Lecture Notes in Computer Science 12251

Founding Editors

Gerhard Goos
Karlsruhe Institute of Technology, Karlsruhe, Germany
Juris Hartmanis
Cornell University, Ithaca, NY, USA

Editorial Board Members

Elisa Bertino
Purdue University, West Lafayette, IN, USA
Wen Gao
Peking University, Beijing, China
Bernhard Steffen
TU Dortmund University, Dortmund, Germany
Gerhard Woeginger
RWTH Aachen, Aachen, Germany
Moti Yung
Columbia University, New York, NY, USA

More information about this series at http://www.springer.com/series/7407

Osvaldo Gervasi · Beniamino Murgante ·
Sanjay Misra · Chiara Garau ·
Ivan Blečić · David Taniar ·
Bernady O. Apduhan · Ana Maria A. C. Rocha ·
Eufemia Tarantino · Carmelo Maria Torre ·
Yeliz Karaca (Eds.)

Computational Science and Its Applications – ICCSA 2020

20th International Conference
Cagliari, Italy, July 1–4, 2020
Proceedings, Part III

 Springer

Editors
Osvaldo Gervasi (iD)
University of Perugia
Perugia, Italy

Sanjay Misra (iD)
Chair- Center of ICT/ICE
Covenant University
Ota, Nigeria

Ivan Blečić (iD)
University of Cagliari
Cagliari, Italy

Bernady O. Apduhan
Department of Information Science
Kyushu Sangyo University
Fukuoka, Japan

Eufemia Tarantino (iD)
Polytechnic University of Bari
Bari, Italy

Yeliz Karaca (iD)
Department of Neurology
University of Massachusetts
Medical School
Worcester, MA, USA

Beniamino Murgante (iD)
University of Basilicata
Potenza, Potenza, Italy

Chiara Garau (iD)
University of Cagliari
Cagliari, Italy

David Taniar (iD)
Clayton School of Information Technology
Monash University
Clayton, VIC, Australia

Ana Maria A. C. Rocha (iD)
University of Minho
Braga, Portugal

Carmelo Maria Torre (iD)
Polytechnic University of Bari
Bari, Italy

ISSN 0302-9743   ISSN 1611-3349 (electronic)
Lecture Notes in Computer Science
ISBN 978-3-030-58807-6   ISBN 978-3-030-58808-3 (eBook)
https://doi.org/10.1007/978-3-030-58808-3

LNCS Sublibrary: SL1 – Theoretical Computer Science and General Issues

This Springer imprint is published by the registered company Springer Nature Switzerland AG
The registered company address is: Gewerbestrasse 11, 6330 Cham, Switzerland

Preface

These seven volumes (LNCS volumes 12249–12255) consist of the peer-reviewed papers from the International Conference on Computational Science and Its Applications (ICCSA 2020) which took place from July 1–4, 2020. Initially the conference was planned to be held in Cagliari, Italy, in collaboration with the University of Cagliari, but due to the COVID-19 pandemic it was organized as an online event.

ICCSA 2020 was a successful event in the conference series, previously held in Saint Petersburg, Russia (2019), Melbourne, Australia (2018), Trieste, Italy (2017), Beijing, China (2016), Banff, Canada (2015), Guimaraes, Portugal (2014), Ho Chi Minh City, Vietnam (2013), Salvador, Brazil (2012), Santander, Spain (2011), Fukuoka, Japan (2010), Suwon, South Korea (2009), Perugia, Italy (2008), Kuala Lumpur, Malaysia (2007), Glasgow, UK (2006), Singapore (2005), Assisi, Italy (2004), Montreal, Canada (2003), and (as ICCS) Amsterdam, The Netherlands (2002) and San Francisco, USA (2001).

Computational science is the main pillar of most of the present research, industrial and commercial applications, and plays a unique role in exploiting ICT innovative technologies. The ICCSA conference series has provided a venue for researchers and industry practitioners to discuss new ideas, to share complex problems and their solutions, and to shape new trends in computational science.

Apart from the general track, ICCSA 2020 also included 52 workshops in various areas of computational science, ranging from computational science technologies to specific areas of computational science, such as software engineering, security, machine learning and artificial intelligence, blockchain technologies, and of applications in many fields. We accepted 498 papers, distributed among 6 conference main tracks, which included 52 in workshops and 32 short papers. We would like to express our appreciation to the workshops chairs and co-chairs for their hard work and dedication.

The success of the ICCSA conference series in general, and of ICCSA 2020 in particular, vitaly depends on the support from many people: authors, presenters, participants, keynote speakers, workshop chairs, session chairs, Organizing Committee members, student volunteers, Program Committee members, Advisory Committee members, international liaison chairs, reviewers, and others in various roles. We take this opportunity to wholeheartedly thank them all.

We also wish to thank our publisher, Springer, for their acceptance to publish the proceedings, for sponsoring part of the Best Papers Awards, and for their kind assistance and cooperation during the editing process.

We cordially invite you to visit the ICCSA website http://www.iccsa.org where you can find all the relevant information about this interesting and exciting event.

July 2020

Osvaldo Gervasi
Beniamino Murgante
Sanjay Misra

Welcome to the Online Conference

The COVID-19 pandemic disrupted our plans for ICCSA 2020, as was the case for the scientific community around the world. Hence, we had to promptly regroup and rush to set in place the organization and the underlying infrastructure of the online event.

We chose to build the technological infrastructure using only open source software. In particular, we used Jitsi (`jitsi.org`) for the videoconferencing, Riot (`riot.im`) together with Matrix (`matrix.org`) for chat and asynchronous communication, and Jibri (`github.com/jitsi/jibri`) for live streaming sessions on YouTube.

Six Jitsi servers were set up, one for each parallel session. The participants of the sessions were helped and assisted by eight volunteer students (from the Universities of Cagliari, Florence, Perugia, and Bari), who assured technical support and smooth running of the conference proceedings.

The implementation of the software infrastructure and the technical coordination of the volunteers was carried out by Damiano Perri and Marco Simonetti.

Our warmest thanks go to all the volunteering students, to the technical coordinators, and to the development communities of Jitsi, Jibri, Riot, and Matrix, who made their terrific platforms available as open source software.

Our heartfelt thanks go to the keynote speakers: Yaneer Bar-Yam, Cecilia Ceccarelli, and Vincenzo Piuri and to the guests of the closing keynote panel: Mike Batty, Denise Pumain, and Alexis Tsoukiàs.

A big thank you goes to all the 454 speakers, many of whom showed an enormous collaborative spirit, sometimes participating and presenting in almost prohibitive times of the day, given that the participants of this year's conference come from 52 countries scattered over many time zones of the globe.

Finally, we would like to thank Google for letting us livestream all the events via YouTube. In addition to lightening the load of our Jitsi servers, that will allow us to keep memory and to be able to review the most exciting moments of the conference.

We all hope to meet in our beautiful Cagliari next year, safe from COVID-19, and finally free to meet in person and enjoy the beauty of the ICCSA community in the enchanting Sardinia.

July 2020

Ivan Blečić
Chiara Garau

Organization

ICCSA 2020 was organized by the University of Cagliari (Italy), University of Perugia (Italy), University of Basilicata (Italy), Monash University (Australia), Kyushu Sangyo University (Japan), and University of Minho (Portugal).

Honorary General Chairs

Antonio Laganà	Master-UP, Italy
Norio Shiratori	Chuo University, Japan
Kenneth C. J. Tan	Sardina Systems, UK
Corrado Zoppi	University of Cagliari, Italy

General Chairs

Osvaldo Gervasi	University of Perugia, Italy
Ivan Blečić	University of Cagliari, Italy
David Taniar	Monash University, Australia

Program Committee Chairs

Beniamino Murgante	University of Basilicata, Italy
Bernady O. Apduhan	Kyushu Sangyo University, Japan
Chiara Garau	University of Cagliari, Italy
Ana Maria A. C. Rocha	University of Minho, Portugal

International Advisory Committee

Jemal Abawajy	Deakin University, Australia
Dharma P. Agarwal	University of Cincinnati, USA
Rajkumar Buyya	The University of Melbourne, Australia
Claudia Bauzer Medeiros	University of Campinas, Brazil
Manfred M. Fisher	Vienna University of Economics and Business, Austria
Marina L. Gavrilova	University of Calgary, Canada
Yee Leung	Chinese University of Hong Kong, China

International Liaison Chairs

Giuseppe Borruso	University of Trieste, Italy
Elise De Donker	Western Michigan University, USA
Maria Irene Falcão	University of Minho, Portugal
Robert C. H. Hsu	Chung Hua University, Taiwan

Tai-Hoon Kim	Beijing Jaotong University, China
Vladimir Korkhov	Saint Petersburg University, Russia
Sanjay Misra	Covenant University, Nigeria
Takashi Naka	Kyushu Sangyo University, Japan
Rafael D. C. Santos	National Institute for Space Research, Brazil
Maribel Yasmina Santos	University of Minho, Portugal
Elena Stankova	Saint Petersburg University, Russia

Workshop and Session Organizing Chairs

Beniamino Murgante	University of Basilicata, Italy
Sanjay Misra	Covenant University, Nigeria
Jorge Gustavo Rocha	University of Minho, Portugal

Award Chair

| Wenny Rahayu | La Trobe University, Australia |

Publicity Committee Chairs

Elmer Dadios	De La Salle University, Philippines
Nataliia Kulabukhova	Saint Petersburg University, Russia
Daisuke Takahashi	Tsukuba University, Japan
Shangwang Wang	Beijing University of Posts and Telecommunications, China

Technology Chairs

| Damiano Perri | University of Florence, Italy |
| Marco Simonetti | University of Florence, Italy |

Local Arrangement Chairs

Ivan Blečić	University of Cagliari, Italy
Chiara Garau	University of Cagliari, Italy
Ginevra Balletto	University of Cagliari, Italy
Giuseppe Borruso	University of Trieste, Italy
Michele Campagna	University of Cagliari, Italy
Mauro Coni	University of Cagliari, Italy
Anna Maria Colavitti	University of Cagliari, Italy
Giulia Desogus	University of Cagliari, Italy
Sabrina Lai	University of Cagliari, Italy
Francesca Maltinti	University of Cagliari, Italy
Pasquale Mistretta	University of Cagliari, Italy
Augusto Montisci	University of Cagliari, Italy
Francesco Pinna	University of Cagliari, Italy

Davide Spano   University of Cagliari, Italy
Roberto Tonelli   University of Cagliari, Italy
Giuseppe A. Trunfio   University of Sassari, Italy
Corrado Zoppi   University of Cagliari, Italy

Program Committee

Vera Afreixo   University of Aveiro, Portugal
Filipe Alvelos   University of Minho, Portugal
Hartmut Asche   University of Potsdam, Germany
Ginevra Balletto   University of Cagliari, Italy
Michela Bertolotto   University College Dublin, Ireland
Sandro Bimonte   CEMAGREF, TSCF, France
Rod Blais   University of Calgary, Canada
Ivan Blečić   University of Sassari, Italy
Giuseppe Borruso   University of Trieste, Italy
Ana Cristina Braga   University of Minho, Portugal
Massimo Cafaro   University of Salento, Italy
Yves Caniou   Lyon University, France
José A. Cardoso e Cunha   Universidade Nova de Lisboa, Portugal
Rui Cardoso   University of Beira Interior, Portugal
Leocadio G. Casado   University of Almeria, Spain
Carlo Cattani   University of Salerno, Italy
Mete Celik   Erciyes University, Turkey
Hyunseung Choo   Sungkyunkwan University, South Korea
Min Young Chung   Sungkyunkwan University, South Korea
Florbela Maria da Cruz   Polytechnic Institute of Viana do Castelo, Portugal
    Domingues Correia
Gilberto Corso Pereira   Federal University of Bahia, Brazil
Alessandro Costantini   INFN, Italy
Carla Dal Sasso Freitas   Universidade Federal do Rio Grande do Sul, Brazil
Pradesh Debba   The Council for Scientific and Industrial Research
                              (CSIR), South Africa
Hendrik Decker   Instituto Tecnológico de Informática, Spain
Frank Devai   London South Bank University, UK
Rodolphe Devillers   Memorial University of Newfoundland, Canada
Joana Matos Dias   University of Coimbra, Portugal
Paolino Di Felice   University of L'Aquila, Italy
Prabu Dorairaj   NetApp, India/USA
M. Irene Falcao   University of Minho, Portugal
Cherry Liu Fang   U.S. DOE Ames Laboratory, USA
Florbela P. Fernandes   Polytechnic Institute of Bragança, Portugal
Jose-Jesus Fernandez   National Centre for Biotechnology, CSIS, Spain
Paula Odete Fernandes   Polytechnic Institute of Bragança, Portugal
Adelaide de Fátima Baptista   University of Aveiro, Portugal
    Valente Freitas

Manuel Carlos Figueiredo	University of Minho, Portugal
Maria Celia Furtado Rocha	PRODEB–PósCultura, UFBA, Brazil
Chiara Garau	University of Cagliari, Italy
Paulino Jose Garcia Nieto	University of Oviedo, Spain
Jerome Gensel	LSR-IMAG, France
Maria Giaoutzi	National Technical University of Athens, Greece
Arminda Manuela Andrade Pereira Gonçalves	University of Minho, Portugal
Andrzej M. Goscinski	Deakin University, Australia
Sevin Gümgüm	Izmir University of Economics, Turkey
Alex Hagen-Zanker	University of Cambridge, UK
Shanmugasundaram Hariharan	B.S. Abdur Rahman University, India
Eligius M. T. Hendrix	University of Malaga, Spain, and Wageningen University, The Netherlands
Hisamoto Hiyoshi	Gunma University, Japan
Mustafa Inceoglu	EGE University, Turkey
Peter Jimack	University of Leeds, UK
Qun Jin	Waseda University, Japan
Farid Karimipour	Vienna University of Technology, Austria
Baris Kazar	Oracle Corp., USA
Maulana Adhinugraha Kiki	Telkom University, Indonesia
DongSeong Kim	University of Canterbury, New Zealand
Taihoon Kim	Hannam University, South Korea
Ivana Kolingerova	University of West Bohemia, Czech Republic
Nataliia Kulabukhova	Saint Petersburg University, Russia
Vladimir Korkhov	Saint Petersburg University, Russia
Rosa Lasaponara	CNR, Italy
Maurizio Lazzari	CNR, Italy
Cheng Siong Lee	Monash University, Australia
Sangyoun Lee	Yonsei University, South Korea
Jongchan Lee	Kunsan National University, South Korea
Chendong Li	University of Connecticut, USA
Gang Li	Deakin University, Australia
Fang Liu	AMES Laboratories, USA
Xin Liu	University of Calgary, Canada
Andrea Lombardi	University of Perugia, Italy
Savino Longo	University of Bari, Italy
Tinghuai Ma	Nanjing University of Information Science and Technology, China
Ernesto Marcheggiani	Katholieke Universiteit Leuven, Belgium
Antonino Marvuglia	Research Centre Henri Tudor, Luxembourg
Nicola Masini	CNR, Italy
Ilaria Matteucci	CNR, Italy
Eric Medvet	University of Trieste, Italy
Nirvana Meratnia	University of Twente, The Netherlands

Noelia Faginas Lago   University of Perugia, Italy
Giuseppe Modica   University of Reggio Calabria, Italy
Josè Luis Montaña   University of Cantabria, Spain
Maria Filipa Mourão   IP from Viana do Castelo, Portugal
Louiza de Macedo Mourelle   State University of Rio de Janeiro, Brazil
Nadia Nedjah   State University of Rio de Janeiro, Brazil
Laszlo Neumann   University of Girona, Spain
Kok-Leong Ong   Deakin University, Australia
Belen Palop   Universidad de Valladolid, Spain
Marcin Paprzycki   Polish Academy of Sciences, Poland
Eric Pardede   La Trobe University, Australia
Kwangjin Park   Wonkwang University, South Korea
Ana Isabel Pereira   Polytechnic Institute of Bragança, Portugal
Massimiliano Petri   University of Pisa, Italy
Maurizio Pollino   Italian National Agency for New Technologies, Energy and Sustainable Economic Development, Italy
Alenka Poplin   University of Hamburg, Germany
Vidyasagar Potdar   Curtin University of Technology, Australia
David C. Prosperi   Florida Atlantic University, USA
Wenny Rahayu   La Trobe University, Australia
Jerzy Respondek   Silesian University of Technology, Poland
Humberto Rocha   INESC-Coimbra, Portugal
Jon Rokne   University of Calgary, Canada
Octavio Roncero   CSIC, Spain
Maytham Safar   Kuwait University, Kuwait
Francesco Santini   University of Perugia, Italy
Chiara Saracino   A.O. Ospedale Niguarda Ca' Granda, Italy
Haiduke Sarafian   Penn State University, USA
Marco Paulo Seabra dos Reis   University of Coimbra, Portugal
Jie Shen   University of Michigan, USA
Qi Shi   Liverpool John Moores University, UK
Dale Shires   U.S. Army Research Laboratory, USA
Inês Soares   University of Coimbra, Portugal
Elena Stankova   Saint Petersburg University, Russia
Takuo Suganuma   Tohoku University, Japan
Eufemia Tarantino   Polytechnic University of Bari, Italy
Sergio Tasso   University of Perugia, Italy
Ana Paula Teixeira   University of Trás-os-Montes and Alto Douro, Portugal
Senhorinha Teixeira   University of Minho, Portugal
M. Filomena Teodoro   Portuguese Naval Academy, University of Lisbon, Portugal
Parimala Thulasiraman   University of Manitoba, Canada
Carmelo Torre   Polytechnic University of Bari, Italy
Javier Martinez Torres   Centro Universitario de la Defensa Zaragoza, Spain
Giuseppe A. Trunfio   University of Sassari, Italy

Pablo Vanegas	University of Cuenca, Ecuador
Marco Vizzari	University of Perugia, Italy
Varun Vohra	Merck Inc., USA
Koichi Wada	University of Tsukuba, Japan
Krzysztof Walkowiak	Wroclaw University of Technology, Poland
Zequn Wang	Intelligent Automation Inc., USA
Robert Weibel	University of Zurich, Switzerland
Frank Westad	Norwegian University of Science and Technology, Norway
Roland Wismüller	Universität Siegen, Germany
Mudasser Wyne	SOET National University, USA
Chung-Huang Yang	National Kaohsiung Normal University, Taiwan
Xin-She Yang	National Physical Laboratory, UK
Salim Zabir	France Telecom Japan Co., Japan
Haifeng Zhao	University of California, Davis, USA
Fabiana Zollo	University of Venice, Italy
Albert Y. Zomaya	The University of Sydney, Australia

Workshop Organizers

Advanced Transport Tools and Methods (A2TM 2020)

Massimiliano Petri	University of Pisa, Italy
Antonio Pratelli	University of Pisa, Italy

Advances in Artificial Intelligence Learning Technologies: Blended Learning, STEM, Computational Thinking and Coding (AAILT 2020)

Valentina Franzoni	University of Perugia, Italy
Alfredo Milani	University of Perugia, Italy
Sergio Tasso	University of Perugia, Italy

Workshop on Advancements in Applied Machine Learning and Data Analytics (AAMDA 2020)

Alessandro Costantini	INFN, Italy
Daniele Cesini	INFN, Italy
Davide Salomoni	INFN, Italy
Doina Cristina Duma	INFN, Italy

Advanced Computational Approaches in Artificial Intelligence and Complex Systems Applications (ACAC 2020)

Yeliz Karaca	University of Massachusetts Medical School, USA
Dumitru Baleanu	Çankaya University, Turkey, and Institute of Space Sciences, Romania
Majaz Moonis	University of Massachusetts Medical School, USA
Yu-Dong Zhang	University of Leicester, UK

Affective Computing and Emotion Recognition (ACER-EMORE 2020)

Valentina Franzoni	University of Perugia, Italy
Alfredo Milani	University of Perugia, Italy
Giulio Biondi	University of Florence, Italy

AI Factory and Smart Manufacturing (AIFACTORY 2020)

Jongpil Jeong	Sungkyunkwan University, South Korea

Air Quality Monitoring and Citizen Science for Smart Urban Management. State of the Art And Perspectives (AirQ&CScience 2020)

Grazie Fattoruso	ENEA CR Portici, Italy
Maurizio Pollino	ENEA CR Casaccia, Italy
Saverio De Vito	ENEA CR Portici, Italy

Automatic Landform Classification: Spatial Methods and Applications (ALCSMA 2020)

Maria Danese	CNR-ISPC, Italy
Dario Gioia	CNR-ISPC, Italy

Advances of Modelling Micromobility in Urban Spaces (AMMUS 2020)

Tiziana Campisi	University of Enna KORE, Italy
Giovanni Tesoriere	University of Enna KORE, Italy
Ioannis Politis	Aristotle University of Thessaloniki, Greece
Socrates Basbas	Aristotle University of Thessaloniki, Greece
Sanja Surdonja	University of Rijeka, Croatia
Marko Rencelj	University of Maribor, Slovenia

Advances in Information Systems and Technologies for Emergency Management, Risk Assessment and Mitigation Based on the Resilience Concepts (ASTER 2020)

Maurizio Pollino	ENEA, Italy
Marco Vona	University of Basilicata, Italy
Amedeo Flora	University of Basilicata, Italy
Chiara Iacovino	University of Basilicata, Italy
Beniamino Murgante	University of Basilicata, Italy

Advances in Web Based Learning (AWBL 2020)

Birol Ciloglugil	Ege University, Turkey
Mustafa Murat Inceoglu	Ege University, Turkey

**Blockchain and Distributed Ledgers: Technologies
and Applications (BDLTA 2020)**

Vladimir Korkhov	Saint Petersburg University, Russia
Elena Stankova	Saint Petersburg University, Russia
Nataliia Kulabukhova	Saint Petersburg University, Russia

Bio and Neuro Inspired Computing and Applications (BIONCA 2020)

Nadia Nedjah	State University of Rio de Janeiro, Brazil
Luiza De Macedo Mourelle	State University of Rio de Janeiro, Brazil

Computer Aided Modeling, Simulation and Analysis (CAMSA 2020)

Jie Shen	University of Michigan, USA

Computational and Applied Statistics (CAS 2020)

Ana Cristina Braga	University of Minho, Portugal

Computerized Evidence Based Decision Making (CEBDEM 2020)

Clarice Bleil de Souza	Cardiff University, UK
Valerio Cuttini	University of Pisa, Italy
Federico Cerutti	Cardiff University, UK
Camilla Pezzica	Cardiff University, UK

Computational Geometry and Applications (CGA 2020)

Marina Gavrilova	University of Calgary, Canada

**Computational Mathematics, Statistics and Information Management
(CMSIM 2020)**

Maria Filomena Teodoro	Portuguese Naval Academy, University of Lisbon, Portugal

Computational Optimization and Applications (COA 2020)

Ana Rocha	University of Minho, Portugal
Humberto Rocha	University of Coimbra, Portugal

Computational Astrochemistry (CompAstro 2020)

Marzio Rosi	University of Perugia, Italy
Cecilia Ceccarelli	University of Grenoble, France
Stefano Falcinelli	University of Perugia, Italy
Dimitrios Skouteris	Master-UP, Italy

Cities, Technologies and Planning (CTP 2020)

Beniamino Murgante	University of Basilicata, Italy
Ljiljana Zivkovic	Ministry of Construction, Transport and Infrastructure and Institute of Architecture and Urban & Spatial Planning of Serbia, Serbia
Giuseppe Borruso	University of Trieste, Italy
Malgorzata Hanzl	University of Łódź, Poland

Data Stream Processing and Applications (DASPA 2020)

Raja Chiky	ISEP, France
Rosanna VERDE	University of Campania, Italy
Marcilio De Souto	Orleans University, France

Data Science for Cyber Security (DS4Cyber 2020)

Hongmei Chi	Florida A&M University, USA

Econometric and Multidimensional Evaluation in Urban Environment (EMEUE 2020)

Carmelo Maria Torre	Polytechnic University of Bari, Italy
Pierluigi Morano	Polytechnic University of Bari, Italy
Maria Cerreta	University of Naples, Italy
Paola Perchinunno	University of Bari, Italy
Francesco Tajani	University of Rome, Italy
Simona Panaro	University of Portsmouth, UK
Francesco Scorza	University of Basilicata, Italy

Frontiers in Machine Learning (FIML 2020)

Massimo Bilancia	University of Bari, Italy
Paola Perchinunno	University of Bari, Italy
Pasquale Lops	University of Bari, Italy
Danilo Di Bona	University of Bari, Italy

Future Computing System Technologies and Applications (FiSTA 2020)

Bernady Apduhan	Kyushu Sangyo University, Japan
Rafael Santos	Brazilian National Institute for Space Research, Brazil

Geodesign in Decision Making: Meta Planning and Collaborative Design for Sustainable and Inclusive Development (GDM 2020)

Francesco Scorza	University of Basilicata, Italy
Michele Campagna	University of Cagliari, Italy
Ana Clara Mourao Moura	Federal University of Minas Gerais, Brazil

Geomatics in Forestry and Agriculture: New Advances and Perspectives (GeoForAgr 2020)

Maurizio Pollino	ENEA, Italy
Giuseppe Modica	University of Reggio Calabria, Italy
Marco Vizzari	University of Perugia, Italy

Geographical Analysis, Urban Modeling, Spatial Statistics (GEOG-AND-MOD 2020)

Beniamino Murgante	University of Basilicata, Italy
Giuseppe Borruso	University of Trieste, Italy
Hartmut Asche	University of Potsdam, Germany

Geomatics for Resource Monitoring and Management (GRMM 2020)

Eufemia Tarantino	Polytechnic University of Bari, Italy
Enrico Borgogno Mondino	University of Torino, Italy
Marco Scaioni	Polytechnic University of Milan, Italy
Alessandra Capolupo	Polytechnic University of Bari, Italy

Software Quality (ISSQ 2020)

Sanjay Misra	Covenant University, Nigeria

Collective, Massive and Evolutionary Systems (IWCES 2020)

Alfredo Milani	University of Perugia, Italy
Rajdeep Niyogi	Indian Institute of Technology, Roorkee, India
Alina Elena Baia	University of Florence, Italy

Large Scale Computational Science (LSCS 2020)

Elise De Doncker	Western Michigan University, USA
Fukuko Yuasa	High Energy Accelerator Research Organization (KEK), Japan
Hideo Matsufuru	High Energy Accelerator Research Organization (KEK), Japan

Land Use Monitoring for Sustainability (LUMS 2020)

Carmelo Maria Torre	Polytechnic University of Bari, Italy
Alessandro Bonifazi	Polytechnic University of Bari, Italy
Pasquale Balena	Polytechnic University of Bari, Italy
Massimiliano Bencardino	University of Salerno, Italy
Francesco Tajani	University of Rome, Italy
Pierluigi Morano	Polytechnic University of Bari, Italy
Maria Cerreta	University of Naples, Italy
Giuliano Poli	University of Naples, Italy

Machine Learning for Space and Earth Observation Data (MALSEOD 2020)

Rafael Santos   INPE, Brazil
Karine Ferreira   INPE, Brazil

Building Multi-dimensional Models for Assessing Complex Environmental Systems (MES 2020)

Marta Dell'Ovo   Polytechnic University of Milan, Italy
Vanessa Assumma   Polytechnic University of Torino, Italy
Caterina Caprioli   Polytechnic University of Torino, Italy
Giulia Datola   Polytechnic University of Torino, Italy
Federico dell'Anna   Polytechnic University of Torino, Italy

Ecosystem Services: Nature's Contribution to People in Practice. Assessment Frameworks, Models, Mapping, and Implications (NC2P 2020)

Francesco Scorza   University of Basilicata, Italy
David Cabana   International Marine Center, Italy
Sabrina Lai   University of Cagliari, Italy
Ana Clara Mourao Moura   Federal University of Minas Gerais, Brazil
Corrado Zoppi   University of Cagliari, Italy

Open Knowledge for Socio-economic Development (OKSED 2020)

Luigi Mundula   University of Cagliari, Italy
Flavia Marzano   Link Campus University, Italy
Maria Paradiso   University of Milan, Italy

Scientific Computing Infrastructure (SCI 2020)

Elena Stankova   Saint Petersburg State University, Russia
Vladimir Korkhov   Saint Petersburg State University, Russia
Natalia Kulabukhova   Saint Petersburg State University, Russia

Computational Studies for Energy and Comfort in Buildings (SECoB 2020)

Senhorinha Teixeira   University of Minho, Portugal
Luís Martins   University of Minho, Portugal
Ana Maria Rocha   University of Minho, Portugal

Software Engineering Processes and Applications (SEPA 2020)

Sanjay Misra   Covenant University, Nigeria

Smart Ports - Technologies and Challenges (SmartPorts 2020)

Gianfranco Fancello   University of Cagliari, Italy
Patrizia Serra   University of Cagliari, Italy
Marco Mazzarino   University of Venice, Italy
Luigi Mundula   University of Cagliari, Italy

| Ginevra Balletto | University of Cagliari, Italy |
| Giuseppe Borruso | University of Trieste, Italy |

Sustainability Performance Assessment: Models, Approaches and Applications Toward Interdisciplinary and Integrated Solutions (SPA 2020)

Francesco Scorza	University of Basilicata, Italy
Valentin Grecu	Lucian Blaga University, Romania
Jolanta Dvarioniene	Kaunas University of Technology, Lithuania
Sabrina Lai	University of Cagliari, Italy
Iole Cerminara	University of Basilicata, Italy
Corrado Zoppi	University of Cagliari, Italy

Smart and Sustainable Island Communities (SSIC 2020)

Chiara Garau	University of Cagliari, Italy
Anastasia Stratigea	National Technical University of Athens, Greece
Paola Zamperlin	University of Pisa, Italy
Francesco Scorza	University of Basilicata, Italy

Science, Technologies and Policies to Innovate Spatial Planning (STP4P 2020)

Chiara Garau	University of Cagliari, Italy
Daniele La Rosa	University of Catania, Italy
Francesco Scorza	University of Basilicata, Italy
Anna Maria Colavitti	University of Cagliari, Italy
Beniamino Murgante	University of Basilicata, Italy
Paolo La Greca	University of Catania, Italy

New Frontiers for Strategic Urban Planning (StrategicUP 2020)

Luigi Mundula	University of Cagliari, Italy
Ginevra Balletto	University of Cagliari, Italy
Giuseppe Borruso	University of Trieste, Italy
Michele Campagna	University of Cagliari, Italy
Beniamino Murgante	University of Basilicata, Italy

Theoretical and Computational Chemistry and its Applications (TCCMA 2020)

| Noelia Faginas-Lago | University of Perugia, Italy |
| Andrea Lombardi | University of Perugia, Italy |

Tools and Techniques in Software Development Process (TTSDP 2020)

| Sanjay Misra | Covenant University, Nigeria |

Urban Form Studies (UForm 2020)

| Malgorzata Hanzl | Łódź University of Technology, Poland |

Urban Space Extended Accessibility (USEaccessibility 2020)

Chiara Garau	University of Cagliari, Italy
Francesco Pinna	University of Cagliari, Italy
Beniamino Murgante	University of Basilicata, Italy
Mauro Coni	University of Cagliari, Italy
Francesca Maltinti	University of Cagliari, Italy
Vincenza Torrisi	University of Catania, Italy
Matteo Ignaccolo	University of Catania, Italy

Virtual and Augmented Reality and Applications (VRA 2020)

Osvaldo Gervasi	University of Perugia, Italy
Damiano Perri	University of Perugia, Italy
Marco Simonetti	University of Perugia, Italy
Sergio Tasso	University of Perugia, Italy

Workshop on Advanced and Computational Methods for Earth Science Applications (WACM4ES 2020)

Luca Piroddi	University of Cagliari, Italy
Laura Foddis	University of Cagliari, Italy
Gian Piero Deidda	University of Cagliari, Italy
Augusto Montisci	University of Cagliari, Italy
Gabriele Uras	University of Cagliari, Italy
Giulio Vignoli	University of Cagliari, Italy

Sponsoring Organizations

ICCSA 2020 would not have been possible without tremendous support of many organizations and institutions, for which all organizers and participants of ICCSA 2020 express their sincere gratitude:

Springer International Publishing AG, Germany
(https://www.springer.com)

Computers Open Access Journal
(https://www.mdpi.com/journal/computers)

IEEE Italy Section, Italy
(https://italy.ieeer8.org/)

Centre-North Italy Chapter IEEE GRSS, Italy
(https://cispio.diet.uniroma1.it/marzano/ieee-grs/
index.html)

Italy Section of the Computer Society, Italy
(https://site.ieee.org/italy-cs/)

University of Cagliari, Italy
(https://unica.it/)

University of Perugia, Italy
(https://www.unipg.it)

University of Basilicata, Italy
(http://www.unibas.it)

Monash University, Australia
(https://www.monash.edu/)

Kyushu Sangyo University, Japan
(https://www.kyusan-u.ac.jp/)

University of Minho, Portugal
(https://www.uminho.pt/)

Scientific Association Transport Infrastructures, Italy
(https://www.stradeeautostrade.it/associazioni-e-organizzazioni/asit-associazione-scientifica-infrastrutture-trasporto/)

Regione Sardegna, Italy
(https://regione.sardegna.it/)

Comune di Cagliari, Italy
(https://www.comune.cagliari.it/)

Referees

A. P. Andrade Marina	ISCTE, Instituto Universitário de Lisboa, Portugal
Addesso Paolo	University of Salerno, Italy
Adewumi Adewole	Algonquin College, Canada
Afolabi Adedeji	Covenant University, Nigeria
Afreixo Vera	University of Aveiro, Portugal
Agrawal Smirti	Freelancer, USA
Agrawal Akshat	Amity University Haryana, India
Ahmad Waseem	Federal University of Technology Minna, Nigeria
Akgun Nurten	Bursa Technical University, Turkey
Alam Tauhidul	Louisiana State University Shreveport, USA
Aleixo Sandra M.	CEAUL, Portugal
Alfa Abraham	Federal University of Technology Minna, Nigeria
Alvelos Filipe	University of Minho, Portugal
Alves Alexandra	University of Minho, Portugal
Amato Federico	University of Lausanne, Switzerland
Andrade Marina Alexandra Pedro	ISCTE-IUL, Portugal
Andrianov Sergey	Saint Petersburg State University, Russia
Anelli Angelo	CNR-IGAG, Italy
Anelli Debora	University of Rome, Italy
Annunziata Alfonso	University of Cagliari, Italy
Antognelli Sara	Agricolus, Italy
Aoyama Tatsumi	High Energy Accelerator Research Organization, Japan
Apduhan Bernady	Kyushu Sangyo University, Japan
Ascenzi Daniela	University of Trento, Italy
Asche Harmut	Hasso-Plattner-Institut für Digital Engineering GmbH, Germany
Aslan Burak Galip	Izmir Insitute of Technology, Turkey
Assumma Vanessa	Polytechnic University of Torino, Italy
Astoga Gino	UV, Chile
Atman Uslu Nilüfer	Manisa Celal Bayar University, Turkey
Behera Ranjan Kumar	National Institute of Technology, Rourkela, India
Badsha Shahriar	University of Nevada, USA
Bai Peng	University of Cagliari, Italy
Baia Alina-Elena	University of Perugia, Italy
Balacco Gabriella	Polytechnic University of Bari, Italy
Balci Birim	Celal Bayar University, Turkey
Balena Pasquale	Polytechnic University of Bari, Italy
Balletto Ginevra	University of Cagliari, Italy
Balucani Nadia	University of Perugia, Italy
Bansal Megha	Delhi University, India
Barazzetti Luigi	Polytechnic University of Milan, Italy
Barreto Jeniffer	Istituto Superior Técnico, Portugal
Basbas Socrates	Aristotle University of Thessaloniki, Greece

Berger Katja   Ludwig-Maximilians-Universität München, Germany
Beyene Asrat Mulatu   Addis Ababa Science and Technology University,
                                  Ethiopia
Bilancia Massimo   University of Bari Aldo Moro, Italy
Biondi Giulio   University of Firenze, Italy
Blanquer Ignacio   Universitat Politècnica de València, Spain
Bleil de Souza Clarice   Cardiff University, UK
Blečić Ivan   University of Cagliari, Italy
Bogdanov Alexander   Saint Petersburg State University, Russia
Bonifazi Alessandro   Polytechnic University of Bari, Italy
Bontchev Boyan   Sofia University, Bulgaria
Borgogno Mondino Enrico   University of Torino, Italy
Borruso Giuseppe   University of Trieste, Italy
Bouaziz Rahma   Taibah University, Saudi Arabia
Bowles Juliana   University of Saint Andrews, UK
Braga Ana Cristina   University of Minho, Portugal
Brambilla Andrea   Polytechnic University of Milan, Italy
Brito Francisco   University of Minho, Portugal
Buele Jorge   Universidad Tecnológica Indoamérica, Ecuador
Buffoni Andrea   TAGES sc, Italy
Cabana David   International Marine Centre, Italy
Calazan Rogerio   IEAPM, Brazil
Calcina Sergio Vincenzo   University of Cagliari, Italy
Camalan Seda   Atilim University, Turkey
Camarero Alberto   Universidad Politécnica de Madrid, Spain
Campisi Tiziana   University of Enna KORE, Italy
Cannatella Daniele   Delft University of Technology, The Netherlands
Capolupo Alessandra   Polytechnic University of Bari, Italy
Cappucci Sergio   ENEA, Italy
Caprioli Caterina   Polytechnic University of Torino, Italy
Carapau Fermando   Universidade de Evora, Portugal
Carcangiu Sara   University of Cagliari, Italy
Carrasqueira Pedro   INESC Coimbra, Portugal
Caselli Nicolás   PUCV Chile, Chile
Castro de Macedo Jose   Universidade do Minho, Portugal
  Nuno
Cavallo Carla   University of Naples, Italy
Cerminara Iole   University of Basilicata, Italy
Cerreta Maria   University of Naples, Italy
Cesini Daniele   INFN-CNAF, Italy
Chang Shi-Kuo   University of Pittsburgh, USA
Chetty Girija   University of Canberra, Australia
Chiky Raja   ISEP, France
Chowdhury Dhiman   University of South Carolina, USA
Ciloglugil Birol   Ege University, Turkey
Coletti Cecilia   Università di Chieti-Pescara, Italy

Coni Mauro	University of Cagliari, Italy
Corcoran Padraig	Cardiff University, UK
Cornelio Antonella	Università degli Studi di Brescia, Italy
Correia Aldina	ESTG-PPorto, Portugal
Correia Elisete	University of Trás-os-Montes and Alto Douro, Portugal
Correia Florbela	Polytechnic Institute of Viana do Castelo, Portugal
Costa Lino	Universidade do Minho, Portugal
Costa e Silva Eliana	ESTG-P Porto, Portugal
Costantini Alessandro	INFN, Italy
Crespi Mattia	University of Roma, Italy
Cuca Branka	Polytechnic University of Milano, Italy
De Doncker Elise	Western Michigan University, USA
De Macedo Mourelle Luiza	State University of Rio de Janeiro, Brazil
Daisaka Hiroshi	Hitotsubashi University, Japan
Daldanise Gaia	CNR, Italy
Danese Maria	CNR-ISPC, Italy
Daniele Bartoli	University of Perugia, Italy
Datola Giulia	Polytechnic University of Torino, Italy
De Luca Giandomenico	University of Reggio Calabria, Italy
De Lucia Caterina	University of Foggia, Italy
De Morais Barroca Filho Itamir	Federal University of Rio Grande do Norte, Brazil
De Petris Samuele	University of Torino, Italy
De Sá Alan	Marinha do Brasil, Brazil
De Souto Marcilio	LIFO, University of Orléans, France
De Vito Saverio	ENEA, Italy
De Wilde Pieter	University of Plymouth, UK
Degtyarev Alexander	Saint Petersburg State University, Russia
Dell'Anna Federico	Polytechnic University of Torino, Italy
Dell'Ovo Marta	Polytechnic University of Milano, Italy
Della Mura Fernanda	University of Naples, Italy
Deluka T. Aleksandra	University of Rijeka, Croatia
Demartino Cristoforo	Zhejiang University, China
Dereli Dursun Ahu	Istanbul Commerce University, Turkey
Desogus Giulia	University of Cagliari, Italy
Dettori Marco	University of Sassari, Italy
Devai Frank	London South Bank University, UK
Di Francesco Massimo	University of Cagliari, Italy
Di Liddo Felicia	Polytechnic University of Bari, Italy
Di Paola Gianluigi	University of Molise, Italy
Di Pietro Antonio	ENEA, Italy
Di Pinto Valerio	University of Naples, Italy
Dias Joana	University of Coimbra, Portugal
Dimas Isabel	University of Coimbra, Portugal
Dirvanauskas Darius	Kaunas University of Technology, Lithuania
Djordjevic Aleksandra	University of Belgrade, Serbia

Duma Doina Cristina	INFN-CNAF, Italy
Dumlu Demircioğlu Emine	Yıldız Technical University, Turkey
Dursun Aziz	Virginia Tech University, USA
Dvarioniene Jolanta	Kaunas University of Technology, Lithuania
Errico Maurizio Francesco	University of Enna KORE, Italy
Ezugwu Absalom	University of KwaZulu-Natal, South Africa
Fattoruso Grazia	ENEA, Italy
Faginas-Lago Noelia	University of Perugia, Italy
Falanga Bolognesi Salvatore	ARIESPACE, Italy
Falcinelli Stefano	University of Perugia, Italy
Farias Marcos	National Nuclear Energy Commission, Brazil
Farina Alessandro	University of Pisa, Italy
Feltynowski Marcin	Lodz University of Technology, Poland
Fernandes Florbela	Instituto Politecnico de Bragança, Portugal
Fernandes Paula Odete	Instituto Politécnico de Bragança, Portugal
Fernandez-Sanz Luis	University of Alcala, Spain
Ferreira Ana Cristina	University of Minho, Portugal
Ferreira Fernanda	Porto, Portugal
Fiorini Lorena	University of L'Aquila, Italy
Flora Amedeo	University of Basilicata, Italy
Florez Hector	Universidad Distrital Francisco Jose de Caldas, Colombia
Foddis Maria Laura	University of Cagliari, Italy
Fogli Daniela	University of Brescia, Italy
Fortunelli Martina	Pragma Engineering, Italy
Fragiacomo Massimo	University of L'Aquila, Italy
Franzoni Valentina	Perugia University, Italy
Fusco Giovanni	University of Cote d'Azur, France
Fyrogenis Ioannis	Aristotle University of Thessaloniki, Greece
Gorbachev Yuriy	Coddan Technologies LLC, Russia
Gabrielli Laura	Università Iuav di Venezia, Italy
Gallanos Theodore	Austrian Institute of Technology, Austria
Gamallo Belmonte Pablo	Universitat de Barcelona, Spain
Gankevich Ivan	Saint Petersburg State University, Russia
Garau Chiara	University of Cagliari, Italy
Garcia Para Ernesto	Universidad del Pais Vasco, EHU, Spain
Gargano Riccardo	Universidade de Brasilia, Brazil
Gavrilova Marina	University of Calgary, Canada
Georgiadis Georgios	Aristotle University of Thessaloniki, Greece
Gervasi Osvaldo	University of Perugia, Italy
Giano Salvatore Ivo	University of Basilicata, Italy
Gil Jorge	Chalmers University, Sweden
Gioia Andrea	Polytechnic University of Bari, Italy
Gioia Dario	ISPC-CNT, Italy

Giordano Ludovica	ENEA, Italy
Giorgi Giacomo	University of Perugia, Italy
Giovene di Girasole Eleonora	CNR-IRISS, Italy
Giovinazzi Sonia	ENEA, Italy
Giresini Linda	University of Pisa, Italy
Giuffrida Salvatore	University of Catania, Italy
Golubchikov Oleg	Cardiff University, UK
Gonçalves A. Manuela	University of Minho, Portugal
Gorgoglione Angela	Universidad de la República, Uruguay
Goyal Rinkaj	IPU, Delhi, India
Grishkin Valery	Saint Petersburg State University, Russia
Guerra Eduardo	Free University of Bozen-Bolzano, Italy
Guerrero Abel	University of Guanajuato, Mexico
Gulseven Osman	American University of The Middle East, Kuwait
Gupta Brij	National Institute of Technology, Kurukshetra, India
Guveyi Elcin	Yildiz Teknik University, Turkey
Gülen Kemal Güven	Namk Kemal University, Turkey
Haddad Sandra	Arab Academy for Science, Technology and Maritime Transport, Egypt
Hanzl Malgorzata	Lodz University of Technology, Poland
Hegedus Peter	University of Szeged, Hungary
Hendrix Eligius M. T.	Universidad de Málaga, Spain
Higaki Hiroaki	Tokyo Denki University, Japan
Hossain Syeda Sumbul	Daffodil International University, Bangladesh
Iacovino Chiara	University of Basilicata, Italy
Iakushkin Oleg	Saint Petersburg State University, Russia
Iannuzzo Antonino	ETH Zurich, Switzerland
Idri Ali	University Mohammed V, Morocco
Ignaccolo Matteo	University of Catania, Italy
Ilovan Oana-Ramona	Babeş-Bolyai University, Romania
Isola Federica	University of Cagliari, Italy
Jankovic Marija	CERTH, Greece
Jorge Ana Maria	Instituto Politécnico de Lisboa, Portugal
Kanamori Issaku	RIKEN Center for Computational Science, Japan
Kapenga John	Western Michigan University, USA
Karabulut Korhan	Yasar University, Turkey
Karaca Yeliz	University of Massachusetts Medical School, USA
Karami Ali	University of Guilan, Iran
Kienhofer Frank	WITS, South Africa
Kim Tai-hoon	Beijing Jiaotong University, China
Kimura Shuhei	Tottori University, Japan
Kirillov Denis	Saint Petersburg State University, Russia
Korkhov Vladimir	Saint Petersburg University, Russia
Koszewski Krzysztof	Warsaw University of Technology, Poland
Krzysztofik Sylwia	Lodz University of Technology, Poland

Kulabukhova Nataliia	Saint Petersburg State University, Russia
Kulkarni Shrinivas B.	SDM College of Engineering and Technology, Dharwad, India
Kwiecinski Krystian	Warsaw University of Technology, Poland
Kyvelou Stella	Panteion University of Social and Political Sciences, Greece
Körting Thales	INPE, Brazil
Lal Niranjan	Mody University of Science and Technology, India
Lazzari Maurizio	CNR-ISPC, Italy
Leon Marcelo	Asociacion de Becarios del Ecuador, Ecuador
La Rocca Ludovica	University of Naples, Italy
La Rosa Daniele	University of Catania, Italy
Lai Sabrina	University of Cagliari, Italy
Lalenis Konstantinos	University of Thessaly, Greece
Lannon Simon	Cardiff University, UK
Lasaponara Rosa	CNR, Italy
Lee Chien-Sing	Sunway University, Malaysia
Lemus-Romani José	Pontificia Universidad Católica de Valparaiso, Chile
Leone Federica	University of Cagliari, Italy
Li Yuanxi	Hong Kong Baptist University, China
Locurcio Marco	Polytechnic University of Bari, Italy
Lombardi Andrea	University of Perugia, Italy
Lopez Gayarre Fernando	University of Oviedo, Spain
Lops Pasquale	University of Bari, Italy
Lourenço Vanda	Universidade Nova de Lisboa, Portugal
Luviano José Luís	University of Guanajuato, Mexico
Maltese Antonino	University of Palermo, Italy
Magni Riccardo	Pragma Engineering, Italy
Maheshwari Anil	Carleton University, Canada
Maja Roberto	Polytechnic University of Milano, Italy
Malik Shaveta	Terna Engineering College, India
Maltinti Francesca	University of Cagliari, Italy
Mandado Marcos	University of Vigo, Spain
Manganelli Benedetto	University of Basilicata, Italy
Mangiameli Michele	University of Catania, Italy
Maraschin Clarice	Universidade Federal do Rio Grande do Sul, Brazil
Marigorta Ana Maria	Universidad de Las Palmas de Gran Canaria, Spain
Markov Krassimir	Institute of Electrical Engineering and Informatics, Bulgaria
Martellozzo Federico	University of Firenze, Italy
Marucci Alessandro	University of L'Aquila, Italy
Masini Nicola	IBAM-CNR, Italy
Matsufuru Hideo	High Energy Accelerator Research Organization (KEK), Japan
Matteucci Ilaria	CNR, Italy
Mauro D'Apuzzo	University of Cassino and Southern Lazio, Italy

Mazzarella Chiara	University of Naples, Italy
Mazzarino Marco	University of Venice, Italy
Mazzoni Augusto	University of Roma, Italy
Mele Roberta	University of Naples, Italy
Menezes Raquel	University of Minho, Portugal
Menghini Antonio	Aarhus Geofisica, Italy
Mengoni Paolo	University of Florence, Italy
Merlino Angelo	Università degli Studi Mediterranea, Italy
Milani Alfredo	University of Perugia, Italy
Milic Vladimir	University of Zagreb, Croatia
Millham Richard	Durban University of Technology, South Africa
Mishra B.	University of Szeged, Hungary
Misra Sanjay	Covenant University, Nigeria
Modica Giuseppe	University of Reggio Calabria, Italy
Mohagheghi Mohammadsadegh	Vali-e-Asr University of Rafsanjan, Iran
Molaei Qelichi Mohamad	University of Tehran, Iran
Molinara Mario	University of Cassino and Southern Lazio, Italy
Momo Evelyn Joan	University of Torino, Italy
Monteiro Vitor	University of Minho, Portugal
Montisci Augusto	University of Cagliari, Italy
Morano Pierluigi	Polytechnic University of Bari, Italy
Morganti Alessandro	Polytechnic University of Milano, Italy
Mosca Erica Isa	Polytechnic University of Milan, Italy
Moura Ricardo	CMA-FCT, New University of Lisbon, Portugal
Mourao Maria	Polytechnic Institute of Viana do Castelo, Portugal
Mourão Moura Ana Clara	Federal University of Minas Gerais, Brazil
Mrak Iva	University of Rijeka, Croatia
Murgante Beniamino	University of Basilicata, Italy
Muñoz Mirna	Centro de Investigacion en Matematicas, Mexico
Nedjah Nadia	State University of Rio de Janeiro, Brazil
Nakasato Naohito	University of Aizu, Japan
Natário Isabel Cristina	Universidade Nova de Lisboa, Portugal
Nesticò Antonio	Università degli Studi di Salerno, Italy
Neto Ana Maria	Universidade Federal do ABC, Brazil
Nicolosi Vittorio	University of Rome, Italy
Nikiforiadis Andreas	Aristotle University of Thessaloniki, Greece
Nocera Fabrizio	University of Illinois at Urbana-Champaign, USA
Nocera Silvio	IUAV, Italy
Nogueira Marcelo	Paulista University, Brazil
Nolè Gabriele	CNR, Italy
Nuno Beirao Jose	University of Lisbon, Portugal
Okewu Emma	University of Alcala, Spain
Oluwasefunmi Arogundade	Academy of Mathematics and System Science, China
Oppio Alessandra	Polytechnic University of Milan, Italy
P. Costa M. Fernanda	University of Minho, Portugal

Parisot Olivier	Luxembourg Institute of Science and Technology, Luxembourg
Paddeu Daniela	UWE, UK
Paio Alexandra	ISCTE-Instituto Universitário de Lisboa, Portugal
Palme Massimo	Catholic University of the North, Chile
Panaro Simona	University of Portsmouth, UK
Pancham Jay	Durban University of Technology, South Africa
Pantazis Dimos	University of West Attica, Greece
Papa Enrica	University of Westminster, UK
Pardede Eric	La Trobe University, Australia
Perchinunno Paola	Uniersity of Cagliari, Italy
Perdicoulis Teresa	UTAD, Portugal
Pereira Ana	Polytechnic Institute of Bragança, Portugal
Perri Damiano	University of Perugia, Italy
Petrelli Marco	University of Rome, Italy
Pierri Francesca	University of Perugia, Italy
Piersanti Antonio	ENEA, Italy
Pilogallo Angela	University of Basilicata, Italy
Pinna Francesco	University of Cagliari, Italy
Pinto Telmo	University of Coimbra, Portugal
Piroddi Luca	University of Cagliari, Italy
Poli Giuliano	University of Naples, Italy
Polidoro Maria João	Polytecnic Institute of Porto, Portugal
Polignano Marco	University of Bari, Italy
Politis Ioannis	Aristotle University of Thessaloniki, Greece
Pollino Maurizio	ENEA, Italy
Popoola Segun	Covenant University, Nigeria
Pratelli Antonio	University of Pisa, Italy
Praticò Salvatore	University of Reggio Calabria, Italy
Previtali Mattia	Polytechnic University of Milan, Italy
Puppio Mario Lucio	University of Pisa, Italy
Puttini Ricardo	Universidade de Brasilia, Brazil
Que Zeli	Nanjing Forestry University, China
Queiroz Gilberto	INPE, Brazil
Regalbuto Stefania	University of Naples, Italy
Ravanelli Roberta	University of Roma, Italy
Recanatesi Fabio	University of Tuscia, Italy
Reis Ferreira Gomes Karine	INPE, Brazil
Reis Marco	University of Coimbra, Portugal
Reitano Maria	University of Naples, Italy
Rencelj Marko	University of Maribor, Slovenia
Respondek Jerzy	Silesian University of Technology, Poland
Rimola Albert	Universitat Autònoma de Barcelona, Spain
Rocha Ana	University of Minho, Portugal
Rocha Humberto	University of Coimbra, Portugal
Rocha Maria Celia	UFBA Bahia, Brazil

Rocha Maria Clara	ESTES Coimbra, Portugal
Rocha Miguel	University of Minho, Portugal
Rodriguez Guillermo	UNICEN, Argentina
Rodríguez González Alejandro	Universidad Carlos III de Madrid, Spain
Ronchieri Elisabetta	INFN, Italy
Rosi Marzio	University of Perugia, Italy
Rotondo Francesco	Università Politecnica delle Marche, Italy
Rusci Simone	University of Pisa, Italy
Saganeiti Lucia	University of Basilicata, Italy
Saiu Valeria	University of Cagliari, Italy
Salas Agustin	UPCV, Chile
Salvo Giuseppe	University of Palermo, Italy
Sarvia Filippo	University of Torino, Italy
Santaga Francesco	University of Perugia, Italy
Santangelo Michele	CNR-IRPI, Italy
Santini Francesco	University of Perugia, Italy
Santos Rafael	INPE, Brazil
Santucci Valentino	Università per Stranieri di Perugia, Italy
Saponaro Mirko	Polytechnic University of Bari, Italy
Sarker Iqbal	CUET, Bangladesh
Scaioni Marco	Politecnico Milano, Italy
Scorza Francesco	University of Basilicata, Italy
Scotto di Perta Ester	University of Naples, Italy
Sebillo Monica	University of Salerno, Italy
Sharma Meera	Swami Shraddhanand College, India
Shen Jie	University of Michigan, USA
Shou Huahao	Zhejiang University of Technology, China
Siavvas Miltiadis	Centre of Research and Technology Hellas (CERTH), Greece
Silva Carina	ESTeSL-IPL, Portugal
Silva Joao Carlos	Polytechnic Institute of Cavado and Ave, Portugal
Silva Junior Luneque	Universidade Federal do ABC, Brazil
Silva Ângela	Instituto Politécnico de Viana do Castelo, Portugal
Simonetti Marco	University of Florence, Italy
Situm Zeljko	University of Zagreb, Croatia
Skouteris Dimitrios	Master-Up, Italy
Solano Francesco	Università degli Studi della Tuscia, Italy
Somma Maria	University of Naples, Italy
Sonnessa Alberico	Polytechnic University of Bari, Italy
Sousa Lisete	University of Lisbon, Portugal
Sousa Nelson	University of Algarve, Portugal
Spaeth Benjamin	Cardiff University, UK
Srinivsan M.	Navodaya Institute of Technology, India
Stankova Elena	Saint Petersburg State University, Russia
Stratigea Anastasia	National Technical University of Athens, Greece

Šurdonja Sanja	University of Rijeka, Croatia
Sviatov Kirill	Ulyanovsk State Technical University, Russia
Sánchez de Merás Alfredo	Universitat de Valencia, Spain
Takahashi Daisuke	University of Tsukuba, Japan
Tanaka Kazuaki	Kyushu Institute of Technology, Japan
Taniar David	Monash University, Australia
Tapia McClung Rodrigo	Centro de Investigación en Ciencias de Información Geoespacial, Mexico
Tarantino Eufemia	Polytechnic University of Bari, Italy
Tasso Sergio	University of Perugia, Italy
Teixeira Ana Paula	University of Trás-os-Montes and Alto Douro, Portugal
Teixeira Senhorinha	University of Minho, Portugal
Tengku Izhar Tengku Adil	Universiti Teknologi MARA, Malaysia
Teodoro Maria Filomena	University of Lisbon, Portuguese Naval Academy, Portugal
Tesoriere Giovanni	University of Enna KORE, Italy
Thangeda Amarendar Rao	Botho University, Botswana
Tonbul Gokchan	Atilim University, Turkey
Toraldo Emanuele	Polytechnic University of Milan, Italy
Torre Carmelo Maria	Polytechnic University of Bari, Italy
Torrieri Francesca	University of Naples, Italy
Torrisi Vincenza	University of Catania, Italy
Toscano Domenico	University of Naples, Italy
Totaro Vincenzo	Polytechnic University of Bari, Italy
Trigo Antonio	Instituto Politécnico de Coimbra, Portugal
Trunfio Giuseppe A.	University of Sassari, Italy
Trung Pham	HCMUT, Vietnam
Tsoukalas Dimitrios	Centre of Research and Technology Hellas (CERTH), Greece
Tucci Biagio	CNR, Italy
Tucker Simon	Liverpool John Moores University, UK
Tuñon Iñaki	Universidad de Valencia, Spain
Tyagi Amit Kumar	Vellore Institute of Technology, India
Uchibayashi Toshihiro	Kyushu University, Japan
Ueda Takahiro	Seikei University, Japan
Ugliengo Piero	University of Torino, Italy
Valente Ettore	University of Naples, Italy
Vallverdu Jordi	University Autonoma Barcelona, Spain
Vanelslander Thierry	University of Antwerp, Belgium
Vasyunin Dmitry	T-Systems RUS, Russia
Vazart Fanny	University of Grenoble Alpes, France
Vecchiocattivi Franco	University of Perugia, Italy
Vekeman Jelle	Vrije Universiteit Brussel (VUB), Belgium
Verde Rosanna	Università degli Studi della Campania, Italy
Vermaseren Jos	Nikhef, The Netherlands

Vignoli Giulio	University of Cagliari, Italy
Vizzari Marco	University of Perugia, Italy
Vodyaho Alexander	Saint Petersburg State Electrotechnical University, Russia
Vona Marco	University of Basilicata, Italy
Waluyo Agustinus Borgy	Monash University, Australia
Wen Min	Xi'an Jiaotong-Liverpool University, China
Westad Frank	Norwegian University of Science and Technology, Norway
Yuasa Fukuko	KEK, Japan
Yadav Rekha	KL University, India
Yamu Claudia	University of Groningen, The Netherlands
Yao Fenghui	Tennessee State University, USA
Yañez Manuel	Universidad Autónoma de Madrid, Spain
Yoki Karl	Daegu Catholic University, South Korea
Zamperlin Paola	University of Pisa, Italy
Zekeng Ndadji Milliam Maxime	University of Dschang, Cameroon
Žemlička Michal	Charles University, Czech Republic
Zita Sampaio Alcinia	Technical University of Lisbon, Portugal
Živković Ljiljana	Ministry of Construction, Transport and Infrastructure and Institute of Architecture and Urban & Spatial Planning of Serbia, Serbia
Zoppi Corrado	University of Cagliari, Italy
Zucca Marco	Polytechnic University of Milan, Italy
Zullo Francesco	University of L'Aquila, Italy

Contents – Part III

International Workshop on Blockchain and Distributed Ledgers: Technologies and Applications (BDLTA 2020)

A Generalization of Bass Equation for Description of Diffusion of Cryptocurrencies and Other Payment Methods and Some Metrics for Cooperation on Market 3
  Victor Dostov and Pavel Shust

Self-sovereign Identity as Trusted Root in Knowledge Based Systems 14
  Nataliia Kulabukhova

Performance of the Secret Electronic Voting Scheme Using Hyperledger Fabric Permissioned Blockchain 25
  Denis Kirillov, Vladimir Korkhov, Vadim Petrunin,
  and Mikhail Makarov

Yet Another E-Voting Scheme Implemented Using Hyperledger Fabric Blockchain 37
  Sergey Kyazhin and Vladimir Popov

Modelling the Interaction of Distributed Service Systems Components 48
  Oleg Iakushkin, Daniil Malevanniy, Ekaterina Pavlova,
  and Anna Fatkina

Data Quality in a Decentralized Environment 58
  Alexander Bogdanov, Alexander Degtyarev, Nadezhda Shchegoleva,
  and Valery Khvatov

A DLT Based Innovative Investment Platform 72
  Alexander Bogdanov, Alexander Degtyarev, Alexey Uteshev,
  Nadezhda Shchegoleva, Valery Khvatov, and Mikhail Zvyagintsev

International Workshop on Bio and Neuro Inspired Computing and Applications (BIONCA 2020)

Application Mapping onto 3D NoCs Using Differential Evolution 89
  Maamar Bougherara, Nadia Nedjah, Djamel Bennouar,
  Rebiha Kemcha, and Luiza de Macedo Mourelle

Identification of Client Profile Using Convolutional Neural Networks 103
  Victor Ribeiro de Azevedo, Nadia Nedjah,
  and Luiza de Macedo Mourelle

International Workshop on Computer Aided Modeling, Simulation and Analysis (CAMSA 2020)

GPU-Based Criticality Analysis Applied to Power System
State Estimation   121
   Ayres Nishio da Silva Junior, Esteban W. G. Clua,
   Milton B. Do Coutto Filho, and Julio C. Stacchini de Souza

Robust Control of the Classic Dynamic Ball and Beam System   134
   Javier Jiménez-Cabas, Farid Meléndez-Pertuz,
   Luis David Díaz-Charris, Carlos Collazos-Morales,
   and Ramón E. R. González

International Workshop on Computational and Applied Statistics (CAS 2020)

Numbers of Served and Lost Customers in Busy-Periods
of $M/M/1/n$ Systems with Balking.   147
   Fátima Ferreira, António Pacheco, and Helena Ribeiro

Simulation Study to Compare the Performance of Signed Klotz
and the Signed Mood Generalized Weighted Coefficients.   157
   Sandra M. Aleixo and Júlia Teles

Impact of OVL Variation on AUC Bias Estimated
by Non-parametric Methods   173
   Carina Silva, Maria Antónia Amaral Turkman, and Lisete Sousa

Adjusting ROC Curve for Covariates with AROC R Package.   185
   Francisco Machado e Costa and Ana Cristina Braga

ROSY Application for Selecting R Packages that Perform ROC Analysis . . .   199
   José Pedro Quintas, Francisco Machado e Costa,
   and Ana Cristina Braga

PLS Visualization Using Biplots: An Application to Team Effectiveness   214
   Alberto Silva, Isabel Dórdio Dimas, Paulo Renato Lourenço,
   Teresa Rebelo, and Adelaide Freitas

Tribological Behavior of 316L Stainless Steel Reinforced with
CuCoBe + Diamond Composites by Laser Sintering and Hot Pressing:
A Comparative Statistical Study   231
   Ângela Cunha, Ana Marques, Francisca Monteiro, José Silva,
   Mariana Silva, Bruno Trindade, Rita Ferreira, Paulo Flores,
   Óscar Carvalho, Filipe Silva, and Ana Cristina Braga

Shiny App to Predict Agricultural Tire Dimensions   247
   Ana Rita Antunes and Ana Cristina Braga

Environmental Performance Assessment of the Transport Sector
in the European Union.   261
   Sarah B. Gruetzmacher, Clara B. Vaz, and Ângela P. Ferreira

Multivariate Analysis to Assist Decision-Making in Many-objective
Engineering Optimization Problems.   274
   Francisco Santos and Lino Costa

**International Workshop on Computerized Evidence-Based Decision
Making (CEBDEM 2020)**

Tuscany Configurational Atlas: A GIS-Based Multiscale Assessment
of Road-Circulation Networks Centralities Hierarchies.   291
   Diego Altafini and Valerio Cutini

The Quali-Quantitative Structure of the City and the Residential Estate
Market: Some Evidences   307
   Valerio Di Pinto and Antonio M. Rinaldi

Assessing the Impact of Temporary Housing Sites on Urban Socio-spatial
Performance: The Case of the Central Italy Earthquake   324
   Camilla Pezzica, Chiara Chioni, Valerio Cutini,
   and Clarice Bleil de Souza

Decision Support Systems Based on Multi-agent Simulation for Spatial
Design and Management of a Built Environment: The Case Study
of Hospitals   340
   Dario Esposito, Davide Schaumann, Domenico Camarda,
   and Yehuda E. Kalay

Future Climate Resilience Through Informed Decision Making
in Retrofitting Projects.   352
   Jonas Manuel Gremmelspacher, Julija Sivolova, Emanuele Naboni,
   and Vahid M. Nik

**International Workshop on Computational Geometry
and Applications (CGA 2020)**

Theoretical Development and Validation of a New 3D Macrotexture Index
Evaluated from Laser Based Profile Measurements   367
   Mauro D'Apuzzo, Azzurra Evangelisti, Daniela Santilli,
   and Vittorio Nicolosi

Calculation of the Differential Geometry Properties of Implicit Parametric
Surfaces Intersection   383
   Judith Keren Jiménez-Vilcherrez

Analytics of the Multifacility Weber Problem   395
    Alexei Yu. Uteshev and Elizaveta A. Semenova

International Workshop on Computational Mathematics, Statistics and Information Management (CMSIM 2020)

Gait Characteristics and Their Discriminative Ability in Patients with Fabry
Disease with and Without White-Matter Lesions   415
    *José Braga, Flora Ferreira, Carlos Fernandes, Miguel F. Gago,
    Olga Azevedo, Nuno Sousa, Wolfram Erlhagen, and Estela Bicho*

Water Meters Inaccuracies Registrations: A First Approach of a Portuguese
Case Study   429
    *M. Filomena Teodoro, Marina A. P. Andrade, Sérgio Fernandes,
    and Nelson Carriço*

Using MDS to Compute the Contribution of the Experts in a Delphi
Forecast Associated to a Naval Operation's DSS   446
    *M. Filomena Teodoro, Mário J. Simões Marques, Isabel Nunes,
    Gabriel Calhamonas, and Marina A. P. Andrade*

Multiscale Finite Element Formulation for the 3D
Diffusion-Convection Equation   455
    Ramoni Z. S. Azevedo and Isaac P. Santos

A Bivariate Multilevel Analysis of Portuguese Students   470
    Susana Faria and Carla Salgado

Impact of Using Excellence Management Models in the Customer
Satisfaction of Brazilian Electricity Distributors - 10 Years of Studies   481
    *Marina A. P. Andrade, Álvaro Rosa, Alexandre Carrasco,
    and M. Filomena Teodoro*

Waste Management and Embarked Staff   492
    M. Filomena Teodoro, José B. Rebelo, and Suzana Lampreia

International Workshop on Computational Optimization and Applications (COA 2020)

On Temperature Variation of the Diabetic Foot   507
    Ana Teixeira and Ana I. Pereira

Impact of the Increase in Electric Vehicles on Energy Consumption
and GHG Emissions in Portugal   521
    Amanda S. Minucci, Ângela P. Ferreira, and Paula O. Fernandes

Penalty-Based Heuristic DIRECT Method for Constrained
Global Optimization 538
   M. Fernanda P. Costa, Ana Maria A. C. Rocha,
   and Edite M. G. P. Fernandes

Comparison of Different Strategies for Arc Therapy Optimization 552
   Humberto Rocha, Joana Dias, Pedro Carrasqueira, Tiago Ventura,
   Brígida Ferreira, and Maria do Carmo Lopes

Locating Emergency Vehicles: Robust Optimization Approaches 564
   José Nelas and Joana Dias

Machine Learning for Customer Churn Prediction in Retail Banking 576
   Joana Dias, Pedro Godinho, and Pedro Torres

A Resource Constraint Approach for One Global Constraint MINLP 590
   Pavlo Muts, Ivo Nowak, and Eligius M. T. Hendrix

Simplified Tabu Search with Random-Based Searches for Bound
Constrained Global Optimization 606
   Ana Maria A. C. Rocha, M. Fernanda P. Costa,
   and Edite M. G. P. Fernandes

A Clustering Approach for Prediction of Diabetic Foot Using Thermal
Images 620
   Vítor Filipe, Pedro Teixeira, and Ana Teixeira

Mixed Integer Linear Programming Models for Scheduling Elective
Surgical Procedures 632
   Hanna Pamplona Hortencio and Débora Pretti Ronconi

PDE Based Dense Depth Estimation for Stereo Fisheye Image
Pair and Uncertainty Quantification 648
   Sandesh Athni Hiremath

Single Screw Extrusion Optimization Using the Tchebycheff
Scalarization Method 664
   Ana Maria A. C. Rocha, Marina A. Matos,
   M. Fernanda P. Costa, A. Gaspar-Cunha, and Edite M. G. P. Fernandes

**International Workshop on Computational
Astrochemistry (CompAstro 2020)**

Binding Energies of N-Bearing Astrochemically-Relevant Molecules
on Water Interstellar Ice Models. A Computational Study 683
   Berta Martínez-Bachs, Stefano Ferrero, and Albert Rimola

Theoretical and Computational Analysis at a Quantum State Level
of Autoionization Processes in Astrochemistry   693
  Stefano Falcinelli, Fernando Pirani, Marzio Rosi,
  and Franco Vecchiocattivi

A Computational Study of the Reaction Cyanoacetylene and Cyano Radical
Leading to 2-Butynedinitrile and Hydrogen Radical.   707
  Emília Valença Ferreira de Aragão, Noelia Faginas-Lago, Marzio Rosi,
  Luca Mancini, Nadia Balucani, and Dimitrios Skouteris

A Theoretical Investigation of the Reactions of N(^{2}D) with Small Alkynes
and Implications for the Prebiotic Chemistry of Titan   717
  Luca Mancini, Emília Valença Ferreira de Aragão, Marzio Rosi,
  Dimitrios Skouteris, and Nadia Balucani

A Theoretical Investigation of the Reaction Between Glycolaldehyde and
H$^+$ and Implications for the Organic Chemistry of Star Forming Regions. . . .   730
  Dimitrios Skouteris, Luca Mancini, Fanny Vazart, Cecilia Ceccarelli,
  Marzio Rosi, and Nadia Balucani

A Computational Study on the Insertion of N(^{2}D) into a C—H or C—C
Bond: The Reactions of N(^{2}D) with Benzene and Toluene
and Their Implications on the Chemistry of Titan   744
  Marzio Rosi, Leonardo Pacifici, Dimitrios Skouteris,
  Adriana Caracciolo, Piergiorgio Casavecchia, Stefano Falcinelli,
  and Nadia Balucani

International Workshop on Cities, Technologies and Planning (CTP 2020)

Territorial Attraction for New Industrial-Productive Plants.
The Case of Pavia Province   759
  Roberto De Lotto, Caterina Pietra, and Elisabetta Maria Venco

A Quantitative Approach in the Ecclesiastical Heritage Appraisal   776
  Francesca Salvo, Manuela De Ruggiero, Daniela Tavano,
  and Francesco Aragona

Assessment of Quality of Life in Residential Estates in Lodz   787
  Małgorzata Hanzl, Jakub Misiak, and Karolina Grela

Situated Emotions. The Role of the Soundscape in a Geo-Based
Multimodal Application in the Field of Cultural Heritage.   805
  Letizia Bollini and Irene Della Fazia

Monitoring Urban Development: National Register of Investment Locations
as a Tool for Sustainable Urban Land Use Management in Serbia.   820
  Ljiljana Živković

Estimation of Risk Levels for Building Construction Projects   836
  Gabriella Maselli, Antonio Nesticò, Gianluigi De Mare,
  Elena Merino Gómez, Maria Macchiaroli, and Luigi Dolores

**International Workshop on Econometric and Multidimensional
Evaluation in Urban Environment (EMEUE 2020)**

City-Port Circular Model: Towards a Methodological Framework
for Indicators Selection   855
  Maria Cerreta, Eugenio Muccio, Giuliano Poli, Stefania Regalbuto,
  and Francesca Romano

The Assessment of Public Buildings with Special Architectural Features
Using the Cost Approach. A Case Study: A Building Sold by EUR S.p.A.
in Rome.   869
  Fabrizio Battisti and Orazio Campo

A Comparison of Short-Term and Long-Term Rental Market in an
Italian City.   884
  Benedetto Manganelli, Sabina Tataranna, and Pierfrancesco De Paola

The Classification of the University for Type of Campus Setting in a World
Sustainability Ranking.   899
  Silvestro Montrone, Paola Perchinunno, and Monica Cazzolle

Real Estate Values, Tree Cover, and Per-Capita Income: An Evaluation
of the Interdependencies in Buffalo City (NY)   913
  Antonio Nesticò, Theodore Endreny, Maria Rosaria Guarini,
  Francesco Sica, and Debora Anelli

COVID 19: Health, Statistical and Constitutional Aspects   927
  Francesco Perchinunno, Luigia Stefania Stucci, and Paola Perchinunno

A Model to Support the Investment Decisions Through Social Impact
Bonds as Effective Financial Instruments for the Enhancement of Social
Welfare Policies   941
  Francesco Tajani, Pierluigi Morano, Debora Anelli,
  and Carmelo Maria Torre

Assessing the Interstitial Rent: The Effects of Touristification
on the Historic Center of Naples (Italy)   952
  Maria Cerreta, Fernanda Della Mura, and Giuliano Poli

Sustainable Redevelopment: The Cost-Revenue Analysis to Support
the Urban Planning Decisions.   968
    Pierluigi Morano, Maria Rosaria Guarini, Francesco Tajani,
    and Debora Anelli

A Procedure to Evaluate the Extra-Charge of Urbanization.   981
    Maria Rosaria Guarini, Pierluigi Morano, and Alessandro Micheli

The Effects of Urban Transformation Projects on the Real Estate Market:
A Case Study in Bari (Italy).   1000
    Pierluigi Morano, Francesco Tajani, Felicia Di Liddo,
    Carmelo Maria Torre, and Marco Locurcio

Circular Enhancement of the Cultural Heritage: An Adaptive Reuse
Strategy for Ercolano Heritagescape   1016
    Maria Cerreta and Valentina Savino

Author Index   1035

International Workshop on Blockchain and Distributed Ledgers: Technologies and Applications (BDLTA 2020)

A Generalization of Bass Equation for Description of Diffusion of Cryptocurrencies and Other Payment Methods and Some Metrics for Cooperation on Market

Victor Dostov[1,2]([✉]) [iD] and Pavel Shust[1,2] [iD]

[1] Saint-Petersburg State University, St Petersburg, Russia
{leonova, shoust}@npaed.ru
[2] Russian Electronic Money and Remittance Association, Moscow, Russia

Abstract. We use case of payment systems to discuss the typology of generalized Bass equation solution for audience growth in systems with cooperative and non-cooperative behavior of users. Based on the models for C2B and P2P payment system models analyzed in our previous papers, we propose an integrated approach. Different types of cooperation are discussed. The paper also proposes some criteria for estimating a degree of cooperation in given real-life system.

Keywords: Bass equation · Payment system · Cryptocurrency · Ricatti equation · Cooperative behavior coefficient

1 Introduction

Bass equation

$$\frac{dx}{dt} = (p + qx)(1 - x) \tag{1}$$

or

$$\frac{dx}{dt} = -qx^2 + (q - p)x + p \tag{2}$$

was proposed by Frank Bass [1] to describe diffusion on innovations, primarily for penetration of consumer durables such as refrigerators and home freezers [2]. It describes how innovations spread in audience with limited size where x is share of maximum audience that already accepted an innovation, p and q are empirical coefficients which, roughly, describe probability for someone to accept innovation as result of advertisement or under the influence of other acceptors, correspondingly. In Bass approach this is one-way process, a person who already uses innovative product or service is supposed to use them forever. This is good assumption for such durables as home freezers but not accurate for wide spectrum of other technologies e.g. CD players or video players, which have relatively short lifecycle. However, Bass model became popular and was later applied to different kinds of innovative goods and services, e.g. in [3, 4].

Bass model is based on the assumption that there are two ways of how users can start using innovative product or service: first type of users are innovators who make independent decisions; second type are imitators, whose decisions depend of those who are already using the product or service. From other perspective, these two modalities are very similar to cooperative and non-cooperative games [5].

2   Generalization of Bass Model to Payment Systems

Initially, Bass model deals with "social" effects, such as advertisement and mutual influence between people. Social effect of mutual influence usually decreases with higher penetration of the technology. However, this is not always a case. In [6, 7] authors demonstrated that even in mature markets cooperation between clients may be important for technological reasons. For example, to make a first phone call, there should be two phone users, not one. This means that in certain technologies cooperation between parties plays an important role.

Interaction in a payment market demonstrates both cooperative and non-cooperative interactions.C2B (consumer to business) systems, like retail card payments are example of non-cooperative game because capacity of given POS terminal to accept payments is virtually infinite. The same is true for cash withdrawal from ATM: cardholder's ability to use the card to withdraw cash does not depend on other cardholders. In other words, client's behavior for such systems does not depend on other people's behavior. This is not the case in P2P (peer to peer) systems, like international money transfers (remittance): if Alice transfers money to Bob, Bob needs to be part of the system too. If Alice wants to transfer her bitcoin from her wallet to Bob, Bob needs to have the wallet as well. All these examples mean that for payment systems (and other systems with "technical" cooperation between client) we can obtain generalization of Bass-like equation for mature market that reflects cooperation behavior driven both by technical specifics and human interaction. Some limited cases of pure C2B and P2P systems and Bass-like equation were considered in [7], in this paper we will draft a general approach from scratch.

Let's consider general type of payment system with number of user x, maximum number of users (maximum audience capacity) N, and, consequently, number of non-users N-x. Then, in growth users rate dx/dt we will have non-cooperative contribution coefficient a_b, describing attractiveness of C2B payment functions and cooperative contribution coefficient a_p for attractiveness of P2P payment functions to describe a probability for non-involved customer to join the payment system at any point of time. For mature payment system or market we also need to introduce fatigue factor b, which describe how often users leave the system, disappointed in its functionality. Thus we obtain a following equation

$$\frac{dx}{dt} = a_p(N - x)x + a_b(N - x) - bx \tag{3}$$

or

$$\frac{dx}{dt} = -a_p x^2 + (a_p N - a_b - b)x + a_b N \tag{4}$$

Then final equation in Ricatti equation [8] with constant coefficients

$$\frac{dx}{dt} = -a_p x^2 + lx + a_b N \tag{5}$$

where

$$l = a_p N - a_b - b \tag{6}$$

Analytical solution of (5) yields

$$x = \frac{1}{2a_p}\left(-D \tanh\left(C - \frac{1}{2}tD\right) + l\right) \tag{7}$$

where D, given by

$$D^2 = l^2 + 4a_b a_p N \tag{8}$$

may be interpreted as reverse time of system evolution.

We can also introduce stationary limit of (7)

$$x_\infty = \frac{1}{2a_p}(D + l) \tag{9}$$

If the initial condition is x(0) = 0, then arbitrary constant C is

$$C = \operatorname{arctanh}(l/D) \tag{10}$$

If we do not have information from beginning of system evolution and in starting point of observation x(0) = x_0, then

$$C = \operatorname{arctanh}\left((l - 2a_p x_0)/D\right) \tag{11}$$

Actually, in practical analysis we have to use (11) with care as it is quite sensitive to errors in x_0.

Here we actually demonstrated, that we can introduce fatigue factor to Bass equation without changing its general appearance, just setting up three independent coefficients instead of two. As seen from (5–6) these two Eqs. (5), one with arbitrary a_p, a_b, N and b and other with the coefficient replaced as following:

$$a_b + b \rightarrow a_b \tag{12}$$

$$N \rightarrow \frac{Na_b}{a_b - b} \tag{13}$$

$$a_p \rightarrow a_p \tag{14}$$

$$b \rightarrow 0 \tag{15}$$

coincide exactly. Such recalibration means that users fatigue b≠0 is qualitatively equivalent to increase of non-cooperative coefficient a_b and decrease of maximal audience N.

Examples of solutions (7) are given in Fig. 1.

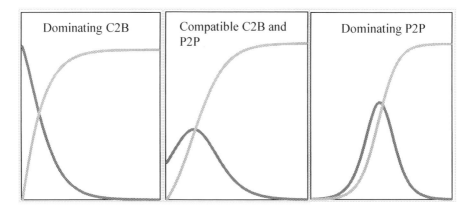

Fig. 1. Transition from C2B-dominating type, $a_b \gg a_p N$ to P2P-dominating type, $a_b \ll a_p N$ for number of users (red) and audience growth rate (blue) (Color figure online).

Also we can make formal but useful observation, that time t is presented in (7) only in combination τ

$$\tau = C - \frac{1}{2} tD \tag{16}$$

This means that, notwithstanding specific parameters, all systems described by (7) are moving along the same trajectory, giving us S-shaped curve like in last graph at Fig. 1, however on different parts of it. The observation period covers only part of this curve, giving us different shapes shown at Fig. 1. This also means that if for current observation period we see C2B type shape with negative d^2x/dt^2, this may mean both pure C2B system or P2P or mixed system in a later stage.

Also, if we start our observation at some point t = 0 which is not point of start of the system operation, we can make extrapolation to negative t, reconstructing previous system behavior. Formally we may find a starting point of system evolution t_0 where x = 0 as

$$t_0 = 2(C - \operatorname{arctanh}\left(\frac{l}{D}\right)/D \tag{17}$$

However, as (7) in part is very sensitive to variation of parameters, the extrapolation back in time leads to fast growing errors and may be irrelevant, so this estimation must be used with big caution.

This leads us to question of algorithm that could be used to obtain system type from practical data. A priori, if we write an Eq. (3) with given coefficients, we can introduce cooperation coefficient

$$K = \frac{a_p N}{a_b} \tag{18}$$

Evidently, $K \gg 1$ yields us cooperative behavior solution and vice versa. Combining K and D we may introduce four type classification for comparing of systems under consideration (Table 1).

Table 1. Typology of payment systems.

	Small D	Big D
Small K	Slow non-cooperative, C2B	Fast non-cooperative, C2B
Big K	Slow cooperative, P2P	Fast cooperative, P2P

Unfortunately, this straightforward criterium (18) cannot be directly applied to practical analysis. In principle, applying LSD or LAD method [9] to real data set, can find best-fitting independent coefficients l, D, a_p for (8), and, correspondingly calculate C, then solve Eqs. (7) and (9) as a system finding N and a_b. From practical standpoint, looking at C is enough. If C is large negative, we get all-time non-cooperative behaviour, if C is a large positive, we get cooperative behaviour, for C around 0 we get mixed type. Therefore, depending on approach, we may use both C and K to estimate degree of cooperation in the system.

3   Practical Examples

To give practical examples, we use data on number payment cards in the European Union (in tens of millions) [10], Russia (in millions) [11, 12], South Korea debit cards (in millions) [13] and Spain bank cards (in millions) [14] at Fig. 2.

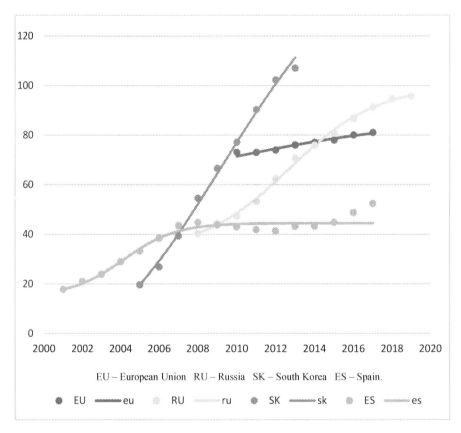

Fig. 2. Dynamics of card number in EU, Russia, South Korea and Spain interpolated by Eq. (8). Circles represent real data, curves represent interpolation by Eq. (7).

In the EU we see the best fit by LAD criterion at $C = 0.0$ with dominating C2B transactions and close to saturation market. In curve for Russia get $C = 1.0$, with both cooperative and non-cooperative contributions and also relatively close to saturation. We also give example of South Korea debit cards [13], for which model work quite well with $C = 0.57$ and far from saturation market and example of Spain bank cards [14], for which market behavior is more complicated. We can see quite a good fit in 2000–2014 with $C = 1,2$ but unpredicted growth later. We may attribute this growth to changes in regulation or market models which resulted in drastic change of model coefficients. We will discuss this situation below.

Similar approach may be used to analyze behavior in cryptocurrency systems: such as bitcoin (abbreviation BTC marks real data) and Ethereum (ETH, correspondingly) using [15] as a source of data. Real-time curve contains a lot of investment-induced quasi-random noise, but we can still assume that average behavior is still described by aforementioned factors. In our previous work [7] we considered cryptocurrencies as pure cooperative systems. In this paper we use (7) and find coefficients on it using LAD criterion. Direct application of LAD to noisy curves does not work pretty well, therefore we first applied it to data averaged by years to get initial estimation: the results are shown on Fig. 3a. Then we applied LAD to full data set to improve this estimation with results shown on Fig. 3b. Evidently, the model is not able to describe hype behavior in 2017–2018 induced by dramatic change of bitcoin/USD exchange rates [16], but we see a good coincidence with average behavior. With C = 1.6 we see domination of cooperative behavior. We also see that cryptocurrency market is definitely reaching its saturation with no significant growth perspective. This fact was demonstrated earlier in [7] and recent data only prove it. The similar behavior is seen in Ethereum platform with C = 3.0. Cryptocurrencies are definitely fast cooperative P2P payment system type in terms of Table 1 and traditional systems are slower and less cooperative C2B system in comparison. However, as role of P2P payments in traditional card system is starting to grow recently [17], we expect a shift to larger values of C in their dynamics.

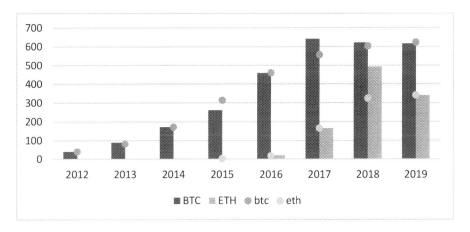

Fig. 3a. Dynamics of cryptocurrency wallets averaged by years for Bitcoin (blue) and Ethereum (green) and their approximation by Eq. (7) (Color figure online).

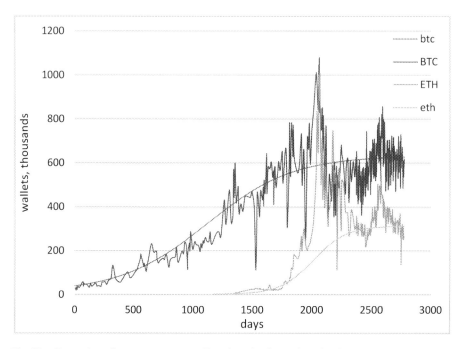

Fig. 3b. Dynamics of cryptocurrency wallets for Bitcoin (red) and Ethereum (green) and their approximation by Eq. (7) (Color figure online).

Table 2. Parameters C,D, x_∞ for different payment instruments.

	EU	SK	RU	ES	BTC	ETH
C	0,0	0,6	1,0	1,2	1.6	3.0
D, 1/year	0,21	0,26	0,44	0,74	1,0	3,0
x_∞, mlns	850	160	99	44	0,63	0,34

We summarize different cases in Table 2, which allows to compare essential parameters for different payment instruments.

We suppose that future studies may provide us with even less-cooperative cases with negative C.

Figure 4 represents graphically Table 2 for card segment with circle size representing audience size. We see clear connection between C, D and x_∞ which will be subject of further studies.

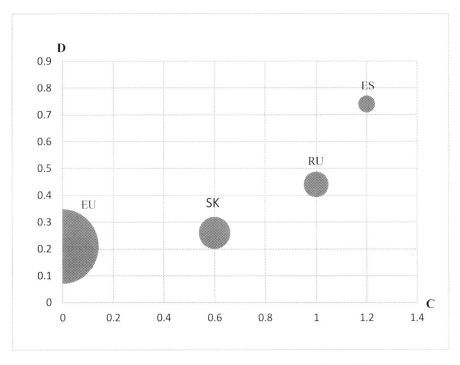

Fig. 4. Representation of significant parameters for card dynamics in different countries.

We also have to underline that that effectiveness of using this approach for longer-term analysis shall not be overestimated. Evidently, coefficients in (3) change slowly but steadily. While typical time of such change is much bigger than 1/D, we can neglect these changes. But sooner or later these changes may become important. E.g. in Figs. 3a and 3b we describe quasi-steady states successfully reached by cryptocurrencies. However, if, e.g. regulatory situation will change and most of countries adopt cryptocurrency regulation, a_p and a_b coefficients will change dramatically and, after transition period, they will switch to new trajectory of development with different coefficients in (7). Evidently, this option is relevant to other types of payment systems as well, including card systems considered above. Very probably, aforementioned card statistics for Spain demonstrates us such situation. However, in short- and middle-time perspective assumption of almost constant coefficients in (3) works well for empirical purposes.

4   Conclusion

We used payment systems to propose the generalization of Bass approach to the case of technology-driven cooperative behavior and possible loss of existing users. In addition to terms of equation describing increase in audience we also introduce a coefficient, responsible for customer outflow and show that this churn shifts effectively system audience to more non-cooperative behavior. We provide solution of Ricatti equation, compare them to the original Bass solution and propose different numerical criteria to describe users' behavior. These results can be applied both to cryptocurrencies audience and "classic" payment system dynamics. We believe that proposed solutions might be used not only for payment systems analysis, but also to describe dynamics of innovations in other fields.

References

1. Bass, F.M.: A dynamic model of market share and sales behavior. In: Proceedings, Winter Conference American Marketing Association, Chicago, IL, USA (1963)
2. Bass, F.M.: A new product growth for model consumer durables. Manage. Sci. **15**(5), 215–227 (1969)
3. Ganjeizadeh, F., Lei, H., Goraya, P., Olivar, E.: Applying looks-like analysis and bass diffusion model techniques to forecast a neurostimulator device with no historical data. In: 27th International Conference on Flexible Automation and Intelligent Manufacturing (FAIM), pp. 1916–1924. Procedia Manufacturing, Modena (2017)
4. Soffer, T., Nachmias, R., Ram, J.: Diffusion of web supported instruction in higher education: The case of tel-aviv university. Educ. Technol. Soc. **13**(3), 212–223 (2010)
5. Takako, F.-G.: Non-Cooperative Game Theory. Springer, Japan (2015)
6. Dostov, V., Shust, P., Popova, E.: Using mathematical models for analysis and prediction of payment systems behavior. In: 34th International Business Information Management Association Conference (IBIMA). IBIMA Publishing, King of Prussia, Madrid (2019)
7. Dostov, V., Shoust, P., Popova, E.: Using mathematical models to describe the dynamics of the spread of traditional and cryptocurrency payment systems. In: 19th International Conference Computational Science and Its Applications – ICCSA, pp. 457–471. Springer, Saint Petersburg (2019)
8. Simmons, G., Robertson, J.: Differential Equations with Applications and Historical Notes, 2nd Edn. International Series in Pure and Applied Mathematics, 640 p. McGraw-Hill, New York (1991)
9. Dodge, Y., Jureckova, J.: Adaptive Regression. Springer, New York (2000)
10. Cherowbrier, J.: Total number of payment cards in the European Union from 2010 to 2018. https://www.statista.com/statistics/444328/number-of-payment-cards-european-union. Accessed 23 Mar 2020
11. Central Bank of Russia Statistics: The number of payment cards issued by credit institutions and the Bank of Russia. http://www.cbr.ru/eng/statistics/nps/, http://www.cbr.ru/Content/Document/File/69593/t13.xlsx. Accessed 23 Mar 2020
12. Odud, A., Platonova, Y.: Analysis of The Bank Cards Market in Russia. Economic vector 11 (2019). http://www.vectoreconomy.ru/images/publications/2019/11/financeandcredit/Odud_Platonova.pdf. Accessed 23 Mar 2020
13. The Korea Bizwire Business Statistics. http://koreabizwire.com/category/feature/kobiz_stats/business-statistics. Accessed 25 Mar 2020
14. CEIC Spain Payment and Cards Statistics. https://www.ceicdata.com/en/spain/payment-and-cards-statistics. Accessed 25 Mar 2020
15. BitInfoChart Bitcoin Active Addresses historical chart. https://bitinfocharts.com/comparison/bitcoin-activeaddresses.html. Accessed 01 Apr 2020

16. Bitnews Today, The annual cryptocurrency market review: what's left after the hype, 2.01.2029. https://bitnewstoday.com/news/the-annual-cryptocurrency-market-review-what-s-left-after-the-hype. Accessed 03 Apr 2020
17. Eremina, A., Yumabaev, D.: Russians make more C2C transactions that card payments in stores (Vedomosti 16.03.2018)

Self-sovereign Identity as Trusted Root in Knowledge Based Systems

Nataliia Kulabukhova$^{(\boxtimes)}$ [iD]

Saint-Petersburg State University, Saint Petersburg, Russia
n.kulabukhova@spbu.ru

Abstract. In this paper we continue to speak about the concept of Self-Sovereign Identity (SSI), but not in the cases of IoT devices as it was in previous works [1]. The main purpose of this research is the usage of digital identity in two cases: a) SSI of a single person in Knowledge based system "Experts Ledger" and b) SSI of a company and candidate in HR matching systems. Though these two systems are developed for different issues, the idea of SSI in both is similar. The overview of these systems is done, and the pros and cons of using SSI with the relation of Zero-Knowledge Proof (ZKP) in each of them is made.

Keywords: Self-sovereign identity · Decentralized Identifiers · Expert systems · Distributed ledgers · Zero-Knowledge Proof

1 Introduction

Self-Sovereign Identity (SSI) become very popular for the last two-three years, because of the growing interest to the blockchain technologies, though the idea of it has its own life before the distributed ledger appears.

That is why, it is more correct to divide the SSI concepts into BlockChain and Non-Blockchain. In this paper, we will focus more on SSI using distributed ledger technologies, because the main stopping point of evaluation of Non-Blockchain approach was the usage of centralized root. The progress of distributed ledger gave the solution of this problem. In this case, the distributed ledger plays the role of a distributed trust root, excluding the possibility of falsifying data about digital identity and its capabilities in the digital world.

In the works [2–4] the survey of rules on which the SSI must be based is done. The main properties of SSI according to these papers are: Existence, Control, Access, Transparency, Persistence, Portability, Interoperability, Consent, Minimalization, Protection, Provable. All these is very important for the applications for SSI. The overview of existing applications [5] (though some of them are just the description of the concept) shown that it is very difficult to follow all these rules at once. Besides, very interesting assumption was made in [6] about the dependence from the type of blockchain: permissionless or permissioned. This makes development of SSI based applications a very comprehensive task, and the main reason for not existing widely used and fully developed open-course one yet.

© Springer Nature Switzerland AG 2020
O. Gervasi et al. (Eds.): ICCSA 2020, LNCS 12251, pp. 14–24, 2020.
https://doi.org/10.1007/978-3-030-58808-3_2

Despite of mentioned problems the concept of digital verified identity of single person, company or even some IoT device, connecting with each other in some environment like knowledge based system is some kind state-of-the art and becoming really necessary in todays digital world.

In this work the idea of using SSI with the ZKP inside in two different knowledge based systems is described. First the overview of the "Experts Ledger" is made, the other is developing for HR purposes. But these two are only the examples of such kind of systems, the list is not limited by them. Further our conversation will show, that the idea is more general.

2   Overview of the Experts Ledger System

The main goal of the system "Experts Ledger" is to create an intellectual knowledge base about researchers in various fields of science, with the ability to rank and compare experts by different relevant criteria. The system also allows the user to browse the information about researchers, their scientific and expert works. Based on all the accumulated information, the system allows you to build the block of analytic data about researchers and their work. Types of work presented in the system are the following:

- Applications for participation in various projects;
- Grants where researchers are either leaders or participant;
- Research publications:
    1. Articles;
    2. Monographs;
    3. Books;
    4. Materials in conference proceedings;
    5. etc.
- Contracts;
- Expertises.

According to this information the system has the following features:

1. Flexible full-text search for people and their works (grants, contracts, articles, expertise);
2. Loading (adding) and uploading data about experts;
3. Graph representation (mapping expert relationships);
4. Analytics based on information available in the database.

2.1   System Architecture

The General scheme of the system interaction is shown on the Fig. 1. The system allows you to load data from several interconnected databases. After comparing and analyzing the data, the system forms its internal database, which is then used for work. In addition to uploading data from external databases, the system also allows to add external files uploaded to the database by system users.

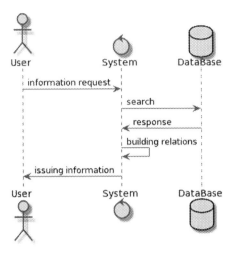

Fig. 1. Interaction diagram of system components

The user communicates with the system via the web interface. Then the system processes the user's requests and interacts with the database (figure), which stores all the information.

2.2 Database Structure

The General scheme of the database structure of the system is shown in the Fig. 2 The information is uploaded from different databases and preprocessing of the received data is made. After that, the database of the knowledge based system is formed and relationships are build.

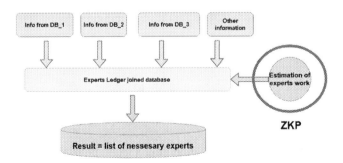

Fig. 2. General view of the system structure

For example, information is being uploaded from one of the bases on the server:

– People (persons);
– Projects (Projects);
– Grants (awards);
– Articles (scientific outputs);
– Requests (applications);
– Organizations (external Organizations);
– Dissertations (student Dissertations).

Updating occurs by adding new information to the existing database, while preserving all dependencies and ratings that were made before. This prevent the possibility of losing data. And all new relationships are made upon the previous iteration of update information, but not constructing from the beginning. This is very important for the SSI, because allows to avoid the reconstruction of Decentralized Identifiers (DID) for each entity of the system.

The database is updated 2 times a week. The system also provides a database update date display to correctly display search data on the date of database upload. Since syncing with Base1 and Base2 databases and then updating the existing database in the system takes several hours, it is not reasonable to update the data more frequently.

2.3   The Technological Stack

The technology stack for backend development includes the following components (the license types under which the software is distributed are given):

– development Language-Python 3 ([7]) (PSFL);
– a web Framework Flask ([8]) (Apache License 2.0);
– expert Database - Elasticsearch ([9]) (Apache License 2.0);
– user Management - SQLAlchemy ([10]) (MIT License).

The technology stack for frontend development includes the following components (the license types that the software is distributed under are shown in parentheses):

– development Language-javascript ES6+/typescript v3.6. 3 ([11]) (Apache License 2.0);
– JavaScript library for creating user interfaces-React V. 16. 0. 0 (React Hooks) ([12]) (MIT License);
– Style primitives-styled-components ([13]) (MIT License);
– to build the project, use parcel V. 1. 12. 3 ([14]) (MIT License).

2.4   The Use of Distributed Ledger

You can use a distributed ledger in this system in two states:

1. as a record of transparency of operations performed in the system (logging information about user actions);
2. anonymization of the expert (confirmation of the reliability of the assessment of expert activity without disclosure the private information about the expert).

In this paper we will discuss more deeply about the second case, though the first is also very interesting to research.

2.5   Estimation Settings Mechanism

In this case, experts are defined as researchers who carry out expert activities in a specific subject area. By evaluating the work of an expert, we mean confirming the expert's skills by specialists. Specialists can be other users of the system who have permission to issue an expert assessment.

Fig. 3. Roles of system users when setting an expert assessment

There are two ways to evaluate the expert's work in the system:

1. Add a comment in the "expert profile";
2. Add a report on the conducted expertise with the results of the expert activity assessment.

How all this work i the concept of SSI and ZKP will be shown in Sect. 4.

3   Overview of Neety System

Neety is the HR platform the main purpose of which is to help employers and job applicants mach each other in an easy and lite way.

Obvious, that currently available applications for hiring employees do not perform the task as desired. The main idea of the solution described here is to create a service that allows user to quickly and objectively find an employer or applicant in accordance with the criteria specified in the resume/vacancy. The goal of development is convenience and reliability in finding employees on the one hand, and work on the other.

The problems of existing HR systems:

1. Job descriptions do not reflect the essence of what you have to work on;
2. There is a lot of superfluous information in job descriptions, but the necessary information is missing;
3. Description summary is too long and unreadable;
4. Criteria for searching for employees and vacancies do not allow you to filter out the right people or companies;
5. The structure of the summary and vacancies does not solve your problem.

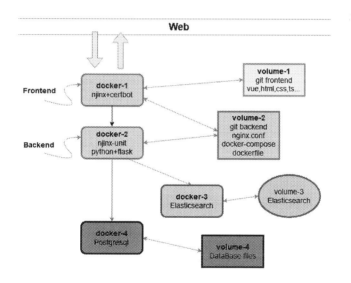

Fig. 4. The general scheme technological stack of the Neety System

Thus, the developed solution should reduce the amount of unnecessary information that hinders both the applicant and the employer when searching. To do this, the system has a built-in matching algorithm based on three main steps:

1. Morphological correspondence;
2. Matching via tags;
3. Neural network.

As you can see, this is a standard approach to natural language processing. The main feature is the method of filling out the questionnaire, which allows you to improve the output of results already at the stage of text processing by the neural network.

The main factors of influence on the results are:

– The relevance of the main skill;
– Relevance for the rest of the skills, taking into account prioritization;
– Relevance for the money (no money → to the end of the issue);
– Rating.

But this algorithm will be discussed in more detail in the next article, because the subject of this article is SSI.

3.1   The Technological Stack

The technologies we use in this case are practically the same as in "Experts Ledger", and they are listed in the Sect. 2.3 and are shown on the Fig. 4.

3.2   The Use of Distributed Ledger

In this case, the distributed ledger will act mainly as a decentralized trusted root, and the usage of ZKP is to evaluate the entities inside the system, based on a trusted, protected voting. Figure 5 shows the idea that using the Decentralized Identifiers based on the blockchain will protects the system from unverified users.

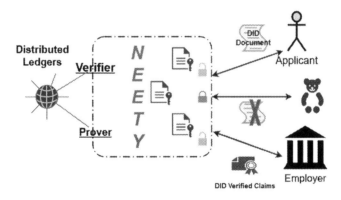

Fig. 5. The general scheme of the Neety System

3.3   Estimation Settings Mechanism

In the case of HR mechanisms ZKP concept as it was mentioned before will work generally as voting.

Factors influencing the rating:

– No confirmation-neutral (zero points added);
– there is a confirmation - plus one point;
– there is a probability that the data does not correspond to reality - minus one point.

4   The SSI Approach

Both systems discussed above are based on working with verified data from both users and organizations. In the case of the first system, we refer only to the reliability of the researchers' data contained in the system. In the second case, the data of both: the individual submitting their resume for a job, and the data of the organization (firm) that placed this vacancy - are important. Let's look at these two cases separately first, and then look at their similarities and differences.

In the first system, when we talk about researchers and their skills and competencies, at the beginning we based only on information collected from various databases. At the same time, this information can be either documented (we will call it verified information), or "written from words" (we will call it unverified information). And when we talk about the SSI concept, we mean just confirmation of data about this unverified information. To do this, the system introduces the concept of "Expert evaluation". What is meant by this concept, let's explain with an example. One of the researchers has information that he is an expert in the field of archaeology, but there is no information in our database about articles, projects and other documents in this field of research. In this case, this is unverified information. And in order to confirm or deny it, the system introduced the concept of "expert". An expert is a Trustee with the right to issue an expert assessment in a particular area.

In this case, the concept of SSI will apply both to the researcher with information that is not verified at the initial stage, and to the Expert who confirms this information. For a complete digital identity of the researcher, it is very important that all information is fully confirmed. To do this, the expert Advisor must also have a fully trusted digital identity. In the context of decentralized identifiers (DIDs), this means that the Expert has the right to rate the researcher. To do this, the expert must have a DID for rating in our example for archaeology. If the Expert has a DID for grading in mathematics, physics, or philosophy, but not in archaeology, the system will not allow the Expert to evaluate the researcher's activities in the field of archaeology.

In the case of the second system, two aspects are also important to us. The first is the confirmed data of the organization that posts vacancies. This will avoid finding "fake" vacancies in the system, and false triggering of the algorithm. In this case, SSI no longer works for one person, but for the entire organization. But on the other hand, the more applicants will also have SSI, the more reliable the information on the other side of the matching algorithm. Ideally, you can consider sending certain decentralized IDs to confirm your skills rather than filling out your competencies and creating a resume. However, this is still not possible in the near future, and it negates the operation of the matching algorithm.

5   Zero-Knowledge Proof in both Cases

First of all, it should be said that the Zero-Knowledge Proof concept (ZKP) as well as SSI existed and was used before the widespread usage of blockchain technology began. The essence of it is that an individual can confirm that he has the right to perform some action, without revealing any information about him- or herself and who gave the right for this action. This principle was widely used in the development of distributed ledger Bitcoin. There it evolved from ZeroCash [16] first to Zcach [17], after it transformed to independent methods of zk-SNARKs [18] and zk-STARKs [19]. The comparison of SNARKs and STARKs is given in the Table 1.

Table 1. Comparison of SNARKs and STARKs main features

$+/-$	ZKP	
	SNARKs	**STARKs**
Pros	– Small proof size; – Fast verification; – Relatively fast proof generation;	– Resistance to hacking by quantum computers – Relatively rapid generation of evidence – Relatively fast verification of evidence – No toxic waste
Cons	– Complex procedure for setting public parameters; – Toxic waste; – Relative complexity of the technology; – Not resistant to hacking using quantum computers	– Complexity of the technology – Large proof size

Currently, these principles are mainly used to track the correct distribution of coins when making transactions in a distributed system. In the described above "Experts ledger" and "Neety" systems, the ZKP ideology is applied not only to the distributed voting, but also to the trustful interaction between the system participants. Let's look at this in more detail.

In this case, in both systems, we can use not only the SSI concept, but also the ZKP concept. When we talk about setting an expert rating for a researcher in the system of the "Experts Ledger", and when the rating of the applicant and the organization is calculated, it is important to correctly calculate the points assigned to all objects participating in both systems. It is important to maintain the reliability and independence of this rating. Therefore, it was decided to perform this calculation based on the blockchain technology and the ZKP concept. The main feature of using these tools allow the corrent user to be sure that this rating can not be forged or changed in anyone's direction. Thus, we get an independent variable rating of all objects in both systems, which on the one hand is based on the opinions of specific people, but on the other hand is independent, devoid of subjectivity.

6   Conclusion

Though, all described above has strong mathematical background, there are still some questions about the security and usability of SSI application itself and those with ZKP either.

Despite this, we are planing to integrate the technologies of ZKP concept in the future work on the projects. The existence of another ZKP application - Bulletproff [20] - should solve some of the problems of SNARKs and STARKs. And the future development of the project will be aimed to research and implement this technology.

Acknowledgment. The author would like to thank the whole team of the project "Experts Ledger": Oleg Jakushkin, Vladimir Korkhov, Artem Shurov, Ivan Marakhovskiy, Alexander Voskresenskiy. The author would like to express gratitude to Evgeniy Abramenko for the opportunity to work with the Neety project.

References

1. Kulabukhova, N.V.: Zero-knowledge proof in self-sovereign identity. CEUR Workshop Proc. **2507**, 381–385 (2019)
2. van Bokkem, D., Hageman, R., Koning, G., Nguyen, L., Zarin, N.: Self-Sovereign Identity Solutions: The Necessity of Blockchain Technology (2019). https://arxiv.org/abs/1904.12816v1
3. Stokkink, Q., Pouwelse, J.: Deployment of a Blockchain-Based Self-Sovereign Identity (2018). https://arxiv.org/pdf/1806.01926.pdf
4. Cameron, K.: The Laws of Identity (2005). https://www.identityblog.com/stories/2005/05/13/TheLawsOfIdentity.pdf
5. Kulabukhova, N., Ivashchenko, A., Tipikin, I., Minin, I.: Self-Sovereign Identity for IoT Devices. In: Misra S. et al. (eds.) Computational Science and Its Applications – ICCSA 2019. ICCSA 2019. Lecture Notes in Computer Science, vol. 11620. Springer, Cham (2019)
6. Panait, A.-E., Olimid, R.F., Stefanescu, A.: Identity Management on Blockchain – Privacy and Security Aspects (2020)
7. https://www.python.org/
8. https://flask.palletsprojects.com/en/1.1.x/
9. https://www.elastic.co/
10. https://www.sqlalchemy.org/
11. https://github.com/microsoft/TypeScript
12. https://ru.reactjs.org/
13. https://www.styled-components.com/
14. https://www.npmjs.com/package/parcel/v/1.12.3
15. Kilian, J.: A note on efficient zero-knowledge proofs and arguments. https://people.csail.mit.edu/vinodv/6892-Fall2013/efficientargs.pdf
16. Bitansky, N., Chiesa, A., Ishai, Y., Ostrovsky, R., Paneth, O.: Succinct non-interactive arguments via linear interactive proofs (2013)
17. Ben-Sasson, E., Chiesa, A., Garman, C., Green, M., Miers, I., Tromer, E., Virza, M.: Zerocash: Decentralized Anonymous Payments from Bitcoin (extended version) (2014)

18. Bunz, B., Fisch, B., Szepieniec, A.: Transparent SNARKs from DARK Compilers (2019)
19. Ben-Sasson, E., Bentov, I., Horesh, Y., Riabzev, M.: Scalable, transparent, and post-quantum secure computational integrity (2018)
20. Bunz, B., Bootle, J., Boneh, D., Poelstra, A., Wuille, P., Maxwell, G.: Bulletproofs: Short Proofs for Confidential Transactions and More. https://eprint.iacr.org/2017/1066.pdf

Performance of the Secret Electronic Voting Scheme Using Hyperledger Fabric Permissioned Blockchain

Denis Kirillov[✉], Vladimir Korkhov, Vadim Petrunin, and Mikhail Makarov

Saint-Petersburg State University, Saint Petersburg, Russia
{d.kirillov,v.korkhov,v.petrunin}@spbu.ru
makarovmma@gmail.com

Abstract. Issues related to reliable electronic voting are still very relevant and not fully resolved. The use of distributed ledger technologies for these purposes has great potential, but even private blockchain solutions often have insufficient bandwidth to satisfy the needs of at least corporate voting, not to mention the public elections. The article studies one of such voting systems, which is based on Hyperledger Fabric permissioned blockchain, where the Identity Mixer technology is used as the anonymizing part.

Keywords: E-voting · Blockchain · Hyperledger fabric · Distributed ledger technologies · Performance

1 Introduction

Attempts to use distributed ledger technologies in various areas of our lives are not always successful. We can observe quite successful cases, for example, the Russian IPChain [1], which allows users to register and perform operations with intellectual property objects. Here, according to the developers, the throughput of about 100 transactions per second is enough, and Hyperledger Fabric [11] can provide this. However, along with this, there are many examples in which the blockchain is used because it is fashionable, despite the fact that there are perfectly functioning examples without using distributed ledger technologies.

The same is observed in the field of electronic voting. Despite the fact that there may be solutions that satisfy many (if not all) of the requirements of an ideal system:

– Eligibility,
– Un-reusability,
– Un-traceability,
– Verifiability,
– Receipt-freeness.

© Springer Nature Switzerland AG 2020
O. Gervasi et al. (Eds.): ICCSA 2020, LNCS 12251, pp. 25–36, 2020.
https://doi.org/10.1007/978-3-030-58808-3_3

there remains one significant performance issue. Most modern systems cannot provide transaction processing speeds of more than a couple of tens of thousands per second (NEO [2], EOS [3]), and the most popular (Bitcoin [4], Ethereum [5]) are only a few tens per second.

In addition to this problem, there are at least two more:

– Complexity,
– Lack of standards.

And if the first one has the consequence of the high cost of the implemented solution, the slowness of development, etc., the second one requires a more global approach, and cannot be resolved momentarily.

The rest of this paper is organized as follows. Section 2 reviews the articles related to this topic. An electronic voting protocol using distributed ledger technologies, as well as a study of its performance, are presented in Sect. 3. Section 4 describes a theoretical modification of the previous protocol, in which the blockchain can be replaced by the more lightweight Identity Mixer technology. Section 5 concludes the paper.

2   Related Work

Historically, electronic voting along with supply chains is one of the most common blockchain cases. Often, electronic voting even appears in tutorials on various platforms of distributed ledger as a basic project for studying technology. This fact strongly affected the modern trends of the scientific developments in the field of blockchain technologies. As a result, today there are a huge number of scientific articles on electronic voting. While some focus on the voting protocol in order to provide basic voting requirements, others set out to explore alternative ways of anonymizing in the system or to study performance.

Most existing solutions are based on the use of well-known DLT platforms, but there are also a large number of works where the authors, inspired by the main advantages provided by the blockchain, create products without using existing platforms. The article [6] describes the principle of operation of the created electronic voting system without using ready-made blockchain platforms. The implemented solution is based on the NodeJS-ReactJS bundle. Distribution is achieved by expanding the various nodes based on Heroku, interacting with PubNub - the implementation of publisher-subscribe. The article describes an example of a system deployment for the case of geographically distributed voting, when each block is deployed on a separate blockchain network node. The described system is a traditional implementation of the public blockchain, the maintenance of which is ensured by supporting the possibility of mining, that is, the basic principles of the PoW algorithm are used. The mathematical PoW problem to be solved is also based on the principle of one-way functions, and the algorithm for adjusting the complexity of the task is similar to the principle of Bitcoin functioning. The authors set a reduction in the amount of stored data as one of the objectives of the study; as a result, the system uses Elliptic-curve cryptography (ECC), which provides a smaller key size, thereby reducing the amount of data storage compared to RSA-based systems.

An example of the implementation of yet another electronic voting system without using DLT platforms was described in [7]. In addition, ECC is also used here. The system is implemented under Ubuntu OS using the Python programming language. Like most systems, the system is designed to meet the basic requirements for electronic voting, including support for the possibility of re-voting until the voting deadline is reached. From an architectural point of view, the implemented system is a modular system and consists of three independent components:

- Synchronized model of voting records based on DLT to avoid forgery of votes,
- User credential model based on elliptic curve cryptography (ECC) to provide authentication and non-repudiation,
- Withdrawal model that allows voters to change their vote before a preset deadline.

For most blockchain applications, the hierarchical principle of maintaining a distributed log is preserved: a log consists of blocks, blocks include transactions. For most votes, this means that each vote is a transaction that will subsequently be combined into blocks based on the block size or timeout and stored in a common journal. In the case of the scheme described in this paper, the block consists exclusively of one bulletin from one user, that is, there is no additional level of the blockchain hierarchy. A similar architecture is typical for Byteball Bytes DAG DLT, where the concept of a block is absent and each transaction contains information about the previous one. The system allows for exclusively public voting without hiding any details of the procedure.

The authors of the article [8] drew attention to another aspect of modern electronic voting protocols - public key infrastructure (PKI). Most existing blockchain voting systems are based on a bunch of cryptographic keys, the reliability of which is guaranteed by number theory for modern computers. However, the rapid development of quantum technologies is a threat to modern cryptographic algorithms, which can become a serious threat to existing systems using cryptographic approaches. This article pays special attention to the issue of anonymization of voters. The scheme proposed in the article combines certificateless ring signature and code-based Niederreiter anonymization approaches. Protection from quantum attacks, which has become the main feature of the described system, is provided by using the modified Niederreiter cryptosystem to distribute Partial Private Keys for each voter. The algorithm is based on code, and its security comes down to the problem of syndrome decoding coding theory. This is an NP-complete problem that is difficult to solve even for quantum computers with powerful computing power. The article provides a formal proof of algorithm security. The scheme proposed in the article is characterized by a linear relationship between security and system efficiency. The optimum value for safety and performance can only be provided for small-scale votings. An increase in the number of voters leads to an increase in the level of security of the system, having a strong influence on the system performance.

To a greater extent, performance issues were addressed in the article [9]. The article talks about blockchain voting based on the Multichain platform. Of course, blockchain is one of the breakthrough technologies of our time, used mainly in the financial sector, but the question of the possibility of comparing financial blockchain systems with traditional payment systems like VISA remains open. The article discusses the various configurations of the blockchain platform for voting in order to conduct a comparative analysis in terms of system performance and scalability. The feasibility of such an analysis becomes possible due to the great flexibility of the Multichain blockchain platform used, which allows managing not only the parameters traditional for such systems, such as block size or transaction, but also the level of access to the blockchain network (commissioned, permissionless).

3   Hyperledger Fabric-Based E-Voting System

The solution based on the Hyperledger Fabric distributed ledger that is examined for performance further is presented in [10]. Here we provide a brief description of the protocol (Fig. 1).

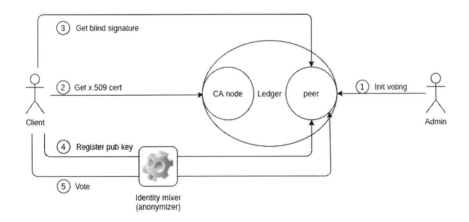

Fig. 1. High-level overview of the Hyperledger Fabric-based e-voting system

The scheme consists of the following steps:

Network Configuration. During this phase organizations, the number of nodes they have, and their rights (read, write) are determined. The network is set up, CA nodes are added, logic (chaincode) is loaded, nodes of the ordering type are determined to achieve consensus.

Voting Configuration. During this phase, the administrator generates a list of questions, determines the list of departments that can participate in this voting. Then each of the administrators of departments forms an additional data (start/end of voting and registration, list of voters) necessary for holding the voting among voters who belong to this department. Also at this stage a key is generated (which is shared between observers). This key will be used to encrypt ballots.

User Registration. This phase is crucial for several purposes: to preserve the possibility of a voter to vote using paper-based ballots (if the user has not been registered he is able to vote only in traditional way); to ensure the choice privacy and maintain eligibility.

- The sub-step of obtaining a blind signature. During the registration each user generates a pair of keys and a random number. Then the voter requests a blind signature from the chaincode. It checks that the current user is allowed to participate in this voting. If it is valid the blind signature is sent back to the user. Then chaincode registers in the ledger that the current user has received a blind signature.
- The sub-step of registering a public key that corresponds to a person is known only by the voter. At this moment the user has a signed public key which he sends it to the chaincode anonymously. Then signature is verified. If it is correct, the public key is recorded to the ledger.
- Ability to cancel registration and vote using the paper-based method. This phase is not anonymous. The voter sends to the chaincode a request to remove the registration key from the list. The chaincode does not completely delete the key but marks it as revoked. This is necessary as if the user changes his mind and wants to register again he has to register a new key. In addition, he is removed from the list of signed users.

Voting. During this phase, users who have registered their private keys for voting can send encrypted ballots to chaincode. Until the end of voting process the voter can change his decision and only the final one is counted.

Result Counting. After the voting is completed observers with private key parts upload them to the chaincode. Thus, it is possible to decrypt the ballots and to get the results that are published on the ledger.

3.1   Performance

The results of calculating the throughput of both the Hyperledger Fabric itself (in the most common configurations) and some of the test cases that were conducted showed disappointing results. For example, an article from the developers themselves [11] shows that out of the box, depending on the parameters, the throughput can vary from 1000 to 3500 tps. Some researchers [12] show that through serious modification of some components, throughput up to 20,000 tps can be achieved. However, at the same time, researchers from Russia (Sberbank) showed that in some cases the speed will be about 1000 tps when using the built-in LevelDB database, and about 300 tps when using CouchDB.

The tests that we conducted showed intermediate results between these values. Several different network configurations were tested:

– Solo (one organization, one node, one orderer, only 5 docker containers),
– Kafka (two organizations, each with 2 nodes, 2 orderers, 3 zookeepers, 4 kafka containers, a total of 20 docker containers).

Each configuration showed similar characteristics: about 2000 tps. After that, some transactions began to fall off by timeout. Clients to interact with the ledger were developed using the following SDKs:

– Java SDK,
– Node SDK,
– Python SDK

Below are the graphs of the average execution time of the voting scenario for one user in ms (for a different number of simultaneous active users. From 100 to 1300 in steps of 200. With more users, the system stopped managing the load and either crashed or some of the users could not receive a response from systems).

The script includes:

– User registration in the HLF system (marked as reg),
– Getting available polls (each user had two polls available, one of which has already ended),
– Getting the results of the ended voting
– Registration in the ongoing vote
– Voting

Testing was carried out on various network configurations (Solo, Kafka) in HLF v1.4.1. The effect of registration was also studied (if the user does not yet have keys for interacting with the ledger, and he needs to contact the CA to get them). At the same time, everything is saved in the ledger: users, voting, bulletins, results etc.

Configuration example: "solo 2 s, 10 msg, 99 MB (reg)" means that the solo network is set up with the following block formation parameters:

– Batch Timeout: The amount of time to wait before creating a batch - 2 s
– Max Message Count: The maximum number of messages to permit in a batch - 10 msg
– Absolute Max Bytes: The absolute maximum number of bytes allowed for the serialized messages in a batch - 99 MB

– Preferred Max Bytes: The preferred maximum number of bytes allowed for the serialized messages in a batch. A message larger than the preferred max bytes will result in a batch larger than preferred max bytes - 512 KB by default

Mark "(reg)" means that users had to register first.

If the fourth parameter is present in the configuration name, then it defines "PreferredMaxBytes" for forming the block.

The plots below illustrate the comparison of operating time in the same configurations depending on the need to initially register in the system (Fig. 2). It can be seen that if it is necessary to register, the execution time of the script increases by 1,5–2 times. This is due to the fact that during registration the user receives a new X.509 certificate, the formation of which is an expensive cryptographic operation. Average voting time of performing the whole voting scenario for a single user is measured.

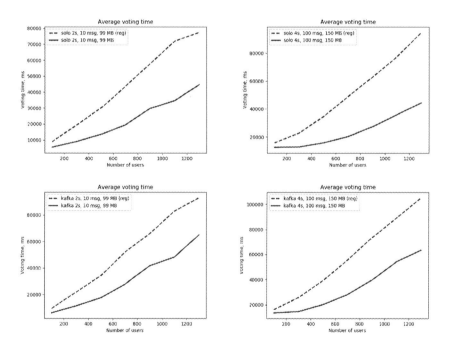

Fig. 2. Comparison of average voting time depending on the need of registration

Comparison of the operating times of different types of consensus (solo and kafka) in the same configurations is presented in Fig. 3. The execution time of the script in the case of Solo was expectedly shorter due to the fact that there is actually only one node that orders transactions, which allows us not to spend resources on consensus between nodes as it happens in Kafka.

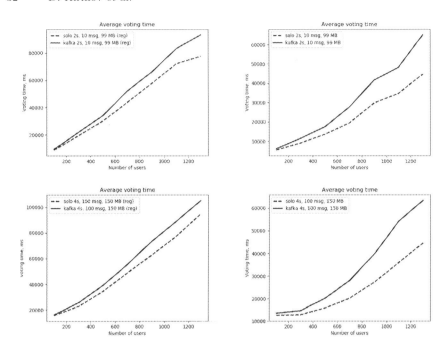

Fig. 3. Comparison of average voting time depending on the type of consensus

Comparison of the operating time with the same type of consensus, but with different parameters of the formation of the block is shown in Fig. 4. It can be seen that an increase in block size when registration was required did not give a performance gain. This is due to the fact that the registration itself takes a comparable time in time to complete the rest of the script. In the case of its absence in the Kafka consensus, we see an increase, especially noticeable after 900 simultaneously active users.

4   Theoretical Alternatives

Identity Mixer (Idemix) technology was developed by IBM and is based on cryptographic techniques, the development of which was presented at the conference and carefully tested by the community. The practical implementation is based on the scheme, which can be found in more detail in [13–15]. Here we provide only a brief description of this technology.

Suppose that there are three participants: the user, issuer (the authority issuing the certificate and confirming the identity of the user) and verifier, which wants to make sure that the user really has some attributes that are necessary to perform some action. In such a scheme, the main problem that Idemix solves is providing to the verifier not all the information about the user, but only certain attributes, or even just proving that the requested attribute falls into a certain range. In addition, it is possible to provide unlinkability, that is, the inability to

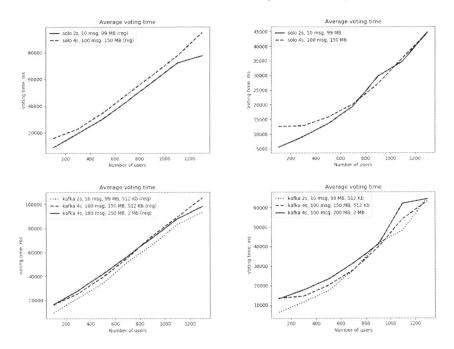

Fig. 4. Comparison of average voting time depending on the block formation parameters

connect two different requests with the same user. In this way, Idemix allows for anonymity and privacy.

Let us briefly describe how Identity Mixer works. Some users want to get Idemix credentials, for this they need to interact with CA:

1. Initialization of receiving credentials.
2. CA returns nonce starting the procedure.
3. The user generates a secret (random number) and on the basis of it, as well as using the public key CA and nonce from the previous step, forms a request, which consists of a commitment's secret and evidence with zero disclosure that the user really knows this secret.
4. CA verifies the evidence and puts its signature on this data along with additional attributes (for example, whether the user belongs to an organization). At the end, this data is returned to the user.
5. The user verifies the CA signature and saves credentials.
6. Now the user can use these credentials to sign messages when communicating with someone. The examiner will be able to make sure that the user really has a signature from the CA, while he will additionally only know the information that the user wants to share.

4.1   Architecture

Registration in the System. To interact with the system, the user must be registered in it. This can be done either simply using a username and password, or using an X.509 certificate.

Create a Vote. When creating a voting, the administrator must specify a list of users who are already registered in the system and who need to access the voting.

Confirmation of Participation in the Vote. After that, users must confirm their participation in this vote. This is achieved with an identity mixer. That is, they make a request for new Idemix credentials, in which the voting identifier acts as an additional attribute on which the signature is placed from the CA.

Voting. Using the credentials received earlier, users send anonymous bulletins (which is one of the main requirements for any voting system). Moreover, verification that this user really has the right to vote is checked on the basis of the signed voting identifier. Users can send their vote several times, only the last one will be taken into account. This makes senseless to enforce to vote in the presence of an evil observer behind.

4.2   Comparison

Let us highlight the main similarities and differences in the two approaches and try to understand whether the absence of a blockchain is critical.

Lack of Additional Cryptographic Operations. In the previous approach, an RSA-based blind signature was used to register participants in each voting. Then, with anonymous access to the ledger, Idemix credentials were used. Now the first part is missing, or rather, is integrated into obtaining Idemix credentials (since the release of this certificate already uses blind signature BSS+).

Mobile Client Support. For the previous solution, there is only one official SDK that supports Idemix, for the Java language. This greatly complicates the development of native client applications for iOS. When using the new solution, it is possible to:

– Writing all-Go native mobile applications,
– Writing SDK applications by generating bindings from a Go package and invoke them from Java (on Android) and Objective-C (on iOS).

Support for Voting Properties. Despite the blockchain dismissal, the security features that were inherent in the previous solution were preserved:

- Eligibility - the user gets access to the vote only when the creator of the vote added it to the list. This is checked when the user confirms the participation in the vote, when he tries to get idemix credential. If he is on the list, he will receive a signed attribute that contains a voting identifier; if not, he will be refused to issue new credentials.
- Un-reusability - when issuing idemix credentials, CA puts a blind signature on the public key, which will be used to determine that the request came from the same anonymous person to exclude double counting
- Un-traceability - it will not work to track the user, since the key under which the ballots are stored in the database is known only to the user (when registering in a vote, he signs with CA blind signature, that is, CA does not know which key is signed), and all interactions are Voting times with the system occur using an identity mixer, that is, anonymously.
- Verifiability - since the user knows his key, he can make sure that in the database in conjunction with this key is exactly the newsletter that he sent.

Absence of Blockchain. The main advantage that a distributed ledger can provide is the lack of centralization and the inability to unilaterally change the data in the database. For state elections this is really important, for corporate voting this might be required to a lesser extent. But let us show what pitfalls may appear. The previous solution allowed connecting external observers by connecting new nodes to the network. However, such observers could not be arbitrarily many (due to scaling problems) and only large organizations became them. In addition, the user alone cannot initially access the network. This happens through authentication in the organization to which this user belongs. And it is the organization (a higher level of hierarchy) that owns the nodes in the network that can take part in reaching consensus. This is a limitation of the Hyperledger Fabric ledger, which is classified as permissioned. Using public blockchains is not suitable again, due to scalability issues, when we have hundreds of thousands or even millions of users, the throughput becomes so low that users cannot send their ballots because the network is congested. That is, the final voter is forced to trust his organization, which issues his identity and within which he joins the distributed network. Thus, one of the main properties of the blockchain is not fully implemented, if not absent at all. It can be argued that the organization cannot change the user's voice, as the transactions are confirmed by other parties. However, a certificate may be revoked or reissued from the user, or the organization may spawn fictitious users in order to achieve the desired voting result.

5 Conclusion

This article conducted a study of the performance of an electronic voting system using distributed ledger technologies. It is shown that in some cases it is advisable to abandon the use of the blockchain without losing the security properties. A theoretical scheme is proposed that can allow anonymous voting using Identity Mixer technology. The proposed solution suggests the possibility of developing native client applications for iOS and Android.

In order to improve the performance of the system, it is proposed to add an additional module that will help ensure receipt-freeness; this can be done using, for example, the approach from [16]. This is the direction of our future work.

References

1. IP Chain association. https://www.ipchain.ru/association/
2. NEO: Open Network for the Smart Economy. https://neo.org/
3. EOS.IO Technical White Paper v2. https://github.com/EOSIO/Documentation/blob/master/TechnicalWhitePaper.md
4. Georgiadis, E.: How many transactions per second can bitcoin really handle? Theoretically. https://eprint.iacr.org/2019/416.pdf
5. A Next-Generation Smart Contract and Decentralized Application Platform. https://github.com/ethereum/wiki/wiki/White-Paper
6. Menon, A., Bhagat, V.: Blockchain based e-voting system. Indian J. Appl. Res. **10**(1), 8–10 (2020)
7. Yi, H.: Securing e-voting based on blockchain in P2P network. EURASIP J. Wirel. Commun. Netw. **2019**(1), 1–9 (2019). https://doi.org/10.1186/s13638-019-1473-6
8. Shiyao, G., Dong, Z., Rui, G., Chunming, J., Chencheng, H.: An anti-quantum e-voting protocol in blockchain with audit function. IEEE Access **7**, 115304–115316 (2019). https://doi.org/10.1109/ACCESS.2019.2935895
9. Khan, K.M., Arshad, J., Khan, M.M.: Investigating performance constraints for blockchain based secure e-voting system. Future Gener. Comput. Syst. **105**, 13–26. ISSN 0167–739X. https://doi.org/10.1016/j.future.2019.11.005
10. Kirillov, D., Korkhov, V., Petrunin, V., Makarov, M., Khamitov, I.M., Dostov, V.: Implementation of an e-voting scheme using hyperledger fabric permissioned blockchain. In: ICCSA 2019. Lecture Notes in Computer Science, vol. 11620. Springer, Cham (2019)
11. Androulaki, E., et al.: Hyperledger fabric: A distributed operating system for permissioned blockchains. In: Proceedings of the Thirteenth EuroSys Conference (EuroSys '18), Article 30, pp. 1–15. Association for Computing Machinery, New York (2018). https://doi.org/10.1145/3190508.3190538
12. Christian, G., Stephen, L., Lukasz, G., Srinivasan, K.: FastFabric: Scaling Hyperledger Fabric to 20,000 Transactions per Second, pp. 455–463 (2019). https://doi.org/10.1109/BLOC.2019.8751452
13. Camenisch, J., Lysyanskaya, A.: Signature schemes and anonymous credentials from bilinear maps. In: Franklin, M. (ed.) CRYPTO 2004. LNCS, vol. 3152, pp. 56–72. Springer, Heidelberg (2004). https://doi.org/10.1007/978-3-540-28628-8_4
14. Au, M.H., Susilo, W., Mu, Y.: Constant-size dynamic k-TAA. In: De Prisco, R., Yung, M. (eds.) Security and cryptography for networks. In: SCN 2006, LNCS, vol. 4116. Springer, Berlin, Heidelberg (2006)
15. Camenisch, J., Drijvers, M., Lehmann, A.: Anonymous attestation using the strong Diffie Hellman assumption revisited. Cryptology ePrint Archive, IACR (2016)
16. Shinsuke, T., Hazim, H., Nazmul, I., Kazi, A.: An incoercible e-voting scheme based on revised simplified verifiable re-encryption mix-nets. Inf. Secur. Comput. Fraud **3**, 32–38 (2015). https://doi.org/10.12691/iscf-3-2-2

Yet Another E-Voting Scheme Implemented Using Hyperledger Fabric Blockchain

Sergey Kyazhin$^{(\boxtimes)}$ and Vladimir Popov

Blockchain Laboratory, Sberbank, Moscow, Russia
`blockchain-research@sberbank.ru`

Abstract. There are many papers whose authors propose various approaches to the construction of electronic voting (e-voting) systems that are resistant to various types of attacks. The two main properties of such systems are anonymity and verifiability. To provide anonymity, a blind signature or a ring signature is often used. To provide verifiability, distributed ledger technologies, in particular blockchain, have recently been used. One of these systems has been presented at ICCSA 2019. This system is implemented using Hyperledger Fabric blockchain platform and uses a blind signature to provide anonymity. One of the disadvantages of this system is that a compromised signer (an organizer generating a blind signature) could independently create valid ballots without detection. In this paper, we modify this system by replacing a blind signature with a linkable ring signature in order to eliminate this disadvantage. As a result, we combine linkable ring signature, Idemix and blockchain technologies to construct a decentralized anonymous e-voting system that does not rely on a trusted signer and can achieve the following properties: eligibility, unreusability, anonymity, and verifiability. In addition, the use of both a linkable ring signature and Idemix allows us to construct a two-stage anonymization, that increases the versatility of the proposed system. This allows us to use a blockchain platform (for example, Hyperledger Fabric) to implement the e-voting system, without making changes to platform standard signature scheme.

Keywords: E-voting · Blockchain · Ring signature · Idemix · Distributed ledger technologies

1 Introduction

The traditional voting procedure, depending on specific circumstances, may have various disadvantages. The main ones are the following:

- non-verifiability the correctness of the result counting;
- voting by the organizer instead of the voter;
- double voting (if the organizer and the voter are in collusion).

© Springer Nature Switzerland AG 2020
O. Gervasi et al. (Eds.): ICCSA 2020, LNCS 12251, pp. 37–47, 2020.
https://doi.org/10.1007/978-3-030-58808-3_4

Therefore, electronic voting (e-voting) is continuously developing. It empowers voters to cast their ballots remotely (through a local or global network), but has some problems. For example, attempts to make voting anonymous often reduce the ability to verify the correctness of the result counting. Therefore, the voting procedure requires public trusted ballot box. To solve this problem, in recent papers, e-voting systems use a distributed ledger technologies to create a public ballot box.

In this paper, we modify one of the existing e-voting systems that uses blockchain technology.

1.1 Security Properties

In various papers, the security requirements that an e-voting system must satisfy are formulated differently. We use the following five properties used in [1].

Eligibility. This property requires that only eligible voters are allowed to vote. In addition, only valid ballots should be counted.

Unreusability or Double Voting Prevention. This property means that voters are allowed to vote only once.

Untraceability or Anonymity. This property requires that no one can reveal the owner of the ballot. In other words, voters are allowed to vote anonymously.

Verifiability. This property means that each voter can verify whether his ballot has been counted correctly.

Receipt-Freeness. This property prevents the voter from proving to others that he has voted for a particular candidate (for example, to sell his vote).

1.2 Related Work

Common cryptographic techniques used to achieve anonymity in e-voting systems are blind signature [2] and linkable ring signature [3].

A blind signature is commonly used in the voter registration phase. Specifically, a legitimate voter receives a blind signature on a random value. The signature value and the random value are then used to prove the authenticity of the voter. This method was first used in the secret e-voting system [4]. However, the use of blind signature requires the signer to be trusted. If the signer is compromised, the attacker will be able to vote as many times as he wants [5].

A linkable ring signature is commonly used as a replacement for a standard signature to provide the authenticity of ballots. With it, you can prevent unauthorized voting while maintain the privacy of legitimate voters. The first e-voting system using a linkable ring signature was proposed in the paper with the first linkable ring signature scheme [3].

In [1], a blind signature is used to construct an e-voting system. User registration includes the step of obtaining a blind signature for a new voting key. Indeed, the system has all the declared security properties. However, there is also a drawback: a compromised organizer can independently create valid ballots.

If the smart contract generates a blind signature, then the secret key for blind signature is available to all peer owners. If the organizer generates a blind signature, then the secret key for blind signature is available to him.

Thus, anyone who has access to a secret key for blind signature can generate and sign as many new voting keys as he wants. Then, having received an anonymous Idemix-certificate (for example, using one of the valid users), he will be able to use each of these keys for voting, since the signatures are valid.

Therefore, we decided to modify the system proposed in [1]. In our construction, we use a linkable ring signature. The described attack is impossible for it. Since neither the organizer nor the smart contract generates any signatures for registration, neither the organizer nor the peer owners will be able to generate additional ballots.

Therefore, when we use a linkable ring signature, we may not require a trusted signer. However, the disadvantage of this approach is that the procedures for generating and verifying the signature have a computation complexity linear to the size of the anonymity set (i. e. the number of legitimate voters).

There are many papers devoted to methods and systems for e-voting, including those that use blockchain technologies and linkable ring signatures, for example, [6–8]. However, we, like [1], also use Idemix technology to anonymize the participants.

1.3   Our Contributions

In this paper, we propose an e-voting system that uses linkable ring signature, Idemix and blockchain technologies. The proposed system is a modification of the system described in [1]. Namely, a blind signature was replaced by a linked ring signature. This modification allows us to improve the scheme, since it eliminates the disadvantage associated with the ability to create valid votes by the compromised organizer.

Thus, we propose a decentralized anonymous e-voting system, which does not rely on a trusted signer, and can achieve the following properties: eligibility, unreusability, anonymity, and verifiability.

Note that the e-voting system described in [3] uses only a linkable ring signature instead of the standard one to provide the anonymity. However, we use both a linkable ring signature and Idemix, because it allows us to construct a two-stage anonymization, that increases the versatility of the proposed system. To implement e-voting system, a blockchain platform (for example, Hyperledger Fabric [9]) can be used without changing the platform standard signature scheme.

1.4   Outline

In the next section, we describe the following used technologies: a linkable ring signature, Idemix, and Hyperledger Fabric blockchain. We propose the e-voting system in Sect. 3. In Sects. 4 and Sect. 5, we provide remarks on the system security and performance properties. We conclude the paper in Sect. 6.

2   Used Technologies

To construct our e-voting system, we use blockchain, Idemix and linkable ring signature technologies.

2.1   Blockchain

Depending on the user access control, all blockchain systems can be divided into two groups: permissionless and permissioned blockchain systems. The presented implementation involves the use of a permissioned system. That is, all voting entities, and their rights are known in advance. Specifically, we use Hyperledger Fabric permissioned blockchain platform.

In the proposed implementation, the blockchain system is used:

– to provide the impossibility of censorship;
– to verify the results by both voters and inspectors.

2.2   Idemix

Suppose there are three parties: the user, the certification authority (can identify the user), and the verifier (wants to verify some user attributes). Idemix technology (details are described in [10–13] and other papers) solves the problem of proving that the user has certain attributes without revealing other information about the user and making it impossible to link two different requests of the same user. The proposed implementation uses the user's entry into the list of system users as an attribute.

2.3   Linkable Ring Signature

Let there be a group of n participants and some message m that needs to be signed. The linkable ring signature scheme consists of the following four algorithms:

– $KGen(1^\lambda)$ is a key generation algorithm that uses a security parameter λ as an input and returns to the i-th participant a secret key sk_i and a public key pk_i, $i = 1, \ldots, n$;
– $Sign(m, pk_1, \ldots, pk_n, i, sk_i)$ is a signature generation algorithm that uses a message m, public keys $pk_1, \ldots, pk_n$ of all group members, a number i of the signer, and his secret key sk_i as inputs, and returns the signature σ;

– $Verify(m, pk_1, \ldots, pk_n, \sigma)$ is a signature verification algorithm that uses a message m, the public keys $pk_1, \ldots, pk_n$ of all group members and the signature σ as inputs, and returns 1 if the signature is valid, 0 otherwise;
– $Link(\sigma_1, \sigma_2)$ is a linking algorithm that uses signatures σ_1, σ_2 as inputs, and returns 1 if these signatures are generated using the same key, 0 otherwise.

Therefore:

– the verifier (the one who runs the algorithm $Verify$) can find out that the signer is a group member, but cannot find out who exactly;
– the linker (the one who runs the algorithm $Link$) can find out that the same signer calculated two signatures on the same or different messages, but cannot find out who exactly.

In the proposed system, a linkable ring signature is used to provide:

– the anonymity of voters (if the user casts a signed ballot, it is known that he is a legitimate voter, but his identity is unknown);
– the impossibility of double voting (if the voter casts the second signed ballot, it is known that he is a double voter, but his identity is unknown).

2.4 Encryption and Secret Sharing

In [1], the voters can send encrypted ballots (a public key cryptosystem is used). After the voting, the decryption key is published, and anyone can decrypt the ballots and verify the results. Secret sharing schemes can be used to prevent intermediate result counting.

For the scheme proposed in this paper we also can use of a public key encryption and secret sharing schemes. However, for simplicity of presentation, we will not describe the corresponding steps, since they are optional.

3 Construction

The described system is a protocol for the interaction of several components.

3.1 System Components

To make it easier to compare our scheme with [1], we do not make changes to the network architecture (see Fig. 1 in [1]). The four types of components involved in the e-voting system is described as follows.

User Applications. The system includes user software applications according to user roles. They are needed for users to interact with the blockchain network. For example:

– the organizer application allows us to initiate a voting;
– the voter application allows us to register, vote and verify the voting results.

Certification Authority. The system includes a certification authority that can operate in two modes: a standard mode (working with user public key certificates) and Idemix mode (working with anonymous Idemix certificates).

Peers. The system includes one or more nodes of the blockchain network (in our implementation, nodes are based on Hyperledger Fabric). This component performs the functions of storing and writing information in the ledger, as well as executing the logic of voting smart contracts (chaincode in Hyperledger Fabric terminology).

Ordering Service. The system includes an ordering service. This component is required to ordering changes in the ledger.

3.2   Entities

The e-voting system contains the following three types of entities, namely, Organizer, Voter and Inspector (optional).

Organizer. This entity makes a list of voting questions and a list of voters, sets the time frames for the registration and voting phases (the entity has nodes of the blockchain network and an organizer application).

Voter. This entity is a voting participant (the entity may have nodes of the blockchain network and has a voter application).

Inspector. This optional entity observes the correct operation of the system (the entity may have nodes of the blockchain network and has an inspector application).

3.3   E-Voting Protocol

In this subsection, we describe the process of e-voting protocol illustrated in Fig. 1. The protocol consists of the following five phases: blockchain network configuration, voting configuration, registration, voting, and result counting.

Blockchain Network Configuration. The administrator creates users in the system, determines their rights in accordance with the roles, configures a certification authority, uploads smart contracts, determines nodes and ordering service.

The certification authority creates key pairs and issues public key certificates for users (1, 2 in Fig. 1). In the paper, it is assumed that:

– each user interaction with the blockchain network involves authentication using the public key certificate;
– writing information to the ledger involves generating requests to change the ledger, its processing on the peers and ordering on the ordering service, and then writing the corresponding information to the ledger.

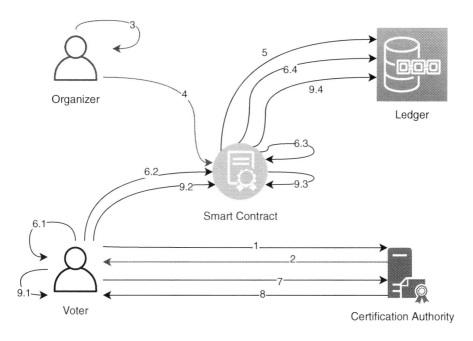

Fig. 1. Overview of our blockchain-based e-voting protocol

Voting Configuration. The organizer:

- creates a voting session, i. e. determines its parameters, including a voting ID, a list of voting questions, a list of voter public keys, a time frames for registration and voting phases (3 in Fig. 1);
- sends this data to the peers that write it in the ledger (4, 5 in Fig. 1).

Registration. Each voter (j is the voter number):

- generates a new voting key pair (public key VP_j and secret key VS_j), where for the linkable ring signature scheme without interaction with a certification authority and despite the existing key pair (6.1 in Fig. 1);
- sends the public key VP_j and the voting ID to the peers, using the public key certificate for authentication (6.2 in Fig. 1).

The smart contract verifies that the current user can participate in this voting by verifying the availability of the corresponding public key in the voting parameters (6.3 in Fig. 1). Then the peers write the key VP_j in the ledger according to the participant's public key certificate (6.4 in Fig. 1).

Voting. Each voter who has registered their voting key:

- receives an anonymous Idemix certificate, interacting with the certification authority, that operates in Idemix mode (7, 8 in Fig. 1);

– downloads the public keys for voting of all voters;
– generates his answers to the questions, calculates the value of the linkable ring signature on these answers, using his private key VS_j and the set of the downloaded public keys for voting (9.1 in Fig. 1);
– sends the voting answers and the linkable ring signature to the peers, using the anonymous Idemix certificate instead of the standard certificate for authentication (9.2 in Fig. 1).

The smart contract:

– verifies that the linkable ring signature is correct by running the algorithm $Verify$ with the voting answers, the public keys for voting of all voters, and the linkable ring signature as inputs (9.3 in Fig. 1);
– verifies that the voter voted earlier (runs the algorithm $Link$ with this signature and each of previously written in the ledger as inputs).

If the signature is valid, the peers write (rewrite if the voter voted earlier) its value and the voter's answers in the ledger (9.4 in Fig. 1).

Before the end of the voting procedure, the voter may revote, having received a new anonymous Idemix certificate and sending new answers. His answers will be rewritten in the ledger because the same voting secret key will be used to calculate the linkable ring signature.

Result Counting. If encryption is not used, the results are counted simultaneously with the voting. The results can be updated after each vote.

4 Remarks on Security Properties

Like the scheme [1], the e-voting system presented in this paper can achieve four of the five properties listed in the introduction, namely, eligibility, unreusability, anonymity, and verifiability.

The **receipt-freeness** property is not fulfilled in this system, because the voter can prove that he has generated a certain linkable ring signature (due to the linkability), thereby proving his answers.

4.1 Eligibility

This property is provided:

– at the registration phase by the security of the authentication procedure in Hyperledger Fabric (an attacker cannot pass the authentication procedure, therefore, cannot add his voting key on the registration phase);
– at the voting phase by the security of the linkable ring signature (an attacker cannot fake the linkable ring signature, therefore, cannot fake the ballot).

4.2   Unreusability

This property is provided:

- at the registration phase by the security of the authentication procedure in Hyperledger Fabric (an attacker cannot register two different voting keys, because the user identity (certificate) is used for authentication);
- at the voting phase by the linkability of the linkable ring signature (an attacker cannot vote twice with the same voting key, because a smart contract can find out that two linkable ring signatures calculated using the same key correspond to the same voter, even if he used different anonymous Idemix certificates).

4.3   Anonymity

This property is provided by the anonymity of the linkable ring signature and the security of Idemix (the identity of the voters cannot be disclosed).

4.4   Verifiability

All votes are written in the ledger according to the linkable ring signatures. Each user can verify whether his ballot has been counted correctly.

5   Remarks on Performance Properties

In this paper, we describe only the e-voting scheme. We don't present the results of experiments evaluating its performance and corresponding comparison with other solutions.

We can give a remark on **scalability**. Compared to the scheme [1], our scheme scales worse because it uses a linkable ring signature. The procedures for its generation and verification have a computational complexity linear to the number of voters.

6   Conclusion and Future Work

In this paper, we modify e-voting scheme [1]. Namely, we replace a blind signature with a linkable ring signature in order to eliminate the disadvantage associated with the ability to cast votes by the compromised organizer.

Thus, we combine linkable ring signature, Idemix and blockchain technologies to construct a decentralized anonymous e-voting system, which does not rely on a trusted signer, and can achieve the following properties: eligibility, unreusability, anonymity, and verifiability.

The use of both a linkable ring signature and Idemix allows us to construct a two-stage anonymization, that increases the versatility of the proposed system. This allows us to use a blockchain platform (for example, Hyperledger Fabric) to implement the e-voting system, without making changes to platform standard signature scheme.

In future work, we would like:

– to carry out experiments for evaluation the performance of our scheme;
– to continue the modification of this scheme to achieve the receipt-freeness.

References

1. Kirillov, D., Korkhov, V., Petrunin, V., Makarov, M., Khamitov, I.M., Dostov, V.: Implementation of an E-voting scheme using hyperledger fabric permissioned blockchain. In: Misra, S., et al. (eds.) ICCSA 2019. LNCS, vol. 11620, pp. 509–521. Springer, Cham (2019). https://doi.org/10.1007/978-3-030-24296-1_40
2. Chaum, D.: Blind signatures for untraceable payments. In: Chaum, D., Rivest, R.L., Sherman, A.T. (eds.) Advances in Cryptology, pp. 199–203. Springer, Boston (1983). https://doi.org/10.1007/978-1-4757-0602-4_18
3. Liu, J.K., Wei, V.K., Wong, D.S.: Linkable Spontaneous anonymous group signature for ad hoc groups. In: Wang, H., Pieprzyk, J., Varadharajan, V. (eds.) ACISP 2004. LNCS, vol. 3108, pp. 325–335. Springer, Heidelberg (2004). https://doi.org/10.1007/978-3-540-27800-9_28
4. Fujioka, A., Okamoto, T., Ohta, K.: A practical secret voting scheme for large scale elections. In: Seberry, J., Zheng, Y. (eds.) AUSCRYPT 1992. LNCS, vol. 718, pp. 244–251. Springer, Heidelberg (1993). https://doi.org/10.1007/3-540-57220-1_66
5. Gong, B., Lu, X., Fat, L.W., Au, M.H.: Blockchain-based threshold electronic voting system. In: Meng, W., Furnell, S. (eds.) SocialSec 2019. LNCS, vol. 1095, pp. 238–250. Springer, Singapore (2019). https://doi.org/10.1007/978-981-15-0758-8_18
6. Lyu, J., Jiang, Z.L., Wang, X., Nong, Z., Au, M.H., Fang, J.: A secure decentralized trustless e-voting system based on smart contract. In: 18th IEEE International Conference on Trust, Security and Privacy in Computing and Communications/ 13th IEEE International Conference on Big Data Science and Engineering (TrustCom/ BigDataSE), pp. 570–577. Rotorua, New Zealand (2019). https://doi.org/10.1109/TrustCom/BigDataSE.2019.00082
7. Yu, B., et al.: Platform-independent secure blockchain-based voting system. In: Chen, L., Manulis, M., Schneider, S. (eds.) ISC 2018. LNCS, vol. 11060, pp. 369–386. Springer, Cham (2018). https://doi.org/10.1007/978-3-319-99136-8_20
8. Hjlmarsson, F., Hreiarsson, G.K., Hamdaqa, M., Hjlmtsson, G.: Blockchain-based E-voting system. In: 11th International Conference on Cloud Computing (CLOUD), pp. 983–986. San Francisco, CA (2018). https://doi.org/10.1109/CLOUD.2018.00151
9. Androulaki, E. et al.: Hyperledger fabric: A distributed operating system for permissioned blockchains. In: Proceedings of the Thirteenth EuroSys Conference (EuroSys18), pp. 1–15. ACM, New York (2018). https://doi.org/10.1145/3190508.3190538
10. Camenisch, J., Van Herreweghen, E.: Design and implementation of the idemix anonymous credential system. In: Proceedings of the 9th ACM conference on Computer and communications security (CCS02), pp. 21–30. ACM, New York (2002). https://doi.org/10.1145/586110.586114
11. Camenisch, J., Lysyanskaya, A.: Signature schemes and anonymous credentials from bilinear maps. In: Franklin, M. (ed.) CRYPTO 2004. LNCS, vol. 3152, pp. 56–72. Springer, Berlin, Heidelberg (2004). https://doi.org/10.1007/978-3-540-28628-8_4

12. Camenisch, J., Drijvers, M., Lehmann, A.: Anonymous attestation using the strong diffie hellman assumption revisited. In: Franz, M., Papadimitratos, P. (eds.) Trust 2016. LNCS, vol. 9824, pp. 1–20. Springer, Cham (2016). https://doi.org/10.1007/978-3-319-45572-3_1
13. Camenisch, J., Drijvers, M., Dubovitskaya, M.: Practical UC-secure delegatable credentials with attributes and their application to blockchain. In: Proceedings of the 2017 ACM SIGSAC Conference on Computer and Communications Security (CCS17), pp. 683–699. ACM, New York (2017). https://doi.org/10.1145/3133956.3134025

Modelling the Interaction of Distributed Service Systems Components

Oleg Iakushkin$^{(\boxtimes)}$, Daniil Malevanniy, Ekaterina Pavlova, and Anna Fatkina

Saint-Petersburg University, 7/9 Universitetskaya nab., St. Petersburg 199034, Russia
o.yakushkin@spbu.ru

Abstract. The development of efficient distributed service systems places particular demands on interaction testing methods. It is required to consider not only the work of individual components at each stage of development but also their interaction as a system. Often, such testing is performed directly at the time of publishing the application with end-users. In this paper, we consider methods of emulating and testing distributed systems on limited resources available to individual developers. What to do with factors beyond the scope of tests of one application? How would the system behave when the network is unstable, and the third-party service does not work correctly? In this work, we present a new tool created to simulate the operation of a full-fledged service system and failure of its specific elements. The main feature of the solution we created is in the provision of abstraction for the developer, which does not require a deep understanding of the work of network infrastructures. It allows us to compare the work of several distributed applications under the same conditions, check the stability of the programme regarding the unstable operation of network elements, study the influence of the network structure on the of services behaviour and detect their vulnerabilities associated with interservice interaction.

Keywords: Network · Modeling systems · Docker · Distributed applications

1 Introduction

Large systems require rigorous testing before commissioning. Let us consider the distributed application, in addition to the correct operation of each of its components, and it is necessary to take into account network interactions. In this work, we focus on testing of such communications.

There are several modelling tools [8] (for example, GNS3 [2], eve-ng [1]) with which you can deploy a network of virtual machines and test the application on it. However, these testing systems require the user to know in the field of system administration.

O. Iakushkin—This research was partially supported by the Russian Foundation for Basic Research grant (project no. 17-29-04288).

ⓒ Springer Nature Switzerland AG 2020
O. Gervasi et al. (Eds.): ICCSA 2020, LNCS 12251, pp. 48–57, 2020.
https://doi.org/10.1007/978-3-030-58808-3_5

Applications can be tested on real users using tools like Pumba [6] or Gremlin [3], but such a solution carries the risk of losing users' interest in the service.

The main goal of our work is to provide an easy way to model a scalable network. An application developer does not require in-depth knowledge of network technology to use the tools described in this paper. We propose to consider a model of a network consisting of Docker image nodes and routers. This approach allows emulation of a distributed system locally without going into the details of network interactions. The model enables reproducing effects observed by developers, such as duplication, loss, delay, or distortion of data packages.

2   Related Work

There are about two dozen emulators of distributed networks [24]. Most of them are free and available in the public domain (including GitHub). Also, HP, Huawei, Cisco have their network emulators oriented to work with equipment of a specific manufacturer. We considered the most popular network modelling environments. We selected some characteristics for comparison (Table 1): ease of use, scalability, and versatility. The ease of use of the tool is evaluated by the following criteria: configuration format, the availability of visualization, automatic deployment, and launch of distributed applications, the ability to access applications on the Internet.

There is no universal solution for emulating large-scale systems. In some cases, preference is given to modelling process details of the network layer L2; in other cases, the interaction is considered at the level of model abstractions. L3NS [4] and MADT [5] solutions allow while preserving the complexity of the services under consideration, to simplify the network abstraction as much as possible, providing the availability of modelling on one or several computing nodes and platforms.

3   Problem Definition

Our goal is to create a universal, scalable solution with which you can quickly simulate the interaction of network components and monitor or change the state of the model in real-time.

To emulate a virtual computer network, the Docker containerization [18] platform is used. Moreover, when the model is run on a cluster WireGuard [15] is used to model IP subnets. Also, a set of FRRouting utilities is used, which allows the implementation of dynamic routing protocols, and the Tcconfig shell, which allows to control [14] the quality of the network dynamically and which is used to improve the realism of the model. The hypergraph separation algorithm KaHyPar is used to organize distributed modelling of virtual [20] computer networks. The jgraph, Flask, and zmq messaging libraries are used to operate the user interface (Fig. 1).

Table 1. Comparative table of existing modeling technologies

Options/ Competitors	L3NS	Madt	Netkit/ Kathara	Mininet	Marionnet	zimulator (VNUML)	Eve-NG	NS3	GNS3	Omnet++ / Omnest
Network configuration format	Python Script	Python Script/ JSON	BASH script / special config	Python API	project file with conf of vm	VNUML Script	labs	c++ script	virual network editor	NED topology description language
Use of lightweight virtualization [7]	+	+	+	- VirtualBox, Qemu, VMware	- VirtualBox, VMware	-	-	+	-	-
Visualization	-	+	+ Netkit Lab Generator	+	+	-	+	+	+	+
Automatically deploy and launch distributed applications	+	+	The application must be correctly installed in the host system	+	+	+	+	+	-	-
The ability to scale the system by using more than 1 physical node for deployment	+	-	-	+	-	-	+ for a fee	+	+	+
Network simulation level	L3	L3	L3	L2	L2	DES*	L2	L2	L2	DES
Connection of external network nodes	+	-	+	+	+	-	+	-	+	+
The ability to access applications on the Internet	+	+	+	+	+	-	+	+	+	+
Platforms on which simulated applications can run	Linux	Linux	all platforms	Unix/ Linux	GNU/ Linux	Linux	VM based Windows, Linux, network devices	Linux, MacOS, FreeBSD	VM based Windows, Linux, host device	all platforms where a modern C++ compiler is available
Open source	MIT	Academic Public License	GPL3	MIT derevative	GPL	GPL3	Academic Public License	GPL2	GPL	Academic Public License

* DES – Discrete-event simulation, not bound to networking level.

Simulation	Additional tools	Monitoring
	 FRRouting suite Tcconfig A tc command wrapper KaHyPar Karlsruhe Hypergraph Partitioning	jgraph An embeddable webGL graph visualization library.  Flask ØMQ

Fig. 1. Technology stack used by MADT and L3NS.

4   Proposed Architecture

The traditional way of modelling computing nodes of a network [12] is the use of hypervisors that manage virtual machines. The kernels of the operating systems are separated; this provides isolation for applications [22]. In case of modelling a network of interacting application services, nodes and applications do not need isolation at the kernel level of the OS, isolation of the network stack that the computing node [11] use is sufficient.

The Linux kernel provides a tool (Namespaces) [13] that allows isolation of the low-level environments [17]. A container is an isolated set of namespaces on the top of which an application is running. The Docker configuration platform allows to create such environments corresponds to predefined templates and run applications in them. Besides, using Docker, we can combine network interfaces

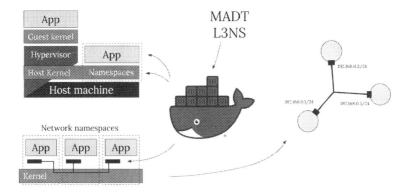

Fig. 2. MADT/L3NS workflow: 1. Commands are sent directly to docker server; 2. Docker server creates containers using Linux namespaces; 3. Docker connects containers with virtual LANs; 4. Virtual LANs and docker containers form a virtual global network.

located in different network namespaces with a virtual bridge into one common IP subnet, the hosts of which are containers. Our solutions allow pre-setting of the network topology and launch parameters of containers. This network is launched by Docker [25] and it turns out a virtual network model that fully works on one computer (Fig. 2). If the simulated application requires a large number of resources (Fig. 3), it is advisable to divide the simulation into several computers. This solves the problem of efficiently sharing the load across hosts to minimize latency in virtual networks connecting containers [16] located on different hosts. In each of the containers launched as part of the model, a Unix socket is mounted. This allows sending messages to the monitoring system [26], even if the network interfaces of the container are blocked or have a long delay, thus ensuring stable monitoring [9]. Using the web interface, you can monitor the state of the model in real-time and interactively manage network parameters for each node, controlling [19] delays, losses, duplicate packets.

5   Modeling Algorithm

The work proceeds according to the following algorithm:

1. Using the API in Python, a "Laboratory" is created — a description of the network model, which includes a list of network nodes, their distribution over subnets, and startup parameters.

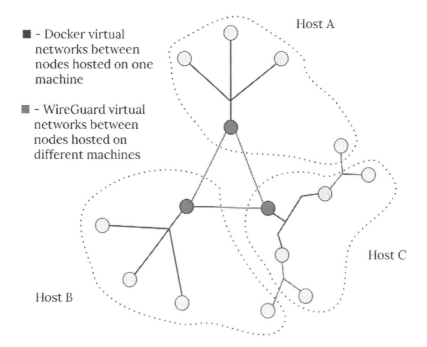

Fig. 3. Modeling a distributed network using the WireGuard tool.

2. "Laboratory" starts by creating network nodes based on Docker containers and network connections between them.
3. Displaying the emulated network is available in the web interface for conducting experiments and monitoring their progress.

6   Implementations of the Proposed Architecture in L3NS and MADT

We have tested several approaches to the implementation of the proposed architecture. The first of them is the MADT program complex, divided into three parts: a library for describing "Laboratories", a server for launching them, and a monitoring interface with zmq sockets. The second approach is implemented in the form of the L3NS library containing all the components necessary for modelling the network in the format of one programmable solution. While MADT allows to familiarize yourself quickly with the concept of the proposed architecture and set the experiment visually, L3NS has the functionality to more quickly and flexibly describe the model of node interaction and implementation in the CI ecosystem.

7   Example: Modeling in Applications with Client-Server Architecture

Let us consider how, using the proposed concept. A network is built with a set of routers and dynamic routing. Such a system can be represented as a tree, the nodes of which are routers and clients. Figure 4 shows how the network looks in case of height of the tree, and the number of leaves is set by 3 and 3.

The construction of such a network is carried out according to the following principle: the graph is expanded using several iterations of adding subnets for child nodes. At the end of each iteration, the array of external nodes is updated. For all nodes, except the external ones, the operation of the dynamic routing protocol is configured. This allows traffic to pass through the tree.

Each child node of the graph is assigned a Docker image that is used to create the client node.

Another step that has to be taken for the correct operation of our virtual network is to define the server address for each client so that they can interact.

When the model starts, it becomes possible to set network problems manually and monitor how they affect the application [21].

8   Modeling Management in the MADT Visual Interface

In MADT, after saving the configuration of the built network, we can go to the graphical interface and monitor the operation of the application. To configure the network status, you have to click on any node and set the interference parameters using the window that opens (Fig. 5a) and then press the "tcset" button. For example, to simulate 50% package loss for a router, we need to set the corresponding value of the loss field.

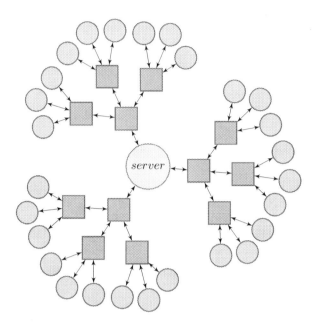

Fig. 4. Network model: server in the middle, blue boxes represent routers, yellow circles depict client nodes. (Color figure online)

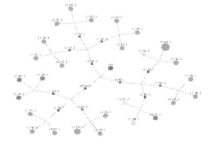

(a) Setting parameters for a single node   (b) Visualization of emulated network

Fig. 5. MADT visual interface. (Color figure online)

After we set the problem for several nodes, we can observe the result on the graph. The colour of the node depends on the status of the incoming message. Green indicates regular operation, and yellow indicates a malfunction, red indicates critical problems, and purple indicates unexpected exceptions (Fig. 5b).

9   Differences in Modelling Capabilities of the L3NS Library and the MADT Software Package

The L3NS code is independent of MADT. The main functional features are the logic of the distribution of IP-addresses. Addresses are distributed according to the description of the "Laboratory", due to which the IP addresses of some nodes can be used to describe others (for example, immediately define the server address for clients) [10]. Also, work with Docker networks has been dramatically simplified in L3NS: network IP addresses, as well as the IP addresses of nodes in them, are set directly, while in MADT a second IP address is given to each interface on the network.

When describing network elements, different strategies for starting and stopping nodes can be used in L3NS. For example, using the local Docker container submodule, the nodes are launched as Docker containers. Meanwhile using cluster submodule, the containers are started on remote machines. L3NS provides the ability to configure remote machines via ssh. In other words, it is possible to access the client on a specific host, and, for example, view a list of Docker images used on it. To run the service system model on several virtual machines, you need to:

– Configure ssh keys,
– Install Docker,
– Install dependencies in python,
– Install WireGuard.

In L3NS support for parallel launching of containers is implemented using ThreadPoolExecutor, which speeds up the launch of "Laboratories" by order of magnitude, as well as networking via Docker Swarm [23] compared to MADT. The approach based on the creation of a single solution in the form of a library for a user programmer showed greater flexibility in system scaling tasks than the MADT program package, which separates the functions of describing, running and configuring "Laboratories".

10   Conclusion

The solutions we developed for testing distributed service systems allow to:

– Create realistic models of large IP networks;
– Deploy distributed applications based on Docker images;
– Manage the quality of work of different parts of the simulated network;

– Visualize the state of a distributed application working in real-time.

Two approaches were implemented — the MADT software package and the self-contained L3NS library. Using the library, it is convenient to test network interaction by emulating the network both on one node and simulating the interaction of nodes in a distributed model. Its use can significantly accelerate the launch of a simulated network without a web interface. While using the complex of programs MADT, it is easy to carry out interactive search experiments. Thus, MADT and L3NS can be used for:

– Comparing the work of several distributed applications under the same conditions;
– Checking application stability regarding unstable network operation;
– Studying the impact of network structure on the operation of each application;
– Checking for network plan vulnerabilities in application interaction.

Acknowledgments. This research was partially supported by the Russian Foundation for Basic Research grant (project no. 17-29-04288).

References

1. EVE-NG webpage. https://www.eve-ng.net/. Last Accessed 20 June 2020
2. GNS3 webpage. https://www.gns3.com/. Last Accessed 20 June 2020
3. Gremlin webpage. https://www.gremlin.com/. Last Accessed 20 June 2020
4. L3NS GitHub repository. https://github.com/rukmarr/l3ns. Last Accessed 20 June 2020
5. MADT GitHub repository. https://github.com/dltcspbu/madt. Last Accessed 20 June 2020
6. Pumba GitHub repository. https://github.com/alexei-led/pumba. Last Accessed 20 June 2020
7. Al-Rakhami, M., et al.: A lightweight and cost effective edge intelligence architecture based on containerization technology. World Wide Web **23**, 1–20 (2019)
8. Barrachina-Muñoz, S., Wilhelmi, F., Selinis, I., Bellalta, B.: Komondor: A wireless network simulator for next-generation high-density wlans. In: 2019 Wireless Days (WD), pp. 1–8. IEEE (2019)
9. Bhushan, K., Gupta, B.B.: Distributed denial of service (ddos) attack mitigation in software defined network (sdn)-based cloud computing environment. J. Ambient Intell. Hum. Comput. **10**(5), 1985–1997 (2019)
10. Huo, C., Yuan, J., Song, G., Shi, Z.: Node reliability based multi-path routing algorithm of high-speed power line communication network. In: 2019 IEEE 4th International Conference on Cloud Computing and Big Data Analysis (ICCCBDA), pp. 570–573. IEEE (2019)
11. Iakushkin, O., Malevanniy, D., Bogdanov, A., Sedova, O.: Adaptation and deployment of panda task management system for a private cloud infrastructure. pp. 438–447. Springer International Publishing (2017)
12. Koganty, R., Alex, N., Su, C.H.: Framework for networking and security services in virtual networks, US Patent 10203972, 12 Feb 2019

13. Lang, D., Jiang, H., Ding, W., Bai, Y.: Research on docker role access control mechanism based on drbac. In: Journal of Physics: Conference Series. vol. 1168, p. 032127. IOP Publishing (2019)
14. Li, T., Gopalan, K., Yang, P.: Containervisor: Customized control of container resources. In: 2019 IEEE International Conference on Cloud Engineering (IC2E), pp. 190–199. IEEE (2019)
15. Lipp, B., Blanchet, B., Bhargavan, K.: A mechanised cryptographic proof of the wireguard virtual private network protocol. In: 2019 IEEE European Symposium on Security and Privacy (EuroS&P), pp. 231–246. IEEE (2019)
16. Lofstead, J., Baker, J., Younge, A.: Data pallets: containerizing storage for reproducibility and traceability. In: International Conference on High Performance Computing, pp. 36–45. Springer (2019)
17. Mahmud, R., Buyya, R.: Modelling and simulation of fog and edge computing environments using ifogsim toolkit. In: Buyya, R., Srirama, S.N. (eds.) Fog and Edge Computing: Principles and Paradigms, pp. 1–35. Wiley, Hoboken (2019)
18. Malevanniy, D., Iakushkin, O., Korkhov, V.: Simulation of distributed applications based on containerization technology. In: Computational Science and Its Applications - ICCSA 2019, pp. 587–595. Springer International Publishing (2019)
19. Malevanniy, D., Sedova, O., Iakushkin, O.: Controlled remote usage of private shared resources via docker and novnc. In: Computational Science and Its Applications - ICCSA 2019, pp. 782–791. Springer International Publishing (2019)
20. Odun-Ayo, I., Geteloma, V., Eweoya, I., Ahuja, R.: Virtualization, containerization, composition, and orchestration of cloud computing services. In: International Conference on Computational Science and Its Applications, pp. 403–417. Springer (2019)
21. Paillisse, J., Subira, J., Lopez, A., Rodriguez-Natal, A., Ermagan, V., Maino, F., Cabellos, A.: Distributed access control with blockchain. In: ICC 2019–2019 IEEE International Conference on Communications (ICC), pp. 1–6. IEEE (2019)
22. Sedayao, J.C., Smith, C.A., Li, H., Yoshii, T.H., Black, C.D., Hassan, V., Stone, D.W.: Virtual core abstraction for cloud computing, US Patent 10176018, 8 Jan 2019
23. Shah, J., Dubaria, D.: Building modern clouds: using docker, kubernetes & google cloud platform. In: 2019 IEEE 9th Annual Computing and Communication Workshop and Conference (CCWC), pp. 0184–0189. IEEE (2019)
24. Son, J., Buyya, R.: Latency-aware virtualized network function provisioning for distributed edge clouds. J. Syst. Softw. **152**, 24–31 (2019)
25. Tarasov, V., Rupprecht, L., Skourtis, D., Li, W., Rangaswami, R., Zhao, M.: Evaluating docker storage performance: from workloads to graph drivers. Clust. Comput. **22**(4), 1159–1172 (2019)
26. Zaman, F.A., Jarray, A., Karmouch, A.: Software defined network-based edge cloud resource allocation framework. IEEE Access **7**, 10672–10690 (2019)

Data Quality in a Decentralized Environment

Alexander Bogdanov[1], Alexander Degtyarev[1,3],
Nadezhda Shchegoleva[1(✉)], and Valery Khvatov[2]

[1] Saint Petersburg State University, 7–9, Universitetskaya Emb.,
St. Petersburg 199034, Russia
{a.v.bogdanov,n.shchegoleva}@spbu.ru, deg@csa.ru
[2] DGT Technologies AG, 15 Wertheim Ct, Richmond Hill,
ON L4B 3H7, Canada
valery.khvatov@gmail.com
[3] Plekhanov Russian University of Economics, 36, Stremyanny per,
Moscow 117997, Russia

Abstract. One of the most important aspects of applying distributed ledger technologies in the field of big data is ensuring the necessary data quality and calculating the corresponding metrics. The paper proposes a conceptual framework for working with Master Data in a decentralized environment. The greatest effect of this framework methods is increasing a real-time integrity for the data segments that have the direct impact on overall data quality. The proposed approach provides the result thanks to a special platform architecture similar to the blockchain and built-in artificial intelligence agents - oracles.

Keywords: Data quality · Distributed ledger technologies · Master data management · F-BFT consensus

1 Introduction

1.1 Motivation

Modern organizations grow their business by forming groups of companies, alliances and by building relationships with subcontractors and partners through the processes of horizontal and vertical integration. Integration allows for the creation of a unified information space within one organization or even along the entire value chain. The effectiveness of integration and business continuity depends on the alignment of core directories, as well as the matching of configurations and metainformation between connected organizations.

Imprecise data is one of the quintessential problems for organizing the consolidation and interaction between various data sources. Ensuring high data precision is especially crucial for the digital economy, since data correctness is the key to making the right and timely business decisions.

A study done by the Gartner Group in 2011 [1] shows that on average organizations lose at least $9.7M per year due to bad data quality, operational ineffectiveness and lost opportunities. According to another study [2], only 12% of organizations use their data's potential to attain strategic advantages.

O. Gervasi et al. (Eds.): ICCSA 2020, LNCS 12251, pp. 58–71, 2020.
https://doi.org/10.1007/978-3-030-58808-3_6

The situation is made worse by the influence of Big Data and by the largely decentralized nature of modern business. However, modern technologies can help overcome the problem of data quality. According to [3], the use of decentralized registry and artificial intelligence technologies will elevate data quality by 50% by 2023. It is true that the inherent qualities of blockchains render them capable of controlling consistency and integrity, which are some of the most important data quality attributes.

However, selecting such solutions in an arbitrary way may lower data accessibility and present issues wish scalability, particularly as a result of inherent limits in organizing storage. The ideas presented below provide a conceptual framework for managing data quality through the integration of master-data using the DGT platform [4], which combines distributed ledger technologies powered by the F-BFT consensus with Machine Learning advancements.

1.2   Data Quality Criteria

Data Quality (DQ) is a characteristic that shows the degree to which data is useful. DQ is an integral characteristic of data, which is presented through its attributes that cover different aspects of data usage. According to the ISO 9000:2015 standards, the key criteria for data quality are: (1) Relevance, (2) Accuracy, (3) Integrity, (4) Completeness, (5) Consistency and (6) Timeliness. In practical cases, these parameters carry different weights for different organizations, so the end model of data quality is heavily dependent on the domain (Fig. 1).

Fig. 1.   The main DQ Attributes by ISO 9000:2015.

Quality attributes are not limited to these core criteria and could feature additional components depending on the specifics of the processing system. Source [5] presents valid quality criteria and their widespread synonyms. The total quality evaluation may be attained through the average of attribute scores within the accepted quality model for a given system.

In terms of decentralized systems, Data Integrity is the most crucial attribute, which we will examine in detail. It is the attribute that is most sensitive to diversions in the object model, terminology and differences in variability of separate measurements.

1.3   Quality Challenges for Big Data

Data of any nature and size is subject to distortions, many of which are connected with missing values, duplicates, contradictions, anomalous values and lack of inclusivity. These problems may be seen from different points of view: violations of structure, format, random or intentional substitution of values, loss of associations and etc.

Big data significantly increases the probability of the occurrence of such errors due to the influence of its characteristics (size, speed, diversity, accuracy and cost), which reflect its distributed nature:

- Unified information frequently includes several data types (structured, half-structured, unstructured), which makes reconciliations difficult even among data that carries the same meanings;
- Semantic differences in definitions may lead to the same positions being filled differently. For example, the term "revenue" may mean different things for various organizations, which means that the tabulated report will lose meaning and lead to incorrect conclusions;
- Differences in formats and syntaxes (for example, time marks) will also lead to internal conflicts and contradictions between subsets of values.

Managing data quality in such conditions makes it impossible to use one metric across the entire set. Moreover, the well-known Brewer's CAP theorem [6] presents an additional limitation: the impossibly of ensuring simultaneous integrity of the complete dataset for a number of distributed systems (in particular for some NoSQL databases, which support only eventual consistency).

Two crucial trends map the direction in which the big data quality framework is growing: decentralization and data virtualization. The first trend illustrates the necessity of adapting distributed ledger technologies to control data quality, while the second shows the necessity of abandoning verification according to a given data structure (since it may vary).

A constructive approach to calculated data quality metrics is the selection of stable information objects and applying validation rules according to them in real-time. In other words, it is proposed to use a two-phase data processing approach:

- Pre-processing of incoming data with the identification of main information objects and validation of their attributes;
- Processing of quality attributes across the entirety of available data, taking into account the discrepancies in versions of transactional information.

Such an approach will prevent "data depletion" as a result of the upload and will adapt a standard mechanism of calculating quality in the decentralized conditions of data processing.

This approach is practically equivalent to building a reliable system for managing master-data in the Big Data paradigm and storing fast transactional information in the maximum number of versions possible.

1.4 Master Data and Identification

The work of any IT-system or a combination of IT-systems comes down to processing, storing and transferring data. Data represents repeatedly interpreted formalized information. Depending on the specifics of the objects reflected in data, it is customary to identify data classes that will indicate the data structure and the objects belonging to a certain type of information.

In terms of any corporate information system that participates in such an informational exchange, we can identify the following information types:

- Transactional (or operational) data – a rapid stream that describes the changes in statuses of information objects, such as money transfers, product shipments, sensor indicators;
- Analytic data – slices of operational data prepared for decision-making;
- Master Data – which is necessary for identifying information objects; these are sets of data with a relatively slow rate of change, including normative-directive information, metadata, parameters and configuration of informational objects.

According to the classic definition stated by Gartner [8], Master Data is defined as "… the consistent and uniform set of identifiers and extended attributes that describes the core entities of the enterprise including customers, prospects, citizens, suppliers, sites, hierarchies and chart of accounts."

Therefore, Master Data is a conditionally constant set of data that defines the composition of the domain being automated and the basis for describing the business logic of the application. Master Data can have a flat, hierarchal or network structure depending on the existing business processes.

Master Data conditionally includes such subgroups as directors, metadata and configurations, depending on the existing models of information management and object life cycles.

Let's take a look at Master Data in a decentralized context. Suppose, several organizations with a common list of counterparties are cooperating with one another. Such a list (directory) is the Master Data and within the framework of a single organization it would be recorded following the approach commonly known as the "golden record". However, synchronizing this list of counterparties for several organizations may be difficult:

- Names of counterparties often come as part of complex unstructured information (such as an invoice) and needs to be selected out;
- Such names may have geographic, language and stylistic differences, which complicate the selection of similar objects from previous entries.

Because of that, the process of evaluating quality (in particular, Integrity) requires an identification procedure that will:

- use probabilistic and statistical methods (in particular, NLP – Natural Language Processing methods);
- rely on a large spectrum of external sources, such as a KYC (Know Your Customer) system or an Open Technical Dictionary (OTD);
- include the process of distributed harmonization and standardization – distributing the information about the decisions made in regards to creating a new object or a link made to an existing object.

Therefore, Master Data has a direct influence on the quality of information and in the context of distributed decentralized systems, it requires a process of agreement/consensus between nodes/hosts that would represent the different sides of the informational exchange.

2   Conceptual DGT Quality Framework

2.1   The Approach

The main strategy is based on the separation of Master Data processing into a separate type of data processing for a distributed environment. Master Data is one of the most important information assets that a modern organization has. With the continuous digitization of processes, the creation of digital twins, and the advent of the fourth industrial revolution, the importance of master data and its management will only grow.

In essence, Master Data refers to all of the static information that is used to identify critical elements of an organization and its business processes. Assigning incoming operational information to objects requires identification, as does the formation of consistent datasets for analytics.

Thus, Master Data, transactional data, and analytic data are interdependent and part of the same context. Errors and discrepancies in Master Data can cause the same or even greater damage as the discrepancies in transaction data. For example, an error in identifying a client during billing may have a much larger impact than an erroneous calculation of the invoice amount for a correct customer.

Master Data maintain the consistency of a common information array between different information systems, divisions and organizations. The most important characteristic of Master Data is the slow rate of change in the informational exchange between several participants. When working with Master Data, the following management styles can be identified:

- Transaction-based. In this style, Master Data is a part of the transmitted transactions; the data is selected directly from the transactions and the management of such Master Data is not a task shared by all of the participants in the information exchange;
- Centralized Master Data. This style features a single system for all participants, which keeps master-data in a common storage. Other systems use it as a reference. Due to a large number of reasons, such a solution is often limited in use by small systems and a small number of participants in the information exchange;

- Shared Master Data. In this style, the management of Master Data is separated in its own distinct stream that is composed either of a series of ETL procedures, which transfer Master Data from one participant to the next, or from an information exchange over distributed ledger technologies.

In a decentralized environment, big data is composed of a large number of independent data sources, additional data streams generated by digital objects, significant flexibility in settings and an overlap of various life cycles. Therefore, the information exchange has additional properties that need to be taken into account:

- The limitations of centralized solutions. Due to the presence of a single point of failure, centralized architectures have a low degree of adaptivity and are subject to risks in managing change processes. They are susceptible to the influence of subjective decisions and cannot support the increasing level of complexity that requires distributed data processing;
- Access to data in real time. Supporting continuity of cooperation between different systems requires an asynchronous access to data – a capability possessed by decentralized registries;
- Smart data processing. Data processes require not only the right integration, but the use of machine learning – artificial intelligence (AI) functions for comparison of complex datasets. New technologies allow for the calculation of the degree of similarity for different datasets, as well as data quality metrics, and other metrics that allow for the improvement of information models;
- Storage of change history. In times of very dense information steams, it is necessary to track not only the changes to parameters, but also connections between these objects. To serve these objectives, there need to be storage systems of a particular type, which would consider hierarchy, correlation, and other specifics (graph databases).

In the framework of the approach being discussed, these problems are solved by utilizing innovational technologies that support great speed of decision-making and reduce losses due to data mismatch:

- The integration layer of the system is built on a high-performance DGT core, which ensures the formation of a unified Master Data registry and its distribution between the participants of an information exchange with a large degree of horizontal scalability and the ability to track the entire history;
- Smart modules (oracles) that track data in real-time and participate in building reconciled datasets while simultaneously measuring quality metrics;
- Developed API that can plug into not only the different corporate systems and analytic instruments, but also to a variety of instruments of data management and profiling.

2.2 Distributed Ledger Technologies Layer

Distributed ledger technologies store shared data in a shared database (registry, ledger) replicated multiple times between several nodes. The rules of inserting data, changing information, and the specifics of registry replication are governed by an algorithm that is commonly called a consensus.

In the DLT approach, the distribution and support of the registry (Master Data registry) is conducted using a network of nodes that are the agents of the distributed MDM structure. A consensus determines which agents may insert data and under which rules. The sum of these rules constitutes the process of validation that prevents entry duplication, while the signature and cryptographic security requirements additionally guarantee the immutability of the registry (its integrity).

Not every consensus will allow to synchronize distributed Master Data. For example, the well-known PoW (Proof-of-Work) that operates in the Bitcoin environment through excessive calculations is not well-suited for validating big data.

The approach below describes the model for working with big data based on the F-BFT consensus [10]. The model's features include:

- Data processing is done in a hybrid consortium-based network built on a federative principle: nodes are grouped in clusters with changing leaders and network access is limited by a set of conditions;
- Registry entry is done as a result of "voting" in a cluster and the subsequent "approvals" of an arbitrator node. Both "voting" and "approval" are a series of checks-validations in the form of calculations with binary results;
- Each network node receives information and identifies informational objects as one of the Master Data classes;
- If an object is new, then there is an attempt to initiate a specialized transaction to insert data into the corresponding registry through a voting mechanism of intermediary nodes (validation process). If the new object is approved by other nodes, the object is added to the registry and the information is disseminated in the network. If a new object is denied, the initializing node receives a link to an existing object;
- The distributed data storage system (registry) takes the form of a graph database (DAG, Directed Acyclic Graph) that allows for coexistence of several transaction families for different object classes, while maintaining the network's horizontal scalability.

This approach allows for the separation of fast streaming data from batch processing of the MDM distributed ledger. Along the way, the quality attributes are calculated in terms of information integrity, as a measure of conflict for a given object. This technology is advantageous in cleaning data in real-time without limiting data availability.

2.3   The Artificial Intelligence Layer

The difficult problem with data quality control is the low efficiency of manual checks with increasing volume and variability of data. In such cases, machine learning modules can help assess quality early in processing, diagnose data absence problems, availability of unforeseen data types, non-standard parameter values, contradictions between different sets, etc.

The use of artificial intelligence (AI) allows for the resolution of several important tasks:

- Clearing text data using Natural Language Processing (NLP) technologies and extract MD from loosely structured texts. NLP modules can determine the degree of correspondence between objects based on context;
- Ensuring compliance against set standards and Master Data management practices; conversion of MD into standard form;
- High-speed comparison of datasets (Entity Resolution) based on closeness metrics (most relevant for configurations);
- Measuring data quality directly based on support vector machine (SVM) algorithm.

AI is based on one of several neural networks that are educated through a labeled model – a prepared dataset. A more advanced approach could be used as well, that being a Generative Adversarial Network (GAN) where one network creates learning templates and the other builds recognition.

Within this article, we will take note of the most in-demand techniques that directly influence the quality of big data and the measurement of quality attributes:

- Advanced technique for information objects discovery and identification;
- Data pattern recognition;
- Prediction analysis;
- Anomaly detection.

In big data, some of the information comes in the form of unstructured information, containing references to people, businesses, geographical locations. ML modules can extract and store such information automatically by fixing the objects themselves and the connections between them. Another problem that is solved automatically is the tracking of duplicate records. Tracking such records can be done using a random forest algorithm. This algorithm not only simplifies frequency analysis, but also allows you to build predictive models, forming the expectations of incoming data flow.

Even a small data entry error can have a significant impact on analytical tools, on the usefulness of the data. Finding anomalies based on machine learning will improve the quality of data, automatically correct found errors, eliminate conflicts of formats, process the inclusion of foreign data.

Algorithms like SVM (support vector machine) allow to categorize texts, images, complex sets of scientific data, search by faces, voices, graphic data.

The AI mechanism is embedded into existing verification mechanisms and works in a distributed approach by using smart agents (oracles) that are directly involved in data validation.

2.4 Data Quality Calculation

The approach formulated above allows for the quantification of data quality in real time with the following assumptions:

- The general quality assessment is conducted on the basis of a weighted indicator, estimated as the number of operations necessary to correct the identified error;

- Some errors may be identified at the data validation stage (ex. differences in object naming that can be fixed, incorrect spelling, etc.), while others only during subsequent analysis. Therefore, the quality attribute for a given dataset is constantly recalculated;
- The attribute's weight depends on the current value of source reliability (in the framework being described that also impacts the number of checks – "votes"), as well as the seriousness and priority of error according to the relevant validation rule;
- Object identification based on fuzzy logic and neural network results;
- Identification of anomalies and correlation with earlier data. When comparing information mechanisms, rules are cartridges (smart-contracts) that are inseparable from the registry and may calculated in accordance with Lowenstein distance, as an example.
- Quality attributes that are incalculable or impossible to evaluate are noted for subsequent analysis.

Algorithm 1. Data Quality Evaluation for Quality Attribute (Integrity)
1:    **procedure** DQE_Attribute
2:        Input try_Object ← Array ();
3:        **for** try_Object **in** Objects () **do**
4:            prove_Object ← IdentificationProcedure(try_Object)
5:            DQ_Assement ← Initialize ()
6:            **for** each Rule **in** Attribute_Validation_Rules **do**
7:                DQ_Assement.Add GetValidResult (Rule, prove_Object)
8:            **end for**
9:            DQ_Assements.Add DQ_Assement
10:        **end for**
11:        **return** DQ_Assements
12:    **end procedure**

The connection with the validation rules is in full accordance with the consensus model – each selected object and its attributes go through checks in the validation rule stack, gradually receiving all of the necessary evaluations. Even the data that is fully rejected may be saved in a separate registry and participate in later evaluations, thus impacting the total quality value.

The identification and validation of quality attributes must be done in accordance with time characteristics – the period given for validation. Checking and rechecking attributes in accordance to time periods is called the alignment of quality processes in the given framework. Such an approach allows for a connection with the life cycle of an information object and for tracking its changes. The general approach to align processes is presented in the Fig. 2 below.

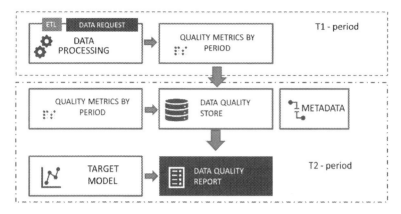

Fig. 2. Quality process aligning.

The total quality ratio can be calculated as weighted average by the following indicators:

- Number of unidentified (unidentified) objects that have been recovered in the future;
- Data inaccessibility statistics based on frequency of requests;
- Processing data-gathering conflicts, including anomalies and going beyond data validation ranges;
- Distance between the initial and final data vectors;
- Coincidence with results from other sources;
- Timeline lengths and data latency;
- Estimates of cleaning time relative to the overall download cycle

Checks are based on historical data directly in validators' rules when "voting" nodes for data insertion.

2.5   Implementation

We used the DGT platform to implement this MDM/DQ platform. Specially adapted transaction families provided MDM information for the sharing ledger on multiple nodes in a cluster of related organizations. The basic architecture of the framework is shown in the Fig. 3 below.

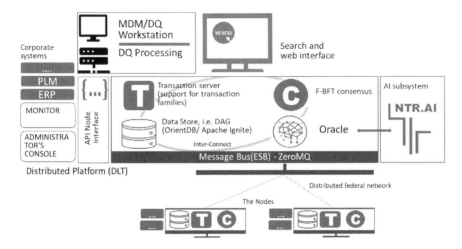

Fig. 3. Framework implementation.

3   Practical Cases

Applying this framework gives special advantages for vertical and horizontal integration of independent economic objects. Below, we describe relevant use scenarios where distributed technologies coupled with artificial intelligence provides the maximum benefit.

3.1   Case A. Data Quality in Aviation

An exchange of data in transport and logistic companies was never a simple issue. In particular, airline companies are subjected to significant expenses from flight cancellations, delays, ground handling overhead, and the search for balance between safety and profitability.

Complex data processing streams combine with the necessity to make decisions in real time, as required by strict industry regulation and by an extensive ecosystem of air travel. This ecosystem includes the airline itself, clients (individual/corporate/freight), airports, partner airlines global distribution systems (GDS), industry associations (IATA), government bodies, suppliers and service suppliers, as well as other participants of the value chain (hotels, car rental companies, tour agents, ad networks).

Even small discrepancies in Master Data may lead to millions of dollars in losses for airlines. Special circumstances affecting the data synchronization methodology is the presence of significant contradictions between the participants of the ecosystem, as well as a large number of groups and alliances between the larger ecosystem, which forms a multi-level federative network where the consensus may take form of intersecting clusters. This will allow for the construction of multi-level consensuses and branched MD registries available in real time.

Perhaps the most pressing problem that unites the interests of the largest players is optimizing ground operations and ensuring timely departures. Directory data and metainformation that coordinates the data streams of on-board crews and support staff (cleaning services, baggage taggers, other ground services) involved in the stages of the flight will optimize expenses and present necessary data for analysis, identify any source of a potential delay, and enable quality decision-making.

A recent study [11] shows that there are currently no universal instruments of managing data exchange in airlines, while existing instruments (ERP, SRM, TMS, WMS) do not allow for full transparency between various participants. Difficulties arise when checking the correctness of paperwork, labeling, and authenticity of information about the origin of transported goods.

The use of the suggested framework will allow to save time on checking paperwork, customs control, and phytosanitary control, as well as enable tracking of goods, simplify the identification of counterfeit products, and improve the quality control of logistics operations.

3.2 Case B. IoT

The Industry 4.0 and greater use of Internet of Things promise significant improvements for managing the environment and industry objects, formulating predictive analytics, improving the quality of equipment and safety levels of industrial facilities.

The large sum of data generated by various sensors, combined with the low cost of sensors, have led to an entire class of IoT platforms that focus on collecting data on the direct physical level. This partially alleviates the issue of heterogeneity of the IoT environment and narrows interactions to a few common platforms/hubs [12].

However, in addition to ensuring data accessibility, there is the issue of overall data integrity between various sub-systems, organizations, and entire industries. Examples of such cluster interactions include:

- Interaction of medical systems, especially concerning sensitive data between different organizations, hospitals, device manufacturers;
- An exchange of data between energy producers, distributive systems, and users;
- Data received in food supply chains, with the necessary integration of the manufacturer's devices with those of logistics companies and food selling organizations.

Significant research, such as [13], has been devoted to the necessity of forming multi-level architecture for distributed devices, where the IoT layer would be supplemented by a DLT layer. A federative network that is capable of processing several transaction families would be highly useful, considering the broad class of devices and their geographic dispersion. In this case, different devices, protocols, and other configurating parameters would be synchronized in the total MDM stream, ensuring the necessary level of data quality (Fig. 4).

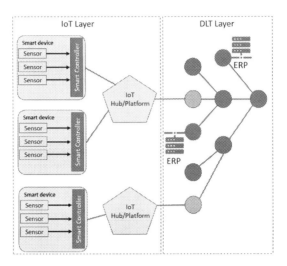

Fig. 4. Two-tier IoT/DLT network.

3.3 Case C. Public Health Care Data

The latest COVID-19 events have shown significant dependence of measures taken by governments to mitigate critical situations on high-quality datasets. The Blockchain Research Institute (BRI) has published an extensive report [14] where it identified five key directions for this technology to be used in fighting the pandemic and its consequences. One of such directions is forming a registry of combined data and statistics for:

- Storing data according to categories, illness statuses;
- Tracking parameters of the measures taken, available hospital beds, ventilators, masks;
- Tracking product deficits and etc.

As everyone witnessed, despite the best efforts of Johns Hopkins University the death rate statistics were quite malleable due to differences in counting methodologies, the consumer market displayed a low level of adaptivity across several goods, while the medical industry of many countries was unprepared for a pandemic. As was shown in the report, countries with good access to data (China, Singapore, Korea) were able to take effective measures in fighting the disease. Countries with limited data capabilities (such as Italy, Spain, the US) experienced significant difficulties.

The presence of a DLT framework as a key element of facilitating data exchange could have saved many lives. We should also consider the following parameters of this necessary solution:

- The presence of horizontal scalability as a mandatory element for working with critical data;
- Elevated security and cryptographic protection according to medical standards;
- Parallel processing of several transaction families to reflect the necessary data variability.

4   Conclusion and Outlook

The framework presented in this study allows us to formulate the requirements for systems necessary to maintain the quality of big data in terms of its integrity. The problems described above allow us to conclude that the distributed ledger and artificial intelligence technologies are in demand for monitoring and assuring quality of big data systems and in a broader sense of the multitude of dynamic systems used in the modern digital world.

Applying distributed ledger technologies to the task of maintaining master data between organizations will provide a united information space for groups of companies integrated horizontally or vertically, allow real-time quality indexes to be calculated and effective information exchange in operational data, improve the quality of analytical data, and ultimately make the decision-making process itself a quality.

References

1. Measuring the Business Value of Data Quality. Gartner Inc. (2011)
2. Assessing Your Customer Intelligence Quotient. Forrester. http://www.forrester.com/Assessing+Your+Customer+Intelligence+Quotient/fulltext/-/E-RES53622?docid=53622
3. Predicts 2020: Data and Analytics Strategies — Invest, Influence and Impact, Gartner Report (2019)
4. DGT Platform Overview. http://dgt.world/docs/DGT_About.pdf
5. List of Conformed Dimensions of Data Quality. CDDQ Open Standard. http://dimensionsofdataquality.com/alldimensions
6. Brewer, E.A.: Towards robust distributed systems. In: (Invited Talk) Principles of Distributed Computing, Portland, Oregon, July 2000
7. ISO/IEC/IEEE 24765-2010: Systems and software engineering – Vocabulary
8. Gartner: Master Data Management (MDM). https://www.gartner.com/en/information-technology/glossary/master-data-management-mdm
9. ECCMA Open Technical Dictionary. https://www.eotd.org/
10. Bogdanov, A., Uteshev, A., Khvatov, V.: Error detection in the decentralized voting protocol. In: Misra, S., Gervasi, O., Murgante, B., Stankova, E., Korkhov, V., Torre, C., Rocha, A.M.A.C., Taniar, D., Apduhan, B.O., Tarantino, E. (eds.) ICCSA 2019. LNCS, vol. 11620, pp. 485–494. Springer, Cham (2019). https://doi.org/10.1007/978-3-030-24296-1_38
11. IATA: Blockchain in aviation. Exploring the fundamentals, use cases, and industry initiatives. White paper|October 2018. https://www.iata.org/contentassets/2d997082f3c84c7cba001f506edd2c2e/blockchain-in-aviation-white-paper.pdf
12. Fernández-Caramés, T.M., Fraga-Lamas, P.: A review on the use of blockchain for the Internet of Things. IEEE Access **6**, 32979–33001 (2018)
13. Casado-Vara, R., de la Prieta, F., Prieto, J., Corchado, J.M.: Blockchain framework for IoT data quality via edge computing. In: The 1st Workshop on Blockchain-enabled Networked Sensor Systems (BlockSys'18), November 4, 2018, Shenzhen, China, p. 6. ACM, New York, NY, USA, Article 39 (2018). https://doi.org/10.1145/3282278.3282282
14. Tapscott, D., Tapscott, A.: Blockchain Solution in Pandemic, BRI, April 2020

A DLT Based Innovative Investment Platform

Alexander Bogdanov[1], Alexander Degtyarev[1,3], Alexey Uteshev[1],
Nadezhda Shchegoleva[1(✉)], Valery Khvatov[2],
and Mikhail Zvyagintsev[2]

[1] Saint Petersburg State University, St. Petersburg, Russia
{a.v.bogdanov, a.uteshev, n.shchegoleva}@spbu.ru,
deg@csa.ru
[2] DGT Technologies AG, Toronto, ON, Canada
valery.khvatov@gmail.com, zma@dgt.world
[3] Plekhanov Russian University of Economics, 36, Stremyanny per,
Moscow 117997, Russia
http://dgt.world/

Abstract. Digital transformation has affected many industries and has become a mega-trend in the information technology systems. However, the venture capital investment process is still rigid. The use of distributed ledger technology and artificial intelligence in the field of finance and investment requires a detailed framework and a rethinking of the interaction model of market participants. The article shows the DLT based approach with attached of F-BFT consensus to improve the efficiency of the investment process in the modern world.

Keywords: Blockchain · Fintech · Distributed ledger technology · Investment platform · F-BFT consensus

1 Introduction

Distributed and decentralized systems are being used with increasing frequency when applied to practical cases of the modern economy. The use of computational algorithms that are resistant to random or intentional distortions of information in the absence of trust between the parties to the information exchange can significantly raise the effectiveness of traditional financial processes. The article [1] describes a consensus search algorithm called F-BFT (Federated Byzantine error tolerance), which can be successfully applied to solve information exchange problems between the parties in the investment process.

Throughout the past five years, crowdfunding has practically transformed from a charity collection method into a purpose-driven vehicle, which allows to conduct investments with the goal of generating returns. Startups received the ability to test the market and attract their target audience, while avoiding due diligence barriers.

Meanwhile, the more "mature" venture capital has not used the market to the full degree. Externally, the absence of transparency and information in the typical 2-paged profiles is baffling. Internally, investors often lack the knowledge to properly evaluate the increasing complexity of truly breakthrough technological projects.

© Springer Nature Switzerland AG 2020
O. Gervasi et al. (Eds.): ICCSA 2020, LNCS 12251, pp. 72–86, 2020.
https://doi.org/10.1007/978-3-030-58808-3_7

Nevertheless, the industry will experience almost a threefold growth from 2018 – 2025, which necessitates a need for a solution [2] (Table 1).

Table 1. Investment statistic.

$65 bn added to global economy	6,455,080 campaigns in one year (2019)
15.98% CAGR in 2018–2025	33.77% rise of funding 2018–2019

1.1 Crowdfunding Trends

Crowdfunding (Crowdfunding Investment, CFI) is a method of attracting relatively small amount of funds from an unlimited number of representatives of the general public through the use of Internet platforms. Essentially, crowdfunding is a part of a more general crowdsourcing process, as well as a component of a common investment model.

The CFI is becoming an increasingly attractive vehicle due to the changes in the business environment as a whole. The barriers to investment are decreasing, as the Internet, mobile payments, fintech and other solutions are simplifying the global transfer of funds. The business models are changing, as globalization and digitization lead to startups becoming more attractive at the expense of non-technological businesses. This also has the effect of increasing the complexity of projects and thus venture capital risks, which complicates access to venture funds. At the same time, venture funds are experiencing pressure from new types of investment vehicles: digitally-focused banks, and venture capital arms of tech companies.

As a result, an increasing number of entrepreneurs seek funding through technological platforms that serve as mediators through algorithms and transparent procedures.

The number of crowdfunding models is increasing as well, with each being distinct in the size of attracted capital, control mechanisms, and investor motivation.

With these underlying needs there is a necessity to create a crowdfunding platform (CFP) of an adaptive type in which the tuning of specific algorithms and models is possible during operation without changing the core of the system.

As a process, CFI reflects not only investments in technology projects, but also into other categories, such as films, art works, social events and charity. According to a number of reports, the crowdfunding market is experiencing significant growth and may reach 25B USD by 2025[1].

CFI is an increasingly important component of the common investment model, which is typically illustrated by a bell curve, describing the size of funding per each successive life stage of a project seeking funds. The stages for an average project are as follows: (1) Idea/Offer, typically served by personal or angel funds; (2) Proof-of-Concept, served by angel capital and CFI Investment of less than $50,000 USD, (3) the Startup or the "window of financing" stage where projects typically collect CFI

---

[1] https://www.statista.com/statistics/1078273/global-crowdfunding-market-size/.

investments of over \$1M USD; (4) Exponential Growth requiring venture capital of over \$1M USD, and finally (5) Business Expansion and attraction of institutional investments.

According to the existing research[2], CFI is most effective in two significant areas of application:

- Early stages of a project: Proof-of-Concept (PoC);
- Bridge financing into existing project structures: startups with minimal viable products (MVP).

The biggest trends converge in the direction of one common reality. Crowdfunding should become more structured, transparent and integrative both for direct participants, regardless of their complexity, and for startups capital as a whole.

Funds raised through CFI reached \$17 200 000 000 per year in North America alone. An increasing number of companies opt for crowdfunding due to it being more efficient than traditional fundraising, its ability to build traction and social validation, the generation of significant media exposure, consistent idea refinement, and an establishing of a customer base of early adopters.

1.2   The Problem in Economy Context

Leading platforms have changed very little in the past few years and have not responded to the industry's clear demand for integration and transparency.

Investors face an increasingly low information environment. Key information about founders and founder history is often missing, as is insight on the products beyond basic profiles. This elevates risk and decreases returns. The issue is exacerbated by the growing complexity of projects. Technological and scientific components of new projects are increasingly harder to valuate without experts, which leaves smaller crowdfunding and angel investors especially vulnerable.

The need for an ecosystem is becoming more apparent. This ecosystem would be integrative to value-building participants, such as experts who could tackle the complexities of projects and service organizations that could tackle investment transparency barriers, including lawyers and accountants.

Entrepreneurs that seek investments are also facing challenges due to a lack of information. Seeking crowdfunding investments at the seed stage places them in the investment shadow zone. It is difficult to gather information about active investors, their project evaluation criteria, and areas of expertise. It becomes a daunting task to attract not only the investors that could provide the most growth opportunities for a specific business, but any investors at all to a breakthrough project with low resources.

The modern platforms are passive and do not provide adequate services for evaluating or advancing the project. They do not curate it nor track its further fate. This makes success difficult for the investors, but also for entrepreneurs, who often repeat the same mistakes as their unsuccessful predecessors. A supportive infrastructure is

---

[2] Crowdfunding and Funding Gap Sources: World Bank (2013).

needed, yet there are no ecosystem approaches migrating from traditional investments into the crowdfunding models yet.

There is a well-known Venture Capital Power Curve heuristic, which states that only 6% of the funds invested drive 60% of the total revenues on a portfolio. There is a need for a solution that would enable VC Funds to be more informed when selecting projects and to gain access to a larger pool of crowdfunding opportunities at an acceptable risk level.

These changes can only come through technological enablement of a seamless data integration between many distinct business entities and omni-sourced information. Such technologies are in the fields of decentralization and deep analytics.

1.3 Existing Platforms Disadvantage

The most popular international investment platforms are based on the principles of crowdfunding. The core problem of their business-model is the bog of subpar projects. Due to the large component of marketing when advancing an investment idea or product, visibility is given to the product that invested the most money into being advanced on the platform, rather than the one that has the highest potential rate of return.

There are many large crowdsourcing campaigns, less crowdfunding, and almost no effective crowd-investing platforms currently in operation. The first two are defined by creative and charity projects, while the latter by investments into businesses.

While crowd-investing platforms do exist, their effectiveness is dampened by a set of serious limitations:

- Projects are often presented in the form of listings, where the only differentiator is the sum paid for visibility, rather than any intrinsic qualities. In fact, in the conditions of limited resources, the projects that generate the most investments are those that sacrificed real gains for visibility on a given platform.
- There is no developed system of expertise to judge the projects. Existing experts quickly turn into for-profit marketing tools.
- There is no mentorship component that would allow a startup to improve their idea or their presentation beyond an automated profile builder.
- There is no ability to follow the post-investment life of the project or the resulting reputation of its founders, ideas, decisions, or investors.

These problems cannot be resolved without the addition of entirely new components into the crowdfunding platform. Only four components are needed to address the quintessential problems of the modern CFI:

1. Matching. The core structure of the system should not be a listing, but a set of projects that is matched to the interest of the investor. The system must have vetting and use intelligence to create an ideal investor-project pair.
2. Crowd due diligence. The project that appears on the platform must go through a filter of an open decentralized community of experts. It is crucial to incentivize the experts to invest in their objective reputations, rather than to exchange positive reviews for money. A decentralized mechanism may ensure that the experts are rewarded for extensive and justified reviews and evaluations.

3. Mentorship. Startups must receive information about the interest level in their projects in order to adjust their quality, without paying their way upwards in the search results. Startups should know how much they are viewed and appear in searches, as well as what should be improved for the growth of these indicators.
4. Edge Computing. The system should be built using edge computing technology. This will allow it to maintain utmost confidentiality while still providing analytics for interested user groups.

2   Decentralized Investment Capital Crowdfunding Platform

2.1   The Approach of Integrated Crowdfunding

In order to implement the required changes, the approach to a Decentralized Investment Capital Crowdfunding Platform must hinge on two ideological pillars. The first is integration and the second one is data maximization.

In the context of CFI, integration is multi-faceted. The platform must create a decentralized cross-IT-environment and a cross-border marketplace that will be able to integrate with enterprise legacy systems and digital mesh technologies. This will enable the system to maximize its information sourcing and analytic capabilities. There must be a technological and business framework for investors, capacity builders, corporations, startups, and professional services to optimize their part of the sourcing process and provide the most synergetic value during the business development phase. The system must not only be decentralized, it must have hierarchal decentralization or federative topology. Instead of one rigid system, it will branch out into many sub-clusters adaptive to its own environment. That will ensure the systems are open, integrative, continuously transformative enough to overcome corporate and environmental borders, as well as economically sensible.

The second pillar is data. There is a necessity of creating an immutable continuous reputation trail that will take into account past information about projects, founders, and investors, as well as the future of these entities beyond the CFI platform. That requires obtaining and analyzing public data for all platform projects, with smart sourcing enabled through artificial intelligence, and decentralization allowing the capacity for such analysis. The aforementioned F-BFT Consensus is optimized for creating cross-organizational and cross-IT-environment data bridges.

In order to securely store the information across organizations a decentralized registry is also needed. A graph registry, such as the Directed Acyclic Graph (DAG) type registry would enable the storage of information to be both horizontally and vertically scalable. Meanwhile, Edge Artificial Intelligence (Edge AI) could provide real-time insight, uncover hidden patterns, and drive effective matching mechanisms. The F-BFT Consensus, the DAG registry, and Edge AI are three core technologies employed by the company DGT [4] in building a CFI platform with a deepened information field. By including publicly sourced data and proper motivation mechanisms for domain experts, DGT can help investors overcome project difficulty, heuristic decision-making, and information gaps.

Integrated Crowdfunding builds investment ecosystems using the hierarchal decentralized network, which contains F-BFT and DAG mechanisms. Projects are openly listed, screened by experts, and become part of a community that maximizes their growth. The platform features strong expert reviews: a network of domain-specific experts, who provide deep reviews and consultations to improve the project vetting process beyond marketing expenditures.

This crowdfunding model integrates Capacity Building Organizations: advisors, organizations and consultants that could maximize business development, utilization of a project's technologies, or provide other benefits at lower costs.

It has enables inter-portfolio synergies. Investors may build portfolios that compliment each other technologically, share data and form cross-product value offerings and also reduces investment friction by integrating legal, forensic accounting, and talent evaluation services for a due diligence process that can be shortened from months to eight (8) days.

2.2 Key Solution Components

A key value point missing from many digital ecosystems is the ability to integrate many data formats, permissions, sources, and uses. DGT uses its decentralized network and deep analytics to overcome such barriers.

It features an immutable ID for tracking reputations of investors, startups, and founders. Its internal economy has a loyalty mechanism that would hold participants within the CFI ecosystem, while also enabling flexible reward-enablement for its growth (such as the motivation of experts). The ecosystem of professional value-adding businesses decreases the friction of the investment process as a whole (Fig. 1).

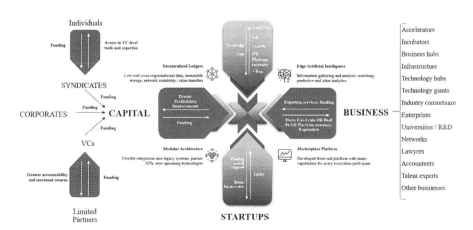

Fig. 1. Intended platform participants and interactions

DGT uses Edge AI that could analyze the Internet for traces of founder history, current industry data, and other information that makes investing less heuristically motivated and founders more committed to their ideas.

A shared network of domain experts decreases complexity-related risk for both small and large institutional investors.

2.3   Expert Reviews in Detail

The network of domain experts is only capable of creating value if expert reviews are objective and plentiful.

Experts are motivated through a variety of mechanisms, including an internal decentralized loyalty currency and/or shares in investments, depending on the structure. Expert reviews follow a unified format, which ensures their accessibility, while still allowing depth.

The unified format allows for an easy comparison of projects based on a single set of criteria:

1. the product, its potential to resolve the problems of the target audience;
2. the business model, its realism, consistency and potential;
3. the technology, the value of innovation and how realistic the claims are, as well as the depth of the work already done;
4. the team, the professional competencies of its participants;
5. the financial model, the accuracy and realism of the monetization, cost structures, ROI, etc.;
6. the risks, an evaluation of where the main risks reside for the company and its investors;
7. PR and marketing, target market quality and reach, communication channels and the spread of the project's information.

Input data is evaluated not only by experts in a specific domain, but also through Edge AI. Through an examination of founder reputation, current trends, competing products, and hidden interdependencies, the Edge AI may provide an estimate for the product's TRL – the Technical Readiness Level. As a result, investors can make decisions that are much more informed without taking on significant costs. Moreover, since this information is openly accessible, each investor does not have to conduct the due diligence work for a project that has already been vetted by others.

2.4   Venture Crowdfunding Ecosystems

The DGT Platform enables the construction of sub-ecosystems of data that brings about further synergetic value. Capital ecosystems share expert insight, AI analytics specialized to a particular domain, and a reduction in total risk. Portfolio ecosystems maximize value by creating data highways between products in themed portfolios, which enables cross-selling, the creation of new services, and an ongoing exchange of customers. Syndicate ecosystems enable angels to create flexible groups and access VC-like analytics and capabilities. Operating Growth Funds (OFGs) are a new investment class that could use the platform to identify top business models and then proceed to launch with or without outside founders from a VC-centric starting point.

DGT uses decentralization to provide valuable input from a greater number of participants without increasing costs, while Edge AI provides an omni-data approach to more successful investments. It is not just a platform – but an entirely new venture capital paradigm.

The platform provides strong data-driven support: fair and full evaluation of start-ups across technological, innovational, business, and financial metrics. The Edge AI module conducts its analysis of revenue drivers, hidden correlation patterns and founder experience at minimal costs. It supports every step of decision-making, taking a portfolio-wide view to improve the standing on the Venture Capital Power Curve. Instead of the reliance on outliers, a transformed investment curve presents a greater percentage of winners and makes the "average performance" one of greater prof-itability, all through a better selective process, integration of more participants into business development, and ongoing omni-data analytics.

DGT ensures that capital may remain smart despite increasing project specialization and subject matter complexity. The platform features domain experts that provide comprehensive, unbiased reviews on each project. Instead of seeking out several experts or relying on the internal analytics teams of venture capital funds that duplicate each other's work, the platform motivates external experts through reputation and economic incentives, while providing them and the investors with the data they need.

An AI-driven predictive optimization engine also solves a typical problem of venture capital: the allocation to underperformers. The AI engine calculates the optimal distribution for follow-on investments, identifies which portfolio companies need more help, and which are a sunk cost. The decentralized Edge AI is smarter and more capable than any non-decentralized artificial intelligence and it grows exponentially so with the platform.

The DGT Platform is highly competitive against any other CFI or VC platform, which are structurally unable to implement the changes necessary to involve as many participants, create cross-organizational value, and insight-driven investments.

2.5   Investment Cycle Advantages

DGT gives an edge on every part of the fund's and investment's life cycle:

1. Fundraising: stronger YoY (year-on-year) performance leading to increased fund reputation; AI-driven connections between funds and LPs;
2. Sourcing: targeted deal flow matched to objectives; ecosystem of vetted projects and cooperation with corporates, industry leaders, other investors;
3. Investing: omni-data driven decision making; deep expert support; identified portfolio synergies; immutable history of past project and founder success;
4. Developing: ecosystem of capability-building organizations with precise domain expertise; AI-driven prediction engine;
5. Exiting: greater number of successful exist; involvement of corporates and alliances early on (mergers & acquisitions – M&A); fully traceable and auditable history for ease of going to (initial public offering – IPO) or further VC (venture capital) investment rounds.

3 Technology and Algorithms

3.1 Conceptual System Framework

The main feature of the system implementation is the adaptation of the classical investment platform to the task of building an open ecosystem of participants who benefit from interaction:

1. under the classical "one-to-many" model, participants interact with one platform backer (e.g. an investment fund). The outlined approach combines these systems into a network with the "many-to-many" relationships. In fact, a data network is being created in which many investors and many projects participate;
2. the composition of participants is expanded by attracting experts, service companies and advertisers. This allows for greater transparency of transactions, elimination of problems of assessing the importance and prospects of projects;
3. unlike classical crowdfunding platforms aimed at collecting small amounts from a large number of participants, the system can be adapted to engage professional investors, and projects can work on more complex fundraising mechanisms (e.g., without restrictions, the kind of "all-or-nothing" specifically tailored to protect small-capital investors);
4. the possibility of implementing a phased payment based on expert assessments reduces the risk of one-time investments.

The Fig. 2 below shows the conceptual framework describing the main modules of the system.

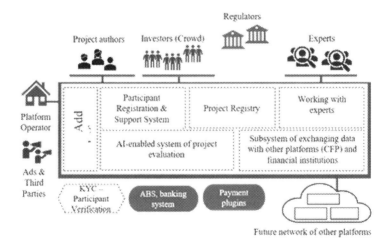

Fig. 2. Conceptual architecture of DGT-based CFI platform

Building a DGT-based investment platform is based on the following assumptions:

1. the solution is a consortium-based network of nodes that process information and participate in the storage of a general data registry. The network can exist even with a single site, but by attracting partners and creating a full-fledged investment network there is an increase in the capacity of investment flows. Nodes receive a commission on each transaction, as well as sell additional financial products (such as deposits of funds);
2. a blockchain-based solution involves decentralization, and the degree of decentralization may vary depending on the extent of the influence of the participants in the information exchange. Joining nodes and setting their topology (e.g., selecting specialized clusters to manage dedicated project portfolios) is based on an anchor mechanism representing a separate system that regulates the growth of the system;
3. each site will feel in checks of financial transactions and formation of ratings of participants. This uses a "reputation calculation" mechanism and project scoring.

Reputation Calculation

Reputation allows you to define projects based on the trust of the team behind the project by the set of rules [5]:

1. Taking into account the experience and fame of the main authors of the project on the Internet/social networks;
2. Overall team structure balance;
3. Technological expertise and formal organizational features (registration of a participant, availability of publications about a participant);
4. Portfolio. Past performance can be considered for teams with experience.

3.2 Security Challenges and Implementation

Attack Prevention

Any CFI system is potentially vulnerable to cyber threats. Some of these threats are addressed by the consensus mechanism, some through an intelligent subsystem. Below are the most widespread attacks specific to CFI:

1. Unfair rating attack:

   - Attempts of participants to change their rating artificially by creating fake profiles, quoting other people's works;
   - Violation of platform rules, use of methods of manipulation in social networks, distortion of delivered materials.

2. Conflict of interest or affiliation between experts and assessed teams;
3. Trying to present similar solutions with slightly different parameters (similar to a double-spending attack for payment systems);
4. White washing/Sybil Attack: identification breach, attempted site capture [7];
5. Creating fake projects to ruin the reputation of competitors.

Hidden Information Problems

In addition to distorting information, another problem is the placement of insufficient information [8]:

1. Screening problems. Checking participants requires third-party involvement and more information;
2. Dynamic behavior. Fundraising may be influenced by the choice of the investment template by the authors of the project. For example, the election model of attracting investments ("All-or-Nothing", "Keep-it-all") or the special qualities of first savers;
3. Herding problem. The behavior of the first investors, the influence of someone's authority or blogger may distort the rating of the project.

Such problems can only be solved with artificial intelligence functions directly embedded in the platform.

3.3   Artificial Intelligence

The investment process described in this article always occurs in an environment of uncertainty and missing information in regards to market conditions, application domain of investment projects, and project authors seeking investments. Despite the efforts of investors to ensure the safety of their investments, there is no procedure that could provide a certain guarantee of minimizing risks and maximizing returns. An environment of fuzziness, inaccuracy, uncertain situations, and free choice to make decisions indicates that this task falls in the realm of artificial intelligence. Hence the artificial intelligence subsystem has an important role on the platform being described.

The first important component of such a system is the domain knowledge base and adaptivity of the intelligence system. In this case, the knowledge base includes a set of rules that are formed both by the experts involved and by the analysis of past projects, including precedents of positive and negative situations, facts, and strategies of finding optimal solutions.

The second component of the system formalizes uncertainties with the goal of identifying the influence of fuzzy factors in making decisions. Factors of uncertainty include the fuzziness of the situation and source data, as well as the fuzziness of goals sought by investors, who don't always formulate their motivators accurately. The resolution of these issues involves fuzzy logic methods with quality and probabilistic uncertainties, inaccuracies and uncertainties in the description of outcomes and so on.

The third component of the system could have a generalized title of procedural. It includes modules based on statistical and machine learning methods. They form the base for the adaptive component and the functioning of the logical inference machine (modus ponens). An example of such a subsystem is one that evaluates the originality of applications and automatically assigns them to one topic or another. For these methods, a well-known comparative text mining mechanism is used [9, 10]. It works with topical collections based on borrowed data, as well as the accumulated documentation.

3.4   Technological Components

In addition to the core F-BFT Consensus, DAG Registry, and Edge AI technologies, the DGT Platform features several important components that differentiate it from the other platforms.

DGT has hierarchal topology, which each cluster of decentralized nodes – enterprises being adaptive to a particular business environment, while remaining part of the total Platform. DGT is a hybrid network, meaning that these clusters may have either public or consortium-based conditions for joining all within one platform. DGT is capable of processing any piece of digital value: financial transactions, IOT data, IDs, digital twins, smart contracts, digital goods and more.

DGT is modular and features flexible divergent architecture. Its network is anchored and secured by pre-existing large economic systems. Its powerful integrative layer features an array of API and other tools to bridge to legacy and digital mesh technologies.

The Architecture

The complex structure of information exchange in the described system is aimed at providing a single information space for different nodes, forming a single reliable copy of data, free from accidental and deliberate errors. Such tasks often arise when a system is functioning without a single center. The logic of checking each transaction, its extension to the entire system (synchronization of registries) is described by a system of rules called consensus. The main components of the system are below:

1. The subsystem is represented by an inherited library and is responsible both for interaction with other nodes and for interaction with other components of the site;
2. Encryption. Library of the formation of hash-functions and the formation of a digital signature. To generate private and public keys, the system use the ECDSA algorithm (elliptic curve) with parameters secp256k1;
3. State. Data Store, Ledger: DAG. Despite the arbitrary positioning of the storage to "keep" the registry, the system uses several data storage solutions at once;
4. Journal is a group of components working on the processing of transactions and their insertion into the registryж
5. Consensus. F-F-BFT Voting Verification and Validation Mechanism;
6. ORACLE. Intelligent component to get additional transaction conditions;
7. Transaction Processor. A module implemented as a separate service(TCP 4004)that supports certain transaction families;
8. REST API. A component that connects customers to the site's core through HTTP/JSON;
9. CLI. A command-line interface that allows you to process information inside a site through a standardized API. It is the primary means of administering the site;
10. DASHBOARD is a system component that reflects the state of the site and the network as a whole;
11. Mobile/Web Client - the end application for the user (Fig. 3).

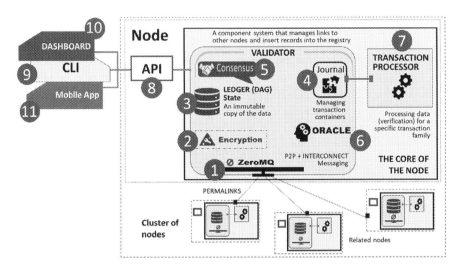

Fig. 3. Conceptual architecture of DGT-based CFI platform

The AI System

The artificial intelligence subsystem is implemented by Apache SystemML infrastructure and supports distributed training data sets. For these purposes, the system supports intelligent agents (Oracles) use two methods: Support Vectors Machines and random forests.

The API Layer

To be able to work with BIG DATA it is necessary to have a set of tools, that we call Ecosystem, out of which the most important is API:

API is a business capability delivered over the Internet to internal or external consumers

1. Network accessible function;
2. Available using standard web protocols;
3. With well-defined interfaces;
4. Designed for access by third-parties.

The key features of API are management tools that make it:

1. Actively advertised and subscribe-able;
2. Available with SLAs;
3. Secured, authenticated, authorized and protected;
4. Monitored and monetized with analytics.

An approach that uses Data API: Unified approach to data integration

1. Conventional APIs: Web, Web Services, REST API – not built for analytics;
2. Database paradigm: SQL, NoSQL, ODBS and JDBC connectors – familiar to analysts;

3. Database Metaphor + API = Data API;
4. Specific API for every type of big data (every "V" (Volume, Variety, Velocity, Veracity) and their combinations) – under a generic paradigm.

A lot of experiments have shown that the Data API is an essential part of the toolkit, as it is an important mechanism for both integration with other systems and access to data and the results of their processing. Therefore, it is necessary to determine the characteristics of data types to select the appropriate software stacks for their effective processing, to establish links between technologies, to develop an interface for convenient user work and to form an effective ecosystem for each data class.

The analysis of efforts in this direction led to the conclusion that the optimal Big Data solution should combine the following technical characteristics:

1. scalability;
2. fault tolerance;
3. high availability;
4. the ability to work securely with data in the public domain;
5. support of analytics, data science and content applications;
6. support for the automation of data processing workflows;
7. ability to integrate with a large number of popular solutions;
8. self - restoring ability.

In addition, during development, the ecosystem should include tools that ensure the following stages of data processing:

1. data collection;
2. data storage (including the possibility of organizing the storage of heterogeneous data), providing real-time access to data, data storage using distributed storages (including the organization of data lakes);
3. data analysis, the formation of a complete picture of the data, data collaboration;
4. data management, cataloging, access control, data quality control and the construction of information processing chains;
5. data production (including building of a data infrastructure, data delivery, integration issues)

Developing an API for a system is an extremely complex, unique process for big data processing systems. Until now, this has been a serious problem that has impeded the widespread use of big data, but the solution proposed by IBM allows us to start solving this problem.

Having a standard, simplified programming model that can simultaneously work on scalar, vector, matrix, and spatial architectures will give developers greater productivity by increasing code reuse and reducing development investment.

OneAPI framework proposed by Intel implements this model. It is based on standards and open specifications and includes the Data Parallel C ++ (DPC ++) language, as well as a set of libraries. This allows equipment suppliers across the industry to develop their own compatible implementations targeting their CPUs and accelerators. Thus, developers can code in only one language using a set of API libraries for different architectures and devices from different manufacturers.

4 Conclusion

The concept of the platform provided above allows for the formation of an integrated investment network that would be attractive both to professional financial institutions (banks and funds) as well as to the participants of the investment market:

1. The platform modernizes the process of collecting investments in the direction of data-driven solutions built with Big Data technologies in mind;
2. Early members of the system (e.g. the bank-owner of the genesis node) receive a tool to create a new financial product and their own ecosystem, allowing the promotion of traditional banking products;
3. The platform facilitates access to capital through direct assessments of the investment attractiveness of projects, allowing the author of projects to focus on the merits of their product, rather than marketing tricks on self-promotion;
4. Intra-VC Marketplace "forks" the decentralized system by allowing a VC to create its own permissioned ecosystem of business support and entrepreneurs. It is monetized through a subscription fee and provides immutable data storage, Edge AI analysis, and the possibility of creating open ecosystems.

References

1. Bogdanov, A., Uteshev, A., Khvatov, V.: Error detection in the decentralized voting protocol. In: Misra, S., Gervasi, O., Murgante, B., Stankova, E., Korkhov, V., Torre, C., Rocha, A.M.A.C., Taniar, D., Apduhan, B.O., Tarantino, E. (eds.) ICCSA 2019. LNCS, vol. 11620, pp. 485–494. Springer, Cham (2019). https://doi.org/10.1007/978-3-030-24296-1_38
2. Statista CrowdFunding Market Size. https://www.statista.com/statistics/1078273/global-crowdfunding-market-size/
3. Crowdfunding and Funding Gap Sources: World Bank (2013). https://www.researchgate.net/figure/Figures-2-Crowdfunding-and-Funding-Gap-Sources-World-Bank2013_fig2_307756562
4. DGT Platform Bluepaper. http://dgt.world/docs/DGT_About.pdf
5. Dennis, R., Owen, G.: Rep on the block: a next generation reputation system based on the blockchain. In: 2015 10th International Conference for Internet Technology and Secured Transactions (ICITST), pp. 131–138. IEEE (2015)
6. Gaikwad, S.N.S., et al.: Boomerang: rebounding the consequences of reputation feedback on crowdsourcing platforms. In: Proceedings of the 29th Annual Symposium on User Interface Software and Technology, pp. 625–637. ACM (2016)
7. Douceur, J.R.: The sybil attack. In: Druschel, P., Kaashoek, F., Rowstron, A. (eds.) IPTPS 2002. LNCS, vol. 2429, pp. 251–260. Springer, Heidelberg (2002). https://doi.org/10.1007/3-540-45748-8_24
8. Bian, S., et al.: IcoRating: a deep-learning system for scam ICO identification (2018)
9. Spinakis, A., Chatzimakri, A.: Comparative study of text mining tools. In: Sirmakessis, S. (ed.) Knowledge Mining. Studies in Fuzziness and Soft Computing, vol. 185. Springer, Berlin, Heidelberg (2005)
10. Krasnov, F., Dimentov, A., Shvartsman, M.: Comparative analysis of scientific papers collections via topic modeling and co-authorship networks. In: Ustalov, D., Filchenkov, A., Pivovarova, L. (eds.) AINL 2019. CCIS, vol. 1119, pp. 77–98. Springer, Cham (2019). https://doi.org/10.1007/978-3-030-34518-1_6

International Workshop on Bio and Neuro Inspired Computing and Applications (BIONCA 2020)

Application Mapping onto 3D NoCs
Using Differential Evolution

Maamar Bougherara[1], Nadia Nedjah[2], Djamel Bennouar[3], Rebiha Kemcha[1],
and Luiza de Macedo Mourelle[4(✉)]

[1] Département d'Informatique, Ecole Normale Supérieure de Kouba, Algiers, Algeria
bougherara.maamar@gmail.com, rebiha.kemcha@gmail.com
[2] Department of Electronics Engineering and Telecommunications, State University
of Rio de Janeiro, Rio de Janeiro, Brazil
nadia@eng.uerj.br
[3] LIMPAF Laboratory, Bouira University, Bouíra, Algeria
djamal.bennouar@univ-bouira.dz
[4] Department of Systems Engineering and Computation,
State University of Rio de Janeiro, Rio de Janeiro, Brazil
ldmm@eng.uerj.br

Abstract. Three-dimensional networks-on-chip appear as a new on-chip communication solution in many-core based systems. An application is implemented by a set of collaborative intellectual property blocks. The mapping of the pre-selected sets of these blocks on three-dimensional networks-on-chip is a NP-complete problem. In this work, we use Differential Evolution to deal with the blocks mapping problem in order to implement efficiently a given application on a three-dimensional network-on-chip. In this sense, Differential Evolution is extended to multi-objective optimization in order to minimize hardware area, execution time and power consumption of the final implementation .

Keywords: Three-dimensional networks-on-chip · Intellectual property mapping · Multi-objective optimization · Differential evolution

1 Introduction

A critical problem in the design of Multi-Processor Systems-on-Chip (MPSoCs) is the on-chip communication, where a Network-on-Chip (NoC) [15] can offer a scalable infrastructure to accelerate the design process. A MPSoC is designed to run a specific application, based on Intellectual Property (IP) blocks. A NoC consists of a set of *resources* (R) and *switches* (S), forming a *tile* [1]. Each resource of the NoC is an IP block, such as processor, memory, Digital Signal Processor (DSP), connected to one switch. Each switch of the NoC implements routing and arbitration, connected by *links*.

The way switches are connected defines the topology, such as the two-dimensional (2D) mesh topology [16]. However, the 2D NoC fails to meet the requirements of SoCs design in performance and area. Three-dimensional (3D) NoCs [2]

O. Gervasi et al. (Eds.): ICCSA 2020, LNCS 12251, pp. 89–102, 2020.
https://doi.org/10.1007/978-3-030-58808-3_8

have proved to be an effective solution to the problem of interconnection complexity in large scale SoCs by using integrated circuit stacking technology. A
3D mesh is implemented by stacking several layers of 2D mesh on top of each
other and providing vertical links for interlayer communication, called Through
Silicon Vias (TSV) [2]. Each switch is connected to up to six other neighbouring
switches through channels, in the same way as 2D mesh does. Figure 1 shows
the architecture of a 3D mesh-based NoC.

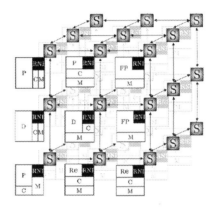

Fig. 1. 3D Mesh-based NoC with 27 resources

Different optimization criteria can be pursued depending on the detailed
information of the application and IP blocks. The application is viewed as a
graph of tasks called Task Graph (TG) [4]. The features of an IP block can be
determined from its companion library [3]. The objectives involved in IP task
assignment and IP mapping are multiple and have to be optimized simultaneously. Some of these objectives are conflicting because of their nature. So, IP
assignment and IP mapping are classified as NP-hard problems [4]. Therefore,
it is mandatory to use a multi-objective optimization strategy, such as Multi-
Objective Evolutionary Algorithms (MOEAs), with specific objective functions.

We use Differential Evolution (DE) as the MOEA [9], modified to suit the
specificities of the assignment and mapping problems in a NoC with mesh topology, and also to guarantee the NoC designer's constraints. In previous work
[12], we applied this strategy to the assignment problem in NoCs. In this paper,
we describe the use of DE to the subsequent problem of mapping onto a 3D
mesh-based NoC.

In Sect. 2, we present some related works, where a multi-objective strategy is
applied in order to optimize some aspects of the design. In Sect. 3, we introduce
the problems of IP assignment and mapping, concerning a SoC design over a
NoC platform. In Sect. 4, we concentrate our attention on IPs mapping using
DE for multi-objective optimization. In Sect. 5, we describe the objective functions for area, power consumption and execution time. In Sect. 6, we show some

performance results, based on the E3S benchmarks suite. In Sect. 7, we draw some conclusions and future work, based on our experiments.

2   Related Work

In some works, the assignment and mapping steps are treated as a single NP-hard problem. The multi-objective nature of these steps is also not taken into account, addressing the problem as a single objective. Since we choose to treat the problem using a multi-objective optimization process, we present here some works that followed the same strategy when dealing with the mapping step.

In [5], the mapping step is treated as a two conflicting objective optimization problem, attempting to minimize the average number of hops and achieve a thermal balance. Every time data cross a switch, before reaching its target, the number of hops is incremented. To deal with this process, they used the multi-objective evolutionary algorithm NSGA.

The problem of mapping IPs/cores onto a mesh-based NoC is addressed by [6] in two systematic steps using NSGA-II. The key problem was to obtain a solution that minimizes energy consumption, considering both computation and communication activities, and also minimizes the link bandwidth requirements.

SPEA-II and NSGA-II are used in [8] for mapping, with some changes in crossover and mutation operators. Energy consumption and thermal balance were the main optimization objectives.

In [17], task mapping on a 3D mesh-based NoC is implemented using fuzzy logic in order to minimize thermal and power consumption.

In [18], the authors propose a multi-objective rank-based genetic algorithm for 3D mesh-based NoCs. Two different models are used for packet latency: under no congestion and with congestion situations.

In [19], a multi-objective immune algorithm is used, where latency and power consumption are considered as the objective functions, constrained by the heating function.

In [20], a centralized 3D mapping (C3Map) is proposed using a new octahedral traversal technology. Combining the C3Map and attractive/repulsive particle swarm optimization, they attempted to reduce energy and latency.

3   IP Assignment and Mapping Problems

The NoC design methodology for SoCs encourages the reuse of components to reduce costs and to reduce the time-to-market of new designs. The designer faces two main problems: selecting the adequate set of IPs (assignment step) and finding the best physical mapping of these IPs (mapping step) onto the NoC infrastructure.

The objective of IP assignment [4,10] is to select, from an IP library (IP repository), a set of IPs, exploiting re-usability and optimizing the implementation of a given application in terms of time, power and area requirements. During this step, no information about physical location of IPs onto the NoC is given. The optimization process must be done based on the Task Graph (TG) and IP features only. Each one of the nodes in the TG is associated with a task type, which corresponds to a processor instruction or a set of instructions. If a given processor is able to execute a given type of instruction, that processor is a candidate to be mapped onto a resource in the NoC structure and will be responsible for the execution of one or more tasks of the TG. The result of this step is a set of IPs that should maximize the NoC performance, *i.e.* minimize power consumption, hardware resources as well as the total execution time of the application. An Application Characterization Graph (ACG) is generated, based on the application's task graph, wherein each task has an IP associated with it.

We structured the used application repository, based on the E3S benchmark suite [21], using XML, both for the TG and the IP repository. Figure 2(a) shows the XML representation of a simple TG of ES3 and Fig. 2(b) shows a simplified XML representation of an IP repository. In previous work [12], we used DE during the assignment step in order to optimize area required, execution time and power consumption.

Given an application, the problem that we are concerned with here is to determine how to topologically map the selected IPs onto the network structure, such that the objectives of interest are optimized [4]. At this stage, a more accurate evaluation can be done taking into account the distance between resources and the number of switches and channels crossed by a data package during a communication session. The result of this process should be an optimal allocation of the prescribed IP assignment to execute the application on the NoC. Figure 3 shows the assignment and the mapping steps.

4   IPs Mapping Using Differential Evolution for Multi-objective Optimization

DE [11] is a simple and efficient Evolutionary Algorithm (EA). It was, initially, used to solve single-objective optimization problems [9]. DE is a population-based global optimization algorithm, starting with a population of NP individuals, of dimension D. Each individual encodes a candidate solution, *i.e* $X_{i,G} = \{X_{i,G}^1, ..., X_{i,G}^D\}$, $i = 1, ..., NP$, where G denotes the generation to which the population belongs [12]. The initial population is generated randomly from the entire search space. The main operations of the DE algorithm are: mutation, crossover and selection.

4.1   Mutation

This operation changes the population with the mutant vector $V_{i,G}$ for each individual $X_{i,G}$ in the population at generation G. The mutation operation can be generated using a specific strategy. In this work, three strategies are used: Rand (Eq. 1); Best (Eq. 2); Current-to-Best (Eq. 3):

$$V_{i,G} = X_{r_1,G} + F.(X_{r_2,G} - X_{r_3,G}),\tag{1}$$

```xml
<?xml version="1.0" encoding="ISO-8859-1"?>
<task_graph>
  <tasks>
    <task code ="0" name="src"    type="45" />
    <task code ="1" name="text"   type="44" />
    <task code ="2" name="sink"   type="45" />
    <task code ="3" name="rotate" type="43"/>
    <task code ="4" name="dith"   type="42" />
  </tasks>
  <EDGES>
    <edge name="a0_0" from="0" to="1" cost="1000"/>
    <edge name="a0_1" from="0" to="3" cost="7070"/>
    <edge name="a0_2" from="3" to="4" cost="7070"/>
    <edge name="a0_3" from="4" to="2" cost="7070"/>
    <edge name="a0_4" from="1" to="2" cost="1000"/>
  </edges>
</task_graph>
```

(a) Example of a TG

```xml
<?xml version="1.0" encoding="ISO-8859-1"?>
<repository>
  <ips>
    <ip procName="AMD_ElanSC520-133_MHz--square"
      price="33.0" taskTime="9e-06" taskPower="1.6"
      area="9.61E-6" taskName="Angle2Time Conversion"
      type="0" procID="0" id="0"
    />
    <ip procName="AMD_ElanSC520-133_MHz--square"
      price="33.0" taskTime="2.3e-05" taskPower="1.6"
      area="9.61E-6" taskName="Basic floating point"
      type="1" procID="0" id="1"
    />
    <ip procName="AMD_ElanSC520-133_MHz--square"
      price="33.0" taskTime="0.00049"taskPower="1.6"
      area="9.61E-6" taskName="Bit Manipulation"
      type="2" procID="0"id="2"
    />
    . . .
  </ips>
</repository>
```

(b) Example of an IP repository

Fig. 2. XML codes

$$V_{i,G} = X_{best,G} + F.(X_{r_1,G} - X_{r_2,G}), \tag{2}$$

$$V_{i,G} = X_{i,G} + F.(X_{best,G} - X_{r_1,G}) + F.(X_{r_2,G} - X_{r_3,G}), \tag{3}$$

where $V_{i,G}$ is the mutant vector to be produced; r_1, r_2, r_3 are integer constants generated randomly in the range of $[1, NP]$, at each iteration; $X_{best,G}$ is the best individual at generation G; F is a scaling factor, which is a real constant usually chosen in the range of $[0, 1]$, controlling the amplification of the difference variation.

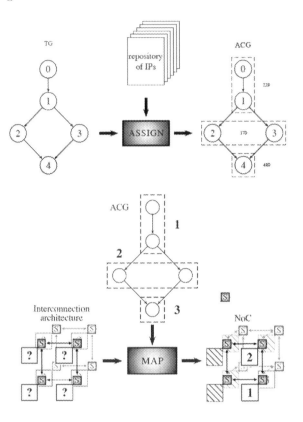

Fig. 3. IP assignment and mapping problems

4.2 Crossover

This operation improves the diversity of the population, being applied after the mutation operation. The crossover operation uses the mutation of the mutant vector $V_{i,G}$ to exchange its components with the target vector $X_{i,G}$, in order to form the trial vector $U_{i,G}$. The crossover operation is defined by Eq. 4:

$$U_{i,G}^{j} = \begin{cases} V_{i,G}^{j} & \text{if}(rand_j[0,1] \le CR)or(j = j_{rand}) \\ \\ X_{i,G}^{j} & \text{otherwise,} \end{cases} \tag{4}$$

where $j = 1, 2, ..., D$; $rand_j$ is the jth evaluation of an uniform random number generator within $[0, 1]$ [11]; the crossover rate CR is an user-specified constant within the range $[0, 1]$; j_{rand} is a randomly chosen integer within the range $[1, D]$ [9].

4.3 Selection

In order to keep the population size constant over subsequent generations, the selection operation is performed. The trial vector is evaluated according to the objective function and compared with its corresponding target vector in the current generation. If the trial vector is better than the target one, the trial vector will replace the target one, otherwise the target vector will remain in the population. The selection operation is represented by Eq. 5:

$$X_{i,G+1} = \begin{cases} U_{i,G} \text{ if} f(U_{i,G}) \leq f(X_{i,G}) \\ \\ X_{i,G} \text{ otherwise.} \end{cases} \quad (5)$$

The mutation, crossover and selection operations are applied for each generation until a termination criterion.

In order to extend the DE algorithm to solve multi-objective optimization problems, we should use the Pareto concept to deal with multiple objectives in order to select the best solution. If the newly generated trial vector dominates the parent vector, then the trial vector will replace the parent one. If the parent dominates the trial, the trial vector will be discarded. Otherwise, when the trial and the parent vectors are not related to each other, the trial vector will be added to the current population for later sorting. Algorithm 1 shows the main steps of the modified DE Multi-Objective (DEMO) algorithm.

Algorithm 1. Modified DEMO

---

    initialize the individuals of the population
    initialize best solutions in archive of leaders
    iteration := 0
    while iteration < max_iteration **do**
      for each individual **do**
        generate a mutated vector using a mutation operation
        generate a trial vector using crossover operation
        evaluate the trial vector
        if the trial vector dominates the individual **then**
          replace individual by the trial vector
        else if the individual dominates the trial vector **then**
          discard the trial vector
        else
          add trial vector to population
        end if
      end for
      update the leaders archive
      iteration := iteration + 1
    end while
    return result from the archive of leaders

---

5 Objective Functions

In this work, we adopted a multi-objective optimization strategy in order to minimize three parameters: area, power consumption and execution time. Here, we describe how to compute each of these parameters in terms of the characteristics of the application and those of the NoC.

5.1   Area

In order to compute the area required by a given mapping, it is necessary to know the area needed for the selected IPs and that required by the used links and switches. The total number of links and switches can be obtained by taking into account all the communication paths between the exploited tiles.

Each communication path between the tiles is stored in the routing table. We adopted an XYZ fixed routing strategy, in which data coming from tile i are sent first to the West or East of the current switch side depending on the target tile position, say j, with respect to i in the 3D mesh NoC, until it reaches the column of tile j. Then, it is sent to the South or North side, depending on the position of tile j with respect to tile i. Finally, it is sent to the Top or Bottom side until it reaches the target tile. The number of links in the described route represents the distance between tiles i and j, corresponding to the *Manhattan distance* [13] as defined by Eq. 6:

$$nLinks(i, j) = |x_i - x_j| + |y_i - y_j| + |z_i - z_j|, \qquad (6)$$

wherein (x_i, y_i, z_i) and (x_j, y_j, z_j) are the coordinates of tiles i and j, respectively.

In order to compute efficiently the area required by all used links and switches, the ACG is associated to a routing table, in which the links and switches necessary interconnect tiles are described. The number of hops between tiles, along a given path, leads to the number of traversal switches. The area is, then, computed summing up the areas required by processors, switches and links involved.

Equation 7 describes the computation involved to obtain the total area for the implementation of a given IP mapping M:

$$Area(M) = \sum_{p \in Proc(ACG_M)} area_p + (A_l + A_s) \times Links(ACG_M) + A_s, \quad (7)$$

wherein function $Proc(.)$ provides the set of distinct processors used in ACG_M and $area_p$ is the required area for processor p; function $Links(.)$ gives the number of distinct links used in ACG_M; A_l is the area of any given link; and A_s is the area of any given switch.

5.2   Power Consumption

The total power consumption of an application NoC-based implementation consists of the power consumption of the processors, while processing the computation performed by each IP, and that due to the data transportation between the tiles, as presented in Eq. 8:

$$Power(M) = Power_p(M) + Power_c(M), \qquad (8)$$

wherein $Power_p(M)$ and $Power_c(M)$ are the power consumption of processing and communication, respectively, as detailed in [13].

The power consumption due to processing is simply obtained summing up attribute $taskPower$ of all nodes in the ACG and is as described in Eq. 9:

$$Power_p(M) = \sum_{t \in ACG_M} power_t. \qquad (9)$$

The total power consumption of sending one bit of data from tile i to tile j can be calculated considering the number of switches and links each bit passes through on its way along the path. It can be estimated as shown in Eq. 10:

$$E_{bit}^{i,j} = nLinks(i,j) \times E_{L_{bit}} + (nLinks(i,j) + 1) \times E_{S_{bit}}, \qquad (10)$$

wherein $E_{S_{bit}}$ and $E_{L_{bit}}$ represent the energy consumed by the switch and link tying the two neighboring tiles, respectively [22]. Function $nLinks(.)$, defined by Eq. 6, provides the number of traversed links (and switches too) considering the routing strategy used in this work and described earlier in this section.

The communication volume $(V(d_{t,t'}))$ is provided by the TG in terms of number of bits sent from the task t to task t' passing through a direct arc $d_{t,t'}$. Let us assume that the tasks t and t' have been mapped onto tiles i and j respectively. Equation 11 defines the total network communication power consumption for a given mapping M:

$$Power_c(M) = \sum_{t \in ACG_M, \forall t' \in Targets_t} V(d_{t,t'}) \times E_{bit}{}^{Tile_t, Tile_{t'}}, \qquad (11)$$

wherin $Targets_t$ provides all tasks that have a direct dependency on data resulted from task t and $Tile_t$ yields the tile number into which task t is mapped.

5.3   Execution Time

The execution time for a given mapping takes into account the execution time of each task, its schedule and the additional time due to data transportation through links and switches along the communication path. $taskTime$ attribute in TG provides the execution time of each task. Each task of the TG needs to be scheduled into its own cycle. Therefore, we used the As-Soon-As-Possible (ASAP) scheduling strategy [10], scheduling tasks in the earliest possible control step.

The routing table allows us to count the number of links and switches required. Two scenarios can lead to the increase in the execution time of the application: (1) when a task in a tile needs to send data to parallel tasks in different tiles through the same initial link, data cannot be sent to both tiles at the same time; (2) when several tasks need to send data to a shared target task, one or more shared links would be needed simultaneously along the partially shared path, which means that the data from both tasks must be pipelined and will not arrive to the target task at the same time.

The overall application execution time takes into account the execution time regarding the underlying task computation and communication for the applied mapping M. It is also necessary to take into account the delay concerning the two aforementioned situations, regarding task scheduling. Therefore, the overall application execution time can be modelled according to Eq. 12:

$$Time(M) = Time_p + Time_c + t_l \times (F_1(M) + F_2(M)), \tag{12}$$

wherein $Time_p$ returns the time necessary to execute the tasks of the TG; $Time_c$ the time spent due to communication among tasks; function F_1 computes the delay caused by the first scenario; function F_2 computes the delay caused the second scenario.

Function F_1 computes the additional time due to parallel tasks that have data dependencies on tasks mapped in the same source tile and yet these share a common initial link in the communication path. Function F_2 computes the additional time due to the fact that parallel tasks producing data for the same target task need to use simultaneously at least a common link along the communication path. Equation 13 defines the time spent with communication between tasks t and t', based on [23]:

$$Time_c = \sum_{t \in APG_M, \forall t' \in Targets_t} \left\lceil \frac{V(d_{t,t'})}{phit} \right\rceil * T_{phit}^{t,}, \tag{13}$$

wherein $V(d_{t,t'})$ is the volume of bits transmitted from task t to task t'. Equation 14 defines the time spent transferring a $phit$:

$$T_{phit}^{t,t'} = nLinks(t,t') \times T_{l_{phit}} + (nLinks(t,t') + 1) \times T_{p_{phit}}, \tag{14}$$

wherein $T_{l_{phit}}$ is the link transmission time and $T_{p_{phit}}$ is the switch processing time used to transfer one phit between two neighboring tiles. A $phit$ represents the physical unit given by the channel width and a $flit$ represents the flow unit, which is a multiple of the $phit$.

6   Results

In order to evaluate the performance of the DEMO algorithm for the mapping step and compare it to that obtained using MOPSO (Multi-Objective Particle Swarm Optimization) algorithm [13], we used the same benchmarks. These are provided by the E3S benchmarks suite, constituted of the characteristics of 17 embedded processors. These characteristics are based on execution times of 16 different tasks, power consumption based on data-sheets, area required based on die size, price and clock frequency. We use 5 random task graphs used in [13], generated by Task Graph For Free (TGFF) [14] to perform experiments and evaluate the performance.

We exploit a population size of 100, with F set to 0.5 and CR set to 0.9. These parameters were set based on simulations. The algorithm was run for

500 iterations. It is noteworthy to emphasize that the same objective functions are used in both works. Besides comparing the algorithms, two topologies are used in the test: 2D mesh with 5×5 and 3D mesh with $3 \times 3 \times 3$. Also, we used As-Soon-As-Possible (ASAP) as the schedulling strategy.

Figure 4 shows the power consumption for the mapping yield by the compared strategies, regarding best results. We can see that the three mutation variants adopted for DEMO offer better results than MOPSO. Among these three, Best shows the best performance.

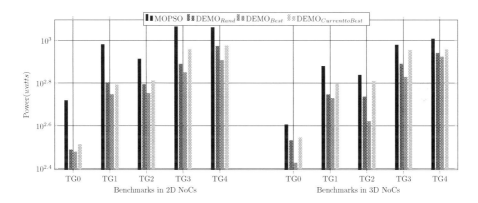

Fig. 4. Comparison of power consumption obtained for each mapping strategy

Also, Fig. 5 provides a comparison of the results regarding execution time, considering the best quality mapping. As can be seen, the performance of DEMO, based on Best mutation strategy, is better than those obtained by MOPSO and the other mutation strategies for DEMO.

Finally, Fig. 6 shows the comparison of the required hardware area related to the mapping obtained by the compared strategies. Here, also, the performance

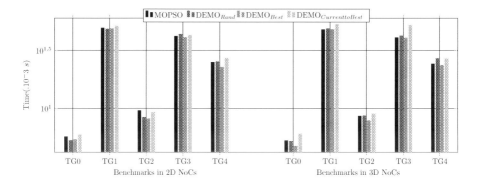

Fig. 5. Comparison of execution time obtained for each mapping strategy

of DEMO is better than that of MOPSO. Among the mutation strategies for DEMO, we can see that Best proves to be the best option.

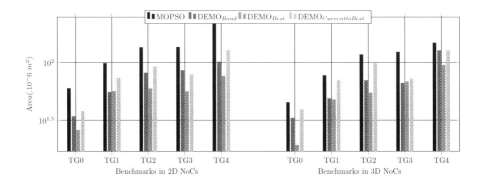

Fig. 6. Comparison of area requirements obtained for each mapping strategy

7   Conclusion

The problem of assigning and mapping IPs into a NoC platform is NP-hard. There are three objectives to be optimized: area required, execution time and power consumption. In this paper, we propose a Multi-Objective algorithm based on Differential Evolution (DEMO) to help NoC designers during the mapping step onto a 3D mesh-based NoC platform, based on pre-selected set of IPs.

We use the same objective functions and the same TGs used in [13], where we applied a Multi-Objective algorithm based on Particle Swarm Optimization (MOPSO), to evaluate the performance of the proposed algorithm. Besides this, the DEMO algorithm offers three strategies, related to variations of the mutation operation: Rand, Best, Current-to-Best. The strategy based on Best presents better performance than the other strategies. We also compare the results obtained by the DEMO algorithm to those obtained by MOPSO algorithm, where DEMO proves to be better than MOPSO. It is interesting to highlight that DEMO requires only two parameters to be set, while MOPSO requires three parameters.

For future work, we plan to experiment with other strategies in the DEMO algorithm and also to use other scheduling algorithms, such as List Schedulling and As Late as Possible.

References

1. Hu, J., Marculescu, R.: Energy-aware mapping for tile-based NoC architectures under performance constraints. In: Proceedings of the 2003 Asia and South Pacific Design Automation Conference, pp. 233–239. ACM, January 2003

2. Davis, W.R., Wilson, J., Mick, S., Xu, J., Hua, H., Mineo, C., Franzon, P.D.: Demystifying 3D ICs: The pros and cons of going vertical. IEEE Des. Test Comput. **22**(6), 498–510 (2005)
3. Ogras, U.Y., Hu, J., Marculescu, R.: Key research problems in NoC design: a holistic perspective. In: Proceedings of the 3rd IEEE/ACM/IFIP International Conference on Hardware/Software Codesign and System Synthesis, pp. 69–74. ACM, September 2005
4. Da Silva, M.V.C., Nedjah, N., de Macedo Mourelle, L.: Evolutionary IP assignment for efficient NoC-based system design using multi-objective optimization. In: 2009 IEEE Congress on Evolutionary Computation, pp. 2257–2264. IEEE, MAy 2009
5. Zhou, W., Zhang, Y., Mao, Z.: Pareto based multiobjective mapping IP cores onto NoC architectures. In: APCCAS, pp. 331–334. IEEE (2006)
6. Jena, R.K., Sharma, G.K.: A multi-objective evolutionary algorithm based optimization model for network-on-chip synthesis. In: ITNG, pp. 977–982. IEEE Computer Society (2007)
7. Radu, C.: Optimized algorithms for network-on-chip application mapping ANoC, Ph.d. thesis, University of Sibiu, engineering Faculty Computer Engineering Department (2011)
8. Radu, C., Mahbub, M.S., Vintan, L.: Developing domain-knowledge evolutionary algorithms for network-on-chip application mapping. Microprocess. Microsyst. Embed. Hardware Des. **37**(1), 65–78 (2013)
9. Robič, T., Filipič, B.: DEMO: differential evolution for multiobjective optimization. In: Coello Coello, C.A., Hernández Aguirre, A., Zitzler, E. (eds.) EMO 2005. LNCS, vol. 3410, pp. 520–533. Springer, Heidelberg (2005). https://doi.org/10.1007/978-3-540-31880-4_36
10. Bougherara, M., Nedjah, N., de Mourelle, L., Rahmoun, R., Sadok, A., Bennouar, D.: IP assignment for efficient NoC-based system design using multi-objective particle swarm optimisation. Int. J. Bio Inspired Comput. **12**(4), 203–213 (2018)
11. Storn, R., Price, K.: Differential evolution-a simple and efficient heuristic for global optimization over continuous spaces. J. Global Optim. **11**(4), 341–359 (1997)
12. Bougherara, M., Kemcha, R, Nedjah, N., Bennouar, D., de Macedo Mourelle, L.: IP assignment optimization for an efficient noc-based system using multi-objective differential evolution. In: International Conference on Metaheuristics and Nature Inspired Computing (META) 2018, pp. 435–444 (2018)
13. Bougherara, M., Nedjah, N., Bennouar, D., Kemcha, R., de Macedo Mourelle, L.: Efficient application mapping onto three-dimensional network-on-chips using multi-objective particle swarm optimization. In: Misra, S., et al. (eds.) ICCSA 2019. LNCS, vol. 11620, pp. 654–670. Springer, Cham (2019). https://doi.org/10.1007/978-3-030-24296-1_53
14. Dick, R.P., Rhodes, D.L., Wolf, W.: TGFF: task graphs for free. In: Proceedings of the Sixth International Workshop on Hardware/Software Codesign (CODES/CASHE 1998), pp. 97–101. IEEE, March 1998
15. Benini, L., De Micheli, G.: Networks on Chip - Technology and Tools. Morgan Kaufmann Publishers, San Francisco (2006)
16. Duato, J., Yalamanchili, S., Ni, L.: Interconnection Networks - An Engineering Approach. Morgan Kaufmann Publishers, San Francisco (2003)
17. Mosayyebzadeh, A., Amiraski, M.M., Hessabi, S.: Thermal and power aware task mapping on 3d network on chip. Comput. Electr. Eng. **51**, 157–167 (2016)
18. Wang, J., Li, L., Pan, H., He, S., Zhang, R.: Latency-aware mapping for 3D NoC using rank-based multi-objective genetic algorithm. In: 2011 9th IEEE International Conference on ASIC, pp. 413–416. IEEE (2011)

19. Sepulveda, J., Gogniat, G., Pires, R., Chau, W., Strum, M.: An evolutive approach for designing thermal and performance-aware heterogeneous 3D-NoCs. In: 2013 26th Symposium on Integrated Circuits and Systems Design (SBCCI), pp. 1–6. IEEE (2013)
20. Bhardwaj, K., Mane, P.S.: C3Map and ARPSO based mapping algorithms for energy-efficient regular 3-d noc architectures. In: Technical papers of 2014 International Symposium on VLSI Design, Automation and Test, pp. 1–4. IEEE (2014)
21. Dick, R.P.: Embedded system synthesis benchmarks suite (E3S). http://ziyang.eecs.umich.edu/~dickrp/e3s/. Accessed 17 Apr 2020
22. Hu, J., Marculescu, R.: Energy-aware mapping for tile-based NoC architectures under performance constraints. In: Proceedings of the 2003 Asia and South Pacific Design Automation Conference, pp. 233–239. ACM (2003)
23. Kreutz, M., Marcon, C.A., Carro, L., et al.: Design space exploration comparing homogeneous and heterogeneous network-on-chip architectures. In: Proceedings of the 18th Annual Symposium on Integrated Circuits and System Design, pp. 190–195. ACM (2005)

Identification of Client Profile Using Convolutional Neural Networks

Victor Ribeiro de Azevedo, Nadia Nedjah, and Luiza de Macedo Mourelle[✉]

Pos-graduation Program in Electronics Engineering,
State University of Rio de Janeiro, Rio de Janeiro, Brazil
v.azevedo9@gmail.com, {nadia,ldmm}@eng.uerj.br

Abstract. In this work, a convolutional neural network is used to predict the interest of social networks users in certain product categories. The goal is to make a multi-class image classification to target social networks users as potential products consumers. In this paper, we compare the performance of several artificial neural network training algorithms using adaptive learning: stochastic gradient descent, adaptive gradient descent, adaptive moment estimation and its version based on infinity norm and root mean square prop. The comparison of the training algorithms shows that the algorithm based on adaptive moment estimation is the most appropriate to predict user's interest and profile, achieving about 99% classification accuracy.

Keywords: Convolutional Neural Networks · Deep learning · Social media image classification · Customer profile identification

1 Introduction

The large volume of information on social networks, such as images and videos, makes it necessary to develop new techniques to extract relevant information from users. Thousands of photos that are being published in daily social networks shows itself to be a relevant process for companies. With a segmentation of customers, it is possible to predict user's interests or create advertisements directed to different publics.

Instagram is one of the most popular social network with more than 1 billion active users monthly, as mentioned by specialists of [2] estimated by the end of 2018. This amount is more than *Twitter* users, but it is behind the amount of users of *WhatsApp* and *Facebook Messenger*. Every day, *Instagram* users post more than 100 million photos as related in [2]. It is important to mention that 72% of the users of *Instagram* say they bought a product they saw in the application as mentioned in [6]. This shows that identifying users by interest is a relevant sales strategy.

Deep learning is a subcategory of machine learning where high complexity neural network models are used to recognize patterns. Recognition of patterns in images and videos can be considered one of the most widely used applications

© Springer Nature Switzerland AG 2020
O. Gervasi et al. (Eds.): ICCSA 2020, LNCS 12251, pp. 103–118, 2020.
https://doi.org/10.1007/978-3-030-58808-3_9

of deep learning. This work contributes to the scientific community, in the field of computational vision and image classification, with exploratory and experimental analysis of the performance of Convolutional Neural Networks (CNNs). Different training algorithms are applied to a group of images in the categories: animals, electronics, games, vehicles and clothes. To improve the performance of the deep learning system, experiments are performed varying the structure of deep learning training algorithms and the initialization of hyperparameters, such as the number of epochs and the rate of learning. In this work, we compare the performance of the following training algorithms: SGD, AdaGrad, Adamax, RMSProp and Adam.

In Sect. 2, we present some related works. In Sect. 3, we introduce the concept of convolutional neural networks. In Sect. 4, we show our proposed solution and the methodology adopted. In Sect. 5, we present the results obtained. In Sect. 6, we drawn some conclusions and future work.

2   Related Works

One of the first projects using CNN was LeNet [7]. The aim was handwriting recognition by implementing the back-propagation algorithm in a Multi Layer Perceptron (MLP), demonstrating the great potential of CNN as extractors of image characteristics.

In [3], an inference of users interests in the social network *Pinterest* is made, based on the classification of texts and images. The work shows that it is possible to recommend products of certain categories based on the classification of images posted by users. The authors report that one of the project's challenges is the interpretation of user's interest based on the image or comment posted on the social network, but the user refers to another subject. In this way, the results were compared using a multi-modal space, combining texts and images, and uni-modal, with texts or images, in order to achieve a better accuracy in the evaluation. The authors use CNNs for image classification, the Bag of Words (BoW) technique for textual representation and the Bag of Visual Words (BoVW) technique for image representation as a feature map. BoW is often used in document and text classification methods, where the occurrence frequency of each word is seen as a characteristic used to train a classifier. Similarly in BoVW, each image is represented as a histogram of occurrence of visual words (small portions of the image, capable of satisfactory describing the image as a whole).

In [14], the binary classification of users of the social network *Pinterest* between masculine and feminine is based on their images. The differential of this project is that it uses the Scalar Invariant Feature (SIF) algorithm to discover local characteristics in each image. To obtain the images for data source of the users, a program is carried out covering 1500 users of the *Pinterest* to obtain the *Twitter* profile. From these profiles, it is possible to obtain the sex of a randomly selected sample with 160 users (80 male and 80 female). Thirty categories are defined from these users.

In [12], the inference of the users' personality of the *Flickr* network is made from images of their galleries. The interaction between the posts, that people leave on social networks, and the traits of their personality are investigated. Experiments with 60.000 images, added to favorites by 300 users, allow them to extract features and segment users.

3 Convolutional Neural Networks

Convolutional Neural Networks are deep learning models usually used for pattern recognition and classification applications in images and videos. A CNN is usually formed by an input layer, which receives the images, a series of layers, which perform image processing operations with characteristics map, and a last step that contains a classification neural network, which receives the characteristics map and outputs the result of the classification.

CNNs are composed of neurons that have weights and bias, which need to be trained. Each neuron receives inputs, to which a scalar product is applied with their respective weights, and provides an output based on a non-linear activation function. A CNN assumes that all inputs are images, which allows us to set properties in the architecture. Traditional artificial neural networks are not scalable to process images, since these require a very high number of weights to be trained.

A CNN consists of a sequence of layers. In addition to the input layer, there are three main layers: convolutional layer, pooling layer and fully connected layer. Besides this, an activation layer is common after the convolutional layer. These layers, when placed in sequence, form the architecture of a CNN, as illustrated in Fig. 1. The convolutional layer is composed of a set of filters that are matrices of real numbers, which can be interpreted as learning weights according to a training. These filters assist in convolution with input data for a feature map. These maps indicate regions in which specific characteristics relative to the filter are found at the entrance. Filter values change throughout the training by having the network learn to identify significant regions to extract features from the data set. The fully connected layer is a common artificial neural network that receives the output of the feature maps and performs the final classification.

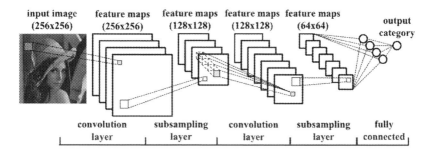

Fig. 1. Architecture of a Convolutional Neural Network

In [1], the pooling layers are mentioned as responsible for reducing the dimensionality of the matrices and the complexity in the neural network. The authors cite the different types of layering aggregation and their advantages. The types of layers cited are: max pooling, average pooling, spatial pyramid pooling, scale dependent pooling.

Activation functions are applied to convolution results matrices. The most used function is the Rectifier Linear Unit - ReLU, used to apply the maximization function to each element of the convolution result. The pooling technique is used after the convolution layer in order to reduce the spatial size of the matrices resulting from the convolution.

The rate of learning is a positive constant, which corresponds to the speed of learning. In the adaptive learning rate (LR) algorithms the value of the learning rate is modified by a term defined in each algorithm. The main adaptive learning algorithms are: AdaGrad, RMSProp and Adam. Besides these, there are others like Adamax, AdaDelta and Nadam.

4   Methodology

For the development of the training model, a virtual environment is created with the container docker. A container docker is a standard encapsulated environment that runs applications through virtualization at the operating system level. An image of docker is a file used to execute code in a docker container. An image is an immutable file that is essentially the representation of a container. Images are created with the compilation command and they will produce a container when started with execution. The image of a docker container is used with the following tools: Jupyter, Matplotlib, Pandas, Tensorflow, Keras and OpenCV [9]. In this work, the Tensorflow and the Keras are used with the language Python to perform the training.

Tensorflow is an open source software library for machine intelligence created by Google in 2015. With Tensorflow, it is possible to define machine learning models, train them with data, and export them. Tensorflow operates with tensors that are multidimensional vectors running through the nodes of a graph.

At the end of the experiments, the performance of each algorithm is analyzed based on the following metrics: accuracy, which is the success rate that the model obtained; precision, which is the rate among those classified as positive; specificity, which is the rate of negative proportion among those that are negative; recall, which is the rate of positive proportion among those who are positive; fmeasure, which is the harmonic mean between precision and recall.

The open source library sklearn [11] is used to measure these performance metrics. According to the documentation of sklearn, it is possible to define the metrics, varying from a minimum of 0 to a maximum of 1 in dimensionless form, as defined in Eq. 1:

$$\begin{cases} precision = \frac{TP}{TP+FP}, \\[2mm] recall = \frac{TP}{TP+FN}, \\[2mm] fmeasure = \frac{2 \times precision}{precision+recall}, \\[2mm] accuracy = \frac{TP+TN}{TP+FN+TN+FP}, \\[2mm] specificity = \frac{TN}{TN+FP}, \end{cases} \tag{1}$$

where TP is the number of true positives, FP are the false positive, TN is the number of true negatives and FN are the false negatives.

The model is tested with 5 learning algorithms: Stochastic Gradient Descent (SGD), AdaGrad, RMSprop, Adam and Adamax. The results of each approach are then compared and ordered according to their performance in relation to accuracy and error.

4.1 Retrieving Data

The selection of training images is based on the *datasets* of the *Kaggle, pyimagesearch* images, on images obtained by *Google Images* and *dataset* [8] as described below. About 1.500 images, organized into 5 categories, are divided into 17 subcategories: domestic animals, such as dogs, cats and birds (264 images) [5]; electronics, such as cell phones, notebooks and TVs (366 images) [8]; games, such as consoles and game scenes (188 pictures); vehicles, such as cars, motorcycles, bikes and planes (386 images) [4]; clothing, such as pants, dresses, blouses, shoes and tennis (298 images) [10].

4.2 Creating the Model

The images go through processing techniques aiming to increase the performance of the classifier model and to reduce the computational effort required to process them. The set of images is divided in 80% for training and 20% for testing. The images are resized to a standard 96px wide and 96px high. The intensity of the pixels is normalized by dividing their scalar values by 255, varying in a range of 0 to 1. The matrices of the images are represented with a dimension of $96 \times 96 \times 3$ with 3 color channels according to the color scale RGB.

The CNN architecture is based on [13]. The network architecture is defined with 7 layers of 2D convolution, with reduction of the dimensionality of the matrices using the max pooling function with dimension 2×2 and 3×3, with 3×3 kernel filters and activation functions ReLU. The fully connected final layer is created using the softmax function for classification. Six layers are used with the sequence of convolution operations, activation function ReLU and pooling. Dropout layers are added to the network to control the overfitting problem and the increased error problem in the training. The Dropout adjusts a certain amount of layer activations based on probabilistic decision making, acting upon activation by a predefined threshold. The network must be able to classify data correctly even without counting on some of its activations at each iteration. This technique becomes important to avoid the overfitting process.

The network then processes the images in the following steps: (1) Obtaining the characteristics: In this first phase the image is processed with the convolutional filters to detect the degree of pertinence of the image with the search characteristics; (2) Positioning the characteristics: In this second phase, the feature maps define what exists in the image and where each characteristic is.

5   Training Results

With the data set prepared, the training routines are established by executing in the *docker* container using the *keras* library with the *python* programming language. The code is run using *Jupyter*, which is a web-based interactive development environment.

5.1   Training Using SGD Algorithm

In the Gradient Descent (GD) algorithm, it is necessary to go through all the samples in the training set to perform parameter updating in a specific iteration. There are three types of Gradient Descent: Batch Gradient Descent, Stochastic Gradient Descent and Mini-batch Gradient Descent.

In the Stochastic Gradient Descent (SGD) algorithm, a training sample is used to update the parameters in a specific iteration. A few samples are selected randomly instead of the whole data set for each iteration. In GD, there is a batch, which denotes the total number of samples from a dataset, that is used for calculating the gradient for each iteration. In typical GD optimization, like Batch Gradient Descent, the batch is taken to be the whole dataset. With large training samples, the use of GD can take a long time. Using SGD, it will be faster, because the performance improvement occurs from the first sample.

Training with the SGD algorithm is carried out using 20 epochs. In Fig. 2, it is possible to observe the decay of the error along the time for the training and testing. Figure 2 shows the error reaching 0.45 for training and 0.46 for test. The Test error curve has some punctual instabilities, but both curves of training and testing have good general relative stability in error decay and in accuracy increasing. In the Fig. 3, it is possible to visualize the growth of training accuracy and validation accuracy reaching performance of 80.80% for training and 82.27% for test. In the Fig. 4, we can see the increase of the hit rate with the visualization of a stable increase of the metrics Fmeasure, precision and recall showing that the relevant results could be correctly classified by the algorithm.

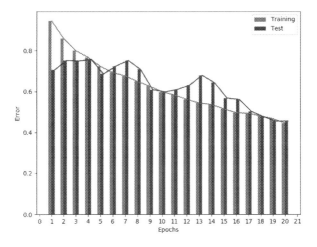

Fig. 2. Error decay during training and test using SGD algorithm

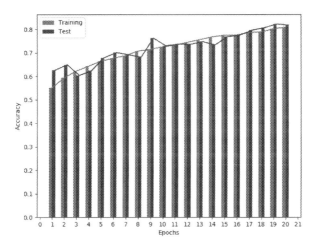

Fig. 3. Accuracy increase during training and test using SGD algorithm

5.2   Training Using AdaGrad Algorithm

The AdaGrad algorithm offers improvements, reducing the learning rate of the weights that receive high gradients and at the same time increasing the rate of learning in the updates of weights. In the AdaGrad algorithm, the learning rate is normalized by the root of the sum of the gradients of each parameter squared.

The results of the trainings and test process are presented using the algorithm AdaGrad with 15 epochs and initial learning rate of 0.01. In Fig. 5, we can see the error decay along the epochs for both training and testing curves. Figure 5 shows some instabilities and peaks for the test curve decay reaching 0.26 and stability for training curve reaching 0.24. Figure 6 shows the growth of training

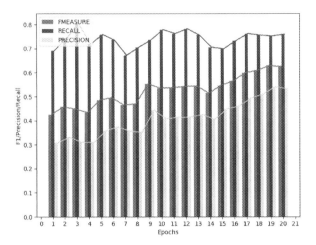

Fig. 4. Metric curves for Fmeasure, Precision, Recall using SGD algorithm

accuracy and validation accuracy reaching performance of 90.21% for training and 89.5% for test offering better results than SGD. In Fig. 7, it is possible to observe an unstable increase of the metrics Fmeasure, Precision and Recall.

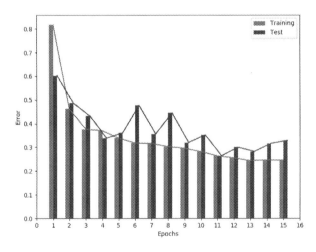

Fig. 5. Error decay during training and test using AdaGrad algorithm

5.3 Training Using RMSprop Algorithm

RMSprop is an algorithm that calculates the magnitudes of the gradients for each parameter and they are used to modify the learning rate individually before applying the gradients. This algorithm modifies the AdaGrad to achieve better

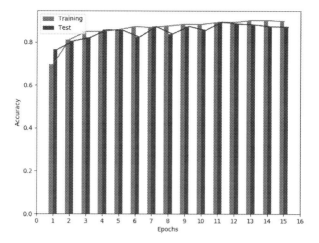

Fig. 6. Accuracy increase during training and test using AdaGrad algorithm

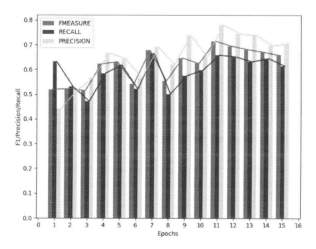

Fig. 7. Metric curves: FMeasure, Precision, Recall using AdaGrad algorithm

performance. The training is performed using the RMSprop with 20 epochs and initial learning rate of 0.001. Figure 8 shows the error decaying to a minimum of 0.20 for training and 0.37 for testing. Both curves of training and testing have good general relative stability in accuracy increasing. In Fig. 9, we can see an unstable increase of the metrics Fmeasure, Precision and Recall. The increase in these metrics along epochs reflects an increase in hit rate. In Fig. 10, the growth in accuracy is shown, reaching a performance of 92.19% for training and 89.16% for testing.

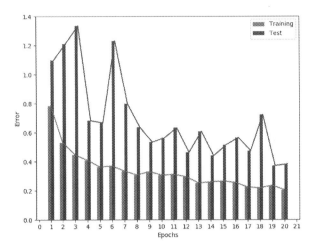

Fig. 8. Error decay during training and test using RMSProp algorithm

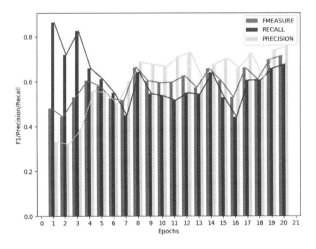

Fig. 9. Metric curves: Fmeasure, Precision, Recall using RMSprop algorithm

5.4   Training Using Adam Algorithm

The Adam training algorithm is a method derived from the RMSprop method
that sets the Adagrad method so that the learning rate does not decrease aggres-
sively. The Adam method's contribution is to add a moment in the upgrade
and smooth out gradient noises before doing this operation. It inherits from
the RMSprop the addition of a decay rate in the sum of the gradients of each
parameter while the learning rate is reduced with each step. This method can
also be defined as a first order gradient-based optimization of stochastic objective
functions.

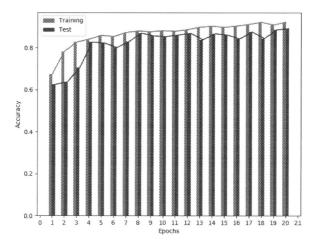

Fig. 10. Accuracy increase during training and test using RMSprop algorithm

The Adam algorithm is run using 50 epochs. Figure 11 shows the error decaying to minimum of 0.12 for training and 0.23 for test showing the good performance of this algorithm. Test error curve has some occurence of instabilities with peaks probability related to some hyperparameter or to some random data chosen in the test step. Figure 12 shows the growth of accuracy, reaching performance of 95.36% for training and 92.12% for test showing a high performance in both steps. In Fig. 13, it is possible to observe some valleys in the increase of the metrics FMeasure, precision and recall. Increasing these metrics along epochs reflects the increase in hit rate. Both curves for training and test have good general relative stability in accuracy increasing.

5.5 Training Using Adamax Algorithm

The Adamax is a variant algorithm of Adam where the second order moment is replaced by the moment of infinite order. Figure 14 shows the error reaching 0.23 for training and 0.36 for test. In Fig. 15, it is possible to visualize the growth of training accuracy and validation accuracy reaching performance of 90.27% for training and 86.15% for test. All curves of error, accuracy, FMeasure, precision and recall are very stables for this algorithm. In Fig. 16, we can observe the increase of the hit rate with the visualization of stable increasing of the metrics Fmeasure, precision and recall.

5.6 Comparative Results

Table 1 shows the maximum accuracy and minimum error for the algorithms used in training and Table 2 shows the corresponding for testing. In addition, data and learning rates are presented for each procedure. In Table 1, the worst accuracy was 80.80% for the SGD algorithm and the two highest accuracy were

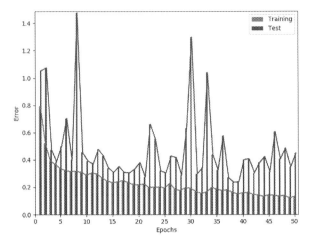

Fig. 11. Error decay during training and test using Adam algorithm

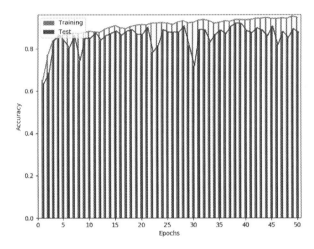

Fig. 12. Accuracy increase during training and test using Adam algorithm

92.19% and 95.36% for the RMSProp and Adam algorithm respectively. The smallest errors observed for training are 0.20 for the RMSProp and 0.12 for the Adam algorithm. Table 2 shows the worst accuracy was 82.27% for the SGD algorithm and the two highest accuracy are 89.50% and 92.10% for the AdaGrad and Adam algorithm respectively. The smallest errors observed for the test are 0.26 for AdaGrad and 0.23 for the Adam algorithm.

Table 3 shows the results of the probabilities of belonging to each category. The tests are presented using the Adam algorithm, which offers the best performance on classifying images of categories animals, electronics, games, vehicles and clothes. It is noteworthy that the performance to classify the categories

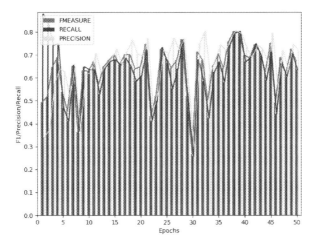

Fig. 13. Metric curves: Fmeasure, Precision, Recall using Adam algorithm

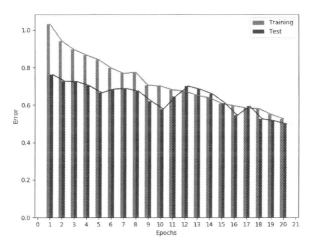

Fig. 14. Error decay during training and test using Adamax algorithm

Table 1. Training performance metrics

Algorithm	Epochs	LR	Max Acc. (%)	Min Error
SGD	20	0.001	80.80	0.45
AdaGrad	15	0.01	90.21	0.24
Adamax	20	0.002	90.27	0.23
RMSProp	20	0.001	92.19	0.20
Adam	50	0.001	95.36	0.12

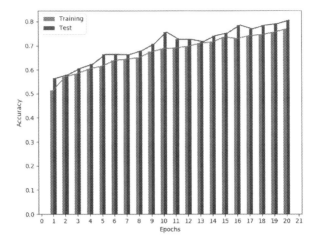

Fig. 15. Accuracy increase during training and test using Adamax algorithm

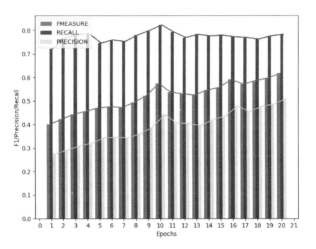

Fig. 16. Metric curves: Fmeasure, Precision, Recall using Adamax algorithm

Table 2. Testing performance metrics

Algorithm	Epochs	L.R.	Max Acc. (%)	Min Error
SGD	20	0.001	82.27	0.46
Adamax	20	0.002	86.15	0.36
RMSProp	20	0.001	89.16	0.37
AdaGrad	15	0.01	89.50	0.26
Adam	50	0.001	92.10	0.23

chair, garden and tower presents maximum probabilities of 66.88%, 0% and 38.42%, respectively, demonstrating that these classes were not used in training. Several images from the trained categories presented performances above 98%, indicating good generalization capacity.

Table 3. Result of images classification in percentage for 5 categories

Image	Anim.	Eletr.	Games	Veic.	Clothes
Dog	99.77	0.00	0.01	0.09	0.00
Cat	99.37	0.01	0.05	0.64	0.07
Bird	99.22	0.05	0.14	0.16	0.07
Notebook	0.12	99.59	0.00	0.08	1.87
TV	0.04	82.62	0.03	66.50	0.14
Cellphone	0.00	99.98	0.00	0.01	0.01
Ps4	0.01	3.07	95.84	0.01	10.32
Car	0.00	0.05	7.16	96.92	0.21
Moto	0.00	0.47	0.80	97.53	3.47
Bike	0.00	0.06	0.08	98.63	4.04
Plane	0.02	0.02	0.05	98.78	1.31
Shoes	0.23	1.33	0.21	55.13	95.92
Dress	1.25	0.00	0.00	0.19	99.70
Pants	12.02	2.34	0.28	0.31	85.49
Chair	0.83	3.17	0.04	6.29	66.68
Garden	0.00	0.00	0.00	0.00	0.00
Tower	0.01	32.76	38.42	1.73	1.78

6   Conclusions

In this work, a convolutional neural network is used as a computational intelligence technique to predict the interest of social network users, in certain categories of products, using image classification. This analysis is valid as a source of knowledge in the segmentation of consumers by interest.

Optimizers and training algorithms are a crucial part of the artificial neural network. Understanding their performance can help choosing the best option for the applications. The knowledge of how hyperparameters can influence the performance of the CNN is very important when training the network. The comparison of the training algorithms SGD, AdaGrad, Adamax, RMSProp and Adam allows us to verify that *Adam* obtains higher accuracy and performances than the others. After completion of all the experiments, when classifying some images of the defined categories, it is possible to obtain satisfactory classification performances as presented in the graphs of performance metrics.

For future work, we suggest to vary the images categories in order to obtain classifications with greater generalization capacity. Another suggestion is to go through the images of users of a certain social network in order to validate on a larger set of images.

References

1. Ajeet Ram Pathak, M.P., Rautaray, S.: Application of deep learning for object detection. Procedia Comput. Sci. **132**, 1706–1717 (2018). https://doi.org/10.1016/j.procs.2018.05.144, http://www.sciencedirect.com/science/article/pii/S18770509 18308767. International Conference on Computational Intelligence and Data Science
2. Aslam, S.: Omnicoreagency: Instagram statistics & facts for 2019 (2019). https://www.websitehostingrating.com/instagram-statistics/. Instagram by the numbers: usage stats, demographics and facts you need to know
3. Cinar, Y.G., Zoghbi, S., Moens, M.: Inferring user interests on social media from text and images. In: 2015 IEEE International Conference on Data Mining Workshop (ICDMW), pp. 1342–1347, November 2015. https://doi.org/10.1109/ICDMW.2015.208
4. Image classification: car, motorbike, bicycle (2012). https://www.kaggle.com/c/image-classification2/data
5. Competition distinguish dogs from cats (2014). https://www.kaggle.com/c/dogs-vs-cats/overview
6. Keyes, D.: Instagram rolls out shoppable posts for more merchants (2017). https://www.businessinsider.com/instagram-rolls-out-shoppable-posts-for-more-merchants-2017-10
7. LeCun, Y., Bottou, L., Bengio, Y., Haffner, P.: Gradient-based learning applied to document recognition. Proc. IEEE **86**(11), 2278–2324 (1998)
8. Li, F.F., Andreetto, M., Ranzato, M.A.: Caltech101 image dataset (2003). http://www.vision.caltech.edu/Image_Datasets/Caltech101/
9. Maksimov, A.: Docker container with jupyter, matplotlib, pandas, tensorflow, keras and opencv (2019). https://hub.docker.com/r/amaksimov/python_data_science/
10. Rosebrock, A.: Multi-label classification with keras (2018). https://www.pyimagesearch.com/2018/05/07/
11. SciKit-learn.org: Machine learn in pyton (2019). https://scikit-learn.org/stable/
12. Segalin, C., Perina, A., Cristani, M., Vinciarelli, A.: The pictures we like are our image: continuous mapping of favorite pictures into self-assessed and attributed personality traits. IEEE Trans. Affect. Comput. **8**(2), 268–285 (2017). https://doi.org/10.1109/TAFFC.2016.2516994
13. Simonyan, K., Zisserman, A.: Very deep convolutional networks for large-scale image recognition. In: International Conference on Learning Representations (2015)
14. You, Q., Bhatia, S., Sun, T., Luo, J.: The eyes of the beholder: gender prediction using images posted in online social networks. In: 2014 IEEE International Conference on Data Mining Workshop, pp. 1026–1030, December 2014. https://doi.org/10.1109/ICDMW.2014.93

International Workshop on Computer Aided Modeling, Simulation and Analysis (CAMSA 2020)

GPU-Based Criticality Analysis Applied to Power System State Estimation

Ayres Nishio da Silva Junior[(✉)], Esteban W. G. Clua,
Milton B. Do Coutto Filho, and Julio C. Stacchini de Souza

Universidade Federal Fluminense, Niteroi, RJ, Brazil
ayresnishio@gmail.com

Abstract. State Estimation (SE) is one of the main tools in the operation of power systems. Its primary role is to filter statistically small errors and to eliminate spurious measurements. SE requires an adequate number of measurements, varied, and strategically distributed in the power grid. Criticality analysis consists of identifying which combinations of measurements, forming tuples of different cardinalities, are essential for observing the power network as a whole. If a tuple of measurements becomes unavailable and, consequently, unobservable, the tuple is considered critical. The observability condition is verified by the factorization of the residual covariance matrix, which usually is a time-consuming computation task. Also, the search for criticalities is costly, being a combinatorial based problem. This paper proposes a parallel approach of the criticality analysis in the SE realm, through multi-threads execution on CPU and GPU environments. To date, no publication reporting the use of GPU for computing critical elements of the SE process is found in the specialized literature. Numerical results from simulations performed on the IEEE 14- and 30-bus test systems showed speed-ups close to $25\times$, when compared with parallel CPU architectures.

Keywords: High performance computing · Supercomputing · Critical analysis · Power system state estimation

1 Introduction

The Energy Management System (EMS) found in control centers encompasses computational tools to monitor, control, and optimize the operation of power networks. Among these tools, the State Estimation (SE) is responsible for processing a set of measurements to obtain the system operating state [1,12]. The SE process depends on the number, type, and distribution of measurements throughout the grid. The observability/criticality analysis reveals the vulnerability of the measuring system to feed SE adequately. This analysis aims to discover the strengths and weaknesses of measurement plans, allowing preventive

© Springer Nature Switzerland AG 2020
O. Gervasi et al. (Eds.): ICCSA 2020, LNCS 12251, pp. 121–133, 2020.
https://doi.org/10.1007/978-3-030-58808-3_10

actions against possible cyber-attacks, and offering different choices of measurement reinforcements. However, due to its combinatorial characteristics, criticality analysis is computationally very expensive.

The word critical is used in the paper to denote extremely important or essential elements of the measuring system devoted to SE. In this sense, the following definitions are stated:

- Critical Measurement (C_{meas}): when not available, it makes the complete system unobservable.
- Critical k-Tuple (C_k): the set of k measurements that, when simultaneously unavailable, leads the system to an unobservable state;

Besides the remarked unobservability cases, the presence of C_{ks} compromises the SE credibility as a function capable of providing a dataset free of gross errors [3]. When there are one or more C_{meas} in the available measurement set, it is impossible to detect any error in the value of these measurements coming from the SE residual analysis. On the other hand, if $(k-2)$ measurements with gross errors belong to a C_k, they can be detected/identified, but if $(k-1)$ or (k) spurious measurements are present in this C_k, they can only be detected.

The criticality analysis becomes computationally costly, particularly when the cardinality of critical tuples k increases, as the consequence of the factorial number of possible combinations to analyze.

This paper presents a parallel programming approach to speed up the process of searching for criticalities (of single measurements or when they are considered forming groups), in which diverse combinations of measurements are simultaneously analyzed. Based on the best knowledge existing so far, there is no publication concerning the use of GPU for identifying critical elements of the SE process is found in the specialized literature. The results (obtained in a multi-core GPU environment) of simulations performed on the IEEE 14- and 30-bus test systems validate the proposed approach.

This paper is organized as follows: Sect. 2 presents some previous works related to criticality analysis. Section 3 presents the residual covariance matrix and its respective application to solve the problem. Section 4 presents our proposed algorithm and it's adaptations to parallel computing. Section 5 shows the results of the implementations in CPU and GPU environments. Finally, Sect. 6 concludes this work.

2   Related Works

One of the first studies on criticality analysis and its relation with the detection/identification of gross measurement errors can be found in [6]. As the identification of all C_{ks} present in a measurement system is a difficult optimization problem [14], most of the published works were limited to study small cardinalities $(k \leq 3)$. This paper adopts the method based on the residual covariance matrix to identify C_{ks} [3].

Criticality analysis is a useful tool to evaluate the robustness of measurement schemes [5,7], as well as to prevent possible cyber-attacks of false data injections, as described in [10,15]. In this type of attack, a malicious attacker intentionally adds gross errors to the available measurements. Based on previous knowledge of the measurement criticalities, it is possible to take preventive actions (e.g., to reinforce the metering plan), so that the effect of an attack can be mitigated.

Few studies have been conducted to address the general problem of searching for C_{ks}. Reference [2] tackles this hard combinatorial problem via the Branch-and-Bound (B&B) approach. B&B is a tree-based exploratory tool for finding implicitly exact solutions.

Parallel programming has been used to speed up the estimation filtering process [9,10]. B&B algorithms in GPUs [4] have also been implemented. The GPU approach to deal with criticality analysis in SE using the residual covariance matrix method can be considered the main contribution of the present paper.

3   The Residual Covariance Matrix Method

The entries of the residual covariance matrix (Ω) represent the degree of interaction between measurements, which can be explored in the criticality analysis [3]. Also, using the definition of a C_k previously introduced, the following properties are useful [6]:

1. The columns of matrix Ω associated with the measurements that form a C_k are linearly dependent.
2. A C_k does not contain a C_j, for $\forall k > j$.

Null elements in Ω indicate the presence of entirely uncorrelated measurements. Therefore, null rows/columns in Ω are associated with C_{meas}.

Now, consider $\tilde{\Omega}$ denoting a submatrix of Ω, composed of the columns/rows associated with a group of k measurements. If $\tilde{\Omega}$ has linearly dependent columns/rows, then the measurements associated with those form a C_k. Gauss elimination can be used to factorize $\tilde{\Omega}$ to check whether there are linearly dependent rows/columns in the matrix.

To gain insight into criticalities, for instance, consider the six-bus system represented in Fig. 1, measured by five power flows and four power injections, for which all C_{ks} present in the set of measurements will be identified.

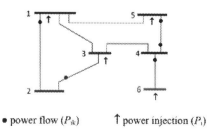

• power flow (P_{ik})   ↑ power injection (P_i)

Fig. 1. Six bus system, with its Measurement set

Symmetric matrix Ω, obtained with the decoupled model for the adopted test system and usually adopted in the classical observability analysis [1] is presented in Table 1.

Table 1. Covariance matrix Ω, obtained from six bus system depicted in Fig. 1

Meas. #	1	2	3	4	5	6	7	8	9
	P_{1-2}	P_{2-3}	P_{4-5}	P_{4-6}	P_{5-4}	P_1	P_3	P_5	P_6
P_{1-2}	0.57	0.38	-0.07	0.00	0.07	-0.23	0.05	-0.19	0.00
P_{2-3}		0.53	0.16	0.00	-0.16	-0.02	0.17	0.15	0.00
P_{4-5}			0.66	0.00	0.34	0.13	0.10	0.23	0.00
P_{4-6}				0.50	0.00	0.00	0.00	0.00	0.50
P_{5-4}					0.66	-0.13	-0.10	-0.23	0.00
P_1						0.17	0.05	0.21	0.00
P_3							0.07	0.12	0.00
P_5								0.34	0.00
P_6									0.50

As one can observe, there is no null columns/rows in Ω; thus, the set of measurements is free from C_{meas}. Now, concerning, for instance, the tuples of cardinality two, a C_2 is identified, involving the measurements #4 and #9, i.e., branch $4-6$ power flow ($P_{4,6}$) and the power injection at bus 6 (P_6). The submatrix $\tilde{\Omega}_{4,9}$ is built with the entries related to the respective rows/columns:

$$(\tilde{\Omega}_{4,9}) = \begin{bmatrix} 0.50 & 0.50 \\ 0.50 & 0.50 \end{bmatrix} \tag{1}$$

$$(\tilde{\Omega}_{4,9}) = \begin{bmatrix} 0.50 & 0.50 \\ 0.00 & 0.00 \end{bmatrix} \tag{2}$$

For tuples of cardinality 3, for instance, rows #2, #7, and #8 are associated with measurements (P_{2-3}, P_3, P_5) that form a C_3. Submatrices $\Omega_{2,7,8}$ in (3) and its equivalent triangular form in (4), support this conclusion.

$$(\tilde{\Omega}_{2,7,8}) = \begin{bmatrix} 0.53 & 0.17 & 0.15 \\ 0.17 & 0.07 & 0.12 \\ 0.15 & 0.12 & 0.33 \end{bmatrix} \tag{3}$$

$$(\tilde{\Omega}_{2,7,8}) = \begin{bmatrix} 0.53 & 0.06 & 0.15 \\ 0.00 & -0.05 & 0.22 \\ 0.00 & 0.00 & 0.00 \end{bmatrix} \tag{4}$$

Together with (P_{2-3}, P_3, P_5) nine more C_3 are identified. They are: (P_{1-2}, P_{2-3}, P_1), (P_{1-2}, P_{2-3}, P_3), (P_{1-2}, P_{2-3}, P_5), (P_{1-2}, P_1, P_3), (P_{1-2}, P_1, P_5), (P_{1-2}, P_3, P_5), (P_{2-3}, P_1, P_3), (P_{2-3}, P_1, P_5), (P_1, P_3, P_5). Table 2 summarizes the results achieved for all C_{ks}. There is an upper limit (k_{lim}) for the cardinality of C_{ks}, which is dependent of the difference between m (number of available measurements) and n (number of state variables), given by: $k_{lim} = m - n + 1$. Thus, for the system under analysis, $k_{lim} = 9 - 6 + 1 = 4$.

Table 2. Number of visited combinations of measurements and C_{ks} identified in the 6-bus system depicted in Fig. 1

Cardinality k	No. of visited combinations of measurements	No. of C_{ks} identified
1	9	0
2	36	1
3	84	10
4	126	10
5	126	0
6	84	0
7	36	0
8	9	0
9	1	0
Total	511	21

4   Proposed Algorithm

The proposed algorithm receives as input Ω, m, and k_{max} and returns as output the solution set (solSet) containing all criticalities C_{ks} in the measurement set. The solution set contains binary vectors, in which the elements that correspond measurements removed from the set are represented by 1. Figure 2 depicts the C_{ks} that were found in the illustrative example of Sect. 4 stored in the solSet. As the number of C_k is previously unknown, solSet was implemented as a stack.

P_{1-2}	P_{2-3}	P_{4-5}	P_{4-6}	P_{5-4}	P_1	P_3	P_5	P_6
0	0	0	1	0	0	0	0	1
0	1	0	0	0	0	1	1	0

Fig. 2. Examples of output representation for criticalities $\{P_{4-6}; P_6\}$ and $\{P_{2-3}; P_3; P_5\}$

The main algorithm adaptation to parallel programming considers the possibility to evaluate all possible k combinations of m measurements simultaneously.

Algorithm 1 shows that each cardinality is analyzed separately within a loop until the adopted k_{max} is reached. Furthermore, each loop consists of four steps, as shown in Algorithm 1. The following subsections detail each one of these steps.

The first step creates a combination matrix ($combMat$) ($\binom{k}{m} \times m$) to store all possible combinations of m measurements taken from k at a time. Steps 2 and 3 are the main focus of the present work, in which CPU and GPU parallel computing are adopted. Step 2 identifies combinations of measurements whose removal compromises the system observability. The third step checks whether or not the solution of the second step is indeed a C_k. Finally, step 4 updates the solution set ($solSet$).

Algorithm 1: Parallel Critcality Analysis

---

    input : Ω, m, k_{max}
    output: $ConjSol$
    for $k \leftarrow 1$ **to** k_{max} **do**
        | *Step 1:* Build Combinaton Matrix
        | *Step 2:* Find out Possible C_ks
        | *Step 3:* Confirm C_ks
        | *Step 4:* Update Solution Set
    end

4.1   *Step 1*: Combination Matrix

The construction of *combMat* is an adaptation of the algorithm presented in [13]. The original algorithm allocates the measurement combinations in a stack so that they can be further successively accessed. However, to allow a parallel implementation, the combinations are represented by rows in a matrix data structure, enabling concurrent access to them.

4.2   *Step 2*: Searching for C_{ks}

At this step, the algorithm builds a submatrix $\tilde{\Omega}$ for each row in *combMat*. Then, the invertibility of each $\tilde{\Omega}$ is tested by using Gauss Elimination, identifying possible C_{ks}.

As can be seen in Algorithm 2, there is an auxiliary binary array *isCrit* in which each element corresponds to a row in *combMat*. If the algorithm finds a possible C_k, it assigns 1 to its corresponding element in *isCrit*; otherwise, it assigns 0.

The multi-thread computing runs each loop in Algorithm 2 simultaneously. The array *isCrit* enables threads to record which row of *combMat* is a possible C_k concurrently. Distinctly, in GPU implementation, this step requires Ω and *combMat* to be transferred to GPU global memory.

Algorithm 2: *Step 2:* Searching for C_{ks}

---

for $i \leftarrow 0$ **to** $\binom{k}{m}$ **do**
    build $\tilde{\Omega}(\Omega, k, combMat)$;
    if $\tilde{\Omega}$ *is invertible* **then**
        isCrit[i]=0;
    else
        isCrit[i]=1;
    end
end

4.3 *Step 3*: Confirmation

Considering C_j $(j < k)$ a criticality solution previously stored in *solSet*, the algorithm compares all C_j with all rows in *combMat*. If a row has its corresponding *isCrit* $= 1$ and contains any C_j, then the measurement combination found in step 2 is not a C_k. Consequently, the algorithm updates its related value in *isCrit* to 0.

The algorithm uses an element-wise subtraction to verify if an array is a subset of another. For instance, considering $C_2 = \{1001\}$ as a critical set previously-stored in *solSet* and $C_3 = \{1011\}$ as possible criticality found in step two. Subtracting each element of those arrays, $(C_3 - C_2) = \{0010\}$, then, since there was no element equal to -1 in such operation, one can conclude that $C_2 \subset C_3$. So C_3 is not considered critical, and its related element in *isCrit* is set to 0.

Algorithm 3: Confirmation

---

for j **in** *solSet* **do**
    for $i=0 : \binom{k}{m}$ **do**
        if *isCrit[i]*==1 **then**
            if *solSet[j]* $\subset$ *matComb[i]* **then**
                isCrit[i]==0;
            end
        end
    end
end

The multi-thread approach executes the inner loop in Algorithm 3 simultaneously. Also, for this implementation on GPU, *solSet* must be initially transferred to the GPU global memory.

4.4 *Step 4*: Solution Set

In this last step, all the rows of *combMat* with the corresponding *isCrit* $= 1$ are added to *solSet*, as shown in Algorithm 4.

Algorithm 4: Solution Set

---

for $i = 0 : \binom{k}{m}$ **do**

    if $isCrit[i]==1$ **then**

        | add($combs[i]$, $solSet$)

    end

end

---

5 Results

This section presents the comparative results between the sequential steps two and three and its parallel implementations in multi-core CPU and GPU environments. For GPU computing, such comparison also considered the data management between host and device.

Simulations carried out on a computer with 8 GB RAM, Intel Core i5-9300H processor, and NVIDIA GeForce GTX 1650 graphics card. The original single-threaded algorithm was implemented in C++. Furthermore, the CPU multi-threading was performed by OpenMP, using 8 threads and GPU computing in NVIDIA CUDA, using 512 threads in each kernel call.

From the performed simulations it could be noted that the use of parallelism did not prove to be advantageous in systems with a few measurements and for small values of k, which resulted in almost any combinations to be analyzed. Therefore, results for $k > 6$ using the IEEE 14-bus and IEEE 30-bus test systems will be presented.

Also, some values of k resulted in a large number of combinations that could not be stored at once in the combination matrix. Then, the k-value analysis (with more than 2^{20} combinations) was partitioned.

5.1 IEEE 14-Bus System

The IEEE 14-bus test case adopts the measuring system presented in [3], in which there are 33 measurements available. Table 3 shows the search space covered by the proposed algorithm and the number of C_{ks} found and confirmed up to $k = 10$.

As can be seen in Figs. 3 and 4, for both steps, better performance has been achieved with the GPU implementation, reaching in Step 2 an average speed-up of 13x when compared to the original serial implementation, as it can be seen in Table 4. Besides, as it is shown in Table 5, in Step 3 we achieved speed-ups of 25x.

5.2 IEEE 30-Bus System

Tests were also performed with the IEEE 30-bus system with a low redundant measurement set, adapted from [2]. Table 6 confirms that the lack of redundant measurements results in more C_{ks} of lower cardinalities. Therefore, Step 3 becomes more expensive, as those C_{ks} are added early to the solution set and are always compared to every possible C_k found in Step 2.

Table 3. IEEE 14-bus system visited combinations and C_{ks} confirmed

Cardinality k	No. of visited combinations of measures	No. of C_{ks} identified
1	33	0
2	528	13
3	5456	0
4	40920	1
5	237336	1
6	1107568	1
7	4272048	13
8	13884156	9
9	13884156	13
10	38567100	11

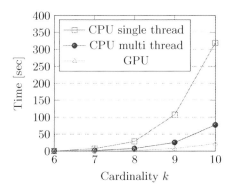

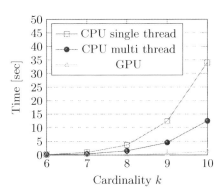

Fig. 3. Performance for Step 2, 14-bus case

Fig. 4. Performance for Step 3, 14-bus case

Table 4. Speed-ups achieved for Step 2 in the IEEE 14-bus system

Cardinality k	Speed-up					Average
	6	7	8	9	10	
CPU Single-Thread × CPU Multi-Thread	4.31	4.32	3.83	4.23	4.16	4.17
CPU Single-Thread × GPU	10.39	13.33	13.69	14.81	14.70	13.37
Multi-Thread × GPU	2.41	3.09	3.56	3.50	3.54	3.22

Figures 5 and 6 show that GPU implementation again presented the best performance. As it is shown in Table 7 and 8, the speedups observed in Steps 2 and 3 were 13x and 18x, respectively. The decrease of the speedup in Step 3,

Table 5. Speed-ups achieved for Step 3 in the IEEE 14-bus system

Cardinality k	Speed-up					Average
	6	7	8	9	10	
CPU Single-Thread × CPU Multi-Thread	2.01	2.40	2.44	2.71	2.72	2.46
CPU Single-Thread × GPU	20.33	25.48	26.50	27.61	26.21	25.23
Multi-Thread × GPU	10.11	10.6	10.83	10.17	9.625	10.27

Table 6. IEEE 30-bus visited combinations and C_{ks} confirmed.

Cardinality k	No. of visited combinations of measurements	No. of C_{ks} identified
1	42	4
2	861	11
3	11480	19
4	111930	96
5	850668	345
6	5245786	1297
7	26978328	2911
8	118030185	8335
9	445891810	39336

concerning the one obtained with the IEEE 14-bus system, occurs because of a larger *solSet* is transferred to GPU global memory in each loop.

The analysis for k = 9 became too expensive for the single thread CPU implementation. In GPU it took around 23 min to perform Step 3.

As mentioned in the introduction section, the specialized literature is sparse in studies related to the problem of searching for critical elements is state

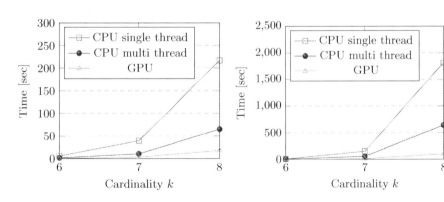

Fig. 5. Performance for Step 2, 30-bus case.

Fig. 6. Performance for Step 3, 30-bus case.

Table 7. Speed-ups achieved for Step 2 in IEEE 30-bus system

Cardinality k	Speed-up			Average
	6	7	8	
CPU Single-Thread × CPU Multi-Thread	3.41	3.75	3.31	3.51
CPU Single-Thread × GPU	11.60	12.92	13.11	12.55
Multi-Thread × GPU	3.05	3.18	3.64	3.29

Table 8. Speed-ups achieved for Step 3 in IEEE 30-bus system

Cardinality k	Speed-up			Average
	6	7	8	
CPU Single-Thread × CPU Multi-Thread	1.32	2.75	2.83	2.30
CPU Single-Thread × GPU	8.12	18.22	18.07	18.03
Multi-Thread × GPU	6.15	5.53	6.38	6.35

estimation. In [2], one can find results of a B&B algorithm implemented in a CPU environment to identify C_{ks} of cardinalities up to five in the IEEE 30-bus system. However, no information on the estimated computing time to perform this task is provided, which could enable comparative studies on the subject. At this point, it is opportune to comment that the results obtained in the paper point to a successful novel application of GPUs to perform the criticality analysis task in power system state estimation studies. As such, the proposed use of GPUs can be seen as a reference implementation, to demonstrate (proof of concept) that it is a feasible option when compared with its CPU counterparts, acting as a starting point for more elaborated (optimized) implementations. Thus, the results obtained here serve as benchmarks for future correlated research works.

6   Conclusion

In this work we developed a novel parallel approach for identification of measurement criticalities. Our solution may affect the quality of power system state estimation, due to its high increase of performance. The use of GPUs proved to be useful to address this difficult combinatorial problem, allowing the determination of cardinalities that would be very costly to achieve in a single-thread CPU. Different test systems and measurement redundancies were tested and the proposed implementation lead to speedups up to 28×, showing its potential for the identification of measurement criticalities, particularly those of high cardinality, whose determination is usually more challenging.

Since managing data transfer between host and device is a bottleneck for all the solutions, in future works, we intend to include steps 1 and 4 at the GPU level. However, adapt those steps to parallel implementation is not a trivial task. In step 1, it is challenging to build the combination matrix in GPU, because it is not straightforward to correlate each combination to each thread identification, according to [11]. In step 4, since we implemented the solution set in a stack, it is not possible to add elements concurrently to set. We believe that exists a better data structure that ensures performance improvements in this implementation. Important improvements in the parallel code are also possible to be made, such as the usage of GPU streams in order to assist large-scale combination arrays. We also believe that we can better explore parallel approaches to assist the Branch and Bound heuristic that were already implemented for criticality analysis solution.

References

1. Abur, A., Gómez Expósito, A.: Power System State Estimation: Theory and Implementation. Dekker, New York (2004)
2. Augusto, A.A., Do Coutto Filho, M.B., de Souza, J.C.S., Guimaraens, M.A.R.: Branch-and-bound guided search for critical elements in state estimation. IEEE Trans. Power Syst. **34**, 2292–2301 (2019)
3. Augusto, A.A., Guimaraens, M.A.R., Do Coutto Filho, M.B., Stacchinni de Souza, J.C.: Assessing strengths and weaknesses of measurement sets for state estimation. In: 2017 IEEE Manchester PowerTech, pp. 1–6. IEEE, Manchester (2017)
4. Boukedjar, A., Lalami, M.E., El-Baz, D.: Parallel branch and bound on a CPU-GPU system. In: 2012 20th Euromicro International Conference on Parallel, Distributed and Network-based Processing, pp. 392–398. IEEE, Munich (2012)
5. Castillo, E., Conejo, A.J., Pruneda, R.E., Solares, C., Menendez, J.M.: $m - k$ Robust observability in state estimation. IEEE Trans. Power Syst. **23**, 296–305 (2008)
6. Clements, K.A., Davis, P.W.: Multiple bad data detectability and identifiability: a geometric approach. IEEE Trans. Power Deliv. **1**, 355–360 (1986)
7. Crainic, E.D., Horisberger, H.P., Do, X.D., Mukhedkar, D.: Power network observability: the assessment of the measurement system strength. IEEE Trans. Power Syst. **5**, 1267–1285 (1990)
8. Ilya, P., Pavel, C., Olga, M., Andrey, P.: The usage of parallel calculations in state estimation algorithms. In: 2017 9th International Conference on Information Technology and Electrical Engineering (ICITEE), pp. 1–5. IEEE, Phuket (2017)
9. Karimipour, H., Dinavahi, V.: Accelerated parallel WLS state estimation for large-scale power systems on GPU. In: 2013 North American Power Symposium (NAPS), pp. 1–6. IEEE, Manhattan (2013)
10. Karimipour, H., Dinavahi, V.: On false data injection attack against dynamic state estimation on smart power grids. In: 2017 IEEE International Conference on Smart Energy Grid Engineering (SEGE), pp. 388–393. IEEE, Oshawa (2017)
11. Luong, T.V., Melab, N., Talbi, E.-G.: Large neighborhood local search optimization on graphics processing units. In: 2010 IEEE International Symposium on Parallel a& Distributed Processing. Workshops and PhD Forum (IPDPSW), pp. 1–8. IEEE, Atlanta (2010)
12. Monticelli, A.: Power System State Estimation: A Generalized Approach. Kluwer, Norwell (1999)
13. Ruskey, F., Williams, A.: The coolest way to generate combinations. Discrete Math. **309**, 5305–5320 (2009)

14. Sou, K.C., Sandberg, H., Johansson, K.H.: Computing critical k-tuples in power networks. IEEE Trans. Power Syst. **27**, 1511–1520 (2012)
15. Xie, L., Mo, Y., Sinopoli, B.: False data injection attacks in electricity markets. In: 2010 First IEEE International Conference on Smart Grid Communications, pp. 226–231. IEEE, Gaithersburg (2010)

Robust Control of the Classic Dynamic Ball and Beam System

Javier Jiménez-Cabas[1], Farid Meléndez-Pertuz[1],
Luis David Díaz-Charris[1], Carlos Collazos-Morales[2(✉)],
and Ramón E. R. González[3(✉)]

[1] Departamento de Ciencias de la Computación y Electrónica,
Universidad de la Costa, Barranquilla, Colombia
fmelende1@cuc.edu.co
[2] Vicerrectoría de Investigaciones, Universidad Manuela Beltrán,
Bogotá, Colombia
cacollazos@gmail.com
[3] Departamento de Física, Universidad Federal de Pernambuco, Recife, Brazil
ramayo_g@yahoo.com.br

Abstract. This article presents the design of robust control system for the Ball and Beam system, as well as the comparison of their performances with classic control techniques. Two controllers were designed based on Algebraic Riccati Equations for the synthesis of H_2 and H_∞ controllers and a third one based on Linear Matrix Inequalities techniques for the design of H_∞ controllers. The results show that H_∞ controllers offer better performance.

Keywords: Robust control H_2 and H_∞ · Uncertainties · Modeling · Ball and beam system

1 Introduction

One of the central topics of discussion in the design of feedback control systems has been the robustness of these systems in the presence of uncertainties, which is the reason of the development of methods to tackle the problem of robust control, has become one of the areas of most active research since the 1970s [1–3].

Several researches have yielded as a result a variety of methods that employ different mathematical techniques [4, 5] extending the field of application of robust control theory to different areas such as the control of chemical processes, control of power systems and the investigation of feedback mechanisms in biological systems, giving great importance to make more contributions in this topic.

In this sense, the ball and beam system has become one of the most popular laboratory models used in the teaching of control engineering [6]. Its main importance is because of is a simple system to understand and is also unstable in open loop, which allows studying both classic and modern control techniques, while analyzing the benefits of each one.

Therefore, this paper shows the synthesis of the robust controllers for the ball and beam system, as well as the verification of their operation and the comparison of their

© Springer Nature Switzerland AG 2020
O. Gervasi et al. (Eds.): ICCSA 2020, LNCS 12251, pp. 134–144, 2020.
https://doi.org/10.1007/978-3-030-58808-3_11

performance of their controls with the classic control techniques. Two devices were designed using solution techniques of Algebraic Riccatis Equations (ARE) for the synthesis of H_2 y H_∞ controls and a third one based on Linear Matrix Inequalities (LMI) techniques for the design of H_∞ controllers. The results shown that the H_∞ results offer better performance.

To verify the operation of the designed systems, the module Ball and Beam from Quanser® was used.

2   Ball and Beam System Description

The module Ball AND Beam from Quanser® (Fig. 1) [7] that was used is made up of a steel bar in parallel with a resistor, which make up the path on which a metal ball travels freely.

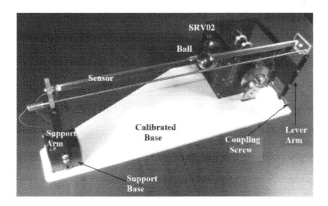

Fig. 1. Ball and beam system attached to the plant SRV02 (Quanser®).

When the ball moves over the path it acts similar to the brush on a potentiometer. The position of the ball is obtained by measuring the voltage on the steel bar. By coupling this system with the SRV02 (from the same manufacturer), a DC motor drives the bar and the motor angle controls the inclination of the steel bar.

The objective is to design a control system that allows the position of the ball to track a reference signal. The manufacturer provides a control system based on two PD loops which is taken as a reference to compare the performances obtained with the robustly designed control systems.

2.1   Mathematical Model

Figure 2 shows the Ball and Beam system [8]. A translational force (due to gravity) and a rotational force (due to the torque produced by its angular acceleration) act on the ball. The first of these forces is given by:

$$F_{tx} = mg \sin \alpha \tag{1}$$

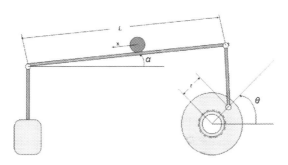

Fig. 2. Ball and Beam System diagram.

The torque produced by the rotational movement of the ball is equal to the radius of the ball multiplied by the rotational force. Using Newton's second law, there is possible to write:

$$T_r = F_{rx}R = Ja = J\frac{\ddot{x}}{R} \tag{2}$$

Taking into account that the moment of inertia of the ball is $J = (2/5)\ mR^2$ (where m is the mass of the ball), it yields:

$$F_{rx} = \frac{2}{5}\ m\ddot{x} \tag{3}$$

Applying Newton's second law to the ball:

$$m\ddot{x} = F_{tx} - F_{rx} = mg \sin \alpha - \frac{2}{5}m\ddot{x} \tag{4}$$

Then,

$$\ddot{x} = \frac{5}{7}g \sin \alpha \tag{5}$$

On the other hand, it has been defined **Arc** as the arc traveled by the lever arm, measuring the angles in radians, then it is possible to obtain:

$$Arc = \theta r = \alpha L \tag{6}$$

Now, there is obtained the model of the position SRV02 [9]. Initially the electrical components of the motor are considered, Fig. 3 shows the schematic diagram of the armature circuit.

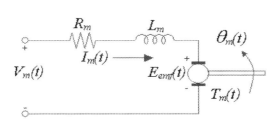

Fig. 3. DC Motor armature circuit.

Applying Kirchhoff's voltage law we obtain:

$$V_m - R_m I_m - L_m \frac{dI_m}{dt} - E_{emf} = 0 \tag{7}$$

Considering that $L_m << R_m$

$$I_m = \frac{V_m - E_{emf}}{R_m} \tag{8}$$

where, E_{emf} is proportional to the motor shaft angular velocity w_m, hence:

$$I_m = \frac{V_m - K_m \dot{\theta}_m}{R_m} \tag{9}$$

Aiming to consider the mechanical aspects of the motor there is important to apply Newton's second law on the motor shaft,

$$J_m \ddot{\theta}_m = T_m - \frac{T_l}{\eta_g K_g} \tag{10}$$

where $T_l/\eta_g K_g$ is the torque of the load seen through the pinions and η_g is the efficiency of the gear box. Now applying Newton's second law to motor load, the next equation emerges:

$$J_l\ddot{\theta}_l = T_l - B_{eq}\dot{\theta}_l \tag{11}$$

where B_{eq} is the viscous coefficient of friction seen at the outlet. Substituting (10) in (11), it becomes

$$J_l\ddot{\theta}_l = \eta_g K_g T_m - \eta_g K_g J_m\ddot{\theta}_m - B_{eq}\dot{\theta}_l \tag{12}$$

Noting that $\theta_m = K_g\theta_l$ y $T_m = \eta_m K_t I_m$ (where η_m is the engine efficiency), can be rewritten (12) as:

$$J_l\ddot{\theta}_l + \eta_g K_g^2 J_m\ddot{\theta}_l + B_{eq}\dot{\theta}_l = \eta_g\eta_m K_g K_t I_m \tag{13}$$

Substituting (8) into (13):

$$\begin{aligned} J_{eq}R_m\ddot{\theta}_l + \left(B_{eq}R_m + \eta_g\eta_m K_m K_t K_g^2\right)\dot{\theta}_l \\ = \eta_g\eta_m K_t K_g V_m \end{aligned} \tag{14}$$

where $J_{eq} = J_l + \eta_g J_m K_g^2$. This can be interpreted as the equivalent moment of inertia of the motor system seen from the output.

The complete model of the system is made up of the Eqs. (5), (6) and (14). Table 1 shows the system parameters [7, 9].

Table 1. System parameters.

Symbol	Description	Value	Unities	Variation
L	Beam length	42.5	cm	
R	Ball radius	2.54	cm	
R	Lever arm offset	2.54	cm	
M	Ball mass	0.064	Kg	
G	Gravitational constant	9.8	m/s^2	
K_t	Torque constant	0.0077	N·m	±12%
K_m	emf constant	0.0077	V/(rad/s)	±12%
K_g	Total Ratio of Pine Nuts	70		
R_m	Armature Resistance	2.6	Ω	±12%
J_{eq}	J engine equivalent	2×10^{-3}	Kg·m^2	±10%
B_{eq}	B equivalent	2×10^{-3}	N·m/(rad/s)	±20%
η_g	Gear box efficiency	0.85		±10%
η_m	Engine efficiency	0.69		±5%

3   System Control Design

To achieve that the ball position tracks a reference signal, a control system was designed implementing two independent control loops [8], an internal loop to control the position of the servo axis and an external loop to control the position of the ball on the bar, manipulating the position of the servo. Figure 4 shows the manufacturer's suggested control configuration for both loops.

The feedback status gains given by the manufacturer for both the internal control loop and the external control loop are Kp = 11.1206 V/rad, Kv = 0.11106 V/rad/s and Kp_bb = 0.0017619 rad/cm, Kv_bb = 0.0022433 rad/cm/s respectively.

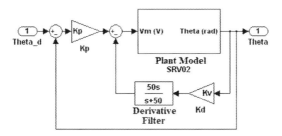

Fig. 4. Control configuration.

For our design, the topology of two control loops was maintained, but using output feedback. The linear models used in each case for the design are as follows.

$$G_m(s) = \left[\begin{array}{cc|c} 0 & 1 & 0 \\ 0 & -35.0427 & 61.6326 \\ \hline 1 & 0 & 0 \end{array}\right] \tag{15}$$

$$G_b(s) = \left[\begin{array}{cc|c} 0 & 1 & 0 \\ 0 & 0 & 0.4179 \\ \hline 100 & 0 & 0 \end{array}\right] \tag{16}$$

3.1   Controllers

To obtain the controllers, unstructured multiplicative uncertainties were considered upon entry into both control loops. Figure 5 shows the block diagram of the closed loop system in which the weights of uncertainties have been included W_i, performance W_d, control signal W_u and desired response W_f.

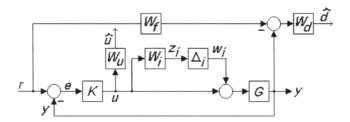

Fig. 5. Close loop system block diagram.

The weight transfer functions used are shown in Table 2.

Table 2. Weights used in the design.

	Internal loop	External loop
W_i	$\dfrac{0.6394s + 6.648}{s + 39.89}$	1.72×10^{-3}
W_d	$\dfrac{0.01s + 0.001}{s + 1 \times 10^{-6}}$	$\dfrac{1 \times 10^{-4}s + 1 \times 10^{-6}}{s + 1 \times 10^{-9}}$
W_f	$\dfrac{492.8}{s^2 + 39.96s + 492.8}$	$\dfrac{9}{s^2 + 5.4s + 9}$
W_u	$\dfrac{s + 157.1}{1 \times 10^{-6}s + 314.2}$	$\dfrac{s + 100}{1 \times 10^{-9}s + 500}$

The system block diagram is rearranged in the form of the configuration Δ–P–K (Fig. 6) for robust controller design [4] and [5].

Below are the controllers obtained for each loop. It should be noted that all of them met the conditions of robust stability and performance in their respective systems.

Controllers H_∞ obtained via method of solution AREs [1–5]. (It should be noted that for instance $1.5e5 = 1.5 \times 10^5$).

$$K_m(s) = \begin{bmatrix}
548.2 & 1 & 0 & 12e\text{-}4 & 0 & -7e\text{-}4 & -7e\text{-}4 & -62e3 \\
15e4 & -35 & 28e3 & 0.3 & -2e\text{-}4 & -0.2 & -0.2 & 17e6 \\
-23.1 & -0.5 & -309.1 & -1 & -2.1 & 0.5 & -0.1 & 0 \\
-79e\text{-}4 & 0 & 0 & 0 & 0 & -0.1 & -0.1 & -403.8 \\
0 & 0 & 81.3 & 0 & -40 & 0 & 0 & 0 \\
-2 & 0 & 0 & 3e\text{-}4 & 0 & -6.6 & -16.5 & -226.4 \\
-5.7 & 0 & 0 & -8e\text{-}4 & 0 & 16.5 & -33.3 & -643.5 \\
0 & 0 & 4 & 0 & 0 & 0 & 0 & 0
\end{bmatrix}$$

$$K_b(s) = \begin{bmatrix}
-136.8 & 1 & 0 & 0 & 0 & 0 & -48e3 \\
-9.4e3 & 0 & 681.1 & 0 & 0 & 0 & -33e5 \\
-0.9 & -0.6 & -104.2 & 22e\text{-}4 & 11e\text{-}4 & -4e\text{-}4 & 0 \\
3e\text{-}4 & 0 & 0 & 0 & 1e\text{-}3 & 1e\text{-}3 & 137.9\ 0 \\
-0.7 & 0 & 0 & 0 & -0.9 & -2.2 & -240.9 \\
-1.2 & 0 & 0 & 0 & 2.2 & -4.5 & -415.6 \\
0 & 0 & 46.2e\text{-}3 & 0 & 0 & 0 & 0
\end{bmatrix}$$

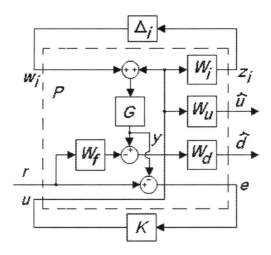

Fig. 6. Configuration Δ–P–K.

Controllers H_∞ based on LMI [10–12].

$$K_m(s)=\left[\begin{array}{ccccccc|c}
-0.2 & 8.5 & -1.2 & 1.4 & -179e\text{-}4 & 2.7 & 0.3 & -5.4 \\
8.8 & -418.8 & 59.3 & -56.1 & 7.2 & -57.7 & -36.4 & 435.8 \\
-1.4 & 55.9 & -41.7 & 8.2 & -16.8 & 13.2 & 5.8 & -64.3 \\
1.2 & -67.7 & 10.4 & -46.7 & 6.9 & 50.9 & -1.9 & 65.6 \\
-0.36 & 21.8 & 13.5 & 1.8 & -10 & -30.2 & 6.2 & -17.9 \\
2 & -41.3 & 4.2 & -14.1 & -31.3 & -400 & 89.7 & 69.8 \\
-51.6 & 2583.1 & -393.8 & 360.3 & -82.1 & 93.6 & -561 & -26e2 \\
1.4 & 60.8 & -10.1 & -9.5 & -33.6 & -389 & 3e\text{-}4 & 0
\end{array}\right]$$

$$K_b(s)=\left[\begin{array}{cccccc|c}
-31e\text{-}3 & -1.7 & -0.8 & 24e\text{-}3 & -55e\text{-}4 & -36e\text{-}4 & -14e\text{-}3 \\
-1.7 & -104.8 & -5 & 1.4 & -0.3 & -1.5 & -1 \\
-0.7 & -1.1 & -107.4 & 0.4 & -81e\text{-}3 & 3.1 & -0.3 \\
32e\text{-}3 & 1.2 & 1.6 & -4.5 & 2.1 & -24e\text{-}3 & 16e\text{-}3\ 0 \\
-0.2 & -8.7 & -2 & -2.1 & -1 & 0.24 & -9e\text{-}2 \\
80.2 & 47e2 & 984.3 & -65.9 & 16.1 & -142.7 & 47.7 \\
3.7 & -395 & 1581.1 & -534e\text{-}4 & 8e\text{-}4 & -19e\text{-}3 & 0
\end{array}\right]$$

Controllers H_2 obtained via method of solution AREs [1–5].

$$K_m(s)=\left[\begin{array}{ccccccc|c}
548.2 & 1 & 0 & 0 & 0 & 0 & 0 & -548.3 \\
15e4 & -35 & 28e3 & 0 & 13e\text{-}4 & 0 & 0 & 15e4 \\
-19.3 & -0.4 & -288.6 & -0.5 & 14.6 & 0.5 & 61e\text{-}4 & 0 \\
-79e\text{-}4 & 0 & 0 & 0 & 0 & -0.1 & -0.1 & -3.6 \\
0 & 0 & 81.3 & 0 & -39.9 & 0 & 0 & 0 \\
-2 & 0 & 0 & 0 & 0 & -6.6 & -16.5 & -2 \\
-5.7 & 0 & 0 & 0 & 0 & 16.5 & -33.3 & -5.7 \\
0 & 0 & 449.4 & 0 & 0 & 0 & 0 & 0
\end{array}\right]$$

$$K_b(s)=\left[\begin{array}{cccccc|c}
-136.8 & 1 & 0 & 0 & 0 & 0 & -1.4 \\
-94e2 & 0 & 681.1 & 0 & 0 & 0 & -93.6 \\
-0.7 & -0.5 & -103.1 & 176e\text{-}4 & 1e\text{-}3 & -3e\text{-}4 & 0 \\
3e\text{-}4 & 0 & 0 & 0 & 1e\text{-}3 & 1e\text{-}3 & 39e\text{-}4\ 0 \\
-0.7 & 0 & 0 & 0 & -0.9 & -2.2 & -68e\text{-}4 \\
-1.2 & 0 & 0 & 0 & 2.2 & -4.5 & -118e\text{-}4 \\
0 & 0 & 1629.8 & 0 & 0 & 0 & 0
\end{array}\right]$$

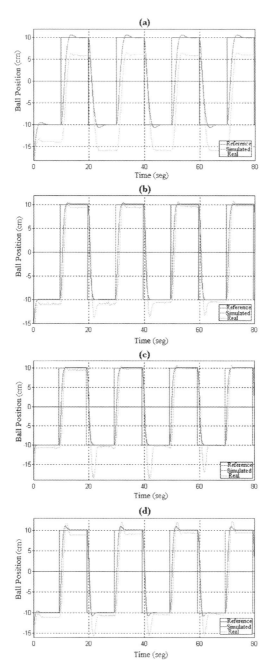

Fig. 7. Response of different closed loop systems. (a). Manufacturer's Suggested Controller RVE, (b). Controller H_∞ ARE, (c). Controller H_∞ LMI and (d). Controller H_2 ARE.

3.2   Results

Figure 7 shows the results obtained with the different controllers mentioned above. Table 3 shows the values of the main parameters that describe the response of the systems as well as the IAE performance index obtained in each case.

Table 3. Results obtained with each controller.

	RVE	H_∞ ARE	H_∞ LMI	H_2
e_{ss} (%)	41.1–65	5.17–10	6.6–5.1	10.8–8.7
t_s (sec)	4.1–4.22	3.18–3.68	3.17–3.9	3.9–3.9
SP (%)	2.3–0	4.2–38.5	2.5–87.3	16.9–58.2
IAE	1130	475.6	490.5	495.5

It is clear that the controllers H_∞ obtained by the AREs Solution Method were, in general, which present the best performance. Furthermore, the manufacturer's suggested controller is widely outperformed by each of the other drivers.

4   Conclusions

The development of this work allowed us to reach the following conclusions:

Based on previous experiences in designing robust controllers for stable plants, it was verified that an unstable plant is more difficult to control compared to a stable plant. This is basically because the presence of an unstable pole in the open-loop transfer function generally imposes a bound below the bandwidth of the closed-loop system. In other words, a rapid closed-loop system response is necessary in order not to compromise performance.

Controllers H_∞ obtained by the Solution Method of AREs offered in this case a better performance since, in general, they achieved the lowest response times, overshoot and errors in steady state, which was reflected in the fact that they presented the lowest value in the IAE performance index.

Generally robust controllers, H_∞ *ARE*, H_∞ *LMI*, H_2 offer better performance than the design based on the classic state variable feedback technique [13, 14].

References

1. Morari, M., Zafiriou, E.: Robust Process Control. Prentice-Hall, Estados Unidos de América (1989)
2. Paganini, F.: A Course in Robust Control Theory. A Convex Approach. Springer, New York (1999)
3. Scherer, C.: Theory of Robust Control. Delft University of Technology, Delft (2001)
4. Skogestad, S., Postlethwaite, I.: Multivariable Feedback Control: Analysis and Design, Second edn. Wiley, Chichester, UK (2001)

5. Zhou, K.: Essentials of Robust Control. Prentice-Hall, New Jersey (1998)
6. Dorf, Richard C. and Bishop, Robert H.: Modern Control Systems, 10th edn. Prentice Hall (2004)
7. Quanser: SRV02-Series. Rotary Experiment # 3. Ball and Beam. User Manual
8. Quanser; SRV02-Series. Rotary Experiment # 3. Ball and Beam. Student Handout
9. Quanser: SRV02-Series. Rotary Experiment # 1. Position Control. Student Handout
10. Boyd, S., Feron, E.: Linear Matrix Inequalities in System and Control Theory. Siam, Stanford, California (1994)
11. Gahinet, P.: Explicit Controller Formulas for LMI-based H_∞ Synthesis. Automatica, vol. 32, no. 7, pp. 1007–1014. Elsevier Science Ltda, Great Britain (1995)
12. Scherer, C., Weiland, S.: Linear Matrix Inequalities in Control, 3ª Version (2000)
13. Jiménez-Cabas, J., et al.: Robust control of an evaporator through algebraic Riccati equations and D-K iteration. In: Misra, S., et al. (eds.) ICCSA 2019. LNCS, vol. 11620, pp. 731–742. Springer, Cham (2019). https://doi.org/10.1007/978-3-030-24296-1_58
14. Misra, S., et al. (eds.): ICCSA 2019. LNCS, vol. 11619. Springer, Cham (2019). https://doi.org/10.1007/978-3-030-24289-3

International Workshop
on Computational and Applied Statistics
(CAS 2020)

Numbers of Served and Lost Customers in Busy-Periods of $M/M/1/n$ Systems with Balking

Fátima Ferreira[1,4] , António Pacheco[2,4] , and Helena Ribeiro[3,4(✉)]

[1] University of Trás-os-Montes and Alto Douro, Vila Real, Portugal
mmferrei@utad.pt
[2] Instituto Superior Técnico, Universidade de Lisboa, Lisbon, Portugal
apacheco@math.tecnico.ulisboa.pt
[3] Escola Superior de Tecnologia e Gestão do Politécnico de Leiria, Leiria, Portugal
helena.ribeiro@ipleiria.pt
[4] CEMAT, Instituto Superior Técnico, Universidade de Lisboa, Lisbon, Portugal

Abstract. In this work we analyze single server Markovian queueing systems with finite capacity and balking, that is $M/M/1/n$ systems with balking. In these systems, the admission of customers is modulated by the state of the system at the instants of customer arrivals. Depending on the size of the queue upon arrival, customers that find place to join the system decide to enter the system with a certain probability. The number of customers in the system amounts to a Markov chain whose transition probabilities incorporate the balking probabilities. Using the Markovian regenerative property of the chain embedded at the instants of arrival or departure of customers, we characterize the joint probability distribution of the number of customers served and the number of customers lost in busy-periods, that is, during continuous occupation periods of the server. This is accomplished implementing a priori a recursive algorithmic procedure for computing the respective probability-generating function. Finally, a numerical illustration of the derived results is presented for different balking policies.

Keywords: Queueing system · Balking · Busy-period · Markov chain

1 Introduction

The perception of the queue length by an arriving customer plays a relevant role in their decision to whether join (or not) a queueing system. In fact, in current rush days there is a growing tendency of customers to balk (i.e., to not join the system) when facing at arrival a long queue size (that exceeds their patience level) or a short queue if a loaded system is perceived as a good condition.

This research was partially supported by Fundação para a Ciência e a Tecnologia (FCT) through project UID/Multi/04621/2019.

As the balking phenomenon may result in significant loss of customers, the analysis of queueing systems with balking is important in practice.

In this work we address the study of finite capacity Markovian queues with a single server under different customer blocking policies when the customers arrive at the system. Using the notation introduced by David G. Kendall, this corresponds to a $M/M/1/n$ system with balking or reverse balking. These systems have many applications because in many real-life situations customers often have the possibility to postpone or give up a given type of service when, on arrival at the system, the level of server congestion is considered unsatisfactory by the same customers.

The notion of customer balking was introduced in the work of Haight [5] that considers an $M/M/1$ queue in which customers do not join the system whenever they face at arrival a queue length larger than a fixed threshold. After this pioneering work, the study of queues under different balking and other types of customer's impatience policies has aroused the interest of several authors (*cf* [2–4, 7–10, 12, 14, 15] and references therein). A review on queueing systems with impatient customers can be found in [13].

The analysis of queues with balking in busy-periods, i.e., in continuous periods of effective use of the server, is relevant from the operator's point of view and provides crucial information for its management. We consider multi busy-periods, *i.e.*, busy-periods initiated with multiple customers in the system. Specifically, by a busy-period initiated by i customers, hereinafter referred to as i-busy-period, we mean the period of time that starts at an arrival instant that makes the system stay with i customers, and ends at the first subsequent time at which the system becomes empty, with a customer initiating service after the arrival instant. This definition is in line with that of remaining busy-period from state i given in Harris [6], of residual busy-period provided in Hanbali [1] and of busy-period initiated with i customers considered in Peköz *et al.* [11].

The aim of this paper is to characterize the joint probability distribution function of the number of customers served and the number of customers lost in busy-periods of $M/M/1/n$ system with balking. In Sect. 2, after describing the system we present the transition probability matrix of the discrete time Markov chain (DTMC) embedded at the moments of arrival or departure of customers. Taking advantage of the Markovian structure of these queues and by using the probability-generating function technique, we obtain in Sect. 3 the joint probability function of the number of customers served and the number of customers lost in busy-periods.

In Sect. 4 we illustrate the results obtained and, finally, we present some concluding remarks in Sect. 5.

2   Model Description

We consider $M/M/1/n$ systems with balking of customers. These are queues with finite capacity, n, at which the customers arrive (one by one) according to a Poisson process with rate λ and are served (one by one) by a single-server, in a first-come first-served discipline. Service times are independent and identically exponential distributed variables with rate μ, and are independent of the arrival process.

These systems differ from the traditional $M/M/1/n$ given that the customer admission is modulated by the state of the system at the time of arrival. As in the traditional $M/M/1/n$ systems, if on arrival a customer finds the system full (with n customers) he is blocked with probability 1. However, if on arrival a customer finds $i < n$ customers in the system, even though there is place in the system to accommodate the customer, he either decides to join the queue with probability e_i or not to enter (balk) with probability $1 - e_i$. For completeness we let $e_n = 0$.

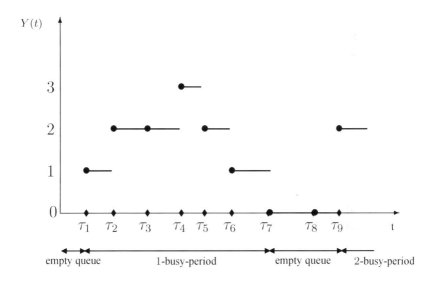

Fig. 1. Typical sample path of $M/M/1/n$ system with balking.

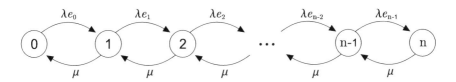

Fig. 2. The transition-rate diagram for an $M/M/1/n$ system with balking.

We let $Y(t)$ denote the number of customers in the system at time t. A typical sample path and the transition-rate diagram of $Y(t)$ are given in Figures 1 and 2, respectively. In Fig. 1, the sequence $(\tau_n)_{n \in \mathbb{N}}$ is the sequence of arrival or departure instants. As the interarrival and the service times have exponential distributions, the process $Y = \{Y(t), t \geq 0\}$ is a continuous time Markov chain with state space $E = \{0, 1, \ldots, n\}$, rate transition matrix $R = (r_{ij})_{i,j \in E}$ with

$$r_{ij} = \begin{cases} \lambda e_i, & j = i+1 \\ \lambda(1 - e_i), & j = i \\ \mu, & j = i-1 \\ 0, & j \in E \setminus \{i-1, i, i+1\} \end{cases},$$

and vector of transition rates out of states $r = (r_i)_{i \in E}$ with $r_i = \lambda + \mu \mathbf{1}_{\{i>0\}}$, where $\mathbf{1}_A$ denotes the indicator function of statement A. As a consequence, the process $\bar{Y} = (Y(\tau_k))_{k \in \mathbb{N}}$ embedded at the sequence $(\tau_k)_{k \in \mathbb{N}}$ of instants of arrival or departure of customers is a DTMC with state space E and transition probability matrix $P = (p_{ij})_{i,j \in E}$, such that

$$p_{ij} = \begin{cases} e_0 & j = 1 \wedge i = 0 \\ 1 - e_0 & j = i = 0 \\ \frac{\lambda e_i}{\lambda + \mu}, & j = i+1 \wedge i \neq 0 \\ \frac{\lambda(1 - e_i)}{\lambda + \mu}, & j = i \neq 0 \\ \frac{\mu}{\lambda + \mu}, & j = i-1 \wedge i \neq 0 \\ 0, & j \in E \setminus \{i-1, i, i+1\} \end{cases}. \tag{1}$$

3   Probability Function of (S_i, L_i)

In this section we derive the joint probability function of the random vector (S_i, L_i) where, for $i \in E \setminus \{0\}$, S_i denotes the number of customers served during an i-busy-period and L_i denotes the number of customers lost during an i-busy-period.

To this purpose, we introduce the probability-generating function of (S_i, L_i),

$$g_i(u, v) = E(u^{S_i} v^{L_i}) = \sum_{s \in \mathbb{N}} \sum_{l \in \mathbb{N}_0} u^s v^l \, P(S_i = s, L_i = l) \tag{2}$$

with $|u| \leq 1$ and $|v| \leq 1$, from which, by derivation, we obtain the probability function of (S_i, L_i),

$$P(S_i = s, L_i = l) = \frac{1}{s! \, l!} \frac{\partial^{s+l} g_i(u, v)}{\partial u^s \partial v^l} \bigg|_{u=0, v=0}, \qquad (s, l) \in \mathbb{N} \times \mathbb{N}_0.$$

The following theorem presents properties of the probability-generating function of (S_i, L_i).

Theorem 1. *The probability-generating function of the number of customers served and the number of customers lost during an i-busy-period of an M/M/1/n system with balking satisfies, for $i \in E\backslash\{0\}$, with the convention that $g_0(u, v) = \theta_{n+1}(u, v) = 1$,*

$$g_i(u, v) = \theta_i(u, v) g_{i-1}(u, v) \tag{3}$$

with

$$\theta_i(u, v) = \frac{u\mu}{\lambda + \mu - v\lambda(1 - e_i) - \lambda e_i \, \theta_{i+1}(u, v)}. \tag{4}$$

Proof. Denoting by X the type of event that occurs at the instant τ_k of the first transition after starting a i-busy-period, we let

$$X = \begin{cases} -1, & \text{if a customer exits the system at } \tau_k \\ 0, & \text{if a customer arrives without entering the system at } \tau_k \\ 1, & \text{if a customer arrives and enters the system at } \tau_k \end{cases}.$$

Conditioning on the event $\{X = x\}$, and letting $=_{st}$ denote equality in distribution, we obtain

$$(S_n, L_n)|_{X=x} =_{st} \begin{cases} (1 + S_{n-1}, L_{n-1}), & x = -1 \\ (S_n, 1 + L_n), & x = 0 \end{cases}$$

and, for $i \in E\backslash\{0, n\}$,

$$(S_i, L_i)|_{X=x} =_{st} \begin{cases} (1 + S_{i-1}, L_{i-1}), & x = -1 \\ (S_i, 1 + L_i), & x = 0 \\ (S_{i+1}, L_{i+1}), & x = 1 \end{cases}.$$

From the total probability law, it follows that, for $(s, l) \in \mathbb{N} \times \mathbb{N}_0$,

$$\begin{aligned} P(S_n = s, L_n = l) =& P(S_{n-1} = s - 1, L_{n-1} = l)P(X = -1) \\ &+ P(S_n = s, L_n = l - 1)P(X = 0) \end{aligned} \tag{5}$$

and, for $i \in E\backslash\{0, n\}$,

$$\begin{aligned} P(S_i = s, L_i = l) =& \mathbf{1}_{\{(i,s,l)=(1,1,0)\}} P(X = -1) \\ &+ \mathbf{1}_{\{i>1\}} P(S_{i-1} = s - 1, L_{i-1} = l)P(X = -1) \\ &+ P(S_i = s, L_i = l - 1)P(X = 0) \\ &+ P(S_{i+1} = s, L_{i+1} = l)P(X = 1). \end{aligned} \tag{6}$$

We will now prove the statement of the theorem using induction, starting with the case $i = n$. Taking into account (1), (2) and (5), we obtain:

$$g_n(u, v) = \frac{u\mu}{\lambda + \mu} g_{n-1}(u, v) + \frac{v\lambda}{\lambda + \mu} g_n(u, v).$$

This implies that

$$g_n(u,v) = \theta_n(u,v)g_{n-1}(u,v)$$

with

$$\theta_n(u,v) = \frac{u\mu}{\lambda + \mu - v\lambda}.$$

Therefore, the Eqs. (3)–(4) are valid for $i = n$.

Let us now assume that the Eqs. (3)–(4) are valid for $i = j + 1$ with j being a fixed number such that $1 < j < n$. Taking into account (1), (2) and (6), we obtain:

$$g_j(u,v) = \frac{u\mu}{\lambda + \mu}g_{j-1}(u,v) + \frac{v\lambda(1 - e_j)}{\lambda + \mu}g_j(u,v) + \frac{\lambda e_j}{\lambda + \mu}g_{j+1}(u,v).$$

Thus,

$$g_j(u,v) = \frac{u\mu}{\lambda + \mu - v\lambda(1 - e_j)}g_{j-1}(u,v) + \frac{\lambda e_j}{\lambda + \mu - v\lambda(1 - e_j)}g_{j+1}(u,v).$$

Now, as the induction hypothesis gives $g_{j+1}(u,v) = \theta_{j+1}(u,v)g_j(u,v)$, we obtain:

$$g_j(u,v) = \frac{u\mu}{\lambda + \mu - v\lambda(1 - e_j)}g_{j-1}(u,v) + \frac{\lambda e_j}{\lambda + \mu - v\lambda(1 - e_j)}\theta_{j+1}(u,v)g_j(u,v).$$

Thus, isolating $g_j(u,v)$, we obtain Eqs. (3)–(4) for $i = j$. Therefore, by induction, the statement of the theorem is valid.

We conclude this section by presenting, in Fig. 3, a recursive procedure to obtain the probability generating function of (S_i, L_i) in $M/M/1/n$ systems with balking, departing from $g_0(u,v) = 1$.

4   Numerical Illustration

In this section, using the above-derived results, we compute the joint probability function of the number of customers served and the number of customers lost in busy-periods of $M/M/1/n$ systems with unit service rate ($\mu = 1$) and the following three customer balking policies.

i. Partial blocking:

$$e^{(1)} = (e_0, e_1, \ldots, e_n) \text{ with } e_i = \begin{cases} 1 & i \in E \backslash \{n\} \\ 0 & i = n \end{cases}$$

that represents the standard partial blocking policy of $M/M/1/n$ systems in which a customer is blocked if and only if the system is full at is arrival to the system.

Input: n, λ, μ, $e = (e_0, e_1, \ldots, e_{n-1}, 0)$

$$\theta_n(u, v) = \frac{u\mu}{\lambda + \mu - v\lambda};$$

For $i = n - 1 : -1 : 1$

$$\theta_i(u, v) = \frac{u\mu}{\lambda + \mu - v\lambda(1 - e_i) - \lambda e_i \theta_{i+1}(u, v)};$$

End for

$g_1(u, v) = \theta_1(u, v);$

For $i = 2 : n$

$$g_i(u, v) = \theta_i(u, v) g_{i-1}(u, v);$$

End for

Output: $(g_i(u, v))_{i=1,2,\ldots,n}$

Fig. 3. Algorithm for computing the probability-generating function of (S_i, L_i) in $M/M/1/n$ systems with balking.

ii. Increasing balking probabilities:

$$e^{(2)} = (e_0, e_1, \ldots, e_n) \text{ with } e_i = \begin{cases} \frac{1}{i+2} & i \in E \backslash \{n\} \\ 0 & i = n \end{cases}$$

in which the balking of customers increases with the size of the queue, that is frequent whenever the customers are in a hurry to be served and tend not to enter the system if they have to wait long periods of time.

iii. Decreasing balking probabilities:

$$e^{(3)} = (e_0, e_1, \ldots, e_n) \text{ with } e_i = \begin{cases} \frac{1}{n+1-i} & i \in E \backslash \{n\} \\ 0 & i = n \end{cases}$$

a reverse balking that favors the entry of customers whenever the system has greater demand and may occur, for instance, in investments in the stock market.

Tables 1 and 2 show the joint probability function of the number of customers served and the number of customers lost in a 1-busy-period of $M/M/1/7$ systems with balking policy $e^{(2)} = (\frac{1}{2}, \frac{1}{3}, \ldots, \frac{1}{8}, 0)$ and arrival rates $\lambda = 0.5$ and $\lambda = 1.1$, respectively. In the system with low traffic intensity (Table 1), as the arrival rate is much lower than the service rate, the queue has a small trend to fill up and to have a high number of losses. Consequently, the higher joint probabilities are associated to the lowest values of the number of customers served and of the number of customers lost. In fact, we observe that $P(S_1 \leq 3, L_1 \leq 3) = 0.9831$.

Table 1. Joint probability function of the number of customers served and the number of customers lost in 1-busy-period of a $M/M/1/7$ system with service rate $\mu = 1$ and arrival rate $\lambda = 0.50$, $P(S_1 = s, L_1 = l)$ for $s = 1, \ldots, 9$ and $l = 0, 1, \ldots, 9$.

$$
\begin{bmatrix}
0.6667 & 0.1481 & 0.0329 & 0.0073 & 0.0016 & 0.0004 & 0.0001 & 0.0000 & 0.0000 & 0.0000 \\
0.0494 & 0.0343 & 0.0159 & 0.0061 & 0.0021 & 0.0007 & 0.0002 & 0.0001 & 0.0000 & 0.0000 \\
0.0064 & 0.0076 & 0.0054 & 0.0030 & 0.0014 & 0.0006 & 0.0002 & 0.0001 & 0.0000 & 0.0000 \\
0.0010 & 0.0016 & 0.0015 & 0.0011 & 0.0007 & 0.0004 & 0.0002 & 0.0001 & 0.0000 & 0.0000 \\
0.0001 & 0.0003 & 0.0004 & 0.0004 & 0.0003 & 0.0002 & 0.0001 & 0.0001 & 0.0000 & 0.0000 \\
0.0000 & 0.0001 & 0.0001 & 0.0001 & 0.0001 & 0.0001 & 0.0000 & 0.0000 & 0.0000 & 0.0000 \\
0.0000 & 0.0000 & 0.0000 & 0.0000 & 0.0000 & 0.0000 & 0.0000 & 0.0000 & 0.0000 & 0.0000 \\
0.0000 & 0.0000 & 0.0000 & 0.0000 & 0.0000 & 0.0000 & 0.0000 & 0.0000 & 0.0000 & 0.0000 \\
0.0000 & 0.0000 & 0.0000 & 0.0000 & 0.0000 & 0.0000 & 0.0000 & 0.0000 & 0.0000 & 0.0000
\end{bmatrix}
$$

In contrast, the system with higher utilization rate (Table 2) has a greater tendency to fill up and to experience more balking and customers blocked. In this case, we observe that $P(S_1 \leq 3, L_1 \leq 3) = 0.8933$ as higher values of the number of customers served and of the number of customers lost still have positive joint probabilities.

Table 2. Joint probability function of the number of customers served and the number of customers lost in 1-busy-period, of a $M/M/1/7$ system with service rate $\mu = 1$ and arrival rate $\lambda = 1.10$, $P(S_1 = s, L_1 = l)$ for $s = 1, \ldots, 9$ and $l = 0, 1, \ldots, 9$.

$$
\begin{bmatrix}
0.4762 & 0.1663 & 0.0581 & 0.0203 & 0.0071 & 0.0025 & 0.0009 & 0.0003 & 0.0001 & 0.0000 \\
0.0396 & 0.0432 & 0.0315 & 0.0191 & 0.0104 & 0.0053 & 0.0026 & 0.0012 & 0.0006 & 0.0003 \\
0.0058 & 0.0107 & 0.0120 & 0.0105 & 0.0078 & 0.0053 & 0.0033 & 0.0019 & 0.0011 & 0.0006 \\
0.0010 & 0.0025 & 0.0039 & 0.0044 & 0.0042 & 0.0035 & 0.0027 & 0.0019 & 0.0013 & 0.0008 \\
0.0002 & 0.0006 & 0.0011 & 0.0016 & 0.0018 & 0.0018 & 0.0016 & 0.0014 & 0.0010 & 0.0008 \\
0.0000 & 0.0001 & 0.0003 & 0.0005 & 0.0007 & 0.0008 & 0.0008 & 0.0008 & 0.0007 & 0.0006 \\
0.0000 & 0.0000 & 0.0001 & 0.0001 & 0.0002 & 0.0003 & 0.0004 & 0.0004 & 0.0004 & 0.0003 \\
0.0000 & 0.0000 & 0.0000 & 0.0000 & 0.0001 & 0.0001 & 0.0001 & 0.0002 & 0.0002 & 0.0002 \\
0.0000 & 0.0000 & 0.0000 & 0.0000 & 0.0000 & 0.0000 & 0.0001 & 0.0001 & 0.0001 & 0.0001
\end{bmatrix}
$$

Figure 4 illustrates the sensitivity of the joint distribution function of (S_1, L_1) at point $(3,3)$ of $M/M/1/10$ systems, as function of the utilization rate $(\rho = \lambda/1)$, considering three customer balking policies: $e^{(1)} = (1, 1, \ldots, 1, 0)$, $e^{(2)} = (\frac{1}{2}, \frac{1}{3}, \ldots, \frac{1}{11}, 0)$ and $e^{(3)} = (\frac{1}{11}, \frac{1}{10}, \ldots, \frac{1}{2}, 0)$. As expected, for the three balking policies, the joint distribution at the considered point decreases as the utilization rate increases. This decrease is more highlighted in the traditional $M/M/1/10$ systems, that is the system with $e^{(1)}$ policy, where customers always enter in the system while it is not full up. Among the $M/M/1/10$ systems considered, the systems with balking, and in particular the system with $e^{(3)}$ policy (reverse balking), show higher joint distribution of (S_1, L_1) at $(3, 3)$ in contrast to the traditional $M/M/1/10$ system.

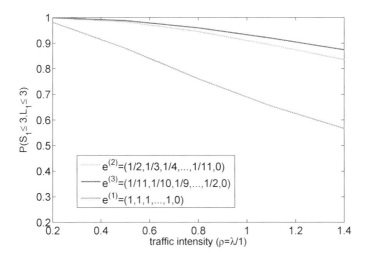

Fig. 4. Joint probability function of the number of customers served and of the number of customers lost in a 1-busy-period of a $M/M/1/10$ system with $\mu = 1$, $P(S_1 \leq 3, L_1 \leq 3)$, as a function of the utilization rate.

5   Conclusion

The number of customers served and the number of customers lost during busy-periods are important queueing performance measures. Although the analysis of their joint behavior constitutes an added value for the characterization of the queueing system, to our knowledge, the joint probability distribution of such measures has not been derived in the literature.

In this work the joint probability function of the number of customers served and the number of customers lost in busy-periods was obtained for single server Markovian queueing systems with finite capacity and balking. In particular, we derived a recursive procedure to compute the probability-generating function of the number of customers served and the number of customers lost in busy-periods that can be initiated by multiple customers in the system.

The derived recursion was applied to numerically compute the joint probability function of the number of customers served and the number of customers lost in busy-periods of $M/M/1/n$ systems under partial blocking and monotonic (increasing/decreasing) customer balking policies.

The approach followed in the paper can be generalized to obtain the joint probability distribution of the number of customers served and the number of customers lost during busy-periods of Markovian queues with batch arrivals and balking and reneging.

References

1. Al Hanbali, A.: Busy period analysis of the level dependent $PH/PH/1/K$ queue. Queueing Syst. **67**(3), 221–249 (2011). https://doi.org/10.1007/s11134-011-9213-6

2. Ancker, C.J., Gafarian, A.V.: Some queuing problems with balking and reneging II. Oper. Res. **11**(6), 928–937 (1963)

3. Ferreira, F., Pacheco, A.: Analysis of $GI^X/M(n)//N$ systems with stochastic customer acceptance policy. Queueing Syst. **58**, 29–55 (2008). https://doi.org/10.1007/s11134-007-9057-2

4. Guha, D., Goswami, V., Banik, A.D.: Algorithmic computation of steady-state probabilities in an almost observable $GI/M/c$ queue with or without vacations under state dependent balking and reneging. Appl. Math. Model. **40**, 4199–4219 (2016)

5. Haight, F.A.: Queuing with balking I. Biometriika **44**(3–4), 360–369 (1957)

6. Harris, T.J.: The remaining busy period of a finite queue. Oper. Res. **19**, 219–223 (1971)

7. Jain, N.K., Kumar, R., Som, B.K.: An $M/M/1/N$ queuing system with reverse balking. Am. J. Oper. Res. **4**(2), 17–20 (2014)

8. Kumar, R., Sharma, S.: Transient analysis of an $M/M/c$ queuing system with balking and retention of reneging customers. Commun. Stat. Theory Methods **47**(6), 1318–1327 (2018)

9. Kumar, R., Som, B.K.: An $M/M/1/N$ queuing system with reverse balking and reverse reneging. AMO-Adv. Model. Optim. **16**(2), 339–353 (2014)

10. Laxmi, P.V., Gupta, U.C.: Analysis of finite-buffer multi-server queues with group arrivals: $GI^X/M/c/N$. Queueing Syst. **36**(1–3), 125–140 (2000). https://doi.org/10.1023/A:1019179119172

11. Peköz, E.A., Righter, R., Xia, C.H.: Characterizing losses during busy periods in finite buffer systems. J. Appl. Probab. **40**(1), 242–249 (2003)

12. Som, B.K., Kumar, R.: A heterogeneous queuing system with reverse balking and reneging. J. Ind. Prod. Eng. **35**(1), 1–5 (2018)

13. Wang, K., Li, N., Jiang, Z.: Queuing system with impatient customers: a review. In: IEEE International Conference on Service Operations and Logistics and Informatics, 15–17 July 2010, Shandong, pp. 82–87 (2010)

14. Wang, Q., Zhang, B.: Analysis of a busy period queuing system with balking, reneging and motivating. Appl. Math. Model. **64**, 480–488 (2018)

15. Yue, D., Zhang, Y., Yue, W.: Optimal performance analysis of an $M/M/1/N$ queue system with balking, reneging and server vacation. Int. J. Pure Appl. Math. **28**, 101–115 (2006)

Simulation Study to Compare the Performance of Signed Klotz and the Signed Mood Generalized Weighted Coefficients

Sandra M. Aleixo[1(✉)] and Júlia Teles[2]

[1] CEAUL and Department of Mathematics, ISEL – Instituto Superior de Engenharia de Lisboa, IPL – Instituto Politécnico de Lisbon,
Rua Conselheiro Em ídio Navarro, 1, 1959-007 Lisbon, Portugal
`sandra.aleixo@isel.pt`
[2] CIPER and Mathematics Unit, Faculdade de Motricidade Humana,
Universidade de Lisbon, Estrada da Costa,
1499-002 Cruz Quebrada - Dafundo, Portugal

Abstract. In this work a Monte Carlo simulation study was carried out to compare the performance of three weighted coefficients that emphasized the top and bottom ranks at the same time, namely the signed Klotz and the signed Mood weighted coefficients previously proposed by the authors [2] and the van der Waerden weighted coefficient [5], with the Kendall's coefficient that assigns equal weights to all rankings. The coefficients for m judges studied in this paper generalize the coefficients previously proposed by the authors for two judges in a preliminary work [1] where a simulation study is also carried out. As the main result of the simulation study, we highlight the best performance of Klotz coefficient in detecting concordance in situations where the agreement is located in a lower proportion of extreme ranks, contrary to the case where the agreement is located in a higher proportion of extreme ranks, in which the Signed Mood and van der Waerden coefficients have the best performance.

Keywords: Monte Carlo simulation · Weighted coefficients · Concordance measures · Signed Klotz coefficient · Signed Mood coefficient · van der Waerden coefficient

1 Introduction

Two weighted correlation coefficients R_S, that allow to give more weight to the lower and upper ranks simultaneously, were proposed by the authors in a previous work [1]. These indexes were obtained computing the Pearson correlation coefficient with a modified Klotz and modified Mood scores. In the sequel of that work, two new generalized weighted coefficients A_S, the signed Klotz A_K and

© Springer Nature Switzerland AG 2020
O. Gervasi et al. (Eds.): ICCSA 2020, LNCS 12251, pp. 157–172, 2020.
https://doi.org/10.1007/978-3-030-58808-3_13

the signed Mood A_M, to measure the concordance among several sets of ranks, putting emphasis on the extreme ranks, were deduced [2]. The van der Waerden weighted coefficient A_W [5] could also be included in these generalized coefficients. In that work, a relationship between the generalized A_S and the average of pairwise R_S was derived. Besides that, the distribution of the generalized weighted agreement coefficients is derived, and an illustrative example is shown. The present paper aims to extend this last work carried out by the authors [2], performing a simulation study to assess the behavior of the new coefficients. So, the goal of this paper is the comparison of the performance of four coefficients: three weighted coefficients to measure agreement among several sets of ranks emphasizing top and bottom ranks at the same time, the signed Klotz, the signed Mood and the van der Waerden coefficients, with the Kendall's coefficient that attributes equal weights to all rankings. This comparison is made through the implementation of a simulation study using the Monte Carlo method, as usual in this kind of studies [3,7,9].

The remainder of paper is organized in five sections. In Sect. 2, the four concordance coefficients compared in the simulation study performed in this work, the Kendall's, the van der Waerden, the Klotz and the Mood coefficient are presented, and treating these coefficients as test statistics, their asymptotic distributions, under the null hypotheses of no agreement among the rankings, are stated. The Monte Carlo simulation study design used in this paper is outlined in Sect. 3. In Sect. 4 the results of the mentioned simulation study are shown and finally, in Sect. 5, conclusions are drawn.

2   Coefficients to Measure Agreement Among Several Sets of Ranks

In the simulation study presented in this paper, it will be used four coefficients to measure agreement among several sets of ranks: the Kendall's coefficient that assigns equal weights to all rankings and three weighted coefficients that emphasized the top and bottom ranks at the same time, the van der Waerden weighted coefficient [5] and the two coefficients previously proposed by the authors, the signed Klotz and the signed Mood [2]. Therefore, this section will briefly present these coefficients, namely the general expression and asymptotic distributions, since they are necessary to carry out the Monte Carlo simulation study.

To present the coefficients described below, it is considered that there are n subjects to be ranked by $m > 2$ independent observers or judges or in m moments, giving rise to m sets of ranks, and also that R_{ij} represents the rank assigned to the jth subject by the ith observer, for $i = 1, 2, \ldots, m$ and $j = 1, 2, \ldots, n$.

2.1   Kendall's Coefficient of Concordance

The Kendall's coefficient of concordance W [6,11] is the ratio between the variance of the sum of the ranks assigned to subjects, represented by S, and the maximum possible value reached for the variance of the sum of the ranks taking into account the values of n and m,

$$W = \frac{S}{\max(S)},$$

where $\max(S) = m^2 n(n^2 - 1)/12$ is the value of S when there is a total agreement among the ranks. Representing by $R_j = \sum_{i=1}^{m} R_{ij}$ the sum of the ranks assigned to the jth subject, for $j = 1, \ldots, n$, the Kendall's coefficient of concordance W is given by

$$W = \frac{n \sum_{j=1}^{n} R_j^2 - \left(\sum_{j=1}^{n} R_j\right)^2}{\dfrac{m^2 n(n^2 - 1)}{12}}.$$

The Kendall's coefficient can take values in the interval $[0, 1]$ [6]. This coefficient attains: the maximum value 1 when there is perfect agreement among the m sets of ranks, only if there are no tied observations; the minimum value zero when there is no agreement, only if $m(n + 1)$ is even.

Although the coefficient W is normally used as a measure of agreement among rankings, it could also be used as a test statistic. There is a relationship between W and the Friedman χ_r^2 statistic, $\chi_r^2 = m(n - 1)W$, which has an approximate chi-squared distribution with $n - 1$ degrees of freedom [4].

2.2 Weighted Coefficients to Measure the Concordance Among Several Sets of Ranks Emphasizing Extreme Ranks at the Same Time

Consider a generic score S_{ij} which is computed based on the rank R_{ij}, for $i = 1, 2, \ldots, m$ and $j = 1, 2, \ldots, n$. The score S_{ij} can be:

– the van der Waerden score [10]

$$W_{ij} = \Phi^{-1}\left(\frac{R_{ij}}{n + 1}\right),$$

– the signed Klotz score [1]

$$SK_{ij} = sign\left(R_{ij} - \frac{n + 1}{2}\right)\left(\Phi^{-1}\left(\frac{R_{ij}}{n + 1}\right)\right)^2,$$

– or the signed Mood score [1]

$$SM_{ij} = sign\left(R_{ij} - \frac{n + 1}{2}\right)\left(R_{ij} - \frac{n + 1}{2}\right)^2.$$

Note that, for the generic score S_{ij} are valid the following equalities:

1. $\sum_{j=1}^{n} S_{1j}^2 = \sum_{j=1}^{n} S_{2j}^2 = \cdots = \sum_{j=1}^{n} S_{mj}^2 = C_s$,
   where C_s is a non null constant that depends on the sample size n and the scores S_{ij}.

2. $\sum_{j=1}^{n} S_{1j} = \sum_{j=1}^{n} S_{2j} = \cdots = \sum_{j=1}^{n} S_{mj} = 0$, that is, $S_{i\bullet} = 0$, for
   $i = 1, 2, \ldots, m$.

These happens because one has:

- in the case of the van der Waerden and signed Klotz scores
  (a) if n odd,

$$\Phi^{-1}\left(\frac{k}{n+1}\right) = -\Phi^{-1}\left(\frac{n+1-k}{n+1}\right),$$

for $k = 1, 2, \ldots, \left[\frac{n}{2}\right]$

(where $[k]$ represents the integer part of k),

and

$$\Phi^{-1}\left(\frac{(n+1)/2}{n+1}\right) = \Phi^{-1}\left(\frac{1}{2}\right) = 0;$$

(b) if n even,

$$\Phi^{-1}\left(\frac{k}{n+1}\right) = -\Phi^{-1}\left(\frac{n+1-k}{n+1}\right),$$

for $k = 1, 2, \ldots, \frac{n}{2}$;

- in the case of signed Mood scores

$$\left(k - \frac{n+1}{2}\right)^2 = \left(n+1-k - \frac{n+1}{2}\right)^2,$$

for $k = 1, 2, \ldots, \left[\frac{n}{2}\right]$.

In a previous work [2], the authors proposed two new weighted concordance coefficients, built similarly to Kendall's coefficient of concordance W defined in Sect. 2.1. These new coefficients, as well as the van der Waerden coefficient, are based on the ratio between the variance of the n sums of m scores awarded to subjects, denoted by S_{ss}^2, and the maximum value that this variance can attain considering the values of m and n. So, the weighted coefficients of agreement to measure the concordance among m sets of ranks, giving more weight to the extremes ones, can be defined by

$$A_s = \frac{S_{ss}^2}{\max S_{ss}^2},$$

with

$$S_{SS}^2 = \frac{1}{n-1} \sum_{j=1}^{n} \left(S_{\bullet j} - \overline{S}_{\bullet\bullet} \right)^2,$$

where $S_{\bullet j}$ is the sum of the scores of m observers for the subject j and $\overline{S}_{\bullet\bullet}$ is the average of all $m \times n$ scores S_{ij}, for $i = 1, 2, \ldots, m$ and $j = 1, 2, \ldots, n$.

Attending that $\overline{S}_{\bullet\bullet} = \sum_{j=1}^{n} S_{\bullet j}/n = 0$ then

$$S_{SS}^2 = \frac{1}{n-1} \sum_{j=1}^{n} S_{\bullet j}^2.$$

Therefore, making some calculus (see [2] for details), the weighted coefficients to measure the concordance among m sets of ranks, putting emphasis on the extremes ones (van der Waerden, Klotz and Mood coefficients), can be given by the general expression

$$A_S = \frac{\sum_{j=1}^{n} \left(\sum_{i=1}^{m} S_{ij} \right)^2}{m^2 \, C_S}. \tag{1}$$

These coefficients take values in the interval $[0, 1]$.

Similarly to the Kendall's coefficient, these three weighted coefficients can be used as test statistics. Their asymptotic distributions were postulated in a theorem, proved in previous work by the authors [2], which states that: under the null hypotheses of no agreement or no association among the rankings, the statistic $m(n-1)A_S$ has an asymptotic chi-square distribution with $n-1$ degrees of freedom.

3   Simulation Study Design

Using R software, version 3.6.2 [8], a Monte Carlo simulation study was carried out to compare the performance of the three weighted coefficients to measure agreement among several sets of ranks emphasizing top and bottom ranks at the same time, presented in Sect. 2.2, with the Kendall's coefficient of concordance that awards equal weights to all rankings, shown in Sect. 2.1. Coefficients performance was estimated by the proportion of rejected null hypotheses, when testing whether the underlying population concordance coefficient is higher than zero. To simplify the writing, from now on, we will refer this proportion as power of coefficient.

The main objective of this simulation study is to contrast the virtue of each one of the three coefficients in identifying agreement among various sets of ranks given emphasis on both lower and upper ranks simultaneously, when compared with the Kendall's coefficient of concordance.

The data generation scheme adopted in this study was similar to the one followed by Legendre [7]. The first group of observations was produced through the generation of a n-dimensional vector of standard random normal deviates. To obtain each one of the others groups of n observations, for some proportion $0 < p < 0.5$, and choosing an appropriate standard deviation σ, it was made the following:

1. random normal deviates with zero mean and standard deviation σ were added to the observations with the most extreme ranks in the first group of observations, that is for $i = 1, \ldots, [np]$ and $i = n - [np] + 1, \ldots, n$;
2. random normal deviates with zero mean and standard deviation σ were generated for the observations with the intermediate ranks in the first sample, that is for $i = [np] + 1, \ldots, n - [np]$.

The m sets of observations were converted into m sets of ranks considering an ascending order. These m sets of ranks are the related samples that were considered.

Simulating the data by this way, the agreement among the groups of ranks is higher in the lower and upper ranks than in the intermediate ones. The six values considered for σ ($\sigma = 0.25, 0.5, 1, 2, 3, 5$) allow to evaluate the performance of coefficients for several intensities of agreement, being that the lower values of σ correspond to higher degrees of agreement. For the proportion of ranks that are in agreement, two values were taken into account ($p = 0.1, 0.25$), which allow to compare the performance of the coefficients in a scenario where the concordance was concentrated on a higher proportion of extreme ranks (scenario 1: $p = 0.25$), with a scenario in which the concordance was focused on a lower proportion of extreme ranks (scenario 2: $p = 0.1$).

Synthesizing, in the simulation study were taken into account:

– three different number of independent observers or judges, $m = 3, 4, 5$;
– five different number of subjects, $n = 20, 30, 50, 100, 200$;
– two values considered for the proportion of ranks that are in agreement, $p = 0.1, 0.25$;
– and six intensities of agreement, $\sigma = 0.25, 0.5, 1, 2, 3, 5$.

Considering all possible combinations of these values, 180 simulated conditions were evaluated. For each simulated condition, 10000 replications were run in order to:

1. calculate the means and standard deviations of Kendall's, van der Waerden, signed Klotz and signed Mood coefficients for the simulated samples;
2. estimate the power of each coefficient by the percentage of rejected null hypotheses, assessed at 5% significance level, when testing whether the underlying population concordance coefficient is higher than zero.

4   Simulation Study Results

The simulation results related to the means and standard deviations of the concordance coefficients are given in Table 1, Table 2 and Table 3 for $m = 3, 4, 5$, respectively, while the corresponding powers are shown for $m = 3$ in Fig. 1 for scenario 1 and in Fig. 2 for scenario 2, for $m = 4$ in Table 4, and for $m = 5$ in Table 5. The tables are in the Appendix. The most relevant results presented in the mentioned figures and tables were analyzed and are exposed below.

To achieve a better understanding of the results of the simulation study, means and standard deviations of the simulated concordance coefficients are presented in Table 1, Table 2 and Table 3. Analyzing these tables, in what concerns the mean estimates for the coefficients under study, we highlight the following results:

- Fixing n and σ, for all values of m considered, the means for all four coefficients estimates are higher in scenario 1 ($p = 0.25$) than the respective values in scenario 2 ($p = 0.1$); this occurs, as expected, because in scenario 1 there is a higher percentage of ranks that are in agreement;
- For both scenarios, for all values of m and n in study, and for all the coefficients, for higher degrees of agreement, corresponding to smaller values of σ, the means of coefficients estimates are higher than for smaller degrees of agreement; that is, as expected, the means of coefficients estimates decrease as σ increases;
- Setting the higher values of σ, for all values of m, for both scenarios, and for each coefficient, the mean estimates are very identical for any sample size n under study;
- Fixing the values of n and σ, for both scenarios and for all coefficients, the mean estimates decrease as m increases;
- For both scenarios, the means of coefficients estimates are higher for the three weighed coefficients A_W, A_K and A_M, when compared with the Kendall's coefficient W, being this difference more evident in scenario 2;
- In both scenarios, the means of van der Waerden coefficient estimates are slightly lower than the correspondent values of the other two weighted coefficients;
- Considering all values of m and n, for higher values of σ, the means of all the four coefficients estimates are similar, for both scenarios.
- Considering all values of m and n, for smaller values of σ, in scenario 1, the means of the Klotz and the Mood coefficients estimates remain practically the same, being slightly higher than the other two coefficients, with Kendall's having the lowest values as already mentioned; whereas in scenario 2, the means of the Klotz coefficient estimates are higher than the values of the other three coefficients, the means of van der Waerden and the Mood coefficients estimates are quite close, being however the values of van der Waerden slightly lower.

Concerning the empirical standard deviations, no pattern was found when comparing the four coefficients, the both scenarios and the six intensities of agreement. The standard deviations of the four coefficients decreased, as n and m increased.

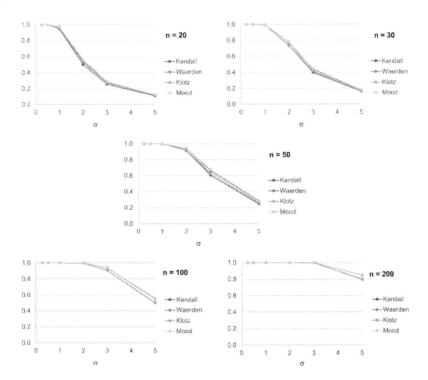

Fig. 1. Powers (%) of Kendall's coefficient of concordance, van der Waerden, signed Klotz, and signed Mood weighted coefficients for $m = 3$ in the scenario 1 ($p = 0.25$), in which the concordance was targeted for a higher proportion of extreme ranks.

Regarding the powers of the four coefficients, analyzing Fig. 1 for scenario 1 and in Fig. 2 for scenario 2, for $m = 3$, and Table 4 and Table 5 for $m = 4$ and $m = 5$ respectively, it can be stated that:

- Fixing n and σ, for all values of m considered, the powers of all indexes are higher in scenario 1 ($p = 0.25$) than the respective values in scenario 2 ($p = 0.1$); this occurs, as expected, because in scenario 1 there is a higher percentage of ranks that are in agreement;
- For both scenarios and for all the number of judges considered ($m = 3, 4, 5$), the powers of all coefficients are higher for smaller values of σ when compared to the higher ones;
- For both scenarios, the three weighed coefficients A_W, A_K and A_M have higher powers than the Kendall's coefficient of concordance W; nevertheless, this difference is more evident in scenario 2 in which the concordance was concentrated on a lower proportion of extreme ranks;
- In scenario 1 (respectively, in scenario 2), for higher values of σ, the Klotz powers are slightly lower (respectively, higher) than the correspondent values of the other two weighted coefficients, what is more evident for lower values of n;

- For all values of m and n, for both scenarios, towards smaller values of σ, the powers attained the maximum value (100%), but for higher values of σ, the powers decrease as σ increases, being this wane more pronounced for small values of n, where the achieved powers are far from been reasonable; for higher values of σ, the powers are not reasonable (because they are much below 70%), and this is worse in scenario 2;
- For all values of m and n, in scenario 1, the powers of all the four coefficients have a similar behavior for each one of the six intensities of agreement in study, whereas, in scenario 2, for higher values of σ, the powers of all coefficients show a distinct behavior, being the powers of the Klotz coefficient the ones that reach always the highest values for the same n and σ, while the van der Waerden and the Mood weighted coefficients have lower and quite similar values;
- For a certain value of n and a fixed σ, the powers of the four coefficients increase as m increases.

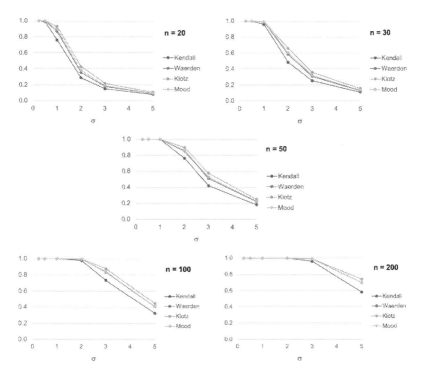

Fig. 2. Powers (%) of Kendall's coefficient of concordance, van der Waerden, signed Klotz, and signed Mood weighted coefficients for $m = 3$ in the scenario 2 ($p = 0.1$) in which the concordance was targeted for a lower proportion of extreme ranks.

5   Conclusions

In this paper, the performance of the three generalized weighted agreement coefficients A_s to measure the concordance among several sets of ranks putting emphasis on the extreme ranks, the two new ones proposed by the authors [2] — the generalized signed Klotz and the signed Mood coefficients— and the van der Waerden coefficient, was compared with the behavior of Kendall's coefficient of concordance W that assigns equal weights to all rankings.

In order to assess whether the generalized weighted agreement coefficients, especially the signed Klotz and the signed Mood coefficients, can be a benefit in assessing the concordance among several sets of ranks, when one intends to place more emphasis on the lower and upper ranks simultaneously, a Monte Carlo simulation study was carried out.

The simulation study allowed to show some interesting results. It can be noticed that all coefficients in study, the three generalized weighted ones and the non-weighted, showed a better performance when the agreement is focused on a higher proportion of extreme ranks than when it is located on a smaller proportion. It can also be stated that the behavior of all coefficients is better for higher degrees of agreement when compared to the smaller ones, for both proportions of extreme ranks under study and for all the number of judges considered. It can still be seen that for a fixed number of subjects and intensity of agreement, the powers of the four coefficients increase as the number of judges increases.

In situations where it is important to give relevance to the agreement on the lower and upper ranks at the same time, we realized that one of the three generalized weighted agreement coefficients should be used instead of Kendall's coefficient of concordance. Besides that, we emphasize that the choice of one of the three types of generalized weighted agreement coefficients depends on the amount of extreme ranks that are the focus of the assessment. It is very interesting to verify that, in cases where the concordance was focused on a lower proportion of extreme ranks, the Klotz coefficient show the best performance when compared with the other two generalized weighted coefficients. But on the contrary, this coefficient is the worst in occurrences where the agreement is located in a higher proportion of extreme ranks, being the Signed Mood and van der Waerden coefficients the ones that reveal the best performance in this case. In both scenarios, the Signed Mood and van der Waerden coefficients present quite similar powers to the several intensities of agreement considered in this study, for a fixed number of subjects and a fixed number of judges.

We believe that this simulation study helps to understand the profits that can result from the use of the generalized weighted coefficients of agreement A_s, instead of the non weighted Kendall's coefficient of concordance, in situations where it is intended to weight the extreme ranks simultaneously.

Acknowledgments. Sandra M. Aleixo was partly supported by the Fundação para a Ciência e Tecnologia, under Grant UIDB/00006/2020 to CEAUL - Centre of Statistics and its Applications. Júlia Teles was partly supported by the Fundação para a Ciência e Tecnologia, under Grant UIDB/00447/2020 to CIPER - Centro Interdisciplinar para o Estudo da Performance Humana (unit 447). *Conflict of Interest*: None declared.

Appendix

Table 1. Mean (standard deviation) of Kendall's W, van der Waerden A_W, signed Klotz A_K, and signed Mood A_M coefficients estimates, with $m = 3$, for two scenarios: on the left, the concordance was targeted for a higher proportion ($p = 0.25$) of extreme ranks, and on the right, the concordance was focused on a lower proportion ($p = 0.1$) of extreme ranks.

n	σ	Scenario 1 ($p = 0.25$)				Scenario 2 ($p = 0.1$)			
		W	A_W	A_K	A_M	W	A_W	A_K	A_M
20	0.25	.87(.03)	.90(.02)	.94(.03)	.94(.02)	.59(.06)	.66(.05)	.80(.03)	.70(.04)
	0.5	.82(.05)	.84(.05)	.87(.06)	.87(.05)	.58(.06)	.64(.06)	.76(.06)	.67(.06)
	1	.68(.08)	.70(.08)	.70(.09)	.71(.08)	.52(.08)	.55(.08)	.62(.10)	.57(.09)
	2	.53(.10)	.54(.10)	.54(.10)	.54(.10)	.43(.09)	.45(.09)	.47(.10)	.46(.10)
	3	.46(.10)	.47(.10)	.47(.10)	.47(.10)	.40(.09)	.41(.09)	.42(.10)	.41(.09)
	5	.41(.09)	.41(.10)	.41(.10)	.41(.10)	.37(.09)	.38(.09)	.39(.10)	.38(.09)
30	0.25	.90(.02)	.92(.02)	.95(.02)	.95(.01)	.61(.05)	.69(.04)	.84(.02)	.73(.03)
	0.5	.84(.04)	.86(.04)	.87(.05)	.88(.04)	.60(.05)	.67(.04)	.80(.04)	.70(.04)
	1	.70(.07)	.72(.06)	.71(.07)	.72(.07)	.54(.07)	.58(.07)	.65(.08)	.59(.07)
	2	.54(.08)	.55(.08)	.54(.08)	.55(.08)	.45(.07)	.47(.08)	.49(.08)	.47(.08)
	3	.47(.08)	.48(.08)	.47(.08)	.48(.08)	.41(.08)	.42(.08)	.43(.08)	.42(.08)
	5	.41(.08)	.42(.08)	.41(.08)	.42(.08)	.38(.07)	.38(.08)	.39(.08)	.39(.08)
50	0.25	.89(.01)	.92(.01)	.95(.02)	.95(.01)	.63(.03)	.72(.03)	.87(.02)	.75(.02)
	0.5	.83(.03)	.86(.03)	.88(.03)	.88(.03)	.62(.04)	.70(.03)	.83(.03)	.73(.03)
	1	.69(.05)	.72(.05)	.72(.06)	.72(.05)	.55(.05)	.60(.05)	.67(.06)	.61(.05)
	2	.54(.06)	.55(.06)	.54(.06)	.55(.06)	.46(.06)	.48(.06)	.50(.07)	.48(.06)
	3	.47(.06)	.48(.06)	.47(.06)	.48(.06)	.41(.06)	.43(.06)	.44(.06)	.43(.06)
	5	.41(.06)	.42(.06)	.41(.06)	.42(.06)	.38(.06)	.39(.06)	.39(.06)	.39(.06)
100	0.25	.90(.01)	.93(.01)	.96(.01)	.96(.01)	.64(.02)	.74(.02)	.89(.01)	.77(.01)
	0.5	.85(.02)	.87(.02)	.88(.02)	.89(.02)	.64(.02)	.72(.02)	.85(.02)	.74(.02)
	1	.70(.03)	.73(.03)	.72(.04)	.73(.03)	.56(.03)	.62(.04)	.69(.05)	.62(.04)
	2	.54(.04)	.55(.04)	.54(.05)	.56(.04)	.46(.04)	.49(.04)	.51(.05)	.49(.04)
	3	.47(.04)	.48(.04)	.47(.04)	.48(.04)	.42(.04)	.43(.04)	.45(.05)	.44(.04)
	5	.41(.04)	.42(.04)	.42(.04)	.42(.04)	.38(.04)	.39(.04)	.40(.04)	.39(.04)
200	0.25	.91(.01)	.94(.00)	.96(.01)	.96(.00)	.65(.02)	.75(.01)	.90(.01)	.77(.01)
	0.5	.85(.01)	.88(.01)	.89(.02)	.90(.01)	.64(.02)	.73(.01)	.86(.01)	.75(.01)
	1	.71(.02)	.73(.02)	.72(.03)	.74(.02)	.57(.02)	.63(.02)	.70(.03)	.63(.03)
	2	.54(.03)	.56(.03)	.55(.03)	.56(.03)	.47(.03)	.49(.03)	.51(.03)	.49(.03)
	3	.47(.03)	.48(.03)	.47(.03)	.48(.03)	.42(.03)	.44(.03)	.45(.03)	.44(.03)
	5	.42(.03)	.42(.03)	.42(.03)	.42(.03)	.38(.03)	.39(.03)	.40(.03)	.39(.03)

Table 2. Mean (standard deviation) of Kendall's W, van der Waerden A_W, signed Klotz A_K, and signed Mood A_M coefficients estimates, with $m = 4$, for two scenarios: on the left, the concordance was targeted for a higher proportion ($p = 0.25$) of extreme ranks, and on the right, the concordance was focused on a lower proportion ($p = 0.1$) of extreme ranks.

n	σ	Scenario 1 ($p = 0.25$)				Scenario 2 ($p = 0.1$)			
		W	A_W	A_K	A_M	W	A_W	A_K	A_M
20	0.25	.86(.03)	.89(.02)	.93(.03)	.93(.02)	.54(.05)	.61(.04)	.78(.03)	.66(.04)
	0.5	.79(.05)	.81(.05)	.84(.06)	.84(.05)	.52(.05)	.59(.05)	.73(.06)	.63(.05)
	1	.62(.08)	.64(.08)	.64(.09)	.65(.09)	.45(.07)	.48(.08)	.55(.10)	.50(.08)
	2	.44(.09)	.45(.09)	.45(.09)	.46(.09)	.35(.08)	.37(.08)	.38(.09)	.37(.08)
	3	.37(.09)	.38(.09)	.38(.09)	.38(.09)	.31(.08)	.32(.08)	.33(.08)	.33(.08)
	5	.32(.08)	.32(.08)	.32(.08)	.32(.08)	.29(.08)	.29(.08)	.30(.08)	.29(.08)
30	0.25	.89(.02)	.91(.02)	.94(.03)	.94(.02)	.57(.04)	.65(.03)	.82(.02)	.69(.03)
	0.5	.81(.04)	.84(.04)	.85(.05)	.86(.04)	.55(.04)	.63(.04)	.77(.04)	.66(.04)
	1	.64(.07)	.66(.07)	.65(.07)	.66(.07)	.47(.06)	.51(.06)	.58(.08)	.53(.07)
	2	.45(.07)	.46(.07)	.46(.08)	.47(.08)	.37(.06)	.38(.07)	.40(.08)	.39(.07)
	3	.38(.07)	.39(.07)	.38(.07)	.39(.07)	.32(.06)	.33(.07)	.34(.07)	.34(.07)
	5	.32(.07)	.33(.07)	.32(.07)	.33(.07)	.29(.06)	.30(.06)	.30(.06)	.30(.06)
50	0.25	.87(.01)	.91(.01)	.94(.02)	.94(.01)	.59(.03)	.68(.02)	.85(.01)	.72(.02)
	0.5	.80(.03)	.84(.03)	.85(.04)	.86(.03)	.57(.03)	.66(.03)	.80(.03)	.69(.03)
	1	.63(.05)	.66(.05)	.66(.06)	.66(.05)	.49(.05)	.54(.05)	.61(.07)	.55(.05)
	2	.45(.06)	.46(.06)	.46(.06)	.47(.06)	.37(.05)	.39(.05)	.41(.06)	.40(.05)
	3	.38(.05)	.39(.05)	.38(.06)	.39(.05)	.33(.05)	.34(.05)	.35(.05)	.34(.05)
	5	.32(.05)	.33(.05)	.32(.05)	.33(.05)	.29(.05)	.30(.05)	.31(.05)	.30(.05)
100	0.25	.89(.01)	.92(.01)	.95(.01)	.95(.01)	.60(.02)	.71(.01)	.87(.01)	.74(.01)
	0.5	.82(.02)	.85(.02)	.86(.03)	.87(.02)	.59(.02)	.68(.02)	.82(.02)	.71(.02)
	1	.65(.04)	.67(.03)	.66(.04)	.67(.04)	.50(.03)	.56(.03)	.63(.05)	.56(.04)
	2	.46(.04)	.47(.04)	.46(.04)	.47(.04)	.38(.04)	.40(.04)	.42(.04)	.40(.04)
	3	.38(.04)	.39(.04)	.38(.04)	.39(.04)	.33(.03)	.35(.04)	.36(.04)	.35(.04)
	5	.33(.04)	.33(.04)	.33(.04)	.33(.04)	.30(.03)	.30(.03)	.31(.04)	.30(.03)
200	0.25	.89(.01)	.93(.00)	.95(.01)	.95(.00)	.61(.01)	.72(.01)	.88(.01)	.74(.01)
	0.5	.82(.01)	.86(.01)	.86(.02)	.87(.01)	.60(.01)	.70(.01)	.83(.01)	.72(.01)
	1	.65(.03)	.68(.02)	.66(.03)	.68(.03)	.51(.02)	.57(.02)	.64(.03)	.57(.03)
	2	.46(.03)	.47(.03)	.46(.03)	.47(.03)	.38(.03)	.41(.03)	.43(.03)	.41(.03)
	3	.38(.03)	.39(.03)	.38(.03)	.39(.03)	.34(.02)	.35(.03)	.36(.03)	.35(.03)
	5	.33(.03)	.33(.03)	.32(.03)	.33(.03)	.30(.02)	.31(.02)	.31(.03)	.31(.02)

Table 3. Mean (standard deviation) of Kendall's W, van der Waerden A_W, signed Klotz A_K, and signed Mood A_M coefficients estimates, with $m = 5$, for two scenarios: on the left, the concordance was targeted for a higher proportion ($p = 0.25$) of extreme ranks, and on the right, the concordance was focused on a lower proportion ($p = 0.1$) of extreme ranks.

n	σ	Scenario 1 ($p = 0.25$)				Scenario 2 ($p = 0.1$)			
		W	A_W	A_K	A_M	W	A_W	A_K	A_M
20	0.25	.85(.03)	.88(.02)	.92(.03)	.92(.02)	.51(.04)	.59(.03)	.76(.02)	.64(.03)
	0.5	.77(.05)	.80(.05)	.82(.06)	.82(.06)	.49(.05)	.56(.05)	.71(.06)	.60(.05)
	1	.58(.08)	.60(.08)	.60(.09)	.61(.09)	.40(.07)	.44(.07)	.51(.10)	.46(.08)
	2	.39(.08)	.40(.08)	.40(.09)	.40(.08)	.30(.07)	.32(.07)	.33(.08)	.32(.08)
	3	.32(.08)	.32(.08)	.32(.08)	.33(.08)	.26(.07)	.27(.07)	.28(.07)	.27(.07)
	5	.26(.07)	.27(.07)	.26(.07)	.27(.07)	.23(.06)	.24(.06)	.24(.07)	.24(.07)
30	0.25	.88(.02)	.91(.02)	.93(.03)	.94(.02)	.54(.03)	.63(.02)	.81(.02)	.67(.02)
	0.5	.79(.04)	.82(.04)	.83(.05)	.84(.04)	.52(.04)	.60(.03)	.75(.05)	.64(.04)
	1	.60(.07)	.62(.07)	.61(.07)	.63(.07)	.43(.05)	.47(.06)	.54(.08)	.49(.07)
	2	.40(.07)	.41(.07)	.40(.07)	.41(.07)	.32(.06)	.33(.06)	.35(.07)	.34(.06)
	3	.33(.06)	.33(.06)	.33(.06)	.33(.06)	.27(.06)	.28(.06)	.29(.06)	.28(.06)
	5	.27(.06)	.27(.06)	.27(.06)	.27(.06)	.24(.05)	.24(.05)	.25(.05)	.24(.05)
50	0.25	.86(.01)	.90(.01)	.94(.02)	.94(.01)	.56(.02)	.66(.02)	.84(.01)	.70(.02)
	0.5	.79(.03)	.82(.03)	.84(.04)	.84(.03)	.54(.03)	.63(.02)	.79(.03)	.67(.03)
	1	.60(.05)	.62(.05)	.62(.06)	.63(.05)	.45(.04)	.50(.05)	.57(.07)	.51(.05)
	2	.40(.05)	.41(.05)	.40(.05)	.41(.05)	.32(.04)	.34(.05)	.36(.06)	.35(.05)
	3	.32(.05)	.33(.05)	.33(.05)	.33(.05)	.28(.04)	.29(.04)	.30(.05)	.29(.05)
	5	.27(.04)	.27(.04)	.27(.04)	.27(.04)	.24(.04)	.25(.04)	.25(.04)	.25(.04)
100	0.25	.88(.01)	.92(.01)	.94(.01)	.95(.01)	.57(.02)	.69(.01)	.86(.01)	.72(.01)
	0.5	.80(.02)	.83(.02)	.84(.03)	.85(.02)	.56(.02)	.66(.02)	.81(.02)	.69(.02)
	1	.61(.04)	.63(.04)	.62(.04)	.64(.04)	.46(.03)	.52(.03)	.59(.05)	.52(.04)
	2	.41(.04)	.42(.04)	.41(.04)	.42(.04)	.33(.03)	.35(.03)	.37(.04)	.35(.03)
	3	.33(.03)	.34(.03)	.33(.03)	.34(.03)	.28(.03)	.29(.03)	.30(.03)	.29(.03)
	5	.27(.03)	.27(.03)	.27(.03)	.27(.03)	.24(.03)	.25(.03)	.25(.03)	.25(.03)
200	0.25	.89(.00)	.92(.00)	.95(.01)	.95(.00)	.58(.01)	.70(.01)	.87(.01)	.73(.01)
	0.5	.81(.02)	.84(.01)	.84(.02)	.86(.01)	.57(.01)	.67(.01)	.82(.02)	.70(.01)
	1	61(.02)	.64(.02)	.62(.03)	.64(.03)	.47(.02)	.53(.02)	.60(.03)	.53(.03)
	2	.41(.03)	.42(.03)	.41(.03)	.42(.03)	.33(.02)	.36(.02)	.37(.03)	.36(.02)
	3	.33(.02)	34(.02)	.33(.02)	.34(.02)	.28(.02)	.30(.02)	.30(.02)	.30(.02)
	5	.27(.02)	.28(.02)	.27(.02)	.28(.02)	.24(.02)	.25(.02)	.25(.02)	.25(.02)

Table 4. Estimated powers (%) of Kendall's W, van der Waerden A_W, signed Klotz A_K, and signed Mood A_M coefficients, with $m = 4$, for two scenarios: on the left, the concordance was targeted for a higher proportion ($p = 0.25$) of extreme ranks, and on the right, the concordance was focused on a lower proportion ($p = 0.1$) of extreme ranks.

n	σ	Scenario 1 ($p = 0.25$)				Scenario 2 ($p = 0.1$)			
		A_W	A_K	A_M	W	A_W	A_K	A_M
20	0.25	100.00	100.00	100.00	100.00	99.97	100.00	100.00	100.00
	0.5	100.00	100.00	100.00	100.00	98.82	99.85	100.00	99.92
	1	99.38	99.61	99.46	99.62	76.02	86.59	92.92	89.32
	2	70.68	73.53	72.11	74.45	28.84	35.44	42.81	38.02
	3	39.23	42.13	40.78	43.04	14.88	17.80	21.85	18.83
	5	17.19	18.43	18.11	18.92	7.89	8.96	10.60	9.41
30	0.25	100.00	100.00	100.00	100.00	100.00	100.00	100.00	100.00
	0.5	100.00	100.00	100.00	100.00	100.00	100.00	100.00	100.00
	1	99.99	100.00	99.99	100.00	95.64	98.63	99.37	98.95
	2	87.98	90.41	88.13	90.78	48.46	58.49	65.87	60.64
	3	58.38	61.89	58.28	62.74	25.37	30.87	36.00	32.20
	5	24.57	25.69	24.37	26.17	11.14	13.16	15.75	13.76
50	0.25	100.00	100.00	100.00	100.00	100.00	100.00	100.00	100.00
	0.5	100.00	100.00	100.00	100.00	100.00	100.00	100.00	100.00
	1	100.00	100.00	100.00	100.00	99.86	99.98	99.99	99.99
	2	97.90	98.60	97.88	98.69	76.43	85.40	89.92	86.50
	3	77.04	80.87	78.19	81.63	41.96	50.91	57.51	52.66
	5	35.67	38.90	37.17	39.94	18.07	21.86	25.11	22.62
100	0.25	100.00	100.00	100.00	100.00	100.00	100.00	100.00	100.00
	0.5	100.00	100.00	100.00	100.00	100.00	100.00	100.00	100.00
	1	100.00	100.00	100.00	100.00	100.00	100.00	100.00	100.00
	2	100.00	100.00	100.00	100.00	97.71	99.39	99.65	99.36
	3	97.48	98.34	97.18	98.39	73.21	83.37	87.64	84.05
	5	64.82	69.66	64.88	69.84	32.52	40.78	44.83	41.14
200	0.25	100.00	100.00	100.00	100.00	100.00	100.00	100.00	100.00
	0.5	100.00	100.00	100.00	100.00	100.00	100.00	100.00	100.00
	1	100.00	100.00	100.00	100.00	100.00	100.00	100.00	100.00
	2	100.00	100.00	100.00	100.00	100.00	100.00	100.00	100.00
	3	99.99	99.99	99.98	100.00	96.07	98.98	99.25	98.91
	5	90.69	93.29	89.34	93.50	58.23	69.97	74.09	69.89

Table 5. Estimated powers (%) of Kendall's W, van der Waerden A_W, signed Klotz A_K, and signed Mood A_M coefficients, with $m = 5$, for two scenarios: on the left, the concordance was targeted for a higher proportion ($p = 0.25$) of extreme ranks, and on the right, the concordance was focused on a lower proportion ($p = 0.1$) of extreme ranks.

n	σ	Scenario 1 ($p = 0.25$)				Scenario 2 ($p = 0.1$)			
		W	A_W	A_K	A_M	W	A_W	A_K	A_M
20	0.25	100.00	100.00	100.00	100.00	100.00	100.00	100.00	100.00
	0.5	100.00	100.00	100.00	100.00	99.85	99.97	99.98	99.97
	1	99.75	99.85	99.84	99.87	89.92	94.90	97.01	95.75
	2	81.48	83.75	82.46	84.69	41.44	48.16	55.11	50.95
	3	49.25	52.06	50.31	53.00	20.28	23.97	28.27	25.49
	5	21.36	22.46	21.79	23.01	9.94	11.19	13.20	11.82
30	0.25	100.00	100.00	100.00	100.00	100.00	100.00	100.00	100.00
	0.5	100.00	100.00	100.00	100.00	100.00	100.00	100.00	100.00
	1	99.99	99.99	99.98	99.99	99.13	99.75	99.89	99.76
	2	94.60	95.62	94.53	95.70	64.25	72.51	78.17	74.40
	3	68.94	71.98	68.54	72.35	32.68	39.11	44.73	40.49
	5	31.42	33.07	31.26	33.66	13.74	16.44	19.78	17.18
50	0.25	100.00	100.00	100.00	100.00	100.00	100.00	100.00	100.00
	0.5	100.00	100.00	100.00	100.00	100.00	100.00	100.00	100.00
	1	100.00	100.00	100.00	100.00	99.98	100.00	100.00	100.00
	2	99.51	99.73	99.49	99.72	88.61	93.48	95.14	93.74
	3	86.89	89.39	86.91	89.77	54.84	63.74	68.83	65.00
	5	44.97	48.75	45.99	49.75	23.71	28.19	31.38	28.94
100	0.25	100.00	100.00	100.00	100.00	100.00	100.00	100.00	100.00
	0.5	100.00	100.00	100.00	100.00	100.00	100.00	100.00	100.00
	1	100.00	100.00	100.00	100.00	100.00	100.00	100.00	100.00
	2	100.00	100.00	100.00	100.00	99.53	99.88	99.96	99.87
	3	99.23	99.61	99.07	99.61	84.57	91.73	93.38	91.95
	5	74.44	78.68	73.60	79.10	41.73	50.47	54.31	51.60
200	0.25	100.00	100.00	100.00	100.00	100.00	100.00	100.00	100.00
	0.5	100.00	100.00	100.00	100.00	100.00	100.00	100.00	100.00
	1	100.00	100.00	100.00	100.00	100.00	100.00	100.00	100.00
	2	100.00	100.00	100.00	100.00	100.00	100.00	100.00	100.00
	3	100.00	100.00	100.00	100.00	98.88	99.69	99.82	99.71
	5	95.35	97.04	94.83	97.02	68.76	79.32	82.43	79.24

References

1. Aleixo, S.M., Teles, J.: Weighting lower and upper ranks simultaneously through rank-order correlation coefficients. In: Gervasi, O., et al. (eds.) ICCSA 2018. LNCS, vol. 10961, pp. 318–334. Springer, Cham (2018). https://doi.org/10.1007/978-3-319-95165-2_23

2. Aleixo, S.M., Teles, J.: Weighted coefficients to measure agreement among several sets of ranks emphasizing top and bottom ranks at the same time. In: Misra, S., et al. (eds.) ICCSA 2019. LNCS, vol. 11620, pp. 23–33. Springer, Cham (2019). https://doi.org/10.1007/978-3-030-24296-1_3

3. Coolen-Maturi, T.: New weighted rank correlation coefficients sensitive to agreement on top and bottom rankings. J. Appl. Stat. **43**(12), 2261–2279 (2016)

4. Friedman, M.: The use of ranks to avoid the assumption of normality implicit in the analysis of variance. J. Am. Stat. Assoc. **32**, 675–701 (1937)

5. Hájek, J., Šidák, Z.: Theory of Rank Tests. Academic Press, New York (1972)

6. Kendall, M.G., Babington-Smith, B.: The problem of m rankings. Ann. Math. Stat. **10**, 275–287 (1939)

7. Legendre, P.: Species associations: the Kendall coefficient of concordance revisited. J. Agric. Biol. Environ. Stat. **10**(2), 226–245 (2005)

8. R Core Team: R: a language and environment for statistical computing. R Foundation for Statistical Computing, Vienna, Austria (2019). https://www.r-project.org

9. Teles, J.: Concordance coefficients to measure the agreement among several sets of ranks. J. Appl. Stat. **39**, 1749–1764 (2012)

10. van der Waerden, B.L.: Order tests for the two-sample problem and their power. Proc. Koninklijke Nederlandse Akademie van Wetenschappen **55**, 453–458 (1952)

11. Wallis, W.A.: The correlation ratio for ranked data. J. Am. Statist. Assoc. **34**, 533–538 (1939)

Impact of OVL Variation on AUC Bias Estimated by Non-parametric Methods

Carina Silva[1,2]([✉]) [iD], Maria Antónia Amaral Turkman[2] [iD], and Lisete Sousa[2,3] [iD]

[1] Lisbon School of Health Technology, Polytechnic Institute of Lisbon,
Lisbon, Portugal
carina.silva@estesl.ipl.pt
[2] Centro de Estatítica e Aplicações, Faculdade de Ciências, Universidade de Lisboa,
Campo Grande, 1749-016 Lisbon, Portugal
{antonia.turkman,lmsousa}@fc.ul.pt
[3] Faculdade de Ciências, Universidade de Lisboa, Campo Grande,
1749-016 Lisbon, Portugal

Abstract. The area under the ROC curve (AUC) is the most commonly used index in the ROC methodology to evaluate the performance of a classifier that discriminates between two mutually exclusive conditions. The AUC can admit values between 0.5 and 1, where values close to 1 indicate that the model of classification has a high discriminative power. The overlap coefficient (OVL) between two density functions is defined as the common area between both functions. This coefficient is used as a measure of agreement between two distributions presenting values between 0 and 1, where values close to 1 reveal total overlapping densities. These two measures were used to construct the *arrow plot* to select differential expressed genes. A simulation study using the bootstrap method is presented in order to estimate AUC bias and standard error using empirical and kernel methods. In order to assess the impact of the OVL variation on the AUC bias, samples from various continuous distributions were simulated considering different values for its parameters and for fixed OVL values between 0 and 1. Samples of dimensions 15, 30, 50 and 100 and 1000 bootstrap replicates for each scenario were considered.

Keywords: AUC · OVL · Arrow-plot · Bias

1 Introduction

Receiver operating curve (ROC) is a widespread methodology to evaluate the accuracy of binary classification systems, particularly in diagnostic tests [1]. The area under the ROC curve (AUC) is the index most commonly used to summarize the accuracy and recently used in genomic studies [2,3]. Another index

This work is partially financed by national funds through FCT Fundação para a Ciência e a Tecnologia under the project UIDB/00006/2020.

O. Gervasi et al. .(Eds.): ICCSA 2020, LNCS 12251, pp. 173–184, 2020.
https://doi.org/10.1007/978-3-030-58808-3_14

that recently has gained attention is the overlapping coefficient (OVL) [3,4], which is a measure of the similarity between two probability distributions. Silva-Fortes et al. (2012) [3] proposed a plot that uses both indices, the *arrow plot* (Fig. 1). This plot displays the overlapping coefficient (OVL) against the area under the ROC curve (AUC) for thousands of genes simultaneously using data from microarray experiments. The *arrow plot* allows to select different types of differentially expressed genes, namely up-regulated, down-regulated and genes with a mixture of both, called *special genes*, who may reveal a presence of different subclasses. Graphic analysis is quite intuitive, as it allows to obtain a global picture of the behavior of the genes and, based on its analysis, the user can choose the cutoff points for AUC and OVL, although this choice is arbitrary. In this approach AUC values near 1 or near 0 will be related by low OVL values, meaning that both probability distributions will not be overlaped.

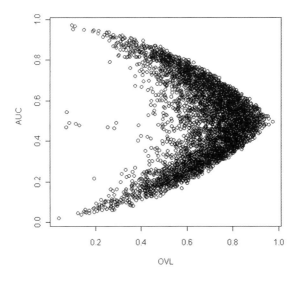

Fig. 1. Arrow plot [3]

In this study we present a simulation analysis to evaluate the impact of the OVL variation in AUC bias when non-parametric methods for their estimation are used. When microarray data analysis is conducted, thousands of AUCs and OVLs are produced, being computationally intensive and time consuming to perform a gene to gene analysis to evaluate if there is a need to make bias corrections. The goal is to understand where in the arrow plot is more likely the need to perform bias adjustments.

2   Methodology

A simulation study using a parametric bootstrap approach was performed using several distributional scenarios in order to reflect a range of distributional behaviours, particularly in genetic studies. AUC bias was analyzed considering non-parametric estimation for fixed OVL values. Consider $X_1, X_2, \ldots, X_{n_1}$ and $Y_1, Y_2, \ldots, Y_{n_2}$ two independent random samples representing some characteristic in population 1 (e.g. control) and population 2 (e.g. cases) respectively. Let F_X be the distribution function of X_i, $i = 1, \ldots, n_1$ and G_Y is the distribution function of Y_j, $j = 1, \ldots, n_2$, and f_X and g_Y their respective density functions. Assume that, without loss of generality, that any cutoff point $c \in \mathbb{R}$, $F_X(c) > G_Y(c)$.

2.1   OVL

OVL is a coefficient used to measure the agreement between two distributions [5, 6], it ranges between 0 and 1, and closer to 1 means higher amount of overlapping area. OVL can detect any differences between two distributions, not only by mean value differences but also by variance differences. One good property of this index is that it is invariant through scale monotone transformations of the variables.

OVL can be expressed under several ways. Weitzman (1970) [5] proposed the expression:

$$\text{OVL} = \int_C \min[f_X(c), g_Y(c)] dc. \tag{1}$$

Results can be extended to discrete distributions, replacing the integral by a summation.

To cover several levels of overlapping areas between the densities, we fixed the OVL values in 0.2, 0.4, 0.6 and 0.8. True OVL values were calculated using (1) for the distributions considered on the different simulated scenarios (see Table 1).

2.2   AUC

A ROC curve φ is defined by the sensitivity $(q(c) = 1 - F_Y(c))$ and specificity $(p(c) = F_X(c))$:

$$\varphi : [0, 1] \longrightarrow [0, 1]$$

$$\varphi(p) = 1 - F_Y(F_X^{-1}(1 - p)), \tag{2}$$

where F_X^{-1} is the inverse function of F_X defined by $F_X^{-1}(1 - p) = \inf\{x \in W(F_X) : F_X(x) \geq 1 - p\}$ and $W(F_X) = \{x \in \mathbb{R} : 0 < F_X(x) < 1\}$ is the support of F_X when continuous distributions are considered.

The AUC is obtained integrating the ROC curve in its domain. It can be proved that it turns out to be equal to $P(X < Y)$, that is to say AUC ranges between 0.5 and 1, where values near 1 indicate a higher classification performance on the discrimination between the two populations. AUC is also invariant under monotone scale transformations of the variables. However, in the *arrow plot* those values range between 0 and 1, because the same classification rule is applied whenever $F_X(c) > G_Y(c)$ or $F_X(c) < G_Y(c)$. This situation may produce ROC curves that are not proper, meaning that they may produce values for the AUC between 0 and 0.5 [3].

True values of the AUC were obtained accordingly with the distributional scenarios (Table 1):

Bi-Normal: When $X \frown N(\mu_1, \sigma_1)$ and $Y \frown N(\mu_2, \sigma_2)$, $\mu_1 < \mu_2$, the AUC is given by [7]:

$$AUC = \Phi \left(\frac{\mu_2 - \mu_1}{\sqrt{\sigma_1^2 + \sigma_2^2}} \right),$$

where $\Phi(.)$ is the standard normal distribution.

Bi-Lognormal: When $X \frown LN(\mu_1, \sigma_1)$ and $Y \frown LN(\mu_2, \sigma_2)$, $\mu_1 < \mu_2$, the AUC is given by [8]:

$$AUC = \Phi \left(\frac{\mu_2 - \mu_1}{\sqrt{\sigma_1^2 + \sigma_2^2}} \right),$$

where $\Phi(.)$ is the standard normal distribution.

Bi-Exponential: When $X \frown Exp(\lambda_1)$ and $Y \frown Exp(\lambda_2)$, $\lambda_1 < \lambda_2$, the AUC is given by [9]:

$$AUC = \frac{\lambda_1}{\lambda_1 + \lambda_2}$$

For non-parametric estimation of the AUC, it was used the empirical and kernel based methods:

Empirical AUC: The empirical estimator of the AUC corresponds to the Mann-Whitney statistic [10]:

$$\widehat{AUC} = \frac{1}{n_1 n_2} \sum_{i=1}^{n_1} \sum_{j=1}^{n_2} \left(I[x_i < y_j] + \frac{1}{2} I[x_i = y_j] \right), \tag{3}$$

where I is the indicator function.

The empirical AUC estimates were obtained using pROC [11] package from R. From now on the notation *emp* is used whenever the empirical estimation of the AUC is mentioned.

Kernel AUC: Several authors have discussed the refinement of the non-parametric approach to produce smooth ROC curves [14]. Among the various non-parametric methodologies, an important method for estimating probability density functions is the kernel estimator. The kernel estimator of a density is given by [15]:

$$\hat{f}(x) = \frac{1}{nh} \sum_{i=1}^{n} K\left(\frac{x - x_i}{h}\right), \forall x \in S, h > 0, \tag{4}$$

where K is the kernel function, h the bandwidth and S the support of X.

Several studies have demonstrated that the quality of the kernel estimator depends more on the choice of the bandwidth h, rather than on the choice of the functional form of the kernel [12]. In this work the Gaussian kernel will be used and three different bandwidth methods choice will be explored.

Lloyd (1997) showed that when a Gaussian kernel is considered, the AUC is estimated as [13]:

$$\widehat{\text{AUC}} = \frac{1}{n_1 n_2} \sum_{i=1}^{n_1} \sum_{j=1}^{n_2} \Phi\left(\frac{y_j - x_i}{\sqrt{h_1^2 + h_2^2}}\right).$$

Silverman (1992) [12] considers the expression (5) as an optimal bandwidth when the kernel is Gaussian.

$$h = \left(\frac{4}{3}\right)^{\frac{1}{5}} \min\left(s, \frac{R}{1.34}\right) n^{-\frac{1}{5}}, \tag{5}$$

where s is the empirical standard deviation and R the interquartile range. Hereinafter, for our convenience, this method will be referred as $nrd0$.

Scott (1992) [16] considers the expression (6) when a Gaussian kernel is used. The notation nrd is used whenever this method is referred to.

$$h = \left(\frac{4}{3}\right)^{\frac{1}{5}} s\, n^{-\frac{1}{5}}. \tag{6}$$

Hall et al. (1991) [17] proposed the plug-in method solve-the-equation for the optimal bandwidth. The notation SJ is used whenever this method is referred to.

2.3   Parametric Bootstrap Estimation

A Monte Carlo simulation study was used to compare bias, standard error (SE) and root mean squared error (RMSE) of the bootstrap AUC estimates obtained from four non-parametric methods ($emp, nrd0, nrd$ and SJ). In order to represent pairs of overlapping distributions with various degrees of separation and skewness (see Table 1), continuous datasets for control and experimental groups, with samples sizes of n: 15, 30, 50 and 100 on both groups, were considered. The bootstrap estimates were obtained from 1000 replicates in each scenario (see Fig. 2).

Table 1. Distributional scenarios for the simulation study.

Scenario	Control	Experimental	True OVL	True AUC
Bi-normal fixed μ	$N(0,1)$	$N(0,0.1)$	0.2	0.5
	$N(0,1)$	$N(0,0.4)$	0.4	0.5
	$N(0,1)$	$N(0,2.4)$	0.6	0.5
	$N(0,1)$	$N(0,1.5)$	0.8	0.5
Bi-normal fixed σ	$N(0,1)$	$N(2.55,1)$	0.2	0.96
	$N(0,1)$	$N(1.65,1)$	0.4	0.89
	$N(0,1)$	$N(1.04,1)$	0.6	0.77
	$N(0,1)$	$N(0.5,1)$	0.8	0.64
Bi-Lognormal	$LN(0,1)$	$LN(0,0.1)$	0.2	0.5
	$LN(0,1)$	$LN(1.65,1)$	0.4	0.87
	$LN(0,1)$	$LN(1.04,1)$	0.6	0.77
	$LN(0,1)$	$LN(0.5,1)$	0.8	0.64
Bi-Exponential	$Exp(1)$	$Exp(0.05)$	0.2	0.95
	$Exp(1)$	$Exp(0.15)$	0.4	0.87
	$Exp(1)$	$Exp(0.32)$	0.6	0.76
	$Exp(1)$	$Exp(0.58)$	0.8	0.63

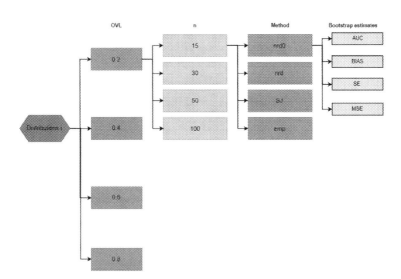

Fig. 2. Scheme of the simulation procedure. Total of 1024 estimates.

The bootstrap estimator of the AUC is:

$$\widehat{\mathrm{AUC}}_B = \frac{1}{1000} \sum_{i=1}^{1000} \widehat{\mathrm{AUC}}_i^*, \tag{7}$$

where $\widehat{\mathrm{AUC}}_i^*$ is the AUC estimate (empirical or kernel) in each bootstrap replicate.

The bootstrap estimator of the standard error of the AUC is given by:

$$\widehat{se}_B(\widehat{\mathrm{AUC}}) = \sqrt{\frac{1}{999} \sum_{i=1}^{1000} (\widehat{\mathrm{AUC}}_i^* - \widehat{\mathrm{AUC}}_B)^2}. \tag{8}$$

The bootstrap estimator of the bias of the bootstrap AUC is:

$$\widehat{\mathrm{bias}}_B(\widehat{\mathrm{AUC}}) = \widehat{\mathrm{AUC}}_B - \mathrm{AUC}, \tag{9}$$

where AUC corresponds to the true AUC value.
The bootstrap estimator of the root mean squared error (RMSE) of the AUC is given by:

$$\widehat{rmse}_B(\widehat{\mathrm{AUC}}) = \sqrt{\frac{1}{1000} \sum_{i=1}^{1000} (\widehat{\mathrm{AUC}}_i^* - \mathrm{AUC})^2}. \tag{10}$$

3   Results and Discussion

Figure 3 and Fig. 4 depict the behaviour of the bootstrap AUC estimates along different OVL values, considering different sample sizes and non-parametric estimation methods for the distributional scenarios presented in Table 1. Analyzing Table 1 it is observed that AUC estimates are influenced by sample sizes and variability in results increases as OVL increase. However, when AUC values are obtained from distributions with equal mean values, which produce not proper ROC curves, the variability of the estimates is constant across different OVL values (Fig. 3a and Fig. 4a).

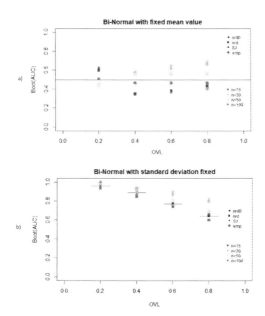

Fig. 3. Comparison of AUC bootstrap estimates against OVL values, considering different sample sizes and non-parametric estimation methods. a) Bi-normal distribution with fixed mean values. b) Bi-normal distribution with fixed standard deviation. True AUC values are represented by the horizontal lines.

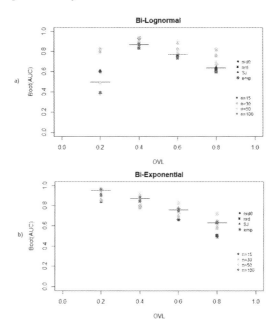

Fig. 4. Comparison of AUC bootstrap estimates against OVL values, considering different sample sizes and non-parametric estimation methods. a) Bi-lognormal distribution. b) Bi-exponential distribution. True AUC values are represented by the horizontal lines.

In general way, the estimated AUC bias is negligible considering all scenarios (|bias| < 0.2) (see Fig. 5).

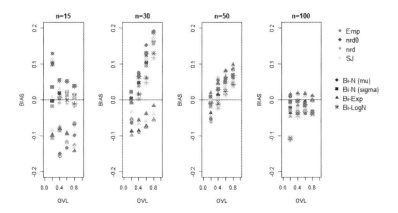

Fig. 5. Comparison of AUC bootstrap bias considering all distributions and non-parametric estimation methods for each sample dimension.

However, as expected, bias is lower for larger samples sizes. Considering small sample sizes, data simulated from bi-exponential distributions tend to underestimate the AUC, contrary to all the other distributions. For high OVL values bias tends to increase.

Precision tends to increase when OVL values decrease, unless if AUC values are obtained from not proper ROC curves (AUC ≈ 0.5), where precision tends to be lower when compared to all other scenarios for all OVL values (see Fig. 6).

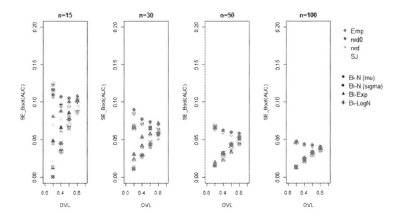

Fig. 6. Comparison of bootstrap estimates of the standard error of the AUC, considering all distributions and non-parametric methods for each sample size.

RMSE tends to be higher in lower samples sizes and increases as OVL increases (see Fig. 7).

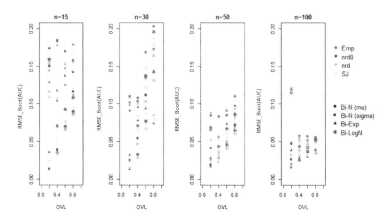

Fig. 7. Comparison of bootstrap estimates of the root of mean squared error of the AUC, considering all distributions and non-parametric methods for each sample size.

In Fig. 8 it is shown the area where bias correction should be considered.

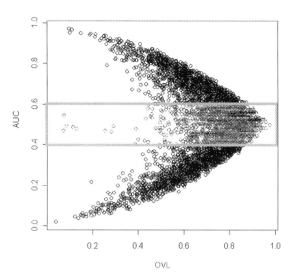

Fig. 8. Area of the Arrow plot where should be considered bias correction.

4   Conclusions

Overall, results of the simulation study suggest that for a broad range of pairs of distributions with several degrees of departure yield close estimates of AUC. Concern about bias or precision of the estimates should not be a major factor in choosing between non-parametric approaches, however there is an advantage in kernel methods since they produce smooth ROC curves. Precision is higher for small values of OVL, however this is not true when AUC values are around 0.5 and are obtained from distributions with the same mean value, leading to not proper ROC curves. This particular situation is related to "special genes" on the *arrow plot*, where greater attention should be paid to the possibility of bias correction.

The main issue with this study is that simulations were performed by using continuous distributions. Future research will include distributions related with count data, where not much research exists concerning the OVL index.

References

1. Lusted, L.: Introduction to Medical Decision Making. Charles C. Thomas, Springfield, Illinois (1968)
2. Parodi, S., Pistoia, V., Musseli, M.: Not proper ROC curves as new tool for the analysis of dierentially expressed genes in microarray experiments. BMC Bioinform. **9**, 410 (2008)
3. Silva-Fortes, C., Amaral Turkman, M.A., Sousa, L.: Arrow plot: a new graphical tool for selecting up and down regulated and genes differentially expressed on sample subgroups. BMC Bioinform. **13**, 147 (2012)
4. Wang, D., Tian, L.: Parametric methods for confidence interval estimation of overlap coefficients. Comput. Stat. Data Anal. **106**, 12–26 (2017)
5. Weitzman, M. S.: Measure of the overlap of income distribution of white and negro families in the United States. Technical report 22, U.S. Department of Commerce, Bureau of the Census, Washington, D.C. (1970)
6. Inman, H.F., Bradley, E.L.: The overlapping coefficient as a measure of agreement between probability distributions and point estimation of the overlap of two normal densities. Commun. Stat. Theor. Meth. **18**(10), 3851–3874 (1989)
7. Faraggi, D., Reiser, B.: Estimation of the area under the ROC curve. Stat. Med. **31**, 3093–3106 (2002)
8. Amala, R., Pundir, S.: Statistical Inference on AUC from a bi-lognormal ROC model for continuos data. Int. J. Eng. Sci. Innov. Technol. **1**(2), 283–295 (2012)
9. Vardhan, R.V., Pundir, S., Sameera, G.: Estimation of area under the ROC curve using Exponential and Weibull distributions. Bonfring Int. J. Data Min. **2**(2), 52–56 (2012)
10. Hanley, J.A., MacNeil, B.J.: A method of comparing the areas under receiver operating characteristic curves derived from the same cases. Radiology **148**(3), 839–843 (1983)
11. Robin, X.: pROC: an open-source package for R and S+ to analyze and compare ROC curves. BMC Bioinform. **12**, 77 (2011)
12. Silverman, B.W.: Density Estimation for Statistics and Data Analysis. Chapman and Hall, London (1998)

13. Lloyd, C.J., Yong, Z.: Kernel estimators of the ROC curve are better than empirical. Stat. Probab. Lett. **44**(3), 221–228 (1999)
14. Zou, K.H., Hall, W.J., Shapiro, D.E.: Smooth non-parametric receiver-operating characteristic (ROC) curves for continuous diagnostic tests. Stat. Med. **16**, 2143–2156 (1997)
15. Rosenblatt, M.: Remarks on some nonparametric estimate of a density function. Ann. Math. Stat. **27**, 832–837 (1956)
16. Scott, D. W.: Kernel Density Estimators. Wiley Series in Probability and Statistics (1992)
17. Hall, P., Sheather, S.J., Jones, M.C., Marron, J.S.: On optimal data-based bandwith selection in kernel density estimation. Biometrika **78**(2), 263–269 (1991)

Adjusting ROC Curve for Covariates with AROC R Package

Francisco Machado e Costa[1]([✉]) [ID] and Ana Cristina Braga[2] [ID]

[1] School of Engineering, University of Minho, Campus de Gualtar, Braga, Portugal
frmachadoecosta@gmail.com
[2] ALGORITMI Centre, University of Minho, Campus de Azurém,
Guimarães, Portugal
acb@dps.uminho.pt

Abstract. The ability of a medical test to differentiate between diseased and non-diseased states is of vital importance and must be screened by statistical analysis for reliability and improvement. The receiver operating characteristic (ROC) curve remains a popular method of marker analysis, disease screening and diagnosis. Covariates in this field related to the subject's characteristics are incorporated in the analysis to avoid bias. The covariate adjusted ROC (AROC) curve was proposed as a method of incorporation. The AROC R-package was recently released and brings various methods of estimation based on multiple authors work.

The aim of this study was to explore the AROC package functionality and usability using real data noting its possible limitations. The main methods of the package were capable of incorporating different and multiple variables, both categorical and continuous, in the AROC curve estimation. When tested for the same data, AROC curves are generated with no statistical differences, regardless of method.

The package offers a variety of methods to estimate the AROC curve complemented with predictive checks and pooled ROC estimation. The package offers a way to conduct a more thorough ROC and AROC analysis, making it available for any R user.

Keywords: Receiver operator characteristic curve · Covariate adjustment · Diagnostic test · Biostatistics · Software tool

1 Introduction

Medical tests meant for use in routine practice must be rigorously scrutinized for their accuracy with statistical analysis before approval.

Differentiating between healthy and diseased populations or, more broadly, control and cases is the fundamental property of any test and of fundamental importance in clinical practice. The receiver operating characteristic (ROC) curve is the most popular statistical tool when dichotomizing test results, particularly with a degree of subjectivity, such as medical imaging [11] and is used

© Springer Nature Switzerland AG 2020
O. Gervasi et al. (Eds.): ICCSA 2020, LNCS 12251, pp. 185–198, 2020.
https://doi.org/10.1007/978-3-030-58808-3_15

broadly in biomedical informatics research to evaluate classification and prediction models for decision support, diagnosis, and prognosis [8].

The ROC curve is defined as a plot of True Positive Fraction (TPF), or sensitivity, and False Positive Fraction (FPF), 1- specificity, pairs obtained by varying threshold c. Defining Y_D and $Y_{\bar{D}}$ as continuous variables for diseased and non-diseased groups respectively with cumulative distribution functions F_D and $F_{\bar{D}}$, we assume all test outcomes greater than c belong to the diseased group, with $c \in \mathbb{R}$. Subsequently, each given c will determine TPF, $TPF(c) = Pr(Y_D \geq c) = 1 - F_D(c)$ and similarly FPF, $FPF(c) = Pr(Y_{\bar{D}} \geq c) = 1 - F_{\bar{D}}(c)$.

The ROC curve is then defined as all TPF-FPF pairs, $ROC(\cdot) = \{(FPF(c), TPF(c)), c \in \mathbb{R}\}$ [13]. By converting FPF at threshold c to t, such as, $t = FPF(c) = 1 - F_{\bar{D}}(c)$, the ROC curve is defined as $\{(t, ROC(t)) : t \in [0, 1]\}$ [16], where

$$ROC(t) = Pr\{Y_D > F_{\bar{D}}^{-1}(1 - t)\} = 1 - F_D\{F_{\bar{D}}^{-1}(1 - t)\}, \quad 0 \leq t \leq 1 \qquad (1)$$

Clinical risk index for babies (CRIB) is a risk assessment tool used in neonatal intensive care units (NICUs) for infants born with less than 31 week gestation or 1500 g and lower birth weight [10, 12]. Along with the Score for Neonatal Acute Physiology (SNAP), these scoring systems and their updates have served as prediction tools more accurate to the previous weight or gestational age univariate predictors.

CRIB score uses six different variables obtained routinely during the first 12 hours of life, namely, birthweight, gestational age, the presence of congenital malformation (excluding inevitably lethal congenital malformations) and indices of physiological status [1] resulting in a score between 0 and 24.

Matching sensitivity and specificity, defining an acceptable trade-off between the two to better characterize test results is, while not its only application, the main focus of ROC curve studies, and CRIB scores have been explored in this regard [1, 2]. However in any diagnostics test a number of factors such as the disease characteristics and specimen features, might affect the marker and its accuracy. In statistics these unaccounted variables are labeled covariates. Incorporating covariate information in ROC analysis is imperative, otherwise biased or oversimplified conclusions might be made about the tests accuracy [7].

In recent years several different mathematical approaches have been introduced to better equate covariates in ROC curve methodology.

By denoting $\mathbf{X}_D$ and $\mathbf{X}_{\bar{D}}$ as diseased and non-diseased vector of covariates of interest and a covariate value $\mathbf{x}$, the covariate-specific ROC curve is built as

$$ROC(t|x) = Pr\{Y_D > F_{\bar{D}}^{-1}(1 - t|\mathbf{X}_{\bar{D}} = \mathbf{x})|\mathbf{X}_D = \mathbf{x}\} \qquad (2)$$

$$1 - F_D\{F_{\bar{D}}^{-1}(1 - t|\mathbf{X}_{\bar{D}} = \mathbf{x})|\mathbf{X}_D = \mathbf{x}\} \qquad (3)$$

The covariate specific ROC curve builds different curves and displays a different test accuracy for each value of $\mathbf{x}$ that, while helpful in understanding optimal and sub optimal populations for the test, lacks a global measurement for the covariate effect. The covariate adjusted ROC curve, AROC, was proposed to mend this issue [7], defined as,

$$AROC(t) = \int ROC(t|\mathbf{x})dH_D(\mathbf{x}) \tag{4}$$

where,

$$H_D(x) = Pr(\mathbf{X}_D \leq \mathbf{x}),$$

the cumulative distribution function of $\mathbf{X}_D$. Therefore the AROC is a weighted average of covariate specific ROC curves according to the distribution of the covariates in the diseased group [16]. Janes and Pepe [7] also demonstrated that the AROC curve can also be expressed as

$$AROC(t) = Pr\{Y_D > F_{\bar{D}}^{-1}(1 - t|\mathbf{X}_D)\} = Pr\{1 - F_{\bar{D}}(Y_D|\mathbf{X}_D)) \leq t\}, \tag{5}$$

proving that the AROC summarizes the covariate-specific performance of the test.

The AROC R-package developed by Rodriguez-Alvarez and Inacio de Carvalho [16] implements different methods of computing covariate information in ROC curve construction and two additional methods of calculating marginal/polled ROC curve. This package has additional value considering the lag on tool development on this sub-field.

This article will explore these different methods on a theoretical and practical basis using CRIB scores of a neonatal database from the Portuguese National Registry on low weight newborns between 2010 and 2012.

2   Methodology

The aim of this study is to explore the AROC package functionality and usability using real data noting its possible limitations and generating preliminary findings on CRIB data that will be explored in the future.

2.1   Dataset

As mentioned the dataset used is part of the Portuguese National Registry on low weight newborns between 2010 and 2012 and was made available for research purposes. The original data included possible confounding observations such as repeated id's and twins that were removed to ensure no unaccounted for variables.

After ensuring all id's were unique these were promptly removed along with any possible identifiable features to abide by EU anonymity and data protection standards. A random sample of 50% was taken resulting in a dataset with 1093 unique observations and 17 columns including gestational time in weeks, the mothers age, biological sex of the infant, CRIB score, survival and other possibly relevant covariates and used for the remainder of the study.

For replication purposes all relevant data was shared to Mendeley Data and is available at data.mendeley.com/datasets/jsmgcmfmdx [9].

2.2   AROC Package Method Exploration

The AROC package for covariate adjusted Receiver Operating Characteristic curve inference offers a comprehensive guide of use present in the CRAN repository. In the remainder of this section we will replicate the examples provided in the documentation with the CRIB dataset, compare the different methods and explore some of the theoretical details of each method.

We begin by importing the dataset, setting a seed for replication purposes and constructing the specnaomit() function to remove NAs on columns under analysis, this is a necessary step since the methods are not built to handle NA values.

```
library(AROC)
set.seed(0824)
neonatal <- read.csv('NeonatalPortugal.csv',
                     sep = ';')

specnaomit <- function(col1, col2, df) {
  newdf <- df

  if (any(is.na(newdf[[col1]]))) {
    newdf <- newdf[!is.na(newdf[[col1]]),]

  }
  if (any(is.na(newdf[[col2]]))) {
    newdf <- newdf[!is.na(newdf[[col2]]),]

  }

  newdf

}

# Crib with Sex as covariate
neonatalCribSex <- specnaomit('sex',
                              'crib',
                              neonatal)

# Crib with Mother's age as covariate
neonatalCribAM <- specnaomit('age_mother',
                             'crib',
                             neonatal)
```

Nonparametric Bayesian. The *AROC.bnp* function estimates the AROC curve using the Bayesian nonparametric method. Working with Eq. 5 we say,

$$AROC(t) = Pr\{Y_D > F_{\bar{D}}^{-1}(1 - t|\mathbf{X}_D)\}$$
$$= Pr\{1 - F_{\bar{D}}(Y_D|\mathbf{X}_D)) \le t\}$$
$$= Pr(U_D \le t), \quad 0 \le t \le 1, \tag{6}$$

where $U_D = 1 - F_{\bar{D}}(Y_D|\mathbf{X}_D)$, a placement value of the test outcome in the diseased population or the standardization of Y_D to the conditional distribution of $Y_{\bar{D}}$ making the AROC a cumulative distribution function of U_D. This method first models the conditional distribution of test outcomes in the nondiseased group, $F_{\bar{D}}$ using a B-splines dependent Dirichlet process mixture of Normals model followed by modeling U_D and it's cumulative distribution using a non parametric regression model through Bayesian bootstrap [16].

For this demonstration we attempt to see the relation between the CRIB score and the sex of the baby, a categorical variable, that could indicate a possible bias on estimating survivability based on sex in the score system. To use this method the only mandatory arguments are *formula.healthy*, an R formula object to define the relation between variables, in this case the 'crib' and 'sex', *group*, that defines the result column, in this case if the infant survived, *tag.healthy*, what identifies the healthy i.e. survival population in the group/result column and *data*, the dataset being used. Further customization is possible with the Bayesian parameters however this will not be explored, the only extra arguments were meant to reduce the default run time.

```
#------ bnp method ------#

bnpCribSex <-
  AROC.bnp(formula.healthy ='crib ~ sex',
                      group = 'status',
                      tag.healthy = 0,
                      data = neonatalCribSex,
                      nsim = 5000,
                      nburn = 1000)

summary(bnpCribSex)

plot(bnpCribSex)
```

The plot functionality allows us to see the computed curve (Fig. 2A) that follow a simplistic visual however the saved object preserves the curve points that allow for any plot customization R programmers use.

Semiparametric Bayesian. The *AROC.bsp* method is very similar to the previous, non parametric method, in both construction and theory, where this models $F_{\bar{D}}$ using a normal linear regression, making it a counterpart to Janes and Pepe [7] frequentist model and its AROC package method - AROC.sp [14].

```
#------ bsp method ------#

bspCribSex <-
  AROC.bsp(formula.healthy ='crib ~ sex',
           group = 'status',
           tag.healthy = 0,
           data = neonatalCribSex)

summary(bspCribSex)

plot(bspCribSex)
```

The object generated by this method can, again, be plotted (Fig. 2B) similarly to the previous method

Semiparametric Frequentist. This semiparametric frequencist method arguments and construction are fairly similar to the previous methods but, as previously hinted, uses a semiparametric location regression model for $Y_{\bar{D}}$ to estimate $F_{\bar{D}}$ and estimates outer probability empirically [6,14] making it less computationally heavy, i.e. faster in comparison. Another advantage is being able to provide direct insight on covariate influence over the test/marker with the fit model's parameters to the crib and sex example we've seen previously.

```
spCribSex <-
  AROC.sp(formula.healthy ='crib ~ sex',
          group = 'status',
          tag.healthy = 0,
          data = neonatalCribSex)

summary(spCribSex)
summary(spCribSex$fit.h)

plot(spCribSex)
```

In this instance, calling the *summary()* function on the *fit.h* element give us an extensive look at test statistics.

Nonparametric Kernel. The *AROC.kernel*, an earlier proposed model by Rodriguez-Alvarez *et al.* [15] is, as the name implies, a kernel based method. Test outcomes for the non-diseased group are modeled with a location-scale regression model where both regression and variance functions are estimated using Nadaraa-Watson local estimators that in turn are used to compute standardised residuals to model U_D and estimate the AROC curve, $\widehat{AROC}(t)$ [14–16].

The package author notes that, for now, this method, unlike the previous, can only handle a single continuous covariate, for this reason the sex variable is unsuitable for testing and is replaced by the age of the mother, *age_mother* to analyze this covariates influence in the CRIB score system and its outcome. Additionally *AROC.sp* is used to offer additional information.

```
#------ kernel method ------#

kernelCribAM <-
   AROC.kernel(marker = 'crib',
               covariate = 'age_mother',
               group = 'status', tag.healthy = 0,
               data = neonatalCribAM,
               B = 10) #time

summary(kernelCribAM)

plot(kernelCribAM)

#------ sp method ------#

spCribAM <-
   AROC.sp(formula.healthy ='crib ~ age_mother',
           group = 'status',
           tag.healthy = 0,
           data = neonatalCribAM)

summary(spCribAM)
summary(spCribAM$fit.h)

plot(spCribAM)
```

Note that for the *AROC.kernel*, not just the default arguments are different from the remaining methods but also that it was necessary to shorten the bootstrap resample process (B) from its default 1000 value, this is due to the execution time of this method being the longest of all the AROC-package functions.

Posterior Predictive Checks (PPC). All previous methods are meant to construct and analyze the AROC curve and the behavior of the incorporated covariates, for the remainder of this section the methods focus on posterior predictive checks and pooled ROC estimation.

Both *predictive.checks.AROC.bnp* and *predictive.checks.AROC.bsp* are implementations of PPCs on their respective Bayesian based method. The premise behind PPCs is evaluating the generated model on how well it is able to generate data similar to the data observed [3] utilizing in this case the B-splines dependent Dirichlet process and Bayesian normal linear regression model for the *AROC.bnp* and *AROC.bsp* objects respectively [14]. To exemplify these methods we use the previously generated objects.

```
predictive.checks.AROC.bnp(bnpCribSex,
                           devnew = FALSE)
predictive.checks.AROC.bsp(bspCribSex,
                           devnew = FALSE)
```

While the mandatory argument is solely the AROC object, the *devnew* was changed from the default *TRUE* argument to display all depicted graphics on the same device. These graphics are histograms of the desired test statistics, by default minimum, maximum, median and skewness and Kernel density estimates showing diagnostic test outcome in the nondiseased group as well.

Pooled ROC. The package also provides two methods of polled ROC estimation, *pooledROC.emp* an implementation of empirical estimation proposed by Hsieh and Turnbull [5,14] and *pooledROC.BB* for Bayesian bootstrap estimation proposed by Gu et al. [4,14]. Both methods are similar to construct and have a similar output structure, mandatory arguments identify diseased and non-diseased groups in the test or marker column which can be achieved with a straightforward indexing in R.

```
# Pooled ROC methods

#------ BB method ------#
BBCrib <-
  pooledROC.BB(
    y0 = neonatal$crib[neonatal$status==0],
    y1 = neonatal$crib[neonatal$status==1])

plot(BBCrib)

#------ emp method ------#

EMPCrib <-
  pooledROC.emp(
    y0 = neonatal$crib[neonatal$status==0],
    y1 = neonatal$crib[neonatal$status==1])

plot(EMPCrib)
```

3   Results

All methods allowed for a visualization of their output with the standard *plot()* function, albeit a simple one, that allows for further customization if necessary.

The *summary()* output for AROC methods all follow a similar template, reporting the requested command with its respective arguments and the AAUC of the resulting AROC curve. While this command offers no additional information for the AROC objects when compared to the *plot()*, that also displays the AAUC, the semiparametric frequentist method generates a specific element, *fit.h* that can be accessed and provides a direct statistical summary on the effect of the covariate to the curve and the fitted regression model (Fig. 1).

```
> summary(spCribSex$fit.h)

Call:
lm(formula = formula.healthy, data = data.h)

Residuals:
    Min      1Q  Median      3Q     Max
-2.0019 -1.9603 -0.9603  0.9981 12.9981

Coefficients:
            Estimate Std. Error t value Pr(>|t|)
(Intercept)  1.96026    0.11918  16.447   <2e-16 ***
sexMale      0.04164    0.16267   0.256    0.798
---
Signif. codes:
0 '***' 0.001 '**' 0.01 '*' 0.05 '.' 0.1 ' ' 1

Residual standard error: 2.537 on 976 degrees of freedom
Multiple R-squared:  6.713e-05,  Adjusted R-squared:  -0.00
09574
F-statistic: 0.06553 on 1 and 976 DF,  p-value: 0.798
```

Fig. 1. The *fit.h summary()* output for semiparametric frequentist adjustment of sex to the CRIB score.

Methods used to adjust the sex covariate to the CRIB ROC curve all displayed AAUC intervals - available through both the *plot()* and *summary()* functions, and curves that overlap each other, inferring no statistical differences between them, an expected yet important attribute, making all methods viable for use in this instance despite a seemingly slight overestimation by the bayesian nonparametric method that holds no statistical relevance. These three curves can be seen in Fig. 2.

The kernel and semiparametric frequentist method shared the same similarities between them when adjusting for the mother's age in the CRIB ROC score, showing both overlapping AAUCs and curves that infer no statistical difference despite the low bootstrap resamples used in the kernel method in the interest of time. Both curves can be seen in Fig. 3. Having established that all AROC adjustment methods display statistically similar curves what follows is to establish the same result for the unadjusted curve methods and then to compare the two. Both bayesian and empirical pooled ROC functions, seen in Fig. 4 displayed similar results just as the adjusted methods, a direct comparison between the two is displayed in Fig. 5C. A comparison between adjusted and unadjusted ROC curves for both the sex of the infant and mother's age can be seen in Fig. 5 A and B respectively. While the AAUC statistics in Fig. 2 through Fig. 4 show a clear overlap between their respective adjustments and the pooled ROC curve and Fig. 5 show several intersections between the curves themselves, an additional source of information on the covariates' influence on their tests is present in the extended summary statistics of the $AROC.sp$ where p-value = 0.798 and p-value = 0.9721 is displayed for biological sex and mother's age respectively a figure well above the 0.05 required to entail any influence over the test for these covariates.

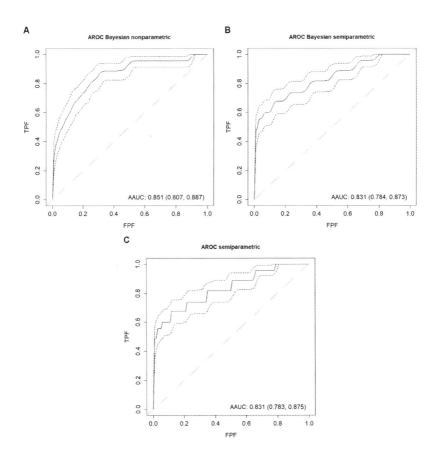

Fig. 2. AROC curve of CRIB score adjusted for the infants sex. **A**) AROC estimation with the nonparametric bayesian method. **B**) AROC estimation with the semiparametric bayesian method. **C**) AROC estimation with the semiparametric bayesian frequentist method.

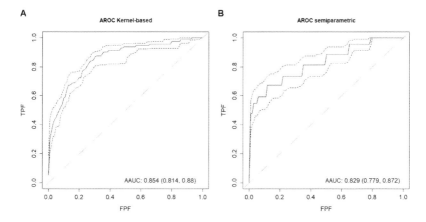

Fig. 3. AROC curve of CRIB score adjusted for the mother's age. **A**) AROC estimation with the kernel based method. **B**) AROC estimation with the semiparametric bayesian frequentist.

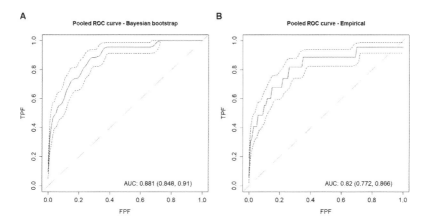

Fig. 4. Pooled ROC curves for CRIB score. **A**) Bayesian bootstrap. **B**) Empirical estimation.

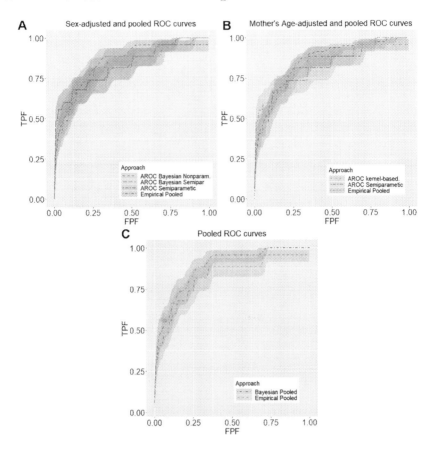

Fig. 5. Visual comparison between ROC and AROC curves. **A)** Comparison between AROC adjustment methods for biological sex and empirical pooled ROC curve. **B)** Comparison between AROC adjustment methods for mother's age and empirical pooled ROC curve. **C)** Visual Comparison between pooled ROC estimation methods

4   Discussion

All AROC estimation methods with the exception of AROC.kernel were capable of handling a categorical covariate rather than the standard continuous covariate. The visualization of the curves Fig. 2 through 5 provide insight on the behavior of the test and its associated covariate.

In efficiency, AROC.sp was the faster and most informative of the methods providing not only a visual representation of the AROC curve, but also descriptive test statistics inferring no statistical relevance to biological sex in the CRIB score that will be explored in the future.

Posterior predictive checks (PPCs) for non parametric bayesian method and for semiparametric bayesian method offer insight on their respective methods behavior and reliability that can lead to the choice of the most reliable method between the two despite, as mentioned, in this case, no real difference being found between methods.

Pooled ROC methods deliver curves with small differences on AUC values due to the nature of their estimators however from AUC alone we can observe the AUC confidence intervals intercepting showing both methods deliver a statistical equivalent curve allowing analysts to choose a preferred method. We can expand this reasoning and see same applies to the all AROC curves further substantiating the AROC.sp statistics.

CRIB scores and biological sex show no clear relation for all methods tested, these findings while preliminary and supported by previous research results, [10,17] are in stark contrast with criticism to the scoring system that lead to the creation of the CRIB II [12]. This data will be further explored to analyze the confounding effect, if any, of this covariate.

5   Conclusion

The AROC package grants multiple methods of analysis and estimation of the AROC curve complemented with predictive checks and pooled ROC estimation. With the exception of the kernel based method, *AROC.kernel*, all AROC estimation methods have proven able to handle both categorical and continuous covariates and are even able to handle several sets of covariates of different natures. All methods generate AROC curves with no statistical differences, passing the choice of which method would be more suitable for the analysis directly to the analyst.

The reliability of this package will allow further research for possible covariates confounding CRIB score.

Overall the package finally brings to the R library a way to conduct a more thorough ROC and AROC analysis making it available for any R user.

Acknowledgements. This work has been supported by FCT - Fundação para a Ciência e Tecnologia within the R&D Units Project Scope: UIDB/00319/2020.

The authors express their gratitude to the Portuguese National Registry for supplying the dataset used in this study.

References

1. Brito, A.S.J.D., Matsuo, T., Gonzalez, M.R.C., de Carvalho, A.B.R., Ferrari, L.S.L.: CRIB score, birth weight and gestational age in neonatal mortality risk evaluation. Revista de saúde pública **37**(5), 597–602 (2003). http://www.ncbi.nlm. nih.gov/pubmed/14569335
2. Ezz- Eldin, Z.M., Abdel Hamid, T.A., Labib Youssef, M.R., Nabil, H.E.D.: Clinical risk index for babies (CRIB II) scoring system in prediction of mortality in premature babies. J. Clin. Diagnostic Res. **9**(6), SC08–SC11 (2015). https://doi. org/10.7860/JCDR/2015/12248.6012

3. Gabry, J., Simpson, D., Vehtari, A., Betancourt, M., Gelman, A.: Visualization in Bayesian workflow. J. Royal Stat. Soc. Ser. A: Stat. Soc. **182**(2), 389–402 (2019). https://doi.org/10.1111/rssa.12378

4. Gu, J., Ghosal, S., Roy, A.: Bayesian bootstrap estimation of ROC curve. Stat. Med. **27**(26), 5407–5420 (2008). https://doi.org/10.1002/sim.3366, http://doi.wiley.com/10.1002/sim.3366

5. Hsieh, F., Turnbull, B.W.: Nonparametric and semiparametric estimation of the receiver operating characteristic curve. Ann. Statist. **24**(1), 25–40 (1996). https://doi.org/10.1214/aos/1033066197

6. Janes, H., Pepe, M.S.: Adjusting for covariates in studies of diagnostic, screening, or prognostic markers: an old concept in a new setting. Am. J. Epidemiol. **168**(1), 89–97 (2008). https://doi.org/10.1093/aje/kwn099

7. Janes, H., Pepe, M.S.: Adjusting for covariate effects on classification accuracy using the covariate-adjusted receiver operating characteristic curve. Biometrika **96**(2), 371–382 (2009). https://doi.org/10.1093/biomet/asp002

8. Lasko, T.A., Bhagwat, J.G., Zou, K.H., Ohno-Machado, L.: The use of receiver operating characteristic curves in biomedical informatics **38**, 404–415 (2005). https://doi.org/10.1016/j.jbi.2005.02.008

9. Machado, E., Costa, F., Braga, A.C.: Neonatalportugal. Mendeley Data (2019). https://doi.org/10.17632/jsmgcmfmdx.1

10. Mourão, M.F., Braga, A.C., Oliveira, P.N.: CRIB conditional on gender: nonparametric ROC curve. Int. J. Health Care Quality Assurance **27**(8), 656–663 (2014). https://doi.org/10.1108/IJHCQA-04-2013-0047

11. Park, S.H., Goo, J.M., Jo, C.H.: Receiver operating characteristic (ROC) curve: practical review for radiologists. Korean J. Radiol. **5**(1), 11 (2004). https://doi.org/10.3348/kjr.2004.5.1.11

12. Parry, G., Tucker, J., Tarnow-Mordi, W.O.: UK neonatal staffing study collaborative: CRIB II : an update of the clinical risk index for babies score For personal use. Only reproduce with permission from The Lancet Publishing Group. Lancet, **361** 1789–1791 (2003). https://doi.org/10.1016/S0140-6736(03)13397-1

13. Pepe, M.S.: The Statistical Evaluation of Medical Tests for Classification and Prediction (2003)

14. Rodriguez-Alvarez, M.X., Inacio de Carvalho, V.: AROC: Covariate-Adjusted Receiver Operating Characteristic Curve Inference (2018). https://CRAN.R-project.org/package=AROC, r package version 1.0

15. Rodríguez-Álvarez, M.X., Roca-Pardiñas, J., Cadarso-Suárez, C.: ROC curve and covariates: extending induced methodology to the non-parametric framework. Stat. Comput. **21**(4), 483–499 (2011). https://doi.org/10.1007/s11222-010-9184-1

16. Rodríguez-Álvarez, M.X., Roca-Pardiñas, J., Cadarso-Suárez, C., Tahoces, P.G.: Bootstrap-based procedures for inference in nonparametric receiver-operating characteristic curve regression analysis. Stat. Methods Med. Res. **27**(3), 740–764 (2018). https://doi.org/10.1177/0962280217742542

17. Terzic, S., Heljić, S.: Assessing mortality risk in very low birth weight infants. Medicinski arhiv **66**, 76–9 (2012). https://doi.org/10.5455/medarh.2012.66.76-79

ROSY Application for Selecting R Packages that Perform ROC Analysis

José Pedro Quintas[1](✉)[ID], Francisco Machado e Costa[1][ID],
and Ana Cristina Braga[2][ID]

[1] School of Engineering, University of Minho, Campus Gualtar, Braga, Portugal
pquintasbcl@gmail.com, frmachadoecosta@gmail.com
[2] ALGORITMI Centre, University of Minho, Campus de Azurem,
Guimarães, Portugal
acb@dps.uminho.pt

Abstract. The empirical ROC curve is a powerful statistical tool to evaluate the precision of tests in several fields of study. This is a two-dimensional plot where the horizontal and vertical axis represent false positive and true positive fraction respectively, also referred to as 1-specificity and sensitivity, where precision is evaluated through a summary index, the area under the curve (AUC). Several computer tools are used to perform this analysis one of which is the R environment, this is an open source and free to use environment that allows the creation of different packages designed to perform the same tasks in distinct ways often resulting in different customization and features often providing similar results. There is a need to explore these different packages to provide an experienced user with the simplest and most robust execution of a needed analysis. This work catalogued the different R packages capable of ROC analysis exploring their performance. A shiny web application is presented that serves as a repository allowing for efficient use of all of these packages.

Keywords: Shiny application · ROC curve · R tools · Packages

1 Introduction

ROC curves are considered one of the best tools to evaluate performance in statistical tests in different areas such as machine learning, data mining, medical diagnosis, experimental psychology, economics and sociology among others. These curves are represented in a bi-dimensional plane where the x- and y-axis represent false positive and true positive fraction respectively, also referred to as 1-specificity and sensitivity, where precision is given by the area under the curve [1].

The current volumes of generated and stored data demand an efficient and swift use leading to the development of software tools in the ROC curve area which in turn popularized the method further. The ubiquitous nature of the

© Springer Nature Switzerland AG 2020
O. Gervasi et al. (Eds.): ICCSA 2020, LNCS 12251, pp. 199–213, 2020.
https://doi.org/10.1007/978-3-030-58808-3_16

ROC curve, being used by such diverse areas, combined with the open source, free to use nature of the R environment allowed the development of several packages, that while performing the same basic analysis present several different and specific features according to the needs of the user, necessitating both quality assessment and cataloguing.

In the field of tool comparison for ROC curves it would be remiss not to mention the work of Stephan *et al.* [16], where eight different computer programs were analysed and compared. While an older article of 2003, making the presented results outdated, its methodology heavily influenced this project. The authors' goal was to select, compare, and evaluate different software using five distinct metrics for the task: data input, data output, program comfort, quality of the user manual and quality of the analysis results. All eight programs were analysed individually by each author receiving a grade for each criterion, conditional on performance.

In a different article by Robin *et al.* [13] the package *pROC* is explored, highlighting some of its features such as, smoothing, pauc (partial area under the curve), confidence intervals (CI), visual output with confidence intervals, statistical tests and CRAN availability, comparing it to *ROCR*, *verification*, *ROC* (Bioconductor) and *pcvsuite*. Considering the author of the paper is one of the package's main creators and current maintainer it is understandable that the focus is on each individual function of the *pROC* rather than the comparison with other packages but it provides useful methodology and information for the current work.

In a more recent study, Cunha and Braga [8] compare the packages *pROC*, *ROCR*, *verification*, *caTools*, *Comp2ROC*, *ROC* (Bioconductor) and *Epi*. In addition to providing a comprehensive guide of each command for all packages, it provides a comprehensive table for each package score regarding different metrics: CRAN availability, ROC processing, ROC smoothing, AUC availability, AUC availability in graphical output and CI intervals on curve comparison. This article's structure and result presentation were a great contribution to the methodology employed in this work.

The Shiny R package allows users the creation of interactive web-based applications. Although a recent addition to the R repository the package and its applications have provided an excellent resource for users to work with complex R packages or perform data analysis through a visual framework requiring only an internet connection. An example of such an application is one created by Wojciechowski, Hopkins and Upton [18] allowing the creation and sharing of pharmaceutical simulations models going as far as granting high school students a study resource for simulated variability of Ibuprofen. Jimmy Doi *et al.* [9] notes the ease of use and of application creation in shiny, emphasising its potential role in education and the ability to create user friendly, interactive and visually stimulating applications, comparing it to Java applets utilizing the R language.

Ebrahim Jahanshiri and Abdul Shariffna [10], using the package in the field of precision agriculture for mapping and data analysis, go even further and mention both R and shiny as a perfect substitute for python in the statistical analysis field.

The common praise for the shiny package in articles of such different fields with its ease of use and versatility, allowing the construction of complex yet appealing applications, served as a motivating factor to this application framework for this ROC curve and ROC package study.

2 Materials and Methods

Several R packages were chosen for comparison in accordance with the referenced papers. Working with the same dataset and criteria these tools were tested and compared. A generic table of contents was produced to enumerate features either present or missing between packages so as to readily inform the user on notable content some packages might have implemented over others.

These packages were implemented into a shiny application allowing the user to interactively perform ROC analysis with different R packages. Observing their differences and distinctiveness, user can choose the more appropriate tool for a specific task based on the checklist represented on application, desired visual output and his goal.

For replication purposes the dataset was shared to Mendeley Data and is available at https://data.mendeley.com/datasets/zx4r8mgn86/1.

2.1 R Packages and Criteria

Appropriate packages were tested, seven of which were selected for the remainder of this study as can be seen in the list below

- *pROC* [13]
- *Comp2ROC* [3]
- *ROCR* [15]
- *ROSE* [11]
- *Epi* [6]
- *ROC* - Bioconductor [5]
- *caTools* [17]

Eleven objective criteria were selected based on the works of Stephan *et al.* [16] and Cunha and Braga [8] to compare features and performance of these seven packages,

- Version
- Last update
- Empirical Curve
- AUC (Area under the curve)
- More than one curve on a chart
- Compare curves
- CI (Confidence Intervals)
- Smooth curve
- Works with inverted scales

- pAUC (Partial AUC)
- Print only one ROC curve

Version and **Last update** are meant as general information regarding the maintenance and frequency of update of a given package while **Empirical Curve** and **AUC**, referencing the ability to create the ROC curve and calculate its respective AUC, are fundamental characteristics necessary to allow a standard ROC study. In ROC tests it is often important a visual representation of different ROC curves using in this work a criterion called **More than one curve on a chart**. This is closely linked with the **Compare curves**, this criterion means ability to compare two (or more) different curves to infer their statistical differences. These were separated because a package might allow visual but not statistical comparison and vice-versa. The **CI** detail the ability of the package to produce CIs in its analysis, for example in the AUC estimation. Smoothing or curve adjustment is often used whether statistically relevant or just to create a more appealing output, therefore the ability to do so was added to the criterion - **Smooth curve**. The **Works with inverted Scales** parameter details a niche but often needed and overlooked detail of many packages. This parameter refers to scales that work inversely in a dataset. In ROC curves analysis, the values of a scale are generally used where a higher score is associated with a positive event (1), but there are scales where the values work in reverse, i.e., a lower score may correspond to the positive event of the result variable, that is equal 1, for instance, in the case of the weight of the newborn, the higher value of the weight corresponding to the greater probability of the event surviving, this is a result equal to zero. However, this is not always the case and different tools might allow a user to work with these non-standard scales or demand a manual conversion of the dataset to fit standard parameters. Another often overlooked requirement is the **pAUC** estimation, referring to a AUC on a given interval of false positive rates, a partial AUC. Finally, the **Print only one ROC curve** details the ability of a package to exclusively work with an isolated ROC curve or a need to compare groups of ROC curves and is meant as a distinguishing factor between packages focused on ROC comparison and the more generalist ROC packages.

After this selection process and package feature assessment, Table 1 was constructed to better guide users on the best packages to fit their needs on the shiny application.

2.2   Dataset

To perform a ROC analysis a dataset must include an indicator, or variable of interest and a binary outcome (negative/positive, healthy/disease, etc). A valid evaluation of all packages requires the use of the same sample of data. To this effect, two distinct paired and independent sample datasets were used.

Table 1. Packages checklist.

	pROC	ROSE	ROCR	Comp2ROC	Epi	ROC (Bioconductor)	caTools
Version	1.16.1	0.0–3	1.0–7	1.1.4	2.40	1.62.0	1.18.0
Last update	2020	2015	2015	2016	2019	2020	2020
Empirical curve	✓	✓	✓	✓	✓	✓	✓
AUC	✓	✓	✓	✓	✓	✓	✓
More than one curve on a chart	✓	✓	✓	✓	✗	✓	✗
Compare curves	✓	✗	✗	✓	✗	✗	✗
CI (Confidence Intervals)	✓	✗	✗	✓	✗	✗	✗
Smooth curve	✓	✗	✗	✗	✗	✗	✗
Works with inverted scales	✓	✓	✗	✓	✗	✗	✓
Partial AUC	✓	✗	✗	✗	✗	✓	✗
Print only one ROC curve	✓	✓	✓	✗	✓	✓	✓

The paired sample dataset, with a total of 169 observations, represents neonatal mortality, where five indicator variables, WEIGHT_AG, CRIB, NTISS, SNAP, SNAPPE access mortality risk and the resulting outcome is represented by the *result* column. Note that, unlike the remaining indicators, WEIGHT_AG is an inverted scale where higher values predict lower mortality. The independent dataset shows different sample sizes of CRIB score measured in four distinct hospitals, *Hospital1–4*, and their resulting mortality, *Res1–4* [4].

Figure 1 and Fig. 2 use the mentioned data format used to test the various packages.

⊿	A	B	C	D	E	F
1	WEIGHT_AG	CRIB	NTISS	SNAP	SNAPPE	result
2	8	6	19	15	15	0
3	9	1	13	22	62	0
4	5	5	22	19	29	0
5	10	0	7	3	3	0

Fig. 1. First rows of the paired sample dataset.

⊿	A	B	C	D	E	F	G	H
1	Hospital1	Res1	Hospital2	Res2	Hospital3	Res3	Hospital4	Res4
2	2	0	2	0	1	0	2	0
3	4	0	12	0	8	0	8	0
4	1	0	2	0	0	0	0	0
5	2	0	3	0	1	0	0	0

Fig. 2. First rows of the independent sample dataset.

2.3   Shiny

Shiny aims to create applications to either be run through a web browser or locally in an R environment. While this tool is mostly aimed at statisticians and data analysts as a way to share their findings, it is mostly available to all areas thanks to its basis on R, making it far more accessible [7].

The package is based around reactive programming, a programming paradigm that facilitates the automatic propagation of change of dataflows. A practical understanding of this concept is, while working with several inputs generating a specific output, be it plots, tables or text, any modification on the input will automatically generate and update the output without requiring a new command or refresh action on the user's side. Another user friendly implementation is the personalization of the shiny interface using widgets and code chunks promoted by RStudio and implemented across several other apps available on their archive [10, 14].

How to Get and Use Shiny. As mentioned, this package was created and made available by RStudio and the easiest way of using it is by installing the RStudio software where all dependencies are included and even a basic template is provided.

When it comes to using created shiny applications, an R interpreter is needed, of which RStudio is the standard. However, the second popular option is using the online platform *shinyapps.io* accessible through an internet browser.

How to Build a Shiny Application. To create the application two components are necessary, a web interface and a web server, as well as an update shiny. A web interface allows users to upload files and supply instructions required for R to calculate the desired output.

To design the structure, either two files, *server.R* and *ui.R* are required or a single *app.R* file, the latter option is merely the former's structure components organized in the same file.

For our purposes, the app must allow a ROC study to be conducted in accordance with user necessity after consulting the checklist for the desired features and selecting the appropriate package.

Shinyapp Flowchart. Initial development of the application saw the creation of a flowchart, that can be seen in Fig. 3, to list possible choices and pathways taken by the user so as to more easily structure the app's development.

The user will initially receive two sources of information regarding each package, the aforementioned checklist followed by each individual and in depth guide of each package.

Afterwards the data must be uploaded, this is the first fork in the flowchart given that *Comp2ROC* has a distinct process for data entry. Following the *Comp2ROC* pathway, the user must select whether the data details paired or independent samples and select the relevant result columns.

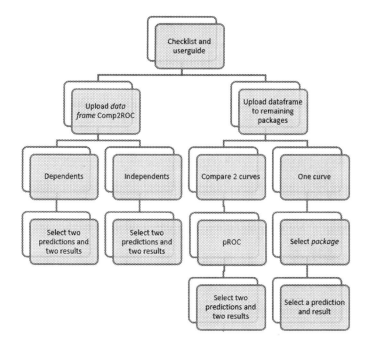

Fig. 3. Flowchart for shiny application creation.

For the remaining packages, data upload and standard ROC analysis are similar with the introduction of a single prediction and result. There is only one exception in the *pROC* package feature when comparing curves that behaves out similarly to *Comp2ROC*, where more than one indicator and result column may be necessary.

3   Results

The developed shiny application is available for use through *shinyapp.io* using the URL https://pquintasbcl.shinyapps.io/ROSY/.

The adopted name, ROSY, came from a mixture of ROC with Shiny through an acronym generator.

This section will detail how each tab of the application works exampling its use with the mentioned datasets. The tool is subject to updates that may result in changes to both visuals and outputs presented at the time of this article's publication.

3.1   Checklist and Userguide

As mentioned, the user is first greeted with the checklist presented in Table 1 implemented into the application as can be seen in Fig. 4, followed by the user guide tab, in Fig. 5 with official documentation for each of the seven packages provided by either CRAN or Bioconductor. Several notes are also provided with some miscellaneous information about each package or criteria to better aid the user in their choice.

Fig. 4. Checklist for R packages within the application.

The package *pROC* is clearly the most feature heavy of the packages however it is still important to provide access to different implementations of these function to ensure a diversity of options and a satisfactory outcome, either with a different visual output or with a swifter computation of results.

Fig. 5. Packages user guide.

3.2   Upload Dataset

Section 2.3 mentions the need to fork the file upload system due to specific requirements of the *Comp2ROC* package. In this section we will explore the upload systems the app uses for the majority of packages as well as the two systems used for *Comp2ROC*.

Firstly, we examine the generic upload system used for the remaining packages, this functionality uses on the standard shiny upload functions and is compatible with .csv and .txt files. The uploaded dataset is displayed on the side. An example of this display and tab functionality can be seen in Fig. 6.

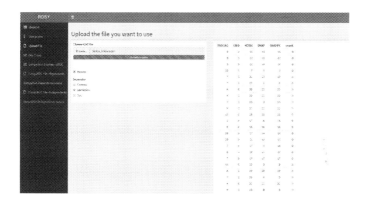

Fig. 6. File upload for all packages except *Comp2ROC*.

This upload function is fairly accessible, needing only the user to press the "Browse" button and search for the relevant files in their directory. After upload the user must specify whether the dataset has headers and how it separates values (comma, semicolon or tab) similarly to the *read.csv* function in R to ensure proper file readability. This tab is used for the *pROC, pROC, Epi, ROC*(Bioconductor) and *caTools* packages.

Dataset upload for *Comp2ROC* was separated between paired and independent samples, creating two distinct tabs for the process to allow simultaneous work on both sections if necessary. As mentioned the specific requirements of data input for this package made it necessary to ease and optimize this process from a user perspective. The code used for this process can be seen below.

```
data2 = reactive({
    req(input$file2)
    inFile2 = input$file2
    df2 = Comp2ROC::read.file(inFile2$datapath, header =
    input$header2 , sep = input$sep2 , ''.'', input$pred1 ,
    input$Direction1, input$pred2,input$Direction2, input$res
    ,T)
    return(df2)})

data3 = reactive({
    input$goPlot
    req(input$file3)
    inFile2 = input$file3
```

```
df3 = Comp2ROC::read.file(inFile2$datapath, header =
input$header3 , sep = input$sep3 , ''.'', input$pred3 ,
input$Direction3 , input$pred4 ,input$Direction4 , input$
res3 ,FALSE ,input$res4)
return(df3)})
```

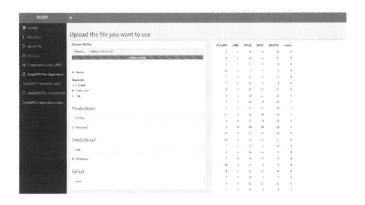

Fig. 7. Data loading for paired samples in *Comp2ROC*.

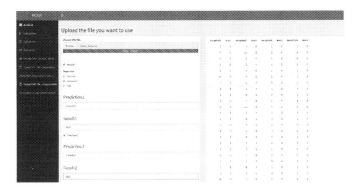

Fig. 8. Data loading for independent samples in *Comp2ROC*.

Much like the previous tab, data uploaded must be supplied with header and separator information. However this data entry must also directly be supplied with information on predictors and result columns. For paired samples these are two predictors for the same result while in independent samples each indicator must have a corresponding result. A "direction" checkbox is also supplied to allow inverted scales without manual inversion. A user must uncheck this box if working with an indicator that does not follow the standard ROC assumption that higher values on indicator corresponding to a negative result, such as the WEIGHT_AG variable in the paired samples dataset.

Figure 7 and Fig. 8 illustrate the use of paired and independent tabs with their respective datasets.

All uploaded data are inserted into reactive objects to allow manipulation in the next tabs of the ROSY application.

3.3 One Curve Section

The data uploaded with the generic system generates its own output in this tab. Outputs from the *Comp2ROC* package are separated to avoid confusion and will be discussed later.

The One Curve tab supplies several options for the user. First, a section of selection boxes asks the user which package should be used, these packages available of this list are, *pROC*, *ROSE*, *pROC*, *caTools*, *Epi* and *ROC* (Biocondutor). Afterwards the users is asked to identify the prediction and response columns (indicator and outcome, respectively), there are default selections already available that can be changed. With this information the app is already able to generate a ROC curve and calculate its AUC in the left-hand side of the screen. Additional options can be selected however, such as smoothing, available if using the *pROC* package, and pAUC selection and calculation, if using *pROC* or Bioconductor.

Finally, users have the possibility to add a second ROC curve to the generated plot. While not a statistical comparison such as in *Comp2ROC*, a visual comparison can be useful in different ROC analysis. This option is available to *ROSE,pROC*, Bioconductor and *pROC* packages. In a different tab *pROC*'s in depth curve comparison will be used.

In Fig. 9 the One Curve tab input options can be seen with their respective output present in Fig. 10.

Fig. 9. Visual appearance of the input values of the "one curve" tab.

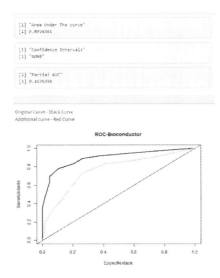

Fig. 10. Output example provided by application.

3.4 Comparison of Two Curves - *pROC*

Since *pROC* allows curve comparison with some statistical parameters a seperate tab was created to highlight this option - Comparison 2 curves - *pROC*.

This tab presents four separate select boxes for input where a user must define predictions and results for both curves, generating an output that represents both curves in the same plot, Z value, p value and AUC.

Figure 11 presents a possible use of this section of the app.

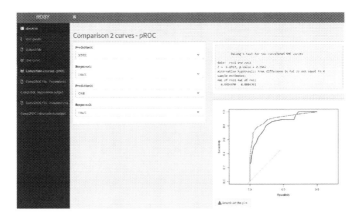

Fig. 11. Example of the output provided on the "Comparison 2 curves - *pROC*" tab.

3.5 Output Comp2ROC - Dependent and Independent Samples

Much like the data input, output for *Comp2ROC*'s analysis is separated in different tabs. After correct data input, "Comp2ROC Dependent output" needs only a press of a button, to display this section's output. Unlike the two previous sections, this tab, due to data entry constraints requires that any change to the inputs on data input must be recalculated with a manual button press, this process takes roughly 30 s to compile all calculations and display them. An example of this output can bee seen in Fig. 12

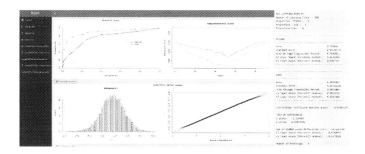

Fig. 12. Example output of roc analysis returned by *Comp2ROC* package on dependent samples tab.

According to the methodology developed by Braga et al. [2] this package is entirely focused on ROC curve comparison therefore, besides the typical plot, provides AUC value comparison, histogram of t and quantiles of standard normal as well as analysis statistics to better compare the two indicators and their performance. These plots can be downloaded using the "Download the plots" button after analysis.

For independent samples, output follows the same format, the only difference being the already discussed input parameters, as can be seen in Fig. 13.

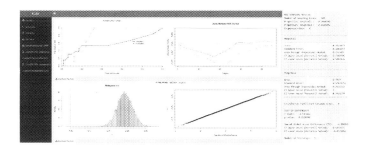

Fig. 13. Example output of roc analysis returned by *Comp2ROC* package on independent samples tab.

4   Conclusions

This work aimed to explore several available packages in R for ROC curve construction. The cataloguing features led to the creation of a checklist to better emphasise the differences between packages in a quick fashion with selected criteria.

To optimize tool selection for ROC analysis a shiny application was created, ROSY, available at https://pquintasbcl.shinyapps.io/ROSY/, giving users the freedom to choose the best package for their desired goal after consulting their features in the checklist and user guide menus.

The application allows any user to use R packages without prior knowledge of programming or scripting replacing command lines and, perhaps, intimidating environments with an easily consumable web based application with select menus and direct outputs.

5   Future Work

This article detailed the first release of the ROSY shiny application and is, as such, subject to change, be it visual improvements or potential bug fixes. Future developments may include an expanding checklist of criteria as well as new packages being implemented to widen the scope of the app and its user appeal.

Acknowledgments. This work has been supported by FCT - Fundação para a Ciência e Tecnologia within the R&D Units Project Scope: UIDB/00319/2020.

References

1. Braga, A.C.: Curvas ROC: Aspectos funcionais e aplicações, Ph.D. thesis, (Portuguese language) (2000)
2. Braga, A.C., Costa, L., Oliveira, P.: An alternative method for global and partial comparison of two diagnostic systems based on ROC curves. J. Stat. Comput. Simulat. **83**(2), 307–325 (2013)
3. Braga, A.C., Frade, H., Carvalho, S., Santiago, A.: Comp2ROC: Compare Two ROC Curves that Intersect. R package version 1.1.4. (2016). https://CRAN.R-project.org/package=Comp2ROC
4. Braga, A.C., Oliveira, P.: Diagnostic analysis based on ROC curves: theory and applications in medicine. Int. J. Health Care Quality Assurance, **16**(4), 191–198 (2003). https://doi.org/10.1108/09526860310479677
5. Carey, V., Enhancements, H.: ROC: utilities for ROC, with uarray focus. R package version 1.60.0. (2019). http://www.bioconductor.org
6. Carstensen, B., Plummer, M., Laara, E., Hills, M.: EPI: A Package for Statistical Analysis in Epidemiology. R package version 2.37 (2019). https://CRAN.R-project.org/package=Epi
7. Cho, W., Lim, Y., Lee, H., Varma, M., Lee, M., Choi, E.: Big data analysis with interactive visualization using r packages. BigDataScience **14**, 103 (2014)

8. Cunha, D., Braga, A.: Receiver Operating Characteristic (ROC) Packages Comparison in R, pp. 545–559 (2017). https://doi.org/10.1007/978-3-319-62395-5_37

9. Doi, J., Gailwong, J., Irvinchi, P.: Web application teaching tools for statistics using R and shiny. Technol. Innov. Stat. Educ. 9, 1–32 (2016)

10. Jahanshiri, E., Shariff, R.: Developing web-based data analysis tools for precision farming using R and shiny. IOP Conf. Ser.: Earth Environ. Sci. 20, 012014 (2014)

11. Lunardon, N., Menardi, G., Torelli, N.: ROSE: a package for binary imbalanced learning. R J. 6(1), 82–92 (2014)

12. Quintas, J.P., Machado e Costa, F.; Braga, A.C.: "Neonatal mortality indicators in hospitals", Mendeley Data, V1, (2020). https://doi.org/10.17632/zx4r8mgn86.1

13. Robin, X., et al.: pROC: an open-source package for R and S+ to analyze and compare ROC curves. BMC Bioinform. 12(1), 77 (2011). https://doi.org/10.1186/1471-2105-12-77

14. RStudio Team.: RStudio: Integrated Development for R. RStudio Inc, Boston, MA (2019). http://www.rstudio.com/

15. Sing, T., Sander, O., Beerenwinkel, N., Lengauer, T.: ROCR: visualizing classifier performance in R. Bioinformatics, 21(20), 7881 (2005). http://rocr.bioinf.mpi-sb.mpg.de

16. Stephan, C., Wesseling, S., Schink, T., Jung, K.: Comparison of eight computer programs for receiver-operating characteristic analysis. Clinical Chemistry 49, 433–9 (2003)

17. Tuszynski, J.: caTools: tools: moving window statistics, GIF, Base64, ROC AUC, etc. R package version 1.17.1.2. (2019). https://CRAN.R-project.org/package=caTools

18. Wojciechowski, J., Hopkins, A., Upton, R.: Interactive pharmacometric applications using R and the shiny package. CPT: Pharmacometrics Syst. Pharmacol. 4, 146–159 (2015)

PLS Visualization Using Biplots: An Application to Team Effectiveness

Alberto Silva[1,2] (ID), Isabel Dórdio Dimas[3,4(✉)] (ID),
Paulo Renato Lourenço[3,5] (ID), Teresa Rebelo[3,5] (ID),
and Adelaide Freitas[1,2] (ID)

[1] Departament of Mathematics, University of Aveiro, 3810-193 Aveiro, Portugal
{albertos, adelaide}@ua.pt
[2] CIDMA, Center for Research & Development in Mathematics
and Applications, University of Aveiro, 3810-193 Aveiro, Portugal
[3] CeBER, Centre for Business and Economics Research, University of Coimbra,
3004-512 Coimbra, Portugal
[4] FEUC, University of Coimbra, 3004-512 Coimbra, Portugal
idimas@fe.uc.pt
[5] FPCEUC, University of Coimbra, 3000-115 Coimbra, Portugal
{prenato, terebelo}@fpce.uc.pt

Abstract. Based on a factorization provided by the Partial Least Square (PLS) methodology, the construction of a biplot for both exploratory and predictive purposes was shown to visually identify patterns among response and explanatory variables in the same graph. An application on a team effectiveness research, collected from 82 teams from 57 Portuguese companies and their respective leaders, containing two effectiveness criteria (team performance and the quality of the group experience as response variables), was considered and interpretation of the biplot was analyzed in detail. Team effectiveness was considered as the result of the role played by thirteen variables: team trust (two dimensions), team psychological capital (four dimensions), collective behavior, transformational leadership, intragroup conflict (two dimensions), team psychological safety, and team cohesion (two dimensions). Results revealed that the biplot approach proposed was able to capture the most critical variables for the model and correctly assigned the signals and the strength of the regression coefficients. Regarding the response variable team performance, the most significant variables to the model were team efficacy, team optimism, and team psychological safety. Concerning the response variable quality of the group experience, intragroup conflict, team-trust, and team cohesion emerged as the most relevant predictors. Overall, the results found are convergent with the literature on team effectiveness.

Keywords: Partial least square · Biplot · Organizational teams · Team effectiveness

© Springer Nature Switzerland AG 2020
O. Gervasi et al. (Eds.): ICCSA 2020, LNCS 12251, pp. 214–230, 2020.
https://doi.org/10.1007/978-3-030-58808-3_17

1 Introduction

Frequently, multivariate data analysis seeks to perceive the existing underlying structure and to understand the relationships established within data. Visual information via graphic displays can be a useful tool to explore the dataset since it summarizes the data more directly and improves its understanding (Koch 2014). Likewise, a graph of the results of a specific statistical method, e.g., the Principal Component Analysis (PCA) biplot, enhances data familiarity. The biplot method permits visual evaluation of the structure of large data matrices through the approximation of a high-rank matrix by one of rank two. The PCA biplot represents observations with points and variables with arrows. Small distances between units can indicate the existence of clusters, while the size of an arrow depicts the standard deviation of the associated variable. Further, the angle between two vectors approximates the linear correlation of the related variables (Gabriel 1971).

When it comes to multivariate regression problems, sometimes one must fix some problems before applying any methodology to estimate parameters and thinking about the graphical representation of its results. This is the case of an ill-posed problem, in which the predictors are many and quasi-collinear, leading to an unstable Ordinary Least Squares (OLS) solution, i.e., the OLS estimates have high variance (Belsley et al. 2004). Under this condition, the Partial Least Squares (PLS) regression gives better results, since it eliminates the quasi-collinearity issue. The PLS method extracts factors that maximize the covariance between the predictors and response variables, and then regresses the response on these latent factors. Based on the outputs of the PLS (scores, loadings, and weights vectors), the variances and correlations of the variables can be revealed by employing an *exploratory PLS biplot*. On the other hand, the PLS biplot can be adapted to provide a visual approximation of the PLS coefficient estimates, hence the reason for naming it *predictive PLS biplot*.

The primary purpose of this article is to provide a straightforward interpretation for the PLS biplot applicable to both exploratory and predictive purposes, illustrating its application in team effectiveness research data. Interest in understanding complex relationships between variables of team effectiveness datasets has been growing in recent years (Mathieu et al. 2019; Ringle et al. 2018) and the PLS biplot method can play a crucial role in the analysis of this kind of data.

In order to achieve the main aim of the present work, the paper is structured in the following sections: Sect. 2 gives a brief overview of how PLS works, describing mathematical details; Sect. 3 presents an application of these methods on a subset of variables of real work teams, exploring the relationships between a set of team effectiveness predictors (team trust, team psychological capital, collective behavior, transformational leadership, intragroup conflict, team psychological safety and team cohesion) and two team effectiveness criteria (team performance and quality of group experience). All the statistical analysis was executed using R software; finally, Sect. 4 includes the discussion of the results found, as well as conclusions and future perspectives.

2 Methods

2.1 Partial Least Squares

Assume a multivariate regression model $\mathbf{Y} = \mathbf{XB} + \mathbf{E}$, in which $\mathbf{Y}$ is an (n × q) response matrix and $\mathbf{X}$ is a (n × m) predictor matrix, and both are column centered, with m and q being respectively the number of predictors and response variables, and n the number of observations. Also, $\mathbf{B}$ is a (m × q) coefficients matrix, and $\mathbf{E}$ is a (n × q) error matrix, such that m > n or the m explanatory variables are highly correlated. In this case, one might use the PLS to estimate the regression coefficients. The PLS model consists of three other models, two external and one internal, as a result of the application of a suitable algorithm, usually the Nonlinear Iterative Partial Least Squares (NIPALS). The method seeks to estimate some underlying factors that decompose $\mathbf{X}$ and $\mathbf{Y}$ simultaneously, maximizing the covariance between them, establishing the so-called outer relations for $\mathbf{X}$ and $\mathbf{Y}$ individually (Geladi and Kowalsky 1986). Considering the extraction of all possible factors, the PLS decomposition results in

$$\mathbf{X} = \mathbf{TP}' \text{ and } \mathbf{Y} = \mathbf{UQ}',$$

where $\mathbf{T}$ contains the scores of the predictors' matrix, $\mathbf{P}$ holds the loadings of $\mathbf{X}$. In turn, $\mathbf{U}$ and $\mathbf{Q}$ are the matrices of scores and loadings relative to the response matrix $\mathbf{Y}$. Additionally, an inner relation links the $\mathbf{X}$-scores and $\mathbf{Y}$-scores matrices as follows:

$$\widehat{\mathbf{u}}_i = a_i \mathbf{t}_i,$$

where

$$a_i = \frac{\mathbf{u}_i' \mathbf{t}_i}{\mathbf{t}_i' \mathbf{t}_i}$$

are the regression coefficients for a given factor. In order to ensure maximum covariance between $\mathbf{Y}$ and $\mathbf{X}$ when extracting PLS components, it is necessary to find two sets of weights $\mathbf{w}$ and $\mathbf{q}$, which allow the vectors $\mathbf{t} = \mathbf{Xw}$ and $\mathbf{u} = \mathbf{Yq}$ to be obtained. It can be done making $\mathbf{t'u}$ maximum and solving the optimization problem

$$\text{argmax}_{w,q}\{\mathbf{w'X'Yq}\},$$
$$\text{subject to: COR } (t_i, t_j) = 0, \forall i \neq j;$$
$$\mathbf{w'w} = 1.$$

2.2 Partial Least Squares Regression

Concisely, the NIPALS algorithm[1] performs the following steps (Abdi 2010):

---

[1] In this context, the symbol $\propto$ means 'to normalize the result of the operation'.

- Step 1. $\mathbf{w} \propto \mathbf{X'u}$ (X-weights).
- Step 2. $\mathbf{t} \propto \mathbf{Xw}$ (X-factor scores).
- Step 3. $\mathbf{q} \propto \mathbf{Y't}$ (Y-weights).
- Step 4. $\mathbf{u} = \mathbf{Yq}$ (Y-scores).

At the i-th iteration of the algorithm, the PLS method estimates a single column $\mathbf{t}_i$ of the matrix $\mathbf{T}$ as a linear combination of the variables X with coefficients $\mathbf{w}$. This vector of weights $\mathbf{w}$ will compose the i-th column of the matrix of weights $\mathbf{W}$. Since in each iteration the matrix $\mathbf{X}$ is deflated, the columns of $\mathbf{W}$ are non-comparable and, hence, $\mathbf{T} \neq \mathbf{XW}$. In contrast, there exists a matrix $\mathbf{R} = \mathbf{W(P'W)}^{-1}$ which allows direct computation of $\mathbf{T}$ by doing $\mathbf{T} = \mathbf{XR}$ (Wold et al. 2004).

The estimated PLS regression equation is:

$$\widehat{\mathbf{Y}} = \mathbf{T}\widehat{\mathbf{B}}, \text{ where } \widehat{\mathbf{B}} = (\mathbf{T'T})^{-1}\mathbf{T'Y}.$$

Moreover, $\widehat{\mathbf{Y}} = \mathbf{XR}\widehat{\mathbf{B}}$ and, thus, $\widehat{\mathbf{B}}_{\text{PLS}} = \mathbf{R}\widehat{\mathbf{B}} = \mathbf{R(T'T)}^{-1}\mathbf{T'Y} = \mathbf{RT'Y} = \mathbf{RQ'}$. Notice that $\mathbf{Q'}$ is the Y-weights matrix composed of the $\mathbf{q}$ vectors estimated in Step 3 of the NIPALS algorithm. Lastly, we can write the predictive model as

$$\widehat{\mathbf{Y}} = \mathbf{X}\widehat{\mathbf{B}}_{\text{PLS}}, \text{ where } \widehat{\mathbf{B}}_{\text{PLS}} = \mathbf{RQ'}.$$

2.3   The Biplot

The term biplot was introduced by Gabriel (1971) and consists of a graphical representation that reveals important characteristics of data structure, e.g., patterns of correlations between variables or similarities between the observations (Greenacre 2010). To achieve this, it uses the decomposition of a $(n \times m)$ target matrix $\mathbf{D}$ into the product of two matrices, such that $\mathbf{D} = \mathbf{GH'}$. The dimension of the $\mathbf{G}$ matrix is $(n \times k)$, and the size of $\mathbf{H}$ matrix is $(m \times k)$. Therefore, each element d_{ij} of the matrix $\mathbf{D}$ can be written as the scalar product of the *i-th* row of the left matrix $\mathbf{G}$ and the *j-th* column of the right matrix $\mathbf{H'}$, as follows:

$$\mathbf{D} = \mathbf{GH'} = \begin{pmatrix} \mathbf{g'_1} \\ \vdots \\ \mathbf{g'_n} \end{pmatrix} (\mathbf{h_1} \quad \dots \quad \mathbf{h_m}) = \begin{pmatrix} \mathbf{g'_1 h_1} & \cdots & \mathbf{g'_1 h_m} \\ \vdots & \ddots & \vdots \\ \mathbf{g'_n h_1} & \cdots & \mathbf{g'_n h_m} \end{pmatrix}.$$

The matrices $\mathbf{G}$ and $\mathbf{H}$ that arise from the decomposition of $\mathbf{D}$ create two sets of points. If these points are two-dimensional (i.e., $k = 2$), then the rows and columns of $\mathbf{D}$ can be represented employing a two-dimensional graph, with the n rows of $\mathbf{G}$ represented by points, and the m columns of $\mathbf{H'}$ reproduced in the form of vectors connected to the origin. In the graph, projecting $\mathbf{g'_i}$ onto the axis determined by $\mathbf{h_j}$ and then multiplying the norm of that projection by the norm of $\mathbf{h_j}$, the result will be equivalent to the geometric definition of the scalar product, which can also be used to represent the element d_{ij} of the target matrix $\mathbf{D}$, that is:

$$d_{ij} = \mathbf{g}'_i \mathbf{h}_j = \|\mathbf{g}_i\| \|\mathbf{h}_j\| \cos\theta,$$

where θ is the angle formed by the vectors $\mathbf{g}_i$ and $\mathbf{h}_j$. Furthermore, each set of coordinates formed by a row of $\mathbf{G}$ (i.e., $\mathbf{g}'_i$) is represented as a *biplot point*, and each column of the transpose of $\mathbf{H}$ (i.e., $\mathbf{h}_j$) is plotted as a *biplot vector*.

2.4   The Exploratory PLS Biplot

Given a rank r data matrix, the PLS allows another matrix to be obtained with rank s that is an approximation of the former, in which $s < r$. The PLS dataset is composed of two centered matrices $\mathbf{X}$ and $\mathbf{Y}$, wherein the matrix of predictors $\mathbf{X}$ has the dimension $(n \times m)$, and the matrix of responses $\mathbf{Y}$ has the size $(n \times q)$. Representing a target matrix $\mathbf{D}$ as a juxtaposition of $\mathbf{X}$ and $\mathbf{Y}$, then it will be $(n \times (m + q))$ and denoted as $\mathbf{D} = [\mathbf{X}\ \mathbf{Y}]$. Considering that the number of PLS components extracted is lower than the rank of $\mathbf{X}$, i.e., $k < r$, thus the matrix product $\mathbf{TP}'$ provides an approximation of $\mathbf{X}$. Similarly, the matrix product $\mathbf{TQ}'$ gives an approximation for $\mathbf{Y}$, instead of $\mathbf{UQ}'$ (Oyedele and Lubbe 2015). As a consequence, $\tilde{\mathbf{D}}$ provides an approximation for $\mathbf{D}$ such that

$$\tilde{\mathbf{D}} = \begin{bmatrix} \tilde{\mathbf{X}} & \tilde{\mathbf{Y}} \end{bmatrix} = [\mathbf{TP}'\ \ \mathbf{TQ}'] = \mathbf{T}[\mathbf{P}\ \ \mathbf{Q}]'.$$

Extracting just two components, the dimension of $\mathbf{T}$ is $(n \times 2)$ and the size of the block matrix $[\mathbf{P}\ \mathbf{Q}]'$ is $(2 \times (m + q))$. So, the rows of $\mathbf{T}$ represent the biplot points in the exploratory PLS biplot, expressing the observations of the sample, while the columns of the block matrix $[\mathbf{P}\ \mathbf{Q}]'$ indicate the biplot vectors and denote the variables, wherein those from column 1 to m refer to the predictors and from column $(m + 1)$ to $(m + q)$ are associated with the responses. Considering each set of biplot vectors separately (predictors and responses), the angle formed by two vectors provides an approximation for the sample correlation coefficient related to the associated variables (Graffelman 2012). Therefore, if $\angle\left(\mathbf{p}'_i, \mathbf{p}'_j\right) \cong 0°$, it means that the associated variables are strongly correlated because the cosine of the angle between the biplot vectors is close to one. On the other hand, when $\angle\left(\mathbf{p}'_i, \mathbf{p}'_j\right) \cong 180°$ and the biplot vectors point to almost opposite directions, then it indicates a negative but substantial correlation. Lastly, a right angle suggests a weak correlation between the related variables. However, the accuracy of this approximation will depend on how much the variables contribute to each of the underlying components estimated (Bassani et al. 2010), as well as the biplot explained variance (Greenacre 2012).

2.5   The Predictive PLS Biplot

As previously seen in Sect. 2.2, the $(m \times q)$ matrix $\widehat{\mathbf{B}}_{PLS} = \mathbf{RQ}'$ contains the PLS coefficient estimates, in which the $\mathbf{R}$ columns are the transformed PLS X-weights, and $\mathbf{Q}$ is the matrix of Y-weights. In the predictive PLS biplot, the rows of the matrix $\mathbf{R}$ denote the biplot points instead of the rows of $\mathbf{T}$. Further, the columns of $\mathbf{Q}'$

symbolize the responses through biplot vectors. Each response can also define a calibrated axis, on which one can project the set of points ($\mathbf{r}_i'$) to get an approximation of the coefficients. Considering a specific response Y_j and a fixed predictor X_i, each element of the matrix $\widehat{\mathbf{B}}_{\text{PLS}}$ is computed as

$$\widehat{b}_{\text{PLS}_{ij}} = \mathbf{r}_i'\mathbf{q}_j = \|\mathbf{r}_i\|\|\mathbf{q}_j\|\cos\theta_{\mathbf{r}_i,\mathbf{q}_j}.$$

Therefore, there are two ways to evaluate an approximation for these estimates in the biplot visually. The first manner consists of the *calibration* of biplot axes (Greenacre 2010; Oyedele and Lubbe 2015) and mentally reading the projection of the biplot point on the biplot axis. In the second mode, the *area biplot* method is applied (Gower et al., 2010; Oyedele and Lubbe 2015), in which the approximation of $\widehat{b}_{\text{PLS}_{ij}}$ is obtained from the area determined by the origin, the rotated biplot point $\mathbf{r}_i'$, and the endpoint of $\mathbf{q}_j$. The area and position of the triangles furnish other relevant information about the PLS regression coefficients, such as the signal and the importance of each predictor to the model.

3   Application

Teams of individuals working together to achieve a common goal are a central part of daily life in modern organizations (Mathieu et al. 2014). By bringing together individuals with different skills and knowledge, teams emerge as a competitive asset in the ever-changing organizational environment. When teams are created, the ultimate goal is to generate value for the organization. Accordingly, studying team effectiveness and the conditions that enable the team to be effective has been a central concern for both research and practice (Kozlowski and Ilgen 2006).

3.1   Variables

In the present research, in line with previous studies (e.g., Hackman 1987), we consider team effectiveness as a multidimensional construct. Thus, in this study, team effectiveness is evaluated through two criteria: team performance and the quality of group experience. *Team performance* (Y_1) refers to the extent to which team outcomes respect the standards set by the organization, in terms of quantity, quality, delivery time and costs (Rousseau and Aubé 2010). The *quality of the group experience* (Y_2) is related to the quality of the social climate within the team (Aubé and Rousseau 2005).

Team effectiveness will be considered, in the present study, as the result of the role played by thirteen variables: team trust (2 dimensions), team psychological capital (4 dimensions), collective behavior, transformational leadership, intragroup conflict (2 dimensions), team psychological safety, and team cohesion (2 dimensions). Each variable will be briefly described as follows.

Team trust refers to the aggregate levels of trust that team members have in their fellow teammates (Langfred 2004) and has been conceptualized as a bidimensional construct: the *affective dimension of team trust* (X_1) is related to the perception of the presence of shared ideas, feelings, and concerns within the team; the *task dimension of team trust* (X_2) has been associated with the recognition by team members of the levels of professionalism and competence of their teammates and on their ability to appropriately perform the tasks (McAllister 1995).

Team psychological capital (PsyCap) can be defined as a team positive psychological state characterized by: having confidence (efficacy) to succeed in challenging tasks; making a positive attribution (optimism) about succeeding now and in the future; persevering, and when necessary, redirecting paths to goals (hope) in order to be effective; and having the ability to bounce back from challenges and setbacks (resilience) (Luthans et al. 2007; Luthans and Youssef-Morgan 2017; Walumbwa et al. 2011). In summary, team PsyCap includes four team psychological resources: *team efficacy* (X_3), *team optimism* (X_4), *team hope* (X_5), and *team resilience* (X_6).

Collective behavior (X_7) refers to the members' tendency to coordinate, evaluate, and utilize task inputs from other team members when performing a group task (Driskell et al. 2010).

Transformational leadership (X_8) can be defined as a leadership style that encourages followers to do more than they originally expected, broadening and changing their interests and leading to conscientiousness and acceptance of the team's purposes (Bass 1990). Carless et al. 2000) described transformational leaders as those who exhibit the following seven behaviors: they 1) communicate a vision; 2) develop staff; 3) provide support for them to work towards their objectives through coordinated team work; 4) empower staff; 5) are innovative by using non-conventional strategies to achieve their goals; 6) lead by example; 7) are charismatic.

Intragroup conflict can be defined as a disagreement that is perceived as creating tension at least by one of the parties involved in an interaction (De Dreu and Weingart 2003). Conflicts in teams may emerge as a result of the presence of different ideas about the tasks performed (X_9) – *task conflict* – or may be related to differences between team members in terms of values or personalities (X_{10}) – *affective conflict* (Jehn 1994).

Team psychological safety (X_{11}) relates to team members' perceptions about what the consequences will be of taking interpersonal risks at the work environment. It means taking beliefs for granted about how others will react when one speaks up or participates. It is a confidence climate that comes from mutual respect and trust between members (Edmondson 1999).

Team cohesion can be defined as the team members' inclination to create social bonds, resulting in the group sticking together, remaining united, and wanting to work together (Carron 1982; Salas et al. 2015). It can be related to the task or the affective system of the team. *Task cohesion* (X_{12}) refers to the shared commitment among members towards achieving a goal that requires the collective efforts of the group. *Social cohesion* (X_{13}) refers to shared liking or attraction to the group and to the nature and quality of the emotional bonds of friendship, liking, caring, and closeness among group members (Chang and Bordia 2001).

3.2 Sample and Data Collection Procedure

Organizations were selected by convenience, using the personal and professional contacts network of the research team. To collect data, key stakeholders in each organization (CEOs or human resources managers) were contacted to explain the purpose and requirements of the study. When the organization agreed to participate, the selection of teams for the survey was based on the following criteria (Cohen and Bailey 1997): teams must be composed of at least three members; should be perceived by themselves and others as a team; they have to regularly interact, interdependently, to accomplish a common goal; and they must have a formal supervisor who is responsible for the actions of the team.

Data was collected following two strategies. In most organizations, questionnaires were filled in during team meetings, in the presence of a member of the research team. When it was not possible to implement this strategy, they were filled in online via an electronic platform. Data was obtained from 104 teams and their respective leaders. After eliminating from the sample teams with a team members' response rate below 50% and participants with more than 10% of missing values, the remaining sample had a total of 82 teams. In this remaining sample, missing values in the questionnaires were replaced by the item average (in case of a random distribution) or by expectation-maximization (EM) method (in case of a non-random distribution).

The 82 teams of the sample are from 57 Portuguese companies. Forty-two per cent of these organizations are small, and the most representative sector is the services sector (73%). Team size ranged from 3 to 18 members, with an average of approximately 6 members (SD = 3.55). Of the team members (N = 353), 67% were female, 63.3% had secondary education or less, with the remaining 36.7% having a higher education background. The mean age was approximately 38 years old (SD = 12.33). The average team tenure was approximately 6 years (SD = 7.25). Regarding team leaders (N = 82), 57% were male, the mean age was about 42 years old (SD = 10.86) and 55.7% had a higher education background. Leaders had, on average, 5 years of experience as leader of the current team (SD = 4.87).

3.3 Measures

Apart from team performance that was assessed by team leaders, all variables were measured by team members. The measures used are identified as follows: *team performance* was measured with a scale developed by Rousseau and Aubé (2010), which has five items; *quality of the group experience* was assessed with the scale developed by Aubé and Rousseau (2005), which is composed of three items; team trust was evaluated with the scale developed by McAllister (1995), which is constituted by 10 items; team psyCap was measured with the scale developed by Luthans et al. (2007), which is composed of 24 items; collective behavior was measured with the scale developed by Driskell et al. (2010), which has 10 items; transformational leadership was measured with the scale developed by Carless et al. (2000), which is composed of seven items; intragroup conflict was evaluated with the scale developed by Dimas and Lourenço (2015), which is composed of nine items; team psychological safety was assessed with the scale developed by Edmonson (1999), which is composed of seven items; team cohesion was measured with the scale developed by Chang and Bordia (2001), which is constituted by eight items. Team trust and team psycap were assessed using 6-point scales, intragroup conflict and team psychological safety were evaluated on 7-point scales and the remaining variables were measured on 5-point scales.

3.4   PLS Biplot Results

In order to reveal a linear relation between the variables describing team effectiveness and the explanatory variables, the PLS was used to construct the external and internal models. First, the predictor matrix $\mathbf{X}_{82 \times 13}$ and the response matrix $\mathbf{Y}_{82 \times 4}$ were centered and scaled. Next, the NIPALS algorithm was used to decompose the data matrices and to extract two PLS components, yielding the matrices $\mathbf{T}_{82 \times 2} = [\mathbf{T}_1 \mathbf{T}_2]$, $\mathbf{P}_{13 \times 2}$, $\mathbf{U}_{82 \times 2}$, $\mathbf{Q}_{4 \times 2}$, $\mathbf{W}_{13 \times 2}$, $\mathbf{R}_{13 \times 2}$, and $\mathbf{B}_{13 \times 2}$. The latter contains the estimates of the PLS regression coefficients, according to Table 1.

The first PLS component $\mathbf{T}_1$ explains 56.5% of the data variability, while the proportion of variance explained by $\mathbf{T}_2$ is 9.5%. Figure 1 shows the exploratory PLS biplot, in which the (black) biplot points (X-scores $\mathbf{t}_i'$) represent the 82 teams, the blue biplot vectors depict the responses (Y-loadings $\mathbf{q}_i'$), and the red biplot vectors symbolize the predictors (X-loadings $\mathbf{p}_i'$).

Figure 1 provides an approximation of the correlation structures of the data, but it must be taken into account that the total proportion of variance explained by the two components $\mathbf{T}_1$ and $\mathbf{T}_2$ is 66%. Table 2 shows some significantly correlated variables evidenced by the biplot (X_2 and X_{12}, X_7 and X_{12}, X_3 and X_{11}, and X_9 and X_{10}), a pair of variables that displayed negative correlation (X_2 and X_9), and others that manifested a weak correlation visually (X_5 and X_{13}, X_6 and X_{13}), all of them flanked by the exact sample correlation coefficients. Table 2 is not exhaustive, and it is possible to pinpoint other exciting examples regarding the correlation structure in Fig. 1, e.g., the weak correlation between the two responses (the correct sample correlation is $\cong 0.28$). Moreover, all of the variables are positively associated with the first PLS component $\mathbf{T}_1$, except *Task conflict* (X_9) and *Affective conflict* (X_{10}), which are negatively associated. Regarding $\mathbf{T}_2$, the predictor *Team trust-affective* (X_1) is negligibly correlated, and the predictors *Team trust-task* (X_2), *Collective behavior* (X_7), *Task cohesion* (X_{12}), and *Social cohesion* (X_{13}) are negatively correlated.

For comparison purposes only, Fig. 2 shows the results of the area biplot method. With respect to response *Team performance* (Y_1) – left biplot, the predictors *Team efficacy* (X_3), *Team optimism* (X_4), and *Team psychological safety* (X_{11}) stand out as the most influential variables to the model, since the triangle related to the regression coefficients b_3, b_4, and b_{11} show the most significant area. On the other side, the variables with the least positive impact on the model are *Team trust-task* (X_2) and *Team task conflict* (X_9), because they are related to the smallest areas. Further, the predictor *Social cohesion* (X_{13}) affects *Team performance* negatively, given that the triangle position is on the right side of the biplot axis. In its turn, regarding the response *Quality of the group experience* (Y_2), the most important predictors are *Task conflict* (X_9), *Affective conflict* (X_{10}), *Team trust-task* (X_2), *Task cohesion* (X_{12}), and *Social cohesion* (X_{13}), with the first two in a negative way. All these findings are following the

Table 1. Punctual estimates of the PLS regression coefficients.

Predictor name	Predictor identification	$\hat{\beta}_1$ related to Team performance (Y_1)	$\hat{\beta}_2$ related to Quality of the group experience (Y_2)
Team trust (affective)	X_1	0.076	0.079
Team trust (task)	X_2	0.005	0.141
Team efficacy	X_3	0.105	0.033
Team optimism	X_4	0.144	−0.027
Team hope	X_5	0.076	0.068
Team resilience	X_6	0.023	0.075
Collective behavior	X_7	0.047	0.095
Transformational leadership	X_8	0.055	0.043
Task conflict	X_9	0.017	−0.124
Affective conflict	X_{10}	0.051	−0.154
Team psychological safety	X_{11}	0.090	0.053
Task cohesion	X_{12}	0.059	0.103
Social cohesion	X_{13}	−0.035	0.109

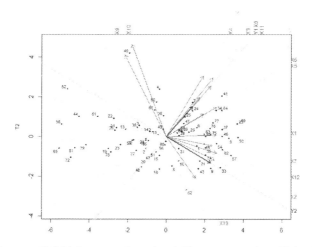

Fig. 1. Exploratory PLS biplot – sample and variables representation. (Color figure online)

PLS results shown in Table 2, but one can easily reach the same conclusions through Fig. 1 (exploratory PLS biplot).

Figure 3 brings a modified version of the exploratory PLS biplot, in which all of the biplot vectors are projected onto the calibrated biplot axis Y_1. One more time, the most significant vector projections refer to *Team efficacy* (X_3), *Team optimism* (X_4), and

Table 2. Correlation approximation by biplot vectors and sample correlation coefficients.

Variables	Correct Correlation Coefficient (r)
X_2 (Team trust - task) and X_{12} (Task cohesion)	0.71
X_7 (Team efficacy) and X_{12} (Task cohesion)	0.70
X_3 (Team efficacy) and X_{11} (Team psychological safety)	0.73
X_9 (Task conflict) and X_{10} (Affective conflict)	0.85
X_2 (Team trust - task) and X_9 (Task conflict)	−0.60
X_5 (Team hope) and X_{13} (Social cohesion)	0.35
X_6 (Team resilience) and X_{13} (Social cohesion)	0.25

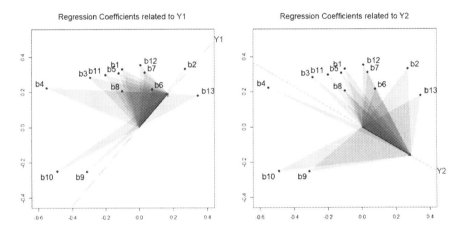

Fig. 2. Area biplot Method applied to the team effectiveness dataset.

Team psychological safety (X_{11}), as well as the less significant projection referring to *Team trust-task* (X_2). Beyond that, only the projection related to the variable *Social cohesion* (X_{13}) falls on the negative part of the biplot axis. The approximation of the regression coefficients related to the dependent variable *Quality of the group experience* is represented in Fig. 4, where the biplot vectors are projected onto the biplot axis Y_2. In this case, similarly to the results of the area biplot method, the largest projections indicate the more relevant variables. On the negative side, the predictors *Task conflict* (X_9) and *Affective conflict* (X_{10}) are the most influential in the model, while the explanatory variables *Team trust-task* (X_2), *Task cohesion* (X_{12}), and *Social cohesion* (X_{13}) have the most significant and positive impact concerning Y_2.

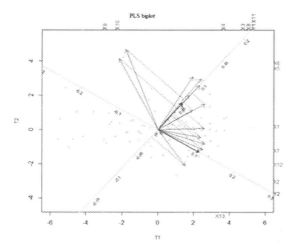

Fig. 3. Visual approximation of the regression coefficients (response Y_1).

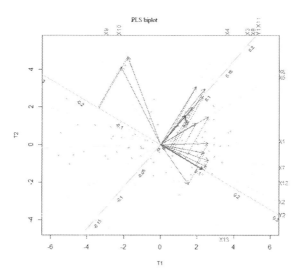

Fig. 4. Visual approximation of the regression coefficients (response Y_2).

4 Discussion and Conclusions

Regarding the application of the method we use in this work, the results point to the "validity" of such an application concerning the relationships found between the group processes and the team output variables considered. In fact, overall, the most significant results found in our study suggest relationships between the predictors and the criteria that are convergent with the literature.

One of the results points out the relevant role of cohesion as a predictor of team outcomes but the different behavior of each one of the team cohesion dimensions.

Indeed, task cohesion (X_{12}) showed a positive relationship with both team effectiveness criteria (although with higher magnitude regarding the quality of the group experience); however, social cohesion (X_{13}), although it emerged as one of the most relevant positive predictors of the quality of group experience, revealed a negative influence on team performance. These results are in line with the literature. Firstly, team cohesion is recognized by researchers as one of the most influential factors on group behavior and, consequently on group outcomes (Carron and Brawley 2000; Dionne et al. 2004). Secondly, and despite that, the literature, namely a meta-analysis conducted by Mullen and Cooper (1994), also suggests that the link between social cohesion and task cohesion with team outcomes can be different. Task cohesion tends to be positively associated with team outcomes, but social cohesion can have a more complex relationship with team outcomes due the fact that social cohesion, although it increases the willingness to help each other and to cooperate, can also lead to uncritical acceptance of solutions and to groupthink (Janis 1972). Thus, social cohesion can both increase the sense of belonging to a group, contributing to a positive perception of the group experience (quality of group experience), and decrease team performance, as suggested by our study.

Another interesting result to highlight is related to the negative relationship of both conflict types – task conflict (X_9) and affective conflict (X_{10}) – with the quality of group experience and the less clear role of task conflict in team performance. Indeed, task conflict revealed a negative relation with the quality of the group experience and a positive (albeit low-level) relation with team performance. These results tend to converge with the literature. On the one hand, the literature points out that conflict is always experienced as a negative experience (e.g., Jehn et al. 2008) and, as a result tends to have a negative influence on the attitudes of team members towards the group. However, on the other hand, studies are not totally consensual with respect to its effects on team performance, especially in what concerns to task conflict (De Wit et al. 2012; Dimas and Lourenço 2015). In fact, most studies found either negative associations between task conflict and team performance (e.g., Janssen et al. 1999) or a nonsignificant relation (e.g., Jordan and Troth 2004), and a meta-analysis conducted by De Dreu and Weingart (2003) supported those findings. However, more recently De Wit et al. (2012) conducted a new meta-analysis and concluded that the effects of task conflict on team outcomes are less negative (or even positive) as compared to affective conflict. Overall, the studies tend to suggest that, in certain circumstances, task conflict may be positively related to group outcomes (e.g., De Wit et al. 2012) emphasizing the role of moderators, such as the conflict-handling strategies used in the team.

It is also interesting to mention the positive role of team trust (X_1 and X_2) in team results and the more significant role of team task trust (X_2) compared to team social trust (X_1). Again, our results are supported by the literature which indicates that trust represents an important determinant of team effectiveness. In this regard, Dirks and Ferrin (2001) pointed out in their meta-analysis that team trust is positively related to performance and team satisfaction (an indicator of the quality of group experience). The fact that, in the present study, task trust has showed to be a more important predictor of performance than social trust can be explained by the fact that our sample is composed of work teams in productive organizations, where trust in the members' skills and their professionalism for the accomplishment of the tasks is more critical.

Finally, it is important to mention the role of team psychological safety (X_{11}), team self-efficacy (X_3) and team optimism (X_4) as positive predictors of team performance. Like the variables that we addressed above, the results obtained are supported by the literature. Regarding the relationship between team psychological safety and team performance, previous studies suggest that team performance can be facilitated, directly or indirectly, by the presence of a psychological security climate (e.g., Edmondson 1999). According to these studies, team performance is increased by a group climate in which team members are encouraged to express themselves without fear of the evaluations of the rest of the group. Regarding team efficacy and team optimism (dimensions of PsyCap), several studies show that collective PsyCap is positively related to team performance (e.g., Norman et al. 2010; Walumbwa et al. 2011). Additionally, previous research suggests that when team members have a collective belief in their ability to be effective, they explore and share knowledge and are more prepared to implement new ways of achieving results, because they believe these behaviors will lead to higher levels of performance (Bandura 1977). Similarly, a team with optimistic beliefs has positive expectations, is usually more actively involved in tasks than a team with a low level of optimism and use more adaptive coping skills when obstacles occur (Avey et al. 2011).

In general, the interpretation method proposed in this work provided excellent results in the application performed in Sect. 3, since it was able to capture the most critical variables for the model, correctly assigned the signals of the regression coefficients and gave an approximation to their values directly in the exploratory PLS biplot. Nevertheless, some inconsistencies were detected. For example, regarding the response Y_1, one can see in Table 2 that $\widehat{\beta}_{41} > \widehat{\beta}_{31}$, but the projections of the biplot vectors corresponding to X_4 and X_3 over the biplot axis yield the opposite result (Fig. 3). In the same sense, the projections of the vectors related to X_6 and X_9 seem to be overestimated considering the associated values ($\widehat{\beta}_{61}$ and $\widehat{\beta}_{91}$) in Table 2. Although with less intensity, the same occurs in Fig. 4, where the projections are made over the biplot axis Y_2.

However, we should keep in mind that biplot is a visualization method whose purpose is to provide a general idea of latent structures in the data, not to mention that the interpretation technique suggested in this paper provides only an approximation of the coefficients, which will be closer to the real values of the estimates, the higher the PLS components' ability to explain the variance.

Acknowledgments. This work was partially supported by The Center for Research and Development in Mathematics and Applications (CIDMA) and by The Center for Business and Economics Research (CeBER) through the Portuguese Foundation for Science and Technology (FCT - Fundação para a Ciência e a Tecnologia), references UIDB/04106/2020, UIDP/04106/2020 and UIDB/05037/2020.

References

Abdi, H.: Partial least squares regression and projection on latent structure regression (PLS regression). WIREs Comput. Stat. **2**, 97–106 (2010)

Avey, J.B., Reichard, R.J., Luthans, F., Mhatre, K.H.: Meta-analysis of the impact of positive psychological capital on employee attitudes, behaviors, and performance. Human Resource Dev. Quarterly **22**(2), 127–152 (2011)

Aubé, C., Rousseau, V.: Team goal commitment and team effectiveness: the role of task interdependence and supportive behaviors. Group Dynamics Theory Res. Practice **9**, 189–204 (2005)

Bandura, A.: Self-efficacy: Toward a unifying theory of behavioral change. Psychol. Rev. **84**(2), 191–215 (1977)

Bass, B.M.: From transactional to transformational leadership: Learning to share the vision. Org. Dyn. **18**(3), 19–31 (1990)

Bassani, N., Ambrogi, F., Coradini, D., Biganzoli, E.: Use of biplots and partial least squares regression in microarray data analysis for assessing association between genes involved in different biological pathways. In: Rizzo, R., Lisboa, Paulo J.G. (eds.) CIBB 2010. LNCS, vol. 6685, pp. 123–134. Springer, Heidelberg (2010). https://doi.org/10.1007/978-3-642-21946-7_10

Belsley, D.A., Kuh, E., Welsch, R.E.: Regression Diagnostics – Identifying Influential Data and Sources of Collinearity. Wiley-Interscience, New Jersey (2004)

Carless, S., Wearing, L., Mann, L.A.: Short measure of transformational leadership. J. Bus. Psychol. **14**(3), 389–405 (2000)

Carron, A.V.: Cohesiveness in Sport Groups: Interpretations and Considerations. J. Sport Psychol. **4**, 123–138 (1982)

Carron, A.V., Brawley, L.R.: Cohesion: Conceptual and measurement issues. Small Group Res. **31**, 89–106 (2000)

Chang, A., Bordia, P.: A multidimensional approach to the group cohesion group performance relationship. Small Group Res. **32**(4), 379–405 (2001)

Cohen, S.G., Bailey, D.E.: What makes teams work: Group effectiveness research from the shop floor to the executive suite. J. Manag. **23**(3), 239–290 (1997)

De Dreu, C.K.W., Weingart, L.R.: Task versus relationship conflict, team performance, and team member satisfaction: A meta-analysis. J. Appl. Psychol. **88**(4), 741–749 (2003)

De Wit, F.R., Greer, L.L., Jehn, K.A.: The paradox of intragroup conflict: A meta-analysis. J. Appl. Psychol. **97**(2), 360–390 (2012)

Dimas, I.D., Lourenço, P.R.: Intragroup conflict and conflict management approaches as determinants of team performance and satisfaction: Two field studies. Negot. Confl. Manage. Res. **8**(3), 174–193 (2015)

Dionne, S.D., Yammarino, F.J., Atwater, L.E., Spangler, W.D.: Transformational leadership and team performance. J. Organ. Change Manage. **17**(2), 177–193 (2004)

Dirks, K.T., Ferrin, D.L.: The role of trust in organizational settings. Organ. Sci. **12**(4), 450–467 (2001)

Driskell, J.E., Salas, E., Hughes, S.: Collective orientation and team performance: Development of an individual differences measure. Hum. Factors **52**(2), 316–328 (2010)

Edmondson, A.: Psychological safety and learning behavior in work teams. Adm. Sci. Q. **44**(2), 350–383 (1999)

Gabriel, K.R.: The biplot graphic display of matrices with application to principal component analysis. Biometrika **58**, 453–467 (1971)

Geladi, P., Kowalsky, B.R.: Partial least squares regression: A tutorial. Anal. Chim. Acta **186**, 1–17 (1986)

Gower, J.C., Groenen, P.J.F., Van de Velden, M.: Area biplots. J. Comput. Graphical Stat. **19**, 46–61 (2010)

Graffelman, J.: Linear-angle correlation plots: new graphs for revealing correlation structure. J. Comput. Graphical Stat. **22**(1), 92–106 (2012)

Greenacre, M.: *Biplots in Practice*. FBBVA, (2010)

Greenacre, M.: Contribution Biplots. J. Comput. Graphical Stat. **22**(1), 107–122 (2012)

Hackman, J.R.: The design of work teams. In: Lorsch, J. (ed.) Handbook of Organizational Behavior, pp. 315–342. Prentice-Hall, Englewood Cliffs (1987)

Janis, I.L.: Victims of groupthink: A psychological study of foreign-policy decisions and fiascoes. Houghton Mifflin (1972)

Janssen, O., Van de Vliert, E., Veenstra, C.: How task and person conflict shape the role of positive interdependence in management teams. J. Manag. **25**(2), 117–141 (1999)

Jehn, K.A.: Enhancing effectiveness: An investigation of advantages and disadvantages of value-based intragroup conflict. Int. J. Conflict Manage. **5**, 223–238 (1994)

Jehn, K.A., Greer, L., Levine, S., Szulanski, G.: The effects of conflict types, dimensions, and emergent states on group outcomes. Group Decis. Negot. **17**(6), 465–495 (2008)

Jordan, P.J., Troth, A.C.: Managing emotions during team problem solving: Emotional intelligence and conflict resolution. Hum. Perform. **17**(2), 195–218 (2004)

Kozlowski, S.W., Ilgen, D.R.: Enhancing the effectiveness of work groups and teams. Psychol. Sci. Public Interest **7**(3), 77–124 (2006)

Koch, I.: Analysis of Multivariate and High-Dimensional Data. Cambridge University Press, New York (2014)

Langfred, C.W.: Too much of a good thing? Negative effects of high trust and individual autonomy in self-managing teams. Acad. Manage. J. **47**(3), 385–399 (2004)

Luthans, F., Avolio, B.J., Avey, J.B., Norman, S.M.: Positive psychological capital: Measurement and relationship with performance and satisfaction. Pers. Psychol. **60**(3), 541–572 (2007)

Luthans, F., Youssef-Morgan, C.M.: Psychological capital: An evidence-based positive approach. Annu. Rev. Organ. Psychol. Organ. Behav. **4**(1), 339–366 (2017)

Mathieu, J.E., Gallagher, P.T., Domingo, M.A., Klock, E.A.: Embracing complexity: Reviewing the past decade of team effectiveness research. Annu. Rev. Organ. Psychol. Organ. Behav. **6**, 17–46 (2019)

Mathieu, J.E., Tannenbaum, S.I., Donsbach, J.S., Alliger, G.M.: A review and integration of team composition models: Moving toward a dynamic and temporal framework. J. Manage. **40**(1), 130–160 (2014)

McAllister, D.: Affect and cognition-based trust as foundations for interpersonal cooperation in organizations. Acad. Manage. J. **38**(1), 24–59 (1995)

Mullen, B., Cooper, C.: The relation between group cohesiveness and performance: An integration. Psychol. Bull. **115**, 210–227 (1994)

Norman, S.M., Avey, J.B., Nimnicht, J.L., Pigeon, N.: The interactive effects of psychological capital and organizational identity on employee organizational citizenship and deviance behaviors. J. Leadership Organ. Stud. **17**(4), 380–391 (2010)

Oyedele, O.F., Lubbe, S.: The construction of a partial least squares biplot. J. Applied Stat. **42**(11), 2449–2460 (2015)

Ringle, C.M., Sarstedt, M., Mitchell, R., Gudergan, S.P.: Partial least squares structural equation modeling in human resource management research. Int. J. Hum. Resour. Manage. **31**, 1617–1643 (2018)

Rousseau, V., Aubé, C.: Team self-managing behaviors and team effectiveness: The moderating effect of task routineness. Group Organ. Manage. **35**(6), 751–781 (2010)

Salas, E., Grossman, R., Hughes, A.M., Coultas, C.W.: Measuring team cohesion: Observations from the science. Hum. Factors **57**(3), 365–374 (2015)

Walumbwa, F.O., Luthans, F., Avey, J.B., Oke, A.: Authentically leading groups: The mediating role of collective psychological capital and trust. J. Organ. Behav. **32**(1), 4–24 (2011)

Wold, S., Eriksson, L., Trygg, J., Kettaneh, N.: The PLS Method - Partial Least Squares Projections to Latent Structures - and Its Applications in Industrial RDP (Research, Development, and Production). Umea University, Umea (2004)

Tribological Behavior of 316L Stainless Steel Reinforced with CuCoBe + Diamond Composites by Laser Sintering and Hot Pressing: A Comparative Statistical Study

Ângela Cunha[1] , Ana Marques[1] , Francisca Monteiro[1(✉)] ,
José Silva[1] , Mariana Silva[1] , Bruno Trindade[2] ,
Rita Ferreira[1] , Paulo Flores[1] , Óscar Carvalho[1] ,
Filipe Silva[1] , and Ana Cristina Braga[3]

[1] CMEMS – Center for Microelectromechanical Systems, University of Minho,
Azurém 4800-058, Guimarães, Portugal
{id8694,id8696,id8700,id8693,
id8692}@alunos.uminho.pt,
rita.mac.ferreira@gmail.com, {pflores,oscar.carvalho,
fsamuel}@dem.uminho.pt
[2] CEMMPRE – Center for Mechanical Engineering, Materials and Processes,
University of Coimbra, Rua Luís Reis Santos, 3030-788 Coimbra, Portugal
bruno.trindade@dem.uc.pt
[3] ALGORITMI Research Centre, University of Minho, Guimarães, Portugal
acb@dps.uminho.pt

Abstract. The aim of this work was to perform a statistical analysis in order to assess how the tribological properties of a laser textured 316L stainless steel reinforced with CuCoBe - diamond composites are affected by diamond particles size, type of technology (laser sintering and hot pressing) and time of tribological test. The analysis started with the description of all response variables. Then, by using IBM® SPSS software, the Friedman's test was used to compare how the coefficient of friction varied among samples in five-time points. From this test, results showed that there was no statistically significant difference in the coefficient of friction mean values over the selected time points. Then, the two-samples Kolmogorov-Smirnov (K-S) test was used to test the effect of the diamond particles size and the type of technology on the mean of the coefficient of friction over time. The results showed that, for both sintering techniques, the size of the diamond particles significantly affected the values of the coefficient of friction, whereas no statistical differences were found between the tested sintering techniques. Also, the two-way ANOVA test was used to evaluate how these factors influence the specific wear rate, which conducted to the same conclusions drawn for the previous test. The main conclusion was that the coefficient of friction and the specific wear rate were statistically affected by the diamond particles size, but not by the sintering techniques used in this work.

Keywords: Statistical analysis · Tribological behavior · Selective Laser Sintering · Hot pressing · Multi-material surface

© Springer Nature Switzerland AG 2020
O. Gervasi et al. (Eds.): ICCSA 2020, LNCS 12251, pp. 231–246, 2020.
https://doi.org/10.1007/978-3-030-58808-3_18

1 Introduction

Austenitic stainless steels are characterized by high applicability in the mechanical, chemical and process industries [1] due to their high resistance to corrosion, high strength and machinability [2–4]. In the automotive industry, the 316L stainless steel is a candidate to substitute the ductile or cast iron in the fabrication of piston rings [5], since it presents similar chemical composition and has an inherent strength that leads to a less chance of ring breakage, with consequently longer service life. The surface of a piston ring must be multifunctional. The reduction of the friction, the retention of oil during operation, the retention of particles from wear and the increase in the thermal conductivity are target properties in the development of automotive piston rings [6, 7]. The 316L stainless steel has low wear resistance, which can be improved by the addition of hard ceramic particles on the surface [8–10]. Different metallic and ceramic material have been used as reinforcements of metallic matrices such as tungsten [11, 12], titanium [12, 13], silicon and niobium carbides [8, 14] and diamond [15, 16] to produce metal matrix composites (MMCs). These composites materials present exceptional combination of properties and performance [11, 17] with improved mechanical and tribological properties when compared with the matrix [13, 18]. The processing temperature of diamond-reinforced MMCs is an important issue because graphitization of diamond into graphite occurs at 900 °C [15, 16, 19]. So, low-temperature processing techniques are required for the production of these materials. In previous work, a laser-textured 316L steel was reinforced with CuCoBe-diamond composites through selective laser sintering (SLS) and hot pressing (HP) [20, 21]. The results showed that the tribological properties of the 316L steel were improved by the reinforcement composite. Selective Laser Sintering (SLS) is an additive manufacturing (AM) process that presents an extraordinary versatility to geometry and materials design, as it is considered a near-net-shape technology [22, 23]. It is also characterized by being a fast technique, which is a potential advantage in mass production. On the other hand, Hot Pressing technique, characterized by the simultaneous application of temperature and pressure, allows obtaining well-consolidated components with high mechanical performance due to the low porosity that it presents [24, 25]. When using pressure and temperature simultaneously, it is possible to compensate temperature with pressure, using lower temperatures. In this case, it is important due to the graphitization of the diamond temperature.

This work aims to study, through statistical analysis, the influence of the two consolidation technologies (SLS and HP) on the tribological behavior of laser-textured 316L samples reinforced with CuCoBe-diamond composites. The size effect of the diamond particles on the tribological properties will also evaluated and discussed.

2 Methodology Procedure

2.1 Materials and Processing Details

Samples of SS316L with 14 mm diameter were textured through an Nd:YAG laser (Sisma Laser) with a wavelength of 1064 nm, laser power of 6 W, scan speed of

128 mm/s, number of passes of 16 and fill spacing of 5 μm. The texture produced (see Fig. 1) was analyzed by scanning electron microscopy (SEM) (Nano-SEM - FEI Nova 200 equipment) and 3D profilometry (InfiniteFocus from Brucker alicona).

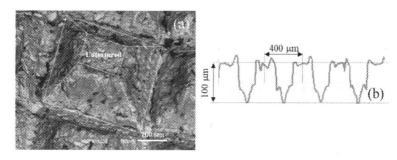

Fig. 1. Texture produced: (a) SEM image of the textured and (b) heights of the profile produced.

The CuCoBe (1.5 wt% cobalt, 0.5 wt% beryllium and the remainder is copper) + 5.8 wt% diamond composites were produced by mechanical alloying (MA) from elemental powders. Two different grades of diamond particles (0.1–0.5 μm and 40–60 μm) were used. The powder size of the CuCoBe alloy was 40–80 μm. Five different samples were produced by laser sintering and hot pressing in this work (Table 1).

Table 1. Samples produced and analyzed in this work.

Sample reference	Description
316L + C_0.1–0.5 (SLS)	Textured 316L stainless steel reinforced with mechanical alloyed CuCoBe + diamond particles (0.1–0.5 μm) produced by SLS
316L + C_40–60 (SLS)	Textured 316L stainless steel reinforced with mechanical alloyed CuCoBe + diamond particles (40–60 μm) produced by SLS
316L + C_0.1–0.5 (HP)	Textured 316L stainless steel reinforced with mechanical alloyed CuCoBe + diamond particles (0.1–0.5 μm) produced by HP
316L + C_40–60 (HP)	Textured 316L stainless steel reinforced with mechanical alloyed CuCoBe + diamond particles (40–0 μm) produced by HP

Figure 2 shows a schematic representation of the two processes used for the reinforcement of the textured 316L stainless steel, SLS and HP.

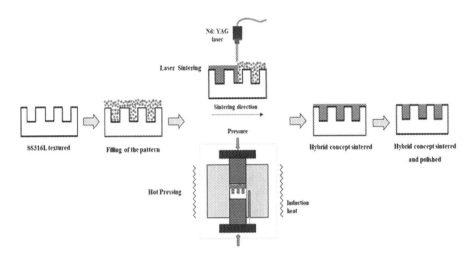

Fig. 2. Schematic representation of the two processes used for the reinforcement of the textured 316L.

For SLS, samples were sintered through the same laser, power and scan speed used for texturing but using just 1 pass and a fill spacing of 20 μm. Under these conditions one line affects a distance of 27 μm, so with the fill spacing used, overlapping of the lines was ensured and consequently sintering of the entire surface was performed. Regarding HP, the process was conducted using a vacuum pressure-assisted sintering system (under a vacuum of 1×10^{-4} Pa) with a high-frequency induction furnace. Samples were heated up to 900 °C, with a heating rate of 100 °C/min, and an applied pressure of 70 MPa, for 30 min. The samples were then cooled down to the room temperature.

After sintering, a polishing operation was necessary to expose the 316L steel surface. The samples were polished with SiC abrasive papers down to a 4000 mesh and ultrasonically cleaned with isopropyl alcohol before tribological tests.

2.2   Tribological Tests

A reciprocating pin-on-plate tribometer (see Fig. 3) was used (Plint TE67-HT) for the tests, which replicates the piston ring-cylinder liner contact. The pin (counter body) consisted of a malleable cast iron cylindrical body with a diameter of 2.25 mm (geometry similar to the engine cylinder body).

The test conditions were defined based on the engine's operating conditions (similar to the in-service conditions) and in accordance with the restrictions of the test equipment. The wear sliding tests were performed dry at 25 N loading (nominal), with a frequency of 1.5 Hz and 7 mm of total stroke length for 4 h. The contact area

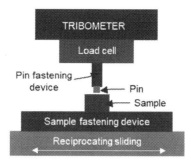

Fig. 3. Schematic representation of the tribological tests.

between two components (pin and sample) was 3.98 mm^2, resulting in contact pressure of 6.29 N/mm^2. Three samples were produced for each processing condition. For each processing condition, three tribological tests were performed and the mean and standard deviation were calculated.

From the tests it was possible to obtain the COF directly and, in addition, the mass loss of the sample was determined (difference between the initial and the final mass). The mass loss and the density of the materials allowed to determine the wear volume (w) in mm^3. So, the specific wear rate (k) of the surfaces was calculated according to the Eq. 1.

$$k = \frac{w}{F_n \cdot s} \tag{1}$$

where F_n represents the normal force in N (25 N) and s is the sliding distance in m ($\cong$284 m).

It should be noted that while the COF variable is measured over time, the variable specific wear rate is a unique value, measured at the end of each test.

2.3 Statistical Analysis

The general procedure followed in this statistical study is schematically presented in Fig. 4. A significance level of 5% was considered for all the statistical analysis.

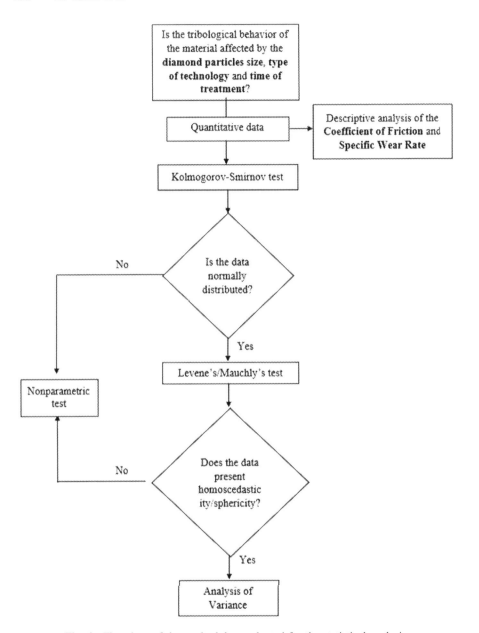

Fig. 4. Flowchart of the methodology adopted for the statistical analysis.

3   Statistical Analysis and Discussion of Results

3.1   Descriptive Analysis of COF

In this section, a statistical study of the dependent variable, coefficient of friction (COF), in each sample [316L + C_0.1–0.5 (SLS), 316L + C_40–60 (SLS), 316L + C_0.1–0.5 (HP) and 316L + C_40–60 (HP)] is presented considering each trial. It was carried out in IBM® SPSS software. Three trials per sample were executed. The obtained results are presented in Table 2.

Table 2. Statistics for each sample considering each trial for COF.

Material		Statistics			
		Mean	Median	Std. Deviation	Skewness
316L + C_0.1–0.5 (SLS)	Trial 1	0.4690	0.4637	0.2112	0.2600
	Trial 2	0.4327	0.4324	0.0087	0.1040
	Trial 3	0.4484	0.4574	0.0242	-0.7920
316L + C_40–60 (SLS)	Trial 1	0.1195	0.1192	0.0029	0.3620
	Trial 2	0.1309	0.1305	0.0033	0.3000
	Trial 3	0.1320	0.1326	0.0035	0.2120
316L + C_0.1–0.5 (HP)	Trial 1	0.5837	0.5939	0.0460	-0.3820
	Trial 2	0.5446	0.5313	0.0286	0.2970
	Trial 3	0.4926	0.4930	0.0115	-0.3280
316L + C_40–60 (HP)	Trial 1	0.1383	0.1375	0.0039	0.6160
	Trial 2	0.1572	0.1567	0.0034	0.5350
	Trial 3	0.1158	0.1145	0.0064	0.8220

The descriptive statistics of each sample considering the three trials performed is discussed. An example is given for Trial 2 of 316L + C_0.1–0.5 (SLS). The same analysis is applicable to Trials 1 and 3, as well as to other samples. Trial 2 presents a mean and median of 0.4327 and 0.4324, respectively. This leads to the conclusion that, for Trial 2, the mean COF value is 0.4327 and half the COF values are less than or equal to 0.4324.

COF's skewness for Trial 2 (0.1040) mirrors an approximately symmetric distribution.

Comparing all histograms from the samples, it is possible to conclude that Trial 1 is the worst regarding to its distribution. This might have due to some aspects of the experiments performed. The same counter body was utilized for all trials. In Trial 3, the counter body has a smoother surface with fewer asperities than in Trials 1 and 2, and fewer rough edges are encountered. Therefore, the contact between the two materials (counter body and sample) becomes more uniform. Due to this fact, the relative movement is more easily maintained when compared to Trials 1 and 2. In Trial 1, the surfaces of the counter body and sample are more irregular at a microscopic level (larger number and size of asperities) and therefore there is higher resistance. This leads to higher difficulty in sliding and consequently, a distribution for COF that is not close to normality. Over time and trials, less rough edges exist, and the track adapts to the counter body geometry, leading to better distribution for COF in Trial 3.

3.2   Descriptive Analysis of Specific Wear Rate

In this section a statistical study of the dependent variable, specific wear rate (k), for each sample [316L + C_0.1–0.5 (SLS), 316L + C_40–60 (SLS), 316L + C_0.1–0.5 (HP) and 316L + C_40–60 (HP)] is presented. The analysis was carried out in IBM® SPSS software, and the obtained results are presented in Table 3.

Table 3. Statistics for each sample considering each trial for k.

Sample	Statistics			
	Mean	Median	Std. Deviation	Skewness
316L + C_0.1–0.5 (LS)	0.00002733	0.00002410	0.00000862	1.451
316L + C_40–60 (LS)	0.00000236	0.00000265	0.00000223	−0.583
316L + C_0.1–0.5 (HP)	0.00001180	0.00001200	0.00000111	−0.782
316L + C_40–60 (HP)	0.00000058	0.00000109	0.00000148	−1.356

From the results obtained, it is possible to conclude that the 316L + C_0.1–0.5 (SLS) sample has the highest mean value of the specific wear rate. Contrarily, the 316L + C_40–60 (HP) sample is the one presenting the lowest specific wear rate. Considering the effect of the particle size, these results show that the samples with higher particles sizes are more resistant to wear than samples with lower particles sizes. Regarding the median value, the 316L + C_40–60 (HP) sample experienced the lowest value for this statistic, meaning that 50% of this sample is subjected to less wear than 50% of the other samples during the performed trials. These results corroborate the above-mentioned conclusion on the mean value for k.

Regarding the skewness statistic for the analysis of k, only one sample [316L + C_0.1–0.5 (SLS)] has a right skewed distribution, with the remaining samples comprising left skewness. Taking the above assumptions in consideration, it can be said that the distribution of 316L + C_0.1–0.5 (SLS) is highly skewed to the right, that is, its right tail is longer than the left tail, and the k distribution is more concentrated on the left side. This is corroborated by the fact that the mean k value for this sample is higher than the median value, which in turn is higher than the mode (0.00002733 > 0.00002410 > 0.00002080). Amongst samples comprising a left skewed distribution and considering the previously referred assumptions, the 316L + C_40–60 (SLS) and 316L + C_0.1–0.5 (HP) ones present a moderately left skewed distribution, and the 316L + C_40–60 (HP) one has a highly left skewed distribution.

3.3   Differences of the Measurement of COF Through Time

Problem. In order to assess if there is a statistically significant difference of COF values of the samples with respect to time, five different time points were selected among the time range. Table 4 presents the five different time points, and the COF values for the four different material conditions, each with three samples, resulting in a total of twelve different conditions.

Table 4. COF values of the four different samples at the five time points.

Sample	Time (s)				
	3600	6300	9000	11700	14397
316L + C_0.1–0.5 (SLS)	0.4473	0.4423	0.4621	0.4837	0.5058
	0.4389	0.4294	0.4389	0.4384	0.4351
	0.4109	0.4252	0.4463	0.4755	0.4654
316L + C_40–60 (SLS)	0.1255	0.1215	0.1192	0.1165	0.1162
	0.1383	0.1383	0.1300	0.1297	0.1250
	0.1389	0.1330	0.1330	0.1280	0.1270
316L + C_0.1–0.5 (HP)	0.4980	0.5643	0.5919	0.6355	0.6406
	0.5144	0.5142	0.5262	0.5825	0.5928
	0.4888	0.4806	0.4905	0.5034	0.4847
316L + C_40–60 (HP)	0.1484	0.1408	0.1344	0.1383	0.1379
	0.1629	0.1607	0.1564	0.1579	0.1556
	0.1288	0.1187	0.1113	0.1087	0.1187

Resolution. In order to study the influence of time in the response variable (COF of the four different samples), the first method attempted was the repeated measures ANOVA. This method eliminates sources of variability between subjects on the experiment error [26]. Therefore, the null hypothesis (H_0) for this test, is given as follows:

H_0: The COF mean values are equal at different time points.

The method implies that the response variable has to be normally distributed within each time points, and sphericity must be guaranteed. Considering the Mauchly's test, $\chi 2$ (4) = 2.760, p-value = 8.66e−09 (p-value < 0.05), the assumption of sphericity is rejected for a significance level of 5%. Corrections could be performed to overcome the violation of sphericity.

Regarding the violation of the normally distributed data assumption (p-value < 0.05), the analysis of variance between time points was again conducted, although with the Friedman test - a non-parametric test [27]. The null-hypothesis and the alternative hypothesis, for this test, are the same as the ones formulated for the repeated measurements ANOVA test. The Friedmann's test is described by Eq. 2 [27].

$$\chi^2 = \frac{12n}{p(p+1)} \sum_{j=1}^{p} \left\{ \bar{r}_j - \frac{1}{2}(p+1) \right\}^2 \tag{2}$$

where $\bar{r}_j$ is the mean rank of the jth time point, p is the number of ranks (time points) and n the number of rows (total number of samples). If χ^2 is too high, then the mean ranks differ significantly [27].

The statistics for this test, $\chi^2(4) = 3.800$, p-value $= 0.434$ (p-value > 0.05), do not reject the null hypothesis of the COF mean values being equal through all the time points. This outcome enables to consider the mean values of COF, through all the time points, for further analysis.

3.4 Effects of Factors on the COF

Problem. The aim of this statistical analysis was to understand how the diamond particles size and the type of technology used in each sample affect the COF. Having this in mind, both levels of each factor were considered and the mean COF over time from the three trials (Table 5) was tested for each combination of factors.

Table 5. Means of COF for each of the three trials over time, for the four samples.

		Diamond particles size (factor A)	
		0.1–0.5 µm	40–60 µm
Type of Technology (factor B)	HP	0.46897; 0.43269; 0.44838	0.11947; 0.13091; 0.13196
	SLS	0.58369; 0.54460; 0.49262	0.13831; 0.15715; 0.11585

Resolution. This two-factor experiment has 12 observations (2 diamond particles sizes x 2 technologies of sintering x 3 trials). In order to test the significance of the effect of each factor, the first attempt was to apply the Analysis of Variance (ANOVA) method with a two-factor factorial design. Firstly, for the application of this method, its assumptions must be satisfied, namely the response variable (mean of COF over time) has to be normally distributed; homoscedasticity has to be verified; as well as randomness of the data [26]. To check the normality of the data, the Kolmogorov-Smirnov (K-S) test was used.

Accordingly, the K-S test points to the same conclusion, as $D(12) = 0.290$, *p-value* $= 0.006$ (p-value < 0.05), therefore rejecting the null hypothesis.

Given that one of the ANOVA assumptions is not fulfilled, it was necessary to resort to another statistical method, in this case, a nonparametric method. Since the response variable presents different variabilities for each combination of factors' levels, a test to compare the distribution of two independent samples was chosen - the two-sample K-S test. The respective statistic, $D_{m,n}$, is calculated as presented in Eq. 3 [28].

$$D_{m,n} = \sup_{|x|<\infty} \left| \hat{F}_{1m}(x) - \hat{F}_{2n}(x) \right| \tag{3}$$

where m and n are the samples sizes, x is the response variable; and $\hat{F}_{1m}(x)$ and $\hat{F}_{2n}(x)$ are the empirical distribution functions obtained from each sample.

Regarding diamond particles size, by applying this method, it was possible to conclude that the distribution of COF across the 0.1–0.5 μm and 40–60 μm diamond particles is not the same, since the software returned the results of $D(12) = 1.732$, p-value (2-sided test) = 0.005. In contrast, the results that concern to the type of technology point to a similar distribution of COF across HP and SLS technologies, as $D(12) = 0.866$, p-value (2-sided test) = 0.441.

Therefore, on the basis of the two-samples K-S test, it is possible to conclude that the diamond particles sizes used in the sintering process produced a statistically significant impact in COF (p-value < 0.05), since the two particles sizes induced a statistically different effect on the response variable for both technologies of sintering. Contrarily, the type of technology did not affect COF in a statistically significant way (p-value > 0.05).

3.5 Effects of Factors on the Specific Wear Rate (k)

Problem. The main goal of this statistical analysis was to study the effects of the diamond particles size and type of technology on the specific wear rate of the sample. Table 6 presents the two levels of each factor considering the combination of factors.

Table 6. k values for each of the three trials, for the four samples.

		Diamond particles size (factor A)	
		0.1–0.5 μm	40–60 μm
Type of Technology	HP	3.71e−05; 2.41e−05; 2.08e−05	4.42e−06;0.00e+00; 2.65e−06
	SLS	1.20e−05; 1.28e−05; 1.06e−05	−1.09e−06; 01.75e−06; 1.09e−06

The negative value of k observed for the SLS samples with particle size of 40–60 μm means that the counter body had transferred mass to sample. The null value of k observed for the same particle size using HP for the same particle size means that there is an equilibrium between the mass transferred from counter body to sample and from to sample to counter body.

Resolution. Again, the ANOVA method was performed in order to test the significance of the effect of particle size and technology type. However, the same assumptions previously described need to be verified. The normality of the data was checked by performing the K-S test.

The null hypothesis is not rejected [$D(12) = 0.198$ and p-value = 0.200 (p-value > 0.05)] and therefore the k data follow a normal distribution. The homoscedasticity was also checked by performing the Levene's test, from which it was possible to confirm the homogeneity of variance, once the obtained p-values were of 0.15 and 0.20 for different technologies [$D(12) = 8.590$] and diamond particles sizes [$D(12) = 7.573$], respectively.

Considering a full factorial model, the analysis of interactions and factors was performed. The effect of each factor indicates a variation in the response variable due to the change in the levels of particle size or technology type. An interaction between the two factors is verified when the effect on one factor depends on the condition of the other factors.

The interaction between particle size and type of technology (H_{03}) was observed since for $F(1,8) = 6.872$ the p-value obtained was 0.031 (p-value < 0.05), and so the null hypothesis is rejected. Therefore, it can be concluded that the interaction between particle size and technology type affects the k. Since an interaction was observed, the analysis of main effects does not explain correctly the effect of factors on the response variable.

The high value of R squared ($R^2 = 0.891$) obtained on the ANOVA test means that the relation between the response variable, the levels of factors and the interaction are well explained by the model. The same conclusion can be taken by observing the corrected value (p-value $\sim 0.00 < 0.05$) on ANOVA table once there is a significative statistical relation. Additionally, the low value of the error (2.067e-11) mirrors the variability of the residuals, which corresponds to the random errors that models cannot explain. Therefore, this allows to conclude that, in fact, there is an interaction.

Figure 5 presents the interaction plot that displays the fitted values of the k variable on the y-axis and the particle size values on the x-axis. The two lines represent the technology types.

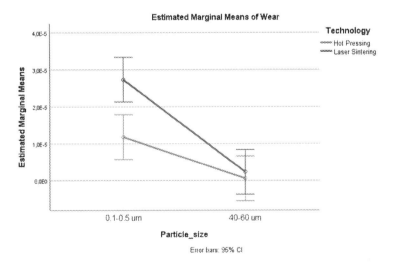

Fig. 5. Interaction effect between particle sizes for two technologies considered.

The different slopes propose that there is an interaction effect, which is confirmed by the p-value for the particle size/technology type. Also, from the distances between the segment edges, it is possible to see that, for smaller particles, there are significative differences between technologies, while for bigger particles this is not verified, suggesting that the behavior of the factor levels changes with type of technology used.

Additionally, the normality of the residuals was checked by performing the K-S test.

The null hypothesis is not rejected $[D(12) = 0.215$ and $p\text{-}value = 0.102$ (p-value > 0.05)] and therefore the k data follow a normal distribution.

The homoscedasticity of the residuals was checked by performing the Levene's test. The plot of residuals against estimated values of k is shown in Fig. 6. The results of Levene's test confirm the homoscedasticity. A p-value of 0.102 for $D(12) = 0.215$ was obtained and therefore the null hypothesis is not rejected. The homogeneity of variance can be also confirmed by a satisfactory pattern in the referred plot. However, the Fig. 6 reveals a funnel pattern for residuals, which means that there are anomalies in the data.

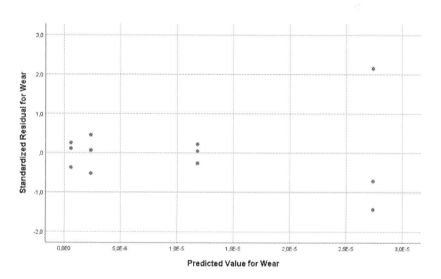

Fig. 6. Plot of residuals against estimated values of k.

In order to understand the anomalies verified in the previous graph, a boxplot of two diamond particles sizes and two type of technologies (Fig. 7) was executed. The graph allowed to conclude that the non-satisfactory pattern might be due to the large dispersion observed for particle size of 0.1–0.5 µm and SLS technology, when compared to the other conditions. No outliers exist in the data.

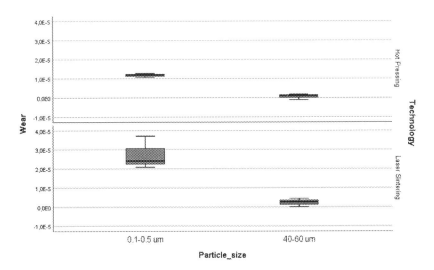

Fig. 7. Boxplot of two diamond particles sizes and two type of technologies.

As mentioned during ANOVA table analysis, the high value of R^2 revealed a well explained model. The corrected model with p-value lower than significance level proves a significative statistical relation and the Levene's a variance homogeneity. So, considering these points and the variability on boxplots, the anomalies found on residuals should not be considered relevant.

4   Conclusions

The modification of the tribological properties of laser textured 316L stainless steel reinforced with a CuCoBe-diamond composites was investigated in this work. After performing the descriptive analysis of the data, five time points were analyzed using the Friedman's test in order to assess the effect of time in the COF. Significant differences were found between the tested time points, suggesting that the reinforcement with diamond particles and the sintering technology affect the COF over time. Then, the ANOVA method with a two-factor factorial design was applied to study how two factor – the diamond particles size and the type of technology – affected the COF. It was found that the diamond particles size significantly impacts the referred parameter, whereas the type of treatment does not produce a significant effect on COF value. Finally, regarding the effects of the factors on the k, a high value of R^2 was obtained in the ANOVA test. This indicates an adjusted model to describe the relation between the response variable, the factors levels and their interaction. Based on the evidences provided in this paper, further statistical assessments must be performed in the topic to investigate the impact of more diamond particle sizes and types of technology on the COF and k, which would allow a more conscious and assertive implementation of such methodologies in specific environments.

Acknowledgements. This work was supported by FCT national funds, under the national support to R&D units grant, through the reference projects UIDB/04436/2020; UIDP/04436/2020; UIDB/00319/2020; and PTDC/CTM-COM/30416/2017.

References

1. Xavior, M.A., Adithan, M.: Determining the influence of cutting fluids on tool wear and surface roughness during turning of AISI 304 austenitic stainless steel. J. Mater. Process. Technol. **209**(2), 900–909 (2009)
2. Mahathanabodee, S., Palathai, T., Raadnui, S., Tongsri, R., Sombatsompop, N.: Dry sliding wear behavior of SS316L composites containing h-BN and MoS 2 solid lubricants. Wear **316**(1–2), 37–48 (2014)
3. Baddoo, N.R.: Stainless steel in construction: A review of research, applications, challenges and opportunities. J. Constr. Steel Res. **64**(11), 1199–1206 (2008)
4. Yan, F.K., Liu, G.Z., Tao, N.R., Lu, K.: Strength and ductility of 316L austenitic stainless steel strengthened by nano-scale twin bundles. Acta Mater. **60**(3), 1059–1071 (2012)
5. Yamagata, H.: The Piston Ring. The Science and Technology of Materials in Automotive Engines, pp. 87–109. Woodhead Publishing (2005)
6. Zabala, B., et al.: Friction and wear of a piston ring/cylinder liner at the top dead centre: Experimental study and modelling. Tribol. Int. **106**, 23–33 (2017)
7. Carvalho, O., Buciumeanu, M., Madeira, S., Soares, D., Silva, F.S., Miranda, G.: Optimization of AlSi – CNTs functionally graded material composites for engine piston rings. Mater. Des. **80**, 163–173 (2015)
8. Kan, W., et al.: A study on novel AISI 304 stainless steel matrix composites reinforced with (Nb0.75, Ti0.25)C. Wear **398–399**, 220–226 (2018)
9. Velasco, F., Lima, W.M., Antón, N., Abenójar, J., Torralba, J.M.: Effect of intermetallic particles on wear behaviour of stainless steel matrix composites. Tribol. Int. **36**(7), 547–551 (2003)
10. Ni, Z., Sun, Y., Xue, F., Bai, J., Lu, Y.: Microstructure and properties of austenitic stainless steel reinforced with in situ TiC particulate. Mater. Des. **32**(3), 1462–1467 (2011)
11. Lou, D., Hellman, J., Luhulima, D., Liimatainen, J., Lindroos, V.: Interactions between tungsten carbide (WC) particulates and metal matrix in WC-reinforced composites. Mater. Sci. Eng. A **340**(1–2), 155–162 (2003)
12. Srivastava, A., Das, K.: The abrasive wear resistance of TIC and (Ti, W)C-reinforced Fe-17Mn austenitic steel matrix composites. Tribol. Int. **43**(5–6), 944–950 (2010)
13. Srivastava, A., Das, K.: Microstructural and mechanical characterization of in situ TiC and (Ti, W)C-reinforced high manganese austenitic steel matrix composites. Mater. Sci. Eng. A **516**(1–2), 1–6 (2009)
14. Knowles, A., Jiang, X., Galano, M., Audebert, F.: Microstructure and mechanical properties of 6061 Al alloy based composites with SiC nanoparticles. J. Alloys Compd. **615**(S1), S401–S405 (2014)
15. Artini, C., Muolo, M.L., Passerone, A.: Diamond-metal interfaces in cutting tools: A review. J. Mater. Sci. **47**(7), 3252–3264 (2012)
16. De Oliveira, L.J., Cabral, S.C., Filgueira, M.: Study hot pressed Fe-diamond composites graphitization. Int. J. Refract. Met. Hard Mater. **35**, 228–234 (2012)
17. Zhang, G., Xing, J., Gao, Y.: Impact wear resistance of WC/Hadfield steel composite and its interfacial characteristics. Wear **260**(7–8), 728–734 (2006)

18. Kumar, V., Venkatesh, C.: Effect of ceramic reinforcement on mechanical properties of aluminum matrix composites produced by stir casting process. Mater. Today Proc. **5**(1), 2466–2473 (2018)
19. Khmelnitsky, R., Gippius, A.: Transformation of diamond to graphite under heat treatment at low pressure. Phase Transit. **87**(2), 175–192 (2013)
20. Cunha, A., Ferreira, R., Trindade, B., Silva, F.S., Carvalho, O.: Reinforcement of a laser-textured 316L steel with CuCoBe-diamond composites through laser sintering. Mater. Manuf. Process. **35**, 1–8 (2020)
21. Cunha, A., Ferreira, R., Trindade, B., Silva, F.S., Carvalho, O.: Production of a laser textured 316L stainless steel reinforced with CuCoBe + diamond composites by hot pressing: Influence of diamond particle size on the hardness and tribological behaviour. Tribol. Int. **146**, 106056 (2020)
22. Liverani, E., Toschi, S., Ceschini, L., Fortunato, A.: Effect of selective laser melting (SLM) process parameters on microstructure and mechanical properties of 316L austenitic stainless steel. J. Mater. Process. Technol. **249**, 255–263 (2017)
23. Bartolomeu, F., et al.: 316L stainless steel mechanical and tribological behavior—A comparison between selective laser melting, hot pressing and conventional casting. Addit. Manuf. **16**, 81–89 (2017)
24. Miranda, G., Buciumeanu, M., Madeira, S., Carvalho, O., Soares, D., Silva, F.S.: Hybrid composites - metallic and ceramic reinforcements influence on mechanical and wear behavior. Compos. Part B **74**, 153–165 (2015)
25. Miranda, G., Buciumeanu, M., Carvalho, O., Soares, D., Silva, F.S.: Interface analysis and wear behavior of Ni particulate reinforced aluminum-silicon composites produced by PM. Compos. Part B **69**, 101–110 (2015)
26. Davis, C.S.: Statistical Methods for the Analysis of Repeated Measurements. Springer, New York (2002)
27. Friedman, M.: The use of ranks to avoid the assumption of normality implicit in the analysis of variance. J. Am. Stat. Assoc. **3**, 1–47 (2012)
28. Bagdonavičius, M.S., Kruopis, V., Nikulin, J.: Nonparametric Tests for Complete Data. ISTE, London; Wiley, Hoboken (2011)

Shiny App to Predict Agricultural Tire Dimensions

Ana Rita Antunes$^{(\boxtimes)}$ and Ana Cristina Braga

ALGORITMI Centre, University of Minho, 4800-058 Guimarães, Portugal
ana_antunes96@hotmail.com, acb@dps.uminho.pt

Abstract. The main objective of this project, carried out in an industrial context, was to apply a multivariate analysis to variables related to the specifications required for the production of an agricultural tire and the dimensional test results. With the exploratory data analysis, it was possible to identify strong correlations between predictor variables and with the response variables of each test. In this project, the principal component analysis (PCA) serves to eliminate the effects of multicollinearity. The use of regression analysis was intended to predict the behavior of the agricultural tire considering the selected variables of each test. In the case of Test 1, when applying the Stepwise methods to select the variables, the model with the lowest value of Akaike Information Criterion (AIC) was achieved with the technique "Both". However, the lowest value of AIC for Test 2 was achieved with "Backward". Regarding the validation of assumptions, both Test 1 and Test 2 were validated. Therefore, all the quantitative variables are important, both in Test 1 and Test 2, because they are a linear combination that determines the principal components. In order to make it easier to compute predictions for future agricultural tires, an application that was developed in Shiny allows the company to know the behavior of the tire before it was produced. Using the application, it is possible to reduce the industrialization time, materials and resources, thus increasing efficiency and profits.

Keywords: Multiple linear regression · Principal component analysis · Siny application · Agricultural tires

1 Introduction

In the industrial process of production of a new tire, it is necessary to consider some specifications. The agricultural tire is constituted with different components like the tread, belt, inner liner, sidewall, bead, among others. In this case, it is important to define the mold, the material and the quantity. After that, the tire has to pass some tests, for example, dimensional and endurance tests, among others. The tire passes the test if the results are in accordance with the legal norms, where the maximum and minimum of dimensional and endurance values are defined. So, when the test result is greater than the maximum defined,

© Springer Nature Switzerland AG 2020
O. Gervasi et al. (Eds.): ICCSA 2020, LNCS 12251, pp. 247–260, 2020.
https://doi.org/10.1007/978-3-030-58808-3_19

the tire doesn't pass and the company has to make changes in the type and/or the quantity of materials.

In this study, the main goal was to apply multivariate analysis to variables related to the specifications required for agricultural tire production and dimensional test results, Test 1 and Test 2. The purpose of this was to understand which variables influence the test results and to predict their values. So, to develop a tire it is important to consider a lot of variables simultaneously and, if it is possible to predict the values for the two tests, it will make it easier for the producers. Multiple Linear Regression (MLR) will help to achieve the results that we want, because the predictors variables are quantitative. MLR has many assumptions to be considered and one of them is multicollinearity effects.

Multicollinearity effects are when two or more predictor variables have a strong correlation among themselves. This can cause problems in MLR. When we estimate regression coefficients and the predictor variables are highly correlated, the coefficients tend to vary widely. Another problem is when we want to make an interpretation of a regression coefficient, the signal can be misleading [5]. One possibility to correct this problem is using Principal Component Regression (PCR), which is a linear regression using principal components. Maxwell et al. [7] wrote an article to tackle with multicollinearity effects and here 5 methodologies were tested: Partial Least Square Regression (PLSR), Ridge Regression (RR), Ordinary Least Square Regression (OLS), Least Absolute Shrinkage and Selector Operator (LASSO) Regression, and the Principal Component Regression (PCR). To compare the 5 methodologies, they used a different number of observations and a number of predictor variables. Root Mean Square Error (RMSE) and AIC were used to compare the performance of each model. With this analysis the authors concluded that PCR has the lowest AMSE and AIC, which means that according to them, PCR is the most efficient in handling critical multicollinearity effects [7]. Lafi and Kaneene used Principal Component Analysis (PCA) to detect and correct multicollinearity effects in a veterinary epidemiological study. In this article were compare OLS and PCR to adjust regression coefficents. The PCR coefficients were more reliable than OLS [6].

After selecting the model for Test 1 and Test 2, a web application was developed to predict the test results before the tire was produced.

2   Methods

2.1   Principal Component Analysis

PCA is a statistic procedure for multivariate problems. It was introduced in 1901 by Pearson and in 1933 it was independently developed by Hotelling [8].

PCA is useful when there are many predictor variables regarding the number of observations in the dataset. It is also used when the predictor variables are highly correlated with each other because it eliminates the effects of multicollinearity. Normally, PCA is used to reduce the dimensionality of a problem and principal component represents most of the information contained in the dataset. This means that the first PC explains the greater proportion of the original variables variation and the second explains the second greater proportion, but it is independent of the first, and so on.

As it is widely known, PCA transforms the variation of a set of variables highly correlated into a new set of variables that are uncorrelated and orthogonal. This new set of variables is a linear combination of the initial p variables. Linear combinations are described as follows (Eq. 1):

$$PC_1 = a_{11}x_1 + a_{12}x_2 + \ldots + a_{1p}x_p$$
$$PC_2 = a_{21}x_1 + a_{22}x_2 + \ldots + a_{2p}x_p$$
$$\vdots$$
$$PC_p = a_{p1}x_1 + a_{p2}x_2 + \ldots + a_{pp}x_p$$

(1)

where a_{ij} are the loadings, x_1, x_2, ..., x_p are the initial variables and PC_1, PC_2, ..., PC_p are the p PCs [4].

After obtaining the linear combinations for each component, and when replacing them with the values of the initial variables, the scores are obtained [5].

2.2 Multiple Linear Regression

With linear regression it is possible to study the linear relationship between response variable (y_i, $i = 1, ..., n$) and one or more predictor variables (x_{ij}, $j = 1, ..., p$), where response variable is a quantitative variable and predictor variables can be quantitative or qualitative. When there is more than one predictor variables it is called Multiple Linear Regression (MLR), (Eq. 2), where β_0 is the constant term and β_p are the coefficients for each variable.

$$y_i = \beta_0 + \beta_1 x_{i1} + \beta_2 x_{i2} + \ldots + \beta_p x_{ip} + \varepsilon_i$$

(2)

To validate the model, it is necessary to verify some assumptions and this can be performed through an explanatory analysis of residuals. Thus, the assumptions to be validated are as follows [3]:

- $E[\varepsilon_i] = 0$, this means the average of the errors must be zero;
- $Var[\varepsilon_i] = \sigma^2$, so the errors variance must be constant;
- $\varepsilon_i \sim N(0, \sigma^2)$, with this, errors must follow a normal distribution;
- Errors are independent.

Another condition to be verified is the existence of multicollinearity and this can be identified by the correlations values and/or considering the Variance Inflation Factor (VIF). When VIF is greater than 10, it means that there are multicollinearity effects in the data. VIF is given by the expression:

$$VIF_j = \frac{1}{1 - R_j^2}$$

(3)

where R_j^2 is the coefficient of determination of x_j relative to the other predictor variables in the model [9].

Variable Selection Method. The stepwise method is used to obtain a model with predictor variables that better explain the variable response and it is possible to consider different criteria, for example, AIC. The "Backward" method builds the regression model using all the predictor variables and removes them considering the chosen criteria. The "Forward" method adds the predictor variables one by one until there are no more candidates that increase the value of the sum of squares in the regression model. However, it is possible to build a regression model with the entry and elimination of the predictor variables, considering the chosen criteria, called "Both" method. The iterative process ends when there are no more variables to be introduced or eliminated according to the criterion adopted [9].

One way to analyze the model that better explains the data in study is the value of AIC. This criterion compares the adequacy of the models when an attempt is made to balance the accuracy of the adjustment and the smallest number of explanatory variables [2]. The AIC value is calculated as follows:

$$AIC_c = -2log(L_p) + 2p \tag{4}$$

where L_p is the maximum value of likelihood function for the model and p is the number of predictor variables present in the model. The models with lowest AIC are the chosen ones [1].

3   Results and Discussion

For this analysis were used 146 experimental agricultural tires, 31 predictor variables, 4 of which are qualitative variables and 27 quantitative variables. We used 2 response variables, y_1 and y_2 for Test 1 and Test 2, respectively. The variables were coded due to a confidentiality agreement. All computations were made in R software using the appropriate packages available to perform the analysis.

Figure 1 represents the correlation coefficients (color intensity and the size of the circle are proportional to the correlation coefficients) and there are strong relationships with variable y_1, variable response for Test 1, as well as with y_2, variable response for Test 2.

Taking into account the values of the correlations of Fig. 1, multicollinearity effects are expected due to the values taken from r between the predictor variables once these variables are correlated with each other. It is also possible to see that x_6, x_7, x_8, x_9, x_{12}, x_{13}, x_{18}, x_{19}, x_{23}, x_{25}, x_{27}, x_{32} and y_1 are correlated (where $r > 0.90$), as well as between x_5, x_{10}, x_{11}, x_{15} and y_2 (where $r > 0.90$).

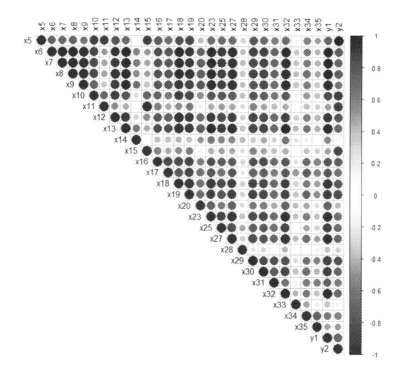

Fig. 1. Pearson's correlations. (Color figure online)

3.1 Principal Component Analysis Results

As referred in Sect. 2.1, PCA can be used to reduce the dimensionality of a problem or to eliminate multicollinearity effects. In this study, it was necessary to prove if multicollinearity effects exist. Regarding this problem, the data were normalized since the variables take different scales of measures.

In Table 1, the VIF values for 19 quantitative predictor variables are presented and the results are the same in Test 1 and Test 2, when an MLR was made for both tests. The VIF values for the other response variables are less than 20. Regarding the results obtained in Table 1, there are multicollinearity effects in the study, because most of the VIFs values are higher then 10.

Since the main objective is to build models that allow predictions for Test 1 and Test 2, the conditions of applicability of MLR models must be guaranteed. For this reason, we opted to use PCA to eliminate the effects of multicollinearity. For this reason, the 27 principal components were used in the models for Test 1 and Test 2 instead of the original variables.

Table 1. VIF value for quantitative predictor variables.

Variable	VIF
X_5	927.10
X_6	608.11
X_7	405.50
X_8	128.60
X_9	843.43
X_{10}	37.90
X_{11}	26452.1
X_{12}	360.31
X_{13}	2707.65
X_{14}	49.92
X_{15}	27120.15
X_{17}	25.91
X_{18}	510.04
X_{19}	77.13
X_{23}	927.66
X_{25}	51.99
X_{27}	41.99
X_{32}	193.62

The graph in Fig. 2 represents the biplot after the rotation varimax for the first two principal components, where the first explains 52.5% and the second 15.6% of the total data variation. It can be seen in Fig. 2 that variables x_5, x_{11} and x_{15} have the greatest positive contribution for the second principal component. Variable x_{14} has the greatest negative contribution for the first component. However, the other variables have the greatest positive contribution for the first component.

In this study, the 27 principal components were used because it was necessary to consider all the information and, for this reason, it was difficult to perform the interpretation of each principal component.

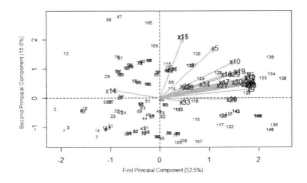

Fig. 2. Biplot for the first and second principal components.

3.2 Multiple Linear Regression Models

After the determination of each PC we proceeded to the construction of MLR models for each tire test. Two models were found using stepwise methods and considering AIC criteria to select the model for Test 1 and Test 2.

In Table 2 the model using "Both" technique has the lowest AIC value, for Test 1, for this reason it was the selected model. For Test 2, the lowest AIC value is using "Backward" and this was the chosen one.

Table 2. AIC values for Test 1 and Test 2.

	Test 1	Test 2
Backward	648.05	764.03
Forward	648.05	766.02
Both	647.45	778.42

Figure 3 shows the set of graphs produced in R using the plot (model) function to validate the assumptions. The first graph, Residuals vs Fitted, proves that the variance of residuals is constant and that residuals are independent because there isn't any pattern or tendency. The second graph shows that the errors follow a normal distribution, since the values are according to the diagonal, except on the extremes, which can indicate the presence of outliers. The Kolmogorov-Smirnov test was used to confirm if the errors follow a normal distribution, considering the following hypotheses: $H_0 : \varepsilon_i \sim N(\mu, \sigma^2)$ VS $H_1 : \varepsilon_i \nsim N(\mu, \sigma^2)$. For this test, the $p - value = 0.615$ reveals that the errors could follow the normal distribution for a significance level $\alpha = 0.05$. The last graph, Residual vs Leverage, shows there are no influence points.

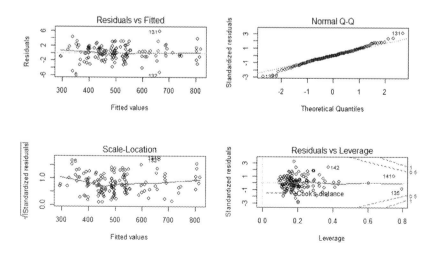

Fig. 3. Validated assumptions for Test 1.

Based on graphs in Fig. 4 it is possible to draw the same conclusions for Test 2. Looking at the Normal Q-Q plot, most of the values are according to the diagonal, except on the extremes, which means there isn't evidence to reject the null hypothesis. Regarding the Kolmogorov-Smirnov test, $p - value = 0.966$, the null hypothesis isn't rejected and the errors could follow the normal distribution for a significance level $\alpha = 0.05$.

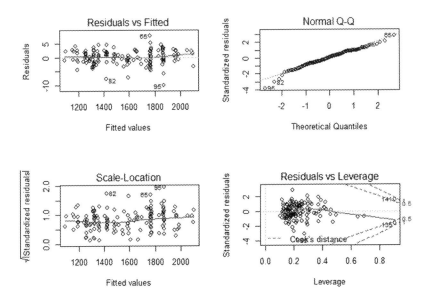

Fig. 4. Validated assumptions for Test 2.

When the extremes in Normal Q-Q plot are straight out it can mean there are outliers. The graphs in Fig. 5 reveal that there are five outliers for Test 1 and four for Test 2. All of them were individually analyzed to understand if it is a process problem or a human error since most of the values are not automatically introduced into company programs.

The entire analysis was repeated, for both models, after removing the outliers and it was found that by using the same criteria the results were not very different and outliers continued to exist. Since all the possible variables were not used in this study and the values of each observation, considered as an outlier, do not appear to be a human error, so it was decided to keep all the observations.

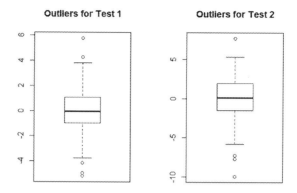

Fig. 5. Box-plots for the residuals for Test 1 and Test 2.

3.3   Shiny Application

The main objective of this study was to predict the results for Test 1 and Test 2 based on the constructed models. For this reason, it was developed a web application using Shiny. In the application it is possible to do two things: upload the dataset and make predictions based on the values of the variables.

Before creating the application it was important to define the necessary steps for its construction, which are represented in Fig. 6.

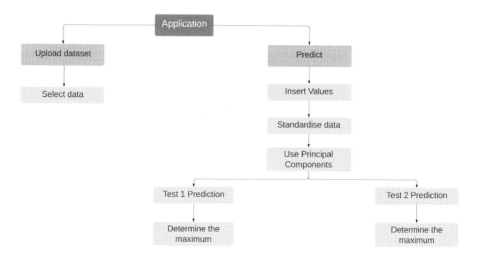

Fig. 6. Flowchart to create the application.

Programming Code. In order to obtain the application interface, a programming code was developed. Shiny is divided into two parts, "ui" and "server". "ui", known as the interface, is used to define how the web application is going to look like. In contrast, "server" is used to define what the application is going to do and this is where the calculations for making predictions for Test 1 and Test 2, are made.

Before starting programming, four Excel documents were added to be used in a later stage. Figure 7 shows the information related to the dataset under study, the coefficients for Test 1 and Test 2, and the loadings for each principal component.

```
 9
10  library(shiny)
11  library(markdown)
12  library(datasets)
13  #  setwd("C:/Users/Ana Antunes/Desktop/DADOS")
14  setwd("C:/Users/Ana Antunes/Desktop/universidade/Tese/Tese/DADOS")
15  dados=read.csv2(file="pesos.csv",sep=";",header=TRUE)
16  teste1=read.csv2(file="teste1.csv",sep=";",header=TRUE)
17  teste2=read.csv2(file="test2.csv",sep=";",header=TRUE)
18  exper=read.csv2(file="experimental2.csv",sep=";",header=TRUE)
19  attach(exper)
20
```

Fig. 7. Information to start the web application.

Firstly, in "ui" the menus were defined as "Upload Dataset" and "Prediction". In lines 24 and 25 is where the user can choose the file to load for the application. Regarding "Prediction", it specifies the quantitative variables, using "numericInput", and the qualitative variables, using "selectInput" (Fig. 8).

```
22
23  ui <- navbarPage("Agricultural Tire",
24              tabPanel("Upload Dataset", fluid=TRUE,
25                      fileInput("fileInput", "Choose file"),
26                      DT::dataTableOutput("table")
27              ),
28              tabPanel("Prediciton", fluid=TRUE,
29                      sidebarPanel("variables",
30                          numericInput(inputId = "obs1",
31                                      label = "x5",
32                                      value = ""),
33                          numericInput(inputId = "obs2",
34                                      label=" x6",
35                                      value=""),
36                          numericInput(inputId = "obs3",
37                                      label="x7",
38                                      value=""),
39                          selectInput(inputId = "obs17",
40                                      label="x21",
41                                      choices = c("1","2")),
```

Fig. 8. Interface code in Shiny.

In order to show the prediction for Test 1 and Test 2, in line 124, was created a button "Go" and the next line is to show the table. On the following lines the colors of the application are defined (Fig. 9).

```
124          actionButton("go2", "Go"),
125          mainPanel(DT::dataTableOutput("tabela"))
126      ),
127
128      tags$style(type = 'text/css', '.navbar { background-color: #04E2FF;
129          font-family: Calibri;
130          font-size: 13px;
131          color: #232426; }',
132
133          '.navbar-dropdown { background-color: #04E2FF;
134          font-family: Calibri;
135          font-size: 13px;
136          color: #232426; }',
137
138          '.navbar-default .navbar-brand {
139          color: #232426;
140          }'
```

Fig. 9. Interface code in Shiny (continuation).

The next step was to define the necessary calculation to predict the value for Test 1 and Test 2. In the first place, the data have different scales and for this reason the data were standardized and the values introduced for each variable were saved (Fig. 10).

```
160 -  observeEvent(input$go2,{
161        a <- ifelse(input$obs1>=1,round((input$obs1-mean(exper$x5))/sd(exper$x5),6),0)
162        b <- ifelse(input$obs2>=1,round((input$obs2-mean(exper$x6))/sd(exper$x6),6),0)
163        q1 <- ifelse(input$obs17=="1",1,0)
164        q2 <- ifelse(input$obs17=="2",1,0)
165        c <- ifelse(input$obs3>=1,round((input$obs3-mean(exper$x7))/sd(exper$x7),6),0)
```

Fig. 10. Server code for the values introduced.

Thereafter, it was important to define which variable is quantitative to determine the principal components for the 27 variables. After that, using the quantitative variables and the loadings obtained before, the principal components were calculated (Fig. 11).

```
206    quant <- c(a,b,c,d,e,f,g,h,i,j,k,l,m,n,o,p,s,u,w,x,y,z,aa,ab,ac,ad,ae)
207
208    # componentes
209    cp1 <- round(quant %*% dados[,1],5)
210    cp2 <- round(quant %*% dados[,2],5)
211    cp3 <- round(quant %*% dados[,3],5)
212    cp4 <- round(quant %*% dados[,4],5)
```

Fig. 11. Server code for creating the principal components.

Finally, the models for Test 1 and Test 2 were calculated using the selected model coefficients for each test and the principal components obtained before. In Fig. 12, n1 and n4 represent the MLR for Test 1 and Test 2, respectively. The maximum is calculated using the expression in n2 and n5. After this, one condition was created to verify if a tire passes the test, represented by n3 and n6. With this information, lines 264, 265 and 266 were used to construct the table with the calculated results. The last line is used to run the application.

```
240    n1 <- teste1[1,2]+teste1[2,2]*cp1+teste1[3,2]*cp2+teste1[4,2]*cp3+teste1[5,2]*cp4+teste1
241
242    # maximum for test 1
243    n2 <- (i1+0.4*(w1*25.4-i1*0.8))*1.05
244
245    # see if the tire passes in test 1
246    n3 <- ifelse(n2<n1,"Not Passed", "Passed")
247
248    # prediction for test 2 using Backward
249    n4 <- teste2[1,2]+teste2[2,2]*cp1+teste2[3,2]*cp2+teste2[4,2]*cp3+teste2[5,2]*cp4+teste2
250
251    # maximum for test 2
252    n5 <- 2*((i1*j1)/100)*1.04+k1*25.4
253
254    # see if the tire passes in test 2
255    n6 <- ifelse(n5<n4, "Not Passed", "Passed")
256
257    output$sum1 <- renderPrint(n1)
258    output$sum2 <- renderPrint(n2)
259    output$sum3 <- renderPrint(n3)
260    output$sum4 <- renderPrint(n4)
261    output$sum5 <- renderPrint(n5)
262    output$sum6 <- renderPrint(n6)
263    }
264    tabela <- data.frame(round(n1,3),n3,round(n4,3),n6)
265    colnames(tabela) <- c("Test 1"," Result", "Test 2", "Result")
266    output$tabela <- DT::renderDataTable({DT::datatable(tabela, options = list(dom = 't'))})
267    }
268    )
269    }
270
271    # Run the application
272    shinyApp(ui = ui, server = server)
```

Fig. 12. Server code for predicting Test 1 and Test 2.

Application Interface. In Upload dataset it is possible to filter the data considering what is necessary to predict the value for Test 1 and Test 2. In Fig. 13 there is an example using a created dataset for an agricultural tire to explain only this functionality.

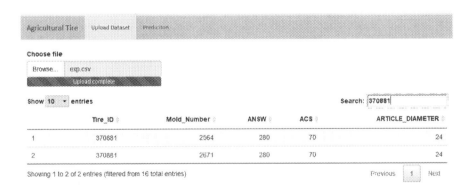

Fig. 13. Upload dataset.

In this case there are five variables and "Search" is an input for what we want to look for: for example, the tire identification number. The data have 15 different tires, where there are 2 tires that contain the number identification 370881 (lower left corner). Whoever wants to use the application for agricultural tires can filter for the tire number identification and its specification appears. This will be necessary for predicting Test 1 and Test 2.

Making predictions was one of the aims for this study and by using the developed application, the results for Test 1 and Test 2 can be predicted before tire production (Fig. 14).

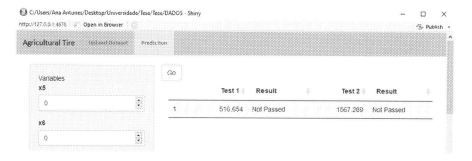

Fig. 14. Making predictions.

The chosen models for Test 1 and Test 2 use principal components that are a linear combination of the initial variables and for this reason it is fundamental to insert the 27 initial variables and the 4 qualitative variables. Therefore, when the variable is quantitative the user has to introduce the value, and when it is qualitative he has to select the pretended level. To make predictions, all the variables have to be filled and after that the results appear when the button "Go" is clicked. The application gives the results for Test 1, y_1, and Test 2, y_2. In addition, the "Result" (Fig. 14) indicates if the tire passed the test. For the production of agricultural tires it is necessary to consider legal norms and both tests have a maximum that cannot be exceeded. When the result is greater than the maximum, the tires do not pass the test, the specification has to be modified and in "Result" appears "Not passed". Otherwise, the agricultural tire passes the test and in "Result" appears "Passed".

4 Conclusion

The main goal was to apply multivariate analysis to variables related to tire production and identify the influences on the two tires tests. In the exploratory analysis it was possible to identify strong correlations between the quantitative variables, including the response variables for each test. With the variance inflation factor, it was possible to identify the existence of multicollinearity between quantity variables and this could be a problem when applying linear regression.

Principal component analysis was used to eliminate multicollinearity effects and to retain as much information as possible to apply to the models. For this reason, it was decided to use the 27 principal components and it was difficult to understand the meaning of each principal component considering the loadings' values.

Multiple linear regression was used to identify the significant variables to improve the agricultural tire production. This was also difficult to identify because we considered the 27 principal components and the qualitative variables. One of the objectives of this study was to find a multiple linear regression for the two tests. For the selection of variables we used Stepwise methods and the choice of the model to be considered was made taking into account the AIC value.

After obtaining the models for the two tests, an application was developed in Shiny in order to quickly and efficiently determine the test results for future agricultural tires. By using the application it is possible to reduce the quantity of materials and resources as there is an increase in efficiency and profits since this application can predict the performance of the tire before starting its production. In addition, reducing the industrialization time is also an advantage, because some specifications can be canceled before the production phase. It also helps to preserve the environment by reducing the destruction of tires with bad performances. Therefore, this application helps the users to select the best specification for the agricultural tire, thus generating more security in the specification to be used and enabling a reduction of errors by the research and development department.

Acknowledgments. This work has been supported by FCT – Fundação para a Ciência e Tecnologia within the R&D Units Project Scope: UIDB/00319/2020.

References

1. Burnham, K.P., Anderson, D.R.: Model Selection and Multimodel Inference: A Practical Information-Theoretic Approach, 2nd edn. Springer, New York (2002). https://doi.org/10.1007/b97636
2. Chattefuee, S., Hadi, A.S.: Regression Analysis by Example. 4. edn. (2006)
3. Chatterjee, S., Simonoff, J.S.: Handbook of Regression Analysis (2013)
4. Johnson, R., Wichern, D.: Applied Multivariate Statistical Analysis, 5th edn. Prentice Hall, Upper Saddle River (2002)
5. Kutner, M.H., Nachtsheim, C.J., Neter, J., Li, W.: Applied Linear Statistical Models, 5th edn. McGraw-Hill Irwin, New York (2005)
6. Lafi, S.Q., Kaneene, J.B.: An explanation of the use of principal-components analysis to detect and correct for multicollinearity. Preventive Veterinary Medicine (1992)
7. Maxwell, O., Amaeze Osuji, G., Precious Onyedikachi, I., Obi-Okpala, C.I., Udoka Chinedu, I., Ikenna Frank, O.: Handling critical multicollinearity using parametric approach. Acad. J. Appl. Math. Sci. **5**(11), 150–163 (2019)
8. Mishra, S.P., et al.: Multivariate statistical data analysis- Principal Component Analysis (PCA). Int. J. Livestock Res. **7**(5), 60–78 (2017)
9. Rawlings, J.O., Pantula, S.G., Dickey, D.A.: Applied Regression Analysis: A Research Tool. Springer Texts in Statistics, 2nd edn. Springer, Heidelberg (1998). https://doi.org/10.1007/b98890

Environmental Performance Assessment of the Transport Sector in the European Union

Sarah B. Gruetzmacher[1,2] , Clara B. Vaz[2,3] , and Ângela P. Ferreira[2(✉)]

[1] Universidade Tecnológica Federal do Paraná, Av. Sete de Setembro. 3165,
Curitiba 80230-901, Brazil
sarah.gruetz@gmail.com

[2] Research Center in Digitalization and Intelligent Robotics (CeDRI), Instituto
Politécnico de Bragança, Campus de Santa Apolónia, 5300-253 Bragança, Portugal
{clvaz,apf}@ipb.pt

[3] Center for Management and Industrial Engineering (CEGI/INESC TEC),
Porto, Portugal

Abstract. The European Union (EU) has been promoting diverse initiatives towards sustainable development and environment protection. One of these initiatives is the reduction of the greenhouse gas (GHG) emissions in 60% below their 1990 level, by 2050. As the transport sector is responsible for more than 22% of those emissions some strategies need to be taken towards a more sustainable mobility, as the ones proposed in 2011 White Paper on transport. Under this context, this study aims to evaluate the environmental performance of the transport sector in the 28 EU countries towards these goals, from 2015 to 2017. The transport environmental performance is measured through the composite indicator derived from the Benefit of the Doubt (BoD) model. The country transport environmental performance is assessed through the aggregation of multiple sub-indicators using the composite indicator derived from the Data Envelopment Analysis (DEA) model. The results indicate that the EU countries slightly improved their transport environmental performance, on average 2.8%. The areas where the inefficient countries need more improvement were also identified: reducing the GHG emissions from fossil fuels, increasing the share of transport energy from renewable sources and improving the public transport share of the total passenger transport.

Keywords: Transport environmental performance · Data
Envelopment Analysis · Sustainable development

1 Introduction

Transportation is an important sector in the European economy, it employs more than 11 million people and accounts for about 5% of Europe's Gross Domestic

This work has been supported by FCT – Fundação para a Ciência e Tecnologia within the Project Scope: UIDB/05757/2020.

O. Gervasi et al. (Eds.): ICCSA 2020, LNCS 12251, pp. 261–273, 2020.
https://doi.org/10.1007/978-3-030-58808-3_20

Product (GDP). Between 2010 and 2050, passenger transport activity is expected to grow by 42% and freight transport activity by 60% [1]. However, GHG emissions of the transport sector, in opposition to other sectors such as industry or energy related industries, have increased in the last 25 years and reached 22% of the total European GHG emissions in 2015 [2]. Under this scenario, the European Commission's, 2011, White Paper on transport - Roadmap to a Single European Transport Area – Towards a competitive and resource efficient transport system [3], proposed strategies for deep changes in the European transport sector aiming a more sustainable and efficient system. Some of the targets addressed in the white paper include: achieving a 60% reduction in CO_2 emissions by 2050 with respect to 1990, phasing out conventionally fuelled cars from cities by the same year and improvement of the road safety.

The United Nations' Sustainable Development Goals (SDG) aim to achieve a more sustainable future for everyone. These goals address global challenges in several areas such as poverty, inequality and climate change, in a total of 17 goals [4]. The SDG targets related to transport sustainability are highlighted in [5] being some of them directly related to the transport sector and others to areas where transport has an important impact, such as energy consumption and emissions. The United Nations emphasizes the interconnection of all goals and the importance of achieving them by 2030 [4].

Therefore, it is important to be able to measure and assess the sustainability of present and future transport policies concerning EU Countries. In order to fulfil this objective, this study aims to evaluate the environmental performance of the transport sector in the EU countries, from 2015 till 2017, towards a more sustainable mobility. The methodology used in this paper to evaluate the environmental performance is based on the Composite Indicator (CI) derived from the DEA model, as proposed by Cherchye in [6], the Benefit of the Doubt model. This CI allows to summarize, compare and track the performance of the countries for complex or multi-dimensional issues [7]. The use of CI is increasingly being recognised as a useful tool in policy analysis and public communication, as it can provide simple comparisons that can be used to illustrate complex and sometimes elusive issues in wide-ranging fields, e.g. environment, economy, society or technological development [8]. In a general level the CI consists in a weighted average of sub-indicators. There are different methods for weighting and aggregating the data of the sub-indicators, being some of them presented in this work.

This paper is organized as follows: the second section presents the literature review about composite indicator usefulness and its construction. Section 3 explains the DEA method and the sub-indicators selected to compose the CI. Section 4 presents the data used and the results obtained. Finally, the conclusions from this work are presented in Sect. 5.

2   Literature Review

Composite Indicators have been proven useful for benchmarking countries performance and are becoming a recognized tool for policy analysis, decision makers and public communication as they provide a big picture often making it easier for the general public to interpret CI rather than identifying common trends across many indicators. However, CI must be considered a means to facilitate a discussion and to stimulate public interest, since their "big picture" results often lead users to draw simplistic analytical or policy conclusions [8].

In [9], four main reasons for the usefulness of indicators are identified: they allow the synthesis of masses of data, show the current position in relation to desirable states, demonstrate progress towards goals and objectives and, finally, they communicate current status to stakeholders leading to effective management decisions towards the established targets.

A Composite Indicator comprises several individual indicators compiled into a single index on the basis of an underlying model and much like mathematical and computational models. Moreover, CI construction owes more to the craftsmanship of the modeller than to universally accepted rules [8].

Different aggregation and weighting techniques have been used in the literature for the assessment of the transport environmental impact.

In [5], the Sustainable Urban Transport Index was developed for cities in the Asian-Pacific region. The sub-indicators were chosen based on the literature while incorporating the Sustainable Development Goals related to urban transport planning. Equal weight of 0.10 was given to the ten selected sub-indicators and the index was calculated by applying the geometric mean for the normalized sub-indicators values.

In [7], a comparison among 33 different combinations regarding normalization, weighting and aggregation techniques for the development of a Composite Indicator was made. A set of 16 sub-indicators was selected to estimate the composite indicator of sustainable urban mobility for Italian provincial towns.

The work performed in [10] evaluates the sustainable transport system in 23 Spanish cities, using a three dimensional CI (economic, social and environmental). The sub-indicators i were normalized using the standardized values method and then aggregated into composite indicators related to the dimensions of sustainability using weights. The signs of the different weights are dependent on the meaning of the sub-indicator, being positive for indicators in which an increase in their values contributes to a more sustainable transport system and negative for those contributing conversely.

A composite indicator for transport sustainability in Melbourne local areas is developed in [11]. Nine sub-indicators were chosen and normalized by the min-max method. The sub-indicators were first aggregated into environmental, social and economic sub-indices using the Principal Component Analysis/Factor Analysis (PCA/FA) and then combined into a single CI.

In [12], a standardized set of transport performance indicators is selected to build the Normalized Transport Sustainability Index. The sub-indicators were normalized in a range between 0 and 1 using the max-min method and the index was calculated using the Euclidean distance between the city evaluated and an hypothetical *worst city*. This hypothetical *worst city* assumes the value one or zero when the effect of the indicator is negative or positive, respectively.

An alternative method to compute the CI is using the DEA method. In [13], an index to assess the performance of 112 countries in green transportation and logistics practices is constructed. The composite indicator combines the logistic performance index, CO_2 emissions and oil consumption using the DEA for weighting and aggregation.

Following the DEA approach, this study proposes the measurement of the transport environmental performance of EU countries through their composite indicator. The CI is calculated through the aggregation of multiple sub-indicators using the BoD model. The selection of the sub-indicators should reflect the targets defined by the EU's White Paper and the SDG goals related to transport. The weight attributed to each sub-indicator is derived endogenously from the DEA model.

3   Methodology

3.1   Data Envelopment Analysis

The DEA is a linear programming method that assesses the efficiency score of multiple decision making units in using multiple inputs to produce multiple outputs [14]. The DEA enables to measure the efficiency in terms of Pareto-Koopmans concept which is attained when an increase in any output (or a decrease in any input) requires a decrease in at least another output (or an increase in at least another input; e.g., [15]).

The CI is derived from the DEA model proposed by [6], named the Benefit of the Doubt model which is equivalent to the original DEA input oriented model, with all indicators considered as outputs and a single dummy input equal to one for all countries. The dummy input can be understood intuitively by regarding the model as a tool for aggregating several sub-indicators of performance, without referencing the inputs that are used to obtain this performance [16]. As the BoD model only includes outputs (the indicators), this DEA model measures the performance rather than the efficiency.

One of the best features of DEA is that it does not require any prior knowledge of weight factors as the model optimizes them endogenously. The weights can vary among countries and are determined in a way to show each of them in the best possible way, i.e., maximizing their performance [16]. Thus, DEA is a popular method in the CI literature as it can solve the problem of subjectivity in the weighting procedure. Another well-known property of the original DEA model is its unit invariance. This is very interesting for the construction of CI as its final value is independent of the measurement units of the sub-indicators which in turn makes the normalization stage redundant [17].

As stated before, the objective is to aggregate the individuals sub-indicators (the outputs) for each country into a single composite indicator defined as the weighted average of the m sub-indicators. Given a cross-section of m sub-indicators and n countries, with y_{ij} being the value of sub-indicator (or output) i for the country j, and w_i the weight attributed to the i-th sub-indicator, which is endogenously defined to maximize the CI value for the country under assessment

[17], without *a priori* expert information. The CI is computed for each country j_o, through the BoD model which has the linear programming formulation (1):

$$CI_{j_o} = max \sum_{i=1}^{m} w_i y_{ij_o} \tag{1}$$

subject to:

$$\sum_{i=1}^{m} w_i y_{ij} \leq 1 \qquad \forall j = 1, ..., n$$

$$w_i \geq 0 \qquad \forall i = 1, ..., m$$

Analyzing the objective function, it can be observed that the problem chooses the w_i that maximizes the resulting CI_{j_o} value. This implies that the highest relative weights are assigned to those dimensions in which the country has the best relative performance when compared to the other countries [17]. The core idea is as follows: if a sub-indicator has a good relative performance it suggests that this country views this policy dimension as relatively important, so it deserves a higher weight. The opposite is also valid, i.e., a sub-indicator with a low relative performance indicates a lower importance attached by the country in that context, therefore it receives a lower weight [6].

The formulation above has only two kinds of restrictions: it is imposed that no country can have a CI value greater than one, ensuring an intuitive interpretation of the indicator; also, each weight should be non-negative, which implies that the CI is a non-decreasing function of the sub-indicators. Consequently, the CI value obtained varies between zero and one for each assessed country j_o, where higher values indicate a better relative performance [17].

The BoD model (1) allows the weights to be freely estimated in order to maximize the relative performance of the country. Thus, in some situations. A country may obtain a higher relative performance by assigning zero weights to some indicators which have worst scores. This means that each sub-indicator associated with the zero weight has no influence in the composite indicator value. This situation should be avoided, since the sub-indicators were carefully selected and therefore they are all important in computing the CI [18]. To accomplish these goals, the model (1) should incorporate additional restrictions for each sub-indicator contribution, by adding virtual proportional weight restrictions, as proposed by [19]. Thus, each sub-indicator is required to have a minimum percentage of contribution (α) in the assessed composite indicator given by (2).

$$\frac{w_i y_{ij_o}}{\sum_{i=1}^{m} w_i y_{ij_o}} \geq \alpha \qquad \forall i = 1, ..., m \tag{2}$$

Another issue that is not considered in the original BoD model is the presence of undesirable sub-indicators, i.e. sub-indicators where the increase of their value is not beneficial, as the percentage of GHG emissions, for example. One possible approach to deal with these undesirable indicators is the use of data transformation techniques. Some of these techniques can be the inversion of the value of the undesirable indicator, the subtraction the undesirable factor from a sufficient large number, or the use of the max-min method. Some of these techniques are presented and compared in [18]. After the data transformation, the transformed undesirable sub-indicators are included in the conventional BoD model and treated as the desirable sub-indicators [20].

3.2   Data and Variables

Three pillars are usually mentioned as defining a sustainable transport system: the economic, the environmental, and the social one [7]. The proposed composite indicator has been developed aiming to achieve a balance between what is necessary to support sustainable transport assessment and the available data for EU countries.

As previously stated, the CI consists in the aggregation of several sub-indicators, being of crucial importance the selection of the indicators to compute the overall performance. Some issues were considered in the sub-indicators selection process: they should reflect the Roadmap targets [3] and other sustainability topics of relevance for transport; and finally, each sub-indicator must measure a specific area of the performance, ensuring the minimum number of sub-indicators; each sub-indicator must be of easy interpretation and should be available for all countries in the time span selected.

Several sub-indicators were considered to incorporate important topics related to the Roadmap and SDG targets. Taken into account these topics and the literature review of previous works with similar concepts on sustainable transport and conceptual framework, the CI is constructed based on the following five sub-indicators: share of buses and trains in total passengers transport, people dead in road accidents, share of energy from renewable sources in transport, GHG emissions by fuel combustion in transport and average CO_2 emissions per kilometer from new passengers cars. These sub-indicators are described hereinafter.

The share of buses and trains in total passengers transport (y_1) reflects the SDG goal related to industry, innovation and infrastructure, which requires building resilient and sustainable infrastructure. On the other hand, the SDG involving sustainable cities and communities, aims to renew and plan cities so they offer access to basic services for all. Future mobility should optimise the use of transport, including car sharing and the integration between different modes of collective transports. Also, the necessity to improve the transport quality, accessibility and reliability is one of the subjects discussed in the Roadmap. Capturing these goals, this indicator measures the share of collective transport in total inland transport. Collective transport refers to buses (including coaches and trolley-buses) and trains, while the total inland transport includes this modes and passenger cars. Trams and metros are not included due to the lack of harmonised data.

The people dead in road accidents (y_2) measures the number of fatalities in road accidents per hundred thousand inhabitants. The average population of the reference year (used as denominator) is calculated as the arithmetic mean of the population on 1st January of two consecutive years. The European Commission aims to make EU a world leader in safety and security of all modes of transport. With initiatives in the areas of technology, enforcement and education, EU aims to reduce fatalities close to zero by 2050. This indicator is also aligned with two SDG, aiming at safer cities and health and well-being status.

The share of energy from renewable sources in transport (y_3) contributes to a significant reduction in the greenhouse gas emissions and also reduces the oil dependence, as well as the local air and noise pollution. The Renewable Energy Directive [21] sets a 10% target for renewable energy in transport for 2020. The Roadmap also suggests a regular phase out of conventionally-fuelled vehicles from urban environments by halving their number in 2030 and phasing them out of the cities by 2050. This indicator shows how extensive is the use of renewable energy and how much it has replacing the fossil fuels.

The GHG emissions by fuel combustion in transport (y_4) measures the transport's fuel combustion contribution in the total greenhouse gas emissions. The value is originally given in thousand tonnes and was normalized using the country's population on 1st January of each year, to consider the dimension of the country. Its unit of measure is thousand tonnes per hundred thousand inhabitants.

The average carbon dioxide (CO_2) emissions per kilometer from new passengers cars (y_5) is defined as the average CO_2 emissions per kilometer in a given year for new passenger cars. The Roadmap highlights the importance of the research and innovation on vehicle propulsion technologies and the improvement of energy efficiency performance of vehicles across all modes. The EU sets a mandatory target for emission reduction for new cars of 95 g of CO_2 per kilometer in 2021. This is a target for the average of the manufacturer's overall fleet, meaning that cars above the limit are allowed as long as they are offset by the production of lighter cars.

These five sub-indicators are used to assess the transport environmental performance of EU countries, as presented in the next section.

4   Results and Discussion

4.1   Descriptive Analysis of the Variables

The transport environmental performance was calculated for the 28 EU countries, from 2015 to 2017. Therefore, data was collected for Belgium, Bulgaria, Czechia, Denmark, Germany, Estonia, Ireland, Greece, Spain, France, Croatia, Italy, Cyprus, Latvia, Lithuania, Luxembourg, Hungary, Malta, Netherlands, Austria, Poland, Portugal, Romania, Slovenia, Slovakia, Finland, Sweden and United Kingdom. It was chosen to use the United Kingdom data, since during the time span of the assessment the country still integrated the European Union. All the data used in this work was gathered from the Eurostat database [22].

Table 1 shows two descriptive statistics for the sub-indicators under analysis across countries for each year. Besides the mean, the dispersion coefficient

(DC), given by the ratio between the standard deviation and the mean, was also calculated in order to facilitate the comparison between sub-indicators.

Table 1. Mean and DC of the indicators data used in the construction of the CI.

	2015		2016		2017	
Indicator	Mean	DC	Mean	DC	Mean	DC
Public transport (y_1)	18.175	0.241	18.011	0.238	17.768	0.246
Deaths road accidents (y_2)	5.800	0.366	5.625	0.325	5.325	0.358
Renewable energy (y_3)	6.544	0.795	6.191	0.746	6.884	0.733
GHG emissions (y_4)	208.670	0.771	211.493	0.714	213.696	0.696
New cars emissions (y_5)	120.946	0.078	118.757	0.066	119.168	0.064

Analysing Table 1, it can be seen that the share of public transport in total passenger transport (y_1) has constantly decreased in the time span under study, by 2017 it was more than 2% lower compared to 2015 levels. The mean of deaths in road accidents (y_2) for all countries has decreased more than 9% from 2015 to 2017. The share of energy from renewable sources in transport (y_3) decreased in 2016 but during 2017 it increased more than 5%, when compared with the 2015 value. The mean of GHG emissions (y_4) for all countries has increased more than 2.4% during the time span studied. The mean of CO_2 emissions per kilometer from new passengers cars (y_5) has increased from 2016 to 2017 but still remained 1.5% below 2015 levels.

The highest difference among countries data is observed in the GHG emissions (y_4) and the share of energy from renewable sources in transport (y_3), as some countries are ahead in utilizing renewable energy, such as Sweden with 26.8% in 2017 and Finland with 24.8% in 2015.

The higher scores of variability relative to the mean are observed for the share of energy from renewable sources in transport (y_3) followed by the GHG emissions from fuel combustion (y_4), although both have been decreasing during the time span studied. These outputs translate the differences among countries in available renewable resources and/or different policies for reducing GHG emissions. The lowest variability relative to the mean is observed for the CO_2 emissions per kilometer from new passengers cars (y_5) showing a higher homogeneity in the energy efficiency performance of vehicles between countries.

4.2 Performance Assessment of the Models

The relative transport environmental performance for each country in a given year is computed by aggregating the sub-indicators y_1, y_2, y_3, y_4 and y_5 through the BoD model given by (1), by computing the CI. To avoid using zero weights in the performance assessment of a given country, the previous model should incorporate proportional virtual weights restrictions, as proposed by (2), imposing α

equal to 5% for each sub-indicator share. The relative transport environmental performance for each country in a given year is assessed by comparison to the best practices observed during the period analysed, i.e., from 2015 until 2017. The obtained results are presented in Table 2, where the Model 1 refers to BoD model defined by (1), and Model 2 refers to the previous one but considering restrictions (2).

Table 2. Transport environmental performance results.

Country	2015		2016		2017	
	Model 1	Model 2	Model 1	Model 2	Model 1	Model 2
Belgium	0.918	0.748	0.924	0.831	0.926	0.847
Bulgaria	0.881	0.799	0.901	0.815	0.889	0.810
Czechia	0.944	0.898	0.974	0.924	0.965	0.929
Denmark	**1.000**	0.929	0.999	0.920	0.999	0.927
Germany	0.932	0.803	0.842	0.821	0.841	0.821
Estonia	0.851	0.234	0.847	0.240	0.879	0.242
Ireland	0.924	0.861	0.936	0.834	0.955	0.909
Greece	**1.000**	0.459	0.996	0.563	0.977	0.782
Spain	0.943	0.476	0.941	0.848	0.908	0.851
France	0.965	0.909	0.968	0.915	0.972	0.920
Croatia	0.946	0.743	0.951	0.494	0.935	0.468
Italy	0.944	0.863	0.958	0.887	0.955	0.865
Cyprus	0.872	0.654	0.883	0.677	0.894	0.668
Latvia	0.835	0.716	0.865	0.666	0.852	0.640
Lithuania	0.812	0.695	0.821	0.681	0.807	0.696
Luxembourg	0.852	0.635	0.853	0.651	0.854	0.666
Hungary	**1.000**	**1.000**	**1.000**	**1.000**	0.992	0.983
Malta	**1.000**	0.871	0.978	0.872	0.985	0.926
Netherlands	**1.000**	0.923	0.970	0.874	0.959	0.902
Austria	0.951	0.902	0.965	0.917	0.958	0.909
Poland	0.914	0.856	0.914	0.782	0.894	0.781
Portugal	0.995	0.902	0.999	0.916	0.997	0.915
Romania	**1.000**	**1.000**	**1.000**	0.970	0.996	0.941
Slovenia	0.866	0.612	0.866	0.530	0.861	0.646
Slovakia	0.946	0.929	0.959	0.939	0.951	0.927
Finland	0.980	0.927	0.911	0.877	0.970	0.936
Sweden	0.975	0.965	0.993	0.987	**1.000**	**1.000**
UK	0.918	0.791	0.925	0.814	0.920	0.804
Mean	0.931	0.789	0.933	0.794	0.932	0.811
Std Dev	0.059	0.178	0.054	0.172	0.054	0.165

The mean of the Model 2 presented in Table 2, shows that the transport environmental performance has increased, on average, 2.8% between 2015 and 2017. These results imply that, overall, the countries are slowly improving towards the sustainable goals and if the results for the GHG emissions and the public transport share in passenger transport were better, the overall performance would be higher.

Considering the Model 1, eight units were efficient: Denmark (in 2015), Greece (in 2015), Hungary (in 2015 and 2016), Malta (in 2015), Netherlands (in 2015), Romania (in 2015 and 2016) and Sweden (in 2017). When the proportional virtual weights restrictions are imposed (Model 2), only four units remain efficient: Hungary (in 2015 and 2016), Romania (in 2015) and Sweden (in 2017) taken into account the five sub-indicators. The countries that are efficient only in the BoD model and become inefficient in the BoD model with restrictions (Denmark, Greece, Malta and Netherlands in 2015 and Romania in 2016) probably had a better result in some sub-indicators but a lower performance in overall sub-indicators. Thus, when restrictions are imposed and all sub-indicators are required to contribute to the final CI score, those countries become inefficient. Hereinafter, it is fair to consider the Model 2, i.e., the Model 1 with restrictions (2), to assess the transport environmental performance of EU countries.

Under this methodology, from 2015 to 2017, most of the countries followed a small improvement in the mean of the overall performance. However, Spain, Greece and Belgium showed a higher improvement in their final score, increasing in 2017 by 79%, 70% and 13% above 2015 levels, respectively. Estonia was the most inefficient country in this analysis with an average CI of 24% and it had almost no improvement in the considered years. Croatia, Poland, Latvia, Romania and Hungary decreased their performance during the time period of analysis, with Romania being efficient in 2015 and Hungary in 2015 and 2016. Finland, Slovenia and the Netherlands decreased their performance in 2016, but by 2017 they managed to improve the environmental performance above 2015 levels.

Analysing Table 2 it is possible to notice that, in all three years, the variability in the results presented in Model 2 was higher compared to the variability presented for the results obtained with Model 1. The highest standard deviation value was observed in the CI from 2015 with Model 2 and the lowest was presented by the CI from 2016 and 2017 using Model 1.

Regarding the transport environmental performance computed by the adopted methodology, this study also compares the benchmark countries with the inefficient ones. This analysis is implemented considering as benchmarks the best performing countries, which obtained a CI score above 0.95, i.e., Hungary (in 2015, 2016 and 2017), Sweden (in 2015, 2016 and 2017) and Romania (in 2015 and 2016). The other countries are considered inefficient. The mean for each sub-indicator is calculated for both groups (benchmarks and inefficient countries), using the original data for the undesirable sub-indicators (GHG emissions, deaths in road accidents and new cars emissions), i.e., without transformation. Figure 1 shows a comparison for each sub-indicator between the benchmark countries and the remaining, considered inefficient.

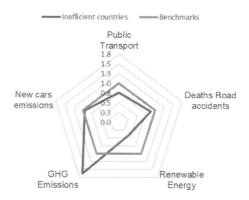

Fig. 1. Comparison between benchmarks and inefficient countries.

Analysing Fig. 1, it is possible to notice that inefficient countries have less than 50% of the share of transport energy from renewable sources observed in the benchmark countries and the GHG emissions from fuel combustion engines are more than 60% higher than the benchmarks. The average CO_2 emissions from new passenger cars is almost the same for both groups. The average share of public transport in the inefficient countries is almost reaching the same level as the benchmark countries. Regarding the number of deaths on road accidents, however, the inefficient countries are slightly better than the benchmarks. This analysis enables to identify the areas where the inefficient countries need to improve by setting out policies and/or redefine output standards, for instance. Most of this work need to be done in drastically reducing the GHG emissions from fossil fuel, increasing the share of transport energy from renewable sources and improving the public transport access and quality to allow a larger share of the total passenger transport.

5   Conclusions

This study assesses the environmental performance of the transport sector in the 28 countries of the European Union towards the targets set in the Roadmap and SDG, from the year 2015 until 2017. The assessment of the transport environmental performance is implemented through the composite indicator derived from the BoD model with virtual proportional weights restrictions. Based on the results achieved, it is possible to conclude that EU countries slightly improved their transport environmental performance, on average 2.8%. This implies that EU countries should develop more efforts to follow the transport environment targets.

Spain, Greece and Belgium showed the highest improvement in their final score during the time frame analysed. Estonia was the most inefficient country and showed almost no improvement over the years. Croatia, Poland, Latvia, Romania and Hungary decreased their performance during this time period. Finland, Slovenia and the Netherlands decreased their performance initially in 2016, but by 2017 they were able to improve above 2015 levels. These results suggest that inefficient countries should improve their practices by emulating the best practices observed on benchmarks.

By using as benchmark units the ones that obtained a performance above 95%, the comparison with the remaining units, considered inefficient, allowed the identification of the areas where policies should greatly impact: the reduction of GHG emissions from fuel combustion engines and the increase of the share of public transport in total passenger transport.

Future works should explore other models for treating undesirable indicators in order to allow results comparison among those different models. Furthermore, some other sub-indicators can be taken into account, to calculate the composite indicator for each country.

References

1. European Commission: EU mobility package: Europe on the move briefing note. http://nws.eurocities.eu/MediaShell/media/EuropeonthemoveBriefingnote.pdf. Accessed 10 Mar 2020
2. European Commission: Proposal for a Regulation of the European Parliament and of the Council setting CO2 emission performance standards for new heavy duty vehicles (2011). https://eur-lex.europa.eu/legal-content/EN/TXT/?uri=SWD:2018:185:FIN
3. European Commission: White Paper on Transport: Roadmap to a Single European Transport Area: Towards a Competitive and Resource-efficient Transport System. Publications Office of the European Union (2011)
4. About the Sustainable Development Goals. https://www.un.org/sustainabledevelopment/sustainable-development-goals/. Accessed 13 Mar 2020
5. Gudmundsson, H., Regmi, M.B.: Developing the sustainable urban transport index. Transport and Communications Bulletin for Asia and the Pacific No. 87 (2017)
6. Cherchye, L., Moesen, W., Rogge, N., Van Puyenbroeck, T.: An introduction to 'benefit of the doubt' composite indicators. Soc. Ind. Res. **82**(1), 111–145 (2007)
7. Danielis, R., Rotaris, L., Monte, A.: Composite indicators of sustainable urban mobility: estimating the rankings frequency distribution combining multiple methodologies. Int. J. Sustain. Transp. **12**, 380–395 (2018)
8. Joint Research Centre-European Commission: Handbook on constructing composite indicators: methodology and user guide. OECD publishing (2008)
9. Mitchell, G., May, A., McDonald, A.: PICABUE: a methodological framework for the development of indicators of sustainable development. Int. J. Sustain. Dev. World Ecol. **2**(2), 104–123 (1995)
10. Alonso, A., Monzón, A., Cascajo, R.: Comparative analysis of passenger transport sustainability in European cities. Eco. Ind. **48**, 578–592 (2015)
11. Reisi, M., Aye, L., Rajabifard, A., Ngo, T.: Transport sustainability index: Melbourne case study. Ecol. Ind. **43**, 288–296 (2014)

12. Zito, P., Salvo, G.: Toward an urban transport sustainability index: an European comparison. Eur. Transp. Res. Rev. **3**(4), 179–195 (2011). https://doi.org/10.1007/s12544-011-0059-0
13. Lu, M., Xie, R., Chen, P., Zou, Y., Tang, J.: Green transportation and logistics performance: an improved composite index. Sustainability **11**(10), 2976 (2019)
14. Charnes, A., Cooper, W.W., Rhodes, E.: Measuring the efficiency of decision making units. Eur. J. Oper. Res. **2**(6), 429–444 (1978)
15. Fried, H.O., Lovell, C.A.K., and Schmidt, S.S.: The measurement of productive efficiency: Techniques and applications, 3–67. Oxford University Press (1993)
16. Chung, W.: Using DEA model without input and with negative input to develop composite indicators. In: International Conference on Industrial Engineering and Engineering Management (IEEM), pp. 2010–2013. IEEE (2017)
17. Cherchye, L., et al.: Creating composite indicators with DEA and robustness analysis: the case of the Technology Achievement Index. J. Oper. Res. Soc. **59**(2), 239–251 (2008)
18. Dyson, R.G., Allen, R., Camanho, A.S., Podinovski, V.V., Sarrico, C.S., Shale, E.A.: Pitfalls and protocols in DEA. Eur. J. Oper. Res. **132**(2), 245–259 (2001)
19. Wong, Y.H., Beasley, J.E.: Restricting weight flexibility in data envelopment analysis. J. Oper. Res. Soc. **41**(9), 829–835 (1990)
20. Färe, R., Karagiannis, G., Hasannasab, M., Margaritis, D.: A benefit-of-the-doubt model with reverse indicators. Eur. J. Oper. Res. **278**(2), 394–400 (2019)
21. European Commission - Renewable energy directive, https://ec.europa.eu/energy/topics/renewable-energy/renewable-energy-directive/overview. Accessed 21 Mar 2020
22. Eurostat Database. https://ec.europa.eu/eurostat/data/database. Accessed 20 Mar 2020

Multivariate Analysis to Assist Decision-Making in Many-objective Engineering Optimization Problems

Francisco Santos🆔 and Lino Costa$^{(\boxtimes)}$🆔

Centro ALGORITMI, Universidade do Minho, 4710-057 Braga, Portugal
francisco_dos_santos@outlook.pt, lac@dps.uminho.pt

Abstract. Data processing (or the transformation of data into knowledge and/or information) has become an indispensable tool for decision-making in many areas of engineering. Engineering optimization problems with many objectives are common. However, the decision-making process for these problems is complicated since there are many trade-offs that are difficult to identify. Thus, in this work, multivariate statistical methods, Principal Component Analysis (PCA) and Cluster Analysis (CA), have been studied and applied to analyze the results of many objective engineering optimization problems. PCA reduces the number of objectives to a very small number, CA through the similarities and dissimilarities, creates groups of solutions, i.e., bringing together in the same group solutions with the same characteristics and behaviors. Two engineering optimization problems with many objectives are solved: a mechanical problem consisting in the optimal design of laminated plates, with four objectives and a problem of optimization of the radar waveform, with nine objectives. For the problem of the design of laminated plates through PCA allowed to reduce to two objectives and through CA it was possible to create three distinct groups of solutions. For the problem of optimization of the radar waveform, it was possible to reduce the objectives from nine to two objectives representing the greatest variability of the data, and CA defined three distinct groups of solutions. These results demonstrate that these tools are effective to assist the decision-making processes in the presence of a large number of solutions and/or objectives.

Keywords: Multivariate analysis · Multi-objective optimization · Dimensionality reduction · Decision-making

1 Introduction

Optimization refers to finding one or more solutions that correspond to extreme values or commitments to one or more objectives. Optimization plays a very important role in the various engineering areas where optimization problems arise, such as, for example, the design of new products, improvement of manufacturing and logistics processes. On the other hand, most of these engineering problems are multi-objective in nature. The simultaneous optimization of conflicting objectives allows to obtain an approximation to the Pareto optimal set. This set contains non-dominated solutions that represent

© Springer Nature Switzerland AG 2020
O. Gervasi et al. (Eds.): ICCSA 2020, LNCS 12251, pp. 274–288, 2020.
https://doi.org/10.1007/978-3-030-58808-3_21

different trade-offs. Then, the decision maker has to choose one of the alternatives according to his preferences. However, when the number of objectives and alternatives is high, the decision process is very difficult. These optimization problems with four or more objectives are, therefore, generally referred to as problems with many objectives.

Evolutionary algorithms are population-based algorithms commonly used in multi-objective optimization, in which one seeks to find approaches to different Pareto optimal solutions [1]. Very recently, these algorithms have been developed and applied to solve problems with a high number of objectives [2]. However, due to the large number of solutions and the high dimension of the objective space, the decision-making process becomes difficult and sometimes almost impractical [3]. Thus, the visualization and analysis of the solutions obtained by the algorithms are crucial for decision-making [4].

In [5], it is proposed a multi-objective evolutionary algorithm to solve problems with many objectives, trying to reduce the dimension of the Pareto front in terms of the number of objectives throughout the search. In practice, it is not always easy to identify the redundancy of objectives. This study highlights the usefulness of dimensionality reduction techniques, in the context of multi-objective optimization with many objectives. The use of PCA of responses in multi-objective problems can also be found in [6].

Brockhoff and Zitzler [7] also identified the difficulties of solving problems with a high number of objectives, both in terms of the quality of the approximations obtained to the optimal set of Pareto, as well as the execution time and also the complexity of the decision-making process. To overcome these difficulties, they proposed the reduction of objectives by identifying those that can be omitted while maintaining the dominance relationship between solutions. Brockhoff and Zitzler [8] defined a generalized notion of conflict between objectives, which specifies the necessary and sufficient conditions for the inclusion of the objectives to be optimized.

Principal component analysis was also used to reduce dimensionality in multi-objective problems by Costa and Oliveira [9] to assist the decision-making process. The use of graphic representations describing the relationship between objectives was also proposed, allowing the simultaneous visualization of all solutions with all objectives, thus facilitating the detection of conflicting objectives. Costa and Oliveira [4] argue that the notion of conflicts between objectives can be misleading, since the hypothesis of the existence of trade-offs between objectives implies that all objectives are important. Thus, even objectives that are not necessary to induce and preserve the dominance relationship between the solutions can contain some useful information to assist the decision-making process.

In this paper, multivariate statistical methods are used to assist the decision-making process for problems with many objectives, in particular, in the engineering area.

In Sect. 2, multivariate analysis techniques and graphical representations in the context of multi-objective optimization are presented. The results of application of the proposed method to two many-objective optimization engineering problems are presented and discussed in Sect. 3. Finally, in Sect. 4, some conclusions and future work are addressed.

2 Multivariate Analysis

In certain multi-objective problems, some objectives are not important to define the Pareto front. In this case, the dimensionality of the problems can be reduced and, consequently, the decision-making process becomes simpler and more treatable. On the other hand, it is interesting to identify the similarities between the solutions belonging to the sets of non-dominated solutions. For this purpose, multivariate statistical methods can be useful in situations where several variables must be measured simultaneously and the connections, similarities and dissimilarities must be identified.

The goal is to summarize, analyze and interpret data in which several response variables are evaluated. In the context of optimization with many objectives, it is intended to interpret and understand the relationships between the objectives as well as the similarities between solutions. For this purpose, multivariate statistical methods such as Principal Component Analysis (PCA) and Cluster Analysis (CA) as well as graphical representation tools such as biplots are of particular use and relevance.

2.1 Principal Component Analysis

Principal Component Analysis is a multivariate statistical method that makes it possible to transform a vector with the variables correlated with each other into a vector of unrelated (orthogonal) variables that are called principal components. In this way, these principal components are calculated in decreasing order of their importance, therefore, the first explains the maximum variance of the original data, the second the maximum variance not yet explained, and so on. Therefore, the last principal component will be the one that least contributes to the explanation of the variation of the initial data. PCA differs from other methods in that its objective is not to find correlation between variables, but to find mathematical functions in the initial variables that can explain as much of the variation that exists in the data and that allows them to describe and reduce them.

The objective of reducing dimensionality is to find a smaller number of objectives that have the relevant information. The reduction in dimension is obtained by transforming the original variables into a new set of variables, called principal components that are not correlated. Ordering is an important property, since it is this that justifies why the first principal components can be considered to have the greatest variability in the data set. However, a disadvantage is that it is difficult to deal with the existence of different types of nonlinearities in data [4].

2.2 Cluster Analysis

Cluster Analysis is used with the objective of grouping objects or individuals classified according to a set of characteristics or variables in groups called clusters. CA can be also useful to reduce the dimensionality by grouping individuals with similarities. In addition to defining a measure of similarity between the objects to be measured, it is also necessary to have a grouping method (hierarchical or non-hierarchical) and an algorithm that groups the objects into clusters according to the measure. Currently, several clustering algorithms are available in the literature [10]. One of the most used is the k-means algorithm because it is, in general, effective and easy to implement. In this algorithm, data points are separated into k distinct groups of equal variation, minimizing the sum of the square of the distance between the points, i.e., the squared sum of the errors. The goal is to group individuals with similar characteristics in the same group, so it is important to define k, the number of clusters. Several methods, such as the elbow method and the gap method can be considered to define the optimal number of clusters.

2.3   Graphical Representations

In general, for optimization problems with many conflicting objectives, decision makers have to compare several alternatives and select the most preferred. The comparison of multidimensional vectors is very difficult, especially without any support tool. In order to support the decision maker to identify and understand the similarities and dissimilarities, several graphical visualization tools can be used. However, visualizing and interpreting graphs is not always simple. On the one hand, the graphs must be easy to understand and must not omit much information, on the other hand, no extra unintended information should be included [11].

Visualization techniques can be classified according to the principles used: representations involving bar diagrams, scatter diagrams or value paths, diagrams using circles and polygons, representations based on hierarchical clustering and projection-based. However, there are other advanced graphic representation techniques that are very useful in large spaces due to the large number of objectives and the number of solutions in the final set of approximations to the Pareto optimal solutions. In the presence of more than two objectives, one of the main difficulties is the visualization of the compromises between different objectives represented by each solution. The usual two-dimensional representation does not provide a better view of how a given solution is positioned in the different projections [4]. Therefore, advanced graphical representations can be used such as biplots and Andrews graphs.

A biplot is a projection-based technique, related to scatter plots. In the context of multi-objective optimization, PCA can be used to find a plan (two dimensions) where the criteria can be projected. In fact, the prefix "bi" refers to the joint display of rows and columns of the original data matrix that contains the alternatives. The biplot is a scatter plot that represents a two-dimensional matrix, both by lines (points) and columns (vectors). This plot allows the simultaneous visualization of all solutions with all objectives, thus facilitating the identification of conflicting objectives and trade-offs [4].

The Andrews graph is one of the solutions for representing and interpreting multivariate data [12]. This graph is a representation that transforms criteria vectors into two-dimensional curves with the aid of Fourier series [11]. All alternatives can be graphically represented in the same coordinate system. The Euclidean distance in the transformation is preserved. Therefore, the criterion vectors next to each other are transformed into curves that are not far from each other.

2.4   Non-dominated Sets Analysis

This section briefly describes the proposed method for analyzing sets of non-dominated solutions. The ultimate goal is to assist the decision-making process. This analysis combines and integrates the multivariate techniques described in the previous sections. Two distinct phases are implemented in order to facilitate the interpretation and to achieve the best results.

In the first phase, the aim is to identify possible redundant objectives. PCA is applied to the set of non-dominated solutions, involving three main steps: preparation of the data, determination of the correlations between objectives, computation of the explicability and the relationship between objectives. Finally, the graphical representation, the objectives and the solutions are presented graphically using, for instance, biplots or the Andrews graph.

In a second phase, CA is applied to the set of non-dominated solutions. This is done considering the results of PCA. The identification of the number of clusters is crucial for the use of the k-means algorithm. The graphic representation of the clusters can be done in terms of the two-dimensional projections of the Pareto front or other multivariate graphics such as biplots, allowing the decision maker to identify trade-offs. The two phases of the proposed method to analyze a set of non-dominated solutions can also be used interactively with the decision maker.

3   Results

This section presents the results of the application of the proposed method to analyze sets of non-dominated solutions for two multi-objective optimization problems of the areas of mechanical engineering and electronic engineering: a laminated plate design optimization problem and a radar performance optimization problem. The methods to analyze these engineering problems were implemented in R language [13].

3.1   Laminated Plate Design Optimization

This multi-objective optimization problem is related to the design of a laminated plate consisting of several layers or laminas of different materials that can have different thickness. Each layer is made of a single material. Thus, the decision variables are the thicknesses, the Young's modulus and the Poisson's ratio. The thicknesses of the layers can take any value from a continuous range and there are a finite number of different materials available. Therefore, this is a mixed integer optimization problem because there are continuous and discrete variables. The problem can be considered an optimization problem with many objectives where the objectives to be minimized are the compliance that is inversely related to the rigidity, the total price, the total mass and the total thickness of the plate. More details about this problem are presented by Costa et al. [14].

The data set comprises a total of 337 non-dominated solutions that are strongly correlated with each other in terms of the four objectives of compliance, total price, total mass and total thickness.

Table 1 shows the correlations between the objectives. There is a negative correlation between complacency and all other goals. On the other hand, there is a positive correlation between price, mass and thickness.

Table 1. Correlations of objectives for the laminated plates design optimization problem.

	Compliance	Price	Mass	Thickness
Compliance	1.0000	−0.5381	−0.5327	−0.3114
Price	−0.5381	1.0000	0.8740	0.6641
Mass	−0.5327	0.8740	1.0000	0.7145
Thickness	−0.3114	0.6641	0.7145	1.0000

The goal of using PCA is to find the objectives that have less variability in the data and that, possibly, can be omitted by losing the minimum quantity of information. Table 2 presents the summary of the principal components (PCs) obtained by PCA. The criteria considered it to retain PCs with variance greater than one and a cumulative proportion of explicability superior to 80%. The variance of PC1 is superior to 1 and the cumulative proportion of PC1 and PC2 is 89.09%. Thus, PC1 and PC2 are retained.

Table 2. Summary of PCs for the laminated plate design optimization problem.

	PC1	PC2	PC3	PC4
Standard deviation	1.6900	0.8411	0.5607	0.3494
Proportion of variance	0.7140	0.1769	0.0786	0.0305
Cumulative proportion	0.7140	0.8909	0.9695	1.0000

Figure 1 shows the correlation cycle between the objectives. In this graph, it is possible to observe that:

- the participation of each objective in PC1 and PC2, where the objectives furthest from the center of the circle are best represented with the contribution value closest to one;
- the correlation between the objectives, where all objectives facing the same direction correlate positively and negatively with the objectives facing the opposite side;
- the contribution of each objective (the red corresponding to the highest contribution and the blue to the lowest contribution);
- the quality of representation of each objective (greater intensity corresponds to greater quality).

Thus, it is possible to observe that the compliance objective has a better contribution and quality in the two PCs. On the other hand, price, mass and thickness objectives are correlated positively with each other and all of them negatively with the compliance. Only the results for PC1 and PC2 are presented as they are the most important for this analysis.

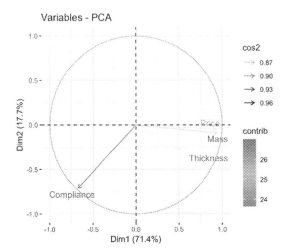

Fig. 1. Correlation cycle between the objectives for the laminated plate design optimization problem (Color figure online).

Figure 2 shows the quality of representation of the solutions in terms of PC1 and PC2 where the red color corresponds to the best represented solutions and the blue one the least well represented. For example, solutions close to the origin of the graph will be less well represented by components. On the other hand, solutions 187 and 20 are better represented in relation to the others, in particular, in the objective compliance (see also Fig. 1). Due to the large number of solutions, CA may play an important role here to identify similarities between solutions and the trade-offs they represent.

The objectives are highly correlated. This fact is important for the analysis of clusters. The purpose of cluster analysis is to group solutions with similar characteristics in the same group. In order to define the optimal number of clusters, the elbow and gap methods were used. The elbow method indicates that the major contributions are found in the first four clusters. On the other hand, the gap method suggests only three clusters. Since the aim is to minimize the number of clusters, the latter solution was adopted and computed using the k-means algorithm. Considering three clusters as indicated by the gap method, it can be observed in Fig. 3 that the clusters are well defined.

The Andrews graph can be represented for each cluster. For instance, observing the Andrews graph for cluster3 (Fig. 4), there are three groups of curves separated into three different colors (red, black and purple). These three colors group solutions with similar characteristics. These solutions are closer together and have very close objective values.

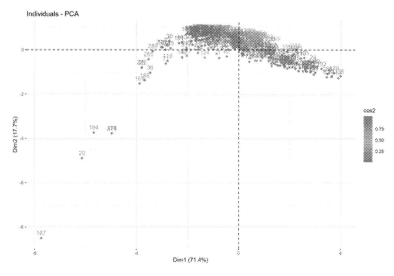

Fig. 2. Quality of representation of the solutions in PC1 and PC2 for the laminated plate design optimization problem (Color figure online).

Figure 5 shows the two-dimensional projections of the Pareto front with the well-defined clusters represented by the colors red, green and black. Among these objectives there are high correlations facilitating the definition of different groups. Considering this information, the decision maker is able to choose which of the clusters best matches his/her preferences.

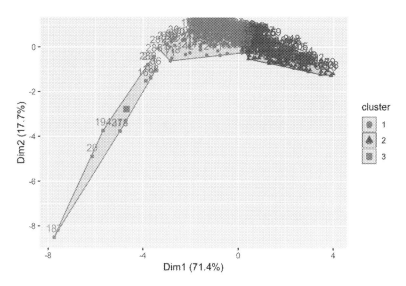

Fig. 3. Representation of the three clusters in terms of PC1 and PC2 for the laminated plate design optimization problem.

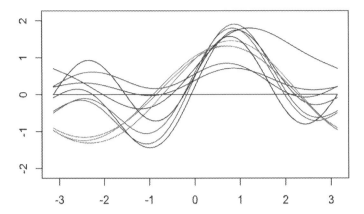

Fig. 4. Andrews graph for cluster3 for the laminated plate design optimization problem (Color figure online).

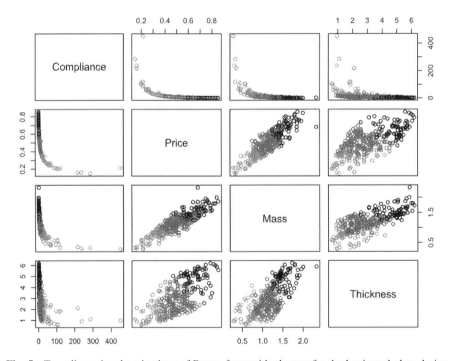

Fig. 5. Two-dimensional projections of Pareto front with clusters for the laminated plate design optimization problem (Color figure online).

3.2   Radar Performance Optimization

This multi-objective optimization problem involves nine objectives [15]. It refers to the design of a waveform for a Pulsed Doppler radar, typical of many aerial combat systems. Radar systems are important for measuring both the distance and the speed of targets. Unfortunately, for very long distances and very high speeds, using a simple waveform it is only possible to measure: i) distance without ambiguity, but ambiguous speed; ii) speed without ambiguity, but with ambiguous distance or; iii) ambiguity in both distance and speed. In order to allow error-free transmissions, a simple set of waves is transmitted, each one subtly different from the other. The results of these multiple waveforms are combined to resolve ambiguities. In fact, the problem is how to choose the optimal set of simple waveforms. More details about this difficult problem are presented by Hughes [15].

The data comprises 11,938 non-dominated solutions. Most of the correlations between the nine objectives are high, which indicate the existence of significant associations. Table 3 summarizes those correlations where the correlations appear ordered according to its value and in bold if greater than 0.5.

Table 3. Correlations of objectives for the radar performance optimization problem.

Objectives	Positive correlations	Negative correlations
f1	f3 > f7 > f5	f9 > f4 > f2 > f8 > f6
f2	f9 > f4 > f8 > f6	f3 > f1 > f7 > f5
f3	f1 > f7 > f5	f9 > f4 > f2 > f8 > f6
f4	f9 > f2 > f8 > f6	f1 > f3 > f7 > f5
f5	f1 > f3 > f7	f9 > f4 > f2 > f8 > f6
f6	f8 > f2 > f9 > f4	f3 > f1 > f5 > f7
f7	f1 > f3 > f5	f4 > f9 > f2 > f8 > f6
f8	f2 > f9 > f4 > f6	f3 > f1 > f5 > f7
f9	f4 > f2 > f8 > f6	f3 > f1 > f7 > f5

The summary of the principal components resulting from PCA is given in Table 4. It can be seen that the first two principal components already accumulate 84% of the variance. Thus, it can be considered that PC1 and PC2 are sufficient to explain the variability of the initial data. In this problem, even without the remaining seven PCs, it is possible to explain most of the data variability.

Table 4. Summary of PCs for the radar performance optimization problem.

	PC1	PC2	PC3	PC4	PC5	PC6	PC7	PC8	PC9
Standard deviation	2.55	1.02	0.81	0.69	0.40	0.32	0.19	0.17	0.04
Proportion of variance	0.72	0.12	0.07	0.05	0.02	0.01	0.00	0.00	0.00
Cumulative proportion	0.72	0.84	0.91	0.96	0.98	0.99	1.00	1.00	1.00

The correlation cycle for the first two principal components PC1 and PC2 is shown in Fig. 6. In addition to the correlations between the objectives, the graph also represents the contribution of the objective to the respective PC (where the red color has the largest representation, intermediate yellow color and blue color the lowest representation). On the other hand, vectors oriented in the same direction indicate that the objectives are positively correlated and are negatively correlated when in opposite directions. So, between objectives f1, f3, f5, f7 there is a positive and very strong correlation, and they are negatively correlated with objectives f2, f4, f8, f9, these being positively correlated between them. The objective f6 can be considered independent of the remaining objectives.

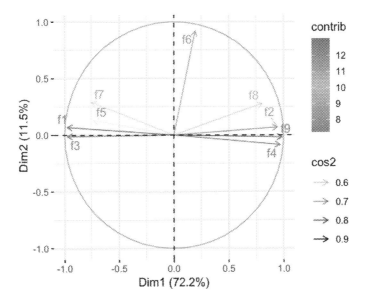

Fig. 6. Correlation cycle between the objectives for the radar performance optimization problem (Color figure online).

Given the very large number of non-dominated solutions, it is not easy to identify trade-offs. In this problem, cluster analysis is crucial. The elbow method indicates three clusters and the gap method suggests nine clusters. Since the aim is to obtain a reduced number of groups of solutions, three clusters were considered and computed by the k-means algorithm. In Fig. 7, the three clusters are represented in terms of PC1 and PC2. Note that points of each cluster almost do not overlap between them, which suggests that the number of clusters is adequate.

Figure 8 shows the two-dimensional projections of the Pareto front for the different combinations of the objectives with the three clusters in different colors (black, red and green). It can be observed that there are some overlaps of the solutions due to the large number of solutions and objectives that difficulties the separation of solutions in groups. For each cluster, it is possible to select one or more representative solutions, for example, those closest to the respective centroids.

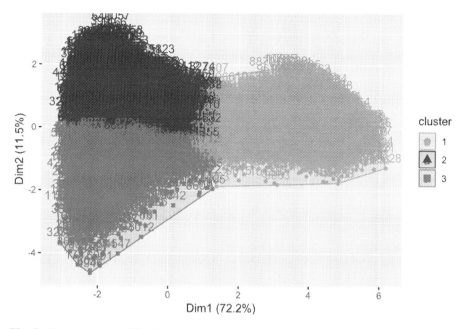

Fig. 7. Representation of the three clusters in terms of PC1 and PC2 for the radar performance optimization problem.

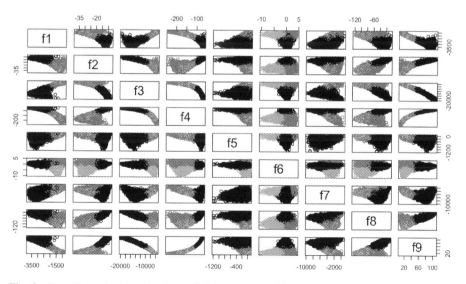

Fig. 8. Two-dimensional projections of Pareto curve with clusters for the radar performance optimization problem (Color figure online).

In order to find representative solutions for each cluster, the solution closest to the centroid was selected. In Fig. 9, it is plotted the Andrews graph where each curve corresponds to the representative solutions of each cluster. The light red curve and the darkest red curve are closer and correspond to more similar representative solutions in terms of objectives. The black curve presents a different behavior in relation to the other curves, taking negative positions when the others are positive.

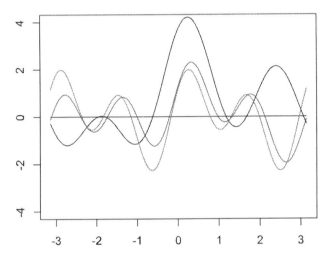

Fig. 9. Andrews graph with the representative solutions for the radar performance optimization problem (Color figure online).

4   Conclusions

In this work, it is proposed to use multivariate statistical methods, such as Principal Components Analysis and Cluster Analysis to help the decision-making process in many-objective optimization problems. When the number of non-dominated solutions and/or objective is large, it is very difficult the visualization of trade-offs as well as the relationships between objectives. Therefore, the proposed method allows to inspect associations between objectives, identify redundant objectives, and similarities between solutions by defining clusters. Clustering is particularly interesting since allow to select one or more representative solutions, for example, those closest to the cluster's centroid. Moreover, graphical tools are critical for the visualization of results. This way, the decision-making process is substantially facilitated.

The proposed method was applied to two difficult many-objective engineering optimization problems. For the problem of the design of laminated plates through PCA allowed to reduce to two objectives and through CA it was possible to create three distinct groups of solutions. For the problem of optimization of the radar waveform, it was possible to reduce the objectives from nine to two objectives representing the greatest variability of the data, and CA defined three distinct groups of solutions. Moreover, it was possible the identification and selection of representative non-dominated solutions.

These results obtained with the engineering problems demonstrate the validity and usefulness of the approach to assist the decision maker.

Future work comprises the development of multiphase strategies to provide interaction with the decision maker as well as the study of other multivariate statistical methods. Strategies to incorporate different levels of importance or preferences of criteria will be also investigated.

Acknowledgements. This work has been supported by FCT – Fundação para a Ciência e Tecnologia within the R&D Units Project Scope: UIDB/00319/2020.

References

1. Deb, K.: Multi-objective Optimization Using Evolutionary Algorithms. Wiley, Chichester (2001)
2. Hughes, E.J.: MSOPS-II: A general-purpose many-objective optimiser. In: 2007 IEEE Congress on Evolutionary Computation, pp. 3944–3951. IEEE (2007)
3. Purshouse, R.C., Fleming, P.J.: Evolutionary many-objective optimisation: An exploratory analysis. In: 2003 Congress on Evolutionary Computation – CEC 2003, vol. 3, pp. 2066–2073. IEEE (2003)
4. Costa, L., Oliveira, P.: Biplots in offline multiobjective reduction. In: 2010 IEEE Congress on Evolutionary Computation - CEC 2010, pp. 18–23. IEEE (2010)
5. Deb, K., Saxena, D.K.: On finding pareto-optimal solutions through dimensionality reduction for certain large-dimensional multi-objective optimization problems. In: 2005 IEEE Congress on Evolutionary Computation – CEC 2005. IEEE (2005)
6. Reis, M.S., Pereira, A.C., Leça, J.M., Rodrigues, P.M., Marques, J.C.: Multi-response and multi-objective latent variable optimization of modern analytical instrumentation for the quantification of chemically related families of compounds: Case study - Solid Phase Microextraction (SPME) applied to the quantification of analytes with impact on wine aroma. J. Chemom. **33**(3), e3103 (2019)
7. Brockhoff, D., Zitzler, E.: Dimensionality reduction in multiobjective optimization with (partial) dominance structure preservation: Generalized minimum objective subset problems. TIK Report, 247 (2006)
8. Brockhoff, D., Zitzler, E.: On objective conflicts and objective reduction in multiple criteria optimization. Peabody J. Educ. 0161956X, **81**(243), 180–202 (2006)
9. Costa, L., Oliveira, P.: Multiobjective optimization: Redundant and informative objectives. In: 2009 IEEE Congress on Evolutionary Computation - CEC 2009, pp. 2008–2015. IEEE (2009)
10. Jain, A.K., Murty, M.N., Flynn, P.J.: Data clustering: A review. ACM Comput. Surv. **31**(3), 264–323 (1999)
11. Miettinen, K.: Survey of methods to visualize alternatives in multiple criteria decision making problems. OR Spectrum **36**(1), 3–37 (2012). https://doi.org/10.1007/s00291-012-0297-0
12. Khattree, R., Naik, D.: Andrews plots for multivariate data: Some new suggestions and applications. J. Stat. Plan. Infer. **100**, 411–425 (2002)
13. R Core Team: R: A Language and Environment for Statistical Computing. R Foundation for Statistical Computing, Vienna, Austria (2019). https://www.R-project.org

14. Costa, L., Fernandes, L., Figueiredo, I., Júdice, J., Leal, R., Oliveira, P.: Multiple- and single-objective approaches to laminate optimization with genetic algorithms. Struct. Multidiscip. Optim. **27**(1–2), 55–65 (2004)
15. Hughes, E.J.: Radar waveform optimisation as a many-objective application benchmark. In: Obayashi, S., Deb, K., Poloni, C., Hiroyasu, T., Murata, T. (eds.) EMO 2007. LNCS, vol. 4403, pp. 700–714. Springer, Heidelberg (2007). https://doi.org/10.1007/978-3-540-70928-2_53

International Workshop
on Computerized Evidence-Based
Decision Making (CEBDEM 2020)

Tuscany Configurational Atlas: A GIS-Based Multiscale Assessment of Road-Circulation Networks Centralities Hierarchies

Diego Altafini$^{(\boxtimes)}$ and Valerio Cutini

Dipartimento di Ingegneria dell'Energia, dei Sistemi, del Territorio e delle Costruzioni, Università di Pisa, Largo Lucio Lazzarino 1, 56122 Pisa, PI, Italy
diego.altafini@phd.unipi.it

Abstract. Digital thematic maps availability has increased with the diffusion of open-source Geographic Information Systems (GIS) suites, which also had important role in urban and regional sciences revamp throughout the late 1990's. These methodological innovations led to the conception of network-based data maps oriented to highlight urban scale road-circulation networks configurational properties, that supported comparative studies regarding cities' morphologies and their representation as complex systems. However, significant hindrances persist for the construction of very large road-circulation network datasets, such as those suitable to regional and supra-regional scale analyses. Owing to their sheer sizes, modelling these expanses require extensive processing times, which impact on research prospects. Data precision is a concern as well, since generalization processes, whereas can reduce computing complexity, oftentimes render comparisons amongst different scales inaccurate, due to certain road structures non-representation. Research requirements for a comparable and accurate multiscale database, suited to evaluate circulation networks configurational properties of centrality, prompted construction of the Tuscany Configurational Atlas as an experiment. Intended as a set of GIS-based digital thematic maps and data repository, it depicts *closeness* and *betweenness* centralities hierarchies of the Tuscan Region road-infrastructure in regional, provincial and municipality scales. This paper summarizes the scope and methodological steps to construct this Configurational Atlas, while reducing regional-wide dataset-related issues. Furthermore, it discusses its contribution as a spatial representation, and evaluates its prospects as an analytical instrument and database. Concluding remarks define forthcoming improvements to be done regarding usability, such as its implementation in a WebGIS suite.

Keywords: Geographic Information Systems · Configurational Atlas · Network analyses

1 Introduction

Geographic Information Systems (GIS) has been throughout continuous development and evolution during the past half-century, in which it made noteworthy contributions to mainstream geographic research [1, 2]. However, amidst much of this period, GIS has been secluded as a somewhat niche instrument, with its usage limited to

O. Gervasi et al. (Eds.): ICCSA 2020, LNCS 12251, pp. 291–306, 2020.
https://doi.org/10.1007/978-3-030-58808-3_22

cartographers and geographers, due to restricted-access software and a steep learning curve, even though, it demonstrated untapped potential applications to other disciplines. In that regard, GIS remained focused on organizing spatial information and establishing geographic rules [3], while seldom employed to explore geo-objects relationships.

Paradigms since then have shifted, as the diffusion of open-source GIS suites has democratized access to geoprocessing, increasing the spatial information acquisition, level of detail and overall production, all circumstances that contributed to the urban and regional sciences revamp throughout the late 1990's [4]. Gradual digitalization of topographical technical charts and the maturation of GIS-based regional and municipal datasets had a most significant role in these developments; further associated to methodological improvements on fields of configurational and network analysis [5], this led to the conception of datamaps oriented to depict urban scale road-circulation networks, supporting comparative studies regarding cities' morphologies and their representation as complex systems [6, 7]. From this standpoint, thematic maps based on network analyses became, over the past decade, important sources of information for urban design, set for describe local movement patterns, as configurational properties of these road networks – their *closeness* and *betweenness* centralities hierarchies – were unveiled as accurate predictors of pedestrian and vehicular movement [8].

Significant hindrances, however, are posed for the construction of very large road-circulation datamaps. Owing to their sheer sizes, modelling regions often require extensive computing times, impairing research prospects [9]. Data precision is a concern as well, as generalization processes, whereas can reduce computing complexity and time, oftentimes render comparisons among different scales inaccurate, due to certain road structures non-representation [8]. This last deterrent regarding data accurateness is not limited to spatial, but also to visual intelligibility. Thematic maps, by definition, should provide visual information regarding the spatial distribution of one or more phenomena, through a clear symbology arranged in hierarchies to illustrate the greater and lesser features about a variable, whilst also presenting specific information about its spatial location [10]. Hence, these maps should retain a *visual metaphor* component capable of evoke the relationships between symbols and the spatial context in which they are inserted [11]. Meeting these requirements may be problematic, above all, when working with interactive datasets with multiple variables set to be depicted both in small and large scales, due to overall data quantity and information overlap.

In that vein, proper organizations of spatial network analyses as thematic maps for open-access databases are quite limited[1], since urban analyses remain isolated as local case studies and regional analyses are not as imparted, due to aforementioned issues. From this perspective, significant urban-regional phenomena dimensions remain rather unexplored and unrepresented. Notwithstanding, research requirements for large configurational datasets, capable of assessing geospatial correlations amid infrastructure and movement, through dissimilar scales tend only to increase. These demands prompted the assemblage of a Configurational Atlas, both as a research phase and as an experiment for

---

[1] The most distinguished initiative in this aspect is the Space Syntax OpenMapping [12]. Created in 2018, it consists of an open data pre-processed spatial network model of Great Britain.

assessing the viability of a future project. Using spatial information from the Tuscan Region (Italy), this Atlas groups a series of thematic maps based on detailed spatial network models from diverse sections of the Tuscan road infrastructure. Their main objective is to illustrate road-circulation network morphologies, as well, to highlight the centralities hierarchies of *closeness* and *betweenness*, in order to represent magnitudes of movement potentials, organizing this information for several territorial extents in a same database.

This paper summarizes the scope and methodological steps to construct the dataset and the models that compose the Configurational Atlas. With this framework in place, it is possible to emphasize characteristics of the GIS interface, as well, the adaptations made for configurational data intelligibility and usability. Furthermore, it will be discussed the Atlas contribution as a spatial representation and data repository, alongside with its prospects as an analytical instrument and future developments regarding data accessibility. Integrating models for geographic simulation and prediction is going beyond GIS' *information system* stage, and towards a dual-core stage, founded on geographic databases sharing and the diffusion of precompiled analysis models [13]. The Tuscany Configurational Atlas is intended as a step on this direction. By making spatial networks datasets available in an open-source geoprocessing-capable instrument, further simulations and correlations can be developed independently, hence, aiding in urban-regional related decision-making processes.

2 Datasets and Configurational Analyses Principles

The Tuscany Configurational Atlas is an outcome of experiments related to ongoing researches in network analyses, focused on the assessment of urban and regional road-infrastructures configurational properties. These were motivated by data requirements for a regional scale depiction of the Tuscan road-circulation network centralities hierarchies, a database to be correlated with other aspects of the built environment. Owing to long processing times entailed in modelling large regional networks, partial analyses using smaller territorial expanses were carried out in the meantime. Results were an extensive collection of original spatial datasets, arranged for consultation as GIS-based thematic maps. Its potential as an analytic instrument, as well as an information broadcaster prompted the Atlas assembly. The Configurational Atlas is organized and hosted as a project in the QGIS 3.10 [14] suite. Since QGIS is an open-source software, all created datasets are integrable to existent WebGIS platforms, therefore, this contributes to this Atlas future availability.

A common database is adopted to construct the Configurational Atlas and produce its network analyses datasets: the Tuscany Road Graph (*Grafo Viario della Toscana*) [15]; a Road-Centre Line (RCL) map that depicts the whole regional road-circulation network topology. This base graph is built based on a pair of regional technical charts

(*Carte Tecniche Regionali – CTR*)[2], scaled 1:10,000 and 1:2,000, hence, able to depict road-infrastructure at an urban quarter level of detail. Its vectoral layout is organized following the typical network structure, being composed by arches, continuous polyline structures dubbed as "road-elements" (*Elemento Stradale*), and node structures, mid-points that establish the linkage between one or more road-elements defined as "road-junctions" (*Giunzione Stradale*) [16].

The base graph is further sectioned, in order to represent the road-infrastructure of the following territorial divisions: region (*regione*), provinces (*province*) and municipalities (*comuni*). Individual datasets were created through an intersection, employing as a frame structure the Tuscan Administrative Limits, which were included in the Atlas as auxiliary maps [17]. From each individual limit, buffers (1 km for provinces; 300 m for municipalities) were used to preserve road-circulation network continuities during sections[3]. Any road-elements that remained disconnected were removed in the graph revamp[4]. A total of 302 unique graph datasets were created[5], corresponding to all Tuscan Region mainland and archipelago[6] continuous road-infrastructure, being subdivided in: region (1), provinces (11) and municipalities (290).

All the Configurational Atlas datasets share an equal coordinate reference system (CRS): Monte Mario/Italy Zone 1 (EPSG:3003); default projection of the Tuscany Road Graph [16] and the Tuscan Administrative Limits maps [17]. It is important to remark that this CRS projection and the WGS 84/Pseudo Mercator (EPSG:3857), often used by satellite imagery repositories, such as Google Maps [18], have parameter values close enough to be considered identical within the accuracy of the transformation. Google Maps satellite imagery was used as a locational reference during the project but will be substituted by the Tuscan Region orthophotos [19] for a future release version.

Prior administrative limits section, the base road-circulation graph was submitted to a generalization process, which employed the Douglas-Peucker simplification algorithm (*v.generalize*) [20] in order to reduce network complexity. Nevertheless, a very

---

[2] While OSM data is available for the Tuscan Region, the CTR data provides a homogeneous dataset that requires less of accessory data generalization. This graph is also used to ensure compatibility with the Tuscan Region Ambiental and Territorial Information System (SITA)

[3] Buffers are set within those threshold radiuses as greater extents collected many discontinuities, due to Tuscany fragmented territorial division, especially at municipality scale.

[4] It is important to remark that sections are not territorially strict, as for network analyses the graphs natural continuities ought to be respected over administrative limits to ensure system wholeness. Hence, road-elements were conserved whenever removal would cause a network gap, even when comprised in a neighboring territorial unit. This adequation was used mainly were road-elements had small segments in another municipality territory.

[5] Some municipalities had to be sectioned in two distinct areas due to network discontinuity, as consequence of the Tuscan fragmented territorial division.

[6] The Tuscan archipelago islands are referred as independent networks, even when partake to another municipality administrative limits. Spatial data regarding some islet's road network, absent in Tuscan Region Graph, were incorporated to the database. Elba Island continuous road-circulation network is represented at provincial scale due to its size, even though it is comprised in Livorno administrative limits. All Tuscan provinces (10) and municipalities (273) are represented, apart from Ca' Raffaello exclave, located in Badia Tedalda municipality (Arezzo Province), which is discontinuous from Tuscan territory, being positioned inside the province of Rimini (Emilia-Romagna Region).

small tolerance value (0.1) was applied during the generalization, with the intent to only diminish excessive numbers of polylines vertices in roundabouts and highway accesses, while preserving their road geometries (Fig. 1).

Fig. 1. Comparison (I.) between non-generalized (II.) and generalized (III.) road-circulation networks from Tuscan Region Road Graph – source: Regione Toscana – Grafo Iternet.

Albeit a minor reduction on the road-elements absolute number (1.06%) (Table 1), mostly due to post-generalization clean-up, the process had substantial impact in vertices count. Hence, this led to a substantial decrease in processing times (Table 2) for when RCL graphs are converted to angular segment maps through DepthMap X 0.5 [21], a network analysis software. This conversion step is required to assess *closeness* and *betweenness centralities* hierarchies in road-circulation networks through Space Syntax' Angular Analyses [22].

Table 1. Comparison between non-generalized and generalized Tuscan Road Graph number of road-elements prior and after angular segmentation.

	Road-elements prior Angular Segmentation	Road-elements after Angular Segmentation	Δ% Road-elements
Non-generalized	393.660	4.657.114	1083,02%
Generalized	389.477	1.251.610	221,35%
Δ% Generalization	1,06%	73,12%	–

Table 2. Average approximated modelling processing times of region, provinces and municipalities for non-generalized and generalized networks converted in Angular segment maps[7].

	Region	Province	Municipalities
Non-generalized	≈7,5 Months	≈23–72 h	≈2–8 min
Generalized	≈2,5 Months	≈7–18 h	≈0,5–3 min

Developed by Turner during the 2000's [22–24], Angular Analyses comprehend a set of methods conceived for evaluating spatial network properties and predict movement potentials on road-circulation networks constructed as dual graphs, such as RCL datasets. These methods require conversion of RCL graphs into angular segment maps,

---

[7] Processing time values for an overclocked Intel i7–8700 k (4.7 GHz); 16 GB of RAM.

in which the network *j-graph* is weighted according to the angle (in radians) amid each connected pair of vertices (in dual graphs, the road-elements), attributing to them a correspondent angular coefficient. System depth is obtained from the shortest *angular* path among all origin-destination pairs of vertices in the network, therefore, this angular coefficient corresponds to a weighted topological step, permitting measurement of *closeness* and *betweenness centralities* hierarchies. Whenever there are angle variations amid a road-elements pair – including any t-intersections or crossings between two roads – the RCL original continuous polyline will be segmented in two (angular segmentation), and the attribution of a corresponding angular value to each individual road-element will ensure, thus, curvatures with many nodes will result in many segments. Continuities will perdure when no interruption or direction change happens.

Closeness centralities analyses are drawn from Space Syntax' Normalized Angular Integration (NAIN) [25]. An algorithm equivalent to mathematical *closeness*, angular integration calculates (sums) the average angular costs of travelling over the shortest path, from each road segment i to all other possible destinations in the network. Hence, it depicts the to-movement potential of a road segment i, or its *relative accessibility* – how central its position is when related to all other positions in the network. The normalization of angular integration measures proposed by Hillier, Yang and Turner [25] ensure that systems' *integration* values are related to the urban average, through the division between node count (NC) and angular total depth (ATD) values regarding a determined segment i, formalized as following:

$$NAIN = (NC)^{\wedge}1,2/ATD \tag{1}$$

This adequation intend to standardize overall angular integration values, reducing the sums absolute number to proper established ranges[8]. This allows better comparisons amongst centralities hierarchies of different sized cities (with non-identical depths) or of diverse scaled road-circulation networks within a same urban settlement.

Betweenness centralities analyses derive from Space Syntax' normalized angular *choice* (NACH), an algorithm based on Angular Analysis [25]. Angular *choice* corresponds to mathematical *betweenness*, as it counts (sums) the number of times each individual road segment i is traversed when travelling, crossing through the overall shortest path, towards all potential destination segments, from all possible origin-destination pairs within the network. In this regard, this measure represents the road segment i through-movement potential, or the probability of its use as a system *preferential route*. Without normalization, *betweenness centralities* absolute values are conditioned to the network size – thus, to its depth. Concerning this, the quantity of *betweenness* acquired by a road segment i is related to its *integration/segregation* and its relative position within the system, with *segregated* designs adding more total and average *betweenness* to the network than *integrated* ones. Hence, angular *choice* measures highlight only the segments with the highest absolute *betweenness centrality* values – the main *preferential route* – while disregarding the remainder of road-circulation

---

[8] Hillier et al. [25] theorical tests demonstrate that NAIN ranges innately differ, depending on node count and total depth proportions. Therefore, there are no exact ranges for NAIN comparisons, but only approximations between top and bottom values depending on the urban settlement structure.

network *preferential routes* hierarchies' due to a high amplitude amid top and bottom *betweenness* sums, resulting in an insufficient representation of through-movement dynamics. Hereof, angular choice is not suited to comparisons amongst different systems or scales. The normalization method devised by Hillier et al. [25] mitigates this purported "*paradox of choice*" by weighting the calculated centralities values (angular *choice* – ACH) by the corresponding angular total depth (ATD) of a segment *i*, as following:

$$NACH = \log(ACHi + 1)/\log(ATDi + 3) \tag{2}$$

Through this relation, centralities values are adjusted according to the angular total depth of its correspondent segment *i*, hence, the depth component is distributed along the network. In this sense, greater the network *segregation*, the more reduced angular *choice* absolute values will be, by being divided by a higher angular total depth value. Normalization will then render visible the lesser *preferential routes*, by attributing to them a relative positional weight within the network hierarchy. The inclusion of a logarithm base standardizes NACH relations in a power law, restricted between 0 and 1, 5+[9] deeming possible comparisons between road-circulation networks with different depths, thus, non-identical sizes and scales, as absolute *betweenness centrality* values are now brought to an equivalence.

Angular Analyses models are exported from DepthMap X 0.5 as MapInfo datafiles (.mif), compatible with QGIS 3.10, then, converted in GIS to ESRI shapefiles (.shp). These models do not include normalized measures, though. While NAIN and NACH could be calculated for each individual graph directly through DepthMap X 0.5, due to the substantial amount of datasets, the use of a Microsoft Excel [26] macro was opted to accelerate the process. This macro imputes the normalization functions in the Excel database sheets (.xlsx) obtained from shapefiles, by creating two columns and normalizing each road-segment (rows). Another macro was used to reconvert these database sheets to a comma-separated values (.csv) file for further remerge with their respective original shapefiles in the GIS-based dataset. This procedure was made for 301 datasets, apart from the regional map, calculated through DepthMap X 0.5 since its road-elements quantity exceeded the Microsoft Excel memory for rows.

3   Tuscany Configurational Atlas Interface, Thematic Maps Data Discussion and Their Usage Prospects as Analytical Instruments

At this stage, datasets and spatial models approaches set, it is possible to emphasize the organizational fundaments applied in the course of the Tuscany Configurational Atlas construction, such as its GIS interface features, and thematic maps' visual adaptations made with data intelligibility in mind.

---

[9] Hillier et al. [25] theoretical tests state that *betweenness centralities* values can sometimes surpass the 1.5 threshold in specific cases, such as systems that exhibit ample differences between mean and total depth.

As mentioned, two parameters – NAIN and NACH – are employed for depicting the Tuscan road-circulation networks' movement potential distributions. Since these parameters are modelled for the 302 road-graphs, 604 individual datasets[10] were produced. To reduce overall files' quantity, both models were imbedded in a same shapefile, which was then classified in GIS to exhibit the desired parameter. Those datasets are organized in a multi-group hierarchical structure, labeled according to the administrative limit that is correspondent to the model extent, therefore, the municipalities (*Comuni*) are set inside their respective provinces (*Provincia*), that are, in their turn, placed in the regional group (*Regione*)[11]. All models can be accessed, shown or hidden through a drop-down list. NAIN and NACH parameters have their values discriminated in graduated legends. The Atlas interface also contains the depiction of the Tuscan road graphs (*Grafi Stradali*), divided in continental network and archipelago networks and corresponding administrative limits (*Aree Amministrative*) (Fig. 2).

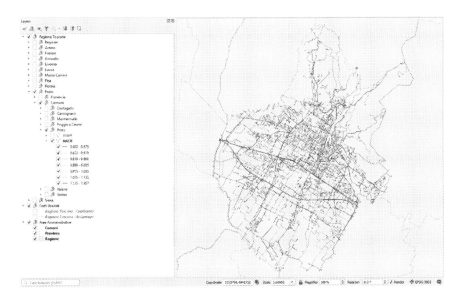

Fig. 2. Tuscany Configurational Atlas GIS project interface and thematic maps organization structure – source: Regione Toscana – Grafo Iternet

Groups were set to be mutually exclusive, thus, when one parameter is set active, the other will be automatically hidden, circumventing thematic maps data superimposition, undesirable from a visual coherence standpoint. Superimposition among different

---

[10] Total size of the Tuscany Configurational Atlas thematic maps datasets, auxiliary maps and original road graphs is 23 GBs.

[11] Groups were created as such to facilitate further incorporation of other regions' datasets in the Atlas database.

scales, however, is made possible, as this can be used to compare how movement patterns and centralities hierarchies are set over dissimilar territorial dimensions.

From a data provider and cartographer perspective, rendering spatial information intelligible in territorial expanses and scales as dissimilar as those intended for a Configurational Atlas ought to be a major concern. In this sense, proper adaptations had to be made to map symbology regarding the vectors' representation and, above all, to the color ranges that individuate *closeness* and *betweenness* centralities hierarchies in the datasets. While issues regarding vector representation where simple to solve, by setting variable line widths in proportion to scale dimension, color range issues were not so trivial, requiring further adjustments. DepthMap X 0.5 [21], where models are first processed, differentiates spatial data through a spectral blue-cyan-yellow-red (bcyr) color band, split in seven gradient intervals that span from dark-blue (lesser values) to dark-red (greater values). Tuned to highlight cyan-yellow mid-range transitions, this spectral gradient distinguishes, through contrast, gradual variations in road-circulation network centralities hierarchies' values. Although QGIS 3.10 [14] possesses a bcyr color range (ctp-city), comparable to DepthMap X 0.5, its blue-cyan and orange-red gradients are predominant over cyan-yellow transitions. Hence, the green gradient remains in a narrow range, in such form that cyan appears to transition directly to yellow (Fig. 3, II.), rendering mid-range centralities data difficult to read in the map.

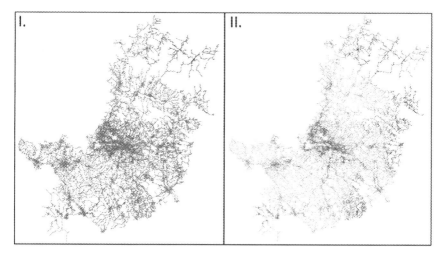

Fig. 3. Comparison between DepthMap X 0.5 (I.) and QGIS 3.10 (ctp-city) (II.) bcyr color ranges for NAIN in the Firenze Province dataset – source: Regione Toscana – Grafo Iternet (Color figure online)

This issue prompted a revision of the QGIS 3.10 [14] bcyr ctp-city color range, and the development of a suitable gradient, capable of highlighting the cyan-green-yellow transition without squandering thematic maps' data intelligibility concerning highest and lowest values for centralities hierarchies (Fig. 4).

I. ▮▮▮▮▮▮▮▮▮▮▮▮▮▮▮▮▮▮▮   II. ▮▮▮▮▮▮▮▮▮▮▮▮▮▮▮▮▮▮▮

Fig. 4. Default QGIS 3.10 bcyr color range (ctp-city) (I.) and revised bcyr color range (II.) used to highlight *closeness* (NAIN) and *betweenness* (NACH) centralities hierarchies (Color figure online)

Comparisons with DepthMap X 0.5 [21] color range demonstrate that QGIS 3.10 revised bcyr can individuate clearer transitions among mid-range values (cyan-green-yellow), which was the main problem regarding QGIS 3.10 ctp-city bcyr [14], and, as well, remark subtle transitions between mid-range and the top-bottom values for centralities hierarchies, absent in the DepthMap X 0.5 gradients (Fig. 5).

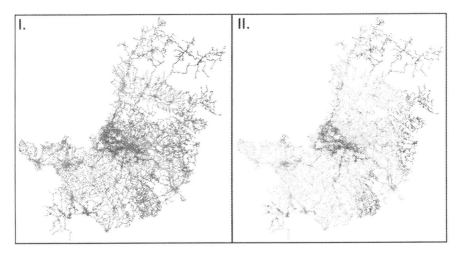

Fig. 5. Comparison between DepthMap X 0.5 (I.) and revised (II.) bcyr color ranges for NAIN in the Firenze Province dataset. – source: Regione Toscana – Grafo Iternet (Color figure online)

Data visualization in the Configurational Atlas interface is further improved through changing the NAIN and NACH color ranges rendering order, so higher valued road-elements (centrality cores) are rendered on top. While this setting is enabled by default on DepthMap X 0.5 [21], it must be manually set on QGIS 3.10.2 [14].

Another issue that had to be taken in account was the innate bcyr issues concerning data intelligibility in grayscale. Even though a minor problem in digital environments, since color bands may be unrestrictedly reproduced, this can be a hindrance when paperback publication or reproduction is considered. While not ideal, the revised bcyr tries to minimize this issue by adopting a darker hue for all gradients, with emphasis in the green gradients that, when converted into grayscale, can mark the point where the transitions from cyan to green and from green to yellow happen (Fig. 6), individuating top middle and bottom centralities hierarchies.

I. ![black bar] II. ![black bar]

Fig. 6. Default QGIS 3.10 bcyr color range (ctp-city) (I.) and revised bcyr color range (II.) when converted to grayscale. (Color figure online)

Regardless that bcyr color ranges are considered far from ideal representations for scientific data [27], requirements of a broad range of graduation that denotes configurational models' values hierarchies make its usage inevitable. These proposed revisions acknowledge and try to minimize bcyr problems and can be used as guidelines for other GIS-based datasets with similar requirements concerning data distribution.

Visual metaphors issues circumvented, the digital framework in which the Tuscany Configurational Atlas datasets are included, and how being in this context can have influence on the Atlas overall prospects as an analytical instrument may be discussed.

Set in a GIS-based environment, the Atlas project permits the apposition of several spatial information layers to its precompiled thematic maps. Being in this context, grant that the configurational models, which depict road-circulation networks movement potentials dimensions, can be visualized alongside other variables that may or may not be affected or interact with these dimensions. From a visual standpoint, this simple juxtaposition can already evoke some aspects about a determined spatial relation that could not be perceived otherwise, therefore, indicate possible correlations. A practical example of this prospect can be observed when a productive built-structures density [28] datamap is layered to the Pisa Municipality NAIN model thematic map, focused in its central urban area (Fig. 7):

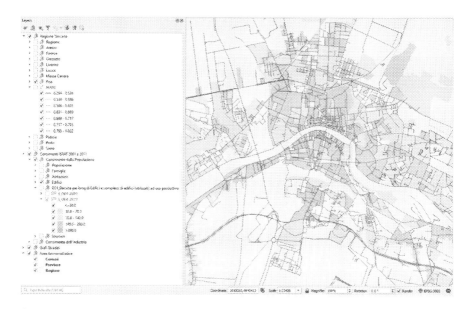

Fig. 7. Productive built-structures density for km2 apposed to Pisa Municipality NAIN dataset – source: ISTAT – Censiamento Generale 2011; Regione Toscana – Grafo Iternet.

The ISTAT datamap [28] depicts the density of buildings or building complexes that have productive – retail, industrial and services – usage throughout the territory. It can be observed that, the productive structures density tends to be generally higher on the areas close to the city center (centered on the map). Outside this case, however, the productive structures density tends to be greater near areas with high to-movement potentials (indicated by NAIN orange and red lines), which are predisposed to gather more pedestrians and vehicles. From a locational economics perspective, this visualized relation explains a known empirical logic, since retail placement tend to be associated to pedestrian and vehicular movement [29] due to customer attraction and presence demands that these activities have.

Even though data juxtaposition can already reveal some aspects about geo-objects spatial relations, these are generally restricted to the initial purpose and characteristics of the precompiled datamaps. However, GIS-based digital environments enable these visual comparisons to be taken one step further, as the datasets can be manipulated. In this sense, GIS' geoprocessing instruments allow the creation of new sets of variables based on the existing data and enable proper geospatial correlations to be set, expanding the possibilities regarding the spatial analyses. This prospect can be exemplified in the following instance, that juxtaposes productive areas data, generated from the Tuscan Land Use and Cover 2007-2016 dataset [30], to the Livorno Municipality NAIN model thematic map (Fig. 8):

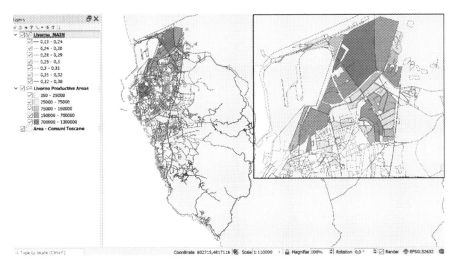

Fig. 8. Productive areas data apposed to the Livorno Municipality NAIN dataset – source: Regione Toscana – Grafo Iternet; Regione Toscana – Uso e Copertura del Suolo 2007 – 2016

Producing original spatial information from a precompiled dataset opens quite a few possibilities regarding spatial analysis. In the example, enacting a sphere of influence associated to density estimation values provided by the heatmap enable statistical correlations between built-structures position, agglomeration and their nearness to network centralities hierarchies. This can yield more detailed and robust results when compared to the previous example (Fig. 7), which is based only in a visual analysis.

Outside shown applications, oriented to juxtapose the spatial datasets to the road-circulation network thematic maps, proper analyses of Configurational Atlas models themselves can, as well, reveal intriguing network dynamics. These can be considered as research topics for further studies to be developed in the fields of theorical and applied mathematics, informatics and cartography.

A group of network phenomena that was observed concerns the *visual similitude* of *betweenness centralities* (NACH) throughout the scales. Regarding this, the found logic is that centralities hierarchies tend to remain largely unaltered in the same places along the network, despite variations of network size (scale), that ought, at least, to change its general hierarchies' distribution. Hence, the Regional model for *betweenness centralities* hierarchies tend to remain substantially simile to Provincial models and Municipality models, which in their turn display similarities between themselves, when considered the same stretches of the network, as highlighted in the example (Fig. 9) between Firenze Province (I.) and Firenze Municipality (II.)

Fig. 9. Visual similitude between Firenze Province (I.) and Firenze Municipality (II.) NACH models - source: Regione Toscana – Grafo Iternet

Without the conception of the Configurational Atlas models and thematic maps, these network phenomena might not have been observed, nor their repetition pattern amid different scales.

Overall, the Tuscany Configurational Atlas has demonstrated stimulating potentials as an analytic instrument. The versatility of the GIS-based digital environment and the availability of geoprocessing instruments enables diverse kinds of data to be apposed and analyses to be developed. Data completeness is an important aspect in decision-making processes, as it helps policymakers to think beyond the single dimensions of the territory, empowering a broader vision of the territorial potentialities.

4   Concluding Remarks

Continuous GIS development implies a transcendence from the *informational systems* stage, towards an instrument that combines geographic databases sharing and analysis models diffusion. Concerning this matter, GIScience progress ought to be oriented for shortening the gap amid informational demands and datasets availability that perdures in several disciplines, urban and regional sciences included. From this perspective, it is expected that, for the near future, digital data repositories constructed in GIS-based environments will have significant roles in setting these transformations in motion, functioning as outsets for the integration of databases and analysis models.

This paper summarized the principles behind a Configurational Atlas construction, describing its data sources, technicalities of its GIS project, and methods used to conceive the configurational models responsible for outline the road-circulation networks movement patterns, main parameters to be exhibited on the repository. In the discussion, some aspects were addressed regarding the Atlas map sets organization on GIS interface, and the amendments devised with intent to avoid information overlap when manipulating the datasets. Another step discussed were the adequations made to circumvent issues concerning the Atlas' *visual metaphors*, and the improvements made to symbology with thematic maps data intelligibility in mind required as part of the incorporation of DepthMapX 0.5 models into GIS. These considerations can be used as construction guidelines for other GIS-based interactive configurational datasets.

Even though constructed as an experimental project, the Tuscany Configurational Atlas has demonstrated potential to go beyond its original purpose as a data repository for consultation. Set in a GIS-based environment, the Atlas permits the apposition of several kinds of spatial information, such as the presented examples from census and built-structures datasets, to the precompiled thematic maps representing the configurational models. From a visual standpoint, this can undoubtedly evoke some aspects about geo-objects spatial relations, yet, since the Atlas' thematic maps can be manipulated, these visual comparisons can be taken one step further, as the configurational datasets may serve as sources for formal spatial correlations using GIS' geo-processing instruments. As the Atlas allows configurational models representation through several scales with the same level of detail, this consented to visualize intriguing network logics that are reproduced in different scales, and ought to be explored in further studies, such as the mentioned *visual similitude* of *betweenness centralities*.

Forthcoming improvements are in course regarding the Configurational Atlas availability and usability, the most important, being its integration to a WebGIS suite. This is intended to be done through the Tuscan Region Ambiental and Territorial Information System, (SITA - GEOScopio) in hope that more comparative studies are developed, but also, to improve governmental access to spatial information, giving further dimensions to be considered in the decision-making processes. Another possible research course consists in expanding the Configurational Atlas to include other Italian regions, in order to make a comprehensive analysis of the Italy's movement potentials logics throughout the territory.

References

1. Goodchild, M.F.: Twenty years of progress: GIScience in 2010. J. Spat. Inf. Sci. (2010)
2. Longley, P.A., Goodchild, M.F., Maguire, D.J., Rhind, D.W.: Geographic Information Science and Systems, 4th edn. Wiley, New York (2015)
3. Lin, H., Chen, M., Lü, G.N.: Virtual geographic environment: a work-space for computer-aided geographic experiments. Ann. Assoc. Am. Geographers **103**(3), 465–482 (2013)
4. Jiang, B., Claramunt, C.: Integration of space syntax into GIS: new perspectives for urban morphology. Trans. GIS **6**, 295–309 (2002)
5. Hillier, B.: Space is the Machine: A Configurational Theory of Architecture, Space is the Machine: A Configurational Theory of Architecture. Space Syntax Ltd., London (2007)
6. Jiang, B., Liu, C.: Street-based topological representations and analyses for predicting traffic flow in GIS. Int. J. Geogr. Inf. Sci. **23**, 9 (2009)
7. Jiang, B., Claramunt, C., Klarqvist, B.: An integration of space syntax into GIS for modelling urban spaces. Int. J. Appl. Earth Obs. Geoinf. **2**, 161–171 (2013)
8. Serra, M., Hillier, B.: Angular and metric distance in road network analysis: a nationwide correlation study. Comput. Environ. Urban Syst. **74**, 194–207 (2019). Elsevier
9. Serra, M., Hillier, B., Karimi, K.: Exploring countrywide spatial systems: spatio-structural correlates at the regional and national scales. In: Karimi, K (ed.) Proceedings of the 10th International Space Syntax Symposium, pp. 1–18. University College of London, London (2015)
10. Tyner, J.A.: Principles of Map Design. Guilford Publications, New York (2014)
11. Roth, R., Kelly, M., Underwood, N., Lally, N., Vincent, K., Sack, C.: Interactive & multiscale thematic maps: a preliminary study. In: Fujita, H. (ed.) Abstracts of the ICA - 29th International Cartographic Conference, 1, 315 Tokyo (2019)
12. Space Syntax OpenMapping Homepage. https://spacesyntax.com/openmapping/. Accessed 02 Feb 2020
13. Lü, G.N., Batty, M., Strobl, J., Lin, H., Zhu, A., Chen, M.: Reflections and speculations on the progress in Geographic Information Systems (GIS): a geographic perspective. Int. J. Geogr. Inf. Sci. **33**(2), 346–367 (2019)
14. QGIS 3.10.2.: 'A Coruña'. https://www.qgis.org/en/site/forusers/download.html. Accessed 02 Feb 2020
15. Regione Toscana, Direzione Urbanistica e Politiche Abitative - Sistema Informativo Territoriale e Ambientale – SITA.: Grafo stradario e ferroviario della Regione Toscana. http://www502.regione.toscana.it/geoscopio/cartoteca.html. Accessed 02 Feb 2020
16. Regione Toscana, Dipartimento Generale delle. Politiche Territoriali, Ambientali e per la Mobilita - Settore Pianificazione del Sistema Integrato della Mobilità e della Logistica.: Specifiche tecniche per la gestione e l'aggiornamento del grafo viario e della numerazione civica. Release 4.6
17. ve - Sistema Informativo Territoriale e Ambientale – SITA.: Limiti Amministrativi. http://www502.regione.toscana.it/geoscopio/cartoteca.html. Accessed 02 Feb 2020
18. Google Maps.: Tuscan Region Map. n.d. https://www.google.it/maps/. Accessed 02 Feb 2020
19. Regione Toscana, Direzione Urbanistica e Politiche Abitative - Sistema Informativo Territoriale e Ambientale – SITA.: Ortofoto. http://www502.regione.toscana.it/geoscopio/ortofoto.html. Accessed 02 Feb 2020
20. Bundala, D., Begenheim, W., Metz, M.: v.generalize source code. https://github.com/OSGeo/grass/tree/master/vector/v.generalize. Accessed 02 Feb 2020

21. Varoudis, T.: DepthmapX 0.5. https://varoudis.github.io/depthmapX/. Accessed 02 Feb 2020
22. Turner, A.: Angular analysis. In: Peponis, J., Wineman, J.D., Bafna, S. (eds.) Proceedings of the 3rd International Symposium on Space Syntax, pp 7–11. Georgia Institute of Technology, Atlanta (2001)
23. Turner, A.: From axial to road-centre lines: a new representation for space syntax and a new model of route choice for transport network analysis. Environ. Plan. **34**, 539–555 (2007)
24. Turner, A.: Getting serious with DepthMap: segment analysis and scripting. In: UCL, pp. 18–25 (2008)
25. Hillier, B., Yang, T., Turner, A.: Normalising least angle choice in Depthmap - and how it opens up new perspectives on the global and local analysis of city space. J. Space Syntax **3** (2), 155–193 (2012)
26. Microsoft Excel 365 ProPlus. Microsoft (2019)
27. Moreland, K.: Diverging color maps for scientific visualization. In: Bebis, G., et al. (eds.) ISVC 2009. LNCS, vol. 5876, pp. 92–103. Springer, Heidelberg (2009). https://doi.org/10.1007/978-3-642-10520-3_9
28. ISTAT: Censimento generale, della popolazione e delle abitazioni – periodo di riferimento 2001–2011, Roma (2016)
29. Hillier, B., Penn, A.: Cities as movement economies. Urban Des. Int. **1**(1), 41–60 (1996)
30. Regione Toscana, Direzione Urbanistica e Politiche Abitative - Sistema Informativo Territoriale e Ambientale – SITA.: Uso e Copertura del Suolo 2007–2016. Accessed 02 Feb 2020

The Quali-Quantitative Structure of the City and the Residential Estate Market: Some Evidences

Valerio Di Pinto[1] and Antonio M. Rinaldi[2,3(✉)]

[1] Dipartimento di Ingegneria Civile, Edile e Ambientale, Universitá di Napoli Federico II, 80125 Via Claudio, 21, Naples, Italy
valerio.dipinto@unina.it
[2] Dipartimento di Ingegneria Elettrica e delle Tecnologie dell'Informazione, Universitá di Napoli Federico II, 80125 Via Claudio, 21, Naples, Italy
antoniomaria.rinaldi@unina.it
[3] IKNOS-LAB - Intelligent and Knowledge Systems - LUPT, Universitá di Napoli Federico II, 80134 Via Toledo, 4, Naples, Italy

Abstract. In the collective imagination there are masterminds speculating on real estate and so defining the "price of the buildings" in a sort of transposition in reality of *Hands over the city* plot. Although it is not the case, the distribution of prices in the contemporary city is a big deal for policy-makers and urban planners as well. Especially in metropolitan areas, where polarization of services is needed, the high complexity of urban scenarios makes the prediction of price variation due to change in the morphology or in the offer of services (for example in transportation and accessibility) very unreliable, despite of the success of policy and planning actions is strictly related to its control. This paper, moving in this perspective, points out some evidences on the relationships between the quali-quantitative structure of the city and the fluctuation of real estate prices with reference to house market, under the idea of investigating what is the role of urban railway stations.

As a matter of fact, railway stations are nodes of the transport network as well places of the urban environment whose big impact on accessibility in the city and the corresponding contribution in the definition of property values is very intuitive. The findings on the effects of railway stations on property values are mixed, since they vary on magnitude and direction.

This paper focuses on the comparison of the impact of urban railway stations on the residential estate property values as well on the structure of the city in itself, grasped in terms of configurational state. To this aim, the paper proposes a case study, evaluating how the correlation among configurational indexes and residential values distribution varies with and without the presence of the railway system.

The main findings of the work make to arise a marginal or negative impact of the railway system, so as to promote the idea that other factors dominates the change in local property values around railway stations. Finally, the authors proposes the integration of ontologies and configu-

© Springer Nature Switzerland AG 2020
O. Gervasi et al. (Eds.): ICCSA 2020, LNCS 12251, pp. 307–323, 2020.
https://doi.org/10.1007/978-3-030-58808-3_23

rational analysis to define a methodological framework able to explain the state of residential estate market.

1   Introduction

The distribution of housing prices in the city is a long lasting theme in urban planning, still actual in the light of crisis that has been affecting cities and urban areas in the last ten years. It is a widespread opinion that the lack in knowledge of the mechanisms behind the dynamics of house prices in the city affects decision and policy making, as well as urban planning choices. If the amount of literature in the field is impressive, there is not a definitive approach that seems to prevail and it leaded to large disagreement in results. The question have been traditionally approached from specific points of view, probably neglecting the needed comprehensive vision: site and real estate peculiarities as well as the human behavior have been modeled to give reproducible results in the frame of an objective environment more than to give reliable outcomes. As an instance, environmental psychology and social sciences proved that the simple optimization of variables don not completely characterize the human behavior [31], taking sometimes priority other elements that are related to the users/environment correlation.

In this context, dominates the hypothesis of monocentricity (Central Business District - CBD) [34], which proposes that market values essentially depend on the accessibility of a location to the most central areas. In this sense, an element of the city pertains the distribution of real estate prices only if it directly impacts on the accessibility or at least in so far as it contributes to define the related patterns. As a result, the real estate market is generally described by a large set of variables that are exogenous to the model of the city and it provokes misunderstanding and distortions [18].

In the last years, the configurational approach to the analysis of the city opened new perspective to the understanding of urban phenomena, assuming the primary role of the space. As a matter of fact, also price distribution has been interpreted recurring to configurational logic and it leaded to significant results raging from housing to office prices [13,16,18,33]. Under the configurational hypothesis, the CBD model is completely integrated into the model of the city, thus resulting in the primary use of endogenous variables.

Stressing the configurational logic, this paper will investigate on the relationship among railway network and accessibility in the distribution of house prices in the city. Since the system of railway connection directly affects the topological structure of urban areas, the analysis will be carried out evaluating the impact of such a network on the correlation among the distribution of house prices and configurational indexes, following a methodological approach described in the following section. Moreover, moving from the common idea that a railway plays in two different domains, which are the accessibility yet introduced (the topological impact of the railway as a network) and the urban form and other externalities (the impact of the railway as a set of stations on the urban area where they lean), the paper will explore also the hypothesis of expanding the purely quantitative approach with a quali-quantitative approach based on the integration of ontology and configurational analysis [7, 9, 19]. This hypothesis will be discussed in the last section.

More in details, in Sect. 2 is provided an analysis and a brief discussion of scientific literature in the domain of interest; in Sect. 3 there is the description of the methodological approach proposed and its application on case study is presented in Sect. 4, as well discussed in Sect. 5 along with general conclusions.

2   Backgrounds

In this section we give a comprehensive overview of existing literature in our domain of interest, focusing on configurational analysis, railway stations and networks, and knowledge representation through ontologies as well.

2.1   The Configurational State of the City

Cities are complex entities that are not prone to be precisely defined, since they defy formal descriptions. It is mainly due to the large amount of variables involved in its dynamic as well as to the overlapping of many competing entities that seem to challenge to dominate its scene. Planners, urban thinkers and designers then have been always trying to simplify the nature of the city, in order to grasp the fundamental nuclei of the urbanity, letting to emphasize hierarchies rather then geometries, or separation of parts from whole as well.

In the last forty years, thanks to the impressive development of automated calculus, a new "urban science" have been developing, stressing the idea to model complexity in cities recurring to the science of complex networks. In the early 80', Bill Hillier pioneered the notion of Space Syntax [31], fostering the use of centrality to define the configurational state of cities and open spaces. Such an approach is based on the extension to the city of measures and algorithms developed in structural sociology some decades before [24], under the idea that a city is made of two main layer that could be read separately, but that work together [48]. In the Space Syntax view, a city is therefore a complex system that behind a physical structure, which is composed by building connected by public spaces (streets, squares, ...), has a social structure, which is made of a large set of activities related by interactions [48]. This is what we call the duality of the city, which feeds a long-lasting debate on what is the dominant dimension and whether they are connected or not. Space Syntax proposes a theoretical position where they are one thing, so as to rethink the city as a "unique entity" that could be considered under a "unique theory" of the city as a whole, assuming the urban open space as *"the common ground for physical and societal cities"* [48].

The dominant elements and dynamics of the city are therefore expressed by the so-called "configurational state" of the urban system, which is expressed by the analysis of the set of *"relations between all various spaces of a system"* [48].

Under this view, the urban layout is completely expressible by a purely topological network, modeling the open space and their mutual connections. Space Syntax rethinks also the way the urban network is built, proposing a "dual" process that leads to a graph where points represent open spaces (typically streets and squares), while edges stand for their connection [30,38]. On this graph, centrality measures could be performed, so as to provide urban elements of numeric values representing its position in many urban hierarchies that define the configurational state of the city, directly related to the urban phenomena. There are many "configurational indexes", each one attributable to a correspondent index in structural sociology [38], although Space Syntax is mainly pivoted on the so-called "Integration Index" and "Choice Index". The first one refers to the concept of closeness in graph theory [41], expressing the idea that is "central" a node that is topologically close to all other in the system. The second one is related to the betweenness centrality [24], under the idea that the centrality of a node pertains its potential to intercept fluxes on the net being on as many geodesics path between any pair of nodes on the graph.

Betweenness centrality index C^b [24] and closeness index C^c [41] of a point p_i on a network of n nodes are:

$$C^b = \sum_{i=1}^{n} \sum_{j>i}^{n} \left(\frac{g_{ij}(p_k)}{g_{ij}}\right)[24]; \qquad C^c = \frac{n-1}{\sum_{i=1}^{n} d(p_i, p_k)}[41]$$

where: $d(p_i, p_k)$ = the number of edges in the geodesic path that links p_i and p_k
g_{ij} = the number of geodesic paths that link p_i and p_k
$g_{ij}(p_k)$ = the number of geodesic paths that links p_i and p_k containing p_k

The role of space in the city is precisely defined by Space Syntax, since it relates the dynamics of urban environment to the rates of movement in the city, assuming the existence of the so-called "natural movement", which is the part entirely and only depending on the grid configuration [29]. It de facto assigns to the urban layout a generative role on urban phenomena, as well as a primary role as catalyst in urban dynamics, since activities located in the city amplify the movement rates, playing as multipliers, generating new movement. It is the so-called "movement economy" [29], which governs the what happens in the city and drive its evolution. For further information, in the city there are also non-configurational attractors, which generate movement rates regardless of the layout. They are generally related to the planned expansion or regeneration of the city and often lead to what we call "urban pathologies" [30].

2.2   Housing Market, Railway Stations and Configurational State

In order to evaluate the impact of railway stations on property values, many authors suggest to see the stations as nodes of the transport network and places in an urban area [5]. The first characteristic relates to accessibility, which generally has a positive effect. The latter, on the other hand, is connected to the externalities of the stations and could have negative or positive effects case by case. In the last 15 years, many empirical studies are pivoted on this approach [17]. As a result, the accessibility effects of a railway station is modeled in terms of distance between houses and the station in itself, as well by considering the quality of the railway network system as a whole, so as to primarily account for reachable destinations and eventually for other qualities like the frequency of the services, the facilities and many others. In this view, higher the accessibility, more positive the effect of the railway station is expected to be on house prices.

The idea that land/house prices and accessibility are strictly related is long-standing, since it roots in the work of Von Thünen, who explained the difference in values of similar farmlands due to diversity in accessibility [49]. Basing on these assumptions, further economists proposed the so-called bid-rent theory, which is pivoted on the relationship among the willingness to pay of an agent and the location of the land, so as to create a gradient declining with distance from a central area, generally named as Central Business District (CBD) [2,50]. The bid-rent model de facto considers as critical for values formation the accessibility to the CBD and therefore the transportation costs to reach it. A location proximal to CBD is considered attractive since there activities trend to polarize so as to give added value.

Investments in transports control prices by limiting friction around the CBD and attract household near to the stations [23]. It is therefore expected that house prices decrease moving away from stations. CBD is therefore an element that is exogenous both to the city in itself and to the price model [13].

In the last years, other researcher proposed models that interpret the distance from the CBD as a local property of real estate directly deriving from its position in the urban structure, so as to make CBD to turn in an endogenous element. Many of that researches analyze the effect of global and local accessibility on office and housing prices recurring to Space Syntax in the frame of a hedonic model of real estate prices analysis. The results show a good predictive power in interpreting the log of the rent, relating it to integration and choice indexes [13,33]. In this view, the railway system is integrated in the topological structure of the city, providing connection between nodes of the urban graph associated to the railway stations in reality [25,26]. It implies that only accessibility provided by the railway is taken into account, while externalities are missing if they not affect the relational state of the city.

Generally, researchers refer to this kind of model as "quantitative" and consider them useful since they assure a sufficient level of simplification not so deeply affecting the capability to read and interpret urban phenomena. However, some formal element of the city seem to be very important to have a deep understanding of complex mechanism acting in the city.

2.3   Representing Urban Knowledge Through Ontologies

Understanding the space in the city is paramount to identify actions for its management, so identifying homogeneous contexts containing highly related and characterizing elements is a necessary task. Therefore, acquire deep knowledge on a built environment is pivoted on the recognition of its peculiarities, interpreted in the context of their physical and cultural environment.

Conceptual misunderstandings and semantic vagueness should be face prior to define a reasonable and reliable description of "urban knowledge", starting by the definition of a shared formalized language. Among the new available techniques to address the issue, ontologies seek to create a shared representation of information. That is intrinsically independent of the information itself, so that it could be isolated, recovered, organized and integrated only depending on its content [40].

With reference to spatial knowledge, ontologies have been gaining momentum in the last fifteen years, since they are used to represent its properties, as well as to harmonize and inter-operate data from different sources [4,32]. Ontologies could be used also to represent the spatial meaning of a community in time, such as physical and cultural landscapes and to give a formal representation of any element of the city.

The most widespread and used approach to ontology starts from the *modeling view* of knowledge acquisition [14]. In this frame, the model relates knowledge to the behavior of a generalized agent (the problem-solving expertise) as well as to its own environment (the problem domain). This approach contrasts the *transfer view*, which consider a knowledge base to be something like a repository that stays in the mind of an expert. Under the modeling view, knowledge is therefore strictly related to the real world, and much less dependent on the way it is used.

Knowledge representation is a pillar of the knowledge engineering process as a whole, in the frame of AI research, although much more interest have been payed to the nature of reasoning, so as to generate a real dichotomy between reasoning and representation, which seems to pertain to the philosophical debate among epistemology and ontology. As defined by Nutter, epistemology is "the field of philosophy which deals with the nature and sources of knowledge" [36], while ontology is much more oriented to the organization of the world, independently form of the related knowledge in itself. As a result, an ontology is *"a formal and explicit specification of a shared conceptualization"* [27], where the term *conceptualization* is to be referred to the model of the reality where the concepts are defined; *explicit* is related to what kind of constrains and concepts are used; *formal* is connected to the "machine-readableness" of the ontology; *shared* relates to the spread consensus of the modeled knowledge. Many other definition of ontologies have been developed, stressing some concepts more then others. Neches proposes a definition providing also a methodological approach to define an ontology: a) identification of terms and their mutual relations; b) definition of arranging rules; c) definition of terms and relations connecting concepts together. Under this view, an ontology is made of the explicit terms as well as of all those that can be derived by the implementation of rules and properties, so as to be a set of "terms" and their mutual "relations" [35].

Three basic questions are directly related to the reading and interpretation of urban environment. The first refers to the context, since the city has several signifiers directly connected to the nature of the ontology. The second is semiotic, in order to discuss the city that is by nature made up of signs in the real world. The third is epistemological and refers to the meanings of the concepts related to the city as an entity. Each question reflects an operational procedure, since upon signifiers, signs and meanings are based almost all the representation sciences, each of them proposing a model of the reality to be analyzed and evaluated to conclude interventions in the reality. Moreover, they are mutually connected by the matter of fact that a sign generates models intervention, goals and values to which planners refers in the definition of meanings that, in turn, affect approaches and criteria of intervention, directly affecting reality. When this latter change, perhaps, also induces changes in meanings, which imply new signs and then a new representation of reality and so on. Simplifying and lowering the complexity of urban environment is therefore mandatory to properly manage and share a knowledge representation of the city, although it is not a easy task.

As reported by Eco, giving a meaning to a signifier includes denotative and connotative elements [21]. To "denote" an object refers to give it a meaning, so as to have a real-time communication, since there are not ideologies or meta-discussions involved. Conversely, if a meaning expresses an ideology (also potentially as well as in hidden way) it is connotative, and it is related to values, symbols and intangible cultural elements. Through forms it is therefore possible to recognize the relation among objects and societies, in diachronic perspective also, giving momentum to the preservation of traces and evidences. When an element in the city is recognized, it become a sign and should be interpreted, not only in reference to itself (i.e. by decomposition) but also in the frame of the spatial and cultural context it belongs to.

In this frame, acquire interest also a formal definition of the concept of relevance information. It could be split in *objective* (system-based) and *subjective* (human (user)-based) relevance classes [28,42]. The objective relevance is a topicality measure, since it pertains to the matching among query and retrieved resources. Diversely, human relevance refers to many criteria [3,37,45], also related to the user's intellectual interpretations of the concepts of *aboutness* and *appropriateness* of the information that is retrieved.

On the basis of such a background, this paper would investigate the role played by railway system in defining the values of house market, under the idea of first evaluating its impact as a network providing accessibility (the set of connections created by the railway system as a whole) and eventually proposing a framework to approach its impact as a set of urban element in surrounding places (railway stations). The approach will be therefore based on a two-step strategy. Firstly, the urban system will be modeled in two scenarios representing the city with and without the railway system. For each scenario the configurational state will be calculated by means of configurational indexes and therefore the correlation among them and house price values will be evaluated, making the impact of railway network to arise. Secondarily, if results of first-step analysis can not clearly state the railway impact, a framework to integrate qualitative (connected to the concept of railway stations as places) and quantitative (connected to the impact of railway network on the configurational state of the city) analysis will be proposed, stressing the notion of ontology.

3   Methodology

In order to evaluate the effect induced by the railway system on the residential estate market, the changes in correlation among the distribution of Space Syntax Integration index and the distribution of residential estate values will be analyzed.

The configurational state of the urban system will be provided recurring to Space Syntax Segment Analysis [43,44], that is an operational technique directly derived from the one pioneered by Hillier and Hanson under the notion of *Axial Analysis* [30,31]. Both the techniques are structurally based upon the building of the so-called *dual network* [38], which proposes the fragmentation of freely accessible spaces into the minimal set of the largest convex spaces, the subsequent tracing of the visual lines connecting them in a system, and, finally, the creation of the urban graph by the inversion of lines into nodes and junction into edges. The process is completely automated into *DepthmapX*, a software by University College of London [46], as well into GIS add-on and plug-in based on DepthmapXnet [47].

The resulting urban graph could be analyzed in order to calculate configurational indexes. This work proposes the use of a tailor-made ArcGIS plug-in, developed to integrate Space Syntax techniques into ArcGIS software [22], using the interface of DepthmapXnet [47]. Configurational data are therefore stored as numerical attributes of urban nodes in the frame of a geographical database. It allows to perform comparative analysis based on the location in space of configurational data.

In the same GIS environment will be stored the data on residential market, in terms of homogeneous price areas, deduced from normalized registered transactions. Thanks to the capability of geo-processing tools of ArcGIS, market data will be pushed on configurational data by adding local properties to nodes (intersect tool), so as to achieve a integrated DB where exploratory analysis of correlation will be performed.

As a result, on the dataset of the nodes of the urban graphs could be performed statistical analysis in order to analyze the correlation among configurational indexes and residential prices. Recurring to multiple model of the city, leading to as much urban networks, the impact of railway system will be analyzed by comparing resulting correlation among the said variables in the two different scenarios (Fig. 1).

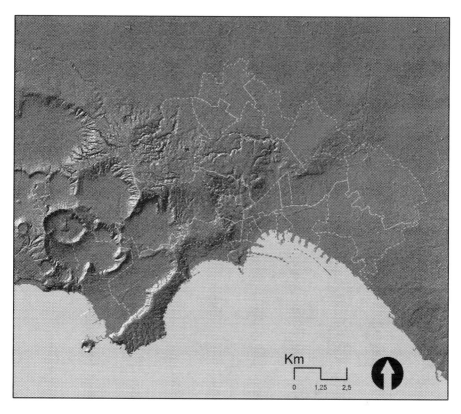

Fig. 1. Digital Terrain Model of the city of Naples. Yellow polygon in transparency highlight the boundary of the city. Dotted lines are the boundary of the administrative neighborhood. The DTM has been retrieved from the official geo-portal of Naples metropolitan city. (http://sit.cittametropolitana.na.it) (Color figure online)

4   Case Study

The proposed approach have been applied in the city of Naples, the third Italian city for magnitude of inhabitants and one of the largest of the Mediterranean basin at all. Despite the city has no more then one million people officially living into its boundaries, the surrounding urban area of the city has about 3.5 million inhabitants, so as to be the eighth urban area of the European Union. Since the territorial surface of the city is of $120\,km^2$, the population density is of $8.000\,inhabitants/km^2$, enclosed in a highly complex morphology, dominated along the east-west direction by the Mount Vesuvius and the Campi Flegrei Volcanic complex, two of the most active volcanoes in Europe. On the normal direction, the city is contained between the sea and a hill line that divide Naples from a large flatten area. Within the city, the morphology is based on two flat areas, divided by a higher hill line sloping toward the coastal line (Fig. 2).

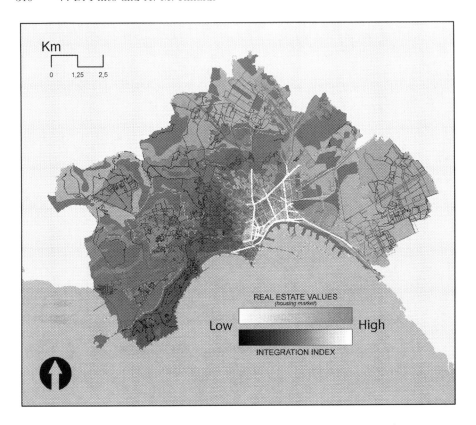

Fig. 2. The distribution of Integration index in SCN0 overlapped to homogeneous prices areas (red scale). Brighter the line, higher the index. Dark gray areas have no prices. Light gray areas are not included in the model, because they are outside of the city of Naples. (Color figure online)

As a result, Naples is a set of pseudo-autonomous area brought together by tunnels, so as to create a complex urban maze, with a long story of use and reuse, where narrow and chaotic spaces are close to neighborhood dominated by a ordered pace, as well as to large and planned boulevards resulting from the heavy urban changes of the nineteenth century. The configurational state of the urban system have been derived defining two different segment maps, in reference to a scenario composed only by the open spaces (SCN0) and a scenario implementing also the topological connection between lines due to the system of railway connections (SCN1). As described elsewhere [18,19], the configurational structure of Naples presents a set of critical lines where the highest values of integration and choice indexes polarize, so as to create what we call a central place. This kind of structure is almost the same in the two scenarios, although the distribution has significant local changes (Fig. 3).

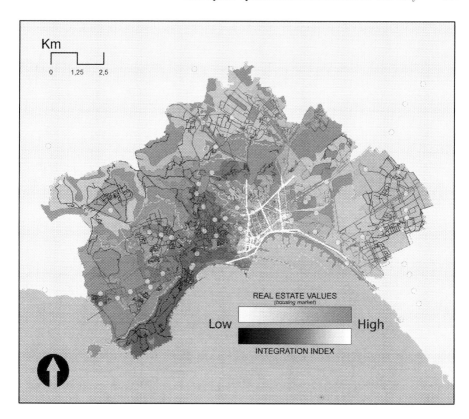

Fig. 3. The distribution of Integration index in SCN1 overlapped to homogeneous prices areas (red scale). Brighter the line, higher the index. Dark gray areas have no prices. Yellow points stand for railway stations. Light gray areas are not included in the model, because they are outside of the city of Naples. (Color figure online)

Recurring to the mapping clustering toolset, the hot spot area registering the higher values of the integration index (the so-called *integration core*) have been defined for SCN0 and SCN1. Figure 4 highlights the differences among them, giving emphasis to the topological effects of the railway system that reinforce the polarization of the integration core.

The residential estate market have been extracted by the official price list of Naples real estate exchange, base on a set of 3402 normalized transactions that have been registered in the 2011 second semester [6]. Using the geospatial tool of GIS environment, the prices have been pushed on the configurational segment, letting to integrate attributes on a singular feature. It leads to perform agile correlation analysis by the implementation of exploratory algorithm. More

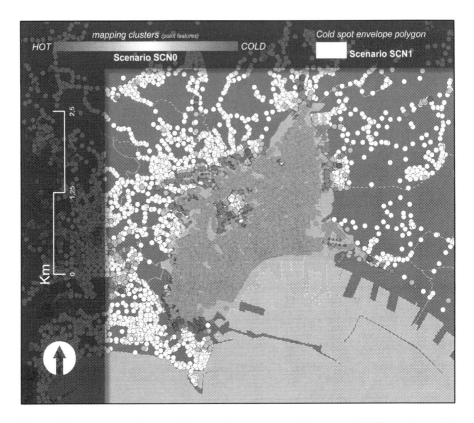

Fig. 4. The envelope polygon of Integration index's cold spots in SCN1 overlapped to the cluster map of Integration index in SCN0. Red dots exceeding the yellow polygon in transparency highlight the higher polarization of the Integration index in SCN1. (Color figure online)

specifically, the residential estate values have been correlated with the three configurational indexes of Integration, Choiche and local integration (400 m radius) [30].

As a result, a correlation matrix has been obtained for the two scenarios so as to let the measure of the impact of railways on such relations. Figure 5 summarizes the change in correlation among the said variables in terms of histogram. Vertical lines above x-axis mean better correlation, while those below mean weaker and the absence of any line means that any change in correlation has been induced. For the largest part of the neighborhoods the correlation is weaker.

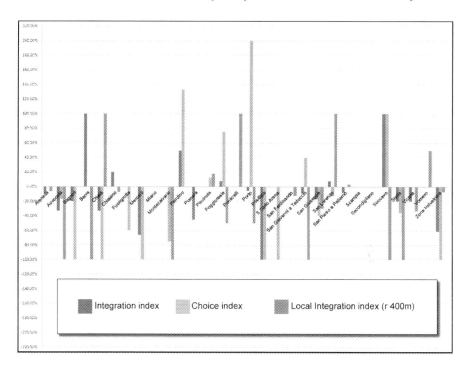

Fig. 5. Histogram summarizing how the correlation among Integration index, Choice index, local Integration index (r 400 m) and housing prices change with the inclusion of the railway system in the configurational model of Naples (y-axis displays the change in correlation expressed as percentage on the original values: if the correlation doubles, the correspondent value on the y-axis is 200%). The graph clearly states that a general lowering is induced.

5   Discussion of Results and Conclusions

The correlation among configurational indexes and housing prices in Naples lowers in reliability if the network model of the city is provided by the set of connections due to the urban railway system. In the view of the authors, it could have at least two different explanations. A first could be an insignificant role of the urban railway system on the effectiveness of accessibility of the city, which, in turn, does not affect the housing prices. Such a possible explanation is however unrealistic for two main reasons. First of all, scientific literature states that almost everywhere railway stations deeply affect the local market, causing a generalized increase in prices with a gradient similar to the CBD logic. Secondarily, only the two most important urban railway routes of Naples (Linea 1 and Linea 2) let 173.146 movements per day [1], which are the 22% of the car movement per day [15] (dominating movement mode in the city).

Such considerations let the authors to get a second explanation to the discussed lowering in the correlation, which refers to the incapability of a purely configurational approach to grasp the nature of the housing market. This latter seems to depend from multiple factors raging from objective (i.e. configurational) to subjective features. Generally speaking, the phenomenon seems to depict the dispute between accessibility and externalities, making the impact of combining benefits and nuisances to arise on real estate values distribution.

Such a complex matter therefore needs to be approached trough an integrated solution, able to formally define both kind of knowledge (objective and subjective). As preliminary proposed elsewhere [9–11,20], a local urban element that is formally defined by a proper ontology (i.e. [12]) in its subjective properties, as well by configurational indexes in its objective qualities, in the frame of an urban environment, could be described by means of the so-called Configurational Ontology [9] as:

$$CO = O(C^c, C^b) \tag{1}$$

where O is the ontology as represented in our model, and C^x are the following network point centrality measures [24]:

$$C^c = \frac{n-1}{\sum_{i=1}^{n} d(p_i, p_k)}; \qquad C^b = \sum_{i=1}^{n} \sum_{j>i}^{n} \left(\frac{g_{ij}(p_k)}{g_{ij}} \right)$$

where:

$d(p_i, p_k)$ = the number of edges in the geodesic path that links p_i and p_k

g_{ij} = the number of geodesic paths that link p_i and p_k

$g_{ij}(p_k)$ = the number of geodesic paths that link p_i and p_k containing p_k

The configurational indexes C^x are attributes of the Configurational Ontology, and they are strictly related to the configurational indexes of Integration (closeness centrality) and Choice (betweenness centrality), as introduced in Sect. 2.

Such an approach combine centrality measures quantitatively influenced by the meaning of urban elements, understood as "local properties" on the nodes composing the city network.

The authors proposes to develop and use such an approach to provide an integrated quali-quantitative knowledge about the city so as to have a clearer view of the relationship among formal properties of the city, its configurational state, and the distribution of housing market values. Under this view, accessibility provided by the railway network and the externalities locally provided by railway stations could be effectively taken into account.

Since the approach is based on the definition of an appropriate ontology of the city, the future developments need to go firstly into that direction, in order to reach an adequate formal description of the city that should be comprehensive of urban element relevant at the scale of the city as a whole as well of their mutual relationships taking into account efficient technique to integrate domain knowledge [8,39].

References

1. Agenzia Campana per la Mobilitá Sostenibile: Piano di riprogrammazione dei servizi di trasporto pubblico locale. Technical report (2013)
2. Alonso, W.: Location and Land Use. Harvard University Press, Cambridge (1964)
3. Barry, C.L.: Document representations and clues to document relevance. J. Am. Soc. Inf. Sci. **49**(14), 1293–1303 (1998). https://doi.org/10.1002/(SICI)1097-4571(1998)49:14⟨1293::AID-ASI7⟩3.3.CO;2-5
4. Bateman, J., Farrar, S.: Towards a generic foundation for spatial ontology. In: Proceedings of the 3rd International Conference on FOIS 2004, pp. 237–248 (2004)
5. Bertolini, L., Spit, T.: Cities on Rails The redevelopment of Railway Station Areas. E & FN Spon, London (1998)
6. Borsa Immobiliare di Napoli: Listino Ufficiale della Borsa Immobiliare di Napoli. Technical report (2011). http://www.binapoli.it/listino/intro.asp
7. Caldarola, E.G., Di Pinto, V., Rinaldi, A.M.: The use of configurational analysis in the evaluation of real estate dynamics. In: Sforza, A., Sterle, C. (eds.) ODS 2017. SPMS, vol. 217, pp. 83–92. Springer, Cham (2017). https://doi.org/10.1007/978-3-319-67308-0_9
8. Caldarola, E.G., Rinaldi, A.M.: A multi-strategy approach for ontology reuse through matching and integration techniques. Adv. Intell. Syst. Comput. **561**, 63–90 (2018)
9. Cataldo, A., Cutini, V., Di Pinto, V., Rinaldi, A.M.: Subjectivity and objectivity in urban knowledge representation. In: Proceedings of the International Conference on Knowledge Discovery and Information Retrieval (KDIR-2014), pp. 411–417. Scitepress (2014)
10. Cataldo, A., Di Pinto, V., Rinaldi, A.M.: A methodological approach to integrate ontology and configurational analysis. In: Murgante, B., et al. (eds.) ICCSA 2014. LNCS, vol. 8580, pp. 693–708. Springer, Cham (2014). https://doi.org/10.1007/978-3-319-09129-7_50
11. Cataldo, A., Di Pinto, V., Rinaldi, A.M.: Representing and sharing spatial knowledge using configurational ontology. Int. J. Bus. Intell. Data Min. **10**(2), 123–151 (2015)
12. Cataldo, A., Rinaldi, A.M.: An ontological approach to represent knowledge in territorial planning science. Comput. Environ. Urban Syst. **34**(2), 117–132 (2010)
13. Chiaradia, A., Hillier, B., Barnes, Y., Schwander, C.: Residential property value patterns. In: Proceedings of the 7th International Space Syntax Symposium, pp. 015:1–015:12. KTH, Stockholm, SVE (2009)
14. Clancey, W.J.: The knowledge level reinterpreted: modelling socio-technical systems. Int. J. Intell. Syst. **8**, 33–49 (1993)
15. Comune di Napoli: Piano generale del traffico urbano. Aggiornamento 2002–2004. Technical report (2002)
16. Cutini, V., Di Pinto, V.: Urban network per il masterplanning strategico nelle grandi aree urbane. configurazione spaziale e rendita fondiaria. Urbanistica Dossier (13), 426–432 (2017)
17. Debrezion, G., Pels, A.J., Rietveld, P.: The impact of rail transport on real estate prices: An empirical analysis of the dutch housing market. Technical report, Rotterdam (2006). https://doi.org/10.2139/ssrn.895270
18. Di Pinto, V.: C come Paesaggio. Analisi configurazionale e paesaggio urbano. Liguori editore, Napoli (2018)

19. Di Pinto, V., Rinaldi, A.M.: A configurational approach based on geographic information systems to support decision-making process in real estate domain. Soft. Comput. **23**(9), 2853–2862 (2018). https://doi.org/10.1007/s00500-018-3142-9

20. Di Pinto, V., Rinaldi, A.M.: Big data. towards an approach to integrate subjectivity and objectivity in representing and managing data to support decision making in urban planning domain. In: Di Pinto, V. (ed.) Decision making in urban and territorial Planning. Maggioli Editore, Santarcangelo di Romagna (Italy) (2020)

21. Eco, U.: La struttura assente. La ricerca semiotica e il metodo strumentale. Bompiani, Milano, Italy (1968)

22. ESRI: Arcgis desktop - version 10.3.1 (2015)

23. Fejarang, R.A.: Impact on property values: A study of the Los Angeles metro rail. In: Public transport planning and operations. Proceedings of seminar h held at the European transport, highways and planning 21st summer annual meeting. pp. 27–41. Transport Research Laboratory (1994)

24. Freeman, L.: Centrality in social networks conceptual clarification. Soci. Networks **1**, 215–239 (1978)

25. Gil, J.: Integrating public transport networks in the axial model. In: Proceedings of the 8th International Space Syntax Symposium. pp. 8103:1–8103:21. PUC, Santiago de Chile, RCH (2012)

26. Gil, J.: Urban Modality - Modelling and evaluating the sustainable mobility of urban areas in the city-region. A+BE — Architecture and the Built Environment (2016)

27. Gruber, T.R.: A translation approach to portable ontology specifications. Knowl. Acquis. **5**(2), 199–220 (1993). https://doi.org/10.1006/knac.1993.1008

28. Harter, S.P.: Psychological relevance and information science. J. Am. Soc. Inf. Sci. **43**(9), 602–615 (1992)

29. Hillier, B., Penn, A., Hanson, J., Grajewski, T., Xu, J.: Natural movement: or, configuration and attraction in urban pedestrian movement. Environ. Plan. **20**(1), 29–66 (1993). https://doi.org/10.1068/b200029

30. Hillier, B.: Space is the Machine. Cambridge University Press, Cambridge (1996)

31. Hillier, B., Hanson, J.: The Social Logic of Space. Cambridge University Press, Cambridge (1984)

32. Janowicz, K., Scheider, S., Adams, B.: A geo-semantics flyby. In: Rudolph, S., Gottlob, G., Horrocks, I., van Harmelen, F. (eds.) Reasoning Web 2013. LNCS, vol. 8067, pp. 230–250. Springer, Heidelberg (2013). https://doi.org/10.1007/978-3-642-39784-4_6

33. Matthews, J., Turnbull, G.: Neighborhood street layout and property value: the interaction of accessibility and land use mix. J. Real Estate Fin. Econ. **35**, 111–141 (2007)

34. Muth, R.: Cities and Housing. University of Chicago Press, Chicago (1969)

35. Neches, R., et al.: Enabling technology for knowledge sharing. AI Mag. **12**(3), 36–56 (1991)

36. Nutter, J.T.: Epistemology. In: Shapiro, S. (ed.) Encyclopedia of Artificial Intelligence. John WyleyS (1998)

37. Park, T.: The nature of relevance in information retrieval: an empirical study. Library Quarterly **63**(3), 318–351 (1993)

38. Porta, S., Crucitti, P., Latora, V.: The network analysis of urban streets: a dual approach. Phys. A **369**(2), 853–866 (2006)

39. Rinaldi, A.M., Russo, C.: A matching framework for multimedia data integration using semantics and ontologies. In: Proceedings - 12th IEEE International Conference on Semantic Computing, ICSC 2018, vol. 2018-January, pp. 363–368 (2018)

40. Rinaldi, A.M., Russo, C.: A semantic-based model to represent multimedia big data. In: MEDES 2018–10th International Conference on Management of Digital EcoSystems, pp. 31–38 (2018)
41. Sabidussi, G.: The centrality index of a graph. Psychometrika **31**, 581–603 (1966). https://doi.org/10.1007/BF02289527
42. Swanson, D.: Subjective versus objective relevance in bibliographic retrieval systems. Lib. Quart. **56**(4), 389–398 (1986)
43. Turner, A.: Angular analysis: a method for the qualification of space. Technical report (2000)
44. Turner, A.: Angular analysis. In: Proceedings of the 3rd International Space Syntax Symposium (2001)
45. Vakkari, P., Hakala, N.: Changes in relevance criteria and problem stages in task performance. J. Doc. **56**(5), 389–398 (2000)
46. Varoudis, T.: DepthmapX - multi-platform spatial network analyses software - version 0.50 (2015). http://archtech.gr/varoudis/depthmapX/
47. Varoudis, T.: DepthmapXnet - version 0.35. http://archtech.gr/varoudis/depthmapX/?dir=depthmapXnet (2017)
48. Vaughan, L.: The spatial syntax of urban segregation. Prog. Plann. **67**, 205–294 (2007)
49. Von Thünen, J.H.: Der Isolierte staat in Beziehung auf Landwirschaft and nationalökonomie. Pflaum, Münich (1863)
50. Wingo, L.: Transportation and Urban Land. Resources for the Future (1961)

Assessing the Impact of Temporary Housing Sites on Urban Socio-spatial Performance: The Case of the Central Italy Earthquake

Camilla Pezzica[1]([⊠]) ⓘ, Chiara Chioni[1] ⓘ, Valerio Cutini[1] ⓘ,
and Clarice Bleil de Souza[2] ⓘ

[1] Department of Energy, Systems Territory and Construction Engineering,
University of Pisa, Largo Lucio Lazzarino, 56122 Pisa, Italy
Pezzicac@cardiff.ac.uk, c.chioni@studenti.unipi.it,
valerio.cutini@ing.unipi.it
[2] Welsh School of Architecture, Cardiff University,
Bute Building, King Edward VII Avenue, Cardiff CF10 3NB, UK
Bleildesouzac@cardiff.ac.uk

Abstract. This paper advocates a performance-based approach to the planning of temporary housing sites after sudden urban disasters. A "functionally graded" configurational analysis method is used to assess, quantitatively and qualitatively, the socio-spatial impact of government-led housing assistance provision across the regional, urban and neighbourhood scales. To highlight the different outcomes achieved in different urban contexts by apparently similar housing recovery plans, a diachronic comparative study of four epicentral historic towns hit by the 2016–2017 Central Italy earthquakes is performed. The research analyses the configurational properties of these settlements at four critical points in time: before the disaster; right after it (emergency phase); during disaster recovery; after the reconstruction is completed. This paper builds on previous research on rapid urban modelling and economic spatial analysis workflows to respond to potential implementation challenges, which include time constraints and geo-data availability issues after disasters. By using a real case scenario, this study seeks to demonstrate the potential benefits of adopting the proposed multidimensional spatial analysis method to foster the delivery of integrated housing recovery solutions, which contribute to sustainable urban development These encompass informing, timely updating, and coordinating strategic, management, and operational decisions related to the design and planning of temporary housing sites.

Keywords: Temporary housing · Disaster recovery · Multiscale spatial modelling · Space syntax · Design Decision-making · Sustainable urban development

© Springer Nature Switzerland AG 2020
O. Gervasi et al. (Eds.): ICCSA 2020, LNCS 12251, pp. 324–339, 2020.
https://doi.org/10.1007/978-3-030-58808-3_24

1 Introduction

1.1 Temporary Housing Sites

Temporary Housing (TH) sites are commonly built by governments and NGOs, after major and sudden urban disasters, to house the homeless population for the duration of the reconstruction works, if any, once the immediate emergency phase has ended. These solutions can be offered as an alternative to cash grants for housing rental and are similarly meant to allow restoring normal housing and life routines, which requires having access to suitable services and infrastructures as well as to adequate housing and neighbourhood conditions. Concentrating the construction of temporary housing on public sites is common in cities and towns where, possibly due to urban density issues, plots lack the requirements to safely implement transitional housing solutions on private land. The construction of TH sites is often justified by an intention to preserve community bonds by maintaining affected individuals and families in close physical proximity. However, their effectiveness and convenience have been seriously questioned [1–3], due to the negative impact that they can have, sometimes unexpectedly, on the overall disaster recovery process. In certain cases, the implementations of inadequate TH solutions has, in fact, undermined the possibility to improve on the pre-disaster vulnerability conditions, which has negatively affected urban development with respect to economic, social and environmental sustainability goals [4].

Certain key decisions concerning their location, number, dimension, density, and layout need to be taken soon after a disaster strikes, often on the basis of limited official information, with or without the support of a strategic plan. Even when present, the strategic plan needs to be promptly updated to better respond to contextual changes and dynamic legislation adjustments, which may be required for different reasons (e.g., aftershocks, governance based on ad hoc regulations, technical practicalities, etc.). As a matter of fact, because of contextual differences present across a large disaster-affected area, requirements may change in different urban contexts and adopting a homogeneous approach in planning TH sites can therefore not be appropriate. Indeed, there is no universal recipe for planning and designing TH sites as, when resorting to a non-diversified housing recovery strategy, dissimilar socio-spatial outcomes can be achieved in apparently similar settings.

Assuming that the number of people in need of this type of housing assistance is known, choosing a TH site location involves either selecting a higher number of TH areas closer to the original settlement or a smaller number of sites further away [5]. Although strategic in nature, this decision is often constrained by considerations of site capacity, availability of infrastructures and services, cost of land and potential land ownership issues, and often requires operating difficult trade-offs. Under certain conditions, the location and layout of TH sites, can seriously affect the spatial performance of a recovering city and, in some extreme cases, trigger the decentralization of a settlement [6] by perturbating its inner geography and characteristic accessibility distribution. Additionally, selected housing typologies, construction technologies and materials, should align with a strategic vision to satisfy related durability requirements and allow for subsequent adjustments, such as longer lifespans, densification plans, and incremental housing transitions, if necessary.

Therefore, decisions about TH sites need to be made coherently, in multiple interdependent domains, across the regional, urban, and neighbourhood scales, and should be guided by planning foresight and informed deliberation. To tackle these issues, this paper, firstly, proposes a "functionally graded" multiscale configurational analysis method aimed at supporting TH design and planning. Then, it qualitatively tests, in a real disaster scenario, the potential of the proposed method to better inform the assessment of the medium- and long-term impact of TH sites on urban spatial performance.

1.2   The Central Italy Earthquake

Four central Italy regions, 10 *province* (districts), and 138 *comuni* (munic-ipalities), scattered around a territory of 8000 km^2, were hit by a long series of devastating seismic waves between the 25th of August 2016 and the 18th of January 2017. These events caused the direct and indirect death of hundreds of people, the disruption of the local economy and the destruction of several cultural heritage sites, with material damages estimated at 23 billion and 530 million euros [7]. The territory impacted by the disaster is characterized by the presence of small, rather segregated, historical settlements, scattered across a mountainous area, currently undergoing a process of depopulation and inhabited by an aging population, whose livelihood depends on small local businesses.

By the 22nd of August 2018, a total of 49844 displaced people was provided with some form of housing assistance, and hosted in either prefabricates structures, hotels, rural emergency modules, containers, municipal facilities or in rented accommodations. Of these, 7782 people requested to be housed in prefabricated houses of 40, 60 or 80 net square meters, called S.A.E.s, within a total of 228 TH sites [8], (Fig. 1).

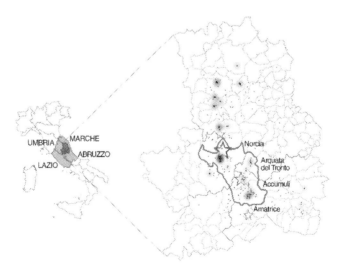

Fig. 1. Location of the disaster affected area in central Italy (which includes portions of Umbria, Abruzzo, Marche and Lazio) and heat map of the number of people hosted in the TH sites. The epicentres of the 2016–2017 earthquakes are marked with stars proportional to their magnitude.

S.A.E.s' Procurement and TH Sites' Construction

The beginning of the period of emergency management was officially declared by the national authorities in August 2016 and its end was then repeatedly postponed until the end of December 2019. During this time numerous ordinances have been issued by the Italian Civil Protection Department (DPC), the body in charge of assisting the affected population, by monitoring and coordinating relief and housing recovery operations.

The design of the S.A.E.s and their furnishing, as well as the conditions for their delivery and installation, had been pre-arranged as part of a strategic framework agreement for prompt disaster response, targeting the entire Italian territory (divided in North, Centre and South), which was signed in 2015 between the DPC and a few suppliers, such as the CNS (the tender's winner in Central Italy) and Consorzio Stabile Arcale (second in the ranking). The CNS's housing modules (Fig. 2) are made of steel and wood and were designed to be comfortable in all Italian climates (although this has been repeatedly contested in the media after the disaster), accessible to disabled people, energy efficient, compliant with the Italian seismic regulations and with safety, hygiene and environmental requirements.

According to the tender documentation, it should be possible to combine the TH units, as well as to remove and recover the S.A.E.s, when they are no longer needed. However, even if around 60% of building components are reusable, their foundation system (consisting of concrete slabs) and urban infrastructure works are hardly reversible, which prevents to easily return the TH sites on which the units are built, to their initial condition and land use. In fact, several of the affected municipalities have decided to expropriate the land occupied by the TH sites, whenever it was not public already. Moreover, further investments will certainly be required in the near future as the CNS's housing units have an intended first lifecycle of only six years [9]. Despite early statements of intentions, the validity of the hypothesis of the temporariness of these TH plans is further challenged by the actual timescale of the sites' urbanisation and housing delivery operations. At the time of writing, almost 4 years after the first quake, the construction works related to TH delivery in central Italy have yet to be totally completed [10].

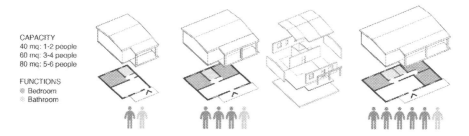

Fig. 2. 2D and 3D schemes of the 3 S.A.E. housing unit types supplied by the CNS.

The ordinance n. 394/2016 identified the four affected regions, namely Umbria, Marche, Lazio, and Abruzzo, as responsible for the construction of the TH sites in their administrative areas. The same document established that the candidate sites for locating the TH structures had to be identified by the municipalities, after a quantitative needs' assessment, considerate of non-housing requirements such as schools, churches, hospitals, etc., and later, had to be assessed for approval by the responsible regional administration. As a general strategy the ordinance required the municipalities to prefer public areas over private ones (to avoid delays due to appeals of landowners to the TAR, the Italian regional administrative court) and to contain the number of TH sites, while considering the needs of people. The strategic agreement stated that the executive urbanisation projects had to be commissioned to the TH suppliers, after approval of the design of the TH site layout by the relevant municipality. However, it did not consider the need for important site preparation works, including site consolidation and safety interventions to mitigate hydrogeologic and hydraulic risk, and works to attach a TH site to the local service grid if the linking point was beyond its boundaries. This required the introduction of a new directive, which highlighted the importance of making informed choices regarding TH sites' selection early on, due to their likely impact on urbanisation and site preparation costs, which could have hardly been sustained in small TH sites with less than 5 units [11]. The same ministerial circular set also a cap for the cost of urbanisation works (of 18000 €), site preparation (7000 €), and a reference limit of dedicated gross surface area (200 m^2) per TH unit. Additionally, it suggested limiting the provision of trees and street furniture in TH sites.

1.3   Scope and Objectives

The examples reported in the scientific literature as well as the case of Central Italy confirm that the level of complexity embedded in the task of planning TH sites after sudden urban disasters, requires a focus on urban systems' performance, in line with long-term urban development objectives. Within a context of dynamic changes, lengthy construction works and bureaucratic procedures, such as that described in Sect. 1.2, it seems crucial to provide humanitarian professionals with instruments that allow a multidimensional understanding of the consequences that morphological transformations have on recovering settlements after a sudden disaster.

Therefore, this study proposes and tests a scenario-based, spatial analysis method, suitable to interrogate aspects of TH sites related to both society and space, and oriented at fostering the implementation of holistic practices in the tactical planning of government-led housing assistance interventions. Specifically, the proposal seeks to augment decision-makers' capacity to align operational planning and design choices, with managerial and strategic decisions concerning TH sites [12]. A statistical correlation analysis between configurational indicators and quantities reflecting the actual planning outcomes in the selected case studies (e.g., distance to POIs, walkability scores, impact of transports etc.) is, however, out of the scope of this paper, as it would inadequately reflect a situation which is still rapidly evolving. Therefore, the validation of the analysis models used in this paper is mainly based on qualitative considerations, supported by official documental evidence, reported news, and in minor part, field observations.

2   Materials and Methods

In line with other research studies [13, 14], this paper suggests adopting a configurational analysis approach based on Space Syntax [15] to look at the problem of locating and designing TH sites. An element of novelty is that this paper uses Space Syntax as an overarching method, which is differently declined to perform the analysis of TH planning scenarios across the macro (regional), meso (urban) and micro (neighbourhood) spatial planning scales, and at multiple significant moments in the disaster management cycle. Performing a "functionally graded" configurational analysis, whose complexity and computational heaviness varies according to the level of information and detail required by the study, simplifies the task of overlapping results of multiple analyses with different theoretical backgrounds and levels of granularity. Furthermore, the multitemporal and multi-scale application of the method, enables disclosing dissimilarities in TH projects, which may otherwise not be obvious nor trivial to identify. The hypothesis is tested through a comparative analysis of four case studies, namely four epicentral historic towns, hit by the 2016–2017 Central Italy earthquakes.

2.1   Space Syntax Approach and Metrics

Space Syntax theory assumes that space in cities (the urban grid) is the enabler of many of the socio-economic interactions that characterise urban life. Despite it being diffused prevalently out of the humanitarian context, its potential in informing and supporting different disaster management tasks has been confirmed by many previous studies. The scientific literature presents examples related to almost all the phases of the disaster management cycle: from disaster prevention and mitigation [16], to emergency response [17], housing relief [18], infrastructure [19] and economic [20] as well as housing recovery [13]. The latter proves that Space Syntax offers novel opportunities for assessing the long-term outcomes of housing recovery plans (i.e., via the quantification of changes in levels of street network resilience and centrality of a given settlement) as it helps understanding the indirect effects of alterations to urban form.

Therefore, this paper exploits different configurational analysis techniques to retrieve quantitative information about the capacity of space to function as intended, so as to enable a nuanced assessment of "what if" situations via a simultaneous study of past, present and possible future scenarios at different spatial scales.

When the focus is on one-dimensional elements (e.g. street networks) it is often recommendable to use an Angular Segment Analysis (ASA) [21]. Two of the most important indices calculated by an ASA are *Normalised Angular Integration* (NAIN) and *Normalised Angular Choice* (NACH), which enable to visualise the to-movement and through-movement patterns generated by an urban grid, respectively. NAIN and NACH can be calculated considering either the entire network system (global analysis, Rn) or just a part of it (local analysis, R3). Past research has found correlations between these metrics and phenomena such as pedestrian movement (NAIN R3), presence of commercial activities (NAIN Rn) and intensity of vehicular flows (NACH Rn).

To establish a link between the configuration of a settlement and the morphology of its open spaces the focus needs to shift from road segments to two-dimensional elements composing a uniform grid. To study the mutual connections of the grid's units it is necessary to perform a Visual Graph Analysis (VGA), [22]; which allows computing several configurational metrics, among which the *Integration Index*, the *Clustering Coefficient* and *Connectivity*. The first indicates the proximity of each point to the most crowded paths and their level of spatial accessibility, the second measures how much a space can be perceived as a spatial unitary entity, whereas the third shows how many nodes of the grid are in direct visual connection. Thus, performing a VGA provides the means to understand the spatial qualities a TH site layout in terms of its specific capacity to: support social interaction, be visually controllable, encourage mutual awareness of people and even accommodate political activism [23]. Arguably, all of these are aspects which contribute to the resilience of displaced communities.

In this study, the first two metrics (NAIN and NACH), are used in the road network analysis at the regional and urban scales, whereas a small selection of VGA indices is used to analyse the TH sites at the neighbourhood scale.

2.2   Cross-scale and Multitemporal Analysis

In order to effectively support the planning of TH sites the analysis method must be flexible enough to accommodate the non-procedural aspects of the design process, thus allowing jumping across scales, without a fixed direction of flow, to iteratively refine the definition of, and assess the response to a design problem. It is suggested that dependencies exist between scales and that decisions made at one level will inevitably constraint those made at the others. Therefore, being able to understand the problem at various scales, opens doors for conflicts' resolution through design and planning innovation or, at the very least, allows recognising opportunities to save important resources by optimising efforts, through better coordination in decision-making.

Despite this paper presents the proposed method as an ordered sequence of analyses going from the regional to the neighbourhood scale, steps can be omitted (or added) as required, making the methodology open-ended. For instance, the same multi-staged scenario-testing approach can be adopted in the analysis of a TH site at just one scale (e.g., to assess the functioning of a certain layout in time as shown in Sect. 3). This type of analysis can be understood as a sort of compressed diachronic study, which includes past and present realities, as well as alternative future scenarios.

Due to their relevance for the objectives of this paper, the selected time steps are:

- T1: corresponding to the pre-earthquake situation.
- T2: immediate aftermath of the disaster, characterising the emergency period when access to the dangerous areas at risk of further damage is denied to the lay public.
- T3: post-earthquake situation, characterising the recovery stage, in which some areas (red zones) are still inaccessible and TH sites are built.
- T4: future scenarios after a reconstruction "as it was, where it was" and with no red zones.

If the analysis considers TH sites to be still present in the T4 scenario, it can assess the impact of their permanency on the spatial geography of a settlement, and help generating possible planning alternatives or regenerative design proposals [24].

Operationally, this paper builds on previous research on rapid configurational analysis workflows using Open Street Map (OSM) data [25] to better respond to potential implementation issues concerning time constraints, geo-data homogeneity across different administrative boundaries and its availability after sudden disasters. To build the T3 and T4 scenarios it was required to gather all the TH site layouts from the municipalities websites and then manually retrace the road centre lines missing in the OSM reference drawings using ®Autocad. The resulting .dxf files where put in the correct reference system before performing the ASA and VGA analyses in Depth-mapX. Finally, results were exported in QGIS for further elaboration. This work was undergone by the research intern, in preparation to a design research project [26].

Section 3 shows how the analysis has been coordinated in practice across scales and illustrates the results obtained in the case of the Central Italy earthquake.

3   Results

Regional Scale

The analysis at the regional scale was conducted after querying the 2018 OSM data-base, to retrieve the required Road Centre Lines (RCL) data, representing the streets accessible to both people and vehicles. The map of the disaster-affected region in Fig. 3 includes the streets, which have been temporarily closed soon after the earthquake.

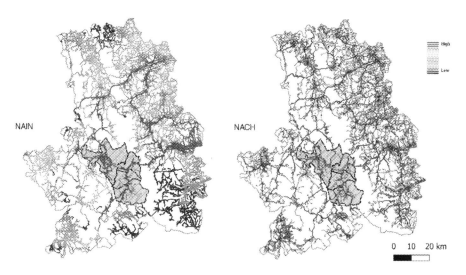

Fig. 3. From the left to the right: NAIN (a) and NACH (b) of the OSM drive network. At the centre, highlighted in grey, the territory of the 4 municipalities analysed in the comparative study. (Color figure online)

High values of NAIN (to-movement, corresponding to the red lines in Fig. 3a) highlight the most accessible and attractive locations of the disaster-impacted region. High values of NACH (through-movement, corresponding to the red lines in Fig. 3b) highlight the streets hosting most of the transports' flow across the disaster-affected region, that is the regional foreground traffic network.

Overall, the results confirm the outcomes of the 2016 ISTAT (Italian National Institute of Statistics) study, done in the same area [27], which describes the territory as characterised by multiple small urban centralities which are rather difficult to reach and whose livelihood depends on a system of local businesses. Additionally, the analysis offers a clear representation of the most and least accessible/segregated locations of the network and highlights the layers of weaker and stronger connections that mark the territory regionally: in the map, NAIN evidences the most important centres while NACH how they are connected to the others.

Urban Scale

Figures 4 and 5 show, with different resolutions, the four neighbouring municipalities that were selected for this study, namely Arquata del Tronto (in Marche), Norcia (in Umbria), Amatrice and Accumoli (in Lazio).

These settlements were chosen considering their proximity to the earthquakes' epicentres. They have suffered the greatest damage, with the consequent displacement of hundreds of people, and host the largest number of TH sites in the region, due to the high devastation caused by the disaster in the surroundings of these towns.

In line with the national political strategic vision, in all four cases, the TH sites were built close to the destroyed settlements. Probably due to geographic constraints, the capacity of these TH sites does not generally exceed a hundred units, except for a few sites in Norcia. This context sets the conditions for the creation of a distributed system of small TH settlements, scattered around the main routes, with two different types of arrangements: a linear distribution in continuity with the existent urban fabric (letters a in Fig. 4) and a nucleated distribution around the historic core (letters b in Fig. 4).

In Fig. 5 the configuration of their street networks has been comparatively analysed at the four points in time described in Sect. 2.1 (T1-T4) to examine differences in terms of short and long-term impact of TH plans on their urban spatial functioning.

Specifically, Fig. 5 shows the multitemporal distribution of NAIN and NACH in the considered cases, which is displayed using the same colour gradient adopted in the analysis at the regional scale (Fig. 3), with red corresponding to high values and blue to low values of the corresponding configurational indices.

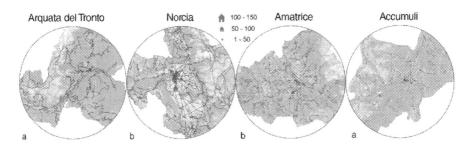

Fig. 4. Location and capacity of the TH sites in Arquata del Tronto, Norcia, Amatrice, Accumuli.

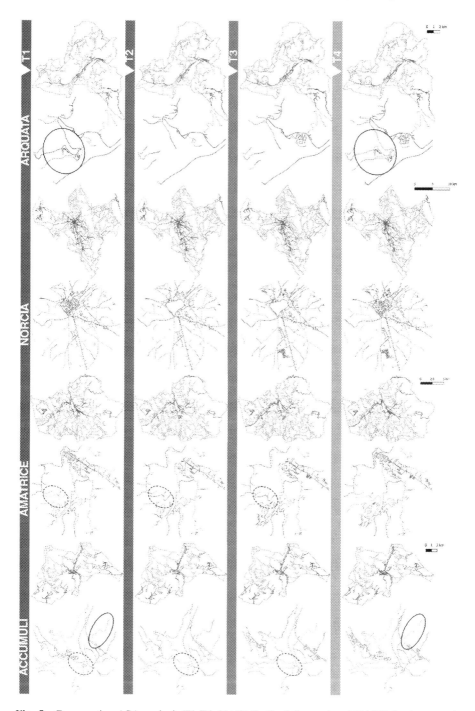

Fig. 5. Comparative ASA analysis T1-T4. NAIN (in the full maps) and NACH (in the zooms).

The analysis of the four towns at T1 highlights that, despite these being all small historic centres developed in the same mountainous area in early Roman times (except for Accumuli which was built in the XII century), the functioning logic of their urban grids differed substantially from each other since before the disaster. For example, it is possible to observe that Norcia and Amatrice have grown more densely and organically around their centres, while Accumuli and Arquata del Tronto have had a more linear development, along their main access roads. This is reflected in the distributed nature of centrality values in the former systems versus the polarised character of the latter.

The earthquake destroyed all the four historic areas, causing the closure of the four town centres for safety and reconstruction purposes: a situation studied in the T2 analysis. Here, the maps show, perhaps not surprisingly, that this partial network closure had a negligible impact on the functioning of the remaining parts of Norcia, which presented a modular "spokes and hubs" structure. Contrarily, Amatrice suffered an important loss of centrality and, although less important, so did Accumuli and Arquata del Tronto. The decrease in centrality of the areas surrounding the original urban core, is further reinforced by the location of the TH sites, built during the disaster recovery phase (T3). Notably, the analysis clearly shows that, Arquata del Tronto will be more significantly affected by the presence of the TH sites than the other towns, should these areas continue to be inhabited in the future (T4). In terms of accessibility, the old centre of Arquata del Tronto will be left behind, while the TH site of *Borgo1* will emerge as the new actual town centre.

This result for Arquata del Tronto confirms the outcomes of a previous study on post disaster recovery and reconstruction processes of small settlements in Italy, which has highlighted the risk of them to trigger an effect of urban decentralisation [6]. However, this study also clarifies why this effect can be far less accentuated in other cases such as Norcia, Amatrice, and notably also Accumuli, which demonstrates the usefulness of adopting the proposed methodology to evaluate housing recovery scenarios. Specifically, the analysis shows that the spatial logic of these three towns will not be majorly affected in the long term by the addition of the new TH sites, even if their physical size is not negligible. Once the reconstruction is completed, it may be possible to smoothly integrate the new temporary settlements in the tissue of the recovered urban system as new neighbourhoods or commercial hubs. On the contrary, in the case of Arquata the new streets of the TH sites are likely to acquire economic value and social desirability, thus competing with the recovered town as duplicate centres, if they are not dismantled.

Neighbourhood Scale

Because of its special interest (see paragraph above), the analysis at the neighbourhood scale has been conducted only in the case of Arquata del Tronto and specifically in the area of *Borgo1*: the most central and biggest TH site of the municipality. Being it the new configurational centre of town, *Borgo1* assumes a key social role and value, which justifies an effort to further strengthen its relational qualities through urban design.

Issues related to the human-centred design and regeneration of TH sites' layouts are explored in this section, in light of their relevance with respect to the municipality's declared priorities which include: the socio-economic revitalization of the territory; the morphological and functional redevelopment of the urban system; and the return of the residents to their permanent homes.

According to the 2015 strategic framework for the provision of TH units, their spatial distribution, i.e. the design of the TH site layout, should consider the geometry, topography and orientation of the plot on which they are located and follow bioclimatic principles. The use of passive design strategies to regulate the units' indoor thermo-hygrometric comfort through TH site design was formally encouraged. For instance, it was suggested to include summer shading by green infrastructure, good natural ventilation, and arrange the units so as to reduce heat loss by minimizing the ratio between their external surfaces and volume. In addition to a set of planning constraints, the suppliers were required to present at least two different planovolumetric layout proposals, comprising different housing arrangement schemes (e.g. detached houses, courtyard or terraced housing distribution), and exploring different density and aggregation options in relation to height, while providing suitable external access to all housing units.

Although certainly relevant and informative for a complete assessment of the design of a TH layout against its original objectives, an environmental analysis of the site has been omitted, because out of the scope of this study. Instead, a VGA analysis is used to respond to the current need to in-vestigate properties of space that inspire a sense of ownership through occupation, which are related to the spatial navigation and visual perception of an existing TH site. The *Integration Index* is used as a proxy for spatial accessibility, while the *Clustering Coefficient* is used to identify pockets with stable visual characteristics, which would allow a comfortable social interaction. *Connectivity* allows assessing the visual capacity of these areas whereas *Control* can be used to identify spaces where people naturally tend to make navigation decisions.

The analysis addresses three points in time, namely T1, T3 and T4 because it assumes that the analysis of T2 would mainly be relevant for deciding the location of the TH site, which in this case is already known and fixed. T3 corresponds to the present layout configuration, while T4 represents the situation in the hypothetical scenario of conservative reconstruction. Notably, other alternative urban reconstruction scenarios could be comparatively tested, as well as a potential T5 situation in which something changes in the present configuration of the TH site as part of a future urban regeneration project, as illustrated in greater detail in [24].

When performing a VGA, a distinction can be made between a visual accessibility analysis (i.e. computing a visibility graph constructed at eye level) and a permeability analysis, which is applied at knee level to reflect how people can move in a space. This paper implements the latter, as it prioritises an understanding of movement around the TH site: Fig. 6 shows that the spatial permeability of the area surrounding the former playground has changed dramatically after the construction of the TH sett-lement (T3).

The resulting local layout modification has, in fact, generated a number of new open spaces with different morphologies, ranging from the little pockets between the TH units to the larger esplanades in various configurationally central locations. Notably, two perpendicular axes emerge with a high value of centrality/accessibility (corresponding to a high value of the *Integration Index*), with the horizontal one crossing the temporary village from side to side. Figure 6 indicates that after the completion of the reconstruction works (T4), if no other change applies, the new road

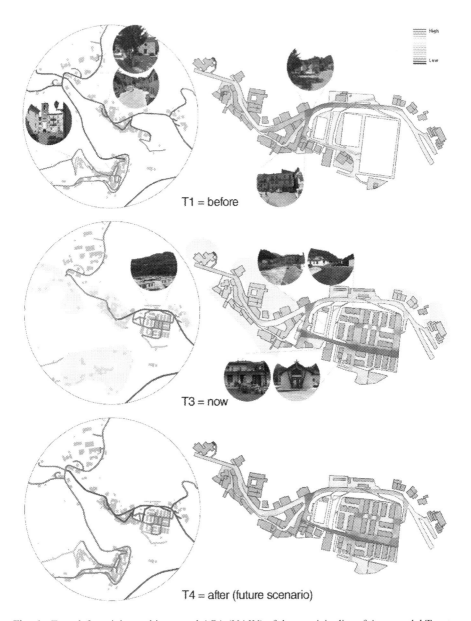

Fig. 6. From left to right: multitemporal ASA (NAIN) of the municipality of Arquata del Tronto and VGA (*Integration Index*) of the area of *Borgo1*, before the earthquake (T1), today (T3) and in the future reconstruction scenario where everything is built "as it was, where it was" and the TH site is not dismantled (T4).

axes will maintain a high configurational importance. This result is confirmed by the distribution of NAIN in the ASA (see zooms in Fig. 6), which shows that a continued presence of the TH sites in the future (the analysis includes all those built in the municipality), is likely to redefine the socio-spatial logic and accessibility pattern of Arquata del Tronto, creating the conditions for local economic activities to move out of the historic town centre. The analysis' prediction is substantiated by the current "temporary" relocation of many services in the TH site, and in its proximity along the SP 89 road, (photos in Fig. 6) and by the data gathered by the research intern during a field visit in 2019 [24].

4 Discussion and Conclusion

The results of this study demonstrate the potential of adopting the configurational analysis as an overarching methodology to support the scenario-based planning and design of spatially performative TH sites. In particular, the proposed "functionally graded" method allowed dealing with different spatial scales at different points in time in a synchronous manner, which could help bridging possible communication gaps between decision-makers and facilitating the sustainable development of TH sites in the long term; both locally and as part of their territory.

Although conducting an extensive empirical validation of all the analysis models, on the one hand, was hardly feasible, and, on the other hand, out of the scope of this research, it was nonetheless possible to find a partial confirmation of the results obtained through the application of the proposed analysis method, in the initial planning outcomes of the TH programme and in some of the policies implemented by the responsible administrations in response to those. For instance, the interventions mentioned in some documents (e.g., Ocdpc n. 510, 518, 538 and 581 among others), involving the addition of services, public transport modifications and road works in the towns, seem to be oriented in the direction shown by the configurational analysis models.

The outputs obtained in the analysis of the selected case studies have practical implications for future housing recovery planning involving the use of prefabricated housing structures. They indicate that special attention should be paid to the choice of the location, as well as to the specific layout, of TH sites, as both can have an important influence on the social and economic functioning of a recovering settlement. This seems particularly relevant for small towns with a linear road network configuration and a limited number of access routes. Understanding the consequences of building TH plans after disasters appears important even in the unlikely event that all TH sites were eventually brought back to their original status, because of their indirect impact on local economic activities. The influence that TH sites exert on pedestrian and vehicular flows during the transitional phase (see circled segments T1–T3 in Fig. 5) can, in fact, last for a potentially very long period, corresponding to the length of reconstruction works. Ultimately, this implies that it is extremely improbable that after a disaster of such a magnitude the post urban geography will correspond exactly to the pre-disaster situation, independently of the actual dismantling of TH sites.

To make the proposed methodology fully operational, future applications should select the most meaningful configurational metrics to answer a specific question, considering the local context. Then, results should be critically interpreted by the experts according to their strategic priorities, acknowledging other non-spatial relevant factors.

Future studies should explore positive synergies between the design of TH units and their spatial arrangement (i.e., the design of the TH site layout), by integrating results of a configurational analysis with those of environmental simulations (e.g., daylighting and wind). This will help designers to respond in a more holistic way to the requirements for bioclimatic solutions, and possibly increase the effectiveness of passive design strategies targeting indoor comfort, with implications for energy efficiency.

References

1. Contreras, D., Blaschke, T., Hodgson, M.E.: Lack of spatial resilience in a recovery process: case L'Aquila, Italy. Technol. Forecast. Soc. Change **121**, 76–88 (2017)
2. Johnson, C.: Impacts of prefabricated temporary housing after disasters: 1999 earthquakes in Turkey. Habitat Int. **31**(1), 36–52 (2007)
3. Wagemann, E.: Need for adaptation: transformation of temporary houses. Disasters **41**(4), 828–851 (2017). https://doi.org/10.1111/disa.12228
4. Alexander, D.: An evaluation of medium-term recovery processes after the 6 April 2009 earthquake in L'Aquila, Central Italy. Environ. Hazards **12**(2013), 60–73 (2013)
5. Hosseini, S.M.A., Pons, O., de la Fuente, A.: A combination of the Knapsack algorithm and MIVES for choosing optimal temporary housing site locations: a case study in Tehran. Int. J. Disaster Risk Reduct. **27**, 265–277 (2018)
6. Alexander, D.: Preserving the identity of small settlements during post-disaster reconstruction in Italy. Disasters **13**, 228–236 (1989)
7. DPC, Terremoto centro Italia: Fondo di Solidarietà dell'Unione Europea, oltre 23 miliardi di euro i costi dell'emergenza e la stima dei danni - Comunicato Stampa (2017). http://www.protezionecivile.gov.it/media-comunicazione/comunicati-stampa/dettaglio/-/asset_publisher/default/content/terremoto-centro-italia-fondo-di-solidarieta-dell-unione-europea-oltre-23-miliardi-di-euro-i-costi-dell-emergenza-e-la-stima-dei-danni. Accessed 17 Jan 2020
8. DPC, I numeri del sisma in Centro Italia (2018)
9. CNS, Le soluzioni abitative in emergenza | CNS (2016). https://www.cnsonline.it/le-soluzioni-abitative-in-emergenza-del-consorzio-nazionale-servizi/. Accessed 17 Jan 2020
10. DPC, Terremoto Centro Italia: le Sae-Soluzioni abitative in emergenza - Dettaglio dossier (2019). http://www.protezionecivile.gov.it/media-comunicazione/dossier/dettaglio/-/asset_publisher/default/content/terremoto-centro-italia-le-sae-soluzioni-abitative-in-emergenza. Accessed 17 Jan 2020
11. DPC, Circolare della Presidenza del Consiglio dei Ministri del 16/01/2017 (2017)
12. Pezzica, C., Cutini, V., Bleil De Souza, C.: Mind the gap: state of the art on decision-making related to post-disaster housing assistance (unpublished)
13. Cutini, V.: The city, when it trembles: earthquake destructions, post-earthquake reconstructions and grid configuration. In: 2013 International Space Syntax Symposium. Sejong University Press (2013)
14. Carpenter, A.: Disaster resilience and the social fabric of space. In: 9th International Space Syntax Symposium, Seoul, Korea (2013)
15. Hillier, B., Hanson, J.: The Social Logic of Space. Cambridge University Press (1984)
16. Srinurak, N., Mishima, N., Kakon, A.N.: Urban morphology and accessibility classification as supportive data for disaster mitigation in Chiang Mai, Thailand. Int. Assoc. Lowl. Technol. **18**, 219–230 (2016)
17. Mohareb, N.I.: Emergency evacuation model: accessibility as a starting point. Proc. Inst. Civ. Eng. - Urban Des. Plan. **164**, 215–224 (2011)

18. Chang, H.-W., Lee, W.-I.: Decoding network patterns for urban disaster prevention by comparing Neihu district of Taipei and Sumida district of Tokyo. MATEC Web Conf. **169**, 01044 (2018). https://doi.org/10.1051/matecconf/201816901044

19. Cutini, V., Pezzica, C.: Street network resilience put to the test: the dramatic crash of genoa and bologna bridges. Sustainability **12**, 4706 (2020). https://doi.org/10.3390/su12114706

20. Penchev, G.: Using space syntax for estimation of potential disaster indirect economic losses. Comp. Econ. Res. **19**(5), 125–142 (2016)

21. Turner, A.: From axial to road-centre lines: a new representation for space syntax and a new model of route choice for transport network analysis. Environ. Plan. B Plan. Des. **34**, 539–555 (2007). https://doi.org/10.1068/b32067

22. Turner, A., Doxa, M., O'Sullivan, D., Penn, A.: From isovists to visibility graphs: a methodology for the analysis of architectural space. Environ. Plan. B Plan. Des. **28**, 103–121 (2001). https://doi.org/10.1068/b2684

23. Cutini, V.: Lines and squares: towards a configurational approach to the morphology of open spaces. In: Proceedings of the 4th International Space Syntax Symposium, London, pp. 49.1–49.14 (2003)

24. Chioni, C., Pezzica, C., Cutini, V., Bleil de Souza, C., Rusci, S.: Multi-scale configurational approach to the design of public open spaces after urban disasters. In: Proceedings of the 5th International Symposium Formal Methods in Architecture. Advances in Science, Technology & Innovation. Springer Nature, Lisbon, n.d

25. Pezzica, C., Cutini, V., Bleil De Souza, C.: Rapid configurational analysis using OSM data: towards the use of Space Syntax to orient post-disaster decision making. In: Proceeding of the 12th International Space Syntax Symposium. Beijing JiaoTong University, Beijing (2019)

26. Chioni, C.: From Temporariness to Permanence. The case of "Borgo di Arquata" after the 2016 central Italy earthquake (2019)

27. ISTAT, Caratteristiche dei territori colpiti dal sisma del 24 agosto 2016 (2016). https://www.istat.it/it/archivio/190370. Accessed 18 Jan 2020

Decision Support Systems Based on Multi-agent Simulation for Spatial Design and Management of a Built Environment: The Case Study of Hospitals

Dario Esposito[1]([⊠]), Davide Schaumann[2], Domenico Camarda[1], and Yehuda E. Kalay[3]

[1] Polytechnic University of Bari, Bari, Italy
{dario.esposito,domenico.camarda}@poliba.it
[2] Jacobs Technion-Cornell Institute at Cornell Tech, New York, USA
davide.schaumann@cornell.edu
[3] Technion, Israel Institute of Technology, Haifa, Israel
kalay@ar.technion.ac.il

Abstract. Social dimension is a fundamental part of sustainable spatial planning, design and management. Envisioning its requirements and consequences is of the utmost importance when implementing solutions. Decision Support Systems aim towards improving decision-making processes in the development of infrastructures and services. They assist decision makers in assessing to what extent the designed places meet the requirements expressed by intended users. However, there is an intrinsic limit in the forecasting of emerging phenomena in spatial complex systems. This is because the use of built environment could differ from its function since it changes as a response to the context and it reflects emergent and dynamic human spatial and social behaviour. This paper proposes a Multi-Agent Simulation approach to support Decision-Making for the spatial design and management of complex systems in risk conditions. A virtual simulation shows a hospital ward in the case of health risk due to Hospital Acquired Infection, with an emphasis on the spatial spread of the risk. It is applied to find out correlations between human characteristics, behaviours and activities influenced by spatial design and distribution. The scenario-building mechanism is designed to improve decision-making by offering a consideration of the simulation outcomes. The visualization of how a building environment is used is suitable in verifying hypotheses and to support operational choices. The proposed framework aims at assessing and forecasting the building's capacity to support user activities and to ensure users safety.

Keywords: Decision support system · Multi-agent simulation · Human spatial behaviour · Hospital acquired infection · Spatial design

© Springer Nature Switzerland AG 2020
O. Gervasi et al. (Eds.): ICCSA 2020, LNCS 12251, pp. 340–351, 2020.
https://doi.org/10.1007/978-3-030-58808-3_25

1   Introduction

In spatial design and management appropriate and rapid decision-making is a critical feature. Nevertheless, decision situations are often complex and multidisciplinary and usually involve multiple agents and types of information. Additionally, the structure and behaviour of a complex situation can raise uncertainty since these are neither well-structured nor clearly defined. As a result, more and more often there is no prescriptive or consistently valid process that can be followed to produce desired solutions.

Decision-making entails the analysis of a real-world system through methods and models and consideration of the context and its variables. The objective of the analysis aims to understand the system of interest and use this knowledge to facilitate decision-making. At the same time, the aim of decision-making influences the way in which decision support systems are designed Fig. 1. The objective of decision-making is to plan policies and measures that oversee and direct the development of the system. Decision-making involves identifying the various possibilities of alternative actions and choosing one or more of these through an evaluation process which should be sensible, rational and based on evidence [1].

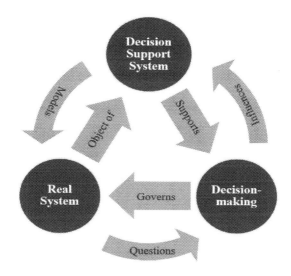

Fig. 1. Decision Support System relational framework.

Decision Support Systems (DSS) help in estimating the implications and consequences of possible decisions before their actual execution, allowing for better decisions [2]. DSSs are fundamental when a decision-making process requires the synthesis of a variety of information produced by multiple sources, when multiple possibilities of choice are involved and when, due to the wide variety of potential outcomes, it is important to justify the decision that is made.

Indeed, DSSs are designed and implemented to address complex decision situations, where it is beyond the capability of individuals to understand and reprocess all the necessary information to significantly describe them. In such cases, distributed data must be collected and organized to support problem analysis. Advanced computer-based tools, such as Multi-Agent simulations, are used to provide a framework for people to explore a problem, learn about it and use the knowledge gained to arrive at conscious decisions. An effective Multi-Agent simulation is a simplified replica of a real-world system which has reached a convenient trade-off between realism and usefulness. DSSs based on Multi-Agent simulations help describe the evolution of the system and its constraints, provide knowledge-based formulation of possible decisions, simulate scenarios of consequences of actions and assist in the formulation of implementation strategies. This decision-making process is often iterative since alternative actions are analysed and information gained from the what-if scenario analysis is used to guide it further.

Decision-making in land engineering, urban planning and architectural design have long-term implications which directly affect the Built Environment, consisting of a network of buildings, infrastructures, open spaces, and its users. They involve the realization of projects and public spaces to develop activities in selected areas. This in turn affects people's experience of cities and life, since built environments support the living and safety needs of their inhabitants. Research in the field of DSS applied to spatial planning and design aims towards improving decision-making processes in public administrations and aiding professionals in the field of infrastructural and service development within cities. Contemporary sustainable planning and design consider social dimensions as a fundamental part of it. Planners, architects and civil engineers should be able to assess to what extent the designed places meet the requirements expressed by their intended users. This is a task inherently oriented to consider human expectations, needs and behaviours. To date, they rarely find formal methods to forecast whether, from a user's perspective, the designed infrastructure will perform before it has been built and tested through use, especially for the most qualitative aspects, such as human satisfaction, productivity, comfort and safety. There is an intrinsic limit to what extent decision makers can use their imagination and experience to forecast emerging phenomena in spatial complex systems. This is because the use of built environment could be different from what the expert may realize, since it could differ from its function. The use changes as a response to the context (cultural, environmental, psychological, and more). It reflects emergent human spatial and social behaviour, and because of this it is more dynamic than functional, e.g. it can change from person to person and in a short time.

However, even if human related aspects are too complex to be predicted accurately and a gap exists between expected and actual behaviour, decisions cannot simply be ignored. In spatial design and management, envisioning the various consequences of implementing specific solutions is of the utmost importance for the decision to succeed [3]. Decision makers constantly search for innovative methods to assess the implications of decisions related to humans in space, as these are crucial in addressing design issues appropriately and as early and thoroughly as possible. Likewise, in everyday hospital life, safety and health risks arise and must be managed through decision-making processes. Here a virtual simulation of a trial case study shows a healthcare system under conditions of risk, with the aim of forecasting and assessing the building's capacity to support user activities and to satisfy users' safety needs. Thus, it is used to illustrate the potential of a Multi-Agent simulation approach to support decisions. All decision-making processes start from the recognition of a problem, and then from its definition to the solution. Specifically, our issue is that of healthcare environment design and management in the case of the health risk due to Hospital Acquired Infection (HAI), i.e. infections contracted during hospitalization, with emphasis on the spatial spread of the risk [4].

2 Background

Practitioners and policymakers usually rely on (and at the same time respect) regulations, design rules and legislative factors. Even though they have at their disposal several computational tools to evaluate quantitative building performances and characteristics such as costs, energy consumption, material features, structural stability and so on [5], analytical approaches in evaluating most qualitative aspects suffer from severe limitations and neither can a real-size prototype be built and tested before construction itself. Therefore, their capacity to fully comprehend the complexity of human-building interaction (e.g. building-use, human spatial behaviour, human satisfaction and safety issues) has shown its limits, mirroring the increasing complexity of building design and variety of human behaviour with all its consequent requirements [6]. Such a real-world system is too complex to be evaluated through analytical solutions, which are not available or are computationally inefficient [7]. Alternatively, it could be replicated and therefore studied by means of computer simulation. A simulation model is preferable to modelling complex systems as it is more appropriate for modelling dynamic and transient effects [8]. It appears to be the best choice for investigating Human Spatial Behaviour. A realistic simulation of a built environment with a Multi-Agent System (MAS) emulates either the behaviour of a single person or collective actions, accounting for how humans act and interact with each other and with the environment [9, 10].

A MAS is defined as a set of agents operating and interacting simultaneously in a shared environment, under different initial conditions and constraints [11]. MAS agents show the ability to solve problems at individual level and to interact in order to reach global objectives. The interaction may take place between agents as well as between agents and their environment. In MAS, the researcher defines the behaviour of the single agent at individual level and the system behaviour emerges from multiple local interactions between them. This leads to a dynamic and unpredictable evolution of the system, which is thus referred to as complex. Multi-Agent Simulation has been recognized by international literature as an efficient method for evaluating the performance of designed systems when the relationships among decision variables are too difficult to be established analytically [12]. To this end, virtual simulation is a valuable approach in investigating "what-if" scenarios. It offers the potential to identify new understanding of how a built environment may operate, providing evidence in support of decision-making processes [13–15].

Recently, following the Multi-Agent paradigm, the Event-Based Modelling and Simulation (EBMS) approach has emerged to address these issues. Developed by Schaumann et al., it is based on the Event notion. Events are designed to coordinate temporal, goal oriented routine activities performed by agents. Rather than describing behaviour from the point of view of each actor, events allow for the description of behaviour from the point of view of the procedures that need to be performed to achieve a task [16]. The event system architecture adds the capacity to manage the coordinated behaviour of multiple agents in a top down fashion to the bottom up structure of the ABM. Its power to simulate complex and realistic scenarios with a flexible user interface allows us to apply it to the simulation of Human Spatial Behaviour in a Built Environment. Such a Multi-Agent Simulation approach incorporates a number of factors such as a description of actors' organizational roles and hierarchies, activities involving more than one actor and adaptability in response to dynamic changes in activity priorities, unplanned events, or non-typical circumstances, such as when an actor is not available to perform a task when expected.

3 Methodology

HAIs are infections caused by microorganisms acquired within a health care facility by a patient who was admitted for a reason other than infection [4]. A prevalence survey conducted under the auspices of the World Health Organisation (WHO) in 55 hospitals of 14 countries representing 4 WHO Regions (Europe, Eastern Mediterranean, South-East Asia and Western Pacific) showed an average of 8.7% of hospital patients who had nosocomial infections [17]. HAIs are a significant burden both for the patient and for public health resources, since treatment is very costly and may not be effective. What is more, organisms causing HAIs can be transmitted to the community through discharged patients, Healthcare Workers (HCW) and visitors, which may cause significant disease in the community.

The contamination propagation phenomenon has multi-factor roots and proceeds through a dynamic transmission mechanism which often leads to outbreaks. It overlaps hospital processes, events and workflows. It is restricted through infection prevention and control procedures and it influences and is influenced by spatial design and distribution. Literature confirms that the conventional ways that hospitals are designed contributes to danger [18]. It shows that the physical environment strongly impacts on hospital acquired infection rates by affecting both airborne and contact transmission routes. Ulrich's research identified more than 120 studies linking infection to the healthcare built environment [19]. A critical challenge for architects is to improve the physical setting design to make hospitals safer, with improved healing procedures and better places to work by reducing risk from HAIs. Nevertheless, guidelines for the design of healthcare facilities are often vague in their formulation [20]. Thus, there is an urgent need for a DSS method which demonstrates how a better hospital ward spatial design could contribute to HAI prevention and control and to support decision makers with choices that could impact on the safety of users.

We propose an event-based decision-support system for hospital design and management, which can be applied to predict, prevent, and manage health risk due to Hospital Acquired Infection (HAI). The model enables the simulation and coordination of multiple actors' behaviours, which affect (and are affected by) the dynamic spatial and social context. The aim is to reveal the mutual interactions between a built environment and the behaviour of its occupants to inform a buildings' design or renovation. In the event-based model, the decision-making authority is stored in Event entities, which direct the collaborative behaviour of a group of actors to perform an activity in a given space. Events afford top-down coordination of actors' scheduled behaviour (e.g. a doctor's patient check round) while accounting for bottom-up adaptations to spatial and social contingencies (e.g., impromptu social interactions between a doctor and a patient), which can delay the performing of scheduled activities. Each event includes a set of (a) *preconditions*, which specify the requirements for an event to be triggered, (b) *performing procedures*, which guide the event's execution, and (c) *postconditions*, which update the state of the entities involved in the event's execution. Events meaningfully combine 3 different components to describe context-dependent behaviour patterns, namely *spaces*, *actors* and *activities*. To reduce events' computational efforts in managing the performing of behaviour patterns, each of the constituents, are endowed with autonomous calculation abilities. Results of such calculations can be communicated to the event to assist its decision-making process. *Spaces* contain geometric and semantic information and can automatically modify their status depending on the actors they host and the activities performed. *Actors* are associated with specific roles in the building organization, knowledge about the space and other actors, and a dynamic status that records the current space where the actor is located, the activity performed, and other additional metrics for evaluation (e.g. the walked distances). *Activities* direct the low-level interactions between actors and their environment. To model human behaviour patterns at increasing levels of complexity, events can be aggregated into tree-like structures, called narratives. A narrative manager coordinates the performing of human behaviour scenarios composed of planned and unplanned narratives. Planned narratives are performed at a specific time, while unplanned narratives emerge in response to social, spatial and environmental conditions.

In the virtual simulation, the behaviour of agents and their conditions and contamination capacity is formulated using a discrete equation formulation. Further details on the weighting of various parameter and their interaction in the system, i.e. likelihood of certain behaviours or events can be found in Esposito [21]. The transmission model and equation were then implemented within the Unity 3D environment where the spatial semantics, actors' profiles, activities, events and the contamination model were coded in C#. This allows for the understanding of the phenomena directly through an infection risk map visualization, which reveals how social interaction and spatial influences affect the spread of HAIs.

4   Case Study Simulation

Decision making is considerably difficult when the objects of interest are complex infrastructures such as hospitals, where performances are related to several functional, typological and organizational requirements. It is also where, among other factors, human considerations such as satisfaction with the quality of care and patient and staff safety are major concerns.

Indeed, human related sciences such as architecture and urban planning tend to work formulating context-dependent methods and developing knowledge models to support and improve the decision-making processes in a specific domain and case [22]. Nevertheless, case study research excels at providing an understanding and explanation of a complex issue and can extend knowledge, add strength or disprove theories [23]. Case study evaluation emphasizes detailed contextual analysis of specific events or conditions and their relationships. Therefore, the case study approach is especially well-suited to produce the kind of knowledge required to support the decision-making process [24]. Indeed, it is useful for both generating and testing hypotheses in carefully planned and crafted researches of real-life situations, issues and problems. The testing of hypotheses through case studies relates directly to the issue of case selection. The strategic choice of case in relation to the research objectives may greatly add to the generalizability of a case study [25].

With this rationale, the proposed simulation illustrates the applications of the EBMS approach to the healthcare environment domain. The simulation displays a building use situation where simulated actors (doctors, nurses, patients, visitors, etc.) perform tasks according to pre-planned schedules (medicine distribution, patient check, patient visiting, etc.), while dynamically responding to social encounters and environmental conditions. In addition, they respond to un-planned events (e.g., "code blue" when a patient is in cardiac distress), and act accordingly. The hypothetical scenario scene shows an HCW workflow interruption situation. To build up a reliable simulation, an observation and analysis of human behaviour in built environments was adopted, based on the POE paradigm. Data was collected on user activities with direct-experience observations, shadowing, tracking people and interviewing medical and administrative staff, patients and visitors. Moreover, references, guidelines and sessions with experienced medical practitioners led us to understand the features of HAIs and the established protocols and best practices to manage them. Further details on this process are provided in Esposito [26]. This phase helped us to accurately represent the complex inter-relations between all the major features involved in the case study narrative, which unfolds as follows: HCWs start from their staff station before moving to the central medicine room to prepare medicines. Afterwards, they move through the patients' rooms to distribute these. During the simulation, a random number of visitors enters the ward to meet their relatives, each one visiting a single patient. They walk through the hallway to reach the patient's room, where a social interaction takes place for a certain amount of time. Afterwards, visitors leave the ward from the same entrance. In the simulation, emergent events could be triggered when specific spatial and social conditions arise. It may occur that when a visitor encounters a HCW, the close proximity between the two drives the visitor to interrupt the HCW scheduled duties to start a social interaction, e.g. visitor asking information about his family member condition, before the HCW returns to his planned activities, as does the visitor. The proposed scene starts with a visitor leaving his relative's room when an actor enters the room to attend to the other patient. This situation forces him to wait in the corridor and then unexpectedly interrupt the HCW workflow while he is performing a round of visits. The visitor who has been in direct contact with his infected relative (making him a carrier), in turn contaminates the HCW. In this situation, the HCW fails to observe Hand Hygiene protocol, therefore he spreads contamination to subsequent patients during the round. In the meanwhile, the space populated by these contaminated actors also becomes contaminated Fig. 2.

The case study reveals the narrative nature of the HAI spread phenomenon, which is well represented through the EBM simulation narrative approach. Moreover, it allows for the visualization of the risk of contamination propagation due to human spatial behaviour and user activities in the built environment, through real-time results displayed with changing colours for actors and spaces, ranging from green to blue for higher contamination risk.

The simulation presented is not focused on predicting the future accurately; instead, our approach is diagnostic, i.e. it is used to understand and explore the MAS model which has been exploited to describe the system [27]. The capacity of the simulation to account for both situations when a clear sequence of observable factors and planned activities can be recognized and when emerging unplanned behaviour complicates situations, demonstrates the capacity of the simulation system to account for changes that are not obvious. The working proof of the "what-if" scenario allows for an understanding of the possible state patterns in the development of the contamination propagation. Moreover, this proves its value as a DSS in the field of hospital management of infection risk, e.g. when employed as a forecasting tool for the evaluation of policies and also on the process of design of hospitals.

In fact, a system user can test the potential risk of different real-life situations by simulating new scenarios with new input conditions and with system parameter tuning, e.g. actors' profiles and behaviour and re-configuration of settings. Although in the simulation the duration of the activities was reduced to condense several hours of activity into a few minutes, the type of each activity and the duration of each contact can be easily modified in the user interface. Likewise, the number of non-colonized, colonized and infected actors can be adjusted to reflect the desired proportion. It is possible to run a scenario in which a percentage of total patients have a predisposition to the acquisition and development of infection, as in the case of the presence of immunocompromised patients or caused by virulent pathogens. Finally, since there can be uncertainty concerning the primary source of transmission, in certain circumstances, HCWs are the cause for transferring bacteria to patients, whereas in others, this could be due to visitors or the environment. If a study into how infection propagates from the flora of a health care environment is required, as in the case of epidemic exogenous environmental infections, it is straightforward at the beginning of the simulation to set a scenario whereby the initial cause of infection spread resides in a contaminated space, adjusting the starting contamination level for the selected spaces.

5   Conclusions

The present study proposes a Multi-Agent simulation functioning as a DSS for hospital management and design, providing evidence in support of underlying decision-making processes. The focus of the paper is on the development and use of a MAS simulation to support decision-making. Indeed, in cases of multiple complexity, its core capability is based on gathering, structuring and presenting knowledge. This involves dealing with the so-called wicked problems for which the decisional approach aims to address

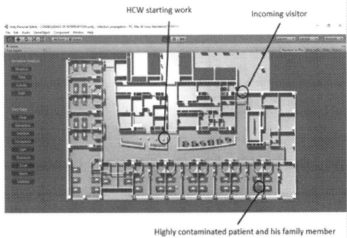

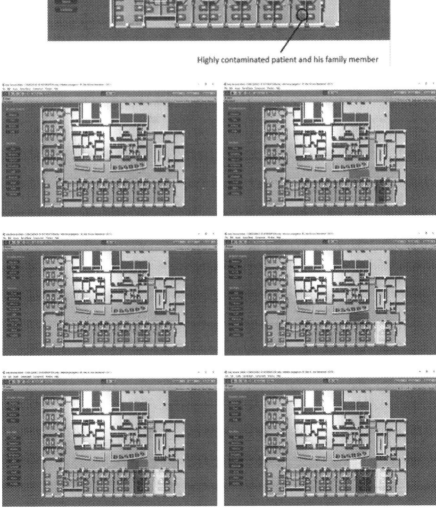

Fig. 2. Contamination risk map for people and spaces.

problems structuring rather than solving them. Furthermore, it means reaching a decision which is satisfactory rather than optimal, avoiding the risk of unintended consequences.

To demonstrate the usability of the proposed approach a virtual scenario is reported. The simulation illustrates the potential applications of the proposed MAS approach through proof-of-concept case studies. The proposed case study interprets the overall course of human spatial behaviour, starting from human states and contextual conditions and ending with activities set in space, allowing for the real-time visualization of contamination transmission under the effect of breaks in prevention measures. It visualizes the pathogen propagation, correlating with the architectural ward layout and workflow organization in a case study of HCW workflow interruption. The scenario-building mechanism is designed to improve decision-making by offering a consideration of the simulation outcomes and their implications. Indeed, the development of the system is fundamental in verifying hypothesis and to support choices. Specifically, this process provides the following features:

- to test "what-if" scenarios in order to explore the effects of possible design solutions for the HAI phenomenon and to define a balanced, satisfactory trade-off with building requirements;
- to estimate the effectiveness of a range of policies aimed at preventing and control the HAI spread and to represent the impact of social and spatial factors on the performance of procedures dealing with the outbreak;
- to evaluate to what extent the EBMS framework can be applied for modelling and simulation of complex human spatial behaviour to support decisions and management for spatial design of a built environment.

The elaboration of the research study can lead to the broader purpose of improving understanding of the potential impact of physical and social settings in a built environment on users. Moreover, it provides insights about human spatial decision-making and actions to help policy makers and experts to interpret the relationship between the organization of places and spatial behaviours. Therefore, the scope of the current research is:

- to provide a visualization of how a building is used and experienced;
- to forecast and assess the building's capacity to support user activities and to satisfy users' functional needs, e.g. safety and satisfaction;
- to examine how well the virtual simulation of human behaviour can be suitable in estimating human-related building performances;
- to support designers while making decisions that could impact on the lives of the users of future buildings;
- to evaluate alternative building project proposals and designers' choices before moving onto the construction phase.

The Multi-Agent simulation is applied to find out correlations between human characteristics, behaviours and activities in case of HAI risk, which give us hints on the role of the space design. Indeed, the principal value of the research is to build a framework to improve the planning and design of a human-cantered built environment, thus enabling a consideration of the use of built infrastructures by user agents.

Providing visualizations of how a building is used in a design phase has the potential of envisioning the various consequences of implementing specific design solutions. Accordingly, an understanding of individual decisions in actions and behavioural processes in space can support professional knowledge and public administrations in their decision-making procedures in the field of urban infrastructures.

A new scenario is to be developed which compares two slightly different spatial configurations, while maintaining all the other fixed variables and conditions. This further experimentation is designed to exploit analysis into whether the architectural layout alone affects, by fostering or hindering, HAI propagation, e.g. understanding how different ward designs can lead to different dynamics of infection diffusion.

The results analysis might enable the evaluation of how an intended design meets infection control and prevention requirements. This application will support the design team and hospital managers in the evaluation of functional design solutions connected with safety requirements, suggesting potential improvements.

Acknowledgments. We wish to thank Professor Jacob Yahav for his meaningful methodological assistance. We gratefully thank the following research group members for their enlightening insights: K. Date, E. Eizenberg, M. Gath Morad, L. Morhayim, N. Pilosof and E. Zinger.

Author Contributions. Conceptualization, investigation, formalization and writing D.E.; software and methodology paragraph D.E. and D.S.; review and editing D.E. and D.S.; supervision and project administration D.C. and Y.K. All authors have read and agreed to the published version of the manuscript.

References

1. Simon, H.A.: The New Science of Management Decision. Prentice-Hall, Upper Saddle River (1977)
2. Furtado, B.A.: Modeling Complex Systems for Public Policies (2015)
3. Borri, D., Camarda, D., Pluchinotta, I., Esposito, D.: Supporting environmental planning: knowledge management through fuzzy cognitive mapping. In: Luo, Y. (ed.) CDVE 2015. LNCS, vol. 9320, pp. 228–235. Springer, Cham (2015). https://doi.org/10.1007/978-3-319-24132-6_29
4. World Health Organization: Prevention of hospital-acquired infections: a practical guide (2002). https://doi.org/WHO/CDS/CSR/EPH/2002.12
5. Schaumann, D., Pilosof, N.P., Date, K., Kalay, Y.E.: A study of human behavior simulation in architectural design for healthcare facilities. Ann. Ist. Super. di Sanità **52**, 24–32 (2016). https://doi.org/10.4415/ANN_16_01_07
6. Simeone, D., Kalay, Y., Schaumann, D., Hong, S.: Modelling and simulating use processes in buildings. In: Proceedings of eCAADe 31, pp. 59–68 (2013)
7. Borshchev, A., Filippov, A.: From system dynamics and discrete event to practical agent based modeling: reasons, techniques, tools. In: 22nd International Conference of the System Dynamics Society, 25–29 July 2004, vol. 45 (2004)
8. Pidd, M.: Computer Simulation in Management Science. Wiley, Chichester (2004)
9. Majid, M.A.: Human Behavior Modeling: An Investigation Using Traditional Discrete Event and Combined Discrete Event and Agent-Based Simulation (2011)

10. Pew, R.W., Mavor, A.S.: Modeling Human and Organizational Behavior. National Academies Press, Washington, DC (1998). https://doi.org/10.17226/6173

11. Ferber, J.: Multi-Agent Systems: An Introduction to Distributed Artificial Intelligence. Addison-Wesley, Harlow (1998)

12. Kalay, Y.E.: Architecture's New Media: Prinicples, Theories and Methods of Computer-Aided Design. MIT Press, Cambridge (2004)

13. Benenson, I., Torrens, P.M.: Modeling Urban Dynamics with Multiagent Systems (2004)

14. Chen, L.: Agent-based modeling in urban and architectural research: a brief literature review. Front. Archit. Res. **1**, 166–177 (2012). https://doi.org/10.1016/j.foar.2012.03.003

15. Batty, M.: Agent-based pedestrian modelling, pp. 81–106 (2003). https://doi.org/10.1068/b2803ed

16. Schaumann, D., Kalay, Y.E., Hong, S.W., Simeone, D.: Simulating human behavior in not-yet built environments by means of event-based narratives, pp. 1047–1054 (2015)

17. World Health Organisation: (WHO) practical guidelines for infection control in health care facilities. World Heal. Organ. **30**, 1–354 (2004). https://doi.org/10.1086/600379

18. Lateef, F.: Hospital design for better infection control. J. Emerg. Trauma Shock **2**, 175–179 (2009). https://doi.org/10.4103/0974-2700.55329

19. Ulrich, R., Quan, X., Systems, H., Architecture, C., Texas, A.: The role of the physical environment in the hospital of the 21st century: a once-in-a-lifetime opportunity. Environment **439**, 69 (2004)

20. Stiller, A., Salm, F., Bischoff, P., Gastmeier, P.: Relationship between hospital ward design and healthcare-associated infection rates: a systematic review and meta-analysis. Antimicrob. Resist. Infect. Control **5**, 51 (2016). https://doi.org/10.1186/s13756-016-0152-1

21. Esposito, D., Schaumann, D., Camarda, D., Kalay, Yehuda E.: Multi-Agent modelling and simulation of hospital acquired infection propagation dynamics by contact transmission in hospital wards. In: Demazeau, Y., Holvoet, T., Corchado, Juan M., Costantini, S. (eds.) PAAMS 2020. LNCS (LNAI), vol. 12092, pp. 118–133. Springer, Cham (2020). https://doi.org/10.1007/978-3-030-49778-1_10

22. Crooks, A.T., Patel, A., Wise, S.: Multi-agent systems for urban planning. In: Technologies for Urban and Spatial Planning: Virtual Cities and Territories, pp. 29–56 (2014). https://doi.org/10.4018/978-1-4666-4349-9.ch003

23. Popper, K.R.: Conjectures and Refutations: The Growth of Scientific Knowledge. Routledge, London (2002)

24. Soy, S.K.: The Case Study as a Research Method. https://www.ischool.utexas.edu/~ssoy/usesusers/l391d1b.htm. Accessed 06 Oct 2017

25. Flyvbjerg, B.: Five misunderstandings about case-study research. Qual. Inq. **12**, 219–245 (2006). https://doi.org/10.1177/1077800405284363

26. Esposito, D., Abbattista, I.: Dynamic network visualization of space use patterns to support agent-based modelling for spatial design. In: Luo, Y. (ed.) Cooperative Design, Visualization, and Engineering. CDVE 2020. Springer, Cham (2020, in press)

27. Saltelli, A., Ratto, M., Andres, T.: Global Sensitivity Analysis: The Primer (2009)

Future Climate Resilience Through Informed Decision Making in Retrofitting Projects

Jonas Manuel Gremmelspacher[1,2]([x]) ⓘ, Julija Sivolova[1],
Emanuele Naboni[2,3] ⓘ, and Vahid M. Nik[1,4,5] ⓘ

[1] The Faculty of Engineering - LTH, Lund University, Box 118,
221 00 Lund, Sweden
jonas@gremmelspacher.net
[2] Schools of Architecture, Conservation and Design, The Royal Danish
Academy of Fine Arts, Philip de Langes Allé 10, 1435 Copenhagen, Denmark
[3] Department of Engineering and Architecture, University of Parma,
Parco Area delle Scienze 181/a, 43124 Parma, Italy
[4] Chalmers University of Technology, 412 96 Gothenburg, Sweden
[5] Queensland University of Technology, 2 George St, Brisbane City,
QLD 4000, Australia

Abstract. High energy use for space conditioning in residential buildings is a significant economic factor for owners and tenants, but also contributes to resource depletion and carbon emissions due to energy generation. Many existing dwellings should thus be retrofitted in order to fulfil the ambitious EU carbon emission mitigation goals by 2050. To investigate how future climate resilience can be implemented in the design process of retrofitting measures, this study concentrates on real case studies that have been retrofitted during the past decade. The performance of retrofitting measures for four case studies in Denmark and Germany were investigated under future climate projections and compared between the non-retrofitted initial stage of the buildings and the retrofitted stage. Building performance simulations were employed to investigate how severe the effects of climate change until the end of the 21st century on the material choice and system design is. Results show that summertime thermal comfort will be a major challenge in the future. Energy use for space heating was seen to decrease for periods in the future, also the severity of cold events decreased, resulting in a decline of heating peak loads. Additionally, not considering extreme events was proven to lead to miss-dimensioning thermal systems. Overall, the study shows that adaptation of informed decisions, accounting for the uncertainties of future climate, can bring a significant benefit for energy-efficient retrofits, potentially promoting adequate passive measures as well as free cooling to prevent overheating and enhance heat removal.

Keywords: Climate change · Resilience · Retrofitting buildings · Residential buildings · Building stock · Building Performance Simulation · Climate action · Future climate

© Springer Nature Switzerland AG 2020
O. Gervasi et al. (Eds.): ICCSA 2020, LNCS 12251, pp. 352–364, 2020.
https://doi.org/10.1007/978-3-030-58808-3_26

1   Introduction

Climate change is a widely discussed topic amongst politicians, the public and media [1]. In order to comply with the Paris Agreement, the global average temperature must not rise more than 2 °C compared to pre-industrial levels and efforts must be made to decrease the number to 1.5 °C [2, 3]. However, the fifth assessment report by the Intergovernmental Panel on Climate Change (IPCC) reveals that climate change will accelerate, rather than decelerate [4]. Furthermore, research shows that weather extremes such as heat-waves will occur more frequently and last over longer periods in the future [5, 6]. The 2030 Agenda for Sustainable Development by the United Nations (UN) set 17 Sustainable Development Goals (SDG) to achieve a better and more sustainable future by 2030 [7, 8], which two of the goals, SDG11 (sustainable cities and communities) and SDG13 (climate action), are addressed in this work by investigating some major energy retrofitting strategies for buildings.

Approximately 75% of the entire building stock is considered energy inefficient [9]. One example is the annual energy use of dwellings in Germany built before 1990 which is almost 50% higher than those built after 1990, when the regulations for mitigating the energy use of buildings were set [10], this can likely be upscaled to other central European countries. Thus, the main focus for retrofitting projects are the pre-1990 buildings in order to save operational energy. This study, however, analysed four buildings that were built before 1970. The bottom step of the Kyoto pyramid [11], which is seen as a general approach to mitigate the energy use in the building stock, is to reduce heat losses and thus, limit the transmission losses by retrofitting their thermal envelope [12]. The concept of energy-efficient retrofitting describes that buildings should be renovated to among others diminish the use of operational energy by lowering the transmission losses and installing systems with high efficiencies [13]. Motivations for building owners and operators were introduced by the EU authorities as well as local governments with a three-step system. Firstly, houses are needed to be classified in an energy labelling system to benchmark the energy-efficiency [14]. Secondly, a renovation roadmap has to be established by property owners to set milestones for energy-efficiency measures [15]. Lastly, the EU member states created funds for subsidies and low-interest loans that are provided to property owners to be able to afford retrofitting measures [15]. Evidently, vast amounts of resources are dedicated to retrofit the European building stock. However, one major challenge is assessing the impact and efficiency of the retrofitting strategies for future climatic conditions. As climatic conditions of the future have a high impact on the performance, thermal comfort, and economic payoff for retrofit measures, considering them when designing retrofit strategies seems essential. Therefore, informed decision making about the appropriate passive and active retrofit strategies of existing dwellings throughout the 21st century was assessed in this work.

This study investigates the energy performance of some common retrofitting strategies for residential buildings in Denmark and Germany on four case studies. Impacts of climate change are thoroughly assessed by studying energy use and indoor thermal comfort under future climate projections. Another major goal is to find how building performance is affected by changing climatic conditions and what indicators can be integrated in the decision-making process at each design stage. Four actual case studies retrofitted in the last decade were assessed in this study by comparing the simulated performance of the non-retrofitted buildings with the retrofitted buildings for three 30-year periods until the end of the 21st century. Two case studies were located in Aarhus and Copenhagen, Denmark and two in Stuttgart, Germany. Thermal comfort in living spaces and energy use of the buildings were evaluated in the study. The impact assessment was performed by applying representative future climate scenarios and modelling the future performance of the buildings with Building Performance Simulations (BPS). The cases and the measures which were carried out throughout the retrofits were modelled and simulated with the limitations that the respective buildings and retrofit designs had.

2 Methodology

2.1 Building Energy Models

Four residential buildings located in Denmark and Germany were modelled; all built before 1970 and undergone major retrofits within the past decade. Case studies were selected to represent a spectrum of constructions that are typical for the countries or regions they are representing. Perspective views of the buildings are displayed in Fig. 1 and some basic information, including the window to wall ratios (WWR) are shown in Table 1. All cases along with their setpoints, parameters and assumptions were based on specifications by the companies owning the buildings and by building physicists involved in the process of certifying the buildings. This represents a limitation of the scope, which was done in order to assess the performance of measures as they were carried out in practice.

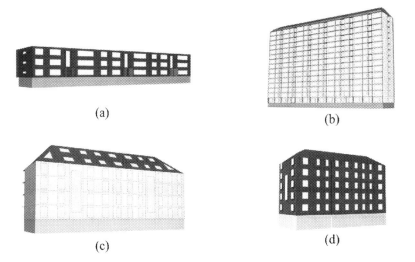

(a)

(b)

(c)

(d)

Fig. 1. Models of the case studies, not to scale: (a) Case 1, Aarhus, Denmark; (b) Case 2, Copenhagen, Denmark; (c) Case 3, Stuttgart, Germany; (d) Case 4, Stuttgart, Germany.

Specifications for HVAC, infiltration and shading were modelled as displayed in Table 2. In all case studies, no active cooling system was applicable; for case 3 and case 4 no active ventilation system was installed. Both Danish projects did not compose any moveable external shading devices, the German buildings were equipped with exterior shutters for windows and skylights in case 3. Setpoints and shading design were modelled according to the specifications of the building owners while infiltration rates were according to on-site measurements before and after the retrofit, or country-specific regulations as further discussed in [16].

Table 1. Model specifications for Cases 1–4.

Case	Heated floor area [m²]	Number of floors	Number of thermal zones	WWR North [%]	WWR East [%]	WWR South [%]	WWR West [%]
Case 1	1 915	3[a]	71	0–5[b]	30–35[b]	0–5[b]	71
Case 2	10 750	15[a]	516	12–15[b]	30–38[b]	15–21[b]	39–48[b]
Case 3	1 225–1 525[b]	4–5[a,b]	50–59[b]	26–34[b]	0	20–43[b]	0
Case 4	1 135	5[a]	60	20	15–22[b]	31	15–22[b]

[a]Basement is excluded in number of floors.
[b]First indication before retrofit, second indication after completion.

Table 2. Heating, Cooling, Infiltration, Ventilation and Shading specifications.

Case	Heating setpoint [°C]	Cooling setpoint [°C]	Infiltration rate (50 Pa) [h⁻¹]	Ventilation rate $\left[\frac{l}{s \times m^2}\right]$	Ventilation type	Heat recovery [%]	Shading type
Case 1	21	N/A	2.1–1.2[a]	0.3	Extraction – balanced[a]	0–82[a]	N/A
Case 2	21	N/A	1.8–0.9[a]	0.4	Extraction – balanced[a]	0–75[a]	N/A
Case 3	22	N/A	5.0–3.0[a]	N/A	N/A	N/A	Ext. Shutters
Case 4	22	N/A	5.0–3.5[a]	N/A	N/A	N/A	Ext. Shutters

[a]First indication before retrofit, second indication after completion.

Table 3. Building envelope area and thermal conductivities.

Case	Area roof [m²]	U-value roof $\left[\frac{W}{m^2 \times K}\right]$	Area floor [m²]	U-value floor $\left[\frac{W}{m^2 \times K}\right]$	Area wall [m²]	U-value wall $\left[\frac{W}{m^2 \times K}\right]$	Area wall gable [m²]	U-value wall gable $\left[\frac{W}{m^2 \times K}\right]$
Case 1	674	0.35[b]–0.08[a,b]	674	1.90–0.26[a,b]	1 354–1 016[a]	0.39[b]–0.18[a,b]	0–3 438[a]	– –0.12[a,b]
Case 2	875	0.23[b]–0.11[a]	875	0.54[b]–0.17[a,b]	5 650	0.54[b]–0.15[a,b]	1 685	0.33[b]–0.16[a,b]
Case 3	442–430[a]	4.05–0.13[a,b]	299	1.11–0.15[a]	854	1.92[b]–0.18[a]	N/A	N/A
Case 4	456	1.59[b]–0.28[a,b]	256	1.59[b]–0.26[a,b]	904	1.46–0.48[a,b]	N/A	N/A

[a]First indication before retrofit, second indication after completion.
[b]Thermally-bridged U-value.

Thermal conductivity specifications for building envelope constructions were set according to Table 3. Areas of roof, floor and walls are gross areas without considering openings of all kinds. U-values including thermal bridges were determined in a pre-study [16]. Case 1 and case 2 were built with different constructions for the gable wall than the oblong façades, therefore, different conductivities were obtained. Table 4 provides an overview of the properties and areas of glazing materials. Case 1, 2 and 4 did not compose skylights; for case 3, skylights were specified separately. G-value of the glazing describes the solar energy transmittance, T_{vis} stands for the visible transmittance and frame ratio defines the ratio between glazing and frame.

Table 4. Glazing area and specifications.

Case	Area glazing [m²]	U-value glazing $\left[\frac{W}{m^2 \times K}\right]$	g-value [%]	T_{vis} [%]	Frame ratio [%]
Case 1	492–524[a]	2.8–1.1[a]	63–50[a]	78–75[a]	20
Case 2	2 105–3 100[a]	2.0–0.9[a]	63–50[a]	78–70[a]	20
Case 3 Windows	140–239[a]	2.7–0.9[a]	63–50[a]	78–60[a]	30
Case 3 Skylights	8,9–71,5[a]	5.4–1.0[a]	90–50[a]	30–65[a]	10–30[a]
Case 4	203–225[a]	2.7–1.3[a]	63–60[a]	78–60[a]	30

[a]First indication before retrofit, second indication after completion.

2.2 Weather Data

For the estimation of location-based building performance, weather data is the most critical input to carry out BPS [17]. The current standard for BPS is the use of weather data composing hourly data, based on historical weather observations typically ranging from 20–30 years and composed into weather files on an hourly resolution over a one-year time frame representing the months with most typical conditions over the assessed timespan [18]. A common format is the typical metrological year (TMY) which includes several meteorological metrics. Selecting the most typical conditions for a multiple-year timeframe based on historical weather measurements also entails two disadvantages. Firstly, extreme weather conditions are underrepresented due to selection of typical months, which leads to averaging [19]. Secondly, basing design decisions in the built environment on historical climate measurements does not represent realistic conditions for the decades in which buildings are meant to withstand weather exposure [20, 21]. Moazami et al. [20] thoroughly discuss the need for climate data that accounts for climate change and its uncertainties in building energy simulations.

TMY files sourcing from the IWEC (International Weather for Energy Calculations) [22] were used in BPS to validate the scripts and models. Future climate conditions were simulated with weather files synthesized according to Nik's method [23, 24]. The weather file sets were retrieved on the basis of RCA4 Regional Climate Models (RCM) [25], developed by the Rossby Centre at the Swedish Meteorological Hydrological Institute (SMHI). A total number of 13 future climate projections over a 90-year period, from 2010 to 2099, that was generated by five global climate models (GCMs) and forced by three Representative Concentration Pathways (RCPs); RCP2.6, RCP4.5 and RCP8.5 were considered. The timeframe until the end of the 21st century was divided into three 30-year periods to be able to compare results for multiple periods: 1) near-term (NT, 2010–2039), 2) medium-term (MT, 2040–2069), and 3) long-term (LT, 2070–2099). Typical Downscaled Year (TDY) weather files were synthesized based on the month representing the most typical temperature within the 30-year period. To account for weather extremes, additional files representing the extreme warm and extreme cold weather conditions within the 30-year and 13 climate predictions matrixes were derived. Extreme Cold Year (ECY) files were hereby representative for the coldest month over the periods. Extreme Warm Year (EWY) files contrarily characterised the warmest months composed into one-year weather files. For more details about applied future weather data sets, the reader is referred to [23].

Figure 2 compares the temperature profile for TDY, ECY and EWY during LT period. The grey background represents all the considered climate scenarios (13 scenarios, each for 30 years) that were used to synthesize TDY, ECY and EWY. The shown temperature graphs are retrieved from Aarhus weather data sets used for case 1. For case 2 located in Copenhagen, weather data retrieved from the RCM for Copenhagen was used. The two last cases, which were in Stuttgart, were simulated with weather files for Munich as it was the closest geographical location for which data was provided.

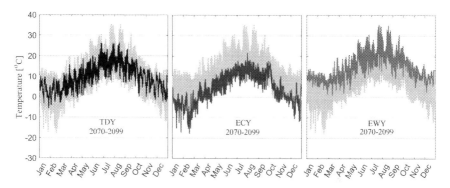

Fig. 2. Temperature variations for TDY, ECY and EWY weather files for the long-term period from 2070-2099 in Aarhus. Grey background is representative for the entire 30-year weather data.

Figure 3 shows the temperature curves in January and February for the NT, MT and LT TDY files of Munich compared with the TMY file for Stuttgart. It can be observed, that, between January and February, the TDY files show very few and only short-term events with temperatures under 0 °C. The TMY file, based on temperature measurements between 1982 and 1992, shows significantly long and frequent cold periods.

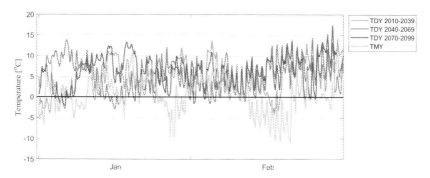

Fig. 3. Temperature variations for TDY NT, MT and LT files for Munich in comparison to the TMY file for Stuttgart based on historical climate measurements.

2.3   Building Performance Simulations

Numerical modelling through BPS was employed to analyse indoor thermal comfort, heating energy use and heating system sizing. Simulations were carried out through the EnergyPlus engine by means of Rhinoceros 3D-modelling [26] and Grasshopper visual scripting [27] using Ladybug Tools plug-ins [28]. The iteration of the workflow was carried out with the TT Toolbox plug-in [29]. The building masses were modelled according to the interior space volumes and glazing as well as frames were assigned parametrically following the window to wall ratios (WWR). Shading by external elements were modelled as opaque surfaces. Moveable shading, presented in some cases, was controlled by a variable schedule, which was enabled during nights when the exterior temperature is under 15 °C, during day and night when the exterior temperature exceeds 24 °C and when the solar radiation incidence was over 400 W/m^2. For all the cases, natural airflow by window ventilation was specified with a variable setpoint controlled by the interior temperature over 24 °C and exterior temperature between 21 °C to 26 °C. 30% of the window area was opened when the setpoint was met. Infiltration, as well as ventilation rates, were case sensitive according to the respective specifications, the same applies for heating setpoint temperatures. Lighting power density was specified as 7 W/m^2 for all cases, density of people was set to be equivalent of four people in 100 m^2 heated floor area. Equipment loads and schedules for occupancy, occupancy activity, lighting, equipment and ventilation were based on ASHRAE standard 90.1-2004 [30]. The indoor thermal comfort was assessed with the adaptive comfort model according to EN 15 251 [31] comfort band II for residential buildings and was plotted as XY-scatters. Validation of the building models and scripts was conducted by comparing energy use data from measurements and calculations received from building owners with simulation data obtained with TMY files.

3   Results and Discussion

3.1   Indoor Thermal Comfort

Results of the indoor comfort study are displayed in Fig. 4 where the presented values indicate the percentage of time exceeding the thermal comfort boundaries. As the measures are based on the adaptive comfort model, the percentage of hours in discomfort displays both time uncomfortable due to cold-stress and time uncomfortable due to heat-stress, relating to the operative temperature being below or above the comfort band temperatures for the entire year, regardless of the occupation. The diagrams in Fig. 4 show the results for TDY periods on the left side, ECY periods in the middle and EWY periods on the right. All measures are averages from zone-specific discomfort values; normalization has been applied according to the floor area. Only lettable spaces were considered in the indoor comfort study. As the extreme cold and warm year scenarios are unlikely to occur continuously, their results can only be used for comparison and drawing (pessimistic) boundaries. By comparing the results retrieved for TDY periods in the initial stage and in the retrofitted stage, an increase of discomfort can be seen over all cases and periods. Case 3 shows an intensive increase of thermal discomfort between initial and retrofitted building. For retrofitted cases 1 and 4, a quantitative increase of hours in discomfort can be seen throughout the TDY periods towards the end of the century. Case 2 shows a slight decrease, whereas for case 3, the outlier is the MT period with an increase in discomfort of 10%. The decrease in case 2 could be explained by the declining number of hours with cold-stress during wintertime, which outweighed the increase of hours in heat-stress. For both cases 1 and 2, the heating setpoints were at 21 °C, leading to a number of hours with cold-stress during wintertime. Especially for zones with large window surfaces this was seen, leading to the conclusion, that the thermostat setpoint will be adjusted by occupants in order to adapt to a comfortable indoor temperature.

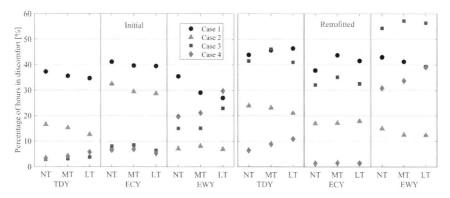

Fig. 4. Percentage of hours in discomfort according to EN 15 251 comfort band II. Non-retrofitted initial buildings on the left, retrofitted buildings on the right.

For the hours of heat-stress and overheating, the possibilities of adaptation were limited. The employed passive measures such as natural window ventilation and shading explored in cases 3 and 4 proved to be insufficient to reduce overheating issues, the setpoints for shading and natural ventilation should thus be reconsidered by the designers. The increase in overheating towards the end of the 21st century indicates that insulation and airtightness measures must be designed in regard to future climate. This entails that, in order to take informed decisions, future climate projections should be considered. Case 3, which showed extraordinarily high percentages of discomfort in the retrofitted building, demonstrates that the addition of 200 mm of thermal insulation outside the thermal mass leads to heat being trapped in the building. Results retrieved for case 4, which is exposed to the same climatic conditions as case 3, were significantly better. Here, the thermal insulation was realized on the inside of the building, resulting in the breaking of thermal mass from the occupied spaces. However, the cases do not allow for comparative conclusions to be drawn, as both airtightness and conductivity of the envelope constructions among others were different.

3.2   Energy Use

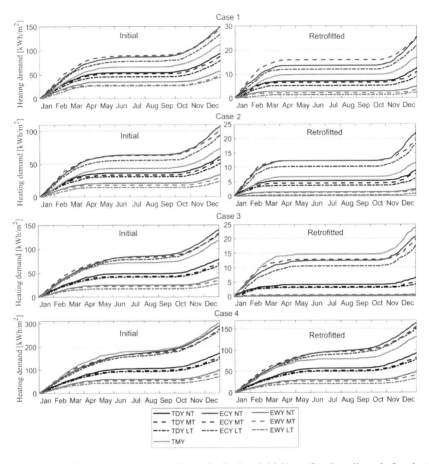

Fig. 5. Cumulative heating energy demand. Scale (initial/retrofitted) adjusted for better visibility.

Figure 5 represents the cumulative heating energy demand for the case studies in both non-retrofitted and retrofitted stages. All diagrams show a total of 10 data sets, obtained using TMY, TDY, ECY and EWY files for NT, MT, and LT periods. Steep periods represent the heating season while the flat periods reflect the times where there is no need for heating. For all four case studies, major reductions of overall heating demand can be seen from the initial to the retrofitted stage. Throughout the case studies, the energy demand for heating was obtained the highest for ECY periods and the lowest for the EWY periods due to the temperature averages which were the lowest for ECY periods and the highest for EWY periods. An exception can be seen in heating demands in cases 3 retrofitted stage and case 4 initial where higher heating energy demands for TMY files were obtained. This inconsistency can be explained by the temperature differences in wintertime as shown in 2.2. For all the periods within the future climate files, a decrease from the NT period towards the end of the 21st century was observed. The differences between the 30-year periods were 5% to 12% for the initial buildings and 5% to 25% for the retrofitted buildings. For the cases 1 and 2, the differences between the results identified for TMY files compared to TDY NT files are between 19% to 35% higher, revealing a larger effect than the one between the TDY periods. For cases 3 and 4, the differences are more significant and reach from 41% to 250%, showing the importance of taking future climate data into account.

3.3   System Sizing

The heating loads of the building were assessed with a statistical analysis seen in Fig. 6. The combination of data sets for TDY, ECY, EWY (called 'Triple') over the considered periods are investigated as suggested by Nik [23] for the initial and retrofitted stages. A comparison with the results for the respective TMY files is given, as the first boxplots for each case. Here it is important to notice that TMY data samples contain 8 760 values representing all hours of one year BPS results, whereas the 'Triple' data sets compose three times as much, representing all hourly data for TDY, ECY and EWY files summed up, thus containing 26 280 values. The reasoning for using 'Triple' data sets is to reflect weather extremes into the system design. Not considering weather extremes in system design but only results from TDY files would result in underestimating loads given during extreme cold or extreme warm periods. The study has shown that especially the cold extremes from BPS with ECY weather files should be considered when designing heating systems. This was in line with previous findings by Moazami et al. [20]. More detailed results showing individual data for TDY, ECY and EWY files can be found in [16]. The findings presented in this article are combining the results for typical and extreme conditions in order to provide adequate building and zone heating systems. This suggests a mild shift in thermal

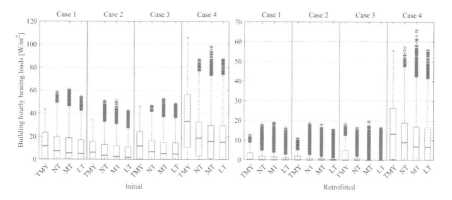

Fig. 6. Hourly heating loads 'triple' data sets and TMY data sets for cases 1–4 from left to right. Non-retrofitted initial building performance (left), retrofitted building performance on the right. TMY: typical metrological year (8 760 values); NT; MT; LT (All 26 280 values). Scale adjusted for better visibility. (Color figure online)

system and device dimensioning strategies which aims for higher resilience in future climate. The discussion is based on results within the whisker range of the boxplot diagrams, the median is shown with a red line within the blue box which is representing the interquartile range. Outliers, plotted as red lines in the upper extrema, are representing a very small fraction of occurrence, so that they were neglected.

Throughout all the cases, the highest heating peak loads were observed for TMY files. This was excepted for case 1 in the initial stage, which showed a slightly lower upper whisker range than throughout the future climate scenarios, which can be explained by not as low winter-temperatures or not as long-lasting cold periods as in the TDY files. Another possible explanation is the location difference between Copenhagen for the TMY file and Aarhus for future weather data. Furthermore, the heating peak loads were seen to be gradually decreasing for the future weather periods for the results obtained for NT to LT. For cases 1–3 in the retrofitted stage, the median is as low as 0 W/m^2, which indicates that for at least 50% of the time on an annual basis, there is no heating need. System design according to TMY files suggests that the heating systems would be over-dimensioned for weather predictions from 2010–2099. This finding emphasises the importance of using future climate projections in the decision-making process, especially to promote resilient design. The decision about dimensioning thermal systems has great influence on economic factors and over or under-dimensioning can have crucial importance on the thermal as well as economic performance of a project.

4   Conclusions

This work investigated the resilience of building retrofits for future climate and the need to consider climate change in the planning and implementation of retrofitting projects. As performance indicators, indoor thermal comfort, heating energy demand and heating system sizing were investigated by means of numerical simulation of the building performance. The impacts of future climate, its long-term changes, and extreme conditions were evaluated for four case studies located in Denmark and Germany. Future climate projections were incorporated in the simulations through weather files derived according to Nik´s method [23]. Each case was studied for the period of 2010–2099, divided into three 30-year periods. Each time period was represented by three weather files to account for the most typical conditions as well as extreme warm and extreme cold conditions.

The study shows that mitigating the energy use for heating was successful through the retrofits of all four case studies. In all cases, the heating demand decreases while approaching the end of the 21st century, indicating the effect of global warming. According to the results, accounting for future climate will provide a more realistic prognose for the energy performance of buildings will be during the usage stage. The severity and frequency of extreme heating loads decreased for the retrofitted buildings, and the heating periods became significantly shorter than the non-retrofitted buildings. Increased thermal discomfort due to heat-stress was seen for all cases when using future climate projections.

Based on the obtained results, it can be concluded that taking future climate projections into account when designing both passive and active measures would be beneficial to evaluate their effectivity, leading to performance-oriented design decisions. The presented method can thus become beneficial when designing according to bioclimatic design principles. Findings from this study suggest that future climate projections and extreme scenarios should be considered for decision-making in the system design to make buildings resilient to climate change and extreme events and to prevent economic and/or performance failures.

Designs that are based on considering future climate conditions enable us to retrofit buildings more sustainably and resource efficiently. Scalability of results was seen between the cases studied, but applicability of the conclusions to other cases in different locations, size and typologies must be studied further.

Acknowledgements. The authors express their gratitude to the companies who supplied data for the real case studies. Claus Fischer from Bau- und WohnungsVerein Stuttgart e.V. and Zeynel Palamutcu from DOMINIA A/S are gratefully acknowledged for their personal effort to provide the case study data. The authors thank the Fraunhofer IBP for supplying educational licenses for WUFI Pro. The authors also express their gratitude to COST Action CA16114 'RESTORE: Rethinking Sustainability towards a Regenerative Economy'.

References

1. Letcher, T.M.: Climate Change. Elsevier (2009). https://doi.org/10.1016/B978-0-444-53301-2.X0001-2
2. United Nations: Paris Agreement (2015). https://unfccc.int/sites/default/files/english_paris_agreement.pdf
3. United Nations Treaty Collection. https://treaties.un.org/pages/ViewDetails.aspx?src=TREATY&mtdsg_no=XXVII-7-d&chapter=27&clang=_en. Accessed 15 Feb 2019
4. IPCC 2013: Climate Change 2013: The Physical Science Basis. Contribution of Working Group I to the Fifth Assessment Report of the Intergovernmental Panel on Climate Change [Stocker, T.F., et al. (eds.)]. Cambridge University Press, Cambridge, United Kingdom and New York, NY, USA, 1535 pp. (2013)
5. Barriopedro, D., Fischer, E.M., Luterbacher, J., Trigo, R.M., García-Herrera, R.: The hot summer of 2010: redrawing the temperature record map of Europe. Science **332**, 220–224 (2011). https://doi.org/10.1126/science.1201224
6. Fischer, E.M., Schär, C.: Consistent geographical patterns of changes in high-impact European heatwaves. Nat. Geosci. **3**, 398–403 (2010). https://doi.org/10.1038/ngeo866
7. The United Nations: Climate Change. https://www.un.org/sustainabledevelopment/climate-change-2/. Accessed 15 Feb 2019
8. Mossin, N., Stilling, S., Chevalier Bøjstrup, T., Grupe Larsen, V., Lotz, M., Blegvad, A.: An Architecture Guide to the UN 17 Sustainable Development Goals. Institute of Architecture and Technology, KADK, Copenhagen (2018)
9. European Commission: Clean energy for all Europeans. https://ec.europa.eu/energy/en/topics/energy-strategy-and-energy-union/clean-energy-all-europeans. Accessed 01 May 2019
10. Deutsche Energie-Agentur GmbH (dena): Statistiken und Analysen zur Energieeffizienz im Gebäudebestand - dena-GEBÄUDEREPORT KOMPAKT2018. https://www.zukunft-haus.info/fileadmin/dena/Dokumente/Pdf/9254_Gebaeudereport_dena_kompakt_2018.pdf. Accessed 22 May 2019

11. Haase, M., Buvik, K., Dokka, T.H., Andresen, I.: Design guidelines for energy efficiency concepts in office buildings in Norway. Des. Principles Pract. Int. J. Annu. Rev. **5**, 667–694 (2011). https://doi.org/10.18848/1833-1874/CGP/v05i04/38156

12. Gustafsson, M.: Energy Efficient Renovation Strategies for Swedish and Other European Residential and Office Buildings (2017)

13. Erbach, G.: Understanding energy efficiency. EPRS | European Parliamentary Research Service. PE 568.361, 10

14. Gram-Hanssen, K.: Existing buildings – users, renovations and energy policy. Renew. Energy **61**, 136–140 (2014). https://doi.org/10.1016/j.renene.2013.05.004

15. Uihlein, A., Eder, P.: Policy options towards an energy efficient residential building stock in the EU-27. Energy Build. **42**, 791–798 (2010). https://doi.org/10.1016/j.enbuild.2009.11.016

16. Sivolova, J., Gremmelspacher, J.: Future Climate Resilience of Energy-Efficient Retrofit Projects in Central Europe (2019). http://lup.lub.lu.se/student-papers/record/8985440

17. Struck, C., de Wilde, P., Evers, J., Hensen, J., Plokker, W.: On Selecting Weather Data Sets to Estimate a Building Design's Robustness to Climate Variations (2009)

18. Finkelstein, J.M., Schafer, R.E.: Improved goodness-of-fit tests. Biometrika **58**, 641–645 (1971). https://doi.org/10.1093/biomet/58.3.641

19. Kershaw, T., Eames, M., Coley, D.: Comparison of multi-year and reference year building simulations. Build. Serv. Eng. Res. Technol. **31**, 357–369 (2010). https://doi.org/10.1177/0143624410374689

20. Moazami, A., Nik, V.M., Carlucci, S., Geving, S.: Impacts of future weather data typology on building energy performance – investigating long-term patterns of climate change and extreme weather conditions. Appl. Energy **238**, 696–720 (2019). https://doi.org/10.1016/j.apenergy.2019.01.085

21. Herrera, M., et al.: A review of current and future weather data for building simulation. Build. Serv. Eng. Res. Technol. **38**, 602–627 (2017). https://doi.org/10.1177/0143624417705937

22. Weather Data by Location | EnergyPlus. https://www.energyplus.net/weather-location/europe_wmo_region_6/DEU//DEU_Stuttgart.107380_IWEC. Accessed 13 May 2019

23. Nik, V.M.: Making energy simulation easier for future climate – synthesizing typical and extreme weather data sets out of regional climate models (RCMs). Appl. Energy **177**, 204–226 (2016). https://doi.org/10.1016/j.apenergy.2016.05.107

24. Nik, V.M.: Application of typical and extreme weather data sets in the hygrothermal simulation of building components for future climate – a case study for a wooden frame wall. Energy Build. **154**, 30–45 (2017). https://doi.org/10.1016/j.enbuild.2017.08.042

25. Swedish Meteorological and Hydrological Institute: Rossby Centre Regional Atmospheric Climate Model (RCA4) (2015)

26. Robert McNeel and Associates: Rhinoceros 3D. Robert McNeel & Associates, USA

27. Algorithmic Modelling for Rhino: Grasshopper (2014)

28. Sadeghipour Roudsari, M., Pak, M.: Ladybug: a parametric environmental plugin for grasshopper to help designers create an environmentally-conscious design (2013)

29. CORE Studio Thornton Tomasetti: TT Toolbox. Thornton Tomasetti, USA (2017)

30. American Society of Heating, Refrigerating and Air-Conditioning Engineers, Inc.: Energy Standard for Buildings Except Low-Rise Residential Buildings (2004)

31. European Committee For Standardization: Indoor environmental input parameters for design and assessment of energy performance of buildings addressing indoor air quality, thermal environment, lighting and acoustics (2007)

International Workshop
on Computational Geometry
and Applications (CGA 2020)

Theoretical Development and Validation of a New 3D Macrotexture Index Evaluated from Laser Based Profile Measurements

Mauro D'Apuzzo[1] , Azzurra Evangelisti[1] , Daniela Santilli[1(✉)],
and Vittorio Nicolosi[2]

[1] University of Cassino and Southern Lazio,
Via G. Di Biasio 43, 03043 Cassino, Italy
{dapuzzo,daniela.santilli}@unicas.it,
aevangelisti.ing@gmail.com
[2] University of Rome "Tor Vergata", via del Politecnico 1, 00133 Rome, Italy
nicolosi@uniroma2.it

Abstract. The Mean Profile Depth (MPD) is the synthetic index worldwide used to describe the macrotexture of road pavement surfaces. MPD is evaluated from two dimensional profiles captured by macrotexture laser-based devices, by means of ISO (International Organization for Standardization) or ASTM (American Society for Testing and Materials International) algorithms.

Several macrotexture laser-based measuring devices are present in the world market and, usually each one provides different MPD values also for the same road pavement. Furthermore the Standard Algorithm application produces MPD values affected by a wide variability. For these reasons the comparison of MPD values deriving from different laser-based macrotexture measuring devices is unsatisfying and it can produces deep misunderstanding on road macrotexture characterization.

In order to reduce the MPD variability and to improve the MPD values comparison a new algorithm to evaluate a more stable macrotexture synthetic index (Estimated Texture Depth) ETD, has been proposed. It has been developed with more than two hundred profiles belonging to virtual pavements and it has been validated on fifteen real road profiles.

The results seems to be promising: the Square Weight Estimated Texture Depth (ETDsw) provides more stable and reliable MPD values and in the same time, it improves the agreement between macrotexture values provided by different laser-based devices.

Keywords: Macrotexture · Profile analysis · Variability · Harmonization issue · Virtual pavement · Square Weight Estimated Texture Depth (ETD$_{SW}$)

1 Introduction

It is worldwide recognized that road pavement macrotexture can deeply influence the safety and environmental performances nowadays required to the road surface. Indeed the macrotexture is directly involved into tire-pavement interaction phenomenon, as it affects the frictional resistance, the water and contaminant drainage, the vehicle riding comfort and the rolling noise and resistance [1].

© Springer Nature Switzerland AG 2020
O. Gervasi et al. (Eds.): ICCSA 2020, LNCS 12251, pp. 367–382, 2020.
https://doi.org/10.1007/978-3-030-58808-3_27

Based on this premise, it is reasonable to believe that, between the road pavement performance requirement classes, the macrotexture is one of the most significant, together with the evaluation of road performance in terms of skid resistance along the lines of the approach adopted in evaluating tire grip class [2]. At the moment, in the international overview, several macrotexture measuring devices and different evaluation algorithms exist.

The volumetric technique is the traditional method to measure the macrotexture of road pavements and provides the Mean Texture Depth (MTD) index, which is a 3-dimensional (3D) measurement of the macrotexture [3]. Lately the use of the macrotexture laser-based devices was becoming predominant and consequently, different algorithms for the profiles analysis and empirical relations for volumetric conversions, have been developed.

According to the ISO 13473 [4] and the ASTM E1845 [5], from the 2D profiles analysis, the Mean Profile Depth (MPD) index can be evaluated and the indirect macrotexture estimation, expressed by means of the Estimated Texture Depth (ETD) index, can be provided by means of the Eq. (1) [5] and (2) [6] if the High Speed Laser (HSL) or Static (Circular Texture Meter, CTM) Devices, have been used, respectively:

$$ETD = 0.8\,MPD + 0.2 \tag{1}$$

$$ETD = 0.947\,MPD + 0.069 \tag{2}$$

Regrettably, the macrotexture values, evaluated on the 2D profiles, are characterized by a wide variability which can be caused by two orders of factors: 1) pavement physically related (as the heterogeneity of used pavement materials [7] and laying techniques) and 2) measurement related (as the devices' technology and operating conditions and the profile analysis algorithms).

This variability can affect the reliability measure of the macrotexture and, in turn, the agreement between ETD estimated on the same pavements, by different devices.

With the aim to investigate the macrotexture variability due to the measurements related factors, a theoretical approach, based on the analysis of more than 200 virtual pavements has been performed. The virtual pavements have simple and repetitive geometric layouts (of which all the features are completely known) and can be considered an extreme simplification of the real pavements but allow to exclude the macrotexture variability caused to pavement physical factors.

Based on this theoretical approach, a new macrotexture descriptor, namely the "square weighted estimated texture depth", ETDsw [8], has been proposed, evaluated and validated on a set of 15 real road pavements belonging to the Virginia Smart Road test track.

2   Theoretical Approach

In order to characterize the macrotexture's variability due to theoretical and technical related factors, as the pavement laying finishing, the laser sampling technology and the algorithms for profiles analysis, and to improve agreement between road macrotexture measurements with different laser-based devices, a theoretical investigation has been performed on virtual pavements. It consists 1) in the generation of a set of virtual pavements, designed on an heterogeneous set of real pavements; 2) in the profiles extraction by simulating different device technology sampling and 3) in the profile analysis, comparing different algorithms for macrotexture evaluation.

2.1   Design and Generation of Virtual Pavements

The outline of the virtual pavements has been inspired by the geometry of the concrete pavements finishing and some examples are reported in the Fig. 1a.

a)

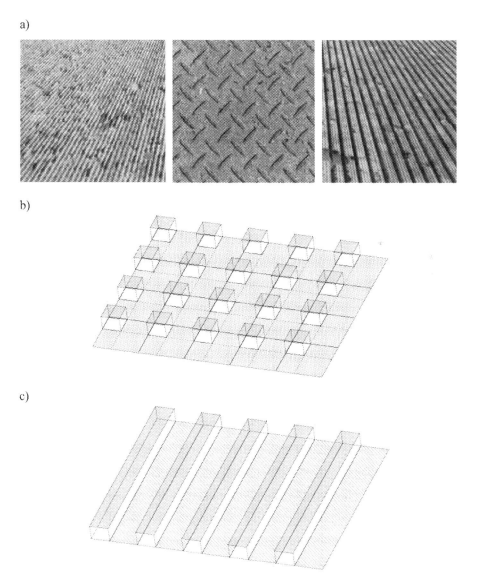

b)

c)

Fig. 1. a) Real concrete pavement finishing geometries; b) isotropic virtual pavement and c) orthotropic virtual pavement.

The generic layouts of virtual pavements have been reported in the Fig. 1b and Fig. 1c. As it is possible to see, they can be described as simple and repetitive geometric entities. If on an hand the layouts can represent a strong simplification, on the other, the most common and used texture synthetic indices, can be totally compute without approximation or uncertainty, as occurs for real road pavements.

According to the standards [4; 5; 6; 9] and to the layout reported in the Fig. 2, the definitions, summarized in the Table 1, can be expressed for both real and virtual pavements, respectively.

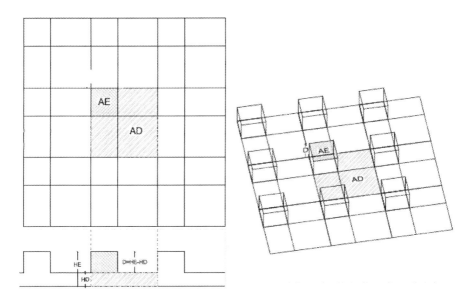

Fig. 2. Generic virtual pavement and nomenclature.

Table 1. Synthetic indices definitions for both real and virtual pavements.

Definition	Real pavement		Virtual pavement	
D = Depth between Peak height (HE) and Valley height (HD)			$HE - HD$	(3)
R = Ratio between Peak area (AE) and Valley area (AD)			AE/AD	(4)
MTD = Mean texture depth [4]	Equation (1) or Eq. (2)	*	$\frac{D}{R+1}$	** (5)
$\mathbf{R_q}$ = Root mean square [9]	$\sqrt{\frac{1}{l}\int_0^l z(x)^2 dx}$	*(6)	$\sqrt{\frac{R}{(R+1)^2} \cdot D^2}$	** (7)
$\mathbf{R_{SK}}$= Skewness [9]	$\frac{1}{l \cdot Rq^3}\int_0^l |z(x)| dx$	*(8)	$\frac{1-R}{\sqrt{R}}$	** (9)

* evaluated on 2D profiles; ** evaluated on 3D surface.
Where:
$z(x)$ = ordinate values of the pavement profile;
l = sampling length.

In order to obtain road pavements macrotexture values, as realistic as possible, a domain modelled on real road pavements has been conceived and used as pattern for the virtual pavement generation. Twelve real heterogeneous pavement mixes, belonging to the Virginia Smart Road track, have been selected: for each single pavement, a moving window, MS (Maximum Aggregate Size of the mixture) wide, has been advanced along the profile and the values of Depth (D) and Skewness (Rsk) [9], for each step, have been evaluated. Then, within a baseline, 100 mm long, the maximum, minimum and mean values, for both D and Rsk, have been computed. Their systematic combination allows to produce a more numerous and significant dataset, that has been used as foundation of the Rsk-D domain (see Fig. 3a).

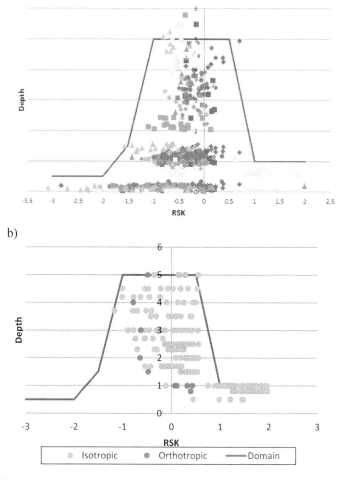

Fig. 3. R_{SK}-D domain from a) real pavements and b) virtual pavements.

Then, based on the domain (red line in the Fig. 3a), more than 200 virtual pavements have been generated (Fig. 3b). According to the real pavement types incidence, the 5% of the generated dataset follows the orthotropic layout (Fig. 1c) and the remaining part the isotropic layout (Fig. 1b).

As it is possible to see in the Fig. 3b, the generated dataset doesn't cover entirely the domain. It is due to the virtual pavements simplified geometry, which in the right-lower part of the domain corresponds to unrealistic layouts (too deep D or too wide AE or AD).

2.2   Extrapolation of Virtual Profiles

In order to simulate the virtual pavement profiles sampled by means of both Static and HSL measuring devices, two different algorithms has been implemented. Their features have been summarized in the Table 2.

Table 2. Features of both static and high speed laser-based measuring devices

Macrotexture laser-based measuring device	Static	High-speed
Length	892.2 mm along a circ.	Along the road pavement
Sample spacing	0.87 mm (approx.)	0.50 mm
Baseline length	111.525 mm	100 mm

In addition, the virtual profiles sampled by the HSL device, are characterized by a straight line, which follows the vehicle path. To better described the variability due to the sampling process, for each virtual pavement, virtual profiles every 5°, from 0° to 90°, have been extracted, as reported in the framework of Fig. 4.

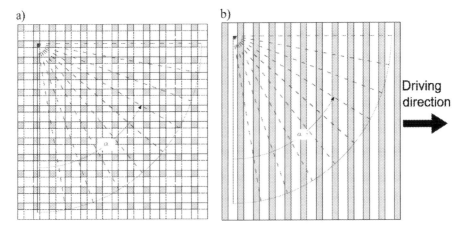

Fig. 4. High-speed laser sampling angles on both a) isotropic and b) orthotropic pavements.

Quite the opposite, the virtual profiles sampled by the Static device, are characterized by a circular line (see Fig. 5), for this reason, the analysis with different angles is not required.

a)   b)

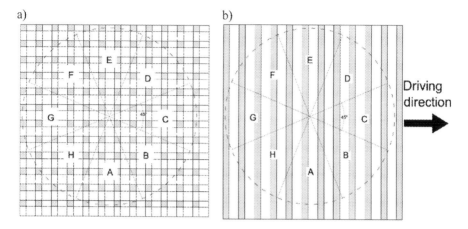

Fig. 5. Static laser sampling line on both a) isotropic and b) orthotropic pavements.

As an example, for a given isotropic pavement, virtual profiles, simulating both HSL and Static devices sampling technique, have been reported in the Fig. 6 and in the Fig. 7, respectively.

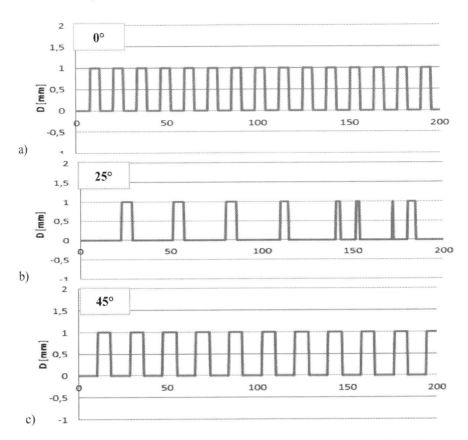

Fig. 6. Example of virtual profiles sampled by high speed laser device with a) 0°; b) 25° and c) 45° incident angle.

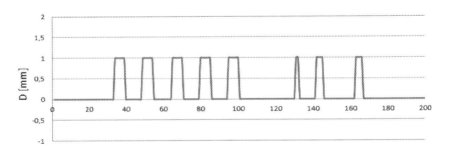

Fig. 7. Example of virtual profile sampled by static device.

2.3   Algorithms for Virtual Profiles Analysis

As far as macrotexture evaluation on 2D profiles has been concerned, according to the standards [4; 5; 6] and to the more widespread custom software, different algorithms have been proposed and their features have been summarized in the following Table 3:

Table 3. Algorithm features

Device	Static	HSL
Algorithm A: standard	A1 [6]	A2 [5]
Sample spacing	0.87 mm	0.5 mm
Baseline (BL) length	111 mm (approx.)	100 mm
Evaluation length for linear regression	55.5 mm	100 mm
Index evaluation frequency	1/111 mm	1/100 mm
ETD evaluation	Equation (2)	Equation (1)
Algorithm B: windowing – half BL	B1	B2
Sample spacing	0.87 mm	0.5 mm
BL length	111 mm (approx.)	100 mm
Evaluation length for linear regression	55.5 mm	50 mm
Index evaluation frequency	1/0.87 mm	1/0.5 mm
ETD evaluation	Equation (2)	Equation (1)
Algorithm C: windowing – whole BL	C1	C2
Sample spacing	0.87 mm	0.5 mm
BL length	111 mm (approx.)	100 mm
Evaluation length for linear regression	111 mm	100 mm
Index evaluation frequency	1/0.87 mm	1/0.5 mm
ETD evaluation	Equation (2)	Equation (1)

For each virtual pavement, by means of the algorithms (Table 3) and the Eq. 5, the macrotexture values have been computed and in the Fig. 8, the comparison of the macrotexture values, evaluated for both Static and HSL devices, has been shown.

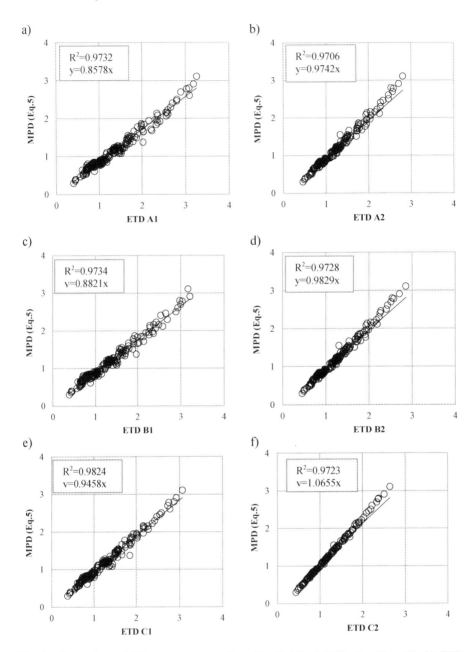

Fig. 8. Comparison of 3D macrotexture values (Eq. 5) VS a) ETD algorithm A1; b) ETD algorithm A2; c) ETD algorithm B1; d) ETD algorithm B2; e) ETD algorithm C1; f) ETD algorithm C2.

In addition, the comparison in terms of Concordance Correlation Coefficient (ρc) [10], Pearson's Coefficient (P), Mean Error and Coefficient of determination (R^2) and angular coefficient of the linear regression (a, b = 0), has been expressed and summarized in the following Table 4:

Table 4. Agreement descriptive coefficient for the Algorithms comparison

Comparison	ρc	Pearson	MeanError	R^2	a
Figure 8a	0,923	0,982	0,148	0,973	0,858
Figure 8b	0,967	0,994	0,063	0,971	0,974
Figure 8c	0,942	0,982	0,120	0,973	0,882
Figure 8d	0,979	0,989	0,074	0,973	0,983
Figure 8e	0,978	0,986	0,066	0,982	0,946
Figure 8f	0,966	0,994	0,062	0,972	1,065

Although the agreement descriptive coefficients seem rather satisfactory (Table 4 and Fig. 8), due to the simple and regular layout of the virtual pavement and to the absence of laser reading errors, it looks evident the necessity to investigate the effectiveness of non-conventional macrotexture synthetic index, to improve the agreement.

3 Proposed 3D Macrotexture Index from 2D Profile Data

Previous efforts have been spent, to evaluate a more stable and effective macrotexture index: the square weighted ETD is based on a sectioning process to identify homogeneous sections in terms of macrotexture and on attributing an only weighted index able to better describe the macrotexture's variability [8]. The theoretical concept of the square weight ETD, expresses by means of the Eq. (10), has been summarized in the Fig. 9:

$$ETD_{SW} = \frac{\sum_{i=1}^{n} MPD_i \cdot L_i^2}{\sum_{i=1}^{n} L_i^2} \tag{10}$$

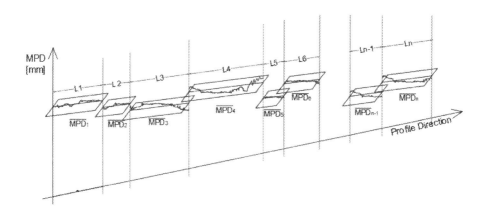

Fig. 9. Graphical representation of square weight ETD evaluation process.

Observing the Fig. 9, it is possible to deduce that, due to the fact that the ETD_{SW} converts MPD values, evaluated on 2D profiles, into areal volumetric macrotexture estimate, the linear transformations (Eq. (1) and Eq. (2)) suggested to transform MPD into ETD are unnecessary.

The algorithm used to evaluate the ETDsw index is summarized in the Table 5:

Table 5. ETDsw algorithm features

Device	Static	HSL
Algorithm D	D1	D2
Sample spacing	0.87 mm	0.5 mm
Baseline length	111 mm (approx.)	100 mm
Evaluation length for linear regression	111 mm	100 mm
Index evaluation frequency	1/0.87 mm	1/0.5 mm
ETD evaluation	Equation (10)	Equation (10)

Then, in the Fig. 10a) and in the Fig. 10b), the comparisons between the effective macrotexture (evaluated by Eq. 5) and the synthetic macrotexture index ETD_{SW}, have been reported, respectively.

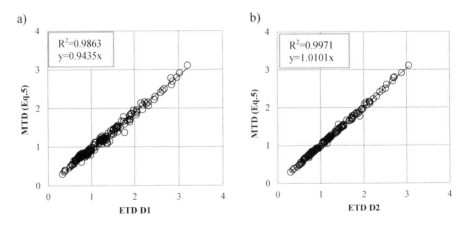

Fig. 10. Comparison of 3D macrotexture values (Eq. 5) VS a) ETD algorithm D1 and b) ETD algorithm D2.

Also in this case the agreement descriptive coefficients in terms of Concordance Correlation Coefficient (ρc), Pearson's Coefficient (P), Mean Error and Coefficient of determination (R^2) and angular coefficient of the linear regression (a, b = 0), have been evaluated and summarized in the following Table 6.

Table 6. Agreement descriptive coefficient for the ETD$_{SW}$ algorithms comparison.

Comparison	ρc	Pearson	MeanError	R^2	a
Figure 10a	0,979	0,989	0,058	0,986	0,943
Figure 10b	0,993	0,994	0,024	0,997	1,01

As far as the harmonization issue is concerned, the comparison between the Static and HSL Devices, for both A and D algorithms, has been performed and the results have been summarized in the following Table 7 and Fig. 11:

Table 7. Descriptive coefficient for the devices comparison

Comparison	ρc	Pearson	MeanError	R^2	a
Figure 11a	0,919	0,986	0,106	0,969	1,131
Figure 11b	0,976	0,988	0,064	0,985	1,067

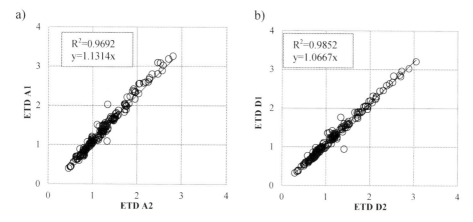

Fig. 11. Comparison between ETD values obtained by a) algorithm A1 VS algorithm A2 and b) algorithm D1 VS algorithm D2.

As it is possible to see, in terms of both, macrotexture evaluation and harmonization issues, the ETD$_{SW}$ provides better results than all the other tested indices and algorithms.

Finally, in order to validate the approach, previously presented, the application of the method to a set of real pavements, has been performed.

4 Validation on Real Pavement

The validation has been performed on a set of 15 real pavements, belonging to the test-track owned by the Virginia Tech Transportation Institute. The entire set has been selected to guarantee a satisfactory heterogeneity in terms of pavement types (asphalt and concrete), used materials and finishing lays.

Both Static and HSL devices, previously introduced (see Table 2 for their main features) have been used to collect 2D profiles.

Two example of real pavement profiles, collected with both Static and HSL devices, have been reported in the Fig. 12a and Fig. 12b, respectively.

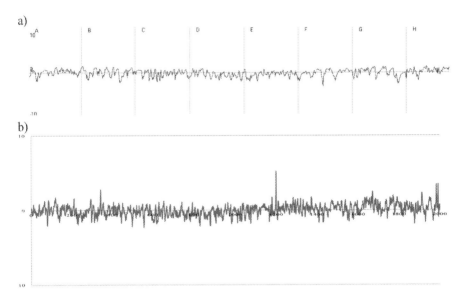

Fig. 12. a) Profile of a generic mix asphalt pavement sampled by a) Static and b) HSL devices.

At this stage, it is important to remember that, as far as real profiles have been concerned, issues related to the invalid laser sensor reading, cannot be neglect and it is highlighted in the Fig. 12b where spikes and drop-outs are present. For this reason, a preliminary filtering process, aim to detect and remove invalid sensor readings is requested [8, 11, 12].

a)   b)

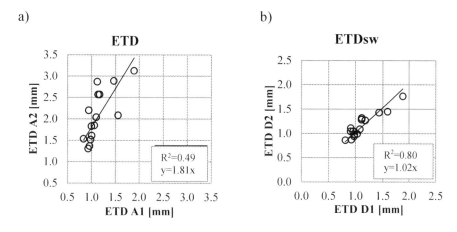

Fig. 13. Comparison between ETD values obtained by a) algorithm A1 VS algorithm A2 and b) algorithm D1 VS algorithm D2.

Table 8. Descriptive coefficient for the devices comparison on real pavements

	ρc	Pearson	MeanError	R^2	a
Figure 13a	0,17	0,67	0,83	0,49	1,81
Figure 13b	0,85	0,865	0,078	0,80	1,02

In the Fig. 13 and in the Table 8 the results of the comparison between the macrotexture values obtained by Static and HSL devices, have been summarized. In particular it is possible to see that the application of the ETD_{SW} (Fig. 13b) improve significantly the agreement respect to the conventional approach (Fig. 13a).

5   Conclusion

In this paper an investigation aimed at improving the reliability of macrotexture measurements has been performed. In order to address the study on the characterization of the macrotexture source of uncertainty only induced by both different operating conditions of measuring devices and different evaluation procedures to compute synthetic indices (pavement physical factors such as mix grading and volumetric properties, energy compaction, finishing laying techniques, have been intentionally excluded), a set of more than 200 virtual pavements have been generated. Two commercial laser-based macrotexture measuring "devices have been selected and used for the analysis of the profiles". Several algorithms have been compared and the ETDsw index has been finally validated on a set of 15 real road pavements, belonging to the Virginia Smart Road facility.

Although further investigation are needed, preliminary results seem to highlight that the ETDsw index reduces the macrotexture variability and provides macrotexture estimate closer to the volumetric macrotexture value of real pavements. In addition, an improvement in term of agreement between different devices, has been demonstrated.

Acknowledgement. The authors wish to acknowledge the Virginia Polytechnic Institute and State University, Professor Gerardo Flintsch, director of the Center for Sustainable Transportation Infrastructure, and the staff of the Virginia Tech Transportation Institute for allowing us to use the data collected on the Virginia Smart Road.

References

1. Mahone, D.C., Sherwood, W.C.: The Effect of Aggregate Type and Mix Design on the Wet Skid Resistance of Bituminous Pavement: Recommendations for Virginia's Wet Accident Reduction Program. Report n°FHWNVA-96-RIO (1995)
2. D'Apuzzo, M., Evangelisti, A., Nicolosi, V.: An exploratory step for a general unified approach to labelling of road surface and tire wet friction. Elsevier Acc. Anal. Prevent. **138**, 105462 (2020). https://doi.org/10.1016/j.aap.2020.105462
3. ASTM E965. Standard Test Method for Measuring Pavement Macrotexture Depth Using a Volumetric Technique. American Society for Testing and Materials (2006)
4. ISO 13473-1. Characterization of pavement texture by use of surface profiles - Part 1: Determination of mean profile depth. International Organization for Standardization (2019)
5. ASTM E1845. Standard Practice for Calculating Pavement Macrotexture Mean Profile Depth. American Society for Testing and Materials (2015)
6. ASTM E2157. Standard Test Method for Measuring Pavement Macrotexture Properties Using the Circular Track Meter. American Society for Testing and Materials (2019)
7. D'Apuzzo, M., Evangelisti, A., Nicolosi, V.: Preliminary findings for a prediction model of road surface macrotexture. Procedia Soc. Behav. Sci. **53**, 1110–1119 (2012). https://doi.org/10.1016/j.sbspro.2012.09.960. ISSN: 1877–0428
8. D'Apuzzo, M., et al.: Evaluation of variability of macrotexture measurement with different laser-based devices. In: Airfield and Highway Pavements 2015: Innovative and Cost-Effective Pavements for a Sustainable Future, pp. 294–305 (2014). https://doi.org/10.1061/9780784479216.027. ISBN (PDF): 9780784479216
9. ISO 4287, Geometrical Product Specifications (GPS) – Surface texture: Profile method – Terms, definitions and surface texture parameters. International Organization for Standardization (ISO) (1997)
10. Lawrence, L.: A concordance correlation coefficient to evaluate reproducibility. Biometrics (International Biometric Society) **45**(1), 255–268 (1989). https://doi.org/10.2307/2532051. JSTOR 2532051. PMID 2720055
11. Katicha, S., Mogrovejo, D., Flintsch, G., de León Izeppi, E.: Latest development in the processing of pavement macrotexture measurements of high speed laser devices. In: 9th International Conference on Managing Pavement Assets (2014)
12. Losa, M., Leandri, P.: The reliability of tests and data processing procedures for pavement macrotexture evaluation. Int. J. Pavement Eng. **12**(1), 59–73 (2011). Taylor and Francis

Calculation of the Differential Geometry Properties of Implicit Parametric Surfaces Intersection

Judith Keren Jiménez-Vilcherrez[1,2]([✉]) [iD]

[1] Universidad Nacional de Piura, Piura, Peru
jjimenezv@unp.edu.pe
[2] Universidad Tecnológica del Perú, Piura, Peru
C19863@utp.edu.pe

Abstract. Generally, to calculate the Frenet-Serret apparatus of a curve, it is necessary to have a parameterization of it; but when it is difficult to obtain a parameterization of the curve, as is the case of the curves obtained by the intersection of two implicit parametric surfaces, it is necessary to develop new methods that make it possible to know the geometric properties of said curve. This paper describes a new Mathematica package, Frenet, with the objective of calculating the properties of the differential geometry of a curve obtained by the intersection of two implicit parametric surfaces. The presented package allows us to visualize the Frenet-Serret mobile trihedron, to know the curvature and torsion at a given point of the curve obtained by the intersection of two implicit parametric surfaces. Package performance is discussed using several illustrative examples. Provide the user with an important tool for visualization and teaching.

Keywords: Frenet-Serret apparatus · Intersection of surfaces · Geometric-differential properties of curves

1 Introduction

In geometry, the study of the differential geometry properties of curves is essential. Generally, to calculate the Frenet-Serret apparatus of a curve, it is necessary to have a parameterization of it [2–4]; but when it is difficult to obtain a parameterization of the curve, as is the case of the curves obtained by the intersection of two implicit parametric surfaces, it is necessary to develop new methods that allow knowing the geometric properties of the curve; methods that have been studied in [1] but this calculation becomes a very cumbersome task due to the amount of mathematical operations that must be carried out for this reason this paper describes a new Mathematica package, Frenet, with objective of calculating the properties of the differential geometry of a curve obtained by the intersection of two implicit parametric surfaces. Enabling the calculation of the Frenet-Serret apparatus of a curve without having a parameterization of it. The outputs obtained

ⓒ Springer Nature Switzerland AG 2020
O. Gervasi et al. (Eds.): ICCSA 2020, LNCS 12251, pp. 383–394, 2020.
https://doi.org/10.1007/978-3-030-58808-3_28

are consistent with Mathematica notation and results. Package performance is discussed using several illustrative examples. The presented package allows us to visualize the Frenet-Serret mobile trihedron, to know the curvature and torsion at a given point of the curve obtained by the intersection of two implicit parametric surfaces, providing the user with a very useful tool for teaching and visualization. The paper is organized as follows: In Sect. 2, the formulas necessary to calculate the properties of the differential geometry of a curve obtained by the intersection of two implicit parametric surfaces are reviewed. In the Sect. 3 introduces the new Mathematica package, Frenet, and describes the implemented commands. Package performance is also analyzed using three explanatory examples. We finish Sect. 4 with the main conclusions of this paper.

2   Mathematical Preliminaries

2.1   Curves

Let $\beta : I \subset \mathbb{R} \to \mathbb{R}^3$ an arbitrary curve with arc-length parametrization, then from the elementary differential geometry, we have

$$\beta'(s) = t$$

$$\beta''(s) = k = \kappa n$$

where t is the unit tangent vector, n is the unit principal normal vector and k is the curvature vector. The curvature is given by

$$k = \sqrt{\kappa \cdot \kappa} = \sqrt{\beta'' \cdot \beta''}$$

the binormal unit vector b is defined by $b = t \times n$. The vectors $\{t, n, b\}$ are called collectively the Frenet- Serret frame, the Frenet–Serret formulas along β are given by

$$t'(s) = \kappa n$$

$$n'(s) = -\kappa t + \tau b$$

$$b'(s) = -\tau n$$

where τ is the torsion of the curve β.

2.2   Implicit Surface Representation

A surface often arises as the locus of a point P which satisfies some restriction, as a consequence of which the coordinates x, y, z of P satisfy a relation of the form

This is called the implicit or constraint equation of the surface, assume that $f(x, y, z) = 0$ is a regular implicit surfaces. In other words $\nabla f \neq 0$, where $\nabla f = (f_1, f_2, f_3)$ is the gradient vector of the surface f, where $f_1 = \frac{\partial f}{\partial x}$, $f_2 = \frac{\partial f}{\partial y}$ and $f_3 = \frac{\partial f}{\partial z}$ denote to partial derivatives of the surface f, then the unit surface normal vector field of the surface f is given by

$$n = \frac{\nabla f}{\|\nabla f\|}$$

2.3 Parametric Surface Representation

A parametric surface in the Euclidean $\mathbb{R}^3$ is defined by a parametric equation with two parameters. Parametric representation is the most general way to specify a surface. In general, if we take the real parameters u and v, then the surfaces can be defined by the vector-value function, $r = r(u, v)$, where $r(u, v) = r(r_1(u, v), r_2(u, v), r_3(u, v))$ and $u_1 \leq u \leq u_2$, $v_1 \leq v \leq v_2$.

2.4 Properties of Parametric Surfaces

Given a parametric surface of the form $r(u, v) = r(r_1(u, v), r_2(u, v), r_3(u, v))$, supposing that $r(u, v)$ is a regular parametric surface. That is to say $r_u \times r_v \neq 0$, where $r_u = \frac{\partial r}{\partial u}$ and $r_v = \frac{\partial r}{\partial v}$ denote to partial derivatives of the surface r. The unit vector normal n at any point on a parametric surfaces is obtained as

$$n = \frac{r_u \times r_v}{|r_u \times r_v|}$$

the first fundamental form given a parametric surface $r(u, v)$, we define the first fundamental form coefficients $E = r_u \cdot r_u$, $F = r_u \cdot r_v$ and $G = r_v \cdot r_v$, then the first fundamental form I of the surface is the quadratic expression defined as,

$$I = E du^2 + 2F du dv + G dv^2$$

the second fundamental form given a parametric surface $r(u, v)$ and its normal vector n, we define the second fundamental form coefficients $L = r_{uu} \cdot n$, $M = r_{uv} \cdot n$ and $N = r_{vv} \cdot n$ then the second fundamental form of the surface is the quadratic expression defined as

$$II = L du^2 + 2M du dv + N dv^2$$

3 Transversal Intersection of Parametric–Implicit Surfaces

Let $r = r(u, v)$ and $f(x, y, z) = 0$ two surfaces regular with unit normal vectors given by

$$n_1 = \frac{r_u \times r_v}{|r_u \times r_v|}$$

$$n_2 = \frac{\nabla f}{\|\nabla f\|}$$

The intersection curve of these surfaces can be seen as a curve of both surfaces as

$$\beta(s) = (x(s), y(s), z(s)), f(x, y, z) = 0$$

$$\beta(s) = r(u(s), v(s)), c_1 < u < c_2, c_3 < v < c_4$$

then you have

$$x(s) = r_1(u(s), v(s))$$
$$y(s) = r_2(u(s), v(s))$$
$$z(s) = r_3(u(s), v(s))$$

where

$$r(u(s), v(s)) = (r_1(u(s), v(s)), r_2(u(s), v(s)), r_3(u(s), v(s)))$$

The surface f can be expressed as

$$h(u, v) = f(r_1, r_2, r_3) = 0$$

Thus the intersection curve is given by

$$\beta(s) = r(u(s), v(s)), h(u, v) = 0, c_1 < u < c_2, c_3 < v < c_4$$

We can easily find the first derivative of the intersection curve

$$\beta(s) = r_u u' + r_v v' \tag{1}$$

the tangent vector of the transversal intersection curve $\beta(s)$ lies on the tangent planes of both surfaces. Therefore it can be obtained as the cross product of the unit normal vectors of the two surfaces at P as

$$t = \frac{n_1 \times n_2}{|n_1 \times n_2|} \tag{2}$$

the surface curvature vector $r(u, v)$ is given by

$$\kappa_1 = L(u')^2 + 2Mu'v' + N(v')^2 \tag{3}$$

since we know the unit tangent vector of the intersection curve from Eq. (2), we can find u' and v' by taking the dot product on both hand sides of Eq. (1) with r_u and r_v, which leads to a linear system

$$Eu' + Fv' = r_u.t$$

$$Fu' + Gv' = r_v.t$$

where E, F, G are the first fundamental form coefficients.
Thus

$$u' = \frac{(r_u \cdot t)G - (r_v \cdot t)F}{EG - F^2}$$

$$v' = \frac{(r_v \cdot t)E - (r_u \cdot t)F}{EG - F^2}$$

similarly

$$\beta'(s) = t = (x'(s), y'(s), z'(s))$$

$$\beta''(s) = k = \kappa n = (x''(s), y''(s), z''(s))$$

where x', y', z' are the three components of t given by Eq. (2) and x'', y'', z'' are the three components of k.

We can calculate the normal curvature of the implicit surface using the equation

$$\kappa_2 = -\frac{f_{xx}(x')^2 + f_{yy}(y')^2 + f_{zz}(z')^2 + 2\left(f_{xy}x'y' + f_{yz}y'z' + f_{xz}x'z'\right)}{\sqrt{f_x^2 + f_y^2 + f_z^2}}$$

consequently, the curvature vector of the intersection curve $\beta(s)$ at P can be calculated as follows:

$$k = \frac{\kappa_1 - \kappa_2 \cos\theta}{\sin^2\theta} n_1 + \frac{\kappa_2 - \kappa_1 \cos\theta}{\sin^2\theta} n_2 \tag{4}$$

the curvature of the intersection curve $\beta(s)$ at P can be calculated using Eq. (4) as follows

$$\kappa = \sqrt{k \cdot k} = \frac{1}{|\sin\theta|}\sqrt{\kappa_1^2 + \kappa_2^2 - 2\kappa_1^2\kappa_2^2 \cos\theta}$$

the unit normal vector and unit binormal vector of the intersection curve $\beta(s)$ given as

$$n = \frac{k}{\kappa}, b = t \times n$$

the torsion of the intersection curve $\beta(s)$ is obtained by

$$\tau = \frac{1}{\kappa \sin^2\theta}\left\{[\lambda_1 - \lambda_2 \cos\theta](b \cdot n_1) + [\lambda_2 - \lambda_1 \cos\theta](b \cdot n_2)\right\}$$

where

$$\lambda_1 = 3\left[Lu'u'' + M(u''v' + u'v'') + Nv'v''\right] + r_{uuu} \cdot n(u')^3 + 3r_{uuv}n(u')^2v'2$$
$$+ 3r_{uvv} \cdot nu'(v')^2 + r_{vvv} \cdot n(v')^3$$

the values of u'', v'' are obtained by solving the following system of equations

$$Eu'' + Fv'' = k \cdot r_u - \frac{E_u}{2}(u')^2 - E_v u'v' - \left(F_v - \frac{G_u}{2}\right)(v')^2$$

$$Fu'' + Gv'' = k \cdot r_v - \left(F_u - \frac{E_v}{2}\right)(u')^2 - G_u u'v' - \frac{G_v}{2}(v')^2$$

similarly

$$\lambda_2 = -\frac{F_1 + F_2 + F_3}{\sqrt{f_x^2 + f_y^2 + f_z^2}}$$

the values of F_1, F_2, F_3 can be calculated by

$$F_1 = f_{xxx}(x')^3 + f_{yyy}(y')^3 + f_{zzz}(z')^3$$

$$F_2 = 3\left[f_{xxy}(x')^2 y' + f_{xxz}(x')^2 z' + f_{xyy}x'(y')^2 + f_{yyz}(y')^2 z' \right.$$
$$\left. + f_{xzz}x'(z')^2 f_{yzz}y'(z')^2 + 2f_{xyz}x'y'z' \right]$$

$$F_3 = 3\left[f_{xx}x'x'' + f_{yy}y'y'' + f_{zz}z'z'' + f_{xy}\left(x''y' + x'y''\right) + f_{yz}\left(y''z' + y'z''\right) \right.$$
$$\left. + f_{xz}\left(x''z' + x'z''\right)\right]$$

4   The Package Frenet: Some Illustrative Examples

In this section we describe the use of the Frenet package, the package works with Mathematica v.11.2 and later versions. Various examples will be used for an introduction to the specific features of the package. First of all, we load the package:

$$<< Frenet.m$$

4.1   Example 01

The first example is given by the curve obtained by the intersection of the implicit surface $f(x, y, z) = z - y^2 - 2 = 0$ and the parametric surface $r(u, v) = (u, uv, v)$.

With the *Frenet* command it is possible to calculate the equations of the Frenet-Serret apparatus at a generic point on the curve.

$$Frenet[z - y^2 - 2, u, u*v, v, x, y, z, u, v]//Simplify$$

Frenet returns the equations of the tangent vector, normal vector, binormal vector as well as curvature and torsion.

$$\left\{\left\{ \frac{2uy-1}{\sqrt{u^2+v^2+1}\sqrt{4y^2+1}\sqrt{\frac{(4y^2+1)v^2+(1-2uy)^2}{(u^2+v^2+1)(4y^2+1)}}}, -\frac{v}{\sqrt{u^2+v^2+1}\sqrt{4y^2+1}\sqrt{\frac{(4y^2+1)v^2+(1-2uy)^2}{(u^2+v^2+1)(4y^2+1)}}}, \right.\right.$$

$$\left.\left. -\frac{2vy}{\sqrt{u^2+v^2+1}\sqrt{4y^2+1}\sqrt{\frac{(4y^2+1)v^2+(1-2uy)^2}{(u^2+v^2+1)(4y^2+1)}}} \right\},\right.$$

$$\left\{ -\frac{v^2\left(2y\left(4y^2+v+1\right)+u\left(v-4\left(4y^4+y^2\right)\right)\right)}{\left((4y^2+1)v^2+(1-2uy)^2\right)^2\sqrt{\frac{v^2\left(v^4+(u^2+1)v^2-4y(u+2y)(2uy-1)v+4y^2(1-2uy)^2(4y^2+1)\right)}{\left((4y^2+1)v^2+(1-2uy)^2\right)^3}}}, \right.$$

$$\frac{v\left(8u^2y^3-8uy^2-2\left(v^3+u^2v-1\right)y+uv\right)}{\left((4y^2+1)v^2+(1-2uy)^2\right)^2\sqrt{\frac{v^2\left(v^4+(u^2+1)v^2-4y(u+2y)(2uy-1)v+4y^2(1-2uy)^2(4y^2+1)\right)}{\left((4y^2+1)v^2+(1-2uy)^2\right)^3}}},$$

$$\left. \frac{v\left(v^3-2uyv+v+4y^2(1-2uy)^2\right)}{\left((4y^2+1)v^2+(1-2uy)^2\right)^2\sqrt{\frac{v^2\left(v^4+(u^2+1)v^2-4y(u+2y)(2uy-1)v+4y^2(1-2uy)^2(4y^2+1)\right)}{\left((4y^2+1)v^2+(1-2uy)^2\right)^3}}} \right\},$$

$$\left\{-\frac{v^3}{(u^2+v^2+1)^{3/2}(4y^2+1)^{3/2}\left(\frac{(4y^2+1)v^2+(1-2uy)^2}{(u^2+v^2+1)(4y^2+1)}\right)^{3/2}\sqrt{\frac{v^2\left(v^4+(u^2+1)v^2-4y(u+2y)(2uy-1)v+4y^2(1-2uy)^2(4y^2+1)\right)}{((4y^2+1)v^2+(1-2uy)^2)^3}}},\right.$$

$$\frac{v\left(4(1-2uy)y^2+v\right)}{(u^2+v^2+1)^{3/2}(4y^2+1)^{3/2}\left(\frac{(4y^2+1)v^2+(1-2uy)^2}{(u^2+v^2+1)(4y^2+1)}\right)^{3/2}\sqrt{\frac{v^2\left(v^4+(u^2+1)v^2-4y(u+2y)(2uy-1)v+4y^2(1-2uy)^2(4y^2+1)\right)}{((4y^2+1)v^2+(1-2uy)^2)^3}}},$$

$$\left.-\frac{v\left(2y+u\left(v-4y^2\right)\right)}{(u^2+v^2+1)^{3/2}(4y^2+1)^{3/2}\left(\frac{(4y^2+1)v^2+(1-2uy)^2}{(u^2+v^2+1)(4y^2+1)}\right)^{3/2}\sqrt{\frac{v^2\left(v^4+(u^2+1)v^2-4y(u+2y)(2uy-1)v+4y^2(1-2uy)^2(4y^2+1)\right)}{((4y^2+1)v^2+(1-2uy)^2)^3}}}\right\},$$

$$2\sqrt{\frac{v^2\left(v^4+(u^2+1)v^2-4y(u+2y)(2uy-1)v+4y^2(1-2uy)^2(4y^2+1)\right)}{((4y^2+1)v^2+(1-2uy)^2)^3}},$$

$$\left(3v\left(-2v(u^2+v^2+1)^{3/2}(u(u+v)+1)y^2(4y^2+1)^{3/2}(2v)\right.\right.$$

$$\frac{2uy-1}{\sqrt{u^2+v^2+1}\sqrt{4y^2+1}\sqrt{\frac{(4y^2+1)v^2+(1-2uy)^2}{(u^2+v^2+1)(4y^2+1)}}}\left(\frac{(4y^2+1)v^2+(1-2uy)^2}{(u^2+v^2+1)(4y^2+1)}\right)^{3/2}$$

$$-4\left(u^2+1\right)y^2(2uy-1)^3+4uv^6y^2(4uy-1)+v^5(4uy-1)\left(4\left(u^2+1\right)y^2-2uy+1\right)-2v^2y(2uy-1)\left(-4yu^3+2u^2+4y^2u+u+8y^3+2y\right)+2uv^4y\left(32u^2\,y^4-24uy^3+4\left(2u\left(u^2+u+1\right)+1\right)y^2-2(3u(u+1)+1)y+2u+1\right)+v^3\left(-32u^2\left(2u^2+1\right)y^5\right.$$
$$+16u\left(5u^2+1\right)y^4\quad+\quad 8u\left(u\left(2u^3+4u-5\right)+2\right)y^3\quad-\quad 4\left(u\left(5u^3+7u-2\right)+1\right)y^2\quad+$$
$$2u\left(5u^2+7\right)y-u^2-2\big)+v(1-2uy)^2\left(u\left(-16uy^3\right.\right.$$
$$\left.\left.+8y^2+4\left(u^2+1\right)y-u\right)-1\right)\big)\big\}\left(v^2+1\right)(u(u\quad+\quad v)\quad+\quad 1)\left(4y^2+1\right)v^2\quad+\quad(1-2uy)^2$$
$$\left(v^4+(u^2+1)v^2-4y(u+2y)(2uy-1)v+4y^2(1-2uy)^2\left(4y^2+1\right)\right\}\big\}$$

The following sentences allow us to obtain the Frenet-Serret apparatus of the curve at one point $(0,0,2)$.

$$\{t,b,n,k,\tau\} = \text{Frenet}\left[\left(-y^2+z-2,\{u,uv,v\},\{x,y,z\},\ \{u,v\},\text{ParVals}\rightarrow\{0,0,2,0,2\}\right]//\text{Simplify}\right.$$

$Frenet$ returns the tangent vector, normal vector, binormal vector as well as curvature and torsion at one point $(0,0,2)$.

$$\left\{\left\{-\frac{1}{\sqrt{5}},\ -\frac{2}{\sqrt{5}},0\right\},\{0,0,1\},\left\{-\frac{2}{\sqrt{5}},\frac{1}{\sqrt{5}},0\right\},8/5,\ -\frac{3}{5}\right\}$$

The following sentences allow us to obtain the graphs of the Frenet-Serret apparatus at the point (0,0,2), the red vector represents the tangent vector, the green vector represents the normal vector and the celestial vector represents the binormal vector.

$pp = \{0,0,2\}; Show[ContourPlot3D[z - y^2 - 2 == 0, \{x,-1,1\},$
$\{y,-2,2\},\{z,1,3\}, ContourStyle \rightarrow Blue],$
$ParametricPlot3D[\{u,uv,v\},\{u,-1,1\},\{v,1,3\},$
$PlotStyle \rightarrow Opacity[0.75]], Graphics3D[\{\{AbsolutePointSize[8], Point[pp]\},$
$\{Red, Arrow[Tube[\{pp,pp+t\}]]\}, \{Green, Arrow[Tube[\{pp,pp+n\}]]\},$
$\{Cyan, Arrow[Tube[\{pp,pp+b\}]]\}\}]]$

See Fig. 1.

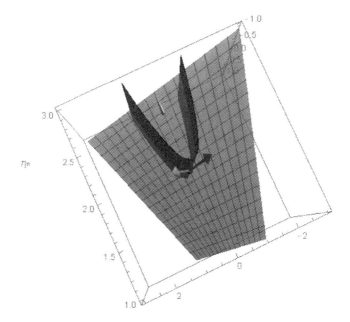

Fig. 1. Graphical visualization of the Frenet-Serret apparatus of the curve of intersection of the implicit $f(x, y, z) = z - y^2 - 2 = 0$ surface and the parametric $r(u, v) = (u, uv, v)$ surface at the point $(0, 0, 2)$ (Color figure online)

4.2   Example 02

The second example is given by the curve obtained by the intersection of the implicit surface $(x - \frac{1}{2})^2 + y^2 = \frac{1}{4}$ and the parametric surface $\{\cos(u)\cos(v),$ $\cos(v)\sin(u), \sin(v)\}$.

The following sentences allow us to obtain the Frenet-Serret apparatus at the point of the curve $\left\{\frac{1}{2}, \frac{1}{2}, \frac{1}{\sqrt{2}}\right\}$.

$$\{t, b, n, k, \tau\} = \text{Simplify}\left[\text{Frenet}\left(\left(x - \frac{1}{2}\right)^2 + y^2 - \frac{1}{4}, \{\cos(u)\cos(v),\right.\right.$$

$$\left.\left.\sin(u)\cos(v), \sin(v)\}, \{x, y, z\}, \{u, v\}, \text{ParVals} \rightarrow \left\{\frac{1}{2}, \frac{1}{2}, \frac{1}{\sqrt{2}}, \frac{\pi}{4}, \frac{\pi}{4}\right\}\right)\right]$$

Frenet returns the tangent vector, the normal vector, the binormal vector, as well as the curvature and torsion.

$$\left\{\left\{-\sqrt{\frac{2}{3}}, 0, \frac{1}{\sqrt{3}}\right\}, \left\{-\frac{1}{\sqrt{39}}, -2\sqrt{\frac{3}{13}}, -\sqrt{\frac{2}{39}}\right\}, \left\{\frac{2}{\sqrt{13}}, -\frac{1}{\sqrt{13}}, 2\sqrt{\frac{2}{13}}\right\},\right.$$

$$\left.\frac{2\sqrt{\frac{13}{3}}}{3}, \frac{6}{13}\sqrt{3}\left(-2\sqrt{\frac{2}{3}} + \sqrt{6}\right)\right\}$$

The following sentences allow us to obtain the graphs of the Frenet-Serret apparatus at the point $\{\frac{1}{2}, \frac{1}{2}, \frac{1}{\sqrt{2}}\}$, the red vector represents the tangent vector, the green vector represents the normal vector and the celestial vector represents the binormal vector.

$$pp = \left\{\frac{1}{2}, \frac{1}{2}, \frac{1}{\sqrt{2}}\right\}; Show\left[ContourPlot3D\left[\left(x - \frac{1}{2}\right)^2 + y^2 == 1/4,\right.\right.$$

$$[\{\cos[u]\cos[v], \cos[v]\sin[u], \sin[v]\}, \{u, -Pi, Pi\}, \{v, -Pi, Pi\},$$

$$PlotStyle \to Opacity[0.5]], Graphics3D[\{\{AbsolutePointSize[8],$$

$$Point[pp]\}, \{Red, Arrow[Tube[\{pp, pp + t\}]]\}, \{Green, Arrow[Tube$$

$$[\{pp, pp + n\}]]\}, \{Cyan, Arrow[Tube[\{pp, pp + b\}]]\}\}], PlotRange \to All]$$

See Fig. 2.

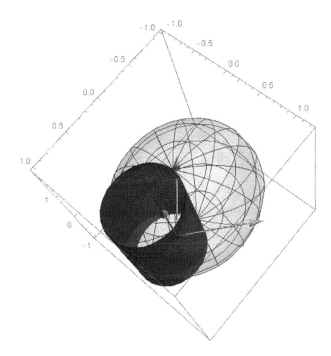

Fig. 2. Graphical visualization of the Frenet-Serret apparatus of the curve of intersection of the implicit $f(x, y, z) = (x - \frac{1}{2})^2 + y^2 - 1/4 = 0$ surface and the parametric $r(u, v) = \{\cos(u)\cos(v), \cos(v)\sin(u), \sin(v)\}$ surface at the point $\{\frac{1}{2}, \frac{1}{2}, \frac{1}{\sqrt{2}}\}$ (Color figure online)

4.3 Example 03

The third example is given by the curve obtained by the intersection of the implicit surface $x^2 + y^2 = 2$ and the parametric surface $\{u^3, v^3, uv\}$. The following sentences allow us to obtain the Frenet-Serret apparatus at the point of the $\{1, 1, 1\}$ curve.

$$\{t, b, n, k, \tau\} = \text{Frenet}\left[x^2 + y^2 - 2, \{u^3, v^3, \text{uv}\}, \{x, y, z\}, \{u, v\}, \text{ParVals} \to \{1, 1, 1, 1, 1\}\right] // FullSimplify$$

Frenet returns the tangent vector, the normal vector, the binormal vector, as well as the curvature and torsion.

$$\left\{\left\{-\frac{1}{\sqrt{2}}, \frac{1}{\sqrt{2}}, 0\right\}, \left\{-\frac{1}{\sqrt{2}}, -\frac{1}{\sqrt{2}}, 0\right\}, \{0, 0, 1\}, \frac{1}{\sqrt{2}}, 0\right\}$$

The following sentences allow us to obtain the graphs of the Frenet-Serret apparatus at the point $\{1, 1, 1\}$, the red vector represents the tangent vector, the green vector represents the normal vector and the celestial vector represents the binormal vector.

$pp = \{1, 1, 1\}; Show[ContourPlot3D[x^2 + y^2 == 2, \{x, -Pi, Pi\}, \{y, -Pi, Pi\},$

$\{z, -2, 2\}, ContourStyle \to Blue], ParametricPlot3D[\{u^3, v^3, uv\},$

$\{u, -1.3, 1.3\}, \{v, -1.3, 1.3\}, PlotStyle \to Opacity[0.5]], Graphics3D$

$[\{\{AbsolutePointSize[8], Point[pp]\}, \{Red, Arrow[Tube[\{pp, pp + t\}]]\},$

$\{Green, Arrow[Tube[\{pp, pp + n\}]]\}, \{Cyan, Arrow[Tube[\{pp, pp$

$+ b\}]]\}\}], PlotRange \to All]$

See Fig. 3.

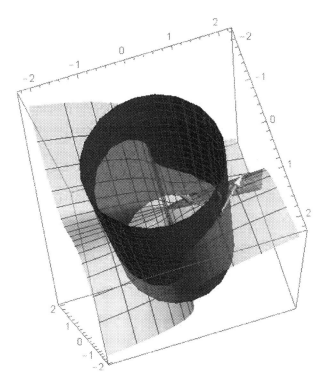

Fig. 3. Graphical visualization of the Frenet-Serret apparatus of the curve of intersection of the implicit $f(x, y, z) = x^2 + y^2 - 2 = 0$ surface and the parametric $r(u, v) = \{u^3, v^3, uv\}$ surface at the point $(1, 1, 1)$ (Color figure online)

5 Conclusions

This paper proposes a program implemented in Mathematica v.11.2 software to calculate the differential geometry properties of curves given by the intersection between two implicit parametric surfaces, based on the results obtained in [1,5] and as a continuation of [6] and [8] whose results coincide with those found in this paper. Demonstrating that it is possible to calculate the Frenet-Serret apparatus of a curve for which it is not necessary to know a parameterization, being of great help when performing operations that can often make said task cumbersome and also provide us with a very useful graphic representation of the problem. Package performance is discussed by means of some illustrative and interesting examples. All of the commands have been implemented in Mathematica v11.2 and are consistent with Mathematica notation and results [7,9,10]. The program is shorter and more efficient from my experience.

Acknowledgement. This research was carried out thanks to the support of the Research Group in Geometry and Symbolic Calculation of the Universidad Nacional de Piura (GIGYCS-UNP) in charge of professors Robert Ipanaqué Chero and Ricardo Velezmoro León expressing our appreciation for their work in the formation of new researchers.

References

1. Ye, X., Maekawa, T.: Differential geometry of intersection curves of two surfaces. Comput. Aided Geom. Des. **16**(8), 767–788 (1999)
2. Ugail, H.: Partial Differential Equations for Geometric Design, 2nd edn. Springer, London (2011). https://doi.org/10.1007/978-0-85729-784-6
3. Willmore, T.J.: An Introduction to Differential Geometry. Reprinted. Courier Corporation, Mineola, New York (2013)
4. Abbena, E., Salamon, S., Gray, A.: Modern Differential Geometry of Curves and Surfaces with Mathematica, 3rd edn. CRC Press, USA (2017)
5. Patrikalakis, N.M., Maekawa, T.: Shape Interrogation for Computer Aided Design and Manufacturing. Illustrated Edition. Springer, New York (2009). https://doi.org/10.1007/978-3-642-04074-0
6. Burgos, G., Jiménez, J., Ascate, Y.: Frenet Serret apparatus calculation of curves given by the intersection of two implicit surfaces in R^3 using Wolfram Mathematica v. 11.2. Selecciones Matemáticas **6**(2), 338–347 (2019)
7. Torrence, B.F., Torrence, E.A.: The Student's Introduction to Mathematica and the Wolfram Language, 3rd edn. Cambridge University Press, Cambridge (2019)
8. Jiménez, J.: Serret: software para el cálculo del tiedro Frenet Serret de curvas dadas por la intersección de dos superficies paramétricas. Revista Científica Pakamuros **8**(1), 69–79 (2020)
9. Wolfram, S.: The Mathematica Book, 5th edn. Wolfram Media Inc., Champaign (2003)
10. Maeder, R.: Programming in Mathematica, 2nd edn. Addison-Wesley, Redwood City (1991)

Analytics of the Multifacility
Weber Problem

Alexei Yu. Uteshev📵 and Elizaveta A. Semenova$^{(\boxtimes)}$📵

St. Petersburg State University,
7-9 Universitetskaya nab., St. Petersburg 199034, Russia
a.uteshev@spbu.ru, semenova.elissaveta@gmail.com

Abstract. For the Weber problem of construction of the minimal cost planar weighted network connecting four terminals with two extra facilities, the solution by radicals is proposed. The conditions for existence of the network in the assumed topology and the explicit formulae for coordinates of the facilities are presented. It is shown that the bifacility network is less costly than the unifacility one. Extension of the results to the general Weber problem is also discussed.

Keywords: Multifacility location problem · Weber problem · Nonlinear optimization

1   Introduction

The classical *Weber* or *generalized Fermat-Torricelli problem* is stated as that of finding the point (facility, junction) $W = (x_*, y_*)$ that minimizes the sum of weighted distances from itself to $n \geq 3$ fixed points (terminals) $\{P_j = (x_j, y_j)\}_{j=1}^n$ in the Euclidean plane:

$$\min_{W \in \mathbb{R}^2} \sum_{j=1}^n m_j |W P_j|. \tag{1}$$

Hereinafter $|\cdot|$ stands for the Euclidean distance and the *weights* $\{m_j\}_{j=1}^n$ are assumed to be positive real numbers.

The treatment of the problem in the case $n = 3$ terminals was first undertaken in 1872 by Launhardt [4] whose interest stemmed from the evident relation to the Economic Geography problem nowadays known as *Optimal Facility Location*. For instance, one can be interested in minimizing transportation costs for a plant manufacturing one ton of the final product from $\{m_j\}_{j=1}^n$ tons of distinct raw materials located at corresponding $\{P_j\}_{j=1}^n$.

Further development of the problem was carried out in 1909 by Alfred Weber [10]. First, he suggested a different economic interpretation for the three-terminal problem. Let P_3 be a place of consumption of m_3 tons of a product produced

Supported by RFBR according to the project No. **17-29-04288**.

O. Gervasi et al. (Eds.): ICCSA 2020, LNCS 12251, pp. 395–411, 2020.
https://doi.org/10.1007/978-3-030-58808-3_29

from two different types of raw materials: m_1 tons of the first type located at P_1 and m_2 tons of the second type located at P_2, let $m_3 < m_1 + m_2$. Where is the optimal location of the production? In the course of the economic background, Weber formulated the following extension of the problem to the case of four terminals[1].

"Let us take a simple case, an enterprise with three material deposits and one which is capable of being split, technologically speaking, into two stages. In the first stage two materials are combined into a half-finished product; in the second stage this half-finished product is combined with the third material into the final product... Let us suppose that possible location of the split production would be in W_1 and W_2; W_1 for the first stage and W_2 for the second stage. What will be the result if the splitting occurs?" [10].

Mathematically the stated problem can be formulated as that of finding the points $W_1 = (x_*, y_*)$ and $W_2 = (x_{**}, y_{**})$ which yield

$$\min_{\{W_1, W_2\} \subset \mathbb{R}^2} F(W_1, W_2) \quad \text{where}$$

$$F(W_1, W_2) = m_1 |W_1 P_1| + m_2 |W_1 P_2| + m_3 |W_2 P_3| + m_4 |W_2 P_4| + m |W_1 W_2| \quad (2)$$

and the weights $\{m_j\}_{j=1}^4$, m are treated as given positive real numbers.

The general *Multifacility Weber problem* is stated as that of location of the given number $\ell \geq 2$ of the facility points (or, simply, facilities) $\{W_i\}_{i=1}^\ell$ in $\mathbb{R}^d$ connected to the terminals $\{P_j\}_{j=1}^n \subset \mathbb{R}^d$ that solve the optimization problem

$$\min_{\{W_1, \ldots, W_\ell\} \subset \mathbb{R}^d} \left\{ \sum_{j=1}^n \sum_{i=1}^\ell m_{ij} |W_i P_j| + \sum_{k=1}^\ell \sum_{i=k+1}^{\ell-1} \widetilde{m}_{ik} |W_i W_k| \right\}; \quad (3)$$

here some of the weights m_{ij} and $\widetilde{m}_{ik}$ might be zero. We will refer to this value as to the **minimal cost of the network**. This problem can be considered as a natural generalization of the celebrated *Steiner minimal tree problem* aimed at construction of the network of minimal length linking the given terminals.

Dozens of papers are devoted to the Weber problem, its ramifications and applications; we refer to [3,5,11] for the reviews. The majority of them are concerned with the problem statement where the objective function (3) is free of the inter-facilities connections. This problem is known as the *Multisource Weber problem* or the *ℓ-median problem*. The present paper is focused on a solution to the Multifacility Weber problem. The mainstream approach in the treatment of this nonlinear optimization problem is the one based on reducing it to an appropriate iterative numerical procedure. For instance, the unifacility version of the problem (1) can be resolved via the modified Weiszfeld algorithm. The main obstacle of this approach consists in the fact that the objective (or cost) function of the Weber problem is non-differentiable at terminal points, and the

---

[1] In the quote we change the original notation of the points.

iterative procedure might diverge if any of the facilities happens to lie close to a terminal (or, in case of the multifacility problem, if two facilities are about to collide).

The present paper is devoted to an alternative approach for the problem, namely an analytical one. We are looking for the conditions for existence of the network and the explicit expressions for the facility coordinates in terms of the problem parameters, i.e. terminal coordinates and weights. This approach has been originated in the recent paper [6] where the unifacility Weber problem for three terminals had been solved *by radicals*. Within the framework of this approach, we will focus here on solution to the planar multifacility Weber problem for the case of $n = 4$ terminals and $\ell = 2$ facilities (i.e. the problem (2)), and also for the case of $n = 5$ terminals and $\ell = 3$ facilities.

Our analytical treatment stems from the geometric solution to the problem originated by Georg Pick and published in the Mathematical Appendix of Weber's book [10]. Nevertheless, Pick did not provide any proof of validity for his algorithm. In the conference paper [8] the present authors have announced without a proof the claim that the Weber problem (2) is solvable by radicals. In a simplified version (and with an extra assumption missed in [8]), this statement is now proved in Sect. 3. In addition, the conditions for the existence of the desired configuration of the network are provided.

In the case of the problem involving variable parameters, analytics provides one with a unique opportunity to evaluate their influence on the solution. In particular, it gives the means to determine the *bifurcation values* for these parameters, i.e. those responsible for the degeneracy of the network topology. We discuss these issues in Sect. 4 via investigation of the facilities dynamics under variation of the terminals location or the value of the involved weights. One may imagine a relevant economic optimization problem with a trawler fishing in the ocean and a floating fish processing facility drifting in anticipation of the catch transferred to it. We also prove here that, in the case of existence, the optimal bifacility network has its cost lower than the unifacility one.

In Sect. 5, we briefly discuss an opportunity for extension of the results to the case of $n \geq 5$ terminals and $\ell \geq 3$ facilities. This extension is based on the reduction of the problem to a similar one with $n - 1$ terminals and $\ell - 1$ facilities via a replacement of a pair of terminals by a suitable auxiliary *phantom* terminal. This trick is just a counterpart of the one utilized in Melzak's algorithm for Steiner tree construction.

2   Unifacility Configuration

Analytical solution for the three-terminal problem, i.e. for finding

$$\min_{W \in \mathbb{R}^2} \left(m_1 |WP_1| + m_2 |WP_2| + m_3 |WP_3| \right) \tag{4}$$

is given in [6]. In the present section we assume the vertices $P_j = (x_j, y_j)$ of the triangle $P_1 P_2 P_3$ be counted counterclockwise. Thus, the value

$$\mathfrak{S} = x_1 y_2 + x_2 y_3 + x_3 y_1 - x_1 y_3 - x_3 y_2 - x_2 y_1$$

is two times the area of this triangle. Denote

$$r_{ij} = |P_iP_j| = \sqrt{(x_i - x_j)^2 + (y_i - y_j)^2} \quad for \ \{i,j\} \subset \{1,2,3\} \ ,$$

by $\alpha_1, \alpha_2, \alpha_3$ the angles of the triangle $P_1P_2P_3$, while by $\beta_1, \beta_2, \beta_3$ the angles of the so-called weight triangle formed by the triple of weights m_1, m_2, m_3. The value $\sqrt{\mathbf{k}}/4$ where

$$\mathbf{k} = (m_1 + m_2 + m_3)(-m_1 + m_2 + m_3)(m_1 - m_2 + m_3)(m_1 + m_2 - m_3) \quad (5)$$

is the Heron formula for the area of the weight triangle.

Theorem 1. *The necessary and sufficient condition for the existence of solution to the problem (4) is that of the following system of inequalities $\{\cos \alpha_j + \cos \beta_j > 0\}_{j=1}^3$. Under this condition, the coordinates of the optimal facility $W = (x_*, y_*)$ are given by the formulae*

$$x_* = \frac{K_1K_2K_3}{2\mathfrak{S}\sqrt{\mathbf{k}d}} \left(\frac{x_1}{K_1} + \frac{x_2}{K_2} + \frac{x_3}{K_3} \right), \ y_* = \frac{K_1K_2K_3}{2\mathfrak{S}\sqrt{\mathbf{k}d}} \left(\frac{y_1}{K_1} + \frac{y_2}{K_2} + \frac{y_3}{K_3} \right)$$

with the cost of the optimal network $\mathfrak{C} = \sqrt{d}$. Here

$$d = \frac{1}{\sqrt{\mathbf{k}}}(m_1^2 K_1 + m_2^2 K_2 + m_3^2 K_3)$$

$$= |\mathfrak{S}|\sqrt{\mathbf{k}} + \frac{1}{2}\left[m_1^2(r_{12}^2 + r_{13}^2 - r_{23}^2) + m_2^2(r_{23}^2 + r_{12}^2 - r_{13}^2) + m_3^2(r_{13}^2 + r_{23}^2 - r_{12}^2) \right],$$

and

$$\begin{cases} K_1 = (r_{12}^2 + r_{13}^2 - r_{23}^2)\sqrt{\mathbf{k}}/2 + (m_2^2 + m_3^2 - m_1^2)\mathfrak{S}, \\ K_2 = (r_{23}^2 + r_{12}^2 - r_{13}^2)\sqrt{\mathbf{k}}/2 + (m_1^2 + m_3^2 - m_2^2)\mathfrak{S}, \\ K_3 = (r_{13}^2 + r_{23}^2 - r_{12}^2)\sqrt{\mathbf{k}}/2 + (m_1^2 + m_2^2 - m_3^2)\mathfrak{S}. \end{cases}$$

The proof consists in formal verification of the equalities

$$m_1 \frac{x_* - x_1}{|WP_1|} + m_2 \frac{x_* - x_2}{|WP_2|} + m_3 \frac{x_* - x_3}{|WP_3|} = 0, \quad (6)$$

$$m_1 \frac{y_* - y_1}{|WP_1|} + m_2 \frac{y_* - y_2}{|WP_2|} + m_3 \frac{y_* - y_3}{|WP_3|} = 0, \quad (7)$$

providing the stationary points of the objective function $\sum_{j=1}^3 m_j|P_jW|$.

The theorem states that the three-terminal Weber problem is solvable by radicals. Geometric meaning of the constants appeared in this theorem is as follows: $\frac{1}{2}|\mathfrak{S}|$ equals the area of the triangle $P_1P_2P_3$ while $\frac{1}{4}\sqrt{\mathbf{k}}$ equals (due to Heron's formula) the area of the weight triangle.

We now formulate some technical results to be exploited later.

Theorem 2. *If the facility W is the solution to the problem (4) for some configuration $\left\{ \begin{smallmatrix} P_1 & P_2 & P_3 \\ m_1 & m_2 & m_3 \end{smallmatrix} \right\}$ then this facility remains unchanged for the configuration $\left\{ \begin{smallmatrix} P_1 & P_2 & \tilde{P}_3 \\ m_1 & m_2 & m_3 \end{smallmatrix} \right\}$ with any position of the terminal $\tilde{P}_3$ in the half-line WP_3.*

Theorem 3. *The facility W lies in the segment P_3Q_1. For any position of the terminal P_3, the facility W lies in the arc of the circle C_1 passing through the points P_1, P_2 and Q_1. Here*

$$Q_1 = \left(\frac{1}{2}(x_1 + x_2) + \frac{(m_1^2 - m_2^2)(x_1 - x_2) - \sqrt{k}(y_1 - y_2)}{2m_3^2}, \right.$$

$$\left. \frac{1}{2}(y_1 + y_2) + \frac{(m_1^2 - m_2^2)(y_1 - y_2) + \sqrt{k}(x_1 - x_2)}{2m_3^2} \right). \quad (8)$$

Theorem 4. *Let the conditions of Theorem 1 be satisfied. Set*

$$\mathfrak{S}_1 = \begin{vmatrix} 1 & 1 & 1 \\ x & x_2 & x_3 \\ y & y_2 & y_3 \end{vmatrix}, \quad \mathfrak{S}_2 = \begin{vmatrix} 1 & 1 & 1 \\ x_1 & x & x_3 \\ y_1 & y & y_3 \end{vmatrix}, \quad \mathfrak{S}_3 = \begin{vmatrix} 1 & 1 & 1 \\ x_1 & x_2 & x \\ y_1 & y_2 & y \end{vmatrix}. \quad (9)$$

For any value of the weight m_3, the optimal facility W lies in the arc of the algebraic curve of the 4th degree given by the equation

$$m_1^2 \mathfrak{S}_2^2 \left[(x - x_2)^2 + (y - y_2)^2 \right] = m_2^2 \mathfrak{S}_1^2 \left[(x - x_1)^2 + (y - y_1)^2 \right]. \quad (10)$$

We next treat the four-terminal case.

Assumption 1. *Hereinafter we will treat the case where the terminals $\{P_j\}_{j=1}^4$, while counted counterclockwise, compose a convex quadrilateral $P_1P_2P_3P_4$.*

Stationary points of the function $\sum_{j=1}^4 m_j|WP_j|$ are given by the system of equations

$$\sum_{j=1}^4 \frac{m_j(x - x_j)}{|WP_j|} = 0, \quad \sum_{j=1}^4 \frac{m_j(y - y_j)}{|WP_j|} = 0. \quad (11)$$

Though this system is not an algebraic one with respect to x, y, it can be reduced to this form via successive squaring of every equation. This permits one to apply the procedure of elimination of a variable via computation of the **resultant**. Thereby, the problem of finding the coordinates of the facility W can be reduced to that of resolving a univariate algebraic equation with coefficients polynomially depending on $\{m_j, (x_j, y_j)\}_{j=1}^4$ [9]. The degree of this equation generically equals 12, it is irreducible over $\mathbb{Z}$, and cannot be expected to be solvable by radicals [1].

3   Bifacility Network for Four Terminals

Assumption 2. *We will assume the weights of the problem (2) to satisfy the restrictions*

$$m < m_1 + m_2, \quad m_1 < m + m_2, \quad m_2 < m + m_1, \quad (12)$$

and

$$m < m_3 + m_4, \quad m_3 < m + m_4, \quad m_4 < m + m_3. \quad (13)$$

From this follows that the values

$$\mathbf{k}_{12} = (m + m_1 + m_2)(m - m_1 + m_2)(m + m_1 - m_2)(-m + m_1 + m_2), \quad (14)$$
$$\mathbf{k}_{34} = (m + m_3 + m_4)(m - m_3 + m_4)(m + m_3 - m_4)(-m + m_3 + m_4) \quad (15)$$

are positive. Additionally we assume the fulfillment of the following inequalities:

$$(m^2 - m_1^2 + m_2^2)/\sqrt{\mathbf{k}_{12}} + (m^2 - m_4^2 + m_3^2)/\sqrt{\mathbf{k}_{34}} > 0, \quad (16)$$
$$(m^2 + m_1^2 - m_2^2)/\sqrt{\mathbf{k}_{12}} + (m^2 + m_4^2 - m_3^2)/\sqrt{\mathbf{k}_{34}} > 0. \quad (17)$$

Theorem 5. *Let Assumptions 1 and 2 be fulfilled. Set*

$$\tau_1 = \sqrt{\mathbf{k}_{12}}\left[\sqrt{\mathbf{k}_{34}}(x_4 - x_3) - (m^2 + m_3^2 - m_4^2)\,y_3 - (m^2 - m_3^2 + m_4^2)\,y_4\right]$$
$$+ 2\,m^2\sqrt{\mathbf{k}_{12}}\,y_2 + \mathbf{k}_{12}(x_1 - x_2) + (m^2 + m_1^2 - m_2^2)\left[\sqrt{\mathbf{k}_{34}}(y_3 - y_4)\right.$$
$$+ (m^2 + m_1^2 - m_2^2)\,x_1 + (m^2 - m_1^2 + m_2^2)\,x_2 - (m^2 + m_3^2 - m_4^2)\,x_3 - (m^2 - m_3^2 + m_4^2)\,x_4\Big],$$

$$\tau_2 = -\sqrt{\mathbf{k}_{12}}\left[\sqrt{\mathbf{k}_{34}}(x_4 - x_3) - (m^2 + m_3^2 - m_4^2)\,y_3 - (m^2 - m_3^2 + m_4^2)y_4\right]$$
$$- 2\,m^2\sqrt{\mathbf{k}_{12}}\,y_1 - \mathbf{k}_{12}(x_1 - x_2) + (m^2 - m_1^2 + m_2^2)\left[\sqrt{\mathbf{k}_{34}}(y_3 - y_4)\right.$$
$$+ (m^2 + m_1^2 - m_2^2)\,x_1 + (m^2 - m_1^2 + m_2^2)\,x_2 - (m^2 + m_3^2 - m_4^2)\,x_3 - (m^2 - m_3^2 + m_4^2)\,x_4\Big],$$

$$\eta_1 = \frac{1}{\sqrt{\mathbf{k}_{12}}}\left[(m^2 - m_1^2 - m_2^2)\tau_1 - 2\,m_1^2\tau_2\right],$$

$$\eta_2 = \frac{1}{\sqrt{\mathbf{k}_{12}}}\left[2\,m_2^2\tau_1 - (m^2 - m_1^2 - m_2^2)\tau_2\right]$$

and set the values for $\tau_3, \tau_4, \eta_3, \eta_4$ via the formulae obtained by the cyclic substitution for subscripts

$$\begin{pmatrix} 1\,2\,3\,4 \\ 3\,4\,1\,2 \end{pmatrix}$$

in the above expressions for $\tau_1, \tau_2, \eta_1, \eta_2$ correspondingly.

If all the values

$$\delta_1 = \eta_2\,(x_1 - x_2) + \tau_2\,(y_2 - y_1), \quad \delta_2 = \eta_1\,(x_1 - x_2) + \tau_1\,(y_2 - y_1), \quad (18)$$
$$\delta_3 = \eta_4\,(x_3 - x_4) + \tau_4\,(y_4 - y_3), \quad \delta_4 = \eta_3\,(x_3 - x_4) + \tau_3\,(y_4 - y_3) \quad (19)$$

and

$$\delta = -\frac{\delta_1\,(m^2 + m_1^2 - m_2^2)}{\sqrt{\mathbf{k}_{12}}} - \frac{\delta_3\,(m^2 + m_3^2 - m_4^2)}{\sqrt{\mathbf{k}_{34}}}$$
$$+ (\eta_1 + \eta_2)\,(y_1 - y_3) + (\tau_1 + \tau_2)\,(x_1 - x_3) \quad (20)$$

are positive then there exists a pair of points W_1 and W_2 lying inside $P_1 P_2 P_3 P_4$ that provides the global minimum value for the function (2). The coordinates of the optimal facility W_1 are as follows:

$$x_* = x_1 - \frac{2\,\delta_1\, m^2\, \tau_1}{\sqrt{k_{12}}\,\Delta}, \tag{21}$$

$$y_* = y_1 - \frac{2\,\delta_1\, m^2\, \eta_1}{\sqrt{k_{12}}\,\Delta}, \tag{22}$$

while those of W_2:

$$x_{**} = x_3 - \frac{2\,\delta_3\, m^2\, \tau_3}{\sqrt{k_{34}}\,\Delta}, \tag{23}$$

$$y_{**} = y_3 - \frac{2\,\delta_3\, m^2\, \eta_3}{\sqrt{k_{34}}\,\Delta}. \tag{24}$$

The corresponding minimum value of the function (2) equals

$$\mathfrak{C} = \frac{\sqrt{\Delta}}{4\,m^3}. \tag{25}$$

Here

$$\Delta = \left[(\eta_1 + \eta_2)^2 + (\tau_1 + \tau_2)^2 \right]. \tag{26}$$

Proof. (I) We first present some directly verified relations between the values τ-s , η-s and δ-s.

$$\tau_1 = \frac{1}{2\,m^2} \left[\sqrt{k_{12}}(\eta_1 + \eta_2) + (m^2 + m_1^2 - m_2^2)(\tau_1 + \tau_2) \right], \tag{27}$$

$$\tau_3 = \frac{1}{2\,m^2} \left[-\sqrt{k_{34}}(\eta_1 + \eta_2) - (m^2 + m_3^2 - m_4^2)(\tau_1 + \tau_2) \right], \tag{28}$$

$$\tau_1 + \tau_2 + \tau_3 + \tau_4 = 0, \quad \eta_1 + \eta_2 + \eta_3 + \eta_4 = 0, \tag{29}$$

$$\sum_{j=1}^{4} (x_j \tau_j + y_j \eta_j) = \frac{\Delta}{4\,m^4}; \tag{30}$$

$$\tau_1^2 + \eta_1^2 = \frac{m_1^2}{m^2}\,\Delta, \tag{31}$$

$$\tau_1 \eta_2 - \tau_2 \eta_1 = \frac{\sqrt{k_{12}}}{2m^2}\,\Delta, \tag{32}$$

$$\tau_2 \eta_3 - \tau_3 \eta_2 = \frac{\sqrt{k_{12}k_{34}}}{4m^4} \left[\frac{m^2 - m_1^2 + m_2^2}{\sqrt{k_{12}}} + \frac{m^2 - m_4^2 + m_3^2}{\sqrt{k_{34}}} \right] \Delta, \tag{33}$$

$$\delta_1 + \delta_3 = (x_1 - x_3)(\eta_1 + \eta_2) - (y_1 - y_3)(\tau_1 + \tau_2), \tag{34}$$

$$2\delta_2 m_2^2 = (m^2 - m_1^2 - m_2^2)\delta_1 - \sqrt{\mathbf{k}_{12}}\left[(y_1 - y_2)\eta_2 + (x_1 - x_2)\tau_2\right], \qquad (35)$$

$$2\delta_4 m_4^2 = (m^2 - m_3^2 - m_4^2)\delta_3 - \sqrt{\mathbf{k}_{34}}\left[(y_3 - y_4)\eta_4 + (x_3 - x_4)\tau_4\right]. \qquad (36)$$

(II) Consider the system of equations for determining stationary points of the objective function (2):

$$\frac{\partial F}{\partial x_*} = m_1 \frac{x_* - x_1}{|W_1 P_1|} + m_2 \frac{x_* - x_2}{|W_1 P_2|} + m \frac{x_* - x_{**}}{|W_1 W_2|} = 0, \qquad (37)$$

$$\frac{\partial F}{\partial y_*} = m_1 \frac{y_* - y_1}{|W_1 P_1|} + m_2 \frac{y_* - y_2}{|W_1 P_2|} + m \frac{y_* - y_{**}}{|W_1 W_2|} = 0, \qquad (38)$$

$$\frac{\partial F}{\partial x_{**}} = m_3 \frac{x_{**} - x_3}{|W_2 P_3|} + m_4 \frac{x_{**} - x_4}{|W_2 P_4|} + m \frac{x_{**} - x_*}{|W_2 W_1|} = 0, \qquad (39)$$

$$\frac{\partial F}{\partial y_{**}} = m_3 \frac{y_{**} - y_3}{|W_2 P_3|} + m_4 \frac{y_{**} - y_4}{|W_2 P_4|} + m \frac{y_{**} - y_*}{|W_2 W_1|} = 0. \qquad (40)$$

Let us verify the validity of (37). First establish the alternative representations for the coordinates (21) and (22):

$$x_* = x_2 - \frac{2\,m^2\,\delta_2\,\tau_2}{\sqrt{\mathbf{k}_{12}}\Delta}, \qquad (41)$$

$$y_* = y_2 - \frac{2\,m^2\,\delta_2\,\eta_2}{\sqrt{\mathbf{k}_{12}}\Delta}. \qquad (42)$$

Indeed, the difference of the right-hand sides of (21) and (41) equals

$$x_1 - x_2 - \frac{2\,m^2\,(\delta_1 \tau_1 - \delta_2 \tau_2)}{\sqrt{\mathbf{k}_{12}}\Delta}$$

and the numerator of the involved fraction can be transformed into

$$\overset{(18)}{=} 2\,m^2 \left[\tau_1 \eta_2(x_1 - x_2) + \tau_1 \tau_2(y_2 - y_1) - \tau_2 \eta_1(x_1 - x_2) - \tau_2 \tau_1(y_2 - y_1)\right]$$

$$= 2\,m^2\,(x_1 - x_2)(\tau_1 \eta_2 - \tau_2 \eta_1) \overset{(32)}{=} (x_1 - x_2)\sqrt{\mathbf{k}_{12}}\Delta.$$

The equivalence of (42) and (22) can be demonstrated in a similar manner. Now express the segment lengths:

$$|W_1 P_1| = \sqrt{(x_1 - x_*)^2 + (y_1 - y_*)^2} \overset{(21),(22)}{=} \frac{2\delta_1 m^2}{\sqrt{\mathbf{k}_{12}}\Delta}\sqrt{\tau_1^2 + \eta_1^2} \overset{(31)}{=} \frac{2\,m\,m_1}{\sqrt{\mathbf{k}_{12}}}\frac{\delta_1}{\sqrt{\Delta}} \qquad (43)$$

and, similarly,

$$|W_1 P_2| \overset{(41),(42)}{=} \frac{2\,m\,m_2}{\sqrt{\mathbf{k}_{12}}}\frac{\delta_2}{\sqrt{\Delta}}. \qquad (44)$$

With the aid of relations (21), (41), (43) and (44) one can represent the first two terms in the left-hand side of the equality (37) as

$$m_1 \frac{x_* - x_1}{|W_1 P_1|} + m_2 \frac{x_* - x_2}{|W_1 P_2|} = -\frac{m}{\sqrt{\Delta}}(\tau_1 + \tau_2). \tag{45}$$

The third summand in the equality (37) needs more laborious manipulations. We first transform its numerator:

$$x_* - x_{**} \overset{(21),(24)}{=} x_1 - x_3 + \frac{2\,m^2}{\Delta} \left[\frac{\delta_3\,\tau_3}{\sqrt{k_{34}}} - \frac{\delta_1\,\tau_1}{\sqrt{k_{12}}} \right].$$

Now write down the following modification:

$$2\,m^2 \left[\frac{\delta_3\,\tau_3}{\sqrt{k_{34}}} - \frac{\delta_1\,\tau_1}{\sqrt{k_{12}}} \right] \overset{(27),(28)}{=} \left[-(\eta_1 + \eta_2) - \frac{m^2 + m_3^2 - m_4^2}{\sqrt{k_{34}}}(\tau_1 + \tau_2) \right] \delta_3$$

$$- \left[(\eta_1 + \eta_2) + \frac{m^2 + m_1^2 - m_2^2}{\sqrt{k_{12}}}(\tau_1 + \tau_2) \right] \delta_1$$

$$= - \Bigg[(\eta_1 + \eta_2)(\delta_1 + \delta_3)$$

$$+ (\tau_1 + \tau_2) \left\{ \frac{\delta_1\left(m^2 + m_1^2 - m_2^2\right)}{\sqrt{k_{12}}} + \frac{\delta_3\left(m^2 + m_3^2 - m_4^2\right)}{\sqrt{k_{34}}} \right\} \Bigg]$$

$$\overset{(20)}{=} - \Bigg[(\eta_1 + \eta_2)(\delta_1 + \delta_3)$$

$$+ (\tau_1 + \tau_2) \left\{ -\delta + (\eta_1 + \eta_2)(y_1 - y_3) + (\tau_1 + \tau_2)(x_1 - x_3) \right\} \Bigg]$$

$$= \delta(\tau_1 + \tau_2) - (\eta_1 + \eta_2) \left[\delta_1 + \delta_3 + (\tau_1 + \tau_2)(y_1 - y_3) - (\eta_1 + \eta_2)(x_1 - x_3) \right] -$$

$$\Delta(x_1 - x_3) \overset{(34)}{=} \delta(\tau_1 + \tau_2) - \Delta(x_1 - x_3).$$

Finally,

$$x_* - x_{**} = x_1 - x_3 + \frac{\delta(\tau_1 + \tau_2) - \Delta(x_1 - x_3)}{\Delta} = \frac{\delta\,(\tau_1 + \tau_2)}{\Delta}.$$

Similarly the following equality can be deduced: $y_* - y_{**} = \frac{\delta\,(\eta_1 + \eta_2)}{\Delta}$, and both formulae yield

$$|W_1 W_2| = \sqrt{(x_* - x_{**})^2 + (y_* - y_{**})^2} = \frac{\delta}{\sqrt{\Delta}}. \tag{46}$$

Therefore, the last summand of equality (37) takes the form

$$mz\frac{x_* - x_{**}}{|W_1W_2|} = m\frac{\delta\,(\tau_1 + \tau_2)\sqrt{\Delta}}{\delta\Delta} = m\frac{\tau_1 + \tau_2}{\sqrt{\Delta}}.$$

Summation this with (45) yields 0 and this completes the proof of (37).

The validity of the remaining equalities (38)–(40) can be established in a similar way.

(III) We now deduce the formula (25) for the network cost. With the aid of the formulae (43), (44), (46) and their counterparts for the segment lengths $|W_2P_3|$ and $|W_2P_4|$, one gets

$$m_1|W_1P_1| + m_2|W_1P_2| + m_3|W_2P_3| + m_4|W_2P_4| + m|W_1W_2|$$

$$= \frac{2m}{\sqrt{\Delta}}\left(\frac{m_1^2\,\delta_1}{\sqrt{k_{12}}} + \frac{m_2^2\,\delta_2}{\sqrt{k_{12}}} + \frac{m_3^2\,\delta_3}{\sqrt{k_{34}}} + \frac{m_4^2\,\delta_4}{\sqrt{k_{34}}} + \frac{\delta}{2}\right)$$

$$\stackrel{(20)}{=} \frac{2m}{\sqrt{\Delta}}\left\{\frac{\delta_1}{2\sqrt{k_{12}}}(-m^2 + m_2^2 + m_1^2) + \frac{\delta_3}{2\sqrt{k_{34}}}(-m^2 + m_3^2 + m_4^2) + \frac{m_2^2\,\delta_2}{\sqrt{k_{12}}} + \frac{m_4^2\,\delta_4}{\sqrt{k_{34}}}\right.$$

$$\left. + \frac{1}{2}\,(\eta_1 + \eta_2)\,(y_1 - y_3) + \frac{1}{2}\,(\tau_1 + \tau_2)\,(x_1 - x_3)\right\}$$

$$\stackrel{(35),(36)}{=} \frac{2m}{\sqrt{\Delta}}\left\{-\frac{1}{2}(y_1 - y_2)\eta_2 - \frac{1}{2}(x_1 - x_2)\tau_2 - \frac{1}{2}(y_3 - y_4)\eta_4 - \frac{1}{2}(x_3 - x_4)\tau_4\right.$$

$$\left. + \frac{1}{2}\,(\eta_1 + \eta_2)\,(y_1 - y_3) + \frac{1}{2}\,(\tau_1 + \tau_2)\,(x_1 - x_3)\right\}$$

$$= \frac{m}{\sqrt{\Delta}}\{y_1\eta_1 + y_2\eta_2 + y_4\eta_4 + x_1\tau_1 + x_2\tau_2 + x_4\tau_4 - x_3(\tau_1 + \tau_2 + \tau_4)$$

$$-y_3(\eta_1 + \eta_2 + \eta_4)\}$$

$$\stackrel{(29)}{=} \frac{m}{\sqrt{\Delta}}\sum_{j=1}^{4}(x_j\tau_j + y_j\eta_j) \stackrel{(30)}{=} \frac{\sqrt{\Delta}}{4\,m^3}.$$

For the proof of the two last statements we refer to [7].

(IV) The facilities W_1 and W_2 providing the solution to the problem (2) lie inside the quadrilateral $P_1P_2P_3P_4$.

(V) The function (2) is strictly convex inside the *convex* (due to Assumption 1) domain given as the Cartesian product $P_1P_2P_3P_4 \times P_1P_2P_3P_4$. Therefore the solution of the system (37)–(40) provides the global minimum value for this function. □

The result of Theorem 5 claims that the bifacility Weber problem for four terminals is solvable by radicals, and thus we get a natural extension of the three-terminal problem solution given in Theorem 1. An additional correlation between these two results can be watched, namely that the denominators of all the formulae for the facilities coordinates contain the explicit expression for the cost of the corresponding network. It looks like every facility "knows" the cost of the network that includes this point.

4 Solution Analysis

Analytical solution obtained in the previous section provides one with an opportunity to analyze the dynamics of the network under variation of the parameters of the configuration and to find the bifurcation values for these parameters, i.e. those responsible for the topology degeneracy. We first treat the cases where either the coordinates of a terminal or the corresponding weight are variated.

Theorem 6. *For any position of the terminal P_3, the facility W_1 lies in the arc of the circle C_1 passing through the points P_1, P_2 and $Q_1 = (q_{1x}, q_{1y})$ given by the formula (8) where substitution $m_3 \to m$ is made. At the same time, the facility W_2 lies in the arc of the circle C_3 passing through the points Q_1, P_4 and $\widetilde{Q}_3$. Here $\widetilde{Q}_3$ is given by (8) where substitution*

$$\begin{array}{c|c|c}
(x_1, y_1) & (x_2, y_2) & \begin{array}{c|c}m_1 & m_2\end{array} \ \ m \\
(x_4, y_4) & (q_{1x}, q_{1y}) & m_4 \ \ \begin{array}{c|c}m & m_3\end{array}
\end{array}$$

is applied to.

Example 1. *For the configuration*

$$\left\{ \begin{array}{c|c|c|c}
P_1 = (1,5) & P_2 = (2,1) & P_3 & P_4 = (6,7) \\
m_1 = 3 & m_2 = 2 & m_3 = 3 & m_4 = 4
\end{array} \ \ m = 4 \right\},$$

find the loci of the facilities W_1 and W_2 under variation of the terminal P_3 moving somehow from the starting position at $(9,2)$ towards P_2.

Solution. The trajectory of P_3 does not influence those of W_1 and W_2, i.e. both facilities do not leave the corresponding arcs for any drift of P_3 until the latter swashes the line $\mathcal{L} = \widetilde{Q}_3 W$ (Fig. 1). At this moment, W_1 collides with W_2 in the point

$$W = \left(\frac{867494143740435 + 114770004066285\sqrt{33} + 14973708000030\sqrt{55} + 19296850969306\sqrt{15}}{435004929875940}, \right.$$
$$\left. \frac{581098602680450 + 10154769229801\sqrt{15} + 9689425113917\sqrt{55} - 18326585102850\sqrt{33}}{145001643291980} \right)$$

$\approx (3.936925, \ 4.048287)$ which stands for the second point of intersection of the circles C_1 and C_3, and yields a solution to the unifacility Weber problem (1) for the configuration $\left\{ \begin{array}{c|c|c|c} P_1 & P_2 & P_3 & P_4 \\ m_1 & m_2 & m_3 & m_4 \end{array} \right\}$. When P_3 crosses the line $\mathcal{L}$, the solution to the bifacility Weber problem (2) does not exist (while the unifacility counterpart (1) still possesses a solution).   □

The following result gives rise to an alternative geometric construction of the facility points W_1 and W_2 for the optimal network.

Theorem 7. *In the notation of Theorem 6, the facility W_2 lies at the point of intersection of the circle C_3 with the line $\widetilde{Q}_3 P_3$. The facility W_1 lies at the point of intersection of the circle C_1 with the line $Q_1 W_2$. The minimal cost of the network equals $\mathfrak{C} = m_3 |\widetilde{Q}_3 P_3|$.*

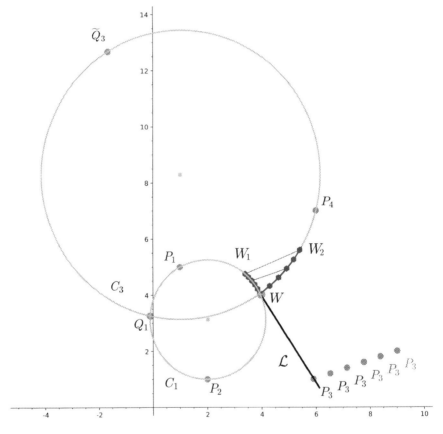

Fig. 1. Example 1. Loci of the facilities W_1 and W_2 under variation of the terminal P_3

Theorem 8. *Let the circle C_1 and the point $Q_1 = (q_{1x}, q_{1y})$ be defined as in Theorem 6. For any value of the weight m_3, the optimal facility W_1 lies in the arc of the circle C_1. At the same time, the facility W_2 lies in the arc of the 4th degree algebraic curve passing through the points P_3, P_4 and Q_1. It is given by the equation*

$$m^2 \begin{vmatrix} 1 & 1 & 1 \\ x & q_{1x} & x_3 \\ y & q_{1y} & y_3 \end{vmatrix}^2 [(x-x_4)^2 + (y-y_4)^2] = m_4^2 \begin{vmatrix} 1 & 1 & 1 \\ x & x_3 & x_4 \\ y & y_3 & y_4 \end{vmatrix}^2 [(x-q_{1x})^2 + (y-q_{1y})^2]. \quad (47)$$

We finally treat the case of the variation of the parameter directly responsible for the inter-facilities connection; the bifurcation equation is now determined by (20).

Example 2. *For the configuration*

$$\left\{ \begin{matrix} P_1 = (1,5) \\ m_1 = 3 \end{matrix} \middle| \begin{matrix} P_2 = (2,1) \\ m_2 = 2 \end{matrix} \middle| \begin{matrix} P_3 = (7,2) \\ m_3 = 3 \end{matrix} \middle| \begin{matrix} P_4 = (6,7) \\ m_4 = 4 \end{matrix} \middle| m \right\},$$

find the loci of the facilities W_1 and W_2 under variation of the weight m within the interval $[2, 4.8]$.

Solution. Due to (46), the trajectories of W_1 and W_2 meet when m coincides with a zero of the equation $\delta(m) = 0$. The latter can be reduced to an algebraic one $24505\, m^{20} - 3675750\, m^{18} + \cdots + 25596924755077 = 0$ with a (closest to $m = 4$) zero $m_{0,1} \approx 4.326092$. The collision point W yields the solution to the unifacility Weber problem (1) for the configuration $\left\{ \frac{P_1}{m_1} \middle| \frac{P_2}{m_2} \middle| \frac{P_3}{m_3} \middle| \frac{P_4}{m_4} \right\}$. Its coordinates $(x_*, y_*) \approx (4.537574, 4.565962)$ satisfy the 10th degree algebraic equations over $\mathbb{Z}$, and this time (as opposed to the variant from Example 1) one cannot expect them to be expressed by radicals.

This scenario demonstrates a paradoxical phenomenon: the weight m increase forces the facilities to a collision, i.e. to a network configuration where its influence disappears completely.

When m decreases from $m = 4$, the facility W_1 moves towards P_1 while W_2 moves towards P_4. The first drift is faster than the second one: W_1 approaches P_1 when m coincides with a zero of the equation $\delta_1(m) = 0$. The latter can be reduced to an algebraic one with a zero $m_{0,2} \approx 3.145546$. □

Theorem 9. *If the optimal bifacility network exists, it is less costly than the unifacility one.*

Proof. If the cost (25) is considered as the function of the configuration parameters then the following identities are valid:

$$\frac{\partial \mathfrak{C}^2}{\partial m_1} \equiv \frac{m_1 \delta_1}{m^2 \sqrt{k_{12}}}, \quad \frac{\partial \mathfrak{C}^2}{\partial m_2} \equiv \frac{m_2 \delta_2}{m^2 \sqrt{k_{12}}}, \quad \frac{\partial \mathfrak{C}^2}{\partial m_3} \equiv \frac{m_3 \delta_3}{m^2 \sqrt{k_{34}}}, \quad \frac{\partial \mathfrak{C}^2}{\partial m_4} \equiv \frac{m_4 \delta_4}{m^2 \sqrt{k_{34}}}$$

$$\text{and} \quad \frac{\partial \mathfrak{C}^2}{\partial m} \equiv \frac{\delta}{2m^3}.$$

The last one results in $\partial \mathfrak{C} / \partial m = \delta / (4m^3 \mathfrak{C})$. Therefore for any specialization of the weights $\{m_j\}_{j=1}^4$, the function $\mathfrak{C}(m)$ increases to its maximal value at the positive zero of $\delta(m)$. □

5   Five Terminals

In order to extend an analytical approach developed in Sect. 3 to the multifacility Weber problem, we first demonstrate here an alternative approach for solution of the four-terminal problem (2). It is based on the reduction of this problem to the pair of the three-terminal Weber problems. We will utilize abbreviations $\{4t2f\}$ and $\{3t1f\}$ for the corresponding problems.

Assume that solution for the $\{4t2f\}$-Weber problem (2) exists. Then the system of equations (37)–(40) providing the coordinates of the facilities could be split into two subsystems. Comparing equations (37) and (38) with (6) and (7) permits one to claim that the optimal facility W_1 coincides with its counterpart for the $\{3t1f\}$-Weber problem for the configuration $\left\{ \begin{smallmatrix} P_1 & P_2 & W_2 \\ m_1 & m_2 & m \end{smallmatrix} \right\}$. A similar statement is also valid for the facility W_2, i.e. it is the solution to the Weber problem for the configuration $\left\{ \begin{smallmatrix} P_3 & P_4 & W_1 \\ m_3 & m_4 & m \end{smallmatrix} \right\}$. From this point of view, it looks like the four-terminal Weber problem can be reduced to the pair of the three-terminal ones. However, this reduction should be modified since the loci of the facilities W_2 or W_1 remain still undetermined. The result of Theorems 2 and 3 permits one to replace these facilities by those with known positions.

Theorem 10. *If the solution to the $\{4t2f\}$-Weber problem (2) exists then the facility W_2 coincides with the solution to the $\{3t1f\}$-Weber problem for the configuration $\left\{ \begin{smallmatrix} P_3 & P_4 & Q_1 \\ m_3 & m_4 & m \end{smallmatrix} \right\}$. Here Q_1 is the point defined by (8) with the substitution $m_3 \to m$. A similar statement is valid for the terminal W_1: it coincides with the solution to the $\{3t1f\}$-Weber problem for the configuration $\left\{ \begin{smallmatrix} P_1 & P_2 & Q_2 \\ m_1 & m_2 & m \end{smallmatrix} \right\}$ where the coordinates for Q_2 are obtained via (8) where the substitution for the indices $1 \to 3, 2 \to 4$ is made together with $m_3 \to m$.*

This theorem claims that the four-terminal Weber problem can be solved by its reduction to the three-terminal counterpart via a formal replacement of a pair of the *real* terminals, say P_3 and P_4, by a single *phantom* terminal Q_2. This reduction algorithm is similar to that used for construction of the Steiner minimal tree (firstly introduced by Gergonne as early as in 1810, and 150 years later rediscovered by Melzak [2]). The approach can be evidently extended to the general case of $n \geq 5$ terminals as will be clarified by the following example.

Example 3. *Find the coordinates of the facilities W_1, W_2, W_3 that minimize the cost*

$$m_1|P_1W_1| + m_2|P_2W_1| + m_3|P_3W_2| + m_4|P_4W_2| + m_5|P_5W_3| \qquad (48)$$
$$+ \tilde{m}_{1,3}|W_1W_3| + \tilde{m}_{2,3}|W_2W_3|$$

for the following configuration:

$$\left\{ \begin{matrix} P_1 = (1,6) & P_2 = (5,1) & P_3 = (11,1) & P_4 = (15,3) & P_5 = (7,11) & \tilde{m}_{1,3} = 10 \\ m_1 = 10 & m_2 = 9 & m_3 = 8 & m_4 = 7 & m_5 = 13 & \tilde{m}_{2,3} = 12 \end{matrix} \right\}.$$

Solution. **(I)** To reduce the problem to the $\{4t2f\}$-case, replace a pair of the terminals P_1 and P_2 by the point Q_1 defined by the formula (8) where the substitution $m_3 \to \widetilde{m}_{1,3}$ is made.

$$Q_1 = \left(-\frac{9}{40}\sqrt{319} + \frac{131}{50}, -\frac{9}{50}\sqrt{319} + \frac{159}{40} \right) \approx (-1.398628, 0.760097).$$

(II) Solve the $\{4t2f\}$-problem for the configuration $\left\{ \begin{array}{c|c|c|c|c} P_5 & Q_1 & P_3 & P_4 & \\ m_5 & \widetilde{m}_{1,3} & m_3 & m_4 & \end{array} \widetilde{m}_{2,3} \right\}$ via formulae (21)–(24) and obtain the coordinates for the facilities

$$W_2 \approx (10.441211, 3.084533) \quad \text{and} \quad W_3 \approx (7.191843, 5.899268).$$

(III) Return P_1 and P_2 instead of Q_1 and solve the $\{3t1f\}$-Weber problem for the configuration $\left\{ \begin{array}{c|c|c} P_1 & P_2 & W_3 \\ m_1 & m_2 & \widetilde{m}_{1,3} \end{array} \right\}$ by the formulae of Theorem 1: $W_1 \approx (4.750727, 4.438893)$ (Fig. 2). We emphasize, that the coordinates of the facilities can be expressed by radicals similar to the following expression for the minimal cost of the network

$$\mathfrak{C} = \frac{\sqrt{10}}{80} \left(4158\sqrt{87087} + 773402\sqrt{231} + 271890\sqrt{319} + 247470\sqrt{143} \right.$$
$$\left. + 326403\sqrt{609} + 104181\sqrt{273} - 4455\sqrt{377} + 15216515 \right)^{1/2} \approx 267.229644 \,.$$

$\square$

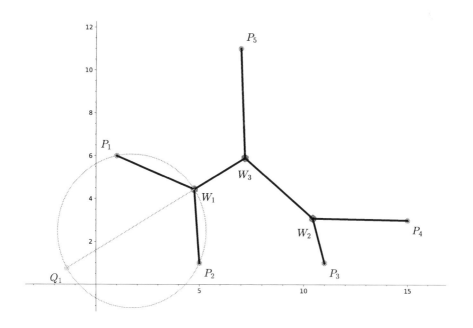

Fig. 2. Example 3. Weber network construction for five terminals

The reduction procedure illuminated in the previous example, in the general case should be accompanied by the conditions similar to those from Theorem 5.

We conclude this section with formulation of two problems for further research. The first one, for simplicity, is given in terms of the last example:

Find the pair of the weights $(\widetilde{m}_{1,3}, \widetilde{m}_{2,3})$ with the minimal possible sum $\widetilde{m}_{1,3} + \widetilde{m}_{2,3}$ such that the corresponding optimal network contains a single facility.

The second problem consists in proving (or disproving) of the following

Conjecture. The {n **terminals** ℓ **facilities**}-Weber problem (3) is solvable by radicals if $\ell = n - 2$ and the valency of every facility in the network equals 3.

6   Conclusions

We provide an analytical solution to the bifacility Weber problem (2) approving thereby the geometric solution by G. Pick. We also formulate the conditions for the existence of the network in a prescribed topology and construct the solution for five terminals.

Several problems for further investigations are mentioned in Section 5. One extra problem concerns the treatment of distance depending functions like $F_L(P) = \sum_{j=1}^{n} m_j |PP_j|^L$ with different exponents $L \in \mathbb{Q} \setminus 0$. The choice $L = -1$ corresponds to Newton or Coulomb potential. It turns out that the stationary point sets of all the functions $\{F_L\}$ can be treated in the universal manner [9]. We hope to discuss these issues in forthcoming papers.

References

1. Bajaj, C.: The algebraic degree of geometric optimization problems. Discrete Comput. Geom. **3**(2), 177–191 (1988). https://doi.org/10.1007/BF02187906
2. Brazil, M., Graham, R.L., Thomas, D.A., Zachariasen, M.: On the history of the Euclidean Steiner tree problem. Arch. Hist. Exact Sci. **68**(3), 327–354 (2014). https://doi.org/10.1007/s00407-013-0127-z
3. Drezner, Z., Klamroth, K., Schöbel, A., Wesolowsky, G.O.: The Weber problem. In: Facility Location: Applications and Theory, pp. 1–36 (2002)
4. Launhardt, W.: Kommercielle Tracirung der Verkehrswege. Architekten-und Ingenieurverein (1872)
5. ReVelle, C.S., Eiselt, H.A.: Location analysis: a synthesis and survey. Eur. J. Oper. Res. **165**(1), 1–19 (2005)
6. Uteshev, A.Y.: Analytical solution for the generalized Fermat-Torricelli problem. Am. Math. Mon. **121**(4), 318–331 (2014)
7. Uteshev, A.Y., Semenova, E.A.: On the multifacility Weber problem for four terminals. In: Proceedings of the 2nd International Conference on Applications in Information Technology, ICAIT-2016, pp. 82–85. University of Aizu Press (2016)
8. Uteshev, A.Y., Semenova, E.A.: Geometry and analytics of the multifacility Weber problem. arXiv preprint arXiv:1912.12973 (2019)
9. Uteshev, A.Y., Yashina, M.V.: Stationary points for the family of Fermat–Torricelli–Coulomb-like potential functions. In: Gerdt, V.P., Koepf, W., Mayr, E.W., Vorozhtsov, E.V. (eds.) CASC 2013. LNCS, vol. 8136, pp. 412–426. Springer, Cham (2013). https://doi.org/10.1007/978-3-319-02297-0_34

10. Weber, A.: Über den Standort der industrien, Tübingen. English Translation: The Theory of the Location of Industries (1909)
11. Xue, G., Wang, C.: The Euclidean facilities location problem. In: Du, D.Z., Sun, J. (eds.) Advances in Optimization and Approximation. NOIA, vol. 1, pp. 313–331. Springer, Boston (1994). https://doi.org/10.1007/978-1-4613-3629-7_17

International Workshop
on Computational Mathematics,
Statistics and Information Management
(CMSIM 2020)

Gait Characteristics and Their Discriminative Ability in Patients with Fabry Disease with and Without White-Matter Lesions

José Braga[1], Flora Ferreira[1,2]([✉]), Carlos Fernandes[1], Miguel F. Gago[3,4], Olga Azevedo[3,4], Nuno Sousa[4], Wolfram Erlhagen[2], and Estela Bicho[1]

[1] Center Algoritmi, University of Minho, Guimarães, Portugal
estela.bicho@dei.uminho.pt
[2] Center of Mathematics, University of Minho, Guimarães, Portugal
{fjferreira,wolfram.erlhagen}@math.uminho.pt
[3] Neurology and Cardiology Services,
Hospital Senhora da Oliveira, Guimarães, Portugal
miguelfgago@yahoo.com
[4] ICVS, School of Medicine, University of Minho, Braga, Portugal

Abstract. Fabry disease (FD) is a rare disease commonly complicated with white matter lesions (WMLs). WMLs, which have extensively been associated with gait impairment, justify further investigation of its implication in FD. This study aims to identify a set of gait characteristics to discriminate FD patients with/without WMLs and healthy controls. Seventy-six subjects walked through a predefined circuit using gait sensors that continuously acquired different stride features. Data were normalized using multiple regression normalization taking into account the subject physical properties, with the assessment of 32 kinematic gait variables. A filter method (Mann Whitney U test and Pearson correlation) followed by a wrapper method (recursive feature elimination (RFE) for Logistic Regression (LR) and Support Vector Machine (SVM) and information gain for Random Forest (RF)) were used for feature selection. Then, five different classifiers (LR, SVM Linear and RBF kernel, RF, and K-Nearest Neighbors (KNN)) based on different selected set features were evaluated. For FD patients with WMLs versus controls the highest accuracy of 72% was obtained using LR based on 3 gait variables: pushing, foot flat, and maximum toe clearance 2. For FD patients without WMLs versus controls, the best performance was observed using LR and SVM RBF kernel based on loading, foot flat, minimum toe clearance, stride length variability, loading variability, and lift-off angle variability with an accuracy of 83%. These findings are the first step to demonstrate the potential of machine learning techniques based on gait variables as a complementary tool to understand the role of WMLs in the gait impairment of FD.

This work has been supported by FCT – Fundação para a Ciência e Tecnologia within the R&D Units Project Scope: UIDB/00319/2020 and UIDB/00013/2020.

Keywords: Fabry Disease · Feature selection · Machine learning · Gait

1   Introduction

Fabry Disease (FD) is a rare genetic X-linked lysosomal disorder caused by the deficiency or absent activity of the enzyme α-galactosidase A, resulting in the accumulation of globotriaosylceramide (GL-3) in several organs, including the kidney, heart and brain [10]. Brain manifestations in FD include progressive white matter lesions (WMLs) [4,14]. Brain WMLs were an early manifestation, affecting 11.1% of males and 26.9% of females under 30 years of age, even without cerebrovascular risk factor outside FD [3]. In [14], WMLs were found in 46% of 1276 patients which tend to occur earlier in males and their prevalence revealed to increase with patients' age. White matter is the brain region responsible for the transmission of nerve signals and for communication between different parts of the brain. WMLs have been associated with gait impairment [23] and the risk of falls [22,27]. Gait abnormalities, such as slower gait and postural instability, have been reported in FD [16]. However, not much is known regarding the impact of WMLs on gait performance in patients with FD. Specific location and distribution of WMLs suggest a specific underlying disease [14] and may reflate in a different gait profile. In fact, gait evaluation can be useful to differentiate different pathologies even in the presence of highly overlapping phenotypes, such as the differences found between two types of Parkinsonism (Vascular Parkinsonism versus Idiopathic Parkinson's Disease) in [9]. Furthermore, gait assessment has been revealing as a good complementary clinical tool to discriminate adults with and without a pathology such as Parkinson's disease [8,9,15], Huntington's disease [17] and recently FD [7].

Gait is usually described by its spatio-temporal and foot clearance characteristics such as speed, stride length, stride time, minimum toe clearance (mean gait characteristics), and their respective variability (given by the coefficient of variation or the standard deviation) [9,15,21]. Different machine learning (ML) techniques have been used to select the best combination of relevant gait characteristics for gait classification [6,8,21]. Recent work [21] shows that a subset of gait characteristics selected using random forest with information gain and recursive features elimination (RFE) technique with Support Vector Machine (SVM) and Logistic Regression improves the classification accuracy of Parkinson's Disease. Widely used machine learning models for gait classification are SVM [1,7,17,20,26], Random Forest (RF) [1,7,26] and K-Nearest Neighbor (KNN) [20,26]. The outcomes of these studies show good performance accuracy in the classification of pathological gait. In particular, the outcomes of previous work [7] show promising results in the use of gait characteristics to discriminate FD patients and healthy adults. However, the implication of the presence or absence of WMLs in the gait performance has not yet been investigated. Hence, the purpose of this study is to identify relevant gait characteristics to discriminate FD patients with and without WMLs from healthy subjects (controls).

Our hypothesis is that from the evaluation of different gait characteristics (e.g., stride length, stride time variability) one can (1) identify unique gait characteristics in the two groups of FD and (2) a selection of relevant gait characteristics can accurately discriminate FD patients with WMLs versus healthy adults, and FD patients without WMLs versus healthy adults.

2 Materials and Methods

2.1 Experimental Protocol

Data from 39 patients with FD (25 with WMLs and 14 without WMLs) and 37 healthy controls were collected. From the control group was constructed two groups of controls aged-matched with the groups of FD patients with WMLs and FD patients without WMLs. The subject demographics of these four groups are summarized in Table 1. For all FD patients, the exclusion criteria were: less than eighteen years of age, the presence of resting tremor, moderate-severe dementia (CDR > 2), depression, extensive intracranial lesions or neurodegenerative disorders, musculoskeletal disease, and rheumatological disorders. Local hospital ethics committee approved the protocol of the study, submitted by ICVS/UM and Center Algoritmi/UM. Written consent was obtained from all subjects or their guardians.

Table 1. Demographic variables for FD patients with and without WMLs and Controls.

Demographics	Fabry with WMLs (n = 25)	Controls (n = 25)	p-value	Fabry without WMLs (n = 14)	Controls (n = 14)	p-value
Age (years)	57.92 (16.75)	60.52 (22.04)	0.174[*]	37.79 (10.48)	38.07 (12.57)	0.946[*]
Male (%)	32%	32%	1[†]	43%	43%	1[†]
Weight (kg)	65.94 (10.05)	64.68 (8.46)	.648[*]	64.75 (6.67)	68.89 (13.25)	.482[*]
Height (m)	1.59 (0.07)	1.63 (0.06)	.052[*]	1.66 (0.09)	1.69 (0.08)	.246[*]

Data is presented as mean (standard deviation). [*]Mann-Whitney U test. [†]Fisher Exact T Test

Two Physilog® sensors (Gait Up®, Switzerland) positioned on both feet were used to measure different gait variables of each stride. This study consists of walking a 60-m continuous course (30 m corridor with one turn) in a self-selected walking while the sensors are acquiring the data. This data consists in the arithmetic mean calculated for all subjects stride time series for 11 spatial-temporal variables: speed (velocity of forward walking for one cycle), cycle duration (duration of one cycle), cadence (number of cycles in a minute), stride length (distance between successive footprints on the ground for the same foot), stance (time in which the foot touches the ground), swing (time in which the foot is in the air and does not touch the ground), loading (percent of stance between the heel strike and the foot being entirely on the ground), foot flat (percent of stance where the foot is entirely on the ground), pushing (percent of stance between the foot being entirely on the ground and toe leaving the ground), double support (percent of the cycle where both feet are touching the ground), peak swing (maximum angular velocity during swing) and 6 foot clearance variables strike angle (angle between the foot and ground when the heel touches the ground), lift-off angle (angle between the foot and the ground at take-off), maximum heel clearance (maximum height reached by the heel), maximum toe clearance 1 (maximum height reached by the toes just after maximum heel clearance), minimum toe clearance (minimum height of the toes during swing) and maximum toe clearance 2 (maximum height reached by the toes just before heel strike) in a total of 17 gait variables that contain the full step data. The average walking speed of FD patients and controls was 1.335 ± 0.174 m/s and 1.333 ± 0.201 m/s, respectively.

2.2   Data Normalization

Gait characteristics of a subject are affected by his demographics properties including height, weight, age, and gender, as well as by walking speed [25,26] or stride length [2,7,9]. To normalize the data regression models according to Wahid et al.'s method [25,26] were used. Comparing to other methods, such as dimensionless equations and detrending methods, MR normalization revealed better results on reducing the interference of subject-specific physical characteristics and gait variables [18,25,26], thereby improving gait classification accuracy using machine learning methods [7,26].

First, to control the multicollinearity among predictor variables within this multiple regression, Variance Inflation Factor (VIF) was calculated [24]. This test measures the colinearity of the physical characteristics (age, weight, height, and gender), speed, and stride length, being a value of VIF greater than 5 an indicator of this strong correlation. In Table 2 are the results of VIF tests. Since, when testing VIF with all the variables, both Speed and Stride Length had a VIF higher than 5 (an indicator of strong correlation), they can not be used simultaneously. When used separately all VIF values are less than 5, so there is no evidence of multicollinearity.

Table 2. Variance inflation factor for physical characteristics (age, weight, height and gender), speed and stride length.

	Age	Weight	Height	Gender	Speed	Stride length
VIF	2.24	1.97	3.76	1.28	6.05	9.89
	1.99	1.93	3.37	1.27	–	2.06
	1.82	1.97	2.9	1.28	1.26	–

Spatio-temporal gait variables and foot clearance variables were normalized as follows:

$$\hat{y}_i = \beta_0 + \sum_{j=1}^{p} \beta_j x_{ij} + \varepsilon_i \tag{1}$$

where $\hat{y}_i$ represents the prediction of the dependent variable for the ith observation; x_{ij} represents the jth physical property of the ith observation including age, weight, height, gender, speed or stride length, β_0 represents the intercept term, β_j represents the coefficient for the jth physical property and ε_i represents the residual error for the ith observation. The model's coefficients are estimated using the physical properties and the mean values of the gait variables of the 37 healthy controls. Although at least 20 subjects per independent variable are recommended in multiple linear regression [13], based on similar studies [18,25] with higher sample sizes, MR models were computed for all combinations with 1, 2, and 3 independent variables using a bisquare weight function. For the models with all significant independent variables ($p < 0.05$) Akaike's information criterion (AIC) [5] and R-squared metrics were used to select the best-fitted model. Statistical assumptions of a linear regression including linearity, normality, and homoscedasticity were verified.

The models created for each feature are summarized in Table 3. These are similar to the ones found in [7] with 34 controls. Left foot models were also computed but no major differences were found among both feet. So, in this paper, we just report the results on the right foot.

Table 3. Multiple linear regression models using only significative independent variables for each gait variable on the right foot

	MR normalization						AIC	Adjusted R^2
	Age	Weight	Height	Speed	Gender	Stride length		
Spatial-temporal variables								
Cycle duration	= 1.083	-0.0011		0.218	-0.274		-124.52	0.658
Cadence	= 110.429	0.133		-24.805	31.282		220.24	0.697
Stride length	= 0.255	-0.0019		0.310	0.506		-114.77	0.889
Stance	= 65.538			-4.291	-0.362		156.56	0.130
Swing	= 34.462			4.291	0.362		156.56	0.130
Loading	= 17.464		-0.131	2.214	2.427		157.45	0.371
Foot flat	= 59.319	0.0851	0.166	-15.311			194.08	0.677
Pushing	= 24.281	-0.075	-0.076	13.336			186.14	0.629
Double support	= 32			-8.804	-1.454		197.81	0.207
Peakswing	= 284.564			-68.139	162.419	16.1879	344.59	0.637
Foot clearance variables								
Strike angle	= 17.135		-0.196		4.811	14.806	184.33	0.640
Lift-off angle	= -41.883	0.209				-27.099	219.29	0.770
Maximum heel clearance	= 0.219	-0.0006			0.039	0.039	-150.06	0.435
Maximum toe clearance 1	= 0.056	0.00021			0.0085		-171.88	0.018
Minimum toe clearance	= 0.012	0.00032					-248.47	0.441
Maximum toe clearance 2	= 0.060	-0.00061			0.027	0.088	-183.44	0.711

In each subject group, the best fitted MR models are used to normalize each stride gait variable by dividing the original value y_i by the predicted gait variable $\hat{y}_i$ from (1), as follow:

$$y_i^n = \frac{y_i}{\hat{y}_i} \tag{2}$$

where y_i^n represents the normalized value for the ith observation.

After normalizing all strides of each of the 16 gait variables, the mean and the standard deviation (SD) of each variable (each gait time series) for all subjects were calculated. In this work, the SD value is used to measure the variability of each gait variable.

2.3   Feature Selection

Due to the high number of gait variables (a total of 32) and a small number of samples a hybrid method (filter method followed by a wrapper method) was employed to select the most relevant gait characteristics. First, a filter method based on Mann Whitney U tests and Spearman's correlation between the variables was used to selected 12 gait characteristics. Mann Whitney U Tests were used to examine the difference between groups (FD with WMLs vs. Controls and FD without WMLs vs. Controls) for each gait characteristics and Spearman's correlation to evaluate the independence and redundancy between them. Therefore, the selected 12 gait characteristics are the ones that present higher U-value and do not present a high correlation between them ($\rho < |0.9|$). Before applying a wrapper method, the selected 12 gait variables were scaled to have zero mean and unit variance. Based on previous work [21], the wrapper was developed using the Recursive Feature Elimination (RFE) technique with three different ML classifiers: Logistic Regression (LR), SVM with Linear kernel and RF. RFE has some advantages over other filter methods [12]. RFE is an iterative method where features are removed one by one without affecting the training error. The selection of the optimal number of features for each model is based on the evaluation metric F1 Score evaluated through 5-fold cross-validation. F1 score is defined as the harmonic mean between precision and recall whose range goes from 0 to 1 and tells how precise the classifier is [11]. The gait characteristics' importance was quantified using the model itself (feature importance for LR and SVM with linear kernel and information gain for RF). The F1 score was used to assess the performance of the different gait characteristics combinations.

2.4   Machine Learning Classifiers

Based on the literature [1,7,17,20,21,26] four different types of classifiers were employed for the purpose of distinguishing FD patients with and without WMLs versus controls: LR, SVM, RF and KNN.

All classifiers hyperparameters were tuned using randomized search and grid search method with 5-fold cross-validation: LR regularization strength constraint and type of penalty according to regularization strength constraint; SVM regularization parameter, gamma and different types of kernel (with different degrees); RF maximum depth, minimum samples leaf, minimum samples split and the number of estimators; finally, KNN number of neighbors, weights and metric. All classifiers were implemented in Python programming language using Scikit-learn library [19].

To evaluate the performance of the different classification models the accuracy metric (the ratio of correct predictions) of the training and validation set was used.

3   Results

3.1   FD Patients with WMLs Versus Controls

Recursive feature elimination technique was performed on the 12 remaining features from filter method: cycle duration, cadence, swing, foot flat, pushing, double support, maximum heel clearance, maximum toe clearance 1, maximum toe clearance 2, loading variability, foot flat variability, and maximum heel clearance variability. The results are summarized in Table 4 and Fig. 1. Five gait characteristics were selected by LR with an F1 Score of 63.96%, while SVM selected 4 features with an F1 score of 63.96%, and RF selected the larger number of features, 9 with an F1 score of 60.21%. In Table 4, are the training and validation accuracies for the optimal models of each algorithm. RF had the higher accuracy in the training but the lower accuracy in validation. LR and SVM show similar accuracies in training and validation.

Table 4. Classification model accuracy for FD patients with WMLs versus controls

	Training accuracy % (Mean ± SD)	Validation accuracy % (Mean ± SD)
Logistic Regression	70.50 ± 4.85	70 ± 14.14
SVM (Linear Kernel)	74.50 ± 3.32	70 ± 10.95
Random Forest	86 ± 2.55	66 ± 13.56

Figure 1 (right side) shows the contribution of each gait characteristic in the classification model. The common gait characteristics among the selected features by each model were three (Top 3): foot flat, pushing, and maximum toe clearance 2. The common gait characteristics from LR and SVM were four (Top 4): foot flat, pushing, maximum toe clearance 2, and foot flat variability. These Top 3 and Top 4 were evaluated with five classification models (LR, SVM Linear kernel, SVM RBF kernel, RF, and KNN) to identify the optimal combination of gait characteristics and the classification model with better performance. Results are presented in Table 5. LR performance increases slightly by

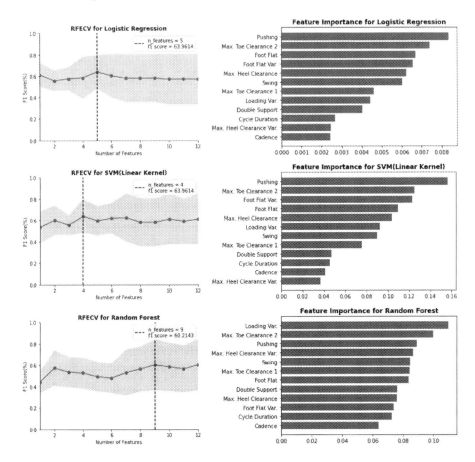

Fig. 1. F1 score versus number of features for selection of optimal numbers of gait characteristics (left) and feature importance results (right) obtained based on Logistic Regression, SVM Linear kernel and Random Forest in FD patients with WMLs vs. controls. Recursive features elimination was used through the 5-fold cross-validation (RFECV). Max.: maximum; Var.: variability.

reducing the number of features to 3. RF shows similar training and validation results with the selected features by correspondent ML model (Table 4), Top 3, and Top 4 (Table 5). With Top 4, higher accuracies are observed in SVM Linear kernel and SVM RBF kernel related to Top 3. In fact, the Top 4 corresponds to the features selected with SVM Linear Kernel. KNN shows equal validation accuracy with Top 3 and Top 4, although it displays higher training accuracy with Top 4. The better performance is observed for LR with Top 3 achieving a validation accuracy of 72% and a training accuracy of 71%, followed by both SVM Linear kernel and SVM RBF kernel with Top 4 achieving around 70% validation accuracy and 74% training accuracy.

Table 5. Classification model accuracy for top common gait characteristics in FD patients with WMLs versus controls

	Top 3 common features (LR, SVM and RF)		Top 4 common features (LR and SVM)	
	Training accuracy (Mean ± SD)	Validation accuracy (Mean ± SD)	Training accuracy (Mean ± SD)	Validation accuracy (Mean ± SD)
LR	71 ± 3.74	72 ± 13.27	74 ± 4.64	70 ± 14.14
SVM (Linear Kernel)	71.50 ± 4.90	66 ± 14.97	74.50 ± 3.32	70 ± 10.95
SVM (RBF kernel)	72 ± 4.85	66 ± 14.97	74 ± 3.39	70 ± 10.95
RF	85 ± 2.74	62 ± 13.27	86 ± 4.06	62 ± 13.27
KNN	68.75 ± 4.84	65 ± 9.35	74.38 ± 3.64	65 ± 9.35

LR: Logistic Regression, RBF: Radial basis function, RF: Random Forest, KNN: K-Nearest Neighbour, SD: Standard deviation

3.2   FD Patients Without WMLs Versus Controls

Recursive feature elimination algorithm was performed on the 12 remaining features from filter method: cycle duration, cadence, loading, foot flat, peak swing, strike angle, minimum toe clearance, stride length variability, loading variability, lift-off angle variability, maximum heel clearance variability, and minimum toe clearance variability. Results are stated in Table 6 and Fig. 2. LR selected 8 features as the optimal number of gait characteristics with an F1 Score of 86.29%, SVM selected 6 features with an F1 Score of 74.29% and RF selected 10 features with an F1 Score of 68.88%. Table 6 presents the training and validation accuracies for the optimal models of each algorithm. LR presents the higher training and validation accuracies.

Table 6. Classification model accuracy for FD patients without WMLs versus controls

	Training accuracy % (Mean ± SD)	Validation accuracy % (Mean ± SD)
LR	96.44 ± 1.79	86.67 ± 19.44
SVM (Linear Kernel)	89.47 ± 6.79	78.33 ± 16.33
RF	95.53 ± 0.15	70 ± 2.67

Taking into account the contribution of each gait characteristic in the classification model (Fig. 2, right side), the common features were 5 (Top 5): loading, foot flat, minimum toe clearance, loading variability, and lift-off angle variability. The common gait characteristics from LR and SVM were 6 (Top 6): loading, foot flat, minimum toe clearance, stride length variability, loading variability, and lift-off angle variability. These Top 5 and Top 6, as well as the Top 3 from LR and SVM (stride length variability, loading variability, and lift-off angle variability), were evaluated with five classification models (LR, SVM Linear kernel, SVM RBF kernel, RF and KNN) to identify the optimal combination of gait

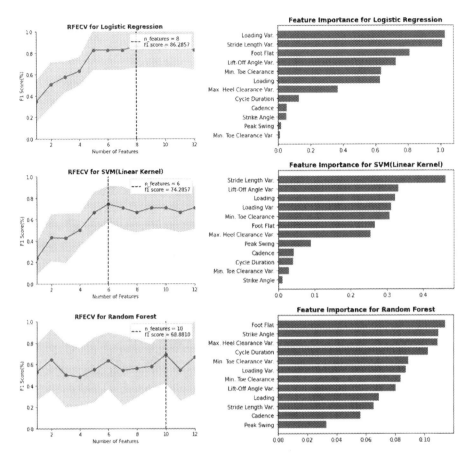

Fig. 2. F1 score versus number of features for selection of optimal numbers of gait characteristics (left) and feature importance results (right) obtained based on Logistic Regression, SVM Linear kernel and Random Forest in FD patients without WMLs vs. controls. Recursive features elimination was used through the 5-fold cross-validation (RFECV). Max.: maximum; Var.: variability; Min.: minimum.

characteristics and the classification model with better performance. Results are displayed in Table 7. With Top 5 SVM RBF kernel, RF and KNN achieved the highest validation accuracy of 76.67%. Looking at the Top 6, LR and SVM RBF kernel showed the highest validation accuracy of 83.33%. Using the Top 3 from LR and SVM, RF showed 78.33% validation accuracy, followed by the SVM Linear kernel, RF, and KNN with 75% validation accuracy, which KNN had the lower standard deviation. The validation accuracy of RF and KNN increased by reducing the feature set from 6 to 3, while the validation accuracy of LR, SVM Linear kernel, and SVM RBF kernel decrease. Overall, LR showed higher mean validation accuracy but also a higher standard deviation.

Table 7. Classification model accuracy for top common gait characteristics in FD patients without WMLs versus controls

	Top 5 common features (LR, SVM and RF)		Top 6 common features (LR and SVM)		Top 3 common features (LR and SVM)	
	Training accuracy (Mean ± SD)	Validation accuracy (Mean ± SD)	Training accuracy (Mean ± SD)	Validation accuracy (Mean ± SD)	Training accuracy (Mean ± SD)	Validation accuracy (Mean ± SD)
LR	85.76 ± 1.21	73.33 ± 17	97.35 ± 2.17	83.33 ± 18.26	79.47 ± 3.52	78.33 ± 12.47
SVM (Linear Kernel)	84.85 ± 1.92	70 ± 19.44	89.47 ± 6.79	78.33 ± 16.33	83.94 ± 4.56	75 ± 16.67
SVM (RBF kernel)	79.55 ± 4.07	76.67 ± 17	84.77 ± 4.75	83.33 ± 14.91	81.29 ± 5.09	70 ± 12.47
RF	89.32 ± 3.49	76.67 ± 13.33	91.91 ± 3.37	73.33 ± 17	90.08 ± 3.52	75 ± 16.67
KNN	82.27 ± 4.17	76.67 ± 13.33	83.11 ± 9.45	70 ± 16.33	84.85 ± 3.46	75 ± 7.45

LR: Logistic Regression, RBF: radial basis function, RF: Random Forest, KNN: K-Nearest Neighbour, SD: Standard deviation

4   Discussion

The present work aimed to identify discriminatory gait characteristics to distinguish FD with WMLs from controls and FD without WMLs from controls. Based on previous literature [1,7,17,20,21,26] different classification models were evaluated with different set of gait characteristics selected using a filter method follow by recursive feature elimination wrapper method with RF, SVM Linear kernel, and LR. Sixteen gait time series were obtained by two wearable sensors. All strides were normalized before developing any ML model according to previous studies [7,18,25]. Then, for each gait time series the mean and the standard deviation (as variability measure) were calculated obtaining 32 gait characteristics. From the feature selection analysis, foot flat, pushing, and maximum toe clearance 2 were identified as important characteristics to classify FD with WMLs. While stride length variability, loading variability, and lift-off angle variability, followed by loading, foot flat, and minimum toe clearance were identified as important gait characteristics to distinguish FD without WMLs from aged-matched healthy adults. Previous work [7] reveals that FD patients (with and without WMLs together) present lower percentages in foot flat and higher in pushing comparing with healthy adults.

For FD patients with WMLs versus controls, validation accuracy of 62–72% and a similar training accuracy of 69–85% was achieved thought the five selected classification models based on Top 3 gait characteristics, showing LR classifier the best performance with validation and training accuracy of 72% and 71%, respectively. With one more feature (foot flat variability) SVM with both Linear or RBF kernel also reveal good performance with an accuracy of 70% for validation and 74% for training. These results corroborated the hypothesis that the gait characteristics can be used to distinguish FD patients with WMLs from controls. This goes in line with the premise that gait is a final outcome of WMLs [22,23,27].

Surprisingly, in the FD patients without WMLs versus controls classification higher training and validation accuracies of 79–97% and 70–83% , respectively, were obtained based on Top 6, Top 5, or Top 3 selected features. LR and SVM RBF kernel classifier displayed the best performance based on Top 6, with an accuracy of 83% for validation and of 97% and 85% for training, respectively. By reducing the feature set for Top 3, overall validation accuracy of 70–78% was achieved, where for the RF and KNN classifiers the accuracy slightly increased to 75%. Similarly, in [21] an increase of the model accuracy was observed with feature reduction. Further, feature selection (reduction) plays an important role to deal with the problem of model overfitting, reduces training time, enhancing the overall ML performance and implementation. These results suggest that selected gait characteristics could be used as clinical features for supporting diagnoses of FD patients even without WMLs from younger ages since the mean age of these patients is 37.786 ± 10.48 years.

Due to the number of subjects involved in this study, all dataset was used in the training and validation of the models and any independent (external) dataset was used for checking the model performances. To test the robustness of classification models based on the selected gait characteristics further research with independent datasets is needed.

5   Conclusion

To the best of our knowledge, this is the first study that explores gait characteristics and their discriminate power in FD patients with WMLs and without WMLs from controls. For the discrimination of FD patients with WMLs the best model was built using LR with the variables foot flat, pushing, and maximum toe clearance 2, with an accuracy of 72%. In contrast, for the discrimination of FD patients without WMLs the best suited model was achieved using LR and SVM with RBF kernel based on the variables loading, foot flat, and minimum toe clearance, stride length variability, loading variability, and lift-off angle variability with an accuracy of 83%. The implications of WMLs on gait compromise in FD or predictive value of each kinematic gait variable remains still elusive, warranting further investigation with a more enriched cohort. Still, our findings are the first step to demonstrate the potential of machine learning techniques based on gait variables as a complementary tool to understand the role of WMLs in the gait impairment of FD. For future research, a larger sample size will be used to confirm and extend these findings.

References

1. Aich, S., Choi, K.W., Pradhan, P.M., Park, J., Kim, H.C.: A performance comparison based on machine learning approaches to distinguish Parkinson's disease from Alzheimer disease using spatiotemporal gait signals. Adv. Sci. Lett. **24**(3), 2058–2062 (2018)

2. Alcock, L., Galna, B., Perkins, R., Lord, S., Rochester, L.: Step length determines minimum toe clearance in older adults and people with Parkinson's disease. J. Biomech. **71**, 30–36 (2018)
3. Azevedo, O., et al.: Natural history of the late-onset phenotype of Fabry disease due to the p. F113L mutation. Mol. Genet. Metab. Rep. **22**, 100565 (2020)
4. Buechner, S., et al.: Central nervous system involvement in Anderson-Fabry disease: a clinical and MRI retrospective study. J. Neurol. Neurosurg. Psychiatry **79**(11), 1249–1254 (2008)
5. Burnham, K.P., Anderson, D.R.: Multimodel inference: understanding AIC and BIC in model selection. Sociol. Methods Res. **33**(2), 261–304 (2004)
6. Caramia, C., et al.: IMU-based classification of Parkinson's disease from gait: a sensitivity analysis on sensor location and feature selection. IEEE J. Biomed. Health Inform. **22**(6), 1765–1774 (2018)
7. Fernandes, C., et al.: Gait classification of patients with Fabry's disease based on normalized gait features obtained using multiple regression models. In: 2019 IEEE International Conference on Bioinformatics and Biomedicine (BIBM), pp. 2288–2295. IEEE (2019)
8. Fernandes, C., et al.: Artificial neural networks classification of patients with Parkinsonism based on gait. In: 2018 IEEE International Conference on Bioinformatics and Biomedicine (BIBM), pp. 2024–2030. IEEE (2018)
9. Ferreira, F., et al.: Gait stride-to-stride variability and foot clearance pattern analysis in idiopathic Parkinson's disease and vascular parkinsonism. J. Biomech. **92**, 98–104 (2019)
10. Giugliani, R., et al.: A 15-year perspective of the Fabry outcome survey. J. Inborn Errors Metab. Screen. **4**, 1–12 (2016)
11. Gu, Q., Zhu, L., Cai, Z.: Evaluation measures of the classification performance of imbalanced data sets. In: Cai, Z., Li, Z., Kang, Z., Liu, Y. (eds.) ISICA 2009. CCIS, vol. 51, pp. 461–471. Springer, Heidelberg (2009). https://doi.org/10.1007/978-3-642-04962-0_53
12. Guyon, I., Elisseeff, A.: An introduction to variable and feature selection. J. Mach. Learn. Res. **3**(Mar), 1157–1182 (2003)
13. Katz, M.H.: Multivariable Analysis: A Practical Guide for Clinicians and Public Health Researchers. Cambridge University Press, Cambridge (2011)
14. Körver, S., Vergouwe, M., Hollak, C.E., van Schaik, I.N., Langeveld, M.: Development and clinical consequences of white matter lesions in Fabry disease: a systematic review. Mol. Genet. Metab. **125**(3), 205–216 (2018)
15. Kubota, K.J., Chen, J.A., Little, M.A.: Machine learning for large-scale wearable sensor data in Parkinson's disease: concepts, promises, pitfalls, and futures. Mov. Disord. **31**(9), 1314–1326 (2016)
16. Löhle, M., et al.: Clinical prodromes of neurodegeneration in Anderson-Fabry disease. Neurology **84**(14), 1454–1464 (2015)
17. Mannini, A., Trojaniello, D., Cereatti, A., Sabatini, A.M.: A machine learning framework for gait classification using inertial sensors: application to elderly, post-stroke and Huntington's disease patients. Sensors **16**(1), 134 (2016)
18. Mikos, V., et al.: Regression analysis of gait parameters and mobility measures in a healthy cohort for subject-specific normative values. PloS One **13**(6), 1–11 (2018)
19. Pedregosa, F., et al.: Scikit-learn: machine learning in Python. J. Mach. Learn. Res. **12**, 2825–2830 (2011)
20. Pradhan, C., et al.: Automated classification of neurological disorders of gait using spatio-temporal gait parameters. J. Electromyogr. Kinesiol. **25**(2), 413–422 (2015)

21. Rehman, R.Z.U., Del Din, S., Guan, Y., Yarnall, A.J., Shi, J.Q., Rochester, L.: Selecting clinically relevant gait characteristics for classification of early Parkinson's disease: a comprehensive machine learning approach. Sci. Rep. **9**(1), 1–12 (2019)

22. Snir, J.A., Bartha, R., Montero-Odasso, M.: White matter integrity is associated with gait impairment and falls in mild cognitive impairment. Results from the gait and brain study. NeuroImage: Clin. **24**, 101975 (2019)

23. Starr, J.M., et al.: Brain white matter lesions detected by magnetic resosnance imaging are associated with balance and gait speed. J. Neurol. Neurosurg. Psychiatry **74**(1), 94–98 (2003)

24. Thompson, C.G., Kim, R.S., Aloe, A.M., Becker, B.J.: Extracting the variance inflation factor and other multicollinearity diagnostics from typical regression results. Basic Appl. Soc. Psychol. **39**(2), 81–90 (2017)

25. Wahid, F., Begg, R., Lythgo, N., Hass, C.J., Halgamuge, S., Ackland, D.C.: A multiple regression approach to normalization of spatiotemporal gait features. J. Appl. Biomech. **32**(2), 128–139 (2016)

26. Wahid, F., Begg, R.K., Hass, C.J., Halgamuge, S., Ackland, D.C.: Classification of Parkinson's disease gait using spatial-temporal gait features. IEEE J. Biomed. Health Inform. **19**(6), 1794–1802 (2015)

27. Zheng, J.J., et al.: Brain white matter hyperintensities, executive dysfunction, instability, and falls in older people: a prospective cohort study. J. Gerontol. Ser. A: Biomed. Sci. Med. Sci. **67**(10), 1085–1091 (2012)

Water Meters Inaccuracies Registrations: A First Approach of a Portuguese Case Study

M. Filomena Teodoro[1,2], Marina A. P. Andrade[3,4(✉)], Sérgio Fernandes[5], and Nelson Carriço[6]

[1] CINAV, Center of Naval Research, Naval Academy, Portuguese Navy,
2810-001 Almada, Portugal
[2] CEMAT, Center for Computational and Stochastic Mathematics,
Instituto Superior Técnico, Lisbon University,
1048-001 Lisboa, Portugal
maria.alves.teodoro@marinha.pt
[3] ISCTE -IUL, Instituto Universitário de Lisboa, 1649-026 Lisboa, Portugal
[4] ISTAR -IUL, Instituto Universitário de Lisboa, 1649-026 Lisboa, Portugal
marina.andrade@iscte.pt
[5] ESTSetúbal, Escola Superior de Tecnologia de Setúbal,
Instituto Politécnico de Setúbal,
Campus do IPS Estefanilha, 2914-508 Setúbal, Portugal
[6] ESTBarreiro, Escola Superior de Tecnologia do Barreiro,
Instituto Politécnico de Setúbal,
Rua Américo da Silva Marinho, 2839-001 Lavradio, Portugal

Abstract. The work described in this article results from a problem proposed by a water utility company in the framework of ESGI 140th, during June 2018. The objective is to evaluate water meters performance using historical data, knowing that, being a mechanical device, the water meters suffer a deterioration with time and use, losing accuracy throughout its cycle of use. We intend to approach a problem capable of identifying anomalies on water consumption pattern. In present work, ARIMA modeling was considered to obtain a predictive model. The results show that in the time series traditional framework revealed significant and adequate in the different estimated models. The in-sample forecast is promising, conducting to adequate measures of performance.

Keywords: Water meter performance · Anomalies identification · ARIMA models

1 Introduction

The study of the behavior of time series data is considered one of the current challenges in data mining [1,2]. A wide number of methods for water demand Forecasting can be found in literature, for example, in [3] we have a good description of such methods which are labeled in five categories. Data Mining refers the extraction of knowledge by analyzing the data from different perspectives and accumulates them to form useful information which could help the decision makers to take appropriate decisions. Due to the unique behavior of time series data, we can find several references about time series data mining in various application domains, namely survey articles [4, 5], PhD thesis where detailed techniques and case studies are analyzed [6], private communications [7], which takes into consideration how Artificial Neural Networks may assist in formulating a GLM, chapters in books [8] where the authors define the major tasks considered by the time series data mining community. Besides that, the existing research is still considered not enough. In [1] time series data is considered one of the 10 challenges in data mining. In particular the discovery of an interesting pattern, also called motif discovery, is a non-trivial task which has become one of the most important data mining tasks. In fact motif discovery can be applied to many domains [2].

Under the 140th European Study Group with Industry, Infraquinta submitted the mathematical challenge: they would like to have an algorithm for evaluating water meters performance by using historical data (hourly water consumption).

In Sect. 2 is described a brief review about the enormous quantity of available methods, analyzing their strengths and weaknesses. Such methods can be clsaasified in four categories: multi-agent models, fundamental models, reduced-form models, statistical models and computational intelligence models.

In present work, usual techniques in time series modeling are used. The ARMA, ARIMA, ARIMAX, SARIMAX are considered useful techniques to obtain a predictive model where its predictive power is discussed.

The outline of this article is developed in six sections. In Sect. 2 is presented some background about short term forecast techniques applied to this kind of problems. Section 3 makes a brief summary about the time series modeling approach. Section 4 provides more details about the challenge proposed by Infraquinta and on the provided data. In Sect. 5 is displayed a short summary about exploratory analysis of the data sets provided and continues with the study on the co-variables that may predict the time series breakpoint. The results of our ARIMA approach are presented. Finally in Sect. 6 some conclusions are drawn and suggestions for future work are pointed.

2   Preliminaries

Water utility companies need to control the supply network system so they can detect undesirable occurrences such as damages, leaks or others issues, so the water distribution network can be considered by the users as reliable and adequate. Another issue that contributes largely to this aim is that the water loss shall be reduced, by one hand due the trend of increasing water cost, by another, due the repair of damages that is usually very expensive. In [9, 10] is evidenced that supervision of network water supply is mandatory to the improvement of the performance of water supply.

With the aim of a contribution to the improvement of the reliability of water distribution networks, the authors of [11] propose a fault measurement detection considering a model-based approach. Considering a fuzzy concept, the diagnosis procedure proposed in [11] takes in consideration all available data and knowledge, revealing that it should be considered in water meters faults management.

In [12] is presented a good contribution to the validation and reconstruction of the flow meter data from a complex water system supply - the Barcelona water distribution network. It is processed a signal analysis to validate (detect) and reconstruct the missing and false data from a large dataset of flow meters in the tele-control system of a water distribution network. It considers two time scale in the proposed models: the daily data is considered in an ARIMA time series approach; the 10 min scale models uses a 10− min demand pattern using correlation analysis and a fuzzy logic classification.

The study presented in [13] concerns about inefficient water metering together with the practice of low tariffs, problem that contributes negatively to the financial sustainability of utilities. Many water utilities in developing countries face this problem. In this work, the performance of 3 m models is analyzed, also the influence of sub-metering is studied.

An interesting approach is done in [14], where the estimated value of not measured water volume in residential connections by loss of accuracy in hydrometers. The authors present the study developed in a Brazilian city from Bahia region, with the intention of estimating the sub-measurement index (percentage of the volume supplied to the consumer that is not measured) in residential consumers, due to the progressive loss of accuracy of the water meters installed due to the installation time. The work was developed based on direct measurements of the consumption profile of typical residential consumers and the evaluation of the real conditions of the meters installed in the city. The work was developed under the aim of the National Program to Combat Water Waste.

A study to find the important explanatory factors of residential consumption [15], spread in several levels categories, built a data base designed with detailed qualitative and quantitative data taking into consideration smart water metering technology, questionnaire surveys, diaries, and household water stock inventory audits. The water demand forecasting models were built using a wide number of several statistical techniques per each user category: cluster analysis, dummy coding, independent t-test, independent one-way ANOVA, bootstrapped regression models. The authors of [15] concluded that socio-demographic-economical and physical characteristics are most significant factors of water demand and that should be taken into consideration in water supply management.

The water demand management is giving more and more importance to smart metering. In last years, we can see that, for example in Australasian water utilities, where customer service, labor optimization, and operational efficiency are important keywords. In [16] are presented surveys and in-depth interviews that are accomplished of smart metering and intelligent water network, projects implemented is last years that got an important feedback. It is evidenced that digital metering combined with data analytics can be used to increase the utility efficiency and costumer service quality inducing an improvement of costumer satisfaction.

In [17], the authors propose the application of the Poisson generation models where the intensity and duration of residential water demand pulse generation are considered. The models can use the readings from households using smart metering technologies, aimed to preserve the mean and cumulative trend of water demand. When several case studies are considered, the models estimated the water demand with adequate quality, even considering multiple time aggregation scales (from 1–15 min until 1 day).

The work presented in [18] follows the same idea: the authors used the data acquired from smart metering of a big number of householders and the detailed information per individual consumer. This work describes a mixed methods study with a certain detail. They have modeled the profile behavior of householder water consumers information together with the infrastructure report details, fact that allowed to improve the efficiency of water use and to promote water conservation. This improvement was evidenced by a home water updates detailed feedback provided by a group of selected households.

3   ARIMA Approach

The identification of an ARIMA model to model the data can be considered one of the most critical phases when using ARIMA approach. For a stationary time series[1] the selection of the model to be used is mainly based on the estimated auto-correlations and partial auto-correlations, which we will use to compare with the theoretical quantities and identify a possible model for the data. Some useful references about ARIMA approach can be found in [19–21].

The auto-regressive models with order p AR(p), the moving average with order q MA(q) and their combination, ARMA(p,q) models have their auto-correlation functions (ACF) with a certain specific feature, similarly to a finger print:

The ACF of an autoregressive process with order p is infinite in extent that decays according to a damped exponential/sinusoidal;

The ACF of a moving average with order q process is finite, i.e. presents a cut after the lag q;

The ACP of ARMA process (p,q) is like a mixture of the processes described in previous items, the ACF has infinite that decays according to exponents/damped sinusoidals after the lag $q - p$.

The idea is to identify a pattern that behaves with the same profile that some theoretical model. In particular, the it is useful to identify MA models but it is not so simple to identify other kind of models. As a possible solution, we can compute the partial auto-correlation function (PACF). This function corresponds to the correlation of X_t, X_{t-k+1} removing the effect of the observations $X_{t-1}, X_{t-2}, \ldots, X_{t-k-1}$ and is denoted by ϕ_{kk}. In the case a stationary time series we can use the Yule-Walker equations to compute the PACF. Again, the PACF have a specific profile for each process like a proper finger print:

The PACF of a MA(d) is infinite in extent that decays according to a damped exponential/sinusoidal (similarly to the behaviour of an ACP from a AR(d) process;

---

[1] A time series is classified as stationary when it is developed in time around a constant mean.

The PACF of AR(p) the process is finite, i.e. presents a cut after the lag q, like the behaviour of an ACP from a MA(p) process;;
The PACF of ARMA process (p,q) is similar to an ACF from a MA(q) process.

A general method for finding a.c. for a stationary process with f.a. is using the Yule-Walker equations.

This method seems to fail in the case of non-stationary time series (the irregular component is significant). To solve this issue we differentiate the non stationary series so many times (d times) as necessary to get a stationary series. After these differences of successive terms of the chain, applied d times, we can apply the same technique: identify which model(s) are identified from ACF and PACF. A model that represents a non-stationary series, differenciated d times, with an auto-regressive component with order p and a moving average component with order q is represents as an ARIMA(p,d,q).

To estimate the best models between several proposals, we usually apply the information criteria AIC, BIC: the best models have the lowest values of AIC and BIC. Also the log of likelihood function is a good statistic to evaluate the quality of the estimated models: the lowest value means a better model.

After selection, the models need to be validated. One of the rules is to analyze the residuals (the ith residual is the difference between the ith obervation and its estimate). Residuals are supposed to be Gaussian and non-correlated. To verify this can be use several statistical tests procedures and other techniques (Ljung-Box test, box-Pierce test, Kolmogorov-Smirnov test, Bera and Jarque test, some graphics, e.g. boxplots, qq plots).

The estimates precision evaluation is another step to include in all process. For that we can compute the usual measures: MAPE, MADE, etc.

In general the procedure of all process is composed by the following iterates:

1. Models formulation: use of ACF and PACF;
2. Models adjustment: estimation of model parameters, application of suitability measures of estimates;
3. Validation of models: selection of variables, diagnostics, residual analysis and interpretation.
4. Analysis of precision and updating the models.

4   Problem Description and Data Available

Infraquinta, E.M. is the water utility that manages the water and the waste water services of a well-known tourist place in Algarve (Portugal), known as Quinta do Lago. Tourism increase trend and climate change scenarios can be identified as two predominant drivers which will strongly influence Infraquinta, E.M. as a water utility. Due to a forecasted decrease in precipitation, especially in warm season, water supply at peak demand times could be reduced and water revenue limited. In this context Infraquinta, E.M. needs to reinforce mechanisms for predictive planning based on data analysis.

They would like to have an algorithm for evaluating water meters performance by using historical data (hourly water consumption). The main purpose of the algorithm should be to find out the meter performance breakpoint and hence where should water meters be replaced.

The main challenge of this task is segmenting the hourly water consumption data into different contributions: seasonality, trend and noise. The trend component is the one related to meter performance.

The available data is composed by:

1. Monthly data on water consumption, including water meter replacement (from 2006 to 2018), of one hotel;
2. Hourly data from 2 water meters, different models, installed in serie (from 2014 to 2018);
3. Hourly data from one water meter (from 2014 to 2018), replaced at 19/08/2014.

It should be noted that the information regarding the second water meter model, from (ii) dataset was not analyzed since it only regards information from 2017 until 2018, hence scarce information to be implemented a robust time series analyzes.

5   Empirical Application

5.1   Preliminary Approach

As it was referred in Sect. 4 data available in this problem have different registration options: hourly, daily, monthly. This is something that should not be ignored. Naturally, the hourly may be aggregated in days and, of course, days may be aggregated in months. But, one shall not forget that each time we group data, necessarily, some information is lost. Therefore, there is the need to, at least have a look on the 'pure' data, where 'pure' is used to mention the data as they were recorded.

5.2   A Brief Descriptive Analysis

Starting with a graphical inspection of the data representation, and considering the thinner representation presented in the data, the choice was to obtain the box-plot representation for the different hours of the day. From this we tried to observe similarities and dissimilarities, for the water meters (dataset (ii)).

Given one set of data has the hourly consumption of two water meters in a serie disposal, the same representation was considered for the other water meter.

As one can observe the graphical representations highlight the different behavior consumption, during day and during night. That is for the period after midnight until 8 a.m. the mean consumption is reduced but there are lots of anomaly observations with huge values for consumption. Even though the two water meters are in a serial disposal it looks like the consumptions are of different

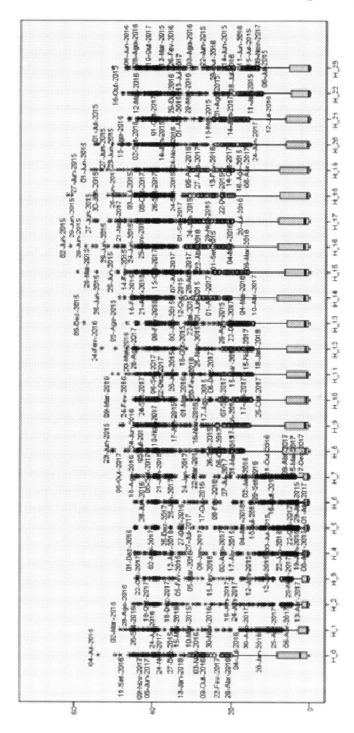

Fig. 1. Boxplot diagrams – water meter 1440.

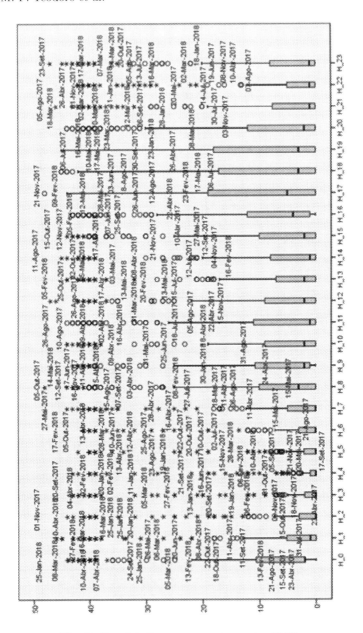

Fig. 2. Boxplot diagrams – water meter 2180.

populations. For water meter 1440 during night, about 75% of the consumption is very reduced but appears a large number of huge values, which should correspond to garden watering. Still the whole consumption seems more homogeneous in comparison to water meter 2180. In what concerns water meter 2180 there are also anomalies but the water meter seems much more sensitive to the different hours, during a day consumption.

For the other set of data expressed in hours (dataset (iii)) has a huge number of zeros at several hours of the day (in many days), which is strange, and therefore no graphical representation was possible, that is, it is not possible to obtain the box-plot. The hourly box-plot representation was not possible to obtain for one data set since there were only monthly registrations of the consumption in one hotel (Figs. 1 and 2).

5.3   The Serial Disposal Water Meters 1440 and 2180 – Some Details

For the serial disposal two water meters, given the disposition of the equipment, what should be expected was, if things were working properly, the same measurements (or at least very close) for the consumption since they were measuring the same. Thus, after observing the data it was considered to obtain the daily aggregation and then to watch what was happening to the overlapping of both commonly days interval registrations, for which a representation is presented in Fig. 3.

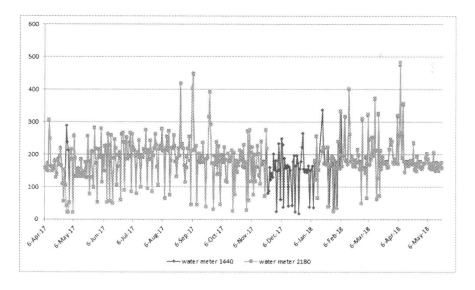

Fig. 3. Water meters series for both 1440 and 2180.

After that, and realizing that the measurement should be the same, it seemed natural to consider the difference between the measurements of each of the water meters on the same days of observation. And from that it was possible to check some irregularities.

When considered all the data registration of data consumption for the two serial disposal water meters was noticed that there is an interval time, from November 22 2017 until January 8 2018 for which water meter 2180 has no registered data. That is clearly identified in Fig. 4. Also, there are two peaks in the graph observed in distinct moments of time.

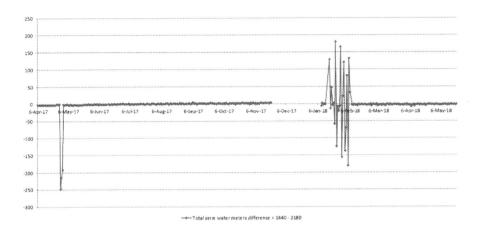

Fig. 4. Water meters difference series.

The first one, may even be an irregular occurrence lonely of a day. But the second peak of irregularities spreads for several days, see in more detail the zoom presented in Fig. 5. In fact, for a whole monthly the difference exhibits big discrepancies whether those are positive or negative.

When observed the whole series of differences, for a certain period it occurs that the difference is positive, that is, the water meter 1440 registers more consumption. It is only after a certain time (last measurements of 2018) that the difference between the measurements is negative, see again Fig. 5, that is, the water meter 2180 begins to measure more accurately, so it was expressed when that observation came up.

5.4   Water Meter 4773 – The Time Series (traditional) Approach

Considering the monthly data on water consumption, including a water meter replacement (dataset (i)), given these are monthly data and some external factors may influence the water consumption. For instance winter/summer seasons consumption or weather variables related with more or less rain, that may constraint the water consumption and it is not possible to have a detailed information on the daily consumption.

Therefore, the first differences, see Fig. 6, and the homologous differences were considered, see Fig. 7, which allows to remove partial auto–correlation issues an seasonality issues.

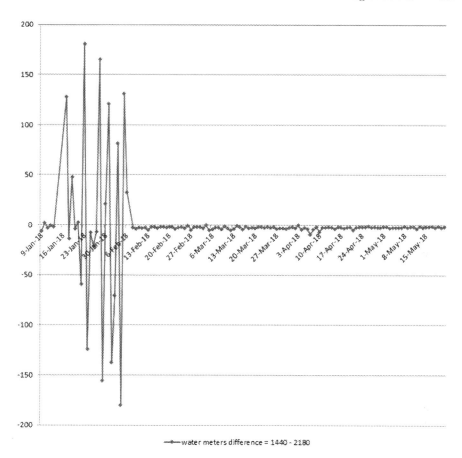

Fig. 5. Water meters difference zoom.

It is easily identified when the water meter was replaced (June 15, 2014) where a peak is clearly exhibited in the graph, in Fig. 6. In the same figure it is possible to identify as areas which may indicate the breakpoints – the slope is higher, in absolute terms, than the pattern or it is subsequently decreasing or subsequently increasing. To consider whether those are or not a breakpoints depends on the company definitions.

When the homologous differences are considered it is also possible to identify the water meter replacement, as already mentioned. In this, it is possible to observe that from January 2014 until June 2014 the consumption decreases.

Then the water meter is replaced (June 2014). From the replacement moment until June 2015 (a year) the water meter shows irregular consumption, which is probably the time needed to the water meter to stabilize, usual time necessary for this kind of equipment to work properly.

By last, were tried and obtained models using the traditional techniques, [25]. In Fig. 8 we have an additive model which decomposes the time series in:

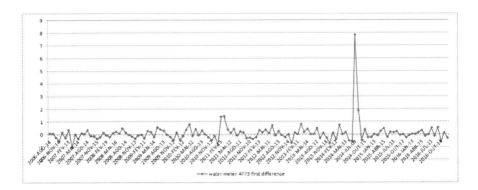

Fig. 6. First difference observation for 4773 water meter.

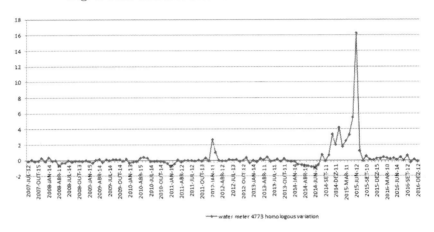

Fig. 7. Homologuos variation for 4773 water meter.

the consumption; the error for consumption from season; the seasonal adjusted series for consumption from season; the seasonal factors for consumption from season; the trend-cycle for consumption from season; with a length of seasonal period of 12 considering an additive effect.

The results obtained by this model are similar to the ones obtained by the non-parametric approach.

For the same data set, another model was obtained. In this case, some times series models were considered and tested. An ARIMA model was obtained presenting adequate goodness measures of fit and prediction.

Again, graphically, as it can be seen in Fig. 9, it is very similar to the non-parametric models.

These models may be considered as a preliminary approach of the non-parametric approach.

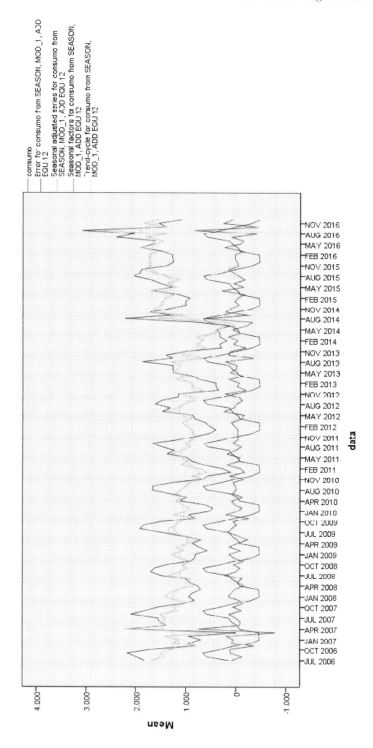

Fig. 8. Water meter 4773 adjusted series with additive effect.

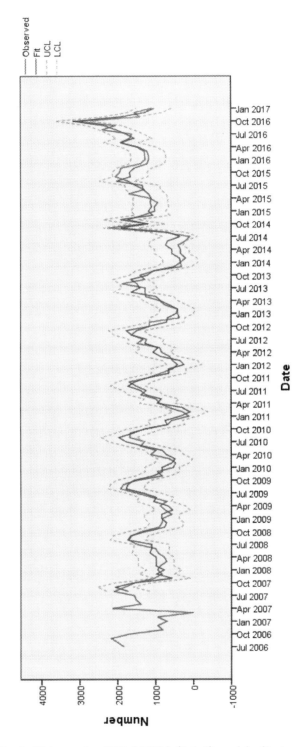

Fig. 9. Water meter 4773 ARIMA $(1, 1, 0)$ model adjustment.

6   Final Comments

In terms of the data analysis of the available data it is important to proceed with a preliminary descriptive analysis and also to consider the traditional time series analysis in a previous data analysis, in a way that the results may work as indicative measures of the non-parametric procedure following the work presented in [22–25] or a more recent approach [26,27] where a breaking point is identified using a recent Package Software [28] developped for time series structural changes. Both ways are complementary, and therefore, both can contribute to the problem solving. However, more work is required to optimize approaches to enable a significant contribution towards more sustainable urban water management. As future work, we can propose consider at least the parametric and non-parametric approach, using the multi-criteria algorithm proposed in [29].

Acknowledgements. This work was supported by Portuguese funds through the *Center of Naval Research* (CINAV), Portuguese Naval Academy, Portugal and *The Portuguese Foundation for Science and Technology* (FCT), through the *Center for Computational and Stochastic Mathematics* (CEMAT), University of Lisbon, Portugal, project UID/Multi/04621/2019.

References

1. Yang, Q., Wu, X.: 10 challenging problems in data mining research. Int. J. Inf. Technol. Decis. Making **5**(4), 597–604 (2006)
2. Fu, T.: A review on time series data mining. Eng. Appl. Artif. Intell. **24**(1), 164–181 (2011)
3. Weron, R.: Electricity price forecasting: a review of the state-of-the-art with a look into the future. Int. J. Forecast. **30**(4), 1030–1081 (2014)
4. Vasimalla, K.: A survey on time series data mining. Int. J. Innovative Res. Comput. Commun. Eng. **2**(5), 170–179 (2014)
5. Pal, S.M., Palet, J.N.: Time series data mining: a review. Binary J. Data Min. Network. **5**, 01–04 (2015)
6. Zhu, Y.: High performance data mining in time series. Techniques and Case Studies. Ph.D. Thesis. Department of Computer Science. University of New York (2004)
7. Mulquiney, P.: Combining GLM and data- mining techniques for modelling accident compensation data. In: XIth Accident Compensation Seminar. Actuaries Institute, 2nd-4th April 2007, Australia. https://actuaries.asn.au/Library/3.a_ ACS07_paper_Mulquiney_Combining
8. Ratanamahatana, C.A., Lin, J., Gunopulos, D., Keogh, E., Vlachos, M., Das, G.: Mining time series data. In: Maimon, O., Rokach, L. (eds.) Data Mining and Knowledge Discovery Handbook. Springer, Boston, MA (2009). https://doi.org/10.1007/978-0-387-09823-4_56
9. Watson, T.G., Christian, C.D., Mason, A.J., Smith, M.H., Meyer, R.: Bayesian-based pipe failure model. J. Hydroinformatics **6**, 259–264 (2004)
10. Kapelan, Z., Babayan, A.V., Savic, D.A., Walters, G.A., Khu, S.T.: Two new approaches for the stochastic least cost design of water distribution system. Water Supply **4**(5–6), 355–363 (2005)

11. Ragot, J., Maquin, D.: Fault measurement detection in an urban water supply network. J. Process Control **16**, 887–902 (2006)
12. Quevedo, J., et al.: Validation and reconstruction of flow meter data in the Barcelona water distribution network. Control Eng. Pract. **18**(6), 640–651 (2010). https://doi.org/10.1016/j.conengprac.2010.03.003
13. Mutikanga, H.E., Sharma, S.K., Vairavamoorthy, K.: Investigating water meter performance in developing countries: a case study of Kampala, Uganda. Water SA **37**(4), 567–574 (2011). https://doi.org/10.4314/wsa.v37i4.18
14. Sanchez, J. G., Motta, L. A., Alves, W. A.: Estimativa de volume de água não medido em ligações residenciais por perde de exactidão dos hidrômetros, na cidade de Juazeiro - BA. In: Congresso Interamericano de Engenharia Sanitária e Ambiental, 27ª, 2000, Porto Alegre. Anais eletrônicos. Porto Alegre-RS: ABES (Associação Brasileira de Engenharia Sanitária e Ambiental), II-051, pp 1–14 (2000)
15. Makki, A.A., Stewart, R.A., Beala, C.D., Panuwatwanich, K.: Novel bottom-up urban water demand forecasting model: revealing the determinants, drivers and predictors of residential indoor end-use consumption. Resour. Conserv. Recycl. **95**, 15–37 (2015)
16. Beal, C.D., Flynn, J.: Toward the digital water age: Survey and case studies of Australian water utility smart-metering programs. Utilities Policy **32**, 29–37 (2015)
17. Creaco, E., Kossieris, P., Vamvakeridou-Lyroudia, V., Makropoulos, C., Kapelan, Z., Savic, D.: Parameterizing residential water demand pulse models through smart meter readings. Environ. Model. Softw. **80**, 33–40 (2016)
18. Liu, A., Giurco, D., Mukheibir, P.: Urban water conservation through customised water and end-use information. J. Cleaner Prod. **112**(44), 3164–3175 (2016). https://doi.org/10.1016/j.jclepro.2015.10.002
19. Box, G.E., Jenkins, G.M., Reinsel, G.C.: Time Series: Forecasting and Control, 3rd edn. Prentice Hall, New Jersey (1994)
20. Chatfield, C.: The Analysis of Time Series: An Introduction, 6th edn. Chapman and Hall/CRC, Boca Raton (2004)
21. Hamilton, J.D.: Time Series Analysis. Princeton University Press, New Jersey (1994)
22. Cleveland, R.B., Cleveland, W.S., McRae, J.E., Terpenning, I.: STL: a seasonal-trend decomposition procedure based on loess. J. Official Stat. **6**(1), 3–73 (1990)
23. Hyndman, R.J., Khandakar, Y.: Automatic time series forecasting: the forecast package for R. J. Stat. Softw. **26**(3), 1–22 (2008). http://www.jstatsoft.org/article/view/v027i03
24. Zeileis, A., Kleiber, C., Kramer, W., Hornik, K.: Testing and dating of structural changes in practice. Computat. Stat. Data Anal. **44**(109–123), 6 (2013). https://doi.org/10.1016/S0167-9473(03)00030-
25. Gonçalves, E., Lopes, N.M.: Séries temporais - Modelações lineares e não lineares, 2nd edn. Sociedade Portuguesa de Estatística, Lisboa, Portugal (2008)
26. Cristina, S., Cordeiro, C., Lavender, S., Costa Goela, P., Icely, J., Newton, A.: MERIS phytoplankton time series products from the SW Iberian Peninsula (Sagres) using seasonal-trend decomposition based on loess. Remote Sens. **8**(6), 449 (2016)
27. Goela, P.C., Cordeiro, C., Danchenko, S., Icely, J., Cristina, S., Newton, A.: Time series analysis of data for sea surface temperature and upwelling components from the southwest coast of Portugal. J. Mar. Syst. **163**, 12–22 (2016)

28. Zeileis, A., Leisch, F., Hornik, K., Kleiber, C.: Strucchange: an R package for testing for structural change in linear regression models. R Package version 1.5 (2015)
29. Fernandes, S., Clímaco, J. and Captivo, M. E.: A cooperative multicriteria group decision aiding tool - a guided tour of the desktop application. In: Proceedings of Group Decision and Negotiation - GDN 2017, Stuttgart-Hohenheim, Germany, pp. 361–368 (2017)

Using MDS to Compute the Contribution of the Experts in a Delphi Forecast Associated to a Naval Operation's DSS

M. Filomena Teodoro[1,2(✉)] ⓘ, Mário J. Simões Marques[1], Isabel Nunes[3,4], Gabriel Calhamonas[4], and Marina A. P. Andrade[5] ⓘ

[1] CINAV, Center of Naval Research, Naval Academy, Portuguese Navy, 2810-001 Almada, Portugal
maria.alves.teodoro@marinha.pt
[2] CEMAT, Center for Computational and Stochastic Mathematics, Instituto Superior Técnico, Lisbon University, 1048-001 Lisboa, Portugal
[3] UNIDEMI, Department of Mechanical and Industrial Engineering, Faculty of Sciences and Technology, New Lisbon University, 2829-516 Caparica, Portugal
[4] FCT, Faculty of Sciences and Technology, New Lisbon University, 2829-516 Caparica, Portugal
[5] ISTAR - ISCTE, Instituto Universitário de Lisboa, Lisboa Portugal, 1649-026 Lisboa, Portugal

Abstract. The Portuguese Navy gave financial support to THEMIS project under the aim of the development of a decision support system to get optimal decisions in short time in a disaster context, optimizing the decision chain, allowing to get a better performance of tasks execution allowing a reduction of costs.

In [14,17], the authors have considered the facilities and high qualified staff of Portuguese Navy and proposed a variant of the Delphi method, a method that is exceptionally useful where the judgments of individuals are considered as an important information source. They proposed a system that prioritize certain teams for specific incidents taking into account the importance of each team that acts in case of emergency.

In the present work we propose a distinct method of computing the weights that represent the importance given to experts opinion in the Delphi method used in [14,17] under the idea that shall not depend on the years of experience of each expert exclusively but also shall be considered the kind of expert experience. To justify this suggestion we have used hierarchical classification, allowing to identify different padrons for experts with the "same experience". Also discriminant analysis and multidimensional scaling revealed to be adequate techniques for this issue. We can classify the experience of each expert evaluating the similarity/distance between the individuals in the group of proposed experts and compare with the number of consensus presented.

In this manuscript we propose an alternative way of weighting the experts experience that contributes to a decision support system capable to prioritize a set of teams for certain disaster incidents involving maritime issues. The decision support system is still been tested but, with this work, we hope to have given an improvement to its optimization.

ⓒ Springer Nature Switzerland AG 2020
O. Gervasi et al. (Eds.): ICCSA 2020, LNCS 12251, pp. 446–454, 2020.
https://doi.org/10.1007/978-3-030-58808-3_32

Keywords: Decision support system · Expert · Delphi method · Questionnaire · Catastrophe · Hierarchical classification · Discriminant analysis · Multidimensional scaling

1 Introduction

The decision support systems (DSS) appear as consequence of the new technologies and the urgency of optimal decisions in a short interval of time. We can observe DSS where there exists a decision chain so it can be optimized allowing to improve the performance of tasks, for example, in medicine, teaching, engineering, transportation [6, 15, 16]. Also it can be found in a maritime environment. In the case of a catastrophe, a DSS for naval operations would allow to improve all stages of intervention: reception of information on the incidents, tasks to execute, navigation orientation to incidents and advice how to perform the tasks.

The project THEMIS, promoted by Portuguese Navy has as purpose the construction and implementation of such DSS that will allow to handle with disaster relief operations, under naval context. Some contributions to THEMIS project can be found in [10–13] where the approach was performed some recent techniques such as augmented reality or user experience design.

We are particularly interested to build and implement a DSS with the ability to prioritize certain teams for specific incidents taking into account the importance of each team that acts in case of emergency, the sequence of tasks that should perform all possible orders to be given.

As the author of [15] says, the Delphi method is a method that is exceptionally useful where the opinions of individuals are needed to dilute the lack of agreement or incomplete state of knowledge. This method is particularly important due its ability to structure and organize group communication. In [5, 8, 18] we can find numerous examples of Delphi method applications. Some examples that illustrate the importance of Delphi method: transportation [6], in paper pulp production [7], the construction of an index in automotive supply chain [2, 3], education [5], natural sciences [15], social policy [1].

We have applied the Delphi method to facilities and high qualified staff of Portuguese Navy [14, 17].

Data collection is already complete. The obtained results depend on the weighting of experts experience. In the [14, 17], the considered weights depend on the time of the expert service. In the present work, we evidence that these weights shall take into account not only the time of experience but the similarity of the experts opinion.

In Sects. 2 and 3 we describe some details about the proposed methodology and the previous work. The evidence of the improvement issue need is provided in Sect. 4. Also are done some conclusions and suggestions.

2 Multidimensional Scaling

The Multidimensional Scaling (MDS) is a multivariate technique that explores a set of multivariate measurements of individuals attributes, representing them in a reduced dimensional system, identifying distinct latent variables associated to the individuals perception [4, 9]. The main objective of MDS is the representation of attributes or individuals in a map of proximities (similarities) with a reduced dimension. We can notice that the data in MDS are similarities/dissimilarities, either measured or perceived or else computed from a set of attributes. The more general application of MDS is the classical MDS. When the data measure some attributes or characteristics, it is necessary to determine a distance between these attributes. When we have quantitative measures the usual distance is the Minkovski distance. In other case, it is usual to use the chi-square statistics for the contingency tables or the correlation coefficient.

Summarizing, the basic steps of MDS are: Assign a number of points to coordinates in a n-dimensional space (can be a 2-dimensional space, a 3-dimensional space, etc); Computation of distances for all pair of points (can be the Euclidean distance - a particular case of the Minkovski distance); Comparison of the similarity matrix with the original input matrix by evaluating the stress function (statistics that evaluates the differences of the actual distances and the estimated distances; The adjutment of coordinates, when necessary so can be minimized the stress.

3   Delphi Method - Round I

In this section, we will describe shortly the first round of Delphi method approach performed in [14, 17].

To proceed with Delphi method, in a preliminary stage to first round of questionnaires, the potential group of experts were inquired about some individual characteristics of their profile: age, gender, professional rank, training class, type of experience in response to disasters (real versus train exercise) and the total ship boarding time (less than 1 year, 1–3 years, 3–5 years, more than 5 years).

With the objective of identifying the degree of priority that which each team should carry out each task, the same group also performed a questionnaire classified in a Likert scale of importance from 1 (Not Important) to 6 (Extremely Important) all possible tasks to be carried out during a humanitarian disaster relief operation for each existing team that can provide its service in the concerned operation (consult all tasks and all possible teams described below).

The questionnaire considered 52 tasks described in Table 1.

Also, the questionnaire included eleven teams as displayed in Table 2.

A total of 572 questions were classified on a Likert scale of 1 to 6. Each expert should answer to questions, indicating the degree of confidence of the given answer and how much experience had with each team previously.

In the first round of questionnaires, 12 experts were considered, all males with at least 5 years on board, all aged between 35 and 54 years, with positions of Captain-lieutenant, Captain-of-sea-and-war and Captain-of-frigate (the most common). Between all experts only 25% have real past experience of disaster response. The remaining 75% experts are training experienced.

To identify the tasks that reached consensus is necessary to determine the Inter-Quartile Range (IQR), that is, the difference between the 1st Quartile and the 3rd Quartile. IQR represents 50% of the observations closest to the median. When IQR is less

Table 1. Collected tasks. Source [14].

Identify incidents	Repair electrical power system
Screen survivors / homeless and injured	Repair communications system
Provide 1st aid	Repair of lighting system [point of interest]
Census individuals	Repair mechanical energy production system
Identify location for [point of interest]	Repair power distribution system
Transport equipment to install [point of interest]	Recover basic sanitation
Mount [point of interest]	Create safety perimeter
Carrying severe injuries	Ensuring perimeter safety
Stabilize serious injury	Impose order and safety
Carry minor injured	Carry out order and safeguard of goodies and property
Rescue imprisoned victim	Carry out flight operations to transport material
Rescue victim from altitude	Perform flight operations for medical evacuations
Rescue victim in collapsed structure	Transport material [type] from [local] to [local]
Rescue isolated victim by land	Shift escort
Rescue single victim by air	Convey distribute food for the wounded
Rescue an isolated victim via water	Convey and Distribute food
Stabilize structures	Dead transport to morgue
Clear paths	Support the funeral ceremony
Build support structures for rescue	Status report
Fire fighting	Evacuate equipment to [point of interest]
Fighting floods	Evacuating population to [point of interest]
Carrying out shoring	Diving for minor repairs
Diving drinking water	Diving for rescue
Restoration of water supply	Evacuation of animals
Control of water leaks	Distribution of animal feed
Repair electric pumping system	Burying dead animal

than 1 it means that more than 50% of the responses obtained are within 1 point of the Likert scale [5].

To process the questionnaires data and classify each one as consensual/no consensual, was necessary to compute a weighted mean (WM), where the weights are based on the experience of each expert responded to with each team. These weights evaluate the individual time of service. Notice that the opinion of a more experienced expert has a more significant weight in the weighted mean of the importance of a certain team and vice-verse. See Table 3.

For the tasks with an IQR greater than 1, was assumed that the experts have not reached a consensus on how important a certain team is to perform a specif task. In this way it will be considered in the next round of questionnaires.

When a task has an IQR less or equal than one, is necessary to analyze its distribution of frequencies. Can be seen that the answer distribution per task is, come times, non consensual. In this case, it was proposed a new criterion: there is no consensus on all questions when there are 2 or more non-adjacent levels of importance that account for at least 20% of the sum of weighted levels of experience.

By the proposed rule, the first round of questionnaires, between the 572 questions, there was no consensus on 290 questions (50.7%), from which 282 have an IQR higher than 1; between the tasks with IQR less or equal to one, only 8 did not meet the requirements of the proposed methodology. These total of 290 questions are included in next round of questionnaires.

Table 2. Available teams. Source [14].

Reconnaissance	Water and Sanitation Technique Brigade
Search and Rescue - SAR	Mechanical Engineering Brigade
Search and Rescue Urban - SAR URB	Technical Brigade of Electricity
Search and Rescue Structures - SAR EST	Supply
Brigade Firefighting Technique	Food
Medical	

Table 3. Weights as function of level of experience of each expert.

Level of experience	Weight
1	0.1
2	0.5
3	0.9
4	1.1
5	1.5
6	1.9

4   Proposal

Here, we present the motivation of improving the table of weights Table 3 in the first round to get the WM. The initial weights depend on the individual time of service (see the Table 3). Notice that the opinion of a more experienced expert has a more significant weight in the weighted mean of the importance of a certain team and vice-versa. But

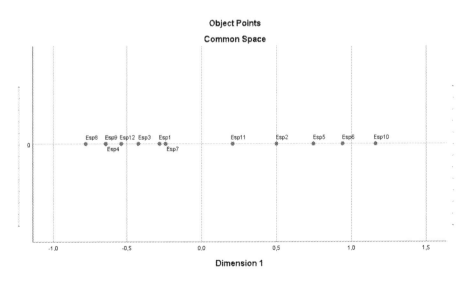

Fig. 1. Global position, dimension 1. The different experts appears in a different order for dimension 1. SAR-REC.team.

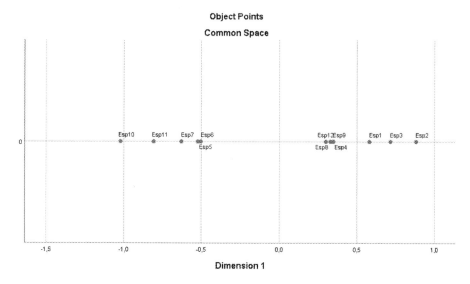

Fig. 2. Global position, dimension 1. The different experts appears in a different order for dimension 1. SAR-URB team.

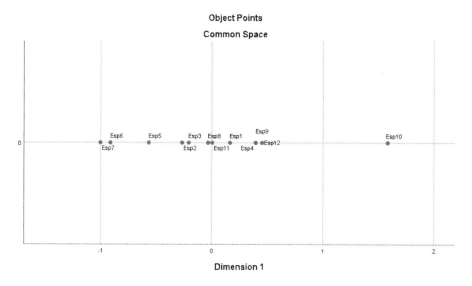

Fig. 3. Global position, dimension 1. The different experts appears in a different order for dimension 1. SAR-ST team.

the experts details, in what specific area shall have a bigger weight or a lower weight is not contemplated in the initial phase. As a first glance, we have performed a hierarchical classification of the experts per task/team. The associated dendograms presented different profiles for distinct tasks. The clusters of individuals were distinct from task to task. The discriminant analysis corroborated with that: the profile of the experts were distinct per task/team. We propose to analyze the proximity of experts per specific team. We have tested some teams and studied the measures of proximity between the experts using multidimensional scaling, algorithm PROXSCAL that minimizes the normalized raw stress. We can give as example, the proximity of the twelve experts has a distinct tip when we consider the SAR-URB, SAR-REC or SAR-EST as we can see in Figs. 1, 2, 3, where the different experts appears in a different order for dimension 1 (we have considered a 2D space, but, for simplicity we represent the 1D space). Also, the measure of distance between experts about the same teams has distinct patterns (see Figs. 4, 5, 6). These facts suggest that the Table 3 can be reformulated taking into account the similarity of the experts per each team. The details about the analysis of results per round using this proposal will be available in an extended version of the this article.

Distances

Dimensionality: 1

	Esp1	Esp2	Esp3	Esp4	Esp5	Esp6	Esp7	Esp8	Esp9	Esp10	Esp11	Esp12
Esp1	.000											
Esp2	1.164	.000										
Esp3	1.810	.646	.000									
Esp4	.411	.753	1.399	.000								
Esp5	1.357	.193	.453	.946	.000							
Esp6	.129	1.035	1.681	.282	1.228	.000						
Esp7	.867	.296	.943	.456	.490	.738	.000					
Esp8	.687	.477	1.123	.276	.670	.558	.180	.000				
Esp9	.411	.753	1.399	.000	.946	.282	.456	.276	.000			
Esp10	1.033	.131	.777	.622	.324	.904	.166	.346	.622	.000		
Esp11	1.523	.359	.287	1.112	.166	1.394	.656	.836	1.112	.490	.000	
Esp12	1.674	.510	.136	1.263	.317	1.545	.807	.987	1.263	.641	.151	.000

Fig. 4. Distance between experts. The different experts appears distinct distances with each other when we consider different teams. SAR-REC team.

Distances

Dimensionality: 1

	Esp1	Esp2	Esp3	Esp4	Esp5	Esp6	Esp7	Esp8	Esp9	Esp10	Esp11	Esp12
Esp1	.000											
Esp2	.301	.000										
Esp3	.134	.167	.000									
Esp4	.235	.536	.369	.000								
Esp5	1.102	1.403	1.236	.867	.000							
Esp6	1.083	1.384	1.217	.848	.019	.000						
Esp7	1.208	1.509	1.342	.973	.106	.126	.000					
Esp8	.253	.555	.387	.019	.849	.829	.955	.000				
Esp9	.235	.536	.369	.000	.867	.848	.973	.019	.000			
Esp10	1.600	1.901	1.734	1.366	.498	.518	.392	1.347	1.366	.000		
Esp11	1.385	1.687	1.520	1.151	.284	.303	.177	1.132	1.151	.215	.000	
Esp12	.282	.583	.416	.047	.820	.800	.926	.029	.047	1.318	1.103	.000

Fig. 5. Distance between experts. The different experts appears distinct distances with each other when we consider different teams. SAR-URB team.

Distances

Dimensionality: 1

	Esp1	Esp2	Esp3	Esp4	Esp5	Esp6	Esp7	Esp8	Esp9	Esp10	Esp11	Esp12
Esp1	.000											
Esp2	1.164	.000										
Esp3	1.810	.646	.000									
Esp4	.411	.753	1.399	.000								
Esp5	1.357	.193	.453	.946	.000							
Esp6	.129	1.035	1.681	.282	1.228	.000						
Esp7	.867	.296	.943	.456	.490	.738	.000					
Esp8	.687	.477	1.123	.276	.670	.558	.180	.000				
Esp9	.411	.753	1.399	.000	.946	.282	.456	.276	.000			
Esp10	1.033	.131	.777	.622	.324	.904	.166	.346	.622	.000		
Esp11	1.523	.359	.287	1.112	.165	1.394	.656	.836	1.112	.490	.000	
Esp12	1.674	.510	.136	1.263	.317	1.545	.807	.987	1.263	.641	.151	.000

Fig. 6. Distance between experts. The different experts appears distinct distances with each other when we consider different teams. SAR-ST team.

Acknowledgements. This work was supported by Portuguese funds through the *Center of Naval Research* (CINAV), Portuguese Naval Academy, Portugal and *The Portuguese Foundation for Science and Technology* (FCT), through the *Center for Computational and Stochastic Mathematics* (CEMAT), University of Lisbon, Portugal, project UID/Multi/04621/2019.

References

1. Adler, M., Ziglio, E.: Gazing Into the Oracle: The Delphi Method and Its Application to Social Policy and Public Health. Kingsley Publishers, London (1996)
2. Azevedo, S.G., Carvalho, H., Machado, V.C.: Agile index: automotive supply chain. World Acad. Sci. Eng. Technol. **79**, 784–790 (2011)
3. Azevedo, S.G., Govindan, K., Carvalho, H., Machado, V.C.: Ecosilient index to assess the grenness and resilience of the up stream automotive supply chain. J. Cleaner Prod. **56**, 131–146 (2013). https://doi.org/10.1016/j.jclepro.2012.04.011
4. Borg, I., et al.: Appl. MDS. Springer Science & Business Media, Heidelberg (2012). https://doi.org/10.1007/978-3-642-31848-1
5. De Vet, E., Brug, J., De Nooijer, J., Dijkstra, A., De Vries, N.K.: Determinants of forward stage transitions: a Delphi study. Health Educ. Res. **20**(2), 195–205 (2005)
6. Duuvarci, Y., Selvi, O., Gunaydin, O., GÜR, G.: Impacts of transportation projects on urban trends in İzmir. Teknik Dergi **19**(1), 4293–4318 (2008)
7. Fraga, M.: A economia circular na indústria portuguesa de pasta, papel e cartão, Master dissertation, FCT, Universidade Nova de Lisboa, Almada (2017)
8. Gunaydin, H.M.: Impact of information technologies on project management functions. Ph.D. dissertation, Chicago University, USA (1999)
9. Kruskal J., Wish, M.: Multidimensional Scaling. SAGE (1978)
10. Simões-Marques, M., Correia, A., Teodoro, M.F., Nunes, I.L.: Empirical studies in user experience of an emergency management system. In: Nunes, I. (eds.) Advances in Human Factors and Systems Interaction. AHFE 2017. Advances in Intelligent Systems and Computing, vol. 592, pp. 97–108. Springer, Cham (2018). https://doi.org/10.1007/978-3-319-60366-7_10
11. Marques, M., Elvas, F., Nunes, I.L., Lobo, V., Correia, A.: Augmented reality in the context of naval operations. In: Ahram, T., Karwowski, W., Taiar, R. (eds.) IHSED 2018. AISC, vol. 876, pp. 307–313. Springer, Cham (2019). https://doi.org/10.1007/978-3-030-02053-8_47

12. Nunes, I.L., Lucas, R., Simões-Marques, M., Correia, N.: Augmented reality in support of disaster response. In: Nunes, I. (eds.) Advances in Human Factors and Systems Interaction. AHFE 2017. Advances in Intelligent Systems and Computing, vol. 592, pp. 155–167. Springer, Cham (2018). https://doi.org/10.1007/978-3-319-60366-7_15

13. Nunes, I.L., Lucas, R., Simões-Marques, M., Correia, N.: An augmented reality application to support deployed emergency teams. In: Bagnara, S., Tartaglia, R., Albolino, S., Alexander, T., Fujita, Y. (eds.) IEA 2018. AISC, vol. 822, pp. 195–204. Springer, Cham (2019). https://doi.org/10.1007/978-3-319-96077-7_21

14. Nunes, I.L., Calhamonas, G., Marques, M., Teodoro, M.F.: Building a decision support system to handle teams in disaster situations - a preliminary approach. In: Madureira, A.M., Abraham, A., Gandhi, N., Varela, M.L. (eds.) HIS 2018. AISC, vol. 923, pp. 551–559. Springer, Cham (2020). https://doi.org/10.1007/978-3-030-14347-3_54

15. Powell, C.: The Delphi technique: myths and realities. Methodological Issues Nurs. Res. **41**(4), 376–382 (2003)

16. Reuven, L., Dongchui, H.: Choosing a technological forecasting method. Ind. Manag. **37**(1), 14–22 (1995)

17. Simões-Marques, M., Filomena Teodoro, M., Calhamonas, G., Nunes, I.L.: Applying a variation of Delphi method for knowledge elicitation in the context of an intelligent system design. In: Nunes, I.L. (ed.) AHFE 2019. AISC, vol. 959, pp. 386–398. Springer, Cham (2020). https://doi.org/10.1007/978-3-030-20040-4_35

18. Wissema, J.G.: Trends in technology forecasting. Res. Dev. Manag. **12**(1), 27–36 (1982)

Multiscale Finite Element Formulation for the 3D Diffusion-Convection Equation

Ramoni Z. S. Azevedo[1](✉) and Isaac P. Santos[1,2](✉)

[1] Optimization and Computational Modeling Lab,
Federal University of Espírito Santo, Vitória, ES, Brazil
rsedano@inf.ufes.br, isaac.santos@ufes.br
[2] Department of Applied Mathematics, Federal University of Espírito Santo,
São Mateus, ES, Brazil

Abstract. In this work we present a numerical study of the Dynamic Diffusion (DD) method for solving diffusion-convection problems with dominant convection on three dimensional domains. The DD method is a two-scale nonlinear model for convection-dominated transport problems, obtained by incorporating to the multiscale formulation a nonlinear dissipative operator acting isotropically in both scales of the discretization. The standard finite element spaces are enriched with bubble functions in order to add stability properties to the numerical model. The DD method for two dimensional problems results in good solutions compared to other known methods. We analyze the impact of this methodology on three dimensional domains comparing its numerical results with those obtained using the Consistent Approximate Upwind (CAU) method.

Keywords: Dynamic diffusion method · Multiscale methods · Diffusion-convection equation

1 Introduction

The diffusive-convective transport equation plays an important role in numerous physical phenomena. It models several problems in engineering, such as the process of pollution on rivers, seas and the atmosphere; oil reservoir flow; heat transfer, and so forth [1]. However, the accurate and reliable numerical simulation of this problem is still a challenging task [2–4].

It is well known that the standard Galerkin Finite Element method (FEM) is not appropriate for solving convection-dominated problems [2], since it experiences spurious oscillations in the numerical solution, that grow as the diffusion coefficient tends to zero. The problem is caused by its loss of stability with respect to the standard H^1-norm [5]. Since the early 1980s, various stabilization techniques have been developed to overcome the weaknesses of the FEM to solve such problems.

One of the most known stabilized formulation to solve convection-dominated transport problem is the Streamline Upwind Petrov-Galerkin (SUPG) method

O. Gervasi et al. (Eds.): ICCSA 2020, LNCS 12251, pp. 455–469, 2020.
https://doi.org/10.1007/978-3-030-58808-3_33

(also known as the Streamline Diffusion Finite Element method (SDFEM)), proposed in [6] and analyzed in [7]. This methodology consists of adding residual-based terms to the FEM, preserving the consistency of the variational formulation. Other methodologies were developed to circumvent the difficulty of the FEM to solve problems of this nature, such as, the Galerkin-Least Square method (GLS) [8] (a variant of the SUPG), edge-based stabilization (the Continuous Interior Penalty (CIP) method [9]), and local projections (the Local Projection Stabilization (LPS) [4,10]). All these numerical formulations are capable to stabilize most of the unphysical oscillations from the FEM, caused by dominating convection. However, localized oscillations may remain in the neighborhood of sharp layers for nonregular solutions [11]. As a remedy, a number of so-called Spurious Oscillations at Layers Diminishing (SOLD) schemes have been proposed; see the excellent reviews [12,13]. SOLD schemes add additional, in general nonlinear, discontinuity (or shock-capturing) terms to these formulations in order to reduce the localized oscillations.

Residual-based stabilized methods have been reinterpreted within the context of the Variational Multiscale (VMS) framework by Hughes and collaborators in the 1990s [14–17]. The starting point of the VMS framework is the splitting of the finite element spaces into two parts: resolved (or coarse) and fine (or subgrid) subspaces. Consequently, the variational problem is decomposed into two subproblems, one associated to the resolved scales and the other associated to the fine ones. The fine scales are approximated or modeled, and then, incorporeted into the coarse scale problem, yielding in an enriched problem for the resolved scales. Many classical stabilization formulation for convection-dominated problems, such as SUPG and GLS can be reinterpreted as residual-based fine-scale models [16,18]. Also, other known multiscale methods, such as the residual-free bubbles method [19,20] and subgrid viscosity methods [21,22], can be placed within this framework.

Over the last years, some nonlinear two-scale variational methods have been developed, such as the Nonlinear Subgrid Stabilization (NSGS) [23,24] and Dynamic Diffusion (DD) methods [11,25] to solve convection-dominated problems; the Nonlinear Multiscale Viscosity (NMV) method [26,27] to solve the system of compressible Euler equations. In [28] the NSGS method was applied to solve incompressible Navier-Stokes equations. In this class of methods, bubble functions are used to enrich the standard finite element spaces and residual-based nonlinear artificial diffusion terms are added to the resulting formulation, in order to improve the stability properties of the numerical model. In particular, the DD method is obtained by incorporating to the VMS formulation, a nonlinear dissipative operator acting isotropically in both scales of the discretization. This method results in good solutions for two dimensional problems, compared to other known stabilized methods, as can be seen in [11,25].

The aim of this work is to present a numerical study of the DD method for solving steady-state diffusion-convection problems with dominant convection on three dimensional domains. We compare its computational results with the SOLD method, Consistent Approximate Upwind Petrov-Galerkin (CAU) [29], in the solution of two convection dominated problems with internal and external layers.

The remainder of this work is organized as follows. Section 2 briefly addresses the governing equation, the numerical formulations of the Galerkin finite element, CAU and DD methods. The numerical experiments are conducted in Sect. 3 and Sect. 4 concludes this paper.

2   Governing Equation and Numerical Formulation

Let $\Omega \subset \mathbb{R}^3$ be a bounded domain with a Lipschitz boundary Γ with an outward unit normal $\boldsymbol{n}$. The linear and steady-state diffusion-convection equation, with a Dirichlet boundary condition, is modeled by

$$- \kappa \Delta u + \boldsymbol{\beta} \cdot \boldsymbol{\nabla} u = f \ \text{ in } \Omega, \tag{1}$$

$$u = g \ \text{ on } \Gamma, \tag{2}$$

where u represents the quantity being transported, $\kappa > 0$ is the diffusivity coefficient, $\boldsymbol{\beta} \in [L^\infty(\Omega)]^3$ is the velocity field such that $\boldsymbol{\nabla} \cdot \boldsymbol{\beta} = 0$, $f \in L^2(\Omega)$ is a given source term and $g \in L^2(\Gamma)$ is a given function.

Consider the set of all admissible functions S and the space of test functions V defined as

$$S = \{u \in H^1(\Omega) : u = g \text{ on } \Gamma\}, \tag{3}$$

$$V = \{v \in H^1(\Omega) : v = 0 \text{ on } \Gamma\}. \tag{4}$$

The continuous variational problem associated to the boundary value problem (1) and (2) can be stated as: find $u \in S$ such that

$$A(u, v) = F(v), \quad \forall v \in V, \tag{5}$$

with

$$A(u, v) = \int_\Omega \left(\kappa \boldsymbol{\nabla} u \cdot \boldsymbol{\nabla} v + (\boldsymbol{\beta} \cdot \boldsymbol{\nabla} u)v \right) d\Omega, \tag{6}$$

$$F(v) = \int_\Omega fv \, d\Omega. \tag{7}$$

2.1   The Galerkin Finite Element Method

We consider a regular partition $\mathcal{T}_h = \{\Omega_1, \Omega_2, \dots, \Omega_{nel}\}$ of the domain Ω into tetrahedral elements. The discrete spaces associated to S and V are defined as

$$S_h = \{u_h \in H^1(\Omega) : u_h|_{\Omega_e} \in \mathbb{P}_1(\Omega_e), \forall \Omega_e \in \mathcal{T}_h, u_h = g \in \Gamma\}, \tag{8}$$

$$V_h = \{v_h \in H^1(\Omega) : v_h|_{\Omega_e} \in \mathbb{P}_1(\Omega_e), \forall \Omega_e \in \mathcal{T}_h, v_h = 0 \in \Gamma\}, \tag{9}$$

in which $\mathbb{P}_1(\Omega_e)$ is the set of first order polynomials on local coordinates. The Galerkin formulation is obtained by restricting the problem (5) to the discrete spaces $S_h \subset S$ and $V_h \subset V$ [2,30]. In other words, the variational statement of the approximate problem consists of finding $u_h \in S_h$, such that,

$$A_G(u_h, v_h) = F_G(v_h), \quad \forall v_h \in V_h, \tag{10}$$

where

$$A_G(u_h, v_h) = \sum_{e=1}^{nel} \int_{\Omega_e} \left(\kappa \nabla u_h \cdot \nabla v_h + (\boldsymbol{\beta} \cdot \nabla u_h) v_h \right) d\Omega_e, \tag{11}$$

$$F_G(v_h) = \sum_{e=1}^{nel} \int_{\Omega_e} f v_h \, d\Omega_e. \tag{12}$$

The Galerkin finite element approximation for the variable u_h is given at the linear tetrahedral element level as follows,

$$u_h = \sum_{j=1}^{4} N_j u_j^e, \tag{13}$$

where N_j is the local interpolation function associated with the nodal point j of the element Ω_e. These functions expressed in global variables $\boldsymbol{x} = (x, y, z)$ are of the form

$$N_1 = \frac{1}{6|\Omega_e|}(a_1 + b_1 x + c_1 y + d_1 z), \quad N_2 = \frac{1}{6|\Omega_e|}(a_2 + b_2 x + c_2 y + d_2 z),$$

$$N_3 = \frac{1}{6|\Omega_e|}(a_3 + b_3 x + c_3 y + d_3 z), \quad N_4 = \frac{1}{6|\Omega_e|}(a_4 + b_4 x + c_4 y + d_4 z),$$

where $|\Omega_e|$ is the volume of the element, calculated by

$$|\Omega_e| = \frac{1}{6} \begin{vmatrix} 1 & x_1 & y_1 & z_1 \\ 1 & x_2 & y_2 & z_2 \\ 1 & x_3 & y_3 & z_3 \\ 1 & x_4 & y_4 & z_4 \end{vmatrix},$$

whereas the coefficients of the shape functions are given by

$$a_1 = \begin{vmatrix} x_2 & y_2 & z_2 \\ x_3 & y_3 & z_3 \\ x_4 & y_4 & z_4 \end{vmatrix}, \quad a_2 = -\begin{vmatrix} x_1 & y_1 & z_1 \\ x_3 & y_3 & z_3 \\ x_4 & y_4 & z_4 \end{vmatrix}, \quad a_3 = \begin{vmatrix} x_1 & y_1 & z_1 \\ x_2 & y_2 & z_2 \\ x_4 & y_4 & z_4 \end{vmatrix}, \quad a_4 = -\begin{vmatrix} x_1 & y_1 & z_1 \\ x_2 & y_2 & z_2 \\ x_3 & y_3 & z_3 \end{vmatrix},$$

$$b_1 = -\begin{vmatrix} 1 & y_2 & z_2 \\ 1 & y_3 & z_3 \\ 1 & y_4 & z_4 \end{vmatrix}, \quad b_2 = \begin{vmatrix} 1 & y_1 & z_1 \\ 1 & y_3 & z_3 \\ 1 & y_4 & z_4 \end{vmatrix}, \quad b_3 = -\begin{vmatrix} 1 & y_1 & z_1 \\ 1 & y_2 & z_2 \\ 1 & y_4 & z_4 \end{vmatrix}, \quad b_4 = \begin{vmatrix} 1 & y_1 & z_1 \\ 1 & y_2 & z_2 \\ 1 & y_3 & z_3 \end{vmatrix},$$

$$c_1 = \begin{vmatrix} 1 & x_2 & z_2 \\ 1 & x_3 & z_3 \\ 1 & x_4 & z_4 \end{vmatrix}, \quad c_2 = -\begin{vmatrix} 1 & x_1 & z_1 \\ 1 & x_3 & z_3 \\ 1 & x_4 & z_4 \end{vmatrix}, \quad c_3 = \begin{vmatrix} 1 & x_1 & z_1 \\ 1 & x_2 & z_2 \\ 1 & x_4 & z_4 \end{vmatrix}, \quad c_4 = -\begin{vmatrix} 1 & x_1 & z_1 \\ 1 & x_2 & z_2 \\ 1 & x_3 & z_3 \end{vmatrix},$$

$$d_1 = -\begin{vmatrix} 1 & x_2 & y_2 \\ 1 & x_3 & y_3 \\ 1 & x_4 & y_4 \end{vmatrix}, \quad d_2 = \begin{vmatrix} 1 & x_1 & y_1 \\ 1 & x_3 & y_3 \\ 1 & x_4 & y_4 \end{vmatrix}, \quad d_3 = -\begin{vmatrix} 1 & x_1 & y_1 \\ 1 & x_2 & y_2 \\ 1 & x_4 & y_4 \end{vmatrix}, \quad d_4 = \begin{vmatrix} 1 & x_1 & y_1 \\ 1 & x_2 & y_2 \\ 1 & x_3 & y_3 \end{vmatrix}.$$

The transformation from global variables $\boldsymbol{x} = (x, y, z)$ to local variables $\boldsymbol{\xi} = (\xi, \eta, \zeta)$, represented in Fig. 1, results in the following interpolation functions [31],

$$N_1 = 1 - \xi - \eta - \zeta, \quad N_2 = \xi, \quad N_3 = \eta, \quad N_4 = \zeta.$$

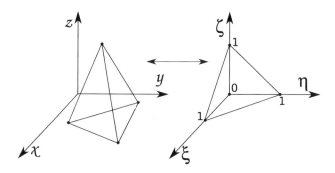

Fig. 1. Transformation from global variables (x, y, z) to local variables (ξ, η, ζ).

The matrix of transformation from global coordinates to local coordinates is defined as

$$\frac{d\boldsymbol{x}}{d\boldsymbol{\xi}} = \begin{bmatrix} \frac{\partial x}{\partial \xi} & \frac{\partial x}{\partial \eta} & \frac{\partial x}{\partial \zeta} \\ \frac{\partial y}{\partial \xi} & \frac{\partial y}{\partial \eta} & \frac{\partial y}{\partial \zeta} \\ \frac{\partial z}{\partial \xi} & \frac{\partial z}{\partial \eta} & \frac{\partial z}{\partial \zeta} \end{bmatrix} = \begin{bmatrix} x_2 - x_1 & x_3 - x_1 & x_4 - x_1 \\ y_2 - y_1 & y_3 - y_1 & y_4 - y_1 \\ z_2 - z_1 & z_3 - z_1 & z_4 - z_1 \end{bmatrix}, \quad (14)$$

and the inverse transformation matrix is determined by

$$\frac{d\boldsymbol{\xi}}{d\boldsymbol{x}} = \begin{bmatrix} \frac{\partial \xi}{\partial x} & \frac{\partial \xi}{\partial y} & \frac{\partial \xi}{\partial z} \\ \frac{\partial \eta}{\partial x} & \frac{\partial \eta}{\partial y} & \frac{\partial \eta}{\partial z} \\ \frac{\partial \zeta}{\partial x} & \frac{\partial \zeta}{\partial y} & \frac{\partial \zeta}{\partial z} \end{bmatrix} = \frac{1}{6|\Omega_e|} \begin{bmatrix} b_2 & c_2 & d_2 \\ b_3 & c_3 & d_3 \\ b_4 & c_4 & d_4 \end{bmatrix}. \quad (15)$$

The integrals present in the formulations can be easily solved analytically, using the formula for integrating products of powers of linear interpolation functions over tetrahedral elements [31],

$$\int_{\Omega_e} N_1^i N_2^j N_3^k N_4^l \, d\Omega_e = \frac{i!\,j!\,k!\,l!}{(i + j + k + l + 3)!} 6|\Omega_e|.$$

The FEM method results in stable and accurate approximate solutions when the transport equation is diffusive dominant [6, 12]. However, when the problem is convective dominant, the numerical solution may presents spurious oscillations, mainly when boundary layers (narrow regions with steep gradients) appear in the solution, requiring the use of stabilized methods [6, 12, 23, 32]. In the following subsections we describe two different stabilization methods used to control the oscillations caused by the dominant convection.

2.2   The Consistent Approximate Upwind Method

The CAU is a stabilized method proposed in [29] for solving convection-dominated problems. It is a SOLD (discontinuity-capturing) model that consists of splitting the stabilization formulation into two parts, where the first is a linear stabilization term, the standard SUPG operator [6], and the second one is a nonlinear operator aiming to control the solution gradient, preventing localized oscillations around boundary layers. In this work we follow the CAU formulation described in [33]. The CAU formulation consists of finding $u_h \in S_h$ such that,

$$A_G(u_h, v_h) + A_{SUPG}(u_h, v_h) + A_{CAU}(u_h; u_h, v_h) = F_G(v_h) + F_{SUPG}(v_h), \quad (16)$$

$\forall v_h \in V_h$, where $A_G(u_h, v_h)$ is given by (11), and

$$A_{SUPG}(u_h, v_h) = \sum_{e=1}^{nel} \int_{\Omega_e} (-\kappa \Delta u_h + \boldsymbol{\beta} \cdot \boldsymbol{\nabla} u_h) \tau_{SUPG} (\boldsymbol{\beta} \cdot \boldsymbol{\nabla} v_h) \, d\Omega_e, \quad (17)$$

$$A_{CAU}(u_h; u_h, v_h) = \sum_{e=1}^{nel} \int_{\Omega_e} \tau_{CAU} \boldsymbol{\nabla} u_h \cdot \boldsymbol{\nabla} v_h \, d\Omega_e, \quad (18)$$

$$F_{SUPG}(v_h) = \sum_{e=1}^{nel} \int_{\Omega_e} f \tau_{SUPG} (\boldsymbol{\beta} \cdot \boldsymbol{\nabla} v_h) \, d\Omega_e. \quad (19)$$

The stabilization parameters τ_{SUPG} and τ_{CAU} are defined as follow,

$$\tau_{SUPG} = \frac{h \alpha(Pe)}{2\|\boldsymbol{\beta}\|},$$

where $\| \cdot \|$ is the Euclidian norm in $\mathbb{R}^3$, Pe is the local Péclet number and h is the characteristic length of mesh:

$$Pe = \frac{h\|\boldsymbol{\beta}\|}{2\kappa}, \quad \alpha(Pe) = max \left\{ 0; 1 - \frac{1}{Pe} \right\},$$

$$h = 2 \frac{\|\boldsymbol{\beta}\|}{\|\boldsymbol{\gamma}\|}, \quad \gamma_i = \sum_{j=1}^{3} \frac{\partial \xi_i}{\partial x_j} \beta_j$$

and

$$\tau_{CAU} = \begin{cases} 0, & \text{if } \frac{|R(u_h)|}{\|\boldsymbol{\beta}\| \, \|\boldsymbol{\nabla} u_h\|} \geq \frac{\bar{\alpha}(\bar{Pe})\bar{h}}{\alpha(Pe)h}, \\[2ex] \frac{\alpha(Pe)h}{2} \left[\frac{\bar{\alpha}(\bar{Pe})\bar{h}}{\alpha(Pe)h} - \frac{|R(u_h)|}{\|\boldsymbol{\beta}\| \, \|\boldsymbol{\nabla} u_h\|} \right] \frac{|R(u_h)|}{\|\boldsymbol{\nabla} u_h\|}, & \text{if } \frac{|R(u_h)|}{\|\boldsymbol{\beta}\| \, \|\boldsymbol{\nabla} u_h\|} < \frac{\bar{\alpha}(\bar{Pe})\bar{h}}{\alpha(Pe)h}, \end{cases}$$

where $R(u_h) = -\kappa \Delta u_h + \boldsymbol{\beta} \cdot \boldsymbol{\nabla} u_h - f$ is the residual of Eq. (1) on Ω_e, $\bar{Pe}$ is the local Péclet number and $\bar{h}$ the caracteristic length of mesh modifieds in termos of an auxiliary vector field $\bar{\boldsymbol{\beta}}$,

$$\bar{h} = 2 \frac{\|\boldsymbol{\beta} - \bar{\boldsymbol{\beta}}\|}{\bar{\gamma}}, \quad \bar{\gamma}_i = \sum_{j=1}^{3} \frac{\partial \xi_i}{\partial x_j} (\beta_j - \bar{\beta}_j),$$

$$\bar{Pe} = \frac{\bar{h}\|\boldsymbol{\beta} - \bar{\boldsymbol{\beta}}\|}{2\kappa}, \quad \bar{\alpha}(\bar{Pe}) = max \left\{ 0; 1 - \frac{1}{\bar{Pe}} \right\}.$$

The velocity field $\bar{\beta}$, given by

$$\bar{\beta} = \begin{cases} \beta, & \text{if } \|\nabla u_h\| = 0, \\ \beta - \dfrac{|R(u_h)|}{\|\nabla u_h\|^2} \nabla u_h, & \text{if } \|\nabla u_h\| > 0, \end{cases}$$

determines a new upwind direction, $\beta - \bar{\beta}$, calculated in each iteration in terms of the approximate solution of the previous iteration. It is important to note that the $\beta - \bar{\beta}$ is in the direction of $\|\nabla u_h\|$, allowing the CAU method to possess greater stability than the SUPG method in regions near to internal and/or external boundary layers [33].

The method is solved by using an iterative procedure defined as: given u_h^n, we find u_h^{n+1} satisfying,

$$A_G(u_h^{n+1}, v_h) + A_{SUPG}(u_h^{n+1}, v_h) + A_{CAU}(u_h^n; u_h^{n+1}, v_h) = F_G(v_h) + F_{SUPG}(v_h),$$

$\forall v_h \in V_h$, where the initial solution u_h^0 is obtained by the standard Galerkin finite element method (10). The nonlinear convergence is checked under a given prescribed relative error with tolerance tol_{CAU} and a maximum number of iterations ($itmax_{CAU}$).

2.3   The Dynamic Diffusion Method

As discussed in the introduction, the DD method is a nonlinear multiscale method that consists of adding a nonlinear artificial diffusion operator acting on both scales of the discretization [11,25]. To present the DD method, we define the enriched spaces S_E and V_E, spaces of admissible functions and test functions, as

$$S_E = S_h \oplus S_B \quad \text{and} \quad V_E = V_h \oplus S_B, \tag{20}$$

where S_B is the space spanned by bubble functions. By denoting $N_b \in H_0^1(\Omega^e)$ as the bubble function defined in each element Ω_e, we define $S_B|_{\Omega_e} = span(N_b)$ and $S_B = \oplus|_{\Omega_e} S_B|_{\Omega_e}$ for all Ω_e in $\mathcal{T}_h$. Here, we use the following function [34],

$$N_b = 256 N_1(x, y, z) N_2(x, y, z) N_3(x, y, z) N_4(x, y, z),$$

where N_j is the local interpolation function associated with the nodal point j of the element Ω_e.

The discrete variational formulation of the DD method can be statement as: find $u = u_h + u_b \in S_E$, with $u_h \in S_h$ and $u_b \in S_B$, such that,

$$A_G(u, v) + A_{DD}(u_h; u, v) = F_G(v), \quad \forall v = v_h + b_b \in V_E, \tag{21}$$

with $v_h \in V_h$ and $v_b \in S_B$, where $A_G(u, v)$ and $F_G(v)$ are given by (11) and (12), respectivelly. The nonlinear operator $A_{DD}(u_h; u, v)$ in (21) is defined as

$$A_{DD}(u_h; u, v) = \sum_{e=1}^{nel} \int_{\Omega_e} \tau_{DD}(u_h) \nabla u \cdot \nabla v \, d\Omega_e, \tag{22}$$

where $\tau_{DD}(u_h) = \mu(h)C_b(u_h)$ represents the amount of artificial diffusion, with

$$C_b(u_h) = \begin{cases} \frac{|R(u_h)|}{2\|\nabla u_h\|}, & \text{if} \quad \|\nabla u_h\| \left(\frac{\|\beta\|\mu(h)}{2\kappa} - 1 \right) > 0, \\ 0, & \text{otherwise}, \end{cases} \tag{23}$$

where $R(u_h) = \beta \cdot \nabla u_h - f$ on Ω_e, and the caracteristic length of mesh,

$$\mu(h) = \begin{cases} 3\sqrt[3]{|\Omega_e|}, & \text{if } \Omega_e \cap \Gamma_+ \neq \emptyset, \\ \sqrt[3]{|\Omega_e|}, & \text{otherwise}, \end{cases}$$

In the last expression, $|\Omega_e|$ is the volume of the element Ω_e, $\Gamma_+ = \{x \in \Omega_e; \beta \cdot n > 0\}$ is the outflow part of Γ, and n is the unit outward normal vector to the boundary Γ.

The method is solved by using an iterative procedure defined as: given u^n, we find u^{n+1} satisfying

$$A_G(u^{n+1}, v) + A_{DD}(u_h^n; u^{n+1}, v) = F_G(v), \quad \forall v \in V_E,$$

where the initial solution is $u_h^0 = 0$. To improve convergence, the following weighting rule to determine (23) is used with $w = 0.5$:

$$C_b = wC_b(u_h^{n+1}) + (1 - w)2C_b(u_h^n).$$

Using the multiscale decomposition process, the Eq. (21) results in two equations (subproblems) in each element $\Omega_e \in \mathcal{T}_h$,

(i) $\int_{\Omega_e} \kappa \nabla u_h \cdot \nabla v_h \, d\Omega_e + \int_{\Omega_e} (\beta \cdot \nabla u_h)v_h \, d\Omega_e + \int_{\Omega_e} (\beta \cdot \nabla u_b)v_h \, d\Omega_e$

$$+ \int_{\Omega_e} \tau_{DD}(u_h)\nabla u_h \cdot \nabla v_h \, d\Omega_e = \int_{\Omega_e} f v_h \, d\Omega_e, \tag{24}$$

(ii) $\int_{\Omega_e} (\beta \cdot \nabla u_h)v_b \, d\Omega_e + \int_{\Omega_e} \kappa \nabla u_b \cdot \nabla v_b \, d\Omega_e$

$$+ \int_{\Omega_e} \tau_{DD}(u_h)\nabla u_b \cdot \nabla v_b \, d\Omega_e = \int_{\Omega_e} f v_b \, d\Omega_e, \tag{25}$$

where we have used the known results

$$\int_{\Omega_e} (\beta \cdot \nabla u_b)v_b \, d\Omega_e = 0 \text{ and } \int_{\Omega_e} \nabla u_b \cdot \nabla v_h \, d\Omega_e = \int_{\Omega_e} \nabla u_h \cdot \nabla v_b \, d\Omega_e = 0,$$

since $u_h|_{\Omega_e}, v_h|_{\Omega_e} \in \mathbb{P}_1(\Omega_e)$ and $u_b, v_b \in H_0^1(\Omega_e)$.

The Eqs. (24) and (25) result in the following local system of algebraic equations, associated with each element Ω_e,

$$\begin{bmatrix} A_{hh}^e & A_{bh}^e \\ A_{hb}^e & A_{bb}^e \end{bmatrix} \begin{bmatrix} U_h^e \\ U_b^e \end{bmatrix} = \begin{bmatrix} F_h^e \\ F_b^e \end{bmatrix}, \tag{26}$$

where the local matrices and vectors in Eq. (26) are defined as follow,

$$A_{hh}^e : \int_{\Omega_e} \left(\kappa \nabla u_h \cdot \nabla v_h + (\boldsymbol{\beta} \cdot \nabla u_h) v_h + \tau_{DD}(u_h) \nabla u_h \cdot \nabla v_h \right) d\Omega_e,$$

$$A_{bh}^e : \int_{\Omega_e} \left((\boldsymbol{\beta} \cdot \nabla u_b) v_h \right) d\Omega_e, \quad A_{hb}^e : \int_{\Omega_e} \left((\boldsymbol{\beta} \cdot \nabla u_h) v_b \right) d\Omega_e,$$

$$A_{bb}^e : \int_{\Omega_e} \left(\kappa \nabla u_b \cdot \nabla v_b + \tau_{DD}(u_h) \nabla u_b \cdot \nabla v_b \right) d\Omega_e,$$

$$F_h^e : \int_{\Omega_e} f v_h \, d\Omega_e, \quad F_b^e : \int_{\Omega_e} f v_b \, d\Omega_e,$$

and U_h^e and U_b^e are the vectors that store the variables $u_h|_{\Omega_e}$ and u_b, respectively. Performing a static condensation to eliminate the unknowns U_b^e at each element, the system (26) can be written in terms of the macro solution U_h^e, as follows,

$$\left(A_{hh}^e - A_{bh}^e (A_{bb}^e)^{-1} A_{hb}^e \right) U_h^e = F_h^e - A_{bh}^e (A_{bb}^e)^{-1} F_b^e. \tag{27}$$

Remark 1. Since $\kappa > 0$, the local matrix A_{bb}^e, associated to the term

$$\int_{\Omega_e} (\kappa + \tau_{DD}(u_h)) \nabla u_b \cdot \nabla v_b \, d\Omega_e,$$

is always nonsingular.

The global system is obtained by assembling all local systems (27) calculated on each element Ω_e,

$$[\boldsymbol{A}_h(\boldsymbol{U}_h^n)] \boldsymbol{U}_h^{n+1} = \boldsymbol{F}_h, \tag{28}$$

$$\boldsymbol{U}_h^0 = \boldsymbol{0}. \tag{29}$$

In each iteration of (28)–(29), the linear system resulting is solved by the Generalized Minimal Residual (GMRES) method [35]. The nonlinear convergence is checked under a given prescribed relative error,

$$\frac{\|\boldsymbol{U}_n^{n+1} - \boldsymbol{U}_h^n\|_\infty}{\|\boldsymbol{U}_n^{n+1}\|_\infty} \leq tol_{DD},$$

and a maximum number of iterations $(itmax_{DD})$.

3   Numerical Experiments

In this section we present two numerical experiments in order to evaluate the quality of the numerical solution obtained with the DD method in three dimensional domains. The computational code was developed in C language in a computer with the following characteristics: Intel Corel i5-2450M, CPU 2.50 GHz × 4 and 8 GB of RAM. The linear systems are solved with the GMRES method considering 150 Krylov vectors in the base, the maximum number of 100 cycles and a tolerance of 10^{-7}. For the nonlinearity of the CAU and DD methods, we use the following tolerance parameters: $tol_{CAU} = tol_{DD} = 10^{-2}$; and the following maximum number of iterations: $itermax_{CAU} = itermaxDD = 50$.

3.1   Problem 1

This problem simulates a three dimensional convection-dominated diffusion-convection problem, with internal and external layers, in a unit cube $\Omega =]0, 1[^3$ with the following coefficients: $\kappa = 10^{-9}$, $\beta = (0, 1, 0)^T$ and $f = 1$. The Dirichlet boundary conditions are defined as $u = 0$ on all faces except in the plane $y = 0$, where a jump discontinuity is introduced in the inflow boundary,

$$u(x, 0, z) = \begin{cases} 1, & \text{for } 0 \le x \le 0.5, \quad z \in [0, 1], \\ 0, & \text{for } 0.5 < x \le 1, \quad z \in [0, 1]. \end{cases}$$

The domain Ω was partitioned in 32 divisions in each direction, so the mesh has 35937 nodes and 196608 elements.

Due to the difficulty in representing results in 3D domains, the solutions are exhibited at the xy plane located at $z = 0.5$. Figure 2 shows the solutions obtained by the CAU and DD methods. The CAU solution (Fig. 2(a)) appears more diffusive in the external boundary layers (narrow regions close to Γ where $\|\nabla u\|$ is large) at $x = 0$, $x = 1$ and $y = 1$, whereas the DD method (Fig. 2(b)) presents some nonphysical under/overshoots in the external layer, at $y = 1$, mainly for $x > 0.5$. In the internal layer, both methods present some diffusion in the region next to the outflow boundary.

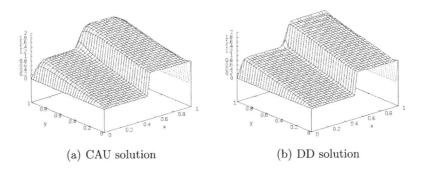

(a) CAU solution   (b) DD solution

Fig. 2. Solutions in plane $z = 0.5$.

For a better understanding of the behavior of the CAU and DD methods, we made three cuts (cross sections) in the solutions described in Fig. 2, at $x = 0.25$, $x = 0.75$ and $y = 0.9$. Figure 3(a) shows the cross section at $x = 0.25$ (lower graph) and at $x = 0.75$ (upper graph). We can observe the diffusive behavior of the CAU method in the external boundary layer (at $y = 1$). In this region the DD solution is better, despite the slight under/overshoots. Figure 3(b) shows the cross section at $y = 0.9$. The CAU method is more diffusive in the three layers at $x = 0$, $x = 0.5$ and $x = 1$.

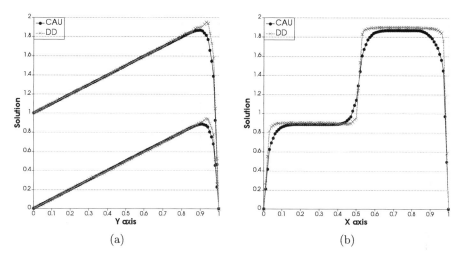

(a)   (b)

Fig. 3. (a) Solutions at $z = 0.5$ and $x = 0.25$ (top), $x = 0.75$ (bottom). (b) Solutions at $z = 0.5$ and $y = 0.9$.

3.2   Problem 2

In this experiment we consider a convection-dominated diffusion-convection problem defined on $\Omega =]0,1[^3$ with the following coefficients: $\kappa = 10^{-6}$, $\beta(1,1,1)^T$ and $f = 1$. The Dirichlet boundary conditions are defined as $u = 0$ on faces $x = 0$, $y = 0$ and $z = 0$ and $u = 1$ on faces $x = 1$, $y = 1$ and $z = 1$. The mesh used in this experiment is the same one as the Problem 1.

The solutions are exposed at the xy plane located at $z = 0.5$. Figure 4 shows the solutions obtained by the CAU and DD methods. The solution obtained with the CAU method appears to be more diffusive at the external boundary layer at $x = 1$ and at $y = 1$ than that of Fig. 4(b). However, some non physical under/overshoots localized at the external layer remain in the DD solution.

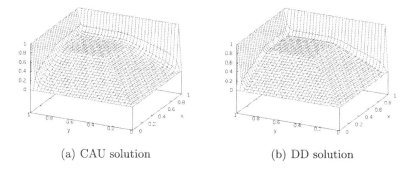

(a) CAU solution   (b) DD solution

Fig. 4. Solutions in plane $z = 0.5$.

These numerical solutions are better understood, analyzing the cuts (cross section) of them shown in the Fig. 5, Fig. 5(a) shows the cross section at $x = 0.7$ and Fig. 5(b) shows the cross section at $y = 0.9$, and Fig. 6 shows the cross section along the diagonal of the xy plane. All of them show the more diffusive behavior of the CAU method in the boundary layer regions and the small oscillations presented by the DD method solution.

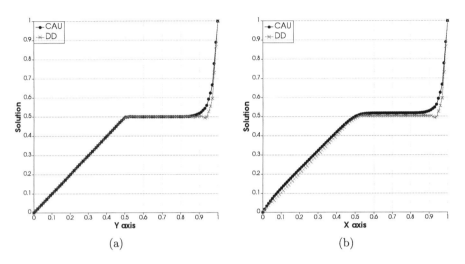

(a)   (b)

Fig. 5. (a) Solutions at $z = 0.5$ and $x = 0.7$. (b) Solutions at $z = 0.5$ and $y = 0.9$.

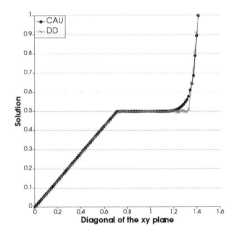

Fig. 6. Solutions along the diagonal of the xy plane at $z = 0.5$.

4   Conclusions

This work deals with the evaluation of the DD method in the solution of diffusion-convection problems with dominant convection on three dimensional domains. We studied two problems with internal and external layers, comparing the results obtained with those from the CAU method. In both experiments, the DD method obtained solutions less diffusive than the CAU, but with small under/overshoot in the external layers.

As in two dimensional problems, the DD method proved to be a promising methodology for solving convection-dominated problems on three dimensional domains. However, more numerical experiments need to be carried out in order to strengthen our conclusions. As future work, we plan to apply this methodology for solving 3D unsteady transport problems, including the Navier-Stokes equations, and also to investigate the convergence rates for the errors in norms $L^2(\Omega)$ and $H^1(\Omega)$.

Acknowledgments. The authors would like to thank the support through the Espírito Santo State Research Support Foundations (FAPES), under Grant Term 181/2017, and the Coordenação de Aperfeiçoamento de Pessoal de Nível Superior - Brasil (CAPES) - Finance Code 001.

References

1. Dehghan, M.: On the numerical solution of the one-dimensional convection-diffusion equation. Math. Probl. Eng. **2005**(1), 61–74 (2005)
2. Donea, J., Huerta, A.: Finite Element Method for Flow Problems. Wiley, Hoboken (2003)
3. Bause, M.: Stabilized finite element methods with shock-capturing for nonlinear convection-diffusion-reaction models. In: Kreiss, G., Lötstedt, P., Målqvist, A., Neytcheva, M. (eds.) ENUMATH 2009, Numerical Mathematics and Advanced Applications 2009, pp. 125–133. Springer, Heidelberg (2010). https://doi.org/10.1007/978-3-642-11795-4_12
4. Barrenechea, G.R., John, V., Knobloch, P.: A local projection stabilization finite element method with nonlinear crosswind diffusion for convection-diffusion-reaction equations. ESAIM: Math. Model. Numer. Anal. **47**(5), 1335–1366 (2013)
5. Roos, H., Stynes, M., Tobiska, L.: Robust Numerical Methods for Singularly Perturbed Differential Equations: Convection-Diffusion-Reaction and Flow Problems, 2nd edn. Springer Science & Business Media, Berlin (2008). https://doi.org/10.1007/978-3-540-34467-4
6. Brooks, A.N., Hughes, T.J.R.: Streamline upwind Petrov-Galerkin formulations for convection dominated flows with particular emphasis on the incompressible Navier-Stokes equations. Comput. Methods Appl. Mech. Eng. **32**(1–3), 199–259 (1982)
7. Johnson, C., Nävert, U., Pitkäranta, J.: Finite element methods for linear hyperbolic problems. Comput. Methods Appl. Mech. Eng. **45**(1), 285–312 (1984)
8. Hughes, T.J.R., Franca, L.P., Hulbert, G.M.: A new finite element formulation for computational fluid dynamics: VIII. the Galerkin/Least-Squares method for advective-diffusive equations. Comput. Methods Appl. Mech. Eng. **73**(2), 173–189 (1989)

9. Burman, E., Hansbo, P.: Edge stabilization for Galerkin approximations of convection-diffusion-reaction problems. Comput. Methods Appl. Mech. Eng. **193**(15), 1437–1453 (2004)
10. Barrenechea, G.R., Burman, E., Karakatsani, F.: Blending low-order stabilised finite element methods: a positivity-preserving local projection method for the convection-diffusion equation. Comput. Methods Appl. Mech. Eng. **317**, 1169–1193 (2017)
11. Valli, A.M.P., Almeida, R.C., Santos, I.P., Catabriga, L., Malta, S.M.C., Coutinho, A.L.G.A.: A parameter-free dynamic diffusion method for advection-diffusion-reaction problems. Comput. Math. Appl. **75**(1), 307–321 (2018)
12. John, V., Knobloch, P.: On spurious oscillations at layers diminishing (SOLD) methods for convection-diffusion equations: part I - a review. Comput. Methods Appl. Mech. Eng. **196**(17), 2197–2215 (2007)
13. John, V., Knobloch, P.: On spurious oscillations at layers diminishing (SOLD) methods for convection-diffusion equations: part II - analysis for P1 and Q1 finite elements. Comput. Methods Appl. Mech. Eng. **197**(21–24), 1997–2014 (2008)
14. Hughes, T.J.R.: Multiscale phenomena: Green's functions, the Dirichlet-to-Neumann formulation, subgrid scale models, bubbles and the origins of stabilized methods. Comput. Methods Appl. Mech. Eng. **127**(1–4), 387–401 (1995)
15. Hughes, T.J.R., Stewart, J.R.: A space-time formulation for multiscale phenomena. J. Comput. Appl. Math. **74**(1–2), 217–229 (1996)
16. Brezzi, F., Franca, L.P., Hughes, T.J.R., Russo, A.: b = ∫g. Comput. Methods Appl. Mech. Eng. **145**, 329–339 (1997)
17. Hughes, T.J.R., Feijóo, G.R., Mazzei, L., Quincy, J.-B.: The variational multiscale method - a paradigm for computational mechanics. Comput. Methods Appl. Mech. Eng. **166**(1–2), 3–24 (1998)
18. Hughes, T.J.R., Scovazzi, G., Franca, L.P.: Multiscale and stabilized methods. In: Encyclopedia of Computational Mechanics, 2nd edn, pp. 1–64 (2018)
19. Brezzi, F., Russo, A.: Choosing bubbles for advection-diffusion problems. Math. Models Methods Appl. Sci. **4**(4), 571–587 (1994)
20. Franca, L.P., Nesliturk, A., Stynes, M.: On the stability of residual-free bubbles for convection-diffusion problems and their approximation by a two-level finite element method. Comput. Methods Appl. Mech. Eng. **166**(1–2), 35–49 (1998)
21. Guermond, J.-L.: Stabilization of Galerkin approximations of transport equation by subgrid modeling. Math. Model. Numer. Anal. **33**, 1293–1316 (1999)
22. Guermond, J.-L.: Subgrid stabilization of Galerkin approximations of linear monotone operators. IMA J. Numer. Anal. **21**, 165–197 (2001)
23. Santos, I.P., Almeida, R.C.: A nonlinear subgrid method for advection-diffusion problems. Comput. Methods Appl. Mech. Eng. **196**(45–48), 4771–4778 (2007)
24. Santos, I.P., Almeida, R.C., Malta, S.M.C.: Numerical analysis of the nonlinear subgrid scale method. Comput. Appl. Math. **31**(3), 473–503 (2012)
25. Arruda, N. C. B., Almeida, R. C., Dutra do Carmo, E. G.: Dynamic diffusion formulations for advection dominated transport problems. In: Dvorkin, E., Goldschmit, M., Storti, M. (eds.) Mecánica Computacional. vol. 29, pp. 2011–2025. Asociación Argentina de Mecánica Computacional, Buenos Aires (2010)
26. Bento, S.S., de Lima, L.M., Sedano, R.Z., Catabriga, L., Santos, I.P.: A nonlinear multiscale viscosity method to solve compressible flow problems. In: Gervasi, O., et al. (eds.) ICCSA 2016. LNCS, vol. 9786, pp. 3–17. Springer, Cham (2016). https://doi.org/10.1007/978-3-319-42085-1_1

27. Bento, S. S., Sedano, R. Z., Lima, L. M., Catabriga, L., Santos, I. P.: Comparative studies to solve compressible problems by multiscale finite element methods. Proc. Ser. Braz. Soc. Comput. Appl. Math. **5**(1) (2017). https://doi.org/10.5540/03.2017.005.01.0317

28. Baptista, R., Bento, S.S., Santos, I.P., Lima, L.M., Valli, A.M.P., Catabriga, L.: A multiscale finite element formulation for the incompressible Navier-Stokes equations. In: Gervasi, O., et al. (eds.) ICCSA 2018. LNCS, vol. 10961, pp. 253–267. Springer, Cham (2018). https://doi.org/10.1007/978-3-319-95165-2_18

29. Galeão, A.C., Dutra do Carmo, E.G.: A consistent approximate upwind Petrov-Galerkin method for convection-dominated problems. Comput. Methods Appl. Mech. Eng. **68**(1), 83–95 (1988)

30. Raviart, P.A., Thomas, J.M.: Introduction à l'analyse numérique des équations aux dérivées partielles. Masson, Paris (1992)

31. Hughes, T. J. R.: The Finite Element Method: Linear Static and Dymanic Finite Element Analysis. Prentice-Hall International (1987)

32. Galeão, A.C., Almeida, R.C., Malta, S.M.C., Loula, A.F.D.: Finite element analysis of convection dominated reaction-diffusion problems. Appl. Numer. Math. **48**(2), 205–222 (2004)

33. Carmo, E.G.D., Alvarez, G.B.: A new stabilized finite element formulation for scalar convection-diffusion problems: the streamline and approximate upwind/Petrov-Galerkin method. Comput. Methods Appl. Mech. Eng. **192**(31–32), 3379–3396 (2003)

34. Ern, A., Guermond, J.-L.: Theory and Practice of Finite Elements, vol. 159, 1st edn. Springer Science & Business Media, New York (2004). https://doi.org/10.1007/978-1-4757-4355-5

35. Saad, Y., Schultz, H.: GMRES: a generalized minimal residual algorithm for solving nonsymmetric linear systems. SIAM. J. Sci. Stat. Comput. **7**(3), 856–869 (1986)

A Bivariate Multilevel Analysis of Portuguese Students

Susana Faria[1]([✉])[iD] and Carla Salgado[2]

[1] Department of Mathematics, Centre of Molecular and Environmental Biology,
University of Minho, Guimarães, Portugal
sfaria@math.uminho.pt
[2] Department of Mathematics, University of Minho, Guimarães, Portugal
salgcarla@gmail.com

Abstract. In this work, we illustrate how to perform a bivariate multilevel analysis in the complex setting of large-scale assessment surveys. The purpose of this study was to identify a relationship between students' mathematics and science test scores and the characteristics of students and schools themselves. Data on about 7325 Portuguese students and 246 Portuguese schools who participated in PISA-2015 were used to accomplish our objectives. The results obtained by this approach are in line with the existing research: the index of the socioeconomic status of the student, being a male student and the average index of the socioeconomic status of the student in school, positively influence the students' performance in mathematics and science. On the other hand, the grade repetition had a negative influence on the performance of the Portuguese student in mathematics and science.

Keywords: Multilevel models · Bivariate models · PISA data

1 Introduction

The Programme for International Student Assessment (PISA) has attracted the attention of many researchers and educational policy makers over the past few years. PISA provides an international view on the reading, science and mathematics achievement among 15-year students from countries of the Organization of Economic and Cultural Development (OECD) every 3 years since 2000. The PISA survey is a self a self-administered questionnaire that tests student skills and gathers information about student's family, home and school background. Since PISA 2015, different from the previous studies, the assessments of all three domains were mainly conducted on computers [1].

Portugal as a founding member of the OECD participated in all editions of PISA. In the early's 2000s, Portugal's performance in PISA was one of the lowest among OECD countries. However, Portugal is one of the few OECD member countries where there has been a tendency for significant improvement in results in the three assessment areas. In PISA 2015, for the first time, Portuguese

O. Gervasi et al. (Eds.): ICCSA 2020, LNCS 12251, pp. 470–480, 2020.
https://doi.org/10.1007/978-3-030-58808-3_34

students ranked significantly above the OECD average score for reading and scientific literacy, being on the OECD average for mathematical literacy [4].

This study utilized data from the 2015 PISA Portuguese sample to investigate the factors from both student and school perspectives, that impact the mathematics and science achievement of 15-year-old students.

Given the hierarchical nature of data (students nested into schools), a multilevel approach is adopted to investigating the impact of school resources and students' characteristics on performance [5]. Multilevel models simultaneously investigates relationships within and between hierarchical levels of grouped data, thereby making it more efficient at accounting for variance among variables at different levels than other existing analyses [6]. With this type of data, classic methods, such as ordinary least squares regression, would not produce correct standard errors [7].

In the literature there are many papers devoted to the analysis of potential factors that influence on student achievement using PISA data and applying multilevel analysis. Examples of the country specific analysis using PISA results can be found in [8], which examines the relationships among affective characteristics-related variables at the student level, the aggregated school-level variables, and mathematics performance by using 2012 PISA Indonesia, Malaysia, and Thailand sample. Using PISA data, Giambona and Porcu study factors affecting Italian students' achievement paying attention on school size [9]. In [10], a multilevel analysis was applied to the OECD-PISA 2006 data with the aim to compare factors affecting students' achievement across Italy and Spain. Using PISA 2012 data, Karakolidis et al. investigated the factors, both at individual and school level, which were linked to mathematics achievement of Greek students [11]. Wu et al. studied the relationship between principals' leadership and student achievement using PISA 2015 United States data [12]. In [13], the study aimed at investigating to what degree external factors, such as cultural and economic capital, parental pressure, and school choice, are related to 15-year-old students' achievement in digital reading and in overall reading on both the student level and the school level in Norway and Sweden, using PISA data from the two countries.

The paper is organised as follows. In Sect. 2, we provide a brief description of the methodological approach adopted in the paper and in Sect. 3 we present the dataset. In Sect. 4, we present the main results arising from our data analysis, while Sect. 5 contains final remarks.

2   Methodology

The model proposed for the empirical analysis is a bivariate two-level linear model in which students (level 1) are nested in schools (level 2) and the outcome variables are mathematics and science achievements. Several studies are present in literature on the mathematics and sciences achievements, where they are treated as separate, applying univariate multilevel models. Using this bivariate model it is possible to compare the associations between students' character-istics or school resources and their performances in mathematics and science achievements.

Let $N = \sum_{j=1}^{J} n_j$ be the total number of students, where $n_j, j = 1, \ldots, J$ is the total number of students of the $j - th$ school. For each school $j = 1, \ldots, J$ and student $i = 1, \ldots, n_j$

$$\mathbf{y}_{ij} = \beta_0 + \sum_{k=1}^{K} \beta_k x_{kij} + \sum_{l=1}^{L} \alpha_l w_{lj} + \mathbf{b}_j + \epsilon_{ij} \tag{1}$$

where $\mathbf{y}_{ij}$ is the bivariate outcome with mathematics and science achievements of student i in school j; $\beta = (\beta_0, \ldots, \beta_K)$ is the bivariate $(K + 1)$-dimensional vector of parameter; x_{kij} is the value of the $k - th$ predictor variable at student's level; $\alpha = (\alpha_1, \ldots, \alpha_L)$ is the bivariate (L)-dimensional vector of parameter; w_{lj} is the value of the $l - th$ predictor variable at school's level; $\mathbf{b}_j \sim N_2(0, \Sigma)$ is the matrix of the bivariate random effects (mathematics and science) at school level, where the covariance matrix of the random effects at school level is given by

$$\Sigma = \begin{bmatrix} \sigma_{b1}^2 & \sigma_{b12} \\ \sigma_{b12} & \sigma_{b2}^2 \end{bmatrix} \tag{2}$$

and $\epsilon \sim N_2(0, W)$ is the matrix of errors, where the covariance matrix of the errors is given by

$$W = \begin{bmatrix} \sigma_{e1}^2 & \sigma_{e12} \\ \sigma_{e12} & \sigma_{e2}^2 \end{bmatrix}. \tag{3}$$

We assume $\mathbf{b}_j$ independent of ϵ.

The modelling procedure in this study has three steps. In the first step, the null model (with no independent variables) was fitted. This model is statistically equivalent to one-way random effects analysis of variance [7] and it is motivated to partition the total variance in the outcome variable into the different levels in the data. At the second step, a student-level model was developed without variables at the school level. School variables were added to the student model at the third step.

3   Data Set

The sample was drawn from the PISA 2015 data set (see [2]) for an overview).

A complex multi-stage stratified random sampling was used to sample the Portuguese 15-year-old student population. The first stage consisted of sampling individual schools in which 15-year-old students could be enrolled. Schools were sampled systematically with probabilities proportional to size, the measure of size being a function of the estimated number of eligible (15-year-old) students enrolled. The second stage of the selection process sampled students within sampled schools. Once schools were selected, a list of each sampled school's 15-year-old students was prepared. From this list, 42 students were then selected with equal probability (all 15-year-old students were selected if fewer than 42 were enrolled). The number of students to be sampled per school could deviate from 42, but could not be less than 20 (see [3]).

The Portuguese sample of the PISA 2015 includes 7325 students nested in $J = 246$ schools. However, analyzing the data set, it was found that there was a school with a lack of information on many variables, so it was removed from the sample. In addition, all students who had missing values in some of the variables were removed. Thus, the final data includes 6549 students nested in $J = 222$ schools.

Our outcome of interest were mathematics and science achievements which are scaled using item response theory to a mean of 500 and a standard deviation of 100. Because the PISA features complex booklet designs, the methods used to estimate achievement are similar to multiple imputation and result in several plausible values (PV) for each student. Following the recommendations for addressing PVs in international large-scale assessments (see [15]), we consider all 10 PVs simultaneously as the dependent variables for the purpose of obtaining unbiased and stable estimates. We remark that in PISA 2015 data, student performance is represented by 10 plausible values each (compared with only 5 plausible values in the previous studies).

When data are based on complex survey designs, sampling weights have to be incorporate in the analysis and any estimation procedure that does not take into account sampling weights provides biased results. To avoid the bias in parameter estimates and to produce nationally representative findings, sampling weights for students and schools provided by the PISA database were included in our analysis.

Centering is an important issue in multilevel analysis. In this study, all predictors were centered on the grand mean at both the student and school levels. The purpose of this is to reduce the multicollinearity among variables and bias in variance estimates so that a more meaningful interpretation can be made [16].

We report some descriptive statistics for the PVs in Table 1 and Table 2. We can see that, although plausible values fluctuate within persons, the means and standard deviations of the ten distributions are very close to each other, as one would expect given their generation from the same distributions.

Table 1. Descriptive statistics of the plausible values of math scores

	Mean	Median	St Dev (SD)	Minimum	Maximum
PV1MATH	492.217	494.543	96.579	157.556	779.685
PV2MATH	491.270	493.850	96.396	145.684	774.265
PV3MATH	492.105	494.146	94.958	172.010	784.568
PV4MATH	491.995	494.766	95.350	109.887	812.811
PV5MATH	489.957	493.287	94.243	163.986	766.498
PV6MATH	490.709	494.043	94.978	192.044	828.550
PV7MATH	491.710	493.282	95.395	167.525	791.994
PV8MATH	491.454	493.692	94.928	147.641	809.749
PV9MATH	491.067	494.055	95.692	173.154	769.480
PV10MATH	492.075	493.283	93.373	153.631	788.372

Table 2. Descriptive statistics of the plausible values of science scores

	Mean	Median	St Dev (SD)	Minimum	Maximum
PV1SCIE	499.377	499.853	92.143	195.810	779.557
PV2SCIE	500.665	501.043	90.927	216.749	779.335
PV3SCIE	500.515	500.697	92.351	141.502	812.208
PV4SCIE	500.379	502.585	91.866	221.472	800.599
PV5SCIE	501.224	501.609	92.347	199.764	802.879
PV6SCIE	501.249	503.953	91.398	192.971	764.132
PV7SCIE	500.068	503.066	91.063	191.538	784.348
PV8SCIE	500.346	503.068	91.429	164.587	779.158
PV9SCIE	500.330	502.035	91.091	221.427	778.648
PV10SCIE	501.607	502.715	91.746	222.347	770.884

Figure 1 shows the scatter plot mathematics achievement versus science achievement (only for plausible value 1). It is immediately clear that there is a positive correlation between the performances of students in the two results. The correlation coefficients among ten PVs of mathematics performance and science performance were computed and the values ranged between 0.891 and 0.901.

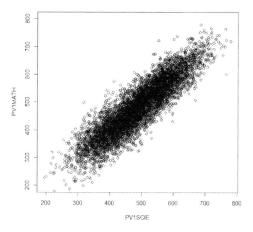

Fig. 1. Scatter plot mathematics achievement vs science achievement (plausible value 1)

With regard to explanatory variables at the student and school level, we include in our model some of the variables that have been most frequently identified as influential in literature.

Students-level variables:

- Age: represents the age of the student using the information drawn from the question referring to the month of birth;
- Gender: dummy variable that takes the value one if the student is male and the value zero if the student is female;
- Rep: dummy variable that takes the value one if the student have retaken at least one year of schooling and zero otherwise;
- Imi: dummy variable that takes the value one if the student is not native and zero otherwise;
- ESCS: student's socio-economic status. This variable is is derived from several variables related to students' family background: parents' education, parents' occupations, a number of home possessions that can be taken as proxies for material wealth, and the number of books and other educational resources available at home. Greater values on ESCS represent a more advantaged social background, while smaller values represent a less advantaged social background. A negative value indicates that the socio-economic status is below the OECD mean.

School-level variables:

- TYPE: dummy variable that takes the value one if the school has private status and takes the value zero if the school is public;
- LOC: dummy variable that takes the value one if the school is located in a town and zero if the school is located in a village;
- RAP: students per teachers ratio;

– PRAP: proportion of girls at school;
– MESCS: average of student's socio-economic status.

Some related descriptive statistics are presented in Table 3 and Table 4.

Table 3. Descriptive statistics for categorical variables

	Variable	Category	N°students	% students	N°schools	% schools
Student	Gender	Male	3288	50.2%		
		Female	3261	49.8%		
	Rep	No	4126	63.0%		
		Yes	2423	37.0%		
	Imi	No	6179	94.4%		
		yes	370	5.6%		
School	TYPE	Public	6264	95.7%	202	91.0%
		Private	285	4.4%	20	9.0%
	LOC	Village	2638	40.3%	99	44.6%
		Town	3911	59.7%	123	55.4%

Regarding students' demographic characteristics, Table 3 shows that 50.2% of the 15-year old students were male, only 5.6% of the students were immigrants and the percentage of students who reported grade repetition was 37.0%. In fact, Portugal is one of the OECD countries with a high repetition rate before the age of 15 years old.

Regarding for school-level variables, 91.0% of the schools are public corresponding to 95.7% of students (in Portugal the number of public schools is higher than number of private schools). Table 3 also shows that 55.4% of the schools were located in towns corresponding to 59.7% of students.

Table 4. Descriptive statistics for continuous variables

Variable	Mean	Median	St Dev (SD)	Minimum	Maximum	Coef. Variation
Age	15.78	15.75	0.28	15.33	16.33	1.27%
ESCS	−0.40	−0.49	1.15	−4.15	3.08	287.76%
PRAP	49.30	49.20	4.81	26.83	68.54	9.81%
RAP	10.42	10.36	4.22	1.98	41.42	40.50%
MESCS	−0.64	−0.71	0.68	−2.28	1.22	107.99%

Table 4 shows that the average age of the students was 15.78 years (SD = 0.28, minimum = 15.33, maximum = 16.33). The average student's economic, social and cultural index was approximately −0.40 (SD = 1.15, minimum = −4.15, maximum $= 3.08$), which means that Portuguese students have a lower economic, social and cultural index than the average of the students across all participating OECD countries of the PISA Program and a greater variability. In general, students with ESCS equal to or greater than two are very socially and culturally advantaged. We see that the average proportion of girls in schools is close to 50% and ranged between 26.83% and 68.54%. The average number of students per teacher is 10.42 students per teacher but there is a great variability of number of students per teacher in Portuguese schools. In fact, there are schools with only 1.98 students per teacher and schools with 41.42 students per teacher. The school PISA index of the economic, social and cultural status values ranged between -2.28 and 1.22 (Mean$= -0.64$, SD $= 0.68$, minimum $= -2.28$, maximum $= 1.22$).

4   Results

This section contains the main results obtained by estimating the bivariate two-level multilevel model explained above to our dataset. All statistical analyses were performed considering sample weights to ensure that the sampled students adequately represent the analyzed total population. We report average results obtained from the use of each one of the PVs.

Firstly, the null model was estimated. This model allows us to explore the correlation structure of the two outcomes. Table 5 reports the estimates of regression coefficients and variance-covariance parameters of the null model. Table 5 shows that $53.53/(53.53 + 37.08) = 59.07\%$ of the total variance in maths achievement was account for by school-level and $51.80/(51.80 + 37.08) = 58.28\%$ of the total variance in science achievement was account for by school-level. These results show that the school explain a relevant portion of the variability in achievement. Note that two scores are highly correlated.

Table 5. Parameter estimates of null model. Asterisks denote different levels of significance $**p - value < 0.05$, $***p - value < 0.01$

	Mathematics	Science
Fixed effects	Estimate(se)	Estimate (se)
Intercept	476.00***(3.87)	482.51***(3.76)
Variance matrix of random effects	$\Sigma = \begin{bmatrix} 53.53 & 18.25 \\ 18.25 & 51.80 \end{bmatrix}$	
Variance matrix of errors	$W = \begin{bmatrix} 37.08 & 20.10 \\ 20.10 & 37.08 \end{bmatrix}$	

Table 6 reports the estimates of regression coefficients and variance-covariance parameters, after removing non-significant variables at the student and school level.

Table 6. Parameter estimates of final model. Asterisks denote different levels of significance $**p-value < 0.05$, $***p-value < 0.01$

	Mathematics	Science
Fixed effects	Estimate(se)	Estimate (se)
Intercept	519.46***(2.71)	525.08***(2.70)
Gender (Female)		
Male	16.31***(1.68)	15.51***(1.68)
Rep (No)		
Yes	−95.06***(1.96)	−89.37***(1.96)
ESCS	14.89***(0.89)	13.46***(0.90)
MESCS	21.09***(2.75)	22.52***(2.72)
Variance matrix of random effects	$\Sigma = \begin{bmatrix} 18.16 & 9.16 \\ 9.16 & 17.76 \end{bmatrix}$	
Variance matrix of errors	$W = \begin{bmatrix} 31.40 & 15.50 \\ 15.50 & 31.79 \end{bmatrix}$	

These models, presented theoretically in Sect. 2, are developed using the R package nlme (see [14]).

With reference to the considered explanatory variables, the student level covariates Age and Imi were not significant, that is, there is not a relevant correlation between scores and age and immigrant and non-immigrant students scored equally well in mathematics and sciences. At school level, the only significant variable was MESCS (school-level average of student socio-economic status). In Table 6, the intercepts represent the average scores for the baseline student: female, not retook at least one year of schooling, and all the other covariates set at mean values. The performance of the baseline student is beyond the international mean of 500 in two outcomes, though the average score in Math is lower than the average score in Science.

In general, the coefficients have the expected signs. Considering the results of existing studies with the same purpose as this work, being a male is positively associated with better results in Maths and Science. Students who have retaken at least one year of schooling have a lower performance in Maths and Science, meaning that these students have more difficulties than students who have not retaken, especially in mathematics. The ESCS is positively associated with the achievements and have similar coefficients in both the fields, suggesting that students with a high socio-economical level are educationally advantaged. Also, at school level, attending a school with higher mean ESCS helps to reach a higher score.

Looking at the variance/covariance matrix of the random effects, we can see that the variability of the random effects are similar (18.16 vs 17.76), therefore attending a specific school has the same influence on the results in mathematics and in science. The two effects are positively correlated.

5   Final Remarks

The research was conducted to find out the relationship between students' achievements and the characteristics of students and schools themselves. The analysis relies on a bivariate multilevel model, thus accounting for both the bivariate nature of the outcome and the hierarchical structure of the data.

Based on an analysis of PISA 2015 data, our findings are in line with the previous studies in which univariate multilevel models had been applied.

The use of bivariate multilevel model allow us to test for differences in the regression coefficients, thus pointing out differential effects of the covariates on the two outcomes. As expected, the characteristics of students and schools are associated in the same way with the two outcomes: mathematics and science achievements.

Acknowledgements. This work was supported by the strategic programme UID-BIA-04050-2019 funded by national funds through the FCT.

References

1. OECD: "What is PISA?" In: PISA 2015 Assessment and Analytical Framework: Science, Reading, Mathematic and Financial Literacy, OECD Publishing, Paris (2016). https://doi.org/10.1787/9789264255425-2-en
2. OECD: PISA databases 2015. OECD Publishing, Paris (2015). http://www.oecd.org/pisa/data/2015database/
3. OECD: PISA 2015 Technical report. OECD Publishing, Paris (2016)
4. Marôco, J., Gonçalves, C., Lourenço, V., Mendes, R.: PISA 2015 - PORTU-GAL.Volume I: Literacia Científica, Literacia de Leitura e Literacia Matemática, Instituto de Avaliação Educativa, I. P. (2016)
5. Goldstein, H.: Hierarchical data modeling in the social sciences. J. Educ. Behav. Stat. **20**, 201–204 (1995)
6. Woltman, H., Feldstain, A., MacKay, J.C., Rocchi, M.: An introduction to hierarchical linear modeling. Tutorials Quant. Methods Psychol. **8**(1), 52–69 (2012)
7. Raudenbush, S.W., Bryk, A.S.: Hierarchical Linear Models: Applications and Data Analysis Methods, 2nd edn. Sage, Newbury Park (2002)
8. Thien, L.M., Darmawan, I.G.N., Ong, M.Y.: Affective characteristics and mathematics performance in Indonesia, Malaysia, and Thailand: what can PISA 2012 data tell us? Large-scale Assessments Educ. **3**(1), 1–16 (2015). https://doi.org/10.1186/s40536-015-0013-z
9. Giambona, F., Porcu, M.: School size and students' achievement. Empirical evidences from PISA survey data. Socio-Econ. Plann. Sci. **64**, 66–77 (2018)
10. Agasisti, T., Cordero-Ferrera, J.M.: Educational disparities across regions: a multilevel analysis for Italy and Spain. J. Policy Model. **35**(6), 1079–1102 (2013)
11. Karakolidis, A., Pitsia, V., Emvalotis, A.: Mathematics low achievement in Greece: a multilevel analysis of the Programme for International Student Assessment (PISA) 2012. Themes Sci. Technol. Educ. **9**(1), 3–24 (2016)
12. Wu, H., Gao, X., Shen, J.: Principal leadership effects on student achievement: a multilevel analysis using Programme for International Student Assessment 2015 data. Educ. Stud. **46**(3), 316–336 (2020)

13. Rasmusson, M.A.: A multilevel analysis of Swedish and Norwegian students' overall and digital reading performance with a focus on equity aspects of education. Large-scale Assessments Educ. **4**(1), 1–25 (2016). https://doi.org/10.1186/s40536-016-0021-7

14. Pinheiro J., Bates D., DebRoy S., Sarkar D.: R Core Team. nlme: linear and Non-linear Mixed effects models. R package version 3.1-147 (2020). https://CRAN.R-project.org/package=nlme

15. Rutkowski, L., Gonzalez, E., Joncas, M., von Davier, M.: International large-scale assessment data: issues in secondary analysis and reporting. Educ. Researcher **39**(2), 142–151 (2010)

16. Kreft, I., de Leeuw, J.: Introducing Statistical Methods. Introducing Multivlevel Modeling. Sage Publications, Inc (1998)

Impact of Using Excellence Management Models in the Customer Satisfaction of Brazilian Electricity Distributors - 10 Years of Studies

Marina A. P. Andrade[1,2]([⊠]) , Álvaro Rosa[1] ,
Alexandre Carrasco[1], and M. Filomena Teodoro[3,4]

[1] ISCTE, Instituto Universitário de Lisboa, 1649-026 Lisbon, Portugal
marina.andrade@iscte.pt
[2] ISTAR, Instituto Universitário de Lisboa, 1649-026 Lisbon, Portugal
[3] CEMAT, Center for Computational and Stochastic Mathematics, Instituto
Superior Técnico, Lisbon University, 1048-001 Lisbon, Portugal
[4] CINAV, Center of Naval Research, Naval Academy, Portuguese Navy,
2810-001 Almada, Portugal

Abstract. In this work we evaluate the impact of the use of the model of excellence in Brazilian management by electricity distribution companies and their impact on customer satisfaction. It was evaluated 10 years of use of the model in groups of companies with different levels of implementation of the model MEG (users, indifferent, engaged and winning) using a statistical approach, firstly using descriptive statistics and by last applying the nonparametric kruskal-Wallis test. It is evidenced the existence of differences between the identified groups revealing the correct decision using the described model. We could attest the positive effects evidenced by the use of reference models of the electric energy sector, on a large scale, during the period analyzed. The results here obtained can be used, at least as guidance, by similar organizations or other industries.

Keywords: Excellence models · Customer satisfaction · Energy distribution · MEG · IASC · Satisfaction survey · Nonparametric tests

1 Introduction

In the recent years, the Brazilian energy sector has been undergoing constant transformations, provoked by dynamic scenarios, a more restrictive regulatory environment and specific performed by industry associations like ABRADEE (Association of Electric Power Distribution Companies).

The challenges imposed on the distributers, due to availability, the progressive improvement in the quality of services rendered, competitive prices, efficiency and cost-effectiveness (Baltazar 2007), besides balancing the interests of several interested parties (Brito 2007), has been leading industry associations like ABRADEE, as well as regulatory agencies to propose actions that enable the companies to face these challenges.

© Springer Nature Switzerland AG 2020
O. Gervasi et al. (Eds.): ICCSA 2020, LNCS 12251, pp. 481–491, 2020.
https://doi.org/10.1007/978-3-030-58808-3_35

Among the actions implemented are the index of satisfaction evaluation of consumer (IASC) and the ABRADEE awards, promoted by Brazil's regulatory agency (ANEEL) and ABRADEE, respectively. In the case of the IASC prize, ANEEL annually conducts a satisfaction survey on services provided to residential consumers, which enables the IASC prize to cover the entire national territory. The ABRADEE prize, in turn, more specifically in the Management Quality category serves as an incentive to the adoption of the MEG (Excellence in Brazilian Management Model) provided by the distribution companies. In this category, participation in the PNQ (National Quality Prize) awards a grade that is subsequently used in the assessment of the ABRADEE Quality in Management Award.

As an effect of the actions mentioned above, at present the sector has participated significantly in the National Quality Prize (PNQ) and in similar awards, both regional and sectoral. For Boutler et al. (2013) and Corredor and Goñi (2010), the organizations that obtain these awards in recognition for their achievements are those that have the best results. Thus, the objective is consistent with the initial aim of the award: to encourage improvements that produce results.

Hence, we are able to evaluate, between the period of 2007 to 2016, if there are differences in performance of the IASC among the companies that regularly use the model, including the award winning companies and the indifferent ones.

The outline of this article is developed in four sections. In Sect. 2 is presented some background about models of excellence in management.Section 3 makes a brief summary about the methodology, the selection of variables is presented, is displayed a short summary about exploratory analysis of the data and it is performed the hypothesis tests. The results of statistical approach are presented. Finally in Sect. 4 some conclusions are drawn and some suggestions are pointed.

2   The MEG Model

According to the National Quality Foundation (FNQ 2016), the MEG is a world-class system of business management or a model for excellence in management. For Puay et al. (1998) and Miguel et al. (2004), these models represent efforts devoted by countries to improve their international reputation in the world market.

In the Brazilian context, the MEG model deserves special attention, seeing that it has become one of the most significant guidelines directed at competitiveness in the country (Cardoso et al. 2012). Furthermore, its importance corroborates ABRADEE's endeavor to promote actions that bring about the best results for the electric energy distribution companies and their interested parts, such as the ABRADEE Award.

The universality of the model and the possibilities of use by any type of organization (Calvo-Mora et al. 2015), as well as the slight differences in scope (Bucelli and Costa Neto 2013) also offer possibilities for the exchange of practices across industries, maximizing the benefits from its use.

Knowledge-exchange can also take place in relation to results obtained and/or practices adopted in similar awards, such as the Deming Prize (Japan), the Malcolm Baldrige National Quality Award (USA) and the European Quality Award (Europe), all of which are among the most eminent awards according to considered by Kholl and Tan (1952), Tan (2002) and Puay et al. (1998).

For this study, it is relevant to emphasize that the versions of the MEG model employed between 2007 and 2016 were composed of two evaluation groups, namely processes and results. Both were aligned by way of fundamentals of excellence.

The criteria related to processes are established through the analysis of detailed information on how the organizations implement their management processes without pre-defined methodologies. Information concerning Leadership, Strategies and Plans, Clients, Social Responsibility, Information and Knowledge, People and Business Processes are requested from each organization. In turn, the results item calls for results that demonstrate the implementation of practices, the attainment of strategic results and the fulfillment of requirements set out by interested parties in a period of at least 3 years, besides benchmarking studies that show evidence of leadership (FNQ 2018). As a result, the evaluated organization receives a written remark about their strong points and areas for improvement, which can be used to improve management, as well as a score ranging from 0 to 1000 points, awarded according to pre-established criteria. The benefits of its use, however, are still controversial, and must be further analyzed individually for each sector or performance attribute. For Doelema et al. (2014), success in the implementation of this model is not assured, despite the dissemination of its use. For Boutler et al. (2013) and Corredor and Goñi (2010), organizations that adopt reference models demonstrate superior results.

In the context of this study, Puay et al. (1998) points out that the users of excellence models not only improve their quality, but also enhance other performance attributes, including client satisfaction, which is represented in the electric energy sector in the form of the IASC. Thus, the relation between GS (Global Score) and the IASC, and the difference in IASC performance among different groups of companies involved (users vs. indifferent, award winning vs. engaged) will be our object of study.

2.1 Measuring the Level of Satisfaction

The evaluation of customers' satisfaction level can be verified in relation to a product or a process. It can be defined as the result of a consumption experience, or as the consumer's response to balancing their expectations concerning a product or service, and the result obtained (Engel et al. 1993).

Used as a comparative tool with external benchmarks, and for gauging the quality of services rendered by the concessionaires of public services of electric energy, for problem-solving and for improving regulation, ANEEL conducts surveys based on studies of methods published by Marchetti and Prado (2001, 2004). Since 2000, ANEEL promotes the IASC prize, which recognizes the organizations that obtained the highest scores in the IASC customer satisfaction survey (ANEEL 2017).

At present, all distribution companies must undertake the survey and are held accountable to its results, which are made available for public consultation on the agency's website. The survey thus serves as a mechanism of popular pressure, among others, employed by the regulatory agency to encourage the improvement of services provided.

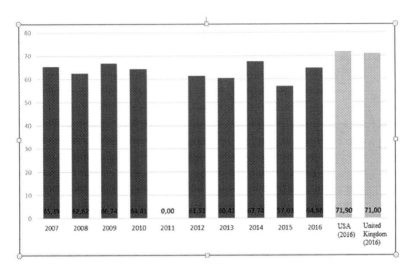

Fig. 1. Evolution of IASC Brazil (2007 to 2016). Evaluation of the quality of public services in Brazil (2016) and comparison with USA (2016) and United Kingdom (2016).

2.2 Satisfaction of Brazilian Customers with Respect to Electric Power Delivery Services

For Oliver (1981, 1997), satisfaction can be understood as the evaluation of the (user's) surprise inherent in a consumption experience. It is, hence, a relevant managerial tool for organizations that seek to improve the provision of their services.

For Marchetti and Prado (2004), in the case of public services such as the supply of electric energy, this type of evaluation has the role of enhancing the process of monitoring the results of the distribution companies, as well as correcting public policies and directing sectoral efforts towards satisfying the needs of consumers. In this way, having been designed for the energy distribution sector, the IASC represents an index of great social and managerial relevance.

The national results (see Fig. 1), however, do not show substantial improvement (in recent years). Weighed up against the comparative benchmarks, they indicate the existence of ample scope for attaining higher levels of improvement.

Compared to other public services (see Fig. 2), the measurement of satisfaction with the electric energy supply rates is the best among the services evaluated. The results presented Fig. 2 are in contrast with the need for improvement so that they can be classified as positive in the same evaluation (50%) (CNI 2016).

Hence, it is evident that the even though there have been advances in the improvement of indicators, sustained efforts are still required to reach a minimum threshold of satisfaction compatible with the standards demanded by the Brazilian population.

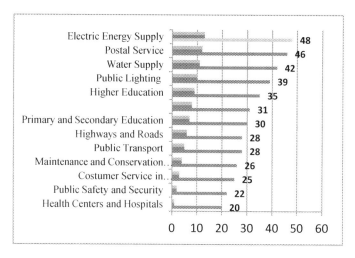

Fig. 2. Evaluation of the quality of public services in Brazil (2016) in contrast with the need for improvement so that they can be classified as positive.

3 Methodology

3.1 Sample and Population

The sample is composed by 31 organizations containing public data made available through sustainability reports, imparting a necessary level of maturity for the process of collating the information, which enables it to be made publicly available while ensuring an adequate degree of transparency. Together, the chosen organizations correspond to 96.1% of the total number of consumers and 95% of the electric energy distributed nationally.

The period of analysis comprehends the years between 2007 and 2016, where the initial year of the series is related with the fact that the PNQ had started to be used as criteria for awarding the ABRADEE Management Quality Prize, in a correlated way.

3.2 Variables Investigated and Data Collection Procedure

For the study in question, the GS and IASC variables were investigated. The GS variable derives from the score obtained by companies in the PNQ, which can vary between 0 to 1000 points, indicating the level of maturity of their management practices. This variable's collection process reflects the judgment of a multidisciplinary board with training and experience for evaluations of models of excellence. In addition, the data used to classify the organizations according to their participation was provided by the FNQ, where the award winners were made public through press releases and announcements to the market and posted on the entity's website.

The IASC scheme, in its turn, is annually organized by ANEEL using a standard methodology (Marchetti and Prado 2001, 2004). Its values can vary from 0 to 100% and are made available for public consultation on the entity's website from the first year an application for the award was submitted. The set of data provided is the very same used for the IASC award (ANEEL 2016).

3.3   Strategies for the Data Analysis

The information was organized and tabulated in such a way as to be analyzed and distilled. In all cases, the organizations have sufficient data for the intended analysis to be carried out.

The organizations were divided into two groups: (1) users; and (2) indifferent. Subsequently, the engaged organizations were subdivided into: (3) prize winners and (4) engaged.

We thus considered:

(1) users - that is, the organizations that have over 3 participations or attained 30% of recognition level in the period of analysis, which characterizes regular participation;

(2) indifferent - that is, those organizations that participated up to 3 times or 30% of the number of times during the analyzed period, sequentially or not;

(3) prize winners - that is, the user organizations that obtained greater recognition in the period of analysis;

(4) engaged – that is, the user organizations that did not receive awards.

Once the groups were formed, we analyzed the variables, initially through descriptive statistics using pure variables, and then through the same analyses of the grouped data. As for the choice of tests, the assumption of normality necessary for some types of analysis was gauged using the Anderson-Darling test (Pino 2014).

The tests used indicated, at a five percent level of significance, that the IASC variable does not present a normal distribution. The option selected for dealing with non-normality, as suggested by Pino (2014), was the use of non-parametric tests. Among the potential tests that could be used, the Kruskal-Wallis (KW) test was applied, seeing as it is suitable for this purpose (Kruskal and Wallis 1952). The KW test is a non parametric test that uses ranks to compare if several independent samples have their origin from the same distribution (one way analysis of variance by ranks). The parametric similar approach of KW Test is the ANOVA (parametric one way analysis of variance).

3.4   Treatment of Data

Once the graphic analyses deriving from descriptive statistics were performed, the database content was analyzed individually, with special attention to the missing data and to the atypical data (outliers) and corrected as necessary.

With reference to the GS, the missing data deriving from the quarantine periods was complemented by the results of the last participation in the PNQ. When the organization ceased to participate, we understood that the practice was interrupted and the gaps were maintained.

As for the IASC, the data from 2011 were not validated by ANEEL, which did not the disseminate indexes obtained (Carrasco 2018). Hence we opted for repeating the values of 2010 for the period of 2011. No correction of the atypical data (outliers) was necessary.

Data Analysis

Descriptive Statistics

The first analysis, deriving from descriptive statistics, allows us to understand the IASC behavior in the various groups of companies (users, indifferent, award winning and engaged). The descriptive statistics can be found in Table 1.

Table 1. Descriptive statistics

Companies	Minimum	Mean	Median	Maximum
Users	48.1	64.3	64.4	79.0
Awarded	51.7	66.5	67.8	79.0
Engaged	48.1	63.1	63.0	77.9
Indifferents	35.5	55.8	56.3	71.8

Best results

Worst results

When evaluating the measures of position, it can be observed that the best IASC results can be found in the group of award winning companies, whereas the worst relate to the indifferent ones.

On the basis of the results, as referred, it was of interest to test the existence of differences in performance of IASC amongst the companies that use regularly the model MEG, including the award winning companies and the indifferent ones. In both cases the null hypothesis, H_0, was conceived as assuming the equality between the groups and the alternative hypothesis, H_1, their difference. These hypotheses, which have been substantiated in the main analysis, are described in Table 2.

Table 2. Proposed hypotheses during the study

HA_0: Users organizations deliver the same level of performance in IASC as indifferent organizations	HA_1: Users organizations deliver a higher level of performance in IASC than indifferent organizations
HB_0: Award winning organizations deliver the same lev of performance in IASC as engaged organizations	HB_1: Award winning organizations deliver a higher level of performance in IASC than engaged organizations

The dispersion graph (see Fig. 3) also corroborates this analysis. Even though the correlation between the indexes is low, it can be noted that the award winning organizations are concentrated in the upper half of the graphs, and the indifferent companies in the lower quadrants.

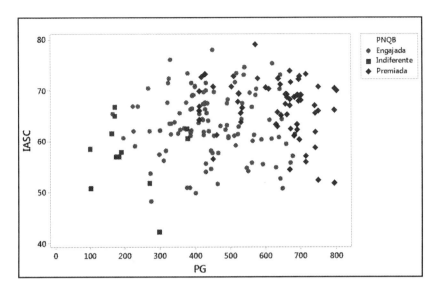

Fig. 3. Scatter analysis GS versus IASC.

The results of the dispersion graph reinforce the low correlation established in the Pearson correlation test (0.18803). It also occurs due to the fact that the IASC represents the satisfaction of consumers, therefore providing an external view of the organization, while the GS indicates internal improvements, as well as those related to the management of the indicators that will reflect the highest satisfaction indexes. On that account, they are complementary, even though they do not explain each other only by themselves.

Testing the Hypothesis

For each hypothesis being tested, we defined two alternatives. The null hypothesis (H_0) consists in admitting that there are no statistical differences among the samples presented. The alternative hypothesis (H_1) consists in a finding of differences. In the case in question, rejecting the null hypothesis (H_0), applying the test Kruskal-Wallis (Martins 2001), is equivalent to confirm the suppositions in this study.

In this way, by obtaining p-value results lower to 0.05 (see Table 3) we consider that there is statistical evidence for the hypotheses HA_1 and HB_1 (shown in Table 1) at a significance level of 5%.

Table 3. P-value results (Kruskal Wallis test).

Groups	$\propto$	n	IASC p-value
Engaged vs Awarded	5%	220	<0.001
Awarded + Engaged vs Indifferent	5%	307	<0.001

4   Conclusions

In this article, we have evaluated the performance difference in IASC of Brazilian electric energy distribution, in a period of ten years (2007 to 2016), vis-à-vis their performance level in the adoption of the MEG.

In our first evaluation, we demonstrated the performance difference between the user companies and the indifferent ones through validation of Hypothesis HA_1. This confirms studies such as that of Puay et al. (1998), where the best satisfaction results are found in companies using excellence models.

The validation of the second hypothesis, HB_1, leads us to confirm that the award winning organizations have better IASC results than the engaged organizations. Hence, these findings corroborate studies like those by Escrig and Menezes (2015), who point to an increase in the correlation between managerial maturity and results in the organizations using this model.

Both conclusions are aligned with the findings presented by Boutler et al. (2013) and Corredor and Goñi (2010), wherein such recognitions are obtained by companies that achieved more positive results, and not by the organizations whose results were not as good.

By adopting a combined and holistic approach, we also identified that the greater managerial maturity of companies, the better their results in the IASC index. Thus we can attest the positive effects evidenced by the use of reference models of the electric energy sector, on a large scale, during the period analyzed.

Acknowledgements. Marina A. P. Andrade acknowledges the financial support by Portuguese funds through the ISTAR -ISCTE, Instituto Universitário de Lisboa.

M. Filomena Teodoro acknowledges the financial support by Portuguese funds through the Center for Computational and Stochastic Mathematics (CEMAT), The Portuguese Foundation for Science and Technology (FCT), University of Lisbon, Portugal, project UID/Multi/04621/2019, and Center of Naval Research (CINAV), Naval Academy, Portuguese Navy, Portugal.

References

ANEEL. Agência Nacional de Energia Elétrica. Evolução Iasc e Benchmarks Internacionais (2016). http://www.aneel.gov.br/metodologia-iasc/-/asset_publisher/ri7lpr3r2ykt/content/evol ucao-iasc-e-benchmarks-internacionais/655804?inheritredirect=false&redirect=http%3a%2f% 2fwww.aneel.gov.br%2fmetodologia. Accessed 10 Oct 2018

ANEEL. Agência Nacional de Energia Elétrica. Regulamento Prêmio Iasc. Despacho N. 2.502 (2017)

Baltazar, A.C.: Qualidade da energia no contexto da reestruturação do setor elétrico brasileiro. Dissertação de Mestrado. Escola Politécnica. Faculdade de Economia e Admistração (2007)

Boulter, L., Bendell, T., Dahlgaard, J.J.: Total quality beyond North America: a comparative analysis of the performance of European excellence award winners. Int. J. Oper. Prod. Manag. 33(2), 197–215 (2013)

Brito, M.P.: Geração distribuída: Critérios e impactos na rede elétrica. Engenharia Elétrica - Universidade Federal do Espírito Santo, Dissertação de graduação (2007)

Buccelli, D.O., Costa Neto, P.L.O.: Prêmio Nacional da Qualidade: Gestão da qualidade ou qualidade da gestão? Trabalho apresentado no XXXIII Encontro Nacional de Engenharia de Produção. A Gestão dos Processos de Produção e as Parcerias Globais para o Desenvolvimento Sustentável dos Sistemas Produtivos Bahia, Brasil 08 a 11 de outubro (2013)

Calvo-Mora, A., Navarro-García, A., Perianez-Cristobal, R.: Project to improve knowledge management and key business results through the EFQM. Int. J. Project Manag. **33**(8), 1638–1651 (2015)

Cardoso, R., Cormack, A.M., Delesposte, J.E., Nascimento, M.K., Boechat, A.S.: O uso da ferramenta "metamodelo de gestão" na integração de múltiplos modelos de referência na modelagem da gestão organizacional. In: Trabalho apresentado no XIX Simpósio de Engenharia de Produção Sustentabilidade na Cadeia de Suprimentos, São Paulo, Brasil, 5 a 7 de Novembro (2012)

Carrasco, A.: Dez anos de estudos sobre o impacto do uso modelos de excelência na qualidade do fornecimento (DEC e FEC) e satisfação de clientes no setor de distribuição de energia elétrica brasileiro. Dissertação de Mestrado. Gestão. ISCTE – Instituto Universitário de Lisboa (2018)

CNI. Confederação Nacional da Indústria: Retratos da sociedade brasileira – serviços públicos, tributação e gasto do governo. Indicadores CNI **5**(33), 1–14 (2016)

Corredor, P., Goñi, S.: Quality awards and performance: is there a relationship? TQM J. **22**(5), 529–538 (2010)

Doeleman, H.J., Ten Have, S., Ahaus, C.T.B.: Empirical evidence on applying the European foundation for quality management excellence model, a literature review. Total Qual. Manag. Bus. Excellence **25**(5–6), 439–460 (2014)

Engel, J.F., Blackwell, R.D., Miniard, P.W.: Consumer Behavior. Dryden Press, Forth Worth (1993)

Escrig, A.B., Menezes, L.M.: What characterizes leading companies within business excellence models? An analysis of "EFQM Recognized for Excellence" recipients in Spain. Int. J. Prod. Econ. **169**, 362–375 (2015)

FNQ. Fundação Nacional da Qualidade Critérios de Excelência. Critérios de Excelência. Avaliação e Diagnóstico da Gestão Organizacional. 19a Edição, São Paulo (2011)

FNQ. Fundação Nacional da Qualidade Critérios de Excelência. Melhores em Gestão. Instruções para candidatura 2018, São Paulo (2018)

FNQ. Fundação Nacional da Qualidade. Melhores em gestão (2016). http://www.fnq.org.br/avalie-se/melhores_em_gestao. Accessed 10 Oct 2018

Khoo, H.H., Tan, K.C.: Managing for quality in the USA and Japan: differences between the MBNQA, DP, and JQA. TQM Mag. **15**(1), 14–24 (2003)

Kruskal, W.H., Wallis, W.A.: Use of ranks in one-criterion variance analysis. J. Am. Stat. Assoc. **47**(260), 583–621 (1952). https://doi.org/10.1080/01621459.1952.10483441

Marchetti, R., Prado, P.H.M.: Um tour pelas medidas de satisfação do consumidor. Rev. Adm. Empresas **41**(4), 56–67 (2001)

Marchetti, R., Prado, P.H.M.: Avaliação da satisfação do consumidor utilizando o método de equações estruturais: Um modelo aplicado ao setor elétrico brasileiro. Rac **8**(4), 9–32 (2004)

Martins, G.D.: Estatística Geral e Aplicada. Atlas, São Paulo (2001)

Miguel, P.A.C., Morini, C., Pires, S.R.I.: Um caso de aplicação do Prêmio Nacional da Qualidade. TQM Mag. **16**(3), 186–193 (2004)

Oliver, R.L.: Measurement and evaluation of satisfaction processes in retailing settings. J. Retail. **57**(3), 25–48 (1981)

Oliver, R.L.: Satisfaction: A Behavioral Approach. Mcgraw-Hill, Nova York (1997)

Pino, A.F.: A questão da normalidade: uma revisão. Rev. Econ. Agrícola **61**(2), 17–33 (2014)

Puay, S.H., Tan, K.C., Xie, M., Goh, T.N.: A comparative study of nine national quality awards. TQM Mag. **10**(1), 30–39 (1998)

Tan, K.C.: A comparative study of 16 national quality awards. TQM Mag. **14**, 165–171 (2002)

Vergara, S.C.: Projetos e Relatórios de Pesquisa em Administração. Atlas, São Paulo (1998)

Waste Management and Embarked Staff

M. Filomena Teodoro[1,2]($\boxtimes$) (iD), José B. Rebelo[1], and Suzana Lampreia[1] (iD)

[1] CINAV, Center of Naval Research, Naval Academy, Portuguese Navy,
2810-001 Almada, Portugal
`maria.alves.teodoro@marinha.pt`
[2] CEMAT, Center for Computational and Stochastic Mathematics, Instituto Superior Técnico,
Lisbon University, 1048-001 Lisboa, Portugal

Abstract. This case study intends to carry out an analysis of the waste management of the Portuguese Navy ships, limiting the study to the residues referring in Annexes I, IV and V of Marpol 73/78 [4], where the pollution by hydrocarbons, sewage and all types of garbage is approached. Methods and forms are analyzed of how the storage and treatment of ship waste is carried out, checking the existing equipment, its operational status and whether there is an on-board waste management plan. However, it is not enough to assess the materials and procedures. The knowledge and cooperation of the military on board for the environment are also determinants. It was built and implemented a questionnaire to the garrison of some selected NRP ships after permission of the Commander of Portuguese Surface Fleet. We have applied an exploratory factorial analysis (EFA) so we could identify the major questions contributions to the latent variables that explain literacy about waste management in ships. An analysis of variance was applied so we could get significant independent variables that contribute to explain the selected factors.

Keywords: Waste management · Environmental guidelines · NRP ships · Literacy · Questionnaire · Statistical analysis · Factorial analysis · ANOVA

1 Introduction

As is well known, the surface earth is covered mostly by water in the liquid state, representing about 71% [1], becoming a very important medium that helps regulate the balance of the entire climate system of our planet. The sea contains many other important factors for human survival, such as its high biodiversity and natural resources, which contribute to global economic development [2,7,8]. Human activity has been intensifying with the growth of the global economy, and as a direct consequence the pollution of the sea is a problem that has increased over the last centuries with the rapid economic development and consequently the population increase. This high growth has created intense pressure on the environment, in this particular case at sea, where marine litter is a major cause of intense human activity [5]. Approximately 80% of the world trade volume is transported by sea [16], causing heavy traffic of ships to be one of the main sources of pollution, which generates solid wastes, sewage and wastes from hydrocarbons, forgetting that they are also an atmospheric pollutant source [5].

© Springer Nature Switzerland AG 2020
O. Gervasi et al. (Eds.): ICCSA 2020, LNCS 12251, pp. 492–503, 2020.
https://doi.org/10.1007/978-3-030-58808-3_36

In this article the objectives are to characterize the profile of waste management in NRP ships. This work was started in [12] where was compiled specific regulation about the theme. Also, in the same monography was implemented a questionnaire to evaluate Knowledge, Attitudes and Practice (KAP) about the waste management during the boarded period in the NRP ships. It was performed a preliminary statistical analysis in [14, 15] using a non complete sample of boarded population, being used as reference in the present work. In this manuscript, we have continued the statistical approach proposed in [13].

This article is comprised of an introduction and results and final remarks Sects., a Sect. 2 containing the description of methodology. The empirical application, discussion and final remarks can be found in Sect. 3.3.

2   Methodology

2.1   Factorial Analysis

Factor analysis (FA) is a technique often used to reduce data. The purpose is to get a reduced number of variables (frequently denominated latent variables) from an initial big set of variables and get easier interpretations [6]. The FA computes indexes with variables that measures similar things. There are two types of factor analysis: exploratory factorial analysis (EFA) and confirmatory factorial analysis (CFA) [17]. It is called EFA when there is no idea about the structure or the dimension of the set of variables. When we test some specific structure or dimension number of certain data set we name this technique the CFA.

There are various extraction algorithms such as principal axis factors, principal components analysis or maximum likelihood (see [3] for example). There are numerous criteria to decide about the number of factors and theirs significance. For example, the Kaiser criterion proposes to keep the factors that correspond to eigenvalues greater or equal to one. In the classical model, the original set contains p variables $(X_1, X_2, \ldots, X_p)$ and m factors $(F_1, F_2, \ldots, F_m)$ are obtained. Each observable variable X_j, $j = 1, \ldots, p$ is a linear combination of these factors:

$$X_j = \alpha_{j1}F_1 + \alpha_{j2}F_2 + \cdots + \alpha_{jm}F_m + e_j, \qquad j = 1, \ldots, p, \qquad (1)$$

where e_j is the residual. The factor loading α_{jk} provides an idea of the contribution of the variable X_j, $j = 1, \ldots, p$, contributes to the factor F_k, $k = 1, \ldots, m$. The factor loadings represents the measure of association between the variable and the factor [6, 17].

FA uses variances to get the communalities between variables. Mainly, the extraction issue is to remove the largest possible amount of variance in the first factor. The variance in observed variables X_j which contribute to a common factor is defined by communality h_j^2 and is given by

$$h_j^2 = \alpha_{j1}{}^2 + \alpha_{j2}{}^2 + \cdots + \alpha_{jm}{}^2, \qquad j = 1, \ldots, p. \qquad (2)$$

According with the author of [9], the observable variables with low communalities are often dropped off once the basic idea of FA is to explain the variance by the common factors. The theoretical common factor model assumes that observables depend on the common factors and the unique factors being mandatory to determine the correlation patterns. With such objective the factors/components are successively extracted until a large quantity of variance is explained. After the extraction technique be applied, it is needed to proceed with the rotation of factors/components maximizing the number of high loadings on each observable variable and minimizing the number of factors. In this way, there is a bigger probability of an easier interpretation of factors 'meaning'.

2.2   Analysis of Variance

Experimental design is primarily due to *Sir* Fisher in designing a methodology for agricultural experiments. The main purpose of these methods is to

– assess how a set of qualitative explanatory variables, factors, affect the answer variable,
– discern the most important factors,
– select the best combination of factors to optimize the response,
– fit a model that can make predictions and/or adjust controllable factors to maintain the response variable in the proposed objective.

Noise factors (uncontrollable) that condition the response variable will not be considered. The different values of a factor are called levels. A combination of levels of different factors is denominated treatment. If there is only one factor, each level is a treatment. This work considers only two distinct cases: one-factor and two-factor experimental design [10, 11].

Experimental Design One Factor.   The purpose of these techniques boils down to comparing k treatments ($k \geq 2$). Suppose there are k groups of individuals chosen at random. Each group is subject to treatment, i, $i = 1, \ldots, k$. Each group does not necessarily have the same group of individuals. Consider n_i the number of individuals in group i. If in each group the number of individuals is equal, the design is denominated as balanced. When two independent ($k = 2$) random samples are available, t-tests can be established to compare means, when there are $k > 2$ independent samples there is no way to establish this test to proceed with their analysis. It is necessary to resort to a completely different technique known as analysis of variance. The data of k samples are generally presented as y_{ij}, the response of individual j in sample i.

Theoretical Model and Parameter Estimation.   Formal inference to compare means of different treatments implies the definition of probabilistic models. It is assumed that the relative data on the $i - th$ treatment have a normal distribution of mean μ_i and variance σ^2. If Y_{ij} is a random variable (rv) associated to the observed value y_{ij} the theoretical model can be represented by (3)

$$Y_{ij} = \mu_i + \varepsilon_{ij}, \ (j = 1, \ldots, n_i, i = 1, \ldots, k), \tag{3}$$

with ε_{ij} rv's independents and Gaussian

$$\varepsilon_{ij} \cap N(0, \sigma^2) \tag{4}$$

Treatment mean μ_i and error variance σ^2 are unknown parameters (to be estimated). Notice that the model (3) (4) is a generalization of models for two independent random samples with equal variance. It is common to write as (5) and (6)

$$\mu = \mu_i + \alpha_i, \tag{5}$$

where μ is designated the global mean given by

$$\mu = \frac{\sum_{i=1}^{k} n_i \mu_i}{\sum_{i=1}^{k} n_i} = \frac{\sum_{i=1}^{k} n_i \mu_i}{N} \tag{6}$$

with $N = \sum_{i=1}^{k} n_i$, the total number of observations. The deviation α_i, from the global mean of the $i-th$ treatment, denominated the treatment effect, is given by (7)

$$\alpha_i = \mu_i - \mu. \tag{7}$$

It is important to note that α_i are subject to the restriction $\sum_{i=1}^{k} n_i \alpha_i = 0$ with only $k - 1$ linearly independent effects. Notice that $\sum_{i=1}^{k} n_i \alpha_i = 0$.

The model can be rewritten as (8)

$$Y_{ij} = \mu + \alpha_i + \varepsilon_{ij}, \ (i = 1, \ldots, k, \ j = 1, \ldots, n_i). \tag{8}$$

The estimate of μ is given by the global sample mean (9) considering all values of k samples, being no more than a weighted average of the estimates of the mean μ_i

$$\widehat{\mu} = \bar{y} = \frac{\sum_{i=1}^{k} \sum_{j=1}^{n_i} y_{ij}}{\sum_{i=1}^{k} n_i} = \frac{\sum_{i=1}^{k} n_i \bar{y}_i}{N}. \tag{9}$$

The estimates of μ_i are the treatment sample averages given by (10)

$$\widehat{\mu}_i = \bar{y}_i \ i = 1, \ldots, k. \tag{10}$$

The estimate $\widehat{\sigma}^2$ of σ^2, the pooled sample variance, it is a weighted average of the (corrected) sample variances per treatment, being the weights the respective degrees of freedom $n_i - 1$, and is given by (11)

$$s^2 = \frac{\sum_{i=1}^{k} \sum_{j=1}^{n_i} (y_{ij} - \bar{y}_i)^2}{\sum_{i=1}^{k} n_i - k} = \frac{\sum_{i=1}^{k} (n_i - 1) s_i^2}{N - k}. \tag{11}$$

The estimate of the effect α_i of treatment i is given by the difference between the mean estimates per treatment i with the global average

$$\widehat{\alpha}_i = \bar{y}_i - \bar{y}. \tag{12}$$

We can verify that the restrictions on the effects of the treatment are maintained for the estimates, i.e.

$$\sum_{i=1}^{k} n_i \widehat{\alpha}_i = 0.$$

3   Empirical Application

3.1   The Questionnaire

After authorization by the Portuguese Surface Fleet Commander, the data collection was carried out through questionnaires and some successive visits to the ships, in which the person responsible for waste management was boarded on board each ship and the questionnaires were distributed to the military belonging to each garrison. The questionnaire is divided into two parts, the first part includes socio-demographic variables and personal details, the second part is composed with questions that allow to evaluate KAP. The initial part concerns the socio-demographic information about each participant in general:

– $Q1_1$ - *"Gender."*
– $Q1_2$ - *"Age."*
– $Q1_3$ - *"Grade."*
– $Q1_4$ - *"Have you ever attended an environmental training course?"*
– $Q1_{41}$ *"If you answered "Yes" in the previous question, it was in Navy."*
– $Q1_5$ *"Do you recycle at home?"*

The second part consists in questions of open or closed response, with the possibility of choosing more than one answer in each question, in the form of Likert scale with four levels from 1 to 4 (1 - Totally Disagree, 2 - Partially Disagree, 3 - Partially agree, 4 - Totally Agree; also some questions have a "yes" or "No" answer; one question has an open answer, it is required to identify factors that contribute for a bad waste management, this question will have a qualitative, not a quantitative analysis. The second part of the questionnaire aimed at evaluating participants' knowledge, attitudes and practices regarding waste management, comprising questions about knowledge issues, other about attitudes and some questions that consider practice details. After filling the questionnaire, the participant should give his participation as finished and submit the questionnaire to the researcher. Follows the list of second set of questions:

– $Q2_1$ - *"The environmental concern on board is always present in my daily life."*
– $Q2_2$ - *"I consider good waste management practice on board ships important."*
– $Q2_3$ - *"There are regular lectures on board on waste management ."*
– $Q2_4$ *"Sometimes I dump small waste into the sea."*
– $Q2_5$ - *"I think there is a good waste management policy on board ships."*
– $Q2_6$ - *" There are some types of waste that we can discharge into the sea."*
– $Q2_7$ - *" The glass can be discharged into the sea, as it ends up in the bottom of the sea, having no interaction with the environment."*
– $Q2_8$ - *"Paper and cardboard can be discharged at sea because they easily degrade."*
– $Q2_9$ *"Proper packaging of waste contributes to the welfare, hygiene and safety of the trim."*
– $Q2_{10}$ - *"Waste storage space is adequate."*
– $Q2_{11}$ - *"The conditions of shipboard equipment allow for the treatment of different types of waste."*
– $Q2_{12}$ - *"Even if conditions are not adequate, there is an effort and concern from the trim to minimize the environmental impact of the ship."*

- $Q2_{13}$ - *"The educational offer of the Navy in the environment preservation is sufficient."*
- $Q2_{14}$ - *"The Navy promotes, with its military staff, the preservation of the environment."*
- $Q2_{15}$ - *"There has been an increase in people's awareness of environmental preservation ."*
- $Q2_{16}$ - *" I know the Navy Environmental Policy and I know where I can consult it."*
- $Q2_{17}$ - *"I am aware of national and international regulations for reducing environmental impact."*
- $Q2_{18}$ - *"Sometimes on board, environmentally harmful acts are performed due to lack of waste treatment conditions."*
- $Q2_{19}$ - *"Feels that their role in minimizing waste generation on board is important for good waste management in the organization."*
- $Q2_{20}$ - *"On board are used environmentally friendly consumables."*
- $Q3$ - *"Has the waste generated on board ever compromised your well-being?."*
- $Q4$ - *"elect from 1 to 2 factors that undermine the proper functioning of onboard waste management."*
- $Q5$ - *"As the Navy is a military organization, do you consider your concern about the ecological footprint at sea important?"*

3.2   Sample Characterization

Firstly, was performed a descriptive analysis of the questionnaire output taking into account the quantitative and non-quantitative character of some variables.

In Fig. 1 we can find the summary about individual characterists of the respondents. On left we can observe the histogram of the of age and gender distributions. In sample, about $4/5$ are men and $1/5$ are women. Almost 45% of participants are aged until 30 years old and 38% are between 30 and 40 years old. The maximum age is 51 years.

Figure 2 evidences that, between the participants, $2/3$ do reclycing at home and $1/3$ have environmental education training.

Several tests were performed, some non-parametric correlations were computed, namely nonparametric Spearman correlation coefficient, non-parametric test of Friedman for paired samples, etc.

3.3   Results and Final Remarks

The first step is to verify the questionnaire inter consistence and homogeneity. to evaluate the internal consistence. The most common measure of questionnaire internal reliability, the alpha-Cronbach coefficient, has given a good internal consistency; also, this index indicates that when some of questions are let out of study, the internal consistence can be improved. This detail is confirmed when we perform some questions distribution comparison tests (for the set of questions asociated to knownledgement). The paired T-test, McNemar's test for frequencies comparison, Crochan's Q test comparison (where the aggregation of the distinct levels per answer as Yes/No took place). Also the Friedman test ($p - value < 0.001$) and the concordance test using the kendals coefficient ($p - value < 0.001$) were applied. The results were consensual and significant:

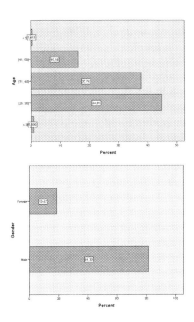

Fig. 1. On left: Sample age distribution. Considered classes: $< 20, [20, 30], [30, 40], [41, 50]$ and > 50. On right: Sample distribution per gender (male/female).

globally the questions conduced to different distribution of answers. The Spearmann correlation coefficient reveals significant relations between some questions, Friedman's test supports such association ($p - value < 0.001$). These preliminary analysys can be found in [14] Table 1 summarizes 1st, 2nd and 3rd quartiles associated to each question. In last three columns of Table 1, are displayed the median tests $p - value$ (runs test and wilcoxon test) and decision. Using such information we can say that there is enidence that m ore than 50% of participants declare that there exists the daily environmental care, consider it an important procedure, also consider that some waste can be left in sea, the waste storage contributes to welfare, security and hygiene of staff. Also embarked staff considers that the existent equipment to process waste is not enough.

The staff declares to know the internal and external rules but claims that there is not a good offer of formation in environmental education. Besides this the environmental awareness is increasing.

With the idea of simplify a high dimensional system, was applied a technique from multivariate Statistics, the EFA, reducing a large number of correlated variables to factors, establishing the correlation of observable variables and organizes them into factors, which in themselves are unobservable variables. The factors communality was computed, the R-matrix gave a significant test, the multi-collinearity or singularity was evaluated. The Bartlett's sphericity test provided a strongly significant level $p < 0.001$, confirming that there exists important patterned relations between the variables. Also, the Kaiser-Meyer-Olkin measure of sampling adequacy evidences confirmed that is appropriate to apply an EFA.

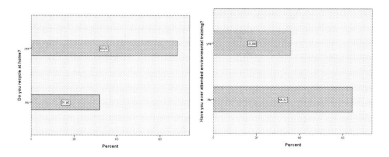

Fig. 2. On left: Sample distribution of training on environmental issues (no/yes). On right: Sample distribution of recycling activity at home (no/yes).

Table 1. Questionnaires answers. Percentiles: 25th, 50th, 75th, Wilcoxon test $p - value$, sign test $p - value$, tests decision.

Question	1st quartile	2nd quartile	3rd quartile	Wilcoxon	Sign	Decision
$Q2_1$	2.2500	3.0000	3.7500	0.000	0.000	$med > 2.5$
$Q2_2$	3.2500	4.0000	4.0000	0.001	0.001	$med > 2.5$
$Q2_3$	1.0000	2.0000	2.0000	0.000	0.000	$med < 2.5$
$Q2_4$	1.0000	1.0000	2.7500	0.000	0.000	$med < 2.5$
$Q2_5$	1.2500	20000	3.0000	0.000	0.000	$med < 2.5$
$Q2_6$	2.0000	3.0000	4.0000	0.000	0.000	$med > 2.5$
$Q2_7$	1.0000	1.0000	1.0000	0.000	0.000	$med < 2.5$
$Q2_8$	1.0000	1.0000	3.0000	0.000	0.000	$med < 2.5$
$Q2_9$	4.0000	4.0000	4.0000	0.000	0.000	$med > 2.5$
$Q2_{10}$	2.0000	2.0000	3.0000	0.000	0.005	$med < 2.5$
$Q2_{11}$	1.0000	2.0000	3.0000	0.000	0.000	$med < 2.5$
$Q2_{12}$	3.0000	3.0000	3.7500	0.000	0.000	$med > 2.5$
$Q2_{13}$	1.0000	2.0000	2.0000	0.000	0.000	$med > 2.5$
$Q2_{14}$	1.0000	2.0000	3.0000	0.000	0.000	$med < 2.5$
$Q2_{15}$	3.0000	3.0000	3.0000	0.000	0.000	$med > 2.5$
$Q2_{16}$	2.0000	3.0000	3.0000	0.000	0.000	$med > 2.5$
$Q2_{17}$	2.0000	3.0000	3.0000	0.317	0.165	$med = 2.5$
$Q2_{18}$	2.2500	3.0000	3.0000	0.000	0.000	$med > 2.5$
$Q2_{19}$	3.0000	3.0000	4.0000	0.000	0.000	$med > 2.5$
$Q2_{20}$	2.0000	3.0000	3.0000	0.000	0.001	$med > 2.5$
$Q3$	0.0000	0.0000	0.7500	0.000	0.000	$med < 0.5$

In Table 2 are displayed the eigenvalues associated to each factor and correspondent explained variance before extraction and after extraction considereding raw data and scaled data. Is also displayed the cumulative variance percentage explained by the first i factors, we have. When we consider raw data, the first 4 factorsexplain almost 50% of variance. By opposite, when we use scaled data, we need to consider 6 factors to explain the same percentage of variance (almost %50).

Table 2. Total variance explained per factor. Top:raw data; bottom: rescaled data.

Total Variance Explained

	Component	Initial Eigenvalues[a]			Extraction Sums of Squared Loadings		
		Total	% of Variance	Cumulative %	Total	% of Variance	Cumulative %
Raw	1	2,912	19,351	19,351	2,912	19,351	19,351
	2	1,545	10,265	29,616	1,545	10,265	29,616
	3	1,482	9,851	39,467	1,482	9,851	39,467
	4	1,104	7,337	46,805	1,104	7,337	46,805
	5	,860	5,716	52,520	,860	5,716	52,520
	6	,823	5,468	57,989	,823	5,468	57,989
	7	,730	4,849	62,838	,730	4,849	62,838
Rescaled	1	2,912	19,351	19,351	4,230	18,392	18,392
	2	1,545	10,265	29,616	1,466	6,374	24,766
	3	1,482	9,851	39,467	1,628	7,079	31,845
	4	1,104	7,337	46,805	1,510	6,566	38,411
	5	,860	5,716	52,520	1,040	4,524	42,935
	6	,823	5,468	57,989	1,228	5,340	48,275
	7	,730	4,849	62,838	,940	4,087	52,362

Extraction Method: Principal Component Analysis.

a. When analyzing a covariance matrix, the initial eigenvalues are the same across the raw and rescaled solution

Table 3. Analysis of variance. Dependent variable:factor 2 (*Hygiene and Safety*). Explanatory variable: Kind of ship.

ANOVA

REGR factor score 2 for analysis 1

	Sum of Squares	df	Mean Square	F	Sig.
Between Groups	16,176	6	2,696	2,819	,011
Within Groups	223,824	234	,957		
Total	240,000	240			

When we use the Kaiser criterion, we select the factors whose eigenvalues are great or equal to one. We have kept the first 4 factors. We can use distinct techniques to select the "best" factors, e.g. the scree plot or the average of extracted communalities can determine the eigenvalue cutt-off (see Fig. 3).

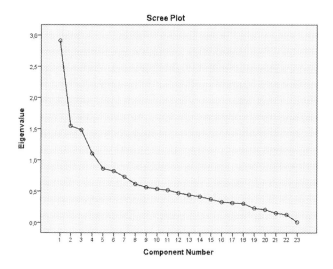

Fig. 3. Scree plot.

After the extraction process, was applied the Varimax approach,allowing to get orthogonal factors. This usefuul algorithm is often applied to identidy variables that can contribute to build indexes or new non correlated variables. We consider the case using raw data, taking the first 4 factors. In this study, we could associate a 'meaning' for the first 3 factors. The interpretation of such meaning is done analyzing the rotated factors scores. We can identify a meaning for each: F_1 combines variables that usually are associated to *Awareness*, F_2 considers variables from *Hygiene and Safety*, F_3 combines variables from *Practice*.

The selected factors (factors that have a higher variance explanation) can be considered as explanatory variables in a predictive model.

With such purpose, we have used the ANOVA technique to investigate if the *new* variables identified as important (the selected factors in the EFA) to describe the problem are related with kind of ship, the attendance of training courses, the military hierarchical posts (*praças, sargentos, oficiais*).

The kind of ship revealed significant differences in the second factor F_2 (see Table 3 when we consider different king of ships. The F test conduced to a $p - value = 0.001$ We can find such differences in the Fig. 4 where is displayed the difference of the global mean of the factor F_2 relatively to the mean per kind of ship. This difference is usually denominated an effect. Notice that from Fig. 4 we can evidence that the *hidrográfico* and *lancha* are the ships with greater positive effects on $F2$, by opposite, the *corveta* is the kind of ship that has the effect with bigger effect with negative sign. The *lancha hidrográfica,veleiro, fragata* e *patrulha oceânica* have smaller effects on factor $F2$. The military hierarchical posts (category) have no significant effects in factors $F1$, $F2$ and $F3$.

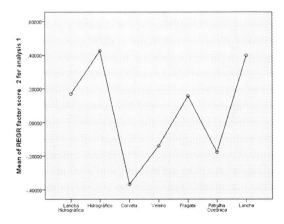

Fig. 4. Effects of the kind of ship in factor 2: difference between the mean estimates per treatment *i* with the global average (12).

The statistical evidence for different means of distinct kind of ships using Scheffé simultaneous intervals for their difference was determined, the difference of distinct means of $F2$ for the several kind of ships was considered, where the $p - value$ associated to F test performed to evaluate the hypothesis was obtained. We can conclude that there is significant statistical difference between the *lancha hidrográfica* and *fragata, hidrográfico, patrulha oceânica*; between *hidrográfico* and *corveta, veleiro*; between *patrulha oceânica* and *corveta, veleiro*; and between *lancha* and *corveta, veleiro*. In general, *corveta, veleiro* have a distinct padron of *Hygiene and Safety* relatively to the most of other ships.

Also were studied another kind of relations withe the 3 factors obtained by EFA, but the study is too detailed and will appear in a continuation of the present article. Some related some questions (attendance of an environmental training course, or the space to store waste) with the different qualitative variables were considered. Also was built and indicator of good practices combining the information of some questions. The results appear to be adequate and in accordance with what is expected.

Acknowledgements. This work was supported by Portuguese funds through the *Center of Naval Research* (CINAV), Portuguese Naval Academy, Portugal and *The Portuguese Foundation for Science and Technology* (FCT), through the *Center for Computational and Stochastic Mathematics* (CEMAT), University of Lisbon, Portugal, project UID/Multi/04621/2019.

References

1. A água no planeta (2019). https://www.sobiologia.com.br/conteudos/Agua/. Accessed 24 Feb 2020
2. APA: Monitorização do lixo marinho (2019). https://www.apambiente.pt/index.php?ref=17&subref=1249. Accessed 24 Feb 2020
3. Child, D.: The Essentials of Factor Analysis. Continuum International Publishing Group, New York (2006)

4. Edition, C.: Marpol 73/78. regulation (2002)
5. Griffin, A.: Marpol 73/78 and vessel pollution: a glass half full or half empty?. Indiana J. Glob. Legal Stud. **1**(2), 489–513 (1994)
6. Harman, H.: Modern Factor Analysis. University of Chicago Press, Chicago (1976)
7. IMO: Focus on imo marpol-25 years (1999). natasha Brown - External Relations Officer. http://www.imo.org/en/KnowledgeCentre/ReferencesAndArchives/FocusOnIMO(Archives)/Documents/FocusonIMO-MARPOL-25years(October1998). pdf. Accessed 24 Feb 2020
8. IMO: International convention for the prevention of pollution from ships (marpol) (2019). http://www.imo.org/en/About/Conventions/ListOfConventions/Pages/International-Convention-for-the-Prevention-of-Pollution-from-Ships-(MARPOL).aspx. Accessed 24 Feb 2020
9. Maroco, J.: Análise Estatística com o SPSS Statistics. Report Number, Pero Pinheiro (2014)
10. Montgomery, D.: Design and Analysis of Experiments, 5th edn. Wiley, New York (2001)
11. Moore, D., McCabe, G.: Introduction to the Practice of Statistics, 4th edn. W. H. Freeman and Company, New York (2003)
12. Rebelo, J.: Impacto Ambiental da Marinha Portuguesa. Análise e resolução da Gestão de Resíduos no mar. Master thesis, Escola Naval (2019)
13. Rebelo, J., Jerónimo, J., Teodoro, M., Lampreia, S.: Literacy about the waste management in boarded staff. In: Proceedings of EUROGEN 2019, 12nd-14th September, Guimarães (2019)
14. Rebelo, J., Jerónimo, J., Teodoro, M., Lampreia, S.: Modeling the waste management in NRP ships. In: Entrepreneurial Ecosystems and Sustainability. Proceedings of Regional Helix 2019, pp. 183–188. AIP (2019). ISBN 978-989-98447-7-3
15. Rebelo, J., Jerónimo, J., Teodoro, M., Lampreia, S.: Preliminary reflexion about waste management plan in NRP ships. In: Simos, T., et al. (eds.) Computational Methods in Science and Engineering, vol. 2186, AIP (2019). https://doi.org/10.1063/1.5138006
16. UN: United nations conference on trade and development (2018). https://unctad.org/en/PublicationsLibrary/dtl2018d1_en.pdf. Accessed 24 Feb 2020
17. Young, A., Pearce, S.: 10 challenging problems in data mining research. Tutorial Quantitative Methods Psychol. **9**(2), 79–94 (2013)

International Workshop
on Computational Optimization
and Applications (COA 2020)

On Temperature Variation of the Diabetic Foot

Ana Teixeira[1]([⊠]) [iD] and Ana I. Pereira[2] [iD]

[1] Mathematics Centre CMAT, Pole CMAT - UTAD, Sciences and Technology School, University of Trás-os-Montes e Alto Douro, Quinta dos Prados, Vila Real, Portugal
ateixeir@utad.pt
[2] Research Centre in Digitalization and Intelligent Robotics (CeDRI), Instituto Politécnico de Bragança, Campus de Santa Apolónia, Bragança, Portugal
apereira@ipb.pt

Abstract. This work aims to give an additional contribute to the development of an alternative diagnostic method to be applied to early detection of foot pathology in Diabetes Mellitus individuals. In this work, the main concepts related to the topic under study are introduced and a framework concerning the use of thermography to evaluate the temperature distribution in the feet is presented. Additionally, in this work, a mathematical model to characterise the plantar temperature distribution variation is presented and an optimization programming problem based on the nonlinear least squares approach is proposed. Some considerations about the two global non-linear optimization metaheuristic methods used to solve this model, namely a Hybrid Genetic Algorithm and a Hybrid Simulated Annealing, are also described. Thermal plantar images of non diabetic and diabetic individuals are used to test the approach. The numerical results obtained with both methods for the different regions of each foot are presented and analysed; the best results were obtained with the Hybrid Genetic Algorithm. Some preliminary conclusions were made.

Keywords: Least squares model · Diabetic foot · Thermography · Genetic Algorithm · Simulated Annealing Algorithm

1 Introduction

The human being is homoeothermic, thus being able to maintain the body temperature constant, regardless of the changes that happen in the environment. This feature is vital for the preservation of a constant environment within the human body with regard to functions and composition of fluids and tissues [1,2]. There are several pathologies that alter the body temperature [2,3] and, in particular, the temperature of the plantar region; Diabetes Mellitus, usually known as diabetes, is one of these diseases.

Diabetes Mellitus is a chronic illness that affects many people and it is estimated that its global cost will increase from 1.2 trillion euros in 2015 to between

© Springer Nature Switzerland AG 2020
O. Gervasi et al. (Eds.): ICCSA 2020, LNCS 12251, pp. 507–520, 2020.
https://doi.org/10.1007/978-3-030-58808-3_37

1.9 and 2.3 trillion euros in 2030 [4]. Additionally, as the foot is the support for the locomotion of human beings, it is the most vulnerable part of a diabetic's body because it becomes the most common region for occurrence of complicated lesions, decreasing the quality of life of the individuals [5]. Thus, prevention and early diagnosis of diabetic foot in Diabetes Mellitus patients is a very pertinent and important subject of study.

Diabetic foot is a pathology developed by individuals with diabetes and it is characterized by a variety of foot injuries, like infection, ulceration and/or destruction of deep tissues that may be associated with neuropathy or vascular disease [6]. Although not all diabetics develop this pathology, those who develop it loose quality of life. In addition, ulceration of the diabetic foot causes serious medical consequences for the patient; for example, the development of ulcers can lead to infections, since ulcers can harbor bacteria and fungi. These infections, when severe, can lead to amputation of the lower limb [7]. Therefore, early identification and effective preventive methods for diabetic foot are essential. A significant number of studies have demonstrated that temperature variations in the plantar feet region may be related to diabetic foot problems [8–10]. Thus, the analysis and characterization of the temperature distribution on the plant of the foot can contribute to the early diagnosis of the onset of certain diseases, namely the diabetic foot, consequently preventing their appearance and increasing the probability of cure.

The use of thermography to examine the feet of diabetic individuals captured the interest of several researchers for some time now, e.g. [8–15]. Nevertheless, to the extent of our knowledge, Bento et al. [16,17], in 2011, were the first ones to apply mathematical optimization techniques and models to this topic, with the aim of finding a characterization of the temperature distribution of plant of the foot of healthy subjects. In a second phase, Carvalho et al. [18,19] continued the work of Bento et al., initiating the characterization of the temperature of the foot of healthy and diabetic individuals and comparing both of them; these two works used mathematical models composed of trigonometric functions. More recently, Contreras et al. [20] proposed a characterization of the plantar temperature distribution based on a probabilistic approach.

In the sequel of the work of Carvalho et al., the objective of this study continues to be the analysis and characterization of the temperature distribution variation of the plant of the foot of non diabetic and diabetic individuals, using thermographic images, as well as the presentation of a mathematical model that approximates these temperature distributions variation; thus, giving an additional contribute to the development of an alternative diagnostic method, using a painless and non-invasive tool, to be applied to diabetic individuals in order to early detect foot pathology.

The structure of this paper considers: in Sect. 2, some background information on the concepts: diabetes mellitus, diabetic foot and thermography is presented; in Sect. 3, the used mathematical model is explained; in Sect. 4, the optimization methods used to solve the model are described; in Sect. 5, the numerical results are presented and analysed; and in the last section the main conclusions are drawn and future work presented.

2   Thermography and Diabetic Foot

2.1   Diabetes Mellitus

Insulin is a hormone produced in the pancreas and it is responsible for transporting the glucose present in the blood to the cells, where it is used as energy. Diabetes is a chronic disease, characterized by the occurrence of abnormalities in glucose metabolism, resulting in defects in insulin secretion, insulin action or both. There are three main types of diabetes: Type 1, Type 2 and Gestational diabetes. Type 1 is associated with autoimmune disease, that is, the human system attacks the beta cells of the pancreas that are responsible for the insulin production. Thus, the body becomes unable to produce the sufficient quantity of insulin and the individual needs daily doses of insulin to regulate blood glucose levels. Type 2 is associated with the most common form of diabetes. The body produces insulin but, due to the resistance developed by the body to this hormone, it does not act effectively. Over time, it leads to high blood glucose levels. The majority of the patients do not require daily insulin treatment, they only take oral medication that helps control blood glucose levels. Gestational diabetes happens to women when the hyperglycemia state is diagnosed during pregnancy and extra care is needed as there is the possibility of complications with the baby.

Diabetes leads to chronic hyperglycemia, that is diagnosed by elevated levels of glucose in the blood. Over time, hyperglycemia causes damage to various tissues of the body, which leads to the development of serious health problems that can compromise the life of the individual [21–23]. Thus, this disease requires early diagnosis and ongoing treatment to prevent serious complications, such as: heart, blood vessel, eyes, nerves and kidneys problems, as well as a high risk of developing infections that can lead to limb amputation.

2.2   Diabetic Foot

Neuropathy is one of the complications of diabetes; it affects the nerves, causing difficulty with sensations, movements and other aspects, depending on the affected nerve. It leads to the loss of foot sensitivity, sometimes resulting in deformation of the foot; a minor trauma (poorly adjusted footwear, walking barefoot or acute injury) suffered by an individual with this disease may precipitate chronic ulceration. Peripheral vascular disease is another consequence of diabetes, provoking obstructive atherosclerotic of the extremities of the body; as in the previous case, minor trauma in patients with this pathology may result in painful and chronic ulceration of the foot [24,25].

Diabetic foot appears in diabetic patients as a result of damaging nerves and blood vessels problems that may increase the risk of ulceration, infection and amputation [22]. In fact, the biggest problem associated with diabetic foot is ulceration [26], which represents the biggest medical, social and economic problem worldwide [5]. Inflammation is one of the first signs of foot ulceration, being characterized by redness, pain, swelling, loss of function and heat [9]. Diabetic foot ulceration occurs due to a variety of factors such as: neuropathy, peripheral vascular disease, foot deformity, arterial insufficiency, trauma, and decreased resistance to infection [23]. Usually, it results from a combination of two or more risk factors, being neuropathy and peripheral vascular disease the main factors that lead to the development of diabetic foot ulcers [5,24].

The diabetic foot is classified according to the type of ulceration that it presents. Diabetic foot ulcers can be classified as neuropathic, ischemic or neuro-ischemic [27,28]. Neuropathic ulcers are usually located in the metatarsal, but are also found on the fingertips, the dorsum of the foot and the plantar zone. They are the least painful for the individual. The surrounding skin shows loss of sensitivity and calluses. This type of ulcer comes from bone deformation caused by neuropathy. Ulcers that develop on the plantar surface are usually circular and perforate in appearance, often deep. The foot that presents this type of ulceration is considered warm. Ischemic ulcers are located in the plantar and dorsal zones of the feet, in the heel, in the big toe, in the medial surface of the head of the first metatarsus and in the lateral surface of the fifth metatarsus. They are black in color and the surrounding skin is pale, shiny and cold. This pathology is not associated with callus or bone deformation, instead it is associated with peripheral vascular disease, for example, atherosclerosis (occlusion of blood vessels). Ischemic foot is considered cold.

2.3   Thermography

All the bodies with temperature above absolute zero emit infrared radiation, also called thermal radiation. The total energy radiated by a surface is known as its emissive power. This emission rate is comprised between 0 and 1, and is obtained as the ratio of the energy radiated from the body's surface and the energy radiated by a blackbody for the same wavelength. The higher the value of the emissivity coefficient, the closer the emissivity of the body is to that of the blackbody, that is, the greater is its energy emission capacity. The emissivity of human skin is almost constant and equal to $0,98 \pm 0,01$ for wavelength range of 2 to 14, thus, being almost an ideal blackbody [2,29].

Medical thermography can be divided into four categories: contact electrical thermometry, cutaneous temperature discrimination, liquid crystal thermography and infrared thermography (IRT) [30]. This work will focus on IRT.

In IRT, the radiation emitted by the body is detected by a thermographic camera and the intensity of the radiation emitted is converted into a temperature. The image generated by the radiation emitted by the body is called a thermogram [31]. One of the great advantages of IRT is that it allows the acquisition of a large number of pixels in a short time and each pixel corresponds to a temperature of a specific point, that is, it has a high resolution [32].

IRT is increasingly being accepted by the medical community because it is a fast, painless, non-contact, non-invasive method, that enables the simultaneous monitoring of a large area. Additionally, it is a non-ionizing technique, without any repetitive use side effects and the color code of the thermograms is easy to interpret [3,31,33]. However, to obtain a quality thermographic image that allows a reliable analysis, certain requirements must be fulfilled, namely: the basic standards of the exam room must be guaranteed; the imaging system, the image acquisition and the image processing must follow a protocol; and, finally, the analysis of results has to follow some criteria in order to enable us to decide about the viability of the data. Moreover, some care should be taken with the patient as there are several factors that can affect body temperature [34].

Another reason that justifies the progressive increase of the use of IRT in the medical area (mainly for diagnostics) is the fact that in many pathological processes changes in body temperature occur [31]. Among others, IRT has applications in: diabetes mellitus; vascular disorder; breast cancer detection; muscular pain; and rheumatoid arthritis [35–40].

For patients with diabetes, IRT has been used to diagnose diabetic neuropathy, assess changes in body temperature and, in particular, to obtain a diabetic foot diagnose, through the analysis of the temperature distribution of the sole of the foot [31,33,41–44]. Although biochemical tests, as blood tests, are the most usual techniques to diagnose diabetes, Sivanandam et al. [45] concluded that IRT diagnosis have higher potential than the previous mentioned tests to achieve this goal.

Some studies using IRT have shown that patients with diabetes and neuropathy have higher plantar temperature than individuals without this pathology [46]; others suggest that thermographic foot patterns analysis allows diabetic patients to be screened for the risk of ulceration and that high temperatures are indicative of ulceration [10]. In most cases, the thermogram of a healthy individual shows a symmetrical butterfly pattern (one wing on each foot), where the highest temperatures are located.

The readers that are interested in deepening their knowledge in this topic should read the reviews presented by [42,47], where more than one hundred references can be found. Both papers classify the studies on the use of IRT in four different categories: i) independent limb temperature analysis, where the studies that perform a temperature analysis on each limb separately fit; ii) asymmetric analysis, that include the works that use the fact that in healthy subjects there is a contralateral symmetry in the skin temperature distribution and consider that an asymmetry in this distribution can be an indicator of an abnormality; iii) temperature distribution analysis, when the studies observe similar skin temperature distribution on the feet of healthy individuals comparing to varying temperature variation on diabetic individuals; and iv) external stress analysis, when the aim is the study of the reaction of the body thermoregulation system under the application of thermal and/or physical stress, such as putting the feet into cold water or running.

3   Mathematical Model

Bento et al. [16,17] were the first to apply optimization techniques and mathematical models to characterize the distribution of the temperature of the feet of healthy subjects, as well as to confirm the symmetric behavior of their temperature. With this objective, ThermaCamResearch program was used to convert each thermographic image into Matlab format file, obtaining the temperature matrix of each feet. In order to analyse the foot temperature in more detail, each foot was divided in three regions, like in Fig. 1; thus, the original temperature matrix was used to obtain six new temperature matrices, that represent the temperature in each one of the six regions presented in Fig. 1.

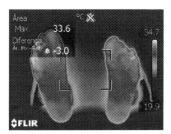 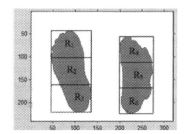

Fig. 1. Feet images

In [16], the author tried to identify the model that best fits the distribution of the temperature for each one of these six regions, minimizing the sum of the quadratic errors of the temperature distribution and the mathematical model. Initially, experiences with quadratic functions were performed, but after having noticed the existence of a wave pattern in the temperature matrix, the authors decided to use functions based on the basic trigonometric functions, concluding that the best model was based on the sum of squares of trigonometric functions.

Carvalho et al. [18,19] continued the work of Bento et al. and initiated the characterization of the temperature distribution of the feet of both healthy and diabetic individuals. Since the temperature values in the plantar foot differ from individual to individual, only the temperature variations presented by each individual are relevant. So, each one of the six different matrices corresponding to each one of the regions in Fig. 1 concern to the temperature variation and were normalized, considering zero as the minimum temperature value. Numerical experiences with models that use variants of trigonometric functions combinations were performed, by minimizing the sum of the quadratic errors between the temperature variation and the mathematical model; the best results were obtained with the models involving the sum of trigonometric functions and/or its squares.

The present study is a sequel of the one developed by Carvalho et al. and the analysis of the distribution of the temperature variation of (non)diabetic feet was carried out by using a mathematical model based in [18] and defined as

$$f(x, i, j) = x_1 \sin^2(ix_2 + jx_3 + x_4) + x_5 \cos^2(ix_6 + jx_7 + x_8) + $$
$$+ x_9 \sin(ix_{10} + jx_{11} + x_{12}) + x_{13} \cos(ix_{14} + jx_{15} + x_{16}) + x_{17}, \quad (1)$$

where $x = (x_1, \ldots, x_{17})$ is the vector of the variables that represents the weights assigned to the different model components, (i, j) represents the pixels positions and $f(x, i, j)$ represents the temperature variation at the (i, j) pixel position.

To characterize the temperature variation matrix, R_k^l, for each image l and each region k, the following nonlinear optimization problem is solved.

$$\min_{x \in \mathbb{R}^{17}} \sum_{i=1}^{M_k^l} \sum_{j=1}^{N_k^l} (r_{ij}^{k,l} - f(x, i, j))^2, \quad (2)$$

where l represents the image to be analyzed; $k \in \{1, \ldots, 6\}$; $M_k^l \times N_k^l$ is the dimension of matrix R_k^l; $r_{ij}^{k,l}$ represents the temperature variation value in region k of image l in the pixels positions (i, j); and f is the function presented in (1).

4   Optimization Methods

In the works of Bento et al. and Soraia et al. different optimization strategies were applied and the best results were obtained with the Genetic Algorithm [16,18,19]. In the current work, the Hybrid Genetic Algorithm and the Hybrid Simulated Annealing Algorithm, described below, were used to find the minimum of the sum of the quadratic errors between the mathematical model and the foot temperature variation matrix.

4.1   Hybrid Genetic Algorithm

The Hybrid Genetic Algorithm (HGA) is based on the Genetic Algorithm (GA) method. The GA is a metaheuristic that simulates the behavior of nature based on an analogy with the mechanisms of natural genetics and is a random search technique with the aim of generating high-quality solutions to an optimization problem [48]. This method is implemented in MatLab software through the pre-defined ga function. This predefined function combines the GA with different strategies with the aim of finding an approximation of a global minimum of a (un)constrained problem. The ga predefined function has a large number of options, e.g., is possible to limit the number of iterations or the number of function evaluations, as well as to define different strategies to crossover and mutation procedures. These options are created by the command $gaoptimset$.

In this work a HGA was considered, by applying a local search after the GA procedure to improve the accuracy of the current solution; the Nelder-Mead method was considered as local search procedure.

4.2   Hybrid Simulated Annealing

The Hybrid Simulated Annealing (HSA) method is based on the Simulated Annealing (SA) method and uses a local procedure to improve the approximate solution provided by the SA. The SA method is a probabilistic technique for approximating the global optimum of a given optimization problem. The process develops in a set of iterations and, at each iteration, a new random point is generated. The distance from the current point to the new point is based on a probability distribution with a scale proportional to a given parameter [49]. This method is also implemented in MatLab software for unconstrained optimization and *simulannealbnd* is the function that defines it; the Nelder-Mead method was used as local search procedure.

5   Numerical Results

In the present work, problem (2), with f defined as in (1), was considered to analyse the distribution of the temperature variation of (non)diabetic feet. Two global optimization metaheuristic algorithms, a Hybrid Genetic Algorithm (HGA) and a Hybrid Simulated Annealing (HSA) were used to solve this problem, in order to analyse which method provides a better approximation of the foot temperature variation.

Additionally, the four non diabetic feet images (H 01, H02, H03, H04) and the four diabetic feet images (D 05, D 06, D 07, D 08), presented in Fig. 2, were used. These images were collected from Polytechnic Institute of Bragança and Centro de Saúde de Santa Maria - Bragança.

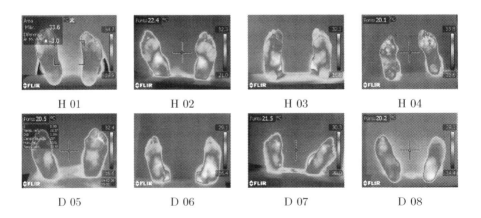

Fig. 2. Non diabetic and diabetic feet images

The numerical results were obtained using an Intel(R) Pentium(R) CPU, with 2.0 GHz, and 4 GB of RAM.

Since both the HGA and the HSA are stochastic and the starting point was generated randomly, the code was executed ten times for each region. Thus, ten possible solutions for the optimization problem (2) are obtained for each region. The computational results, obtained by both optimization methods for each one of the six regions of the eight analysed thermographic images (four of non diabetic individuals and four of diabetic individuals), are presented in Table 1. This table presents the information concerning the minimum value of the objective function of problem (2), Min, as well as the mean value, Mean, and the standard deviation, Std Dev, of the ten obtained solutions, for each one of the six regions.

As can be seen in Table 1, the HGA method obtained better results than the HSA method, for both diabetics and non-diabetic individuals. Not only the minimum values (Min) are lower with the HGA method, but also the same happens with the mean values. So, it is possible to conclude that HGA gives better results than HSA method. This conclusion is corroborated by the values of the standard deviations, which indicate that the ten objective values obtained by the HGA are more similar to those obtained by the HSA.

Furthermore, HGA is always faster than HSA, providing to the solution in a couple of minutes.

As the HGA performed better than HSA, some global considerations about the numerical results obtained using this method are presented in Table 2. This table presents the minimum (Min), maximum (Max) and the average (Mean) values in Table 1 for all regions; the range is also presented (Δ) as well as the relation between the minimum values obtained in all regions, where $\#Ri$ represented the minimum value obtained by HGA method in region i.

Observing Table 2 is possible to conclude that the lowest values of minimum (Min), maximum (Max) and average (Mean) are obtained for diabetic images, which means that in these cases this model fits better in the pixel network of the thermographic image. With the obtained computational results, it is also possible to verify the symmetric behavior for the non diabetic images. This means that when it is obtained the best fit in Region 1 (comparing the results of R1, R2 and R3), then it is expected that the best fit will happen in Region 4 (when comparing R4, R5 and R6). This situation is described in reflection column, for example $\#R3 < \#R2 < \#R1$ means that the best value in Region 3 is smaller than the value in the Region 2, and the last one is smaller than the value in the Region 1. Similar situation occurs in the right foot ($\#R6 < \#R5 < \#R4$). This situation occurs in images H01, H02, H03 and D06. So, this preliminary results indicate that this symmetric behavior occurs more often in non diabetic foot images than in foot images of diabetic individuals. This situation is corroborated by health professionals that considered the temperature variation usually is similar in both feet.

Table 1. Numerical results for the eight thermographic images using HGA and HSA methods

Images	Methods		Regions					
			R1	R2	R3	R4	R5	R6
H 01	HGA	Min	4.0E3	1.3E3	6.8E2	3.0E3	1.1E3	6.7E02
		Mean	4.6E3	1.9E3	8.2E2	3.5E3	1.5E3	9.1E02
		Std Dev	1.2E3	6.5E2	1.1E2	6.2E2	9.8E2	4.8E02
	HSA	Min	1.1E4	4.6E3	2.0E3	6.7E3	3.4E3	2.5E03
		Mean	1.9E4	9.2E3	3.3E3	9.9E3	6.0E3	4.1E03
		Std Dev	7.0E3	3.1E3	1.6E3	5.1E3	1.5E3	1.1E03
H 02	HGA	Min	1.5E3	1.5E3	2.4E3	1.2E3	1.3E3	1.5E03
		Mean	1.8E3	1.6E3	2.9E3	1.3E3	1.5E3	1.9E03
		Std Dev	5.6E2	1.0E2	6.6E2	1.7E2	1.0E2	5.0E02
	HSA	Min	3.4E3	5.0E3	5.4E3	3.8E3	2.3E3	4.6E03
		Mean	4.8E3	7.4E3	8.3E3	4.5E3	6.5E3	8.4E03
		Std Dev	2.2E3	5.1E3	5.9E3	1.2E3	2.7E3	4.8E03
H 03	HGA	Min	2.4E3	5.9E3	4.6E3	1.1E3	7.1E3	3.4E3
		Mean	2.8E3	6.4E3	5.1E3	1.2E3	7.9E3	4.1E3
		Std Dev	5.2E2	8.5E2	4.9E2	4.2E2	8.8E2	6.7E2
	HSA	Min	5.0E3	5.1E3	9.4E3	3.2E3	6.5E3	8.9E3
		Mean	7.7E3	5.9E3	1.1E4	4.3E3	1.3E4	1.0E4
		Std Dev	4.4E3	3.1E2	3.4E3	1.8E3	1.4E4	1.7E3
H 04	HGA	Min	2.7E3	5.9E2	9.3E2	2.1E3	6.7E2	6.5E2
		Mean	3.0E3	7.0E2	1.1E3	2.7E3	8.9E2	6.9E2
		Std Dev	8.6E2	1.3E2	1.1E2	7.0E2	1.8E2	3.8E1
	HSA	Min	8.0E3	2.7E3	3.1E3	4.7E3	2.2E3	1.9E3
		Mean	9.5E3	3.7E3	4.1E3	6.4E3	3.6E3	3.1E3
		Std Dev	3.2E3	1.6E3	2.0E3	3.7E3	2.3E3	2.3E3
D 05	HGA	Min	1.4E3	1.1E3	7.9E2	1.5E3	4.3E2	5.9E2
		Mean	1.7E3	1.2E3	1.1E3	1.6E3	4.9E2	6.6E2
		Std Dev	5.4E2	1.2E2	2.0E2	1.1E2	4.7E1	9.9E1
	HSA	Min	4.4E3	1.3E3	3.6E3	5.5E3	2.5E3	3.0E3
		Mean	6.6E3	5.2E3	5.0E3	7.0E3	3.9E3	4.7E3
		Std Dev	3.1E3	5.1E3	2.1E3	2.4E3	3.8E3	3.6E3
D 06	HGA	Min	1.4E2	3.2E2	9.7E2	7.0E1	3.1E2	9.8E2
		Mean	1.8E2	4.7E2	1.2E3	1.7E2	4.5E2	1.2E3
		Std Dev	3.2E1	2.8E2	2.3E2	1.1E2	2.3E2	2.2E2
	HSA	Min	1.1E3	4.6E2	8.3E2	2.7E2	1.2E3	1.1E3
		Mean	2.4E3	1.2E3	1.4E3	4.2E2	1.7E3	2.2E3
		Std Dev	1.6E3	1.2E3	6.8E2	1.9E2	1.2E3	1.5E3
D 07	HGA	Min	7.3E2	3.9E2	1.1E3	1.0E3	1.5E3	1.3E3
		Mean	1.2E3	5.5E2	1.1E3	1.1E3	1.7E3	1.4E3
		Std Dev	1.2E3	2.6E2	4.3E1	1.6E2	1.4E2	2.0E2
	HSA	Min	4.6E3	3.5E3	3.5E3	7.0E3	1.1E3	2.3E3
		Mean	8.7E3	7.6E3	5.0E3	8.8E3	3.1E3	2.8E3
		Std Dev	3.6E3	5.2E3	3.3E3	2.9E3	2.6E3	1.7E3
D 08	HGA	Min	2.8E2	2.0E2	1.9E2	3.4E1	2.1E2	4.5E2
		Mean	3.3E2	2.8E2	2.7E2	1.2E2	2.2E2	5.9E2
		Std Dev	1.5E2	8.4E1	1.5E2	2.0E2	1.6E1	2.6E2
	HSA	Min	2.2E2	1.6E3	2.5E3	2.0E2	1.7E3	2.8E3
		Mean	3.7E2	1.7E3	4.2E3	3.5E2	4.3E3	6.3E3
		Std Dev	7.6E1	6.0E0	2.1E3	5.1E1	3.2E3	2.4E3

Table 2. Numerical analysis using HGA method

Images	Min	Max	Mean	Δ	Reflection
H 01	6.74E02	3.95E03	1.78E03	3.28E03	$\#R3 < \#R2 < \#R1$ $\#R6 < \#R5 < \#R4$
H 02	1.23E03	2.37E03	1.57E03	1.14E03	$\#R1 = \#R2 < \#R3$ $\#R4 < \#R5 < \#R6$
H 03	1.05E03	7.06E03	4.07E03	6.01E03	$\#R1 < \#R2 < \#R3$ $\#R4 < \#R5 < \#R6$
H 04	5.94E02	2.68E03	1.27E03	2.09E03	$\#R2 < \#R3 < \#R1$ $\#R6 < \#R5 < \#R4$
D 05	4.32E02	1.52E03	9.70E02	1.09E03	$\#R3 < \#R2 < \#R1$ $\#R5 < \#R6 < \#R4$
D 06	7.04E01	9.77E02	4.65E02	9.06E02	$\#R1 < \#R2 < \#R3$ $\#R4 < \#R5 < \#R6$
D 07	3.93E02	1.52E03	1.00E03	1.13E03	$\#R2 < \#R1 < \#R3$ $\#R4 < \#R6 < \#R5$
D 08	3.39E01	4.47E02	2.28E02	4.13E02	$\#R3 < \#R2 < \#R1$ $\#R4 < \#R5 < \#R6$

6 Conclusions and Future Work

In this work, eight thermal plantar feet images (four images of non diabetic individuals and four images of diabetic individuals) were used to test the proposed mathematical model, which describes the distribution of the variation of the temperature of the feet. Two optimization techniques were used to minimize the sum of the square error between the results produced by the mathematical model and the observed temperature.

The numerical results indicate that, in general, the proposed mathematical model fits well the distribution of the temperature variation for images of diabetic individuals, when comparing with the results obtained with images from non diabetic individuals. Furthermore, is possible to confirm the symmetric behavior of the temperature variation on the images from non diabetic individuals. The results also indicate that the Hybrid Genetic Algorithm obtains a better solution when compared with the Hybrid Simulated Annealing method and is faster.

As future work, the authors intent to study more plantar feet images from non diabetic and diabetic individuals and test different mathematical models to approximate the feet temperature variation on both cases. This study can also be expanded to patients with other types of diseases that affect directly their feet.

Acknowledgments. The authors thank University of Trás-os-Montes and Alto Douro, Polytechnic Institute of Bragança and Centro de Saúde de Santa Maria, Bragança, Portugal. This work has been supported by FCT – Fundação para a Ciência e Tecnologia within the Projects Scope UIDB/05757/2020 and UIDB/00013/2020.

References

1. Gilman, S.: Neurobiology of Disease. Academic Press, Cambridge (2011)
2. Jones, B.F.: A reappraisal of the use of infrared thermal image analysis in medicine. IEEE Trans. Med. Imaging **17**(6), 1019–1027 (1998)
3. Lahiri, B.B., Bagavathiappan, S., Jayakumar, T., Philip, J.: Medical applications of infrared thermography: a review. Infrared Phys. Technol. **55**(4), 221–235 (2012)
4. Riddle, M.C., Herman, W.H.: The cost of diabetes care-an elephant in the room. Diabetes Care **41**(5), 929–932 (2018)
5. Sinwar, P.D.: The diabetic foot management-recent advance. Int. J. Surg. **15**, 27–30 (2015)
6. International Working Group on the Diabetic Foot, et al.: International consensus on the diabetic foot: Amsterdam 1999. Technical report. Amsterdam, The Netherlands (1999). ISBN 90-9012716-X
7. Apelqvist, J., Larsson, J.: What is the most effective way to reduce incidence of amputation in the diabetic foot? Diab. Metab. Res. Rev. **16**(S1), S75–S83 (2000)
8. Armstrong, D.G., Holtz-Neiderer, K., Wendel, C., Mohler, M.J., Kimbriel, H.R., Lavery, L.A.: Skin temperature monitoring reduces the risk for diabetic foot ulceration in high-risk patients. Am. J. Med. **120**(12), 1042–1046 (2007)
9. Lavery, L.A., et al.: Home monitoring of foot skin temperatures to prevent ulceration. Diabetes Care **27**(11), 2642–2647 (2004)
10. Lavery, L.A., et al.: Preventing diabetic foot ulcer recurrence in high-risk patients. Diabetes Care **30**(1), 14–20 (2007)
11. Hernandez-Contreras, D., Peregrina-Barreto, H., Rangel-Magdaleno, J.: Similarity measures to identify changes in plantar temperature distribution in diabetic subjects. In: 2018 IEEE International Autumn Meeting on Power, Electronics and Computing (ROPEC), pp. 1–6. IEEE (2018)
12. Hernandez-Contreras, D., Peregrina-Barreto, H., Rangel-Magdaleno, J., Gonzalez-Bernal, J., Altamirano-Robles, L.: A quantitative index for classification of plantar thermal changes in the diabetic foot. Infrared Phys. Technol. **81**, 242–249 (2017)
13. Hernandez-Contreras, D., Peregrina-Barreto, H., Rangel-Magdaleno, J., Orihuela-Espina, F., Ramirez-Cortes, J.: Measuring changes in the plantar temperature distribution in diabetic patients. In: 2017 IEEE International Instrumentation and Measurement Technology Conference (I2MTC), pp. 1–6. IEEE (2017)
14. Peregrina-Barreto, H., Morales-Hernandez, L., Rangel-Magdaleno, J., Vazquez-Rodriguez, P.: Thermal image processing for quantitative determination of temperature variations in plantar angiosomes. In: 2013 IEEE International Instrumentation and Measurement Technology Conference (I2MTC), pp. 816–820. IEEE (2013)
15. Peregrina-Barreto, H., Morales-Hernandez, L.A., Rangel-Magdaleno, J., Avina-Cervantes, J.G., Ramirez-Cortes, J.M., Morales-Caporal, R.: Quantitative estimation of temperature variations in plantar angiosomes: a study case for diabetic foot. In: Computational and Mathematical Methods in Medicine, vol. 2014 (2014)
16. Bento, D.: Modelação matemática da variação da temperatura no pé. Master's thesis. Instituto Politécnico de Bragança, Escola Superior de Tecnologia e Gestão (2011)

17. Bento, D., Pereira, A.I., Monteiro, F.: Mathematical model of feet temperature. In: AIP Conference Proceedings, vol. 1389, pp. 787–790. American Institute of Physics (2011)
18. Carvalho, S.: Caracterização da distribuição da temperatura na planta do pé. Master's thesis. Instituto Politécnico de Bragança, Escola Superior de Tecnologia e Gestão (2014)
19. Carvalho, S., Pereira, A.: Characterization of feet temperature. In: Numerical Analysis and Applied Mathematics, Book Series: AIP Conference Proceedings, vol. I–III (1558), pp. 574–577 (2013)
20. Hernandez-Contreras, D.A., Peregrina-Barreto, H., Rangel-Magdaleno, J.D.J., Orihuela-Espina, F.: Statistical approximation of plantar temperature distribution on diabetic subjects based on beta mixture model. IEEE Access **7**, 28383–28391 (2019)
21. Technical report
22. International Diabetes Federation: International diabetes federation: Brussels. Technical report, International Diabetes Federation (2015)
23. Noor, S., Zubair, M., Ahmad, J.: Diabetic foot ulcer-a review on pathophysiology, classification and microbial etiology. Diab. Metab. Syndr. Clin. Res. Rev. **9**(3), 192–199 (2015)
24. Bakker, K., Apelqvist, J., Schaper, N.C.: Practical guidelines on the management and prevention of the diabetic foot 2011. Diab. Metab. Res. Rev. **28**(S1), 225–231 (2012)
25. Lobmann, R., Rümenapf, G., Lawall, H., Kersken, J.: Der Diabetologe **13**(1), 8–13 (2017). https://doi.org/10.1007/s11428-016-0173-7
26. Glaudemans, A., Uçkay, I., Lipsky, B.A.: Challenges in diagnosing infection in the diabetic foot. Diabet. Med. **32**(6), 748–759 (2015)
27. Alavi, A., et al.: Diabetic foot ulcers: Part I. Pathophysiology and prevention. J. Am. Acad. Dermatol. **70**(1), 1–12 (2014)
28. Clayton, W., Elasy, T.A.: A review of the pathophysiology, classification, and treatment of foot ulcers in diabetic patients. Clin. Diab. **27**(2), 52–58 (2009)
29. Modest, M.F.: Radiative Heat Transfer. Academic Press, Cambridge (2013)
30. Bharara, M., Cobb, J.E., Claremont, D.J.: Thermography and thermometry in the assessment of diabetic neuropathic foot: a case for furthering the role of thermal techniques. Int. J. Lower Extremity Wounds **5**(4), 250–260 (2006)
31. Bagavathiappan, S., et al.: Correlation between plantar foot temperature and diabetic neuropathy: a case study by using an infrared thermal imaging technique. J. Diab. Sci. Technol. **4**(6), 1386–1392 (2010)
32. Bouzida, N., Bendada, A., Maldague, X.P.: Visualization of body thermoregulation by infrared imaging. J. Therm. Biol. **34**(3), 120–126 (2009)
33. Ring, F.: Thermal imaging today and its relevance to diabetes. J. Diab. Sci. Technol. **4**(4), 857–862 (2010)
34. Ring, F., Ammer, K.: The technique of infrared imaging in medicine. Thermol. Int. **10**(1), 7–14 (2000)
35. Bagavathiappan, S., et al.: Infrared thermal imaging for detection of peripheral vascular disorders. J. Med. Phys. **34**(1), 43 (2009)
36. Boquete, L., Ortega, S., Miguel-Jiménez, J.M., Rodríguez-Ascariz, J.M., Blanco, R.: Automated detection of breast cancer in thermal infrared images, based on independent component analysis. J. Med. Syst. **36**(1), 103–111 (2012)

37. Dibai Filho, A.V., Packer, A.C., de Souza Costa, K.C., Berni-Schwarzenbeck, A.C., Rodrigues-Bigaton, D.: Assessment of the upper trapezius muscle temperature in women with and without neck pain. J. Manipulative Physiol. Ther. **35**(5), 413–417 (2012)

38. Huang, C.-L., et al.: The application of infrared thermography in evaluation of patients at high risk for lower extremity peripheral arterial disease. J. Vasc. Surg. **54**(4), 1074–1080 (2011)

39. Ng, E.Y.-K.: A review of thermography as promising non-invasive detection modality for breast tumor. Int. J. Therm. Sci. **48**(5), 849–859 (2009)

40. Snekhalatha, U., Anburajan, M., Teena, T., Venkatraman, B., Menaka, M., Raj, B.: Thermal image analysis and segmentation of hand in evaluation of rheumatoid arthritis. In: Computer Communication and Informatics (ICCCI), pp. 1–6. IEEE (2012)

41. Anburajan, M., Sivanandam, S., Bidyarasmi, S., Venkatraman, B., Menaka, M., Raj, B.: Changes of skin temperature of parts of the body and serum asymmetric dimethylarginine (ADMA) in type-2 diabetes mellitus Indian patients. In: Engineering in Medicine and Biology Society, EMBC, 2011 Annual International Conference of the IEEE, pp. 6254–6259. IEEE (2011)

42. Hernandez-Contreras, D., Peregrina-Barreto, H., Rangel-Magdaleno, J., Gonzalez-Bernal, J.: Narrative review: diabetic foot and infrared thermography. Infrared Phys. Technol. **78**, 105–117 (2016)

43. Sejling, A.-S., Lange, K.H.W., Frandsen, C.S., Diemar, S.S., Tarnow, L., Faber, J., Holst, J.J., Hartmann, B., Hilsted, L., Kjaer, T.W., Juhl, C.B., Thorsteinsson, B., Pedersen-Bjergaard, U.: Infrared thermographic assessment of changes in skin temperature during hypoglycaemia in patients with type 1 diabetes. Diabetologia **58**(8), 1898–1906 (2015). https://doi.org/10.1007/s00125-015-3616-6

44. Sivanandam, S., Anburajan, M., Venkatraman, B., Menaka, M., Sharath, D.: Estimation of blood glucose by non-invasive infrared thermography for diagnosis of type 2 diabetes: an alternative for blood sample extraction. Mol. Cell. Endocrinol. **367**(1), 57–63 (2013)

45. Sivanandam, S., Anburajan, M., Venkatraman, B., Menaka, M., Sharath, D.: Medical thermography: a diagnostic approach for type 2 diabetes based on non-contact infrared thermal imaging. Endocrine **42**(2), 343–351 (2012)

46. StessStess, R.M., et al.: Use of liquid crystal thermography in the evaluation of the diabetic foot. Diabetes Care **9**(3), 267–272 (1986)

47. Adam, M., Ng, E.Y., Tan, J.H., Heng, M.L., Tong, J.W., Acharya, U.R.: Computer aided diagnosis of diabetic foot using infrared thermography: a review. Comput. Biol. Med. **91**, 326–336 (2017)

48. Chambers, L.D.: Practical Handbook of Genetic Algorithms: Complex Coding Systems, vol. 3. CRC Press, Boca Raton (1998)

49. Van Laarhoven, P.J.M., Aarts, E.H.L.: Simulated annealin. In: Simulated Annealing: Theory and Applications, pp. 7–15, Springer, Dordrecht (1987). https://doi.org/10.1007/978-94-015-7744-1_2

Impact of the Increase in Electric Vehicles on Energy Consumption and GHG Emissions in Portugal

Amanda S. Minucci[1] [iD], Ângela P. Ferreira[2]([⊠]) [iD],
and Paula O. Fernandes[3] [iD]

[1] Universidade Tecnológica Federal do Paraná (UTFPR), Ponta Grossa, Brazil
asminucci@hotmail.com
[2] Research Centre in Digitalization and Intelligent Robotics (CeDRI),
Instituto Politécnico de Bragança, Campus de Santa Apolónia,
5300-253 Bragança, Portugal
apf@ipb.pt
[3] Applied Management Research Unit (UNIAG), Instituto Politécnico
de Bragança, Campus de Santa Apolónia, 5300-253 Bragança, Portugal
pof@ipb.pt

Abstract. The sector with the higher weight in final energy consumption is the transport sector, reflecting the one responsible for most greenhouse gas (GHG) emissions. In this context, and given the increasing penetration of zero or low GHG emissions vehicles with high energy efficiency, this work intends to contribute to the future identification of the impact of the transport sector on energy consumption and resulting emissions. It aims at identifying the energy consumption in the sector and quantify the GHG levels from the comparative analysis of the increase of electric based vehicle fleet in detriment of those based in internal combustion engines, considering five scenarios until the year 2030. The study is applied in mainland Portugal considering the fleet of light passenger and commercial vehicles, which comprise the vast majority of the Portuguese fleet. The Bottom-up model, where the hierarchical tree detailing is constructed from detail to the whole, is applied to the study to determine the energy consumption and GHG emissions variables. The analysis is performed through the application of the simulation tool Long-range Energy Alternatives Planning system (LEAP) of scenario-based and integrated modelling, herein utilized to determine energy consumption and account for GHG emission sources. Results show that the increase of electric vehicles directly influences the reduction of GHG emissions and final energy consumption while electric energy consumption increases.

Keywords: Electric vehicles · Energy consumption · Greenhouse gas emissions · Bottom-up model

This work has been supported by FCT - Fundação para a Ciência e Tecnologia within the Project Scope: UIDB/05757/2020.

O. Gervasi et al. (Eds.): ICCSA 2020, LNCS 12251, pp. 521–537, 2020.
https://doi.org/10.1007/978-3-030-58808-3_38

1   Introduction

Energy consumption and associated greenhouse gas (GHG) emissions have, in recent years, become the target of major worldwide investments so that the medium and long term impacts are mitigated through various measures, such as, for example, increasing equipment efficiency, whether for residential applications, transport or industry.

This work aims to analyse and forecast the impacts on energy consumption and GHG emissions, following the increase in the penetration of electrically based vehicles in the Portuguese mobility system, in replacement of internal combustion engines based vehicles. The study considers five different scenarios, with different levels of penetration of light (passenger and commercial), electric, hybrid, and plug-in vehicles.

The model developed to analyse energy consumption in the light vehicle transport sector, corresponds to a hybrid approach, combining characteristics of bottom-up and top-down models. The projection of energy consumption in the light vehicle segment is determined from disaggregated data in order to describe the energy end users and technological options in detail, according to the requirements of the bottom-up model [1–5]. The forecast of the total car fleet has been implemented exogenously, seeking to increase the robustness of the model. By adopting this approach, the model gathers a top-down characteristic, in a hybrid approach. This model focuses on the economy and on various relationships inherent to it, addressing the simulation of future supply and demand for a given product.

The analysis of GHG emissions is based on the bottom-up model, consolidating the method adopted by the Portuguese National Inventory Report on Greenhouse Gases, which considers the recommendations of the Intergovernmental Panel on Climate Change (IPCC) regarding the use of methodologies based on disaggregated levels in the estimates of carbon dioxide (CO_2) and non-CO_2 gas emissions [4].

Several studies have assessed the growth of the fleet with the increase in electric vehicles, based on the bottom-up model. The study [2] estimated the future energy demand and GHG emissions from China's road transport until 2050. The work developed in [1] determined the energy consumption and GHG emissions in the transport sector of São Paulo city, Brazil, considering three different scenarios until 2035. Moreover, [6] estimated the composition of the vehicle market and calculated the energy consumption and CO_2 emissions of light passenger vehicles in Japan, in a long-term basis.

This paper is organized in three major sections, the first of which concerns the presentation of the current scenario of the transport sector in Portugal. Then, the methodology used to determine the energy consumption and GHG emissions of vehicles, the object of this study, is developed. Finally, section four presents and analyses the emissions levels and the energy consumption for the different scenarios under consideration, as well as the main consequences in the Portuguese energy system.

2   Theoretical Framework

2.1   Portuguese Transport Sector

According to the Portuguese Private Business Association (ACAP), responsible for publishing the Automobile Sector Statistics (ESA) yearbook, the vehicle fleet in Portugal is divided into two major categories, light and heavy, being predominantly composed of light vehicles, with approximately 97.5% of the total [7], as indicated in Table 1. Thus, the study is limited to this category, given that it is the most likely to change from the technological point of view, i.e., replacement of vehicles based on combustion engines for vehicles with electric traction, totally or partially.

Table 1. Vehicle fleet in Portugal (2011) [7]

Vehicle fleet		Units	(%)
Light	Passenger	4 522 000	77
	Commercial	1 206 000	20.5
Heavy		145 000	2.5

According to data provided by ACAP [7], it can be seen that Portuguese fleet of light vehicles had a significant increase in last years. Figure 1 shows the evolution of the fleet of light vehicles from 1984 until 2017, where there is a significant growth trend between the years 1988 and 2006. Between 2006 and 2011 there was a slow-down, but the trend was still of increasing. In recent years, 2012 until 2016, the fleet stabilized and increased again in 2017 to levels close to 6 000k.

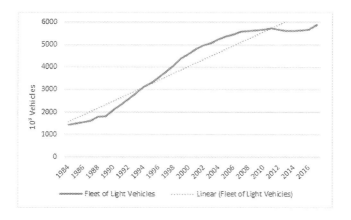

Fig. 1. Fleet of light vehicles (passengers and commercial) in Portugal (1984–2017) [7].

Figure 2 presents the data related to the sales of light vehicles (passenger and commercial) in Portuguese territory for the same period, observing a downward trend in sales when considering a 35-year time horizon [7].

The combination of information related to the data above allows identifying some important factors, such as the high average age of the circulating fleet in the country. The insertion of new vehicles does not increase at the same way the existing fleet, being this hypothesis corroborated from the automotive statistics sector data at 2011 [7], identifying that 63.2% of the fleet is over 10 years old, with an average age of 12.6 years (passenger) and 13.7 years (commercial) [7].

When conducting the previous analysis in a smaller sampling period, the last five years, the scenario changes, with the sales growing considerably. This growth may be explained as the result of the exponential technological development in the last decade, reflected in consumer goods; the conditions offered to the consumer are more advantageous and attractive, with more efficient and modern vehicles. In addition, there are incentives that promote the insertion of electric vehicles in the current vehicle fleet, since it relates to one of the country's goals for future reduction in GHG emissions.

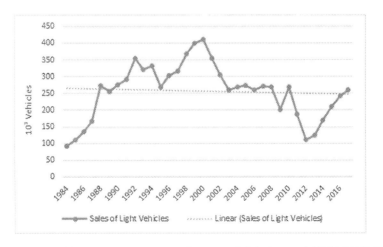

Fig. 2. Sales of light cars (passengers and commercial) in Portugal (1984–2017) [7].

2.2   Portuguese Energy Balance and Greenhouse Gas Emissions

The Directorate-General for Energy and Geology (DGEG), in a study carried out in 2017 pointed out that the transport sector continues to be the main final energy consumer, accounting for 37.2% of the energy consumed. The industry sector accounts with 31.3% of the final energy consumption, the domestic sector with 16.4%, the services sector with 12.2% and the agriculture and fisheries sector with 2.9% [8], as shown in Fig. 3. Thus, the transport sector is one of the main sectors with a higher impact on the economy, tracking the progress made towards reducing the environmental impacts associated with energy production.

The technological development of the sectors, including transport, combined with the change in the lifestyle of the population, with the increase in the Gross Domestic Product (GDP) per capita, are directly reflected in the increase in energy consumption worldwide [9].

The demand for energy and the increase in mobility from 1990 onwards reflect Portuguese economic evolution, which in turn triggered the increase in GHG emissions. According to the Report of the State of the Environment from the Portuguese Environment Agency (APA) [10], the stabilization of the levels of GHG emissions started with the technological development of pollution control systems and energy efficiency. In addition, the emergence of fuels with lower polluting rates and the increase of energy based in renewable sources from the 2000s, contributed to the GHG emissions actual state. However, according to [10], the transport sector presents itself as one of the main emitters during the historical series. In 2017, the percentage of emissions was 24.3% in the country, with a growth of 4.53% compared to 2016.

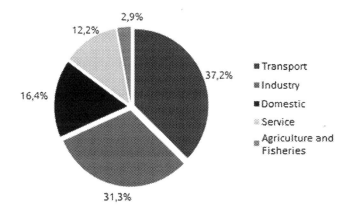

Fig. 3. Final energy consumption by sector (%) in Portugal (2017) [8].

2.3 Goals Related to Energy and Climate 2021–2030

The Regulation on the Governance of the Energy Union and Climate Action [Regulation (EU) 2018/1999] [11], established that all Member States must prepare and submit to the European Commission an integrated National Energy and Climate Plan (NECP), with a medium-term perspective (horizon 2021–2030) [12]. The European Council approved targets related to energy and climate, which linked an internal reduction of −40% in GHG emissions related to 1990 levels. It was also set the following energy targets: increasing energy efficiency by at least 32.5%, increasing the share of renewable energy to at least 32%, with 14% of renewables in transport, and guaranteeing at least 15% electricity inter-connection levels between neighbouring Member States [12].

Portugal's targets for 2030 GHG emissions are comprised between −45% and −55%, compared to 2005 levels, and the renewable energy contribution is estimated in 47% of the national gross final consumption of energy, with 20% of renewables in the transport sector. Regarding energy efficiency, the proposed contribution is more modest, with projected values of -2.3 Mtoe (million tonnes of oil equivalent) of primary energy consumption due to energy efficiency [13].

3 Methodological Approach

The methodology for forecasting the impact on energy consumption due to the increased penetration of electric vehicles, is performed using the bottom-up and top-down hybrid model. To estimate the levels of GHG emissions, from the consumption forecast, the bottom-up model is applied.

The bottom-up model seeks to perform a structural detailing of the technology used in the energy conversion and usage, using disaggregated data aiming at describing the final energy utilization and technological options in detail. In the model methodology, technologies and replacement indexes of the analysed equipment can be identified, in this case, vehicles. On the other hand, the top-down model takes into account only carbon dioxide (CO_2) emissions from energy production and consumption data, without detailing how this energy is consumed.

For the development of the present research, light vehicle fleet, commercial and passenger typologies were classified separately, since the characteristic parameters of each one of those variants presented slightly different indicators.

The determination of energy consumption of any class in the transport sector uses three factors: fleet, average annual mileage and average specific consumption. The average distance travelled is directly related to the need of society to transport people and goods, according to economic development, which in turn may be related with the frequency of use of the vehicle and the age of the existing fleet [14]. The average consumption of these vehicles, whether passenger or commercial, depends mainly on the underlying technology used.

The model used in the present study is developed using the environment and energy modelling tool based on scenarios, the Long-range Energy Alternatives Planning system (LEAP) [15], developed by the Stockholm Institute for the Environment with the aim of conducting energy policy analyses and assessing climate change mitigation.

3.1 Variable Modelling

For the application of the proposed model, the energy consumption in the transport sector is defined based on the product of the related variables, also implemented in [2], given by:

$$
\begin{aligned}
Energy\ Consumption_{f,v,y} = {} & Stock\ of\ Vehicles_{f,v,y} \times Annual\ Vehicle\ Mileage_{f,v,y} \\
& \times\ Fuel\ Economy_{f,v,y}
\end{aligned} \tag{1}
$$

where v is the type of vehicle considered (passenger or commercial), f is the type of fuel (gasoline, diesel, electricity or liquefied petroleum gas) and y is the calendar year in which the energy consumption is to be determined. The variables stated in Eq. (1) are detailed as follows.

The stock of vehicles is given by

$$Stock\ of\ Vehicles_{f,v,y} = \sum_{s} Sales_{v,s} \times Survival_{v,y-s} \qquad (2)$$

where subscript s relates to the vintage (i.e., the year in which a vehicle starts its use). The stock of vehicles of a specific type is, therefore, given by the total sum of the sales times the survival index, which represents the fraction of the vehicles of the considered typology in the year y minus the vintage year, s.

Regarding the vehicle's average annual mileage, it is estimated through

$$Annual\ Vehicle\ Mileage_{f,v,y} = Mileage_{v,y} \times MIDegradation_{y,y-s} \qquad (3)$$

from which the average miles driven by vehicle type v, with fuel type f, in the year y is expressed by the vehicle mileage of the type v into use in the year s, times the degradation of the vehicle mileage of the type v, considering $y - s$. By this way, the degradation is modelled through an exponential function.

The average consumption is estimated through the distance travelled, in kilometres, by vehicles of type v in the year y and it is considered constant throughout the life span of the vehicle, on the assumption that maintenance is carried out with the correct periodicity.

With regard to GHG emissions, their determination is carried out using the emission factor, EF, times $Activity$, which translates the amount of energy consumed or distance covered by a given mobile source, i.e.,

$$Emissions = EF_{i,f,v} \times Activity_{f,v} \qquad (4)$$

being i the gas (CO_2, CO, NO_X, CH_4, etc.), f the type of fuel and v the type of vehicle.

The above procedure is applied by LEAP and is in accordance with the guidelines of the Intergovernmental Panel on Climate Change (IPCC) [16]. In detail, it translates as follows:

- Determine the amount of energy consumed in terajoules (TJ), by type of fuel, for each sector and sub-sector;
- Multiply the amount of energy consumed by the specific emission factor for each fuel, for each sector and sub-sector, in tonnes/TJ. The GHG emissions in Eq. (4) are presented in tonnes (t).

To convert the emissions as given by Eq. (4) to tonnes of carbon dioxide equivalent (tCO_2eq), the following applies:

$$tCO_2eq = tonnes(gas) \times GWP(gas) \qquad (5)$$

being GWP the Global Warming Power which corresponds to one of the five IPCC Assessment Reports (AR1-AR5) [16]. These reports are established by the relative importance of each gas related to carbon dioxide, in the production of an amount of energy per unit volume several years after an emission impulse, i.e., the GWP values measure the heating potential of one tonne of each gas in relation to one tonne of CO_2.

3.2   Historical Data and Projections

From the construction of the base scenario, the relevant data for the construction of the hierarchical tree are gathered, where the main data structure is displayed. The base year is 2011 and the forecast is made until 2030, considering only mainland Portugal. The tree is structured as follows:

- Passenger vehicles
  - Internal combustion engine
  - Purely electrical, hybrid and Plug-in
- Commercial vehicles
  - Internal combustion engine
  - Purely electrical

Hybrid and plug-in vehicles are not considered in the light commercial fleet because, according to Automotive Sector Statistics from ACAP 2018 yearbook [7], there are no data in the sales history till 2017. The construction of the base scenario is dependent on the historical data in the base year. The determination of future energy consumption considers various indicators related to the projection of the amount of energy in the transportation sector and they are presented in the following sections. It should be noted that the order of insertion of the data does not interfere with the final result.

Fleet
Historical fleet data were obtained from the automotive sector statistics from ACAP yearbooks [7] where it is possible to characterize the vehicle fleet presumably in circulation by vehicle type and according to the main fuel. As introduced in the previous Sect. 2.1, this study is limited to light passenger and commercial vehicles. Total number of vehicles in these two categories in 2011 (the base year) is presented in Table 1.

Survival
The survival of the vehicles under study was estimated according to the Gompertz curve [17], given by:

$$S(t) = e^{-e^{-(a + bt)}} \tag{6}$$

where $S(t)$ is the function that describes scrapped vehicles, t is the age of the vehicle and a and b are curve adjustment parameters, variable according to each type of vehicle, assuming, in this work, the values presented in the Table 2. The adopted parameters were determined from the calibration of the model using historical data presented in the reference report of the GHG emissions by mobile sources in the energetic sector from the first Brazilian inventory of anthropogenic GHG emissions [18].

Fleet Profile of Existing Vehicles
The existing fleet in the years under study consists of vehicles of different ages and, as such, the identification of the fleet' profile reflects directly in the algorithm of the future

Table 2. Adjustment parameters of Gompertz model.

	a	b
Passengers	1,798	−0,137
Commercials	1,618	−0,141

fleet. Historical data were obtained from the Automotive Sector Statistics from ACAP [7] and are shown in Table 3. The fleet life cycle profile was determined as 25 years old and differentiated between the two categories under analysis.

Table 3. Fleet of road passenger vehicles presumably in circulation, by age (2011) [7].

Category	Passenger	Commercial
Vehicle age	%	%
1 year	5.54	3.93
2 years	4.08	3.32
3 years	5.72	4.70
4 years	5.58	5.62
5 years	5.35	5.38
6 a 10 years	26.79	30.59
11 a 15 years	27.67	29.39
16 a 20 years	15.32	13.06
21 a 25 years	3.95	4.03

Sales

Historical sales data were obtained from ACAP data (Automotive Sector Statistics) [7], where it is possible to survey sales for the base year. In 2011, 153 404 passenger cars and 34 963 light commercial vehicles were sold, each type subdivided into two other categories according to the propulsion system: internal combustion engine and propelled on electrical energy using at least one electric motor, purely electric, hybrid or plug-in, as shown in Table 4.

Table 4. Sales of light vehicles by category and type of fuel (2011) [7].

Propulsion		Passenger	Commercial
Internal combustion	Gasoline	44 544	3
	Diesel	106 832	34 955
	Liquefied petroleum gas (LPG)	839	0
Electric	Purely electric	203	5
	Hybrid	932	0
	Plug-in	54	0

Fleet Forecast

Although there is a wide range of different models for forecasting and several ways to build a model, there might be a best way to build a particular model for a specific purpose. In the present study and for the specific problem, to determine the future fleet of both groups of vehicles, the simple linear regression model will be applied. The simple linear regression is a statistical method that allows summarising and studying relationships between two continuous variables [19].

Equation (7) represents the simple linear regression model. The dependent variable is designated by Y and the independent variable by X. The unknown parameters in the model are the intercept β_0 and the slope β_1, being $\in$ the random or stochastic error and the variance σ^2 [19].

$$Y_i = \beta_0 + \beta_1 X_i + \in_i, \quad i = 1, 2, 3, \ldots, n \quad Y_i = \beta_0 + \beta_1 X_i + \in_i, \quad i = 1, 2, 3, \ldots, n \tag{7}$$

where $E(\in_i | X_i) = 0$ and $V(\in_i | X_i) = \sigma^2$.

In the current problem, after several trials, GDP per capita is found as the most suitable independent variable (X) to forecast the number of light vehicles (passenger and commercial vehicles) as dependent variable (Y).

It is important to evaluate the suitability of the regression model for making forecasts. To do so, it will be necessary to calculate the coefficient of determination, r^2, to analyse the fit quality of the model and to know the correlations between variables, which is fundamental to use the value of one variable to predict the value of another. To achieve correlations between variables, Pearson's correlation coefficient, r, will be used.

For the problem under analysis an r^2 of 0.933 was obtained, which means that 93.33% of the variability stems from the independent variable, X (in this case, GDP per capita), of the light passenger fleet. For the light commercial vehicle fleet, an r^2 equal to 0.8472 was found, from which GDP per capita explains 84.72% of this category of vehicle fleet.

It is also possible to verify that there is a strong positive correlation between the variables GDP per capita and light passenger vehicle fleet and also with light commercial vehicle fleet, being the correlation higher in the first case than is the second one, $r = 0.966$ and 0.92, respectively.

Annual Average Vehicle Mileage

The annual average vehicle mileage, or the intensity of use, is understood as the estimated average distance travelled by each vehicle in the circulating fleet in the time unit under analysis (yearly basis).

The values used for this projection were obtained from the Portuguese National Inventory Report on Greenhouse Gases, 1990–2012 [20].

This report proposes a degradation rate of usage as a function of the age, given by a model developed based on data from vehicle inspection centres [20]. The distance travelled (in kilometres) in a yearly basis is estimated through:

$$A_2 + \frac{(A_1 - A_2)}{1 + (age/x_0)^p} \tag{8}$$

being A_1, A_2, x_0 and p parameters defined by vehicle technology as presented in Table 5.

Table 5. Usage curve parameters [20].

	A_1	A_2	x_0	p
Gasoline, LPG, hybrid passenger vehicles	13354.668	737.09264	19.69152	2.4209
Diesel-powered passenger vehicles	19241.066	6603.86725	17.45625	2.53695
Diesel-powered commercial vehicles	20800.215	2597.42606	15.44257	2.32592

Fuel Consumption

Energy consumption is directly linked to the type of fuel used. These values were estimated using the LEAP tool database [15], where there are pre-determined values for the two main propulsion systems, those based on internal combustion and the ones using electric propulsion, including pure electric, hybrid and plug-in vehicles. The consumption of the vehicles under analysis adopted are shown in Table 6.

Table 6. Consumption by category and type of fuel [15].

Propulsion	Fuel		Consumption
Internal combustion	Gasoline (l/100 km)		8.4
	Diesel (l/100 km)		9.4
	LPG (km/m^3)		12 754.3
Electric (full or partial)	Electric (kWh/100 km)		19.88
	Hybrid	Gasoline (l/100 km)	4.9
		Diesel (l/100 km)	4.3
	Plug in	Gasoline (l/100 km)	4.9
		Electric (kWh/100 km)	18.64

Emission Factor

The emission factor parameters used are shown in Table 7 and were obtained through the fifth assessment report (AR5) from IPCC [16].

Table 7. Emission factors

Fuel	tCO$_2$eq/TJ
Diesel	73.28
Gasoline	68.56
LPG	62.71

3.3 Characterization of the Scenarios

For the study proposed in this work, four scenarios were defined from the insertion of electric, hybrid and plug-in vehicles and a theoretical scenario where there is no insertion of this class of vehicles, in the period under analysis, as outlined below:

- Base Scenario: the rate of sales made in the base year (2011) are maintained in the time frame of the study for all types of vehicles disaggregated by type of propulsion;
- Scenario 10%: penetration of electric, hybrid and plug-in vehicles in detriment of sales of internal combustion engines based vehicles, with a rate of 10%;
- Scenario 30%: penetration of electric, hybrid and plug-in vehicles in detriment of sales of internal combustion engines based vehicles, with a rate of 30%;
- Scenario 60%: penetration of electric, hybrid and plug-in vehicles in detriment of sales of internal combustion engines based vehicles, with a rate of 60%;
- Scenario 100%: penetration of electric, hybrid and plug-in vehicles in detriment of sales of internal combustion engines based vehicles, with a rate of 100%.

The scenarios were chosen in order to enable the establishment of a comparative analysis between them. The 100% scenario presents a purely theoretical configuration with the objective of establishing a maximum value for the insertion of electric powered (fully or partially) light vehicles.

The study considers that electric based vehicles started to be introduced in the Portuguese market in the base scenario (2011) according to a linear model until the year 2030.

4 Results and Analysis

4.1 Fleet Forecast

From the forecast proposed by the methodology described in Sect. 3, the results regarding energy consumption and GHG emissions for the five proposed scenarios can be viewed from the projections obtained through the LEAP tool.

Figure 4 presents the projection of the fleet of purely electric, hybrid and plug-in vehicles from the year 2011 until 2030, within the proposed scenarios. The fleet for each year is composed of vehicles of different ages (vintage) and thus varies from year to year.

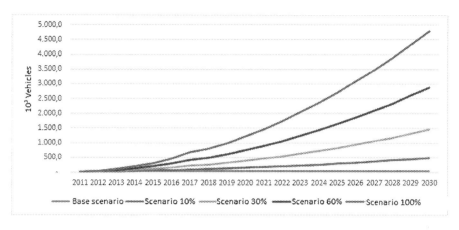

Fig. 4. Projection of the fleet of electric, hybrid and plug-in vehicles.

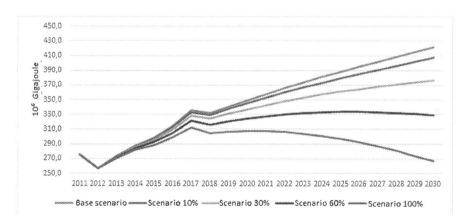

Fig. 5. Comparison of energy consumption between the proposed scenarios.

4.2   Energy Consumption Forecast

The total energy consumption for the different scenarios under consideration are presented in a comparative way in Fig. 5.

The behaviour of the predicted energy consumption is similar for the base, 10% and 30% scenarios, while for the other two scenarios, the behaviour is slightly different. The former ones present a total energy consumption with an increasing tendency, while the energy consumption remains almost constant for 60% scenario and decreases under the 100% scenario.

This behaviour can be explained in view of the higher energy efficiency of electric propulsion based vehicles, which have an increased penetration in these two scenarios, able to maintain or decrease the predicted total energy consumption, for the 60% and 100% scenarios, respectively. In light of these results, the penetration of electric propulsion based vehicles favours the reduction of the country's energy dependence.

The energy consumption mentioned above refers to total energy consumption, including fossil fuel and electric.

In order to preview the impact the fleet change may have in the electrical grid infrastructure and electric energy demand, an analysis based on the forecasting of the electric energy consumption for the different scenarios under consideration is also performed (Fig. 6). From the estimates, it is possible to conclude that the penetration of electric propulsion based vehicles will translate in an increase in electrical energy demand, as would be expected. In the year 2030, the analysis forecasts that for each 10% increased penetration of electric, hybrid and plug-in vehicles, the electricity consumption will increase about 700 MWh (megawatt hours).

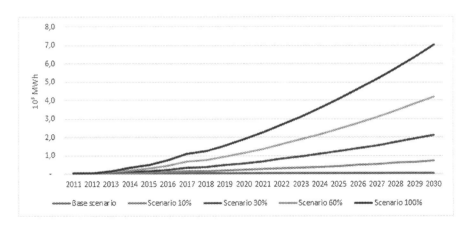

Fig. 6. Comparison of electric energy consumption between the proposed scenarios.

4.3   Greenhouse Gas Emissions Forecast

The GHG emissions forecast for the considered scenarios is shown in Fig. 7. As would be expected, the higher the penetration of electrically-powered vehicles in the commercial and passenger fleet, the lower are GHG emissions levels. With a considerable penetration of these vehicles (60% and 100% scenarios), it can be seen a decrease in the GHG emissions forecast in the transport sector.

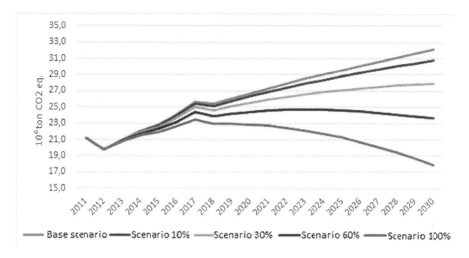

Fig. 7. Comparison of GHG emissions between the proposed scenarios.

5 Conclusions

Forecast analyses involving energy consumption and GHG emissions are of paramount importance for issues involving global sustainability. In this way, the international community has been discussing objectives and targets for sectors to seek technological development with greater energy efficiency and effectiveness with the objective to decrease environmental impacts.

From a macro perspective, this work indicates that the modernization of the transport sector for personal use reflects on positive rates in relation to the reduction of national energy dependency. In addition, it contributes to the control of GHG emissions from internal combustion vehicles that comprise the vast majority of the actual country's transport fleet.

Concerning total energy consumption, with a considerable penetration in sales of electric propulsion based vehicles in detriment of internal combustion engines based vehicles, the energy demand decreases. This is because electric vehicles present higher energy efficiency in the transport sector, specifically in light passenger and commercial vehicles.

The analysis between the scenarios shows that for every 10% increase in pure electric, hybrid and plug-in vehicles in the fleet, in 2030 there will be a 3.84% reduction in total energy consumption, in comparison to the base scenario. However, the penetration of electric vehicle propulsion technologies will shift the energy source used in the vehicles. The electricity demand will increase, from which may be required an upgrade of the electrical grid infrastructure including energy sources, as renewable ones, for instance.

Following the behaviour of total energy consumption, electric propulsion based vehicles are one of the main alternatives for clean transport, in the viewpoint of GHG emissions with reduced impacts on the environment. For every 10% increase in purely electric, hybrid and plug-in vehicles in the fleet, in detriment of internal combustion engine vehicles, there is a reduction of 4.46% of GHG emissions, when comparing the levels reached in the year 2030 to the values of the base scenario.

References

1. Dias, M., et al.: The impact on electricity demand and emissions due to the introduction of electric cars in the São Paulo Power System. Energy Policy **65**, 298–304 (2014)
2. Zhou, N., et al.: China's energy and emissions outlook to 2050: perspectives from bottom-up energy end-use model. Energy Policy **53**, 51–62 (2013)
3. Du Can, S., et al.: Modeling India's energy future using a bottom-up approach. Appl. Energy **238**, 1108–1125 (2019)
4. APA Agência Portuguesa do Ambiente: Portuguese National Inventory Report on Greenhouse Gases, 1990–2017, Amadora (2019). http://www.apambiente.pt
5. van Beeck, N.: Classification of Energy Models. Operations research. Tilburg University & Eindhoven University of Technology (1999). https://pure.uvt.nl/ws/portalfiles/portal/532108/777.pdf
6. González Palencia, J., Otsuka, Y., Araki, M., Shiga, S.: Scenario analysis of lightweight and electric-drive vehicle market penetration in the long-term and impact on the light-duty vehicle fleet. Appl. Energy **204**(C), 1444–1462 (2017). Elsevier
7. ACAP Associação Automóvel de Portugal, Estatística do Sector Automóvel. https://acap.pt/pt/estatisticas. Accessed 15 Oct 2019
8. DGEG Direção-Geral de Energia e Geologia, Caraterização Energética Nacional 2019. http://www.dgeg.gov.pt/. Accessed 15 Oct 2019
9. Murat, Y., Ceylan, H.: Use of artificial neural networks for transport energy demand modeling. Energy Policy **34**(17), 3165–3172 (2006)
10. APA Agência Portuguesa do Ambiente, Relatório do Estado do Ambiente. Pegada Energética e Carbónica dos Transportes. https://rea.apambiente.pt/. Accessed 15 Oct 2019
11. EU European Union, Regulation (EU) 2018/1999 of the European Parliament and of the Council of 11 December 2018, on the Governance of the Energy Union and Climate Action. https://eur-lex.europa.eu/legal-content/EN/TXT/?uri=uriserv:OJ.L_.2018.328.01.0001.01.ENG&toc=OJ:L:2018:328:FULL. Accessed 15 Oct 2019
12. EU European Commission, EU climate action and the European Green Deal. https://ec.europa.eu/clima/policies/eu-climate-action_en. Accessed 20 Sept 2019
13. DGEG Direção-Geral de Energia e Geologia: National Energy and Climate Plan, Portugal, 2021–2030 (PNEC 2030) (2019). https://ec.europa.eu/energy/sites/ener/files/documents/pt_final_necp_main_pt.pdf
14. Stone, A., Merven, B., Maseela, T., Moonsamy, R.: Providing a foundation for road transport energy demand analysis: a vehicle pare model for South Africa. J. Energy South. Afr. **29**(2), 29–42 (2018)
15. SIE Stockholm Institute for the Environment: Long-range Energy Alternatives Planning (LEAP) System (2007). http://www.energycommunity.org
16. IPCC Intergovernamental Panel on Climate Change, Assessment Reports. https://www.ipcc.ch/. Accessed 21 Sept 2019

17. Gompertz, B.: On the nature of the function expressive of the law of human mortality and on the new mode of determining the value of life contingencies. Philos. Trans. R. Soc. Lond. **115**, 513–583 (1825)
18. MCT Ministério da Ciência e Tecnologia: Emissão de gases de efeito estufa por fontes móveis, no setor energético. Primeiro Inventário Brasileiro de Emissões Antrópicas de Gases de Efeito Estufa, Brasília (2006). https://cetesb.sp.gov.br/inventario-gee-sp/wp-content/uploads/sites/34/2014/04/3.pdf
19. Wasserman, L.: All of Statistics. STS. Springer, New York (2004). https://doi.org/10.1007/978-0-387-21736-9
20. APA Agência Portuguesa do Ambiente: Portuguese National Inventory Report on Greenhouse Gases, 1990–2012, Amadora (2014). http://www.apambiente.pt

Penalty-Based Heuristic DIRECT Method for Constrained Global Optimization

M. Fernanda P. Costa[1]($\boxtimes$), Ana Maria A. C. Rocha[2],
and Edite M. G. P. Fernandes[2]

[1] Centre of Mathematics, University of Minho, 4710-057 Braga, Portugal
mfc@math.uminho.pt
[2] ALGORITMI Center, University of Minho, 4710-057 Braga, Portugal
{arocha,emgpf}@dps.uminho.pt

Abstract. This paper is concerned with an extension of the heuristic DIRECT method, presented in [8], to solve nonlinear constrained global optimization (CGO) problems. Using a penalty strategy based on a penalty auxiliary function, the CGO problem is transformed into a bound constrained problem. We have analyzed the performance of the proposed algorithm using fixed values of the penalty parameter, and we may conclude that the algorithm competes favourably with other DIRECT-type algorithms in the literature.

Keywords: Global optimization · DIRECT method · Heuristic · Penalty auxiliary function

1 Introduction

In this paper, we aim to find the global solution of a non-smooth and non-convex constrained optimization problem using a non-differentiable penalty function and the DIRECT method [1]. The constrained global optimization (CGO) problem has the form:

$$\min_{x \in \Omega} f(x)$$
$$\text{subject to} \quad h(x) = 0 \tag{1}$$
$$g(x) \leq 0,$$

where $f : \mathbb{R}^n \to \mathbb{R}$, $h : \mathbb{R}^n \to \mathbb{R}^m$ and $g : \mathbb{R}^n \to \mathbb{R}^p$ are nonlinear continuous functions and $\Omega = \{x \in \mathbb{R}^n : -\infty < l_i \leq x_i \leq u_i < \infty, i = 1, \dots, n\}$. Denoting the feasible set of problem (1) by $\mathcal{F} = \{x \in \Omega : h(x) = 0, g(x) \leq 0\}$, we define a non-negative function

$$\theta(x) = \sum_{i=1}^{m} |h_j(x)| + \sum_{i=1}^{p} \max\{g_i(x), 0\}, \tag{2}$$

© Springer Nature Switzerland AG 2020
O. Gervasi et al. (Eds.): ICCSA 2020, LNCS 12251, pp. 538–551, 2020.
https://doi.org/10.1007/978-3-030-58808-3_39

where $\theta(x) = 0$ if $x \in \mathcal{F}$. Since convexity is not assumed, many local minima may exist in the feasible region, although we require only a global solution. For non-smooth problems, the derivative-free methods are the most appropriate. Deterministic and stochastic methods have been proposed to solve CGO problems [2,3]. Using deterministic methods, the convergence to a global optimal solution can be guaranteed and a solution with a required accuracy is obtained in a finite number of steps. On the other hand, stochastic methods are not guaranteed to find a global optimal solution although they are often able to find very good solutions after a (moderate) large number of steps. Stochastic convergence may be established using probability theory.

From the class of deterministic methods, the DIRECT method [1] has proven to be quite effective in converging to the global solution while avoiding to be trapped in a local solution, as far as bound constrained global optimization problems are concerned. The method has attracted considerable interest from the research community and several strategies have been incorporated into DIRECT, including the local search reinforcement [4,5], the improvement of the global search [6], new ideas for the selection of potentially optimal hyperrectangles [7,8] and new partition schemes [9–11].

The most popular methods to solve the problem (1) combine the objective function with a penalty term that aims to penalize constraint violation. Penalty functions within a DIRECT-type framework are proposed in [12–14]. An auxiliary function that combines in a special manner information on the objective and constraints is presented in [15]. Other techniques that involve the handling of the objective function and constraints violation separately can be found in [5,16,17].

The main contribution of this paper is the following. The two-phase heuristic DIRECT algorithm, proposed by the authors in [8], is extended to solve CGO problems using an auxiliary penalty function. The auxiliary function proposed in [15] is redefined to transform the CGO problem (1) into a bound constrained global optimization (BCGO) problem.

The paper is organized as follows. Section 2 briefly presents some ideas and the main steps of the DIRECT method. Section 3 describes a heuristic incorporated into the DIRECT algorithm to reduce the number of identified potentially optimal hyperrectangle, and the corresponding proposed extension to handle CGO problems, in particular, the use of a non-differentiable auxiliary penalty function. Finally, Sect. 4 contains the results of our preliminary numerical experiments and we conclude the paper with the Sect. 5.

2 DIRECT Method

The DIRECT (DIviding RECTangles) algorithm [1], originally proposed to solve BCGO problems of the form

$$\min_{x \in \Omega} f(x), \tag{3}$$

assumes that the objective function f is a continuous function, and iteratively produces finer and finer partitions of the hyperrectangles generated from Ω (see also [18]). The algorithm is a modification of the standard Lipschitzian approach, where f is assumed to satisfy the Lipschitz condition,

$$|f(x_1) - f(x_2)| \leq K\|x_1 - x_2\| \text{ for all } x_1, x_2 \in \Omega,$$

and the Lipschitz constant $K > 0$ is viewed as a weighting parameter that indicates how much emphasis to place on global versus local search. DIRECT is a deterministic method that does not require any analytical or numerical derivative information and searches (locally and globally) the feasible region Ω for hyperrectangles that are known as potentially optimal hyperrectangle (POH). These POH satisfy the two conditions established in the following definition:

Definition 1. *Given the partition $\{\mathcal{H}^i : i \in H\}$ of Ω, let ϵ be a positive constant, and let $f_{\min}$ be the current best function value. A hyperrectangle j is said to be potentially optimal if there exists some rate-of-change constant $\hat{K}^j > 0$ such that*

$$
\begin{aligned}
f(c_j) - \frac{\hat{K}^j}{2}\|u^j - l^j\| &\leq f(c_i) - \frac{\hat{K}^i}{2}\|u^i - l^i\|, \text{ for all } i \in H \\
f(c_j) - \frac{\hat{K}^j}{2}\|u^j - l^j\| &\leq f_{\min} - \epsilon|f_{\min}|
\end{aligned}
\tag{4}
$$

where c_j (resp. c_i) is the center and $\|u^j - l^j\|/2$ (resp. $\|u^i - l^i\|/2$) represents the size of hyperrectangle $j \in H$ (resp. i), and H is the set of indices of the hyperrectangles at the current iteration [1, 15].

The use of $\hat{K}^j$ in the definition intends to show that it is not the Lipschitz constant. The second condition in (4) aims to prevent the algorithm from identifying as POH the hyperrectangle with center that corresponds to $f_{\min}$. This way, small hyperrectangles where very small improvements may be obtained are skipped to be further divided.

The most important step in the DIRECT algorithm is the identification of POH since it determines the search along the feasible set. Each identified hyperrectangle is trisected along its longest sides and two new points in the hyperrectangle are sampled and remain center points of the other hyperrectangles (of the trisection).

A global search driven strategy would identify POH from the biggest hyperrectangles. On the other hand, a local search driven strategy would identify POH whose center point corresponds to $f_{\min}$. Good solutions are found rather quick but the hyperrectangle that contains the global solution may be missed if its center point has a bad function value. The main steps of the DIRECT algorithm are shown in Algorithm 1.

Algorithm 1. DIRECT algorithm

Require: η, f^*, $Nfe_{\max}$;

  1: Set $Nfe = 0$;
  2: **repeat**
  3:    Identification procedure for POH (Selection) according to Definition 1;
  4:    Selection procedure for division along dimensions (Sampling);
  5:    Division procedure;
  6:    Update index sets; Update Nfe;
  7: **until** $Nfe \geq Nfe_{\max}$ or $|f_{\min} - f^*| \leq \eta \max\{1, |f^*|\}$

3 Heuristic DIRECT Method Based on Penalties

This section presents the extension of a heuristic DIRECT algorithm [8] to handle nonlinear equality and inequality constraints.

3.1 Heuristic DIRECT Method

Firstly, we briefly describe a heuristic that can be incorporated into the DIRECT algorithm [8] aiming

- to divide a promising search region into three subregions, so that the number of hyperrectangles that are candidate to be potentially optimal is reduced;
- to choose between a global search driven phase or a local search driven phase.

Since avoiding the identification of POH that were mostly divided can enhance the global search capabilities of DIRECT [6] and identifying POH that are close to the hyperrectangle which corresponds to $f_{\min}$ may improve the local search process, the heuristic incorporated into the DIRECT method divides the region of the hyperrectangles with least function values in each *size* group - denoted by *candidate* hyperrectangles - into three subregions.

Each subregion is defined by the indices based on *size* of the hyperrectangles. The larger the *size* the smaller the index. The leftmost subregion includes hyperrectangles whose indices are larger than $i_l = \lfloor 2/3i_{\min} \rfloor$, where $i_{\min}$ is the index of the hyperrectangle that corresponds to $f_{\min}$. The rightmost subregion contains the hyperrectangles with indices that are smaller than $i_u = \lfloor 1/3i_{\min} \rfloor$ and the middle subregion contains hyperrectangles with indices between i_l and i_u (including these limits).

To be able to guarantee convergence to the global solution while avoiding the stagnation in a local solution, the algorithm cycles between global and local search phases. It starts with a global driven search, where $G_{\max}$ iterations are performed using all *candidate* hyperrectangles from the rightmost subregion, 50% of the *candidate* hyperrectangles from the middle subregion (randomly selected) and 10% of the *candidate* hyperrectangles from the leftmost subregion (randomly selected). This choice of percentages is hereinafter denoted by $(10/50/100)\%$. At each iteration, the set of POH are identified among these selected hyperrectangles. Afterwards, a local driven search is implemented for $L_{\max}$ iterations with the percentages of selected *candidate* hyperrectangles in the leftmost and rightmost subregions changed, denoted by $(100/50/10)\%$. This cycling process is repeated until convergence.

3.2   Penalty Auxiliary Function

We now extend this heuristic DIRECT method to handle nonlinear equality and inequality constraints. We use an auxiliary function that takes into consideration the violation of inequality constraints by combining information of the objective and constraint functions [15]. This function penalizes any deviation of the function value at the center c_j of a hyperrectangle above the global optimal value f^*:

$$P(c_j) = \max\{f(c_j) - f^*, 0\} + \sum_{i=1}^{p} \mu_i \max\{g_i(c_j), 0\} \tag{5}$$

where μ_i are positive weighting coefficients. Note that when the hyperrectangle has a feasible center point c_j, the second term is zero, and when it is infeasible, the second term is positive and the first term only counts for the cases where $f(c_j)$ is above f^*. Since f^* is unknown in general, but satisfies $f^* \leq f_{\min} - \varepsilon$, for a small tolerance $\varepsilon > 0$, we redefine the following variant of the auxiliary function

$$P(x; \mu) = \max\{f(x) - (f_{\min} - \varepsilon), 0\} + \mu \left(\sum_{i=1}^{m} |h_i(x)| + \sum_{i=1}^{p} \max\{g_i(x), 0\} \right) \tag{6}$$

where $f_{\min}$ is the current best function value found so far among all feasible center points. Although different weights might prove to be useful for some problems, we consider only one constant weighting coefficient for all the constraints, and extend the penalized constraint violation term to the equality constraints, since in our formulation they are treated separately from the inequality constraints. We remark that if no feasible point has been found so far, the function $P(x; \mu)$ is reduced to the second term alone in (6).

The definition of POH (recall Definition 1 above) is now adapted to the strategy that aims to find a global minimum solution of the problem

$$\min_{x \in \Omega} P(x; \mu) \tag{7}$$

for a fixed $\mu > 0$ value, in the sense that the sequence of approximations $x_{\min}^k$ (resp. $f_{\min}^k$) converges to x^* (resp. f^*), the global optimal solution of problem (1), as k increases. In this context, the new algorithm searches (locally and globally) the feasible region Ω to identify hyperrectangles that are known as POH with respect to $P(x; \mu)$ and satisfy:

Definition 2. *Given the partition* $\{\mathcal{H}^i : i \in H\}$ *of* Ω, *let* $\epsilon > 0$ *and* $\mu > 0$ *be constants and let* $f_{\min}$ *be the current best function value among feasible center* points. A hyperrectangle j is said to be potentially optimal with respect to $P(x; \mu)$ if there exists some rate-of-change constant $\hat{K}^j > 0$ such that

$$
\begin{aligned}
P(c_j; \mu) - \frac{\hat{K}^j}{2} \|u^j - l^j\| &\leq P(c_i; \mu) - \frac{\hat{K}^i}{2} \|u^i - l^i\|, \quad \text{for all } i \in H \\
P(c_j; \mu) - \frac{\hat{K}^j}{2} \|u^j - l^j\| &\leq P_{\min} - \epsilon |P_{\min}|
\end{aligned}
\tag{8}
$$

where $P_{\min}$ is the current best penalty function value and H is the set of indices of the selected candidate hyperrectangles at the current iteration.

The main steps of the proposed penalty-based heuristic DIRECT algorithm are presented in Algorithm 2.

Algorithm 2. Penalty-based heuristic DIRECT algorithm

Require: $\eta_1, \eta_2, G_{\max}, L_{\max}, f^*, Nfe_{\max}$;
1: Set $Nfe = 0$, $flag = G$, $it = 0$;
2: **repeat**
3:      Set $it = it + 1$;
4:      **if** $flag = G$ **then**
5:          Based on $i_{\min}$ and function P, randomly select the *candidate* hyperrectangles from the 3 subregions of indices based on the percentages $(10/50/100)\%$;
6:      **else**
7:          Based on $i_{\min}$ and function P, randomly select the *candidate* hyperrectangles from the 3 subregions of indices based on the percentages $(100/50/10)\%$;
8:      **end if**
9:      Identification procedure for POH according to Definition 2, among those selected *candidate* hyperrectangles (Selection);
10:     Selection procedure for division along dimensions (Sampling);
11:     Division procedure;
12:     Update index sets; Update Nfe;
13:     **if** $flag = L$ and $it \geq L_{max}$ **then**
14:         Set $flag = G$, $it = 0$;
15:     **end if**
16:     **if** $flag = G$ and $it \geq G_{max}$ **then**
17:         Set $flag = L$, $it = 0$;
18:     **end if**
19: **until** $Nfe \geq Nfe_{\max}$ or $(\theta(x_{\min}) \leq \eta_1$ and $|f_{\min} - f^*| \leq \eta_2 \max\{1, |f^*|\})$

Unless otherwise stated, the stopping conditions for the algorithm are the following. We consider that a good approximate solution x^k, at iteration k, is found, if the conditions

$$
\theta(x^k_{\min}) \leq \eta_1 \quad \text{and} \quad \frac{|f^k_{\min} - f^*|}{\max\{1, |f^*|\}} \leq \eta_2
\tag{9}
$$

are satisfied, for sufficiently small tolerances $\eta_1, \eta_2 > 0$, where $x^k_{\min}$ is the best computed solution to the problem, i.e., is the feasible center point of the hyperrectangle that has the least function value $f^k_{\min}$. However, if conditions (9) are not satisfied, the algorithm runs until a maximum number of function evaluations, $Nfe_{\max}$, is reached.

4   Numerical Experiments

In these preliminary numerical experiments, a set of seven benchmark problems with $n \leq 5$ is used. The MATLAB® (MATLAB is a registered trademark of the MathWorks, Inc.) programming language is used to code the algorithm and the tested problems. The parameter values for the algorithm are set as follows: $\epsilon = 1\text{E-}04$, $G_{\max} = 10$, $L_{\max} = 10$, $\varepsilon = 1\text{E-}06$, $\eta_1 = 1\text{E-}04$, $\eta_2 = 1\text{E-}04$ and $Nfe_{\max} = 1\text{E+}05$. Due to the random issue present in the algorithm, namely the selection of the *candidate* hyperrectangles, each problem is run five times. The reported results in the subsequent tables correspond to average values obtained in the five runs.

First, we consider two problems, one has 2 variables and 2 inequality constraints and the other 3 variables and 3 equality constraints, to show the effectiveness of the proposed strategy based on the penalty auxiliary function (6) when compared to the more usual L1 penalty-based technique.

Problem 1. (Problem 8 in [19])

$$\min_{x \in \Omega} x_1^4 - 14x_1^2 + 24x_1 - x_2^2$$
$$\text{s. t. } x_2 - x_1^2 - 2x_1 \leq -2$$
$$-x_1 + x_2 \leq 8$$

with $\Omega = \{x \in \mathbb{R}^2 : -8 \leq x_1 \leq 10, 0 \leq x_2 \leq 10\}$ and $f^* = -118.70$.

Problem 2. (Problem 5 in [19])

$$\min_{x \in \Omega} x_3$$
$$\text{s. t. } 30x_1 - 6x_1^2 - x_3 = -250$$
$$20x_2 - 12x_2^2 - x_3 = -300$$
$$0.5(x_1 + x_2)^2 - x_3 = -150$$

with $\Omega = \{x \in \mathbb{R}^3 : 0 \leq x_1 \leq 9.422, 0 \leq x_2 \leq 5.903, 0 \leq x_3 \leq 267.42\}$ and $f^* = 201.16$.

To analyze the gain in efficiency of Algorithm 2, we report in Table 1 the average values of f ($f_{\min}$), θ ($\theta(x_{\min})$), number of iterations (k) and number of function evaluations (Nfe) obtained after the 5 runs using the stopping conditions in (9) (or a maximum of 1E+05 function evaluations is reached). The standard deviation of the obtained f values ($St.D.$) is also reported. From the results in Table 1 we can conclude that the algorithm with the penalty (6) gives significantly better results than with the penalty L1. Similarly, from Table 2 we conclude that penalty (6) performs much better compared to L1. The algorithm

Table 1. Comparison between penalty functions, when solving Problem 1.

Algorithm 2	μ	$f_{\min}$	$St.D.$	$\theta(x_{\min})$	k	Nfe
Penalty (6)	0.5	-118.691829	2.26E$-$03	0.00E+00	109	2175
	1	-118.692153	2.01E$-$03	0.00E+00	86	1457
	10	-118.688994	2.65E$-$04	0.00E+00	44	526
	100	-118.688657	1.69E$-$04	0.00E+00	77	1121
L1 penalty	0.5	(-217)	(6.15E$-$05)	(4.5E+00)	(4848)	>1E+05
	1	(-217)	(6.15E$-$05)	(4.5E+00)	(4852)	>1E+05
	10	(-215)	(5.02E$-$05)	(4.0E+00)	(4828)	>1E+05
	100	-118.689247	6.48E$-$04	0.00E+00	72	865

In parentheses, achieved values when the algorithm stops due to $Nfe > 1\text{E}+05$.

Table 2. Comparison between penalty functions, when solving Problem 2.

Algorithm 2	μ	$f_{\min}$	$St.D.$	$\theta(x_{\min})$	k	Nfe
Penalty (6)	0.5	201.159343	0.00E+00	7.83E$-$05	43	577
	1	201.159343	0.00E+00	7.83E$-$05	43	543
	10	201.159343	0.00E+00	7.83E$-$05	45	624
	100	201.159343	0.00E+00	7.83E$-$05	41	531
Penalty L1	0.5	(201)	(0.00E+00)	(7.6E$-$04)	(8803)	>1E+05
	1	(201)	(0.00E+00)	(1.4E$-$04)	(7612)	>1E+05
	10	201.159343	0.00E+00	7.83E$-$05	320	5864
	100	201.159343	0.00E+00	7.83E$-$05	45	577

In parentheses, achieved values when the algorithm stops due to $Nfe > 1\text{E}+05$.

with the penalty L1 works better with the larger values of the weighing parameter while the performance of the algorithm with penalty (6) is not too much affected by the value of μ.

In Table 3 we compare the results obtained by Algorithm 2 based on the penalty auxiliary function (6) for two values of the weighting parameter (that provide the best results among the four tested) with those obtained by previous DIRECT-type strategies that rely on the filter methodology to reduce both the constraint violation and objective function [8,17]. The results are also compared to those obtained by variants DIRECT-GL and DIRECT-GLce reported in [14]. We note that the reported $f_{\min}$, $\theta(x_{\min})$, k and Nfe selected from [8] correspond also to average values, while the values from the other papers in comparison correspond to just a single solution (one run of deterministic methods). A slight gain in efficiency of the proposed penalty-based heuristic DIRECT algorithm has been detected.

To analyze the performance of the Algorithm 2 when compared to the strategy proposed in [15] and two filter-based DIRECT algorithms (in [8,17]), we consider the problem Gomez #3 (available in [15]):

Table 3. Comparative results for Problems 1 and 2.

	μ	$f_{\min}$	$\theta(x_{\min})$	k	Nfe	f^*
Problem 1						
Algorithm 2 (penalty (6))	10	-118.688994	0.00E+00	44	526	-118.70
	100	-118.688657	0.00E+00	77	1121	
DIRECT-type^a [8]		-118.700976	0.00E+00	19	823	
DIRECT-type^b [8]		-118.700976	0.00E+00	19	797	
DIRECT-type^c [8]		-118.692210	0.00E+00	23	689	
filter-based DIRECT [17]		-118.700976	0.00E+00	23	881	
DIRECT-GLc in [14]		-118.6892	–	–	1197	
DIRECT-GLce in [14]		-118.6898	–	–	1947	
Problem 2						
Algorithm 2 (penalty (6))	1	201.159343	7.83E−05	43	543	201.16
	100	201.159343	7.83E−05	41	531	
DIRECT-type^a [8]		201.159343	7.83E−05	30	1015	
DIRECT-type^b [8]		201.159343	7.83E−05	30	883	
DIRECT-type^c [8]		201.159343	7.83E−05	30	769	
filter-based DIRECT [17]		201.159343	7.83E−05	30	1009	
DIRECT-GLc in [14]		201.1593	–	–	819	
DIRECT-GLce in [14]		201.1593	–	–	819	

^{a}with filter; ^{b}with filter and upper bounds on f and θ;
^{c}with filter and upper bounds on f and θ as well as a heuristic.
"–" information not available.

Problem 3.

$$\min_{x \in \Omega} \left(4 - 2.1x_1^2 + \frac{x_1^4}{3}\right) x_1^2 + x_1 x_2 + (-4 + 4x_2^2)x_2^2$$
$$\text{s. t. } -\sin(4\pi x_1) + 2\sin^2(2\pi x_2) \leq 0$$

with $\Omega = \{x \in \mathbb{R}^2 : -1 \leq x_i \leq 1, \, i = 1, 2\}$.

Table 4 compares the performance of the tested algorithms. Our Algorithm 2 was tested with four different values of the fixed weighting parameter. When solving the Problem 3, our algorithm reports a considerable sensitivity to the selected μ value, with a better performance achieved when small values are used.

The following problem, known as T1 is tested with 3 different values of n.

Problem 4.

$$\min_{x \in \Omega} \sum_{i=1}^{n} x_i$$
$$\text{s. t. } \sum_{i=1}^{n} x_i^2 \leq 6$$

with $\Omega = \{x \in \mathbb{R}^n : -1 \leq x_i \leq 1, \, i = 1, \ldots, n\}$.

Table 4. Comparison results when solving Problem 3.

	μ	f_{min}	$\theta(x_{min})$	k	Nfe	f^*
Algorithm 2 (penalty (6))	0.5	−0.971021	2.45E−05	51	606	−0.9711
	1	−0.971018	1.34E−05	74	983	
	10	−0.971018	1.34E−05	455	12952	
	100	(−0.97)	(4.4E−05)	(2625)	>1E+05	
DIRECT-type[a] [8]		−0.971006	6.00E−05	17	615	
DIRECT-type[b] [8]		−0.971006	6.00E−05	17	683	
DIRECT-type[c] [8]		−0.971041	3.17E−05	18	555	
filter-based DIRECT [17]		−	−	18	733	
DIRECT in [15]		−	−	−	513	

In parentheses, the achieved values when the algorithm stops due to $Nfe > 1E+05$.
[a]with filter; [b]with filter and upper bounds on f and θ;
[c]with filter and upper bounds on f and θ as well as a heuristic.
"−" information not available.

The results obtained by Algorithm 2, and those in [14] (variants DIRECT-GLc and DIRECT-GLce), as well as the results obtained by the variant DIRECT-GL and the original DIRECT (when they are implemented in a penalty-based strategy with penalty function L1) are shown in Table 5 for comparison. Our algorithm with the penalty (6) works much better with the smaller values of the fixed weighting parameter and those results outperform in general the other results in comparison, for the same solution quality accuracy, as far as function evaluations are concerned.

Finally, the last 3 problems, known as g04, g06 and g08 in [16], have inequality constraints.

Problem 5.

$$\min_{x \in \Omega} 5.3578547x_3^2 + 0.8356891x_1x_5 + 37.293239x_1 - 40792.141$$
$$\text{s. t. } 0 \le 85.334407 + 0.0056858x_2x_5 + 0.0006262x_1x_4 - 0.0022053x_3x_5 \le 92$$
$$90 \le 80.51249 + 0.0071317x_2x_5 + 0.0029955x_1x_2 + 0.0021813x_3^2 \le 110$$
$$20 \le 9.300961 + 0.0047026x_3x_5 + 0.0012547x_1x_3 + 0.0019085x_3x_4 \le 25$$

with $\Omega = \{x \in \mathbb{R}^5 : 78 \le x_1 \le 102, 33 \le x_2 \le 45, 27 \le x_i \le 45, i = 3, 4, 5\}$.

Problem 6.

$$\min_{x \in \Omega} (x_1 - 10)^3 + (x_2 - 20)^3$$
$$\text{s. t. } -(x_1 - 5)^2 - (x_2 - 5)^2 \le -100$$
$$(x_1 - 6)^2 + (x_2 - 5)^2 \le 82.81$$

with $\Omega = \{x \in \mathbb{R}^2 : 13 \le x_1 \le 100, 0 \le x_2 \le 100\}$.

Table 5. Comparison results when solving Problem 4.

	μ	$f_{\min}$	$\theta(x_{\min})$	k	Nfe	f^*
$n = 2$						
Algorithm 2 (penalty (6))	0.5	-3.464079	3.60E$-$05	26	383	-3.4641
	1	-3.464052	3.30E$-$05	26	370	
	10	-3.464106	9.29E$-$05	40	723	
	100	-3.464106	7.68E$-$05	85	3927	
DIRECT-type[a] [8]		-3.464106	9.29E$-$05	14	1395	
DIRECT-type[b] [8]		-3.464106	9.29E$-$05	14	893	
DIRECT-type[c] [8]		-3.464106	5.72E$-$05	13	335	
DIRECT-L1 in [14]	10	$-$	$-$	$-$	3345	
	100	$-$	$-$	$-$	8229	
DIRECT-GL-L1 in [14]	10	$-$	$-$	$-$	1221	
	100	$-$	$-$	$-$	1921	
DIRECT-GLc in [14]		$-$	$-$	$-$	1373	
DIRECT-GLce in [14]		$-$	$-$	$-$	2933	
$n = 3$						
Algorithm 2 (penalty (6))	0.5	-4.242687	7.25E$-$05	266	29187	-4.2426
	1	-4.242443	4.38E$-$05	104	6989	
	10	-4.242443	0.00E+00	110	8577	
	100[d]	-4.242443	2.30E$-$05	260	85472	
DIRECT-type[a] [8]		-4.242443	0.00E+00	28	16885	
DIRECT-type[b] [8]		-4.242443	0.00E+00	35	37977	
DIRECT-type[c] [8]		-4.242443	9.17E$-$05	29	3233	
DIRECT-L1 in [14]	10	$-$	$-$	$-$	66137	
	100	$-$	$-$	$-$	>1E+06	
DIRECT-GL-L1 in [14]	10	$-$	$-$	$-$	75105	
	100	$-$	$-$	$-$	16625	
DIRECT-GLc in [14]		$-$	$-$	$-$	26643	
DIRECT-GLce in [14]		$-$	$-$	$-$	8297	
$n = 4$						
Algorithm 2 (penalty (6))	0.5	-4.898440	0.00E+00	74	9514	-4.899
	1	-4.898440	0.00E+00	62	6201	
	10[e]	-4.898440	3.42E$-$05	133	54981	
	100[d]	-4.898440	5.80E$-$05	98	31440	
DIRECT-type[a] [8]		-4.898847	0.00E+00	42	151753	
DIRECT-type[b] [8]		-4.898847	3.42E$-$05	39	78859	
DIRECT-type[c] [8]		-4.898440	3.30E$-$05	51	36219	
DIRECT-L1 in [14]	10	$-$	$-$	$-$	127087	
	100	$-$	$-$	$-$	>1E+06	
DIRECT-GL-L1 in [14]	10	$-$	$-$	$-$	180383	
	100	$-$	$-$	$-$	189595	
DIRECT-GLc in [14]		$-$	$-$	$-$	192951	
DIRECT-GLce in [14]		$-$	$-$	$-$	47431	

[a]with filter; [b]with filter and upper bounds on f and θ;
[c]with filter and upper bounds on f and θ as well as a heuristic.
[d]80% successful runs; [e]60% successful runs.
"$-$" information not available.

Problem 7.

$$\min_{x \in \Omega} -\frac{\sin^3(2\pi x_1)\sin(2\pi x_2)}{x_1^3(x_1+x_2)}$$
$$\text{s. t. } x_1^2 - x_2 + 1 \leq 0$$
$$1 - x_1 - (x_2 - 4)^2 \leq 0$$

with $\Omega = \{x \in \mathbb{R}^2 : 0 \leq x_i \leq 10, i = 1, 2\}$.

Table 6. Comparison results when solving Problems 5, 6 and 7.

	μ	$f_{\min}$	$\theta(x_{\min})$	k	Nfe	f^*
Problem 5[a]						
Algorithm 2 (penalty (6))	0.5^b	-30665.2339	9.99E$-$05	387	36277	-30665.53867
	1^c	-30665.4237	9.99E$-$05	377	40331	
	10	-30665.2450	9.96E$-$05	132	5119	
	100	-30665.2329	9.79E$-$05	133	5247	
	1000	-30665.2329	9.79E$-$05	135	5746	
DIRECT-GL-L1 in [14]	1000	–	–	–	1799	
DIRECT-GLc in [14]		–	–	–	5907	
DIRECT-GLce in [14]		-30663.5708	–	–	21355	
eDIRECT-C in [16]		-30665.5385	–	–	65	
Problem 6[a]						
Algorithm 2 (penalty (6))	0.5	-6961.9092	6.86E$-$05	120	2815	-6961.81387558
	1	-6961.8763	4.80E$-$05	118	2699	
	10	-6961.8088	2.18E$-$05	114	2758	
	100	-6961.7868	1.66E$-$05	121	2939	
	1000	-6961.8150	2.77E$-$05	186	4941	
DIRECT-GL-L1 in [14]	1000^d	–	–	–	289	
DIRECT-GLc in [14]		–	–	–	3461	
DIRECT-GLce in [14]		-6961.1798	–	–	6017	
eDIRECT-C in [16]		-6961.8137	–	–	35	
Problem 7						
Algorithm 2 (penalty (6))	0.5	-0.095825	0.00E+00	18	174	-0.095825
	1	-0.095825	0.00E+00	16	158	
	10	-0.095825	0.00E+00	16	152	
	100	-0.095825	0.00E+00	16	164	
	1000	-0.095825	0.00E+00	15	153	
DIRECT-GL-L1 in [14]	1000	–	–	–	471	
DIRECT-GLc in [14]		–	–	–	471	
DIRECT-GLce in [14]		-0.0958	–	–	1507	
eDIRECT-C in [16]		-0.0958	–	–	154	

[a] results for $\eta_2 = 1\text{E}-05$; [b] 20% successful runs;
[c] 80% successful runs; [d] final solution outside the feasible region.
"–" information not available.

The results obtained by our Algorithm 2 for five values of μ are compared to those of the variants DIRECT-GL-L1, DIRECT-GLc and DIRECT-GLce in [14], and eDIRECT-C from [16]. We remark that eDIRECT-C incorporates a local minimization search (MATLAB `fmincon`). From the results in Table 6, we may conclude that contrary to Problem 5 which presents significantly better results with the larger fixed μ values, the other two problems report reasonable good performances with all the tested μ values, competing with the other algorithms is comparison.

5   Conclusions

In this paper, we present an extension of the heuristic DIRECT method (available in [8]) to solve nonlinear CGO problems. The herein proposed extension transforms the CGO problem (1) into a BCGO one, using a penalty strategy based on the penalty auxiliary function (6). We have analyzed the performance of the penalty-based heuristic DIRECT algorithm for a set of fixed penalty parameter values, using well-known benchmark CGO problems. Neither too small nor too large parameter values (1, 10 and 100) have produced results that show the robustness and efficiency of proposed algorithm hereby competing favourably with other available DIRECT-type algorithms.

Although, for now, we have considered a fixed value for the parameter ε (in the definition of the penalty (6)), we feel that a sequence of decreased values may further improve the efficiency of the algorithm. This will be an issue for future research.

Acknowledgments. The authors wish to thank two anonymous referees for their comments and suggestions to improve the paper.

This work has been supported by FCT – Fundação para a Ciência e Tecnologia within the R&D Units Project Scope: UIDB/00319/2020, UIDB/00013/2020 and UIDP/00013/2020 of CMAT-UM.

References

1. Jones, D.R., Perttunen, C.D., Stuckman, B.E.: Lipschitzian optimization without the Lipschitz constant. J. Optim. Theory Appl. **79**(1), 157–181 (1993). Based on exact penalty functions
2. Hendrix, E.M.T., G.-Tóth, B.: Introduction to Nonlinear and Global Optimization. Optimization and its Applications. Springer, New York (2010). https://doi.org/10.1007/978-0-387-88670-1
3. Sergeyev, Y.D., Kvasov, D.E.: Deterministic Global Optimization. SO. Springer, New York (2017). https://doi.org/10.1007/978-1-4939-7199-2
4. Gablonsky, J.M., Kelley, C.T.: A locally-biased form of the DIRECT algorithm. J. Glob. Optim. **21**(1), 27–37 (2001)
5. Di Pillo, G., Liuzzi, G., Lucidi, S., Piccialli, V., Rinaldi, F.: A DIRECT-type approach for derivative-free constrained global optimization. Comput. Optim. Appl. **65**(2), 361–397 (2016). https://doi.org/10.1007/s10589-016-9876-3
6. Paulavičius, R., Sergeyev, Y.D., Kvasov, D.E., Žilinskas, J.: Globally-biased DIS-IMPL algorithm for expensive global optimization. J. Glob. Optim. **59**(2–3), 545–567 (2014)

7. Stripinis, L., Paulavičius, R., Žilinskas, J.: Improved scheme for selection of potentially optimal hyper-rectangles in DIRECT. Optim. Lett. **12**(7), 1699–1712 (2018)
8. Costa, M.F.P., Fernandes, E.M.G.P., Rocha, A.M.A.C.: A two-phase heuristic coupled DIRECT method for bound constrained global optimization. In: EUROGEN 2019 Conference Proceedings, 8 p. Guimarães, Portugal (2019)
9. Paulavičius, R., Žilinskas. J.: Simplicial Lipschitz optimization without the Lipschitz constant. J. Glob. Optim. **59**, 23–40 (2014)
10. Liu, H., Xu, S., Wang, X., Wu, J., Song, Y.: A global optimization algorithm for simulation-based problems via the extended DIRECT scheme. Eng. Optim. **47**(11), 1441–1458 (2015)
11. Paulavičius, R., Sergeyev, Y.D., Kvasov, D.E., Žilinskas, J.: Globally-biased BIRECT algorithm with local accelerators for expensive global optimization. Expert. Syst. Appl. **144**, 113052 (2020)
12. Di Pillo, G., Lucidi, S., Rinaldi, F.: An approach to constrained global optimization based on exact penalty functions. J. Glob. Optim. **54**(2), 251–260 (2012)
13. Di Pillo, G., Lucidi, S., Rinaldi, F.: A derivative-free algorithm for constrained global optimization based on exact penalty functions. J. Optim. Theory Appl. **164**, 862–882 (2015)
14. Stripinis, L., Paulavičius, R., Žilinskas, J.: Penalty functions and two-step selection procedure based DIRECT-type algorithm for constrained global optimization. Struct. Multidisc. Optim. **59**, 2155–2175 (2019)
15. Jones, D.R.: The DIRECT global optimization algorithm. In: Floudas, C., Pardalos, P. (eds.) Encyclopedia of Optimization, pp. 431–440. Kluwer Academic Publisher, Boston (2001)
16. Liu, H., Xu, S., Chen, X., Wang, X., Ma, Q.: Constrained global optimization via a DIRECT-type constraint-handling technique and an adaptive metamodeling strategy. Struct. Multi. Optim. **55**(1), 155–177 (2016). https://doi.org/10.1007/s00158-016-1482-6
17. Costa, M.F.P., Rocha, A.M.A.C., Fernandes, E.M.G.P.: Filter-based DIRECT method for constrained global optimization. J. Glob. Optim. **71**(3), 517–536 (2017). https://doi.org/10.1007/s10898-017-0596-8
18. Gablonsky, J.M.: DIRECT version 2.0 user guide. Technical report CRSC-TR-01-08, Center for Research in Scientific Computation, North Carolina State University (2001)
19. Birgin, E.G., Floudas, C.A., Martínez, J.M.: Global minimization using an Augmented Lagrangian method with variable lower-level constraints. Technical report MCDO121206, Department of Computer Science IME-USP, University of São Paulo (2007). http://www.ime.usp.br/~egbirgin/

Comparison of Different Strategies for Arc Therapy Optimization

Humberto Rocha[1,2](✉)⬤, Joana Dias[1,2]⬤, Pedro Carrasqueira[2]⬤,
Tiago Ventura[2,3]⬤, Brígida Ferreira[2,4]⬤, and Maria do Carmo Lopes[2,3]⬤

[1] CeBER, Faculty of Economics, University of Coimbra, 3004-512 Coimbra, Portugal
{hrocha,joana}@fe.uc.pt
[2] INESC-Coimbra, 3030-290 Coimbra, Portugal
pedro.carrasqueira@deec.uc.pt
[3] Medical Physics Department, IPOC-FG, EPE, 3000-075 Coimbra, Portugal
{tiagoventura,mclopes}@ipocoimbra.min-saude.pt
[4] Health School, Polytechnic of Porto, 4200–072 Porto, Portugal
bcf@ess.ipp.pt

Abstract. Radiotherapy (RT) has seen considerable changes in the last decades, offering an increased range of treatment modalities to cancer patients. Volumetric Modulated Arc Therapy (VMAT), one of the most efficient RT arc techniques, particularly with respect to dose delivery time, has recently considered noncoplanar arc trajectories while irradiating the patient, thanks to technological advances in the most recent generation of RT systems. In a preliminary work we have proposed a two-step optimization approach for noncoplanar arc trajectory optimization (ATO). In this paper, treatment plans considering 5-, 7-, 9-, 11- and 21-beam ensembles in the first step (optimal selection of noncoplanar irradiation directions) of the proposed approach are compared to assess the trade-offs between treatment quality and computational time required to obtain a noncoplanar VMAT plan. The different strategies were tested resorting to a head-and-neck tumor case already treated at the Portuguese Institute of Oncology of Coimbra (IPOC). Results obtained presented similar treatment quality for the different strategies. However, strategies that consider a reduced number of beams in the first step clearly outperform the other strategies in terms of computational times. Results show that for a similar tumor coverage, treatment plans with optimal beam irradiation directions obtained an enhanced organ sparing.

Keywords: Noncoplanar optimization · IMRT · Arc therapy · Treatment planning

1 Introduction

Cancer incidence will continue to rise worldwide, with an expected increase of 63.1% of cancer cases by 2040, compared with 2018 [1]. Radiotherapy (RT) is

© Springer Nature Switzerland AG 2020
O. Gervasi et al. (Eds.): ICCSA 2020, LNCS 12251, pp. 552–563, 2020.
https://doi.org/10.1007/978-3-030-58808-3_40

used in more than 50% of all cancer cases in high-income countries [2]. In Europe, currently, less than 75% of the patients who should be treated with RT actually are [3].

Radiation oncology relies on cutting edge technology to provide the best possible treatments. There are different treatment modalities available, taking advantage of technological advances that allow an increased control over the shape and intensity of the radiation. In external RT, radiation is generated by a linear accelerator mounted on a gantry that can rotate around the patient that lays immobilized in a treatment couch that can also rotate. RT systems use a multileaf collimator (MLC) to modulate the radiation beam into a discrete set of small beamlets whose individual intensities can be optimized – fluence map optimization (FMO) problem. In static intensity-modulated RT (IMRT), the nonuniform radiation fields obtained can be delivered while the gantry is halted at given beam irradiation directions that can be optimally selected – beam angle optimization (BAO). In rotational (or arc) IMRT, the patient is continuously irradiated with the treatment beam always on while the gantry is rotating around the patient.

Volumetric modulated arc therapy (VMAT) is a modern IMRT arc technique, particularly efficient with respect to dose delivery time [4]. Typically, VMAT uses beam trajectories that lay in the plane of rotation of the linear accelerator for a 0° couch angle (coplanar trajectories). The most recent RT systems allow the simultaneous movement of the gantry and the couch while continuously irradiating the patient. The highly noncoplanar arc trajectories obtained combine the short treatment times of VMAT [4], with the improved organ sparing of noncoplanar IMRT treatment plans [5]. We have proposed an optimization approach composed of two steps for arc trajectory optimization (ATO) in a preliminary work [6]. A set of (seven) optimal noncoplanar beam irradiation directions is initially calculated in a first step. Then, anchored in these beam irradiation directions (anchor points), additional anchor points are iteratively calculated until 21 anchor points are obtained defining the noncoplanar arc trajectory. In this paper, the trade-offs between the computational time needed to find an optimal noncoplanar beam ensemble – fastest if a small number of beams are considered or slowest when more beams are included in the beam orientation optimization – and the overall quality of the treatment plans obtained are investigated. A head-and-neck tumor case treated previously at the Portuguese Institute of Oncology of Coimbra (IPOC) is used to compare VMAT treatment plans considering 5-, 7-, 9-, 11- and 21-beam ensembles in the first step of the proposed approach. The rest of the paper is organized as follows. The head-and-neck cancer case is described in section two. In the following section, the noncoplanar ATO strategy is briefly described. In section four, the computational results are presented. The last section is devoted to the conclusions.

2   Head-and-Neck Cancer Case

A complex head-and-neck tumor case already treated at IPOC is used in the computational tests. The planing target volume (PTV) is composed of both the tumor and the lymph nodes. Two different dose prescription levels were considered for each patient. A 70.0 Gy radiation dose was prescribed to the tumor (PTV_{70}) while a 59.4 Gy radiation dose was prescribed to the lymph nodes ($PTV_{59.4}$).

Treatment planning of head-and-neck cancer cases is difficult due to the large number of organs-at-risk (OARs) surrounding or even overlapping both the tumor and the lymph nodes. The list of OARs considered in our tests is composed of spinal cord, brainstem, oral cavity, and parotid glands. Spinal cord and brainstem are serial organs, i.e. organs that may see their functionality impaired even if only a small part is damaged while the oral cavity and the parotid glands are parallel organs, i.e. organs whose functionality is not impaired if only a small part is damaged. Thus, maximum-dose constraints are considered for serial organs while mean-dose constraints are considered for parallel organs. The remaining normal tissues, called Body, is also included in the optimization procedures to prevent dose accumulation elsewhere. Table 1 depicts the doses prescribed for the PTVs and the tolerance doses for the OARs included in the optimization.

Table 1. Doses prescribed for the PTVs and tolerance doses for the OARs included in the optimization.

Structure	Tolerance dose		Prescribed dose
	Mean	Max	
PTV_{70}	–	–	70.0 Gy
$PTV_{59.4}$	–	–	59.4 Gy
Right parotid	26 Gy	–	–
Left parotid	26 Gy	–	–
Oral cavity	45 Gy	–	–
Spinal cord	–	45 Gy	–
Brainstem	–	54 Gy	–
Body	–	80 Gy	–

3   Arc Trajectory Optimization

The ATO framework evolves in two steps. An optimal noncoplanar beam irradiation ensemble is calculated in the first step, using a previously developed BAO algorithm [7]. Then, anchored in these beam irradiation directions, additional beam directions are iteratively calculated in order to define the trajectory of the noncoplanar arc. The output of the FMO problem is the measure considered to guide both optimization procedures in each of the steps. Aiming at minimizing the possible discrepancies to fully deliverable VMAT plans, direct aperture optimization (DAO) is used in this work for fluence optimization rather than the conventional beamlet-based FMO. The DAO approach used in this work is presented next followed by the description of the two steps that compose the ATO approach.

3.1   Fluence Map Optimization – DAO

Direct aperture optimization produces a deliverable plan by calculating aperture shapes instead of beamlet intensities that need to be converted to aperture shapes. The use of DAO during treatment planning can thus decrease possible discrepancies to fully deliverable VMAT plans.

The head-and-neck case considered in this work was assessed in matRad [8], an open source RT treatment planning system written in Matlab. matRad provides an experimental DAO implementation that can be customized by selecting, from a set of options available, objectives, constraints or weights assigned to each structure. matRad uses a gradient-based DAO algorithm [9] that starts with a good initial solution obtained by conventional beamlet-based FMO including sequencing [10].

Considering the appropriate options in matRad, fluence optimization can be formulated as a convex voxel-based nonlinear model [11]:

$$\min_w \left[\underline{\lambda}_i \left(T_i - \sum_{j=1}^N D_{ij} w_j \right)_+^2 + \overline{\lambda}_i \left(\sum_{j=1}^N D_{ij} w_j - T_i \right)_+^2 \right]$$

$$s.t. \quad 0 \leq w_j \leq w^{max}, \; j = 1, \ldots, N,$$

where D_{ij} is the dose delivered by beamlet j to voxel i, w_j is the intensity of beamlet j, T_i is the prescribed/tolerated dose for voxel i, $\overline{\lambda}_i$ and $\underline{\lambda}_i$ are overdose and underdose penalties, respectively, w^{max} is the maximum weight (intensity) of a beamlet and $(\cdot)_+ = \max\{0, \cdot\}$. This FMO formulation implies that overdose or underdose may be clinically accepted at reduced levels, but are decreasingly acceptable for increased deviations from the prescribed/tolerated doses [11].

In matRad, FMO is addressed using IPOPT [12], an interior point optimizer solver developed by the COIN-OR initiative.

3.2   First Step – BAO

The BAO problem has been formulated by us considering all possible continuous beam irradiation directions rather than a discretized set of irradiation directions around the tumor. Thus, instead of a combinatorial optimization problem we tackle a continuous global optimization problem [7,13–17]. If n is the number of irradiation directions previously defined, θ denotes a gantry angle and ϕ denotes a couch angle, the mathematical formulation of the continuous BAO problem is

$$\min f\Big((\theta_1,\phi_1),\ldots,(\theta_n,\phi_n)\Big)$$

$$s.t. \ \Big(\theta_1,\ldots,\theta_n,\phi_1,\ldots,\phi_n\Big) \in \mathbb{R}^{2n},$$

with f an objective function where the best beam angle ensemble is attained at the function's minimum. The optimal FMO value is the measure used as the objective function, f, to provide guidance for the BAO procedure. Collision between the linear accelerator gantry and the patient may occur for some pairs of couch and gantry angles. In order to consider only feasible directions while maintaining an unbounded formulation, the following penalization is considered:

$$f\Big((\theta_1,\phi_1),\ldots,(\theta_n,\phi_n)\Big) = \begin{cases} +\infty & \text{if collisions occur} \\ \text{optimal FMO value} & \text{otherwise.} \end{cases}$$

The BAO continuous search space presents a property of symmetry due to the fact that irradiation order is irrelevant. If the beam directions are kept sorted, only a small portion of the entire BAO search space needs to be explored [7]. The continuous BAO formulation is suited for the use of derivative-free optimization algorithms. Pattern search methods (PSM) are a class of derivative-free optimization algorithms that need a reduced number of function evaluations to converge making them an excellent option to address the highly non-convex and time-consuming BAO problem [18,19]. For a thorough description of the complete BAO framework used in this first step see, e.g., Ref. [7].

3.3   Second Step – ATO

For this second step, a discrete set of beams spaced $10°$ apart for both the couch and the gantry angles is considered. Figure 1 displays, both in 2D (Fig. 1(a)) and in 3D (Fig. 1(b)), the resulting candidate beams. Note that pairs of gantry-couch directions that would cause collision between gantry and patient for head-and-neck cancer cases were removed. The ATO strategy proposed is anchored in beam directions (anchor points) calculated in the first step, adding novel anchor points based on optimal FMO values. For illustration purposes, Fig. 2 displays in red the 5-beam ensemble obtained in the first step. ATO iterative procedure starts with these 5 initial anchor points and halts when 21 beam directions are obtained, which corresponds to the number of anchor beams that is typically used to define the arc trajectory [20,21]. The ATO strategy is based on dosimetric considerations, similarly to the BAO approach, being guided by the optimal FMO values. In order to enhance the short delivery times characteristics of VMAT plans, the gantry/couch movements are constrained according to the following conditions:

- The starting position of the couch and the gantry corresponds to the beam with lowest value of gantry angle of the noncoplanar BAO solution, i.e. is the leftmost beam in Fig. 2;

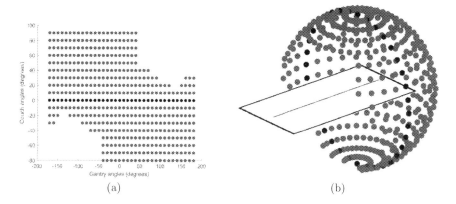

(a)   (b)

Fig. 1. Feasible equispaced candidate beams represented in 2D – (a) and in 3D – (b). Coplanar beams for a fixed 0° couch angle are displayed in black while noncoplanar candidate beams are displayed in blue. (Color figure online)

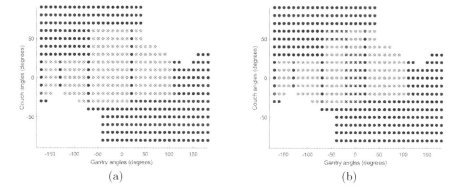

(a)   (b)

Fig. 2. The 5-beam ensemble, solution of the noncoplanar BAO problem is displayed in red while the possible beams considered in the iterative search of a novel anchor beam are displayed in green – (a). New anchor beam calculated for the largest set of candidate green beams is added while candidate green beams that fail the gantry/couch movement restrictions are removed – (b). (Color figure online)

- The anchor beam to visit next has the lowest value of gantry angle among the beams that are yet to be visited;
- The last position of the couch and the gantry corresponds to the beam with highest value of gantry angle of the noncoplanar BAO solution, i.e. is the rightmost beam in Fig. 2;
- The movement of the gantry must always be towards the next beam while the couch can be halted of move towards the next beam.

Figure 2(a) displays in green the candidates for novel anchor point to add when considering a 5-beam ensemble obtained in the first step as initial anchor points and following the definition of the possible gantry and coach movements. The most populated set is selected for searching the new anchor beam to add to the current arc trajectory for two reasons: to add anchor beams where more degrees of freedom exist and to reduce the computational time by reducing as much as possible the overall number of green points. Iteratively, each candidate beam of the most populated set is temporarily inserted in the trajectory and the optimal FMO value for the corresponding beam ensemble is calculated. The candidate beam that lead to the best beam ensemble in terms of optimal FMO value is selected, the candidate green points that become infeasible due to the gantry/couch movement restrictions are removed and a new iteration can proceed. Figure 2(b) illustrates one iteration of this arc trajectory optimization approach. This iterative procedure ends when the number of anchor beams is 21. The process of obtaining the optimized arc trajectory has been completely automated in order to get the required solution without additional human intervention. The pseudocode of the ATO algorithm is presented in Algorithm 1.

Algorithm 1. Iterative arc trajectory algorithm

Initialization:

- Calculate the initial anchor beams resorting to the noncoplanar BAO algorithm;
- Identify the possible candidate green beams to iteratively calculate new anchor beams;

Iteration:

While less than 21 anchor beams exist **do**

1. Choose the set of candidate beams between two anchor beams with larger cardinality;
2. Add each beam of the previous set to the current set of anchor beams and calculate the optimal FMO value of the resulting beam ensemble;
3. Select as new anchor point the candidate green beam that corresponds to the beam ensemble with best optimal FMO value in the previous step;
4. The candidate beams that fail the gantry/couch movement restrictions are removed.

---

4 Computational Results

A personal computer with an Intel i7-6700 processor @ 2.60 GHz was used for the computational tests. All tests were performed in a sequential way to allow a better comparison of computational times. Nevertheless, all algorithms are

Table 2. Optimal FMO results and computational time.

#BAO beams	Plan					BAO time	ATO time	Total time
	$IMRT_5$	$IMRT_7$	$IMRT_9$	$IMRT_{11}$	VMAT			
0 (Equi)	199.4	179.9	171.7	169.6	163.4	–	–	–
5	**178.7**	170.4	165.9	162.2	157.7	10.3	20.7	31.0
7	–	**166.9**	165.1	162.5	158.6	21.5	10.2	31.7
9	–	–	**164.4**	163.1	158.3	30.4	11.9	42.3
11	–	–	–	**161.6**	158.1	47.1	13.5	60.6
21	–	–	–	–	**158.4**	121.4	–	121.4

Table 3. Tumor coverage achieved by the different plans.

	$VMAT_0$	$VMAT_5$	$IMRT_{11}$
PTV_{70} D_{95}	66.66	66.74	66.63
$PTV_{59.4}$ D_{95}	58.13	58.12	57.98

implemented in parallel and computational times are typically divided by 12 – the maximum number of cores that our Matlab license allows us to use.

VMAT treatment plans considering 5-, 7-, 9-, 11- and 21-beam ensembles in the first step of the proposed approach were obtained as well as IMRT plans with optimal 5-, 7-, 9- and 11-beam ensembles. Table 2 depicts the optimal FMO value obtained for each of these VMAT and IMRT plans as well as the computational times required for BAO and/or ATO. As expected, the optimal FMO value improves for treatment plans with more beam directions but at the cost of larger computational times required to obtain optimal irradiation directions using BAO (depicted in bold). For instance, computing an optimal 21-beam ensemble using BAO takes twice the time of computing an optimal 11-beam ensemble with only a small gain in terms of optimal FMO value. The optimal FMO value obtained for the VMAT plans following the different strategies – less beam directions in the first step up to all directions in the first step – present only small differences. However, in terms of computational times, the strategies that consider less beams in the first step clearly outperform the other strategies in terms of computational times.

The quality of the treatment plans is also acknowledged by a set of dose metrics. These dose metrics are compared for the noncoplanar VMAT plan with best optimal FMO value (and overall computational time) denoted $VMAT_5$, the coplanar VMAT plan denoted $VMAT_0$ and the best IMRT plan denoted $IMRT_{11}$, whose noncoplanar arc trajectory, coplanar arc trajectory and noncoplanar beams, respectively, are displayed in Fig. 3. Target coverage is one of the metrics typically used for the tumors, i.e. the amount of PTV that receives 95% of the dose prescribed (D_{95}). Typically, 95% to 98% of the PTV volume is

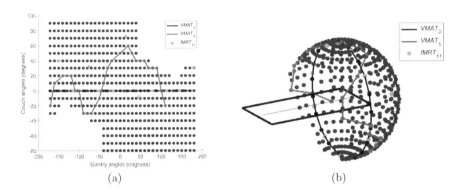

(a)   (b)

Fig. 3. Trajectories obtained by $VMAT_0$, the coplanar VMAT plan, and $VMAT_5$, the noncoplanar VMAT plan with best optimal FMO value (and overall computational time), and noncoplanar beams of $IMRT_{11}$, the best IMRT plan, in 2D – (a) and in 3D – (b).

Table 4. Organ's sparing achieved by the different plans.

OAR	Mean dose (Gy)			Max dose (Gy)		
	$VMAT_0$	$VMAT_5$	$IMRT_{11}$	$VMAT_0$	$VMAT_5$	$IMRT_{11}$
Brainstem (BS)	–	–	–	45.94	44.99	46.12
Spinal cord (SC)	–	–	–	36.92	35.97	36.56
Right parotid (RPrt)	28.34	27.17	26.16	–	–	–
Left parotid (LPrt)	22.57	21.39	20.85	–	–	–
Oral Cavity (OCav)	29.42	27.72	27.65	–	–	–

required. Table 3 reports the tumor coverage metrics. We can observe that tumor coverage is very similar for the three plans with a small advantage for the $VMAT$ plans. OARs sparing results are depicted in Table 4. Results obtained by the different plans fulfill most of the times the tolerance doses with no clear advantage of one approach for all structures. For instance, for serial organs, brainstem and spinal cord, $VMAT_5$ treatment plan obtained the best sparing while for parallel organs, oral cavity and parotids, $IMRT_{11}$ treatment plan obtained the best sparing results. Target coverage and organ sparing for the different structures are displayed in Fig. 4 allowing a more comprehensive view of the results.

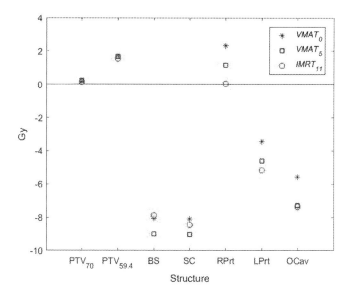

Fig. 4. Comparison of tumor coverage and organ sparing metrics obtained by $VMAT_0$, $VMAT_5$ and $IMRT_{11}$ treatment plans. The horizontal line displayed represents the difference between actual dose metric and the prescribed or tolerance (mean or maximum) dose metric for each structure. Target volume values should be above the line while organs values should be under the line.

5   Conclusions

Radiation oncology has seen considerable changes in the last decades, offering an increased range of treatment modalities to cancer patients. Efforts dedicated to beam angle optimization in IMRT enable the conclusion that optimizing the beam directions will always result in plans that are at least as good as the ones planned by a human, but can be much better for some patients in terms of tumor target coverage and normal tissue sparing. Since performing beam angle optimization will not waste any human resources, it can be assessed as being a valuable tool in clinical routine [22]. More recently, VMAT treatment plan optimization for photon therapy has also been researched, namely deciding the best trajectory of arcs [6].

In this work, treatment plans considering 5-, 7-, 9-, 11- and 21-beam ensembles in the first step (BAO) of a two-step ATO approach are compared to assess the trade-offs between treatment quality and computational time required to obtain a noncoplanar VMAT plan. The different strategies tested obtained similar treatment quality as measured by the optimal FMO value. However, strategies that consider less beams in the first step clearly outperform the other strategies in terms of computational times. Generally, results show that for a similar tumor coverage, treatment plans with optimal beam directions obtained an enhanced organ sparing.

As future work, further test should be performed with more patients and considering different tumor sites in order to validate the conclusions withdrawn using a single head-and-neck cancer case.

Acknowledgments. Support for this work was provided by FCT under project grants UIDB/05037/2020 and UIDB/00308/2020 and by project grant POCI-01-0145-FEDER-028030.

References

1. Cancer tomorrow. http://gco.iarc.fr/tomorrow/home. Accessed 03 Mar 2020
2. Atun, R., et al.: Expanding global access to radiotherapy. Lancet Oncol. **16**, 1153–1186 (2015)
3. Lievens, Y., et al.: Radiation oncology. Optimal health for all together. ESTRO vision, 2030. Radiother. Oncol. **136**, 86–97 (2019)
4. Otto, K.: Volumetric modulated arc therapy: IMRT in a single gantry arc. Med. Phys. **35**, 310–317 (2008)
5. Bangert, M., Ziegenhein, P., Oelfke, U.: Comparison of beam angle selection strategies for intracranial IMRT. Med. Phys. **40**, 011716 (2013)
6. Rocha, H., Dias, J., Ventura, T., Ferreira, B.C., Lopes, M.C.: An optimization approach for noncoplanar intensity-modulated arc therapy trajectories. In: Gervasi, O., et al. (eds.) ICCSA 2018. LNCS, vol. 11621, pp. 199–214. Springer, Cham (2019)
7. Rocha, H., Dias, J., Ventura, T., Ferreira, B.C., Lopes, M.C.: Beam angle optimization in IMRT: are we really optimizing what matters? Int. Trans. Oper. Res. **26**, 908–928 (2019)
8. Wieser, H.-P., et al.: Development of the open-source dose calculation and optimization toolkit matRad. Med. Phys. **44**, 2556–2568 (2017)
9. Cassioli, A., Unkelbach, J.: Aperture shape optimization for IMRT treatment planning. Phys. Med. Biol. **58**, 301–318 (2013)
10. Xia, P., Verhey, L.J.: Multileaf collimator leaf sequencing algorithm for intensity modulated beams with multiple static segments. Med. Phys. **25**, 1424–1434 (1998)
11. Aleman, D.M., Kumar, A., Ahuja, R.K., Romeijn, H.E., Dempsey, J.F.: Neighborhood search approaches to beam orientation optimization in intensity modulated radiation therapy treatment planning. J. Global Optim. **42**, 587–607 (2008)
12. Wächter, A., Biegler, L.T.: On the implementation of an interior-point filter linesearch algorithm for large-scale nonlinear programming. Math. Program. **106**, 25–57 (2006)
13. Rocha, H., Dias, J., Ferreira, B.C., Lopes, M.C.: Selection of intensity modulated radiation therapy treatment beam directions using radial basis functions within a pattern search methods framework. J. Glob. Optim. **57**, 1065–1089 (2013)
14. Rocha, H., Dias, J., Ferreira, B.C., Lopes, M.C.: Beam angle optimization for intensity-modulated radiation therapy using a guided pattern search method. Phys. Med. Biol. **58**, 2939–2953 (2013)
15. Rocha, H., Dias, J., Ferreira, B.C., Lopes, M.C.: Pattern search methods framework for beam angle optimization in radiotherapy design. Appl. Math. Comput. **219**, 10853–10865 (2013)
16. Rocha, H., Dias, J., Ferreira, B.C., Lopes, M.C.: Noncoplanar beam angle optimization in IMRT treatment planning using pattern search methods. J. Phys. Conf. Ser. **616**, 012014 (2015)

17. Rocha, H., Dias, J., Ventura, T., Ferreira, B.C., Lopes, M.C.: A derivative-free multistart framework for an automated noncoplanar beam angle optimization in IMRT. Med. Phys. **43**, 5514–5526 (2016)
18. Alberto, P., Nogueira, F., Rocha, H., Vicente, L.N.: Pattern search methods for user-provided points: application to molecular geometry problems. SIAM J. Optim. **14**, 1216–1236 (2004)
19. Custódio, A.L., Vicente, L.N.: Using sampling and simplex derivatives in pattern search methods. SIAM J. Optim. **18**, 537–555 (2007)
20. Papp, D., Bortfeld, T., Unkelbach, J.: A modular approach to intensity-modulated arc therapy optimization with noncoplanar trajectories. Phys. Med. Biol. **60**, 5179–5198 (2015)
21. Wild, E., Bangert, M., Nill, S., Oelfke, U.: Noncoplanar VMAT for nasopharyngeal tumors: plan quality versus treatment time. Med. Phys. **42**, 2157–2168 (2015)
22. Ventura, T., Rocha, H., Ferreira, B., Dias, J., Lopes, M.C.: Comparison of two beam angular optimization algorithms guided by automated multicriterial IMRT. Phys. Med. **64**, 210–221 (2019)

Locating Emergency Vehicles: Robust Optimization Approaches

José Nelas[1,2] and Joana Dias[1,3(✉)] (iD)

[1] Universidade de Coimbra, CeBER, Faculdade de Economia,
Coimbra, Portugal
joana@fe.uc.pt
[2] Hospital Pediátrico - CHUC, Coimbra, Portugal
[3] Universidade de Coimbra, INESC-Coimbra, Coimbra, Portugal

Abstract. The location of emergency vehicles is crucial for guaranteeing that populations have access to emergency services and that the provided care is adequate. These location decisions can have an important impact on the mortality and morbidity resulting from emergency episodes occurrence. In this work two robust optimization models are described, that explicitly consider the uncertainty that is inherent in these problems, since it is not possible to know in advance how many will be and where will the emergency occurrences take place. These models consider the minimization of the maximum regret and the maximization of the minimum coverage. They are based on a previous work from the same authors, where they develop a model with innovative features like the possibility of vehicle substitution and the explicit consideration of vehicle unavailability by also representing the dispatching of the vehicles. The developed robust stochastic models have been applied to a dataset composed of Monte Carlo simulation scenarios that were generated from the analysis of real data patterns. Computational results are presented and discussed.

Keywords: Emergency vehicles · Location · Optimization · Regret

1 Introduction

The optimal location of emergency vehicles has been largely studied in the past years [1]. Initially, most of the optimization models developed were deterministic, and the uncertainty that is inherent to emergency episodes occurrence was not explicitly considered. Historically, it is usually assigned to Toregas et al. [2] the first publication where a set covering model is used for determining the location of emergency services [3, 4]. The main objective was to guarantee total coverage, minimizing the total number of located services. More recently, other models have been developed, considering in an explicit way uncertainty [5] or survival related features [6, 7].

When considering the location of emergency vehicles, it is important to take into consideration that a vehicle will not be always available to be assigned to an emergency episode. If, for instance, a vehicle is already assigned to an emergency episode that is still taking place, then this vehicle cannot be used for any other episodes simultaneously occurring. One way of dealing with this situation is to consider double coverage.

© Springer Nature Switzerland AG 2020
O. Gervasi et al. (Eds.): ICCSA 2020, LNCS 12251, pp. 564–575, 2020.
https://doi.org/10.1007/978-3-030-58808-3_41

This means that a demand point is considered covered only if there are at least two vehicles located within the maximum defined time limit (see, for instance, [8–10]). Another way of dealing with the possibility of several simultaneous emergency calls is to use queue models. Yoon and Albert [11] present a coverage model with different priorities assigned to emergency occurrences, and considers explicitly the existence of queues. It is also possible to consider a given probability of a vehicle being unavailable [12]. McLay [13] considers two different types of vehicles. The demand is classified into three different priority levels, and congestion is explicitly considered. The objective is to maximize the coverage of the more severe occurrences, to maximize survival possibilities.

The occurrence of emergency episodes is inherently uncertain, and it is important to take this uncertainty explicitly into play when deciding where to locate emergency resources. Sung and Lee [14], for instance, develop a model based on a set of scenarios that represent the uncertainty of emergency calls in different time periods. Bertsimas and Ng [15] present a stochastic formulation, minimizing the non-assisted emergency calls. Beraldi, Bruni and Conforti [16] present a stochastic model using probabilistic restrictions that guarantee a determined level of service. In a following work they explicitly consider the equipment's congestion problem [17].

Ingolfsson et al. [18] maximize coverage considering uncertainty in the means availability and travel times. The uncertainty regarding the response time is also considered in Zhang et al. [19]. Boujemaa et al. [20] consider different scenarios and different levels of assistance.

In a previous work [21], we have developed a coverage stochastic optimization model that brings several new advances compared with the existing models. One of the contributes of this model is the explicit consideration of different vehicle types, with different assistance levels, and the substitutability possibilities between these different types. In a real setting, if one given vehicle type is not available, then a vehicle or set of vehicles of different types are sent. The model also allows for the explicit consideration of emergency episodes that need more than one vehicle. The unavailability of a vehicle, if it is already assigned to an occurring episode, is also taken into account by using an incompatibility matrix that defines whether a given vehicle can or cannot be simultaneously assigned to two different emergency episodes, taking into account the time periods of occurrence of these episodes. This model is based on coverage location models, and it maximizes the expected total number of episodes covered. An episode is considered covered if it receives all the needed vehicles (or adequate substitutes) and if they all arrive within a defined time window. The consideration of an objective function of this type does not accommodate the point of view of robustness, that is of great importance in the field of emergency assistance. Actually, when looking at an expected coverage value, it is possible that we are not taking into account what is happening in some of the most complicated scenarios, that can end up having low levels of coverage even if the average value is acceptable. A robust solution will be found looking at other objectives that can better represent the idea that the system must respond as well as possible even in more challenging scenarios.

In this work we present two models where robust location decisions are considered. One of the models maximizes the total number of covered episodes in the worst case.

The other model minimizes the maximum regret. These models were applied to a set of scenarios, that were built based on real data, and using Monte Carlo simulation.

This manuscript is organized as follows: in the next section the characteristics of the emergency vehicles location problem that we have considered are briefly described. Section 3 describes the main features of the mathematical model developed. Section 4 presents the results obtained by applying the two robust approaches, and these results are then discussed in Sect. 5. Section 6 presents the conclusions and some paths for future work.

2   Description of the Problem

In this problem we consider a given geographical area, with already existing bases where emergency vehicles can be located. These are the potential locations that can be chosen. There is a fixed fleet of emergency vehicles. We have to decide on which base to locate each one of the existing vehicles. Emergency occurrences can occur anywhere within the geographical area. An emergency episode is considered covered if and only if it receives all the needed vehicles within an acceptable time window.

The geographical area of interest was the Coimbra district, located at the central part of Portugal. At the present moment there are 35 vehicles available that are distributed among 34 existing bases. We also consider this number of vehicles and their corresponding types, and the already existing bases.

The uncertainty that is inherent to the occurrence of emergency episodes is represented by a set of scenarios. The generated scenarios are meant to represent the characteristics of real data. The real data that was used includes information about all the emergency occurrences that required the assignment of advanced life support vehicles during one year, and data published by the National Medical Emergency Institute INEM for the last 3 years, that allowed the estimation of the total number of occurrences for the same period.

The model considers the dispatching decisions also: when an emergency episode occurs, the model will decide which are the vehicles that are sent to that call. Including these dispatching decisions is very important in this kind of model since it will allow a more realistic representation of the unavailability of vehicles when they are already assigned to other emergency calls.

The objective will be to find a robust solution: the location of the vehicles should guarantee the best possible coverage under different scenarios.

2.1   Emergency Vehicles

We consider the existence of five different types of emergency vehicles. Medical Emergency Motorcycles (MEM) are vehicles that have an emergency technician. They are able to provide Basic Life Support (BLS) and are equipped with external automatic defibrillation. They cannot transport victims when they need to receive further assistance in health institutions. Assistance Ambulances (AA) are usually drove by volunteers or professionals with specific training in prehospital emergency techniques. They provide BLS only but have transportation capacity. Medical Emergency

Ambulances (MEA) have, usually, two emergency technicians and have transportation capacity. Immediate Life Support (ILS) Ambulances (ILSA) guarantee more differentiated health care than the previous means, such as resuscitation maneuvers. The crew consists of a nurse and a pre-hospital emergency technician, providing Advanced Life Support (ALS). Medical Emergency and Resuscitation Vehicles (MERV) have a crew composed by a nurse and a medical doctor with competence and equipment for ALS. They do not have the ability to transport victims, and are many times used in conjunction with other vehicles with this capacity.

Considering the type of care that each one of these vehicle types can provide (namely BLS and ALS, and also taking into account the composition of the team), and whether they are or are not able to transport victims, it is possible to define substitutability possibilities between these vehicles. Table 1 presents this information. Each vehicle type presented in a given row can be directly substituted by one single vehicle of another type represented in the columns whenever a 1 is in the corresponding cell.

Table 1. Substitutability between different vehicle types

Vehicle type	AA	MEM	MEA	ILSA	MERV
AA	1	0	1	1	0
MEM	1	1	1	1	0
MEA	1	0	1	1	0
ILSA	0	0	0	1	0
MERV	0	0	0	0	1

There are also other possible substitutions, where a single vehicle can be substitute by more than one vehicle. A vehicle of type ILSA, for instance, can also be substituted by sending simultaneously an AA vehicle and a MERV or a MEM vehicle. These more intricated substitutions are also considered in the optimization model.

2.2 Emergency Episodes

Each emergency episode is characterized by the place where it occurs, the number of vehicles of each type that it will need, the time period in which it takes place (from the moment it begins until the moment when the vehicles that were assigned to it are ready to be assigned to other occurrences).

Each scenario will consider all the episodes that occur in a 24-h period. The generation of these scenarios took into consideration the information gathered from real data.

The region of interest considered in this work presents different emergency occurrence patterns in different subregions, either because these subregions correspond to urban centres or because they are traversed by heavy traffic roads, for instance. These differences were also taken into account, by splitting the region into different polygons with different characteristics considering the number and severity of the emergency occurrences (Fig. 1).

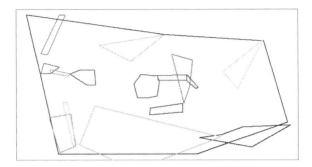

Fig. 1. Definition of different areas with different patterns regarding the occurrence of emergency episodes. Each polygon defines a subregion with specific patterns regarding the occurrence of emergency episodes.

To generate the emergency episodes that constitute one given scenario, the number of episodes that will occur is randomly generated, considering a Poisson distribution. Each episode is then allocated to one of the defined subregions, considering the emergency episodes occurrence probabilities for each one. Within this subregion, a particular location is randomly generated (with latitude and longitude coordinates). Then, the number and types of vehicles that will be needed for that episode are also randomly generated. The road driving times between each one of the existing bases and the location of the emergency episodes are calculated using Google Maps API. This allows the creation of a binary coverage matrix, considering a defined time limit. The time limits were defined as 30 min for rural areas and 15 min for urban centres. This matrix will dictate which vehicles are within the coverage radius of each emergency episode. The occurrence time period associated with each episode is also randomly generated (from the moment the episode begins to the moment where all the assigned vehicles are already operational and can assist other episodes). More detailed information regarding the generation of scenarios can be found in [21].

3   Modelling the Problem

The mathematical model is a two-stage stochastic model, with uncertainty included in the model using a set of different scenarios There will be decisions that are scenario independent, and others that are scenario dependent. The location variables will not be scenario dependent (these decisions are first level decisions, since they must be made without knowing what the future will bring). Variables related to the dispatching of vehicles to emergency episodes will depend on each particular scenario.

The mathematical model has to consider a very large number of constraints, so that it represents, as faithfully as possible, the real context of emergency vehicles dispatching. The constraints considered guarantee that:

- An episode belonging to a given scenario is considered "covered" if and only if it receives all the necessary vehicles of all the needed types in that scenario.

- An emergency vehicle can only be assigned to an episode if this episode is within the coverage radius of the vehicle, and it is available when the episode occurs.
- An emergency vehicle can only be assigned to two different episodes if their occurrence time periods do not intersect.
- It is not possible to anticipate the future in each scenario, considering the assignment decisions. This means that the dispatching decisions will be made based on the emergency occurrences that have already taken place, not considering the ones that will happen in the future. The inclusion of non-anticipative constraints is very important to guarantee the least biased results in terms of coverage.
- There is a maximum number of vehicles that can be located at each base, and vehicles can only be assigned to a base that has been prepared.
- Each emergency vehicle is located in one, and only one base.

The implementation of these conceptual restrictions gives rise to a high-dimensional mathematical programming problem. Details can be found in [21]. The developed model considered an objective function of maximization of the expected total number of covered episodes.

In the emergency location setting it can be better to look for robust solutions other than looking for the best expected coverage values. There are many interpretations for the concept of robustness. It is usually related with finding solutions that perform well under different scenarios. Performing well can be related with admissibility (finding solutions that are admissible under most of the scenarios) or with optimality (finding solutions that are near-optimal under most scenarios). In the current setting, the defined restrictions have to be always fulfilled, because they guarantee that the calculated solution makes sense. The robustness, in the current context, has to do with optimality.

Many stochastic robust models consider an objective function of the type max-min: they are concerned with finding the solution that is the optimal one for the worst case possible. In this case we would like to locate the emergency vehicles in order to maximize the coverage in the scenario where coverage is the worst among all the scenarios. The drawback of this approach is that, sometimes, this point of view is too conservative. Actually, the worst scenario is usually not the one with the highest probability of occurring and looking only at this scenario can ignore solutions that could behave better for the majority of the future situations.

This drawback can be overcome if another measure of robustness is considered, namely the minimization of the maximum regret. Regret has already been used in the location optimization field [22, 23]. It measures how much we are losing for not being capable of predicting the future and choosing the optimal solution for what will occur. We must make a decision without being able to anticipate what the future will be. This means that the chosen solution will not, with a very high probability, be the best one for that particular situation.

If we have a set of different scenarios, we can calculate the best solution for each one. This will give us the maximum possible coverage for that scenario. Then we want to find the solution that minimizes the maximum regret: our greatest loss for not doing what would be the best thing if we could anticipate the future.

These two approaches were considered and the model presented in [21] was changed accordingly. As the max-min coverage and min-max regret objective functions are non-linear, it is necessary to linearize them using auxiliary variables and constraints, increasing the dimension of the problem.

In the next section we present some computational results and discuss the obtained results.

4   Computational Experiments

We have solved different model instances. First, we have fixed the vehicles' locations to their current locations to see how the current solution behaves under the two different objective functions. Then we have solved both the min-max regret and the max-min coverage problems and we have compared the obtained solutions. Each instance considered 30 scenarios, corresponding to 1978 emergency occurrences in total. The instances were solved by Cplex, version 12.7, using Intel Xeon Silver 4116, 2.1 GHz, 12-core processor, 128 GB RAM. These instances took around 3 h of computational time. When the location decisions are fixed, the computational time is of few minutes only.

Table 2 presents the current vehicles' locations and the ones calculated by the two different modeling approaches. Table 3 presents the coverage results obtained.

Using the same scenario set to calculate the optimal vehicles location and to assess their true adequacy to the real setting can be misleading. Actually, we are comparing solutions that are definitely the best, for that particular objective function and for that particular set of scenarios. But what we really want to assess is their ability to perform well under different and unseen situations. This is why we have generated an extra set of 15 scenarios, with a total of 951 emergency episodes, and tested all the solutions under this unseen set. Table 4 presents the results for this out-of-sample set. Table 3 and Table 4 present the expected coverage, the worst coverage and the worst regret for four different solutions: the current one (where we have fixed all the vehicles' location variables in their current positions), the max-min coverage, the min-max regret and the maximization of the expected coverage (these last results consider the original model applied in [21]).

Looking at the differences on the vehicles' locations between the current and the calculated solutions, and knowing the characteristics of each of the considered regions, it is possible to observe that some vehicles were relocated to areas with a higher population density, with higher rates of emergency occurrences. These changes try to correct some situations were high level of occurrences were not being properly covered. There are also some changes considering the location of vehicles with transportation capacity, positioning them in bases where they can be more useful.

These changes increase the response capacity in both the 30 scenario and 15 scenario sets. The current locations present worse results than the coverage results presented by the other three alternative solutions.

Table 2. Current vehicles' locations and the ones calculated by the two different modeling approaches

Locations	Current locations					Minimizing the maximum regret					Max the minimum coverage				
	AA	MEM	MEA	ILSA	MERV	AA	MEM	MEA	ILSA	MERV	AA	MEM	MEA	ILSA	MERV
Base 1	1	0	0	0	0	0	0	1	0	0	2	0	1	0	0
Base 2	1	0	0	0	0	0	1	0	1	0	1	0	0	0	0
Base 3	1	0	0	0	0	1	0	0	0	0	0	0	1	0	0
Base 4	0	0	0	0	1	1	0	0	0	0	1	0	0	0	1
Base 5	0	0	0	0	1	1	0	0	0	1	1	0	0	0	0
Base 6	0	0	0	0	0	1	0	0	0	0	1	0	0	0	0
Base 7	0	0	0	0	0	0	0	0	0	0	0	0	0	0	0
Base 8	0	1	3	0	0	0	0	1	0	1	1	0	0	0	1
Base 9	1	0	0	0	0	0	0	0	0	0	0	1	0	0	0
Base 10	1	0	0	0	0	1	0	0	0	0	1	0	0	0	0
Base 11	1	0	0	0	0	0	0	0	1	0	0	0	0	0	0
Base 12	0	0	0	0	0	1	0	0	0	0	0	0	0	0	0
Base 13	1	0	0	0	0	1	0	0	0	0	0	0	0	0	0
Base 14	0	0	0	1	1	1	0	0	0	1	0	0	1	0	1
Base 15	1	0	0	0	0	0	0	1	0	0	0	0	0	0	0
Base 16	0	0	0	0	0	1	0	0	0	0	1	0	0	0	0
Base 17	0	0	0	0	0	1	0	0	0	0	1	0	0	0	0
Base 18	0	0	0	0	0	0	0	0	0	0	0	0	0	0	0
Base 19	1	0	0	0	0	1	0	0	0	0	1	0	0	0	0
Base 20	1	0	0	0	0	1	0	0	0	0	1	0	0	1	0
Base 21	1	0	0	0	0	1	0	0	0	0	1	0	0	0	0
Base 22	1	0	0	0	0	1	0	0	0	0	1	0	0	0	0
Base 23	1	0	0	0	0	1	0	0	0	0	1	0	0	0	0
Base 24	1	0	0	0	0	1	0	0	0	0	0	0	0	0	0
Base 25	1	0	0	0	0	0	0	1	0	0	1	0	0	0	0
Base 26	1	0	0	0	0	1	0	0	0	0	1	0	0	0	0
Base 27	1	0	0	0	0	1	0	0	0	0	1	0	0	0	0
Base 28	1	0	0	0	0	0	0	0	0	0	1	0	0	0	0
Base 29	1	0	0	0	0	0	0	0	0	0	1	0	0	0	0
Base 30	1	0	0	0	0	1	0	0	0	0	1	0	0	0	0
Base 31	1	0	0	0	0	1	0	0	0	0	0	0	0	1	0
Base 32	1	0	0	0	0	1	0	0	0	0	1	0	0	0	0
Base 33	1	0	0	0	0	1	0	0	0	0	1	0	0	0	0
Base 34	1	0	1	1	1	2	0	0	0	1	1	0	1	0	1

Table 3. Summary of the results considering the 30 scenarios set

	Average coverage	Worst case coverage	Maximum regret
Current location	84.20%	69.60%	10
Max min coverage	84.70%	83.10%	10
Min max regret	87.70%	78.80%	3
Expected coverage	89.50%	78.79%	4

Table 4. Summary of the results considering the 15 scenarios set

	Average coverage	Worst case coverage	Maximum regret
Current location	84.40%	78.33%	6
Max min coverage	87.42%	80.85%	6
Min max regret	88.52%	81.67%	6
Expected coverage	87.95%	81.67%	6

As expected, in the 30-scenario set, each solution is the best one among the four considering the specific objective function that was used in the calculation of the respective solution. The best value for the maximum regret is obtained with the min-max regret solution, the best expected coverage is obtained with the optimal solution of this model, and so on. Looking at the values in Table 3, it is difficult to decide on what solution to choose. Should we consider as most important the expected coverage, or is it more important to guarantee that we are better off in the worst situation? This question has not got a simple answer. But since each of these solutions is being tested with the same data that was used for the calculation of the optimal solution of each of the objective functions, this analysis can be biased.

Looking at Table 4, it is possible to conclude that one solution stands out as the best one when compared with the others. The solution that minimizes the maximum regret obtains the best average coverage and also the best coverage in the worst case. The maximum regret is the same for all the solutions. Assessing the solutions in these independent scenario set allows for an unbiased analysis that gives a better idea of how the solutions would behave in the real setting.

5   Discussion

When looking at the implementation of changes in the vehicle's locations, there are several issues that have to be taken into consideration and that can make these decisions hard to put into practice.

One of these barriers is the fact that different vehicles belong to different institutions (volunteer firemen, red cross, national emergency institute, and so on). Some institutions will not feel comfortable in letting their vehicles being positioned somewhere else. This is particularly true for vehicles of the AA type, that are usually located at firemen departments. This observation motivated us to analyse the impact of restricting the locations of these vehicles to their current positions. We have run all of the models, fixing the locations of all the AA vehicles. Table 5 presents the obtained results for the 30-scenarios set and Table 6 for the out-of-sample scenario set.

Table 5. Summary of the results considering the 30 scenarios set and fixing the location of all the AA vehicles to their current positions

	Average coverage	Worst case coverage	Maximum regret
Current location	84.20%	69.60%	10
Max min coverage	83.59%	82.76%	10
Min max regret	86.20%	77.27%	4
Expected coverage	88.90%	78.79%	5

Table 6. Summary of the results considering the 15 scenarios set and fixing the location of all the AA vehicles to their current positions

	Average coverage	Worst case coverage	Maximum regret
Current location	84.40%	78.33%	6
Max min coverage	84.34%	83.33%	9
Min max regret	69.31%	57.58%	21
Expected coverage	89.28%	83.33%	5

It is interesting to observe the behaviour of the min-max regret solution in the out-of-sample set. This solution is clearly more dependent on particular changes that are done in the location of some vehicles. Not being able to go through with those needed changes, namely fixing the location of all the AA vehicles, makes the solution present very bad coverage metrics in the out-of-sample set. The model that seems to behave better when all the AA vehicles' locations are fixed is the maximization of the expected coverage model. The results in out-of-sample, for this model where we are considering these AA vehicles fixed, are even better than the solution where these restrictions were not imposed. Actually, the location of AA vehicles is the result of a long time-period adaptation of different institutions to the specificities of their regions. These locations may implicitly consider an empirical knowledge about the emergency occurrences characteristics. However, the results are much better than the ones obtained with the current solution, meaning that the change in the location of some vehicles can have an important impact in terms of coverage.

These results motivate further studies, namely considering enlarged scenario sets.

Another issue that is not being considered in these models has to do with the episodes that are not being covered. These episodes will not receive any emergency vehicles. This would never happen in a real setting. In a real situation all emergency episodes will receive assistance, but probably later than they should or resorting to less adequate vehicles. It is possible that the impact of this limitation is not very significant, since the model is not also accounting for vehicles that are positioned in the boundaries of the region of interest, and that can cover emergency occurrences that take place in the frontier of this region. However, it will be important to take these features explicitly into account in future works.

The placement of vehicles nearer some health institutions can also increase the affluence of victims to these institutions, that may not be prepared to this increased demand. This can jeopardize the quality of care, or increase the costs in order to cope with this increase.

The generation of scenarios did not take into consideration differences in the emergency occurrence patterns in different days of the week, or different times of the year. The models do not also consider the possibility of dynamically changing the vehicles' location during the day. This is a possibility that could increase coverage, since the number and location of occurrences have different patterns during the day. However, such dynamic changes are usually not well accepted by the emergency technicians and medical doctors, so these solutions are even harder to put into practice.

6 Conclusions

In this work we extend the model presented in [21], by considering two different objective functions that allow the calculation of robust solutions: one model considers the maximization of the worst case coverage; the other considers the minimization of the maximum regret.

Assessing the different solutions considering an unseen set of 15 scenarios, it is possible to conclude that the solution that minimizes the maximum regret seems to be the best one under all the different criteria. From a decision-making point of view, it is also a decision that is supported by the reasoning of being as close as possible to the optimal solution in each situation using the available resources. However, when we fix all the AA type vehicles to their current locations, this is the model that suffers the most in terms of worsening the coverage metrics.

The presented models suffer from some limitations that have been identified. These limitations motivate the pursuing of new developments. It is important, for instance, to consider the possibility of partial coverage: a situation where an emergency occurrence does not receive the most adequate resources and/or they do not arrive within the defined time window. These situations are not being considered, but they should be as they consume resources that cannot be assigned to other occurrences and can thus somehow bias the coverage results.

We are not considering the possibility of acquiring new vehicles, or preparing new potential locations to receive emergency vehicles. This is also an interesting line of research.

Acknowledgments. This study has been funded by national funds, through FCT, Portuguese Science Foundation, under project UIDB/00308/2020 and with the collaboration of Coimbra Pediatric Hospital – Coimbra Hospital and University Center.

References

1. Brotcorne, L., Laporte, G., Semet, F.: Ambulance location and relocation models. Eur. J. Oper. Res. **147**(3), 451–463 (2003)
2. Toregas, C., Swain, R., ReVelle, C., Bergman, L.: The location of emergency service facilities. Oper. Res. **19**(6), 1363–1373 (1971)

3. Daskin, M.S., Dean, L.K.: Location of health care facilities. In: Brandeau, M.L., Sainfort, F., Pierskalla, W.P. (eds.) Operations Research and Health Care. International Series in Operations Research & Management Science, vol. 70, pp. 43–76. Springer, Boston (2005)

4. Li, X., Zhao, Z., Zhu, X., Wyatt, T.: Covering models and optimization techniques for emergency response facility location and planning: a review. Math. Meth. Oper. Res. **74**(3), 281–310 (2011)

5. Zhou, Y., Liu, J., Zhang, Y., Gan, X.: A multi-objective evolutionary algorithm for multi-period dynamic emergency resource scheduling problems. Transp. Res. Part E Logist. Transp. Rev. **99**, 77–95 (2017)

6. Knight, V.A., Harper, P.R., Smith, L.: Ambulance allocation for maximal survival with heterogeneous outcome measures. Omega **40**(6), 918–926 (2012)

7. Leknes, H., Aartun, E.S., Andersson, H., Christiansen, M., Granberg, T.A.: Strategic ambulance location for heterogeneous regions. Eur. J. Oper. Res. **260**(1), 122–133 (2017)

8. Moeini, M., Jemai, Z., Sahin, E.: Location and relocation problems in the context of the emergency medical service systems: a case study. CEJOR **23**(3), 641–658 (2014). https://doi.org/10.1007/s10100-014-0374-3

9. Moeini, M., Jemai, Z., Sahin, E.: An integer programming model for the dynamic location and relocation of emergency vehicles: a case study. In: International Symposium on Operational Research in Slovenia, SOR, pp. 343–350 (2013)

10. Pulver, A., Wei, R.: Optimizing the spatial location of medical drones. Appl. Geogr. **90**, 9–16 (2018)

11. Yoon, S., Albert, L.A.: An expected coverage model with a cutoff priority queue. Health Care Manage. Sci. **21**(4), 517–533 (2017). https://doi.org/10.1007/s10729-017-9409-3

12. Daskin, M.S.: A maximum expected covering location model: formulation, properties and heuristic solution. Transp. Sci. **17**(1), 48–70 (1983)

13. McLay, L.A.: A maximum expected covering location model with two types of servers. IIE Trans. **41**(8), 730–741 (2009)

14. Sung, I., Lee, T.: Scenario-based approach for the ambulance location problem with stochastic call arrivals under a dispatching policy. Flex. Serv. Manuf. J. **30**, 1–18 (2016). https://doi.org/10.1007/s10696-016-9271-5

15. Bertsimas, D., Ng, Y.: Robust and stochastic formulations for ambulance deployment and dispatch. Eur. J. Oper. Res. **279**(2), 557–571 (2019)

16. Beraldi, P., Bruni, M.E., Conforti, D.: Designing robust emergency medical service via stochastic programming. Eur. J. Oper. Res. **158**(1), 183–193 (2004)

17. Beraldi, P., Bruni, M.E.: A probabilistic model applied to emergency service vehicle location. Eur. J. Oper. Res. **196**(1), 323–331 (2009)

18. Ingolfsson, A., Budge, S., Erkut, E.: Optimal ambulance location with random delays and travel times. Health Care Manage. Sci. **11**(3), 262–274 (2008)

19. Zhang, B., Peng, J., Li, S.: Covering location problem of emergency service facilities in an uncertain environment. Appl. Math. Model. **51**, 429–447 (2017)

20. Boujemaa, R., Jebali, A., Hammami, S., Ruiz, A., Bouchriha, H.: A stochastic approach for designing two-tiered emergency medical service systems. Flex. Serv. Manuf. J. **30**(1), 123–152 (2017). https://doi.org/10.1007/s10696-017-9286-6

21. Nelas, J., Dias, J.: Optimal emergency vehicles location: an approach considering the hierarchy and substitutability of resources. Eur. J. Oper. Res. **287**, 583–599 (2020)

22. Marques, M.C., Dias, J.: Dynamic location problem under uncertainty with a regret-based measure of robustness. Int. Trans. Oper. Res. **25**, 1361–1381 (2015)

23. Dias, J.M., do Céu Marques, M.: A multiobjective approach for a dynamic simple plant location problem under uncertainty. In: Murgante, B., et al. (eds.) ICCSA 2014. LNCS, vol. 8580, pp. 60–75. Springer, Cham (2014). https://doi.org/10.1007/978-3-319-09129-7_5

Machine Learning for Customer Churn Prediction in Retail Banking

Joana Dias[1,2] (ID), Pedro Godinho[2(✉)] (ID), and Pedro Torres[2] (ID)

[1] INESC-Coimbra, University of Coimbra, Coimbra, Portugal
joana@fe.uc.pt
[2] CeBER, Faculty of Economics, University of Coimbra, Coimbra, Portugal
{joana,pgodinho}@fe.uc.pt, pedro.torres@uc.pt

Abstract. Predicting in advance whether a given customer will end his relationship with a company has an undeniable added value for all organizations, since targeted campaigns can be prepared to promote customer retention. In this work, six different methods using machine learning have been investigated on the retail banking customer churn prediction problem, considering predictions up to 6 months in advance. Different approaches are tested and compared using real data. Out of sample results are very good, even with very challenging out-of-sample sets composed only of churners, that truly test the ability to predict when a customer will churn. The best results are obtained by stochastic boosting, and the most important variables for predicting churn in a 1–2 months horizon are the total value of bank products held in recent months and the existence of debit or credit cards in another bank. For a 3–4 months horizon, the number of transactions in recent months and the existence of a mortgage loan outside the bank are the most important variables.

Keywords: Churn · Retail banking · Machine learning · Validation

1 Introduction

Customer churn prediction is an important component of customer relationship management since it is less profitable to attract new customers than to prevent customers to abandon the company [1, 2]. Companies that establish long term relationships with their customers can focus on customers' requirements, which is more profitable than looking for new customers [3]. This may also enable companies to reduce service costs [2] and enhances opportunities to cross and up sale [1]. Long-term customers are also less likely to be influenced by competitors' marketing campaigns [4], and tend to buy more and to spread positive word-of-mouth [2]. Customers that abandon the company may influence others to do the same [5]. Actually, customers that churn because of "social contagion" may be harder to retain [6]. Thus, building predictive models that enable the identification of customers showing high propensity to abandon are of crucial importance, since they can provide guidance for designing campaigns aiming to persuade customers to stay [7].

Machine learning (ML) techniques have been used for churn prediction in several domains. For an overview of the literature after 2011 see [1, 7]. Few publications

© Springer Nature Switzerland AG 2020
O. Gervasi et al. (Eds.): ICCSA 2020, LNCS 12251, pp. 576–589, 2020.
https://doi.org/10.1007/978-3-030-58808-3_42

consider churn prediction in the financial sector or retail banking. In the work presented in [8], only 6 papers considered the financial sector. In another literature analysis focusing on applications of business intelligence in banking, the authors conclude that credit banking is the main application trend [9]. Words like "retention" or "customer relationship management" are not the most relevant term frequencies found.

Most of the referenced work consider applications in the telecommunication sector (see, for instance, [10–15]). Applications of churn prediction methods in other sectors can also be found. Coussement et al. [16] consider churn prediction in a newspaper publishing company. Miguéis et al. [17] consider the retailing context. Buckinx and Van den Poel [18] predict partial defection in fast-moving consumer goods retailing. Predicting churn in the logistics industry is considered in [19].

As aforementioned, research focusing on retail banking customer churn prediction has been scarce. The prediction of churn in this context has some differentiating features. There is a strong competitive market and a single customer can have several different retail banking service providers at the same time, which offer more or less the same type of products with similar costs. The relationship between the customer and the service provider is, in general, of the non-contractual type, meaning that the customer has control over the future duration and type of the relationship with the company. The customer has control over the services he wants to acquire. A customer can, most of the time, leave the bank without informing it of this intention, making it harder to distinguish between churners or simply inactive customers.

In a non-contractual setting, it is often difficult to know what is the proper churn definition that should be considered and different definitions can have different economic impacts for the companies, namely considering the cost and impact of marketing campaigns against churn [20]. In the retail banking sector, the situation is even more complicated, since the cost of terminating the relationship with a service provider can be very different for different types of customers: it is negligible for many, but not for all. Customers that have signed loans for home purchase, for instance, can incur in significant costs for terminating their relationship with the bank, and this relationship is much more similar to a contractual type one than for other types of customers.

Churn prediction in the banking sector has been addressed not only by machine learning approaches but also by survival analysis models. Mavri and Ioannou [21] use a proportional hazard model to determine the risk of churn behavior, considering data from Greek banking. Their objective is not to predict whether a specific customer will churn or not, but rather to understand what features influence the switching behavior of bank customers. Survival analysis is also used by Larivière and Van den Poel [22], who consider Kaplan-Meier estimates to determine the timing of churn. The authors use data from a large Belgian financial service provider. The authors also study the influence of product cross-selling on the customer propensity to churn. The same authors apply machine learning approaches, namely random forests, to the same dataset and compare its performance with logistic regression for customer retention and profitability prediction. They conclude that the best results are achieved by random forests [23].

Glady et al. [24] propose the use of customer lifetime value as a measure for classifying churners in the context of a retail financial service company. A churner is defined as a customer with a decreasing customer lifetime value. The authors develop a new loss function for misclassification of customers, based in the customer lifetime value. The authors apply decision trees, neural networks and logistic regression models as classifiers. The predictor variables used are of the type RFM (recency, frequency and monetary). Xie et al. [25] apply random forests to churn prediction considering banks in China, integrating sampling techniques and cost-sensitive learning, penalizing more the misclassification of the minority class and achieving very high accuracy rates. De Bock and Van den Poel [26] develop a model able to predict partial churn for a European bank (the closing of a checking account by a customer, not taking into consideration whether that customer has other accounts or services in the bank) in the next 12 months. The authors conclude that the most important features influencing churn prediction are related to RFM variables. Verbeke et al. [7] consider the importance of the comprehensibility and justifiability of churn prediction models, and use AntMiner+ and Active Learning Based Approach for Support Vector Machine rule extraction. The concern with interpretability is also shared by [26], that use an ensemble classifier based upon generalized additive models and apply it to a set of six datasets. The model interpretability is illustrated by considering a case study with a European bank. De Bock and Poel [27] apply two different forest-based models, Rotation Forest and RotBoost, to four of these six real-life datasets. Gür Ali and Arıtürk [8] describe a dynamic churn prediction framework in the context of private banking including environmental variables in addition to customers' behavior attributes. Churn prediction in a defined future time window is tackled using binary classifiers, and the performance is compared with Cox Regression. Shirazi and Mohammadin [28] focus on the customer retiree segment of the Canadian retail banking market. They develop a model that considers not only data directly related with the customers' interactions with the bank but also with the customers' behavior (tracking of customers' behavior on various websites, for instance).

Churn prediction for bank credit card customers is addressed by several authors. Farquad et al. [29] apply support vector machine in conjunction with Naïve Bayes Trees for rule generation. This hybrid approach outperforms other approaches in the computational tests performed using a Latin American bank dataset. Lin et al. [30] use rough set theory to extract rules that explain customer churn. Nie et al. [31] apply logistic regression and decision trees to a dataset from a Chinese bank, reaching the conclusion that logistic regression slightly outperforms decision trees.

In this work, six machine learning techniques are investigated and compared to predict churn considering real data from a retail bank. Individual results obtained by each methodology are also compared with the results obtained by applying survival analysis tools. The objective is to develop a methodological framework capable of predicting not only which customers will churn but also when they will churn, considering a future horizon of six months. The models aim at predicting which will be the customers churning in the next month, two months from now, and so on.

By employing different machine learning tools and evaluating the models using a large retail banking dataset, this study makes several contributions to the literature:

- It is focused on retail banking, an industry in which research has been scarce, and shows how retail banking data can be leveraged by structuring data in new ways.
- It defines of a global framework for churn prediction that not only identifies the customers that are more likely to churn, but also predicts when they are going to

churn, allowing a continuous reassessment of the churn probabilities for different time horizons. It can therefore be a useful tool for supporting the management of resources and scheduling of campaigns in retail banks.

- It proposes a new validation methodology that truly assesses the prediction accuracy, in a highly-biased dataset. This methodology considers the use of balanced out-of-sample sets that are only composed of churners, with different churning times.
- It presents computational results using real data, showing that it is possible to achieve high levels of predictive accuracy. This means that the model is useful in a real-life context.

Although a large dataset is used, the time horizon of the sample dataset (two years) is an important constraint. This limitation was overcome by creating different records based on different rolling time windows, but a dataset covering a longer time period might provide more insights.

This paper is organized as follows: Sect. 2 describes the methodology applied. Section 3 presents the main computational results. Section 4 presents a discussion of the results and some ideas for future research.

2   Exploratory Data Analysis

The data used in this study considers information related to more than 130 000 customers of a retail bank, including all the monthly customer interactions with the bank for two years: product acquisition, product balances, use of bank services, in a total of 63 attributes for each month. All the data is anonymized, preventing the identification of customers. Personal data considers age, location, marital status, employment situation, date of bank account opening and the way in which the customer opened the account (using the bank online platform, in a bank branch or any other way). Considering the large number of attributes characterizing the customers' interactions with the bank, some of these attributes were consolidated. The final set of attributes considered were: total value of bank products held by the customer, total value of personal loans, total value of mortgage loans, number of insurance policies held, total number of transactions of any kind (debit or credit cards, bank transfers, deposits, etc.), binary values stating whether the customer had a mortgage loan or personal loan in another bank, and whether he has debit or credit cards in another bank.

Regarding the sociodemographic characterization of customers, most of them are single, followed by those that are married, as shown in Fig. 1. Most customers were born between 1980 and 1990 (Fig. 2).

Most of the customers opened an account using a bank branch (Fig. 3), but the oldest ones have mostly opened their accounts using the web platform (Fig. 4).

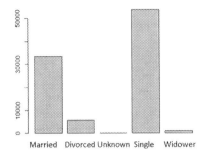

Fig. 1. Distribution of customers by marital status

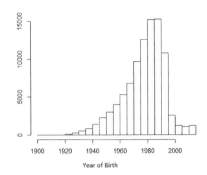

Fig. 2. Distribution of customers by year of birth

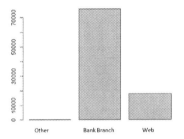

Fig. 3. Distribution of customers by acquisition channel

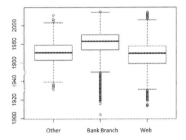

Fig. 4. Distribution of customers by acquisition channel and year of birth

Less than 1% of the customers have churned during the time horizon corresponding to the available data. In order to better understand what characterizes a churner, a survival analysis was performed, using Cox Regression. The acquisition channel, marital status, age, level of academic education and number of descendants showed to be statistically significant attributes ($p < 0.01$) in explaining churn. Survival curves were plotted considering different customer related attributes. Figures 5 and 6 illustrate two of those survival plots. It was possible to conclude that customers that begin their relationship with the bank using the web are less prone to churn. Customers that are single present a higher probability of churning in earlier stages of their involvement with the bank. Older customers are less prone to churn. Regarding the academic education, less educated customers are the ones that churn earlier. These results are fully aligned with other similar results found in the literature. The results in [32], for instance, show that older people are more likely to stay with the bank, and that more educated people are more likely to be loyal.

All the attributes were pre-processed in the following way: all quantitative attributes were centred and scaled and the natural logarithm was applied, due to the skewness that the histograms of most attributes presented.

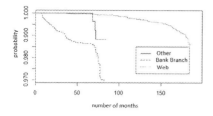

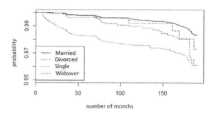

Fig. 5. Survival curves considering acquisition channel

Fig. 6. Survival curve considering marital status

3 Prediction Models and Methodology

The bank has its own definition of churn: it wants to be able to predict customers that fall below a certain threshold in terms of the relationship with the bank: a customer is a churner if he does not interact in any way with the bank for 6 consecutive months, the balance of assets in the bank is smaller than or equal to 25 € and the balance of debts is also smaller than or equal to 25 €. It is considered that the time of churn is the first time these conditions are simultaneously met.

The problem of predicting up to six months in advance the customers that will churn is treated as a classification problem with two classes (churners and non-churners). The dataset is highly unbalanced since the minority class (churners) constitutes less than 1% of the total dataset.

The objective is to predict which customers will churn and when they will churn, considering a future time horizon of 6 months. Therefore, six different models are built. One model will predict the customers that will churn in the next month. Another model will predict which customers will churn two months from now, and so on. These predictions will be based on the data corresponding to the past six months, along with the known customer attributes.

To train the machine learning models, different datasets considering rolling time windows dependent on the prediction horizon were created. For example, consider that the model will make predictions for the next month. As there are 24 months of data available, the first time window will consider the initial seven months of data: the first six months will originate the values of the explanatory variables (attributes). The information of whether a customer has churned or not in the seventh month will be the output variable (to be predicted). Then this time window is rolled one time period forward. Next, all records with explanatory variables considering months 2 to 7 and the output variable calculated for month 8 are added to the dataset. This is performed until there are no more data available. In the case of one month ahead prediction, the last time window includes months 18–24. Building the datasets in this way means that a given customer will contribute more than one record for the global dataset. Actually, a customer that will churn at a given month will contribute with records classified as "non-churner" until the churn actually occurs, and one record classified as "churner" corresponding to the month in which he churns.

Figures 7 and 8 illustrate the use of these rolling time windows for predictions made one and two months ahead, respectively.

Due to the highly-unbalanced feature of the data, it is important to balance the dataset before feeding the data into the machine learning models. An undersampling strategy was chosen. All the "churner" samples were considered and a subset of the "non-churner" samples, composed by a number of samples that matched the "churner" ones, was randomly selected. Burez and Van den Poel [33] study the problem of class imbalance that is often present in churn prediction datasets, since the churners are usually the minority class. The authors state that undersampling can be useful, improving prediction accuracy, and they find no advantages in using advanced sampling techniques. Other authors also state that undersampling seems to present advantages when compared with oversampling [34, 35].

To use the churners data as most as possible, leave-one-out cross validation was chosen to test the models and to calculate accuracy metrics. From a given data set, all data related to a given customer is removed. The model is trained with the remaining data, and then it tries to predict what happens with the removed customer. For each model, average accuracy (number of correct predictions divided by the total number of predictions) and average area under the curve (AUC, defined as the area below a plot of the true positive rate against the false positive rate for the different thresholds) are calculated. In a balanced dataset, a perfect forecast will produce a 100% value for the accuracy and for the AUC, whereas a completely random prediction will produce a value of about 50% for each of these metrics.

When predicting churn, one possible criticism is that customers that churn and that do not churn have different features that facilitate the classification process. To test even further the machine learning models applied, the creation of the datasets has been slightly changed in a way that emphasizes the ability of the methods to accurately predict the time of churn. As, for churners, several records will usually be available (one record classified as "churner", for the month when churn occurred, and all other records classified as "non-churners"), sample sets created exclusively with customers that have churned during the 24 months considered are created. The methods will have to find out if they have churned in the next month, next two months, and so on.

Six different methods were used, using the indicated R packages: random forests (RF-randomForest), support vector machine (SVM-kernlab), stochastic boosting (SB-ADA), logistic regression (LR-Rpart), classification and regression trees (CART-Rpart), multivariate adaptive regression splines (MARS-Earth).

Months

1	2	3	4	5	6	7	8	9	10	...	24
1	2	3	4	5	6	7	8	9	10	...	24
1	2	3	4	5	6	7	8	9	10	...	24

Fig. 7. Rolling window for building the dataset considering predictions one month in advance (values from the explanatory variables will come from the months in grey background, output variables will be calculated using data from the underlined months).

Months

1	2	3	4	5	6	7	_8_	9	10	...	24
1	2	3	4	5	6	7	8	_9_	10	...	24
1	2	3	4	5	6	7	8	9	_10_	...	24

Fig. 8. Rolling window for building the dataset considering predictions two months in advance (values from the explanatory variables will come from the months in grey background, output variables will be calculated using data from the underlined months).

RF are a combination of tree predictors [36], each one built considering a random subset of the available data, and contributing with one vote to the final result, calculated by majority voting. RF with 1000 trees were created. SVM are non-probabilistic supervised classifiers that find the hyperplane such that the nearest training data points belonging to different classes are as distant as possible. Radial basis gaussian functions were used. SB considers the application of simple classifiers, each one using a different version of the training data with different weights associated [37]. CART does a recursive partitioning for classification, building a binary tree by deciding, for each node, what is the best splitting variable to choose (minimizing a classification error). MARS creates a piecewise linear model using the product of spline basis functions [38].

The methodological framework used is described in pseudo-code:

1. $n \leftarrow$ forecasting horizon
2. Build the dataset for this forecasting horizon:
   a. Initialize the rolling time window with $t \leftarrow 1$. *Dataset* $\leftarrow \{\}$
   b. For each customer, build a sample considering the explanatory variables retrieved from data from month t to $t + 5$, and determine the value of the output variable (churner/non-churner) by looking at month $t + 5+n$. Include samples from all available customers in the *Dataset*.
   c. If $t + 5+n < 24$ then $t \leftarrow t + 1$ and go to 2.b. Else go to 3.
3. Repeat for 5 times:
   a. Create a set with all the samples that correspond to churners.
   b. Include in this set an identical number of samples randomly retrieved from the "non-churner" records.
   c. For each existing customer i in this set:
      (1) Remove customer i from the set.
      (2) Train the machine learning method with the remaining samples.
      (3) For each record belonging to customer i predict the corresponding output value.

Repeating five times the third step is important for two reasons: on one hand, some of the machine learning methods that are going to be applied have a random behaviour, so it is important to assess average behaviour instead of reaching conclusions based on a single run; on the other hand, as the sampling set is built using a random sampling procedure, it is also important to see if the results are sensitive to this sampling.

4 Computational Results

All methods were assessed considering classification accuracy and AUC, for each forecast horizon, and considering the two situations: datasets built considering churners and non-churners, and datasets built considering churners only.

These methods were also compared with the use of Cox Regression and the corresponding survival functions.

Survival analysis is able to predict whether a customer will churn or not during the considered time horizon (24 months). Considering a threshold of 0.5, so that if the survival probability is less than this threshold it predicts churn, survival analysis has an average accuracy of 80.73% and an average AUC of 89% (considering balanced sets). However, it is not capable of accurately predicting when a customer will churn. The average error is 6 months.

Table 1 presents the accuracy obtained when the dataset considers both churners and non-churners. A threshold of 0.5 was considered for all methods. Average values and standard deviations are shown for all the six methods tested, and for all the forecasting horizons. These values are related with out-of-sample testing only. As expected, with the increase of the forecasting horizon the precision deteriorates. The method that presents the best results in general is SB. The values of the standard deviation show that the methodology is not very sensitive to the random components of the procedures that were employed.

Table 2 presents the results for AUC. These results are also very good for SB. The methods are being tested considering balanced datasets, which means that a random classifier would have an AUC near 50%.

When the datasets were exclusively composed of churners that did churn sometime during the 24 months period, the quality of the results deteriorates, as expected. However, the accuracy and AUC are still very good (Table 3 and Table 4). SB continues to be the method presenting the best results for AUC. CART is slightly better than SB when it comes to accuracy. These results show that it is not only possible to accurately classify churners and non-churners, but it is also possible to accurately predict when they are going to churn.

It is possible to better understand the attributes that define whether a customer will churn or not by looking at the importance of each variable in the predicted outcome. For SB, the importance of each variable is related with the number of times it is selected for boosting.

Considering SB models predicting churn one and two months ahead, the most important variables are the total value of bank products held by the customer in the past 3 months, along with the existence of debit or credit cards in another bank. Considering models that predict churn 3 to 4 months ahead, the preceding attributes continue to be important, but the most important ones are the number of transactions in the past 2 months and the information of whether the customer has a mortgage loan outside the bank. These variables are mostly the same considering the two different types of sample sets (with churners and non-churners and churners only).

Table 1. Accuracy considering a dataset with churners and non-churners (grey cells- the best value for the corresponding forecasting horizon; italic and bold values – the worst results for the corresponding forecasting horizon).

Future	RF		SVM		SB		LR	
	Average	Standard Deviation	Average	Standard Deviation	Average	Standard Deviation	Average	Standard Deviation
1 month	88,39%	0,59%	*82,39%*	0,42%	89,07%	0,67%	84,50%	0,72%
2 months	89,37%	0,41%	*80,39%*	0,98%	88,46%	1,23%	82,28%	0,65%
3 months	87,88%	0,87%	77,85%	1,08%	88,33%	0,87%	*76,53%*	5,54%
4 months	87,67%	1,31%	78,50%	0,84%	88,06%	0,72%	*75,99%*	5,95%
5 months	86,15%	1,25%	78,12%	0,53%	87,43%	0,67%	*75,94%*	5,55%
6 months	85,68%	0,95%	*75,55%*	2,18%	86,05%	0,76%	78,74%	1,44%

Future	CART		MARS	
	Average	Standard Deviation	Average	Standard Deviation
1 month	86,55%	2,12%	85,52%	0,91%
2 months	85,59%	2,46%	85,55%	1,12%
3 months	87,88%	0,89%	85,84%	1,37%
4 months	85,60%	1,96%	86,11%	1,49%
5 months	85,32%	2,98%	83,80%	2,47%
6 months	85,83%	0,65%	82,96%	1,84%

Table 2. AUC considering a dataset with churners and non-churners (grey cells- the best value for the corresponding forecasting horizon; italic and bold values – the worst results for the corresponding forecasting horizon).

Future	RF		SVM		SB		LR	
	Average	Standard Deviation	Average	Standard Deviation	Average	Standard Deviation	Average	Standard Deviation
1 month	96,19%	0,27%	*88,32%*	0,99%	96,53%	0,31%	91,90%	0,60%
2 months	96,29%	0,17%	*86,80%*	0,55%	96,14%	0,39%	89,92%	0,46%
3 months	95,52%	0,35%	84,99%	0,44%	95,78%	0,34%	*81,32%*	9,51%
4 months	95,38%	0,61%	85,41%	1,19%	95,70%	0,44%	*80,74%*	9,55%
5 months	94,03%	0,48%	85,62%	1,53%	95,20%	0,52%	*80,51%*	9,43%
6 months	93,38%	0,41%	*82,82%*	1,52%	94,45%	0,40%	86,10%	1,26%

Future	CART		MARS	
	Average	Standard Deviation	Average	Standard Deviation
1 month	90,59%	1,18%	94,79%	0,36%
2 months	90,06%	1,82%	94,70%	0,68%
3 months	85,25%	4,76%	94,71%	0,69%
4 months	81,87%	7,91%	94,36%	0,76%
5 months	86,25%	6,72%	93,65%	1,20%
6 months	87,15%	1,97%	92,35%	1,21%

Table 3. Accuracy considering a dataset with churners only (grey cells- the best value for the corresponding forecasting horizon; italic and bold values – the worst results for the corresponding forecasting horizon).

Future	RF		SVM		SB		LR	
	Average	Standard Deviation	Average	Standard Deviation	Average	Standard Deviation	Average	Standard Deviation
1 month	82.97%	0.53%	70.46%	3.14%	82.48%	0.56%	*67,36%*	2.46%
2 months	79.62%	0.91%	70.53%	3.90%	81.19%	0.76%	*64,42%*	5.09%
3 months	78.27%	1.44%	63.82%	5.29%	79.85%	1.15%	*63,62%*	3.62%
4 months	77.25%	0.90%	*60,35%*	5.44%	78.13%	0.92%	62.23%	10.02%
5 months	76.56%	1.40%	*60,67%*	4.26%	76.63%	1.13%	62.61%	2.38%
6 months	73.19%	1.48%	*56,71%*	6.05%	75.61%	0.91%	57.93%	2.71%

Future	CART		MARS	
	Average	Standard Deviation	Average	Standard Deviation
1 month	78.26%	3.90%	80.30%	1.59%
2 months	79.31%	1.31%	80.23%	1.34%
3 months	81.13%	1.06%	79.97%	1.74%
4 months	80.48%	0.88%	78.02%	1.17%
5 months	78.70%	1.22%	77.27%	3.06%
6 months	78.66%	0.88%	75.39%	3.25%

Table 4. AUC considering a dataset with churners only (grey cells- the best value for the corresponding forecasting horizon; italic and bold values – the worst results for the corresponding forecasting horizon).

Future	RF		SVM		SB		LR	
	Average	Standard Deviation	Average	Standard Deviation	Average	Standard Deviation	Average	Standard Deviation
1 month	89.73%	0.31%	78.24%	3.14%	90.49%	0.70%	*72,79%*	1.87%
2 months	87.37%	0.72%	77.02%	3.90%	89.25%	0.58%	*69,59%*	6.98%
3 months	85.93%	1.21%	71.75%	5.29%	88.31%	0.71%	*68,82%*	3.78%
4 months	84.69%	0.26%	68.14%	5.44%	86.84%	0.52%	65.25%	12.29%
5 months	84.02%	0.48%	*66,06%*	4.26%	85.85%	0.72%	67.06%	2.50%
6 months	79.38%	1.04%	*58,48%*	6.05%	84.00%	0.82%	59.88%	2.48%

Future	CART		MARS	
	Average	Standard Deviation	Average	Standard Deviation
1 month	78.76%	3.90%	87.62%	0.62%
2 months	80.31%	1.31%	87.61%	0.41%
3 months	78.90%	1.06%	86.79%	1.31%
4 months	78.19%	0.88%	85.05%	1.14%
5 months	77.82%	1.22%	84.50%	0.83%
6 months	75.71%	0.88%	82.10%	1.37%

5   Conclusions

In this study, a methodology is investigated that allows a retail bank to produce each month a list of customers that are more likely to churn in the next six months, detailing which ones are likely to churn in each month ahead. This tool can be an important asset for determining focused retaining campaigns that will only be targeting customers with a high probability of leaving, and that the bank wants to keep. There is also the possibility that some of these customers are not interesting to the bank, so churning should not be avoided in these cases. The importance of not only correctly identifying churners but also deciding which of them to include in retaining campaigns, with the objective of maximizing profit, is addressed in [39], for instance.

The classification threshold considered was 0.5, so that no bias was introduced in the computational results shown. However, this value can be optimized to account for different costs associated with incorrectly predicting a churner or a non-churner. Predicting that a given customer will churn, when in fact he is not going to, can imply the cost of an unnecessary retaining action. Predicting that a customer is not a churner when indeed he is will imply the cost of losing future profits associated with this customer, which can be more significant to the bank. This analysis should be made to decide on the best threshold value to consider.

In the computational tests, SB had the best overall performance. Attributes related to the relationship that the customer has with other bank institutions are important for churn prediction.

In future work, an integrated methodology between churner prediction and the prediction of the next product to buy can be developed. This would help a company understanding what kind of product should offer to a possible churner in order to retain him as a customer. Future studies could also use other evaluation metrics that include profits [1].

This study only superficially addresses the motivations to churn. Computational results show that it is somehow related to the relationship of each customer with other banks. Inspired by Óskarsdóttir et al. [10], it would be interesting to better understand the motivations to churn, including "social contagion".

Acknowledgments. This study has been funded by national funds, through FCT, Portuguese Science Foundation, under project UIDB/00308/2020.

References

1. De Caigny, A., Coussement, K., De Bock, K.W.: A new hybrid classification algorithm for customer churn prediction based on logistic regression and decision trees. Eur. J. Oper. Res. **269**, 760–772 (2018)
2. Ganesh, J., Arnold, M.J., Reynolds, K.E.: Understanding the customer base of service providers: an examination of the differences between switchers and stayers. J. Mark. **64**, 65–87 (2000)
3. Reinartz, W.J., Kumar, V.: The impact of customer relationship characteristics on profitable lifetime duration. J. Mark. **67**, 77–99 (2003)

4. Colgate, M., Stewart, K., Kinsella, R.: Customer defection: a study of the student market in Ireland. Int. J. Bank Mark. **14**, 23–29 (1996)
5. Nitzan, I., Libai, B.: Social effects on customer retention. J. Mark. **75**, 24–38 (2011)
6. Verbraken, T., Bravo, C., Weber, R., Baesens, B.: Development and application of consumer credit scoring models using profit-based classification measures. Eur. J. Oper. Res. **238**, 505–513 (2014)
7. Verbeke, W., Martens, D., Mues, C., Baesens, B.: Building comprehensible customer churn prediction models with advanced rule induction techniques. Expert Syst. Appl. **38**, 2354–2364 (2011)
8. Gür Ali, Ö., Arıtürk, U.: Dynamic churn prediction framework with more effective use of rare event data: The case of private banking. Expert Syst. Appl. **41**, 7889–7903 (2014)
9. Moro, S., Cortez, P., Rita, P.: Business intelligence in banking: A literature analysis from 2002 to 2013 using text mining and latent Dirichlet allocation. Expert Syst. Appl. **42**, 1314–1324 (2015)
10. Óskarsdóttir, M., Van Calster, T., Baesens, B., Lemahieu, W., Vanthienen, J.: Time series for early churn detection: Using similarity based classification for dynamic networks. Expert Syst. Appl. **106**, 55–65 (2018)
11. Tsai, C.-F., Chen, M.-Y.: Variable selection by association rules for customer churn prediction of multimedia on demand. Expert Syst. Appl. **37**, 2006–2015 (2010)
12. Tsai, C.-F., Lu, Y.-H.: Customer churn prediction by hybrid neural networks. Expert Syst. Appl. **36**, 12547–12553 (2009)
13. Huang, B., Kechadi, M.T., Buckley, B.: Customer churn prediction in telecommunications. Expert Syst. Appl. **39**, 1414–1425 (2012)
14. Idris, A., Rizwan, M., Khan, A.: Churn prediction in telecom using Random Forest and PSO based data balancing in combination with various feature selection strategies. Comput. Electr. Eng. **38**, 1808–1819 (2012)
15. Vafeiadis, T., Diamantaras, K.I., Sarigiannidis, G., Chatzisavvas, KCh.: A comparison of machine learning techniques for customer churn prediction. Simul. Model. Pract. Theor. **55**, 1–9 (2015)
16. Coussement, K., Benoit, D.F., Van den Poel, D.: Improved marketing decision making in a customer churn prediction context using generalized additive models. Expert Syst. Appl. **37**, 2132–2143 (2010)
17. Miguéis, V.L., Camanho, A., Falcão e Cunha, J.: Customer attrition in retailing: an application of multivariate adaptive regression splines. Expert Syst. Appl. **40**, 6225–6232 (2013)
18. Buckinx, W., Van den Poel, D.: Customer base analysis: partial defection of behaviourally loyal clients in a non-contractual FMCG retail setting. Eur. J. Oper. Res. **164**, 252–268 (2005)
19. Chen, Kuanchin., Hu, Ya-Han, Hsieh, Yi-Cheng: Predicting customer churn from valuable B2B customers in the logistics industry: a case study. Inf. Syst. e-Bus. Manage. **13**(3), 475–494 (2014). https://doi.org/10.1007/s10257-014-0264-1
20. Clemente-Císcar, M., San Matías, S., Giner-Bosch, V.: A methodology based on profitability criteria for defining the partial defection of customers in non-contractual settings. Eur. J. Oper. Res. **239**, 276–285 (2014)
21. Mavri, M., Ioannou, G.: Customer switching behaviour in Greek banking services using survival analysis. Manag. Financ. **34**, 186–197 (2008)
22. Larivière, B., Van den Poel, D.: Investigating the role of product features in preventing customer churn, by using survival analysis and choice modeling: the case of financial services. Expert Syst. Appl. **27**, 277–285 (2004)

23. Larivière, B., Van den Poel, D.: Predicting customer retention and profitability by using random forests and regression forests techniques. Expert Syst. Appl. **29**, 472–484 (2005)
24. Glady, N., Baesens, B., Croux, C.: Modeling churn using customer lifetime value. Eur. J. Oper. Res. **197**, 402–411 (2009)
25. Xie, Y., Li, X., Ngai, E.W.T., Ying, W.: Customer churn prediction using improved balanced random forests. Expert Syst. Appl. **36**, 5445–5449 (2009)
26. De Bock, K.W., Van den Poel, D.: Reconciling performance and interpretability in customer churn prediction using ensemble learning based on generalized additive models. Expert Syst. Appl. **39**, 6816–6826 (2012)
27. De Bock, K.W., den Poel, D.V.: An empirical evaluation of rotation-based ensemble classifiers for customer churn prediction. Expert Syst. Appl. **38**, 12293–12301 (2011)
28. Shirazi, F., Mohammadi, M.: A big data analytics model for customer churn prediction in the retiree segment. Int. J. Inf. Manage. **48**, 238–253 (2018)
29. Farquad, M.A.H., Ravi, V., Raju, S.B.: Churn prediction using comprehensible support vector machine: an analytical CRM application. Appl. Soft Comput. **19**, 31–40 (2014)
30. Lin, C.-S., Tzeng, G.-H., Chin, Y.-C.: Combined rough set theory and flow network graph to predict customer churn in credit card accounts. Expert Syst. Appl. **38**, 8–15 (2011)
31. Nie, G., Rowe, W., Zhang, L., Tian, Y., Shi, Y.: Credit card churn forecasting by logistic regression and decision tree. Expert Syst. Appl. **38**, 15273–15285 (2011)
32. Van den Poel, D., Larivière, B.: Customer attrition analysis for financial services using proportional hazard models. Eur. J. Oper. Res. **157**, 196–217 (2004). https://doi.org/10.1016/S0377-2217(03)00069-9
33. Burez, J., Van den Poel, D.: Handling class imbalance in customer churn prediction. Expert Syst. Appl. **36**, 4626–4636 (2009)
34. Chen, C.: Using Random Forest to Learn Imbalanced Data. University of California, Berkeley (2004)
35. Weiss, G.M.: Mining with rarity: a unifying framework. ACM SIGKDD Explor. Newslett. **6**, 7 (2004)
36. Breiman, L.: Random forests. Mach. Learn. **45**, 5–32 (2001)
37. Friedman, J., Hastie, T., Tibshirani, R.: Additive logistic regression: a statistical view of boosting. Ann. Stat. **28**, 337–407 (2000)
38. Friedman, J.H.: Multivariate adaptive regression splines. Ann. Stat. 1–67 (1991)
39. Verbeke, W., Dejaeger, K., Martens, D., Hur, J., Baesens, B.: New insights into churn prediction in the telecommunication sector: A profit driven data mining approach. Eur. J. Oper. Res. **218**, 211–229 (2012)

A Resource Constraint Approach for One Global Constraint MINLP

Pavlo Muts[1], Ivo Nowak[1], and Eligius M. T. Hendrix[2(✉)]

[1] Hamburg University of Applied Sciences, Hamburg, Germany
{pavlo.muts,ivo.nowak}@haw-hamburg.de
[2] Computer Architecture, Universidad de Málaga, 29080 Málaga, Spain
eligius@uma.es

Abstract. Many industrial optimization problems are sparse and can be formulated as block-separable mixed-integer nonlinear programming (MINLP) problems, where low-dimensional sub-problems are linked by a (linear) knapsack-like coupling constraint. This paper investigates exploiting this structure using decomposition and a resource constraint formulation of the problem. The idea is that one outer approximation master problem handles sub-problems that can be solved in parallel. The steps of the algorithm are illustrated with numerical examples which shows that convergence to the optimal solution requires a few steps of solving sub-problems in lower dimension.

Keywords: Decomposition · Parallel computing · Column generation · Global optimization · Mixed-integer nonlinear programming

1 Introduction

Mixed Integer Nonlinear Programming (MINLP) is a paradigm to optimize systems with both discrete variables and nonlinear constraints. Many real-world problems are large-scale, coming from areas like machine learning, computer vision, supply chain and manufacturing problems, etc. A large collection of real-world MINLP problems can be found in MINLPLib [13].

In general, the problems are hard to solve. Many approaches have been implemented over the last decades. Most of the nonconvex MINLP deterministic solvers apply one global branch-and-bound (BB) search tree [3,4], and compute a lower bound by a polyhedral relaxation, like ANTIGONE [8], BARON [11], Couenne [1], Lindo API [7] and SCIP [12]. The challenge for these methods is to handle a rapidly growing global search tree, which fills up the computer memory. An alternative to BB is successive approximation. Such methods solve an optimization problem without handling a single global search tree. The Outer-Approximation (OA) method [5,6,9,15] and the Extended Cutting Plane method [14] solve convex MINLPs by successive linearization of nonlinear constraints. One of the challenges we are dealing with is how to handle this for nonconvex MINLP problems.

© Springer Nature Switzerland AG 2020
O. Gervasi et al. (Eds.): ICCSA 2020, LNCS 12251, pp. 590–605, 2020.
https://doi.org/10.1007/978-3-030-58808-3_43

In this paper, we focus on practical potentially high dimension problems, which in fact consist of sub-problems that are linked by one coupling constraint. This opens the opportunity to reformulate the problem as a resource constraint bi-objective problem similar to the multi-objective view used in [2] for integer programming problems. We investigate the potential of this approach combining it with Decomposition-based Inner- and Outer-Refinement (DIOR), see [10].

To investigate the question, we first write the general problem formulation with one global constraint and show that this can be approached by a resource constrained formulation in Sect. 2. Section 3 presents a possible algorithm to solve such problems aiming at a guaranteed accuracy. The procedure is illustrated stepwise in Sect. 4. Section 5 summarizes our findings.

2   Problem Formulation and Resource Constraint Formulation

We consider *block-separable* (or *quasi-separable*) MINLP problems where the set of decision variables $x \in \mathbb{R}^n$ is partitioned into $|K|$ blocks

$$\min c^T x \quad \text{s.t.} \quad a^T x \le b, \ x_k \in X_k, \ k \in K \tag{1}$$

with

$$X_k := \{y \in [\underline{x}_k, \overline{x}_k] \subset \mathbb{R}^{n_k} | y_j \in \mathbb{Z}, j \in J_k, g_{ki}(y) \le 0, \ i \in I_k\}. \tag{2}$$

The dimension of the variables $x_k \in \mathbb{R}^{n_k}$ in block k is n_k such that $n = \sum_{k \in K} n_k$. The vectors $\underline{x}, \overline{x} \in \mathbb{R}^n$ denote lower and upper bounds on the variables. The linear constraint $a^T x \le b$ is called the resource constraint and is the only global link between the sub-problems. We assume that the part a_k corresponding to block k has $a_k \ne 0$, otherwise the corresponding block can be solved independently. The constraints defining feasible set X_k are called *local*. Set X_k is defined by linear and nonlinear local constraint functions, $g_{ki} : \mathbb{R}^{n_k} \to \mathbb{R}$, which are assumed to be bounded on the set $[\underline{x}_k, \overline{x}_k]$. The linear objective function is defined by $c^T x := \sum_{k \in K} c_k^T x_k, c_k \in \mathbb{R}^{n_k}$. Furthermore, we define set $X := \prod_{k \in K} X_k$.

The Multi-objective approach of [2] is based on focusing on the lower dimensional space of the global constraints of the sub-problems rather than on the full $n-$dimensional space. We will outline how they relate to the so-called Bi-Objective Programming (BOP) sub-problems based on a resource-constrained reformulation of the MINLP.

2.1   Resource-Constrained Reformulation

If the MINLP (1) has a huge number of variables, it can be difficult to solve it in reasonable time. In particular, if the MINLP is defined by discretization of some infinitely dimensional variables, like in stochastic programming or in differential equations. For such problems, a resource-constrained perspective can be promising.

The idea is to view the original problem (1) in n–dimensional space from two objectives; the objective function and the global resource constraint. Define the $2 \times n_k$ matrix C_k by

$$C_k = \begin{bmatrix} c_k^T \\ a_k^T \end{bmatrix} \tag{3}$$

and consider the transformed feasible set:

$$V_k := \{v_k := C_k x_k \ : \ x_k \in X_k\} \subset \mathbb{R}^2. \tag{4}$$

The *resource-constrained formulation* of (1) is

$$\min \sum_{k \in K} v_{k0} \quad \text{s.t.} \quad \sum_{k \in K} v_{k1} \leq b, \tag{5}$$
$$v_k \in V_k, \quad k \in K.$$

The approach to be developed uses the following property.

Proposition 1. *Problem* (5) *is equivalent to the two-level program*

$$\min \sum_{k \in K} v_{k0}^*$$
$$\text{s.t.} \quad \sum_{k \in K} v_{k1} \leq b, \tag{6}$$

of finding the appropriate values of v_{k1} where v_{k0}^* is the optimal value of sub-problem RCP_k given by

$$v_{k0}^* := \min c_k^T x_k \quad \text{s.t.} \quad x_k \in X_k, C_{k1} x_k \leq v_{k1}. \tag{7}$$

Proof. From the definition it follows that the optimum of (6) coincides with

$$\min c^T x \quad \text{s.t.} \quad x \in X, \quad a^T x \leq b.$$

$\square$

This illustrates the idea, that by looking for the right assignment v_{k1} of the resource, we can solve the lower dimensional sub-problems in order to obtain the complete solution. This provokes considering the problem as a potentially continuous knapsack problem. Approaching the problem as such would lead having to solve the sub-problems many times to fill a grid of state values with a value function and using interpolation. Instead, we investigate the idea of considering the problem as a bi-objective one, where we minimize both, v_{k0} and v_{k1}.

2.2   Bi-objective Approach

A bi-objective consideration of (5) changes the focus from the complete image set V_k to the relevant set of Pareto optimal points. Consider the sub-problem BOP_k of block k as

$$\min C_k x_k \quad \text{s.t.} \quad x_k \in X_k. \tag{8}$$

The Pareto front of BOP$_k$ (8) is defined by a set of vectors, $v_k = C_k x_k$ with $x_k \in X_k$ with the property that there does not exist another feasible solution $w = C_k y_k$ with $y_k \in X_k$, which dominates v_k, i.e., for which $w_i \leq v_{ki}$ for $i = 0, 1$, and $w_i < v_{ki}$ for $i = 0$ or $i = 1$. We will call an element of the Pareto front a nondominated point (NDP). In other words, a NDP is a feasible objective vector for which none of its components can be improved without making at least one of its other components worse. A feasible solution $x_k \in X_k$ is called efficient (or Pareto optimal) if its image $v_k = C_k x_k$ is a NDP, i.e. it is nondominated. Let us denote the Pareto front of NDPs of (8) as

$$V_k^* := \{v \in V_k : v \text{ is a NDP of (8)}\}.$$

Proposition 2. *The solution of problem* (5) *is attained at* $v^* \in V^*$.

$$\min \sum_{k \in K} v_{k0} \quad \text{s.t.} \quad \sum_{k \in K} v_{k1} \leq b, \tag{9}$$
$$v_k \in V_k^*, \quad k \in K.$$

Proof. Assume there exist parts of $\hat{v}_k^* \notin V^*$ of an optimal solution v^*, i.e the parts are dominated. This implies $\exists w_k \in V_k^*$ which dominates v_k^*, i.e. $w_{ki} \leq v_{ki}^*$ for $i = 0, 1$. Consider $\hat{v}$ the corresponding solution where in v^* the parts v_k^* are replaced by w_k. Then $\hat{v}$ is feasible for RCP given $\sum_{k \in K} \hat{v}_{k1} \leq \sum_{k \in K} v_{k1}^* \leq b$ and its objective value is at least as good, $\sum_{k \in K} \hat{v}_{k0} \leq \sum_{k \in K} v_{k0}^*$, which means that the optimum is attained at NDP point $\hat{v} \in V^*$.   □

In bi-objective optimization, a NDP can be computed by optimizing a weighted sum of the objectives

$$\min d^T C_k x_k \quad \text{s.t.} \quad x_k \in X_k. \tag{10}$$

For a positive weight vector $d \in \mathbb{R}^2$, an optimal solution of (10) is an efficient solution of (8), i.e., its image is nondominated. Such a solution and its image are called a *supported efficient solution* and a *supported NDP*, respectively. Thus, an efficient solution x_k is supported if there exists a positive vector d for which x_k is an optimal solution of (10), otherwise x_k is called unsupported.

Example 1. To illustrate the concepts, we use a simple numerical example which can be used as well to follow the steps of the algorithms to be introduced. Consider $n = 4$, $K = \{1, 2\}$, $c = (-1, -2, -1, -1)$, $a = (2, 1, 2, 1)$, $b = 10$, $\underline{x} = (0, 0, 2, 1)$ and $\overline{x} = (5, 1.5, 5, 3)$. Integer variables are $J_1 = \{1\}, J_2 = \{3\}$ and the local constraints are given by $g_{11}(x_1, x_2) = 3x_2 - x_1^3 + 6x_1^2 - 8x_1 - 3$ and $g_{21}(x_3, x_4) = x_4 - \frac{5}{x_3} - 5$. The optimal solution is $x = (1, 1.5, 2, 2.5)$ with objective value -8.5. The corresponding points in the resource space are $v_1 = (-4, 3.5)$ and $v_2 = (-4.5, 6.5)$.

Figure 1 sketches the resource spaces V_1 and V_2 with the corresponding Pareto front. In blue, now the dominated part of V_k not covered by the Pareto front V_k^* is visible. The number of supported Pareto points is limited.

The example suggests that we should look for an optimal resource v_k in a two-dimensional space, which seems more attractive than solving an n-dimensional problem. Meanwhile, sub-problems should be solved as few times as possible.

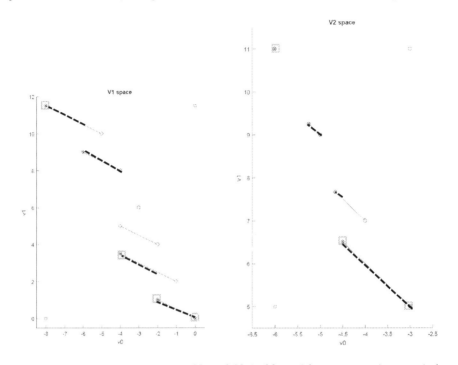

Fig. 1. Resource constraint spaces V_1 and V_2 in blue with extreme points as circles. The Pareto front is in black. Supported NDPs are marked with a green square. The red squares represent the Ideal $\underline{v}_k$ (left under) and the Nadir point $\overline{v}_k$ (right-up).

3   Algorithmic Approach

The optimization of problem (1) according to Proposition 1 reduces to finding v_k^* in the search space which according to Proposition 2 can be found in the box $\mathcal{W}_k = [\underline{v}_k, \overline{v}_k]$. First of all, if the global constraint is not binding, dual variable $\mu = 0$ and the sub-problems can be solved individually. However, usually $\mu > 0$ gives us a lead of where to look for v_k^*. Notice that if v_k^* is a supported NDP, it can be found by optimizing sub-problem

$$y_k(\beta) \in \operatorname{argmin}\{c_k^T x + \beta a_k^T x, \ x \in X_k\}, \tag{11}$$

for $\beta = \mu$. In that case, $v_k^* = C_k^T y_k$. However, we do not know the dual value μ beforehand and moreover, v_k^* can be a nonsupported NDP. Notice that the resulting solution y_k is an efficient point and $C_k y_k$ is a supported NDP for $\beta \geq 0$.

To look for the optimum, we first create an outer approximation (OA) $\mathcal{W}_k$ of V^* by adding cuts using an LP master problem to estimate the dual value and iteratively solving 11. Second, we will use the refinement of $\mathcal{W}_k$ in a MIP OA, which may also generate nonsupported NDPs.

3.1 LP Outer Approximation

Initially, we compute the Ideal and Nadir point $\underline{v}_k$ and $\overline{v}_k$ for each block k. This is done by solving (11) with $\beta = \varepsilon, \beta = \frac{1}{\varepsilon}$ respectively. Let $r_1 = C_k y_k(\varepsilon)$ and $r_2 = C_k y_k(\frac{1}{\varepsilon})$, then $\underline{v}_k = (r_{1,0}, r_{2,1})^T$ and $\overline{v}_k = (r_{2,0}, r_{1,1})^T$. These vectors bound the search space for each block and initiate a set P_k of outer cuts. We use set $R_k = \{(r, \beta)\}$ of supported NDPs with a corresponding weight β, to define local cut sets

$$P_k = \{v \in [\underline{v}_k, \overline{v}_k](1, \beta)v \geq (1, \beta)r, \forall (r, \beta) \in R_k\}.$$

An initial cut is generated using the orthogonal vector of the plane between r_1 and r_2, i.e. find $y_k(\beta_{k0})$ in (11) with

$$\beta_{k0} = \frac{\overline{v}_{k0} - \underline{v}_{k0}}{\overline{v}_{k1} - \underline{v}_{k1}}. \tag{12}$$

Notice that if r_1 is also a solution of the problem, then apparently there does not exist a (supported) NDP v at the left side of the line through r_1 and r_2, i.e. $(1, \beta)v < (1, \beta)r_1$. The hyperplane is the lower left part of the convex hull of the Pareto front. Basically, we can stop the search for supported NDPs for the corresponding block.

An LP outer approximation of (5) is given by

$$w = \operatorname{argmin} \sum_{k \in K} v_{k0} \quad \text{s.t.} \quad \sum_{k \in K} v_{k1} \leq b, \tag{13}$$
$$v_k \in P_k, \quad k \in K.$$

It generates a relaxed solution w and an estimate β of the optimal dual μ. Then β is used to generate more support points by solving problem (11). Notice that for several values of β the same support point $C_k y_k(\beta)$ may be generated. However, each value leads to another cut in P_k. Moreover, the solution w of (13) will be used later for refinement of the outer approximation. A sketch of the algorithm is given in Algorithm 1.

3.2 MIP Outer Approximation

An outer approximation $\mathcal{W}_k$ of the Pareto front V_k^* is given by the observation that the cone $\{v \in \mathbb{R}^2, v \geq \underline{v}\}$ contains the Pareto front. Basically, we refine this set as a union of cones based on all found NDPs. We keep a list $\mathcal{P}_k = \{p_{k0}, p_{k1}, \ldots, p_{k|\mathcal{P}_k|}\}$ of NDPs of what we will define as the extended Pareto

---

Algorithm 1. Generate OA

1: **function** INITOA($\varepsilon, q_{\max}$)
2:     **for** $k \in K$ **do**
3:         stop$_k \leftarrow$ false
4:         Use (8) to determine $\underline{v}_k$ and $\overline{v}_k$ via r_1 and r_2
5:         Determine β_{k0} of (12). Solve (11) with β_{k0}
6:         **if** $r_1 \in$ argmin of (11) **then**
7:             stop$_k \leftarrow$ **true**, stop searching for supported NDPs
8:         **else** $R_k \leftarrow \{(C_k y_k(\beta_{k0}), \beta_{k0})\}$
9:     $q \leftarrow 1$
10:     $(w, \beta_q) \leftarrow$ (primal, dual) solution (13)
11:     **repeat**
12:         **for** $k \in K$ **do**
13:             **if** not stop$_k$ and $\beta_q \neq \beta_{k0}$ **then**
14:                 $y_k(\beta_q) \leftarrow$ solution (11), Store $(C_k y_k(\beta_q), \beta_q)$ in R_k
15:         $q \leftarrow q + 1$, $(w, \beta_q) \leftarrow$ (prima, dual) solution (13) using R_k to define P_k
16:     **until** $(\exists j = 1, \ldots, q-1, |\beta_q - \beta_j| < \varepsilon)$ or $(q = q_{\max})$ or $(\forall k \in K,$ stop$_k)$
17:     **return** $R_k, w,$ stop$_k$

---

front ordered according to the objective value $p_{k00} < p_{k10} < \cdots < p_{k|\mathcal{P}_k|0}$. Initially, the list $\mathcal{P}_k$ has the supported NDPs we found in R_k. However, we will generate more potentially not supported NDPs in a second phase. Using list $\mathcal{P}_k$, the front V_k^* is (outer) approximated by $\mathcal{W}_k = \cup W_{ki}$ with

$$W_{ki} = \{v \in \mathbb{R}^2 | v_0 \geq p_{k(i-1)0}, v_1 \geq p_{ki1}\}.$$

We solve a MIP master problem (starting with the last solution found by the LP-OA) to generate a solution w, for which potentially not for all blocks k we have $w_k \in V_k^*$. Based on the found points, we generate new NDPs v_k to refine set $\mathcal{W}_k$, up to convergence takes place. Moreover, we check whether v_k is supported NDP, or generate a new supported NDP in order to add a cut to P_k.

The master MIP problem is given by

$$w = \operatorname{argmin} \sum_{k \in K} v_{k0} \quad \text{s.t.} \quad \sum_{k \in K} v_{k1} \leq b,$$
$$v_k \in P_k \cap \mathcal{W}_k, \quad k \in K, \tag{14}$$

which can be implemented as

$$w = \operatorname*{argmin}_{k \in K} \sum v_{k0}$$

$$\text{s.t.} \quad \sum_{k \in K} v_{k1} \leq b,$$

$$v_k \in P_k, \quad k \in K,$$

$$v_{k0} \geq \underline{v}_{k0} + \sum_{i=1}^{|\mathcal{P}_k|} (p_{k(i-1)0} - \underline{v}_0) \delta_{ki}, \quad k \in K,$$

$$v_{k1} \geq \underline{v}_{k1} + \sum_{i=1}^{|\mathcal{P}_k|} (p_{i0} - v_{k1}) \delta_{ki}, \quad k \in K$$

$$\sum_{i=1}^{|\mathcal{P}_k|} \delta_{ki} = 1, \quad k \in K,$$

$$\delta_{ki} \in \{0, 1\}, \quad i = 1, \ldots, |\mathcal{P}_k|, k \in K.$$

(15)

Basically, if the solution w of (14) corresponds to NDPs, i.e. $w_k \in V_k^*$ for all blocks k, we are ready and solved the problem according to Proposition 2. If $w_k \notin V_k^*$, we refine $\mathcal{W}_k$ such that w_k is excluded in the next iteration. In order to obtain refinement points p, we introduce the extended Pareto front, which also covers the gaps. Define the extended search area as

$$\overline{V}_k = [\underline{v}_k, \overline{v}_k] \setminus \{w \in \mathbb{R}^2 | \exists v \in V_k^*, w < v\}. \tag{16}$$

Then the extended Pareto front is given by

$$\overline{V}_k^* = \{v \in \mathbb{R}^2 | v \text{ is NDP of } \min v \quad \text{s.t.} \quad v \in \overline{V}_k\}. \tag{17}$$

To eliminate w_k in the refinement, we perform line-search $v = w_k + \lambda(1, \beta)^T, \lambda \geq 0$ in the direction $(1, \beta)^T$, with

$$\beta = \frac{\overline{v}_1 - w_1}{\overline{v}_0 - w_0}. \tag{18}$$

A possible MINLP sub-problem line search is the following:

$$\lambda_k = \operatorname*{argmin} \lambda$$

$$\text{s.t.} \quad Cx \leq w_k + \lambda(1, \beta)^T, \tag{19}$$

$$\lambda \geq 0, x \in X_k$$

and taking $v_k = w_k + \lambda_k(1, \beta)^T$ as NDP of the (extended) Pareto front. Notice that if $\lambda_k = 0$, apparently $w_k \in V_k^*$.

Algorithm 2. MIPOA decomposition

1: **function** OASUBS(ε)
2:　　Take the results from Algorithm 1
3:　　For all k, initiate list $\mathcal{P}_k$ with the NDPs in R_k
4:　　$w \leftarrow$ solution LP OA master (13)
5:　　**repeat**
6:　　　**for** $k \in K$ **do**
7:　　　　**if** $w_k \notin \mathcal{P}_k$ **then**
8:　　　　　find v_k solving (19)
9:　　　　　Insert v_k in list $\mathcal{P}_k$ and update $\mathcal{W}_k$
10:　　　　　**if** $(w_k \neq v_k)$ and (not $\mathtt{stop}_k$) **then**
11:　　　　　　$\beta \leftarrow \frac{\bar{v}_{k1} - w_{k1}}{\bar{v}_{k0} - w_{k0}}$
12:　　　　　　Solve (11) and add $(C_k y_k(\beta), \beta)$ to R_k, update P_k
13:　　　$w \leftarrow$ solution MIP master (14)
14:　　**until** $\forall k \in K, \exists p \in \mathcal{P}_k, \|w_k - p\| < \varepsilon$

Moreover, we try to generate an additional cut by having a supported NDP. The complete idea of the algorithm is sketched in Algorithm 2. Let $z_k = w_{k0}$, if $w_k \in \mathcal{P}_k$ and $z_k = v_{k0}$ else. Then $\sum z_k$ provides an upper bound on the optimal function value. In this way, an implementation can trace the convergence towards the optimum. Based on the concept that in each iteration the MIP-OA iterate w is cut out due to the refinement in v_k, it can be proven that the algorithm converges to an optimal resource allocation v^*.

4   Numerical Illustration

In this section, we use several instances to illustrate the steps of the presented algorithms. All instances use an accuracy $\varepsilon = 0.001$ and we focus on the number of times OA-MIP (14) is solved and sub-problem (11) and line-search sub-problem (19). First we go stepwise through example problem. Second, we focus on a concave optimization problem known as ex_2_1_1 of the MINLPlib [13]. At last, we go through the convex and relatively easy version of ex_2_1_1 in order to illustrate the difference in behaviour.

4.1   Behaviour for the Example Problem

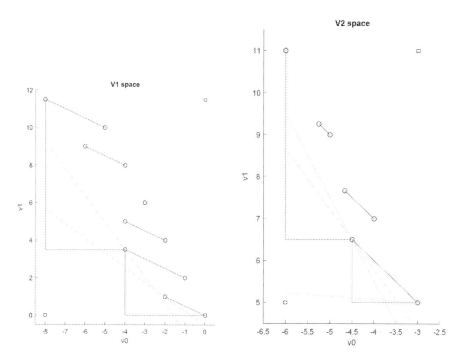

Fig. 2. Generated cuts (green) and refinement (mangenta) define the outer approximation after one iteration for both search areas. (Color figure online)

First of all, we build OA W_k of the Pareto front following Algorithm 1. Ideal and Nadir point for each block, $W_1 = \left[\begin{pmatrix} -8 \\ 0 \end{pmatrix}, \begin{pmatrix} 0 \\ 11.5 \end{pmatrix} \right]$ and $W_2 = \left[\begin{pmatrix} -6 \\ 5 \end{pmatrix}, \begin{pmatrix} -3 \\ 11 \end{pmatrix} \right]$ are found by minimizing cost and resource. Based on these extreme values, we run sub-problem (11) for step 8 in Algorithm 1 using direction vectors $\beta_{1,0} = 0.6957$ and $\beta_{2,0} = 0.5$ according to (12). For this specific example, we reach the optimal solution corresponding to $(v_{1,0}, v_{1,1}, v_{2,0}, v_{2,1}) = (-4, 3.5, -4.5, 6.5)^T$ which still has to be proven to be optimal.

One can observe the first corresponding cuts in Fig. 2. We run the LP OA which generates a dual value of $\beta_{k1} = 0.6957$, which corresponds to the angle of the first cut in the V_1 space. This means that sub-problem (11) does not have to be run for the first block, as it will generate the same cut. For the second block, it finds the same support point $v_2 = (-4.5, 6.5)^T$, but adds a cut with a different angle to P_2, as can be observed in the figure.

Algorithm 2 first stores the found extreme points and optimal sub-problem solutions in point sets $\mathcal{P}_k$. The used `matlab` script provides an LP-OA solution of

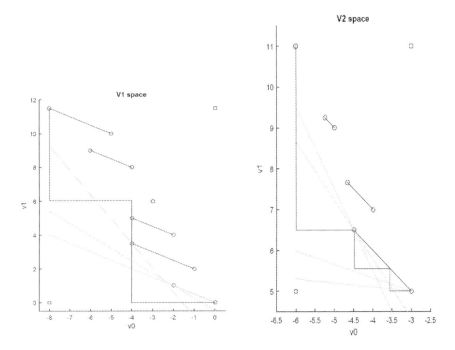

Fig. 3. Outer approximation of the Pareto front given by generated cuts (green) and refinement (mangenta) after convergence of the algorithm to the optimal resource allocation. (Color figure online)

$w = (-4.0138, 3.5198, -4.4862, 6.4802)$, which is similar to the optimal solution with rounding errors. Solving the line search (19) provides a small step into the feasible area of $\lambda_1 = 0.014$ and $\lambda_2 = 0.0015$ in the direction of the Nadir point.

Due to the errors, both blocks do not consider $v_k = w_k$ and add a cut according to step 12 in the algorithm. Interesting enough is that the MIP OA in the space as drawn in Fig. 2 reaches an infeasible point w, i.e. $w_k \in \mathcal{W}_k$, but $w_k \notin V_k$ further away from the optimum. This provides the incentive for the second block to find v_1 in the extended Pareto set as sketched in Fig. 3. This helps to reach the optimum with an exact accuracy.

For this instance, the MIP OA was solved twice and in the end contains 6 binary variables. Both blocks solve two sub-problems to reach the Ideal and 2 times the line search problem (19). The first block solves sub-problem (11) 3 times and the second block once more to generate an additional cut. In general, in each iteration at most two sub-problems are solved for each block. The idea is that this can be done in parallel. Notice that the refinement causes the MIP OA to have in each iteration at most $|K|$ additional binary variables.

4.2 Behaviour for ex_2_1_1

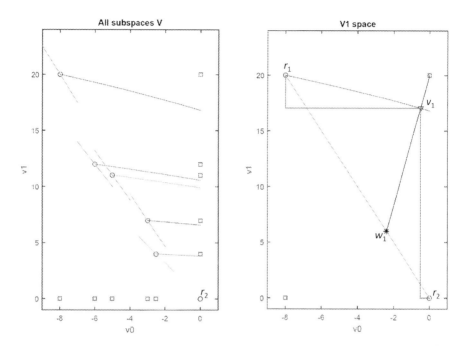

Fig. 4. At the left, the Pareto front of all sub-problems combined. The minimum resource point $r_2 = (0,0)^T$ is common for all sub-problems and the minimum cost solution is given by a blue circle. The only cut (line between r_1 and r_2) is sketched in green. At the right, OA for V_1 after one iteration. The line search between w_1 and Nadir point providing v_1 and first refinement of the OA (mangenta) are sketched. (Color figure online)

This instance can be characterised as a worst-case type of knapsack problem, where the usual heuristic to select the items with the best benefit-weight ratio first provides the wrong solution. As we will observe, a similar behaviour can be found using the OA relaxations. All variables are continuous and we are implicitly minimizing a concave quadratic function. In our description $n = 10$, $c = (0, 1, 0, 1, 0, 1, 0, 1, 0, 1)$, $a = (20, 0, 12, 0, 11, 0, 7, 0, 4, 0)$, $b = 40$, $K = \{1, 2, 3, 4, 5\}$ and divides vector x into 5 equal blocks $x_k = (x_{2k-1}, x_{2k})$, $k \in K$. The local constraints are given by $g_k(y, z) = q_k y - 50y^2 - z$, where $q = (42, 44, 45, 47, 47.5)$. The bounds on the variables are given by $[\underline{x}_{k1}, \overline{x}_{k1}] = [0, 1]$ and $[\underline{x}_{k2}, \overline{x}_{k2}] = [q_k - 50, 0]$, $k \in K$.

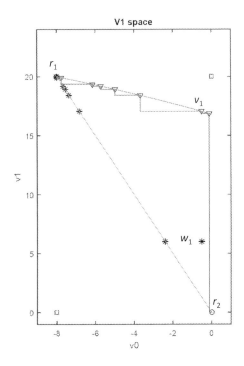

Fig. 5. Final refinement to reach the accuracy of $\varepsilon = 0.001$ of the approximation w of the optimal solution; w_1 given by a black star. In each iteration a new refinement point v_1 is generated, indicated by a black triangle.

The optimal solution is $x = (1, -8, 1, -6, 0, 0, 1, -3, 0, 0)$ with objective value -17. However, the LP OA relaxation provides a first underestimate of the objective of -18.9, where all subproblems take point r_1 apart from the first one, where $w_1 = (-2.4, 6)$. The solution and its consequence for the first iteration is sketched in Fig. 4. One can also observe the resulting line-search to find the first solution v_1 to come to the first refinement. The bad approximation of the first solution corresponds to a result of a greedy knapsack algorithm. For this instance, an iterative refinement is necessary that generates the points v_k up to convergence is reached.

Another typical characteristic is the concave shape of the Pareto front. This implies that the first cut is in fact the green line between r_1 and r_2. This is detected in step 6 of Algorithm 1 and implies that there are no more supported NDPs than r_1 and r_2, so that one does not have run sub-problem (11) anymore. However, the algorithm requires to solve for more and more sub-problems the line search problem (19) in each MIP OA iteration.

In total, the algorithm requires 9 steps of the MIP OA algorithm to reach the accuracy. In the end, the problem contains 26 binary variables δ over all subproblems. The intensity is best depicted by the refinement of $\mathcal{W}_1$ in Fig. 5.

The algorithm requires 3 times solving sub-problem (11) to generate r_1, r_2 and to find the cut for each block. In total it solved the line search (19) 21 times.

4.3   Behaviour for a Convex Variant of ex_2_1_1

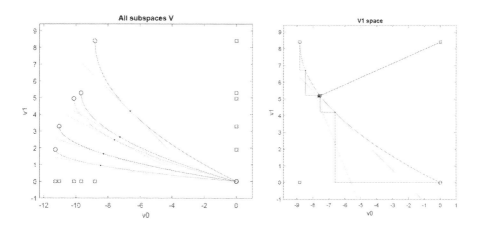

Fig. 6. At the left, the Pareto front of all sub-problems combined. The minimum resource point $r_2 = (0,0)^T$ is common for all sub-problems and the minimum cost solution is given by a blue circle. The first cut is illustrated by a green line. At the right, OA for V_1 after one iteration. Actually, it stays the same during future iterations, as w_1 coincides numerically with a found NDP. (Color figure online)

Convex problems are usually considered easy to solve. The idea of having a so-called zero duality gap is captured by solving the problem already by the LP OA using the dual value to generate cuts. We use the data of the former instance, but are now implicitly minimizing a convex quadratic function. Again $n = 10$, $c = (0, 1, 0, 1, 0, 1, 0, 1, 0, 1)$, $a = (20, 0, 12, 0, 11, 0, 7, 0, 4, 0)$, but now taking $b = 15$ in order to have a binding global constraint. The local constraints are given by $g_k(y, z) = -q_k y + 50y^2 - z$, where $q = (42, 44, 45, 47, 47.5)$. The bounds on the variables are given by $[\underline{x}_{k1}, \overline{x}_{k1}] = [0, 1]$ and $[\underline{x}_{k2}, \overline{x}_{k2}] = [0, -\frac{q_k^2}{200}]$, $k \in K$.

The optimal objective function value is -45.62. The Pareto fronts do not exhibit gaps, but are smooth quadratic curves. This means that solving the weighted problem with different values of β provides different optima. In Fig. 6, all Pareto fronts are depicted together with the first generated cut. However, the LP OA relaxation of Algorithm 1 provides a first relatively sharp underestimate of the objective of -45.7. This illustrates the idea of decomposition for convex problems where a good value for the dual provides the optimum for the primal problem.

The big difference with the concave case is that now in each iteration in principle new cuts can be generated. Considering the first block V_1, this does not really happen, as the optimum w_1 coincides with a point v_1 on the Pareto front found by the line search. Figure 6 illustrates at the right the MIP OA based on the points in set $\mathcal{P}_1$ on the Pareto front. Further refinement does not improve the bound, as the optimum has already been approximated in the first iteration.

5   Conclusions

Mixed Integer Nonlinear Programming (MINLP) is a strong concept for formulating practical optimization problems. Solvers based on the branch and bound concept, usually suffer from the number of variables n of the problem. For instances having one global inequality, we investigated the potential of using decomposition requiring sub-problems of a smaller size $n_k, k \in K$ to be solved using a MIP master problem exploiting the concept of resource constraint programming and a Pareto front.

The result of our investigation is a decomposition algorithm aiming at convergence up to a guaranteed accuracy. In each iteration, a MIP master problem is solved and at most $2|K|$ sub-problems, that can be solved in parallel. The process is illustrated with graphical examples which are solved up to an accuracy of $\varepsilon = 0.001$ after a few iterations.

At the moment, we are working on designing algorithms for instances with more than one global constraint, resulting in a Pareto set in higher dimension. The approach presented here can be extended and proven to converge. However, to obtain an algorithm with reasonable practical performance requires rethinking the cut generation in a resource space of higher dimension. We will report on this topic in future papers.

Acknowledgments. This paper has been supported by The Spanish Ministry (RTI2018-095993-B-100) in part financed by the European Regional Development Fund (ERDF) and by Grant 03ET4053B of the German Federal Ministry for Economic Affairs and Energy.

References

1. Belotti, P., Lee, J., Liberti, L., Margot, F., Wächter, A.: Branching and bounds tightening techniques for non-convex MINLP. Optim. Methods Softw. **24**(4–5), 597–634 (2009)
2. Bodur, M., Ahmed, S., Boland, N., Nemhauser, G.L.: Decomposition of loosely coupled integer programs: a multiobjective perspective (2016). http://www.optimization-online.org/DB_FILE/2016/08/5599.pdf
3. Burer, S., Letchford, A.: Non-convex mixed-integer nonlinear programming: a survey. Surv. Oper. Res. Manage. Sci. **17**(2), 97–106 (2012)
4. Bussieck, M.R., Vigerske, S.: MINLP Solver Software (2014). www.math.hu-berlin.de/~stefan/minlpsoft.pdf
5. Duran, M., Grossmann, I.: An outer-approximation algorithm for a class of mixed-integer nonlinear programs. Math. Program. **36**, 307–339 (1986)

6. Fletcher, R., Leyffer, S.: Solving mixed integer nonlinear programs by outer approximation. Math. Program. **66**(3(A)), 327–349 (1994)
7. Lin, Y., Schrage, L.: The global solver in the LINDO API. Optim. Methods Softw. **24**(4–5), 657–668 (2009)
8. Misener, R., Floudas, C.: ANTIGONE: Algorithms for coNTinuous/Integer Global Optimization of Nonlinear Equations. J. Glob. Optim. **59**(2–3), 503–526 (2014)
9. Muts, P., Nowak, I., Hendrix, E.M.T.: The decomposition-based outer approximation algorithm for convex mixed-integer nonlinear programming. J. Glob. Optim. **77**(1), 75–96 (2020). https://doi.org/10.1007/s10898-020-00888-x
10. Nowak, I., Breitfeld, N., Hendrix, E.M.T., Njacheun-Njanzoua, G.: Decomposition-based Inner- and Outer-Refinement Algorithms for Global Optimization. J. Glob. Optim. **72**(2), 305–321 (2018). https://doi.org/10.1007/s10898-018-0633-2
11. Tawarmalani, M., Sahinidis, N.: A polyhedral branch-and-cut approach to global optimization. Math. Program. **103**(2), 225–249 (2005)
12. Vigerske, S.: Decomposition in multistage stochastic programming and a constraint integer programming approach to mixed-integer nonlinear programming. Ph.D. thesis, Humboldt-Universität zu Berlin (2012)
13. Vigerske, S.: MINLPLib (2018). http://minlplib.org/index.html
14. Westerlund, T., Petterson, F.: An extended cutting plane method for solving convex MINLP problems. Comput. Chem. Eng. **21**, 131–136 (1995)
15. Yuan, X., Zhang, S., Piboleau, L., Domenech, S.: Une methode d'optimisation nonlineare en variables mixtes pour la conception de procedes. RAIRO **22**(4), 331–346 (1988)

Simplified Tabu Search with Random-Based Searches for Bound Constrained Global Optimization

Ana Maria A. C. Rocha[1(✉)], M. Fernanda P. Costa[2],
and Edite M. G. P. Fernandes[1]

[1] ALGORITMI Center, University of Minho, 4710-057 Braga, Portugal
{arocha,emgpf}@dps.uminho.pt
[2] Centre of Mathematics, University of Minho, 4710-057 Braga, Portugal
mfc@math.uminho.pt

Abstract. This paper proposes a simplified version of the tabu search algorithm that solely uses randomly generated direction vectors in the exploration and intensification search procedures, in order to define a set of trial points while searching in the neighborhood of a given point. In the diversification procedure, points that are inside any already visited region with a relative small visited frequency may be accepted, apart from those that are outside the visited regions.

The produced numerical results show the robustness of the proposed method. Its efficiency when compared to other known metaheuristics available in the literature is encouraging.

Keywords: Global optimization · Tabu search · Random searches

1 Introduction

This paper aims to present a simplified tabu search algorithm, in line of the Directed Tabu Search (DTS) [1], that uses random exploration and intensification local search procedures. This means that the exploration in the neighborhood of a current solution, as well as the final local intensification search are based only on randomly generated vectors. This proposal aims to contribute to the research area of nonlinear bound constrained global optimization (BCGO). The problem is formulated as follows:

$$\begin{aligned} \min \ & f(x) \\ \text{subject to} \ & x \in \Omega, \end{aligned} \tag{1}$$

where $f : \mathbb{R}^n \to \mathbb{R}$ is a nonlinear function and $\Omega = \{x \in \mathbb{R}^n : -\infty < l_i \leq x_i \leq u_i < \infty, i = 1, \ldots, n\}$ is a bounded feasible region. We do not assume that the objective function f is differentiable and convex and many local minima may exist in Ω. The optimal set X^* of the problem (1) is assumed to be nonempty

© Springer Nature Switzerland AG 2020
O. Gervasi et al. (Eds.): ICCSA 2020, LNCS 12251, pp. 606–619, 2020.
https://doi.org/10.1007/978-3-030-58808-3_44

and bounded. The global minimizer is represented by x^* and the global optimal value by f^*.

Solution methods for global optimization can be classified into two classes: the exact methods and the approximate methods [2,3]. Exact methods for global optimization are guaranteed to find an optimal solution within a finite run time that is problem-dependent. Thus, they solve a problem with a required accuracy in a finite number of steps. With approximate methods, it is not possible to guarantee an optimal solution. However, very good solutions can be obtained within a reasonable run time. Most of the approximate methods are stochastic. These methods differ from the deterministic ones in that they rely on random variables and random searches that use selected heuristics to search for good solutions. Metaheuristics are heuristics that can be applied to any general and complex optimization problem. Their search capabilities are not problem-dependent [4].

The Tabu Search (TS) is a trajectory-based metaheuristic that was primary developed in 1986 for combinatorial problems [5], and later on extended to solve continuous optimization problems, e.g., [6–8]. TS for continuous optimization guides the search out of local optima and continues the exploration of new regions for the global optimum. Important aspects of the TS are the definition of a neighborhood of the current solution and the management of a tabu list (a list of the already computed solutions). Along the iterative process, TS maintains a list of the most recent movements, which is used to avoid subsequent movements that may lead to solutions that have already been visited.

To overcome the slow convergence of TS, a local search strategy has been applied to the promising areas found by the TS algorithm, e.g., the simplex method of Nelder-Mead [9] or the Hooke-and-Jeeves [10]. Details can be seen in [1,11–13]. In general, this type of hybridization occurs in the final stage of the iterative process when the current solution is in the vicinity of the global optimum. The DTS method proposed in [1] bases its searches for the global optimum on three main procedures:

- the "neighborhood search procedure" - that aims to generate trial points - is based on local search strategies, like the Nelder-Mead [9] or the adaptive pattern search strategy [14], it generates a tabu list (TL) of already visited solutions and defines tabu regions (TRg) to prevent anti-cycling;
- the "diversification search procedure" goals the random generation of new points outside the already visited regions (VRg) and it also stores a list of the VRg, with their visited frequencies (VRf);
- the "intensification search procedure" aims to refine the best solution found so far and is applied at the final stage of the process.

Hybrid strategies that incorporate other popular metaheuristics into the TS have been proposed so far, but they are mostly applied in solving combinatorial problems. The diversity of TS applications - mostly in the combinatorial optimization area - are illustrated in [15]. The work also gives a detailed description of some of the TS structures and attributes, e.g., tabu list, neighborhood, move, threshold value, short-term memory, recency-based memory, frequency-based memory. For continuous problems, the paper [16] presents a hybrid by the integration of the Scatter Search into the TS.

1.1   The Contribution

The herein presented simplification of the TS algorithm, to solve bound constrained global optimization problems, is denoted by "Simplified Tabu Search (with random-based searches)" (S-TS), and focuses on the random generation of direction vectors to search in the neighborhood of a central point x. These random searches of trial points are implemented in the "exploration search procedure" and in the 'intensification search procedure":

- in the "exploration search procedure", n trial points around the central point x are computed, where each point is obtained by adding a randomly generated number to each component of x;
- In the "intensification search procedure", n normalized direction vectors are randomly generated, one different vector for each trial point computed around the central x.

Furthermore, during the "diversification procedure", our proposal allows the acceptance of a randomly generated point in $[l, u]$ that is inside an already VRg as long as its correspondent VRf is relatively small, apart from the acceptance of a point that is outside all the already VRg.

1.2   Notation

For quick reference, a summary of the notation used in this paper follows:

- $\mathcal{L}$: set of indices of points in the TL,
- $|\mathcal{L}|$: cardinal of $\mathcal{L}$,
- L_{edge}: basic edge length,
- R_{TRg}: radius of the TRg (fixed, for all TRg),
- $NTL_{\max}$: maximum number of points in the TL,
- N_{VRg}: number of VRg,
- VR^i: the center point of the VRg i, where $i = 1, \ldots, N_{VRg}$,
- VRf^i: the frequency of the VRg i (number of generated points that are inside the VRg i, including the center),
- R_{VRg}: radius of the VRg (fixed, for all VRg),
- $Perc$: percentage that allows that a point inside at least one of the already VRg is accepted as a new diversification point,
- $NnoI_{\max}$: allowed maximum number of consecutive iterations where an improved point is not found,
- δ: step size of the search along each random direction,
- ϵ: tolerance to analyze vicinity of f^* (the best known global minimum),
- $rand$: random number uniformly distributed in $[0, 1]$.

1.3   Organization

This paper is organized as follows. In Sect. 2, the most important parts of the new algorithm are presented, namely the new "exploration search procedures", "diversification procedure", "neighborhood search procedure" and the "intensification search procedure". The comparative experiments are shown in Sect. 3 and the paper is concluded in Sect. 4.

2   Simplified Tabu Search

This section describes the S-TS method. We remark that the bound constraints of the problem are always satisfied by projecting any component i of a computed trial point that lies outside the interval $[l_i, u_i]$ to the bound l_i or u_i or, alternatively, randomly projecting into the interval $[l_i, u_i]$ (as it is shown in (2)).

Initially, this variant of the TS algorithm randomly generates a point $x \in [l, u]$

$$x_j = l_j + rand\,(u_j - l_j)\,, j = 1, \ldots, n, \tag{2}$$

and the TL is initialized with x. Using x, the "exploration search procedure" explores randomly in the neighborhood of x to find a set of n trial points, denoted by $s^i, i = 1, \ldots, n$, using randomly generated numbers in $[-1, 1]$ and a positive step size $\delta \in \mathbb{R}^+$,

$$s_j^i = x_j + \delta\,rand_j^{[-1,1]}, j = 1, \ldots, n \tag{3}$$

for $i = 1, \ldots, n$, where $rand_j^{[-1,1]} = -1 + 2\,rand$ represents a random number in $[-1, 1]$ generated from a uniform distribution. (See Algorithm 1 below.) The best of the trial solutions (the one with least objective function value, $f_{\min}$) is saved at $x_{\min}$. If this point has not improved relative to x, the step size δ is reduced and a new set of points $s^i, i = 1, \ldots, n$ are randomly generated centered at x (in line of (3)). If there is an improvement relative to x, all the points $s^i, i = 1, \ldots, n$ are shifted by the same amount. The amount is $x_{\min} - 2x$ if the number of improved iterations so far exceeds n, otherwise is $x_{\min} - x$. This exploration algorithm terminates when the number of iterations exceeds $It_{\max}$, δ falls under $1E - 06$, or the difference between the worst and the best of the trial points is considered small. See Algorithm 2 below.

Algorithm 1. Random exploration trial points algorithm

Require: n, x, l, u, δ;
  1: **for** $i = 1, \ldots, n$ **do**
  2:    **for** $j = 1, \ldots, n$ **do**
  3:       Compute $s_j^i = x_j + \delta\,rand_j^{[-1,1]}$;
  4:    **end for**
  5:    Project s^i componentwise into $[l, u]$;
  6: **end for**

Algorithm 2. Exploration search algorithm

Require: n, x, f, l, u, δ, ϵ, $It_{\max}$;
 1: Set $Nf_{exp} = 0$, $It = 1$, $NImpr = 0$;
 2: Based on x and δ, generate s^i, $i = 1, \ldots, n$ using Algorithm 1;
 3: **repeat**
 4:     Compute $f(s^i)$, $i = 1, \ldots, n$, identify $x_{\min}$, $f_{\min}$ and $x_{\max}$; Update Nf_{exp};
 5:     **if** $f_{\min} < f$ **then**
 6:         Set $NImpr = NImpr + 1$;
 7:         **if** $NImpr > n$ **then**
 8:             Set $d_{\min} = x_{\min} - 2x$;
 9:         **else**
10:             Set $d_{\min} = x_{\min} - x$;
11:         **end if**
12:         Set $x = x_{\min}$, $f = f_{\min}$;
13:         Compute $s^i = s^i + d_{\min}$, project s^i componentwise into $[l, u]$, $i = 1, \ldots, n$;
14:     **else**
15:         Set $\delta = 0.5\delta$;
16:         Based on x and δ, generate s^i, $i = 1, \ldots, n$ using Algorithm 1;
17:         Set $x_{\min} = x$, $f_{\min} = f$;
18:     **end if**
19:     Set $It = It + 1$;
20: **until** $\|x_{\max} - x_{\min}\| \leq 10^2 \epsilon$ or $It > It_{\max}$ or $\delta \leq 1E - 06$

The formal description of the implemented S-TS algorithm for solving the BCGO problem (1), based on random searches to define a set of trial points in the neighborhood of a specified point is presented in Algorithm 3.

This S-TS algorithm aims to find a global optimal solution of a BCGO problem, $f(x_{better})$, within an error of $100\,\epsilon\%$ relative to f^*, i.e., the algorithm stops if the following condition holds:

$$|f(x_{better}) - f^*| \leq \epsilon \max\{1, |f^*|\}, \tag{4}$$

where ϵ is a small positive tolerance. However, if the above condition is not satisfied and the number of function evaluations required by the algorithm to reach the current approximation, Nfe, exceeds the target $Nfe_{\max}$, the algorithm also stops.

2.1   Diversification Procedure

The main loop of the S-TS algorithm (from line 9 to 23 in Algorithm 3) invokes a "diversification procedure" (see line 12 of the algorithm). Here, a randomly generated point y that is not inside any of the already VRg is accepted. Each VRg is defined by its center VR^i, the radius R_{VRg} and its frequency VRf^i, $i = 1, \ldots, N_{VRg}$, being N_{VRg} the number of visited regions defined so far. In this case, a new VRg is created centered at $VR^j = y$, VRf^j is set to 1 and N_{VRg} is updated. However, a point y that is inside any of the VRg may also be accepted. Let k be the index of the VRg that has its center VR^k closest to y,

Algorithm 3. Simplified Tabu Search algorithm

Require: n, l, u, L_{edge}, ϵ, f^*, $Nfe_{\max}$, ϵ_{TS}, Nfe_{TS}, $NnoI_{\max}$;
1: Set $NnoI = 0$, $\mathcal{L} = \emptyset$,
2: Randomly generate $x \in [l, u]$ according to (2), compute $f = f(x)$;
3: Set $x_{better} = x$, $f_{better} = f$, $Nfe = 1$, $\delta = 2L_{edge}$, $success = No$;
4: Based on x, use Algorithm 2 to provide $x_{\min}$ (after Nf_{exp} function evaluations);
5: Set $Nfe = Nfe + Nf_{exp}$;
6: Set $x_{better} = x_{\min}$, $f_{better} = f(x_{\min})$;
7: Initialize TL, $z^1 = x$ $(\mathcal{L} = \{1\})$, set $VR^1 = z^1$, $VRf^1 = 1$, $N_{VRg} = 1$;
8: Set $x_{old} = x = x_{\min}$, $f_{old} = f = f(x_{\min})$;
9: **repeat**
10:     Using x and Algorithm 5, provide $x_{\min}$ (after Nf_{nei} function evaluations);
11:     Set $Nfe = Nfe + Nf_{nei}$;
12:     Generate y using Algorithm 4;
13:     Set $x = y$, compute $f = f(x)$, $Nfe = Nfe + 1$; Identify z^W in TL;
14:     Update TL: set $z^W = x$, $f^W = f$;
15:     **if** $f(x_{\min}) < f_{old}$ **then**
16:         Set $NnoI = 0$, $x_{old} = x_{\min}$, $f_{old} = f(x_{\min})$;
17:     **else**
18:         Set $NnoI = NnoI + 1$;
19:     **end if**
20:     **if** $f(x_{\min}) < f_{better}$ **then**
21:         Set $x_{better} = x_{\min}$, $f_{better} = f(x_{\min})$;
22:     **end if**
23: **until** $NnoI \geq NnoI_{\max}$ or $Nfe \geq Nfe_{TS}$ or $|f_{better} - f^*| \leq \epsilon_{TS} \max\{1, |f^*|\}$
24: Set $\delta = 2L_{edge}$, $Nf^{int}_{\max} = Nfe_{\max} - Nfe$;
25: Using x_{better} and Algorithm 6, provide $x_{\min}$ (after Nf_{int} function evaluations);
26: Set $Nfe = Nfe + Nf_{int}$;
27: Set $x_{better} = x_{\min}$, $f_{better} = f(x_{\min})$;
28: **if** $|f_{better} - f^*| \leq \epsilon \max\{1, |f^*|\}$ **then**
29:     Set $success = Yes$;
30: **end if**

among all VRg that contain y. If its frequency VRf^k is small (relative to all the frequencies of the other VRg), i.e., if

$$\frac{VRf^k}{\sum_{i=1}^{N_{VRg}} VRf^i} < Perc,$$

the point is accepted, where $Perc$ is the percentage for the acceptance of a point y inside an already VRg. In this case the corresponding frequency is updated. The details concerning the acceptance issue and the updating of VRg are shown in Algorithm 4.

The main loop terminates when the number of consecutive iterations with non-improved points exceeds $NnoI_{\max}$, the number of function evaluations exceeds a target value, Nfe_{TS}, or the best function value in the TL has an error of $100\,\epsilon_{TS}\%$ relative to f^*.

Algorithm 4. Diversification search algorithm

Require: $VR^i, VRf^i, i = 1, \ldots, N_{VRg}, R_{VRg}, Perc$;

1: Set $accept = 0$;
2: **while** $accept = 0$ **do**
3:     Randomly generate $y \in [l, u]$ according to (2);
4:     Compute $d_{min} = \min_{i=1,\ldots,N_{VRg}}\{\|y - VR^i\|/R_{VRg}\}$; Set $k = \arg\min_i\{\|y - VR^i\|/R_{VRg}\}$;
5:     **if** $d_{min} > 1$ **then**
6:         Set $accept = 1$, $N_{VRg} = N_{VRg} + 1$, $VR^j = y$ with $VRf^j = 1$ where $j = N_{VRg}$;
7:     **else**
8:         **if** $(VRf^k/\sum_i VRf^i) < Perc$ **then**
9:             Set $accept = 1$, $VRf^k = VRf^k + 1$;
10:        **end if**
11:    **end if**
12: **end while**

2.2   Neighborhood Search Procedure

Based on a starting point x, each iteration of the main loop of the S-TS algorithm also tries to compute a point better than the x, by invoking the "neighborhood search procedure" (see line 10 of the Algorithm 3). Firstly, the randomly generated set of trial points should be preferentially far away from the other points already in the TL. This is accomplished by selecting a specific value for the step size δ. The points in the TL, $z^i, i \in \mathcal{L}$, are the centers of the TRg, and the radius, R_{TRg}, is a fraction of L_{edge}. This means that if x is inside any of the TRg, δ is chosen to be a value greater than the maximum distance from x to the centers of those TRg that contain x, i.e., $\|x - z^i\| \leq R_{TRg}$ (for those i) holds. Otherwise, δ is greater than L_{edge}. The trial points around x are generated by the "exploration search procedure" that provides x_{min}.

The point x is then added to the TL if the number of points in the TL is smaller than the threshold value NTL_{max}. Otherwise, our proposal here is to replace the worst point in the TL, z^W, by x. Parallel to that, the frequencies of the VRg that contain x are updated or, a new VRg is defined centered at x with frequency set to 1, if none of the VRg contains x.

The point x_{min} is compared to x. If an improvement has been obtained (in terms of f), the counter for consecutive iterations with no improvement is set to 0, otherwise, the counter is updated. This iterative procedure terminates when the number of iterations exceeds a target, It_{max}^N, or the number of consecutive iterations with no improvement computed points reaches a threshold value $NnoI_{max}$. For the output of the procedure, x_{min} is identified as the best point (with least objective function value) among the points in TL. See the details in Algorithm 5.

Algorithm 5. Neighborhood search procedure

Require: n, x, f, L_{edge}, $z^i, i \in \mathcal{L}$ (points in the TL), $VR^i, VRf^i, i = 1, \ldots, N_{VRg}$,
$\quad NnoI_{max}$, It^N_{max}, NTL_{max}, R_{TRg}, R_{VRg};
1: Set $NnoI = 0$, $Nf_{nei} = 0$, $It = 0$;
2: **repeat**
3: $\quad$ Set $It = It + 1$, $\mathcal{L}_< = \emptyset$, $flag_{VRg} = 0$;
4: $\quad$ **for all** $i \in \mathcal{L}$ such that $\|x - z^i\| \leq R_{TRg}$ **do**
5: $\quad\quad$ Set $\mathcal{L}_< = \mathcal{L}_< \cup \{i\}$;
6: $\quad$ **end for**
7: $\quad$ **if** $\mathcal{L}_< \neq \emptyset$ **then**
8: $\quad\quad$ Compute $d_{max} = \max\{\|x - z^i\|\}$ for all $i \in \mathcal{L}_<$;
9: $\quad\quad$ $\delta = (1 + rand)d_{max}$;
10: $\quad$ **else**
11: $\quad\quad$ $\delta = (1 + rand)L_{edge}$;
12: $\quad$ **end if**
13: $\quad$ Using x and Algorithm 2, provide x_{min} (after Nf_{exp} function evaluations);
14: $\quad$ Set $Nf_{nei} = Nf_{nei} + Nf_{exp}$;
15: $\quad$ **if** $|\mathcal{L}| < NTL_{max}$ **then**
16: $\quad\quad$ Update TL: set $|\mathcal{L}| = |\mathcal{L}| + 1$, $z^i = x$, $f^i = f$ where $i = |\mathcal{L}|$;
17: $\quad$ **else**
18: $\quad\quad$ Update TL: set $z^W = x$, $f^W = f$;
19: $\quad$ **end if**
20: $\quad$ **for all** i such that $\|x - VR^i\| < R_{VRg}$ **do**
21: $\quad\quad$ Set $VRf^i = VRf^i + 1$; Set $flag_{VRg} = 1$;
22: $\quad$ **end for**
23: $\quad$ **if** $flag_{VRg} = 0$ **then**
24: $\quad\quad$ Set $N_{VRg} = N_{VRg} + 1$, $VR^j = x$ and $VRf^j = 1$ where $j = N_{VRg}$;
25: $\quad$ **end if**
26: $\quad$ **if** $f(x_{min}) < f$ **then**
27: $\quad\quad$ Set $NnoI = 0$;
28: $\quad$ **else**
29: $\quad\quad$ Set $NnoI = NnoI + 1$;
30: $\quad$ **end if**
31: $\quad$ Set $x = x_{min}$, $f = f(x_{min})$;
32: **until** $NnoI \geq NnoI_{max}$ or $It \geq It^N_{max}$
33: Identify x_{min}, f_{min} in TL;

2.3 Intensification Search Procedure

Finally, the "intensification search procedure" is used to intensify the search around the best point found from the main loop of S-TS algorithm (see line 25 in Algorithm 3). It is simple to implement and it does not require any derivative information. Details of this procedure are shown in the Algorithm 6.

The search begins with a central point x (which on entry is the best point found so far, x_{better}) and a set of n trial approximations

$$s^i = x + \delta v^i, \tag{5}$$

where $v^i \in \mathbb{R}^n$ is a normalized vector with random components in $[-1, 1]$, for each $i = 1, \ldots, n$.

This procedure follows a strategy similar to the "exploration search procedure" although with important differences. For each point s^i a random direction vector v^i is generated and the reduction of the step size (when the best of the trial points has not improved over the central point x) is more moderate and δ is not allowed to fall below $1E - 06$. We remark that when the best of the trial points improves relative to x, the algorithm resets δ to a fraction of the value on entry only if there was no improvement in the previous iteration. Furthermore, the termination of the algorithm is activated only when the best of the n trial approximations, denoted by $x_{\min}$, satisfies the stopping conditions as shown in (4), or when the number of function evaluations required by the algorithm exceeds the target $Nf_{\max}^{int} = Nfe_{\max} - Nfe$ (the remaining function evaluations until the maximum $Nfe_{\max}$ is attained, where Nfe is the function evaluations required until "intensification search procedure" is invoked).

Algorithm 6. Intensification search algorithm

Require: n, x, f, l, u, δ, $Nf_{\max}^{int}$, ϵ, f^*;

1: Set $Nf_{int} = 0$, $\delta_0 = \delta$, $flag_{noMove} = 0$;
2: Based on x and δ, generate s^i, $i = 1, \ldots, n$ using Algorithm 7;
3: **repeat**
4:     Compute $f(s^i)$, $i = 1, \ldots, n$; Update Nf_{int}; Identify $x_{\min}$, $f_{\min}$;
5:     **if** $f_{\min} < f$ **then**
6:         Set $d_{\min} = x_{\min} - x$, $x = x_{\min}$, $f = f_{\min}$;
7:         Compute $s^i = s^i + d_{\min}$, project s^i componentwise into $[l, u]$, $i = 1, \ldots, n$;
8:         **if** $flag_{noMove} = 1$ **then**
9:             Set $\delta = 0.95\delta_0$, $\delta_0 = \delta$, $flag_{noMove} = 0$;
10:        **end if**
11:    **else**
12:        Set $\delta = \max\{0.75\delta, 1E - 06\}$, $flag_{noMove} = 1$;
13:        Based on x and δ, generate s^i, $i = 1, \ldots, n$ using Algorithm 7;
14:        Set $x_{\min} = x$, $f_{\min} = f$;
15:    **end if**
16: **until** $|f_{\min} - f^*| \le \epsilon \max\{1, |f^*|\}$ or $Nf_{int} \ge Nf_{\max}^{int}$

Algorithm 7. Random intensification trial points algorithm

Require: n, x, l, u, δ;

1: **for** $i = 1, \ldots, n$ **do**
2:     Generate $v^i \in \mathbb{R}^n$ with random components in $[-1, 1]$ and $\|v^i\| = 1$;
3:     Compute $s^i = x + \delta v^i$; Project s^i componentwise into $[l, u]$;
4: **end for**

---

3   Numerical Results

To analyze the performance of our Algorithm 3, we use two sets of benchmark problems. The first set contains 9 problems and is known as Jones set: Branin (BR) with $n = 2$, Camel Six-Hump (C6) with $n = 2$, Goldstein & Price (GP) with $n = 2$, Hartman 3 (H3) with $n = 3$, Hartman 6 (H6) with $n = 6$, Shekel 5 (S5) with $n = 4$, Shekel 7 (S7) with $n = 4$, Shekel 10 (S10) with $n = 4$, Schubert (SHU) with $n = 2$ (see the full description in [1]).

The second set contains sixteen problems: Booth (BO) with $n = 2$, Branin (BR) with $n = 2$, Camel Six-Hump (C6) with $n = 2$, Dekkers & Aarts (DA) with $n = 2$, Goldstein & Price (GP) with $n = 2$, Hosaki (HSK) with $n = 2$ Matyas (MT) with $n = 2$, McCormick (MC) with $n = 2$, Modified Himmelblau (MHB) with $n = 2$, Neumaier2 (NF2) with $n = 4$, Powell Quadratic (PWQ) with $n = 4$, Rastrigin with $n = 2$, $n = 5$, $n = 10$ (RG-2, RG-5, RG-10), Rosenbrock (RB) with $n = 2$ and Wood (WF) with $n = 4$, see the full description in [17].

The MATLAB® (MATLAB is a registered trademark of the MathWorks, Inc.) programming language is used to code the algorithm and the tested problems.

The values for the parameters are set as follows: $L_{edge} = 0.1 \min_{i=1,\ldots,n}(u_i - l_i)$, $R_{V\,Rg} = 2L_{edge}$, $R_{T\,Rg} = 0.2L_{edge}$, $NnoI_{max} = 2n$. Unless otherwise stated, in the stopping condition (4), the tolerance is set to $\epsilon = 1E - 04$, and $Nfe_{max} = 50000$. We also set $\epsilon_{TS} = 1E + 02\,\epsilon$, $Nfe_{TS} = 0.2\,Nfe_{max}$ and $Perc = 0.25$. In the "neighborhood search procedure", we set the number of iterations with consecutive no improvement points $NnoI_{max} = 2n$, and $It^{N}_{max} = 3n$, $NTL_{max} = 5n$. The parameter for the "exploration search procedure" is $It_{max} = 2n$.

Because there are elements of randomness in the algorithm, each problem was solved several times (100 and 30 depending on the set of problems) by the algorithm using different starting seed for the pseudo-random number generator.

The subsequent tables report the average number of function evaluations required to achieve the stopping condition (4). To test the robustness of the algorithm, the rate of success, i.e., the percentage of runs in which the algorithm obtains a solution satisfying (4), is also shown (inside parentheses). The average number of the objective function evaluations is evaluated only in relation to the successful runs.

With the Tables 1 and 2, we aim to compare our results to those of the Enhanced Continuous TS (ECTS) in [7], the DTS method [1], the Continuous greedy randomized adaptive search procedure (C-GRASP) [18], the Hybrid Scatter TS (H-STS) [16], a simulated annealing hybridized with a heuristic pattern search (SAHPS) [14], a mutation-based artificial fish swarm algorithm (m-AFS) [19], and an improved TS with the Nelder-Mead local search (iTS-NM) available in [12]. The results obtained from our algorithm emanate from the two implemented scenarios when a component of a point lies outside $[l_j, u_j]$:

Case 1 - projecting randomly into $[l_j, u_j]$;
Case 2 - projecting to the bound l_j or u_j.

A comparison with the results of the opposition-based differential evolution (ODE) [20], where the error tolerance in (4) is reduced to $\epsilon = 1E - 08$ and $Nfe_{\max} = 1E + 06$ (only for the Case 1 scenario) is also shown. The results of the listed algorithms are taken from the original papers.

Table 1. Average number of function evaluations and rate of success over 100 runs

	Algorithm 3[†]	ECTS [†,a]	DTS [†,b]	C-GRASP [†,c]	H-STS [†,d]
	Case 1				
BR	242 (100)	245 (100)	212 (100)	10090 (100)	1248 (100)
C6	199 (100)	–	–	–	–
GP	338 (100)	231 (100)	230 (100)	53 (100)	809 (100)
H3	708 (100)	548 (100)	438 (100)	1719 (100)	298 (100)
H6	1015 (100)	1520 (100)	1787 (83)	29894 (100)	1263 (100)
S5	1445 (99)	–	819 (75)	9274 (100)	9524 (100)
S7	1586 (97)	–	812 (65)	11766 (100)	3818 (100)
S10	1742 (96)	–	828 (52)	17612 (100)	3917 (100)
SHU	456 (100)	370 (100)	274 (92)	18608 (100)	1245 (100)

[†] Results based on $\epsilon = 1E - 04$ in (4) and $Nfe_{\max} = 50000$.
[a] Results reported in [7]; [b] results reported in [1];
[c] Results reported in [18]; [d] results reported in [16] (25 runs).
"–" Information not available.

First, when the results of Case 1 are compared to those of Case 2, it is possible to conclude that Case 1 is slightly more robust and efficient in general. The first comparison is with the results of DTS. Our S-TS algorithm wins clearly as far as robustness is concerned. When we compare our results to those of C-GRASP and H-STS - algorithms with similar robustness level - we conclude that Algorithm 3 wins on efficiency. From Table 2, it is possible to conclude that Algorithm 3 wins on robustness - although requires more function evaluations - when compared to SAHPS and iTS-NM. The comparison with ODE is also favourable to our algorithm.

Table 3 contains the results obtained with the second set of problems and aims to compare our Algorithm 3 to the stochastic coordinate descent method (St-CS) available in [21]. A comparison with the results of the ODE [20], where the error tolerance is reduced to $\epsilon = 1E - 08$ and $Nfe_{\max} = 1E + 06$ (only for the Case 1 scenario) is also shown. We note that Case 2 is now slightly more efficient than Case 1. The comparison with St-CS is favourable to the Algorithm 3 as far as robustness and efficiency are concerned. From the comparison with ODE, we may conclude that Algorithm 3 was not able to converge to the solution, with the required accuracy, on 4 of the 16 tested problems, but produced very good results (in terms of robustness and efficiency) to the other problems.

Table 2. More comparisons of function evaluations and rate of success over 100 runs

	Algorithm 3[†]	SAHPS [†,a]	m-AFS [†,b]	iTS-NM [†,c]	Algorithm 3[§]	ODE [§,d]
	Case 2				Case 1	
BR	255 (100)	318 (100)	475 (−)	178 (100)	656 (100)	2804 (100)
C6	192 (100)	−	247 (−)	−	n.a. (0)	2366 (100)
GP	354 (100)	311 (100)	417 (−)	165 (100)	871 (100)	2370 (100)
H3	751 (100)	517 (95)	1891 (−)	212 (100)	1573 (100)	1796 (100)
H6	1143 (100)	997 (72)	2580 (−)	880 (66)	3276 (100)	n.a. (0)
S5	1940 (84)	1073 (48)	1183 (−)	777 (75)	3067 (100)	n.a. (0)
S7	1686 (94)	1059 (57)	1103 (−)	751 (89)	2594 (97)	n.a. (0)
S10	1778 (82)	1031 (48)	1586 (−)	751 (89)	2841 (93)	2316 (100)
SHU	504 (100)	450 (86)	523 (−)	402 (100)	908 (100)	−

[†] Results based on $\epsilon = 1E - 04$ in (4) and $Nfe_{max} = 50000$;
[§] results based on $\epsilon = 1E - 08$ in (4) and $Nfe_{max} = 1E + 06$.
[a] Results reported in [14]; [b] results reported in [19] (30 runs);
[c] results reported in [12]; [d] results reported in [20] (50 runs).
"−" Information not available; "n.a." not applicable.

Table 3. Average number of function evaluations and rate of success over 30 runs

	Algorithm 3[†]		St-CS [†,a]	Algorithm 3[§]	ODE [§,b]
	Case 1	Case 2		Case 1	
BO	201 (100)	244 (100)	1555 (100)	686 (100)	−
BR	244 (100)	232 (100)	239 (100)	650 (100)	2804 (100)
C6	175 (100)	224 (100)	512 (100)	n.a. (0)	2366 (100)
DA	518 (100)	485 (100)	1020 (100)	n.a. (0)	1116 (100)
GP	326 (100)	353 (100)	1564 (100)	790 (100)	2370 (100)
HSK	134 (100)	175 (100)	110 (100)	n.a. (0)	1654 (100)
MT	128 (100)	120 (100)	2159 (100)	436 (100)	1782 (100)
MC	134 (100)	128 (100)	172 (100)	n.a. (0)	1528 (100)
MHB	450 (100)	449 (93)	1450 (100)	937 (83)	−
NF2	10236 (33)	8962 (40)	n.a. (0)	13497 (3)	364300 (8)
PWQ	1230 (100)	1130 (100)	n.a. (0)	52084 (100)	3998 (100)
RG-2	544 (97)	515 (87)	2074 (100)	1023 (93)	−
RG-5	1532 (100)	1356 (100)	6981 (100)	4080 (100)	−
RG-10	4241 (100)	4107 (100)	20202 (100)	12068 (100)	170200 (24)
RB	905 (100)	939 (100)	n.a. (0)	2324 (100)	−
WF	13034 (100)	11856 (100)	n.a. (0)	23305 (100)	54136 (84)

[†] Results based on $\epsilon = 1E - 04$ in (4) and $Nfe_{max} = 50000$;
[§] results based on $\epsilon = 1E - 08$ in (4) and $Nfe_{max} = 1E + 06$.
[a] Results reported in [21]; [b] results reported in [20] (50 runs).
"−" Information not available; "n.a." not applicable.

4   Conclusions

A simplified version of the TS metaheuristic, denoted by S-TS, is presented. The simplifications are concerned mainly with the diversification, exploration and intensification procedures. When searching in the neighborhood of a given central point, both "exploration search procedure" and "intensification search procedure" rely solely on randomly direction vectors. In the "diversification procedure", our S-TS algorithm also allows the acceptance of a randomly generated point that falls inside any already visited region, as long as its visited frequency is small. The new algorithm has been tested and compared to other well-known metaheuristics in the literature.

The testified robustness and efficiency when solving BCGO problems are encouraging and induce us to extend this simplified TS algorithm to solving general nonlinear constrained global optimization problems. The idea is to handle constraint violation and objective function values separately although giving priority to trial points that are feasible with respect to the general constraints.

Acknowledgments. The authors wish to thank two anonymous referees for their comments and suggestions to improve the paper.

This work has been supported by FCT – Fundação para a Ciência e Tecnologia within the R&D Units Project Scope: UIDB/00319/2020, UIDB/00013/2020 and UIDP/00013/2020 of CMAT-UM.

References

1. Hedar, A.-R., Fukushima, M.: Tabu Search directed by direct search methods for nonlinear global optimization. Eur. J. Oper. Res. **170**, 329–349 (2006)
2. Hendrix, E.M.T., Boglárka, G.T.: Introduction to Nonlinear and Global Optimization. In: Optimization and its Applications, vol. 37. Springer, New York (2010). https://doi.org/10.1007/978-0-387-88670-1
3. Stork, J., Eiben, A.E., Bartz-Beielstein, T.: A new taxonomy of continuous global optimization algorithms, 27 August 2018. arXiv:1808.08818v1
4. Sörensen, K.: Metaheuristics - the metaphor exposed. Int. Trans. Oper. Res. **22**, 3–18 (2015)
5. Glover, F.W.: Future paths for integer programming and links to artificial intelligence. Comput. Oper. Res. **13**(5), 533–549 (1986)
6. Siarry, P., Berthau, G.: Fitting of tabu search to optimize functions of continuous variables. Int. J. Num. Meth. Eng. **40**(13), 2449–2457 (1997)
7. Chelouah, R., Siarry, P.: Tabu search applied to global optimization. Eur. J. Oper. Res. **123**, 256–270 (2000)
8. Franzè, F., Speciale, N.: A tabu-search-based algorithm for continuous multiminima problems. Int. J. Numer. Meth. Eng. **50**, 665–680 (2001)
9. Nelder, J.A., Mead, R.: A simplex method for function minimization. Comput. J. **7**, 308–313 (1965)
10. Hooke, R., Jeeves, T.A.: Direct search solution of numerical and statistical problems. J. ACM **8**, 212–229 (1961)

11. Chelouah, R., Siarry, P.: A hybrid method combining continuous tabu search and Nelder-Mead simplex algorithms for the global optimization of multiminima functions. Eur. J. Oper. Res. **161**, 636–654 (2005)
12. Mashinchi, M.H., Orgun, M.A., Pedrycz, W.: Hybrid optimization with improved tabu search. Appl. Soft. Comput. **11**, 1993–2006 (2011)
13. Ramadas, G.C.V., Fernandes, E.M.G.P.: Self-adaptive combination of global tabu search and local search for nonlinear equations. Int. J. Comput. Math. **89**(13–14), 1847–1864 (2012)
14. Hedar, A.-R., Fukushima, M.: Heuristic pattern search and its hybridization with simulated annealing for nonlinear global optimization. Optim. Method. Softw. **19**(3–4), 291–308 (2004)
15. Glover, F., Laguna, M., Martí, R.: Principles and strategies of tabu search. In: Gonzalez, T.F. (ed.) Handbook of Approximation Algorithms and Metaheuristics: Methologies and Traditional Applications, 2nd edn., vol. 1, pp. 361–375. Chapman and Hall, London (2018)
16. Duarte, A., Martí, R., Glover, F., Gortázar, F.: Hybrid scatter tabu search for unconstrained global optimization. Ann. Oper. Res. **183**(1), 95–123 (2011)
17. Ali, M.M., Khompatraporn, C., Zabinsky, Z.B.: A numerical evaluation of several stochastic algorithms on selected continuous global optimization test problems. J. Glob. Optim. **31**(4), 635–672 (2005). https://doi.org/10.1007/s10898-004-9972-2
18. Hirsch, M.J., Meneses, C.N., Pardalos, P.M., Resende, M.G.C.: Global optimization by continuous GRASP. Optim. Lett. **1**(2), 201–212 (2007). https://doi.org/10.1007/s11590-006-0021-6
19. Rocha, A.M.A.C., Costa, M.F.P., Fernandes, E.M.G.P.: Mutation-based artificial fish swarm algorithm for bound constrained global optimization. In: AIP Conference Proceedings, vol. 1389, pp. 751–754 (2011)
20. Rahnamayan, S., Tizhoosh, H.R., Salama, M.M.A.: Opposition-based differential evolution. IEEE Trans. Evol. Comput. **12**(1), 64–79 (2008)
21. Rocha, A.M.A.C., Costa, M.F.P., Fernandes, E.M.G.P.: A population-based stochastic coordinate descent method. In: Le Thi, H.A., Le, H.M., Pham Dinh, T. (eds.) WCGO 2019. AISC, vol. 991, pp. 16–25. Springer, Cham (2020). https://doi.org/10.1007/978-3-030-21803-4_2

A Clustering Approach for Prediction of Diabetic Foot Using Thermal Images

Vítor Filipe[1,2] ⓘ, Pedro Teixeira[1] ⓘ, and Ana Teixeira[1,3(✉)] ⓘ

[1] University of Trás-os-Montes e Alto Douro, Quinta de Prados,
5001-801 Vila Real, Portugal
{vfilipe,ateixeir}@utad.pt,
miguelbento1997@hotmail.com
[2] INESC TEC - INESC Technology and Science, 4200-465 Porto, Portugal
[3] Mathematics Centre CMAT, Pole CMAT - UTAD, Vila Real, Portugal

Abstract. Diabetes Mellitus (DM) is one of the most predominant diseases in the world, causing a high number of deaths. Diabetic foot is one of the main complications observed in diabetic patients, which can lead to the development of ulcers. As the risk of ulceration is directly linked to an increase of the temperature in the plantar region, several studies use thermography as a method for automatic identification of problems in diabetic foot. As the distribution of plantar temperature of diabetic patients do not follow a specific pattern, it is difficult to measure temperature changes and, therefore, there is an interest in the development of methods that allow the detection of these abnormal changes.

The objective of this work is to develop a methodology that uses thermograms of the feet of diabetic and healthy individuals and analyzes the thermal changes diversity in the plantar region, classifying each foot as belonging to a DM or a healthy individual. Based on the concept of clustering, a binary classifier to predict diabetic foot is presented; both a quantitative indicator and a classification thresholder (evaluated and validated by several performance metrics) are presented.

To measure the binary classifier performance, experiments were conducted on a public dataset (with 122 images of DM individuals and 45 of healthy ones), being obtained the following metrics: Sensitivity = 0.73, Fmeasure = 0.81 and AUC = 0.84.

Keywords: Diabetic foot · Thermography · Clustering · Prediction · Binary classification

1 Introduction

Diabetes Mellitus (DM) or diabetes is a chronic disease characterized by the inability of our body to use its main source of energy, glucose (sugar), resulting in increased blood glucose levels (glycemia) [1]. Diabetes is a disease for life, which can have serious consequences if not well controlled [2].

The International Diabetes Federation (IDF) estimated that in 2015 there were about 415 million people with diabetes, and by 2040 the number would increase to

© Springer Nature Switzerland AG 2020
O. Gervasi et al. (Eds.): ICCSA 2020, LNCS 12251, pp. 620–631, 2020.
https://doi.org/10.1007/978-3-030-58808-3_45

642 million, representing 1 in 10 adults worldwide [3]. In 2017, diabetes resulted in approximately 3.2 to 5.0 million deaths [4].

Diabetic foot is one of the main complications observed in diabetic patients, and can be defined as infection, ulceration and/ or destruction of deep tissues [5, 6]. Diabetic patients have between 12% and 25% of risk of developing foot ulcers during life, which is mainly related to peripheral neuropathy, and often with the peripheral vascular disease [7]. Peripheral vascular disease is a significant complication of diabetes and can produce changes in blood flow that will influence a change in skin temperature. An increase in skin temperature can also indicate tissue damage or inflammation associated with some type of trauma or excessive pressure [8]. The risk of ulceration is directly linked to an increase of the temperature in the plantar region; thus, there is a growing interest in monitoring plantar temperature frequently [9, 10].

Infrared thermography (IRT) is a fast, non-invasive and non-contact method that allows visualizing the foot temperature distribution and analyzing the thermal changes that occur [11]. Temperature analysis involves the identification of characteristic patterns and the measurement of thermal changes. It has been shown that in healthy individuals, there is a specific spatial pattern, called the butterfly pattern; while, in the DM groups there is a wide variety of spatial patterns [12–14].

The use of thermal images for the detection of complications in diabetic foot assumes that a variation in plantar temperature is associated with these types of complications. Several research works on the application of thermography to automatic identification of problems in diabetic feet can be found in the literature, being possible to classify those works in four categories, based on the type of the analyzes carried out: independent analysis of the temperature of the limbs, asymmetric analysis, analysis of the distribution of temperature and analysis of external stress [10, 15, 16]. Some brief considerations on these four classifications are presented.

The analysis of an external stress consists in the application of external stimulus to the patient, e.g. immersing the feet in cold water or walking for a while, and analyzing the behavior of the plantar temperature to that stimulus. The main drawback of this type of analysis is that it can be uncomfortable for some people [15, 17, 18]. The independent analysis of temperature allows to obtain representative temperature ranges between the different study groups, however, it is not possible to detect specific areas with some risk related to problems in diabetic foot [10, 15]. Asymmetric temperature analysis consists in comparing the temperature of the foot with the contralateral, in order to define a limit that enables to detect risk areas. This approach has shown good results in several studies, however, it has some limitations; for example, when the patient has similar complications in both feet, because it is not possible to detect the risk areas, and when the patient has a partial or total amputation of the foot, as there is no area to compare with [15, 19, 20]. The analysis of temperature distribution is a method whose main advantage is that it does not use the comparison of the temperature of the feet, allowing to analyze each patient's foot separately. This approach makes it possible to measure changes by calculating a representative value for each foot in the DM group. Therefore, the measurement depends on the temperature distribution and not on a spatial pattern [12, 21, 22].

Several works present a thermal analysis based on the observation of specific points, or specific regions. Considering the temperature of the entire foot it is possible to carry out a complete analysis of its general condition. However, as the foot does not have an uniform temperature, it is important to consider a regional division [23]. For example, in [24, 25], temperatures were recorded in thirty-three regions of interest (ROIs) on both feet (considering the points of the foot that are most likely to ulcer) before the analysis. Nevertheless, as one of the main causes of diabetic foot ulceration is the decrease in blood supply, the division that has been mostly used and discussed in recent years [12–14, 26, 27], divides the feet into four regions according to the concept of angiosome, which is a region of tissue that has blood supply through a single artery. Some of the studies that use a partition of the feet based on the angiosomes propose quantitative indices to estimate the temperature variation; for example, in [26] quantitative information on the distribution of plantar temperature is provided, by identifying differences between the corresponding areas of the right and left feet, while the index presented in [12], takes into account not only the temperature difference between the regions, but also the temperature interval of the control group (constituted by healthy subjects), in order to avoid the limitations of the asymmetric approach.

The aim of this study is to develop a methodology to analyze the diversity of thermal changes in the plantar region of diabetic and healthy individuals, classifying each foot as belonging to a DM or to a healthy individual; thermographic photos of the plant of the feet are used. With this in mind, a binary classifier, based on the clustering concept, was developed to predict the diabetic foot, using both, a quantitative temperature index (CTI) and a classification temperature threshold (CTT).

The calculation of the CTI is based on the division of the foot in clusters and uses the average temperature of each region; thus, low ranges of temperature values within the same region give a better index. The clustering method consists in grouping approximate values of the temperature in a cluster; therefore, dividing the foot in different areas (clusters) that present similar values of temperature [28]. When clusters are used to divide the foot in regions, the temperature values within each region are similar and the range of temperatures within each one of the clusters is low; thus, it is expected to obtain an index capable of measuring thermal variations.

To measure the classifier's performance, binary experiments were performed using a public data set (with 122 images of individuals with DM and another 45 of healthy ones); the performance metrics Sensitivity, Specificity, Precision, Accuracy, F-measure, and Area Under the Curve were used to evaluate and validate the proposed CTT.

This paper is organized as follows: In Sect. 2 the proposed binary classifier is described. In Sect. 3 the used database is introduced and the obtained results are presented and analyzed. Finally, conclusions and guidelines for future work are presented in Sect. 4.

2   Methodology

The proposed method has three stages of processing (Fig. 1): temperature clustering, index computation and classification.

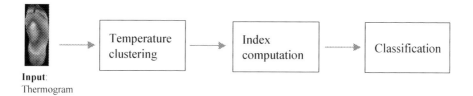

Input:
Thermogram

Fig. 1. Method overview.

In the first stage the foot is divided in regions based on a temperature clustering algorithm. For each cluster, regional parameters of temperature are calculated. Next, an index of temperature variation in relation to the reference values is computed. Finally, applying a thresholding procedure on the index, results concerning the subject classification as healthy or diabetic are presented.

2.1 Clustering

Clustering is the process of identifying natural clusters within multidimensional data based on some similarity measure.

The global objective is to group data points with similar characteristics and assign them to the same cluster. This method is used in several fields, including machine learning, pattern recognition, image analysis, information retrieval, bioinformatics, or computer graphics [28]. One of the most used clustering algorithms is k-means clustering, due to its simplicity and being computationally fast, when operated with a large data set. This algorithm divides a set of data into a k number of clusters, being an iterative process in which it minimizes the sum of the distances of each object to the cluster centroid, in all clusters [29].

Considering an image $I(x,y)$, the objective is to group the pixels in k number of clusters ($Cl_i, i = 1, \ldots k$). Let $p(x,y)$ be an input pixel and c_i the centroid of each cluster (a centroid is the center of a cluster). The k-means algorithm is composed of 5 steps:

1. Initialize the number of clusters, k, and randomly choose k pixels from the image as initial centroids.
2. For each centroid, calculate the Euclidean distance between the pixel and the centroid using (1).

$$d_i = \|p(x, y) - c_i\|, \quad i = 1..k \tag{1}$$

3. Assign the pixel to the nearest center based on the distances computed in (1).
4. After all pixels have been assigned, recalculate the new centroid's value, using (2).

$$c_i = \frac{1}{k} \sum_{y \in Cl_k} \sum_{x \in Cl_k} p(x, y), \quad i = 1, \ldots, k \tag{2}$$

5. Repeat steps 2, 3 and 4 until the centroids stop moving, i.e. k-means algorithm has converged.

6. Reshape the cluster pixels into image.

Although k-means has the great advantage of being easy to implement, it has some disadvantages, being the main one the fact that the quality of the result depends on the arbitrary selection of the initial centroid; thus, being possible to achieve to different results for each cluster in the same image [29].

In this case, as we are dealing with foot temperatures, the goal will be to assign the pixels to different groups, according to its temperature, in order to guarantee that all pixels in the same group has the closest temperature values.

In Fig. 2 two sets of images, resulting from the application of the k-means algorithm to the thermograms (a) and (e), are presented. The first set ((a) to (d)) represents a non-diabetic individual, while the second one ((e) to (h)) a diabetic individual; the second, third and fourth images of each set corresponds to the division of the plant of the foot in three, four and five clusters, respectively.

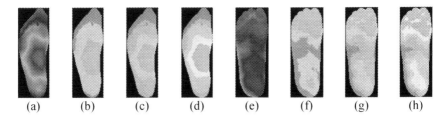

| (a) | (b) | (c) | (d) | (e) | (f) | (g) | (h) |

Fig. 2. Application of the k-means algorithm using different numbers of clusters (a) Original thermogram of a non diabetic individual (b) Obtained from (a) using 3 clusters (c) Obtained from (a) using 4 clusters (d) Obtained from (a) using 5 clusters (e) Original thermogram of a diabetic subject (f) Obtained from (e) using 3 clusters (g) Obtained from (e) using 4 clusters (h) Obtained from (e) using 5 clusters.

After dividing the image in k clusters, for each one them, regional parameters of the temperature are calculated, namely, average, standard deviation, maximum and minimum values.

2.2   The Cluster Temperature Index

Based on the concept of clustering and in order to provide a quantitative estimate of the thermal changes in the foot caused by DM, a new quantitative index is proposed. This index is based on the temperature variation of each cluster in relation to a reference temperature obtained from the healthy individuals (control group). To obtain the reference value of the temperature for each region (cluster), the average of the temperature of the corresponding region (cluster) of the control group is computed.

For each subject, the value of the CTI is calculated as the average of the positive differences between the temperatures of the clusters of an individual (IND) and the correspondent reference values, obtained from the control group, as is (3), where k represents the number of used clusters, n the number of feet of the individuals, $\overline{Tc_i}$ is the reference temperature value of cluster i and IND_{ij} is the average temperature value of cluster i of the jth foot, respectively.

$$CTI_j = \frac{\sum_i^k |\overline{Tc_i} - IND_{ij}|}{k}, \quad j = 1, \ldots, n \tag{3}$$

This index is used to classify each foot as belonging to a healthy or diabetic subject.

2.3 Classification

The CTI index measures the differences in the distribution of a subject's plantar temperature regarding to the reference temperatures, obtained using the control group. Thus, taking into account the reference values and the 334 CTI values obtained using (3), it can be observed that higher index values correspond to greater values of temperature variation; therefore, implying higher risk of the individual developing foot lesion.

Based on the CTI values, in the last stage of the method the selection of an appropriate threshold (CTT), to classify the thermograms in one of the categories: healthy or diabetic is proposed. The success of this classification depends on the CTT used.

2.4 Performance Evaluation and Classification Scheme

To measure the classifier's performance, several experiments were carried out using the metrics: Sensitivity, Specificity, Precision, Accuracy, F-measure, and Area Under the Curve (AUC), defined by (4) to (8), where TP, FP, TN and FN represent the number of cases of True Positive, False Positive, True Negative and False Negative, respectively. The AUC summarizes the entire location of the Receiver Operating Characteristic (ROC) curve and, rather than depending on a specific operating point, is an effective and combined measure of sensitivity and specificity that describes the inherent validity of diagnostic tests [30]. As the classifier is being used to detect the presence of a disease, Sensitivity and Specificity are the most relevant metrics to assess the integrity of the classifier [31]:

$$Sensitivity = \frac{TP}{TP + FN} \tag{4}$$

$$Specificity = \frac{TN}{FP + TN} \tag{5}$$

$$Precision = \frac{TP}{TP + FP} \tag{6}$$

$$Accuracy = \frac{TP + FN}{TP + TN + FP + FN} \tag{7}$$

$$F_{measure} = \frac{2 * TP}{2 * TP + FP + FN} \tag{8}$$

3   Results

To test the binary classifier, thermograms of individuals with and without DM were obtained from a public database that contains 334 individual plantar thermograms, corresponding to 167 individuals (122 diabetic and 45 healthy) [23] (Table 1). Each thermogram, that corresponds to the plant of the left or right feet, is already segmented and vertically adjusted.

Table 1. Characterization of both groups in the database. (Adapted from [23])

	Control group	DM group
Volunteers	45	122
Female	16	89
Male	29	33
Age(years)	27.76 ± 8.09	55.98 ± 10.57

In order to determine the number of clusters that should be used to obtain the best classification, several experiments were carried out, starting with three clusters and increasing this number until six. For each one of these cases, the reference values were computed using the control group temperature values and the CTI was calculated individually for each foot of both groups. From these experiments it was possible to observe that worse metric results are obtained with three clusters than with four or five. But, although the metric values improve when the number of clusters increases from three to four and from four to five, the same does not happen when six clusters are considered; therefore, in this work, five clusters are used. It is worth mentioning that when six or more clusters are used, the values of the metrics do not improve because the temperature variations between each cluster are low and, so, different clusters present very close temperature values; thus, affecting the calculation of the CTI index. For the five clusters case, the reference average values of the temperature per cluster ($\overline{Tc}$) and the correspondent standard deviation (Std. dev.), both measured in centigrade degrees (°C), are the ones in Table 2 and the reference CTI value, obtained with the control group data, is 1.23 ± 0.91 °C.

Table 2. Mean and standard deviation of the temperature (°C) for the control group, per cluster.

Cluster	1	2	3	4	5
$\overline{Tc}$	23.93	25.52	26.59	27.61	28.88
Std. dev.	1.76	1.54	1.57	1.51	0.92

Table 3 illustrates the index values obtained for seven thermograms from the dataset (CG027, CG002, DM102, DM004, DM032, DM027, DM006 in [23]), the first two are from the control group and the remaining five from the DM group.

In the columns of the table there is information concerning: the subject's identification, the thermographic image, the image obtained after clustering, the CTI value, and the measures (mean, standard deviation, minimum and maximum values) of the entire foot and of each one of the five clusters, respectively.

During the feet thermal analysis, it could also be observed that when the clusters of a foot have average temperature values close to the ones of the control reference (Table 2), low values of the CTI are obtained and the butterfly pattern can be observed; this is what happens with the subjects in the control group, e.g. (Table 3 (CG027 and CG002)). Nevertheless, notice that the foot of the example Table 3 (CG002) presents, in general, average values of the temperature clusters more distant from the reference values than the ones of example Table 3 (CG027) and presents a CTI value more than 1 °C higher than the one of Table 3 (CG027) (although still concordant with the reference CTI value); this shows that the TCI index detects small temperature variations.

For the feet of the DM group wide variations of the values of the CTI index can be observed and thermal changes can vary from slightly different from the butterfly pattern to a completely different pattern and this reflects in the CTI value. As shown in Table 3, the CTI is closer to the reference CTI value when slight temperature variations occur, as in Table 3 (DM102) example. As variations start to be more evident (Table 3 (DM102 and DM004)), the CTI value increases, moving further and further away from the reference CTI value. It was also observed that, when the hot spots evolve, they not only become wider, but also warmer; thus, higher CTI values are obtained, e.g. (Table 3 (DM032 and DM027)). The hot spots may even cover the entire plantar region, reaching very high temperatures and presenting very high CTI values, e.g. (Table 3 (DM006)).

Based on the obtained CTI values, a new classification methodology is proposed. As the success of the classification methodology depends on the threshold value used to determine whether the individual is diabetic or not, several threshold values were tested, considering the reference CTI value. In Table 4 the performance measures used to evaluate the classification, for each one of the tested CTT, are presented.

In order to visualize the tradeoffs between the true positive rate (sensitivity) and false positive rate (1-specificity) for different threshold values, we plot in Fig. 3 the Receiver Operating Characteristic (ROC) curve. Calculating the area under the curve (AUC) we obtained value the value 0.838.

Given that the problem under study concerns a disease, special attention must be payed to the values of the sensitivity and specificity metrics [31]. In order to obtain a CTT that balances these two performance metrics, the value of 1.9 is proposed as the limit to classify the individuals as diabetic or healthy. This value was chosen using an approach known as the point closest-to-(0,1) corner in the ROC plane, which defines the optimal cut-point as the point that minimizes the Euclidean distance between the ROC curve and the (0,1) point [32].

Table 3. CTI and temperature measure values, per cluster, for some feet thermograms of the control and the DM groups in [23].

Sub-ject	Ther-mal	Clus-ters	CTI	Meas-ures	Gen-eral	C_1	C_2	C_3	C_4	C_5
CG0 27			0.15	Min	18.01	18.01	24.85	26.15	27.22	28.41
				Max	30.24	24.84	26.15	27.21	28.40	30.24
				Mean	26.75	24.24	25.51	26.72	27.72	29.07
				Std dev	1.48	0.66	0.37	0.29	0.33	0.41
CG0 02			1.23	Min	23.85	23.85	25.97	27.29	28.37	29.46
				Max	30.59	25.97	27.29	28.37	29.45	30.59
				Mean	28.08	25.20	26.76	27.86	28.87	30.00
				Std dev	1.38	0.51	0.35	0.28	0.31	0.29
DM1 02			2.33	Min	25.51	25.51	27.35	28.34	29.33	30.33
				Max	31.62	27.34	28.33	29.31	30.32	31.62
				Mean	28.52	26.88	27.84	28.81	29.84	30.82
				Std dev	1.22	0.36	0.28	0.26	0.29	0.39
DM0 04			3.55	Min	25.79	25.79	28.68	29.70	30.60	31.37
				Max	32.71	28.67	29.70	30.59	31.36	32.71
				Mean	30.71	28.04	29.30	30.15	30.99	31.77
				Std dev	1.08	0.50	0.29	0.26	0.22	0.26
DM0 32			4.10	Min	27.24	27.24	28.74	29.94	31.28	32.49
				Max	34.16	28.74	29.94	31.28	32.49	34.16
				Mean	31.13	28.08	29.36	30.57	32.01	32.99
				Std dev	1.74	0.38	0.33	0.37	0.35	0.35
DM02 7			5.17	Min	26.67	26.67	30.58	31.39	32.06	32.73
				Max	34.28	30.58	31.38	32.05	32.73	34.28
				Mean	31.91	30.08	31.11	31.66	32.43	33.08
				Std dev	0.82	0.55	0.21	0.18	0.19	0.33
DM0 06			7.11	Min	31.46	31.46	32.89	33.32	33.74	34.44
				Max	35.77	32.89	33.31	33.73	34.43	35.77
				Mean	33.61	32.60	33.14	33.49	34.04	34.78
				Std dev	0.63	0.28	0.12	0.12	0.18	0.26

Table 4. Performance measures of the classifier for the different CTT

CTT	Sensitivity	Specificity	Precision	Accuracy	F-measure
2.5	0.631	0.911	0.951	0.707	0.759
2.4	0.648	0.900	0.946	0.716	0.769
2.3	0.664	0.867	0.931	0.719	0.775
2.2	0.672	0.844	0.921	0.719	0.777
2.1	0.697	0.822	0.914	0.731	0.791
2.0	0.705	0.800	0.905	0.731	0.793
1.9	0.730	0.778	0.899	0.743	0.805
1.8	0.734	0.744	0.886	0.737	0.803
1.7	0.750	0.733	0.884	0.756	0.812
1.6	0.766	0.689	0.870	0.746	0.815
1.5	0.795	0.678	0.870	0.764	0.831
1.4	0.816	0.678	0.873	0.778	0.843
1.3	0.828	0.644	0.863	0.778	0.845
1.2	0.853	0.600	0.853	0.784	0.853

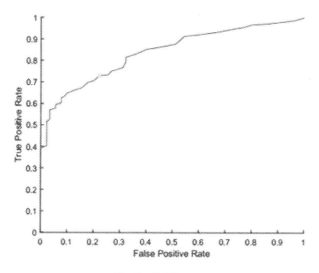

Fig. 3. ROC curve

4   Conclusions

In this work, a methodology that is capable of analyzing a diversity of thermal variations in the plantar region, classifying each foot as belonging to a DM or to a healthy individual is presented. In this approach, the cluster concept is used to obtain an index, CTI, that measures the temperature variations of the plant of the foot. The classification was evaluated by a threshold, that balances the metrics sensitivity and specificity metrics.

Based on the presented results, the concept of cluster proved to be an effective approach to help measuring temperature variations in the plant of the feet and the presented index is able to detect those variations, indicating that higher index values correspond to greater values of temperature variation, therefore, implying a higher risk of the individual developing foot lesion. Additionally, the threshold that was determined allows to classify an individual as having DM or being healthy.

Therefore, this work contributes for health professionals to have access to a classification instrument, that can assist in medical diagnosis, early detection of injury risk and helping to prevent ulceration.

As future work we intend to classify the thermograms of the DM individuals in multiple categories. We also intend to extend the dataset, in order to balance the number of elements of healthy and DM individuals.

Acknowledgments. This work is financed by National Funds through the Portuguese funding agency, FCT – Fundação para a Ciência e a Tecnologia within project UIDB/50014/2020.

References

1. Brison, D.W.: Definition, diagnosis, and classification. In: Baumeister, A.A., (ed.) Menial Retardation, pp. 1–19 (2017). Amelior. Ment. Disabil. Quest. Retard
2. Clark, Jr., C.M., Lee, D.A.: Prevention and treatment of the complications of diabetes mellitus. New Engl. J. Med. **332**(18), 1210–1217 (1995)
3. S. Edition, "IDF Diabetes Atlas," Int. Diabetes Fed (2015)
4. Diabetes. https://www.who.int/en/news-room/fact-sheets/detail/diabetes. Accessed 04 May 2020
5. Apelqvist, J., Larsson, J.: What is the most effective way to reduce incidence of amputation in the diabetic foot? Diab. Metab. Res. Rev. **16**(SUPPL. 1), S75–S83 (2000)
6. Leung, P.: Diabetic foot ulcers - a comprehensive review. Surgeon **5**, 219–231 (2007)
7. Glaudemans, A.W.J.M., Uçkay, I., Lipsky, B.A.: Challenges in diagnosing infection in the diabetic foot. Diab. Med. **32**(6), 748–759 (2015)
8. Frykberg, R.G., et al.: Diabetic foot disorders: a clinical practice guideline (2006 revision). J. Foot Ankle Surg. **45**(5 SUPPL), S1–S66 (2006)
9. Ring, F.: The Herschel heritage to medical thermography. J. Imaging **2**(2), 13 (2016)
10. Hernandez-Contreras, D., Peregrina-Barreto, H., Rangel-Magdaleno, J., Gonzalez-Bernal, J.: Narrative review: Diabetic foot and infrared thermography. Infrared Phys. Technol. **78**, 105–117 (2016)
11. Pereira, C.B., Yu, X., Dahlmanns, S., Blazek, V., Leonhardt, S., Teichmann, D.: InfrCHed thermography. In: Multi-Modality Imaging: Applications and Computational Techniques, pp. 1–30. Springer International Publishing (2018)
12. Hernandez-Contreras, D., Peregrina-Barreto, H., Rangel-Magdaleno, J., Gonzalez-Bernal, J. A., Altamirano-Robles, L.: A quantitative index for classification of plantar thermal changes in the diabetic foot. Infrared Phys. Technol. **81**, 242–249 (2017)
13. Nagase, T., et al.: Variations of plantar thermographic patterns in normal controls and non-ulcer diabetic patients: Novel classification using angiosome concept. J. Plast. Reconstr. Aesthetic Surg. **64**(7), 860–866 (2011)
14. Mori, T., et al.: Morphological pattern classification system for plantar thermography of patients with diabetes. J. Diab. Sci. Technol. **7**(5), 1102–1112 (2013)

15. Adam, M., Ng, E.Y.K., Tan, J.H., Heng, M.L., Tong, J.W.K., Acharya, U.R.: Computer aided diagnosis of diabetic foot using infrared thermography: a review. Comput. Biol. Med. **91**, 326–336 (2017)
16. Madhava Prabhu, S., Verma, S.: A systematic literature review for early detection of type II diabetes. In: 2019 5th International Conference on Advanced Computing and Communication Systems, ICACCS 2019, pp. 220–224 (2019)
17. Balbinot, L.F., Robinson, C.C., Achaval, M., Zaro, M.A., Brioschi, M.L.: Repeatability of infrared plantar thermography in diabetes patients: a pilot study. J. Diab. Sci. Technol. **7**(5), 1130–1137 (2013)
18. Agurto, C., Barriga, S., Burge, M., Soliz, P.: Characterization of diabetic peripheral neuropathy in infrared video sequences using independent component analysis. In: 2015 IEEE 25th International Workshop on Machine Learning for Signal Processing (MLSP), vol. 2015, pp. 1–6 (2015)
19. Kaabouch, N., Chen, Y., Anderson, J., Ames, F., Paulson, R.: Asymmetry analysis based on genetic algorithms for the prediction of foot ulcers. Vis. Data Anal. **7243**, 724304 (2009)
20. Liu, C., Van Baal, J.G., Bus, S.A.: Automatic detection of diabetic foot complications with infrared thermography by asymmetric analysis with infrared thermography by asymmetric analysis. J. Biomed. Opt. **20**(2), 026003 (2015)
21. Hernandez-Contreras, D.A., Peregrina-Barreto, H., De Jesus Rangel-Magdaleno, J., Orihuela-Espina, F.: Statistical approximation of plantar temperature distribution on diabetic subjects based on beta mixture model. IEEE Access **7**, 28383–28391 (2019)
22. Hernandez-Contreras, D., Peregrina-Barreto, H., Rangel-Magdaleno, J., Ramirez-Cortes, J., Renero-Carrillo, F.: Automatic classification of thermal patterns in diabetic foot based on morphological pattern spectrum. Infrared Phys. Technol. **73**, 149–157 (2015)
23. Hernandez-Contreras, D.A., Peregrina-Barreto, H., Rangel-Magdaleno, J.D.J., Renero-Carrillo, F.J.: Plantar thermogram database for the study of diabetic foot complications. IEEE Access **7**, 161296–161307 (2019)
24. Macdonald, A., et al.: Thermal symmetry of healthy feet: a precursor to a thermal study of diabetic feet prior to skin breakdown. Physiol. Meas. **38**(1), 33–44 (2017)
25. Macdonald, A., et al.: Between visit variability of thermal imaging of feet in people attending podiatric clinics with diabetic neuropathy at high risk of developing foot ulcers. Physiol. Meas. **40**(8), 084004. (2019)
26. Peregrina-Barreto, H., Morales-Hernandez, L.A., Rangel-Magdaleno, J.J., Avina-Cervantes, J.G., Ramirez-Cortes, J.M., Morales-Caporal, R.: Quantitative estimation of temperature variations in plantar angiosomes: A study case for diabetic foot. In: Computational and Mathematical Methods in Medicine (2014)
27. Peregrina-Barreto, H., Morales-Hernandez, L.A., Rangel-Magdaleno, J.J., Vazquez-Rodriguez, P.D.: Thermal image processing for quantitative determination of temperature variations in plantar angiosomes. In: 2013 IEEE International Instrumentation and Measurement Technology Conference, pp. 816–820 (2013)
28. Omran, M.G.H., Engelbrecht, A.P., Salman, A.: An overview of clustering methods. Intell. Data Anal. **11**(6), 583–605 (2007)
29. Dhanachandra, N., Manglem, K., Chanu, Y.J.: Image segmentation using k-means clustering algorithm and subtractive clustering algorithm. Procedia Comput. Sci. **54**, 764–771 (2015)
30. Hajian-Tilaki, K.: Receiver operating characteristic (ROC) curve analysis for medical diagnostic test evaluation. Casp. J. Int. Med. **4**(2), 627–635 (2013)
31. Taha, A.A., Hanbury, A.: Metrics for evaluating 3D medical image segmentation: analysis, selection, and tool. BMC Med. Imaging **15**(1), 29 (2015)
32. Unal, I.: Defining an optimal cut-point value in ROC analysis: an alternative approach. In: Computational and Mathematical Methods in Medicine, vol. 2017 (2017)

Mixed Integer Linear Programming Models for Scheduling Elective Surgical Procedures

Hanna Pamplona Hortencio$^{(\boxtimes)}$ and Débora Pretti Ronconi

Department of Production Engineering, EP-USP, University of São Paulo,
Av. Prof. Almeida Prado, 128, Cidade Universitária, São Paulo 05508-900, Brazil
hannapamplona@gmail.com

Abstract. The problem of scheduling surgeries consists of allocating patients and resources to each surgical stage, considering the patient's needs, as well as sequencing and timing constraints. This problem is classified as NP-hard and has been widely discussed in the literature for the past 60 years. Nevertheless, many authors do not take into account the multiple stages and resources required to address the complex aspects of operating room management. The general goal of this paper is to propose a mathematical model to represent and solve this problem. Computational tests were also performed to compare the proposed model with a similar model from the literature, with a 64% average reduction in computational time.

Keywords: Scheduling · Surgery · Mixed integer linear programming · Multiple resources · Multiple stages

1 Introduction

The aging population, a higher incidence of chronic diseases and reduced health care budgets are some of the reasons why hospitals need to improve their productivity [1,2]. In a hospital, the surgery department accounts for the highest revenue (around 70%), although it also represents the highest costs (approximately 40%) [3]. A survey conducted by the Brazilian Federal Council of Medicine and Data Folha Institute [4] on the Brazilian public health care system (SUS) revealed that 45% of patients wait longer than six months to book a surgery. Thus, it is extremely important that public and private hospitals have an efficient management system to reduce patients' waiting time and earn good profits.

Higher productivity in surgery departments depends on some factors, like efficient scheduling of activities, fast setups, and punctuality [5]. Operation research tools used in other service sectors like hotels or restaurants can be used in hospitals to find good results [6]. In this paper, we propose a mixed integer linear

This work has been partially supported by the Brazilian funding agencies FAPESP (grants 2013/07375-0 and 2016/01860-1), CNPq (grants 306083/2016-7), and CAPES.

© Springer Nature Switzerland AG 2020
O. Gervasi et al. (Eds.): ICCSA 2020, LNCS 12251, pp. 632–647, 2020.
https://doi.org/10.1007/978-3-030-58808-3_46

programming (MILP) model to solve the Problem of Scheduling Multiple Surgery Resources.

1.1 Problem Definition

The surgery scheduling problem consists of allocating resources to meet patients' needs, when and where these resources are required [7]. This work focuses on the offline operational hierarchical level, because its goal is to define the scheduling of known patients, considering all the necessary resources and the sequence of activities.

Surgical processes addressed in this paper include three stages: preoperative, perioperative, and postoperative. Many resources may be required to perform each stage and a resource can only be used by one patient at a time. The resource is released when the patient moves to the next stage; this constraint is called a blocking constraint [9]. Figure 1 shows a Gantt chart with an example of scheduling involving three patients. The first column shows patient stages, while the second presents the resources required for each stage. After determining which resources will be used by each patient at each stage, the start and end times of each stage are established, taking into account time constraints.

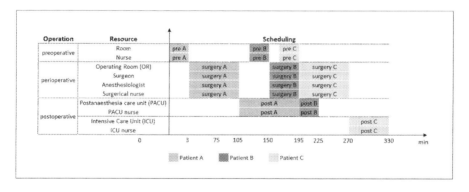

Fig. 1. Gantt chart: example with three patients.

This paper encompasses four sections. Section 1.1 brings a literature review on surgery scheduling at the offline hierarchical level. Section 2 presents two models - one from the literature and the proposed model. Section 3 reports computational experiments to validate and evaluate our model. Final remarks are given in Sect. 4.

1.2 Literature Review

This section consists of a brief literature review of the procedures used to schedule elective surgeries at the offline operational level. These papers have various degrees of complexity, ranging from one single operation and resource to multiple operations and resources. Different solution strategies were found, such as mathematical modeling, exact methods, heuristic approaches, and simulation. Furthermore, distinct objective functions can be defined for the surgery scheduling problem. A literature review shows that 78% of the papers published between 2000 and 2017 use objective functions to efficiently manage resources [8]. Weinbroum, Ekstein and Ezri [9] investigated the reasons that lead to low efficiency in operating rooms and those authors concluded that the unavailability of operating rooms and staff accounts for approximately 66% of delays, while other causes are related to activities that take place in the operating room before or after the surgery, such as the cleaning activities and preoperative and postoperative stages.

Conforti, Guerriero and Guido [10] presented a multi objective model that determines how medical specialties are distributed in surgical rooms and schedules surgeries for elective patients; therefore, it simultaneously addresses problems at the tactical and offline operational levels. Surgery scheduling is based on patients' clinical priorities and considers one stage and one resource. In order to minimize total costs, Fei *et al.* [11] presented a branch-and-price approach to solve a surgery scheduling problem with just one stage and one resource. In a subsequent publication [12], those authors developed a heuristic to define the tactical plan for the surgery department and an exact algorithm to define the patients' schedule. Two resources and just one stage were considered in [12].

Testi and Tafani [13] developed an MILP model that not only schedules surgeries at the operational level but also determines the tactical distribution of specialties. The model works with two stages - perioperative and postoperative -, and a single resource is used in each stage. The objective function seeks to maximize patients' well-being, based on waiting time indicators. Cardoen, Demeulemeester and Beliën [14] developed a model to solve a multiple-objective surgery sequencing problem, where two stages (perioperative and postoperative) are considered with one resource each. The same authors [15] proposed a branch-and-price approach model to solve a day-care scheduling problem. In addition to the perioperative and postoperative stages, the cleaning procedure was also considered. Both studies used data from a Belgian hospital. Perdomo, Augusto and Xie [16] formulated a surgery scheduling problem model that uses two stages (perioperative and postoperative) and two resources (operating room and PACU (Post-Anesthesia Care Unit)). In a subsequent work [17], those authors consider the same stages and resources, but allow a fraction of the postoperative stage to be performed in an operating room, if necessary.

To maximize the marginal financial contribution of each patient, Gartner and Kolisch [18] developed a model to solve the surgery scheduling problem. They described the problem in a Flexible Job Shop machine environment and took into account several stages. Riise, Mannino and Burke [19] also proposed a multi-resource, multi-stage model. The multi-objective function has four terms that are normalized and weighted according to the hospital's needs, therefore being a priority multi-objective optimization method. As pointed out by T'Kindt and

Billaut [20], this method is sensitive and can lead to different results depending on the weighting values used. Pham and Klinkert [21] introduced an MILP model which, in addition to preoperative, perioperative and postoperative stages, also considers cleaning and setup stages, thus addressing around 88% of the causes of delay (see [9]). The model also allows the inclusion of several resources in each stage to better represent an actual hospital setting.

It can be observed from the literature review presented previously in this paper that several surgery scheduling research studies do not consider multiple stages and multiple resources. This paper analyzes the model of Pham and Klinkert [21] and presents suggestions for improvements. This paper also proposes an alternative MILP model for the problem.

2 Mixed Integer Linear Programming Models

This section presents and discusses two MILP models. The first one was developed by Pham and Klinkert [21] and uses the concept of resource modules to simplify resource allocation. The second model was developed by the authors of this paper and aims to improve and expand Pham and Klinkert's model.

2.1 Model Proposed by Pham and Klinkert (2008)

The model presented in [21] for the elective surgery scheduling problem is called Multi-Mode Blocking Job Shop (MMBJS). The authors described the problem as a Flexible Job Shop machine environment, where multiple resources are required to process each task (job). The Job Shop is an environment with m machines and n tasks, where each task has its own route (sequence of operations) to follow. When an operation can be performed by more than one machine, this environment is called the Flexible Job Shop (FJS).

For the problem at hand, patients are equivalent to tasks in the FJS problem and operations correspond to stages in a surgical process. For the execution of each stage, a set of different resources is needed and they are grouped into modules, similarly to machines in the FJS environment. Once a stage is started, all the resources that make up the module cannot be allocated to another patient until the stage is completed and the patient is transferred to the next stage or leaves the system. This restriction is called blocking constraint and aims to eliminate patient waiting time between two stages of the surgical process without being allocated to any module. The MMBJS model and the notations used (Table 1) will be presented next.

Set $\Im$ contains all patients. For each patient $J \in \Im$ there is a set of processing stages (operations); the present work defines three stages for each patient: preoperative, perioperative, postoperative. The set of operations for all patients is called I. The set of two consecutive operations for the same patient J is defined as O_J. The patient to which operation i belongs is called J_i. Set M represents all the resources available in the system. Each operation requires a different set of resources for its execution. A module is a set of resources that meet the requirements for an operation and are available in the same period of time. Modules

Table 1. Notation of the MMBJS.

Sets	
$\mathfrak{I}$	Patients
I	Operations
O_j	Consecutive pairs of operations of patient $J \in \mathfrak{I}$
M	Resources
M_i^r	Resources contained in module r that serves operation $i \in I$, $r \in R_i$, $M_i^r \subseteq M$
R	Module indexes, $R = \cup_{i \in I} R_i$
R_i	Possible module indexes for operation $i \in I$

Parameters	
p_i^r	Processing time of operation i in module r, $i \in I$, $r \in R_i$
p_i^{su}	Setup time before operation $i \in I$
p_i^{cl}	Cleanup time after operation $i \in I$
b_i	Maximum waiting time allowed for operation $i \in I$ before the patient is moved to the next stage
e^r	Start time of the availability range of module $r \in R$
f^r	End time of the availability range of module $r \in R$
H	Big number: $H = \sum_{i \in I} \max_{r \in R_i} \{p_i^r\} + \max_{r \in R_i, i \in I} \{p_i^r\} + \max_{i \in I} \{p^s u_i\} + \max_{i \in I} \{p^c l_i\}$
α	Small weight factor
J^i	Patient which operation $i \in I$ belongs to

Decision variables	
z_i^r	1, if module r is assigned to operation $i \in I$; 0, otherwise
y_{ij}	1, if operation j is processed after operation i; 0, otherwise $i, j \in I$, $i < j$
x_i	Start time of operation $i \in I$
l_i	End time of operation $i \in I$
τ	Makespan

that share the same resource cannot be used by two patients simultaneously. The set of resources that can process operation i is defined as $M_i^r \in M$, where r belongs to the set of indexes of the modules that can process operation i. A fictional module is available at the end of the time horizon. Unscheduled patients are assigned to an artificial module that does not consume any system resources, which is called fictitious module. The set of all indexes of the modules is defined by R. Each operation $i \in I$ has the following parameters: processing time (p_i^r), which is dependent on the module to be used, setup time (p_i^{su}), cleaning (p_i^{cl}), and maximum waiting time (b_i). Modules are available in interval indicated by $[e^r, f^r]$, in which $0 \leq e^r \leq f^r$. Parameters H and *alpha* are a large number and a small number, respectively.

Decision binary variables z_i^r and y_{ij} indicate whether module r will execute operation i and determine the precedence between two operations, respectively. Integer variables x_i and l_i determine the beginning and end of each operation. The makespan is represented by τ and indicates the moment when the last patient leaves the system.

MMBJS model:

$$\min \quad \tau + \alpha \sum_{i \in I} x_i \tag{1}$$

subject to:

$$\sum_{r \in R_i} z_i^r = 1 \quad i \in I \tag{2}$$

$$l_i - x_i - p_i^r z_i^r \geq 0 \quad i \in I, r \in R_i \tag{3}$$

$$l_i - x_i - \sum_{r \in R_i} p_i^r z_i^r \leq b_i \quad i \in I \tag{4}$$

$$l_i - x_j = 0 \quad (i,j) \in O_j, J \in \Im \tag{5}$$

$$x_j - l_i + H(2 - z_i^r - z_j^s) + H(1 - y_{ij}) \geq p_i^{cl} + p_j^{su} \quad i,j \in I, i < j, J^i \neq J^j, \tag{6}$$
$$r \in R_i, s \in R_j, M_i^r \cap M_j^s \neq 0$$

$$x_i - l_j + H(2 - z_i^r - z_j^s) + Hy_{ij} \geq p_j^{cl} + p_i^{su} \quad i,j \in I, i < j, J^i \neq J^j, \tag{7}$$
$$r \in R_i, s \in R_j, M_i^r \cap M_j^s \neq 0$$

$$x_i - e^r z_i^r \geq p_i^{su} \quad i \in I, r \in R_i \tag{8}$$

$$\sum_{r \in R_i} f^r z_i^r - l_i \geq p_i^{cl} \quad i \in I \tag{9}$$

$$\tau - l_i \geq p_i^{cl} \quad i \in I \tag{10}$$

$$x_i, l_i, \tau \geq 0 \quad i \in I \tag{11}$$

$$y_{ij} \in \{0,1\} \quad i,j \in I, i < j \tag{12}$$

$$z_i^r \in \{0,1\} \quad i \in I, r \in R_i \tag{13}$$

The objective function (1) minimizes the makespan, and then the sum of all start times, weighted by factor *alpha*. Constraints (2) assign a single module to each operation. Restrictions (3) and (4) determine the time of entry and exit of an operation. Restrictions (5) determine that there is no time interval between two subsequent operations. Restrictions (6) and (7) prevent two modules that share the same resource from being scheduled simultaneously. Restrictions (8) and (9) guarantee that setup and cleaning times will be carried out within the module's availability interval. Constraints (10) define the makespan value. Restrictions (11), (12) and (13) indicate the domain of decision variables.

Objective Function Analysis. The makespan is defined as $max(C_1, C_2, ..., C_k , ..., C_n)$, where C_k indicates the end time of operation k, that is, the makespan is when the last operation leaves the system. In the model analyzed in this paper, the makespan is calculated based on the set of restrictions (10) and (11) and on the minimization of the objective function. Restrictions (10) indicate that all operations are considered for the evaluation of the makespan, including those that are in the fictional module which does not represent real resources.

Therefore, patients allocated to this module should not be considered in the calculation of the makespan. Moreover, as the fictitious module is at the end of the time horizon, whenever an operation is rejected, the makespan is associated with an unscheduled patient. The following set of constraints can calculate the actual makespan value:

$$\tau \geq l_i + p_i^{cl} - G \sum_{r \in R_i^r} z_i^r \quad i \in I \tag{14}$$

where R_i^r is the set of all non-fictitious modules and G is a sufficiently large number.

The sum of start times of operations aims to reduce the number of patients allocated in the fictitious module. The definition of α is a fundamental task, since it is intended to weigh the values of the two terms of the objective function and guarantee an optimal schedule. By analyzing the objective function, it can be noted that, when the value of α is large, the generated solution can anticipate the beginning of operations without worrying about finishing them as early as possible, worsening the makespan. On the other hand, considering that the makespan calculation is performed in the operations allocated in the real modules, when the α factor is very small, many patients may not be scheduled, i.e. they are allocated in the fictitious module.

Constraint Analysis. The set of constraints (3) and (4) determines the start and end times of operations. Although both constraints have similar functions, they have different structures. Note that the device used in the second expression reduces the number of restrictions. The same occurs for the set of restrictions (8) and (9) that defines that the activities must be performed within the time window of the module. Thus, restrictions (3) and (8) can be replaced with restrictions (15) and (16), respectively.

$$l_i - x_i - \sum_{r \in R_i} p_i^r z_i^r \geq 0 \qquad i \in I \tag{15}$$

$$x_i - \sum_{r \in R_i} e^r z_i^r \geq p_i^{su} \qquad i \in I \tag{16}$$

These alternative constraints may lead to a gain in computational time while solving this problem.

Additional Remarks. By analyzing the objective function (1), it can be seen that there is no prioritization rule for patients. This fact can benefit patient scheduling with shorter processing times. In a hospital environment, the lack of a patient prioritization rule may be undesirable, since patients who require more complicated and therefore longer procedures may not be scheduled. Pham and Klinkert [21] presented a solution to this problem by establishing patient priorities i by fixing variable $y_{ij} = 1$ for all j. By predetermining these variables, it is possible to schedule priority patients when they compete with other patients for the same resource, ensuring that they are scheduled. However, previously determining the sequence of patients reduces the search space and can exclude solutions that include priority patients and have better makespan.

2.2 Proposed Model

The proposed model considers all the analyses presented in the previous section to improve the MMBJS model. First, in order to guarantee the correct makespan evaluation, the dummy module was excluded from the model and an alternative strategy to select the patients was implemented. Furthermore, considering that the definition of the parameter α is not a trivial task and has a great influence on the quality of the solution, a new objective function is proposed. This function also minimizes the number of unscheduled patients but does not use parameter α. A weighting factor associated with each patient was also included in the objective function to consider priority patients.

The set of constraints (3) and (8) can be reduced if replaced with restrictions (15) and (16), respectively. As observed in preliminary tests, fewer restrictions reduced the model's execution time to find an optimal solution. Therefore, this replacement was carried out in the proposed model.

Table 1 and Table 2 show all the notations used in the proposed model. Table 2 presents notations that are exclusive to the proposed model. The makespan value is indicated by mkp to differentiate it from the makespan value of the MMBJS model.

Proposed model:

$$min \quad mkp - G \sum_{i \in I} \sum_{r \in R_i} w_i z_i^r \tag{17}$$

Table 2. Notation of the proposed model.

Parameters	
G	Big number used in the objective function $G = \max_{r \in R}\{f^r\}$
w_i	Weighted priority factor for operation $i \in I$
Decision variables	
mkp	Makespan

subject to:

$$\sum_{r \in R_i} z_i^r \leq 1 \quad i \in I$$

(18)

$$l_i - x_i - \sum_{r \in R_i} p_i^r z_i^r \geq 0 \quad i \in I$$

(19)

$$l_i - x_i - \sum_{r \in R_i} p_i^r z_i^r \leq b_i \quad i \in I$$

(20)

$$l_i - x_j = 0 \quad (i,j) \in O_j, J \in \Im$$

(21)

$$x_j - l_i + H(2 - z_i^r - z_j^s) + H(1 - y_{ij}) \geq p_i^{cl} + p_j^{su} \quad i,j \in I, i < j, J^i \neq J^j,$$

(22)

$$r \in R_i, s \in R_j, M_i^r \cap M_j^s \neq 0$$

$$x_i - l_j + H(2 - z_i^r - z_j^s) + H y_{ij} \geq p_j^{cl} + p_i^{su} \quad i,j \in I, i < j, J^i \neq J^j,$$

(23)

$$r \in R_i, s \in R_j, M_i^r \cap M_j^s \neq 0$$

$$x_i - \sum_{r \in R_i} e^r z_i^r \geq p_i^{su} \sum_{r \in R_i} z_i^r \quad i \in I$$

(24)

$$\sum_{r \in R_i} f^r z_i^r - l_i \geq p_i^{cl} \sum_{r \in R_i} z_i^r \quad i \in I$$

(25)

$$mkp - l_i \geq p_i^{cl} \quad i \in I$$

(26)

$$x_i, l_i, mkp \geq 0 \quad i \in I$$

(27)

$$y_{ij} \in \{0,1\} \quad i,j \in I, i < j$$

(28)

$$z_i^r \in \{0,1\} \quad i \in I, r \in R_i$$

(29)

In the objective function (17), the number of scheduled patient activities, weighted by the priority factor, is calculated by the term $\sum_{i \in I} \sum_{r \in R_i} w_i z_i^r$. This part is multiplied by constant G to ensure that this term will be the problem at the top level, while makespan remains at the bottom level. Note that, in the proposed model, patients' priority is established through factor (w_i). Thus, it is possible to schedule a higher priority patient in the time horizon without necessarily getting this patient to be the first to be scheduled.

Observe that the set of constraints (18) was converted into inequalities when compared with an analogous set of constraints of the MMBJS model. Given the absence of the fictitious module, these restrictions guarantee that each operation should only be scheduled by no more than one module. Constraints (19) and (20) determine the time at which an operation starts and ends, respecting the maximum allowed waiting time. As previously discussed, the set of restrictions (19) was reduced in relation to the MMBJS model. The set of constraints (21) guarantees that there will be no waiting time for the patient between two consecutive operations. Constraints (22) and (23) ensure that modules that share the same resource are not scheduled simultaneously. Restriction sets (24) and (25) ensure that activities will be allocated to modules within their availability range. Two changes were made to these sets with respect to the MMBJS model. The set of restrictions (24) was reduced and the term $\sum_{r \in R_i} z_i^r$ added to the end of the restrictions, thus eliminating the fictitious module of the model. Constraints (26) determine that the makespan is the longest time to end all system activities. These restrictions do not need require any changes in relation to the MMBJS model since there is no longer a fictitious module at the end of the time horizon. Finally, restrictions (27), (28) and (29) determine the domain of decision variables.

Objective Function Analysis. The objective function of the proposed model maximizes the number of scheduled patients and, among those solutions with the largest number of patients, it chooses one that minimizes makespan. It is worth noticing that the modeled problem is not equivalent to the problem of finding a single schedule that maximizes the number of scheduled patients while, in a second stage and preserving the same sequence, minimizes the makespan.

Here is an example to illustrate the reason to use this function. Two situations are analyzed: (i) using the proposed model to maximize only the number of scheduled patients and, subsequently, making an adjustment to reduce the makespan and (ii) analyzing the problem in two levels using the objective function (17). In the example, there are five patients available at the beginning of the horizon and five modules: 1 preoperative, 2 perioperative, and 2 postoperative modules. Figure 2(a) shows the Gantt chart that represents the solution obtained by solving the proposed model to maximize only the number of scheduled patients. In this solution, four patients are scheduled and the makespan value is 1050. Figure 2(b) presents the same sequence of patients' activities adjusted to reduce idle times and, consequently, the makespan. The value of the makespan of this schedule is 980 units of time. Figure 2(c) shows the solution obtained through the resolution of the complete proposed model. The number of scheduled patients remains four, as expected, and the makespan value is 940 units of time. The makespan value is reduced because this solution was found using the proposed MILP model which, among all schedules for 4 patients, seeks one that minimizes makespan.

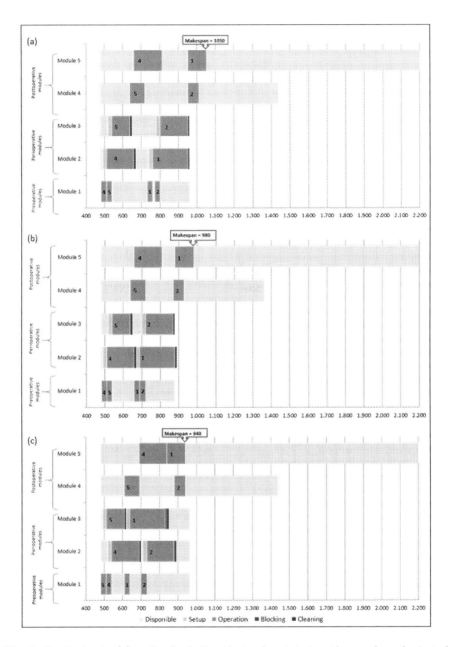

Fig. 2. Gantt charts: (a) optimal solution that only minimizes the number of rejected patients, (b) solution with the same sequence presented in a Gantt chart of (a) without idle times, and (c) optimal solution found by solving the proposed MILP model.

3   Numerical Experiments

The MILP models were implemented in CPLEX ILOG version 12.6. The default configuration parameters have not been changed, except for a CPU time limit of two hours per instance and a suboptimality tolerance level of 0.005%. The experiments were carried out with an Intel (R) Xeon (R) processor, 6 cores, 4.20 20 GHz, 512 GB of RAM, and on a Linux operating system.

Since the problem of scheduling elective surgeries with multiple resources and multiple stages has not yet been widely studied in the literature, new instance sets were generated for this problem based on instances from the literature. Ozkarahan [22] proposed instances with surgery processing time divided by specialty. Pham and Klinkert [21] completed the data from Ozkarahan [22], adding the processing times of preoperative and postoperative stages, maximum waiting time, and setup and cleaning times. Since full instances have not been made available, the present work also includes similar instances that were randomly generated. The proposed set of instances considers three medical specialties in which patients are distributed. The random number generator and the seeds proposed by Taillard [23] were used to generate processing, setup, cleaning, and maximum waiting times, as well as a weighted priority factor for patients using a discrete uniform distribution. The resources required for each stage were determined as suggested by Pham and Klinkert [21] and the time horizon considered was five days.

In the first part of the numerical experiments, a small set of instances was generated to conduct a comparative assessment of the models. Next, a sensitivity analysis was carried out to investigate the performance of the proposed model in different scenarios. With this purpose, 245 instances were generated and the obtained results will be briefly discussed in Sect. 3.2.

3.1   Comparative Evaluation

In this section, numerical experiments were carried out to compare the proposed model and the MMBJS model. Since the MMBJS model cannot properly calculate the makespan value if a patient is rejected, instances with enough resources were generated to avoid unscheduled patients. Moreover, in the experiments with this model, each instance was tested several times in order to find the appropriate value of α.

Table 3 shows the results obtained using the MILP models presented in the previous section. In the instances considered, there are 10 patients, 16 resources, and 47 modules distributed in 5 days. The MMBJS* model corresponds to the MMBJS model with the replacement of constraints (3) and (8) with constraints (15) and (16), respectively. The first column identifies the instance. The following columns show the running time required to solve each instance and the column named GAP (%) indicates the percentage difference between the incumbent solution and the lower bound determined by CPLEX.

Figures in the table show that the solver execution time using the MMBJS model is greater than the corresponding time of the proposed model for all instances. The use of the proposed model led to a 64% reduction in the average computational time. Furthermore, the proposed model was able to find the optimal solution for all instances, while with the MMBJS model, three optimal solutions could not be found within the limit time, with an average gap of 22.34%. By comparing models MMBJS* and MMBJS, it can be seen that, in most instances, the computational time of CPLEX is shorter using the MMBJS* model. The average reduction in computational time was 54%, indicating that a reduction in the restriction set size was beneficial. On the other hand, when comparing this model with the proposed model, it can be noted that, in general, the proposed model is solved in less time and, unlike the MMBJS* model, all the optimal solutions were found.

Table 3. Performance of the analysed models.

Instance	MMBJS		MMBJS*		Proposed	
	Time (s)	GAP (%)	Time (s)	GAP (%)	Time (s)	GAP (%)
1	247.97	00.00	196.59	00.00	169.19	00.00
2	338.99	00.00	90.38	00.00	256.82	00.00
3	7214.19	29.90	7200.37	21.25	144.08	00.00
4	7206.98	13.25	1200.00	00.00	120.19	00.00
5	146.58	00.00	117.15	00.00	84.03	00.00
6	922.54	00.00	1188.33	00.00	407.17	00.00
7	7238.41	23.88	749.04	00.00	154.46	00.00

3.2 Sensitivity Analysis

In this section, numerical experiments were performed to analyze the results obtained by CPLEX when using the proposed model for the resolution of instances with different characteristics. Three characteristics of the base group of instances ($ID = 0$) were modified: availability interval of modules, except for the ICU modules, number of ICU postoperative modules, and number of preoperative modules. For each characteristic, two groups of instances were created. In the first group, modules were added or the availability of modules was increased in relation to the base problem configuration, making the system more flexible. In the second group, the availability of modules was reduced or modules were excluded, consequently, the system became more restricted. In each group, five different amounts of available patients were considered (5, 7, 10, 12, 15). For each combination of characteristic, group, and number of patients, 7 instances were generated, i.e, $3 \times 2 \times 5 \times 7 = 210$ instances were solved. Additionally, 35 instances (5 sets of patients $\times$ 7 instances) were generated to compose the base group.

The solver was able to achieve the optimal solution within the allowed execution time in approximately 86% of instances. In the remaining instances, the average gap was around 9%. As expected, the solutions found by the solver for more restricted instances contain results (number of scheduled patients) equal to or worse than the results of the base group. Conversely, in instances where idleness was added to the proposed model, the solver found similar or better results. This pattern was verified in the 210 instances tested. In order to illustrate the solver's behavior using the proposed model, results obtained for instances with 10 patients and different features are detailed in Table 4. In this table, the first column (*ID*) identifies the instance and its attributes (characteristic and group), while the column named GAP (%) indicates the percentage difference between the incumbent solution and the lower bound determined by CPLEX. Details about the numerical experiment described in this section can be found in [24].

Table 4. Results for an instance with 10 patients.

ID	Makespan	# scheduled patients	Unscheduled patients	GAP (%)	CPU time (s)
0	3 d 13 h 43 min	8	6, 9	0.0*	203.40
1.1	4 d 16 h 41 min	10	–	0.0*	169.19
1.2	4 d 13 h 25 min	8	6, 9	0.0*	144.30
2.1	4 d 13 h 25 min	10	–	0.0*	2369.01
2.2	3 d 14 h 52 min	6	1, 6, 9, 10	0.0	85.61
3.1	3 d 13 h 43 min	8	6, 9	0.0*	176.45
3.2	4 d 13 h 16 min	8	6, 9	0.0*	238.49

*indicates that the suboptimality tolerance level has been reached

4   Final Remarks

This research addressed the scheduling of elective surgical procedures with multiple stages and limited resources. According to the literature review, it was possible to observe that most prior studies have not considered multiple stages and resources in their mathematical models. The present work aimed to study and improve the MILP model presented in [21], called Multi-Mode Blocking Job Shop. First, the MMBJS model was analyzed and modifications were suggested to improve the addressed model, such as a reduction in the number of constraints and an effective way to evaluate the makespan of the final schedule. Then, a mixed integer linear programming model based on MMBJS was proposed. The benefits of this objective function were discussed through the analysis of an example.

Numerical experiments were performed. A comparison between the MMBJS model and the proposed model showed a significant improvement in execution time. A sensitivity analysis was then performed to interpret the model's behavior. As a suggestion for future work, heuristics and metaheuristics should be developed to solve large instances seeking to reduce the computational effort.

Acknowledgments. The authors would like to thank the careful reading of the reviewers, whose comments helped to improve the quality of the manuscript.

References

1. Demeulemeester, E., Belien, J., Cardoen, B., Samudra, M.: Operating room planning and scheduling. In: Denton, B. (ed.) Handbook of Healthcare Operations Management: Methods and Applications, pp. 121–152. Springer, New York (2013). https://doi.org/10.1007/978-1-4614-5885-2_5
2. Samut, P., Cafri, R.: Analysis of the efficiency determinants of health systems. Soc. Ind. Res. **129**, 113–132 (2016)
3. Molina-Parente, J.; Framian, J.: Testing planning policies for solving the Elective Case Scheduling phase: a real application. In: 35th International Conference on Operational Research Applied to Health Services, pp. 1–10. ORAHS, Leuven (2009)
4. Federal Council of Medicine Homepage. http://portal.cfm.org.br/. Accessed 15 Mar 2020
5. Kaye, A., McDowell, J., Diaz, J., Buras, J., Young, A., Urman, R.: Effective strategies in improving operating room case delays and cancellations at an academic medical center. J. Med. Pract. Manage. **30**(6), 24–29 (2015)
6. Gupta, D., Denton, B.: Appointment scheduling in health care: challenges and opportunities. IIE Trans. **40**(9), 800–8019 (2008)
7. Hans, E., Vanberkel, P.: Operating theatre planning and scheduling. In: Hall, R. (ed.) Handbook of Heathcare System Scheduling, pp. 105–130. Springer, New York (2013). https://doi.org/10.1007/978-1-4614-1734-7_5
8. Hortencio, H., Ronconi, D., Miyake, D.: Produtividade na programaç ao de centros cirúrgicos: revisão de literatura. XX Simpósio de Administração da Produção. Logística e Operações Internacionais, pp. 1–16. FGV-EAESP, São Paulo (2017)
9. Weinbroum, A., Ekstein, P., Ezri, T.: Efficiency of the operating room suite. Am. J. Surg. **185**(3), 244–250 (2003)
10. Conforti, D., Guerriero, F., Guido, R.: A multi-objective block scheduling model for the management of surgical operating rooms: new solution approaches via genetic algorithms. In: 2010 IEEE Workshop on Health Care Management, Venice, pp. 1–5. IEEE (2010)
11. Fei, H., Chu, C., Meskens, N., Artibac, A.: Solving surgical cases assignment problem by a branch-and-price approach. Int. J. Prod. Econ. **112**(1), 96–108 (2008)
12. Fei, H., Meskens, N., Combes, C., Chu, C.: The endoscopy scheduling problem: a case study with two specialised operating rooms. Int. J. Prod. Econ. **120**(2), 452–462 (2009)
13. Testi, A., Tànfani, E.: Tactical and operational decisions for operating room planning: efficiency and welfare implications. Health Care Manage. Sci. **12**(363), 363–373 (2009)
14. Cardoen, B., Demeulemeester, E., Beliën, J.: Optimizing a multiple objective surgical case sequencing problem. Int. J. Prod. Econ. **119**(2), 354–366 (2009)
15. Cardoen, B., Demeulemeester, E., Beliën, J.: Sequencing surgical cases in a day-care environment: an exact branch-and-price approach. Comput. Oper. Res. **36**(9), 2660–2669 (2009)
16. Perdomo, V., Augusto, V., Xie, X.: Operating theatre scheduling using Lagrangian relaxation. Eur. J. Ind. Eng. **2**(2), 172–189 (2008)

17. Augusto, V., Xie, X., Perdomo, V.: Operating theatre scheduling with patient recovery in both operating rooms and recovery beds. Comput. Ind. Eng. **58**(5), 231–238 (2010)
18. Gartner, D., Kolisch, R.: Scheduling the hospital-wide flow of elective patients. Eur. J. Oper. Res. **233**(3), 689–699 (2014)
19. Riise, A., Mannini, C., Burke, E.: Modelling and solving generalised operational surgery scheduling problems. Comput. Oper. Res. **66**, 1–11 (2016)
20. T'Kindt, V., Billaut, J.: Multicriteria Scheduling - Theory, Models and Algorithms, 2nd edn. Springer, Heidelberg (2006). https://doi.org/10.1007/b106275
21. Pham, D., Klinkert, A.: Surgical case scheduling as a generalized job shop scheduling problem. Eur. J. Oper. Res. **185**(3), 1011–1025 (2008)
22. Ozkarahan, I.: Allocation of surgeries to operating rooms by goal programing. J. Med. Syst. **24**(6), 339–378 (2000)
23. Taillard, E.: Benchmarks for basic scheduling problems. Eur. J. Oper. Res. **64**(2), 278–285 (1993)
24. Hortencio, H.: Técnicas de pesquisa operacional aplicadas ao problema de programação de cirurgias eletivas. São Paulo (2019). www.teses.usp.br. Accessed 23 Apr 2020

PDE Based Dense Depth Estimation for Stereo Fisheye Image Pair and Uncertainty Quantification

Sandesh Athni Hiremath$^{(\boxtimes)}$

Valeo Schalter und Sensoren, Hummendorfer Straße 76, 96317 Kronach-Neuses,
Germany
sandesh.athni-hiremath@valeo.com

Abstract. Dense estimation of depth from stereo pinhole image pair is
a well known problem and seems to be easily solvable by both classical
as well as machine learning algorithms. However, dense depth estimation
from stereo fisheye image pair is apparently a more challenging problem.
There are two main factors adding to the complexity of the problem: (I)
owing to the wider field of view, the fisheye lenses are inherently nonlinear thus making the estimation problem harder. (II) because we want
to estimate depth for every pixel coordinate from just two images, i.e.
a calibrated stereo pair, the data is statistically insufficient thus greatly
complexifying the problem. To alleviate (II) many depth estimation algorithms enlarge their parameter space by incorporating pose parameters
and work on image sequence (esp. videos). Here we stick to the stereo
camera setting and provide a novel estimation technique and quantify
its uncertainty. We use the technique of variational calculus to derive
a (noisy) pde for which wellposedness is proved and then solved in the
least square sense using proximal algorithms. This setting enables us
to invoke the standard machinery of nonlinear least squares for generating the covariance estimates. Most state of the art algorithms that
are based on deep learning (nonlinear regression) technique fail to say
anything about the uncertainty of the outputs which is of paramount
importance in safety critical applications for e.g. arising in automotive domain. Lastly, our methodology has the advantage of being mas-
sively parallelizable for hardware implementation thereby offering lesser
runtime.

Keywords: Stereo fisheye correspondence · Nonlinear optimization ·
Uncertainty quantification

1 Introduction

Tremendous research in image processing algorithms has given rise to a wide
variety of applications across all domains ranging from photography, entertain-
ment, surveillance, medicine, robotics, etc. to name a few. Similarly, even the

© Springer Nature Switzerland AG 2020
O. Gervasi et al. (Eds.): ICCSA 2020, LNCS 12251, pp. 648–663, 2020.
https://doi.org/10.1007/978-3-030-58808-3_47

automotive industry has seen a surge of vision based applications especially for semi or fully automated driving. The range of applications include lane detection, traffic sign recognition, object/obstacle detection, parking assistance and so on. Consequently, cameras are becoming one of the mandatory sensor for an automobile. Due to limitation on the number of installed cameras and the need for incorporating more number of safety features, the choice for a wider angle cameras become quite obvious [7,8]. Owing to this necessity fisheye cameras, whose FOV is larger than 180°, are becoming popular for automotive applications. As a result, many of the computer vision algorithms are being adapted for wide-angle lenses [1–4,9,10,22].

Stereo image correspondence is a well known classical computer vision problem which involves finding the location of the objects of the first image in the second image. For a pinhole cameras this problem simplifies to finding the shift in pixels i.e. disparity between the images. The disparity is then used to calculate the depth of the observed objects, thereby enabling a dense 3D reconstruction. Many well know algorithms, both classical as well as deep learning based approaches, exists for this problem [5,6,14,17,23] and some of them serve as a standard in industry. However, the same correspondence problem for fisheye images is still a open problem. The main challenge when dealing with fisheye image is the distortion introduced by the lens (see [1]). This distortion is modeled by a nonlinear camera projection model. A simple workaround is to use the camera projection model to remove the distortion and produce a rectified stereo image pair for which pinhole stereo algorithms can be applied. This approach was followed in [20] which heavily relied of the stable disparity outputs of the standard stereo matching algorithms such as [5,6].

However, there is a growing demand for algorithms that directly work on raw distorted images mainly due to growth in automotive, robotics and drone applications. In [16], the authors used conventional stereo matching algorithms using directly a nonlinear fisheye projection model and in [13] PDE based technique was used to find image correspondences. Analogous to [14], in [19] the authors have used variational methods to directly match fisheye images obtained at different poses by locally linearizing the error term. In contrast to these classical algorithms, authors in [15] adapted deep learning techniques, where they modified/adapted the existing network for pinhole models and trained it on fisheye images. One of the main aspect lacking in all these methods is that they fail to quantify the uncertainty of their estimates and (one of the) goal of this paper is to fill this gap by providing a probabilistic modeling of the problem and to quantify the uncertainty of the estimates.

The paper is organized in the following manner, first we formulate the problem and propose a model in Sect. 2, in Sect. 3 we provide a probabilistic interpretation of the model and propose a method for estimating the variance, in Sect. 4 we propose a numerical method for solving the model and provide results. Finally in Sect. 5 we discuss the reults and make some concluding remarks.

2   Modelling

The problem we want to solve is the following: given a pair of images I_L and I_R obtained from two calibrated fisheye cameras forming a stereo setup, estimate the depth densely i.e. for every pixel of the image grid.

2.1   Problem Formulation

Mathematically, this can be written in the following precise way- given two calibrated fisheye images $I_L, I_R : \mathfrak{D} \to \mathfrak{R}$, with compact domain $\mathfrak{D} \subset \mathbb{R}^2$ and range $\mathfrak{R} \subset \mathbb{R}^3$, find the mapping $\lambda : \mathfrak{D} \to \mathbb{R}^+$ that indicates the depth of objects represented in the image I_L(or I_R)[1].

To solve this problem we need to appeal to the underlying epipolar constraints. According to that, if the baseline of the stereo camera is sufficiently wide, images I_L and I_R will have significant overlap and could be viewed as shifted versions of each other. However, for fisheye cameras, unlike in pinhole cameras, the shift is along a curve[2] instead of along a line. The curve is obtained by projecting the curve on the right camera (resp. left camera) surface, traced by a light ray incident on the left camera (resp. right cam), onto the right image (resp. left image) plane. For fisheye and other omnidirectional lenses, images are distorted more on the edges as compared to the middle. Based on the coordinate position of the left image (resp. right image) the curves traced on I_R is of varied degree of curvature, this again adds to the complexity.

Different models for fisheye cameras have been proposed [1,11,21] which have varied degree of accuracy and simplicity. For our work, we use the *enhanced unified camera model* (EUCM) [12] mainly because of its simplicity and reasonable accuracy.

In Figs. 1a–c we illustrate the validity of the EUCM projection model. Firstly, for a given pair of fisheye stereo images of a scene, a set of prominent matching features are selected. The ray passing through a feature point of left image would trace a curve (as a function of depth) on the right image. For the correct calibration parameters, the curve passes through the exact same feature point corresponding to the left image. The point where the curve hits the matching feature point represents the true depth value. The feature points are marked in black dots in left image and the curves are marked in cyan on the right image. The curves pass through the feature points (nearly) exactly. Thus the correctness of projection model and the calibration of the involved parameters is validated.

2.2   Optimization Model

As per the problem formulation Sect. 2.1, the aim is to estimate a positive valued scalar function λ. To this end, we adopt the framework of 'calculus of variations'.

---

[1] Here we always associate λ as a depth mapping for the left image I_L.

[2] Based on vertical or horizontal stereo setup, the curve extends either vertically or horizontally.

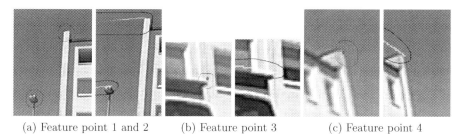

(a) Feature point 1 and 2   (b) Feature point 3   (c) Feature point 4

Fig. 1. The Figs. 1a–c show the magnification of the feature points (shown as red dot) of the left image and curve (cyan color) that passes through those respective features on the right image. The pairs are the subimages of the corresponding fisheye stereo image pair. (Color figure online)

Before we proceed, let us make some basic assumptions on the functions and their spaces. We assume that I_L, I_R are the elements of the Hilbert space $L^2(\mathfrak{D})$ and we expect λ also be an element of $L^2(\mathfrak{D})$.

In order to invoke the tools variational calculus, we first need to define a loss functional whose minimizing element represents the solution to the problem reasonably well.

$$E[\lambda(u,v)] = \frac{1}{2} \int_{\mathfrak{D}} \overbrace{\|I_L(u,v) - I_R(\hat{u}(\lambda), \hat{v}(\lambda))\|_{L^2}^2}^{\text{Data fidelity term}} + \overbrace{s^2(\lambda, \partial_u\lambda, \partial_v\lambda)}^{\text{regularizer term}} \, d\mu \qquad (1)$$

1. $(u,v) \in \mathfrak{D} \subset \mathbb{R}^2$
2. $\lambda : \mathfrak{D} \to \mathbb{R}$ is the depth map which needs to be estimated
3. $I_L : \mathfrak{D} \to \mathbb{R}$ and $I_R : \mathfrak{D} \to \mathbb{R}$ are the left and right fisheye images forming a stereo pair.
4. $(\hat{u}, \hat{v}) \in \mathfrak{D}$ is the pixel coordinate of the right image for the corresponding pixel coordinate $(u,v) \in \mathfrak{D}$ of the left image. More precisely we have that $(u,v) \mapsto \hat{u}(\lambda(u,v))$ and $(u,v) \mapsto \hat{v}(\lambda(u,v))$
5. $(u,v) \mapsto s(u,v)$ is the magnitude of the normal vector to the tangent plane, at the position (u,v) on the surface generated by the 3D point $Z(\lambda(u,v))$.

The **data fidelity** term is based on the brightness constancy assumption and is required to minimize the error in reconstruction of the first image after sampling from the second image.

The **regularizer** term is required to ensure structural consistency as well as to ensure solvability of the minimization problem and the candidacy the solution in the appropriate function space. The intuitive meaning of the term used here is that $(u,v) \mapsto s(u,v)$ represents the infinitesimal area of the 3D surface generated by the mapping

$$(u,v) \mapsto Z(\lambda(u,v)) := \lambda(u,v)\, X(u,v).$$

Here Z represents the 3d world point which can also be seen as a function of the depth map $(u,v) \mapsto \lambda(u,v)$. More precisely, we also have that $\lambda \mapsto Z(\lambda) := \lambda X$.

Note that, since $X(u,v)$ is obtained after unprojecting the image coordinates (using the projection model), it is of unit norm and remains constant for a given set of camera intrinsics. Consequently, Z varies as per the depth map $\lambda(u,v)$. Based on this, the area of the surface spanned by Z can be computed in the following way:

$$S = \iint s(u,v) \, du \, dv := \iint |Z_u \times Z_v| \, du \, dv$$

where, Z_u and Z_v are the partial derivatives of Z w.r.t u and v respectively. Altogether, we have that

$$S = \iint s(u,v) \, du \, dv := \iint |Z_u \times Z_v| \, du \, dv := \iint |a\lambda_u \lambda + b\lambda\lambda_v + c\lambda^2| \, du \, dv.$$

where, $(u,v) \mapsto a(u,v) \in \mathbb{R}^3$, $(u,v) \mapsto b(u,v) \in \mathbb{R}^3$, and $(u,v) \mapsto c(u,v) \in \mathbb{R}^3$ are constant vector functions given as

$$a := X \times X_v = (a_1, a_2, a_3)^T, \, b := X_u \times X = (b_1, b_2, b_3)^T, \, c := X_u \times X_v = (c_1, c_2, c_3)^T.$$

From the above formula it is clear that the regularizer term s^2 in (1) is directly inducing a smoothness constraint on the depth map.

To estimate the depth map for either one of the stereo fisheye image pair (I_L, I_R), we need to solve the following variational problem:

$$\hat{\lambda}(u,v) = \operatorname*{argmin}_{\lambda \in L^2(\mathfrak{D})} E[\lambda(u,v)]. \tag{2}$$

A necessary condition for $\hat{\lambda}$ to be the minimizer of $E[\lambda]$ is that it must satisfy the corresponding Euler-Lagrange equation which is given as:

$$\left(I_L(u,v) - I_R(\hat{u}(\hat{\lambda}), \hat{v}(\hat{\lambda})) \right) \left(\nabla I_R|_{\hat{\lambda}} \cdot \nabla \hat{\mathbf{u}}|_{\hat{\lambda}} \right) - \partial_\lambda s^2 + \nabla \cdot (\nabla_{\lambda_u, \lambda_v} s^2) = 0 \tag{3}$$

The partial derivatives of s^2 w.r.t λ are computed in the following way: let

$$s(\lambda, \partial_u \lambda, \partial_v \lambda) := |Z_u \times Z_v| = |a\lambda_u \lambda + b\lambda\lambda_v + c\lambda^2|, \tag{4}$$

then

$$\partial_\lambda s^2 = 2\left[\sum_{k=1}^{3} \lambda(\lambda_u a_k + \lambda_v b_k + \lambda c_k)^2 + \sum_{k=1}^{3} \lambda^2 c_k(\lambda_u a_k + \lambda_v b_k + \lambda c_k) \right],$$

$$\partial_{\lambda_u} s^2 = 2\left[\lambda^2 (\lambda_u \|a\|_2^2 + \lambda_v(a,b)_2 + \lambda(c,a)_2) \right],$$

$$\partial_{\lambda_v} s^2 = 2\left[\lambda^2 (\lambda_u(a,b)_2 + \lambda_v \|b\|_2^2 + \lambda(c,b)_2) \right].$$

Next we make an important simplification, namely we want to convexify the regularizer term. Note that, from Eq. (4) it is clear that the mapping $\lambda \mapsto s^2(\lambda, \cdot, \cdot)$ is a quadric (4th-order) polynomial, consequently its first order derivatives will be at most a cubic polynomial which in general is non-convex. In order to account for this we simply define a new mapping $\lambda \mapsto \hat{s}(\lambda, \cdot, \cdot) := \frac{1}{\lambda} s(\lambda, \cdot, \cdot)$ and take $\hat{s}$ as the regularizer term. As per this, we have the following modified loss functional

$$\hat{E}[\lambda(u,v)] = \frac{1}{2} \int_{\mathfrak{D}} \overbrace{\|I_L(u,v) - I_R(\hat{u}(\lambda), \hat{v}(\lambda))\|_{L^2}^2}^{\text{Data fidelity term}} + \overbrace{\hat{s}^2(\lambda, \partial_u \lambda, \partial_v \lambda)}^{\text{convexified regularizer term}} \quad d\mu. \quad (5)$$

Its corresponding Euler-Lagrange equation reads as

$$\left(I_L(u,v) - I_R(\hat{u}(\hat{\lambda}), \hat{v}(\hat{\lambda}))\right)\left(\nabla I_R|_{\hat{\lambda}} \cdot \nabla \hat{\mathbf{u}}|_{\hat{\lambda}}\right) - \partial_\lambda \hat{s}^2 + \nabla \cdot (\nabla_{\lambda_u, \lambda_v} \hat{s}^2) = 0, \quad (6)$$

where,

$$\hat{s}(\lambda, \partial_u \lambda, \partial_v \lambda) := \frac{1}{\lambda} |Z_u \times Z_v| = |a\lambda_u + b\lambda_v + c\lambda|$$

$$\partial_\lambda \hat{s}^2 = 2\left(\lambda_u(a,c) + \lambda_v(b,c) + \lambda \|c\|_2^2\right),$$

$$\partial_{\lambda_u} \hat{s}^2 = 2\left(\lambda_u \|a\|_2^2 + \lambda_v(a,b)_2 + \lambda(c,a)_2\right),$$

$$\partial_{\lambda_v} \hat{s}^2 = 2\left(\lambda_u(a,b)_2 + \lambda_v \|b\|_2^2 + \lambda(c,b)_2\right).$$

Equation (6) can be abstractly written in the following form:

$$A(u,v)\lambda = f(\lambda), \quad (7)$$

where,

1. the mapping $(u,v) \mapsto A(u,v) \in L(D(A), L^2(\mathfrak{D}))$ with

$$A(u,v) := \nabla \cdot \left(K_1(u,v) \nabla \lambda\right) + K_2(u,v) \cdot \nabla \lambda - r(u,v)\lambda,$$

where

$$K_1(u,v) := \kappa_1 \begin{pmatrix} p_1 & p_2 \\ p_2 & p_4 \end{pmatrix} := \kappa_1 \begin{pmatrix} \|a\|_2^2 & (a,b)_2 \\ (a,b)_2 & \|b\|_2^2 \end{pmatrix}$$

$$K_2(u,v) := \kappa_2 \begin{pmatrix} p_3 \\ p_5 \end{pmatrix} := \kappa_2 \begin{pmatrix} (a,c)_2 \\ (b,c)_2 \end{pmatrix}$$

$$r(u,v) := q(u,v) + \kappa_2 \begin{pmatrix} p_3 \\ p_5 \end{pmatrix} \cdot \nabla$$

$$q(u,v) := \kappa_3 \|c\|_2^2$$

2. the mapping $\lambda \mapsto f(\lambda) \in L^2(\mathfrak{D})$ with

$$f(\lambda) := -\left(I_L(u,v) - I_R(\hat{u}(\lambda), \hat{v}(\lambda))\right)\left(\nabla I_R|_{\hat{\lambda}} \cdot \nabla \hat{\mathbf{u}}|_{\hat{\lambda}}\right) - \kappa_4 \left(\lambda \nabla \cdot K_2(u,v)\right)$$

3. $\kappa := (\kappa_1, \kappa_2, \kappa_3, \kappa_4)$ is a vector of model parameters and is a positive constant.

In order to solve (7) we still need to enforce boundary condition for which we make the obvious choice of Neumann boundary condition.

Theorem 1. *Let the images I_L, $I_R \in H^1(\mathfrak{D}) \cap L^\infty(\mathfrak{D})$, the functions $a(u,v)$, $b(u,v)$, $c(u,v) \in C^2(\mathfrak{D}) \cap L^\infty(\mathfrak{D})$, the coordinate functions $\hat{u}$ and $\hat{v} \in C^2(\mathfrak{D}) \cap L^\infty(\mathfrak{D})$. Additionally, let $K_1(u,v)$ is uniformly positive definite for each $u, v \in \mathfrak{D}$, i.e. $\xi^T K_1 \xi \geq \theta|\xi|^2$, $\theta > 0$, $\forall \xi \in \mathbb{R}^2, (u,v) \in \mathfrak{D}$ and $q(u,v) \geq q_{min} > 0$. Then the pde given in (7) has a unique solution $\lambda \in H^1(\mathfrak{D})$.*

Proof. Consider the bilinear form $B[\phi, \psi] := \int_\mathfrak{D} K_1 \nabla\phi \cdot \nabla\psi + \int_\mathfrak{D} q\phi\psi$

$$B[\phi, \psi] = \int_\mathfrak{D} p_1 \partial_u\phi\partial_u\psi + 2(p_2\partial_v\phi\partial_u\psi) + p_4\partial_v\phi\partial_v\psi + \int_\mathfrak{D} q\psi\phi$$

$$\Rightarrow |B[\phi, \psi]| \leq C\|\phi\|_{H^1}\|\psi\|_{H^1}$$

$$B[\phi, \phi] = \int_\mathfrak{D} (\partial_u\phi a, \partial_u\phi a)_2 + 2(\partial_v\phi a, \partial_u\phi b)_2 + (\partial_v\phi b, \partial_v\phi b)_2 + \int_\mathfrak{D} q\phi\phi$$

$$\geq \theta\|\phi\|_{H^1}^2$$

Thus by Lax-Milgram we have an unique solution to $A\lambda = g$ in $H^1(\lambda)$ for $g := f(\lambda) \in L^2(\mathfrak{D})$. This implies the resolvent operator $A^{-1} : L^2 \to H^1$ maps every $g \in L^2$ to a corresponding $\lambda \in H^1$.

Since $\|\lambda\|_{H^1}^2 \leq \frac{1}{\theta}B[\lambda, \lambda] \leq \frac{1}{\theta}\|g\|_2\|\lambda\|_{H^1}$ we get that $\|A^{-1}g\|. = \|\lambda\|. \leq \|\lambda\|_{H^1} \leq \frac{1}{\theta}\|g\|_{L^2}$, we get that

$$\|A^{-1}g\|_{H^1} \leq \frac{1}{\theta}\|g\|_{L^2}, \quad \text{as well as} \quad \|A^{-1}g\|_{L^2} \leq \frac{1}{\theta}\|g\|_{L^2}.$$

Hence, both $A^{-1} : L^2 \to H^1$ and $A^{-1} : L^2 \to L^2$ are continuous linear operators. Howerver, since $H^1 \hookrightarrow\hookrightarrow L^2$ the operator $A^{-1} : L^2 \to L^2$ is compact.

Now the weak formulation of the original problem is equivalent to finding a fixed point of $T\lambda := A^{-1}f(\lambda)$. Following as above, we have $\|T\lambda\|. = \|A^{-1}f(\lambda)\|. \leq \frac{1}{\theta}\|f\|_{L^2} \leq \frac{1}{\theta}\|f\|_{L^\infty}$. Thus maps bounded sets in L^2 to bounded sets in H^1 and since $H^1 \hookrightarrow\hookrightarrow L^2$ we get that $T : L^2 \to L^2$ is a compact operator. To show that $T : L^2 \to L^2$ is also continuous, consider a sequence $(\lambda_m)_{m\in\mathbb{N}}$ such that $\lambda_m \to \lambda$ in $L^2(\mathfrak{D})$, then there exists a subsequence $(\lambda_{m_k})_{k\in\mathbb{N}}$ such that $\lambda_{m_k}(\cdot) \to \lambda(\cdot)$ a.e. in $\mathfrak{D}$. Thus, by continuity of A^{-1} we have that

$$\|T\lambda_{m_k} - T\lambda\|_{L^2} = \|A^{-1}f(\lambda_{m_k}) - A^{-1}f(\lambda)\|_{L^2} = \|A^{-1}(f(\lambda_{m_k}) - f(\lambda))\|_{L^2}$$

$$\leq \frac{1}{\theta}\|f(\lambda_{m_k}) - f(\lambda)\|_{L^2}$$

Since, $f \in C^2$, $\lambda_{m_k} \to \lambda$ implies $\|f(\lambda_{m_k}) - f(\lambda)\|_{L^2} \to 0$ as $k \to \infty$. Thus we also get that $\|T\lambda_{m_k} - T\lambda\|_{L^2} \to 0$ as $k \to \infty$. Finally, we have that $T : L^2 \to L^2$ is compact and continuous in L^2 with $T(H^1) \subset H^1$ thus by Shauder's fixed point theorem we get that $\exists \lambda \in L^2$ such that $\lambda = T\lambda$ in $L^2(\mathfrak{D})$. Moreover, since $\|T\lambda\|_{H^1} = \|T\lambda\|_{H^1} \leq C < \infty$ implies that $\lambda \in H^1(\mathfrak{D})$.

Due to the boundedness of the $\mathfrak{D} \subset \mathbb{R}^2$, I_L, I_R clearly belong to $L^2(\mathfrak{D})$. Because of the continuity and boundedness of the EUCM projection model, the coordinate functions $\hat{u}, \hat{v}$ also smooth and bounded. Similarly, even the cross product functions a, b, c can be verified to be smooth and bounded. To ensure uniform ellipticity of the diffusion coefficient and edge aware diffusion we modify $K_1(u, v)$ as:

$$K_1(u, v) := \kappa_1 \begin{pmatrix} \exp\left[-\kappa_5|\partial_u I_L| + \kappa_6(1 + p_1)\right] & \exp\left[-\frac{\kappa_5}{2}(|\partial_u I_L| + |\partial_v I_L|) + \kappa_6(1 + p_2)\right] \\ \exp\left[-\frac{\kappa_5}{2}(|\partial_u I_L| + |\partial_v I_L|) + \kappa_6(1 + p_2)\right] & \exp\left[-\kappa_5|\partial_v I_L| + \kappa_6(1 + p_4)\right] \end{pmatrix} \quad (8)$$

Since $|\partial_x I_L|$ and $|\partial_y I_L|$ are uniformly bounded, K_1 is uniformly elliptic if and only if $p_1 + p_4 > 2p_2$ i.e. $\|a\|^2 + \|b\|^2 - (a, b)_2 > 0$ i.e. $\|a - b\|^2 > 0$. If necessary, a sufficiently small constant is added to the diagonal elements, (especially where $a(u, v) = b(u, v)$ or $a = b = 0$), of the K_1 in order to make it strictly positive definite.

3 Uncertainty Quantification

3.1 Noise Model

The necessary condition for finding the minimum of the Functional (5) is to solve the Eq. (7). In order to consider the errors induced in the image formation we incorporate noise in the model which takes the following form:

$$A(u, v)\lambda = f(\lambda) + W, \quad (9)$$

where $W \sim N(0, \Sigma(\lambda))$ is a $L^2(\mathfrak{D})$-valued Gaussian random variable with covariance operator $\Sigma \in L(L^2(\mathfrak{D}); L^2(\mathfrak{D}))$. The noise term depends on the depth function which facilitates positivity of the solution. It is important to note here that, even though we have introduced the noise at the level of Euler-Lagrange equation (7), one may also, equivalently, introduce at the level of functionals (5). Intuitively, what this means is that, the data fidelity term reads as $\|I_L(u, v) - I_R(u'(\lambda), v'(\lambda)) + W\|_{L^2}^2$, where $W \sim N(0, \Sigma(\lambda))$ is a $L^2(\mathfrak{D})$ valued Gaussian random variable.

The noisy model is now solved in the L^2 sense i.e.

$$\underset{\lambda \in L^2(\mathfrak{D})}{\operatorname{argmin}} \frac{1}{2}\|A\lambda - f(\lambda)\|_{\Sigma^{\frac{1}{2}}(L^2)}^2. \quad (10)$$

We note here that the solution to a least squares problem is given by solving the normal equation.

Theorem 2 (Least squares solution). *Let H_1 and H_2 be two Hilbert spaces, let $A : H_1 \to H_2$ be a bounded linear operator. Then the least squares solution to $Ax = y$ for $x \in H_1$ and $y \in H_2$ is given as $\hat{x} = \operatorname{argmin}_{x \in H_1} \frac{1}{2}\|Ax - y\|_{H_2}^2$ and is obtained by solving the normal equation*

$$A^*Ax = A^*y \quad i.e \quad \hat{x} := (A^*A)^{\dagger}A^*y,$$

where $(A^*A)^{\dagger}$ denotes the pseudo inverse of A^*A.

This is a nonlinear least squares problem which we solve following the techniques provided in the next section. In order to numerically solve (10) we invoke the machinery of convex optimization esp. techniques involving the use of proximal operators. Additionally, it can be supplemented by constraints such as positivity, smoothness, shearlet, wavelet and so on. To facilitate this we consider the following generic minimization formulation:

$$\underset{x,y}{\operatorname{argmin}} \; \frac{1}{2}\|Ax - f\|_2^2 + \iota_S(y) \quad \text{s.t.} \quad Qx = y \qquad (11)$$

Letting ι_S be the indicator function of the set S and Q a bounded linear operator. We now mention a primal-dual algorithm that is used to solve the minimization problems of the type (11).

Algorithm 1: PDHGMx

Data: $\kappa > 0$, $\tau_x, \tau_p > 0$, $\vartheta \in (0,1]$, x^0, p^0

1   $\bar{x}^0 := x^0$
2   $\tilde{x}^0 := x^0$
3   **for** $n = 0, 1 \ldots$ **do**
4     $p^{n+1} = p^n + \tau_p Q\bar{x}^n - \tau_p P_S\left(\frac{1}{\tau_p}p^n + Q\bar{x}^n\right)$
5     $\tilde{x}^{n+1} = x^n - \tau_x Q^* p^{n+1}$
6     $x^{n+1} = (A^*A + \frac{1}{\tau_x})^{-1}(A^*f + \frac{1}{\tau_x}\tilde{x}^{n+1})$
7     $\bar{x}^{n+1} = x^{n+1} + \vartheta(x^{n+1} - x^n)$
8   **end**

3.2 Covariance Computation

Based on the Algorithm 1 we can compute the uncertainty of the estimate x^{n+1} in the following way. Let $\hat{A} = [(A,0),(0,\frac{1}{\sqrt{\tau_x}})]$ and $\hat{f} = [f, \frac{1}{\sqrt{\tau_x}}\tilde{x}^{n+1}]$. Furthermore, let $\hat{f} \sim N(\mu_{\hat{f}}, \Sigma_{\hat{f}})$, then line-6 can be seen as a solution to the following problem

$$x^{n+1} = \underset{x}{\operatorname{argmin}} \; \frac{1}{2}\|\hat{A}x - \hat{f}\|^2_{\Sigma_{\hat{f}}^{\frac{1}{2}}}.$$

The solution to this is given as $x^{n+1} \in N(\mu_x, \Sigma_x)$ with

$$\mu_x = (\hat{A}^*\Sigma_{\hat{f}}^{-1}\hat{A})^{-1}\hat{A}^*\Sigma_{\hat{f}}^{-1}\mu_{\hat{f}} \text{ and } \Sigma_x = (\hat{A}^*\Sigma_{\hat{f}}^{-1}\hat{A})^{-1}.$$

For a special case when f and $\tilde{x}^{n+1}$ are independent of each other, Σ_f and Σ_f^{-1} can be written as

$$\Sigma_f = \begin{bmatrix} \Sigma_1 & 0 \\ 0 & \Sigma_2 \end{bmatrix} \text{ and } \Sigma_f^{-1} = \begin{bmatrix} \Sigma_1^{-1} & 0 \\ 0 & \Sigma_2^{-1} \end{bmatrix}$$

where $f \sim N(\mu_1, \Sigma_1)$ and $\tilde{x}^{n+1} \sim N(\mu_2, \Sigma_2)$. Correspondingly, we have that

$$\hat{A}^*\Sigma_{\hat{f}}^{-1}\hat{A} = A^*\Sigma_1^{-1}A + \frac{1}{\tau_x}\Sigma_2^{-1}, \quad \hat{A}^*\Sigma_{\hat{f}}^{-1}\mu_{\hat{f}} = A^*\Sigma_1^{-1}\mu_1 + \frac{1}{\tau_x}\Sigma_2^{-1}\mu_2$$

and

$$\Sigma_x = \left(A^* \Sigma_1^{-1} A + \frac{1}{\tau_x} \Sigma_2^{-1}\right)^{-1}. \tag{12}$$

4   Implementation

As mentioned above (Sect. 2) we are interested in estimating the depth map corresponding to an raw image obtained by fisheye cameras in a stereo setup. Firstly, the intrinsic camera parameters K_L and K_R are determined and kept fixed. Similarly, the relative transformation R and T between the two cameras are also determined and kept fixed. The operators R and T is used to transform the origin of left camera into the origin of the right camera as explained above in Sect. 2. The images I_L and I_R obtained are converted to grayscale images and are normalized to $[0, 1]$. The raw images are of resolution 800×1200 which are cropped to 512×1024 in order to consider only the main overlapping regions of the images. Finally, the 512×1024 images are downscaled to 128×256 and fed for optimization. The obtained depth map of size 125×512 which is then upscaled to the actual size. Since (10) is nonlinear we solve it locally and iteratively, i.e.

$$\lambda_{m+1} = \underset{\lambda \in L^2(\mathfrak{D})}{\operatorname{argmin}} \frac{1}{2} \|A\lambda - f(\lambda_m)\|^2_{\Sigma^{\frac{1}{2}}(L^2)}. \tag{13}$$

The constraints $a > 0, b > 0, c > 0$ and $a \neq b$, mentioned below formula (8), were numerically verified for the parameters in Table 1. This ensures uniform ellipticity of K_1 as a result of which we get existence of a weak solution. Consequently, the least square solution is equivalent to the weak solution. Moreover, to ensure positivity of the solution, we enforce box-constraints i.e. $0 < \lambda_{min} \leq \lambda \leq \lambda_{max}$ with $\lambda_{min} := .5$ and $\lambda_{max} := 100$. By setting $Q := I$ and $P_S := P_{[\lambda_{min}, \lambda_{max}]}$ in Algorithm 1 we get the Algorithm 2.

For evaluating I_R at non integer pixel locations $(\hat{u}, \hat{v})$ simple bilinear interpolation was used. For solving the resulting matrix equations, algebraic multigrid solvers were used [18]. For the covariance estimate we use the formula (12) with

$$\Sigma_1 := \left(\frac{\lambda}{(1+\lambda)} (\nabla I_R|_\lambda \cdot \nabla \hat{\mathbf{u}}|_\lambda)\right)^2 \text{ and } \Sigma_2 := I. \tag{14}$$

Since we are solving the quadratic problem (Eq. (13)) locally, numerical values of Σ_1 are obtained by evaluating (14) at λ_m^n. Based on this we have that $\Sigma_{\lambda_m^{n+1}} = \left(A^*(\Sigma_1(\lambda_m^n))^{-1} A + \frac{1}{\tau_x}\right)^{-1}$. As $\lambda_m^n \to \hat{\lambda}$ in $L^2(\mathfrak{D})$ we have $\Sigma_1(\lambda_m^n) \to \Sigma_1(\hat{\lambda})$, consequently we also get that $\Sigma_{\lambda_m^n} \to \Sigma_{\hat{\lambda}}$. Since covariance computation involves matrix inversion it is a relatively expensive operation. In order to make the computation practically tractable, we make the following simplification: (i) during covariance computation we downscale the image to the

---

Algorithm 2: PDHGMx

Data: $\kappa > 0$, τ_x, $\tau_p > 0$, $\vartheta \in (0,1]$, λ^0, $p^0 = 0$, tol $= 10^{-15}$

1 **Initialization:** $\bar{\lambda}_0^0 := \lambda^0$, $\tilde{\lambda}_0^0 := \lambda^0$, $\Sigma_{\lambda_0^0} := I$

2 **for** $m = 0, 1 \ldots$ **do**

3 $\quad$ $f_m := f(\tilde{\lambda}_m^0)$

4 $\quad$ **for** $n = 0, 1 \ldots$ **do**

5 $\quad\quad$ $p_m^{n+1} = p_m^n + \tau_p \bar{\lambda}_m^n - \tau_p P_S \left(\frac{1}{\tau_p} p_m^n + \bar{\lambda}_m^n \right)$

6 $\quad\quad$ $\tilde{\lambda}_m^{n+1} = \lambda_m^n - \tau_x p_m^{n+1}$

7 $\quad\quad$ $\lambda_m^{n+1} = (A^* A + \frac{1}{\tau_x})^{-1}(A^* f_m + \frac{1}{\tau_x}\tilde{\lambda}^{n+1})$

8 $\quad\quad$ $\bar{\lambda}_m^{n+1} = \lambda_m^{n+1} + \vartheta(\lambda_m^{n+1} - \lambda_m^n)$

9 $\quad\quad$ $\Sigma_{\lambda_m^{n+1}} = \left(A^*(\Sigma_1(\lambda_m^n))^{-1} A + \frac{1}{\tau_x} \right)^{-1}$

10 $\quad$ **end**

11 $\quad$ $\bar{\lambda}_{m+1}^0 = \bar{\lambda}_m^{n+1}$

12 $\quad$ $\tilde{\lambda}_{m+1}^0 = \tilde{\lambda}_m^{n+1}$

13 $\quad$ **if** $\|\tilde{\lambda}_m^{n+1} - \lambda_m^{n+1}\| < tol$

14 $\quad\quad$ break;

15 $\quad$ **endif**

16 **end**

Table 1. Simulation parameters

	Model parameters						Numerical parameters					
	κ_1	κ_2	κ_3	κ_4	κ_5	κ_6	ϑ	τ_x	τ_p	m	n	Tol
Scene1	.01, .001	1	1	.001	.05	.1	1	1	.001	35	1	10^{-15}
Scene2	.01, .005	1	1	.001	.05	.1	1	1	.001	35	1	10^{-15}
Scene3	.01	1	1	.1	.05	.1	1	1	.001	25	1	10^{-15}
Scene4	.01	1	1	.1	.05	.1	1	1	.001	25	1	10^{-15}

size 32×64, and (ii) compute only the variance for these pixels. The obtained variance image is then upscaled to the original resolution of 512×1024.

Due to the nonlinearity of the problem the choice of initial condition is highly important. In order to generate a good initial estimate one may use any available algorithms ranging from naive exhaustive global search based method or SGM mehtod or an estimate generated from DNN. We take the output of the FisheyeDistanceNet [15] and use the semantic segmentation output to assign max values to the regions of the image that represent sky. It is specially important for outdoor scenes where sky occupies majority of the scene. This strategy motivates one to supplement an offline algorithm, such as multi-task (trained to infer semantic and depth values) DNN, with an online estimation algorithm. The combination provides two fold advantage: (i) on the one hand it provides a more realistic result by improving the initial estimate based on the actual data and (ii) on the other hand improves safety of the system by rectifying grossly erroneous DNN outputs. We use of this philosophy to estimate depths for different outdoor scenes. Their results are as shown below in Fig. 2. Figures 2b–q show the estimated depths. The use of semantically segmented Figs. 2a–p has enabled easy identification of the sky region resulting in a much clear depth output. Figures 3a and 3b depict the convergence of the numerical algorithm for all four scenes considered. Scene3 and Scene4 are taken from a video, for which we used more than two images to obtain image correspondences. Here we have used 6 pairs, the spikes in the convergence plot indicative of the use of a new neighboring frame as the sampling image. This was done purely to indicate the plausible improvement by incorporating multiple views that serves to improves statistical sufficiency of the data. This improvement was validated by comparing the estimate with the ground-truth generated by the Valeodyne sensors.

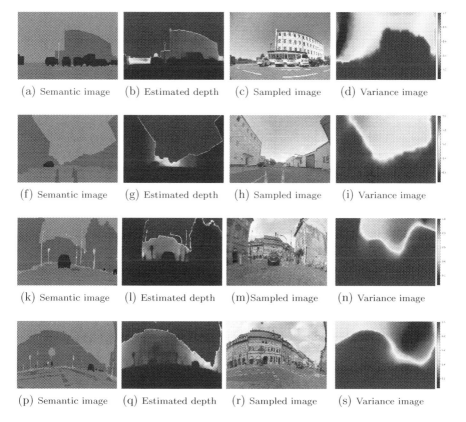

(a) Semantic image   (b) Estimated depth   (c) Sampled image   (d) Variance image

(f) Semantic image   (g) Estimated depth   (h) Sampled image   (i) Variance image

(k) Semantic image   (l) Estimated depth   (m) Sampled image   (n) Variance image

(p) Semantic image   (q) Estimated depth   (r) Sampled image   (s) Variance image

Fig. 2. Simulation results for 4 different scenes: Scene1-Scene4. Each row depicts the result of one particular scene. The left most column is the semantic segmentation of the scene taken as an input for sky identification. The second (from left) column is the estimated depth image. Third (from left) column is the reconstructed target image generated by sampling the right image. The rightmost column is the variance of the estimated depth.

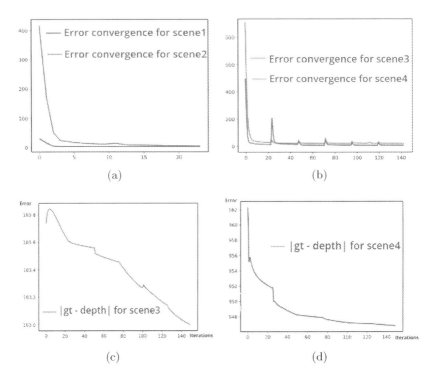

(a)

(b)

(c)

(d)

Fig. 3. Plots of convergence errors and ground truth errors. Figure 3a depict the convergence of the Algorithm 2 for Scene1 and Scene2. Figure 3b depict the convergence result for Scene3 and Scene4. Figures 3c and 3d shows the difference of the predicted depth with the available ground truth (gt) values.

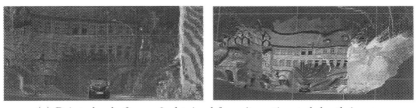

(a) Point cloud of scene3 obtained from its estimated depth image

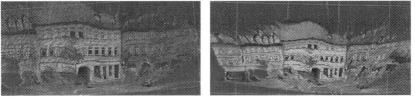

(b) Point cloud of scene4 obtained from its estimated depth image

Fig. 4. Reconstructed 3d point cloud from the estimated depth

This is highlighted in Figs. 3c and 3d. Finally, the Figs. 4b show the point clouds corresponding to Scene3 and Scene4 which are generated using the estimated depth values. This facilitates the usage of point processing libraries to refine the depth estimate by probabilistically fitting ground planes and building surfaces. Furthermore, such point processing algorithms can be incorporated in a closed loop fashion to enhance the structural consistency of the depth maps.

5 Discussion

In this paper we have presented a framework to estimate depths from two or more fisheye images. We employ variational technique to estimate the entire depth map at once. A novel surface area based term is used as a regularizer for the energy functional. As a result we get an anisotropic diffusion term which we believe facilitates in a more realistic surface reconstruction through the means of realistic depth estimates. The regularizer term also assists in establishing the Lax-Milgram property which is essential in proving existence of a unique weak solution in $H^1(\mathfrak{D})$. In order to ensure positivity of the solution and to facilitate the use of additional regularizing constraints we use primal-dual methods to solve (9) in least squares sense. The least-squares interpretation enables us to infer the 2nd order moment i.e. the covariance operator of the depth estimate. Since we assume Gaussian noise, covariance (along with the mean) is sufficient to estimate the entire distribution of the depth map. It should be noted that a least-squares interpretation is not a necessary step to have an uncertainty estimation, alternatively one can look at (9) as a stochastic pde which also enables one to estimate the uncertainty of the solution. We believe that uncertainty quantification is of paramount importance for many safety critical applications (e.g. robotics, autonomous driving). Complex applications requires interaction of various modules that work with different types of data. In such cases covariance of the estimate is a crucial information that facilitates in fusing heterogeneous data. Thereby, seamlessly integrating different higher level modules. Furthermore, an entire range of Bayesian algorithms become applicable for image based inferences. Other popular approaches (esp. deep learning based techniques) lack to provide any kind of inference about the distribution of the estimate, thus incorporation of these techniques into complex real-time systems puts the entire system at great risk. However, as mentioned above, such offline algorithms, that offer quick initial estimates, can be combined with our online algorithm to improve the overall reliability of the system.

References

1. Bettonvil, F.: Fisheye lenses. WGN J. Int. Meteor Organ. **33**(1), 9–14 (2005)
2. Caruso, D., Engel, J., Cremers, D.: Large-scale direct slam for omnidirectional cameras, pp. 141–148, September 2015. https://doi.org/10.1109/IROS.2015.7353366
3. Drulea, M., Szakats, I., Vatavu, A., Nedevschi, S.: Omnidirectional stereo vision using fisheye lenses, pp. 251–258, October 2014. https://doi.org/10.1109/ICCP.2014.6937005

4. Friel, M., Hughes, C., Denny, P., Jones, E., Glavin, M.: Automatic calibration of fish-eye cameras from automotive video sequences. Intell. Transp. Syst. IET **4**, 136–148 (2010). https://doi.org/10.1049/iet-its.2009.0052

5. Geiger, A., Roser, M., Urtasun, R.: Efficient large-scale stereo matching. In: Kimmel, R., Klette, R., Sugimoto, A. (eds.) ACCV 2010. LNCS, vol. 6492, pp. 25–38. Springer, Heidelberg (2011). https://doi.org/10.1007/978-3-642-19315-6_3

6. Hirschmüller, H.: Stereo processing by semi-global matching and mutual information. IEEE Trans. Pattern Anal. Mach. Intell. **30**, 328–341 (2008)

7. Hughes, C., Glavin, M., Jones, E., Denny, P.: Wide-angle camera technology for automotive applications: a review. Intell. Transp. Syst. IET **3**, 19–31 (2009). https://doi.org/10.1049/iet-its:20080017

8. Hughes, C., O'Malley, R., O'Cualain, D., Glavin, M., Jones, E.: Trends towards automotive electronic vision systems for mitigation of accidents in safety critical situations, January 2011. https://doi.org/10.5772/12907

9. Häne, C., et al.: 3d visual perception for self-driving cars using a multi-camera system: calibration, mapping, localization, and obstacle detection. Image Vis. Comput. (2017). https://doi.org/10.1016/j.imavis.2017.07.003

10. Kakani, V., Kim, H., Kumbham, M., Park, D., Jin, C., Nguyen, V.: Feasible self-calibration of larger field-of-view (FOV) camera sensors for the advanced driver-assistance system (ADAS). Sensors **19**, 3369 (2019). https://doi.org/10.3390/s19153369

11. Keselman, L.: What's the best lens for stereo? An exploration of lens models and their stereoscopic depth performance. Stanford (2016). https://pdfs.semanticscholar.org/39cf/7c007e33725552d6242eef2f1c6713830812.pdf?_ga=2.238984768.1758024819.1591704773-1508129036.1591704773

12. Khomutenko, B., Garcia, G., Martinet, P.: An enhanced unified camera model. IEEE Robot. Autom. Lett. **1**, 1–1 (2015). https://doi.org/10.1109/LRA.2015.2502921

13. Kim, H., Hilton, A.: 3d scene reconstruction from multiple spherical stereo pairs. Int. J. Comput. Vis. **104** (2013). https://doi.org/10.1007/s11263-013-0616-1

14. Kosov, S., Thormählen, T., Seidel, H.-P.: Accurate real-time disparity estimation with variational methods. In: Bebis, G., et al. (eds.) ISVC 2009. LNCS, vol. 5875, pp. 796–807. Springer, Heidelberg (2009). https://doi.org/10.1007/978-3-642-10331-5_74

15. Kumar, V.R., et al.: FisheyeDistanceNet: self-supervised scale-aware distance estimation using monocular fisheye camera for autonomous driving (2019)

16. Li, S.: Binocular spherical stereo. IEEE Trans. Intell. Transp. Syst. **9**, 589–600 (2009). https://doi.org/10.1109/TITS.2008.2006736

17. Newcombe, R., Lovegrove, S., Davison, A.: DTAM: dense tracking and mapping in real-time, pp. 2320–2327, November 2011. https://doi.org/10.1109/ICCV.2011.6126513

18. Olson, L.N., Schroder, J.B.: PyAMG: algebraic multigrid solvers in Python v4.0 (2018). https://github.com/pyamg/pyamg. Release 4.0

19. Roxas, M., Oishi, T.: Variational fisheye stereo. IEEE Robot. Autom. Lett. **PP**, 1–1 (2020). https://doi.org/10.1109/LRA.2020.2967657

20. Schneider, J., Stachniss, C., Förstner, W.: On the accuracy of dense fisheye stereo. IEEE Robot. Autom. Lett. **1**(1), 227–234 (2016). https://doi.org/10.1109/LRA.2016.2516509

21. Schwalbe, E.: Geometric modelling and calibration of fisheye lens camera systems. In: International Archives of the Photogrammetry, Remote Sensing and Spatial Information Sciences, vol. 36, May 2012

22. Shah, S., Aggarwal, J.K.: Depth estimation using stereo fish-eye lenses. In: Proceedings of 1st International Conference on Image Processing, vol. 2, pp. 740–744, November 1994. https://doi.org/10.1109/ICIP.1994.413669
23. Vieira, G., et al.: Stereo vision methods: from development to the evaluation of disparity maps, pp. 132–137, October 2017. https://doi.org/10.1109/WVC.2017.00030

Single Screw Extrusion Optimization Using the Tchebycheff Scalarization Method

Ana Maria A. C. Rocha[1(✉)], Marina A. Matos[1], M. Fernanda P. Costa[2], A. Gaspar-Cunha[3], and Edite M. G. P. Fernandes[1]

[1] ALGORITMI Center, University of Minho, 4710-057 Braga, Portugal
{arocha,emgpf}@dps.uminho.pt, aniram@live.com.pt
[2] Centre of Mathematics, University of Minho, 4710-057 Braga, Portugal
mfc@math.uminho.pt
[3] Institute for Polymer and Composites, University of Minho, 4800-058 Guimarães, Portugal
agc@dep.uminho.pt

Abstract. The optimal design of a single screw extrusion (SSE) is a very difficult task since it deals with several conflicting performance indices. Past research to find the optimal SSE design has been successfully conducted by optimization procedures, in particular by multi-objective optimization. Problems with two or more objectives have been addressed by multi-objective evolutionary algorithms that search for the whole set of promising solutions in a single run. Our approach has been guided by the bi-objective optimization problems, using a methodology based on the weighted Tchebycheff scalarization function. The numerical results show that the proposed methodology is able to produce satisfactory results with physical meaning.

Keywords: Single screw extrusion · Multi-objective optimization · Tchebycheff scalarization · Simulated annealing

1 Introduction

In the context of the polymer extrusion, the optimal design of a single screw extrusion (SSE) is a very difficult task since it deals with several performance indices (the objective functions) that are conflicting [1–4]. That is, the improvement of one performance index leads to another performance index degradation. This SSE design is concerned with the definition of the optimal screw geometry/configuration for obtaining the best performance indices, while manufacturing a certain product. Frequently, the screw geometry is established based on empirical knowledge, combined with a trial-and-error approach until the desirable performance indices are achieved. However, a more efficient approach is to handle the SSE design as an optimization problem, where the several conflicting objective functions are optimized simultaneously.

© Springer Nature Switzerland AG 2020
O. Gervasi et al. (Eds.): ICCSA 2020, LNCS 12251, pp. 664–679, 2020.
https://doi.org/10.1007/978-3-030-58808-3_48

Previous studies concerning with this SSE process have addressed the multi-objective optimization (MOO) problem by adopting a methodology based on multi-objective evolutionary algorithms (MOEA), namely the reduced Pareto set genetic algorithm (RPSGA) [2,3,5]. Most MOEA treat the MOO problem as a whole and find the entire set of promising and desirable solutions in a single run of the algorithm. They are mainly population-based stochastic using well-known and established meta-heuristic, e.g., genetic algorithm, particle swarm optimization, differential evolution, ant colony optimization, to name a few [6–11]. Simulated annealing and tabu search are point-to-point based stochastic algorithms that have also been used in this context [12,13].

Multi-objective evolutionary algorithms based on decomposition (recognized in the scientific literature as MOEA/D) decompose a MOO problem into a number of scalar optimization subproblems (using a weight-dependent scalar aggregation function) and optimize them simultaneously [14–16]. The subproblems are simultaneously solved by handling a population of solutions that comprise the best solutions found so far for each subproblem.

Methods for constructing a scalar aggregation function, combining the multi-objective functions into a weighted scalar objective function, that is used in a single objective optimization (SOO) context, thus producing a single optimal solution, are the easiest to understand and implement. Furthermore, well-established and known either deterministic or stochastic SOO algorithms can be used to find one optimal solution [6,7]. However, to obtain a set of promising and desirable solutions of the MOO problem, the SOO method must be run as many times as the desired number of points using different sets of weights. The most popular scalar aggregation functions include the weighted sum and the weighted Tchebycheff approaches.

The contribution of this paper is to show that a methodology based on the weighted Tchebycheff scalarization function, when used to optimize bi-objective optimization problems, and based on an evenly distributed set of weight vectors, provides acceptable and reasonably good approximations to the solutions of the SSE optimization problem. This study clearly shows that the used methodology is an effective tool to the SSE design optimization. The achieved optimal solutions are meaningful in physical terms.

This paper is organized as follows. Section 2 describes the single screw extrusion problem exhibiting the objective functions to be optimized and the decision variables of the problem. Section 3 presents some basic concepts concerning with MOO and the proposed methodology based on the weighted Tchebycheff approach. Finally, Sect. 4 exposes the carried out numerical experiments and we conclude the paper in Sect. 5.

2   Single Screw Extrusion Optimization

The most relevant performance indices in the SSE design are the mass output (Q), the length of the screw required for melting the polymer (Z_t), the melt temperature at die entrance (T_{melt}), the mechanical power consumption $(Power)$, the weighted-average total strain ($WATS$) and the viscous dissipation ($Visco$). These objective functions vary and depend on the values of two sets of parameters: the geometric and the operating parameters. Their values are obtained using numerical modelling routines that describe the plasticizing SSE process [1]. The geometric (or/and the operating) parameters are the inputs of the *computerized simulator* and the objective values Q, Z_t, T_{melt}, $Power$, $WATS$, $Visco$ are the output. Usually, the best design is attained by maximizing the objectives Q and $WATS$, and minimizing Z_t, T_{melt}, $Power$ and $Visco$.

The geometric parameters are related to the internal screw diameter of the feed zone (D_1) and metering zone (D_3), the axial lengths of the feed (L_1), compression (L_2) and metering (L_3) zones, the flight thickness (e) and the screw pitch (p). See Fig. 1. The operating parameters that correspond to the operating conditions of the extruder are: the screw speed (N) and the temperature profile of the heater bands in the barrel (Tb_1, Tb_2, Tb_3). In this paper, it is assumed that the operating parameters are previously fixed. The aim is to find the optimal values for the geometric parameters - the decision variables of the MOO problem - while optimizing the objectives. This is a crucial issue since, for example, if the compression zone is too short, the rate of decreasing channel depth downstream could become higher than the melting rate, resulting in material clogging. Furthermore, since the shallower the screw channel the higher the melting rate, a very long compression zone will result in an unnecessarily long melting stage.

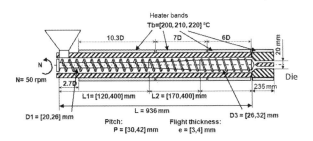

Fig. 1. Operating and geometric parameters of the SSE

For easy of notation, and since the total length of the screw (L) is to be minimized (corresponds to the objective Z_t), the length of the metering zone can be computed by the equation $L_3 = L - L_1 - L_2$. Thus, the set of decision variables of the herein analyzed MOO problems is represented by the vector (L_1, L_2, D_1, D_3, p, e). Usually, the ranges of variation of the geometric decision variables are defined based on empirical knowledge. Thus, taking into account the extruder illustrated in Fig. 1 and the polymer material (High Density Poyethylene-HDPE), the lower and upper bound values for the decision variables vector are $(120, 170, 20, 26, 30, 3)$ and $(400, 400, 26, 32, 42, 4)$, respectively.

The SSE MOO problem can be formulated as follows:

> Find a set of values for the vector $(L_1, L_2, D_1, D_3, p, e) \in \Omega \subset \mathbb{R}^6$
> such that the vector $(Q, Z_t, T_{melt}, Power, WATS, Visco)$ is optimized,   (1)

where the set Ω of feasible solutions is defined as $\Omega = \{(L_1, L_2, D_1, D_3, p, e) :$
$120 \leq L_1 \leq 400, 170 \leq L_2 \leq 400, 20 \leq D_1 \leq 26, 26 \leq D_3 \leq 32, 30 \leq p \leq$
$42, 3 \leq e \leq 4\}$. However, due to the complexity of problem (1) that involves a
vector of six objective functions, the design of the SSE process is optimized
considering the optimization of only two objective functions simultaneously.
Since the mass output Q is the most relevant performance index in the polymer
extruder, Q is present in all of the five alternative formulated bi-objective optimization problems, while the geometric parameters are optimized. The analyzed
bi-objective optimization problems are formulated as:

> Find values for the vector

1		(Q, Z_t)	is optimized
2		(Q, T_{melt})	is optimized
3	$(L_1, L_2, D_1, D_3, p, e) \in \Omega \subset \mathbb{R}^6$ such that	$(Q, Power)$	is optimized
4		$(Q, WATS)$	is optimized
5		$(Q, Visco)$	is optimized.

(2)

Thus, in this study, we deal with bi-objective optimization problems. They
are easier to solve and to check if the solutions produced by the optimization
algorithm are suitable for the extrusion process. Moreover, it is simpler to identify and to visualize the trade-offs between the solutions.

3   Multi-objective Optimization Approach

Nowadays many design problems faced by decision-makers are tackled by a multi-
objective approach. This means that more than one objective may be measured
and are expected to be optimized. Frequently, there is a conflict while optimizing more than one objective, i.e., does not exist one single solution that optimizes simultaneously all the objectives. However, a compromise solution can be
selected among a set of promising and desirable solutions - further ahead denoted
by Pareto optimal set - according to the preferences of the decision-maker.

3.1   Multi-objective Optimization

In general terms, the MOO problem can be formally defined as:

> Find $x^* \in \Omega \subseteq \mathbb{R}^n$ that optimizes the functions vector $(f_1(x), \dots, f_r(x))$,
> (3)

where $x \in \mathbb{R}^n$ is the vector of the decision variables, n is the number of decision
variables, Ω is the feasible search region and $f_1, \dots, f_r$ denote the r functions to

be optimized. In the minimization context, a vector $f = (f_1, \ldots, f_r)$ is said to dominate $\bar{f} = (\bar{f}_1, \ldots, \bar{f}_r)$ if and only if

$$\forall i \in \{1, \ldots, r\} \ f_i \leq \bar{f}_i \ \text{and} \ \exists i \in \{1, \ldots, r\} \ \text{such that} \ f_i < \bar{f}_i. \qquad (4)$$

Thus, when two solutions $f^1 = f(x^1)$ and $f^2 = f(x^2)$, $x^1, x^2 \in \Omega \subseteq \mathbb{R}^n$ are compared, one of these three cases holds: i) f^1 dominates f^2, ii) f^1 is dominated by f^2, iii) f^1 and f^2 are non-dominated.

Two other important definitions in the MOO context are the following.

Definition 1. *Let $f \in \mathbb{R}^r$ be the objective functions vector. A solution $x^1 \in \Omega$ is said to be Pareto optimal if and only if there is no other solution $x^2 \in \Omega$ for which $f(x^2)$ dominates $f(x^1)$.*

Assuming that the optimization problem involves the minimization of the functions in f, the Definition 1 says that x^1 is Pareto optimal if there is no other feasible solution x^2 which would decrease some objective f_i without causing a simultaneous increase in at least one other objective. That is, does not exist a single solution, but a set of solutions called Pareto optimal set (in the space of the decision variables). Their corresponding function vectors are said to be non-dominated.

Definition 2. *Given a MOO problem with objective function vector $f \in \mathbb{R}^r$ and the Pareto optimal set X^*, the Pareto optimal front (PF^*) is defined as:*

$$PF^* = \{f = (f_1(x), \ldots, f_r(x)) \ \text{such that} \ x \in X^*\}.$$

The algorithms for MOO aim to find a good and balanced approximation to the Pareto optimal set (and Pareto front PF^*). That is, the goal is to find a manageable number of Pareto optimal (function) vectors which are evenly distributed along the Pareto optimal front [14]. The goal is to support the decision-maker to formulate his/her preferences and identify the best (or compromise) solutions. The most popular methods to solve a MOO problem are based on:

   i) the aggregation of the objectives,
  ii) the ϵ-constraint strategy,
 iii) producing an approximation to the PF^* directly.

The aggregation of the objectives that requires weighting coefficients is used as an *a priori* approach since a total order is defined on the objective space (by defining a scalar function) and a statement of additional preferences, e.g., a weights vector for the objectives, should be provided by the decision-maker prior to the optimization. The aggregation method combines the objective functions into a scalar objective function that is used in a SOO context, thus producing one single compromise solution. To obtain an approximation to the PF^*, the SOO method must be run as many times as the desired number of points using different sets of weights vector [16].

In the ϵ-constraint method, one objective is selected to be minimized and all the other objective functions are converted into inequality constraints by setting upper bound values to each one, $\epsilon_1, \ldots, \epsilon_{r-1}$ [6]. Again, by varying and properly choosing the values set to the bounds $\epsilon_1, \ldots, \epsilon_{r-1}$, one can obtain different points on the Pareto optimal front [7]. Further, the method is able to find Pareto optimal solutions in convex and non-convex regions of the Pareto optimal front.

Methods to compute an approximation to the PF^* in a single run are in general stochastic population-based search techniques. They belong to the class of *a posterior* approaches. Popular MOEA rely on meta-heuristics and work reasonably well on difficult problems. In general, they are based on generating iteratively a population of solutions which are manipulated through operations like crossover, mutation and selection. Population-based meta-heuristics are naturally prepared to produce many solutions from which the set of Pareto optimal solutions can be emanated. Known examples with industry applications are NSGA-II [8], SPEA-2 [17], MOEA/D [14] and RPSGA [3]. The reader is referred to [6,7,18] for more details.

3.2   Weighted Tchebycheff Approach

Our proposal to solve the SSE problem, as formulated in (1), is to use the weighted Tchebycheff approach, a rather efficient aggregation method [19]. The aggregation method combines the objective functions into a scalar objective function that is used in a SOO context, thus producing one single compromise solution. However, defining the scalar objective function requires specific knowledge about search domain and function ranges that are in general not available. The most used aggregation method is the weighted sum approach that assigns to each objective function f_i, of the vector f, a non-negative weight w_i in a way that $\sum_{i=1}^{r} w_i = 1$, and minimizing the function that is the weighted sum of the objectives. This approach could not be able to find certain Pareto optimal solutions in non-convex regions of the Pareto optimal front.

The weighted Tchebycheff approach also assigns a weights vector to the objectives to form a single objective [15,19]. As opposed to the linear aggregation of the weighted sum approach, the Tchebycheff approach relies on a nonlinear weighted aggregation of the functions f_i, that is why it can deal with non-convex Pareto front [7]. In the minimization context, the SOO problem of the Tchebycheff approach has the form

$$\begin{aligned} \text{minimize} \quad & g(x; w) \equiv \max \left\{ w_1 \left(f_1(x) - z_1^* \right), \ldots, w_r \left(f_r(x) - z_r^* \right) \right\} \\ \text{subject to} \quad & x \in \Omega \end{aligned} \tag{5}$$

where $w = (w_1, \ldots, w_r)$ is the vector of non-negative weights and $z^* = (z_1^*, \ldots, z_r^*)$ is the ideal point in the objective space, i.e., $z_i^* = \min\{f_i(x)$ such that $x \in \Omega\}$ for $i = 1, \ldots, r$. The Tchebycheff approach guarantees finding all Pareto optimal solution with ideal solution z^*. In practice, this approach requires finding z^* by independently optimizing each objective function. We also note that problem (5) is non-differentiable (function $g(x; w)$ is not smooth at some points), although this disadvantage can be easily overcome implementing a derivative-free optimization method. We remark the following:

Remark 1. Under some mild conditions, for each Pareto optimal $x^* \in X^*$ there exists a weight vector w such that x^* is the optimal solution of problem (5), and each optimal solution of problem (5) (associated with a weights vector w) is a Pareto optimal solution to problem (3) [6,14,15].

It is advisable to rescale or normalize the objective functions so that their objective values are approximately of the same magnitude. Based on the ideal objective vector z^* and on the nadir objective vector, z^{nad}, each objective f_i is replaced by

$$F_i(x) = \frac{f_i(x) - z_i^*}{z_i^{nad} - z_i^*} \tag{6}$$

and, consequently, the range of the normalized function is $[0, 1]$. The nadir point is constructed with the worst objective function values in the complete Pareto optimal set X^*, i.e., $z_i^{nad} = \max\{f_i(x)$ such that $x \in X^*\}$ for $i = 1, \dots, r$, which makes the accurate estimation of the nadir objective values a difficult task [20]. An estimate of z^{nad}, herein denoted by the vector $f^{\max}$, obtained from a payoff table may be used instead. The function $g(x; w)$ in this Tchebycheff approach is then replaced by

$$g(x; w) \equiv \max\left\{w_1\left(\frac{f_1(x) - z_1^*}{f_1^{\max} - z_1^*}\right), \dots, w_r\left(\frac{f_r(x) - z_r^*}{f_r^{\max} - z_r^*}\right)\right\}. \tag{7}$$

Remark 2. For normalized objectives, the maximization of $f_i(x)$ can be reformulated as a minimization objective as follows:

$$\arg\max f_i(x) = \arg\min\left(1 - \frac{f_i(x) - z_i^*}{z_i^{nad} - z_i^*}\right).$$

3.3   Simulated Annealing Method

Nowadays, a large number of complex and difficult to solve real-world optimization problems are solved in an approximate way by meta-heuristics. These algorithmic frameworks combine heuristics with local search and population-based strategies to explore the feasible region and escape from local optimal solutions. Frequently, they are greedy methods. Meta-heuristics provide a good quality solution to the problem in a reasonable amount of time (instead of guaranteeing convergence to an optimal solution like the exact methods do). Meta-heuristics are general-purpose optimizers and mostly problem-independent, making them better suited to real world optimization problems [21].

The simulated annealing (SA) method is a single solution-based meta-heuristic with origins in statistical mechanics. The method models the physical process of heating a material and, to minimize the system energy, it carefully controls the reduction of the temperature - the cooling process - in order to reduce defects [22]. This process is known as annealing. It is adequate to solve unconstrained and bound constrained global optimization problems.

Since the feasible region Ω of each formulated bi-objective problem (2) is defined by box constraints alone, our proposal to solve the resulting problem (5) - with the objective function (7) (with $r = 2$) - in the Tchebycheff approach context, is to use a simulated annealing algorithm [23]. This is a well-known meta-heuristic for global optimization. At each iteration, of the SA algorithm, a new point is randomly generated. The distance between this new point and the current point, or the extent of the search, is based on a probability distribution with a scale proportional to the temperature. The algorithm accepts a new point if it improves the objective function, but also accepts, with a certain probability, a new point that deteriorates the objective. Using the temperature parameter, the SA algorithm controls the search for the global solution, e.g., a higher temperature allows more new points to be accepted which lead to the exploration of different regions of the search space. On the other hand, a lower temperature favors the acceptance of improving new points which result in the local exploitation of a promising region. Along the iterative process, the temperature is systematically decreased through a cooling schedule [24]. The main steps of the SA algorithm are presented in Algorithm 1. For the sake of brevity, details of the SA algorithm are not included and the reader is referred to the literature.

Algorithm 1. SA algorithm

Require: T_0, $It_{\max}$
1: Set a starting point x and evaluate the objective using (7)
2: Set $T = T_0$ and a cooling rate $0 < \kappa < 1$
3: Set N_t the number of trials per temperature
4: Set $It = 0$
5: **repeat**
6:     **for** $j = 1$ to N_t **do**
7:         Generate a new point $\bar{x}$ (in the neighborhood of x and according to a generating probability) and evaluate the objective using (7)
8:         **if** $\bar{x}$ is accepted according to the probability $P(x, \bar{x}; T)$ **then**
9:             Set $x = \bar{x}$
10:         **end if**
11:     **end for**
12:     Set $T = \kappa T$
13:     Set $It = It + 1$
14: **until** $It \geq It_{\max}$

3.4 Weighted Tchebycheff Scalarization Methodology

First, the issue related to the computation of approximations to the z_i^* and $f_i^{\max}$ (see previous Subsect. 3.2), $i = 1, \ldots, r$ in the SSE context has been addressed. The approximation to each z_i^*, hereinafter denoted by $f_i^{\min}$, has been defined based on the practical experience with this SSE and the polymer material. Similarly, the maximum value $f_i^{\max}$ is defined based on the practical experience with this particular extruder and the polymer material.

In the sequence of the strategy referred to in Sect. 2, Table 1 exposes the objective functions and their $f^{\min}$ and $f^{\max}$ values to be used in each of the five previously mentioned bi-objective optimization problems.

Table 1. Objective functions and their $f^{\min}$ and $f^{\max}$ values, for the five bi-objective problems

Problem	Objectives		$f_1^{\min}$	$f_1^{\max}$	$f_2^{\min}$	$f_2^{\max}$
	f_1	f_2				
1	Q	Z_t			0.2	0.9
2	Q	T_{melt}			150.0	220.0
3	Q	$Power$	1.0	20.0	0.0	9200.0
4	Q	$WATS$			0.0	1300.0
5	Q	$Visco$			0.9	1.2

Let $w^1, \ldots, w^P$ be the set of evenly spread weight vectors, where P represents the chosen number of tested weight vectors. According to the above referred Remark 1, for each set of weights, w^i $(i = 1, \ldots, P)$, the computed approximation to the optimal solution of problem (5), $x^*(w^i)$, is an approximation to a Pareto optimal solution (of the set X^*) to the problem (3).

Thus, our methodology to obtain an approximation to the PF^* is as follows. For each weights vector w^i, Algorithm 1 is used to compute an approximation to $x^*(w^i)$ and the corresponding functions vector, approximation to $f(x^*(w^i))$. This process is repeated N_{times} independent times. From the N_{times} sets of function vectors (approximations to the Pareto optimal front), the non-dominated function vectors are selected to better represent the trade-off between the objectives. From there on the decision-maker may identify a set of compromise solutions. Algorithm 2 describes the main steps of the methodology.

4   Results

The weighted Tchebycheff scalarization algorithm was coded in MATLAB® (MATLAB is a registered trademark of the MathWorks, Inc.). The code invokes the simulated annealing algorithm from the Global Optimization Toolbox (simulannealbnd function) as the optimization solver to compute the optimal solution of each MOO problem under study, throughout the minimization of the objective in (7). The solver simulannealbnd resorts the *computerized simulator of the SSE process* that provides the objective function values (output) given a set of values of the decision variables (input). This simulator is a computer program that simulates the SSE process. It comprises a dynamic model of the

---

Algorithm 2. Weighted Tchebycheff scalarization algorithm

Require: N_{times}, P;
1: Set $step = 1/(P - 1)$
2: Randomly generate $y \in \Omega$
3: **for** $N = 1$ to N_{times} **do**
4:      Set $w_1^1 = 0$
5:      **for** $i = 1$ to P **do**
6:          Set $w_2^i = 1 - w_1^i$
7:          Based on y, use Algorithm 1 to provide $x(w^i)$, an approximation to $x^*(w^i)$
8:          Set $PF^{N,i} = f(x(w^i))$
9:          Set $w_1^{i+1} = w_1^i + step$
10:     **end for**
11: **end for**
12: Select the non-dominated function vectors among the vectors $PF^{N,i}$, $i = 1, \ldots, P$,
      $N = 1, \ldots, N_{times}$.

SSE process in action predicting the interaction of all the components in the process. The use of a simulator is far and wide recommendable as an alternative to running a pilot plant or a full scale production equipment since these are quite costly and time consuming. With the simulator, one is able to provide changes in the screw geometry (or/and in the operating conditions), search for the optimum processing conditions and see immediately the results.

The experimental results are obtained with the following parameters setting: $It_{\max} = 50$ (adopted stopping criterion for `simulannealbnd`), $T_0 = 100$ (default value), $\kappa = 0.95$ (default value), $N_{times} = 5$ and a value of $P = 11$ is adopted with the weight vectors $(0, 1), (0.1, 0.9), (0.2, 0.8), \ldots, (0.8, 0.2), (0.9, 0.1), (1, 0)$. The fixed values assigned to the operating parameters are: $N = 50$, $Tb_1 = 200$, $Tb_2 = 210$ and $Tb_3 = 220$.

Problem 1. With the objectives Q and Z_t optimized simultaneously, the algorithm converges to a curve, the Pareto front, that defines the trade-offs between the objectives. See Fig. 2. The (blue) small full circles represent the solutions obtained for all the sets of weight vectors, over 5 runs, and the (red) large circles are the non-dominated solutions among the whole set. It is possible to see that the higher the mass output Q the higher is the length of screw for melting. Table 2 shows the decision variables and the corresponding objective functions values for four selected solutions from the Pareto front. The solution D provides the maximum value of the mass output, while the solution A gives the minimum value of the length of the screw required for melting. Two different trade-offs between the objectives are shown in the table that correspond to points B and C in Fig. 2. We note that from solution A to B a small increase in Q leads to a large degradation in Z_t. While Q improves from solution C to D, a large degradation is noted in Z_t.

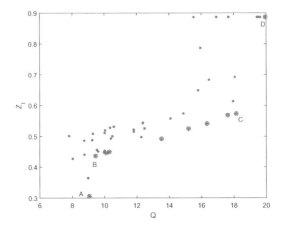

Fig. 2. Pareto front for Problem 1 (Color figure online)

Table 2. Solutions with the best values of the objectives Q and Z_t for Problem 1

	Geometric parameters						Objectives		Other performance indices			
	L_1	L_2	D_1	D_3	p	e	Q	Z_t	T_{melt}	$Power$	$WATS$	$Visco$
A	156	316	25.3	31.2	37.8	3.5	9.07	**0.305**	219	1507	373	1.02
B	270	325	25.2	31.3	35.1	3.1	9.46	0.435	219	1287	304	1.03
C	327	380	25.4	28.9	40.4	3.4	18.15	0.573	212	1468	150	0.99
D	350	318	24.5	26.1	41.7	3.2	**19.96**	0.886	203	1218	2	0.92

Problem 2. Among the whole set of Pareto front approximations, only four points are non-dominated solutions. It is possible to see from Fig. 3 that the higher the mass output the higher is the melt temperature, as the viscous dissipation becomes more important. Table 3 reports the values of the decision variables, Q and T_{melt}, and the other objective values obtained from the non-dominated points A and B. Solutions A and B are the optimal solutions in terms of T_{melt} and Q respectively.

Table 3. Solutions with the best values of the objectives Q and T_{melt} for Problem 2

	Geometric parameters						Objectives		Other performance indices			
	L_1	L_2	D_1	D_3	p	e	Q	T_{melt}	Z_t	$Power$	$WATS$	$Visco$
A	284	298	23.9	26.0	35.3	3.9	18.42	**200**	0.886	1159	3	0.90
B	222	276	26.0	26.2	35.9	4.0	**19.99**	206	0.886	1218	2	0.93

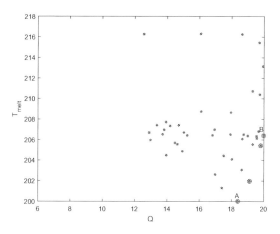

Fig. 3. Pareto front for Problem 2

Problem 3. Relative to Problem 3, we show in Fig. 4 the Pareto front computed by our proposed approach. Table 4 reports four non-dominated solutions. At solution A, we note that Q has the worst value but *Power* attains the best value, and at solution D, Q achieves the best value while *Power* has its worst value. The remaining solutions are compromise solutions between the two objectives. When comparing solutions A and B, B is clearly preferable since the loss in *Power* is negligible but the gain in Q is reasonable. When solution C is analyzed, and compared to D, we observe that a significant reduction in *Power* leads to a small degradation in Q.

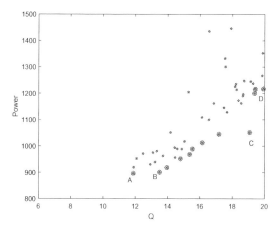

Fig. 4. Pareto front for Problem 3

Table 4. Solutions with the best values of the objectives Q and *Power* for Problem 3

	Geometric parameters						Objectives		Other performance indices			
	L_1	L_2	D_1	D_3	p	e	Q	*Power*	Z_t	T_{melt}	*WATS*	*Visco*
A	282	305	20.0	26.0	30.0	4.0	11.89	**896**	0.870	214	9	1.00
B	277	209	21.8	26.6	33.2	3.3	13.52	900	0.886	207	3	0.94
C	302	202	23.7	26.1	41.6	3.0	19.10	1051	0.886	215	2	0.98
D	172	246	25.9	26.0	31.7	3.0	**19.93**	1217	0.886	211	2	0.96

Problem 4. When both Q and *WATS* are maximized by the weighted Tcheby-cheff scalarization algorithm, a set of 17 non-dominated solutions are obtained. Figure 5 shows the Pareto front computed by the proposed approach. The maximum value of *WATS* is attained at solution A and the best solution in terms of Q is the solution D. The other solutions show the trade-offs between the objectives. See Table 5. An emphasis is placed on solutions B and C. C is clearly preferred when compared to B since a very small degradation on *WATS* implies a large improvement on Q.

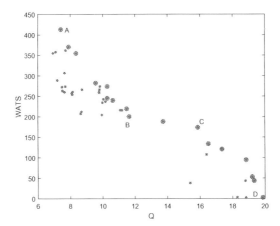

Fig. 5. Pareto front for Problem 4

Problem 5. Figure 6 shows the Pareto front obtained when maximizing Q and minimizing *Visco*. From the overall 5 non-dominated solutions, we stress points A and B. The best value in terms of *Visco* is attained at solution A and the maximum value for Q is reported with the solution B. The reader is referred to Table 6 for details concerning the values of the other objectives and the optimized parameter values.

After these experimental studies, we may conclude that the weighted Tcheby-cheff scalarization approach to solve the bi-objective problems formulated for the SSE optimal design has supplied a valuable procedure to identify good trade-offs between conflicting objectives.

Table 5. Solutions with the best values of the objectives Q and $WATS$ for Problem 4

	Geometric parameters						Objectives		Other performance indices			
	L_1	L_2	D_1	D_3	p	e	Q	$WATS$	Z_t	T_{melt}	$Power$	$Visco$
A	331	223	25.9	32.0	31.5	3.0	7.42	**412**	0.563	219	1127	1.03
B	400	363	25.6	31.7	41.4	3.1	11.65	200	0.555	220	1200	1.02
C	316	336	26.0	29.0	40.4	3.0	15.87	173	0.562	218	1278	1.03
D	372	368	25.2	26.6	35.3	3.0	**19.87**	2	0.886	207	1188	0.94

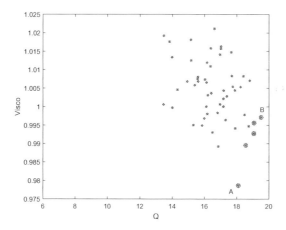

Fig. 6. Pareto front for Problem 5

Table 6. Solutions with the best values of the objectives Q and $Visco$ for Problem 5

	Geometric parameters						Objectives		Other performance indices			
	L_1	L_2	D_1	D_3	p	e	Q	$Visco$	Z_t	T_{melt}	$Power$	$WATS$
A	326	360	25.0	27.5	39.0	3.8	18.11	**0.98**	0.617	210	1569	122
B	377	399	22.6	29.3	38.1	4.0	**19.53**	1.00	0.721	214	1250	84

5   Conclusions

The weighted Tchebycheff scalarization is a simple and easy to understand methodology that provides a viable approach to solve the MOO problems that emerge from the SSE design optimization. In particular, the direct visualization of the trade-offs through the solutions of the approximate Pareto front assists the decision-maker in the selection of crucial SSE performance index and geometric parameter values. From the experimental studies on the five bi-objective problems, good trade-offs between conflicting objectives have been identified. We may also conclude that the proposed methodology provides optimal solutions that are meaningful in physical terms.

Acknowledgments. The authors wish to thank two anonymous referees for their comments and suggestions to improve the paper.

This work has been supported by FCT – Fundação para a Ciência e Tecnologia within the R&D Units Project Scope: UIDB/00319/2020, UIDB/05256/2020 and UIDP/05256/2020, UIDB/00013/2020 and UIDP/00013/2020 of CMAT-UM, and the European project MSCA-RISE-2015, NEWEX, Reference 734205.

References

1. Covas, J.A., Gaspar-Cunha, A., Oliveira, P.: An optimization approach to practical problems in plasticating single screw extrusion. Polym. Eng. Sci. **39**, 443–456 (1999)
2. Covas, J.A., Gaspar-Cunha, A.: Optimisation-based design of extruders. Plast. Rubber Compos. **33**(9–10), 416–425 (2004)
3. Gaspar-Cunha, A., Covas, J.A.: RPSGAe - reduced pareto set genetic algorithm: application to polymer extrusion. In: Gandibleux, X., Sevaux, M., Sörensen, K., Tkindt, V. (eds.) Metaheuristics for Multiobjective Optimisation. Lecture Notes in Economics and Mathematical Systems, vol. 535, pp. 221–249. Springer, Heidelberg (2004). https://doi.org/10.1007/978-3-642-17144-4_9
4. Covas, J.A., Gaspar-Cunha, A.: Polymer extrusion - setting the operating conditions and defining the screw geometry. In: Gaspar-Cunha, A., Covas, J.A. (eds.) Optimization in Polymer Processing, pp. 87–113. Nova Science Publishers (2011)
5. Gaspar-Cunha, A., Covas, J.A., Costa, M.F.P., Costa, L.: Optimization of single screw extrusion. In: Sikora, J.W., Dulebová, L. (eds.) Technological and Design Aspects of the Processing of Composites and Nanocomposites, vol. I, Scientific-Practical International Workshop (NewEX H2020-MSCA-RISE-2017) Technical University of Košice (2018)
6. Miettinen, K.M.: A posteriori methods. In: Nonlinear Multiobjective Optimization. International Series in Operations Research & Management Science, vol. 12, pp. 77–113. Springer, Boston (1998). https://doi.org/10.1007/978-1-4615-5563-6_4
7. Emmerich, M.T.M., Deutz, A.H.: A tutorial on multiobjective optimization: fundamentals and evolutionary methods. Nat. Comput. **17**(3), 585–609 (2018). https://doi.org/10.1007/s11047-018-9685-y
8. Deb, K., Pratap, A., Agarwal, S., Meyarivan, T.: A fast and elitist multiobjective genetic algorithm: NSGA-II. IEEE Trans. Evolut. Comput. **6**(2), 182–197 (2002)
9. Coello, C.A.C., Lechuga, M.S.: MOPSO: a proposal for multiple objective particle swarm optimization. In: Proceedings of the Congress on Evolutionary Computation (CEC 2002) pp. 1051–1056 (2002)
10. Santana-Quintero, L.V., Coello, C.A.C.: An algorithm based on differential evolution for multi-objective problems. Int. J. Comput. Intell. Res. **1**(2), 151–169 (2005)
11. Angelo, J.S., Barbosa, H.J.C.: On ant colony optimization algorithms for multiobjective problems. In: Ostfeld, A. (ed.) Ant Colony Optimization - Methods and Application, InTech Europe, pp. 53–74 (2011)
12. Bandyopadhyay, S., Saha, S., Maulik, U., Deb, K.: A simulated annealing-based multiobjective optimization algorithm: AMOSA. IEEE Trans. Evol. Comput. **12**(3), 269–283 (2008)
13. Molina, J., Laguna, M., Marti, R., Caballero, R.: SSPMO: a scatter tabu search procedure for non-linear multiobjective optimization. INFORMS J. Comput. **19**(1), 91–100 (2007)

14. Zhang, Q., Li, H.: MOEA/D: a multiobjective evolutionary algorithm based on decomposition. IEEE Trans. Evol. Comput. **11**(6), 712–731 (2007)

15. Zhang, Q., Liu, W., Tsang, E., Virginas, B.: Expensive multiobjective optimization by MOEA/D with Gaussian process model. IEEE Trans. Evol. Comput. **14**(3), 456–474 (2010)

16. Feng, Z., Zhang, Q., Zhang, Q., Tang, Q., Yang, T., Ma, Y.: A multiobjective optimization based framework to balance the global and local exploitation in expensive optimization. J. Glob. Optim. **61**, 677–694 (2015)

17. Zitzler, E., Laumanns, M., Thiele, L.: SPEA2: improving the strength Pareto evolutionary algorithm. TIK-Report 103, Computer Engineering and Networks Laboratory (TIK), Department of Electrical Engineering, Swiss Federal Institute of Technology (ETH), Zurich ETH Zentrum, Zurich (2001)

18. Deb, K.: Multi-Objective Optimization Using Evolutionary Algorithms. Wiley, Hoboken (2001)

19. Steuer, R.E., Choo, E.U.: An interactive weighted Tchebycheff procedure for multiple objective programming. Math. Program. **26**, 326–344 (1983)

20. Deb, K., Miettinen, K., Chaudhuri, S.: Toward an estimation of nadir objective vector using a hybrid of evolutionary and local search approaches. IEEE Trans. Evol. Comput. **14**(6), 821–841 (2010)

21. Gogna, A., Tayal, A.: Metaheuristics: review and application. J. Exp. Theor. Artif. Intell. **25**(4), 503–526 (2013)

22. Kirkpatrick, S., Gelatt, C., Vecchi, M.: Optimization by simulated annealing. Science **220**, 671–680 (1983)

23. Ingber, L.: Adaptive simulated annealing (ASA): lessons learned. Control Cybern. **25**(1), 33–54 (1996)

24. Wah, B.W., Chen, Y., Wang, T.: Simulated annealing with asymptotic convergence for nonlinear constrained optimization. J. Glob. Optim. **39**, 1–37 (2007)

International Workshop
on Computational Astrochemistry
(CompAstro 2020)

Binding Energies of N-Bearing Astrochemically-Relevant Molecules on Water Interstellar Ice Models. A Computational Study

Berta Martínez-Bachs[(✉)], Stefano Ferrero, and Albert Rimola

Departament de Química, Universitat Autònoma de Barcelona, 08193 Bellaterra,
Catalonia, Spain
Berta.martinezi@e-campus.uab.cat

Abstract. Different molecules and radical species have been detected and identified in the interstellar medium (ISM) through rotational and infrared techniques. The presence of dust grains in ISM have also been observed, which in the denser regions are covered by thick ice mantles consisting mainly of water molecules. Quantifying the interaction of gas phase species with water ice is essential to accurately model the evolution of the ISM. Hence, the importance to obtain accurate binding energies. In this contribution, the binding energies (BEs) of different N-containing interstellar species on water ice surfaces are presented. They have been calculated by means of reliable periodic quantum chemical simulations. Two different methods have been used: i) hybrid DFT methods, i.e., B3LYP for closed-shells species and M06-2X for open-shell radicals, and ii) the cost-effective DFT//HF-3C, i.e., geometry optimization at HF-3c followed by a single point energy refinement at the DFT level. As a first step, BEs have been computed on a crystalline water ice surface model of a thickness of 12 atomic layers, in which different super cells have been adopted (involving 36 or 72 water molecules per unit cell), depending on the size of the adsorbate. Moreover, an ONIOM-like local approximation combining CCSD(T) with DFT functional as high and low theory levels, respectively, has been adopted to correct BEs.

1 Introduction

Interstellar medium (ISM) is the matter and radiation present in the regions between stars. Interstellar matter can be found as ionic, atomic and molecular gas forms but also as tiny dust grains. Interstellar matter is not homogeneous and its temperature ranges between 10 K–10^6 K and atomic densities between 10^{-4} cm^{-3}–10^8 cm^{-3} depending on the environment and region [1].

In the coldest and densest regions of the ISM, cloud-like environments are found. These are regions of the Universe with extension between 100 parsecs and 300 light years, in which gaseous atoms and molecules and dust grains are mixed together [2]. In these regions, dust grains are covered by thick ice mantles, which are mainly composed by water (H_2O) but also by other minor volatile species, such as carbon monoxide (CO), carbon dioxide (CO_2), ammonia (NH_3), methane (CH_4) and methanol (CH_3OH) [3].

© Springer Nature Switzerland AG 2020
O. Gervasi et al. (Eds.): ICCSA 2020, LNCS 12251, pp. 683–692, 2020.
https://doi.org/10.1007/978-3-030-58808-3_49

Infrared-based astronomical observations reveal that these ices are in an amorphous structural state. They consist of different ice mantles coating the core of the dust grains. Ice mantles can be in the form of (i) water-rich, so-called polar phases, which mainly contain H_2O but also CH_3OH and NH_3 and where hydrogenation reactions dominate, or (ii) water-poor mantles, so-called apolar phases, which contain mainly CO but also CO_2 and HCOOH and where the reactions involving CO are dominating. Furthermore, molecules from the gas phase can adsorb on the ice surfaces, diffuse on them and react to form other, more complex species [3].

Astrochemical models are computational tools the purpose of which is to reproduce and represent the physical and chemical behaviour of interstellar clouds. Initially, astrochemical models were built to represent the abundances of early gas phase species [4, 5]. However, after noticing the fundamental role of dust grains in the increase of the molecular complexity in the ISM, novel astrochemical models have been developed that account for the chemistry on the dust grains [6–8].

To obtain astrochemical modelling results, a set of physical and energetic parameters must be introduced as input data. Binding energy (BE) values are input parameters of paramount importance in the gas-grain astrochemical models as they dictate ice accretion, surface diffusion and desorption events, which are fundamental elementary physico-chemical processes in the grain surface chemistry. One of the major current problems is the lack of accurate molecule/ice BE values [8]. Temperature programmed desorption is the usual technique to determine BEs experimentally, but laboratory experiments present severe limitations as they are not capable to reproduce neither the actual interstellar conditions nor realistic ice grain analogues [10]. This leads to obtain unreliable BEs, which is a dramatic drawback in astrochemical modelling, and hence the importance to accurately quantify the interaction between gas-phase species and ice surfaces.

In this work, results on the binding energies (BEs) of different N-containing species of interstellar interest on water ice surfaces modelled by a crystalline system adopting a periodic slab approach are presented.

2   Methods

BEs have been calculated by means of robust periodic quantum chemical simulations adopting different methods and density functional theory (DFT) functionals. One quantum chemical method is the novel cost-effective HF-3c, in which the electronic energy is grounded on the Hartree-Fock framework, but three corrections are added in a posteriori way [11]. Two DFT functionals have also been used: B3LYP for closed-shell species and M06-2X for radicals open-shell species. Both methods are hybrid DFT functionals as they incorporate a certain amount of exact exchange, 20% and 45%, respectively [12–14]. For B3LYP, the D3 empirical correction has been added to account for dispersive forces [15]. For both DFT functionals, the Ahlrichs TZV basis with polarization functions were used, while for HF-3c the MINIX basis set was used. HF-3c values do not suffer from the basis set superposition error (BSSE) but DFT values do. Accordingly, for these cases, single point energy calculations to obtain BSSE-corrected BE values have been performed [16].

Truncated Coupled Cluster (CC) methods are one of the most accurate electronic structure methods but due to its high computational cost they can only be applied for small chemical systems. In this work the CCSD(T), a method that treats the single and double excitations with a CC formalism and the triple ones perturbatively, has also been used to refine the DFT BEs adopting an ONIOM-like approach [17], hereafter referred to as O[CCSD(T)//DFT]. In this apporach, a molecular system including the adsorbate species and the binding site region has been extracted from the periodic system and calculated at CCSD(T). The O[CCSD(T)//DFT] BEs are obtained as:

$$BE(O[CCSD(T)//DFT]) = BE(low; real) + BE(high; model) - BE(low; model)$$

where the model system is the molecular "adsorbate+binding site" region and the real system is the actual periodic system, and the high method is CCSD(T) and the low method the DFT ones. For these calculations, we used the DFT optimized geometries.

The surface mimicking the icy water dust grain has been modelled adopting a periodic approach. The water ice surface is based on the crystalline P-ice structure. P-ice is the proton ordered analogue to the hexagonal (proton-disordered) water ice, which has been demonstrated to reproduce fairly well the physico-chemical features of crystalline water ice [18]. The surface model has been generated by cutting out the P-ice 3D periodic bulk structure perpendicular to the [010] direction, resulting with the (010) slab surface model. We chose the (010) slab model because it is one of the most stable planes of the P-ice and because it has no dipole component perpendicular to them [19]. The slab model consists of twelve atomic layers and consists of 72 atoms (24 water molecules) per unit cell. It is worth mentioning that a structurally proper model based on the hexagonal ice (Ih) requires a large unit cell as all the proton arrangements must be covered. Indeed, as mentioned above Ih is a proton-disordered system. Since the position of the protons in the ice follows the Bernal-Fowler-Pauling (BFP) rules [20, 21], i.e., i) each water molecule is oriented such that its two hydrogen atoms are directed approximately toward two of the four surrounding oxygen atoms arranged almost in a tetrahedron, ii) only one hydrogen atom is present on each O···O linkage and iii) each oxygen atom has two nearest neighbouring hydrogen atoms such that the water molecule structure is preserved, a structural model for Ih envisaging all the hypothetical proton-ordered configurations accounting the BFP rules would be very large and, accordingly, costly to treat at a quantum mechanical level. Because of that, using P-ice is a good compromise between accuracy and computational cost.

3 Results

The BEs of eight N-containing species, i.e., HNO, N_2O, NO, NO_2, N, NH, NH_2 and HNC, on the water crystalline slab model have been computed. These species can be classified in two groups, according to their capability to establish hydrogen bonds (H-bonds): molecules that can only act as H-bond acceptors (N, NO, NO_2 and N_2O, hereafter referred to as Hb-A), and molecules that can act as both H-bond donors and acceptors (HNO, NH, NH_2 and HNC, hereafter referred to as Hb-AD).

As mentioned above, three different quantum chemical methodologies have been used: i) a full DFT treatment, using B3LYP for closed-shell species and M06-2X functional for open-shell species, ii) the cost-effective DFT//HF-3c treatment, in which geometry optimizations were carried out at a HF-3c level followed by a single point energy calculation at the corresponding DFT level, and iii) an ONIOM-like local approximation with CCSD(T) as high level of theory and the DFT functionals as low level of theory, from here on referred as O[CCSD(T)//DFT]. The purpose is to obtain accurate and robust BE values as well as to carry out a calibration study to assess the accuracy of the cost-effective DFT//HF-3c method, that is, if the computed BE values are in fair agreement with the full DFT (DFT//DFT) ones and the O[CCSD(T)//DFT] ones.

The structures of the complexes formed by adsorption of the Hb-A and Hb-AD species on the water ice surface model are shown in Figs. 1 and 2, respectively. Table 1 reports the BE values obtained at both DFT//DFT, DFT//HF-3c and O[CCSD(T)//DFT] methods.

As expected, for all cases, H-bond interactions and the dispersion forces are the main intermolecular forces that dictate the adsorption of the species on the water ice surface. Optimized complexes at full DFT theory and at the HF-3c method present differences in the H-bond distance, which in most of the cases lay within 0.2 Å. In addition to the intrinsic differences in the definition of each method, the fact that HF-3c does not suffer from BSSE in the optimization processes while this is indeed the case for the DFT methods also has an influence in the observed H-bond differences.

In relation to the calculated BE values, the general trend is that Hb-AD species present larger BEs than Hb-A species. Taking the O[CCSD(T)//DFT] BE values, the former group have BEs spanning the 70.4–35.1 kJ mol^{-1} range while the latter group the 29.2–13.8 kJ mol^{-1} range. This is because the Hb-AD species can establish more and more efficient H-bond interactions with the surface than the Hb-A ones. More H-bonds because for all the Hb-AD cases two H-bonds are established, i.e., one as H-bond donor and one as H-bond acceptor. This is not the case for Hb-A species, as they can only establish one or two H-bonds, in this later case the acceptor atom being the same. And more efficient H-bonds because the fact to act as both H-bond donor and acceptor enables H-bond cooperativity, which reinforces the H-bond network in which the adsorbate species is involved. This is reflected by the H-bond distances identified in both groups of species: Hb-AD species establish H-bond interactions of between 1.6–2.0 Å, while Hb-A species of between 2.2–3.0 Å (at HF-3c theory level), meaning that this later group present weak H-bonds and accordingly dispersion is a major contributor to the BEs.

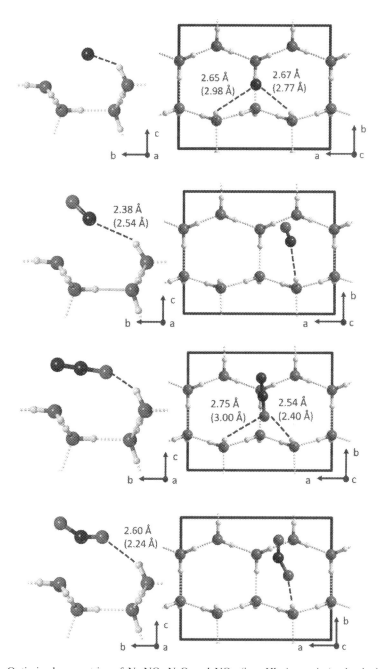

Fig. 1. Optimized geometries of N, NO, N$_2$O and NO$_2$ (i.e., Hb-A species) adsorbed on the crystalline slab (010) water ice surface model, as lateral (right panel) and top (left panel) views. Bare distances are values corresponding to the full DFT optimizations while in parenthesis to those corresponding to the HF-3c optimizations. Color legend: red, O; blue, N; white, H. (Color figure online)

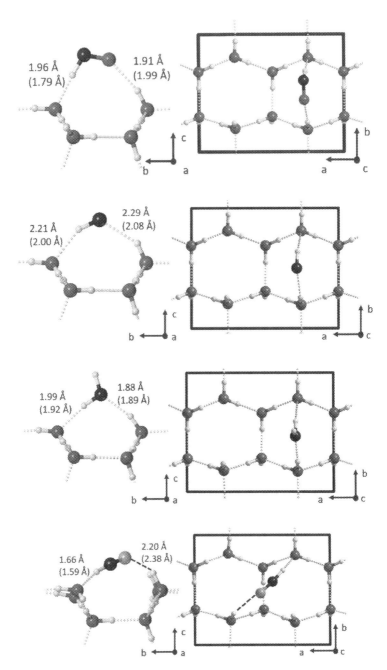

Fig. 2. Optimized geometries of HNO, NH, NH$_2$ and HNC (i.e., Hb-AD species) adsorbed on the crystalline slab (010) water ice surface model, as lateral (right panel) and top (left panel) views. Bare distances are values corresponding to the full DFT optimizations while in parenthesis to those corresponding to the HF-3c optimizations. Color legend: red, O; blue, N; white, H; brown, C. (Color figure online)

Table 1. Calculated BE values obtained at the DFT//HF-3c, DFT//DFT and O[CCSD(T)//DFT] theory levels. Units are in kJ mol^{-1}.

Species		DFT//HF-3c	DFT//DFT	O[CCSD(T)//DFT]
Hb-A	N	15.9	18.5	13.8
	NO	11.4	13.6	14.5
	NO$_2$	18.3	22.3	23.8
	N$_2$O	25.7	27.3	29.2
Hb-AD	NH	34.7	38.6	35.1
	HNO	38.3	42.8	41.3
	NH$_2$	50.2	49.4	50.3
	HNC	70.5	72.8	70.4

In the Hb-AD group, HNC is the species presenting the largest BE value because it presents the strongest H-bond interaction (1.59 Å at HF-3c). In contrast, NH is the species with the lowest Hb-AD BE because their H-bonds are relatively weak (about 2.0 Å at HF-3c). In the Hb-A group, NO$_2$ presents the largest BE value because, being one of the largest molecules (and hence presenting large dispersion contributions), it exhibits the "strongest" H-bond (2.24 Å) among the species of its group. In contrast, N has the lowest BE because the H-bonds are actually very weak (2.77 and 2.98 Å at HF-3c) and dispersion contribution is very minor as it is a single atom.

An identified general trend is that DFT//DFT BE values are slightly more favourable than the DFT/HF-3c ones. This is due to the effects of the BSSE in the optimization at either full DFT or HF-3c theory levels. As mentioned above, optimized HF-3c geometries, which do not suffer from BSSE, generally present H-bond distances larger than the DFT optimized ones. Because of these larger H-bonds in the HF-3c geometries, the DFT single point energies provide lower BE energies than the DFT// DFT ones. Despite this fact, BE values obtained with the two methods are in almost perfect correlation, as one can see in Fig. 3. Moreover, both methods also correlate very well with the BE values obtained through the O[CCSD(T)//DFT] approach (see Fig. 3), allowing us to conclude that the cost-effective DFT//HF-3c is a suitable method to obtain robust and accurate enough BE values. As a final interesting comment, we would like to highlight that B3LYP and M06-2X functional methods, based on the comparison with the O[CCSD(T)//DFT] approach, describe well the electronic structure of respective closed-shell and open-shell systems.

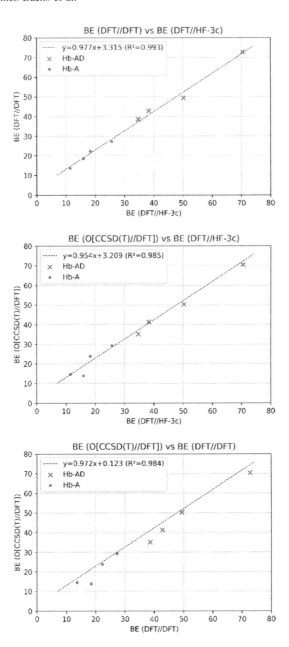

Fig. 3. Correlation between the BE values obtained with the DFT//DFT, DFT//HF-3c and O[CCSD(T)//DFT] methods. The units of the BEs are kJ mol^{-1}. The linear regressions are also shown.

4 Conclusion

In this contribution, results on the binding energy (BE) values of eight N-containing species of astrochemical relevance (i.e., N, NO, NO_2, N_2O, NH, HNO, NH_2, HNC) on water icy grain surfaces, in this case modelled by a periodic (010) crystalline water ice slab model of P-ice, are presented. The main purpose is to assess the robustness of the cost-effective DFT//HF-3c approach (i.e., optimization at the composite HF-3c theory level followed by a single point energy calculation at DFT theory level) to calculate accurate BEs. To this end, DFT//HF-3c BEs have been compared to those provided at DFT//DFT as well as to those obtained through a CCSD(T) correction adopting an O [CCSD(T)//DFT] approach.

The N-containing molecules considered here, in their adsorption on the water ice surface can act either as just H-bond acceptors (Hb-A, N, NO, NO_2 and N_2O) or as H-bond acceptors and donors (Hb-AD, NH, HNO, NH_2 and HNC). Consequently, the latter group presents more favorable BE values than the former one. Indeed, while Hb-AD species establish relatively strong H-bond interactions with the surfaces, in Hb-A-based complexes H-bond interactions are relatively weak, leaving the major contribution to the total BE in the dispersive forces.

Despite the small differences between the optimized geometries at HF-3c and at full DFT levels, BE energies computed at DFT//HF-3c and at DFT//DFT are in almost perfect agreement. Moreover, BE values obtained with the two methods also agree very well with those obtained adopting the O[CCSD(T)//DFT] approach. These results indicate that DFT//HF-3c is a reliable cost-effective method to obtain accurate BEs, and accordingly it can be used in more realistic interstellar water ices, such as amorphous surfaces, which require the use of larger unit cells since the complex surface heterogeneity and binding sites diversities have to be accounted for.

References

1. Herbst, E., Yates, J.T.: Introduction: astrochemistry. Chem. Rev. **113**(12), 8707–8709 (2013)
2. Herbst, E., van Dishoeck, E.F.: Complex organic interstellar molecules. Annu. Rev. Astron. Astrophys. **47**(1), 427–480 (2009)
3. Van Dishoeck, E.F.: Astrochemistry of dust, ice and gas: introduction and overview. Faraday Discuss. **168**, 9–47 (2014)
4. Bates, D.R., Spitzer, L.J.: The density of molecules in interstellar space. Astrophys. J. **113**, 441 (1951)
5. Herbst, E., Klemperer, W.: The formation and depletion of molecules in dense interstellar clouds. Astrophys. J. **185**, 505 (1973)
6. Tielens, A.G.G.M., Tokunaga, A.T., Geballe, T.R., Baas, F.: Molecular dynamics simulation of the H2 recombination on a graphite surface. Astrophys. J. **381**, 181–199 (1991)
7. Hasegawa, T.I., Herbst, E., Leukng, C.M.: Models of gas-grain chemistry in dense interstellar clouds with complex organic molecules. Astrophys. J. Suppl. Ser. **82**, 167 (1992)
8. Garrod, R.T.: A three-phase chemical model of hot cores: the formation of glycine. Astrophys. J. **765**(1), 60 (2013)

9. Cuppen, H.M., Walsh, C., Lamberts, T., Semenov, D., Garrod, R.T., Penteado, E.M., Ioppolo, S.: Grain surface models and data for astrochemistry. Space Sci. Rev. **212**(1), 1–58 (2017). https://doi.org/10.1007/s11214-016-0319-3

10. Smith, I.W.M., Cockell, C.S., Leach, S.: Astrochemistry and Astrobiology. Springer, Berlin, Heidelberg (2013)

11. Sure, R., Grimme, S.: Corrected small basis set Hartree-Fock method for large systems. J. Comput. Chem. **34**(19), 1672–1685 (2013)

12. Becke, A.D.: Density-functional exchange-energy approximation with correct asymptotic behavior (1988)

13. Becke, A.D.: A new mixing of Hartree-Fock and local density-functional theories. J. Chem. Phys. **98**(2), 1372–1377 (1993)

14. Zhao, Y., Truhlar, D.G., Zhao, Y., Truhlar, D.G.: The M06 suite of density functionals for main group thermochemistry, thermochemical kinetics, noncovalent interactions, excited states, and transition elements: two new functionals and systematic testing of four M06-class functionals and 12 other function. Theor. Chem. Acc. **120**, 215–241 (2008)

15. Grimme, S.: Accurate description of van der Waals complexes by density functional theory including empirical corrections. J. Comput. Chem. **25**(12), 1463–1473 (2004)

16. Boys, S.F., Bernardi, F.: The calculation of small molecular interactions by the differences of separate total energies. Some procedures with reduced errors. Mol. Phys. **19**(4), 553–566 (1970)

17. Tsuneda, T.: Density Functional Theory in Quantum Chemistry, vol. 9784431548256. Springer, Japan (2013)

18. Pisani, C., Casassa, S., Ugliengo, P.: Proton-ordered ice structures at zero pressure. A quantum-mechanical investigation. Chem. Phys. Lett. **253**(3–4), 201–208 (1996)

19. Zamirri, L., Casassa, S., Rimola, A., Segado-Centellas, M., Ceccarelli, C., Ugliengo, P.: IR spectral fingerprint of carbon monoxide in interstellar water-ice models. MNRAS **480**, 1427–1444 (2018)

20. Bernal, J.D., Fowler, R.H.: A theory of water and ionic solution, with particular reference to hydrogen and hydroxyl ions. J. Chem. Phys. **1**(8), 515–548 (1933)

21. Pauling, L.: The structure and entropy of ice and of other crystals with some randomness of atomic arrangement. J. Am. Chem. Soc. **57**(12), 2680–2684 (1935)

Theoretical and Computational Analysis at a Quantum State Level of Autoionization Processes in Astrochemistry

Stefano Falcinelli[1]([⊠]) [iD], Fernando Pirani[2,3], Marzio Rosi[1,3], and Franco Vecchiocattivi[1]

[1] Department of Civil and Environmental Engineering, University of Perugia, Via G. Duranti 93, 06125 Perugia, Italy
{stefano.falcinelli,marzio.rosi}@unipg.it,
franco@vecchio.it
[2] Department of Chemistry, Biology and Biotechnologies, University of Perugia, Via Elce di Sotto 8, 06100 Perugia, Italy
fernando.pirani@unipg.it
[3] ISTM-CNR, 06123 Perugia, Italy

Abstract. Changes in atomic-molecular alignment and/or in molecular orientation, can affect strongly the fate of basic collision events. However, a deep knowledge of these phenomena is still today not fully understood, although it is of general relevance for the control of the stereo-dynamics of elementary chemical-physical processes, occurring under a variety of conditions, both in gas phase and at surface. In particular, understanding the mode-specificity in reaction dynamics of open-shell atoms, free radicals, molecules, atomic and molecular ions, under hyper-thermal, thermal and sub-thermal conditions is of fundamental importance for catalysis, plasmas, photo-dynamics as well as interstellar and low-temperature chemistry. In this paper recent results on the role of atomic alignment effects on the stereo-dynamics of autoionization reactions are presented and discussed.

Keywords: Autoionization · Stereo-dynamics · State to state · Charge transfer · Transition state · Electron spectroscopy · Astrochemistry

1 Introduction

The knowledge of strength, anisotropy and radial dependence of the leading intermolecular interaction components is a crucial objective to assess the selectivity of the molecular dynamics under a variety of conditions. The charge transfer (CT), a basic component of the intermolecular interaction, whose role is often not fully understood, affects myriad of phenomena, including also the formation of intermolecular halogen and hydrogen bonds [1–3].

© Springer Nature Switzerland AG 2020
O. Gervasi et al. (Eds.): ICCSA 2020, LNCS 12251, pp. 693–706, 2020.
https://doi.org/10.1007/978-3-030-58808-3_50

The focus of this paper is to point out a theoretical and computational methodology based on the selective role of anisotropic CT on the stereo-dynamics of autoionization processes, also known as Penning or chemi-ionization phenomena [4, 5]. Such elementary reactions can be represented according to Eq. (1) below:

$$X^* + Y \rightarrow [X \text{ - - - } Y]^* \rightarrow [X \text{ - - - } Y]^+ + e^- \rightarrow \text{ion products.} \tag{1}$$

In Eq. (1), X^* is a metastable excited species inducing ionization of a collisional target Y that can be an atom or a molecule; $[X \text{ - - - } Y]^*$ and $[X \text{ - - - } Y]^+$ are the intermediate neutral and ionic collisional complexes, respectively, being the first the transition state of the reaction [6, 7]. In general, X^* species that can be used to induce autoionization are the rare gas atoms in their first excited electronic level which is characterized by a high excitation energy and a considerably long lifetime (see Table 1 below [4, 8]).

Table 1. Metastable noble-gas atoms and their fundamental properties [4, 8].

Metastable species	Electronic configuration	Lifetime (s)	Energy (eV)
$He^*(2^1S_0)$	$1s\ 2s$	0.0196	20.6158
$He^*(2^3S_1)$	$1s\ 2s$	9000	19.8196
$Ne^*(^3P_0)$	$2p^5\ 3s$	430	16.7154
$Ne^*(^3P_2)$	$2p^5\ 3s$	24.4	16.6191
$Ar^*(^3P_0)$	$3p^5\ 4s$	44.9	11.7232
$Ar^*(^3P_2)$	$3p^5\ 4s$	55.9	11.5484
$Kr^*(^3P_0)$	$4p^5\ 5s$	0.49	10.5624
$Kr^*(^3P_2)$	$4p^5\ 5s$	85.1	9.9152
$Xe^*(^3P_0)$	$5p^5\ 6s$	0.078	9.4472
$Xe^*(^3P_2)$	$5p^5\ 6s$	150	8.3153

As clearly indicated by the data reported in Table 1, the autoionization reactions are promoted by collisions of an open shell atom, electronically excited in a high energetic metastable state, with another atomic/molecular partner giving rise to the spontaneous emission of electrons accompanied by the formation of parent ions, aggregate ions and fragmentation ions. Therefore, the measure of the energy dependence of emitted electrons, also known as Penning Ionization Electron Spectra (PIES), provides direct information on the electronic rearrangements occurring within the collision complex, which directly correlates with the transition state of the autoionization processes. Moreover, molecular ionization probability and emitted PIES are strongly dependent on symmetry and energy of the atomic or molecular orbital from which the electron is extracted and then on their spatial orientation within the collision complex [7, 9].

Autoionization processes represent also barrier-less reactions, driven by an aniso-tropic optical potential (see next section), whose real part controls approach of reactants and removing of products, while the imaginary part triggers the passage from neutral reactants to ionic products through an electronic rearrangement. The investigation of these reactions is important for fundamental researches, to assess the coherent control of reactive events at low temperature and then to explore the quantum nature of matter [10]. Autoionization reactions are of interest in radiation and plasma chemistry [11, 12], for the development of excimer lasers and in combustion chemistry [13–15]. More recently, they attracted the attention of the scientific community since they can be involved in the chemistry of planetary ionospheres as well as in astrochemistry [16–24]. These processes can be studied in laboratory using various techniques employing the molecular beam method in high vacuum apparatuses able to detect reactive events in a single-collision condition [25–27]. The main diagnostic experi-mental procedures are the mass spectrometry [28–31] and the electron spectroscopy [32, 33] coupled with either ion-imaging [34, 35] and coincidence techniques [36, 37]. All these techniques are available in our laboratory.

The Fig. 1 shows a schematic view of the molecular beam apparatus operating in our laboratory (briefly presented in the Sect. 3) and depicts a metastable Ne* atom, whose external electron is excited in the 3s orbital and its ionic core exhibits the same electronic configuration of the high electron affinity fluorine atom, that approaches to a water molecule [33]. This event originates a collision complex where the spontaneous electron jumps from one of HOMO orbitals of water to the ionic core of Ne* and releases enough energy to eject the 3s electron with a defined kinetic energy. Therefore, the measure of the energy dependence of emitted electrons by PIES experiments provides a direct characterization of the transition state of the autoionization reaction (i.e. symmetry and energy of the involved molecular orbitals and the spatial molecular orientation within the collision complex, as depicted in Fig. 1). We characterized in details such features for important hydrogenated molecules, as water, ammonia and hydrogen sulfide, and obtained results have been reported in previous published papers [7, 9, 33, 38].

However, the dependence of the reaction probability on the valence orbital align-ment of open-shell atoms still represents a basic open question. Recently [39, 40], our investigation has been focused on some prototype atom-atom reactions, as that involving Ne* - Kr, in order to obtain information on this basic target. New insights on the stereo-dynamics of elementary processes involved are presented and discussed in the next section.

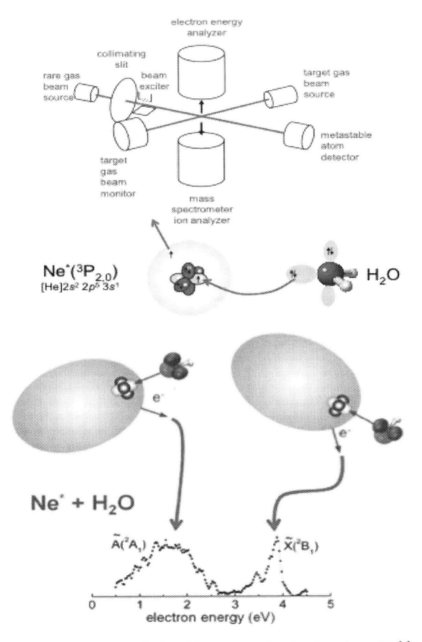

Fig. 1. - *Upper panel*: A schematic view of the apparatus, where the primary beam of Ne*($^3P_{2,0}$) atoms, emerging from an electron bombardment supersonic seeded source, crosses at right angles the secondary beam of H_2O molecules. PIES have been measured exploiting a hemispherical electron energy analyzer, while total, partial cross sections and branching ratios have been determined by the mass spectrometry technique. - *Middle panel*: A scheme of the atomic and molecular orbitals involved in the electron exchange. - *Lower panel*: A scheme of two different transition states leading to the formation of water ion in ground and excited states with the associated PIES [33, 38].

2   The Theoretical and Computational Approach

From a theoretical point of view, autoionization reactions are represented employing the so called optical potential model first introduced by Bethe in 1940 [41], which is based on a complex potential W, according to Eq. (2):

$$W = V - \frac{i}{2}\Gamma \tag{2}$$

Where V is the real part V describing the $X^* + Y$ collision and the relative interaction while Γ represents the imaginary part of such a complex potential accounting for the probability of the electron ejection from the intermediate collisional complex $[X - - - Y]^*$, i.e. the transition state of the reaction (see Eq. (1)) [42–44].

In our theoretical approach, the two V and Γ components are, for the first time, considered as interdependent [5, 45], whose magnitude is related to the internuclear distance, R, between X and Y, their relative orientation and the manifold of quantum states attainable for the system along the collision.

In the description of the autoionization event induced by the X^*+Y collision, the lifetime, τ, of the intermediate autoionizing complex $[X - - - Y]^*$ is depending on the internuclear distance R, and is given by the following simple equation:

$$\tau(R) = \frac{\hbar}{\Gamma(R)} \tag{3}$$

Along the element dR, at the distance R, during a collision with asymptotic speed g and a collision energy E, with impact parameter b, the probability that the system ionizes is given by

$$P(R)dR = \frac{\Gamma(R)}{\hbar g \left[1 - \frac{V(R)}{E} - \frac{b^2}{R^2} \right]^{\frac{1}{2}}} dR \tag{4}$$

During a complete collision, the probability that the system survives (i.e. does not give rise to ionization) from infinite distance to the turning point R_c is:

$$F_{R_c,\infty}(b,g) = exp\left[-\int_{R_c}^{\infty} P(R)dR \right] = exp\left[-\int_{R_c}^{\infty} \frac{\Gamma(R)}{\hbar g \left[1 - \frac{V(R)}{E} - \frac{b^2}{R^2} \right]^{\frac{1}{2}}} dR \right] \tag{5}$$

The total ionization cross section will be given by:

$$\sigma_{tot}(g) = 2\pi \int_0^{\infty} P(b)bdb = 2\pi \int_0^{\infty} \left[1 - F_{R_c,\infty}^2(b,g) \right] bdb \tag{6}$$

For associative ionization, it is necessary to define the limit distance, R_{ass}, above which associative ionization (i.e. the formation of the XY^+ molecular ion) cannot take place. If $V_{in}(R)$ is the input potential (describing the incoming collision between the neutral $X^* + Y$ partners) and $V_{out}(R)$ is the output potential (accounting for the interaction between the separating products $X + Y^+$), R_{ass} is defined by the condition:

$$V_{in}(R_{ass}) - E = V_{out}(R_{ass}) - H \tag{7}$$

where H is the height of the centrifugal barrier, i.e. the maximum of the following function:

$$\frac{Eb^2}{R_{ass}^2} \tag{8}$$

The probability of associative ionization is given by:

$$P_{ass}(b, g) = F_{R_{ass}, \infty} \left[1 - F_{R_c, ass}^2(b, g) \right] \tag{9}$$

from which the cross section for associative ionization is obtained

$$\sigma_{ass}(g) = 2\pi \int_0^\infty P_{ass}(b) b db = 2\pi \int_0^\infty F_{R_{ass}, \infty} \left[1 - F_{R_c, ass}^2(b, g) \right] b db \tag{10}$$

3   Results and Discussion

The experimental devices used to investigate the dynamics of the autoionization reaction involving metastable atoms and using mass spectrometry, electron spectroscopy, ion-imaging and coincidence techniques has been presented in details in previous papers [46–50] and a schematic view is given in the upper part of Fig. 1.

Shortly, it is a molecular beam (MB) apparatus, formed by a noble gas beam source from which the emerging Ne atoms are electronically excited by collisions with energetic electrons. The Ne* MB crosses at right angles the MB of target species, which in recent experiments are noble gas atoms. PIES are measured by exploiting an hemispherical electron energy analyzer, while total, partial ionization cross sections and branching ratios (BRs), with their collision energy dependence, are obtained exploiting the mass spectrometry technique. In the present case, BRs of relevance are also those associated to the relative formation probability of the *parent ion* Kr^+ and of the ionic adduct Kr^+ - Ne, also indicated as *associated ion*.

Results of recent experiments [39, 40, 46], performed at low collision energy and under high resolution conditions, permitted us to separate in measured PIES (left panel in Fig. 2) the contributions of entrance and exit channels referred to specific spin-orbit levels of Ne* reagent and of Kr^+ product, which are both open shell species. Moreover, PIES have been also measured as a function of the collision energy and obtained results are plotted in the right panel of Fig. 2.

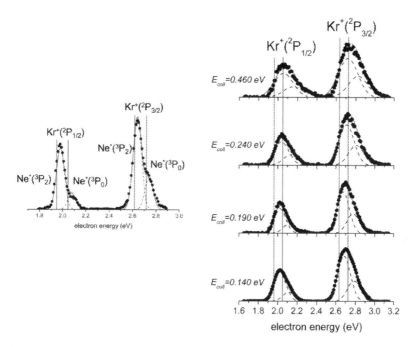

Fig. 2. *Left panel*: A Ne* - Kr PIES measured at low collision energy (50 meV), where the contributions of four reaction channels, associated to two different spin-orbit states *J* of Ne* neutral reactant (J_i = 2,0) and of Kr⁺ ionic product (J_f = 3/2, 1/2), have been resolved. Vertical continuous lines represent the peak positions as predicted for the ionization of Kr by Ne(I) photons, and the shift of observed maxima in measured PIES relates to structure and stability of the reaction transition state [39]. The peak area ratios, defining the relative reaction yields of the four channels, have been also evaluated through the analysis performed adopting four independent Gaussian functions with the same width [39, 40]; *Right panel*: PIESs measured as a function of the collision energy (vertical lines as in the left lower panel). Their analysis emphasizes the dependence of the peak position and of the peak area ratios on the collision energy.

Their analysis provided the dependences on the collision energy of the peak positions, related to the change of the transition state stability, and of the peak area ratios, determined by the change of relative reaction yields. These experimental findings, coupled with the ample phenomenology achieved in our laboratory on the anisotropic dynamical behavior of open-shell atoms [51, 52], suggested us that electronic rearrangements driving the reaction directly arise from polarization of external-floppy cloud of the 3s electron, charge transfer and modifications of angular momentum couplings of other valence electrons within the collision complex. Such rearrangements are accompanied by *adiabatic* and *non-adiabatic effects*, which play a crucial role in the control of the collision dynamics.

In particular, *adiabatic effects* mostly arise from strength and selectivity of configuration interaction which couples entrance, V_{in}, and exit, V_{out}, channels, having the same molecular symmetry and differing for one electron exchange. They affect the anisotropic behavior of the real part, V, of the optical potential (see Eq. (2)) and account for the adiabatic conversion of atomic states, defined in terms of $|J, \Omega >$ quantum numbers, where J represents the total electronic angular momentum of open shell species and Ω describes its projection along the interatomic direction, into molecular states of Σ and Π symmetry. The latter are emerging only at short separation distances, while the atomic states are representative of the system at large distances. Moreover, this conversion involves both entrance and exit channels and obtained interaction components.

The imaginary Γ components (see Eq. (2)), defined in terms of quantum numbers proper of entrance and exit channels, are controlled by strength and radial dependence of *non-adiabatic effects*. They arise again from polarization, selective configuration interactions, changes in electron angular momentum couplings, spin-orbit and Coriolis contributions.

The nature of the *non-adiabatic effects* suggests that the autoionization reaction occurs through two complementary microscopic mechanisms, as discussed in recent papers [39, 40]:

They are classified as:

i) *direct mechanism* - it is triggered by a homogeneous electron exchange, with coupling terms, between entrance and exit channels, called $A_{\Sigma-\Sigma}$ and $A_{\Pi-\Pi}$ on the basis of molecular character (Σ or Π) of initial and final state;

ii) *indirect mechanism* - it is stimulated by a heterogeneous electron exchange and it is accompanied by mixing/exchange between initial and final states of different symmetry. Such mechanism is basically promoted by spin-orbit and Coriolis coupling effects.

It is important to note that the two mechanisms show a different radial dependence and therefore their relative role varies with the collision energy as previously discussed [39, 40].

The interaction components, formulated as summarized above and published in recent papers [45, 53], permitted us to calculate within the semiclassical method outlined in Sect. 3, the total ionization cross sections in a wide collision energy range for three investigated Ne* - Ng (where Ng = Ar, Kr and Xe) autoionization systems. Obtained results are plotted in Fig. 3. Using this methodology, we were able also to calculate branching ratios (BRs) between selected reaction channels (see Eq. (11) below). This provided an internally consistent rationalization of most relevant experimental findings, characterized with different techniques in various laboratories, that includes total and partial ionization cross sections, BRs and PIES measurements [53–56]. The energy dependence of BRs between selected reaction channels must also relate with the PIES spectra for Ne* - Ng systems measured in our laboratory [5, 39, 40] as those reported in Fig. 2 where, as an example, are showed only those for Ne* - Kr.

The PIES spectra of Ne* - Ng are composed of 4 peaks, and each of relative area R_A, R_B, R_C, R_D are evaluated according to the scheme of Fig. 4, and according to the following equations:

$$R_A = \frac{Q_{2-3/2}}{Q_{0-3/2}}; \; R_B = \frac{Q_{0-1/2}}{Q_{0-3/2}}; \; R_C = \frac{Q_{2-1/2}}{Q_{0-3/2}}; \; R_D = \frac{Q_{2-1/2}}{Q_{2-3/2}} \qquad (11)$$

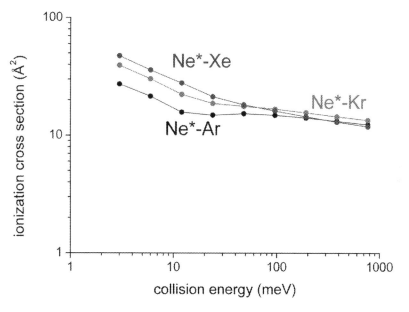

Fig. 3. The total ionization cross sections calculated by the semiclassical method outlined in Sect. 3 for Ne* - Xe (points and blue line), Ne* - Kr (points and red line) and Ne* - Ar (points and black line). (Color figure online)

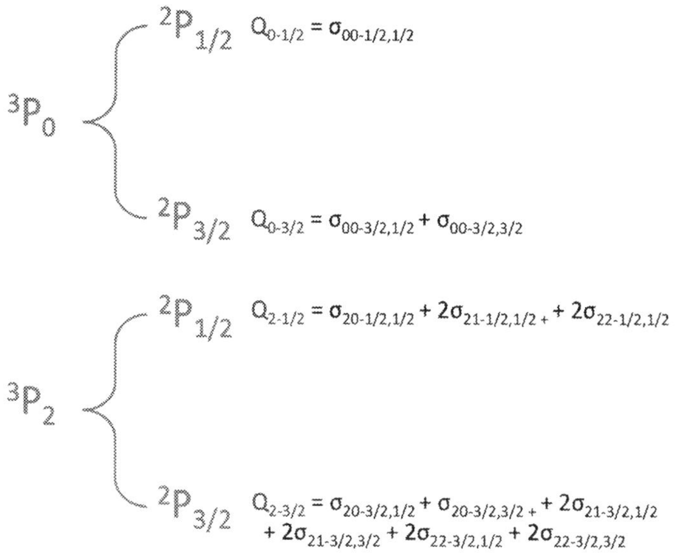

$^3P_0$
$\begin{cases} \\ \\ \\ \end{cases}$

$^2P_{1/2}$ $\quad Q_{0\text{-}1/2} = \sigma_{00\text{-}1/2,1/2}$

$^2P_{3/2}$ $\quad Q_{0\text{-}3/2} = \sigma_{00\text{-}3/2,1/2} + \sigma_{00\text{-}3/2,3/2}$

$^3P_2$
$\begin{cases} \\ \\ \\ \end{cases}$

$^2P_{1/2}$ $\quad Q_{2\text{-}1/2} = \sigma_{20\text{-}1/2,1/2} + 2\sigma_{21\text{-}1/2,1/2} + + 2\sigma_{22\text{-}1/2,1/2}$

$^2P_{3/2}$ $\quad Q_{2\text{-}3/2} = \sigma_{20\text{-}3/2,1/2} + \sigma_{20\text{-}3/2,3/2} + + 2\sigma_{21\text{-}3/2,1/2}$
$\qquad \qquad \quad + 2\sigma_{21\text{-}3/2,3/2} + 2\sigma_{22\text{-}3/2,1/2} + 2\sigma_{22\text{-}3/2,3/2}$

Fig. 4. The scheme used for the evaluation of the of relative area R_A, R_B, R_C, R_D in the observed peaks of Ne*-Ng PIES spectra (see text).

Where the used parameters are defined in Fig. 5.

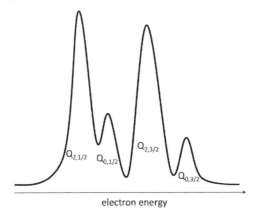

electron energy

Fig. 5. The typical Ne* - Ng PIES spectrum with the definition of parameters used in Eq. (11) (see text).

By analyzing the PIES spectra previously measured in our laboratory (see for example those reported in Fig. 2) [39, 46] we performed cross section ratios calculations of different channels $|J_i \rightarrow J_f\rangle$ for all Ne* - Ng (Ng = Ar, Kr, and Xe) systems on the basis of the semiclassical treatment discussed in Sect. 3. The obtained data are reported in Figs. 6, 7 and 8 where continuous lines of different colors are the results of the present treatment carried out assuming a $^3P_2/^3P_0$ population ratio of about 3, as found for a Ne* beam generated by electron impact [33].

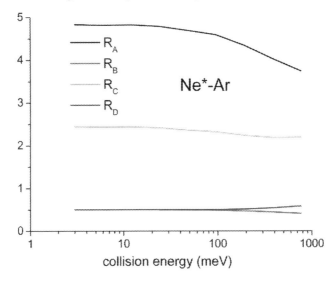

collision energy (meV)

Fig. 6. Cross section ratios of different channels $|J_i \rightarrow J_f\rangle$ calculated for Ne* - Ar system as a function of the collision energy, and analyzing the peak areas obtained in PIES measurements according to Eq. (10) and Figs. 4 and 5. The continuous lines are the results of the present treatment carried out assuming a $^3P_2/^3P_0$ population ratio of about 3, as found for a Ne* beam generated by electron impact (see text).

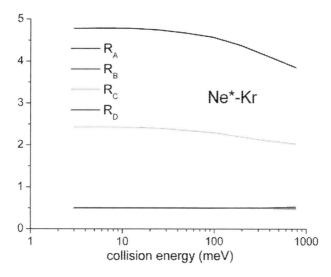

Fig. 7. Cross section ratios of different channels $|J_i \rightarrow J_f\rangle$ calculated for Ne* - Kr system as a function of the collision energy, and analyzing the peak areas obtained in PIES measurements according to Eq. (10) and Figs. 4 and 5. The continuous lines are the results of the present treatment carried out assuming a $^3P_2/^3P_0$ population ratio of about 3, as found for a Ne* beam generated by electron impact (see text). In the figure the red line of R_B cross section ratio is not visible since superimposed by the blue ones related to R_D. (Color figure online)

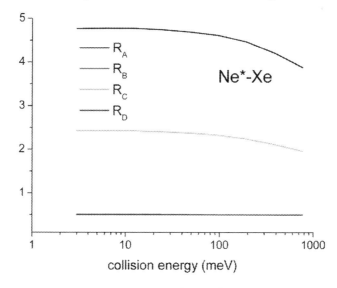

Fig. 8. Cross section ratios of different channels $|J_i \rightarrow J_f\rangle$ calculated for Ne* - Xe system as a function of the collision energy, and analyzing the peak areas obtained in PIES measurements according to Eq. (10) and Figs. 4 and 5. The continuous lines are the results of the present treatment carried out assuming a $^3P_2/^3P_0$ population ratio of about 3, as found for a Ne* beam generated by electron impact (see text). In the figure the red line of R_B cross section ratio is not visible since superimposed by the blue ones related to R_D. (Color figure online)

4   Conclusions

In conclusion, the discussed methodology provides unique information on the stereo-dynamics of autoionization reactions. In this paper it is applied to prototype $Ne^* - Ng$ atom-atom systems, and allows to rationalize in a unifying picture most of the available experimental findings from our and other laboratories. This opened the possibility to obtain *state to state* cross sections which is of great interest for the investigation of quantum effects in the coherent control of collision processes in astrochemistry, promoting Penning and associative ionization, from ultra-cold up to thermal reactive collisions [10, 57, 58]. Obtained results suggest also how to extend the methodology to autoionization reactions involving molecules [7, 9, 59], which are of great interest in several fields, including the balance of phenomena occurring in interstellar environments and planetary atmospheres. In addition, also electron-molecule impacts are of crucial relevance in many applications of molecular plasmas [5].

Acknowledgments. This work was supported and financed with the "Fondo Ricerca di Base, 2018, dell'Università degli Studi di Perugia" (Project Titled: Indagini teoriche e sperimentali sulla reattività di sistemi di interesse astrochimico). Support from Italian MIUR and University of Perugia (Italy) is acknowledged within the program "Dipartimenti di Eccellenza 2018–2022".

References

1. Bartocci, A., Belpassi, L., Cappelletti, D., Falcinelli, S., et al.: Catching the role of anisotropic electronic distribution and charge transfer in halogen bonded complexes of noble gases. J. Chem. Phys. **142**(18), 184304 (2015)
2. Cappelletti, D., Bartocci, A., Grandinetti, F., Falcinelli, S., et al.: Chem. Eur. J. **21**(16), 6234–6240 (2015)
3. Pirani, F., Cappelletti, D., Falcinelli, S., Cesario, D., Nunzi, F., Belpassi, L., Tarantelli, F.: Angew. Chem. Int. Ed. **58**(13), 4195–4199 (2019)
4. Siska, P.E.: Rev. Mod. Phys. **65**, 337–412 (1993)
5. Falcinelli, S., Pirani, F., Candori, P., Brunetti, B.G., Farrar, J.M., Vecchiocattivi, F.: Front. Chem. **7**, 445 (2019)
6. Benz, A., Morgner, H.: Mol. Phys. **57**, 319–336 (1986)
7. Falcinelli, S., Bartocci, A., Cavalli, S., Pirani, F., Vecchiocattivi, F.: Chem. Eur. J. **22**(2), 764–771 (2016)
8. Falcinelli, S., et al.: Modeling the intermolecular interactions and characterization of the dynamics of collisional autoionization processes. In: Murgante, B., et al. (eds.) ICCSA 2013. LNCS, vol. 7971, pp. 69–83. Springer, Heidelberg (2013). https://doi.org/10.1007/978-3-642-39637-3_6
9. Falcinelli, S., Rosi, M., Cavalli, S., Pirani, F., Vecchiocattivi, F.: Chem. Eur. J. **22**(35), 12518–12526 (2016)
10. Arango, C.A., Shapiro, M., Brumer, P.: Phys. Rev. Lett. **97**, 193202 (2006)
11. Falcinelli, S., Capriccioli, A., Pirani, F., Vecchiocattivi, F., Stranges, S., Martì, C., et al.: Fuel **209**, 802–811 (2017)
12. Falcinelli, S.: Catal. Today **348**, 95–101 (2020)
13. Cavallotti, C., Leonori, F., Balucani, N., Nevrly, V., Bergeat, A., et al.: J. Phys. Chem. Lett. **5**, 4213–4218 (2014)

14. Leonori, F., Balucani, N., Nevrly, V., Bergeat, A., et al.: J. Phys. Chem. C **119**(26), 14632–14652 (2015)
15. Leonori, F., Petrucci, R., Balucani, N., Casavecchia, P., Rosi, M., et al.: Phys. Chem. Chem. Phys. **11**(23), 4701–4706 (2009)
16. Rosi, M., Falcinelli, S., Balucani, N., Casavecchia, P., Leonori, F., Skouteris, D.: theoretical study of reactions relevant for atmospheric models of titan: interaction of excited nitrogen atoms with small hydrocarbons. In: Murgante, B., et al. (eds.) ICCSA 2012. LNCS, vol. 7333, pp. 331–344. Springer, Heidelberg (2012). https://doi.org/10.1007/978-3-642-31125-3_26
17. Alagia, M., Candori, P., Falcinelli, S., Lavollée, M., Pirani, F., Richter, R., Stranges, S., Vecchiocattivi, F.: Chem. Phys. Lett. **432**, 398–402 (2006)
18. Alagia, M., Candori, P., Falcinelli, S., Lavollée, M., Pirani, F., Richter, R., Stranges, S., Vecchiocattivi, F.: J. Phys. Chem. A **113**, 14755–14759 (2009)
19. Alagia, M., Candori, P., Falcinelli, S., Lavollèe, M., Pirani, F., Richter, R., Stranges, S., Vecchiocattivi, F.: Phys. Chem. Chem. Phys. **12**, 5389–5395 (2010)
20. Falcinelli, S., Pirani, F., Alagia, M., Schio, L., Richter, R., et al.: Chem. Phys. Lett. **666**, 1–6 (2016)
21. Falcinelli, S., Pirani, F., Vecchiocattivi, F.: Atmosphere **6**(3), 299–317 (2015)
22. Biondini, F., Brunetti, B.G., Candori, P., De Angelis, F., et al.: J. Chem. Phys. **122**(16), 164307 (2005)
23. Biondini, F., Brunetti, B.G., Candori, P., De Angelis, F., et al.: J. Chem. Phys. **122**(16), 164308 (2005)
24. Podio, L., Codella, C., Lefloch, B., Balucani, N., Ceccarelli, C., et al.: MNRAS **470**(1), L16–L20 (2017)
25. Bartolomei, M., Cappelletti, D., De Petris, G., Rosi, M., et al.: Phys. Chem. Chem. Phys. **10**(39), 5993–6001 (2008)
26. Leonori, F., Petrucci, R., Balucani, N., Casavecchia, P., Rosi, et al.: J. Phys. Chem. A **113**(16), 4330–4339 (2009)
27. De Petris, G., Cartoni, A., Rosi, M., Barone, V., Puzzarini, C., Troiani, A.: Chem. Phys. Chem. **12**(1), 112–115 (2011)
28. Falcinelli, S., Fernandez-Alonso, F., Kalogerakis, K., Zare, R.N.: Mol. Phys. **88**(3), 663–672 (1996)
29. Alagia, M., Balucani, N., Candori, P., Falcinelli, S., Richter, et al.: Rend. Lincei Sci. Fisiche e Naturali **24**, 53–65 (2013)
30. Ben Arfa, M., Lescop, B., Cherid, M., Brunetti, B., Candori, P., et al.: Chem. Phys. Lett. **308**, 71–77 (1999)
31. Brunetti, B.G., Candori, P., Ferramosche, R., Falcinelli, S., et al.: Chem. Phys. Lett. **294**, 584–592 (1998)
32. Hotop, H., Illenberger, E., Morgner, H., Niehaus, A.: Chem. Phys. Lett. **10**(5), 493–497 (1971)
33. Brunetti, B.G., Candori, P., Cappelletti, D., Falcinelli, S., et al.: Chem. Phys. Lett. **539–540**, 19–23 (2012)
34. Falcinelli, S., Rosi, M., Candori, P., Farrar, J.M., Vecchiocattivi, F., et al.: Planet. Space Sci. **99**, 149–157 (2014)
35. Pei, L., Carrascosa, E., Yang, N., Falcinelli, S., Farrar, J.M.: J. Phys. Chem. Lett. **6**(9), 1684–1689 (2015)
36. Alagia, M., Candori, P., Falcinelli, S., Pirani, F., et al.: Phys. Chem. Chem. Phys. **13**(18), 8245–8250 (2011)
37. Alagia, M., Candori, P., Falcinelli, S., Lavollée, M., et al.: J. Chem. Phys. **126**(20), 201101 (2007)

38. Balucani, N., Bartocci, A., Brunetti, B., Candori, P., et al.: Chem. Phys. Lett. **546**, 34–39 (2012)
39. Falcinelli, S., Vecchiocattivi, F., Pirani, F.: Phys. Rev. Lett. **121**(16), 163403 (2018)
40. Falcinelli, S., Vecchiocattivi, F., Pirani, F.: J. Chem. Phys. **150**(4), 044305 (2019)
41. Bethe, H.A.: Phys. Rev. **57**, 1125–1144 (1940)
42. Miller, W.H., Morgner, H.: J. Chem. Phys. **67**, 4923–4930 (1977)
43. Brunetti, B., Candori, P., Falcinelli, S., Pirani, F., Vecchiocattivi, F.: J. Chem. Phys. **139**(16), 164305 (2013)
44. Falcinelli, S., Candori, P., Pirani, F., Vecchiocattivi, F.: Phys. Chem. Chem. Phys. **19**(10), 6933–6944 (2017)
45. Falcinelli, S., Rosi, M., Vecchiocattivi, F., Pirani, F.: Analytical potential energy formulation for a new theoretical approach in penning ionization. In: Misra, S., et al. (eds.) ICCSA 2019. LNCS, vol. 11621, pp. 291–305. Springer, Cham (2019). https://doi.org/10.1007/978-3-030-24302-9_21
46. Brunetti, B., Candori, P., Falcinelli, S., Lescop, B., et al.: Eur. Phys. J. D **38**(1), 21–27 (2006)
47. Brunetti, B., Candori, P., De Andres, J., Pirani, F., Rosi, M., et al.: J. Phys. Chem. A **101**(41), 7505–7512 (1997)
48. Alagia, M., Boustimi, M., Brunetti, B.G., Candori, P., et al.: J. Chem. Phys. **117**(3), 1098–1102 (2002)
49. Alagia, M., Candori, P., Falcinelli, S., Mundim, M.S.P., Pirani, F., et al.: J. Chem. Phys. **135**(14), 144304 (2011)
50. Alagia, M., Bodo, E., Decleva, P., Falcinelli, S., et al.: Phys. Chem. Chem. Phys. **15**(4), 1310–1318 (2013)
51. Pirani, F., Maciel, G.S., Cappelletti, D., Aquilanti, V.: Int. Rev. Phys. Chem. **25**, 165–199 (2006)
52. Candori, P., Falcinelli, S., Pirani, F., Tarantelli, F., Vecchiocattivi, F.: Chem. Phys. Lett. **436**, 322–326 (2007)
53. Falcinelli, S., Vecchiocattivi, F., Pirani, F.: Commun. Chem. **3**(1), 64 (2020)
54. Gordon, S.D.S., Omiste, J.J., Zou, J., Tanteri, S., Brumer, P., Osterwalder, A.: Nat. Chem. **10**, 1190–1195 (2018)
55. Gordon, S.D.S., Zou, J., Tanteri, S., Jankunas, J., Osterwalder, A.: Phys. Rev. Lett. **119**, 053001 (2017)
56. Gregor, R.W., Siska, P.E.: J. Chem. Phys. **74**, 1078–1092 (1981)
57. Skouteris, D., Balucani, N., Faginas-Lago, N., et al.: A&A **584**, A76 (2015)
58. Skouteris, D., Balucani, N., Ceccarelli, C., Faginas Lago, N., et al.: MNRAS **482**, 3567–3575 (2019)
59. Jankunas, J., Bertsche, B., Jachymski, K., Hapka, M., Osterwalder, A.: J. Chem. Phys. **140**, 244302 (2014)

A Computational Study of the Reaction Cyanoacetylene and Cyano Radical Leading to 2-Butynedinitrile and Hydrogen Radical

Emília Valença Ferreira de Aragão[1,2]([⊠]) (ORCID), Noelia Faginas-Lago[1] (ORCID),
Marzio Rosi[3] (ORCID), Luca Mancini[1], Nadia Balucani[1] (ORCID), and Dimitrios Skouteris[2]

[1] Dipartimento di Chimica, Biologia e Biotecnologie, Università degli Studi
di Perugia, 06123 Perugia, Italy
{emilia.dearagao,luca.mancini2}@studenti.unipg.it,
{noelia.faginaslago,nadia.balucani}@unipg.it
[2] Master-up srl, Via Sicilia 41, 06128 Perugia, Italy
{emilia.dearagao,d.skouteris}@master-up.it
[3] Dipartimento di Ingegneria Civile ed Ambientale, Università degli Studi di Perugia,
06125 Perugia, Italy
marzio.rosi@unipg.it

Abstract. The present work focuses on the characterization of the reaction between cyanoacetylene and cyano radical by electronic structure calculations of the stationary points along the minimum energy path. One channel, leading to C_4N_2 (2-Butynedinitrile) + H, was selected due to the importance of its products. Using different ab initio methods, a number of stationary points of the potential energy surface were characterized. The energy values of these minima were compared in order to weight the computational costs in relation to chemical accuracy. The results of this works suggests that B2PLYP (and B2PLYPD3) gave a better description of the saddle point geometry, while B3LYP works better for minima.

Keywords: Ab initio calculations · Titan atmosphere · Astrochemistry · Chemistry of the interstellar medium

1 Introduction

Cyanopolyynes are a family of carbon-chain molecules that have been detected in numerous objects of the Interstellar medium (ISM), such as hot cores, star forming regions and cold clouds [1–4]. They are all linear molecules, with alternating carbon-carbon triple and single bonds. The simplest cyanopolyyne, HC_3N, has been among the first organic molecules to be detected in the ISM [5] and up to date also HC_5N, HC_7N, HC_9N and $HC_{11}N$ have been detected at least once in the ISM [6,7] (the detection of $HC_{11}N$, however, has been recently disputed by

O. Gervasi et al. (Eds.): ICCSA 2020, LNCS 12251, pp. 707–716, 2020.
https://doi.org/10.1007/978-3-030-58808-3_51

Loomis et al. [8] and Cordiner et al. [9]). HC_3N and HC_5N are also abundant in solar-type protostars (see for instance a recent work on IRAS 16293-2422 by Jaber Al-Edhari et al. [10]).

The shortest and most abundant member of the cyanopolyyne family, HC_3N (cyanoacetylene), is a precursor of chain elongation reactions: the successive addition of C_2H molecules generates the other members of its family, as summarised by Cheikh et al. [11]. According to the same authors, however, addition of CN radical instead of C_2H would result in a chain termination reaction by the formation of dicyanopolyynes, $NC-(CC)_n-C N$. These species have not been observed in the ISM so far because they lack a permanent electric dipole moment and cannot be detected through their rotational spectrum. However, it has been suggested that they are abundant in interstellar and circumstellar clouds [12]. The recent detection of NCCN [13] via its protonated form $NCCNH^+$ seems to corroborate the suggestion by Petrie et al. [12].

In 2004, Petrie and Osamura had explored through computational means the formation of C_4N_2 (2-Butynedinitrile) in Titan's atmosphere through many pathways [14]. They had found in particular that the cyano radical addition to cyanoacetylene leads to C_4N_2. In order to characterize all the stationary points of the potential energy surface, the authors had used at the time the hybrid density functional method B3LYP in conjunction with triple-split valence Gaussian basis set 6-311G** for geometry optimizations and vibrational frequency calculations. Moreover, they had also performed single-point calculations with CCSD(T) in conjunction with aug-cc-pVDZ basis-set, a choice made at the time due to computational costs. As computational power has risen in the last 15 years and new methods have been implemented into quantum chemistry software, more accurate results for the geometries and energies can be obtained.

In our laboratory we have already investigated several reactions of astrochemical interest providing insightful results for the understanding of processes observed in the ISM [15,16] including those involving or leading N-bearing organic molecules [17–19]. Recently, we have focused on studying the reaction between HC_3N and CN in collaboration with the experimental part of the Perugia group. The preliminary investigation of the potential energy surface of the system HC_3N + CN showed the presence of a number of reactive channels. The search for intermediate and product species always involves the computation of four carbon atoms, two nitrogen atoms and one hydrogen atom, i.e. 39 electrons in total. Ab initio calculations can become prohibitively expensive with a rising number of electrons, therefore a balance between chemical accuracy and computational cost must be reached. The focus of this paper is the exit channel of the reaction between cyanoacetylene and cyano radical that leads to the formation of 2-Butynedinitrile and hydrogen. Four different computational methods were benchmarked in order to check if relatively cheap methods can get accurate results for this system. A comparison to Petrie and Osamura's results was also done.

The paper is organized as follows: in Sect. 2, the methods and the construction of the potential energy surface are outlined. Preliminary results are reported in Sect. 3 and in Sect. 4 concluding remarks are given.

2   Methods

The Potential Energy Surface (PES) of the system was investigated through the optimization of the most stable stationary points. The overall multiplicity of the system was 2 (doublet) and calculations were performed adopting an unrestricted formalism. Following a well established computational scheme [20–26], we optimized the geometry of the stationary points, both minima and saddle points, using a less expensive method with respect to the one employed in order to get accurate energies. Calculations for geometries were performed in order to benchmark three methods: Unrestricted-Hartree-Fock (UHF) [27,28], density functional theory (DFT), with the Becke-3-parameter exchange and Lee-Yang-Parr correlation (B3LYP) [29,30], and double-hybrid DFT method B2PLYP [31] combined or not with Grimme's D3BJ dispersion [32,33]. All methods were used in conjunction with the correlation consistent valence polarized basis set aug-cc-pVTZ [34]. In each level of theory, a vibrational frequency analysis was done to confirm the nature of the stationary points: a minimum in the absence of imaginary frequencies and a saddle point if one and only one frequency is imaginary. The assignment of the saddle points was performed using intrinsic reaction coordinate (IRC) calculations [35,36]. Then for each stationary point for all methods, the energy was computed with coupled cluster including single and double excitations and a perturbative estimate of connected triples (CCSD(T)) [37–39]. CCSD(T) is a more accurate method than the ones used for the optimizations, but prohibitive computational costs restricts the use of the method in this particular system to fixed geometry calculations. Finally, zero-point energy correction obtained through the frequency calculations were added to energies obtained from all methods to correct them to 0 K. All calculations were performed using the Gaussian 09 code [40].

3   Results and Discussion

The calculation of the different electronic structures shows that the attack of cyano radical on the cyanoacetylene is an energetically favorable process that leads to the formation of an adduct intermediate. Figure 1 gather the geometries at the minimum energy path, starting from the HC$_3$N and CN reactants and leading to 2-Butynedinitrile and hydrogen. These geometries were optimized at UHF/aug-cc-pVTZ, B3LYP/aug-cc-pVTZ, B2PLYP/aug-cc-pVTZ and B2PLYPD3/aug-cc-pVTZ levels. The reported interatomic distances are in red for UHF, blue for B3LYP and green for both B2PLYP and B2PYLP with Grimme's D3 dispersion, since no difference in their geometries was recorded. The geometries obtained with the methods above are very similar. The differences between bond lengths and angles are small, with the exception of the

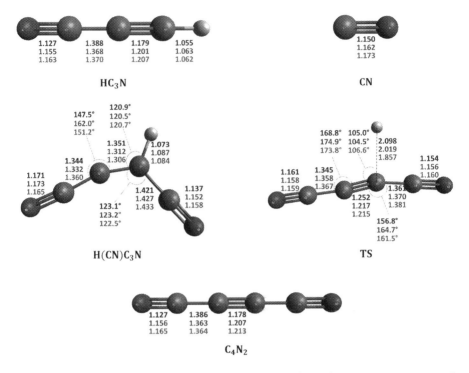

Fig. 1. Optimized geometries of the stationary points along the minimum energy path leading cyanoacetylene and cyano radical to 2-Butynedinitrile and hydrogen. Bond lengths are shown in Å and bond angles are displayed in degrees. UHF values in red, B3LYP values in blue, B2PLYP values in green. (Color figure online)

transition state geometries. In the saddle point structure, B2PLYP seems to provide a more reasonable distance for weak interaction as the one between carbon and hydrogen: it is 0.162 Å shorter than the distance obtained with B3LYP and 0.241 Å shorter than the one proposed by UHF. B2PLYP is a double-hybrid density functional that combines Becke exchange and Lee, Young and Parr correlation with Hartree-Fock exchange and a perturbative second-order correlation obtained from Kohn-Sham orbitals [31]. According to the author of the method, B2PLYP reports very good results for transition state geometries.

Most of the geometries might be similar, but this does not necessarily means that all energies also are. Table 1 gathers the relative energies of all geometries computed with every method. The reactants were taken as the reference for the energy. Within the B2PLYP method, the account of Grimme's D3 dispersion changes slightly the energy value, even for an identical geometry. As every method returns different absolute energy values, a direct comparison is not pertinent. However, the trend in the evolution of the relative energies can be compared. First, it can be observed that all stationary points are below the energy of the reactants in all the methods that have been employed.

Table 1. Energies ($kJ.mol^{-1}$, $0 K$) of the different geometries relative to the reactants. The energies are computed at UHF/aug-cc-pVTZ, B3LYP/aug-cc-pVTZ, B2PLYP/aug-cc-pVTZ, and B2PLYPD3/aug-cc-pVTZ levels of theory. In parentheses are the values for the same geometry computed at CCSD(T)/aug-cc-pVTZ level of theory.

	ΔH_0^0 (kJ.mol^{-1})			
	UHF (CCSD(T))	B3LYP (CCSD(T))	B2PLYP (CCSD(T))	B2PLYPD3 (CCSD(T))
$HC_3N + CN$	0.0 (0.0)	0.0 (0.0)	0.0 (0.0)	0.0 (0.0)
MIN 1	-276.6 (-240.8)	-259.0 (-223.8)	-231.7 (-221.0)	-236.2 (-221.0)
TS (MIN 1 → C_4N_2 + H)	-62.8 (-50.2)	-65.3 (-44.9)	-48.9 (-25.3)	-53.8 (-25.2)
C_4N_2 + H	-50.3 (-58.7)	-82.7 (-68.1)	-85.6 (-68.1)	-88.0 (-67.9)

Second, in all methods the adduct intermediate MIN 1 is the most energetically stable structure. In addition, a barrier between the adduct and the products is well characterized for B3LYP and both B2PLYP methods, but it is not the case for the UHF method. Though the saddle point was identified with UHF, the energy of the products returned a higher value. It will be shown, however, that when a single-point energy calculation with these same geometries optimized with UHF is done with the CCSD(T)/aug-cc-pVTZ method, the energy of the products is below the energy of the saddle-point.

In order to compare the geometries optimized with all those methods, single-point CCSD(T) calculations were also performed. The results are reported again in Table 1, in parenthesis. Firstly, it can be observed that the energies of all stationary points, i.e. minima and transition state, are below the energy of the reactants in all the methods used. Secondly, as B2PLYP and B2PLYPD3 optimized geometries were mostly identical, the energies computed with CCSD(T)/aug-cc-pVTZ were also the same. Thirdly, it can be noticed that the energies of the B3LYP optimized geometries are very close to the ones of B2PLYP methods, with the exception of the transition state geometry: for this saddle point, energy difference is around 19.6 $kJ.mol^{-1}$. At last, the energies of the UHF optimized geometries are the most different from the others methods. In particular, the energies of the intermediate and transition state gave the lowest energy at CCSD(T) level. In contrast, the energy of the products was the highest one using the UHF geometry. Nevertheless, the energy barrier between the intermediate and the products is now well characterized for all methods.

It is also interesting to look at the barrier height values obtained for each method. Table 2 reports the energy changes and barrier heights, computed at the same levels of theory, for the process leading to 2-Butynedinitrile and hydrogen. The interaction of HC_3N and CN gives rise to the adduct MIN 1 (or $H(CN)C_3N$), more stable than the reactants in all levels of theory. This adduct evolves, through a barrier leading to the transition state, to the products of the reactive channel C_4N_2 and radical H. In all levels of theory, the products are less stable than the adduct. Making a comparison between methods, UHF estimates the largest enthalpy changes and barrier height. UHF is followed by B3LYP, B2PLYPD3 and B2PLYP in this order. In relation to CCSD(T), a higher-level method, UHF and B3LYP energies are systematically overestimated. B2PLYP and B2PLYPD3 overestimates the CCSD(T) in the enthalpy variation attributed to the formation of the adduct, but underestimate both the barrier and enthalpy change for the reaction that leads the intermediate to the product.

Table 2. Enthalpy changes and barrier height (kJ.mol^{-1}, 0 K) computed at UHF/aug-cc-pVTZ, B3LYP/aug-cc-pVTZ, B2PLYP/aug-cc-pVTZ, and B2PLYPD3/aug-cc-pVTZ levels of theory. In parentheses are the values for the same geometry computed at CCSD(T)/aug-cc-pVTZ level of theory.

	ΔH_0^0 (kJ.mol^{-1})				Barrier height			
	UHF (CCSD(T))	B3LYP (CCSD(T))	B2PLYP (CCSD(T))	B2PLYPD3 (CCSD(T))	UHF (CCSD(T))	B3LYP (CCSD(T))	B2PLYP (CCSD(T))	B2PLYPD3 (CCSD(T))
HC$_3$N + CN → MIN 1	-276.6 (-240.8)	-259.0 (-223.8)	-231.7 (-221.0)	-236.2 (-221.0)				
MIN 1 → C$_4$N$_2$ + H	226.4 (182.1)	176.2 (155.7)	146.0 (152.9)	148.2 (153.1)	213.9 (190.6)	193.6 (178.9)	182.7 (195.7)	182.3 (195.8)

While in this work the aug-cc-pVTZ basis set was employed for every method, Petrie and Osamura [14] carried geometry optimizations at B3LYP level in conjunction with the 6-311G** basis-set and computed the single-point energies at CCSD(T) level in conjunction with aug-cc-pVDZ basis-set. In respect to the geometries obtained, only the transition state and the intermediate geometries were published in their paper. The bond distance and angle values are similar to the ones in the B3LYP/aug-cc-pVTZ optimized geometries. The difference is smaller than when the B3LYP geometries were compared to the ones obtained with other methods. The energies computed at CCSD(T)/aug-cc-pVDZ level are listed in Table 2 of that same paper. Since the authors provided the values of total energy for all stationary points, it was possible to calculate the energy of each point in relation to HC$_3$N + CN. The energy at 0 K is -231.1 kJ.mol^{-1} for the adduct intermediate, -39.5 kJ.mol^{-1} for the transition state and -58.6 kJ.mol^{-1} for the products. The energies of the stationary points are once again below the energy of the reactants. CCSD(T)/aug-cc-pVDZ provides here a lower energy value for the adduct intermediate and a higher energy value for the transition state and the products than CCSD(T)/aug-cc-pVTZ. At 191.6 kJ.mol^{-1}, the height of the barrier between the adduct intermediate and the products is larger in comparison to CCSD(T)/aug-cc-pVTZ. On the other hand, at 172.5 kJ.mol^{-1} the enthalpy change between the adduct and the product is very close to the one showing on Table 2.

4   Conclusions

UHF, DFT (with B3LYP functional) and double-hybrid DFT (with B2PLYP and B2PLYPD3 functionals) were benchmarked on four points of the PES of the astrochemically relevant reaction between cyanoacetylene (HC$_3$N) and cyano radical. As far as optimized geometries for stationary points are concerned, the UHF method seems to be inadequate, while DFT methods seem to be more reliable. In particular, B3LYP functional seems to work well for minima, while better functionals like the B2PLYP seem to be necessary for transition state geometries when van der Waals interactions are present. As far as energies are concerned, however, only more correlated methods like CCSD(T) seems to provide accurate results.

A more general conclusion concerns the reaction mechanism suggested by our calculations, which is also in line with the previous determination by Petrie and Osamura. Similarly to the case of other reactions involving CN and other species holding a triple C−C bond (such as ethyne, propyne and 2-butyne [41–43]), the CN radical interacts with the electron density of triple bond to form an addition intermediate without an activation energy. This is in line with the large value of the rate coefficient derived for the reactions also at very low temperature [11] and makes this process a feasible route of dicyano-acetylene (2-Butynedinitrile) even under the harsh conditions of the interstellar medium.

After completing the derivation of the potential energy surface for the title reaction, we will run kinetic calculations to derive the rate coefficient and product branching ratio. The calculated rate coefficient will be compared with the experimental values derived by Cheikh et al. [11] while the reaction mechanism and product branching ratio will be compared with those inferred by the crossed molecular beam experiments which are now in progress in our laboratory. A thorough characterization of the title reaction will allow us to establish its role in the nitrogen chemistry of the interstellar medium. To be noted that both cyano- and dicyano-acetylene have a strong prebiotic potential [44].

Acknowledgements. This project has received funding from the European Union's Horizon 2020 research and innovation programme under the Marie Skłodowska-Curie grant agreement No 811312 for the project "Astro-Chemical Origins" (ACO). E.V.F.A. thanks the Dipartimento di Ingegneria Civile ed Ambientale of University of Perugia for allocated computing time. N.F.L. thanks Perugia University for financial support through the AMIS project ("Dipartimenti di Eccellenza-2018–2022"), also thanks the Dipartimento di Chimica, Biologia e Biotecnologie for funding under the program Fondo Ricerca di Base 2017. M.R. acknowledges the project "Indagini teoriche e sperimentali sulla reattività di sistemi di interesse astrochimico" funded with Fondo Ricerca di Base 2018 of the University of Perugia.

References

1. Wyrowski, F., Schilke, P., Walmsley, C.: Vibrationally excited HC_3N toward hot cores. Astron. Astrophys. **341**, 882–895 (1999)
2. Taniguchi, K., Saito, M., Sridharan, T., Minamidani, T.: Survey observations to study chemical evolution from high-mass starless cores to high-mass protostellar objects I: HC_3N and HC_5N. Astrophys. J. **854**(2), 133 (2018)
3. Mendoza, E., et al.: A search for cyanopolyynes in L1157–B1. Monthly Not. R. Astron. Soc. **475**(4), 5501–5512 (2018)
4. Takano, S., et al.: Observations of ^{13}C isotopomers of HC_3N and HC_5N in TMC-1: evidence for isotopic fractionation. Astron. Astrophys. **329**, 1156–1169 (1998)
5. Turner, B.E.: Detection of interstellar cyanoacetylene. Astrophys. J. **163**, L35–L39 (1971)
6. Broten, N.W., Oka, T., Avery, L.W., MacLeod, J.M., Kroto, H.W.: The detection of HC_9N in interstellar space. **223**, L105–L107 (1978). https://doi.org/10.1086/182739
7. Bell, M., Feldman, P., Travers, M., McCarthy, M., Gottlieb, C., Thaddeus, P.: Detection of $HC_{11}N$ in the cold dust cloud TMC-1. Astrophys. J. Lett. **483**(1), L61–L64 (1997)
8. Loomis, R.A., et al.: Non-detection of $HC_{11}N$ towards TMC-1: constraining the chemistry of large carbon-chain molecules. Monthly Not. R. Astron. Soc. **463**(4), 4175–4183 (2016)
9. Cordiner, M.A., Charnley, S.B., Kisiel, Z., McGuire, B.A., Kuan, Y.J.: Deep K-band observations of TMC-1 with the green bank telescope: detection of HC_7O, nondetection of $HC_{11}N$, and a search for new organic molecules. Astrophys. J. **850**(2), 187 (2017). https://doi.org/10.3847/1538-4357/aa970c
10. Jaber Al-Edhari, A., et al.: History of the solar-type protostar IRAS 16293–2422 as told by the cyanopolyynes. A&A **597**, A40 (2017). https://doi.org/10.1051/0004-6361/201629506
11. Cheikh Sid, S., Morales, S.B., Guillemin, J.C., Klippenstein, S.J., Sims, I.R.: Low temperature rate coefficients for the reaction $CN + HC_3N$. J. Phys. Chem. A **117**(46), 12155–12164 (2013). https://doi.org/10.1021/jp406842q
12. Petrie, S., Millar, T., Markwick, A.: NCCN in TMC-1 and IRC+ 10216. Monthly Not. R. Astron. Soc. **341**(2), 609–616 (2003)
13. Agúndez, M., et al.: Probing non-polar interstellar molecules through their protonated form: detection of protonated cyanogen ($NCCNH^+$). Astron. Astrophys. **579**, L10 (2015)
14. Petrie, S., Osamura, Y.: NCCN and NCCCCN formation in Titan's atmosphere: 2. HNC as a viable precursor. J. Phys. Chem. A **108**(16), 3623–3631 (2004)
15. Podio, L., et al.: Silicon-bearing molecules in the shock L1157–B1: first detection of SiS around a Sun-like protostar. Monthly Not. R. Astron. Soc. Lett. **470**(1), L16–L20 (2017)
16. Skouteris, D., et al.: Interstellar dimethyl ether gas-phase formation: a quantum chemistry and kinetics study. Monthly Not. R. Astron. Soc. **482**(3), 3567–3575 (2019)
17. Skouteris, D., Balucani, N., Faginas-Lago, N., Falcinelli, S., Rosi, M.: Dimerization of methanimine and its charged species in the atmosphere of Titan and interstellar/cometary ice analogs. Astron. Astrophys. **584**, A76 (2015)

18. Balucani, N., Skouteris, D., Ceccarelli, C., Codella, C., Falcinelli, S., Rosi, M.: A theoretical investigation of the reaction between the amidogen, NH, and the ethyl, C_2H_5, radicals: a possible gas-phase formation route of interstellar and planetary ethanimine. Mol. Astrophys. **13**, 30–37 (2018)

19. Sleiman, C., El Dib, G., Rosi, M., Skouteris, D., Balucani, N., Canosa, A.: Low temperature kinetics and theoretical studies of the reaction $CN + CH_3NH_2$: a potential source of cyanamide and methyl cyanamide in the interstellar medium. Phys. Chem. Chem. Phys. **20**(8), 5478–5489 (2018)

20. Falcinelli, S., Rosi, M., Cavalli, S., Pirani, F., Vecchiocattivi, F.: Stereoselectivity in autoionization reactions of hydrogenated molecules by metastable noble gas atoms: the role of electronic couplings. Chem. A Eur. J. **22**(35), 12518–12526 (2016)

21. Leonori, F., et al.: Crossed-beam and theoretical studies of the $S(^1D) + C_2H_2$ reaction. J. Phys. Chem. A **113**(16), 4330–4339 (2009)

22. Bartolomei, M., et al.: The intermolecular potential in $NO-N_2$ and $(NO-N_2)^+$ systems: implications for the neutralization of ionic molecular aggregates. Phys. Chem. Chem. Phys. **10**(39), 5993–6001 (2008)

23. de Petris, G., Cartoni, A., Rosi, M., Barone, V., Puzzarini, C., Troiani, A.: The proton affinity and gas-phase basicity of sulfur dioxide. ChemPhysChem **12**(1), 112–115 (2011)

24. Leonori, F., et al.: Observation of organosulfur products (thiovinoxy, thioketene and thioformyl) in crossed-beam experiments and low temperature rate coefficients for the reaction $S(^1D) + C_2H_4$. Phys. Chem. Chem. Phys. **11**(23), 4701–4706 (2009)

25. de Petris, G., Rosi, M., Troiani, A.: SSOH and HSSO radicals: an experimental and theoretical study of $[S_2OH]^{0/+/-}$ species. J. Phys. Chem. A **111**(28), 6526–6533 (2007)

26. Rosi, M., Falcinelli, S., Balucani, N., Casavecchia, P., Leonori, F., Skouteris, D.: Theoretical study of reactions relevant for atmospheric models of Titan: interaction of excited nitrogen atoms with small hydrocarbons. In: Murgante, B., et al. (eds.) ICCSA 2012. LNCS, vol. 7333, pp. 331–344. Springer, Heidelberg (2012). https://doi.org/10.1007/978-3-642-31125-3_26

27. Roothaan, C.C.J.: New developments in molecular orbital theory. Rev. Mod. Phys. **23**(2), 69 (1951)

28. Pople, J.A., Nesbet, R.K.: Self-consistent orbitals for radicals. J. Chem. Phys. **22**(3), 571–572 (1954)

29. Becke, A.D.: Density functional thermochemistry. III. The role of exact exchange. J. Chem. Phys. **98**(7), 5648–5652 (1993). https://doi.org/10.1063/1.464913

30. Stephens, P.J., Devlin, F.J., Chabalowski, C.F., Frisch, M.J.: *Ab Initio* calculation of vibrational absorption and circular dichroism spectra using density functional force fields. J. Phys. Chem. **98**(45), 11623–11627 (1994)

31. Grimme, S.: Semiempirical hybrid density functional with perturbative second-order correlation. J. Chem. Phys. **124**(3), 034108 (2006)

32. Grimme, S., Ehrlich, S., Goerigk, L.: Effect of the damping function in dispersion corrected density functional theory. J. Comput. Chem. **32**(7), 1456–1465 (2011)

33. Goerigk, L., Grimme, S.: Efficient and accurate double-hybrid-meta-GGA density functionals- evaluation with the extended GMTKN30 database for general main group thermochemistry, kinetics, and noncovalent interactions. J. Chem. Theory Comput. **7**(2), 291–309 (2011)

34. Dunning Jr., T.H.: Gaussian basis sets for use in correlated molecular calculations. I. The atoms boron through neon and hydrogen. J. Chem. Phys. **90**(2), 1007–1023 (1989)

35. Gonzalez, C., Schlegel, H.B.: An improved algorithm for reaction path following. J. Chem. Phys. **90**(4), 2154–2161 (1989)

36. Gonzalez, C., Schlegel, H.B.: Reaction path following in mass-weighted internal coordinates. J. Phys. Chem. **94**(14), 5523–5527 (1990)

37. Bartlett, R.J.: Many-body perturbation theory and coupled cluster theory for electron correlation in molecules. Ann. Rev. Phys. Chem. **32**(1), 359–401 (1981)

38. Raghavachari, K., Trucks, G.W., Pople, J.A., Head-Gordon, M.: A fifth-order perturbation comparison of electron correlation theories. Chem. Phys. Lett. **157**(6), 479–483 (1989)

39. Olsen, J., Jørgensen, P., Koch, H., Balkova, A., Bartlett, R.J.: Full configuration-interaction and state of the art correlation calculations on water in a valence double-zeta basis with polarization functions. J. Chem. Phys. **104**(20), 8007–8015 (1996)

40. Frisch, M., et al.: Gaussian 09, Revision A. 02, 2009. Gaussian. Inc., Wallingford CT (2009)

41. Huang, L.C.L., Balucani, N., Lee, Y.T., Kaiser, R.I., Osamura, Y.: Crossed beam reaction of the cyano radical, CN (X $^2\Sigma^+$), with methylacetylene, CH_3CCH (X $^1A_1$): observation of cyanopropyne, CH_3CCCN (X $^1A_1$), and cyanoallene, $H_2CCCHCN$ (X $^1A'$). J. Chem. Phys. **111**(7), 2857–2860 (1999)

42. Balucani, N., et al.: Crossed beam reaction of cyano radicals with hydrocarbon molecules. II. Chemical dynamics of 1-cyano-1-methylallene ($CNCH_3CCCH_2$; X $^1A'$) formation from reaction of CN (X $^2\Sigma^+$) with dimethylacetylene CH_3CCCH_3 (X $^1A_1'$). J. Chem. Phys. **111**(16), 7472–7479 (1999)

43. Huang, L.C.L., et al.: Crossed beam reaction of cyano radicals with hydrocarbon molecules. IV. Chemical dynamics of cyanoacetylene (HCCCN; X $^1\Sigma^+$) formation from reaction of CN (X $^2\Sigma^+$) with acetylene, C_2H_2 (X $^1\Sigma_g^+$). J. Chem. Phys. **113**(19), 8656–8666 (2000)

44. Balucani, N.: Elementary reactions of N atoms with hydrocarbons: first steps towards the formation of prebiotic N-containing molecules in planetary atmospheres. Chem. Soc. Rev. **41**(16), 5473–5483 (2012)

A Theoretical Investigation of the Reactions of N(^{2}D) with Small Alkynes and Implications for the Prebiotic Chemistry of Titan

Luca Mancini[1]($^{(\boxtimes)}$), Emília Valença Ferreira de Aragão[1,3], Marzio Rosi[2], Dimitrios Skouteris[3], and Nadia Balucani[1]

[1] Dipartimento di Chimica, Biologia e Biotecnologie, Università degli Studi di Perugia, 06123 Perugia, Italy
{luca.mancini2, emilia.dearagao}@studenti.unipg.it,
nadia.balucani@unipg.it
[2] Dipartimento di Ingegneria Civile e Ambientale, Università degli Studi di Perugia, 06125 Perugia, Italy
marzio.rosi@unipg.it
[3] Master-Up, Via Sicilia 41, 06128 Perugia, Italy
d.skouteris@master-up.it

Abstract. The reactions between atomic nitrogen, in its first electronically excited state (^{2}D), and two different hydrocarbons, methyl acetylene and acetylene, have been analyzed by performing electronic structure calculations of the potential energy surface. For each reaction, H-displacement channels leading to cyclic molecular products have been identified, together with an H_2 elimination channel for the reaction N(^{2}D) + acetylene and a methyl (CH_3) loss channel for the N(^{2}D) + methylacetylene reaction. Since both reactions have been found to be exothermic and without an entrance barrier, we suggest that they are fast and efficient under the conditions of the upper atmosphere of Titan. These data will be used to perform kinetic calculations and will be compared with detailed experimental results in future work.

Keywords: Ab initio calculations · Titan atmosphere · Astrochemistry

1 Introduction

The chemistry of nitrogen atoms in their first electronically excited state ^{2}D is of great relevance in several extraterrestrial environments, such as the upper atmosphere of Titan (the largest moon of Saturn) and Pluto [1–5]. One of the few moons of the Solar System which possess thick atmospheres, Titan has attracted the attention of the scientific community because of possible similarities with primitive Earth [3]. Due to the fact that the appearance of life on Earth has drastically modified our planet, it is impossible to reconstruct the first steps in the evolution of prebiotic chemistry and, thus, the study of planets or moons which share similarities with primitive Earth has become fundamental for its understanding [3–5]. Interesting data about Titan became available starting from the Voyager missions, which reported molecular nitrogen (N_2) as the major component of its atmosphere [6], with minor constituents being methane

© Springer Nature Switzerland AG 2020
O. Gervasi et al. (Eds.): ICCSA 2020, LNCS 12251, pp. 717–729, 2020.
https://doi.org/10.1007/978-3-030-58808-3_52

and higher hydrocarbons (e.g. ethane, ethylene, acetylene, etc.) [7]. Methylacetylene was also observed [8] while the presence of its structural isomer allene, predicted to be present by photochemical models, was confirmed only very recently [9].

Our understanding of the chemistry of the atmosphere of Titan relies on a multidisciplinary approach. First of all, observation, performed with the support of both space mission and ground-based instruments, allows one to derive the abundances of trace compounds which are indicators of an active atmospheric chemistry. As a complement, laboratory experiments aiming at reproducing the conditions of Titan can furnish a first look into the dominant chemical processes (see, for instance, Refs. [10–12] and references therein). Finally, a more refined approach is based on photochemical models that include physical and chemical parameters (see, for instance, Refs. [13–16]), with the latter being determined in theoretical and/or experimental investigations. To build a photochemical model able to reproduce the abundance of the chemical species present in trace amounts (which are also those with a strong prebiotic potential) it is imperative to know the reaction rates of all relevant elementary reactions [16].

Since the first photochemical model developed by Yung et al. [13] after the Voyager mission, one of the starting steps that initiate the chemistry of nitrogen is considered to be the dissociation of molecular nitrogen in the thermosphere induced by free electrons:

$$N_2 + e^- \rightarrow N(^4S) + N(^2D) + e^- \tag{1}$$

The dissociation of molecular nitrogen leads to the formation of atomic nitrogen in the ground, ^4S, and first electronically excited, $^2D_{5/2,3/2}$, states. The ^2D state is metastable with a long radiative lifetime of ca 48 h [1], has an energy content of 230.0 kJ/mol with respect to the ground state and is produced in a large amount also by the EUV photodissociation of N_2 as well as by the electron recombination of the N_2^+ ions [1]. Nitrogen atoms can then react with the hydrocarbons present in the upper atmosphere leading to the formation of N-containing organic compounds, some of which were detected at the time of the Voyager 1 mission [7, 8]. Remarkably, while atomic nitrogen in the ground state is not a very reactive species (only its reactions with open-shell radicals are fast enough to be of any relevance in the conditions of the upper atmosphere of Titan), the reactions of $N(^2D)$ have been invoked to explain the formation of molecules containing a novel C-N bond [13]. Numerous reactions involving $N(^2D)$ and the most common hydrocarbons in the atmosphere of Titan, as well as molecular hydrogen and water, have been already investigated from both an experimental and a theoretical point of view (namely, $N(^2D) + C_2H_2$, C_2H_4, C_2H_6, C_6H_6 [17–26]). More specifically, on the side of theoretical calculations, a combined approach has been used in which the overall potential energy surface has been determined via electronic structure calculations at the DFT and CCSD(T) levels of theory. The resulting stationary points along the minimum energy path have been used to perform kinetic calculations, using a Rice-Ramsperger-Kassel-Marcus (RRKM) code implemented for this purpose [17]. The comparison with the experimental data, coming from the crossed molecular beams (CMB) experiments, allows the derivation of the reaction mechanism.

In this preliminary account, we report on dedicated electronic structure calculations of the stationary points of the potential energy surfaces (PESs) for the reactions involving the two simplest alkynes, that is, ethyne (or acetylene) and propyne (or methylacetylene):

$$N(^2D) + C_2H_2 \rightarrow HCCN + H \tag{2a}$$

$$\rightarrow H_2 + C_2N \tag{2b}$$

$$N(^2D) + CH_3CCH \rightarrow H + C_3H_3N \tag{3a}$$

$$\rightarrow CH_3 + C_2NH \tag{3b}$$

Previous theoretical investigations at a different level of theory are available for the reaction (2) [18, 27, 28], while, to the best of our knowledge, no data are available for reaction (3). The present results will be used to perform kinetic calculations to obtain rate coefficients and product branching ratios to be compared with the experimental results obtained in the CMB experiments with mass spectrometric detection in a future work.

2 Computational Details

The $[C_2H_2 + N]$ and $[C_3H_4 + N]$ reactions have been analyzed by exclusively considering the lowest doublet electronic state for both systems. The PESs for the two systems have been characterized through optimization of the most stable stationary points, adopting a computational strategy which has already been utilized successfully in several cases [17, 20, 26, 29–35]. In particular, density functional calculations have been performed at the B3LYP [36, 37] level of theory in conjunction with the correlation consistent valence polarized basis set aug-cc-pVTZ [38–40]. The harmonic vibrational frequencies have been computed at the same level of theory in order to determine the nature of each stationary point, i.e. minimum if all the frequencies are real and saddle point if there is one, and only one, imaginary frequency. Intrinsic reaction coordinates (IRC) calculations [41, 42] have been performed in order to assign the saddle points. Then, the energy of each stationary point has been computed with the more accurate coupled cluster theory including both single and double excitations and using a perturbative estimate of the effect of the triple excitations (the CCSD(T) level) [43–45], with the same basis set aug-cc-pVTZ. The zero-point correction (computed using the scaled harmonic vibrational frequencies obtained at the B3LYP/aug-cc-pVTZ level) has been added in order to correct the energies at 0 K. The experimental [46] separation N(^{4}S) – N(^{2}D) of 230.0 kJ/mol has been added to the energy of N(^{4}S) at all the levels of calculations in order to estimate the energy of N(^{2}D). All calculations have been carried out using GAUSSIAN 09 [47] while the analysis of the vibrational frequencies has been done using MOLDEN [48, 49].

3 Results

3.1 N(^{2}D) + Acetylene

In the PES for the system N(^{2}D) + acetylene, shown below in Fig. 1, three cyclic minima have been located (MIN1AC, MIN2AC, MIN3AC), connected by three transition states (TS1AC, connecting MIN1AC and MIN2AC, TS2AC, connecting MIN1AC and MIN3AC, and TS3AC, connecting MIN2AC and MIN3AC). The same stationary points were previously investigated by Takayanagi et al. in 1998 at the MP2/cc-pVTZ level for geometry optimization and MP4(full, SDTQ)/cc-pVTZ level for the energy evaluation [27]. We can notice a reasonable agreement between the calculations, considering the different levels of theory, except for the presence of a barrier of 2.9 kcal/mol at the MP4(full, SDTQ)/cc-pVTZ level, which has not been identified at the CCSD(T) level. We investigated the initial attack of N(^{2}D) also at the B2PLYP/aug-cc-pVTZ level of theory in order to include the second order perturbation theory in the calculations. Neither method allows us to locate the barrier. In Fig. 2 a comparison of the geometries of the reactant and the first intermediate optimized at B3LYP/aug-cc-pVTZ and B2PLYP/aug-cc-pVTZ level of theory is shown.

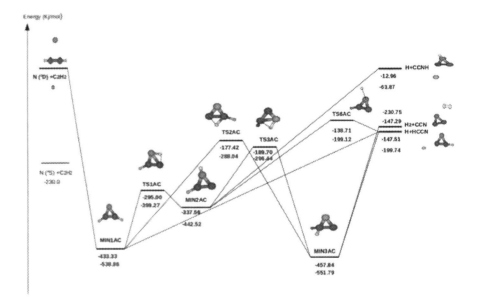

Fig. 1. Schematic representation of the PES obtained for the system N(^{2}D) + acetylene, with the relative energies (kJ/mol) evaluated at the CCSD(T)/aug-cc-pVTZ (in black) and B3LYP/aug-cc-pVTZ (in blue) levels of theory. (Colour figure online)

The absence of an entrance barrier was also verified by very accurate calculations by Nuñez-Reyes et al. [28], that accompanied an experimental determination of the reaction rate coefficient. Notably, the value of the rate coefficient derived by the kinetic experiments carried out by Nuñez-Reyes et al. [28] is very large and, together with the temperature dependence, points to a barrierless reaction.

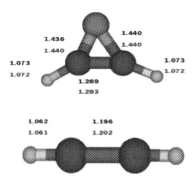

Fig. 2. Geometries of the reactant and the first intermediate (distances in angstroms) optimized at B3LYP/aug-cc-pVTZ (in black) and B2PLYP/aug-cc-pVTZ levels of theory (in red). (Color figure online)

In summary, we can conclude that the reaction starts with the barrierless addition of N(^{2}D) to the triple bond of acetylene. The addition leads to the formation of a cyclic intermediate MIN1AC, in which two C-N bonds are present. The MIN1AC intermediate can then isomerize to MIN2AC by overcoming the transition state TS1AC, which is characterized by the shift of an H atom from one of the C atoms to the N atom, with the consequent formation of a new N-H bond. Finally, two different sequences can lead to the formation of the third intermediate MIN3AC, which is the most stable species in the C_2NH_2 PES: *a)* the isomerization of MIN1AC through the transition state TS2AC (overcoming a barrier of 243.63 kJ/mol), and *b)* the isomerization of MIN2AC through the transition state TS3AC (overcoming a barrier of 147.86 kJ/mol).

In addition to the isomerization processes, all the three intermediates can lead to the formation of different products by losing an H atom. In particular, MIN1AC can directly decompose into H and the cyclic fragment HCCN (which shows a relative energy, with respect to the reactant asymptote, of −147.51 kJ/mol). MIN2AC can lose an H atom and form either the cyclic C(NH)C radical (−12.96 kJ/mol with respect to the reactant asymptote) and cyclic HC(N)C (−147.51 kJ/mol with respect to the reactant asymptote), through a transition state (TS6AC, −138.71 kJ/mol) which clearly shows the breaking of the N-H bond. Finally, MIN3AC can also form both C(NH)C and HC(N)C by losing an H atom or the cyclic fragment CCN by an H_2-elimination mechanism (this channel is located at −147.29 kJ/mol with respect to the reactant asymptote). All identified stationary points lie under the reactant energy asymptote.

The structures of the minima and transition states optimized at the B3LYP/aug-cc-pVTZ level of theory are shown in Fig. 3. Table 1 reports on the enthalpy changes and barrier heights for each step, computed at the CCSD(T)/aug-cc-pVTZ level of theory.

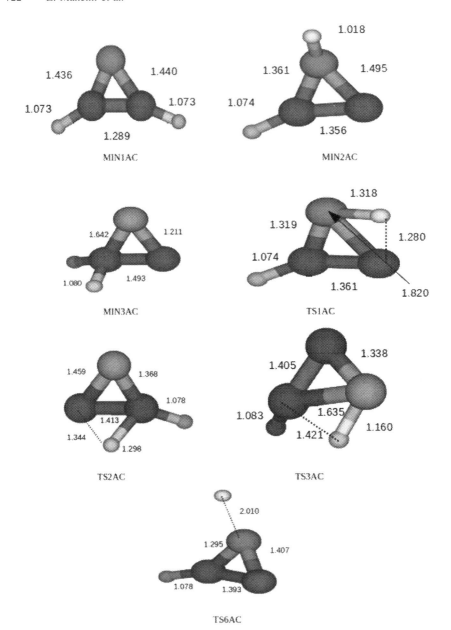

Fig. 3. Optimized structures calculated at the B3LYP/aug-cc-pVTZ level of theory of the minima and transition states for the $N(^2D)$ + acetylene reaction (distances in angstroms).

Table 1. Enthalpy changes (kJ/mol) and barrier heights (kJ/mol) computed at the CCSD(T)/aug-cc-pVTZ level of theory for the reaction of the system N(^{2}D) + acetylene.

	ΔH_0^0	Barrier heights
N(^{2}D)+C$_2$H$_2$ → MIN1AC	−433.33	
MIN1AC → H+HCCN	285.82	
MIN1AC → MIN2AC	95.77	138.33
MIN1AC → MIN3AC	−24.51	255.91
MIN2AC → H+CCNH	324.6	
MIN2AC → H+HCCN	190.05	198.85
MIN2AC → MIN3AC	−120.28	147.86
MIN3AC → H+HCCN	310.33	
MIN3AC → H$_2$+CCN	310.55	

3.2   N(^{2}D) + Methylacetylene

Figure 4 shows the schematic PES for the reaction N(^{2}D) + methylacetylene, in which three cyclic minima can be identified (MIN1MA, MIN2MA and MIN3MA), linked by two transition states (TS1MA, connecting MIN1MA and MIN2MA, and TS4MA, connecting MIN2MA and MIN3MA). As in the previous system, the reaction starts with a barrierless addition of N(^{2}D) to the triple bond of the methylacetylene molecule, leading to the formation of the cyclic intermediate MIN1MA (located at −446.15 kJ/mol with respect to the reactant energy asymptote). This structure can generate different products depending on which bond undergoes fission. Two different barrierless channels can lead to the loss of atomic hydrogen, one leading to the isomer H$_3$CC(N)C (the resulting exothermicity of this channel is −166.18 kJ/mol) by the fission of the acetylenic CH bond, and one leading to the isomer H$_2$CC(N)CH (the resulting exothermicity of this channel is −220.15 kJ/mol) by the fission of one of the CH bonds of the methyl group. Moreover, the intermediate MIN1MA can undergo a C-C bond breaking with the formation of CH$_3$ + cyclic-C(N)CH. MIN2MA can be easily formed via isomerization of MIN1MA overcoming a barrier of 140.6 kJ/mol. The related transition state clearly exhibits the breaking of a C-H bond in the methyl group and simultaneous formation of a N-H bond. The new intermediate can, in turn, dissociate forming atomic hydrogen and the cyclic molecule H$_2$CC(NH)C (the resulting exothermicity of this channel is −90.17 kJ/mol) or the isomer H$_3$CC(N)C, also formed by the dissociation of MIN1MA. MIN2MA can also undergo an isomerization process to produce the third intermediate MIN3MA, which is the lowest minimum along the PES. MIN3MA can undergo a decomposition with the formation of atomic hydrogen and H$_3$CC(N)C or CH$_3$ and C(N)CH. All identified stationary points lie under the reactant energy asymptote. In Fig. 5 the optimized structure of the minima and transition states identified in the potential energy surface at the B3LYP/aug-cc-pVTZ level of theory are reported, while in Table 2 the enthalpy changes and barrier heights for each step of the reaction are shown.

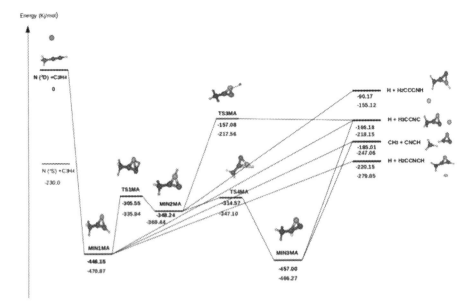

Fig. 4. Schematic representation of the PES obtained for the N(^{2}D) + methylacetylene reaction, with the relative energies (kJ/mol) evaluated at the CCSD(T)/aug-cc-pVTZ (in black) and B3LYP/aug-cc-pVTZ (in blue) levels of theory. (Color figure online)

4   Discussion

As expected, the two reactions N(^{2}D) + C$_2$H$_2$ and N(^{2}D) + CH$_3$CCH share several similarities as far as their initial steps are concerned. The electrophilic nature of N(^{2}D) drives its attack toward the electron density of the triple bond present in both C$_2$H$_2$ and CH$_3$CCH molecules. As a result, a first cyclic intermediate is formed in both cases, where the nitrogen atom is bound to the two acetylenic carbon atoms. When the co-reactant is methylacetylene, the presence of a -CH$_3$ group makes the addition inter-mediate asymmetric and more channels become open because of the possibility of losing a CH$_3$ moiety. It is also possible to lose an H atom from the methyl group with the formation of the cyclic product H$_2$CC(NH)C. As a consequence, the reaction with methylacetylene is much more complex than the reaction with acetylene, as already observed in many other cases (for instance, see the reactions of atomic oxygen, CN radicals or BO radicals with acetylene as opposed to methylacetylene [50–54]).

From previous investigations on the reaction N(^{2}D) + C$_2$H$_2$ [18] we know that additional reaction pathways are possible following the ring-opening of the MIN2AC cyclic intermediate with the formation of (almost) linear HCCN and/or HNCC + H. These pathways and their analogues for the N(^{2}D) + CH$_3$CCH system have not been investigated in the present study, but will be characterized in a future work.

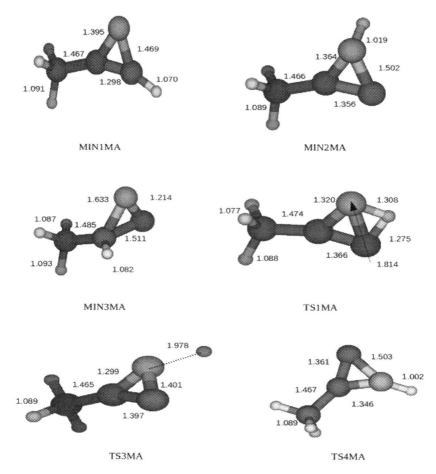

Fig. 5. Optimized structures calculated at the B3LYP/aug-cc-pVTZ level of theory of the minima and transition states for the N(^{2}D)+methyl acetylene reaction (distances in angstroms).

Table 2. Enthalpy changes (kJ/mol) and barrier heights (kJ/mol) computed at the CCSD(T)/aug-cc-pVTZ level of theory for the reaction of the system N(^{2}D) + methylacetylene.

	ΔH_0^0	Barrier heights
$C_3H_4 \rightarrow$ MIN1MA	−446.15	
MIN1MA $\rightarrow$ H+H$_2$CCNCH	226.00	
MIN1MA $\rightarrow$ CH$_3$+CNCH	261.14	
MIN1MA$\rightarrow$ H+H$_3$CCNC	279.97	
MIN1MA$\rightarrow$ MIN2MA	97.91	140.6
MIN2MA $\rightarrow$ H+H$_2$CCCNH	258.07	
MIN2MA $\rightarrow$ H+H$_3$CCNC	182.06	191.16
MIN2MA $\rightarrow$ MIN3MA	−108.76	33.67
MIN3MA $\rightarrow$ CH$_3$+CNCH	271.99	
MIN3MA $\rightarrow$ H+H$_3$CCNC	290.82	

5 Conclusions

In the present work, electronic structure calculations have been employed in order to derive the potential energy surface for two reactions of relevance in different natural environments, such as terrestrial and planetary atmospheres. The $N(^2D)$ addition processes have been found to be barrierless for the analyzed systems at all the employed levels of theory. Furthermore, the identified stationary points for the two processes lie below the reactant energy asymptote, therefore the global reactions are feasible under the low temperature conditions of Titan. Kinetic calculations are required to derive the rate constants and branching ratios of the two processes. The derived kinetic parameters will allow us to include the title reactions in the chemical models of the atmosphere of Titan and of other bodies of the Solar System where the molecules involved are present. The present study can be combined with the results coming from experimental investigations, such as CMB experiments, in order to understand the overall reaction mechanism.

Acknowledgments. This research was supported by the Italian Space Agency (ASI, DC-VUM-2017-034, Grant no 2019-3 U.O Life in Space).

References

1. Dutuit, O., et al.: Critical review of N, N^+, N_2^+, N^{++} and N_2^{++} main production processes and reactions of relevance to Titan's atmosphere. Astrophys. J. Suppl. Ser. **204**, 20 (2013)
2. Vuitton, V., Dutuit, O., Smith, M.A., Balucani, N.: Chemistry of Titan's atmosphere. In: Mueller-Wodarg, I., Griffith, C., Lellouch, E., Cravens, T. (eds.) Titan: Surface, Atmosphere and Magnetosphere. Cambridge University Press, Cambridge (2013)
3. Balucani, N.: Nitrogen fixation by photochemistry in the atmosphere of titan and implications for prebiotic chemistry. In: Trigo-Rodriguez, J., Raulin, F., Muller, C., Nixon, C. (eds.) The Early Evolution of the Atmospheres of Terrestrial Planets. Astrophysics and Space Science Proceedings, vol. 35, pp. 155–164. Springer, New York (2013). https://doi.org/10.1007/978-1-4614-5191-4_12
4. Balucani, N.: Elementary reactions of N atoms with hydrocarbons: first steps towards the formation of prebiotic N-containing molecules in planetary atmospheres. Chem. Soc. Rev. **41**, 5473–5483 (2012)
5. Balucani, N.: Elementary reactions and their role in gas-phase prebiotic chemistry. Int. J. Mol. Sci. **10**, 2304–2335 (2009)
6. Lindal, G.F., Wood, G.E., Hotz, H.B., Sweetnam, D.N. Eshleman, V.R., Tyler, G.L.: The atmosphere of Titan. An analysis of the Voyager 1 radio occultation measurements. Icarus **2** (53), 348–363 (1983)
7. Hanel, R., et al.: Infrared Observations of the Saturnian System from Voyager 1. Science **212** (4491), 192–200 (1981)
8. Maguire, W.C., Hanel, R.A., Jennings, D.E., Kunde, V.G., Samuelson, R.E.: C_3H_8 and C_3H_4 in Titan's atmosphere. Nature **292**(5825), 683–686 (1981)
9. Lombardo, N.A., et al.: Detection of propadiene on Titan. Astrophys. J. Lett. **881**, L33 (2019). (6 pp)
10. Imanaka, H., Smith, M.A.: Formation of nitrogenated organic aerosols in the Titan upper atmosphere. PNAS **107**, 12423–12428 (2010)

11. Somogyi, A., Oh, C.H., Smith, M.A., Lunine, J.I.: Organic environments on Saturn's moon, Titan: simulating chemical reactions and analyzing products by FT-ICR and ion-trap mass spectrometry. J. Am. Soc. Mass Spectrom. **16**(6), 850–859 (2005)
12. Cassidy, T., et al.: Radiolysis and photolysis of icy satellite surface: experiments and theory. Space Sci. Rev. **153**, 299–315 (2010)
13. Yung, K.L., Allen, M., Pinto, J.P.: Photochemistry of the atmosphere of Titan: comparison between models and observations. Astrophys. J. Suppl. Ser. **55**, 465–506 (1984)
14. Lavvas, P., Coustenis, A., Vardavas, I.: Coupling photochemistry with haze formation in Titan's atmosphere, Part I: model description. Planet. Space Sci. **56**, 27 (2008)
15. Lavvas, P., Coustenis, A., Vardavas, I.: Coupling photochemistry with haze formation in Titan's atmosphere, Part II: results and validation with Cassini/Huygens data. Planet. Space Sci. **56**, 67 (2008)
16. Hébrard, E., Dobrijevic, M., Bénilan, Y., Raulin, F.: Photochemical kinetics uncertainties in modeling Titan's atmosphere: first consequences. Planet. Space Sci. **55**, 1470–1489 (2007)
17. Balucani, N., Bergeat, A., Cartechini, L., Volpi, G.G., Casavecchia, P., Skouteris, D., Rosi, M.: Combined crossed molecular beam and theoretical studies of the N(^{2}D) + CH$_4$ reaction and implications for atmospheric models of Titan. J. Phys. Chem. A **113**, 11138–11152 (2009)
18. Balucani, N., et al.: Cyanomethylene formation from the reaction of excited nitrogen atoms with acetylene: a crossed beam and ab initio study. J. Am. Chem. Soc. **122**, 4443–4450 (2000)
19. Balucani, N., Leonori, F., Petrucci, R., Stazi, M., Skouteris, D., Rosi, M., Casavecchia, P.: Formation of nitriles and imines in the atmosphere of Titan: combined crossed-beam and theoretical studies on the reaction dynamics of excited nitrogen atoms N(^{2}D) with ethane. Faraday Discuss. **147**, 189–216 (2010)
20. Balucani, N., et al.: Combined crossed beam and theoretical studies of the N(^{2}D) + C$_2$H$_4$ reaction and implications for atmospheric models of Titan. J. Phys. Chem. A **116**, 10467–10479 (2012)
21. Balucani, N., Cartechini, L., Alagia, M., Casavecchia, P., Volpi, G.G.: Observation of nitrogen-bearing organic molecules from reactions of nitrogen atoms with hydrocarbons: a crossed beam study of N(^{2}D) + ethylene. J. Phys. Chem. A **104**, 5655–5659 (2000)
22. Balucani, N., et al.: Dynamics of the N(^{2}D) + D$_2$ reaction from crossed-beam and quasiclassical trajectory studies. J. Phys. Chem. A **105**, 2414–2422 (2001)
23. Balucani, N., et al.: Experimental and theoretical differential cross sections for the N(^{2}D)+H$_2$ reaction. J. Phys. Chem. A **110**, 817–829 (2006)
24. Homayoon, Z., Bowman, J.M., Balucani, N., Casavecchia, P.: Quasiclassical trajectory calculations of the N(^{2}D) + H$_2$O reaction elucidating the formation mechanism of HNO and HON seen in molecular beam experiments. J. Phys. Chem. Lett. **5**, 3508–3513 (2014)
25. Balucani, N., Cartechini, L., Casavecchia, P., Homayoon, Z., Bowman, J.M.: A combined crossed molecular beam and quasiclassical trajectory study of the Titan-relevant N(^{2}D) + D$_2$O reaction. Mol. Phys. **113**, 2296–2301 (2015)
26. Balucani, N., Pacifici, L., Skouteris, D., Caracciolo, A., Casavecchia, P., Rosi, M.: A theoretical investigation of the reaction N(^{2}D) + C$_6$H$_6$ and implications for the upper atmosphere of Titan. In: Gervasi, O., et al. (eds.) ICCSA 2018. LNCS, vol. 10961, pp. 763–772. Springer, Cham (2018). https://doi.org/10.1007/978-3-319-95165-2_53
27. Takayanagi, T., et al.: Measurements of thermal rate constants and theoretical calculations for the N(^{2}D, ^{2}P) + C$_2$H$_2$ and C$_2$D$_2$ reactions. J. Phys. Chem. A **102**, 6251–6258 (1998)
28. Nuñez-Reyes, D., Loison, J.-C., Hickson, K.M., Dobrijevic, M.: Rate constants for the N(^{2}D) + C$_2$H$_2$ reaction over the 50–296 k temperature range. Phys. Chem. Chem. Phys. **21**, 22230–22237 (2019)

29. Falcinelli, S., Rosi, M., Cavalli, S., Pirani, F., Vecchiocattivi, F.: Stereoselectivity in autoionization reactions of hydrogenated molecules by metastable gas atoms: the role of electronic couplings. Chem. Eur. J. **22**, 12518–12526 (2016)

30. Leonori, F., et al.: Crossed-beam and theoretical studies of the $S(^1D)$ + C_2H_2 reaction. J. Phys. Chem. A **113**, 4330–4339 (2009)

31. Bartolomei, M., Cappelletti, D., De Petris, G., Moix Teixidor, M., Pirani, F., Rosi, M., Vecchiocattivi, F.: The intermolecular potential in $NO-N_2$ and $(NO-N_2)^+$ systems: implications for the neutralization of ionic molecular aggregates. Phys. Chem. Chem. Phys. **10**, 5993–6001 (2008)

32. De Petris, G., Cartoni, A., Rosi, M., Barone, V., Puzzarini, C., Troiani, A.: The proton affinity and gas-phase basicity of sulfur dioxide. ChemPhysChem **12**, 112–115 (2011)

33. Leonori, F., et al.: Observation of organosulfur products (thiovinoxy, thioketene and thioformyl) in crossed-beam experiments and low temperature rate coefficients for the reaction $S(^1D)$ + C_2H_4. Phys. Chem. Chem. Phys. **11**, 4701–4706 (2009)

34. De Petris, G., Rosi, M., Troiani, A.: SSOH and HSSO radicals: an experimental and theoretical study of $[S_2OH]^{0/+/-}$ species. J. Phys. Chem. A **111**, 6526–6533 (2007)

35. Rosi, M., Falcinelli, S., Balucani, N., Casavecchia, P., Leonori, F., Skouteris, D.: Theoretical study of reactions relevant for atmospheric models of titan: interaction of excited nitrogen atoms with small hydrocarbons. In: Murgante, B., et al. (eds.) ICCSA 2012. LNCS, vol. 7333, pp. 331–344. Springer, Heidelberg (2012). https://doi.org/10.1007/978-3-642-31125-3_26

36. Becke, A.D.: A new mixing of Hartree-Fock and local density-functional theories. J. Chem. Phys. **98**(2), 1372–1377 (1998)

37. Stephens, P.J., Devlin, F.J., Chablowski, C.F., Frisch, M.J.: Ab initio calculation of vibrational absorption and circular dichroism spectra using density functional force fields. J. Phys. Chem. **98**(45), 11623–11627 (1994)

38. Dunning, T.H., Jr.: Gaussian basis sets for use in correlated molecular calculations. I. The atoms boron through neon and hydrogen. J. Chem. Phys. **90**, 1007–1023 (1989)

39. Woon, D.E., Dunning, T.H., Jr.: Gaussian basis sets for use in correlated molecular calculations. III. The atoms aluminum through argon. J. Chem. Phys. **98**, 1358–1371 (1993)

40. Kendall, R.A., Dunning Jr., T.H., Harrison, J.R.: Electron affinities of the first-row atoms revisited. Systematic basis sets and wave functions. J. Chem. Phys. **96**, 6796–6806 (1992)

41. Gonzalez, C., Schlegel, H.B.: An improved algorithm for reaction path following. J. Chem. Phys. **90**, 2154–2161 (1989)

42. Gonzalez, C., Schlegel, H.B.: Reaction path following in mass-weighted internal coordinates. J. Phys. Chem. **94**, 5523–5527 (1990)

43. Bartlett, R.J.: Many-body perturbation theory and coupled cluster theory for electron correlation in molecules. Annu. Rev. Phys. Chem. **32**, 359–401 (1981)

44. Raghavachari, K., Trucks, G.W., Pople, J.A., Head-Gordon, M.: Quadratic configuration interaction. A general technique for determining electron correlation energies. Chem. Phys. Lett. **157**, 479–483 (1989)

45. Olsen, J., Jorgensen, P., Koch, H., Balkova, A., Bartlett, R.J.: Full configuration–interaction and state of the art correlation calculations on water in a valence double-zeta basis with polarization functions. J. Chem. Phys. **104**, 8007–8015 (1996)

46. Schofield, K.: Critically evaluated rate constants for gaseous reactions of several electronically excited species. J. Phys. Chem. Ref. Data **8**, 723–798 (1979)

47. Frisch, M.J., et al.: Gaussian 09, Revision A.02. Gaussian, Inc., Wallingford CT (2009)

48. Schaftenaar, G., Noordik, J.H.: Molden: a pre- and post-processing program for molecular and electronic structures. J. Comput. Aided Mol. Des. **14**, 123–134 (2000)

49. Schaftenaar, G., Vlieg, E., Vriend, G.: Molden 2.0: quantum chemistry meets proteins. J. Comput. Aided Mol. Des. **31**(9), 789–800 (2017). https://doi.org/10.1007/s10822-017-0042-5
50. Leonori, F., Balucani, N., Capozza, G., Segoloni, E., Volpi, G.G., Casavecchia, P.: Dynamics of the O(^{3}P) + C$_2$H$_2$ reaction from crossed molecular beam experiments with soft electron ionization detection. Phys. Chem. Chem. Physiscs **16**, 10008–10022 (2014)
51. Gimondi, I., Cavallotti, C., Vanuzzo, G., Balucani, N., Casavecchia, P.: Reaction dynamics of O(^{3}P) + Propyne: II. Primary products, branching ratios, and role of intersystem crossing from ab initio coupled triplet/singlet potential energy surfaces and statistical calculations. J. Physycal Chem. A **120**(27), 4619–4633 (2016)
52. Kaiser, R.I., Balucani, N.: Exploring the gas phase synthesis of the elusive class of boronyls and the mechanism of boronyl radical reactions under single collision conditions. Acc. Chem. Res. **50**, 1154–1162 (2017)
53. Huang, L.C.L., et al.: Crossed beam reaction of the cyano radical with hydrocarbons molecule. IV. Chemical dynamics of cyanoacetylene (HCCCN; X$^1\Sigma^+$) formation from reaction of CN(X$^2\Sigma^+$) with acetylene, C$_2$H$_2$ (X$^1\Sigma_g^+$). J. Chem. Phys. **113**, 8656 (2000)
54. Balucani, N., Asvany, O., Kaiser, R.I., Osamura, Y.: Formation of three C4H3N isomers from the reaction of CN (X$^2\Sigma^+$) with Allene, H2CCCH2 (X^{1}A^1), and methylacetylene, CH3CCH (X^{1}A^1): a combined crossed beam and ab initio study. J. Phys. Chem. A **106**, 4301–4311 (2002)

A Theoretical Investigation of the Reaction Between Glycolaldehyde and H⁺ and Implications for the Organic Chemistry of Star Forming Regions

Dimitrios Skouteris[1(✉)], Luca Mancini[2], Fanny Vazart[3],
Cecilia Ceccarelli[3], Marzio Rosi[4], and Nadia Balucani[2,3]

[1] Master-Up, Via Sicilia 41, 06128 Perugia, Italy
d.skouteris@master-up.it
[2] Dipartimento di Chimica, Biologia e Biotecnologie, Università degli Studi di Perugia, 06123 Perugia, Italy
luca.mancini2@studenti.unipg.it,
nadia.balucani@unipg.it
[3] Université Grenoble Alpes, IPAG, 38000 Grenoble, France
{fanny.vazart,cecilia.ceccarelli}@univ-grenoble-alpes.fr
[4] Dipartimento di Ingegneria Civile e Ambientale, Università degli Studi di Perugia, 06125 Perugia, Italy
marzio.rosi@unipg.it

Abstract. The characterization of one of the possible pathways in the reaction between H⁺ and glycolaldehyde (the channel leading to COH + CH_3OH^+) has been carried out by performing electronic structure calculations of the stationary points along the minimum energy path. We have employed different theoretical methods verifying that, while geometry optimizations can be performed with a relatively low level of theory, quantitative results for the energies require higher level calculations. The same methodology will be applied to the complete scheme of the title reaction as well as similar processes which are needed to characterize the destruction routes of interstellar complex organic molecules.

Keywords: Computational chemistry methods · Astrochemistry · Electronic structure calculations

1 Introduction

Interstellar medium (ISM) is the non-condensed matter in between stars and their planetary system objects in our Galaxy. It is composed of gas (with the main component being hydrogen) and dust distributed in an inhomogeneous way. On average the number density is extremely low (as low as ~ 0.1 particles per cm^3 in the diffuse warm and ionized regions), but there are regions where baryonic matter becomes denser, the so-called molecular clouds ($\sim 10^4$ particles per cm^3). In some of those regions, new solar systems are formed [1]. The observation of organic molecules of some complexity

© Springer Nature Switzerland AG 2020
O. Gervasi et al. (Eds.): ICCSA 2020, LNCS 12251, pp. 730–743, 2020.
https://doi.org/10.1007/978-3-030-58808-3_53

(hereafter iCOMs for *interstellar complex organic molecules*) in star-forming regions (for a recent survey see [2]) has aroused plenty of interest, since those organic molecules can be the legacy of interstellar chemistry to newly formed planets, thus seeding them with relatively complex organic molecules that behave as the building blocks of life [1, 3–7]. Among the most interesting iCOMS, glycolaldehyde ($HCOCH_2OH$) is the smallest molecule containing both a hydroxyl group and a carbonyl group, thus resembling a sugar. The formation of glycolaldehyde is an intermediate step towards sugars in the formose reaction and, therefore, it is believed to have a strong prebiotic potential. Its detection in high mass and low mass star forming regions [8–12] has posed the question of how a molecule with this degree of complexity can be formed in the harsh conditions of the interstellar medium. Various formation mechanisms have been envisaged involving either gas-phase chemistry or reactions occurring on the icy mantles of interstellar dust grains [13–17]. In particular, the sequence of two neutral-neutral reactions

$$OH + CH_3CH_2OH \rightarrow CH_2CH_2OH + H_2O \tag{1}$$

$$O + CH_2CH_2OH \rightarrow HCOCH_2OH + H \tag{2}$$

seems to be very promising since it is able to reproduce the relative abundance of ethanol (the "mother" molecule) and glycolaldehyde in low mass [13] and high mass [12] star forming regions.

In both cases, whether released in the gas phase from the sublimating icy mantles of the interstellar dust grains or formed directly in the gas phase, to build up realistic astrochemical models and account for the observed abundances it is necessary to characterize the destruction routes of each species in the gas phase. Unfortunately, there are no data available in the literature concerning the possible destruction routes of glycolaldehyde. In the conditions of the ISM, most molecules are consumed by their reactions with energetic ions, such as HCO^+, H_3^+, He^+ and H^+ (see for instance [18]), which are abundant in the ISM. While in the case of the reactions with the ions HCO^+ and H_3^+ the most probable outcome is a proton transfer mechanism and in the case of He^+ only a charge transfer mechanism is possible, the collisions of interstellar molecules with H^+ can occur either by charge transfer (in this case the molecule is ionized and can also undergo dissociative ionization) or by a real chemical reaction, in which H^+ adds to the molecule and a reaction pathway starts.

Given the paucity of information on the processes following the interaction of iCOMs with H^+, in our laboratory we have decided to start a systematic investigation of these processes by resorting to electronic structure calculations of the stationary points along the relevant potential energy surfaces (PESs) and kinetics estimates. In this contribution, we present the first results concerning the reactive pathway

$$HCOCH_2OH + H^+ \rightarrow COH + CH_3OH^+ \tag{3}$$

Different theoretical methods have been tested and compared to assess their performance and select the best method for our planned systematic investigation of iCOMs reactions.

2 Computational Details

The characterization of the PES for the reaction (3) has been performed by locating the lowest stationary points with different computational strategies listed in Table 1.

Table 1. List of combinations method/basis set used.

Method	Basis set
MP2	6-31G(d,p)
B3LYP	STO-3G
B3LYP	aug-cc-pVDZ
B3LYPD3	aug-cc-pVDZ
B2PLYP	aug-cc-pVTZ
B2PLYPD3	aug-cc-pVTZ

Glycolaldehyde is characterized by four possible conformers, among which the *cis-cis* structure is the most stable one because of an intramolecular hydrogen bond. Therefore, we have performed our calculations starting from the *cis-cis* conformer (see Fig. 2). As an example, the structures of this conformer of glycolaldehyde as obtained with the different methods are reported in Fig. 1. A few other stationary points have been also characterized using the methods listed in Table 1.

Subsequently, the geometries of all the stationary points, including reactants and products, have been optimized only at the B3LYP level of theory [19, 20], in conjunction with the correlation consistent aug-cc-pVTZ basis set [21–23]. Harmonic vibrational frequencies have been computed at the same level of theory, in order to define the nature of the identified stationary points, *i.e.*, minimum if all the frequencies are real, saddle point if there is one and only one imaginary frequency. The identified saddle point has been assigned through Intrinsic Reaction Coordinates (IRC) calculations [24, 25]. The zero-point-energy correction (computed using the scaled harmonic vibrational frequencies) has been added in order to correct the energies at 0 K. All calculations have been repeated (to assess the accuracy of the obtained data) with the double hybrid B2PLYP functional in addition to the contribution of the semiempirical dispersion considered in the D3BJ model (B2PLYPD3) [26, 27], with the same aug-cc-pVTZ basis set. In the end, coupled-cluster single and double excitations augmented by a perturbative treatment of the triple excitations (CCSD(T)) [28–30] calculations have been performed with the same basis set and the B2PLYPD3/aug-cc-pVTZ optimized geometries, following a computational scheme which has been used in previous cases [31–38]. In addition, computational work has been carried out in order to establish which combination of method and basis set would be the most suitable one for the analyzed reaction. All calculations have been performed using Gaussian 09 [39] while the analysis of the vibrational frequencies has been done using Avogadro [40, 41].

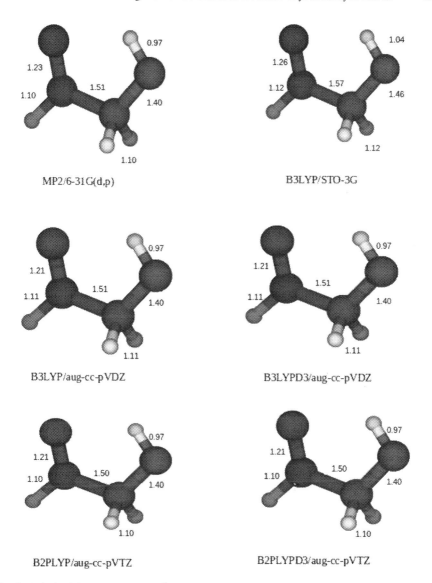

Fig. 1. Principal bond lengths (in Å) of the molecule glycolaldehyde calculated at different levels of theory.

3   Results and Discussion

The complete potential energy surface obtained for the reaction (3) is reported in Fig. 2, with the geometries optimized at the B3LYP/aug-cc-pVTZ level of theory. In particular, in the figure the energies have been reported at the B3LYP/aug-cc-pVTZ (in red), B2PLYPD3/aug-cc-pVTZ (in black) and CCSD(T)/aug-cc-pVTZ (in black, in parentheses) levels of theory. In order to have an analysis of the minimum energy path, the more accurate CCSD(T) energies are considered.

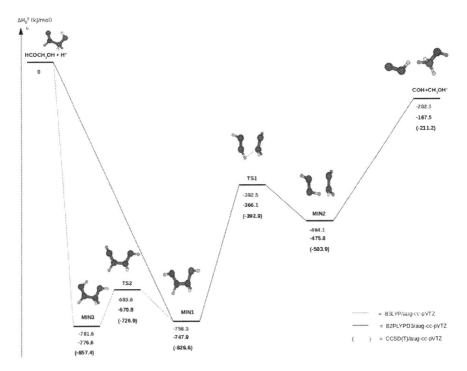

Fig. 2. Potential energy surface obtained, with the geometries optimized at the B3LYP/aug-cc-pVTZ level of theory. The energies have been reported at the B3LYP/aug-cc-pVTZ (in red), B2PLYPD3/aug-cc-pVTZ (in black) and CCSD(T)/aug-cc-pVTZ (in black, in parentheses) levels of theory. (Color figure online)

There are two possible attack sites for the incoming free proton.

In the initial approach analyzed in this work, the reaction starts with a barrierless addition of the proton to the oxygen of the carbonyl group of the molecule, leading to the formation of MIN1, in which a new O-H bond is formed at a distance of 0.977 Å, while the value of the C-O bond length goes from 1.206 Å to 1.261 Å, indicating a double-to-single bond variation. The energy of the stationary point, with respect to the reactant energy asymptote, taken as a reference, is −826.6 kJ/mol. The alternative additional initial approach can feature, instead, the addition of H⁺ on the oxygen of the hydroxyl group (the investigation of this pathway is currently under way).

Starting from MIN1, it is possible to obtain the second minimum in the PES, named MIN2. In this case the C-C bond is broken and two fragments are formed: HCOH and CH$_2$OH. In this structure, the hydrogen atom linked to the carbon in the HCOH fragment strongly interacts with the carbon atom in the CH$_2$OH fragment at a distance of 1.351 Å and the total charge is distributed over the complex. The relative energy of MIN2, with respect to the reactant energy asymptote, is −503.9 kJ/mol. The transition from MIN1 to MIN2 is possible through the overcoming of a barrier of 433.7 kJ/mol. The related transition state, TS1, clearly shows the breaking of the C-C bond. The distance between the two C atoms, indeed, is increased up to 2.146 Å. The analysis of the vibrational frequencies for TS1 allows us to notice the interaction of the H in the HCOH fragment with the carbon atom in the CH$_2$OH moiety. The distance between the two mentioned atoms is 1.924 Å. Once formed, MIN2 directly leads to the formation of the two products (HCO and CH$_3$OH$^+$), located at 211.2 kJ/mol below the reactant energy asymptote. No transition states have been found for this process. In general, it is possible to notice that all the identified stationary points lie below the reactant energy asymptote, so the reaction is globally exothermic and can take place in the harsh conditions of the interstellar medium.

The analysis of the stationary points in the PES has been performed at different levels of theory; in Figs. 3, 4 are reported the values of the bond lengths (in Å) resulting from the geometry optimized at B3LYP/aug-cc-pVTZ (in red) and B2PLYPD3/aug-cc-pVTZ (in black) level of theory. As can be seen from the reported data, there are no particular differences in the bond lengths evaluated with the two different computational strategies. The only appreciable difference is shown in the C-H interaction in TS1, in which the chemical bond is not completely formed. In this case the B2PLYPD3/aug-cc-pVTZ combination provides a length which is 0.031 Å lower than the one resulting from the B3LYP/aug-cc-pVTZ calculation. A different behavior can be seen from the analysis of the energies. In Tables 2, 3, 4 are reported the enthalpy changes and barrier heights at 0 K, computed at the three previously mentioned levels of theory.

Since the geometries of the different stationary points are very similar, it is possible to conclude that the differences in energies are related to the differences in the two methods. In particular the B2PLYPD3 functional, in addition to the D3BJ dispersion, includes a contribution related to second order perturbation theory, which can be considered the origin of the noticed differences. The same conclusion can be reached considering the results coming from the use of different methods and basis sets. As can be seen in the results shown in Fig. 1, in the case of glycolaldehyde there are no particular differences in the values of the bond lengths evaluated at different levels of theory. The results are, instead, more varied for the MIN2 and TS1 structures. The main differences can be noticed when the B3LYP functional is used in conjunction with the

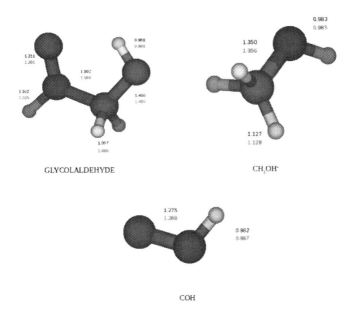

Fig. 3. Values of the bond lengths (in Å) resulting from the geometry optimized at B3LYP/aug-cc-pVTZ (in red) and B2PLYPD3/aug-cc-pVTZ (in black) levels of theory for the reactant and products of the reaction. (Color figure online)

STO-3G basis set. From a comparison with the cases in which the same functional is used together with other different basis sets, it can be concluded that the differences in the bond lengths derive mainly from the use of the STO-3G basis set, which is less accurate since it is a minimum basis set. From an energetic point of view, it can be noticed that the combination that most differs from the reference (which in this case is the CCSD(T) energy calculation) is once again due to the use of the STO-3G basis set. The values of the energy obtained at different levels of theory are shown in Table 5. In some cases, the results obtained at the B3LYP level of theory seem to be closer to the CCSD(T) values (which is considered the best approach) than the ones obtained at the B2PLYPD3 level. However, the B2PLYPD3/aug-cc-pVTZ approach tends to give a more accurate estimate for the barrier heights, a crucial aspect when it comes to deriving rate coefficients and product branching ratios (Figs. 5 and 6).

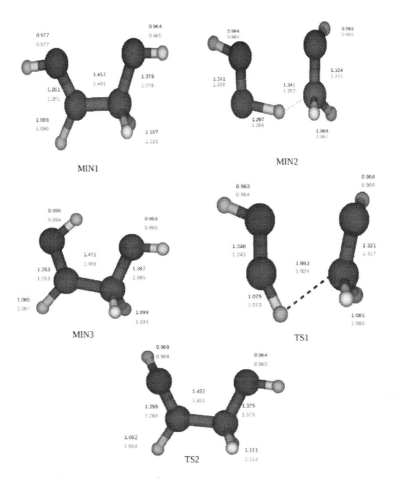

Fig. 4. Values of the bond lengths (in Å) resulting from the geometry optimized at B3LYP/aug-cc-pVTZ (in red) and B2PLYPD3/aug-cc-pVTZ (in black) levels of theory for the identified minima and transition states. (Color figure online)

Table 2. Enthalpy changes and barrier heights (in kJ/mol) for the system, evaluated at the B3LYP/aug-cc-pVTZ level of theory.

	ΔH°_0	Barrier heights
H$^+$ + HCOCH$_2$OH → MIN1	−756.3	
MIN1 → MIN2	262.2	373.8
MIN1 → MIN3	−25.3	72.7
MIN2 → COH + CH$_3$OH$^+$	292.0	

Table 3. Enthalpy changes and barrier heights (in kJ/mol) for the system, evaluated at the B2PLYPD3/aug-cc-pVT level of theory.

	$\Delta H°_0$	Barrier heights
$H^+ + HCOCH_2OH \rightarrow MIN1$	−747.9	
$MIN1 \rightarrow MIN2$	272.1	381.8
$MIN1 \rightarrow MIN3$	−28.7	77.1
$MIN2 \rightarrow COH + CH_3OH^+$	308.3	

Table 4. Enthalpy changes and barrier heights (in kJ/mol) for the system, evaluated at the CCSD(T)/aug-cc-pVTZ level of theory considering the geometry optimized at the B2PLYPD3/aug-cc-pVTZ level of theory.

	$\Delta H°_0$	Barrier heights
$H^+ + HCOCH_2OH \rightarrow MIN1$	−826.6	
$MIN1 \rightarrow MIN2$	322.7	433.7
$MIN1 \rightarrow MIN3$	−30.8	99.7
$MIN2 \rightarrow COH + CH_3OH^+$	292.7	

Table 5. Energy values (in kJ/mol) of the different stationary points with respect to the reactant energy asymptote at different levels of theory

Level of theory	TS1	MIN2
MP2/6-31G(d,p)	−339.2	−462.1
B3LYP/STO-3G	−538.0	−680.8
B3LYP/aug-cc-pVDZ	−376.9	−490.8
B3LYPD3/aug-cc-pVDZ	−379.6	−492.1
B2PLYP/aug-cc-pVTZ	−365.0	−475.3
B2PLYPD3/aug-cc-pVTZ	−366.1	−476.0

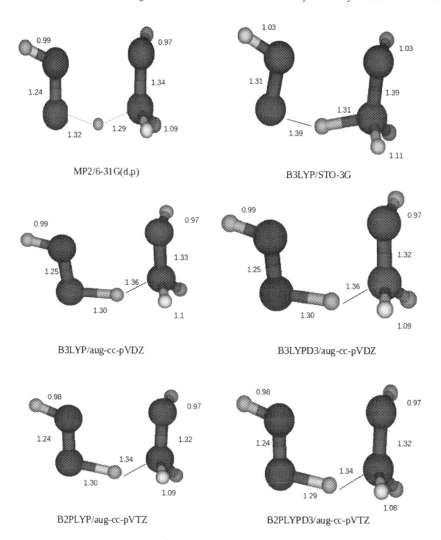

Fig. 5. Principal bond lengths (in Å) of the minimum MIN2 calculated at different levels of theory

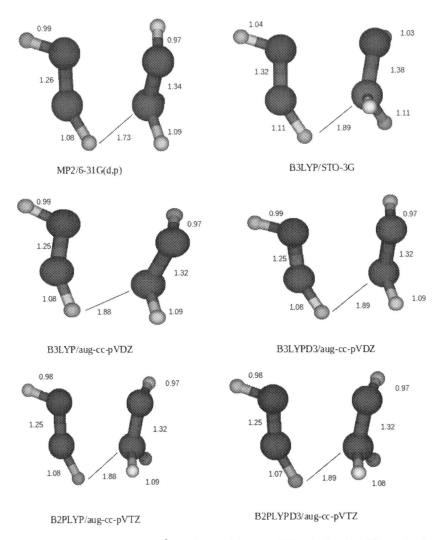

Fig. 6. Principal bond lengths (in Å) of the transition state TS1 calculated at different levels of theory.

4 Conclusion

In this work, different ab initio methods have been employed for the study of the reaction between H$^+$ and glycolaldehyde with the addition site on the oxygen of the carbonyl group and with the formation of COH and CH$_3$OH$^+$. The aim was to define the best compromise between chemical accuracy and computational costs. The best approach will be later employed in our systematic investigation of this and similar reactions. From the data analyzed we can conclude that relatively low-level calculations, such as B3LYP/aug-cc-pVTZ, can be used for geometry optimization and harmonic vibrational frequencies calculations. However, the energy must be evaluated at a higher level of theory, *e.g.* CCSD(T), in order to obtain more accurate values and construct a realistic potential energy surface. This conclusion is in agreement with previous work.

The reaction channel investigated in this contribution is only part of the PES of the glycolaldehyde-H$^+$ system. The global PES is expected to be much more complex with other possible exit channels (namely, those leading to the products HCOH$^+$ + H$_2$COH, HCOH + H$_2$COH$^+$, H$_2$ + HCOHCHO$^+$, H$_2$O + CH$_2$COH$^+$, H$_2$ + OCCH$_2$OH$^+$, H$_3$O$^+$ + OCCH$_2$ and H$_2$O + CH$_3$CO$^+$). Another complication is due to the presence of a second site for H$^+$ addition. Having verified that a combined B3LYP/aug-cc-pVTZ and CCSD(T) approach is a viable method to characterize the PES of this system, further investigations complemented by kinetics calculations are needed. We aim to complete those calculations to provide reliable estimates of the rate coefficients to be employed in astrochemical models.

Acknowledgments. This project has received funding from the European Research Council (ERC) under the European Union's Horizon 2020 research and innovation programme, for the Project "The Dawn of Organic Chemistry" (DOC), grant agreement No 741002 and from the European Union's Horizon 2020 research and innovation programme under the Marie Sklodowska-Curie grant agreement No 811312 for the project "Astro-Chemical Origins" (ACO). M.R. acknowledges the project "Indagini teoriche e sperimentali sulla reattività di sistemi di interesse astrochimico" funded by Fondo Ricerca di Base 2018 of the University of Perugia. DS wishes to thank "Master-Up" for funding.

References

1. Caselli, P., Ceccarelli, C.: Our astrochemical heritage. Astron. Astrophys. Rev. **20**, 56 (2012)
2. McGuire, B.A.: 2018 Census of interstellar, circumstellar, extragalactic, protoplanetary disk, and exoplanetary molecules. Astrophys. J. Suppl. Ser. **239**, 2 (2018)
3. Ceccarelli, C., et al.: Seeds Of Life In Space (SOLIS): The Organic Composition Diversity at 300–1000 au Scale in Solar-type Star-forming Regions. ApJ, 850 (2017)
4. Herbst, E., van Dishoeck, E.F.: Complex organic interstellar molecules. Annu. Rev. Astron. Astrophys. **47**(1), 427–480 (2009)
5. Balucani, N.: Elementary reactions and their role in gas-phase prebiotic chemistry. Int. J. Mol. Sci. **10**, 2304–2335 (2009)

6. Balucani, N.: Elementary reactions of N atoms with hydrocarbons: first steps towards the formation of prebiotic N-containing molecules in planetary atmospheres. Chem. Soc. Rev. **41**, 5473–5483 (2012)

7. Ehrenfreund, P., Charnley, S.B.: Organic molecules in the interstellar medium, comets, and meteorites: a voyage from dark clouds to the early earth. Ann. Rev. Astron. Astrophys. **38**, 427–483 (2000)

8. Hollis, J.M., Lovas, F.J., Jewell, P.R. Interstellar Glycolaldehyde: The First Sugar. ApJ, 540, 2, L107–L110 (2000)

9. Lefloch, B., et al.: L1157-B1, a factory of complex organic molecules in a solar-type star-forming region. MNRAS **469**, L73 (2017)

10. Jørgensen, J.K., et al.: The ALMA Protostellar Interferometric Line Survey (PILS). First results from an unbiased submillimeter wavelength line survey of the Class 0 protostellar binary IRAS 16293–2422 with ALMA. Astron. Astrophys. 595, A117 (2016)

11. Bouchez, A., et al.: The submillimeter spectrum of deuterated glycolaldehydes. Astron. Astrophys. **540**, A51 (2012)

12. Xue, C., Remijan, A.J., Burkhardt, A.M., Herbst, E.: ALMA observations of the spatial distribution of three $C_2H_4O_2$ isomers toward Sgr B_2(N). ApJ **871**, 1 (2019)

13. Skouteris, D., et al.: The genealogical tree of ethanol: gas-phase formation of glycolaldehyde, acetic acid, and formic acid. ApJ **854**, 2 (2018)

14. Coutens, A., et al.: Chemical modelling of glycolaldehyde and ethylene glycol in star-forming regions. MNRAS **475**(2), 2016–2026 (2018)

15. Chuang, K.-J., Fedoseev, G., Ioppolo, S., van Dishoeck, E.F., Linnartz, H.: H-atom addition and abstraction reactions in mixed CO, H_2CO and CH_3OH ices - an extended view on complex organic molecule formation. MNRAS **455**(2), 1702–1712 (2016)

16. Garrod, R.T., Widicus, S.L.V., Herbst, E.: Complex chemistry in star-forming regions: an expanded gas-grain warm-up chemical model. ApJ **682**(1), 283–302 (2008)

17. Holdship, J., Rawlings, J., Viti, S., Balucani, N., Skouteris, D., Williams, D.: Investigating the efficiency of explosion chemistry as a source of complex organic molecules in TMC-1. ApJ **878**, 65 (2019)

18. Ascenzi, D., et al.: Destruction of dimethyl ether and methyl formate by collisions with He^+. Astron. Astrophys. **625**, A72 (2019)

19. Becke, A.D.: A new mixing of Hartree-Fock and local density functional theories. J. Chem. Phys. **98**(2), 1372–1377 (1998)

20. Stephens, P.J., Devlin, F.J., Chablowski, C.F., Frisch, M.J.: Ab initio calculation of vibrational absorption and circular dichroism spectra using density functional force fields. J. Phys. Chem. **98**(45), 11623–11627 (1994)

21. Dunning Jr., T.H.: Gaussian basis sets for use in correlated molecular calculations. I. The atoms boron through neon and hydrogen. J. Chem. Phys. **90**, 1007–1023 (1989)

22. Woon, D.E., Dunning Jr., T.H.: Gaussian basis sets for use in correlated molecular calculations. III. The atoms aluminum through argon. J. Chem. Phys. **98**, 1358–1371 (1993)

23. Kendall, R.A., Dunning Jr., T.H., Harrison, J.R.: Electron affinities of the first row atoms revisited. Systematic basis sets and wave functions. J. Chem. Phys. **96**, 6796–6806 (1992)

24. Gonzalez, C., Schlegel, H.B.: An improved algorithm for reaction path following. J. Chem. Phys. **90**, 2154–2161 (1989)

25. Gonzalez, C., Schlegel, H.B.: Reaction path following in mass-weighted internal coordinates. J. Phys. Chem. **94**, 5523–5527 (1990)

26. Goerigk, L., Grimme, S.: Efficient and accurate double-hybrid-meta-GGA density functionals- evaluation with the extended GMTKN30 database for general main group thermochemistry, kinetics and noncovalent interactions. J. Chem. Theory Comput. **7**, 291–309 (2011)

27. Grimme, S., Ehrlich, S., Goerigk, L.: Effect of the damping function in dispersion corrected density functional theory. J. Comput. Chem. **32**, 1456–1465 (2011)
28. Bartlett, R.J.: Many-body perturbation theory and coupled cluster theory for electron correlation in molecules. Ann. Rev. Phys. Chem. **32**, 359–401 (1981)
29. Raghavachari, K., Trucks, G.W., Pople, J.A., Head-Gordon, M.: Quadratic configuration interaction. A general technique for determining electron correlation energies. Chem. Phys. Lett. **157**, 479–483 (1989)
30. Olsen, J., Jorgensen, P., Koch, H., Balkova, A., Bartlett, R.J.: Full configuration–interaction and state of the art correlation calculations on water in a valence double-zeta basis with polarization functions. J. Chem. Phys. **104**, 8007–8015 (1996)
31. Vazart, F., Calderini, D., Puzzarini, C., Skouteris, D., Barone, V.: State-of-the-art thermochemical and kinetic computations for astrochemical complex organic molecules: formamide formation in cold interstellar clouds as a case study. J. Chem. Theory Comput. **12**, 5385–5397 (2016)
32. Falcinelli, S., Rosi, M., Cavalli, S., Pirani, F., Vecchiocattivi, F.: Stereoselectivity in autoionization reactions of hydrogenated molecules by metastable noble gas atoms: the role of electronic couplings. Chem. Eur. Jour. **22**(35), 12518–12526 (2016)
33. Leonori, F., et al.: Crossed-beam and theoretical studies of the S(^{1}D) + C$_2$H$_2$ reaction. J. Phys. Chem. A **113**(16), 4330–4339 (2009)
34. Bartolomei, M., et al.: The intermolecular potential in NO-N2 and (NO-N2) + systems: implications for the neutralization of ionic molecular aggregates. PCCP **10**(39), 5993–6001 (2008)
35. de Petris, G., Cartoni, A., Rosi, M., Barone, V., Puzzarini, C., Troiani, A.: The proton affinity and gas-phase basicity of sulfur dioxide. ChemPhysChem **12**(1), 112–115 (2011)
36. Leonori, F., et al.: Observation of organosulfur products (thiovinoxy, thioketene and thioformyl) in crossed-beam experiments and low temperature rate coefficients for the reaction S(^{1}D) + C$_2$H$_4$. Phys. Chem. Chem. Phys. **11**(23), 4701–4706 (2009)
37. de Petris, G., Rosi, M., Troiani, A.: SSOH and HSSO radicals: an experimental and theoretical study of [S2OH]0/± species. J. Phys. Chem. A **111**(28), 6526–6533 (2007)
38. Rosi, M., Falcinelli, S., Balucani, N., Casavecchia, P., Leonori, F., Skouteris, D.: Theoretical study of reactions relevant for atmospheric models of titan: interaction of excited nitrogen atoms with small hydrocarbons. In: Murgante, B., Gervasi, O., Misra, S., Nedjah, N., Rocha, A.M.A.C., Taniar, D., Apduhan, B.O. (eds.) ICCSA 2012. LNCS, vol. 7333, pp. 331–344. Springer, Heidelberg (2012). https://doi.org/10.1007/978-3-642-31125-3_26
39. Gaussian 09, Revision A.02, Frisch, M.J., et al.: Gaussian, Inc., Wallingford CT (2009)
40. Avogadro: an open-source molecular builder and visualization tool. Version 1.XX. (http://avogadro.cc)
41. Hanwell, M.D., Curtis, D.E., Lonie, D.C., Vandermeersch, T., Zurek, E., Hutchison, R.: Avogadro: an advanced semantic chemical editor, visualization, and analysis platform. J. Cheminformatics **4**, 17 (2012)

A Computational Study on the Insertion of N(^{2}D) into a C—H or C—C Bond: The Reactions of N(^{2}D) with Benzene and Toluene and Their Implications on the Chemistry of Titan

Marzio Rosi[1(✉)], Leonardo Pacifici[2], Dimitrios Skouteris[3],
Adriana Caracciolo[2], Piergiorgio Casavecchia[2], Stefano Falcinelli[1],
and Nadia Balucani[2]

[1] Dipartimento di Ingegneria Civile e Ambientale, Università degli Studi di Perugia, 06125 Perugia, Italy
{marzio.rosi,stefano.falcinelli}@unipg.it

[2] Dipartimento di Chimica, Biologia e Biotecnologie, Università degli Studi di Perugia, 06123 Perugia, Italy
leo.pacifici1@gmail.com, adriana.caracciolo@yahoo.it,
{piergiorgio.casavecchia,nadia.balucani}@unipg.it

[3] Master-Up, Via Sicilia 41, 06128 Perugia, Italy
d.skouteris@master-up.it

Abstract. The reactions between nitrogen atoms in their first electronically excited state ^{2}D with benzene and toluene have been characterized by electronic structure calculations of the stationary points along the minimum energy path. We focused our attention, in particular, to the channels leading to the imidogen radical for the first reaction implying the insertion of nitrogen into a C—H bond and to the NCH$_3$ radical for the second reaction implying the insertion of nitrogen into a C—C bond. The minima along these reaction paths have been characterized using different ab initio methods in order to find a reasonable compromise between chemical accuracy and computational costs. Our results suggest that, while for geometry optimizations even relatively low level calculations are adequate, for energies higher level of calculations are necessary in order to obtain accurate quantitative results, in particular when strong correlation effects are present.

Keywords: Ab initio calculations · Titan atmosphere · Chemistry of planetary atmospheres · Prebiotic chemistry · Astrochemistry · Astrobiology

1 Introduction

Titan, the largest moon of Saturn, is the only solar system object with a thick dinitrogen atmosphere, besides the Earth [1, 2]. The other components of its atmosphere are mainly simple hydrocarbons like CH$_4$, C$_2$H$_6$, C$_2$H$_4$ and C$_2$H$_2$, simple nitriles like HCN and HCCCN, H$_2$ and Ar. This atmosphere is reminiscent of the primeval atmosphere of Earth [3, 4]. The reactivity of these species is very relevant as suggested by the

© Springer Nature Switzerland AG 2020
O. Gervasi et al. (Eds.): ICCSA 2020, LNCS 12251, pp. 744–755, 2020.
https://doi.org/10.1007/978-3-030-58808-3_54

information provided by the NASA/ESA/ASI Cassini-Huygens mission [5] and the observations performed with the ALMA interferometer [6]. Among the species identified by Cassini Ion Neutral Spectrometer (INMS), benzene shows a relatively important mole fraction, being 1.3×10^{-6} at 950 km [7]. Solid benzene has been identified also on the surface of Titan [8]. Toluene is efficiently produced in Titan's atmosphere since C_6H_5, the main product of the photodissociation of benzene, reacts with the radicals most abundant in Titan's atmosphere, which are H and CH_3 [9]. However, molecular dinitrogen, in its closed shell electronic configuration, cannot react with hydrocarbons in the atmosphere of Titan because of the presence of relatively high activation energy barriers. In the upper atmosphere of Titan, however, molecular dinitrogen can be converted into atomic nitrogen or ions by energetic processes [10] or by the interaction with Extreme Ultra-Violet (EUV) radiation. The dissociation of molecular dinitrogen induced by dissociative photoionization, galactic cosmic ray absorption, N_2^+ dissociative recombination, or dissociation induced by EUV photons produces atomic nitrogen in its electronic ground state ^{4}S and in the first excited ^{2}D state in similar amounts [10]. While atomic nitrogen in its ^{4}S ground state exhibits very low reactivity with closed shell molecules, atomic nitrogen in its first electronically excited ^{2}D state shows a significant reactivity with several molecules identified in the atmosphere of Titan. Atomic nitrogen in its excited ^{2}D state is metastable but it shows a radiative lifetime long enough to react in binary collisions with other constituents of the upper atmosphere of Titan (6.1×10^4 s and 1.4×10^5 s for the ^{2}D$_{3/2}$ and ^{2}D$_{5/2}$ state, respectively) [11–19]. We have already investigated the reactions of atomic nitrogen in its excited ^{2}D state with aliphatic hydrocarbons, like CH_4, C_2H_2, C_2H_4, C_2H_6, in laboratory experiments by the crossed molecular beam technique and by ab initio and kinetics calculations [20–25]. More recently, with the same approach, we started the investigation of the reaction between N(^{2}D) and aromatic hydrocarbons like benzene [26]. In this study we will report preliminary calculations on the interaction of N(^{2}D) with benzene and toluene paying attention to the computational requirements. Indeed, the ab initio study of these systems is computationally very challenging and a compromise between chemical accuracy and computational resources is necessary.

2   Computational Details

The reactive channels of the N(^{2}D) + C_6H_6 and $C_6H_5CH_3$ systems leading to the insertion of atomic nitrogen into a C—H or C—C bond were investigated by locating the lowest stationary points at the B3LYP level of theory [27, 28], in conjunction with the $6 - 311 + G^{**}$ basis set [29, 30], on the doublet ground state potential energy surface. The basis set superposition error (BSSE) has been estimated using the counterpoise method [31, 32]. At the same level of theory we have computed the harmonic vibrational frequencies in order to check the nature of the stationary points, *i.e.* minimum if all the frequencies are real, saddle point if there is only one imaginary frequency. The assignment of the saddle points was performed using intrinsic reaction coordinate (IRC) calculations [33, 34] The geometry of all the species was optimized without any symmetry constraints considering for all the investigated species the electronic ground state. In order to check the accuracy of the computed geometries, we have optimized all the minima using a more extended basis set. We have optimized all the minima at the B3LYP level [27, 28] with the correlation consistent aug-cc-pVTZ basis set [35]. For all the stationary points, the energy was computed also at the higher level of calculation CCSD(T) [36–38] using the same correlation consistent aug-cc-pVTZ basis set [35] and the B3LYP/aug-cc-pVTZ optimized geometries, following a well established computational scheme [39–49]. The accuracy of the employed approach, in particular as far as basis set completeness is concerned, has been recently investigated [50]. Both the B3LYP and the CCSD(T) energies were corrected to 0 K by adding the zero point energy correction computed using the scaled harmonic vibrational frequencies evaluated at B3LYP level. The energy of $N(^2D)$ was estimated by adding the experimental [51] separation $N(^4S) - N(^2D)$ of 55 kcal mol^{-1} to the energy of $N(^4S)$ at all levels of calculation. All calculations were done using Gaussian 09 [52] while the analysis of the vibrational frequencies was performed using Molekel [53, 54].

3   Results and Discussion

Figure 1 reports the optimized geometries of the stationary points localized along the reactive channel leading from $N(^2D)$ + benzene to the insertion product of $N(^2D)$ into a C—H bond and finally to the formation of the imidogen radical NH. We have found four minima along this reaction channel connected by three transition states: the first minimum (1) is only an adduct where N is slightly interacting with the benzene ring; the second one (2) shows a C atom of the ring bonded to an H and the N atoms; the third minimum (3) shows the nitrogen atom bonded to two adjacent carbon atoms and then the fourth minimum (4) shows the insertion of the nitrogen atom into the C—H bond with the formation of an NH group. Figure 1 shows the main geometrical parameters optimized at B3LYP level using both the $6 - 311 + G^{**}$ and the aug-cc-pVTZ basis set. The last results are reported in parentheses. Some of these optimized structures were previously reported [55]. We can notice that there is a good agreement between the geometries optimized with the $6 - 311 + G^{**}$ and the aug-cc-pVTZ basis set, the differences being within 0.01Å. There is only one relevant difference represented by the N—C distance in species 1: this distance indeed is 3.044 Å with the $6 - 311 + G^{**}$ basis set and 3.186 Å with the aug-cc-pVTZ basis set. However, this is an adduct and the energy dependence from the distance in this case is weak, so it is very difficult to compute the correct bond length. In Fig. 2 we have reported the minimum energy path for the channel leading to species 4 and then to the formation of the imidogen radical. The approach of the nitrogen to the benzene ring leads to the adduct 1 which is more stable than the reactants by 30.1 kcal/mol. This species, through the transition state **TS12** which is still well under the reactants gives rise to species **2** where we have a true C—N bond. Species **2**, through a very small barrier of only 0.1 kcal/mol evolves towards species **3** which shows the nitrogen bridging two adjacent carbon atoms. It is necessary to overcome a barrier of 48.4 kcal/mol for species **3** to evolve to the more stable species **4** which represents the lowest minimum in this path. Species **4** shows the insertion of the nitrogen into a C—H bond. The dissociation of the C—N bond leads to the formation of the imidogen radical in a process endothermic by 101.2 kcal/mol. The imidogen radical can be formed also by direct abstraction of the hydrogen by the N(^{2}D). If the nitrogen approaches the benzene ring from the hydrogen side, the adduct **5** is formed which is 30.0 kcal/mol more stable than the reactants. This adduct, through a transition state which is 22.7 kcal/mol under the reactants gives the products.

The direct abstraction of hydrogen implies only one step for the formation of the imidogen radical, while the reaction path through species **4** implies several steps which lie however at lower energies. In order to state which one is the preferred path, kinetic calculations are necessary and they will be performed in the future.

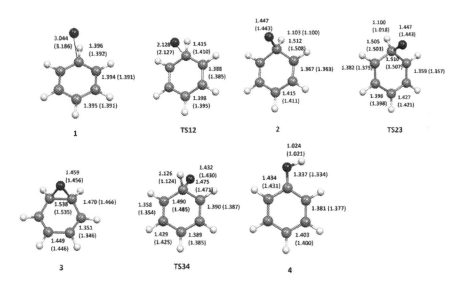

Fig. 1. Optimized geometries at B3LYP/6 − 311 + G** and B3LYP/aug-cc-pVTZ (in parentheses) level of the stationary points considered along the minimum energy paths leading from N(^{2}D) + C$_6$H$_6$ to C$_6$H$_5$NH. Bond lengths in Å.

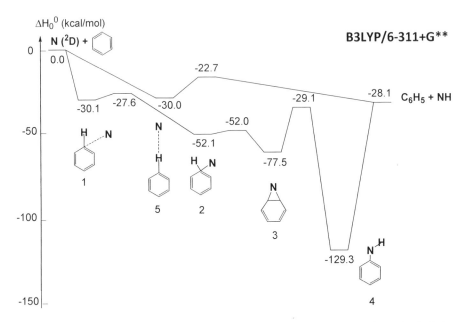

Fig. 2. Main steps along the minimum energy path for the channel leading to C_6H_5NH. Relative energies (kcal/mol, 0 K) with respect to the reactants $N(^2D) + C_6H_6$ computed at B3LYP/6 − 311 + G** level.

Figure 3 reports the optimized geometries of the stationary points localized along the reactive channel leading from $N(^2D)$ + toluene to the insertion product of $N(^2D)$ into a C—C bond and finally to the formation of the NCH_3 radical. We have found three minima along this reaction channel connected by two transition states: the first minimum **1** is only an adduct where N is slightly interacting with the toluene ring, being the C—N distance as high as 3.027 Å; the second one **2** shows the nitrogen atom bonded to two adjacent carbon atoms and then the third minimum **3** shows the insertion of the nitrogen atom into the C—C bond with the formation of an NCH_3 group. Figure 3 shows the main geometrical parameters optimized at B3LYP level using both the 6 − 311 + G** and the aug-cc-pVTZ basis set. The last results are reported in parentheses. We were not able to optimize the initial adduct at the B3LYP/aug-cc-pVTZ level. From Fig. 3 we can notice that there is a good agreement between the geometries optimized with the 6 − 311 + G** and the aug-cc-pVTZ basis set, the differences being within 0.01Å. We can also notice that the presence of the methyl substituent in toluene, with respect to benzene, changes the electronic distribution around the carbon atom, preventing the formation of the equivalent of species **2** reported in Fig. 1. In Fig. 4 we have reported the minimum energy path for the channel leading to species **3** and then to the formation of the NCH_3 radical. The approach of the nitrogen to the ipso carbon of the toluene ring leads to the adduct **1** which is more stable than the reactants by 30.2 kcal/mol. This species, through the transition state **TS12** which is still well under the reactants gives rise to species **2** which shows the nitrogen bridging the ipso and the ortho carbon atoms. It is necessary to overcome a barrier of 47.5 kcal/mol for species **2** to evolve to the more stable species **3** which represents the lowest minimum in this path. Species **3** shows the insertion of the nitrogen into a C—C bond. The dissociation of the C—N bond leads to the formation of the NCH$_3$ radical in a process endothermic by 88.5 kcal/mol. Comparing Fig. 2 and Fig. 4 we can notice that, except for the presence of minimum **2** in Fig. 2 which, however, can evolve to minimum **3** through an almost negligible barrier height, the insertions of N(^{2}D) into a C—H or a C—C bond are very similar processes. Moreover, this seems to be a realistic path to the formation of a imidogen or a NCH$_3$ radical.

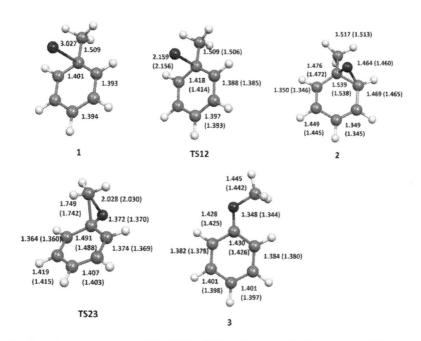

Fig. 3. Optimized geometries at B3LYP/6 − 311 + G** and B3LYP/aug-cc-pVTZ (in parentheses) level of the stationary points considered along the minimum energy path leading from N(^{2}D) + toluene to C$_6$H$_5$NCH$_3$. Bond lengths in Å.

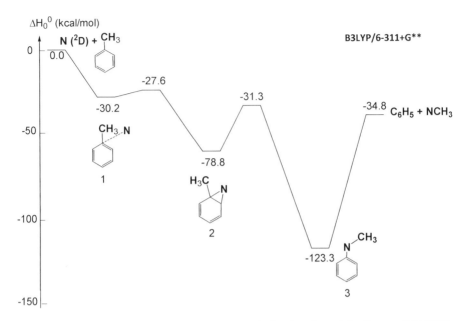

Fig. 4. Main steps along the minimum energy path for the channel leading to $C_6H_5NCH_3$. Relative energies (kcal/mol, 0 K) with respect to the reactants $N(^2D) + C_6H_5CH_3$ computed at B3LYP/6 − 311 + G** level.

Although, both in Fig. 2 and Fig. 4 all the species, including the transition states, are in energy well under the reactants, some doubts could emerge due to the low level of accuracy of the performed calculations. For this reason, we performed more accurate calculations using the CCSD(T) method in conjunction with a more extended basis set as the aug-cc-pVTZ. We optimized all the stationary points (minima and saddle points) at the B3LYP/aug-cc-pVTZ level and at these optimized geometries we performed CCSD(T)/aug-cc-pVTZ calculations. The energies of main steps along the minimum energy path for the reaction $N(^2D) + $ benzene $\rightarrow C_6H_5 + NH$ and $N(^2D) + $ toluene $\rightarrow C_6H_5 + NCH_3$ are reported in Table 1. For the main processes we reported both the enthalpy differences and the barrier heights computed at 0 K. In Table 1, the first entry is the energy computed at B3LYP/6 − 311 + G**, the second entry the relative energy computed at B3LYP/aug-cc-pVTZ level and the third entry the energy computed at CCSD(T)/aug-cc-pVTZ//B3LYP/aug-cc-pVTZ level. We were not able to optimize the geometry of species **1** for the toluene reaction at the B3LYP/aug-cc-pVTZ level. For this reason, in Table 1 we have reported also the reaction from the reactants to species **2**.

Table 1. Enthalpy changes and barrier heights (kcal/mol, 0 K) computed at the B3LYP/6 − 311 + G**, B3LYP/aug-cc-pVTZ and CCSD(T)/aug-cc-pVTZ//B3LYP/aug-cc-pVTZ level of theory for the minimum energy path for the reactions N(^{2}D) + C$_6$H$_6$ → C$_6$H$_5$ + NH and N(^{2}D) + C$_6$H$_5$CH$_3$ → C$_6$H$_5$ + NCH$_3$. First entry B3LYP/6 − 311 + G**, second entry B3LYP/aug-cc-pVTZ and third entry CCSD(T)/aug-cc-pVTZ//B3LYP/aug-cc-pVTZ

	ΔH_0^0	Barrier height
Benzene		
N(^{2}D) + C$_6$H$_6$ → 1	−30.1, −31.2, −39.7	
1 → 2	−22.0, −21.4, −4.1	2.5, 2.7, 18.2
N(^{2}D) + C$_6$H$_6$ → 2	−52.1, −52.6, −43.8	−27.6, −28.5, −21.5
2 → 3	−25.4, −26.1, −30.2	0.1, 0, 0.3
3 → 4	−51,8, −52.2, −46.5	48.4, 48.7, 48.1
4 → C$_6$H$_5$ +NH	101.2, 101.8, 101.7	
Toluene		
N(^{2}D) + C$_6$H$_5$CH$_3$ → 1	−30.2	
1 → 2	−48.6	2.6
N(^{2}D) + C$_6$H$_5$CH$_3$ → 2	−78.8, −79.8, −75.9	−27.6, −28.3, −22.3
2→ 3	−44.5, −45.2, −37.4	47.5, 47.4, 54.0
3 → C$_6$H$_5$ + NCH$_3$	88.5, 89.1, 93.7	

Comparing the values of Table 1 we can notice that the B3LYP/6 − 311 + G** and the B3LYP/aug-cc-pVTZ energies differ always by a small amount of energy (around 1 kcal/mol). This difference is essentially due to the basis set superposition error computed with the counterpoise method which is 1.1 kcal/mol [31, 32]. Therefore, comparing the geometries and the energies, we can conclude that it is worthless to use a more extended basis set, with respect to the 6 − 311 + G**, to optimize the geometries at B3LYP level. Comparing the B3LYP and CCSD(T) energies we have a complete different situation, being the difference sometimes higher than ten kcal/mol. In particular, the difference between B3LYP and CCSD(T) energies is relevant for reactions involving species **1**: this is an expected result since species **1** is an adduct which is very difficult to describe accurately at B3LYP level; moreover, in species **1** there is a N(^{2}D) atom almost unperturbed and it is well known that the B3LYP method cannot describe correctly the nitrogen excited ^{2}D state.

For the reactions considered in this context, the situation however does not change dramatically going from B3LYP to CCSD(T) results because the energies of all the species investigated are still well under the reactants, but for other reaction this point could be very relevant.

4   Conclusions

The study at ab initio level of the reactions N(^{2}D) + C$_6$H$_6$ → C$_6$H$_5$ + NH and N(^{2}D) + C$_6$H$_5$CH$_3$ → C$_6$H$_5$ + NCH$_3$, performed using different methods in order to find a reasonable compromise between chemical accuracy and computational costs suggests that, while for geometry optimizations even relatively low level calculations are adequate, for energies, calculations of higher level are necessary in order to obtain accurate quantitative results.

Acknowledgments. This work has been supported by MIUR "PRIN 2015" funds, project "STARS in the CAOS (Simulation Tools for Astrochemical Reactivity and Spectroscopy in the Cyberinfrastructure for Astrochemical Organic Species)", Grant Number 2015F59J3R. We acknowledge the Italian Space Agency for co-funding the Life in Space Project (ASI N. 2019-3-U.O). SF and MR acknowledge the project "Indagini teoriche e sperimentali sulla reattività di sistemi di interesse astrochimico" funded with Fondo Ricerca di Base 2018 of the University of Perugia. M.R. thanks the Dipartimento di Ingegneria Civile e Ambientale of the University of Perugia for allocated computing time within the project "Dipartimenti di Eccellenza 2018-2022".

References

1. Hörst, S.M.: Titan's atmosphere and climate. J. Geophys. Res. Planets **122**, 432–482 (2017)
2. Vuitton, V., Yelle, R.V., Anicich, V.G.: The nitrogen chemistry of Titan's upper atmosphere revealed. Astrophys. J. **647**, L175–L178 (2006)
3. Vuitton, V., Dutuit, O., Smith, M.A., Balucani, N.: Chemistry of Titan's atmosphere. In: Mueller-Wodarg, I., Griffith, C., Lellouch, E., Cravens, T. (eds.) Titan: Surface, Atmosphere and Magnetosphere. Cambridge University Press, Cambridge (2013)
4. Balucani, N.: Elementary reactions of N atoms with hydrocarbons: first steps towards the formation of prebiotic N-containing molecules in planetary atmospheres. Chem. Soc. Rev. **41**, 5473–5483 (2012)
5. Brown, R., Lebreton, J.P., Waite, J. (eds.): Titan from Cassini-Huygens. Springer, Netherlands (2010). https://doi.org/10.1007/978-1-4020-9215-2
6. Lai, J.C.-Y., et al.: Mapping vinyl cyanide and other nitriles in Titan's atmosphere using ALMA. Astron. J. **154**(206), 1–10 (2017)
7. Vuitton, V., Yelle, R.V., Cui, J.: Formation and distribution of benzene on Titan. J. Geophys. Res. **113**, E05007 (2008)
8. Clark, R.N., et al.: Detection and mapping of hydrocarbon deposits on Titan. J. Geophys. Res. **115**, E10005 (2010)
9. Loison, J.C., Dobrijevic, M., Hickson, K.M.: The photochemical production of aromatics in the atmosphere of Titan. Icarus **329**, 55–71 (2019)
10. Lavvas, P., et al.: Energy deposition and primary chemical products in Titan's upper atmosphere. Icarus **213**, 233–251 (2011)
11. Dutuit, O., et al.: Critical review of N, N^+, N_2^+, N^{++} and N_2^{++} main production processes and reactions of relevance to Titan's atmosphere. Astrophys. J. Suppl. Ser. **204**, 20 (2013)
12. Balucani, N.: Nitrogen fixation by photochemistry in the atmosphere of Titan and implications for prebiotic chemistry. In: Trigo-Rodriguez, J.M., Raulin, F., Muller, C., Nixon, C. (eds.) The Early Evolution of the Atmospheres of Terrestrial Planets, edited by Springer Series in Astrophysics and Space Science Proceedings, vol. 35, pp. 155–164 (2013). https://doi.org/10.1007/978-1-4614-5191-4_12
13. Balucani, N.: Elementary reactions and their role in gas-phase prebiotic chemistry. Int. J. Mol. Sci. **10**, 2304–2335 (2009)
14. Imanaka, H., Smith, M.A.: Formation of nitrogenated organic aerosols in the Titan upper atmosphere. PNAS **107**, 12423–12428 (2010)

15. Balucani, N., et al.: Dynamics of the N(^{2}D) + D$_2$ reaction from crossed-beam and quasiclassical trajectory studies. J. Phys. Chem. A **105**, 2414–2422 (2001)
16. Balucani, N., et al.: Experimental and theoretical differential cross sections for the N(^{2}D) + H$_2$ reaction. J. Phys. Chem. A **110**, 817–829 (2006)
17. Homayoon, Z., Bowman, J.M., Balucani, N., Casavecchia, P.: Quasiclassical trajectory calculations of the N(^{2}D) + H$_2$O reaction elucidating the formation mechanism of HNO and HON seen in molecular beam experiments. J. Phys. Chem. Lett. **5**, 3508–3513 (2014)
18. Balucani, N., Cartechini, L., Casavecchia, P., Homayoon, Z., Bowman, J.M.: A combined crossed molecular beam and quasiclassical trajectory study of the Titan-relevant N(^{2}D) + D$_2$O reaction. Mol. Phys. **113**, 2296–2301 (2015)
19. Israel, G., et al.: Complex organic matter in Titan's atmospheric aerosols from in situ pyrolysis and analysis. Nature **438**, 796 (2005)
20. Balucani, N., et al.: Combined crossed molecular beam and theoretical studies of the N(^{2}D) + CH$_4$ reaction and implications for atmospheric models of Titan. J. Phys. Chem. A **113**, 11138–11152 (2009)
21. Balucani, N., et al.: Cyanomethylene formation from the reaction of excited Nitrogen Atoms with Acetylene: A crossed beam and ab initio study. J. Am. Chem. Soc. **122**, 4443–4450 (2000)
22. Balucani, N., Cartechini, L., Alagia, M., Casavecchia, P., Volpi, G.G.: Observation of nitrogen-bearing organic molecules from reactions of nitrogen atoms with hydrocarbons: a crossed beam study of N(^{2}D) + ethylene. J. Phys. Chem. A **104**, 5655–5659 (2000)
23. Balucani, N.: Formation of nitriles and imines in the atmosphere of Titan: combined crossed-beam and theoretical studies on the reaction dynamics of excited nitrogen atoms N(^{2}D) with ethane. Faraday Discuss. **147**, 189–216 (2010)
24. Balucani, N., et al.: Combined crossed beam and theoretical studies of the N(^{2}D) + C$_2$H$_4$ reaction and implications for atmospheric models of Titan. J. Phys. Chem. A **116**, 10467–10479 (2012)
25. Rosi, M., Falcinelli, S., Balucani, N., Casavecchia, P., Skouteris, D.: A theoretical study of formation routes and dimerization of methanimine and implications for the aerosols formation in the upper atmosphere of Titan. In: Murgante, B., et al. (eds.) ICCSA 2013. LNCS, vol. 7971, pp. 47–56. Springer, Heidelberg (2013). https://doi.org/10.1007/978-3-642-39637-3_4
26. Balucani, N., Pacifici, L., Skouteris, D., Caracciolo, A., Casavecchia, P., Rosi, M.: A theoretical investigation of the reaction N(^{2}D) + C$_6$H$_6$ and implications for the upper atmosphere of Titan. In: Gervasi, O., et al. (eds.) ICCSA 2018. LNCS, vol. 10961, pp. 763–772. Springer, Cham (2018). https://doi.org/10.1007/978-3-319-95165-2_53
27. Becke, A.D.: Density functional thermochemistry. III. The role of exact exchange. J. Chem. Phys. **98**, 5648–5652 (1993)
28. Stephens, P.J., Devlin, F.J., Chablowski, C.F., Frisch, M.J.: Ab initio calculation of vibrational absorption and circular dichroism spectra using density functional force fields. J. Phys. Chem. **98**, 11623–11627 (1994)
29. Krishnan, R., Binkley, J.S., Seeger, R., Pople, J.A.: Self-consistent molecular orbital methods. XX. A basis set for correlated wave functions. J. Chem. Phys. **72**, 650–654 (1980)
30. Frisch, M.J., Pople, J.A., Binkley, J.S.: Self-consistent molecular orbital methods 25 Supplementary functions for Gaussian basis sets. J. Chem. Phys. **80**, 3265–3269 (1984)
31. Simon, S., Duran, M., Dannesberg, J.J.: How does basis set superposition error change the potential surfaces for hydrogen-bonded dimers? J. Chem. Phys. **105**, 11024–11031 (1996)
32. Boys, S.F., Bernardi, F.: The calculation of small molecular interactions by the differences of separate total energies Some procedures with reduced errors. Mol. Phys. **19**, 553–566 (1970)

33. Gonzalez, C., Schlegel, H.B.: An improved algorithm for reaction path following. J. Chem. Phys. **90**, 2154–2161 (1989)
34. Gonzalez, C., Schlegel, H.B.: Reaction path following in mass-weighted internal coordinates. J. Phys. Chem. **94**, 5523–5527 (1990)
35. Dunning Jr., T.H.: Gaussian basis sets for use in correlated molecular calculations. I. The atoms boron through neon and hydrogen. J. Chem. Phys. **90**, 1007–1023 (1989)
36. Bartlett, R.J.: Many-body perturbation theory and coupled cluster theory for electron correlation in molecules. Annu. Rev. Phys. Chem. **32**, 359–401 (1981)
37. Raghavachari, K., Trucks, G.W., Pople, J.A., Head-Gordon, M.: Quadratic configuration interaction. A general technique for determining electron correlation energies. Chem. Phys. Lett. **157**, 479–483 (1989)
38. Olsen, J., Jorgensen, P., Koch, H., Balkova, A., Bartlett, R.J.: Full configuration–interaction and state of the art correlation calculations on water in a valence double-zeta basis with polarization functions. J. Chem. Phys. **104**, 8007–8015 (1996)
39. de Petris, G., Cacace, F., Cipollini, R., Cartoni, A., Rosi, M., Troiani, A.: Experimental detection of theoretically predicted N_2CO. Angew. Chem. **117**, 466–469 (2005)
40. De Petris, G., Rosi, M., Troiani, A.: SSOH and HSSO radicals: an experimental and theoretical study of $[S_2OH]^{0/+/-}$ species. J. Phys. Chem. A **111**, 6526–6533 (2007)
41. Bartolomei, M., et al.: The intermolecular potential in $NO-N_2$ and $(NO-N_2)^+$ systems: implications for the neutralization of ionic molecular aggregates. PCCP **10**, 5993–6001 (2008)
42. Leonori, F., et al.: Observation of organosulfur products (thiovinoxy, thioketene and thioformyl) in crossed-beam experiments and low temperature rate coefficients for the reaction $S(^1D) + C_2H_4$. PCCP **11**, 4701–4706 (2009)
43. Leonori, F., et al.: Crossed-beam and theoretical studies of the $S(^1D) + C_2H_2$ reaction. J. Phys. Chem. A **113**, 4330–4339 (2009)
44. De Petris, G., Cartoni, A., Rosi, M., Barone, V., Puzzarini, C., Troiani, A.: The proton affinity and gas-phase basicity of sulfur dioxide. ChemPhysChem **12**, 112–115 (2011)
45. Berteloite, C., et al.: Low temperature kinetics, crossed beam dynamics and theoretical studies of the reaction $S(^1D) + CH_4$ and low temperature kinetics of $S(^1D) + C_2H_2$. Phys. Chem. Chem. Phys. **13**, 8485 (2011)
46. Rosi, M., Falcinelli, S., Balucani, N., Casavecchia, P., Leonori, F., Skouteris, D.: Theoretical study of reactions relevant for atmospheric models of Titan: rocarbons. In: Murgante, B., et al. (eds.) ICCSA 2012. LNCS, vol. 7333, pp. 331–344. Springer, Heidelberg (2012). https://doi.org/10.1007/978-3-642-31125-3_26
47. Skouteris, D., Balucani, N., Faginas-Lago, N., Falcinelli, S., Rosi, M.: Dimerization of methanimine and its charged species in the atmosphere of Titan and interstellar/cometary ice analogs. Astron. Astrophys. **584**, A76 (2015)
48. Falcinelli, S., Rosi, M., Cavalli, S., Pirani, F., Vecchiocattivi, F.: Stereoselectivity in autoionization reactions of hydrogenated molecules by metastable gas atoms: the role of electronic couplings. Chemistry Eur. J. **22**, 12518–12526 (2016)
49. Troiani, A., Rosi, M., Garzoli, S., Salvitti, C., de Petris, G.: Vanadium hydroxide cluster ions in the gas phase: bond-forming reactions of doubly-charged negative ions by SO_2-promoted V-O activation. Chem. Eur. J. **23**, 11752–11756 (2017)
50. Rosi, M., et al.: An experimental and theoretical investigation of 1-butanol pyrolysis. Front. Chem. **7**, 326 (2019). https://doi.org/10.3389/fchem.2019.00326
51. Moore, C.E.: Atomic Energy Levels, Natl. Bur. Stand. (U.S.) Circ. N. 467 (U.S., GPO, Washington, DC, 1949)
52. Gaussian 09, Revision A.02, Frisch, M.J., et al.: Gaussian, Inc., Wallingford CT (2009)

53. Flükiger, P., Lüthi, H. P., Portmann, S., Weber, J.: MOLEKEL 4.3, Swiss Center for Scientific Computing, Manno (Switzerland), 2000–2002

54. Portmann, S., Lüthi, H.P.: MOLEKEL: an interactive molecular graphics tool. Chimia **54**, 766–769 (2000)

55. Balucani, N., et al.: A computational study of the reaction N(^{2}D) + C_6H_6 leading to pyridine and phenylnitrene. In: Misra, S., et al. (eds.) ICCSA 2019. LNCS, vol. 11621, pp. 316–324. Springer, Cham (2019). https://doi.org/10.1007/978-3-030-24302-9_23

**International Workshop on Cities,
Technologies and Planning (CTP 2020)**

Territorial Attraction for New Industrial-Productive Plants. The Case of Pavia Province

Roberto De Lotto[✉], Caterina Pietra, and Elisabetta Maria Venco

DICAr – University of Pavia, via Ferrata 3, 27100 Pavia, Italy
uplab@unipv.it

Abstract. Territorial attractiveness to production activities is a determining factor for the revitalization of degraded, abandoned, and impoverished territories and urban areas from an economic, social and environmental point of view. A competitive territory is capable of producing wealth and economic prosperity for its citizens and, at the same time, enhancing the environment, guaranteeing the protection of natural resources and cultural heritage and encourage the joint intervention of different subjects and institutional levels. New experiences in industrial plants planning concretely demonstrate how it is possible to create production plants capable of integrating sustainable principles in planning, design and management phases.

It is well recognized the possibility to pursue (according to principles and sustainability objectives) the competitiveness of the production system through localization choices capable of guaranteeing economic development, environmental protection and the involvement of different public and private subjects. Authors, in accordance to the new Lombardy Region Law (LR 18/2019) on urban regeneration, develop a Planning Support System (PSS) that leads to the study and definition of key elements for the planning of complex production-industrial system. In particular, the logical framework summarizes every steps of the planning process of productive activities on a territory. Moreover, authors present the case study related to the territorial and economic organization of Pavia Province.

Keywords: Planning Support System · Territorial competitiveness · Urban regeneration

1 Introduction

Over the last decades, the theme of territorial development and competitiveness, addressed for a long time exclusively in economic terms [1, 2], has been enriched with new visions and approaches that consider competitiveness as a complex concept, multidimensional and deeply linked to the principles and objectives of sustainable development [3–5]. A competitive territory is capable of producing wealth and economic prosperity for its citizens (economic competitiveness) and, at the same time, enhancing the environment, guaranteeing the protection of natural resources and historical-cultural-anthropological-territorial heritage (environmental competitiveness) and encourage the joint intervention of different subjects and institutional levels (social competitiveness).

© Springer Nature Switzerland AG 2020
O. Gervasi et al. (Eds.): ICCSA 2020, LNCS 12251, pp. 759–775, 2020.
https://doi.org/10.1007/978-3-030-58808-3_55

In a territorial context, the presence of an efficient production system can represent a valid possibility to pursue the concept of competitiveness in all its components. Experiences such as the Ecologically Equipped Production Areas [6] concretely demonstrate how it is possible to create production plants capable of integrating sustainable principles in planning, design and management phases. The use of innovative tools aims at promoting economic growth, the reduction of environmental impacts and synergies between institutions, citizens and businesses [7]. The success of these experiences is partly due to the planning process that identifies the complex elements (in socio-economic, infrastructural, environmental systems) that characterize it and that define its predisposition to be the right place of certain productive activities, clarifying its vocation. The analysis of the territory's vocation is a fundamental requirement to guarantee the protection and enhancement of the territory itself and therefore of its competitiveness.

The main scope (location of productive activities) can be analyzed from the general supply and demand law that in territorial meanings may be translated in two points of view: the first as an activity that wants to establish itself in the territory and therefore expresses a demand (What-Where), the latter as a planning process that guides the development of the territory by defining its offer (Where-What). Defined the production activity type and its specific characteristics, the What-Where model allows to identify the most appropriate areas for their location. Otherwise, given a specific area of intervention, the Where-What model allows to highlight the functional vocation to be attributed, taking into account the area peculiarities and the mutual influence with the territorial context [8, 9]. In the planning choices of both models, a plurality of territorial stakeholders contributes (including public administration bureaus, business companies, monitoring agencies, common citizens, labor unions, research centers, etc.). According to a multi-level governance, they work together in order to identify factors, conditions and enhancement actions for the establishment of new production activities.

2   Main Objectives

Nowadays, it is recognized the possibility to pursue (according to sustainable principles) the competitiveness of the production system through localization choices capable of guaranteeing economic development, environmental protection and the involvement of different public and private subjects. In order to guide the planning process and then the design activities of metropolitan area, authors develop a methodology based on a set of tools and techniques, Planning Support System (PSS), which allows focalizing roles and interests of the different involved stakeholders. The presented methodology leads to the study and definition of key elements for the planning of complex production-industrial system. For this purpose, authors develop a partially automatic GIS tool based on the processual method that combines the characteristics and typical elements of a DSS (Decision Support System), a KDDM (Knowledge Discovery Data Mining) and an ES (Expert System) [10, 11].

The logical framework summarizes the planning process of productive activities on a territory in order to pursue the objectives established a priori starting from the reference model (What-Where or Where-What). In particular, the tool supports users (professionals, administrative staff, and stakeholders) in making decisions, in evaluating the effects (direct and indirect) of actions and in monitoring results.

Authors also present the case study related to the territorial and economic organization of Pavia Province. In particular, it is analyzed: the territorial development with specific reference to some municipal realities; the widespread and specific territorial marketing; the territorial attraction to industrial/productive activities; the definition of new relationships between territorial and economic systems; the urban regeneration of dismissed areas and the definition of optimal scenarios for the regeneration of territorial, economic and social elements.

3 Methodology: Parameters and Analysis Techniques

The PSS, based on an analytical, multidisciplinary and trans-scalar approach of the territory, is divided into logical and coherent steps (punctually described in the next paragraphs):

- Analysis of territory characteristics and GIS tool construction;
- Multicriteria analysis;
- GIS tool query and spatial analysis techniques.

Table 1 summarize the main involved subjects, with specific roles and interests in the different process phases.

Table 1. PSS stakeholder.

Phase	Role	Stakeholder
Analysis of territory characteristics	Collecting data	Administrative bureau (Region, Province, Municipality) Business enterprises Business organization (Confindustria) Labor union Citizen committee Monitoring agencies (ARPA, ASST) Environmental groups Research institutes (University, Technological centers)
	GIS tool development	Professionals

(continued)

Table 1. (*continued*)

Phase	Role	Stakeholder
Multicriteria analysis	Qualitative and quantitative criteria definition	Administrative bureau (Region, Province, Municipality) Business enterprises Business organization (Confindustria) Professionals
	Qualitative and quantitative criteria evaluation	Administrative bureau (Region, Province, Municipality) Business enterprises Business organization (Confindustria) Labor union Citizen committee Monitoring agencies (ARPA, ASST) Environmental groups Research institutes (University, Technological centers) Professionals
GIS tool query	Spatial analysis	Professionals

3.1   Analysis of Territory Characteristics and GIS Tool Construction

The territorial analysis aims to provide a general overview of landscape, environmental, anthropogenic and socio-economic assets in the area in order to identify potentialities, fragilities, environmental sensibilities, anthropic pressure and local community needs. Authors start the analysis from the collection of the huge amount of available territorial data [12]. Furthermore, different thematic maps are drawn up (by overlapping shape-files). They represent synthetic graphic tools showing the current state of affairs of the territory at the different planning scales:

- Infrastructure charter: it is analyzed the existing structure and the potential of territory infrastructures (at national, regional and provincial level) related to railway, main road and intermodal centers;
- Charter of natural and heritage assets and environmental restrictions: the databases of environmental heritage, the plains environmental bases, the Regional Ecological Network (REN), database of municipality plans and protected areas give all the information relating to environmental and landscape elements and sensitivities;
- Charter of the settlement system: it is analyzed the presence of strategic structures as driving force for the economic, cultural and social growth of the territory. These structures often perform as polarization function towards the surrounding areas for

trade, work, leisure, education and health, becoming attracting poles (main production areas, technological centers, research institutes, universities, hospitals, colleges);

- Charter of industrial districts and meta-districts: the system of industrial districts and meta-districts defines the state of manufacturing activities present in a territory, in particular the cluster of companies specialized in a complex production process and the supply chains with high technological potential (Lombardy Region recognizes 16 districts and 6 meta-districts).

3.2 Multicriteria Analysis

In order to evaluate and establish a priority of importance among the different aspects highlighted in the analysis phase, and in order to define the optimal localization of the productive activities' establishment, it is necessary to define qualitative and quantitative indicators.

In particular, the indicators belonging to the infrastructural system assess the efficiency of the local road system, measured in terms of accessibility, travel time, proximity to the main infrastructures (highway, pan-European freight transit corridors). The indicators that belong to the demographic field consider the local population as a source of customers and potential workforce. The economic sector refers to factors capable of attracting industrial settlements such as capital, the price of land and the presence of investments. Lastly, those relating to the environmental system evaluate the ability of the interventions to reduce the impact on the environment by limiting the soil consumption and protecting the areas with the greatest environmental value through adequate physical distance.

For the indicators' assessment, authors use the pairwise comparison matrix (attribution of weight from 1 – equivalence of criteria – to 6 – strong prevalence of one over the other); the weights are added together and normalized, in order to identify the most relevant and to facilitate the ranking. Expert evaluators supported by a group of public-private subjects (Table 1) assign the weights considering their knowledge and the specific needs of the type of industrial activity selected.

3.3 GIS Tool Query and Spatial Analysis Techniques

The final phase of the PSS foresees the graphic representation of the indicators and the spatial analysis allowing the processing of the obtained results, their visualization and the creation of future development scenarios [8, 9]. In particular:

- Querying: it allows making a selection from a group of items;
- Proximity analysis: it allows analyzing the vector objects and their context, identifying areas of spatial influence. It is based on two functions: spatial property selection and buffering. The selection by spatial properties sets a choice based on the spatial relationship with other objects belonging or not to the same layer. Buffering operations create polygons whose perimeter has a constant and determined distance from existing objects;
- Overlay analysis: it allows intersection, union, difference, fade and cut operations between overlapping themes.

4   Case Study: Territorial Attraction for Industrial Activities in Pavia Province

In 2014, the regional law LR 31/2014 [13] introduced new provisions aimed at limiting the consumption of soil and promoting the regeneration of urbanized areas. Since that the soil is a non-renewable resource, the priority objective of reducing soil consumption takes the form of re-directing urban-building transformation activities not on virgin areas but on already urbanized, degraded or dismissed areas, in order to activate an urban renewal and regeneration process. It establishes also that it is a task of Territorial Government Plan – PGT [14] to identify the areas where urban and territorial regeneration processes start. To gain a better urban and territorial regeneration in the regional area, the Regional Territorial Plan intends to systematize transversal policies (territorial planning, economic, social and environmental policies) and activate direct collaboration and co-planning with local administrations (to avoid the fragmentation of interventions). The Regional Territorial Plan intends also to systematize all the policies creating a systemic vision, capable of attracting and directing investment and innovation (respecting the functional vocation and the peculiarities of each contexts).

Lombardy Region, through RL 18/2019 – art.1, pursuing the objective of sustainable development, recognizes the interventions aimed to urban and territorial regeneration, concerning areas (with different dimensions) or buildings, as priority actions to reduce soil consumption, improve functional, environmental and landscape quality of territories and settlements, as well as the socio-economic conditions of the population [15].

The RL 18/2019 brings significant changes to RL 12/2005 and RL 31/2014 by increasing the importance of interventions aimed at urban and territorial regeneration. In particular, in the new art. 4 of LR 12/2005, it defines criteria for the reduction of urbanization costs and the contribution on the construction cost up to 50%. Moreover, it defines also an increasing (up to 50%) of the contribution on the construction cost for interventions that consume agricultural land included or not in the consolidated urban fabric and for logistics or road haulage operations that do not incident on the regeneration areas. It is evident how the law plays an important role as driver of the sustainable development of the territory and pushes private stakeholder to develop targeted urban regeneration actions.

The Province of Pavia has over 4 sq. km of dismissed area (former industrial/productive/storage areas) spread on its territory. The city of Pavia, alone, has over 800.000 sq. m of former disused industrial areas to which must be added the areas that need effective regeneration: for decades, they have represented not only disqualifying urban voids but also areas of strong environmental and social pressure. They represent a reality that is dimensionally and temporally strategic in relation to urban and territorial development: they are actual opportunity for economic and entrepreneurial development.

The basic need is to identify the best location for a mechanical industry with a regional reference market (What-Where model), within the context of Pavia Province.

The main analyzed requirements are:

- Efficient infrastructure system, in particular regional level transport lines (fast-link road network - highways and main roads, railway network);
- Dismissed (industrial) areas, in order to start a regeneration process and to obtain the regional bonus;
- Highly skilled workforce (number of persons – 14–65 years of age – with a specific educational and technical college degree);
- Differentiated and specialized industrial cluster;
- Capital for scientific research in order to struggle competition and bring innovations;
- Availability of raw materials.

Based on these considerations, the specific stakeholders define the qualitative and quantitative parameters (Table 2) and, then, evaluate them.

Table 2. Qualitative and quantitative indicators related to the different fields of interest.

		Medium-long term indicator
Infrastructural system	a	Proximity to the fast link road system Proximity to the nearest motorway exit (5 km sphere of influence)
	b	Proximity to the main road system Proximity to main roads (500 m sphere of influence)
	c	Proximity to railway areas for loading-unloading goods Proximity to the nearest freight yard (1 km sphere of influence)
	d	Proximity to freight sorting and distribution site Proximity to the nearest intermodal pole (5 km sphere of influence)
	e	Proximity to the railway network Proximity (on foot) to the nearest train station (1 km sphere of influence)
Demographic field	f	Potentially customer base Average real-estate price (€/sq. m)
	g	Potential workforce Proximity to technological centers, universities, research institutes (10 km sphere of influence)
Economic sector	h	Real-estate values in the context Average real-estate price (€/sq. m)
	i	Proximity to strategic infrastructures Proximity to technological centers, universities, research institutes (10 km sphere of influence)
	j	Specialized industrial cluster Existence of other similar enterprises within a 15-min isochronous (15 km sphere of influence)
	k	Availability of economic investments and capital Ability to take advantage of capital-financing aimed at localization in certain areas
	l	Availability of resources Proximity to raw materials resources (water, energy, wood, etc.)

(*continued*)

Table 2. (*continued*)

		Medium-long term indicator
Environmental system	m	Distance from protected natural areas
		Distance from protected natural areas (more than 1 km)
	n	Distance from protected cultural heritage
		Distance from protected cultural heritage (more than 1 km)
	o	Distance from sensitive structures
		Distance from health and social care structures (500 m), and school (300 m)
	p	Distance from sensitive urban functions
		Distance from city center and residential areas (3 km)

Subsequently, authors assess the parameters using the pairwise comparison matrix and then define the classification of the parameters relating to the defined production activity. Due to this type of activity needs, the most relevant parameters concern the proximity to the main roads and fast link road system (a, b); while those who obtained a lower score are related to potential customers and environmental issues (f, m, n, o, p). The remaining parameters (g, i, j, k) and (c, d, e, h, l) obtained an average and sufficient evaluation respectively.

At the same time, experts build the GIS tool through the union of different thematic maps, representative of the territory (Table 3).

Table 3. Main data used to create the territory representation and to make the spatial analysis.

Thematic maps	Data type
Infrastructure charter	Railway network, road network (main and secondary), motorway exit, railway station, railway freight yard, intermodal, bridges
Protected cultural heritage and natural sites	General beauties, individual beauties, areas of particular environmental interest, rivers, streams, public waterways and their banks, Po river embankment area, natural monuments, local parks, natural parks, regional-national parks, SCI (Sites of Community Importance), Special Protection Areas (SPAs), active springs, relevance items protected by law 1089/39 (now Legislative Decree 22 January 2004, n° 42), primary ecological corridors REN, first level REN areas, second level REN areas
Settlement system charter	Dismissed areas, industrial-production area
Industrial districts and meta-districts charter	Industrial districts and meta-districts

The GIS assessment phase involves the use of different spatial analysis techniques: the buffering technique for displaying the parameters defined by creating specific sphere of influence (buffer) and the intersection between dismissed areas and industrial-production areas (existing and planned) and different systems buffers. Specifically, for the infrastructure and economic system, the intersection has a positive value; for the environmental system, negative. For the demographic system, there are problems in the visualization of data as these are obtained only from statistical processing (Istat) with no available shapefile. In particular, Table 4 shows the information related to the data availability and the possibility of displaying it in GIS environment.

Table 4. Data availability and data visualization on GIS tool.

		Medium-long term indicator	Data availability	Visualization on GIS tool
Infrastructural system	a	Proximity to the fast link road system	X	X
	b	Proximity to the main road system	X	X
	c	Proximity to railway areas for loading-unloading goods	X	X
	d	Proximity to freight sorting and distribution site	X	X
	e	Proximity to the railway network	X	X
Demographic field	f	Potentially customer base	X	–
	g	Potential workforce	X	–
Economic sector	h	Real-estate values in the context	X	–
	i	Proximity to strategic infrastructures	X	X
	j	Specialized industrial cluster	–	–
	k	Availability of economic investments and capital	–	–
	l	Availability of resources	–	–
Environmental system	m	Distance from protected natural areas	X	X
	n	Distance from protected cultural heritage	X	X
	o	Distance from sensitive structures	X	X
	p	Distance from sensitive urban functions	X	X

5 Results

The 2 selected territorial contexts in Pavia Province (Pavia, Broni-Stradella) are analyzed below, considering:

- Infrastructural system;
- Environmental system: green landscape and cultural heritage [16];
- Settlement system: sensitive facilities to noise, car traffic, and air-water pollution;
- Economic system.

5.1 Infrastructural System

The intersection between the brownfield/production areas and the buffers of the infrastructural system identifies the best location of the production activity (Figs. 1 and 2).

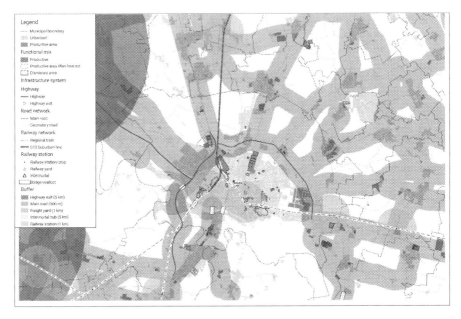

Fig. 1. Pavia infrastructural system analysis, scale 1:50.000.

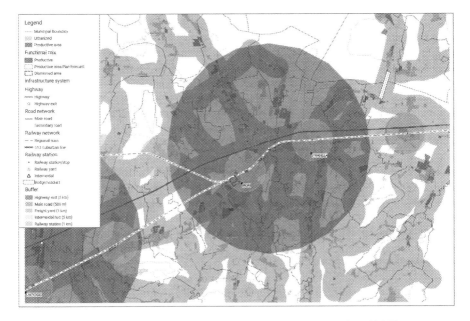

Fig. 2. Broni-Stradella infrastructural system analysis, scale 1:50.000.

5.2   Environmental System

The difference between the brownfield/production areas and the buffers of the environmental system identifies the best location of the production activity (Figs. 3, 4, 5 and 6).

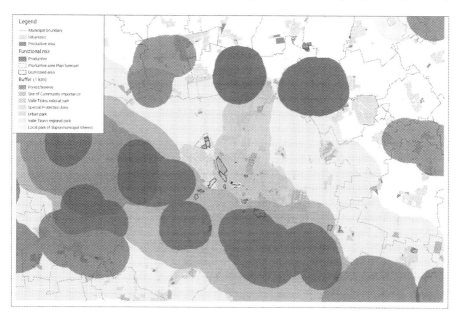

Fig. 3. Pavia environmental system analysis, scale 1:50.000.

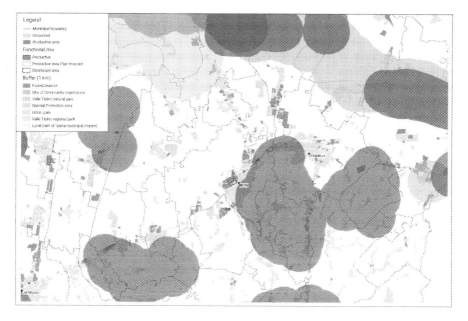

Fig. 4. Broni-Stradella environmental system analysis, scale 1:50.000.

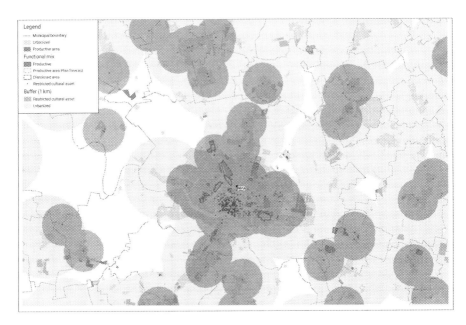

Fig. 5. Pavia cultural heritage analysis, scale 1:50.000.

Fig. 6. Broni-Stradella cultural heritage analysis, scale 1:50.000.

5.3   Settlement System

The difference between the brownfield/production areas and the buffers of the sensitive facilities identifies the best location of the production activity (Figs. 7 and 8).

Fig. 7. Pavia settlement system analysis, scale 1:50.000.

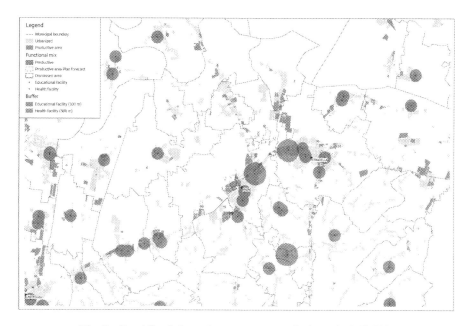

Fig. 8. Broni-Stradella settlement system analysis, scale 1:50.000.

5.4   Economic System

The intersection between the brownfield/production areas and the buffers of the economic system identifies the best location of the production activity (Fig. 9).

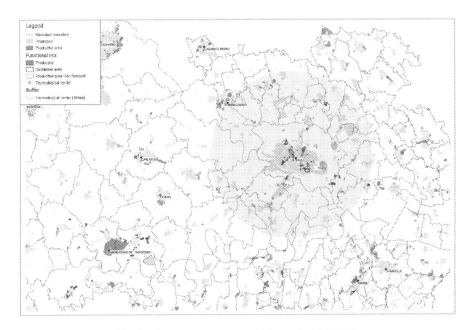

Fig. 9. Economic system analysis, scale 1:125.000.

5.5   Final Localization Ranking

The final step is the attribution of a different color to the intersections created: considering the relevance of each parameter, authors define a chromatic scale (from the darkest blue to the lightest blue) at the various localization levels (Fig. 12–14). In particular, the optimal, good, discrete and insufficient localization is obtained as an intersection between abandoned areas and industrial-productive areas (existing and planned) and the buffers relating respectively to parameters (a, b), (g, i, j, k), (c, d, e, h, l) and (f, m, n, o, p) (Figs. 10 and 11).

Fig. 10. Pavia area localization ranking, scale 1:50.000.

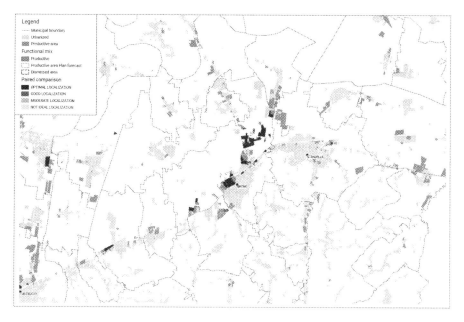

Fig. 11. Broni-Stradella area localization ranking, scale 1:50.000.

6   Conclusions

Each planning activity can be seen as an interactive and iterative process applied to the complex system in which it is not always easy to define the relationships that exist among the different elements and therefore the correct set of procedures to develop at a single decision-making process. In particular, in the territorial planning of industrial settlements it becomes fundamental to consider a set of factors that define the main peculiarities of these areas, the organization of which must facilitate, economically and technically, the individual business companies established.

The presented PSS represents contains tools and techniques suitable to help all the subjects involved in the production sites planning process. In particular, the stakeholders coordinated intervention to define the correct localization strategies allows determining the needs (socio-economic and environmental) of a specific territory by sharing knowledge and interests and by guaranteeing the protection and enhancement of the territory itself hence its competitiveness.

The critical issues found in the presented application case (above all the lack of updated spatial shape-form data availability) make necessary to strengthen these relationships among the subjects (public and private) in order to improve the efficiency of the tool and therefore of the entire planning process.

References

1. Porter, M.E.: The Competitive Advantage of Nations. Macmillan, Londra (1990)
2. Storper, M.: Competitiveness policy options: the tecnologicy-regions connection. Growth Change **26**, 285–308 (1995)
3. Kitson, M., Martin, R., Tyler, P.: Regional Competitiveness: an Elusive yet Key Concept? Reg. Stud. **38**(9), 991–999 (2004)
4. Ciciotti, E., Dallara, A., Rizzi, P.: Una geografia della sostenibilità dei sistemi locali italiani. In: XXVII Conferenza di Scienze Regionali (2006)
5. Gemmiti, R: Competitività territoriale in sostenibilità. L'interpretazione alla base della ricerca. In: Prezioso, M., Bencardino, F. (eds.) Competitività in sostenibilità: la dimensione territoriale nell'attuazione dei processi di Lisbona/Gothenburg nelle regioni e nelle province italiane. Geotema, Special Issue, pp. 31–32 (2007)
6. Borsari, L., Stacchini, V. (eds.): Linee guida per la realizzazione delle Aree produttive Ecologicamente Attrezzate della provincia di Bologna (2006)
7. Conticelli, E., Tondelli, S.: La pianificazione delle aree produttive per lo sviluppo sostenibile del territorio. Alinea, Firenze (2009)
8. Cattaneo, T., De Lotto, R., Venco, E.M.: Functional reuse and intensification of rural-urban context: rural architectural urbanism. Int. J. Agric. Environ. Inf. Syst. **7**(1), 1–27 (2016)
9. Cattaneo, T., De Lotto, R., Venco, E.M.: Methodology and applications for rurban sustainable development. In: Mastorakis, N.E., Corbi, O., Corbi, I. (eds.) Environmental and Agricultural Science, Atti della IV Conferenza Internazionale Energia, Ambiente, Imprese e Innovazione (ICESEEI15), pp. 111–120. WSEAS Press, Dubai (2015)
10. Densham, P.: Spatial Decision Support Systems. In: Maguire, D.J., Goodchild, M.F., Rhind, D.W. (eds.) Geographical Information Systems: Principles and Applications, pp. 403–412. John Wiley and Sons, New York (1991)

11. Brail, R.K.: Planning Support Systems for Cities and Regions. Lincoln Institute of Land Policy, Cambridge (2008)
12. Regione Lombardia, Geoportale della Lombardia. http://www.geoportale.regione.lombardia. it. Accessed 11 Feb 2020
13. Legge Regionale 28 novembre 2014, n. 31 Disposizioni per la riduzione del consumo di suolo e la riqualificazione del suolo degradato
14. Legge Regionale 11 marzo 2005, n. 12 Legge per il Governo del Territorio
15. Legge Regionale 26 novembre 2019, n. 18 Misure di semplificazione e incentivazione per la rigenerazione urbana e territoriale, nonché per il recupero del patrimonio edilizio esistente. Modifiche e integrazioni alla legge regionale 11 marzo 2005, n. 12 (Legge per il governo del territorio) e ad altre leggi regionali (BURL n. 48, suppl. del 29 Novembre 2019)
16. Decreto Legislativo 22 gennaio 2004, n. 42 Codice dei beni culturali e del paesaggio

A Quantitative Approach in the Ecclesiastical Heritage Appraisal

Francesca Salvo[1]([✉]), Manuela De Ruggiero[1], Daniela Tavano[1],
and Francesco Aragona[2]

[1] Department of Environmental Engineering, University of Calabria,
Via Pietro Bucci Cubo 46b, 87036 Rende, Italy
francesca.salvo@unical.it,
manueladeruggiero@gmail.com, daniela_tavano@libero.it
[2] New York, USA

Abstract. Cultural, environmental and religious resources are a repository of values participating in the process of recognizing the environmental semantic roots. The presence and use of these resources transform an anonymous place into a place of art, contemplation, meeting and cultural exchange; they contribute to psychological well-being, to the realization of social equity and to the economic development by attracting new investments and strengthening existing ones.

Ecclesiastical goods represent a conspicuous part, in terms of quantity and quality, of the historical and artistic heritage. Religious heritage is also an economic asset, having the characteristics of utility, usability and limited availability that distinguish economic assets (increasingly rare) from non-economic and natural ones.

The ecclesiastical cultural heritage, given its widespread distribution, is certainly a testimony, like other cultural heritage, of history, civilization and culture of a country, but at the same time also of people religiosity as used for liturgical or devotionals purposes.

Given the particular nature of these assets, the valuation approach aimed at their appraisal is also particular and targeted.

This contribution aims to increase knowledge about the evaluation of ecclesiastical cultural heritage, for which at present there is still a few attention, due to its marked atypical nature, uniqueness and complexity.

Keywords: Ecclesiastical heritage · Real estate appraisal · Contingent valuation

1 Introduction

The concept of ecclesiastical heritage, as well as that of cultural heritage, is not clear; it is not easy to find a definition able to make agreement on an economic and legal level.

F. Aragona—Freelance engineer.

© Springer Nature Switzerland AG 2020
O. Gervasi et al. (Eds.): ICCSA 2020, LNCS 12251, pp. 776–786, 2020.
https://doi.org/10.1007/978-3-030-58808-3_56

All assets belonging to a public juridical person referable to the Church constitute the category of ecclesiastical goods. All other assets are excluded, whether they belong to faithful individually or to private legal people.

As López Alarcón states, the ecclesiastical condition of a good derives from the condition of the subject who owns it [1]. No property has the universal Church as direct owner, even if each of them, as managed by public juridical persons, belongs to the Church.

All the ecclesiastical assets constitute the ecclesiastical heritage, which is subject to administrative law [2].

Ecclesiastical goods represent a very conspicuous part, both from a quantitative and qualitative point of view, of the country's historical-artistic heritage and in recent decades the sensitivity for the protection of this heritage has greatly increased, both in society and in the civil and ecclesiastical institutions. Ecclesiastical cultural goods have a double significance. On the one hand they are evidence of history, civilization and culture of a country, on the other they retain a religious function, as they are intended for liturgical or devotional purposes [3].

The economic analysis of ecclesiastical cultural goods, regardless of the actual legal nature of existing property rights, presupposes their qualification as common property. The ecclesiastical cultural goods are therefore neither private nor public but they are certainly unique and unproducible. These characteristics, analogous to environmental assets, allow to extend some evaluation methods, elaborated by the environmental economy, to the evaluation of cultural heritage, which do not find in the market the suitable place for the formation of prices/values.

2   Research Background

Assigning an economic value to the heritage of art, history and culture to which ecclesiastical cultural heritage belongs is essential in order to promote conservation and enhancement, through diversified and targeted intervention measures [4, 5].

The first phase to be carried out in the economic evaluation of an ecclesiastical cultural asset consists in identifying a notion of value attributable to the use of the asset itself.

The reasons why an ecclesiastical cultural asset is used are essentially summarized:

1. in the celebration of sacred rites;
2. in the manifestation of a spiritual search;
3. in visiting works of art;
4. in historical research;
5. in anthropological studies.

To attribute a value to ecclesiastical cultural properties, we could consider valuations by the individual protagonists of religious celebrations or of the so-called "cultural/spiritual" tourism could be used. However, there would be biased data, as the sample is not complete but represents an elite of people who find it pleasant to stay in these places. To make this assessment operation more complex is also, to be taken into due account, the beneficial emotional effect that these assets determine for individuals and for which effect there is no market compensation. This beneficial emotional effect attributable to the concept of externality determines in individuals a different conception of the property and therefore a diversified attribution of value. Ecclesiastical cultural goods, unlike public goods in general, are placed in an intermediate stage between consumer goods for private use only and pure public goods or those that can be used by all without particular restrictions. You almost never pay an entrance ticket to enter a structure such as a church or an oratory unlike a museum or a historic building; rarely these goods are protected or sustained in conservation by visitors as are the goods of the State. This is because ecclesiastical cultural goods are intended as consumer goods and not for profit, although they perform functions for the benefit of visitors.

The analysis aimed at identifying the possible values attributable to such an asset, foresees irreproducibility and often also the plurality of functions in addition to uniqueness.

The study of the appraisal aspects of ecclesiastical cultural heritage involves various areas which, each to their own extent, contribute to defining the overall value of these goods. Assigning a value to ecclesiastical cultural heritage necessarily implies attributing a value to the externalities or benefits they produce and to the functions they perform as instruments of growth and as treasures of history and art.

It is necessary to specify that the cultural asset, a category in which ecclesiastical cultural goods are placed, does not differ much from others, such as public or environmental goods, with regard to the policies and to the interventions of restoration or enhancement to be implemented. Traditional economic theory then does not attribute a value to goods without a market price. The need therefore arises to promote, for the purposes of the development and conservation of these assets, experimental evaluation strategies that can offer concrete support for political decisions.

Already in the 1920s, in the USA, there was an experimentation of evaluation methods able of transforming the implicit preferences that people show, with their behavior regarding these goods, into monetary value [6].

The economic evaluation of ecclesiastical cultural goods must aim to evaluate the benefits that current or future consumers obtain from the goods themselves. This evaluation is based, unlike the cultural evaluation that relies on the judgment of critics and experts, on the individual preferences of citizens. Economic theory in fact reveals that collective well-being is determined by the satisfaction of individual preferences since a benefit is defined as anything that increases individual well-being and a cost anything that reduces it [7, 8].

In order to be able to translate the value of an ecclesial cultural asset into monetary terms, the theory of total economic value can be used [9].

It is a theory that takes into consideration two main values that every good possesses: the instrumental value or use value and the non-use value [10].

In summary, we can express the total economic value (TEV) of an ecclesiastical cultural asset through the following identity:

$$TEV = Use\ value + Non\ use\ value \tag{1}$$

in which the non-use value refers to:

- option value, linked to the desire to ensure the availability of the asset in future;
- existence value, linked to the possibility of preserving the good from possible destruction;
- bequest value, which has as its precise reference the possibility of taking advantage of a certain asset by future generations.

This mathematical exemplification of course cannot be exhaustive to valuate the economic value of an ecclesiastical cultural asset; a general analysis of copious literature reveals, in fact, that over time there have been numerous proposals for assessing quality. In principle, the evaluation of the qualitative values of a resource, be it historical, monumental, artistic, environmental or religious, can be carried out according to two fundamental approaches: the economic approach and the extra-economic approach [11, 12]. The first one is usually used in order to express economic convenience judgment in transformation projects, while the second one, because not expressing in monetary terms, is used to drive design choices.

The economic approach expresses the value of resources in terms of willingness to pay and market price and takes place in a context that considers real demand and supply.

The economic approach for the evaluation of the qualitative values of the architectural-environmental and cultural resources is essentially based on three quantitative procedures aimed at estimating the willingness to pay and the market price. This approach carries out monetary measures of qualitative values. In any case, these are "account prices", that is to say, prices set for the purposes of the valuation analysis that takes place on alternative projects and for selection purposes. The account price is a function of the objectives set by the valuation, therefore it increases in terms of revenues and decreases in terms of costs if it pursues the objectives and operates vice versa when it moves away from it. The rational basis of the approach is basically the same as the extra-economic approach, but the advantage of the economic approach consists in the fact that it is expressed in a single numerary and allows an immediate comparison with the costs to carry out the different alternative projects. Furthermore, an immediate comparison can be made between the results of the projects on the one hand and the budget and financing models of the projects themselves on the other; between productive investment projects and intervention projects that invest qualitative resources or qualitative values of resources.

The procedures for evaluating architectural, environmental and cultural resources are: the hedonic price method (HPM), the travel cost method (TVM), the contingent valuation method (CVM). The first two procedures analyze the revealed preferences with the consumption of private goods complementary to the use of the qualitative resource [13, 14], the third procedure instead examines the preferences expressed (stated preferences) in a specially built virtual market [15–17].

The extra-economic approach is independent of this real context and expresses the value through the analysis of the individual attributes. This approach finds its starting point in the theory of multi-attribute utility. This theory, which is the basis of multi-criteria analyzes, is closely connected to the processes of choice and starts from the general idea that in decisions what tends to maximize is not only economic utility but also a set of other elements or criteria. This approach developed for the first time by Lancaster [18] and then by Zeleny [19], Keeney and Raiffa [20], approaches the real way of reasoning and evaluating, when make choices by each individual. In the selection processes, the multicriteria analysis relates in particular to the evaluation aimed at planning. The assessment procedures based on multi-criteria analysis are able to simultaneously take into account several objectives and consequently several judgment criteria, without necessarily providing for monetary quantification.

The purpose of this research is to propose a quantitative approach in order to evaluate the economic convenience in the protection and enhancement of ecclesiastical goods, to be achieved through an economic approach. Although ecclesiastical cultural goods have their own value (sometimes even considerable), they are not considered by market mechanisms, which are typical of private goods. In the absence of a price system, generally, the economic value of a cultural asset can be measured by the quantity of money corresponding to the variation in individual well-being (utility), caused by a change in its availability. A monetary valuation of this variation can be provided by the consumer surplus which is defined as the difference between the maximum amount of money that consumers are willing to pay for a good (gross value) and the amount of money they actually spend on the good itself (financial value) [21, 22]. It can therefore be said that the gross monetary value reflects the total utility of an asset, while the financial value represents the sacrifice or disutility to obtain it. Therefore, by subtracting from the measure of utility that of disutility, an indication of the net utility deductible from the consumption of the good is obtained.

3 Case Study

3.1 Property Description

The case study refers to the conservation and/or transformation project of the Episcopal Seminary in San Marco Argentano (CS-Italy).

The preference granted to the Episcopal Seminary of San Marco Argentano (CS) is generated both by the centrality of the property with respect to the territorial system and by its ability to testify the local and diocesan civilization in its crucial historical phases. It is representative of many buildings in the Italian dioceses. It is, in fact, an ancient building, which plays a significant role in the perception of well-being by the community of reference, and which requires an onerous intervention of protection and enhancement (Fig. 1).

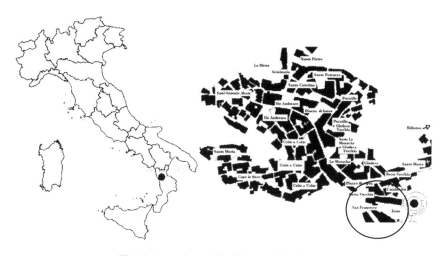

Fig. 1. Location of the Episcopal Seminary

The ownership of the Church and the protective action undertaken by the curia itself, as well as the provision of public benefits that the presence of the seminary implies, reveal the belonging of this property to the category of ecclesiastical goods.

This property represents that huge building complexes of high architectural and cultural value, first used as places of spiritual and cultural training for seminarians, which today are oversized compared to the number of future priests and therefore they are largely unused and subject to physical and structural deterioration. This type of architectural/cultural asset therefore represents a territorial resource almost completely unused. The preference granted to the Episcopal Seminary of San Marco Argentano (CS) is generated both by the centrality of the good with respect to the territorial system and by its characteristics that testifies the local and diocesan civilization in its crucial historical phases.

The territory of the municipality of San Marco Argentano extends for about 80 km^2 between the valley of the Crati river and the Thyrrenian mountain range. The ancient center develops on a hill (416 m s.l.m.) which rises from the valley of the Fullone river; in the surrounding flat areas modern settlements and industrial area develop.

The new Episcopal Seminary appears in the classic form with a church, cloister and overlying rooms.

The Seminary of San Marco Argentano building is still in excellent condition; the exterior of the seminar is welcoming with large green spaces for students' meditation and reflection. In the last period, a sports facility was also built welcoming not only seminarians but also young people from San Marco parishes and the whole diocese.

A reserved area inside the Seminary houses the nuns who generously offer the necessary care for the canteen service and the cleaning of the entire building (Fig. 2).

Fig. 2. The Episcopal Seminary exterior

3.2   Methods

In order to appraise the case study property, we've decided to the Contingent Valuation, proceeding as follows:

- drafting of the survey;
- sampling plan;
- conducting interviews;
- statistical data analysis;
- appraisal of the property value.

The survey has been structured in order to acquire information regarding: personal characteristics of the interviewed people (gender, age, provenance, educational qualifications, interests, income bracket, etc.); information relating to the interview conditions (day of the week, time, weather conditions, etc.); willingness to pay.

In particular, it has been asked willingness to pay for a una tantum voluntary contribution linked to the recovery and refunctionalization of the asset. An informative book containing photos and descriptions of the proposed action intervention is presented to the interviewees.

A pilot survey was first carried out, 20 individuals of both sexes residing in the Municipality of San Marco Argentano participated, after which the definitive survey was launched, held in the months of May–June 2019, conducting 210 direct interviews.

The lack of a place where to receive the interviewees required the preliminary elaboration of an information book, equipped with images and brief descriptions useful for the description of the transformation scenario.

The investigation was administered by the three people, properly educated, who directly approached the respondents in the seminar places, at the workplace, at their home or simply on the street. Given the numerous refusals in the initial phase of the investigation, involvement was favored by the issue of a participation certificate. The duration of the interview was on average 25 min.

The sampling plan adopted to carry out the survey is a proportional stratified type one, and it provided for the stratification of the sample in 6 sub-samples of 35 units.

With regard to the method of investigation for the acquisition of data regarding the willingness to pay, it was decided to use the double bounded dichotomous choice technique, following a 4-level model. Table 1 shows the offer vector used during the survey. The highest level of offer (one-time payment of € 500.00) can be considered a plausible price limit, while the lowest level (one-time payment of € 5.00) can be considered an amount that any respondent can pay.

The vector price varied in the sample. The 210 unit sample is divided into six sub-samples of 35 units each. Each sub/sample is associated with a vector price for which it is asked to express willingness to pay.

Table 1. Price offer vector

First proposed price	Lowest proposed price	Highest proposed price
10,00	5,00	20,00
25,00	10,00	50,00
50,00	25,00	100,00
100,00	50,00	150,00
150,00	100,00	250,00
250,00	150,00	500,00

3.3   Results

The results of the statistical analysis allow us to state that the sample investigated to carry out the investigation has the characteristics established by the prepared sampling plan. The results of the statistical analysis allow to identify the main characteristics of the question:

a) only 16 out of 210 individuals have showed a null willingness to pay;
b) the population is resident in 94% of cases in San Marco Argentano (CS);
c) the respondents are equally divided in males and females and they have an average age of 47;
d) the interviewees only attended compulsory school in more than 27% of cases, 35% have a high school diploma, while 38% have obtained a university degree;
e) the interviewees are largely believers (about 94%)
f) the interviewee is full-time employed in over 30% of the observations, more than 11% are looking for employment while over 15% are retired;
g) the respondents declare that their average family income is around € 1,750.00 per month with a percentage of around 21% declaring an income of around € 1,625.00 and € 2,125.00 and some particular situations whose income reaches € 4,000.00 or just € 37.50;
h) 80% of the survey participants usually read newspapers, and about 64% of them read the cultural and leisure page with interest;
i) 55% of the survey participants watch TV and radio reports and documentaries on cultural and/or religious themes;
j) 45% of the respondents read books on subjects concerning monuments, places of worship, museums and historical centers;
k) the entire sample interviewed is aware of the existence of the Seminar;
l) over 94% of the survey participants visited or viewed it at close range;

m) the protection of the Episcopal Seminary to continue promoting the missionary nature of the Church (religious value) is important for 38% people or very important (62%);

n) for most of the interviewees it is important or very important to keep the Seminar both to allow San Marco Argentano citizens to enjoy it everyday (value of direct use) and that of future generations (legacy value) and to ensure the opportunity of development, cultural growth and employment for the local community (indirect use value).

The demand curve was estimated by regressing the accumulated frequencies of the interviews on the corresponding willingness to pay (Table 2).

Table 2. Detected frequencies in relation to WTP

WTP (€)	Sub-sample dimension	Frequencies (yes)	Frequencies (yes) %
10,00	35	33	27,00
25,00	35	30	24,60
50,00	35	31	25,40
100,00	35	22	18,00
150,00	35	5	4,10
250,00	35	1	0,80
Total	210	122	100

Among the recurring functional forms, the logarithmic one has proven to be able to interpret the data in a better way, providing a value of the determination coefficient R^2 equal to 0.958.

The functional relationship that binds the dependent variable WTP to the independent variable Fc is the following one:

$$WTP = 287,99 - 44,80 lnFc, \tag{2}$$

With reference to the chosen sample, the surplus generated by the protection and enhancement of the Episcopal Seminary is estimated by integrating the truncated demand function between zero and the maximum valid frequency.

$$\int_0^{194}(287,99 - 44,80 lnF)dF = € 18.777,25. \tag{3}$$

The pro capite surplus is given by:

$$\int_0^{194}(287,99 - 44,80 lnF)dF : 194 = € 96,79. \tag{4}$$

The surplus was then extended to the population residing in the catchment area (7,590 units), considering a reduction factor based on the incidence of those who, within the sample of respondents, declared they were not willing to pay.

In this way, the economic value attributed to the protection and enhancement of the asset was estimated at € 678,663.82.

The utility flow supplied by the asset can be considered plausible for the purpose of its inclusion in a neo-classical economic analysis. In fact, subsequently, in order to determine the economic convenience of the public investment, the economic value thus calculated was taken as an ex ante measure of the net social benefits associated with the protection and enhancement project.

The approach linked to the evaluation of the consumer surplus seems to suggest new perspectives as a support to the public decision-maker in the classification of "goods of excellence", thus reconciling the cultural instances expressed by the community with the need for rationalization in the use of limited financial resources.

4 Conclusion

Ecclesiastical assets represent a conspicuous part, in terms of quantity and quality, of the historical and artistic heritage. In recent decades, the issues of conservation and protection of cultural heritage in general and ecclesiastical cultural heritage in particular have been at the center of debates promoted by both civil and ecclesiastical institutions.

Conserving means moving in a multidisciplinary field in which the economic criterion cannot be ignored. The cultural heritage, testimony to historical or artistic or religious values, is also an economic asset, having the characteristics of utility, usability and limited availability that distinguish economic assets from non-economic and natural goods.

Given the particular nature of these assets, the evaluation approach aimed at their valuation is also particular and targeted.

The appraisal of these properties is centered around the search for the monetary evaluation of the qualitative values that are generally found in cultural heritage in general and all other aspects connected to the religious function that ecclesiastical cultural heritage performs. A not negligible aspect for valuation purposes is the lack of a market for ecclesiastical cultural properties, which are usually not freely traded.

The valuation of ecclesiastical cultural heritage can be approached recurring to different procedures reported in literature, mainly implemented in the context of historical-cultural and environmental heritage. Among the existing methods, the Contingent Valuation Method (CVM) appears to be the most suitable for the monetary valuation of the market value of ecclesiastical cultural heritage.

This study just marks a first step in the path of issues concerning the valuation of ecclesiastical cultural heritage by providing scientific-estimative answers to the public decision-making processes of programming which require, with increasing frequency, the availability of suitable tools to provide a measurable sign, in monetary terms, of the collective benefits resulting from the recovery, reuse and management of real estate resources characterized by historical and architectural values.

References

1. Lopez Alacron, M.: La administración de los bienes eclesiásticos. Ius Canonicum **24**(47) (1984)
2. De Paolis, V.: I beni temporali nel Codice di Diritto Canonico. Monitor Ecclesiasticus 111 (1986)
3. Feliciani, G.: I beni culturali ecclesiastici. Dall'Accordo di revisione del Concordato lateranense alla recente Intesa. Vita e pensiero **80**, 493–507 (1997)
4. Mossetto, G., Vecco, M. (eds.): Economia del Patrimonio Monumentale. Franco Angeli, Venezia (2001)
5. Mehrling Perry, G.: The Money Interest and the Public Interest; American Monetary Thought, 1920–1970. Harvard University Press, Cambridge (1997)
6. Morano, P., Locurcio, M., Tajani, F.: Cultural heritage valorization: an application of ahp for the choice of the highest and best use. Procedia Soc. Behav. Sci. **223**, 952–959 (2016)
7. Mazzanti, M.: Cultural Heritage as multi-dimensional, multi-value and multi-attribute economic good: toward a new frame work for economic analysis and valuation. J. Soc.-Econ. **31**, 529–558 (2002)
8. Feliciani, G.: Beni culturali di interesse religioso. Il Mulino, Bologna (1995)
9. Girard, L.F.: Risorse architettoniche e culturali. Valutazioni e strategie di conservazione. FrancoAngeli, Milano (1990)
10. Bergstrom, J.C.: Concepts and measures of the economic value of environmental quality: a review. J. Environ. Manage. **31**, 215–228 (1990)
11. Stellin, G., Rosato, P.: La valutazione economica dei beni ambientali. UTET, Torino (1998)
12. Wright, W.C.C., Eppink, F.V.: Drivers of heritage value: a meta-analysis of monetary valuation studies of cultural heritage. Ecol. Econ. **130**, 277–284 (2016)
13. Campbell, H., Brown, R.: Benefit-Cost Analysis. Financial and Economic Appraisal using Spreadsheets. Cambridge University Press, Cambridge (2003)
14. Garrod, G.D., Willis, K.G.: Valuing goods' characteristics: an application of the hedonic price method to environmental attributes. J. Environ. Manage. **34**(1), 59–76 (1992)
15. Mitchell, R.C., Carson, R.I.: Using surveys to value public good. The contingent valuation method. Resources for the future, Washington (1989)
16. Cummings, R.J., Brookshire, D.S., Schulze, W.D.: Valuing Environmental Goods. An Assessment of the Contingent Valuation Method. Rowman and Littlefield, Totowa (1986)
17. Throsby, D.: Determining the value of cultural goods: How much (or how little) does contingent valuation tell us? J. Cult. Econ. **27**(3-4), 275–285 (2003)
18. Lancaster, K.J.: A new approach to consumer theory. J. Polit. Econ. **74**(2), 132–157 (1966)
19. Zeleney, M.: Multiple Criteria Decision Making. Mc Graw Hill, New York (1982)
20. Keeney, R.L., Raiffa, H.: Decisions with Multiple Objectives: Preferences and Value Trade-Offs. Cambridge University Press, Cambridge (1993)
21. Hicks, J.R.: The four consumer's surpluses. Rev. Econ. Stud. **11**(1), 31–41 (1943)
22. Bedate, A., Herrero, L.C., Sanz, J.Á.: Economic valuation of the cultural heritage: application to four case studies in Spain. J. Cult. Heritage **5**(1), 101–111 (2004)

Assessment of Quality of Life in Residential Estates in Lodz

Małgorzata Hanzl[(✉)] , Jakub Misiak , and Karolina Grela

Institute of Architecture and Town Planning, Lodz University of Technology,
Al. Politechniki 6A, 90-240 Lodz, Poland
mhanzl@p.lodz.pl

Abstract. Quality of life remains a principal concern of contemporary urban planning and design. This keyword repeats in documents of global organisations such as the UN and the policy documents of the European Union. To satisfy this requirement, we need proper evaluation framework which would enable planners and all involved stakeholders to make the appropriate decisions and transform the human living environment in the right directions. In the current paper, we have reviewed several assessment schemes used for urban benchmarking looking for methods which might indicate how to improve the quality of life of residents. The article uses the case studies of three housing estates in Lodz, Poland, which we assess looking for recommendations for future planning. The evaluation scheme addresses several features considered essential for citizens' well-being. We present the method of how to combine them to obtain a consolidated system of indicators. The current study is the first step towards a more developed framework of evaluation for the quality of life in urban planning.

Keywords: Quality of life · Assessment · Urban planning · Indicators

1 Introduction

Increasingly, we begin to pay attention to how, where, and in what conditions we live. First of all, both mental and physical health and care for it have become essential to us. The level of our awareness has increased significantly, and, as a result, we have started, as people, to pay more attention to the space that surrounds us, care for the natural environment or consider the quality of the air we breathe.

The notion of quality of life is also prevalent among researchers and scientists who have devoted many publications and scientific papers to this issue. It gained particular popularity in the area of public health, and in social science - it was from there that it migrated to the language of modern urban planning, but also into humanistic geography or economics. The form of rankings and ways the quality of life is measured reflect these disciplinary associations. The ambiguity of this concept has created many definitions that describe its meaning emphasising different aspects of reality, depending on the approach, but each of them comes down to asking the question of 'how people live?' (Petelewicz and Drabowicz 2016, p. 7).

© Springer Nature Switzerland AG 2020
O. Gervasi et al. (Eds.): ICCSA 2020, LNCS 12251, pp. 787–804, 2020.
https://doi.org/10.1007/978-3-030-58808-3_57

Despite the technological progress, formal aspects of development, such as the distribution of spatial structures, their proportions and dimensions, the nature and form of the relationship between them are critical and require optimisation. We refer here both to the forms of the public realm: squares, streets, parks, informal urban vegetation including the activities they accommodate, and to the built structures - their volume, shape, parcellation of lots and land uses. All these themes belong to the domain of urbanism.

The current study examines the quality of life in selected Lodz housing estates and tries to give recommendations on how to improve the living conditions of their inhabitants by using tools from the field of urban planning and design, depending on the scale of the concerned site and applied measures. After this introduction, we briefly present the research aims and scope and then review the current literature on the topic of quality of life in order to define to principal methods of assessment. Next, we present the available rankings and their methodology. Further, we proceed with the presentation of the three case studies of housing estates in Lodz, Poland. The discussion of the quality of life evaluation conducted for all of them follows along with the recommendations. Conclusions and paths for future research ventures wrap up the article.

2   Aims and Scope

In order to answer the question: how people live in selected Lodz housing estates and how to improve their living conditions by using urban planning and design tools, we have at first looked at the current way of assessment of the quality of life. The topicality of the presented issue and their multithreading requires comprehensive analyses of publications and scientific papers, evaluation standards and assessment schemes regarding the quality of life. The research outcomes served to prepare the evaluation matrix, which we have applied to the analysis of the case studies. A survey conducted among residents of the three selected housing estates completed the quantitative and qualitative assessment.

In scientific terms, the project aims to collect, organise and analyse data, which represent three parts of the city as a structure and as a system, including their evolution.

The more general practical goal of the study is to create a tool that supports improving urban design and rationalisation of functional solutions in Lodz housing estates based on norms and standards for assessing the quality of life. On this basis, the tool shall assist in transforming urban structures to improve living conditions.

The spatial scope of work includes fragments of three housing estates located in the City of Łódź:

- Dąbrowa located in the southern part of the city in the Górna district,
- Montwiłł-Mirecki housing estate in the western part of the city in the Polesie district,
- Radogoszcz - East in the northern part of the city in the Bałuty district.

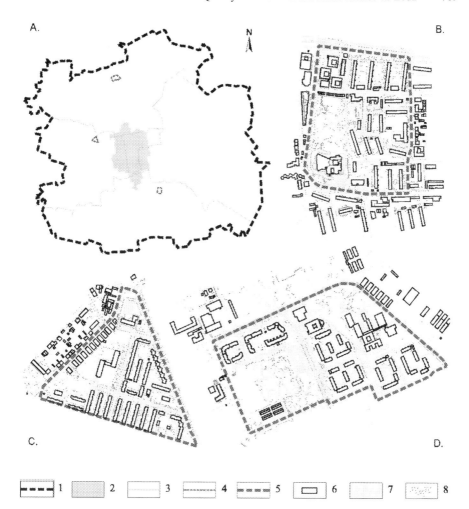

Fig. 1. The research extents: A. the locations of the three housing estates examined in the current study; B. Dąbrowa housing estate; C. Montwiłł-Mirecki housing estate; D. Radogoszcz Wschód housing estate. Symbols: 1. City administrative borders, 2. Metropolitan zone, 3. Districts; 4, 5. Problem areas, 6. Buildings, 7. Roads and path, 8. Existing trees.

Selected fragments analysed in this study are comparable in terms of surface and functional systems. Similar characteristics enable us to compile the results of the study and conduct a comparative analysis. Moreover, we assume that based on the results of the analysis, it shall be possible to determine joint recommendations for housing estates in Lodz. Referring to the temporal scope of work, it covers the period from March 2019 to February 2020 (Grela 2020).

3   Research Background

The topic of quality of life has been extensively researched for the past decades in several fields, of which the most representative are public health (Guyatt GH 1993, Carr 2001, Devlin 2018) and social science (Testa and Simonson 1996, Kruk et al. 2018). As the first step, we conducted an exploratory bibliometric analysis of the SCOPUS database that confirms this observation (Fig. 2A) and shows quantities of yearly publications as high as 38 769 and rising trend lines. The analysis referred to the presence of the string 'quality of life' in article's title or keywords. Contrary to the popularity of the subject in the disciplines mentioned above, the urban planning and design research observe much less abundant development; however, trends are growing too (Fig. 2B). In this case, the measures included asking a question related to the previous condition in conjunction with the presence of expressions' urban design' or 'urban planning' anywhere within the body of the article. The trends prove the increasing interests in the topic among theoreticians and practitioners of city planning.

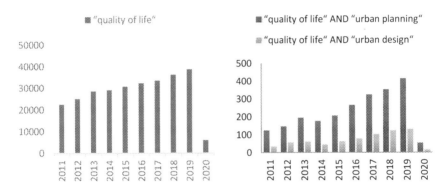

Fig. 2. The results of an exploratory bibliometric analysis of Scopus with regard to A. quantities of publications on quality of life (string 'quality of life' in the title or keywords), B. the conjunction of two research interests: quality of life and urban planning or design.

3.1   Quality of Life – Research Review

The state of knowledge in the subject might be systematised as follows: the first group covers extensive research on the subject, and the second – audits and rankings by various national and international bodies and institutions. Moreover, a robust body of literature includes government, European and global documents, legislation and manuals by UN (2017), UN-Habitat, EU, WHO and other institutions. The European Union's and United Nations' priority is to improve the quality of life. SDG-3 directly focuses on ensuring healthy lives and promoting "well-being for all at all ages" (UN 2016); other goals also indirectly address this topic. Quality of life remains one of the critical issues of the New Urban Agenda 2030 (UN 2017). These goals are to be achieved thanks to, among others, the focus on optimised urban form (UN 2016a).

Moreover, the requirement to achieve the collaboration and support of all concerned stakeholders calls for the explicit assessment framework (UN-Habitat 2014). European Union actively implements the program 'A Europe that protects its citizens and enhances their quality of life' (EU 2019), which directly refers to the environmental conditions.

As researchers emphasise, the socio-economic and environmental conditions of quality of life should be measured, but our understanding should not be limited to just that. In order to fully grasp the complexity of the urban environment, we need to picture specific assessment against the backdrop of other related phenomena, such as, among others, sustainability and climate resilience (Schmitt 2015, Naboni 2019). Quality of life and well-being are the elements which should play a role in a broader integrated system of the parametric data used for regenerative design, next to the vulnerability and physical environment assessment (Naboni 2019).

3.2   Indicators and Measurements

Planners and decision-makers use indicators to evaluate specific phenomena and to assess the ongoing transformations and implementation of defined goals. They need such explicit information to support decisions and scenarios in the political process. Specific measures require established standards in order to become accepted and hence useful. The shared framework enables the common understanding of a concept and hence, communication between all parties.

Indicators stem directly from parameters which serve as input data. Coefficients, by turn, are derived from indicators; they are mathematically combined sets of values. These parameters portray and provide the measures of the phenomena which we study. An established set of indicators offers a normative framework for the assessment of the planning process and the comparison of the outcomes. UNECE (2019) proposes a classification of indicators into the following groups: leading indicators (LI) which use objective values, and sentiment indicators (SI) – based on surveys and evaluation by the concerned citizens.

Based on the environmental conditions, indicators should offer clear information on the current trends which might be understood by all stakeholders (World Bank 1997). This requirement assumes the need for the clarity of synthesis and might involve some simplifications. Verbruggen and Kuik (1991) emphasise the possible need for generalisation in order to increase conciseness. The complexity of urban issues makes the approximation inherent part of indicators system.

In the current elaboration, we analyse the liveability based on the criteria used by quality of life assessment schemes. The most recognised are the following ones:

- Global liveability rankings:
  - Human Development Index (HDI), United Nations Development Programme (UNDP) – since 1990,
  - Global Liveability Index (GLI) by the Economist Intelligence Unit (EIU),
  - Quality of Living Survey (QoLS) by Mercer,
  - Gallup & Healthways Well Being Index (G&H WBI), since 2012,
  - Worldwide Cost of Living Survey (WCoLS) by EIU,
  - Quality of Life Survey by Monocle (the lifestyle Magazine), since 2006,
  - World Development Indicators (WDI) by World Bank,
  - Global Liveable and Smart Cities Index (GLSCI) by the Asia Competitiveness Institute,
  - TomTom Traffic Index;
- European liveability rankings:
  - Urban Audit (UA), since 1997 by the European Statistical System (ESS) and Directorate-General for Regional and Urban Policy of European Commission,
  - Quality of Life in European Cities (QoLiEC) by the European Commission (EC) Eurobarometer,
  - the European Quality of Life Survey (EQLS) by the European Foundation for the Improvement of Living and Working Conditions (Eurofound),
  - EU-SILC – The European Union Statistics on Income and Living Conditions (Eurostat),
  - European green city index (EGCI) by EIU and Siemens,
  - European Cities Monitor (ECM) by Cushman & Wakefield (C&W);
- Polish rankings:
  - Quality of life in Poland. 2017 edition,
  - Public opinion research on quality of life assessment in Lodz and residents' expectations regarding the budget of the City, 2007 (Lodz QoLA),
  - Eurostat's Urban Audit in Kielce in 2009.

The criteria which define liveability according to specific rankings have been listed in Table 1. The official rankings use the coefficients that consist of several criteria and many indicators to define the quality of life in cities (Schmitt et al. 2015). Even though they measure the same phenomena, the approaches differ, which makes explicit the various understanding of the concept.

Table 1. Fields covered in specific rankings; the table taken and adapted from (Grela 2020)

	Public health	Knowledge and science	Standard of living	Culture	Environment	Infrastructure &transport	Politics	Safety	Social interaction	Demography	Labour market
Global liveability rankings											
HDI by UNDP	X	X	X								
GLI by EIU	X	X		X	X	X		X			
QoLS by Mercer	X	X	X	X	X	X	X		X		
G&H WBI	X		X						X		
WCoLS by EIU			X								
QoLS by Monocle	X				X	X	X	X			X
WDI by World Bank	X	X			X	X	X	X		X	X
GLSCI by ACI	X	X	X		X	X	X	X	X	X	X
TomTom TI						X					
European liveability rankings											
UA by ESS	X	X	X	X	X			X	X	X	X
QoLiEC by EC			X		X	X					X
EQLS by Eurofound	X	X	X					X	X		X
EU-SILC by Eurostat	X		X							X	X
EGCI by EIU					X						
ECM by C&W						X					X
Polish liveability rankings											
QoL in Poland 2007	X	X	X		X			X	X		X
Lodz QoLA 2007			X								X
UA Kielce 2009					X						X

ISO norms developed methodologies to measure city services, sustainability and quality of life along with additional criteria and indicators for smart and resilient cities:

- ISO 37101:2017 'Sustainable development in communities – management system for sustainable development for use',
- ISO 37120:2018 'Sustainable cities and communities — Indicators for city services and quality of life',
- ISO 37122:2019 'Sustainable cities and communities – Indicators for smart cities',
- ISO 37123:2019 'Sustainable cities and communities – Indicators for resilient cities'.

4 Methodology – Case Study

In the current research, we applied two methods: the first one based on the established standards and indicators and the second, which uses a questionnaire to capture a subjective assessment by the inhabitants. The former we might classify as a leading indicator (LI), the latter – as sentiment indicator (SI) (UNECE 2019). Having reviewed and summarised the methods to evaluate the quality of life carried both by private and public institutions, we have selected a comprehensive set of indicators for the investigation of parts of the three housing estates. In the first group, principal sources of data include housing associations, internet sites of local government and statistical agencies, mobile applications, and direct measurement of physical structures. At this stage, due to the data availability, assessment based on social research has partially replaced formal data sources.

4.1 Analysis of Indicators

Based on the analysis of indicators above, we have selected the recurring themes, such as security, social and technical infrastructure, mobility, sport and recreation, environment and climate change, and, referring to the research profile, urban planning. Each of these areas gets a set of indicators (Fig. 3). Most of them follow the cited standards; some we have transformed and adapted to the conditions of the analysed housing estates. While referring to security, we propose two indicators: (A1) a sense of security at home and (A2) direct contact with crime expressed by the number of crime events.

The topic of social infrastructure also receives two indicators: (B1) the availability of services and their diversity represented by the Walkscore calculation algorithm and (B2) the percentage of public buildings accessible to people with disabilities. The measures which describe technical infrastructure include (C1) the share of the city's population served by sewage and (C2) number of connections to the gas network per 1 000 people (residential).

Mobility receives six indicators that refer to (D1) number of users using shared transport per 1000 inhabitants, (D2) length (km) of bicycle paths per 1000 inhabitants, (D3) number of bicycles available from bicycle-sharing services within the estate per 1000 inhabitants, (D4) length (km) f public transport system per 1,000 inhabitants, (D5) annual number of public transport journeys per inhabitant and (D6) number of parking spaces.

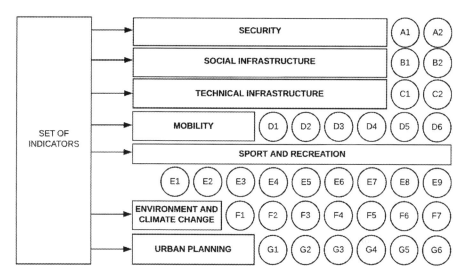

Fig. 3. Categories and indicators describing quality of life assumed in the current study.

We use nine indicators to depict sport and recreation. In this number six address urban vegetation: (E1) green areas (hectares) per 1000 inhabitants, (E2) satisfaction with recreational and green spaces, (E3) the share of people who live closer than 300 m from the park, (E4) the size of urban parks per inhabitant, (E5) the ratio of park area to the built-up area, (E6) housing estate vegetation per each housing estate owner. Remaining three indices feature (E7) housing share in the total estate area, and sites for recreation: (E8) the number of square meters of public indoor recreational space per inhabitant, (E9) the number of square meter of outdoor recreational space per inhabitant.

The next theme environment and climate change features six indicators: (F1) exposure to excessive noise (% of households), (F2) share of biologically active space, (F3) percentage of buildings constructed or renovated in the last five years following the principles of ecological construction, (F4) number of air quality monitoring stations in real-time per square kilometre ($km2$), (F5) share of energy from renewable sources, (F6) particulate matter 2,5 μm (PM 2.5) and (F7) particulate matter 10-μm concentration (PM10).

The last category covers spatial planning and management issues defined by a set of five indicators, including (G1) land use, (G2) coverage with local spatial development plans (%), (G3) yearly number of citizens involved in the planning process per 1000 inhabitants, (G4) cleanliness and aesthetics of the city, (G5) number of decisions on building conditions and decisions on public goal investment location per 1000 inhabitants, and (G6) number of building permits per 1000 inhabitants.

4.2   Social Research

The spatial scope of the questionnaire examining the citizens' opinions on the quality of life in the three housing estates covered the entire Radogoszcz-Wschód estate and two fragments of the two other estates: Montwiłł-Mirecki and Dąbrowa; the extents are shown in Fig. 1. The online survey was running between the first and the 20th of January 2020. The results are exploratory; thirty respondents of a variegated profile in each location answered the questions (Table 2). We used both open and closed questioned; additionally, there was a form for respondents to add their insights.

Moreover, to measure the value of features describing the quality of life, we have asked questions referring to the impact of these features. Respondents answered a question about the most relevant characteristic. We used the number and frequency of indications by respondents to assign weights to the specific indicators and categories. Each indicator, as defined in the previous section, received its weight to show a more objective picture of the housing estate, reflecting the current values and norms which refer to the quality of life of residents.

5   Results and Discussion

5.1   Assessment of the Quality of Life in the Three Housing Estates

The Table 2 includes the primary data on three selected fragments of housing estates. Table 3 presents the results of assessments of the three sites based on the quality of life standards. The research uses the data from the Geostatistics Portal (https://geo.stat.gov. pl/imap/), from the National Census of Population and Housing 2011.

Table 2.   Basic information about the problem areas

Housing estate	Dąbrowa	Montwiłł-Mirecki	Radogoszcz Wschód
Size of the area	17,16 ha	14,25 ha	24,15 ha
Population living in a given area	5253	1948	5754
Built up areas	3,38 ha	2,68 ha	2,86 ha
Total green areas	8,03 ha	5,91 ha	13,92 ha
Vegetation within institutions	1,96 ha	1,39 ha	1,38 ha
Vegetation in the housing estate	4,26 ha	4,52 ha	9,59 ha
Parks area	1,81 ha	–	2,94 ha

Security has been evaluated based on the two indicators: the personal sense of security using the questionnaire responses (A1) and direct contact with crime (A2), based on the National Hazard Safety Map data (mapa.geoportal.gov. pl/iMapLite/KMZBPublic.html). The dataset contains no incident threatening the safety of residents for these housing estates.

Table 3. The results of assessments of the three sites based on the quality of life standards

Theme	No.	Source	Dąbrowa	Montwiłł-Mirecki	Radogoszcz Wschód
Security	A1	GUS, 2017	3,15	3,58	4,2
	A2	Rokicka, 2013	0	0	0
Social infrastructure	B1	PN-ISO 37120	77	72	57
	B2	PN-ISO 37122	0,7	0,67	1
Technical infrastructure	C1	PN-ISO 37120	100%	100%	100%
	C2	PN-ISO 37120	100%	100%	100%
Mobility	D1	PN-ISO 37122	103	238	307
	D2	PN-ISO 37120	0 km	0 km	0 km
	D3	PN-ISO 37122	2,86	0	0
	D4	PN-ISO 37120	0,22 km	0,63 km	0,24 km
	D5	PN-ISO 37120	3,00	5,55	2,24
	D6	QoLS, Lodz 2012	5%	25%	25%
Sport and recreation	E1	PN-ISO 37120	1,53 ha	3,03 ha	2,42 ha
	E2	GUS, 2017	2,55	3,6	4,28
	E3	UA Kielce 2009	10%	0%	20%
	E4	UA Kielce 2009	3,45 m^2	0 m^2*	5,11 m^2
	E5	UA Kielce 2009	0,54	0 m^2*	1,03
	E6	UA Kielce 2009	8,11 m^2	23,18 m^2	16,67 m^2
	E7	UA Kielce 2009	24,82%	31,68%	39,73%
	E8	PN-ISO 37120	0 m^2	0 m^2	0 m^2
	E9	PN-ISO 37120	0,65 m^2	1,79 m^2	0,54 m^2

(*continued*)

Table 3. (*continued*)

Theme	No.	Source	Dąbrowa	Montwiłł-Mirecki	Radogoszcz Wschód
Environment and climate change	F1	PN-ISO 37120	5%	10%	0%
	F2	UA Kielce 2009	47%	41%	57,61%
	F3	ISO 37122	No info	1 building	No info
	F4	ISO 37122	No info	No info	No info
	F5	PN-ISO 37120	No info	No info	No info
	F6	PN-ISO 37120	17 $\mu g/m^3$	64 $\mu g/m^3$	9 $\mu g/m^3$
	F7	PN-ISO 37120	24 $\mu g/m^3$	70 $\mu g/m^3$	14 $\mu g/m^3$
Spatial planning and management	G1	UA Kielce 2009	0,19	0,19	0,12
	G2	UA Kielce 2009	0%	0%	0%
	G3	ISO 37122	No info	No info	No info
	G4	QoLS, Lodz 2012	2,68	2,73	3,73
	G5	UA Kielce 2009	4,19	3,08	2,26
	G6	UA Kielce 2009	4,38	11,29	1,56

Table 4. Walk Score punctation, Source: https://www.walkscore.com/, access: 7.02.2020

Walk score	Description	
90–100	Excellent accessibility	Everyday matters do not require a car
70–89	Good accessibility	Most things might be fixed on foot
50–69	Average accessibility	Some things might be fixed on foot
25–49	Car dependent	Most things require a car
0–24	Total car dependency	Nearly all things require a car to be fixed

Social infrastructure assessment involves two indicators: the availability of services (B1) and accessibility of public buildings for disabled people (B2). We have used the online Walk score tool to calculate the availability of services within walking distance. Table 4 reveals the threshold values of the method. The outcomes of the walk score evaluation are biased for the Radogoszcz Wschód estate due to the assumed research extent, which excludes the primary service hub. The results for the two other estates are similar, with slightly higher values for the Dąbrowa estate. The facilities for disabled are available in all public institutions in Radogoszcz Wschód; both other estates lack sufficient equipment. Both technical infrastructure conditions: sewage (C1) and gas access (C2) are satisfied for all three case study areas.

Six features define mobility. Shared modes of transportations (D1) include shared bikes systems which occur the most popular in Radogoszcz Wschód and the least used in Dąbrowa. The results might stem from the level of awareness of inhabitants, their age profile and the stations' availability. The length of bicycle paths (D2) in all three sites equals zero that offers the possibility for improvements. Numbers of bicycles available for sharing per 1000 inhabitants (D3) might be assessed only for Dąbrowa estate; the two other sites do not accommodate such facilities. Montwiłł-Mirecki housing estate features the highest length of the public transport network per 1000 inhabitants (D4), whereas the Dąbrowa estate is the least equipped. We acquired the two remaining mobility-related indicators from the survey. The annual number of travels by public transport per capita (D5) shows that in Montwiłł-Mirecki citizens use this mode of transportation the most. At the same time, the lowest levels affect housing estate Radogoszcz - East. The results might be affected by the age of respondents as well as the inhabitants' income levels. The question about the parking lots (D6) demonstrates that all three estates lack sufficient parking spaces according to the local citizens. The situation in Dąbrowa has been assessed as the worst. The reasons might be related to the rising numbers of individual vehicles and the inability to meet rapidly growing demands.

Nine indicators fall into sport and recreation category; of this number six address green areas and remaining three housing and recreation availability. The highest share of green space per 1000 residents (E1) occurs in Montwiłł-Mirecki estate, which stems from the direct vicinity of the Zdrowie Park. This parameter is reflected by the presence of the buffer zone shown in Fig. 7, which illustrates the catalogue of features being measured. The residents of Radogoszcz Wschód estate feel the most satisfied with recreation and green areas (E2), while these living in Dąbrowa are the least happy. The next indicator, which refers to the share of people living closer than 300 m from the public park (E3) confirms the above results. Montwiłł-Mirecki estate leads for the same reason as above, followed by Radogoszcz Wschód where 20% of residents live close to the park and Dąbrowa estate with 10% of inhabitants living within the buffer of 300 m from urban greenery. We calculated the share of urban greenery per capita (E4) and the ratio of park areas to built-up areas (E5) for two sites: Dąbrowa and Radogoszcz Wschód, the results repeat these described above, with a higher amount of public greenery in Radogoszcz Wschód estate. Instead, the Montwiłł-Mirecki estate features a vast urban park in the direct vicinity. Moreover, the same estate features the highest share of housing estate green per inhabitant (E6), followed again by Radogoszcz Wschód. The highest proportion of green areas within the estate to its total area (E7) we find in Radogoszcz Wschód and the lowest in the Dąbrowa estate. There is no indoor public recreational space in any of the three housing estates which means that the next indicator (E8) equals zero. The size of outdoor facilities per capita (E9) is the highest in Montwiłł-Mirecki estate and the lowest in Radogoszcz Wschód.

The category of environment and climate change includes seven indicators. We have estimated the exposure to excessive noise expressed by the share of affected households (F1) based on the acoustic maps and have found the highest levels in

Montwiłł-Mirecki estate – due to the proximity of railway, shooting range and a speedway stadium. The Radogoszcz Wschód estate remains the quietest one. The percentage of biologically active land (F2) amounts to over 40% in all sites, with the highest values in Radogoszcz Wschód and the lowest one in Montwiłł-Mirecki estate. The three next indicators (F3 to F5) have not been evaluated due to the limited availability of data. Air pollution results measured as concentrations of PM2,5 (F6) and PM10 (F7) were based on remote real-time air monitoring stations data and on the information made available by social activists – initiative MM and refer to 4th February 2020. The air in Radogoszcz Wschód estate was the least polluted, while in the Monwiłł Mirecki estate the air quality was the worst.

The built-up land ratio (G1) remains at similar levels for all the three sites, with the lowest parameter's values in Radogoszcz Wschód estate. The two next indicators (G2 and G3) are not available as the estates are not covered with local plans of spatial development. For the Monwiłł Mirecki estate, the procedure has recently started, and the citizens' participation in it is not yet possible to assess either. Citizens assess Radogoszcz Wschód estate' s cleanness and aesthetics (G4) the highest, the second rank belongs to Montwiłł-Mirecki estate. The most decisions on construction conditions per 1000 residents (G5) have been issued in Dąbrowa estate, while the least in Radogoszcz Wschód. The highest numbers of building permits (G6) were issued in Montwiłł-Mirecki estate, the lowest to Radogoszcz Wschód.

5.2   A Consolidated Assessment – Method

Each indicator was assigned its weight to learn which of the selected fragments of housing estates features the highest quality of life, both based on criteria derived from standards and social research. The weight of the indicator enables us to show a more objective picture of the housing estate, reflecting the current values and norms which refer to the quality of life of residents (Fig. 4). The weights were assigned based on social research.

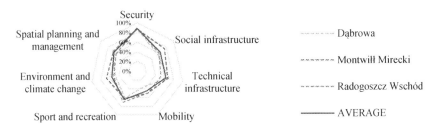

Fig. 4. The chart shows the weights collected through the social survey which reflect citizens' attitudes towards the indicators used in the study.

Respondents pointed out the most relevant indicators based on their preferences. We used the number and frequency of indications by respondents to assign weights to the specific indicators and categories. We summed up the quantities of choices per indicators within each class to obtain more consistent results. Next, we have brought all the values to a common denominator. Therefore, we normalised the results using the total of respondents multiplied by the number of indicators.

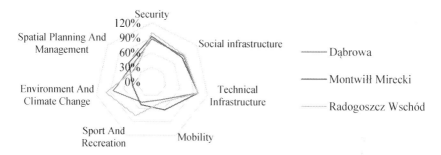

Fig. 5. The assessment results: The chart presents objective values based on the measurements and collected data

The chart (Fig. 5) which presents the assessment results uses percentage units. We normalised every parameter using the maximum for a given indicator. In case when a given category comprises more than one indicator, the diagram shows the arithmetic mean. For the negative values – noise and the concentrations of PM2,5 and PM10, we included the adverse numbers. In the first case, we used the maximum noise level based on the Directive relating to assessment and management of environmental noise (European Commission 2002), in the two other the respective levels of pollutants according to the Directive on ambient air quality and cleaner air for Europe (European Commission 2008). The results, which indicated exceeding the norms, received 0%, and those within the acceptable range – an evaluation proportional to the highest measured value. The chart (Fig. 6) presents the results multiplied by the weights counted in the previous step. Therefore, the last graph shows an assessment that takes into account the normative system established based on the local citizens' preferences.

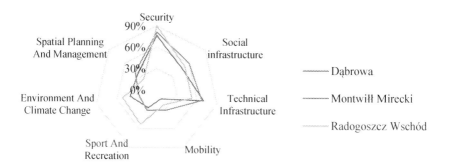

Fig. 6. The chart takes into account the values of weights based on the citizens opinions and values expressed in the survey.

As the last step, we have performed the summative assessment based on the average values of all the categories. The results show that residents of Radogoszcz Wschód estate perceive their quality of life the highest, whereas those who live in Dąbrowa the lowest.

The current research shows that the existing evaluation schemes which are used for benchmarking and rankings of cities from various perspectives address the quality of the physical environment, which is subject of interest in physical planning. We derived the data for the current study either from the social survey among citizens or the measurement of physical features of the environment. In both cases, the questions pertain to the actual qualities of the physical form. Figure 7 shows the elements of the physical environment for which the calculations have been performed.

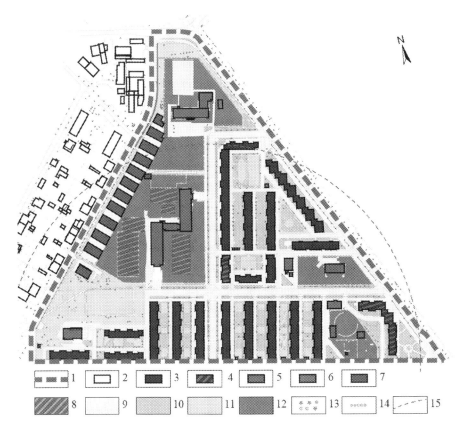

Fig. 7. The plan drawing which presents all the features being evaluated in the current study, using the example of Montwiłł-Mirecki housing estate. Symbols: 1. Study area, 2. Buildings, 3. Multifamily housing, 4. Residential and service development, 5. Services, 6. Technical infrastructure, 7. Garages, 8. Sport and recreation areas, 9. Transportation areas – impervious surfaces, 10. Organised green areas – parks, 11. housing estate vegetation, 12. Public institutions vegetation, 13. Deciduous and coniferous trees, 14. Rows of trees. 15. The 500 m buffer to the nearby park.

6   Conclusions and Future Research Avenues

The contemporary urban planning requires informed decisions which take into account the well-being and quality of life of urban citizens. The parametric tools might support such decisions and help in their optimisation (Naboni 2019). They offer the potential to integrate the directions for the environment transformations with citizens' bottom-up insights. In the current paper, we have reviewed the existing evaluation and benchmarking schemes in order to design a framework for the assessment of the quality of life as currently understood in these various approaches. The assessment included both the quantitative and qualitative elements, based either on the formal features of physical structures and on the residents' opinions.

The method has been tested for the evaluation of three housing estates in Lodz, Poland. We have compiled the assessment and obtained summative results which enabled the overall evaluation both of the objective physical and demographic characteristics of quality of life and the subjective perception by the residents. The latter has been done thanks to the usage of weights based on the survey results.

Furthermore, in the next phase, we envisage the following steps:

- more direct linking of the parameters to the GIS system to measure, analyse and compare the physical, social and economic data on city functioning,
- PPGIS-based surveys which would enable citizens to express their opinions within the GIS systems directly.

Using GIS in spatial planning can provide technical input to measuring the spatial quality of life and improve accuracy in decision-making. The results may contribute to public discussion and help involve residents in the spatial planning process.

References

Carr, A.J.: Measuring quality of life: is quality of life determined by expectations or experience? BMJ 322, 1240–1243 (2001). https://doi.org/10.1136/bmj.322.7296.1240

Devlin, N.J., et al.: Valuing health-related quality of life: an EQ-5D-5L value set for England. Health Econ. 27(1), 7–22 (2018). https://doi.org/10.1002/hec.3564

Eurofound: Challenges and prospects in the EU: quality of life and public services. Publications Office of the European Union, Luxembourg (2019). https://doi.org/10.2806/558895

European Commission: Directive 2002/49/EC of the European Parliament and of the Council of 25th June 2002 relating to assessment and management of environmental noise. OJ L 189, 18 July 2002, p. 12 (2002)

European Commission: Directive 2008/50/EC of the European Parliament and of the Council of 21st May 2008 on ambient air quality and cleaner air for Europe. OJ L 152, 11 June 2008, pp. 1–44 (2008)

European Political Strategy Centre: Sustainable Europe 2030. From Goals to Delivery, Brussels (2019)

Grela, K.: Assessment of quality of life in residential estates in Lodz. Unpublished master's thesis, Lodz University of Technology, Lodz (2020)

Guyatt, G.H.: Measuring health-related quality of life. Ann. Intern. Med. 118(8), 622 (1993). https://doi.org/10.7326/0003-4819-118-8-199304150-00009

ISO 37101: Sustainable development in communities – Management system for sustainable development – Requirements with guidance for use (2016)

ISO 37120: Sustainable cities and communities – Indicators for city services and quality of life (2018)

ISO 37122: Sustainable cities and communities — Indicators for smart cities (2019)

Kruk, M.E., et al.: High-quality health systems in the sustainable development goals era: time for a revolution. Lancet Glob. Health **6**(11), e1196–e1252 (2018). https://doi.org/10.1016/S2214-109X(18)30386-3

Kuik, O., Verbruggen, H.: In Search of Indicators of Sustainable Development. Kluwer Academic Publishers, Dordrecht (1991)

Testa, M.A., Simonson, D.C.: Assessment of quality-of-life outcomes. N. Engl. J. Med. **334**(13), 835–840 (1996). https://doi.org/10.1056/nejm199603283341306

Naboni, E., Havinga, L.: Regenerative Design in Digital Practice. A Handbook for the Built Environment. Eurac Research, Bolzano (2019)

Petelewicz, M., Drabowicz, T.: Jakość życia – globalnie i lokalnie. Pomiar i wizualizacja. University of Lodz, Lodz (2016)

Schmitt, G.: Information Cities, Zurich, Singapore (2015). https://doi.org/10.3929/ethz-a-010403946

UN-Habitat: Planning for Climate Change: A Strategic, Values-based Approach for Urban Planners. Nairobi, Kenya (2014)

UN-Habitat: Hot Cities: Battle-ground for Climate Change, United Nations (2011)

United Nations: Transforming our World: The 2030 Agenda for Sustainable Development. A/RES/70/1, New York, USA (2016)

United Nations: Habitat III New Urban Agenda Draft Outcome Document for Adoption in Quito (2016a)

United Nations: Habitat III New Urban Agenda, Quito, Ecuador (2017)

United Nations: The Sustainable Development Goals Report (2019)

United Nations: Transforming our World: The 2030 Agenda for Sustainable Development. Resolution adopted by the General Assembly on 25th September 2015. A/RES/70/1 (2015)

United Nations Economic Commission For Europe (UNECE): Guidelines On Producing Leading, Composite and Sentiment Indicators, Geneva (2019)

World Bank: World Development Indicators, Washington (1997)

Situated Emotions. The Role of the Soundscape in a Geo-Based Multimodal Application in the Field of Cultural Heritage

Letizia Bollini[1]([⋈]) ⓓ and Irene Della Fazia[2]

[1] Free University of Bozen-Bolzano, piazza Università, 1, 39100 Bolzano, Italy
`letizia.bollini@unibz.it`
[2] University of Milano-Bicocca, piazza dell'Ateneo Nuovo 1,
20126 Milano, Italy

Abstract. Cartography has traditionally been the privileged key to reading an urban territory. The form "map", in fact, translates in two-dimensionally and with an abstract language the complexity of the space that we perceived and interact with through our physicality. In this process of conceptualisation many of our visceral abilities must be distilled into a form of representation that elects vision as a privileged channel. The possibility of using mobile devices has widened the range of expression and the ways in which we orient people in the environment. But, well beyond this instrumental function, the use of the acoustic channel, for example, opens up design scenarios for the construction of an identity of places that involves users from the point of view of experiential richness. Generally used to give procedural indications and free the user in movements, the sound interface has in itself further potential for exploring the communicative design and reading of the anthropic environment. The paper proposes an interpretation of the urban landscape of the city of Milan and its historical transformation through the narrative key of one of the major Italian novels and the construction of soundscapes that gives back the emotional richness of reality. Based on a holophonic recording, the prototype of the mobile app explores the possibility of extending a space-contextual experience through a multimodal storytelling that generates "sound vision".

Keywords: Soundscape design · Holophonic interactions · Interaction and experience design

1 Introduction: Digital, Senses, Emotions

According to the book written in 2005 by Donald Norman [1], emotions are one of the most powerful drivers of our actions in the world. There are three possible levels of intervention of design: *visceral*, *behavioural*, and *reflective*. Most of the digital products and services are conceived to satisfy the second one fully, that means to solve practical problems and processes, giving useful a functional solution to people. The middle level is mainly addressed to the cognitive process and based on relatively simple everyday behaviour patterns. Nevertheless, it seems that this approach is not enough. The richness and complexity of our nature relay on many other assets: we are not just cognitive beings, but rather emotional and cultural ones. Besides, the way we experience the world around us and socially interact with others is profoundly rooted in our senses aimed to feel and respond to the physicality of the environment. On the one hand, the visceral component is how we react to sensory elements, i.e. what can be perceived by our senses: visual, acoustic, tactile, taste or olfactory. The primordial reactions to these stimuli have kept us safe. They let us "decide" in an almost immediate timespan what is good and what is bad, what is safe and what is dangerous for us and our survival. On the other hand, the reflective level pertains to our conscious response based on our thoughts, reflections, and previous experiences. All the three aspects coexist and influence mutually inhibiting or enhancing each other.

Nevertheless, many technologies aren't designed to face all these aspects, although strongly impacting our lives. In particular, the visceral component should be kept in higher consideration. Visual appearance, motion effects, sounds and sensory sensations – real or synesthetic – can play a huge role in digital interactions. Emotions have been often overlapped or confused with the blurred concept of beauty or aesthetic when discussing design issues and opposed to functionality [2]. To design for visceral level "is actually for affect – that is, eliciting the appropriate psychological or emotional response for a particular context – rather than for aesthetic alone." underline Alan Cooper [3]. That not the place to debate the different positions, but empirical and experimental studies conducted in the digital field have definitively shown that people find that what is beautiful is also useful. Specifically, Kurosu and Kashimura [4] introduced the concept of *apparent usability* – opposite to usability [5] – to describe the *pleasantness* that some interfaces were able to convey to users when compared to just *usable* ones. It has been Tratchinsky [6] together with Katz and Ikar [7] who explicitly referred to beautiful as component appreciated by people when using digital systems aimed to be experimentally tested. Emotional or visceral aspects, therefore, are essential and useful to users when facing the challenge to interact with complex systems and, in particular, with digital technologies. They are, in fact, one of the most powerful means of knowledge of the world.

The transition from "fixed position" – that means a desktop or laptop computer managed with a keyboard and a mouse/track-pad – to a *poketable* connection has profoundly changed the role of digital in our daily experience. Mobile devices – namely smartphones – are the mediator of our interaction with the world in an *on-life* [8] ecosystem. But the two-dimensional surface of displays tends to dramatically reduce the potentialities of physical interactions to a singular sense, the vision according to the Graphical User interface pattern and its language and metaphors. Touch devices, as well, use this sense to mediate through a flattened space our experience of the surrounding environment both physical and on-line. Together with the evolution of connection in mobility, new tools are gaining popularity, for instance, smart-glasses and VR visors. The investments made by big tech companies – Facebook in Oculus Drift and Microsoft with HoloLens – reveals a grooving niche opening new design opportunities and challenges in a blended environment where virtual, augmented and mixed reality are promising field to be explored. The revamp of virtual reality after the decline of the last almost 30 years – is opening the opportunity to create richer interactions. The 3D simulation (or based on real photos) let people explore a simulative space where reality can be replay or (re)invented [9]. The point is not if virtual is real or vice versa, but rather that virtuality is, again, involving our senses, proxemic and synesthetic perception. Besides, starting from the second generation of mobile devices up to smart home devices – such as Alexa in 2014 – in recent years voice interfaces [10] have opened the possibility to interact in a verbal and dialogic way, bringing the acoustic component among the project variables.

2 Soundscape

Environments also have their own sound language, i.e. they produce a significant and recognisable identity in the construction of the urban experience. Already in 1969 Southworth introduces the concept of soundscape in his research on the relationship between acoustic and visual perception of space: "Two aspects of the soundscape that appear particularly important in city perception were central to the study. First, we evaluated the identity of the sounds, including (a) the uniqueness or singularity of local sounds in relation to those of other city settings and (b) their informativeness or the extent to which a place's activity and spatial form were communicated by sound." [10]. The experimental study he conducted in those years is even more significant for the type of subjects involved – blind or deaf – which emphasise even more the close relationship between the different factors of urban perception. The theoretical definition of the concept of "soundscape", however, is attributed to Canadian composer Raymond Murray Schafer in the 1970s. It indicates everything that makes up the acoustic environment. The soundscape is made up of all the acoustic, natural and artificial resources within a given area from the environment. A sound "photograph" of the environment. According to Schafer, the soundscape is generally composed of 3 different elements:

- **Keynote sounds:** This is a musical term that identifies the key of a piece, not always audible "The character of the people living there". They are created by nature (geography and climate): wind, water, forests, plains, birds, insects, animals. In many urban areas, traffic has become the keynote sound.
- **Sound signals:** These are foreground sounds, which are listened to consciously; examples would be warning devices, bells, whistles, horns, sirens, etc.
- **Soundmark:** This is derived from the term landmark. A soundmark is a sound that is unique to an area. The elements have been further defined as to essential sources [11].

A soundscape is a sound or a combination of sounds that are formed or come from an immersive environment and therefor based in three elements: *geophony, biophony* and *anthrophony* [12–14]. To better understand the concept these concepts, it is possible to refer to Krause's studies. An American musician and ecologist of the soundscape, who during a TED Global, explains the three music sources that contribute to the composition of a natural soundscape: "The soundscape is made up of three basic sources. The first is the geophony, or the nonbiological sounds that occur in any given habitat, like wind in the trees, water in a stream, waves at the ocean shore, movement of the Earth. The second of these is the *biophony*. The biophony is all the sound that is generated by organisms in a given habitat at one time and in one place. And the third is all of the sound that we humans generate that is called anthrophony. Some of it is controlled, like music or theater, but most of it is chaotic and incoherent, which some of us refer to as noise." [15, 16]. Moreover, in relation to these 3 elements of the soundscape, Krause states that there is an important interaction between biophony and the other sources of sound: geophony and *anthropophony*. For instance, several studies, particularly those in process by Nadia Pieretti at Urbino University, have shown that birds alter their vocalizations to accommodate themselves to urban noise. And killer whales (*Orcinus orca*) do the same with boat noise in their marine environments. According Bernie Krause "The soundscape concept consists of what I call signature sources, meaning that each type of sound, from whatever origin, contains its own unique signature, or quality, one that inherently contains vast stores of information. That individual signature is unlike any other. So, also, is the natural soundscape unique in its collective state, especially as it becomes the voice of an entire habitat. [...] I use the resulting term, soundscape ecology, to describe new ways of evaluating the living landscapes and marine environments off the world, mostly through their collective voices [17]. The premise behind this concept is the fact that the sense of hearing is the experience and knowledge of the surrounding reality. So, the soundscape is part of the context that affects human existence. This relationship between sound and place is a way of thinking about everyone's responsibility in the "composition" of the shared sound landscape [18]. This "reading of the world" through a specific acoustic scenario turns out to be a fundamental element for the construction and success of the project. Through the activation of the sound, the user is transported in time and space in another era, thus allowing a totally new use of that place. In this case the effect that we try to reproduce is narrative: evoking situations and contents extraneous to the user. The intent is to create a plausible environment from an acoustic point of view. This is possible thanks to the creation of a soundscape composition [19].

2.1   Designing Soundscapes

Artistic experimentation has explored the possibilities and influences of environmental sounds on musical compositions since the 1960'. On the one hand we have the recording practices and the innovative effects introduced by Lee "Scratch" Perry: "He buried microphones under trees to get a different sound, ran tapes backwards, used found sounds and techniques which 20 years later would be called sampling, and blew ganja smoke over tapes." [20]. On the other side are the sound installations and ambient music developed by Brian Eno. Although starting from opposite approaches, in both cases the room becomes the element of acoustic production capable of suggesting the sound mood and arousing emotions.

In more recent time, the soundscape composition as a form of electroacoustic music, has been experimented at Simon Fraser University during the *World Sound-scape Project* (WSP). This melody is characterised by the presence of recognisable sounds and environmental contexts, in order to evoke the listener's associations, memories and imagination related to the soundscape [21].

3 Holofony

The word "holophony" from the ancient Greek *holos* means "everything" and *phonia* means "sound" and describes the particular recording technique created by Hugo Zuccarelli, [22] an Argentinian scientist and Umberto Maggi, an Italian musician. They wanted to emulate the idea of holography to a sound level. This, almost hypnotic effect, cannot be perceived with headphones nor by the classical stereo arc, but "out of the mind", almost in the exact spatial coordinates of recording. In this case we can talk about binaural registration. Binaural recording (with two ears) is a three- dimensional recording method of sound that has the purpose of optimising the recording for its listening through headphones, reproducing as faithfully as possible the acoustic perceptions of a listener located in the original environment of sound event, maintaining its 360° spherical directional characteristics.

The holophonic principle develops the process of sound perception as it is performed by the human auditory apparatus. At a technical level, this sound modifies some classic parameters in the audio recording system: instead of using two microphones, one for the right channel and one for the left channel, as in the case of the stereo, a plastic head is used in which are inserted, in correspondence to the auricles, of the microphone capsules called "holophones", capable of capturing the sound coming from any direction, all trying to simulate the auditory capacities of the human head. In this way the brain introduces a slowness in the auditory reception: if the sound is perceived on the right, it will be perceived on the left only after a significant delay, causing the head to act as if it were a resonance box (Fig. 1).

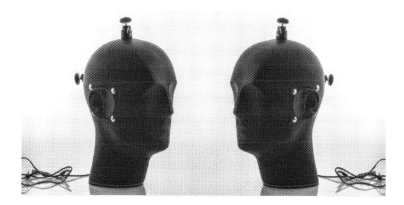

Fig. 1. Example of a holophonic microphone

The first to use this type of technology, commercially, were the Pink Floyd in the song *Get Your Filthy Hands Off My Desert*, in *The final cut*, the twelfth studio album released on 21 March 1983. *The Final Cut* was recorded using "holophonics" - an audio processing technique used to enhance the aural three-dimensional 'feel' of the recording. Holophonics was used to make sound effects on the album appear more three-dimensional to the listener - sound effects, particularly when heard on headphones, appear to pass not just from left to right in the stereo spectrum but also from in front to behind of the listener and around them. Perhaps the most notable use of holophonics on the album is on the song *Get Your Filthy Hands Off My Desert* - during the intro, an airplane is heard to fly swiftly overhead, passing from in front of the listener to behind them, before a huge explosion from the bomb it has dropped is heard surrounding the listener both in front and behind them and to either side of the stereo picture. The use of this technique was in keeping with Pink Floyd's long-standing interest in using atmospheric sound effects combined with advances and innovations in audio technology to enhance the listener's experience of their music. It was also claimed that this process could not be duplicated if one made a copy of the recording i.e., copying from the vinyl record to tape cassette [23]. Despite the great innovation, this type of technology is known by very few and has never reached the apex of success, probably due to system difficulties: the inconvenient use of the plastic head and the obligation to use headphones to warn the effect.

4   The Betrothed Next: A Milan Synesthetic Experience

The Betrothed Next is the evolution of a previous case study developed between 2013 and 2016 [24]. The project was born from a research and analysis work related to the Lombard cultural heritage and more specifically to the Manzoni's novel "I Promessi Sposi". The two previous projects The Betrothed 2.0 [25] and 3.0 explored augmented reality applications and storytelling using geo-referenced information on historical maps [26]. This last phase explores the synesthetic and sound aspects of multisensory interweaving with urban space and historical and cultural heritage [27].

The project is based on the discovery of the Milan urban evolution, thanks to the historical stratification based on a visual *time-machine*. The project, in particular, explores new approaches to the reading of cultural heritage and its "exhibition" [28]. The interpretative key, in fact, wants to reintroduce the emotional and sensory aspects since it is an urban environment that can be actively explored. The key is threefold: the visual [29], spatial and acoustic dimensions experienced both directly and synesthetically [30] through the mediation of digital technology. While the reading plans are two: the narrative of the writing that transposes the contemporary in a metaphorical key and the plot of the novel, which in its historical transposition reconstructs the urban and social antecedent. Following in the footsteps of one of the main characters the user can in fact walk along the streets of Milan discovering, through the use of mobile devices, the cultural and architectural change that occurred through three reference periods: the historical setting of the novel (17th century), the historical period of the author (18th century) and the contemporary city.

In the development of the project, two of the 5 senses have been taken as reference: sight and hearing. In both cases, through the three historical periods, the user has the opportunity to discover and learn about the change in the city. In the first case through Augmented Reality, in the second thanks to the use of special sound effects, created emulating the holophonic recording. The main objective, particularly linked to this latter approach, is to implement is the "sound vision" of each century. Thanks to this expedient, the subject will learn to let him/herself be transported in the various eras not only visually (concept already successfully conceived and developed several times), but also from an acoustic point of view. These two elements joined together through the use of digital storytelling, allow a double immersion (Table 1).

Table 1. Soundscapes of Milan

	XVII Century	XIX Century	XXI Century
Piazza Duomo	Angry mob	Sound in the square: people and music	Turist and street vendors
Porta Venezia	The sound of the countryside	The noise of the wagons and hoofbeats	Friends during an happy hour
Lazzaretto	Hideous screams and complaints	Shops and train whistling	The noise of the street and multi-ethnic music
Porta Nuova	The sound of the countryside	The noise of the wagons and hoofbeats	Business calls

Introductory note	Renzo invites the user to discover Milan

4.1 Methodology

The first step to understand the project's viability has been to assess the developed prototype. The subjects involved in the research were asked to perform a Task Analysis and more specifically a summative test, in order to verify the effectiveness of interaction between the user and the interface. To better understand the user's point of view, the Thinking Aloud method was used (or TA): each user was invited to talk about his/her actions and thoughts while using the app.

Before performing the test, it was evaluated level of knowledge and familiarity with digital technology, throughout an interview. The technologies referred to are that of augmented reality, 3D maps or itinerary and the use of a virtual assistant (e.g. Siri by Apple, Cortana by Microsoft or Amazon Alexa). In the latter case, the user did not use this type of intelligent technology as a support for research, but acoustic interfaced with a sound element.

For the effective research of users and in particular for the number of subjects chosen, we referred to Nielsen's research: "Some people think that usability is very costly and complex and that user tests should be reserved for the rare web design project with a huge budget and a lavish time schedule. Not true. Elaborate usability tests are a waste of resources. The best results come from testing no more than 5 users and running as many small tests as you can afford" [31]. Nielsen explains how there is no need to find large budgets or a large number of users for the success of an effective test, where it is not necessary to analyse several different groups. This is because after a certain number of subjects, the actions will be repeated more or less all in the same way, without making any truly significant intervention. Such a limited number of subjects is useful when testing users belong to different personas, as in the following case.

For the test phase, the selected targets belong to three different archetypes, for this reason it is advisable to find 15 users, 5 for each group:

- Milanese citizens: from subject 1 to subject 5
- Tourists: from subject 6 to subject 10
- Letter teachers: from subject 11 to subject 15

Each of them was subjected to a persona/scenario-based user test with 5 specific tasks to complete:

1. Listen carefully to the audio and explain what the purpose of the app is for you.
2. Find out how Manzoni describes the Bastioni of Porta Venezia.
3. Observe the architectural change of Piazza Duomo and discover the etymology of the word "Rebecchino".
4. Listen to the sounds and noises belonging to the Lazzaretto, starting from the century closer to and going backwards. After each sound you try to explain what feelings you've tried.
5. Where do you think the Lazzaretto is located? Try to give at least a couple of points geographic references (e.g. Piazza Duomo, Galleria). By viewing the maps, starting from the farthest century, you try to read today's map and understand where the hospital is located geographically.

Objective measures were collected for each activity, first of all the success rate or the percentage of activities that each user is able to complete. In addition, for tasks no. 2, 3, 4 and 5 the time it took for each tester to perform the task was verified. The Table 3 shows all the data relating to the achievement of the tasks and the relative timing. As the professor Roberto Polillo affirms in "Facile da usare. Una moderna introduzione all 'ingegneria dell' usabilità", it is possible to calculate the overall success rate, through the formula [31]:

$$[S + (P/2)] / \text{n. tot di task} =$$

S equals the number of successfully executed tasks, P to the number of partially executed tasks, which conventionally will be counted as half of a total success, finally F represents the number of tasks never completed.

$$[71 + (3/2)]/75 =$$

The overall success rate of the system presented is 97%. Through this data it is possible to obtain a significant indication on the usability of the app.

For a better interpretation, however, it is necessary to use some information relating to users. As already explained above, each tester was asked what their relationship with technology was and what, specifically, the apps they use. Most users have shown themselves to be in favour of the use of each proposed app, responding to use each of these functions daily (Fig. 2 and Table 2).

- Have you ever used voice testing (e.g. WhatsApp)?
- Have you ever used a 3D map/route system to reach a place?
- What type of device do you use (smartphone/tablet/kindle/pc/smartwatch)
- What type of operating system do you use? (IOS/Android/OS/Windows)
- Have you ever used some kind of Voice user interface? (Siri/Cortona Alexa/Google Assistant)

Table 2. Task results

	Task 1	Task 2	Task 3	Task 4	Task 5
User 1	S	S - 2'29"	S - 1'28"	S - 0'08"	S - 0'09"
User 2	S	S - 0'58"	S - 2'00"	S - 0'50"	S - 0'12"
User 3	S	S - 1'31"	S - 3'04"	S - 0'28"	S - 1'15"
User 4	S	S - 2'30"	S - 2'20"	S - 0'18"	S - 0'33"
User 5	S	S - 2'00"	P	S - 0'30"	S - 0'11"
User 6	S	S - 1'27"	S - 1'51"	S - 0'15"	S - 0'05"
User 7	S	S - 0'40"	S - 0'25"	S - 0'24"	S - 0'04"
User 8	S	S - 2'00"	S - 1'50"	S - 0'18"	S - 0'08"
User 9	S	S - 3'05"	P	S - 0'30"	S - 0'10"
User 10	S	S - 1'45"	F	S - 0'35"	S - 0'09"
User 11	S	S - 1'38"	S - 1'17"	S - 0'18"	S - 0'10"
User 12	S	S - 2'23"	S - 1'00"	S - 0'15"	S - 0'11"
User 13	S	S - 1'33"	S - 1'37"	S - 0'24"	S - 0'09"
User 14	S	S - 1'47"	P	S - 0'33"	S - 0'07"
User 15	S	S - 1'22"	S - 2'55"	S - 0'25"	S - 0'06"

S = Succes
P = Partial Succes
F = Failure

User 1 - 5 = Milanese citizens
User 6 - 10 = Tourists
User 11 - 15 = Teachers

5   Results

Observing the general data obtained by using the calculation of the response time, it is possible to affirm that the application, as a whole, has a good degree of usability and UX. However, by analysing in more detail the result that emerged from the success rate, we can see that 4 out of 5 tasks were performed by all users without difficulty. Otherwise, task number 3 appears to have been more difficult to complete. Through the collected data it is possible to highlight how 3 out of 15 users have only partially achieved the task, while one user has failed in the task. Through the analysis proposed by Professor Polillo and his formula, we obtain that this task had achieved a success equal to 83%.

Table 3. Questions results

	Question 1	Question 2	Question 3	Question 4	Question 5
User 1	Yes	Yes	Smartphone, pc	Android, Windows	Yes, Cortona
User 2	Yes	Yes	Smartphone, pc, tablet, kindle smartwatch	iOS, Os, Windows	Yes, Siri
User 3	Yes	Yes	Smartphone, pc, tablet, kindle	iOS, Os, Windows	Yes, Siri
User 4	Yes	Yes	Smartphone, pc, tablet	iOS, Os	Yes, Siri
User 5	Yes	Yes	Smartphone, pc	Android, Windows	Yes, Google assistant
User 6	Yes	Yes	Smartphone, pc, tablet, smartwatch	Android, Windows	Yes, Google assistant
User 7	Yes	Yes	Smartphone, pc, tablet	Android, Windows	Yes, Google assistant
User 8	Yes	Yes	Smartphone, pc	iOS, Os	Yes, Siri
User 9	Yes	Yes	Smartphone, pc, tablet	Android, Windows	Yes, Google assistant
User 10	Yes	Yes	Smartphone, pc, tablet	Android, Windows	Yes, Alexa
User 11	Yes	Yes	Smartphone, pc, tablet	iOs, Android, Windows	Yes, Siri
User 12	Yes	Yes	Smartphone, pc, tablet	iOs, Android, Windows	No
User 13	Yes	Yes	Smartphone, pc, tablet, kindle	iOS, Os	Yes, Siri
User 14	Yes	Yes	Smartphone, tablet, kindle	Android, Windows	Yes, Alexa
User 15	Yes	Yes	Smartphone, pc	iOS, Os	Yes, Siri

User 1 - 5 = Milanese citizens
User 6 - 10 = Tourists
User 11 - 15 = Teachers

$$(11 + (3/2)/15 = 83\%$$

This result does not seem to depend on a lack of confidence with the smartphone or with this type of application, because as already analysed, 100% of users use functions similar to those present in the task analysis (Fig. 3).

Fig. 2. Zoom on

The subjects, whose result is equivalent to a partial success, apparently did not consider the element "other" to be important, because when they reached this screen they continued to move between the various centuries without however completing the second part of the task "[. ..] discover the etymology of the word "Rebecchino".

Let's examine the average duration used for the execution of each task.

- Task n.2 = 1'66 (average obtained on 15 users)
- Task n.3 = 1'62 (average obtained on 11 users)
- Task n.4 = 0'24 (average obtained over 15 users)
- Task n.5 = 0'51 (average obtained over 15 users)

Task n.2, despite having been completed by all users, is the most difficult based on the average execution time. This "slowness" of operation was dictated by the type of iconography used. The book, in fact, represented as an element attributable to the Manzoni's novel, does not seem to have created any kind of assonance. The users, as verified after a more detailed conversation, having never seen this type of image, have never developed a precise cognitive psychological connotation. On the other hand, Tasks n.4 and n.5, have been much simpler and more immediate, precisely because the required task was based on two widely recognised iconographies: headphones for task n.4 and the symbol of geolocation for the task n.5. In addition to the analysis relating to the timing, it is interesting to check what the responses related to the second part of task no.2 have been: "[…] After each sound, try to explain what feelings you felt". By analysing these answers, it is possible to observe how in this case what emerges is a subjective and emotional aspect. Although the request is very clear, users are able to express the sensations they experienced only by listening to the latest audio, that relating to the seventeenth century. Each subject, in an unconscious way, passes from defining a clear, defined and concrete situation, to the explanation of totally abstract concepts, such as fear. Listening to the sound belonging to the 21st century (Road noise, multi-ethnic music) users do not express their feelings, but clearly explain what they listened to: multi-ethnic music, sounds of road works, sounds of machines, and so on (Figs. 4 and 5).

Fig. 3. Tag cloud of the 21th century

Fig. 4. Tag cloud of the 19th century

In the case of 19th-century sounds (shop noises and the trains whistle in the distance), users do not define the sounds, but give physical references to each of them: animals, steam train, bells, church, market, etc. In both situations, users are unable to connect a specific feeling because, probably, none of these sounds can transmit a situation extraneous to their daily lives. The two soundscapes, although two centuries apart, do not emerge as such, rather they are considered in the same way: contexts of daily life. The sounds of the seventeenth century are the only ones that manage to convey a strong and predominant emotion, in this case in fact, no more physical

Fig. 5. Tag cloud of the 17th century

elements emerge, but only feelings; sometimes, according to users, even difficult to explain.

The subjects claim to feel emotionally linked to fear, anxiety, loneliness and suffering. Through an analysis of the contexts, it could be affirmed that this exponential estrangement and this trend in moving from concrete to abstract depends on the situations experienced by the user.

In the course of life, in fact, it is assumed that each subject experienced situations similar to those presented both in the 21st century (a busy street, characterised by the noise of road works and ethnic music from some bars) and in the 19th century (a market fruit and vegetables located near a train station). Precisely for this reason, each user is able to view the immersive context presented by the two soundscapes, defining a precise context. Hardly, however, will have experienced a situation like that of the Lazzaretto in the seventeenth century (a field of dying people affected by the plague). Having never experienced situation like that, and therefore difficult for him to visualise and physically connotate, the external user experiences what he feels through emotions. In parallel to this first qualitative evaluation, it would be interesting to expand the test by involving other parameters. Besides, the correlation between the impact of the sound experience relative to the single place was investigated, using the emotions aroused by the different historical thresholds. A wider investigation could explore, vicevesa, the relationship and frequency of words expressed in relation to places.

6   Conclusions

The transition from systems of interaction based mainly on the fixed visual component to a wider and richer world allows a communicative range more similar to our sensorial richness and perception of the world around us. On the one hand, the forms of reality mediated by technology – augmented, virtual, mixed or *blend* – allow an exploration that tries to reproduce the three-dimensionality of the world or to enter into a direct relationship with it. On the other hand, the perspectives opened up by the Voice User Interfaces and, in general, by the use of the acoustic channel offer a range and a language to explode to designers. The combination of these two factors in mobility creates the scenario to transform space-related data into a narrative concept to be explored using one's own senses. An opportunity to discover a spatial identity played on new parameters - the soundscape - and not only on the "image of the city". Besides, the three-dimensionality of the holophonic technique also plays on the role that spatiality and haptic interactions play in the complexity of our knowledge process. The situational context, the plasticity of the experience reconstructs the emotional connection with the territory and its genius loci in a process of transformation of data into experience, of space into a place, of history into a narration.

Acknowledgments. Although the paper is a result of the joint work of all the authors, Letizia Bollini is in particular the author of Sects. 1, 2 and 5; and Irene Della Fazia is the author of Sects. 3 and 4.

References

1. Norman, D.: Emotional Design: Why We Love (or Hate) Everyday Things. Basic Books, New York (2004)
2. Bollini, L.: Beautiful interfaces. From user experience to user interface design. In: Di Lucchio, L., Imbesi, L., Atkinson, P. (eds.) Design for Next. Proceedings of the 12th European Academy of Design Conference. Sapienza University of Rome, 12–14 April 2017. Taylor & Francis Group, Abingdon (2017)
3. Cooper, A., Reimann, R., Cronin, D.: About Face. Hungry Minds Inc., New York (1995)
4. Kurosu, M., Kashimura, K.: Apparent usability vs. inherent usability: experimental analysis on the determinants of the apparent usability. In: Conference Companion on Human Factors in Computing Systems, pp. 292–293 (1995)
5. Nielsen, J.: Usability inspection methods. In: Conference Companion on Human Factors in Computing Systems, pp. 413–414 (1994)
6. Tractinsky, N.: Aesthetics and apparent usability: empirically assessing cultural and methodological issues. In: Proceedings of the ACM SIGCHI Conference on Human Factors in Computing Systems, pp. 115–122 (1997)
7. Tractinsky, N., Katz, A.S., Ikar, D.: What is beautiful is usable. Interact. Comput. **13**(2), 127–145 (2000)
8. Floridi, L.: The Logic of Information: A Theory of Philosophy as Conceptual Design. Oxford University Press, Oxford (2019)
9. Bollini, L.: Virtual & augmented reality representation. Experiencing the cultural heritage of a place. In: Pellegri, G. (ed.) De-Sign Environment Landscape City, pp. 572–583. Genova University Press, Genova (2018)
10. Pearl, C.: Designing Voice User Interfaces. O'Reilly, Sebastopol
11. Southworth, M.: The sonic environment of cities. Environ. Behav. **1**(1), 49–70 (1969). https://doi.org/10.1177/001391656900100104
12. Schafer, R.M.: The soundscape: our sonic environment and the tuning of the world. In: Inner Traditions/Bear & Co., p. 10 (1993)
13. Krause, B.: The great animal orchestra: finding the origins of music in the world's wild places. In: Little Brown, p. 278 (2002)
14. Krause, B.: The anatomy of a soundscape. J. Audio Eng. Soc. (2008)
15. Pijanowski, B.C., et al.: Soundscape ecology: the science of sound in the landscape. In: BioScience, pp. 203–216 (2011)
16. TED: La voce del mondo naturale. https://www.ted.com/talks/bernie_krause_the_voice_of_the_natural_world/transcript?language=it. Accessed 27 Jan 2020

17. The Next Tech: Bernie Krause, l'uomo che da 45 anni registra i paesaggi sonori del mondo. https://thenexttech.startupitalia.eu/54730-20160523-krause-registrare-mondo-suoni-natura. Accessed 27 Jan 2020

18. Biophony Soundscape ecology plunges us into a wilder world beyond the mundane and merely visual. https://www.anthropocenemagazine.org/2017/08/biophony/

19. Il paesaggio sonoro. https://library.weschool.com/lezione/il-paesaggio-sonoro-20696.html. Accessed 27 Jan 2020

20. ErrolNazareth. (2020). https://twitter.com/ErrolNazareth/status/1241324819469254662

21. Soundscape composition. https://www.sfu.ca/~truax/scomp.html. Accessed 27 Jan 2020

22. ABC: Qué es el audio 8D, la última moda sonora del "todo está inventado" que produce "orgasmos sonoros". https://www.abc.es/tecnologia/electronica/sonido/abci-audio-ultima-moda-sonora-todo-esta-inventado-produce-orgasmos-sonoros-201810081257_noticia.html. Accessed 27 Jan 2020

23. https://www.neptunepinkfloyd.co.uk/forum/viewtopic.php?t=21764

24. Bollini, L., De Palma, R., Nota, R.: Walking into the past: design mobile app for the geo-referred and the multimodal user experience in the context of cultural heritage. In: Murgante, B., et al. (eds.) ICCSA 2013. LNCS, vol. 7973, pp. 481–492. Springer, Heidelberg (2013). https://doi.org/10.1007/978-3-642-39646-5_35

25. Bollini, L., De Palma, R., Nota, R., Pietra, R.: User experience & usability for mobile geo-referenced apps. A case study applied to cultural heritage field. In: Murgante, B., et al. (eds.) ICCSA 2014. LNCS, vol. 8580, pp. 652–662. Springer, Cham (2014). https://doi.org/10.1007/978-3-319-09129-7_47

26. Bollini, L., Begotti, D.: The *time machine*. Cultural heritage and the geo-referenced storytelling of urban historical metamorphose. In: Gervasi, O., et al. (eds.) ICCSA 2017. LNCS, vol. 10406, pp. 239–251. Springer, Cham (2017). https://doi.org/10.1007/978-3-319-62398-6_17

27. Bollini, L.: Sensitive environments. Spatial interactive technologies for preserving cultural heritage. Artlab **40**, 44–47 (2011)

28. Bollini, L., Borsotti, M.: Reshaping exhibition & museum design through digital technologies: a multimodal approach. Int. J. Virtual R. **8**(3), 25–31(2009)

29. Bollini, L., Cerletti, V.: Knowledge sharing and management for local community: logical and visual georeferenced information access. In: International Conference on Enterprise Information Systems and Web Technologies (EISWT-09), pp. 92–99 (2009)

30. Riccò, D.: Sentire il design. Sinestesie nel progetto di comunicazione. Carocci (2008)

31. Nielsen Norman Group: Why You Only Need to Test with 5 Users. https://www.nngroup.com/articles/why-you-only-need-to-test-with-5-users/

32. Pollillo, R.: Facile da usare. Apogeo (2010)

Monitoring Urban Development: National Register of Investment Locations as a Tool for Sustainable Urban Land Use Management in Serbia

Ljiljana Živković[(✉)]

Ministry of Construction, Transport and Infrastructure, Belgrade, Serbia
liliana.zivkovic@gmail.com

Abstract. A stalled urbanization process in Serbia is an obstacle to the national economic growth. One of the reasons for that is a poor local economic development, which is caused also by the lack of efficient urban land management system in Serbia. Thus, in order to encourage sustainable urban development, local self-governments in Serbia need a new model of sustainable urban planning and urban land management that should include market-based instruments, like, investments, required for the urban renewal process triggering. Aiming to underpin this strategic ambitious, Serbian Ministry of construction, transport and infrastructure, Republic geodetic authority and Swedish partners are implementing a Project "Improvement of Investment Environment in Serbia" in order to establish a National Register of Investment Locations (NRIL) and related monitoring system, which would improve efficiency and effectiveness of decision-making process for the investment locations and urban land use management in general within 11 pilot local self-governments. Building of this register and its monitoring system would include development of an investment location data model, centralized database and 3 Web-based applications for the investment-related data collection and their interactive visualization using the Serbian National Spatial Data Infrastructure framework and advantages of its geoportal Geosrbija. Also, to increase the role of Serbian cities and strengthen corrective function of the urban plans and urban planning in general, one of the Project objectives is to provide a direct communication channel between the potential investors and local self-governments. Finally, it is expected that the Project results would be implemented in the other local self-governments in Serbia as well, and that they would contribute to the goals achievement of a recently adopted national Sustainable Urban Development Strategy 2030, as well as to some other strategic projects and initiatives launching in Serbia.

Keywords: Urban development · Urban land use management · Monitoring system · National Register of Investment Locations (NRIL) · geoportal Geosrbija

© Springer Nature Switzerland AG 2020
O. Gervasi et al. (Eds.): ICCSA 2020, LNCS 12251, pp. 820–835, 2020.
https://doi.org/10.1007/978-3-030-58808-3_59

1 Introduction

Along with the establishment of a spatial planning monitoring system, as an evidence-base platform for the national territorial development decision-making and policy management in Serbia [6], it has been recognized a need for establishment of the monitoring system on a local –urban planning- level that is struggling today to adjust its urban development and land management instruments and tools to the market-based economy mechanisms [22–24]. Also, after the years of post-socialist transition and delayed land reforms in Serbian cities, identified discrepancy between the current urban land management model and economic model has resulted in an inefficient land use management system, a poor local economic growth and, eventually, stalled urbanization process on national level [12, 22–24]. In order to overcome identified problems, Ministry of construction, transport and infrastructure (MCTI) [15] has adopted the first national Sustainable Urban Development Strategy 2030 (SUDS) [21], but it has also launched together with the Republic Geodetic Authority (RGA) [31] development of a national investment locations register and local level monitoring system for the efficient and effective investment locations and, thus, sustainable urban land use management in general [20]. Once built, this National Register of Investment Locations (NRIL) and related monitoring system, implemented as a part of the Serbian National Spatial Data Infrastructure (NSDI) framework and its geoportal Geosrbija [30], should provide advantages of an interactive visualization tool that would be able to support integrated urban development planning and management, which would stimulate sustainable local economic growth [33, 34].

On the European level, after the adoption of European Spatial Development Perspectives (ESDP) document and establishment of the European Observation Network for Territorial Development and Cohesion (ESPON) program, a number of monitoring systems for various policies and strategic goals implementation has increased, prevailingly on the national and regional levels [3, 8, 9]. However, despite this increase in number, the research and literature about monitoring systems are still somewhat limited [1, 5]. At the same time, a need for these systems adjusted by their contents and designs to the particular territorial development issue(s) is becoming critical, especially on the sub-regional or local level where the decreasing public budgets demand always innovative concepts and approaches for reaching a good quality of life and protection of public interest in general [1, 2, 4, 5]. Finally, since this local level is the one where a corrective role of urban plans and urban planning in general would be directly reflected in the sustainability extent of a market-driven urban development, it is expected that number and research on the urban monitoring systems would increase in the near future [4, 9].

Therefore, the aim of this paper is to present the first results on development of the NRIL register and related monitoring system for the investment locations and sustainable urban land use management in Serbia, as well as to analyze their potential direct and indirect benefits and impacts on the investment level and other areas within the urban planning and development domain in Serbia, both from technological and stakeholder perspectives. Also, this Project is implementing in the cooperation between MCTI, RGA and the Swedish counterpart Lantmäteriet [36], supported by the Swedish

International Development Cooperation Agency (SIDA) [35], within the 2-year Project "Improvement of Investment Environment in Serbia" [29].

After the Introduction section, the paper structure continues with general overview of a recent literature on monitoring systems and land use planning and management in European context. Afterwards, current status of the Serbian urban development and land management system in general are described, followed by the main information about the Project and its methodological approach. Further, the first results from an initial Project phase are described, and then Project results' benefits and potentials are briefly discussed. Paper is closing with the concluding section that, besides general concluding remarks, includes possible future directions of the development.

2   Monitoring Territorial Development in Europe

Regardless of the governance level, monitoring process is an integral part of a public policy management, related decision-making and its implementation management, and it influences thus directly the public policies efficiency and effectiveness in protection of the public interest [1, 3]. Therefore, according to Kleibrink et al. [1], the first and logical step in the public policy management for protecting the particular public interest is to establish the relevant monitoring information platform, which should enable efficient and meaningful decision-making in a continuous way. In order for the needed monitoring continuity to be accomplished, some authors claim the necessity of a monitoring process institutionalization [1, 2], as well as establishment of an appropriate monitoring system design [2, 5, 17, 19]. Regardless of the public policy domain, this monitoring system design should be able to perform the next functions [1–3, 5, 6]:

- Gather information and make them available to the decision-makers, in order to learn about transformation and provide responses accordingly;
- Clarify the purpose and functioning of the particular public policy and public interest, and make these comprehensible to all stakeholders and broader public, in order to build and reinforce mutual trust and cooperation; and
- Support the constructive involvement and participation of all stakeholders through transparent channels, in order to keep them accountable.

2.1   Territorial Development Monitoring Systems: From European to Local Level

Over the past decades, a large number of the territorial development monitoring systems or territorial observatories [5, 9] have emerged in Europe due to the many factors, including: increasing complexity of urban systems and territorial development dynamics in general; development of GIScience; and continually evolving European Union territorial policies [2, 3].

Since the adoption of common policy options for all tiers of administration in Europe with a spatial planning responsibility by the ESDP document in 1999, which has been followed by the ESPON program since 2002, growing number of the different territorial monitoring systems have been covering a range of diversifying development topics and related monitoring activities [3, 8, 9]. These activities include different tasks, like: diagnosing of the territorial trends; development of the new indicators for monitoring urban systems; creation of the relevant monitoring databases; etc. [3].

According to Lindberg and Dubois [2], development of each territory is strongly influenced by its geographical characteristics. Since the territorial development and its management processes are strongly influenced by the types and unique combination of geographical features (cities, sparsely populated areas, mountainous regions, etc.), it is difficult to design an appropriate monitoring system, regardless of a public governance level [3]. Thus, in domain of the spatial and urban planning in Europe, territorial development monitoring systems vary depending on the geographic features and other related statistics and their scale, where the level of details for identified development trends are moving from the general and indicative in macro-regional monitoring systems, to the detailed information in urban monitoring systems (Fig. 1).

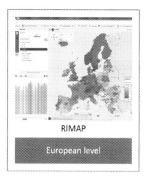
RIMAP
European level

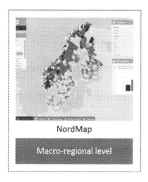
NordMap
Macro-regional level

l'Observatoire des territoires
National level

Geoportale Piemonte
Regional level

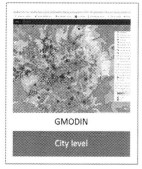
GMODIN
City level

smartdata.wien
Neighbourhood level

Fig. 1. Examples of geovisualization for territorial monitoring systems in Europe on different levels: RIMAP (http://rimap.espon.eu/) (top left), NordMap (http://nordmap.org/) (top middle), l'Observatoire des territoires (http://www.observatoire-des-territoires.gouv.fr/observatoire-des-territoires/en/node) (top right), Geoportale Piemonte (http://www.geoportale.piemonte.it/cms/) (bottom left), GMODIN (https://mappinggm.org.uk/gmodin/) (bottom middle) and Smart City Wien geoportal (https://smartdata.wien/) (bottom right)

Also, if the territorial monitoring system is to be successful and durable, they should include the strategic issues important for the current political ambitions, besides the statistical evidence on past and current territorial dynamics and trends [2]. In this way, monitoring system would enable all relevant stakeholders to ground their decisions and action plans on the same politics-related information on this knowledge-creating platform. For this very reason, an early involvement and a participatory approach to the establishment of territorial development monitoring system are important [1, 2, 5, 6].

Supported by the modern computers and GIS-based applications, geovisualization capabilities of the monitoring systems are becoming critical for their success and usability as well. By definition, geovisualization is more than a cartographic presentation, or communication of the visual thinking results by the static user-customized maps, since it assumes functionalities of an online mapping application with interactive user interfaces, needed for adding and removing data layers during the visual exploration, facilitated by the Web services, digital databases and new functionalities necessary for a dashboard type of presentation [9, 13, 16, 17].

Finally, as sustainable territorial development is a generally accepted objective and principle in spatial and urban planning, as Ostojić and Glažar [7] confirmed, a land use management on the local level is its main instrument for achieving sustainability objectives within the urban development domain [20]. And, while the comprehensive monitoring system for achieving sustainability objectives in the urban development domain is still challenging, these authors claim that optimal approach assumes the stakeholders participation during the urban land use determination and implementation through "simulation of local well-being and economic prosperity, [where defined land uses should] accelerate mutual positive effects, diminish negative effects and increase the economic value of land" [3]. This approach is further supported by Korthals Altes [7] who finds that planning agencies, besides continuing their work on the improvement of urban places quality, must at the same time facilitate investments in the property development, and thus increase the land economic value, because "planning is not only driven by the realization of public works such as streets, parks and arterial access roads, but also by its effect on private investment in housing, industry, retail, etc." This leads to the conclusion that, in situation of the cutting public funds and aiming to do-more-with-less [2], the public interest can be protected today as well through the private investments, where the investment processes should strictly conform to the provisions of plans or planning documents promoting sustainable development and contributing to the preservation of spatial order in general [7, 23].

3   Urban Development and Land Management System in Serbia

On one side, adoption of the ESDP document and Territorial Agenda of the European Union 2020 [8] has triggered general rethinking of the strategic territorial development goals on the European level [2, 3, 6, 13, 17]. On the other side, the recent Urban Agenda for the EU [28], based on the UN-Habitat New Urban Agenda [27] and underpinned by the International Guidelines on Urban and Territorial Planning [26], has launched new generation of the urban visioning and strategic urban development planning cycles for the foreseen challenges of the future cities and urbanization process in general up to 2030.

Correspondingly, on the national level in Serbia, the adopted Spatial Development Strategy (SDS) 2009-2013-2020 [25] and Spatial Plan of the Republic of Serbia 2010–2020 [14] have introduced for the first time some new instruments and tools on the national level, like the monitoring system for continual sustainable territorial development goals management, which are in line with the EUROPE 2020 strategy [18].

Also, pursuing sustainability as a general development principle [7], mentioned national SDS has provided directions for the compatible strategic urban development in Serbia, emphasizing the need for transition to either neoliberal or Scandinavian or some hybrid land management system, as well as to new urban land use and urban planning and development model, which would increase a role of the local authority and the use of urban plans and urban planning in general as main development corrective. Also, this new urban development model should include the new approaches and methods, and contribute to the establishment of a more streamlined and efficient system of urban land management, overcoming existing insufficiencies[1] of the same (system) by including -among other things- market-based instruments and tools within the strategic urban management domain in Serbia, like the private investments for public interest protection [22, 24].

However, despite the strategic urban development directions provided in the national SDS in 2009, only in 2019 the first national urban strategy -SUDS- has been adopted, which relies on the principles and solutions stipulated in the relevant international documents and initiatives [26–28], and includes the vision, goals and measures for steering sustainable development within the Serbian cities up to 2030. Also, in order to create strong commitment to the implementation of identified long-term measures and goals within SUDS, through the local Integrated Urban Development Strategies (IUDS) development, national SDS strategy proposes the general urban planning shift from 'command-and-control' to 'learn-and-adapt' model, as well as building of the monitoring system for the urban development evidence-based decision-making and status reporting [21, 24].

3.1 Monitoring Urban Development and Sustainable Land Use Management

Finally, although the national urban development strategy document –SUDS- has been adopted, it still needs to be translated into the concrete action plans and directives. In the meantime, majority of the current urban land use planning and development issues are regulated by the valid Law on Construction and Planning (LCP) [11]. Besides those basic urban land management-related issues, LCP prescribes also establishment of the

---

[1] Like, for example: inefficient urban land consumption; lack of synergy between the urban land market and urban development planning; an enormous increase of illegal and/or informal construction; poor administrative arrangements for land use management including the transparency of the system; unresolved issue of urban land leasehold rights conversion into urban land property rights; uncontrolled suburban expansion - sprawling, etc.

local information systems in accordance with the INSPIRE principles, which would once built provide comprehensive data platform required for the urban development and land use monitoring, decision-making and steering [10, 11, 22]. And, while development status of these local monitoring systems varies between the local self-governments in Serbia, mostly due to the absence of relevant standards, resources and bylaws, some components –like digital registers- are already in building phase on the national level as part of the NSDI initiative [11, 30], which is becoming major pillar for the e-Government development in Serbia as well [19, 33]. Among those digital registers, NRIL establishment would be also prioritized due to the recognised 1) importance of an efficient urban land use management for an effective urban renewal process in Serbian cities, which are implementing the urban regeneration, revitalization and reconstruction projects, on one side; as well as 2) importance of the private investments attraction, either brownfield, greyfield or greenfield, for protecting the public interest in domain of sustainable urban development when public funds are decreasing.

Therefore, the aim of this article is to present up-to-date status in establishment of the NRIL register and related framework within the 11 local self-governments in Serbia, as well as development of the interactive urban development monitoring system for the investment locations and sustainable urban land use management using the NSDI platform advantages.

4   Project "Improvement of Investment Environment in Serbia"

4.1   Aim and Objectives

In order to create possibilities for improvement of the local economic development and launching of the urban renewal process in Serbia in a systematic way, through the targeted investments into the local strategic projects from the urban revitalization, regeneration and reconstruction domain, MCTI together with RGA has started Project "Improvement of Investment Environment in Serbia" [29] with the Swedish counterpart Lantmäteriet, supported by the Swedish International Development Cooperation Agency (SIDA).

Due to the previously described urban land system inefficiency in Serbia, and its negative effects for the local economic development, urbanization process and quality of life, the **purpose** of this Project is to contribute to a better climate for investments that would improve local socio-economic, environmental and other conditions necessary for a sustainable smart and inclusive urban development in future [32].

Therefore, the **aim** of the Project is to support sustainable local economic development in Serbia by establishing a digital national register of investment locations – NRIL-, and related on-line and interactive monitoring system for the transparent and efficient investment decision-making, locations management and their status monitoring, as well as for a direct communication between the local self-governments and potential domestic and foreign investors. Additionally, combined with the other relevant NSDI registers and datasets, the aim of this monitoring system is to support sustainable urban land use planning and management, as well as integrated urban development monitoring in Serbia.

Based on the defined aim, the **objectives** of the Project include:

- To develop a model, database and digital platform for storing, processing, searching/retrieving, analyzing, geovisualizing and monitoring status and effects of the selected investment locations within the 11 local self-governments in Serbia;
- To collect, visualize and communicate investment locations data to the potential investors and other end-user communities; and
- To build capacities of the Project stakeholders –RGA and pilot local self-governments- on a market-based urban land use principles and tools for the investments attracting and management.

Project duration is 2 years in period 2019–2021, while the total budget of 2 million euros is allocated among the 5 main components that are necessary for the establishment of investment locations database and mentioned monitoring system, as a tool for informed and efficient sustainable urban land use management in Serbia.

4.2   NRIL and Its Monitoring System Concept

The concept of the investment locations database and its communication-oriented monitoring system needs to be designed in such a way to satisfy identified user requirements related to the NRIL data model and technical solution, which main features are listed in Fig. 2.

Data model requirements	Technical solution requirements
To enable a unique identification of each investment location,	Applications to be platform independent,
To include a spatial position of the investment locations,	System to be modular and scalable with independent components integrated by the services,
To establish a uniform national classification of the investment locations,	Applications to be built as centralized systems with Web-based interfaces,
To be user-friendly,	To use Internet and HTTP/HTTPS protocols for enterprise application integration,
To provide potential investors a relevant information on the investment locations,	Technological diversity to be controlled to minimize the cost of maintaining expertise,
To be reliable platform for the investment decision-making by providing quality data,	Data to be protected from unauthorized use and disclosure,
To secure transparency for the investment locations management to citizens, public authorities and third parties,	Data integrity to be maintained in multiuse environment,
To provide possibilities for connecting to the other relevant registers and monitoring systems on national level,	Data modifications to be traceable, and change of data values would be recorded,
To ensure compliance with the ISO19100 standards and INSPIRE principles and rules, and	Data are accessed through services,
To secure for all data modifications to be traceable, and change of data values would be recorded.	All data to be centrally stored and online available in a central data repository, and
	Software and hardware to conform to defined standards that promote interoperability for data, applications, and technology.

Fig. 2. NRIL and its monitoring system requirements

5 Methodology

The recent World Bank study [12] has revealed an uneven spatial distribution of the lagging and leading regions in Serbia, as well as generally stalled urbanization process that has been an obstacle to Serbia's economic growth. Thus, recognizing the importance of achieving sustainable, inclusive and balanced regional development on the national level through stable local economic development, couple of initiatives[2] for the establishment of investment locations register has been launched lately in Serbia. These initiatives failed due to the lack of national standards for the investment locations data collection, spatial presentation and unified management. Also, being identified as one of the registers with strategic importance for the sustainable urban development and urbanization in general in Serbia, it has been decided that planned NRIL database would be part of the already established NSDI database and its geoportal Geosrbija [30]. In this way, it is expected that this new database and planned monitoring system would benefit from the existing technological capabilities as well as combination with the other NSDI datasets, securing thus directly conditions needed for creating the interactive cartographic tool for the informed and efficient investment decision-making and urban development monitoring.

Therefore, the Project methodology for establishment of this tool has been organized around the 5 main Project components and related activities for creation of the outputs needed for the identified aim and objectives achievement (Table 1) [32].

Table 1. Project methodological approach

Component	Activity	Output
National investment location model	Development of a registry model for the investment locations, including database implementation	Data model and database "National Register of Investment Locations"
Data collection tool	Development of tool for collecting data on the investment locations on a local level	Web-based application for data collection
Capacity building of local self-governments	Building capacities of local self-governments in the domains of market-based urban land management and attracting investments by presenting available resources to potential investors	Improved knowledge and skills capacities of 11 Serbian local self-governments
Location pool tool	Establishment of direct and transparent channel of communication between investors and a local self-governments	Web-based user interface for communication "Location pool"
Public monitoring and visibility tool	Development of a tool for public monitoring and promotion of the Project and its results	Web-based user interface for the Project visibility

---

[2] For example, https://ras.gov.rs/podrska-investitorima/baza-investicionih-lokacija.

And, although the Project methodology includes above listed 5 components, final component in a technical sense is the fourth one.

6 Project Results

Since this Project is currently under implementation, only its first results are subject of this article and analysis, and they would be presented and described in the next lines in a following order:

- Proposal for a common data model for the investment locations database (Subsect. 6.1),
- Collection of the initial –pilot- data for investment locations and their publication on the NSDI geoportal Geosrbija (geosrbija.rs) (Subsect. 6.2), and
- Implementation of the first capacity building event for the selected Project stakeholders, i.e. 11 pilot local self-governments[3] (Subsect. 6.3).

6.1 National Investment Location Data Model

According to the above presented requirements from the NRIL data model and solution (Subsect. 4.2), and following the lessons-learnt from the previous similar initiatives in Serbia, as well as results of the first consultations with the selected pilot local self-governments, initial data model for the investment locations management and monitoring in Serbia has been proposed (Fig. 3).

This national investment location data model, in line with the collected user requirements analysis, should include approximately 21 attributes divided into the 7 categories of information that are find to describe each investment location within Serbian cities into the sufficient level of details for the informed investment decision-making. These categories and their attributes with the specified value domains present a good starting point and, in fact, a first national standard for the investment locations data collection and database design, and eventually full digitalization of the NRIL register.

Also, from the technology perspective, digitalization of the NRIL register would open up possibilities for the usage of modern GIS-based and related technologies' advantages for the investment locations data analyses, their (geo)visualization, as well as efficient monitoring of the investments realization and impacts on the neighboring land uses and urban functions, quality of life and urban development in general, as presented in the next section (Subsect. 6.2).

Further, since the digital NRIL register is planned to be implemented as part of the already established NSDI database and its digital platform Geosrbija, which stores and publishes today over 220 various datasets, it would be possible to easily combine and analyze together investment locations data with the other relevant NSDI datasets. This would support integrated decision-making and management of each investment location's value, feasibility and future potential separately.

---

[3] Arilje, Bor, Valjevo, Ćurpija, Knjaževac, Sombor, Vranje, Zrenjanin, Pirot, Požarevac, Čačak.

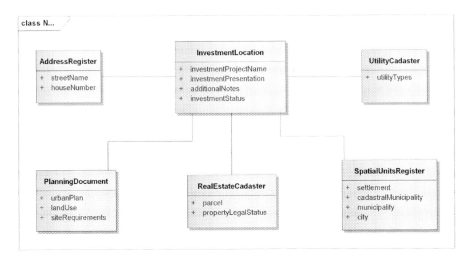

Fig. 3. Conceptual model for the investment location management and monitoring domain (indicative)

Finally, based on the previously stated user requirements and expectations, above presented is a proposed conceptual model for the investment location management and monitoring domain (Fig. 3), which consists of the existing NSDI datasets, like utility, real estate cadaster, spatial units and address registers, as well as the new datasets that still need to be created, like investment locations and planning documents datasets.

6.2   NRIL Applications and Pilot Investment Locations Visualization

Following the predefined technological requirements for the NRIL monitoring system (Subsect. 4.2), and relying on the advantages of above proposed investment location data model (Subsect. 6.1), RGA supported by the Swedish partners has started the initial phase of Project by developing the test versions of NRIL components and monitoring environment. Thus, the results of this initial phase have included development of a temporary Web form for collecting data on a selected set of pilot investment locations, as well as collected data visualization on the NSDI geoportal Geosrbija.

This means that, using this temporary Web form based on the proposed investment location standard data model, those 11 local self-governments have collected and combined geometry data with the attribute data for each pilot investment location (Fig. 4). Collected investment locations data has been stored directly within the NSDI database, while testing of the resulting data and visualization quality, including available geovisualization capabilities, have been performed using the geoportal Geosrbija functionalities (Fig. 5), which is based on the module-oriented map application Adaptive [16].

Following the already identified main Project outputs in the methodology Section (Sect. 5), the NRIL monitoring system would be provided also with the 3 Web-based applications, which should secure this strategic national register sustainability

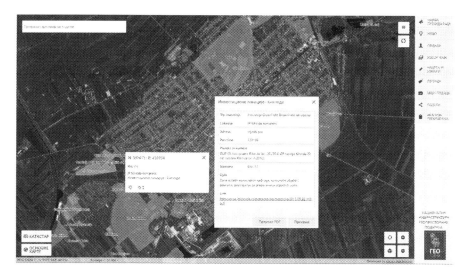

Fig. 4. Visualization of the investment location data on the geoportal Geosrbija; example investment location in Kikinda

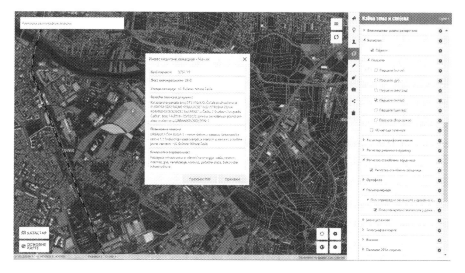

Fig. 5. Interactive cartographical visualization possibilities for the investment location data on the geoportal Geosrbija; example investment location in Čačak

also after the Project ending. Those proposed 3 Web-based applications are further elaborated during this initial Project phase, and it was defined that they would support the next NRIL monitoring system functions:

- **Web-based application for data collection** would support user-friendly and standardized collection and modification (editing, updating, etc.) of the investment locations data directly in the NSDI database, structured according to the proposed standardized data model (Fig. 3);

- **Web-based user interface "Location pool"** would provide public –online- access and transparent selection of the stored investment locations data, based on the criteria set by investors; also, it would allow direct communication between the potential investors and local self-governments. Transparency quality of the solution would be secured by the planned reporting function on the potential investors' queries, as well as through the real-time investment locations status monitoring; and
- **Web-based user interface for the Project visibility** would provide simple and transparent interactive cartographic presentation of the Project results, including monitoring of the investments realization status in real-time.

6.3   Capacity Building of Local Self-government and RGA

Finally, following the first results of the Project in data and technology domains, also first capacity building event for pilot local self-governments was organized by the Swedish partners. Besides the general introduction to the Swedish local government administration level and investments attraction and management tools, provided trainings included presentation of the organizational, legal, financial and technology issues and framework for the similar investment-related initiative in Sweden. During the trainings it was also emphasized the Swedish model for municipal public financing options, as well as state system for the investment support.

It is expected that shared experiences of the Swedish local administration should encourage Serbian local self-governments to change or improve efficiency of the current municipal public financing model and options, as well as to build realistic expectations from the investment locations in Serbia, which are necessary for the local economic growth underpinning and successful implementation of the future local strategic goals identified by the IUDSs, and national –SUDS- ones in general.

7   Discussion

Selected methodological approach and the first results of the Project "Improvement of Investment Environment in Serbia", described in the previous Sections (Sects. 5 and 6), confirm that the investment locations and urban land use management and monitoring system under development would provide necessary technological advantages for the interactive geovisualization. This characteristic of the monitoring system is important not just for the usability and durability of this system, but it's also critical for a better visibility, analysis and understanding of the investment locations potential, including their impacts on the neighboring urban functions and urban land use patterns in general. This means that the outputs of this Project once built should lead to the optimally informed investment decision-making, as well as later efficient management of the investments realization on side of the investors in Serbia.

Also, on side of Serbian local self-governments, emerging investment locations monitoring system should lead to the better understanding of the investors' needs, including possibility for simulating potential effects of each investment decision-making on public interest, besides strengthening the strategic role of local self-governments in urban development process in general. Further, digitalization of the standardized, modular and scalable NRIL register, and its reliance on the other NSDI registers, would support learning process needed for creation of the new model of urban land use planning and management, as well as designing of the new concept of land management system and land use policy in Serbia in general, which would be able to boost local economic growth in the cities.

Also, as part of the NRIL register digitalization, standardization of the land use dataset that is part of this register (following the INSPIRE implementation rules and local specific needs) would also support interactive visual thinking about the new land use model, as well as better adapted urban planning methods and approaches to Serbian city's needs. This would lead to a new urban development knowledge creation, needed for sustainability goals implementation along with the encouraged local economic development.

Once built, this national investment locations' register and urban land use monitoring system would be able also to tackle today stalled urbanization process in Serbia, and secure its sustainable evolving using the controlling mechanisms of the plans and urban planning in general. Additionally, this new urbanization process could in future be supported by the currently developing compatible paradigms and concepts in Serbia and Europe, like Smart City, Open Data, Big Data, EU Green Deal, etc.

Finally, digitalization of the land use dataset from the urban plans and other planning documents would directly support their already confirmed corrective role, while the assumed modularity and scalability of the monitoring system solution would encourage establishment of the spatial and urban planning documents database in Serbia in near future, as well as launching of so-called e-Space geoportal that would be a national gateway for the territorial planning and development related content.

8   Conclusions

In order for the urban planning community and local self-governments in Serbia to understand and respond appropriately to the poor local economic growth, lagging reforms in the land management domain, as well as increasing urban development challenges in general, they need the integrated urban planning and development monitoring system able to support evidence-based decision-making and relevant urban policy management. Relying on the various datasets and registers, this monitoring system should support interactive geovisualization capabilities, needed for the visual thinking and customized user-driven approach to the digital maps creation, proved to be useful tool for the territorial development stakeholders gathering, empowering and development options negotiation.

In order to underpin building of this type of the local monitoring and information system in Serbia, MCTI together with RGA and Swedish partners has launched the Project "Improvement of Investment Environment in Serbia" to establish the digital NRIL register that would support the investment locations decision-making and management, as well as the urban land use monitoring within 11 pilot Serbian cities. This register and related monitoring system would be part of the NSDI initiative and its geoportal Geosrbija, securing thus direct access to the over 220 datasets for combining and analyzing together with the investment locations data. Besides technical components, the planned Project activities include also capacity building events for RGA and local self-governments, which would increase the strategic role of the latter and create the critical opportunities for learning, reinventing and adopting the new land management and urban development models and approaches in Serbia.

Finally, it is expected that the results of this Project would motivate and encourage digitalization of some new registers in Serbia, as well as launching of some new projects of strategic importance within the territorial development domain, including establishment of the spatial and urban planning documents database and e-Space geoportal.

References

1. Kleibrink, A., Gianelle, C., Doussineau, M.: Monitoring innovation and territorial development in Europe: emergent strategic management. Eur. Plan. Stud. **24**(8), 1438–1458 (2016). https://doi.org/10.1080/09654313.2016.1181717
2. Lindberg, G., Dubois, A.: How to Monitor Territorial Dynamics. Nordregio News Publication Issue (2014). http://www.nordregio.se/Publications/Publications-2014/Monitoring-Territorial-Dynamics/
3. Soria-Lara, J.A., Zúñiga-Antón, M., Pérez-Campaña, R.: European spatial planning observatories and maps: merely spatial databases or also effective tools for planning? Environ. Plan. B Plan. Des. **42**(5), 904–929 (2015). http://journals.sagepub.com/doi/abs/10.1068/b130200p
4. Korthals Altes, W.K.: Multiple land use planning for living places and investments spaces. Eur. Plan. Stud. **27**, 1146–1158 (2019). https://doi.org/10.1080/09654313.2019.1598022
5. Segura, S., Pedregal, B.: Monitoring and evaluation framework for spatial plans: a spanish case study. Sustainability **9**(10), 1706 (2017). https://doi.org/10.3390/su9101706
6. Živković, L.: A proposal for the spatial planning monitoring system in Serbia. In: Gervasi, O., et al. (eds.) ICCSA 2017. LNCS, vol. 10407, pp. 555–570. Springer, Cham (2017). https://doi.org/10.1007/978-3-319-62401-3_40
7. Ostojić, I., Glažar, T.: Criteria for evaluation and guidelines for land use planning in terms of sustainable urban development. Creativity Game Theor. Pract. Spat. Plann. **2014**(2), 24–32 (2014). https://doi.org/10.15292/IU-CG.2014.02.024-032. Scientific journal, No. 2/2014
8. Territorial Agenda of the European Union 2020. http://ec.europa.eu/regional_policy/en/information/publications/communications/2011/territorial-agenda-of-the-european-union-2020
9. ESPON Inspire Policy Making by Territorial Evidence. https://www.espon.eu/main/
10. INSPIRE Infrastructure for spatial information in Europe. http://inspire.ec.europa.eu/
11. Law on Construction and Planning, Official Gazette of the Republic of Serbia 9/2020. https://mgsi.gov.rs/cir/dokumenti/zakon-o-planiranju-i-izgradnji
12. World Bank 2019: Western Balkans and Croatia Urbanization and Territorial Review, vol. 2 (English). World Bank Group, Washington, D.C. http://documents.worldbank.org/curated/en/404331565242902772/Western-Balkans-and-Croatia-Urbanization-and-Territorial-Review
13. Dühr, S.: The Visual Language of Spatial Planning: Exploring Cartographic Representation for Spatial Planning in Europe. Routledge, New York (2007)

14. Spatial Plan of the Republic of Serbia 2010–2020 (Abridged version in English language). http://195.222.96.93//media/zakoni/Spatial%20Plan%20of%20the%20Republic%20of%20Serbia_2010-2020_abridged%20(1).pdf
15. Ministry of construction, transport and infrastructure (MCTI). http://mgsi.gov.rs/
16. Adaptive. https://www.avinet.no/index_en.html, https://www.avinet.no/adaptive.html
17. Živkovic, L., Barboric, B.: CO-TAMP: Transnational platform for territorial attractiveness monitoring in the danube region. In: Misra, S., et al. (eds.) Computational Science and its Applications – ICCSA 2019. LNCS, vol. 11621. Springer, Cham (2019). https://doi.org/10.1007/978-3-030-24302-9_27
18. EUROPE 2020: A strategy for smart, sustainable and inclusive growth. http://ec.europa.eu/eu2020/pdf/COMPLET%20EN%20BARROSO%20%20%20007%20-%20Europe%202020%20-%20EN%20version.pdf
19. Portal eUprave Republike Srbije. https://euprava.gov.rs/
20. Izakovičová, Z., Špulerová, J., Petrovič, F.: Integrated approach to sustainable land use management. Environments **5**(3), 37 (2018). https://doi.org/10.3390/environments5030037
21. Sustainable urban development strategy of the Republic of Serbia until 2030. https://www.mgsi.gov.rs/cir/dokumenti/urbani-razvoj
22. Zeković, S.: Evaluation of the current urban land system in Serbia. Spatium **2008**, 55–60 (2008). https://doi.org/10.2298/SPAT0818055Z
23. Zeković, S.: Possibilities for transformation of the urban land management in Serbia. Spatium **2009**(19), 15–19 (2009). https://doi.org/10.2298/SPAT0919015Z
24. Zeković, S., Vujošević, M., Maričić, T.: Spatial regularization, planning instruments and urban land market in a post-socialist society: the case of Belgrade. Habitat Int. **48**, 65–78 (2015). https://doi.org/10.1016/j.habitatint.2015.03.010
25. Spatial Development Strategy of the Republic of Serbia 2009-2013-2020. http://www.apps.org.rs/wp-content/uploads/strategije/Strategija_PROSTORNI%20RAZVOJ%20Republike%20Srbije.pdf
26. International Guidelines on Urban and Territorial Planning. https://unhabitat.org/sites/default/files/download-manager-files/IG-UTP_English.pdf
27. New Urban Agenda. http://habitat3.org/wp-content/uploads/NUA-English.pdf
28. Urban Agenda for the EU. https://ec.europa.eu/regional_policy/en/policy/themes/urban-development/agenda/
29. Improvement of Investment Environment in Serbia. http://rgz.gov.rs/projekti/projekat-unapre%C4%91enje-poslovne-klime-u-srbiji/
30. NSDI geoportal Geosrbija. https://geosrbija.rs/en/
31. Republic geodetic authority (RGA). http://en.rgz.gov.rs/
32. RGA/SIDA/Lantmäteriet: INCEPTION REPORT Improvement of Investment Environment in Serbia 2018–2021. Internal document (2019)
33. Jovanović Milenković, M., Rikanović, S., Teodosijević, B., Đuran, D., Vučetić, D.: Model of investment map by application of the e-Government service. In: Proceedings of the XLVI International Symposium on Operational Research, Kladovo, Serbia, pp. 103–108 (2019)
34. Spatial data standardization model for national registry of investment locations. https://collab.geosrbija.rs/topic/
35. Swedish International Development Cooperation Agency (SIDA). https://www.sida.se/
36. Lantmäteriet. https://www.lantmateriet.se/sv/

Estimation of Risk Levels for Building Construction Projects

Gabriella Maselli[1]([⊳⊲]) (ID), Antonio Nesticò[1] (ID), Gianluigi De Mare[1] (ID),
Elena Merino Gómez[2] (ID), Maria Macchiaroli[3] (ID),
and Luigi Dolores[1] (ID)

[1] Department of Civil Engineering, University of Salerno, Fisciano, SA, Italy
{gmaselli,anestico,gdemare,ldolores}@unisa.it
[2] Nebrija University, Calle Pirineos, 55, 28040 Madrid, Spain
emerino@nebrija.es
[3] Department of Industrial Engineering, University of Salerno,
Fisciano, SA, Italy
mmacchiaroli@unisa.it

Abstract. This paper aims to define an innovative method to estimate the
investment risk thresholds, in order to provide the analyst with essential terms
for the economic evaluation of the projects. The idea is to borrow from the As
Low As Reasonably Practicable (ALARP) principle the concepts of the
acceptability threshold and tolerability threshold of risk. Following this principle, generally used to assess the risk of human life loss, a risk is ALARP: when
the costs for its mitigation appear to be disproportionate to the achievable
benefits; that is when it lies between the unacceptable and the broadly acceptable
region. To estimate the thresholds of acceptability and tolerability, the theoretical reference is the Capital Asset Pricing Model (CAPM), which enables to
compare the investment risk not only to the return of the production sector in
which the project under consideration falls, but also to the return of the market
as a whole. The combined use of CAPM and statistical survey methods makes it
possible to attribute the two risk thresholds to investments in a specific sector. In
this paper, the focus is on risk assessment for building construction projects. For
this production sector, the proposed analysis model is implemented with reference to official data concerning the Italian economy.

Keywords: Risk assessment · Economic evaluation · ALARP logic · Capital
Asset Pricing Model · Building construction projects

1 Introduction

Many factors influence the riskiness of a civil engineering project. First, the difficulty
of expressing with certainty forecasts on the critical variables, namely those that significantly influence the value of the economic performance indicators. Secondly, the

G. Maselli, A. Nesticò, G. De Mare, E.M. Gómez, M. Macchiaroli, L. Dolores—The six authors
contributed equally to this work.

O. Gervasi et al. (Eds.): ICCSA 2020, LNCS 12251, pp. 836–851, 2020.
https://doi.org/10.1007/978-3-030-58808-3_60

need to consider environmental, social and cultural parameters in assessments, which are often difficult to estimate [1–4]. This leads to having to express the outcome of the Cost-Benefit Analysis (CBA) in stochastic terms. Specifically, investment is risky if the probability that the profitability indicator is lower than a threshold value considered critical is high. The greatest difficulty is precisely estimating these limit levels of risk, not suggested either by the sector literature or by the national and EU regulatory guidelines [5, 6].

Thus, the objective of the paper is to establish limit thresholds useful for assessing the acceptability of investment risk. The idea is to establish the risk thresholds based on the As Low As Reasonably Practicable (ALARP) principle. In accordance with this principle, widely established to assess safety in the industrial environment, a risk is defined as ALARP when the costs for its mitigation appear to be disproportionate to the achievable benefits or, in other words, if it is between the acceptability threshold and tolerability threshold [7–9]. From ALARP, therefore, we take up the concepts of acceptability threshold and tolerability threshold, which respectively represent the limit below which the risk is broadly acceptable and the one above which it is intolerable.

Once these two risk thresholds have been defined, an innovative method is proposed to estimate them. In the financial field, these thresholds correspond to the expected return from an investment project with an "acceptable" and "tolerable" risk profile respectively. For this reason, the Capital Asset Pricing Model (CAPM), which explains how the risk of a financial investment affects its expected return, represents the theoretical reference useful for assessing the acceptability and tolerability of risk [10, 11]. The joint use of the CAPM and statistical survey tools also makes it possible to specify risk limit values depending both on the investment sector and on the territorial context in which the project takes place.

The paper is structured into four sections. Section 2 examines the investment risk assessment in support of the CBA, as suggested by the EU regulatory guidelines. In Sect. 3, the ALARP principle and the CAPM are first exposed, and then the methodology for estimating the acceptability and tolerability thresholds of the investment risk is proposed. In Sect. 4, we implement the proposed method to estimate the acceptance limits of the risk related to building construction investments in the Italian territorial context. In the last section, final reflections are analysed.

2 The Investment Risk in the Building Construction Projects

The European regulatory guidelines highlight the importance of risk analysis in the *ex ante* evaluation of investment projects. In this regard, Regulation no. 1303/2013 of the European Commission specifies that risk analysis can be required either by virtue of the complexity or by virtue of the project size and in relation to the availability of data necessary for the evaluation [12]. In particular, it is mandatory in the case of major projects, defined as «works, activities or services intended in itself to accomplish an indivisible task of a precise economic or technical nature which has clearly identified goals and for which the total eligible cost exceeds € 50 000 000».

The Guide to Cost-Benefit Analysis of investment projects of the European Commission (EC) for the 2014–2020 programming period, on the other hand, identifies four main phases in the risk management process [13]:

1) sensitivity analysis;
2) qualitative risk analysis;
3) probabilistic risk analysis;
4) risk prevention and mitigation.

Phase 1. The CBA allows expressing a judgment on the economic performance of the investment, as well as to choose between alternative projects. This technique consists of: forecasting the costs and benefits that the project initiative generates in the analysis period; in the subsequent discounting of the Cash Flows (CF); therefore in the estimate of the performance indicators, Net Present Value (NPV), Internal Rate of Return (IRR), Benefits/Costs ratio, Payback Period [14–19]. If we consider the IRR as an indicator of economic performance, then:

$$\sum_{t=0}^{n} \frac{B_t - C_t}{(1IRR)^t} = 0 \tag{1}$$

Where B_t e C_t represent respectively the costs and benefits at time t.

In the implementation of the CBA, the sensitivity analysis enables to identify the "critical" variables of the project, namely those that have the greatest impact on the result of the evaluation. The sensitivity analysis is carried out by changing the values associated with each variable and evaluating the effect of this change on the profitability indicators of the project. Considering "critical" the variables for which a variation of $\pm 1\%$ of the value adopted in the base case determines a variation of more than 1% of the value of the economic performance indicator can be a guiding criterion for identifying sensitive variables. To study the impact on the project determined by the simultaneous variation of the critical variables, the sensitivity analysis can be completed with the scenario analysis. The estimation of profitability indicators in optimistic and pessimistic scenarios allows expressing a preliminary judgment on the project risks.

Phase 2. The qualitative risk analysis is substantiated: in: the identification of possible events with negative implications on the execution of the project; the consequent definition of a risk matrix for each adverse event, from which it is possible to read the probability of occurrence (P) and severity of the impact (S); in the interpretation of the risk matrix in order to evaluate the risk levels associated with the project (P · S); planning major risk mitigation interventions according to the level of estimated risk.

Phase 3. The probabilistic risk analysis is expressed in the stochastic description of the critical variables of the project and in the subsequent estimate of the probability distribution of the profitability indicator. The transition from the cumulative probability curve of the risky variables to that of the project IRR occurs through the Montecarlo analysis. Briefly, the random extraction of the probable values for each critical variable allows deriving the respective value of the profitability indicator. By repeating the procedure for a sufficiently large number of extractions, the probability distribution of the IRR is derived.

From the reading of the probability distribution of the IRR it is possible to derive important information on the project risk, for example the expected value and the variance of the profitability index E(IRR).

If p(IRR) is a continuous random variable with probability density, the expected value (mean or mathematical expectation) of the variable is defined as the integral extended throughout $\mathbb{R}$ of the product between IRR and the density function p(IRR):

$$E(\text{IRR}) = \int_{-\infty}^{+\infty} \text{IRR} \cdot p(\text{IRR}) \tag{2}$$

Discretizing the probability density function of the IRR, then the expected value of the discrete random variable E(IRR) is the sum of the products between the IRR$_i$ values and the respective probability p(IRR$_i$), that is:

$$E(\text{IRR}) = \sum_{i=1}^{n} \text{IRR}_i \cdot p(\text{IRR}_i) \tag{3}$$

With n number of discretization intervals of the probability distribution of the random variable IRR.

The comparison between E(IRR) and performance limit values makes it possible to express a judgment on the project risk. In this sense, however, it should be noted that the regulatory guidelines do not provide clear indications on the acceptability levels of the project risk.

Phase 4. The definition of an effective risk mitigation and/or prevention strategy is a direct consequence of the results of the previous phases. This phase includes the selection of mitigation measures; the implementation of the prepared plan; the analysis and evaluation of the residual risk, i.e. the risk that remains despite the mitigation strategy undertaken. In other words, it is necessary to re-estimate the probability distributions of the risky variables of the project, deriving from the implementation of the risk containment measures and, consequently, that of the economic performance indicator.

3   The Investment Risk in the Building Construction Projects

It has already been highlighted that for the "building construction" sector, the literature and regulations do not indicate limit values for the acceptance of investment risk. Thus, the aim of the work is twofold: 1) to establish threshold levels for risk; 2) to define a methodology useful for estimating risk acceptance thresholds that can guide the analyst in the economic evaluation of the projects.

As regards the first objective, the reference is the ALARP principle, summarized in Subsect. 3.1. Specifically, the ALARP principle is based on the concepts of risk acceptability threshold and tolerability threshold. These are respectively the limit value below which the risk is broadly acceptable and the limit value above which the risk is acceptable so that mitigation measures must necessarily be planned.

Instead, to achieve the second objective, we use the Capital Asset Pricing Model (CAPM), described in Subsect. 3.2. The CAPM is the theoretical reference for establishing risk tolerability and acceptability thresholds, as it allows to assess the risk-adjusted discount rate $r(\beta)$, which can be interpreted as the minimum return expected from an investment project with a β risk profile [10]. The combined use of CAPM and statistical survey methods makes it possible to calibrate the thresholds of acceptability and tolerability of risk to investments in a specific sector, based on a statistically acceptable return on investment (ROI) panel of companies operating in a given territorial context. The use of these tools allows characterizing an innovative method for estimating the acceptability and tolerability thresholds of investment risk, as detailed in Subsect. 3.3.

3.1 Risk Assessment According to ALARP Criteria

The risk assessment is developed in the safety field, that is, to analyse the risk of loss of human life in the performance of dangerous activities. Since 1960, tragic accidents in installations with dangerous substances have demonstrated the need for criteria to judge the tolerability/acceptability of these activities. The first risk acceptance criteria are born in the nuclear field, where even low probability of accidents can have catastrophic consequences [20]. It is in this context that in the 1960s–70s the probabilistic analyses of accidents made the first quantitative risk assessments possible. In 1992, in response to the need to manage industrial risks, the British Agency Health and Safety Executive (HSE) defined general risk acceptance principles [21–23]. Among these, of absolute importance is the As Low As Reasonably Practicable principle, according to which all risk mitigation measures must be implemented as long as the costs do not appear disproportionate to the benefits that can be achieved. An intervention can be defined as "practicable" as long as its technical feasibility is demonstrated, while "reasonableness" implies the need to also consider extra-monetary aspects, that is, social, cultural and environmental aspects. In other words, any risk reduction intervention is ALARP if "reasonably" feasible and sustainable in the broad sense, that is, it is tolerable if further mitigation interventions have costs disproportionate to the achievable benefits. Thus, a risk is ALARP if it lies between the acceptability threshold and the tolerability threshold. Risks that fall below the tolerability threshold must be necessarily reduced because they are unacceptable; those included between the tolerability and acceptability thresholds are in the ALARP area, that is, they must be mitigated up to reasonable practicability; on the other hand, the risks above the acceptability threshold are "broadly acceptable" or it is not necessary to mitigate them [24–26].

The ALARP approach, which originated in the nuclear sector, is increasingly used to plan land use in the immediate vicinity of industries or dams, to manage landslide risk, and to assess risk in tunnels [27–31]. If so far it has been generally applied to assess the security risk associated with the safeguarding of human life, we believe that the ALARP criterion can also be applied again in the assessment of the riskiness of investments in the civil sector. Even in this case, in fact, the main objective is the triangular balance between risk, mitigation costs and achievable benefits.

3.2   The Capital Asset Pricing Model to Evaluate Risky Projects

Sharpe [32] Lintner [33] and Black [34] introduced the Capital Asset Pricing Model into financial economics. This model extends the market portfolio model introduced initially by Markowitz [35] who argue that investors are risk averse investors and will choose a portfolio by trading off between risk and return for one investment period, [36]. The CAPM a static model of portfolio allocation in conditions of uncertainty and risk aversion, which relates the return r_i of the i-*th* investment with the risk-free return r_f and the market return r_m [37–39] according to the formula:

$$E[r_i] = r_f + \beta \cdot \left(E[r_m] - r_f \right) \tag{4}$$

In (4): E is the expectation operator; r_i and r_m respectively represent the gross return on the investment in question and the return on the market portfolio; r_f, on the other hand, is the return referring to a risk-free investment. The difference between the market return and the risk-free return returns the market risk premium, i.e. the remuneration for the risk assumed by the investor. The β coefficient gives a measure of the systematic, i.e. non-diversifiable, risk of a company and expresses the expected percentage change in the excess return on an investment for a 1% change in the excess return on the market portfolio, so if:

- $\beta = 0$, the investment is risk-free and its return is equal to r_f;
- $\beta = 1$, the investment has the same risk as the market and its return is equal to r_m;
- $\beta < 0$, the investment is risky but its risk level moves "against the trend" to the general average;
- $0 < \beta < 1$, the initiative is risky but less than the market and its risk level moves "in the same direction" as the latter;
- $\beta > 1$, the risk level of the project still moves "in the same direction" as the market but is higher than the average [40–44].

In other words, β summarizes how much the investment in question amplifies the risks affecting the global market. In formula:

$$\beta = \frac{\text{cov}\,(r_i, r_m)}{\text{var}\,r_m} \tag{5}$$

In which the numerator is the covariance between return r_i of the generic investment and market return r_m, while the denominator coincides with the variance of the market return r_m.

To test the model, we write the Eq. (4) in the form:

$$r_i = a_o + a_1 \cdot \beta + u_i \tag{6}$$

where:

$a_o = r_f$;
$a_1 = E[r_m] - r_f$;
r_i = realized return on asset i over our sample;
u_i = expectational error = $r_i - E[r_i]$.

It follows that graphically β corresponds to the inclination of the straight line that best interpolates the excess returns of the investment compared to the excess returns of the market [45–47].

3.3 Estimation of Risk Acceptability Threshold

The *acceptability threshold* T_a, which in the safety field is defined as the limit below which the risk is widely acceptable, in the case study it represents the expected return on an investment project whose risk profile represents the average risk profile of construction companies with "worse" returns, statistically those of first quartiles (0–25%). Therefore:

$$r(\beta_a) = T_a = r_f + \beta_a \cdot (r_m - r_f) \tag{7}$$

In (7) r_f is the risk-free rate, estimated as the average yield on 10-year government bonds. β_a is the "acceptable" systematic risk, a function of the return r_I of the first quartile companies and the return r_m of an ideal "market portfolio" made up of the total manufacturing enterprises of a country. β_a is given by the inclination of the line that best interpolates the average excess returns Y_I of the sector companies represented in the first quartile with respect to the average excess returns X of the market. In formula:

$$X = r_m - r_f \tag{8}$$

$$Y_I = r_I - r_f \tag{9}$$

where r_I is the average *Return of Investment* (ROI) of the sector companies belonging to the lower quartile.

3.4 Estimation of Risk Tolerability Threshold

The *tolerability threshold* T_t, which in the safety environment separates the ALARP area from the unacceptable one, coincides with the expected return of the company statistically representative of the second quartile of data (50–75%). So:

$$r(\beta_t) = T_t = r_f + \beta_t \cdot (r_m - r_f) \tag{10}$$

with β_t "tolerable" systematic risk, a function of both the return r_{II} of the second quartile company and the return r_m of the market. β_t coincides with the inclination of the straight line that best interpolates the excess returns Y_{II} of the average company in the sector with respect to the average excess returns X of the market:

$$Y_{II} = r_{II} - r_f \tag{11}$$

with r_{II} equal to the ROI of the average sector company.

4 Construction Building Projects in Italy: Risk Limit Values

The method described in Sect. 3 is implemented to estimate the tolerability and acceptability thresholds of risk for investments in the "building construction" sector (ATECO 41) in Italy. Once the parameters common to (7) and (10), i.e. the risk-free rate r_f and the market return r_m, have been determined, the "acceptable" systemic risk β_a and the "tolerable" systematic risk β_t are estimated, hence the threshold values T_a and T_t.

4.1 Estimation of r_f and r_m

The risk free rate r_f is equal to the average rate of return of the 10-year Treasury bonds (BTP) in the period 2009–2018 (source: Ministry of Economy and Finance, Department of the Treasury). The elaborations return $r_f = 3.44\%$.

The rate of market return r_m is assumed to be equal to the average ROI of the main 2095 Italian manufacturing companies in the decade 2009–2018 (source: Mediobanca Research Office). From the analyses derives $r_m = 7.76\%$.

The risk premium X, given by the difference between r_m and r_f, is 3.32%.

Table 1 shows the estimates of return rate of the market r_m, the risk-free rate r_f and the risk premium X.

Table 1. Estimate of r_m, r_f and X (sources: Ministry of Economy and Finance, Treasury Department; Mediobanca Studies Office).

Year	2009	2010	2011	2012	2013	2014	2015	2016	2017	2018	Mean
r_f (%)	4.32	4.01	5.25	5.65	4.38	3.00	1.70	1.40	2.14	2.54	**3.44**
r_m (%)	8.10	6.70	8.30	7.30	7.40	7.20	8.40	8.10	7.90	8.20	**7.76**
X (%)											**4.32**

4.2 Estimation of β_a and Acceptability Threshold T_a

The estimation of the systematic risk β_a is necessary to evaluate the acceptability threshold, which separates the ALARP risk region from the "widely acceptable" one. The β_a risk refers to the first quartile Italian companies in the "building construction" sector, which is 25% of the companies studied with the lowest ROI. Therefore, β_a is given by the relationship between two terms: covariance between the return r_I of the first quartile company and the market return r_m; variance of market yield r_m

$$\beta_a = \frac{\text{cov}\,(r_I, r_m)}{\text{var}\,r_m} \tag{12}$$

In a Cartesian plane, β_a corresponds to the inclination of the straight line that best interpolates the average excess returns $Y_I = r_I - r_f$ of the first quartile firms with respect to the average excess returns $X = r_m - r_f$ of the market in the time interval 2009–2018. r_I is estimated as the average ROI of 6949 Italian companies of the first quartile with code ATECO 41. The data are taken from the AIDA database of the Bureau Van Dijk. This database contains financial and commercial information of more than 500,000 companies active on the national territory. The analysis was conducted over a period of only 10 years because the AIDA database provides the values of the financial performance indicators of companies only for the last decade. The elaborations return $\beta_a = 1.26$. The results are summarized in Table 2 and Fig. 1.

The estimate of the determination coefficient R_2 allows evaluating the accuracy of the regression. The value $R^2 = 0.91$ demonstrates a high correlation between the variables. Tables 3 and 4 show the statistical parameters of the regression and the standard error of the terms that make up the regression line, i.e. both the intercept on y and β_a.

Table 2. Estimation of β_a

Year	r_I (%)	r_m (%)	r_f (%)	Y_I (%)	X (%)
2009	1.94	8.10	4.32	−2.38	3.78
2010	2.30	6.70	4.01	−1.71	2.69
2011	2.38	8.30	5.25	−2.87	3.05
2012	2.06	7.30	5.65	−3.59	1.65
2013	1.19	7.40	4.38	−3.19	3.02
2014	2.36	7.20	3.00	−0.64	4.20
2015	3.65	8.40	1.70	1.95	6.70
2016	3.53	8.10	1.40	2.13	6.70
2017	3.76	7.90	2.14	1.62	5.76
2018	4.41	8.20	2.54	1.87	5.66
Mean	2.76	7.76	3.44	−0.68	4.32
COV(X,Y$_I$)	0.00040064	VAR X	0.000317697	β_a	1.261

4.3   Estimation of β_t and Tolerability Threshold T_t

The tolerability threshold T_t, which separates the ALARP risk region from the unacceptable one, is instead a function of the systematic risk β_t. This risk is associated with the second quartile company in Italy. Thus, β_t corresponds to the slope of the line that best interpolates the average excess returns $Y_{II} = r_{II} - r_f$ of the second quartile company with respect to the average excess returns X of the market:

$$\beta_t = \frac{\text{cov}\,(r_{II}, r_m)}{\text{var}\,r_m} \tag{13}$$

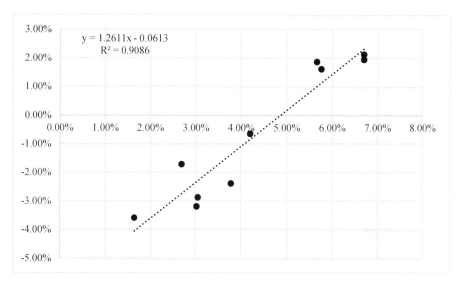

Fig. 1. Regression line for the estimate of β_a.

Table 3. Regression statistics for T_a.

	T_a
R multiple	0.953186
R^2	0.908563
R^2 correct	0.897134
Standard error	0.007563
Observations	10

Table 4. Standard error on β_a and T_a.

	Coefficients	Standard error	Stat t	Significance value	Lower 95%	Higher 95%
T_a	−0.06131	0.006563	−9.34119	1.41E−05	−0.07644	−0.04617
β_a	1.261077	0.141442	8.915839	1.99E−05	0.934911	1.587244

Where r_{II} equals the average ROI of companies with code ATECO 41 of the second quartile. Also for the estimate of β_t the ROI values of the Italian companies of the last decade were used, because they are the only values provided by the AIDA database.

The elaborations, summarized in Table 5 and Fig. 2, return $\beta_t = 0.84$.

Tables 6 and 7 show the statistical parameters of the regression and the standard error of the terms that make up the regression line, i.e. both the intercept on y and the β_t. For the β_t estimation, the analyses reveal a high correlation between the variables, with a value of R^2 equal to 0.93.

Table 5. Estimation of β_t.

Year	r_{II} (%)	r_m (%)	r_f (%)	Y_{II} (%)	X (%)
2009	6.77	8.10	4.32	2.45	3.78
2010	6.20	6.70	4.01	2.19	2.69
2011	6.31	8.30	5.25	1.06	3.05
2012	5.68	7.30	5.65	0.03	1.65
2013	5.79	7.40	4.38	1.41	3.02
2014	5.73	7.20	3.00	2.73	4.20
2015	5.98	8.40	1.70	4.28	6.70
2016	6.08	8.10	1.40	4.68	6.70
2017	6.14	7.90	2.14	4.00	5.76
2018	6.55	8.20	2.54	4.01	5.66
Mean	6.77%	7.76	3.44	2.45	4.32
COV(X,Y_{II})	0.000266225	VAR X	0.000317697	β_a	0.838

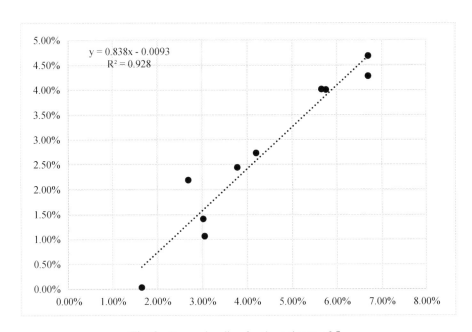

Fig. 2. Regression line for the estimate of β_t.

Table 6. Regression statistics for T_t.

	T_t
R multiple	0.96332
R^2	0.927986
R^2 correct	0.918984
Standard error	0.004413
Observations	10

Table 7. Standard error on β_t and T_t.

	Coefficients	Standard error	Stat t	Significance value	Lower 95%	Higher 95%
T_t	−0.00935	0.00383	−2.44088	0.040509	−0.01818	−0.00052
β_t	0.837984	0.082533	10.15327	7.58E−06	0.647662	1.028307

5 Estimation of Acceptability Threshold T_a and Tolerability Threshold T_t: Discussion of the Results

From the implementation of the formulas (7) and (10), we obtain the value of the acceptability threshold T_a and tolerability threshold T_t:

$$T_a = r_f + \beta_a \cdot (r_m - r_f) = 3.44\% + 1.26 \cdots (7.66\% - 3.44\%) = 8.90\%$$

$$T_t = r_f + \beta_t \cdot (r_m - r_f) = 3.44\% + 0.84 \cdots (7.66\% - 3.44\%) = 7.06\%$$

These thresholds represent limit values useful for assessing the risk and residual risk associated with projects in the "construction of buildings" sector in Italy. Thus, for such investments, the risk is:

- widely acceptable if $E(\text{IRR}) > 8.90\%$, i.e. if the expected internal rate of return is greater than 8.90%;
- unacceptable for the investor if $E(\text{TIR}) < 7.06\%$;
- ALARP 7.06% $< E(\text{TIR}) < 8.90\%$.

It is worth remembering that the European regulatory guidelines do not provide clear indications on the acceptability levels of the project risk. So, the proposed methodology based on the comparison between $E(\text{IRR})$ and performance limit values can be a useful guide for the analyst when judging the economic performance of risky investments.

6 Conclusions

The aim of the paper is to provide the analyst with guidelines for assessing the risk of investments in the "building construction" sector. To this end, risk acceptance thresholds are first defined, then a methodology is outlined to estimate these thresholds.

The theoretical reference useful for establishing the risk thresholds is the ALARP logic. From this logic, widely consolidated in safety risk management in industrial procedures, the concepts of acceptability threshold and risk tolerability threshold are borrowed. According to this logic, investment risk can be defined as low as reasonably practicable if it is included among the above-mentioned thresholds or if the costs to mitigate it appear disproportionate to the benefits that can be pursued [4–8].

To estimate the tolerability and acceptability thresholds of the investment risk, the CAPM is used. Through this approach, it is possible to relate the risk of the project not only to the return of the production sector in which the project falls, but also to the return of the market in its entirety. Indeed, the CAPM allows evaluating the risk-adjusted discount rate $r(\beta)$, which can be interpreted as the expected return from an investment with a risk profile β [9].

Operationally, the joint use of the CAPM and statistical investigation methods enables to calibrate the limit values of risk acceptability and tolerability for investments in a specific sector, based on a statistically significant panel of Return on Investment (ROI) values for companies operating in a specific territorial context. In particular: the acceptability threshold T_a coincides with the expected return of a project with a risk profile that represents the "worst" companies in the sector falling within the survey territory, or those belonging to the lower quartile; the tolerability threshold T_t, on the other hand, is assessed as the expected return of an investment with the risk profile of the "mean" company in the sector.

In the second part of the paper, the analysis of the profitability indices of 6949 companies in the "building construction" sector and 2095 manufacturing companies, all active in Italy in the decade 2009–2018, makes it possible to estimate values of the tolerability thresholds T_t and acceptability thresholds T_a of the investment risk of 7.10% and 8.90% respectively. This means that an investment in building construction in Italy has a risk: a) unacceptable, that is, it is necessary to plan mitigation actions, if E (IRR) < 7.06%; b) which falls within the ALARP area, or is tolerable (in the sense that further mitigation measures have costs disproportionate to the persuasive benefits), if 7.06% < E(IRR) < 8.90%, c) broadly acceptable, therefore there is no need to envisage any containment action, if E(IRR) > 8.90%.

Since it is able to estimate the risk in quantitative terms, the proposed methodology can be of help in the decision-making processes, until it becomes a real economic policy tool.

References

1. Nesticò, A., De Mare, G., Fiore, P., Pipolo, O.: A model for the economic evaluation of energetic requalification projects in buildings. A real case application. In: Murgante, B., et al. (eds.) ICCSA 2014. LNCS, vol. 8580, pp. 563–578. Springer, Cham (2014). https://doi.org/10.1007/978-3-319-09129-7_41
2. Fiore, P., Nesticò, A., Macchiaroli, M.: The energy improvement of monumental buildings. An investigation protocol and case studies. Valori e Valutazioni **16**, 45–55 (2016). ISSN: 2036-2404

3. Nesticò, A., Macchiaroli, M., Pipolo, O.: Historic buildings and energetic requalification a model for the selection of technologically advanced interventions. In: Gervasi, O., et al. (eds.) ICCSA 2015. LNCS, vol. 9157, pp. 61–76. Springer, Cham (2015). https://doi.org/10.1007/978-3-319-21470-2_5

4. Della Spina, L.: A multi-level integrated approach to designing complex urban scenarios in support of strategic planning and urban regeneration. In: Calabrò, F., Della Spina, L., Bevilacqua, C. (eds.) ISHT 2018. SIST, vol. 100, pp. 226–237. Springer, Cham (2019). https://doi.org/10.1007/978-3-319-92099-3_27

5. De Mare, G., Nesticò, A., Benintendi, R., Maselli, G.: ALARP approach for risk assessment of civil engineering projects. In: Gervasi, O., et al. (eds.) ICCSA 2018. LNCS, vol. 10964, pp. 75–86. Springer, Cham (2018). https://doi.org/10.1007/978-3-319-95174-4_6

6. Nesticò, A., He, S., De Mare, G., Benintendi, R., Maselli, G.: The ALARP principle in the cost-benefit analysis for the acceptability of investment risk. Sustainability 10(12), 1–22 (2018). https://doi.org/10.3390/su10124668

7. Aven, T., Abrahamsen, E.B.: On the use of cost-benefit analysis in ALARP processes. Int. J. Perform. Eng. 3(3), 345–353 (2007). https://doi.org/10.23940/ijpe.07.3.p345.mag

8. Jones-Lee, M., Aven, T.: ALARP – what does it really mean? Reliab. Eng. Syst. Saf. 96(8), 877–882 (2011). https://doi.org/10.1016/j.ress.2011.02.006

9. Melchers, R.: On the ALARP approach to risk management. Reliab. Eng. Syst. Saf. 71(2), 201–208 (2001). https://doi.org/10.1016/S0951-8320(00)00096-X

10. Gollier, C.: Pricing the Planet's Future: The Economics of Discounting in an Uncertain World. Princeton University Press, New Jersey, US (2011)

11. Nesticò, A., Maselli, G.: A protocol for the estimate of the social rate of time preference: the case studies of Italy and the USA. J. Econ. Stud. 47(3), 527–545 (2020). https://doi.org/10.1108/JES-02-2019-0081

12. European Commission: Regulation (EU) No 1303/2013 of the European Parliament and of the Council of 17 December 2013. https://eur-lex.europa.eu/legal-content/EN/TXT/PDF/?uri=CELEX:32013R1303&from=it. Accessed 19 Jan 2020

13. Commission, E.: Guide to Cost-Benefit Analysis of Investment Projects: Economic Appraisal Tool for Cohesion Policy 2014–2020. Directorate General for Regional and Urban Policy. European Commission, Brussels, Belgium (2014)

14. De Mare, G., Nesticò, A., Macchiaroli, M.: Significant appraisal issues in value estimate of quarries for the public expropriation. Valori e Valutazioni 18, 17–23 (2017). ISSN: 20362404

15. Dolores, L., Macchiaroli, M., De Mare, G.: A model for defining sponsorship fees in public-private bargaining for the rehabilitation of historical-architectural heritage. In: Calabrò, F., Della Spina, L., Bevilacqua, C. (eds.) ISHT 2018. SIST, vol. 101, pp. 484–492. Springer, Cham (2019). https://doi.org/10.1007/978-3-319-92102-0_51

16. Nesticò, A.: Risk-analysis techniques for the economic evaluation of investment projects. In: Mondini, G., Fattinnanzi, E., Oppio, A., Bottero, M., Stanghellini, S. (eds.) SIEV 2016. GET, pp. 617–629. Springer, Cham (2018). https://doi.org/10.1007/978-3-319-78271-3_49

17. Macchiaroli, M., Pellecchia, V., D'Alpaos, C.: Urban water management in Italy: an innovative model for the selection of water service infrastructures. WSEAS Trans. Environ. Develop. 15, 463–477 (2019). ISSN: 17905079

18. Troisi, R., Alfano, G.: Towns as safety organizational fields: an institutional framework in times of emergency. Sustainability 11(24), 7025 (2019). https://doi.org/10.3390/su11247025

19. Nesticò, A., Maselli, G.: Sustainability indicators for the economic evaluation of tourism investments on islands. J. Clean. Prod. 248, 119217 (2020). https://doi.org/10.1016/j.jclepro.2019.119217

20. Macciotta, R., Lefsrud, L.: Framework for developing risk to life evaluation criteria associated with landslides in Canada. Geoenviron. Disasters **5**(1), 1–14 (2018). https://doi.org/10.1186/s40677-018-0103-7

21. HSE (Health and Safety Executive): The tolerability of risk from nuclear power stations. Her Majesty's Stationery Office, London, UK (1992)

22. HSE (Health and Safety Executive): Reducing Risks, Protecting People. Her Majesty's Stationery Office, London, UK (2001)

23. Health and Safety Executive: Principles and Guidelines to Assist HSE in Its Judgements That Dutyholders have Reduced Risk as Low as Reasonably Practicable (2014). http://www.hse.gov.uk/risk/theory/alarp1.htm. Accessed 20 Dec 2019

24. Benintendi, R., De Mare, G., Nesticò, A.: Upgrade the ALARP model as a holistic approach to project risk and decision management: a case study. Hydrocarbon Process. **97**(7), 77–82 (2018)

25. Vanem, E.: Principles for setting risk acceptance criteria for safety critical activities. In: Bérenguer G., Soares G. (eds.) Advances in Safety, Reliability and Risk Management, pp. 1741–1751 (2012)

26. Aven, T.: Risk assessment and risk management: review of recent advances on their foundation. Eur. J. Oper. Res. **253**(1), 1–13 (2016). https://doi.org/10.1016/j.ejor.2015.12.023

27. Morgenstern, N.R.: Managing risk in geotechnical engineering. In: The 3rd Casagrande Lecture. Proceedings 10th Pan-American Conference on Soil Mechanics and Foundation Engineering, Guadalajara, Mexico, vol. 4, pp. 102–126 (1995)

28. ERM-Hong Kong Ltd.: Landslides and Boulder Falls from Natural Terrain: Interim Risk Guidelines, p. 183. ERM-Hong Kong Ltd., The Government of Hong Kong Special Administrative Region (1998)

29. Ho, K.K.S., Leroi, E., Roberds, W.J.: Quantitative risk assessment application, myths and future direction. In: Proceedings of the International Conference on Geotechnical and Geological Engineering, GeoEng2000, Melbourne, Australia, pp. 269–312 (2000)

30. Leroi, E., Bonnard, C., Fell R., McInnes, R.: Risk assessment and management. In: Hungr, O., Fell, R., Couture, R., Eberhardt, E. (eds.) Landslide Risk Management. Proceedings of the International Conference on Landslide Risk Management, Vancouver, Canada, pp. 159–198 (2005)

31. Porter, M., Jakob, M., Holm, K.: Proposed landslide risk tolerance criteria. In: 62nd Canadian Geotechnical Conference and 10th Joint CGS/IAH-CNC Groundwater Conference, Halifax, Nova Scotia, Canada, pp. 533–541 (2009)

32. Sharpe, W.F.: Capital asset prices: a theory of market equilibrium under conditions of risk. J. Finan. **19**, 425–442 (1964)

33. Lintner, J.: The valuation of risk assets and the selection of risky investments in stock portfolios and capital budgets. Rev. Econ. Stat. **47**(1), 13–37 (1965). https://doi.org/10.2307/1924119

34. Black, F.: Beta and return. J. Portf. Manage. **20**, 8–18 (1993)

35. Markowitz, H.: Portfolio selection. J. Finan. **7**(1), 77–99 (1952). https://doi.org/10.1111/j.1540-6261.1952.tb01525.x

36. Elbannan, M.A.: The capital asset pricing model: an overview of the theory. Int. J. Econ. Finan. **7**(1), 216–228 (2015)

37. Nesticò, A., De Mare, G., Frusciante, B., Dolores, L.: Construction costs estimate for civil works. A model for the analysis during the preliminary stage of the project. In: Gervasi, O., et al. (eds.) ICCSA 2017. LNCS, vol. 10408, pp. 89–105. Springer, Cham (2017). https://doi.org/10.1007/978-3-319-62404-4_7

38. Dolores, L., Macchiaroli, M., De Mare, G.: A dynamic model for the financial sustainability of the restoration sponsorship. Sustainability **12**(4), 1694 (2020). https://doi.org/10.3390/su12041694

39. Tilfani, O., Ferreira, P., El Boukfaoui, M.Y.: Multiscale optimal portfolios using CAPM fractal regression: estimation for emerging stock markets. Economics **32**, 77–112 (2020)

40. Fama, E.F., French, K.R.: Common Risk Factors in the Return on stocks and bonds. J. Finan. Econ. **33**(1), 3–56 (1993). https://doi.org/10.1016/0304-405X(93)90023-5

41. Brealey, R., Stewart, M.: Principles of Corporate Finance. McGraw Hill, New York (1981)

42. Mankiw, N.G., Shapiro, M.D.: Risk and return: consumption beta versus market beta. Rev. Econ. Stat. **68**(3), 452–459 (1986)

43. Fama, E.F., French, K.R.: The capital asset pricing model: theory and evidence. J. Econ. Perspect. **18**(3), 25–46 (2005). https://doi.org/10.2469/dig.v35.n2.1671

44. Wijaya, E., Ferrari, A.: Stocks investment decision making capital asset pricing model (CAPM). Jurnal Manajemen **24**(1) (2020). http://dx.doi.org/10.24912/jm.v24i1.621

45. Rosenberg, B., Guy, J.: Beta and investment fundamentals. Finan. Anal. J. **32**, 60–72 (1976)

46. De Mare, G., Nesticò, A., Macchiaroli, M., Dolores, L.: Market prices and institutional values. In: Gervasi, O., et al. (eds.) ICCSA 2017. LNCS, vol. 10409, pp. 430–440. Springer, Cham (2017). https://doi.org/10.1007/978-3-319-62407-5_30

47. Manganelli, B., De Mare, G., Nesticò, A.: Using genetic algorithms in the housing market analysis. In: Gervasi, O., et al. (eds.) ICCSA 2015. LNCS, vol. 9157, pp. 36–45. Springer, Cham (2015). https://doi.org/10.1007/978-3-319-21470-2_3

International Workshop on Econometric and Multidimensional Evaluation in Urban Environment (EMEUE 2020)

City-Port Circular Model: Towards a Methodological Framework for Indicators Selection

Maria Cerreta[⊠], Eugenio Muccio, Giuliano Poli,
Stefania Regalbuto, and Francesca Romano

Department of Architecture, University of Naples Federico II, Via Toledo 402,
80134 Naples, Italy
{maria.cerreta,giuliano.poli,
stefania.regalbuto}@unina.it,
eugenio.muccio@gmail.com, romano.fra5@gmail.com

Abstract. Maritime transport technologies and infrastructures development fostering global economies since the second half of the 20th century, conditioning the reshaping of the territorial system in terms of social and spatial organisation, detaching port functions from urban ones and weakening their relationships. The reorganisation of the same activities, often within larger areas of the existing urban structure, has emphasised the so-called misalignment of the City-Port, given by dynamics having benefits at the regional scope and localised negative impacts. The European recommendations provided by both the Sustainable Development Goals (SDGs) framework and the Maritime Spatial Planning European Directive 2014/89, within the theoretical and methodological framework provided by Circular City Model (CCM) approach, define the context to reconsider City-Port dynamics, leading to a regeneration of both the port and the city. To identify sustainable design strategies' portfolio for Naples City-Port, in Italy, as pivotal action to trigger the regenerative process, the selection of a suitable set of indicators, within a multidimensional and multi-scale decision support system, has been developed. The study is part of the Italian research Project of Relevant National Interest (PRIN, 2015), "Metropolitan cities: territorial economic strategies, financial constraints, and circular regeneration", coordinated by the Polytechnic of Milan, Italy.

Keywords: Circular Economy · City-Port model · SDGs · Indicators selection · Decision support system

1 Introduction

Strongly influenced by dynamics and urban morphology, port areas, over centuries, have been the economic and social driving force of cities [1], deploying effects both locally and regionally [2]. Nevertheless, the development of maritime transport technologies and infrastructures, fostering global economies since the second half of the 20th century, has been reshaping the territorial system in terms of social and spatial organisation, detaching port functions from urban ones and weakening their relationships. In response

© Springer Nature Switzerland AG 2020
O. Gervasi et al. (Eds.): ICCSA 2020, LNCS 12251, pp. 855–868, 2020.
https://doi.org/10.1007/978-3-030-58808-3_61

to the growing need for autonomy and expansion spaces, in the major European ports, there has been decentralisation and relocation of activities no longer compatible with the urban structure [3, 4]. In Italy, instead, the reorganisation of the same activities, often within more extensive areas of the existing urban structure [5], has emphasised the so-called misalignment of the port-city, given by dynamics having benefits at the regional scope and localised negative impacts. Indeed, if on one hand commercial ports as nodes of the logistic system for international exchange, play a decisive role for the economic development of the territory and the nation, on the other hand, have significant environmental, social and economic impacts on the urban scale [6], influencing well-being and quality of life.

In this perspective, the Circular City Model (CCM) approach [7–9] has been assumed as the theoretical and methodological framework supporting paradigm shift in the context of which to rephrase City-Port development processes [10, 11], following Circular Economy (CE) principles [12, 13]. Furthermore, a significant correlation between the SDGs and the CE indicators has been noticed [14].

Considering long-term horizon where the guidelines for urban planning unfold as well as environmental, economic and socio-cultural impacts that these actions have on the territories, a "place-based" and "use-inspired" approach to transition [15, 16] has been adopted. In this perspective, the City-Port system identifies a specific enabling context [17]. Moreover, it is felt that reconsidering City-Port dynamics within a sustainable approach can support to redefine the relationship between the port and the urban region that surrounds it, leading to a regeneration of both the port and the city [18]. In order to overtake arisen conflicts within City-Port context, the research intends to operationalising the European recommendations provided by both the Sustainable Development Goals (SDGs) framework [11] and the Maritime Spatial Planning European Directive 2014/89, that, coherently with the European principles for integrated urban development [19], aim to implement a multidimensional decision support system working with multi-scale indicators within the theoretical and methodological framework provided by CCM approach.

In the context of literature, a growing body of research related to Sustainable Development Indicators already occurs. Due to the complexity of its context, hence the significant number of considered indicators derives, analysis capabilities of sustainable development indicators system have been often questioned [20].

As part of the decision-making problem in maritime spatial planning, the selection of a consistent set of indicators has been assumed as crucial both for detection of the context features, other than for the evaluation of City-Port system development scenarios.

In order to identify a suitable set of Sustainable Development Indicators, a place-based approach [21] wherein the main characters of the place have been spelt out including both theme-based and field-based indicators, within CCM approach, has been assumed. Therefore, reaping recommendations provided by global policies as the starting point for defining the goal steering transition towards the CE, and considering, at once, local interventions as crucial to turn a goal into reality, multiple dimensions, other than multiple scales have been assumed as part of decision support system. The contribution aims to identify a methodological framework for the selection of suitable indicators, which are useful to describe the City-Port system and evaluate the impacts of likely transformations. The proposed methodological approach has been implemented on the focus area located in the eastern part of Naples City-Port, in the South of Italy, including both the commercial port and the neighbourhood urban settlement. The focus area has been selected based on information, data and issues developed within Master's Degree Course Level II in "Sustainable planning and design of port areas" of the Department of Architecture (DiARC), University of Naples Federico II, Italy.

The paper articulation proceeds as follows: the first part (Sect. 2) highlights the content analysis of selected paper undertaken to identify the most appropriate methodological approach for qualitative indicators' selection; the second part (Sect. 3) shows the methodological approach; the third one (Sect. 4) introduces the case study in order to test the indicators' selection process; and the fourth one (Sect. 5) points out preliminary outcomes, highlighting potentials and shortcomings.

2 Indicators Selection: Approaches and Criteria

Assessing urban sustainability processes for planning new multi-scale strategies, policies and actions means dealing with a wide dashboard of indicators, which are affected by mutual interrelationships at once. As the Sustainable Development Indicators set is still comprehensive, it is necessary to identify criteria for selecting them [22, 23]. As early as 1998 Meadows identified a set of desirable features of sustainable development indicators. Later, other criteria for the selection of sustainability indicators have been proposed [24–26].

Although multiple evaluation methods to foster sustainable development have been experimenting for more than three decades, some conceptual and operational difficulties have to be still addressed [27]. Considering a place-based approach, in an example, the broad availability of indicators at the global scale generally does not correspond to site-specific data and monitoring. Therefore, the availability of information has to be verified at multiple investigation scales.

Moreover, a consistent selection of criteria and indicators at the local scale, within defined theoretical and analytical bounds, is the pivotal step for the effectiveness of sustainability assessments in different contexts [28]. On the one hand, it has been deemed that in social, economic and environmental features of urban systems a certain potential to become circular lies, on the other, CE principles, indeed, had to comply with the peculiarities of the context [29].

According to Munda [30], indeed, the evaluation of a consistent set of indicators can be performed, grouping them into two main categories considering their local and global priorities for sustainability. In this perspective, a methodological approach for indicators' selection has to be chosen in order to manage complex information better, avoiding redundancies. Many authors [31–34] have been tracing back the sustainability indicators selection methods to two main approaches, according to the focused research goals. The approaches mentioned above involve:

- aggregating single indicators into composite indices for judging the sustainability degree;
- managing a system, or dataset, of multidimensional indicators which measure separate features of sustainable development.

Nevertheless, these approaches are not able to solve the selection problem, which, generally, is referred to as internal choices or unilateral decision. At the same time, a partial selection process could lead to misunderstandings and incorrect interpretation of results. In this regard, preliminary operations aiming at indicators' refinement are addressed to check matching with the common criteria of appropriateness, comprehensibility, validity, simplicity, reliability, availability, measurability and sensitivity [35]. In particular, it is crucial evaluating data availability at the local scale, combining soft and hard information at once [36–38]. Evaluating the relationships among SDGs - along with their related indicators - and available local data in qualitative terms is a way to operationalise and internalise the sustainability issues [39].

About the criteria of appropriateness, sensitivity and simplicity, the selection procedures can be performed through qualitative and quantitative methods, according to the grain and quality of available indicators. One of the most recurring methods involves the design of surveys with qualitative evaluation scale to rate indicators respect to their importance on sustainability issues [40]. The respondents can be experts or citizens depending on the type of preferences to be elicited, and the indicators feature to be assessed. Moreover, other authors extend the use of SWOT analysis not only to short-term sustainable operational objectives' evaluation but also to the selection of key indicators of socio-economic monitoring [41]. The primary skills of SWOT analysis, indeed, are inherent to its simplicity and the feasible comparison of internal and external factors of sustainability within groups of people with different expertise. The hybridisation of the methods mentioned above, furthermore, allows the experts to score the indicators and determine their influence degree.

3   The Case Study: East Naples City-Port

Located in the eastern part of Naples City-Port, in the South of Italy, the focus area selected to implement the CCM approach lies on the East coastal zone including both the commercial port and the neighbourhood urban settlement. The area with original agricultural characterisation, as a result of Naples-Portici passenger railway construction, dates back to 1839, experienced a changing pattern in land use. A growing urbanisation process that reached its maximal peak in buildings stock development during the industrial age. Due to its economic and social features, the focus area became a driver for the whole urban settlement. Later, the substantial decrease in demand for industrial activities occurring during the process of de-industrialisation, caused the emergence of abandoned industrial buildings and brownfields, making the coastline even more neglected and inaccessible. Still today there is an unemployment rate of 36,5%. Bordered to the North, to the West and the East by urban landmark (respectively the industrial area, the central railway station of Naples and the Vesuvian municipalities) and bathed to the South by the Tyrrhenian Sea, the port area, localised in San Giovanni a Teduccio neighbourhood has been selected basing on information, data and issues developed within Master's Degree Course Level II in "Sustainable planning and design of port areas" of the Department of Architecture (DiARC), University of Naples Federico II (Fig. 1).

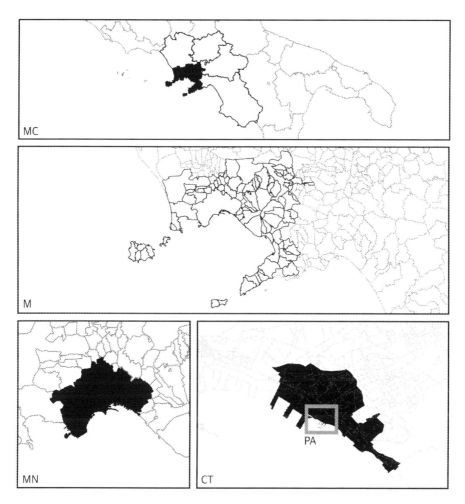

Fig. 1. The East Naples City-Port

Taking into account both the administrative boundaries and the geographical scale covered by the available data, in line with the approach proposed within the Italian research Project of Relevant National Interest (PRIN, 2015), "Metropolitan cities: territorial economic strategies, financial constraints, and circular regeneration", coordinated by the Polytechnic of Milan, the focus area has been furthermore explored according to five levels of analysis corresponding to five different spatial scales related to Metropolitan City (MC), Municipalities (M), Municipality of Naples (MN), Census Tracts (CT) and Port Area (PA). Already recognised as relevant by both Municipality and Port Authority System; indeed, the area is considered strategic for the development of the territorial system that includes the city of Naples, the Metropolitan City and also the Campania Region.

4 Materials and Methods

The research aims to identify a methodological framework for the selection of suitable indicators, which are useful to describe the City-Port system and evaluate the impacts of likely transformations. Therefore, assuming indicators as crucial tools in the evaluative process, a methodological approach (Fig. 2) combining theoretical background and place-based studies has been adopted in order to restore the system of relationships in Naples City-Port, within the perspective provided by CE framework. The study is articulated into four main phases: 1. the identification of a preliminary dashboard of indicators; 2. the decision-making problem structuring; 3. the hybrid qualitative method; 4. Outcome.

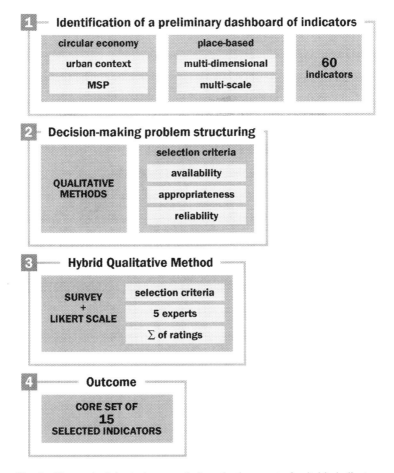

Fig. 2. The methodological approach for selecting a set of suitable indicators

The main research question underlying the decision-making problem leverages on how to choose an appropriate set of indicators for Naples City-Port sustainable transition. In this context, CE principles have been assumed as both theoretical and operative framework in order to reach Sustainable Development. The proposed methodological approach (Fig. 2) leverages on the selection of indicators as a pivotal step in the context of sustainable development assessment. The following paragraphs will specifically show the operating steps of the methodological framework.

4.1 The Identification of a Preliminary Dashboard of Indicators

A preliminary dashboard of indicators has been deduced combining results obtained from research related to the urban context and maritime spatial planning within CE perspective, and the actual data availability found on the ground. Consisting of a long-term strategy aimed at reducing the inefficiency of resources and social exclusion in specific places, the place-based approach, which aims at promoting made-to-measure development policies [16], has been assumed as the operational framework of the indicators' selection process, considering:

- multiple dimensions gathered into the three main categories: environmental, economic and socio-cultural;
- cross-scale perspective consists of five levels of analysis.

A preliminary dashboard of 60 indicators has been thus identified.

4.2 The Decision-Making Problem Structuring

In structuring the decision-making problem, three main dimensions of Naples City-Port corresponding to "Urban and Environmental Quality", "Economic Development" and "Innovation and Culture", has been assumed as strategic fields into which gather indicators. As the preliminary dashboard of indicators related to Sustainable Development is still comprehensive, the definition of selection criteria for structuring a feasible set of indicators has been provided. Considering both topics identified within the scientific literature and peculiarities related to the focus area, three main criteria concerning to "availability", "appropriateness" and "reliability" have been assumed as discriminating factor for indicators' selection.

4.3 The Hybrid Qualitative Method

For the selection of a core set of Sustainable Development Indicators, a combination of survey method and Likert scale evaluation has been employed. The preliminary dashboard including 60 indicators has been discussed within a group of five experts, each of whom, using Likert scale, have assigned a score between 1 and 5 to each of three of them (from 1 = least important, 3 = important and 5 = most important). According to the aforementioned set of selection criteria including "availability", "appropriateness" and "reliability", from the preliminary dashboard of indicators, a core set has been carried out. Considering that the maximum score for each criterion corresponds to 25 and that the minimum score is equal to 5, the threshold for selection was placed at the value of 20. Therefore, only the indicators whose criteria have been assigned a score of at least 20 have been selected. This has allowed the identification of a core set of 15 Sustainable Development Indicators.

5   Outcome

Considering both the theoretical background provided by scientific literature suitably selected and the actual data availability offered by the national database at the different spatial scale of analysis, preliminary dashboard indicators gathered as follows has been obtained:

- environmental dimension, 16 indicators (Fig. 3);
- economic dimension, 17 indicators (Fig. 4);
- socio-cultural dimension, 26 indicators (Fig. 5).

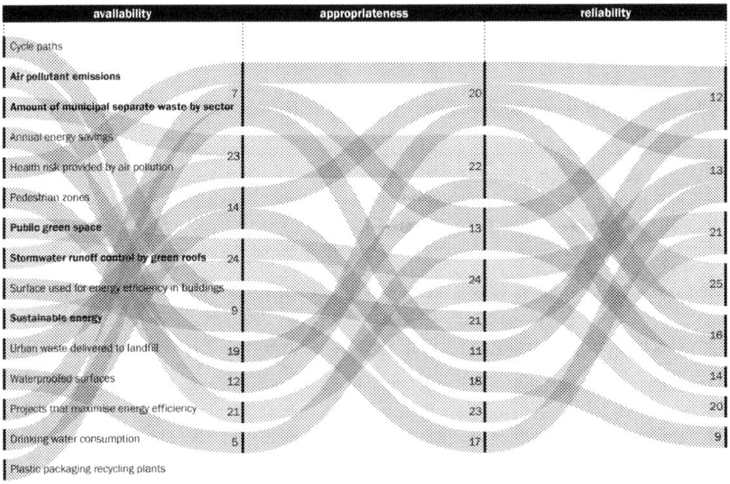

Fig. 3. Environmental indicators selection (16 indicators).

The process of qualitative selection described above defines the 15 selected indicators that have been therefore gathered into three domains corresponding to the strategic objectives already identified in the context of the decision tree (Table 1).

The domains are:

- Urban and Environmental Quality (corresponding to the previous environmental one);
- Economic Development (corresponding to the previous economic one);
- Innovation and Culture (corresponding to the previous socio-cultural one).

The three domains previously identified, have been, indeed, renamed taking into account the features described from the selected indicators. In addition to that, the core set includes indicators related to the spatial scales of analysis previously described, according to a cross-scale perspective. The indicators have been selected both from national databases (ASIA, Borsino Immobiliare, Tagliacarne Institute, Porth System Authority of the Central Tyrrhenian Sea, BES, ISTAT) and scientific literature. Among databases mainly consulted are those of ISTAT and the Port Authority System of the

Central Tyrrhenian Sea (ADSP). In particular, ISTAT - Italian National Institute of Statistics - is the main producer of official statistics in Italy. Its activity includes a census of the population other than social, economic and environmental analyses. Among that, there is BES report aiming to evaluate the quality of well-being in Italy within a multidimensional approach. AdSP, meanwhile, is the public authority dealing with the management of Naples, Salerno and Castellammare di Stabia ports.

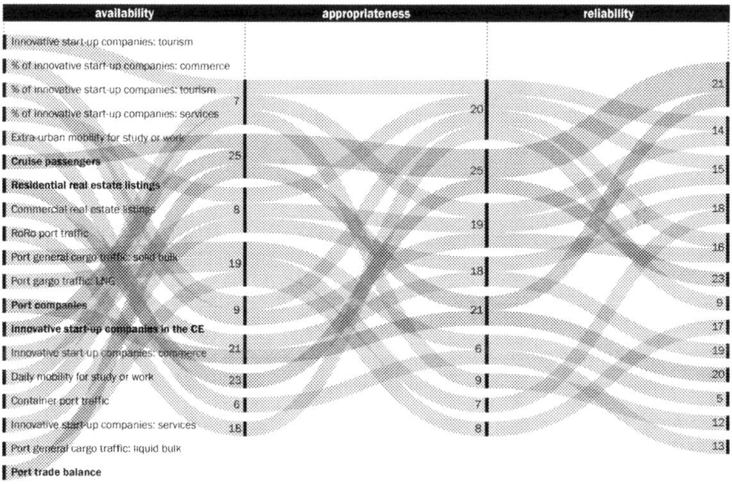

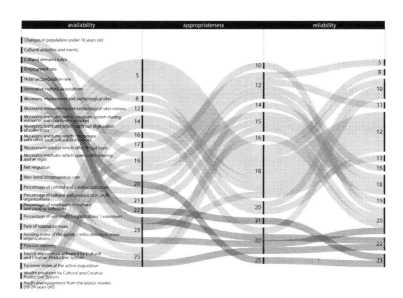

Fig. 4. Economic indicators selection (17 indicators).

Fig. 5. Socio-cultural indicators selection (26 indicators).

It has administrative and financial autonomy, and it is responsible for the entire area belonging to the port. Its statistical studies, therefore, concern the port mentioned above areas mainly in terms of cargo-handling and traffic fleeting. ASIA is a company in the Campania Region that carries out environmental hygiene services such as separate collection, sweeping and transportation of waste to treatment plants.

Table 1. The core set of indicators

Domain	Indicator	UM	Scale	Data source	Year
Urban and Environmental Quality	Sustainable energy	%	Metropolitan City	BES	2016
	Air pollutant emission	Ton/year	Port Area	Authors elaboration	2018
	Public green space	Square metres per inhabitant	Municipality of Naples	BES	2017
	Amount of municipal separate waste by sector	Ton/year	Municipality of Naples	ASIA	2017
	Stormwater runoff control by green roofs	%/year	Port Area	Authors elaboration	2019
Economic Development	Port trade balance	Ton/year	Port Area	AdSP del Mar Tirreno Centrale	2018
	Port companies	n.	Port Area	AdSP del Mar Tirreno Centrale	2019
	Innovative start-up companies in the circular economy	n.	Municipality of Naples	Elaboration Istituto Guglielmo Tagliacarne on Istat data	2018
	Cruise passengers	n./year	Port Area	AdSP del Mar Tirreno Centrale	2018
	Residential real estate listings	€/square metres	Census Tracts	Borsino Immobiliare	2018

(*continued*)

Table 1. (*continued*)

Domain	Indicator	UM	Scale	Data source	Year
Innovation and Culture	Employment rate	%	Municipalities	ISTAT	2011
	Percentage of employees in cultural and creative industries	%	Municipalities	ISTAT	2011
	Innovative cultural associations	n.	Municipality of Naples	Authors elaboration	2019
	Tourism intensity	n.	Metropolitan City	ISTAT	2017
	Demand for cultural heritage	n.	Metropolitan City	ISTAT	2017

The data are published online on its statistical platform. Borsino Immobiliare, a property price list, publishes continuously updated statistical values according to the price fluctuations collected at the local and national level. Finally, Tagliacarne Institute which provides training and research in economics.

6 Conclusions

The methodological process allowed to identify a set of relevant indicators able to analysing the specificities of the context and assessing the possible transformations. Eliciting weights from the preferences of different stakeholders has been representing an open issue within decision-making theories for a long time. This contribution is an attempt to operationalise a simple - even though not comprehensive - methodology for indicators' selection through a scoring technique. The main advantages concerning the proposed approach are inherent to its simplicity, intelligibility and, especially, flexibility in revising the weights faster, whether the stakeholders preference changes. Some limitations of the methodology refer to the linearity of the weighted sum, which supposes independence among variables (criteria) to be assessed. In this regard, it does not include in all the cases the interrelationships among criteria. Moreover, the weights are strongly affected by the expertise of decision-makers, and this makes it complex to elicit different ways of thinking, preferences and trade-offs, so far as the positions are multiple and divergent. Nevertheless, the first shortcoming has been solved by choice of few and separated criteria, e.g. the availability of an indicator does not necessarily exclude its appropriateness or reliability.

The purpose of combining Sustainable Development Goals (SDGs) framework and the Maritime Spatial Planning European Directive 2014/89, within the theoretical and methodological context provided by Circular City Model (CCM) approach, defines the context to reconsider City-Port dynamics and to develop a decision-making process open to complex analysis and description of the City-Port system and to evaluate the impacts of possible transformations. The integration of different domains for indicators identification and selection allows us to consider a plural perspective, in which different specificities converge, including synergistic perspectives and knowledge.

References

1. Wakeman, R.: What is a sustainable port? The relationship between ports and their regions. J. Urban Technol. **3**(2), 65–79 (1996)
2. Geissdoerfer, M., Savaget, P., Bocken, N.M.P., Hultink, E.J.: The circular economy - a new sustainability paradigm? J. Clean **143**, 757–768 (2017)
3. Ducruet, C.: The port city in multidisciplinary analysis. In: Alemany, J., Bruttomesso, R. (eds.) The Port City in the XXIst Century: New Challenges in the Relationship Between Port and City, pp. 32–48. RETE (2011)
4. Hoyle, B.S.: The port-city interface: trends, problems and examples. Geoforum **20**, 429–435 (1989)
5. Cerreta, M., Giovene di Girasole, E., Poli, G., Regalbuto, S.: L'Agenda 2030 per la città-porto: approcci e strumenti per trasformazioni sostenibili. In: XI INU Study Day: Interruptions, Intersections, Sharing and Overlappings. New Perspectives for the Territory. Urbanistica Informazioni, vol. 278, pp. 302–305 (2018)
6. Cerreta, M., De Toro, P. Integrated Spatial Assessment (ISA): a multi-methodological approach for planning choices. In: Burian, J. (ed.) Advances in Spatial Planning, pp. 77–108. IntechOpen, Rijeka, Croatia (2012)
7. ENEL (Ente nazionale per l'energia elettrica): Circular Cities Cities of Tomorrow, 1st edn. (2018). https://www.enel.com/content/dam/enel-com/media/document/cities-of-tomorro_en.pdf
8. ENEL (Ente nazionale per l'energia elettrica): Circular Cities. Cities of Tomorrow, 2nd edn. (2019). https://corporate.enel.it/content/dam/enel-it/media/documenti/circular-cities-cities-of-tomorrow_en.pdf
9. Eurocities: Full Circle, Cities and the Circular Economy. Eurocities, Brussels, Belgium (2017)
10. Prendeville, S., Cherim, E., Bocken, N.: Circular cities: mapping six cities in transition. Environ. Innov. Soc. Trans. **26**, 171–194 (2018)
11. United Nations General Assembly: Implementation of Agenda 21, the Programme for the Further Implementation of Agenda 21 and the Outcomes of the World Summit on Sustainable Development. United Nations (2009)
12. Ayres, R.U., Ayres, L.W.: A Handbook of Industrial Ecology. Edward Elgar, Cheltenham, UK (2002)
13. Ellen MacArthur Foundation: Cities in the Circular Economy: An Initial Exploration (2017)
14. Rodríguez-Antón, J.M., Alonso-Almeida, M.D.M.: The circular economy strategy in hospitality: a multicase approach. Sustainability **11**(20), 5665 (2019)
15. Rosati, F., Faria, L.G.D.: Business contribution to the sustainable development agenda: organisational factors related to early adoption of SDG reporting. Corp. Soc. Responsib. Environ. Manag. **26**(3), 588–597 (2019)
16. Barca, F.: An agenda for a reformed cohesion policy. A place-based approach to meeting EU challenges and expectations. Independent Report (2009). https://www.europarl.europa.eu/meetdocs/2009_2014/documents/regi/dv/barca_report_/barca_report_en.pdf

17. Bonato, D., Orsini, R.: Urban circular economy: the new frontier for European cities'sustainable development, Chap. 12, 2nd edn. In: Clark, W.W. (ed.) Sustainable Cities and Communities Design Handbook, Green Engineering, Architecture, and Technology, pp. 235–245. Elsevier, Amsterdam, The Netherlands (2018)

18. Wakeman, R.: What is a sustainable port? The relationship between ports and their regions. J. Urban Technol. Spec. Issue Port Trends Growth Technol. Sustain. **3**(2), 65–79 (2009)

19. COTER-VI/018: Articolo 307 del TFUE: La riqualificazione delle città portuali e delle aree portuali (2017). https://cor.europa.eu/it/our-work/Pages/OpinionTimeline.aspx?opId=CDR-5650-2016

20. Pissourios, I.A.: An interdisciplinary study on indicators: a comparative review of quality-of-life, macroeconomic, environmental, welfare and sustainability indicators. Ecol. Ind. **34**, 420–427 (2013)

21. Huang, L., Wu, J., Yan, L.: Defining and measuring urban sustainability: a review of indicators. Landscape Ecol. **30**(7), 1175–1193 (2015). https://doi.org/10.1007/s10980-015-0208-2

22. Joumard, R., Nicolas, J.-P.: Transport project assessment methodology within the framework of sustainable development. Ecol. Indicat. **10**, 136–142 (2010)

23. Trans, J., Joumard, R., Gudmundsson, H., Folkeson, L.: Framework for assessing indicators of environmental impacts in the transport sector. Transp. Res. Rec. J. Transp. Res. Board **2242**, 55–63 (2011)

24. OECD (Organisation for Economic Co-operation and Development): Handbook on Constructing Indicators: Methodology and User Guide. OECD and Joint Research Centre (JRC) (2008)

25. UN Indicators of Sustainable Development: Guidelines and Methodologies, 3rd edn. United Nations, New York (2007)

26. da Silva, J., Fernandes, V., Limont, M., Rauen, W.B.: Sustainable development assessment from a capitals perspective: analytical structure and indicator selection criteria. J. Environ. Manage. **260**, 110147 (2020)

27. Fernandes, V., Malheiros, T.F., Philippi Jr., A., Sampaio, C.A.C.: Metodologia de avaliação estratégica de processo de gestão ambiental municipal. Saúde e Sociedade **21**, 128–143 (2012)

28. Pissourios, I.A.: An interdisciplinary study on indicators: a comparative review of quality-of-life, macroeconomic, environmental, welfare and sustainability indicators. Ecol. Indicat. **34**, 420–427 (2013)

29. Levoso, A.S., Gasol, C.M., Martínez-Blanco, J., Durany, X.G., Lehmann, M., Gaya, R.F.: Methodological framework for the implementation of circular economy in urban systems. J. Clean. Prod. **248**, 119227 (2020)

30. Munda, G.: Indicators and Evaluation Tools for the Assessment of Urban Sustainability (2001)

31. World Bank: World Development Indicators 2009. World Development Indicators. World Bank. https://openknowledge.worldbank.org/handle/10986/4367 (2009)

32. Manns, J.: Beyond Brudtland's compromise. Town Country Plan. **2010**, 337–340 (2010)

33. Nguyen, P.T., Wells, S., Nguyen, N.: A systemic indicators framework for sustainable rural community development. Syst. Pract. Action Res. **32**(3), 335–352 (2018). https://doi.org/10.1007/s11213-018-9456-9

34. De Lucia, C., Pazienza, P., Balena, P., Caporale, D.: Exploring local knowledge and socio-economic factors for touristic attractiveness and sustainability. Int. J. Tourism Res. **22**(1), 81–99 (2020)

35. Fischer, T.B., Dalkmann, H., Lowry, M., et al.: The dimensions and context of transport decision making. In: Joumard, R., Gudmundsson, H. (eds.) Indicators of Environmental Sustainability in Transport, pp. 79–102. Les collections de l'Inrets, Paris (2010)
36. Cerreta, M., Panaro, S.: From perceived values to shared values: a multi-stakeholder spatial decision analysis (M-SSDA) for resilient landscapes. Sustainability 9(7), 1113 (2017)
37. Torre, C.M., Morano, P., Taiani, F.: Social balance and economic effectiveness in historic centers rehabilitation. In: Gervasi, O., et al. (eds.) ICCSA 2015. LNCS, vol. 9157, pp. 317–329. Springer, Cham (2015). https://doi.org/10.1007/978-3-319-21470-2_22
38. Murgante, B., Tilio, L., Lanza, V., Scorza, F.: Using participative GIS and e-tools for involving citizens of Marmo Platano-Melandro area in European programming activities. J. Balkan Near East. Stud. 13(01), 97–115 (2011)
39. Cerreta, M., Poli, G., Regalbuto, S., Mazzarella, C.: A multi-dimensional decision-making process for regenerative landscapes: a new harbour for Naples (Italy). In: Misra, S., et al. (eds.) ICCSA 2019. LNCS, vol. 11622, pp. 156–170. Springer, Cham (2019). https://doi.org/10.1007/978-3-030-24305-0_13
40. Xu, Y.: Property-led urban growth, institutional change and urban sustainability: the case of China in an international perspective. In: Caprotti, F., Yu, L. (eds). Sustainable Cities in Asia, p. 322. Routledge, London (2017)
41. Shcherbak, V., et al.: Use of key indicators to monitor sustainable development of rural areas. Glob. J. Environ. Sci. Manag. 6(2), 175–190 (2020)

The Assessment of Public Buildings with Special Architectural Features Using the Cost Approach. A Case Study: A Building Sold by EUR S.p.A. in Rome

Fabrizio Battisti[✉] and Orazio Campo

Department of Planning, Design, Architecture Technology (PDTA), Faculty of Architecture "Sapienza", University of Rome, Via Flaminia 72, 00196 Rome, RM, Italy
{fabrizio.battisti, orazio.campo}@uniroma1.it

Abstract. The growing interest in issues to do with the administration, enhancement and sale of public building stock has increased the need for their valuation and, as a result, has increasingly highlighted the need to identify suitable methods for estimating value that take into account the peculiarities of public buildings. In Italy, public buildings often boast 'special' architectural features (interfloor distance, layout, finishings, type of wiring/heating systems, etc.) that render them 'extra-ordinary' assets; in some cases, these features also confer grand and/or historical importance to such buildings. Thus when carrying out valuation work in Italy, it is necessary to adopt suitable valuation methods designed for buildings boasting special features prior to drafting any administration plans, enhancement projects or even plans for the sale of public building stock. Given the lack of comparable examples or income-based parameters specifically concerning buildings with special features, the Cost Approach is the only method that can be used to estimate the market value of such properties. This paper aims to identify the ways the Cost Approach can be applied when estimating the market value of properties with special features that fall into the 'extra-ordinary' category. The theoretical part will be supported by a practical section where the Cost Approach will be used to estimate the market value of an extra-ordinary landmark building in the EUR district: the "Palazzo della Scienza Universale", previously owned by the EUR S.p.A. company (formerly known as the Ente EUR), which was recently sold (2015).

Keywords: Market value · Cost approach · Appraisal · Public real estate

1 Introduction

A substantial swathe of Italy's public buildings boast 'special' architectural features (interfloor distance, layout, finishings, type of wiring/heating systems, etc.) that render them 'extra-ordinary' assets; in some cases, these features confer a landmark 'monumental' importance to such buildings. According to traditional valuation sciences, such features are considered 'intrinsic positional features' and affect a building's market value [1].

© Springer Nature Switzerland AG 2020
O. Gervasi et al. (Eds.): ICCSA 2020, LNCS 12251, pp. 869–883, 2020.
https://doi.org/10.1007/978-3-030-58808-3_62

Nevertheless, when adopting property valuation procedures, it often proves difficult to estimate the market value of buildings with special features due to the lack of comparable buildings with similar characteristics (and known market prices). Even income data is often unavailable given that in Italy, unlike other countries, local authorities often use public buildings (with special and/or monumental features, for administrative or governmental ends) for free or at a cost that is well below the market price. The growing interest amongst legislators in issues to do with the administration, enhancement and sale of public buildings increasingly highlights the need to identify and establish suitable methods for estimating the market value of buildings that boast special features and are therefore considered 'extra-ordinary' [2, 3]. Such methods could be applied to the valuation of properties prior to drafting administration plans, enhancement projects or even before selling public building stock [4–8].

Given the lack of comparable examples and income-based parameters specifically concerning buildings with special features, the Cost Approach is the only method that can be used to estimate the market value of such properties. This is because the Cost Approach allows us to estimate the market value of a property whilst considering both the value of the land and the cost of its construction and/or reconstruction, taking into account any depreciation [9, 10]. We can be justified in applying the Cost Approach, even to buildings with special features with no comparable examples, because the value of buildable land, understood as being the value of building rights, is in no way influenced by the features (whether they be ordinary or special) found on a building. Moreover, when it comes to the cost, special architectural features can always be linked to technical elements or processes, which usually make it possible to infer data and parameters that render an estimate of the cost of reconstruction feasible. The Cost Approach can also be used to estimate the market value of monuments. In such cases, however, it does not calculate the marginal cost conferred by the historic and/or artistic characteristics of a monument (wherever present). The same can be said for buildings of particular architectural value. In this case as before, the Cost Approach does not calculate the marginal cost due to the particular architectural value of the building.

Generally speaking, the Cost Approach concerns the valuation of special buildings constructed, with either a small market or no market at all [4].

Thus if we want to apply the Cost Approach, we must: i) estimate the cost of the property with special features; ii) estimate the value of the land upon which the building stands.

This paper aims to identify the ways the Cost Approach can be applied when estimating the market value of buildings with special features, i.e. buildings that are deemed 'extra-ordinary'. The method we propose, which is tailored to suit the Italian property market, will be organised in a way that makes it possible to adapt it to a number of different foreign legislative and regulatory circumstances and, in particular, allows it to be used by those local authorities that are called upon to manage, enhance and, above all, sell extra-ordinary public buildings.

The method proposed here involves estimating the value of the land and the cost of reconstruction, including any depreciation, in keeping with established methods of applying the Cost Approach, as codified in international field literature. Land value is estimated in accordance with the most common valuation methods codified by international standards. The most innovative part of the method we propose is the estimate of the cost of reconstruction, including any depreciation, in that by taking as its starting point a sample of the special architectural features most commonly found in public buildings designed for the tertiary sector, it proves possible to estimate the impact that such features could have on the cost, thus investigating the relationship between cost and special features. By identifying the link, in terms of cost, between the ordinary and extra-ordinary features of a building, made clear by suitable correlation coefficients, it proves possible to estimate the cost of reconstructing an extra-ordinary building using a direct method, based on data regarding ordinary buildings; where necessary, it can then be depreciated in line with current international methods [10].

In order to test the operational validity of the model we propose, the Cost Approach will be applied to estimate the value of an extra-ordinary landmark property in the EUR district: the "Palazzo della Scienza Universale", previously owned by the EUR S.p.A. company (formerly known as the Ente EUR), which was recently sold (2015).

2 Context Analysis and 'Evaluative' Background

Among the issues recently debated in the property sector, those concerning the administration, enhancement, use and sale of public buildings have become particularly important. In many countries (the European Union, the USA and Canada for instance), property management models similar to those used in the private sector have been adopted in order to make the most of public assets. Such buildings are often owned by special-purpose entities (SPEs) almost entirely run by public organisations that, however, obey market principles [11]. Public buildings are usually rented to private organisations or local authorities at market prices. Such organisations use measures derived from private legislation in order to generate profit and increase public income, benefiting state coffers. To this end, e.g., the Public Building Service operates at a federal level in the USA, while Canada has Canada Lands Company Limited, Austria has the Bundesimmobiliengesellschaft, Germany has the Bundesanstalt fur Immobilienaufgaben and the UK has the Crown Estate, which manages the British crown's entire estate using trusts [12–15].

This issue has also become a high profile concern in Italy. A succession of new regulations have been drafted over the past 20 years. Articles 58 of legislative decree no. 112 of 2008 and 33 bis of law decree no. 201 of 2011 are particularly important, as they basically state that in order to renovate, manage or enhance the public property owned by regional governments, provincial governments, town councils or other local organisations, each organisation can make a list of the single properties found within its catchment area that are not needed to carry out its institutional functions and that could be enhanced or sold, and can draft a plan for selling or renovating such properties and attach it to its budget, with the support of the "Agenzia del Demanio" state property office [16], if it so wishes, to create a SPE. These new regulations stimulated the need to identify the market value of properties that could be managed, enhanced, used or sold.

Nevertheless, many public buildings boast 'special' architectural features (interfloor distance, layout, finishings, type of wiring/heating systems, etc.) that, *de facto*, render them 'extra-ordinary' assets. Thus it can prove difficult to calculate their market value using a direct methodology (comparison approach), due to the lack of comparables upon which an estimate can be based. Even an income approach to valuation can prove difficult, if not impossible, to carry out due to the lack of comparables. It is only possible in cases where a building with special features makes an income in keeping with market forces. In such cases, the rental income allows us, on the one hand, to measure and assess the real cost of using such an asset (thus aiding the elimination of useless areas), whilst, should a public asset be sold, providing us with definite market-based information – rent paid by the tenant – that can be used to assess market value based on the capitalisation of income, in line with internationally accepted valuation procedures [17–19].

However, local authorities in Italy frequently have the right to use buildings for governmental ends free of charge, or in any case pay leases that are not in line with market rate [20]. In such cases, given the lack of information regarding income, the market value of a public property must be estimated without any information to support its valuation when it is sold.

Given that we cannot use either the comparison approach or the income approach to estimate the market value of an extra-ordinary building, we are forced to identify the features that confer extra-ordinary value and assess their impact on its value, compared to ordinary assets. This principle is also found in RICS valuation protocols, where aspects associated with the valuation of special features must be taken into account by valuers in order to establish what is known as its 'fair value' [19].

In the European Union, valuation processes that take into account the specific characteristics of a building (including unusual features) are codified in documents concerning international valuation standards drafted, e.g., in the UK by the Royal Institution of Chartered Surveyors (the RICS). The RICS has standardised property valuation methods and techniques in its Red Book, which has been the benchmark manual for over 20 years now, and is also used to estimate the value of specific architectural features, including such an estimate within the concept of 'fair value' (see Sect. 4, RICS Valuation - Professional Standard, Red Book). In RICS protocols, the tangible features of buildings, whether they be ordinary or extra-ordinary, taken along with aspects associated with 'productive' features (limits to property rights), have an impact on a building's market value [19].

RICS principles have been adopted in Italian valuation practices as reproduced in Tecnoborsa's Italian Property Valuation Standard (the Codice delle Valutazioni Immobiliari), but there is no systematic treatment of extra-ordinary property valuation [21].

In the light of increasing interest amongst legislators in the sale of public buildings, given the particular nature of Italian public building stock, it becomes essential to establish a suitable valuation procedure that should be adopted before drafting plans to sell public buildings that are often of an extra-ordinary nature, despite the lack of comparables or income data. In such cases, the use of the Cost Approach becomes the only possible way forward.

3 Applying the Cost Approach When Estimating the Market Value of Properties with Special Features

The Cost Approach allows us to estimate the market value (MV) of a property by adding the market value of the land (MVland) to the cost of reconstructing the building, minus any depreciation that may have accumulated at the time of the valuation (CVdep) [22]. Generally speaking, it can be exemplified by the following formula:

$$MV = MV_{land} + CV_{dep}$$

The MVland estimate is included in traditional buildable land valuation methods (Transformation Value Method, Highest and Best Use) [23].

In order to estimate the CV_{dep}, as for the more common cost value, one can either adopt the direct or indirect methodology. However, the indirect methodology, which is carried out by drafting a bill of quantities and their unit prices, can prove difficult due to the complexity of this procedure and may not be possible at all should the special features be the product of unusual construction methods that do not appear on pricelists and thus make it difficult to formulate new prices [24].

When it comes to buildings with special features, valuers must take care when estimating the CV_{dep} parameter. The method we propose below (see Subsect. 3.2) suggests an innovation to the traditional Cost Approach that involves adding a step to the valuation of the CV_{dep} through an 'intermediate' way between direct and indirect method: i.e. detailing a comparison of the cost of reconstruction by applying corrective coefficients that can highlight the differences between an ordinary property whose cost of reconstruction is known and the property with special features that requires valuation. Should a building boast features of monumental, historic, artistic and/or architectural value, we can add a further element to the parameters mentioned above. Estimating such an element is undoubtedly complicated given the extremely variable nature of the concept of monumental, historic, artistic and/or architectural value and the importance of such qualities in properties, so it will not be taken into account below.

So let us move on to establish the methods for estimating the elements that form the Cost Approach when calculating the value of buildings with special features.

3.1 Estimating MV_{land}

The market value of buildable land can be generally estimated in two ways: i) using the direct methodology if information (known prices) regarding similar properties is available (same type of area, similar floor area ratio and purpose, a similar legislative/administrative context); ii) using the indirect methodology with a financial approach or using the Transformation Value Method (TVM) [25, 26], which can be implemented using the following equation:

$$MV_{land} = \frac{MV_{bl} - \Sigma K_{p}}{(1 + r')^{n}}$$

where:

MV_{bl} = is the market value of all properties that have been constructed and can be sold, added together;

ΣK_p = is the total construction cost (with reference to the properties constructed);

r' = is the capitalisation rate of the property transaction;

n = is the number of years required to complete the property transaction.

3.2   Estimating CV_{dep}

The cost of reconstructing an existing building (erected in the past) is calculated by adding together all the costs that a construction company would have to pay at the time of the estimate (present) in order to complete an identical or equivalent building during a hypothetical construction process, linked to market production costs and a particular timescale. In other words, the cost of reconstruction is the cost of constructing an existing building from scratch.

The cost of reconstruction includes the physical cost of reconstructing the building and the other costs that are payable to complete it. To be more precise, the main items included in the cost of reconstruction in the traditional system of organising such projects are the following [27, 28]:

- the physical cost of reconstruction, which in turn includes: a.1) the sum of labour costs, rental costs, materials and transportation; a.2) general costs; a.3) profit (of the construction company);
- bureaucratic/administrative fees;
- legal fees for reconstruction/construction (planning permission/town planning), if required;
- financial costs;
- profit (of the company overseeing the reconstruction project).

When calculating the cost of reconstruction, it may prove necessary to take into account the cost of demolishing, recycling and/or removing the rubble of the existing building, depending on the particular circumstances of each case and the reasons that led to its reconstruction.

Depreciation of the cost of reconstruction is calculated by taking into account, either directly or indirectly, the age of the building compared to its life expectancy.

In order to estimate CV_{dep}, we propose an 'intermediate' way that provides implementation of direct methodology cost estimate plus the application of corrective coefficients to compensate the differences between the assets considered comparable and the building that requires valuation. Special architectural features often make it impossible to find entirely similar buildings where recent prices are known for construction or reconstruction. It is therefore impossible to estimate the cost of reconstructing a building by comprehensively comparing it to assets considered comparable with known, comparable prices.

Thus the valuation of CV_{dep} must be carried out following the next steps:

1. Estimate of a technical unit cost value, reference for the estimate ($CVtu_{rif}$). In this first step must be identified a similar range (to the building being estimated) of recently constructed buildings (that can be considered comparable) whose

construction costs are known [29, 30]; applying the direct methodology we can estimate $CVtu_{rif}$. If the building being assessed is made up of non-homogeneous parts/surfaces, the $CVtu_{rif}$ estimation must be carried out for each of these; in general they are: i) primary areas: serving the building's main purpose; ii) secondary areas: supplementary functions found in lower ground floors or mezzanines; iii) the external area, grouped into the following categories: a) extensions: open spaces within a building (porticoes, balconies, cloisters); appurtenances: open spaces outside a building (courtyards, forecourts, parking spaces).

2. Identifying a specific technical cost value relating to the building requiring evaluation (CVtu). In order to compensate for the differences between the sample range of buildings considered to be comparable enough (due to specific morphological or technological characteristics, etc.) and the building requiring valuation (which could possess features that differ from those found in the comparable sample), and in order to ensure that the overall benchmark valuation figure is consistent with the specific characteristics (the special features) of the building requiring valuation, to extrapolate a CVtu we need to apply corrective coefficients to the abovementioned overall benchmark $CVtu_{rif}$ value considering: Location (ksite), Morphology (kmorph), Structure (kstru), Interior and exterior finishings (kfin), Systems (kimp), Interior and exterior windows and doors (kwind), Ceiling height multiplier (Khm). If multiple types of surfaces are present in the building requiring valuation, this operation must be repeated for each surface. When estimating the CVtu, the above requires the application of the following equation:

$$CVtu = [CVtu_{rif} * (1 + k_{site} + k_{morph} + k_{stru} + k_{fin} + k_{imp} + k_{win})] * K_{hm}$$

3. Calculate the depreciation of the specific technical unit cost value for the building requiring valuation ($CVtu_{dep}$). Having calculated the total physical cost of reconstructing a building with special features (CVtu), we then need to apply the depreciation coefficient, taking into account the building's remaining lifespan in order to estimate the overall depreciated physical cost ($CVtu_{dep}$); the percentage of depreciation therefore corresponds to the percentage of 'operational state' compared to its life expectancy.

4. Estimate the total depreciated cost of reconstruction (CV_{dep}). Having established the technical unit depreciated cost of reconstruction ($CVtu_{dep}$), we can then go on to estimate the physical cost of reconstructing the entire building, multiplying each figure by the single figure estimated for each part of the building. The total amount will be the total depreciated physical cost of reconstruction. Nevertheless, in order to calculate the cost of construction, we need to take into account the sum of the following: the physical cost of construction, bureaucratic fees, any legal costs payable for constructing or reconstructing a building (depending on applicable legislation), financial costs and the profit of the company overseeing the reconstruction.

4   The Case Study: Estimating the Market Value of Palazzo della Scienza, Sold by EUR S.p.A. in December 2015

EUR S.p.A. (90% owned by the Italian Ministry of Economics and Finance and 10% owned by Roma Capitale town council) is a property management and development company, founded in 2000 by the at the time called Treasury Ministes and converted from the "Ente Autonomo Esposizione Universale" di Roma world's fair organisation.

It boasts particularly valuable assets that are extraordinary due to the sheer number of grand architectural landmarks produced by Italy's rationalist movement, unique both in terms of their size and quality. EUR S.p.A.'s property assets also include 70 ha of parks and gardens that are open to the public and considered to be an extraordinary reserve of biodiversity.

Its mission is to manage and enhance its assets in order to maximise their profitability whilst respecting their artistic and historical importance, working with Roma Capitale's heritage department and the Lazio region's office for cultural heritage and landscape.

In keeping with its social purpose and with what is stated in its charter, EUR S.p.A. carries out conservation work safeguarding its historic, artistic and landscape heritage, renting out renovated premises through its property management department. In contrast, its asset management activities involve the implementation of property development projects and urban improvements.

In the early 2000 s, EUR S.p.A. launched its new "La Nuvola" conference centre project; the cost of completing the project (still being defined due to reservations made by the construction company) about €350 million, led to a considerable level of debt, which EUR S.p.A.'s management decided to address by selling four building complexes to INAIL (the national institute for insurance against accidents at work): i) the State Archive buildings; ii) Palazzo delle Arti e Tradizioni Popolari; iii) Palazzo della Scienza Universale; iv) Palazzo della Polizia Scientifica.

The sale of these buildings to INAIL was worth approximately €297.5 million for roughly 92,000 sqm of primary gross floor area and roughly 46,500 sqm of secondary gross floor area. Taking into account the secondary gross floor area with a correlation coefficient of 25%, whilst ignoring the external areas, appurtenances and extensions, the total commercial surface area amounts to approximately 103,000 sqm, from which we can infer a price of roughly €2,900 per commercial square metre. The figures concerning the Palazzo della Scienza Universale building complex's proportions, which allowed us to apply the Cost Approach, were gleaned from publications regarding EUR S.p.A.'s assets (Table 1).

We therefore proceeded to estimate the MVland and the CVdep.

The MVland was estimated using the proposed method, by adopting the direct methodology. Data from 2015 provided us with a market value of buildable land in the EUR district of approximately €950 per sqm of possible gross floor area. Given that secondary areas and porticoes can be constructed without being counted in Floor Area Ratios, if we take the size of the building to be 24,996 sqm (primary and commercial areas), we end up with an MVland rounded up to €23,700,000.

Table 1. The Palazzo della Scienza Universale's proportions

The Palazzo della Scienza Universale building				
Proportions				
ID	Gross floor area	Purpose	Size	
A.1	Primary	Exhibition space, offices, conference hall, apartments	sqm	24,945
A.2	Secondary	Mezzanines, storage rooms, utility rooms	sqm	13,846
A.3a	Exterior: extensions	Porticoes[a]	sqm	2,421
A.3b	Exterior: extensions	Cloisters[a]	sqm	170
A.4	Commercial	Shops	sqm	51
A.5	Exterior: appurtenances	Courtyard	sqm	528
As	Total	(A.1 + A.2 + A.3a + A.3b + A.4 + A.5)	sqm	41,961
Av	Total (volume)		cm	223,000

[a]Approximate unverified data

We went on to estimate the CVdep, in keeping with step 1 of our proposed method, which involved the selection of a sample range of known prices, inferred from publications detailing EUR S.p.A.'s assets, regarding the construction of buildings (Building.n, acronym B.n) that could allow us to infer the basic single figure (CVturif) for our valuation. The single estimated figure of value (2015 data) Cvturif of the Palazzo della Scienza Universale building are: B.1.1. primary and commercial area: €/sqm 1,390; B.1.2 secondary area: €/sqm 1,080; B.2 porticos €/sqm 680. Having inferred the basic single value of each area, we went on to apply corrective coefficients (see step 2 of this proposed method for estimating CVdep); in the current application the corrective coefficients were inferred from EUR S.p.A. records (Table 2); the use of Building Information Modeling (BIM) systems can contribute to the coefficient formation process by highlighting the relationship between type of building work and range.

We then went on to estimate the CVtu of reconstructing the property (Table 3).

In order to apply the depreciation coefficient (step 3 of the proposed method for estimating CVdep) based on EUR S.p.A. records, we were able to infer that the building complex was in excellent condition from a structural point of view. As far as other construction elements were concerned, we calculated that those areas were at 75% of their life expectancy. We therefore proceeded to estimate the CVtudep on the basis of these coefficients.

$$k_{dep} = [(1/1) * 1/3] + [(0.75/1) * 2/3] = 0.84$$

Last but not least, we went on to estimate the CVtdep and other costs in order to come up with the CVdep. Depreciated Cost of Reconstruction (Table 4).

Table 2. Corrective coefficients for estimating CVtu

The Palazzo della Scienza Universale building

a) Primary area

The application of corrective coefficients

Elements	Indicators	Primary areas	Secondary areas	Porticoes	Commercial areas
		Corr. Coeff. (%)	Corr. Coeff. (%)	Corr. Coeff. (%)	Corrective coefficient (%)
Location	Access to construction site	1.50%	1.50%	1.50%	1.50%
	Logistics	0.00%	0.00%	0.00%	0.00%
	Planning restrictions	15.00%	15.00%	15.00%	15.00%
Morphology	Layout	4.00%	0.00%	0.00%	0.00%
Structure	Span between columns	5.00%	2.50%	5.00%	0.00%
	Number of floors	0.00%	0.00%	0.00%	0.00%
	Foundations	0.00%	0.00%	0.00%	0.00%
Finishings	Flooring	1.00%	−1.50%	0.00%	0.00%
	Interior finishings	1.00%	−1.50%	–	0.00%
	Exterior finishings	4.00%	−2.00%	4.00%	3.00%
Systems	Wiring	0.00%	0.00%	–	0.00%
	Plumbing/heating	0.00%	0.00%	–	2.00%
	Mechanical systems	0.00%	0.00%	–	−0.30%
	Air conditioning	3.00%	0.00%	–	0.00%
	Fire prevention	1.50%	1.50%	–	0.00%
	Other (TV, telephone, other cables)	0.70%	0.00%	–	0.00%
Windows and doors	Interior	3.00%	0.00%	–	0.00%
	Exterior	0.00%	0.00%	–	0.00%
C.1a	**Overall corrective coefficient**	**39.70%**	**15.50%**	**25.50%**	**21.20%**
C.2a	**Ceiling height multiplier**	**72.00%** (average h = 7.5 m)	**27.00%** (average h = 5 m)	**72.00%** (average h = 7.5 m)	**27.00%** (average h = 5 m)

By adding the MVland to the CVdep. we were able to calculate the MV of the Palazzo della Scienza Universale building, for an amounts of €129,463,000, which

Table 3. Estimated CVtu

ID	Description	Source	Unit of measure	Value
The Palazzo della Scienza Universale building				
An estimate of the overall cost of physically reconstructing the building				
Primary area (a)				
B.1.1	Overall estimated value inferred from comparables	From Table 2	€/sqm	1,390
C.1.a	Comprehensive corrective coefficient	Source: EUR S.p.A.	%	39.70
	Overall estimated value Surface to Volume Ratio 3.2 lm	B.1.1 * C.1.a	€/sqm	1,942
C.2.a	Ceiling height multiplier	From Table 2	%	72
CVtu.a	Overall physical cost of reconstruction	B.1a * C.2.a	€/sqm	3,340
Secondary areas (b)				
B.1.2	Overall estimated value inferred from comparables	From Table 2	€/sqm	1,080
C.1.b	Comprehensive corrective coefficient	Source: EUR S.p.A.	%	15.50
	Overall estimated value Surface to Volume Ratio 3.2 lm	B.1.2 * C.1.b	€/sqm	1,247
C.2.b	Ceiling height multiplier	Source: EUR S.p.A.	%	27
CVtu.b	Overall physical cost of reconstruction	B.1b * C.2.b	€/sqm	1,584
Porticoes (c)				
B.2	Overall estimated value inferred from comparables	From Table 2	€/sqm	680
C.1.c	Comprehensive corrective coefficient	Source: EUR S.p.A.	%	25.50
	Overall estimated value Surface to Volume Ratio 3.2 lm	B.2 * C.1.c	€/sqm	853
C.2.c	Ceiling height multiplier	Source: EUR S.p.A.	%	72
CVtu.c	Overall physical cost of reconstruction	B.1c * C.2.c	€/sqm	1,468
Commercial area (d)				
B.1.1	Overall estimated value inferred from comparables	From Table 2	€/sqm	1,390
C.1.d	Comprehensive corrective coefficient	Source: EUR S.p.A.	%	21.20
	Overall estimated value Surface to Volume Ratio 3.2 lm	B.1.1 * C.1.d	€/sqm	1,685
C.2.d	Ceiling height multiplier	Source: EUR S.p.A.	%	27
CVtu.d	Overall physical cost of reconstruction	B.1d * C.2.d	€/sqm	2,140

corresponds to an overall estimated figure of €4,550 per sqm, considering a correlation coefficient of 100% for primary areas and 25% for secondary areas. Compared to the actual sale price, from which we can infer an average MV (for the four buildings sold) of €2,900/sqm of commercial area, the application of the Cost Approach provides us with an estimated MV approximately 57% higher than the actual sale value.

Table 4. The CV_{dep} of the Palazzo della Scienza Universale building

The Palazzo della Scienza Universale building – Estimated cost of reconstruction				
Id.		Source	Unit of measure	Value
As	Total gross floor area	From Table 3	sqm	41,961
Av	Total volume		cm	223,000
A.1	Primary areas (exhibition spaces, offices, conference hall, apartments)		sqm	24,945
A.2	Secondary areas (usable mezzanines, storage rooms, utility rooms)		sqm	13,846
A.3a	Extensions (porticoes)		sqm	2,421
A.3b	Extensions (cloisters)		sqm	170
A.4	Commercial areas		sqm	51
A.5	Appurtenances (courtyard)		sqm	528
$CVtu_{dep}.a$	Overall Physical Cost of Reconstruction: primary areas	$CVtu * k_{dep}$	€/sqm	2,800
$CVtu_{dep}.b$	Overall Physical Cost of Reconstruction: secondary areas		€/sqm	1,330
$CVtu_{dep}.c$	Overall Physical Cost of Reconstruction: extensions (porticoes)		€/sqm	1,230
$CVtu_{dep}.d$	Overall Physical Cost of Reconstruction: commercial areas		€/sqm	1,800
Cea	Cost of renovating exteriors	Source: EUR S.p.A.	€/qm	61
ID	Cost entries	Source	Amount	
CVt_{dep}	Physical Cost of Reconstruction	(A.1 * CVtudep. a) + (A.2 * CVtudep. b) + (A.3a * CVtudep. c) + (A.4 * CVtudep. d)	€	91,310,000
K.1	Cost of renovating exteriors	(A.3b + A.5) * Cea	€	43,000

(*continued*)

Table 4. (*continued*)

Id.		Source		Unit of measure	Value
The Palazzo della Scienza Universale building – Estimated cost of reconstruction					
K.2	Bureaucratic fees	5%	of CVtdep + K.1	€	4,570,000
K.3	Financial costs	5%	of CVtdep + K.1 + K.2	€	4,800,000
K.4	Profit of company overseeing the reconstruction	5%	of CVtdep + K.1 + K.2 + K.3	€	5,040,000
CV_{dep}	**Total**		**CVt_{dep} + K.1 + K.2 + K.3 + K.4**	€	**105,763,000**

5   Conclusion

When we lack comparables and/or income data, we can still estimate a property's market value thanks to the Cost Approach. The Cost Approach was adjusted to estimate buildings with special features. The crucial step when using the Cost Approach to carry out a valuation of such buildings is the estimate of its CVdep. Further improvements to our proposed method could therefore aim to identify coefficients concerning differences in the cost of various construction elements that are 'ordinarily' present, based on a sample of relevant data taken as a benchmark, perhaps using multiple regression models or tools for checking the cost of construction work involving BIM processes or patents such as the SISCO [31, 32]. Other developments could involve the inclusion of the element of value linked to historical importance in our proposed procedure.

While we may well believe that this proposed procedure requires further work, the results it produced when applied to the Palazzo della Scienza Universale building complex have been interesting; compared to the actual sale price, the Cost Approach generated a significantly higher market value. Such results support our belief that the Cost Approach for estimating the market value of buildings is particularly useful when selling public buildings that boast special features.

References

1. Forte, F.: Appraisal and evaluation in the production of works for the transformation of the contemporary city (2017). https://doi.org/10.1007/978-3-319-49676-4_10
2. Guarini, M.R., Chiovitti, A., Battisti, F., Morano, P.: An integrated approach for the assessment of urban transformation proposals in historic and consolidated tissues (2017). https://doi.org/10.1007/978-3-319-62398-6_40
3. Manganelli, B.: Real Estate Investing: Market Analysis, Valuation Techniques, and Risk Management, pp. 1–210 (2015). https://doi.org/10.1007/978-3-319-06397-3
4. Yeh, I.C., Hsu, T.K.: Building real estate valuation models with comparative approach through case-based reasoning. Appl. Soft Comput. **65**, 260–271 (2018)

5. Connellan, O.P., Britton, W., Crofts, M.K.: The Cost Approach to Valuation: A Research Report for the RICS (1991)
6. Guo, J., Xu, S., Bi, Z.: An integrated cost-based approach for real estate appraisals. Inf. Technol. Manage. **15**(2), 131–139 (2014)
7. Huber, W.R., Messick, L.P., Pivar, W.H.: Real Estate Appraisal: Principles and Procedures. Educational Textbook Company (2006)
8. Artemenkov, A.I., Mikerin, G.I., Artemenkov, I.L.: Professional valuation and investment-financial valuation: distinctions in valuations for private and public markets. Appraisal J. **76** (4) (2008)
9. Langston, C.: Life-Cost Approach to Building Evaluation. Routledge (2013)
10. Baxter, J., French, N., Gabrielli, L.: Market value and depreciated replacement cost: contradictory or complementary? J. Property Invest. Financ. (2007)
11. Chinloy, P.: Real Estate: Investment and Financial Strategy, vol. 1. Springer Science & Business Media (2012)
12. https://www.big.at/
13. https://www.gsa.gov/about-us/organization/public-buildings-service
14. https://www.bundesimmobilien.de/
15. https://www.thecrownestate.co.uk/
16. https://www.agenziademanio.it/opencms/it/
17. Lundström, S., Lind, H.: Valuation of public real estate: context and concept. J. Property Valuation Invest. (1996)
18. Yeh, I.C., Hsu, T.K.: Building real estate valuation models with comparative approach through case-based reasoning. Appl. Soft Comput. **65**, 260–271 (2018)
19. Royal Institution of Chartered Surveyors: RICS Appraisal and Valuation Standards. RICS, London (2017)
20. Vermiglio, C.: Public property management in Italian municipalities. Property Manag. (2011)
21. Giammaria, V., Bambagioni, G., Simonotti, M., Tecnoborsa & Associazione Bancaria Italiana: Codice Delle Valutazioni Immobiliari: Italian Property Valuation Standard. Tecnoborsa, Roma (2018)
22. Burada, C.O., Demetrescu, T.C.: Historical real estate valuation by cost approach. In: Applied Mechanics and Materials, vol. 880, pp. 371–376. Trans Tech Publications (2018)
23. Morano, P., Tajani, F.: Break even analysis for the financial verification of urban regeneration projects (2013). https://doi.org/10.4028/www.scientific.net/AMM.438-439. 1830
24. Ashworth, A., Perera, S.: Cost Studies of Buildings. Routledge (2015)
25. Acampa, G., Giuffrida, S., Napoli, G.: Appraisals in Italy identity, contents, prospects [La disciplina estimativa in Italia Identità, conoscenza, prospettive]. Valori e Valutazioni (20), 13–32 (2018)
26. Battisti, F., Campo, O.: A methodology for determining the profitability index of real estate initiatives involving public-private partnerships. A case study: the integrated intervention programs in Rome. Sustainability (Switzerland) **11**(5), 1371 (2019). https://doi.org/10.3390/su11051371
27. Battisti, F., Campo, O., Forte, F.: A methodological approach for the assessment of potentially buildable land for tax purposes: the Italian case study. Land **9**(1) (2020). https://doi.org/10.3390/land9010008
28. Campo, O., Rocca, F.: The parameterization of physical quantities in the definition of parametric costs the legislative decree n. 50/2016 on public works design. Valori e Valutazioni **19**, 3–9 (2017)

29. Guarini, M.R., Buccarini, C., Battisti, F.: Technical and economic evaluation of a building recovery by public-private partnership in Rome (Italy) (2017). https://doi.org/10.1007/978-3-319-49676-4_8
30. Guarini, M.R., Battisti, F., Buccarini, C.: Rome: re-qualification program for the street markets in public-private partnership. A further proposal for the flaminio II street market (2014). https://doi.org/10.4028/www.scientific.net/AMR.838-841.2928
31. Giovanna, A., Crespo Cabillo, I., Marino, G.: Representation by drawing against simulation by BIM systems. opportunity or threat for architecture [Representación del dibujo frente a simulación de los sistemas BIM. oportunidad o amenaza para la arquitectura]. Archit. City Environ. 14(40), 111–132 (2019). https://doi.org/10.5821/ace.14.40.6689
32. Acampa, G., García, J.O., Grasso, M., Díaz-López, C.: Project sustainability: criteria to be introduced in BIM. Valori e Valutazioni 2019(23), 119–128 (2019)

A Comparison of Short-Term and Long-Term Rental Market in an Italian City

Benedetto Manganelli[1]([⊠]) ⓘ, Sabina Tataranna[1],
and Pierfrancesco De Paola[2] ⓘ

[1] School of Engineering, University of Basilicata, Viale dell'Ateneo Lucano,
85100 Potenza, Italy
{benedetto.manganelli,sabina.tataranna}@unibas.it
[2] Department of Industrial Engineering, University of Naples Federico II,
Piazzale Vincenzo Tecchio 80, 80125 Naples, Italy
pierfrancesco.depaola@unina.it

Abstract. The increasing trend of the amount of properties entering in the short-term rental market has been involving most of global and European capitals. Latest innovative computer technologies, especially digital platforms, the availability of a housing stock until now underutilized or unused and changing styles of travel, concurrently, are the factors that have determined the success of Airbnb. The opportunity to obtain an extra income makes housing owners behaving as businesspeople, and therefore, entering their properties in the short-term rental market, when the latter is particularly profitable compared to the long-term rental market. However, the extent of this profit, and how it is distributed spatially between different hosts, requires some analysis considering specific conditions about the local real estate market. This research present compares the profitability of the two rental market segments in the city of Bologna. This study integrates in a research field, both at international and at national sphere, currently.

Keywords: Airbnb · Real estate market · Short-term rent · Long-term rent · GIS

1   Introduction

In general, the success of the sharing economy in real estate field is a consequence of many factors. On the one hand, there is the dynamic of existing housing stock, which had sized and designed specifically for lifestyle, needs and family size outdated, at this point, and that instead now is very often underutilized or unused. On the other hand, there is a new style of travel, based, not only, on the deep knowing of the physical place but also on interests which could be defined of anthropological nature. Latest innovative computer technologies, especially digital platforms have allowed these factors to meet according to the law of supply and demand.

In the last decade, it has been observed an increasing trend of the amount of listing available on different digital platforms. Currently, Airbnb is one of the leader companies in the sphere of economic sharing. It is one of the largest digital marketplaces in

© Springer Nature Switzerland AG 2020
O. Gervasi et al. (Eds.): ICCSA 2020, LNCS 12251, pp. 884–898, 2020.
https://doi.org/10.1007/978-3-030-58808-3_63

the world offering about 7 millions of accommodations, 40,000 handcrafted activities, available in over 220 countries and over 100,000 cities. The real estate offer is particularly diversified, including not only flats and tiny-houses, but also castles and tree house.

A study aimed to identify the macroeconomic factors that have contributed to the growth of the listings, has observed a positive correlation with the part GDP resulting from the tourism and trade industry, with the nominal exchange rate and with the technological development index [1].

The phenomenon of functional conversion of housing into hotel "hotelization" so-called [2], b&b for short term rent, alter the balances of the urban zones scheme guaranteed from the rules of urban planning. This has raised the attention of local governments, worried about the potential consequences in terms of environmental and economic impact and social equity, not only in the central areas but also in the others neighbourhoods of the city.

Even though the centrality remains one of the most decisive in the choice to convert the house use, as demonstrated by listings location, some studies have observed, at an international level, a greater dispersion of the same, which have spread over time also to the peripheral residential neighbourhoods, close to the central areas [3–5]. There are some cases, such as the European cities of Berlin, London, and Milan, where the spatial distribution of Airbnb listings is more scattered and it was developed, early, in residential neighbourhoods, often located in suburban area, that turn out to be very attractive and highly recognizable [6, 7].

A study carried out for the main tourism-orientated Italian cities has highlighted that in the inner cities the percentage related to the 'entire apartment' typology is greater than the 'single room' and 'share room' typologies [6]. The same research has shown that this phenomenon is growing and, in some cities, has reached high values, such as 18% in Florence, 25% in Matera, 8% in Rome. This could lead to progressive gentrification process.

About the possible property externalities due to short-term rent in residential neighbourhoods, conflicting views appear from researches. Some researchers assert that the economic affect local companies due to the growing demand for goods and services, increase market prices of properties caused by increasing of demand for retail spaces [8]. The latter dynamic could produce a phenomenon of informal regeneration of the suburbs [9]. Other authors [3] instead, showed that the flow of tourism and thus the presence of strangers inside a building and neighbourhoods, and the consequent reduction of perceived degree of safety, the increasing of noises, the decline of social cohesion of neighbourhood relationships, caused the decrease of the property market values.

A research work carried out for European cities to estimate the effect of Airbnb on hotels highlighted a dual effect. On the one hand a decrease of hotel occupancy rates, and on the other hand an increasing of the average daily rates and the consequence positive effect on total hotel revenue. However, the latter condition has greater likely to be reached by high-end hotels that meet a demand less elastic to price changes [10]. Therefore, the low-cost hotels are the most disadvantaged class. The same findings have been confirmed by a research carried out on the city of Venice, highlighting that

Airbnb accommodations are a business rivals of hotels of intermediate price category rather than luxury and cheaper hotels [11].

This research provides a comparison between the profitability the Short-Term and Long-Term rental market. For this aim, we used Short-Term rent data extracted directly from Airbnb platform by Inside Airbnb, while the Long-term rent data has been extracted from Italian Real Estate Market Observatory.

The second section reports a short review of researches aimed to evaluate possible interactions between Short-Term rental market and the Long-Term rental market. The third section explains the case study, the data, the data analysis methodology and the results. Finally, in the fourth section the conclusions and the future research developments are reported.

2   Airbnb Effects on the Local Long-Term Rental Market

On the supply side, Short-Term rent can be an alternative to the Long-Term rent. Usually, expected profits related to the two rental market segments are different each other. The measure of this difference could make owners move from one to the other, and in special conditions, choose to invest through acquisition of new real estate properties, to be included in the short-term rental market.

In Italy, these conditions could be produced by incipient introduction of fiscal incentives aimed to reduce the costs of refurbishment able to upgrade seismic and energy performances, not only for first residential houses but also for second residential houses. The reduction of these refurbishment's costs would make the entire real estate investment process, refurbishment and subsequent rent more economic and financial convenient. As result, a probable increase of house demands as investment, usually not influenced by the possibility to obtain a profit into long term rental market [12]. However, the validity of this hypothesis requires further investigations about eventual sale prices fluctuations, these being linked not only to the refurbishment and maintenance costs, but also to other factors, intrinsic and extrinsic to the asset [13, 14]. Furthermore, permanence of the above conditions, intended as the time frame in which the fiscal incentives are effective, will can influence on the size of phenomenon, thus it should be considered.

The relationship between the two market segments has been examined in several national and international research works. Moreover, at national level, researches did not explore the issue of regulations, impact of the phenomenon on traditional local tourism companies, prices development, and correlation with macroeconomics factors. A short review of international researches about interaction between two segments has reported, below.

The study carried out by Ayouba K. et al. (2019) [15] on eight French cities has shown that the density of Airbnb rentals puts upward pressure on rents in Lyon, Montpellier, and Paris, respectively 0.3851%, 0.3982% and 0.5242%, whereas it has no significant effects in the other cities (Nantes, Marseille, Nice, Toulouse and Bayonne). An increasing of density of professional Airbnb rentals, by host with at least two lodgings and/or more than 120 days of reservations per years has a greater effect on rents in Paris with 1.2372%, while it no longer affects Lyon and Montpellier.

Analogously, Segù M. (2018) for Barcelona found that presence of Airbnb is responsible for 4% increase in Long-Term rents [16]. Horn et al. (2017) showed that, in Boston, a one standard deviation increase in Airbnb listing is associated with an increase in asking rents of 0.4% [17].

While in other cities, such as Prague the quantification of the percentage attributable to Airbnb rents, is difficult due to numerous factors influencing prices in local rental market. One of these is growing of demand linked to an increasing demographic trend. This latter and the lack of housing, even with a low number of Airbnb listings, could reduce housing supply in the long-term rental market, significantly, and a consequent price increasing [18].

In New York, Coles et al. [3] determined the Break-even Nights, intended as the number of nights that allow reaching in short-term rent market the same revenue of the long-term market, assume lower value in lower-income and middle-income neighborhoods, and in neighbourhoods outside of Manhattan, such as Brooklyn and Queens. On contrary, it assumes higher value in Manhattan. This because the premium that long-term renters are willing to pay to live in the city's most central and highest rent neighborhoods exceeds the premium that short-term renters are willing to pay to live in those same neighbourhoods. Moreover, the most spread Airbnb property typology in less-central neighbourhoods is 'single room' suggesting that most hosts are residents rather than investors or professional hosts. The same study highlighted that over the analysis period, 2011–2016, short-term rentals have become less profitable compared to those of long-term rentals. The risk of conversion of long-term rental property into Airbnb is relatively low, consequently.

Based on the mentioned researches, the profitable of Short-term rental market and effects in terms of prices of long-term rents, are localized: these change from one city to another, and in the same city from one neighbourhood to another. Moreover, it is no simple to evaluate the impact of Airbnb platform on long-term rental market due to several macroeconomic variables that condition it simultaneously, for examples: employment rate, GDP, demographic trend, availability of housing property in the rental market.

Therefore, governments, to control the growing of the Airbnb phenomenon, should consider of the above aspects; hence select the best kind of regulations, such as restriction of nights rental, fiscal measures, authorizations or permits.

3 Case Study

3.1 Bologna and the Touristic Flow

Bologna is the city chosen as the case study. It has been selected because it represents one of Italian art cities with greater tourist vocation. From the report *"Tourism in the metropolitan city of Bologna"* dated November 2019 [19], appears that 2,372,172 tourist arrivals and 4,729,192 tourist presences were recorded over 2018, respectively 8.19% and 2.64% increasing compared to 2017. In comparative terms, the arrivals have increased since 2014 by 44%, against 20.2% of the Italian average. The 'average length of stay' of tourists in accommodations of town is about 2 nights, against 3.35 nights of Italian one (Table 1).

Table 1. Average period of stay

Stay Index (Nights)					
Area	2014	2015	2016	2017	2018
Bologna	1.96	1.89	2.01	2.12	1.98
The rest of the metropolitan city	2.02	1.93	2.03	2.07	2.01
Metropolitan city	1,98	1.90	2.02	2.10	1.99

In Bologna, since May to September, arrivals of foreign visitors exceed those of Italian, while over autumn and winter trends are switch. In fact, foreign tourists do not prefer Bologna as destination to Christmas holidays. The graph (Fig. 1) shows that alternation of the two tourist flows over year, guarantees an almost constant number of attendees during the different seasons.

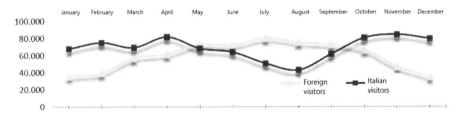

Fig. 1. Attendees per month-2018

The same report highlighted another event: Bologna has involved by a real outbreak of other tourist accommodations than hotels, over 2018, that represents a 20% increasing compared to 2017 (Fig. 2). Simultaneously a stagnation of traditional tourist accommodations, such as hotels, has observed. The growing of other tourist accommodations reached about 70% since 2013 to 2018, greater than 48% Italian average growing. In Bologna, 64.11% of other tourist accommodations are rental housing. The above increasing is likely due to the n. 50/2017 Legislative Decree [20], which had introduced a tax for short-term rental contracts, related to a length of stay less than 30 days, stipulated by natural persons or intermediaries.

The attendees rate, approximately constant over the entire year is a favorable factor to development of Airbnb phenomenon, as an alternative to long-term rents, certainly.

However, the way in which the increasing has spread over neighbourhoods, and possible effects, in terms of spatial spread and development of listings trend caused by introduction of a different kind of rules, such as the restriction of nights rental, have not been investigated yet.

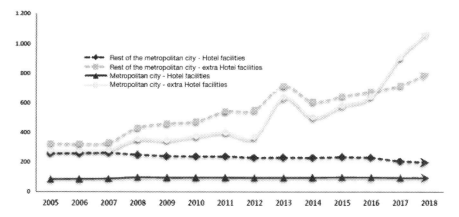

Fig. 2. Time series of the number of accommodation, hotel and non-hotel structures in the city (Bologna) and in the rest of the metropolitan city

3.2 Housing Stock in Bologna

A large percentage of available housing stock, which has been resulting unused, is a decisive factor to spread of short-term rents, as already mentioned.

From a report [21] about the development of housing stock, published in 2005, appears that, in 2001, no inhabitants occupied 21,500 housing. Two-thirds (2/3) of these (14,500 units or 7% of the entire housing real estate) accommodated university students. The remaining one-thirds (1/3), about 7,000 units, were unused in a systematic or occasional way, for example vacant houses or available to owners, or in a bad state of conservation. These latter represent 3–4% of existing real estate.

In 2011 [22] the preliminary data of housing units is about 206,582 in Bologna, and 88.4% of these was occupied by resident families. These data are in line with those provided by a study [23] carried out on Italian art cities, particularly for city of Bologna it provides that the ratio between number of Airbnb listings and available residential asset is 2.7%.

In 2018, the real estate registry consists of 225,000 residential units. The residential asset has grown by over 4,200 units (+2%) over 9 years, and 630 of those over the last year. New units become, not only by new constructions, but also very often by split of property, merger of existing properties, or amendment related to new units entering in the real estate asset registry [24].

All these phenomena are probably linked not only to a decrease of housing space consequently to those of number of members in the household, but also to conversion of housing into studios or mini-apartments to be rented in the short-term rent market.

3.3 Data

Airbnb Data

The data have been extracted from Inside Airbnb, a no commercial web site that collect data related to listings from web site of Airbnb directly. The city of Bologna shows the same trend of other Italian cities (Fig. 3) characterized by a prevalence of listings related to entire apartments compared to private rooms and shared rooms. Only a part of provided information for each listing have been useful to the present research, as following:

- ID codes, associated to each listing and to each host;
- Spatial information: geographical coordinates (latitude, longitude) and neighbour hoods;
- Property typology: single room, share room, entire apartment; number of bathrooms, number of bedrooms, number of guests, description of property provided by hosts;
- Number of reviews in last twelve months;
- Minimum length of stay;
- Availability over year;
- Price per night.

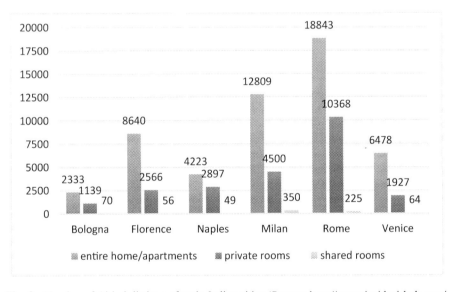

Fig. 3. Number of Airbnb listings of main Italian cities (Source: http://www.insideairbnb.com/, last accessed on 10th June 2020)

In order to compare the profitability of the two segments of market, short-term rent and long-term rent, twelve months, from December 2018 to November 2019, have been chosen as analysis period. Only listings related to property typology of entire apartment and at least one review have been taking into account (Fig. 4). The analyzed sample consist of 830 properties.

The description of property provided by hosts have been undergone text analysis aimed to extract information about the apartment surface area, where host had declared it. Using observed surfaces data related to 384 apartments, a multivariate regression

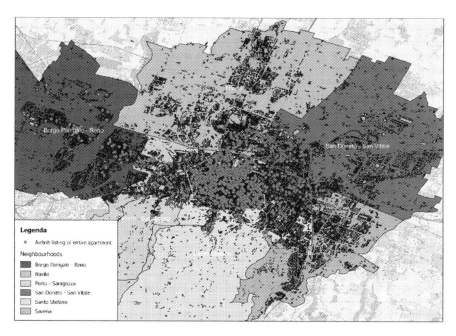

Fig. 4. Listings of Entire apartments

analysis allowed to determinate surface (independent variable) as a function of following dependent variables: number of guests, number of bathrooms and number bedrooms.

The regression Eq. (1) and check parameters of analysis are showed in Tables 2 and 3.

$$sqm = -8.7187 + 4.3192 * guests + 28.4993 * bathrooms + 18.4445 * bedrooms \tag{1}$$

Table 2. Regression statistics

R	0.7820	**R-Squared**	**0.6115**	Adjusted R-Squared	0.6084
MSE	539.727	S	23.2320	MAPE	24.7245
Durbin-Watson	2.0573				
PRESS	210314.8713	PRESS RMSE	23.4028	Predicted R-Squared	0.6016

Table 3. Regression parameters

	Coefficients	Standard error	t Stat	p-value	H0 (5%)
Intercept	−8.7186985	3.8043726	−2.2917573	.0224659	Rejected
Guests	4.3191848	1.1196786	3.8575220	.0001345	Rejected
Bathrooms	28.4992876	3.5281220	8.0777501	.00000	Rejected
Bedrooms	18.4444509	2.2048502	8.3653987	.00000	Rejected

Gross revenues have been calculated based on the occupancy rate, which has been estimated trough the methodology suggested by Inside Airbnb, therefore using a 50% review rate to convert reviews to estimated bookings[1] and then, multiply these latter by the minimum length of stay and by average price. At the end the ratio of gross revenue per month and surface area, has been calculated. This ratio represents a parameter useful to the comparison with the revenues of long-term rental market.

Data of Long-Term Rental Market

The data about the profitability of long-term rental market have been extracted from the database of Real Estate Market Observatory[2]. Based on market values and rental prices, local real estate market is shared into homogeneous areas and for each of these a price range is provided as a function of the state of conservation and building typology. The real estate asset of Bologna consists of 28 areas with a residential destination. An indicator of profitability as the average value of the range's rental prices, with reference to the consultation of first six months 2019, related to the more spread building typology in the area (Fig. 5), has been considered.

3.4   The Comparison Between Long-Term and Short-Term Rental Market

Bringing together as a function of the above-mentioned areas, larger part of Airbnb listings are located in inner neighbourhoods city or those adjacent to these. Particularly, 18% of the sample is in central areas, 43% in those adjacent to these, 37% in peripheral areas, and at the end 2% in suburban areas. The number of Airbnb listings to each area is reported in Fig. 6.

The ratio (Q) of the gross revenues per month and surface area, respectively, of short-term rent and long-term rent allowed for a comparison between the profitability of the two rental market segments. The ratio has been calculated to each individual listing. The obtained values have formed two main classes $Q < 1$; $Q >= 1$. Then, each main class has been divided in subclasses. The cases with a greater Q ratio than one, which have shown economic advantage in the short-term rental market, and the corresponding secondary classification, are reported in Fig. 7. While the cases with a less Q ratio than one, with an economic advantage in the long-term rental market are reported in Fig. 8.

---

[1] http://insideairbnb.com/about.html.

[2] https://wwwt.agenziaentrate.gov.it/geopoi_omi/index.php.

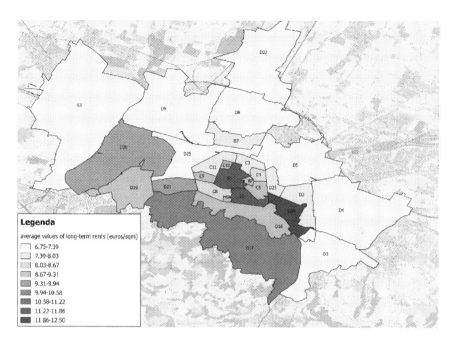

Fig. 5. Own depiction of Average long-term rents of areas based on www.agenziaentrate.gov.it/geopoi_omi/index.php

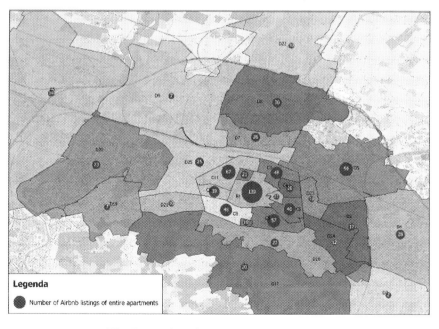

Fig. 6. Number of Airbnb listings to each area

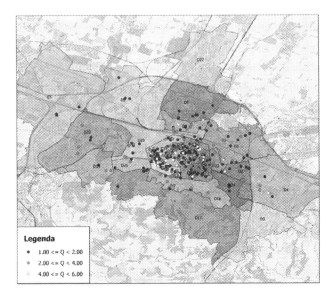

Fig. 7. Airbnb bookings with Q >= 1

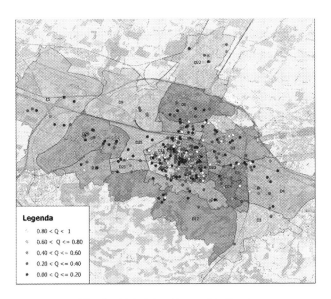

Fig. 8. Airbnb bookings with Q < 1

Generally, 49% of the Airbnb properties reach an advantage economic respect to the long-term rental market, and it is noted that moving away from central areas the convenience in long-term rent increases. Figures 8 and 10 show, respectively, the numerical distributions inside the secondary classifications. In the histogram of the

Fig. 9, it is observed that great part of the properties succeeds to obtain a gross revenue in the short term until four times advanced to the correspondent one in the long term.

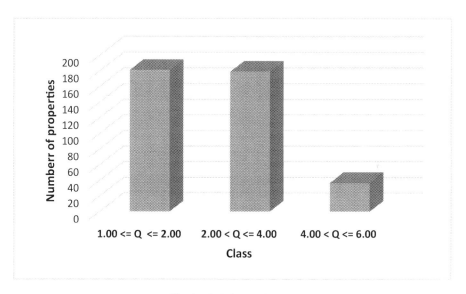

Fig. 9. Subclasses Q >= 1

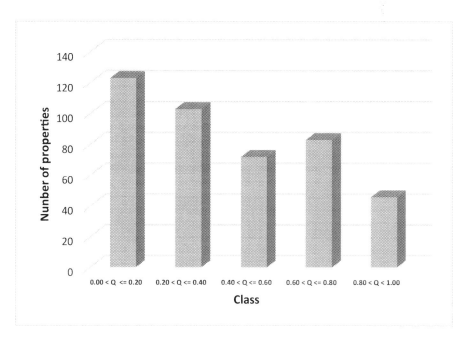

Fig. 10. Subclasses Q < 1

Moreover, the comparison (Fig. 11) reveals that inside the central areas and adjacent to those the properties with an advantage in short term is more numerous than those for which the same advantage does not occur. Meanwhile the opposite situation has been realized in peripheral areas, except for the D25 area.

Fig. 11. Comparison of economic vantage in short-term rent and long-term rent

The revenues of short-term rental market have been determined only based on the estimated bookings of the Airbnb platform. However, most hosts, usually, promote their properties simultaneously on different digital platforms. Therefore, the percentage of owners that achieve economic advantage to rent their properties in short-term, is likely higher than the one reported here.

4 Conclusions and Future Developments

The Airbnb platform allows to housing owners to put easily and flexibly their apartments in the short-term rental market, over time. Owners consider this option as a convenient alternative, in terms of profitability, as compared to traditional long-term rental market.

The present contribute provides the comparison of the profitability of the two options in the city of Bologna. The comparison has been carried out through an indicator as ratio (Q) of the gross revenues per month and surface area, respectively, of short-term rent and long-term rent. For 49% properties, listed on Airbnb platform now, this condition is more profitable than the long-term one. However, this data is underestimated. In fact, most hosts, usually, promote their properties simultaneously on different digital platforms, intercepting a larger demand than one observed here that is limited to Airbnb platform, and thus they can reach greater revenue. The results also showed that properties that reach a larger profitability through short-term lease are mostly concentrated in central neighbourhoods, Instead in peripheral neighbourhoods long-term rental obtain economic advantages. Based on these results, next future researches are going to investigate the spatial heterogeneity of the profitability indicator. Starting from these results future researches will investigate spatial heterogeneity of the indicator of profitability, considering potential factors linked to it, such as: socio-economic factors, capital and income availability, social composition of the inhabitants of neighbourhoods, housing stock of areas.

This could highlight phenomena of informal regeneration or of gentrification occurred in some neighbourhoods, and then, provide useful suggestions to define rules capable to adjust and to control the development those phenomena.

References

1. Heo, C.Y., Blengini, I.: A macroeconomic perspective on Airbnb's global presence. Int. J. Hospitality Manage. **78**, 47–49 (2019). https://doi.org/10.1016/j.ijhm.2018.11.013
2. Lee, D.: How Airbnb short-term rentals exacerbate, Los Angeles's affordable housing crisis: analysis and policy recommendations. Harvard Law Policy Rev. **10**, 229–253 (2016)
3. Coles, P., Egesdal, M., Ellen, I.G., Li, X., Sundararajan, A.: Airbnb Usage Across New York City Neighborhoods: Geographic Patterns and Regulatory Implications. Cambridge Handbook of the Law of the Sharing Economy (forthcoming),12 October 2017. https://ssrn.com/abstract=3048397
4. Quattrone, G., Proserpio, D., Quercia, D., Capra, L., Musolesi, M.: Who Benefits from the Sharing Economy of Airbnb? In: Proceedings of the 26th International ACM Conference on World Wide Web (WWW), pp. 1385–1394 (2016)
5. Rubino, I.: Affitti brevi e trasformazioni nelle aree urbane: il caso di Torino. Territorio Italia **1**, 91–109 (2018). https://doi.org/10.14609/Ti_1_18_5i
6. Picascia, S., Romano A., Teobaldi, M.: The airification of cities: making sense of the impact of peer to peer short term letting on urban functions and economy. In: Lisbona AESOP Annual Congress (2017). https://www.researchgate.net/publication/318707990
7. Schafer, P., Braun, N.: Misuse through short-term rentals on the Berlin housing market. Int. J. Hous. Markets Anal. **9**(2), 287–311 (2016). www.emeraldinsight.com/1753-8270.htm
8. Sheppard, S., Udell, A.: Do Airbnb properties affect house prices? No 2016-03, Department of Economics Working Papers, Department of Economics, Williams College (2016)
9. Balampanidis, D., Maloutas, T., Papatzani, E., Pettas, D.: Informal urban regeneration as a way out of the crisis? Airbnb in Athens and its effects on space and society. Urban Res. Pract. (2019). https://doi.org/10.1080/17535069.2019.1600009
10. Coyle, D., Yeung, T.: Understanding AirBnB in Fourteen European cities. The Jean-Jacques Laffont Digital Chair Working Papers, vol. 7088, pp. 1–33 (2016)
11. Rossi, A.: The sharing economy in the Italian cities: an analysis of Airbnb in Venice. Ph.D. Thesis (2018)
12. Manganelli, B., Morano, P., Tajani, F.: House prices and rents. The Italian experience. WSEAS Trans. Bus. Econ. **11**(1), 219–226 (2014)

13. Morano, P., Tajani, F., Di Liddo, F., Anelli, D.: A feasibility analysis of the refurbishment investments in the Italian residential market. Sustainability **12**, 1–20 (2020). https://doi.org/10.3390/su12062503

14. Morano, P., Rosato, P., Tajani, F., Manganelli, B., Di Liddo, F.: Contextualized property market model vs. generalized mass appraisal: an innovative approach. Sustainability **11**, 4896 (2019). https://doi.org/10.3390/su11184896

15. Ayouba, K., Breuille', M.L., Grivault, C., Le Gallo, J.: Does Airbnb disrupt the private rental market? An empirical analysis for French cities. Int. Reg. Sci. Rev. **43**(1-2), 76–104 (2020)

16. Garcia-López, M.À., Jofre-Monseny, J., Martínez Mazza, R., Segú, M.: Do short-term rent platforms affect housing markets? Evidence from Airbnb in Barcelona. MPRA Paper No. 96131 (2019). https://mpra.ub.uni-muenchen.de/96131/

17. Horn, K., Merante, M.: Is home sharing driving up rents? Evidence from Airbnb in Boston. J. Hous. Econ. **38**, 14–24 (2017). https://doi.org/10.1016/j.jhe.2017.08.002

18. Bestakova, S.: The influence of short-term rental on rental housing prices in Prague. In: 9th Business & Management Conference, Prague, 17 June 2019. ISBN 978-80-87927-64-9. https://doi.org/10.20472/bmc.2019.009.002

19. Il turismo nella città metropolitana di Bologna, Servizio studi e Statistica per la programmazione strategica, Rapporto novembre 2019. https://www.cittàmetropolitana.bo.it/statistica/. Accessed 15 Mar 2020

20. Decreto-legge 24 aprile 2017, n. 50

21. I dati in sintesi per la costruzione del piano strutturale, l'evoluzione del patrimonio abitativo (2005). http://www.comune.bologna.it. Accessed 15 Mar 2020

22. Dipartimento programmazione Settore statistica, 15° Censimento generale della popolazione e delle abitazioni 2011, La condizione abitativa a Bologna (2014)

23. Bisello, A., Vettorato, D., Stephens, R., Elisei, P. (eds.): SSPCR 2015. GET. Springer, Cham (2017). https://doi.org/10.1007/978-3-319-44899-2

24. Il catasto urbano nel comune di Bologna. http://www.inumeridibolognametropolitana.it. Accessed 15 Mar 2020

The Classification of the University for Type of Campus Setting in a World Sustainability Ranking

Silvestro Montrone[1], Paola Perchinunno[1(✉)], and Monica Cazzolle[2]

[1] Department of Economics, Management and Business Law,
University of Bari A. Moro, Largo Santa Scolastica 53, Bari, Italy
{silvestro.montrone,paola.perchinunno}@uniba.it
[2] Statistics Unit of the Institutional Affairs Department,
University of Bari A. Moro, Piazza Umberto I, Bari, Italy
monica.cazzolle@uniba.it

Abstract. Sustainability indicators and integrated indexes are gaining lot of importance and increasingly recognized as a powerful tool for policy making and public communication, in providing information on countries in fields such as environment, economic or social improvement. All over the world a network is being developed to tackle these issues in a homogeneous way. UI GreenMetric world ranking has been recognized globally as the only simple and accessible sustainability ranking and acts as a reference point, in particular to help universities in developing countries to create sustainability, promote research and development technology and accelerate development in all research areas, through various technological improvements. The aim of this work is to study the relationships between the rankings made on the basis of sets of sustainability indicators and the corresponding typology by Campus Setting.

Keywords: International ranking · Sustainability indicators · Campus setting

1 Introduction

Sustainability indicators and integrated indexes are gaining lot of importance and increasingly recognized as a powerful tool for policy making and public communication, in providing information on countries in fields such as environment, economic or social improvement [1, 2].

All over the world a network is being developed to tackle these issues in a homogeneous way. In particular, the UIGWURN (UI GreenMetric World University Ranking Network) network has defined three main thematic priorities for activities: 1) shaping global higher education and sustainability research, 2) creating global sustainability leaders and 3) collaborating on solutions to sustainability challenges.

The contribution is the result of joint reflections by the authors, with the following contributions attributed to S. Montrone (Sect. 1, Subsect. 3.2), M. Cazzolle (Sect. 2, Subsect. 3.1) and to P. Perchinunno (Subsect. 3.3, Sect. 4).

© Springer Nature Switzerland AG 2020
O. Gervasi et al. (Eds.): ICCSA 2020, LNCS 12251, pp. 899–912, 2020.
https://doi.org/10.1007/978-3-030-58808-3_64

The 2030 agenda, with the 17 sustainable development goals (SDGs), also presents an integrated vision of the three different development dimensions. Among the proposed objectives, the following are particularly important and can be related to the indicators used in green metric classification:

GOAL 3 - Ensuring health and well-being conditions for everyone at all ages: reduction of the world mortality rate of maternal, neonatal and children under 5 years of age due to preventable causes, through adequate health care for all and by supporting the research and development of vaccines and medicines for communicable diseases.

GOAL 4 - Offer quality, inclusive and equal education and promote lifelong learning opportunities for all: education can guarantee young people a better future because reading and writing has helped men and women by eliminating all forms of gender discrimination and promoting equal access to all levels of education.

GOAL 6 - Ensure the availability and sustainable management of water and hygiene conditions for everyone: water is a source of life and must be accessible to anyone; universal access to clean and drinking water is essential and hygiene conditions must be adjusted with particular attention to the most vulnerable people.

GOAL 7 - Ensuring access to clean, cheap and sustainable energy for all Energy systems are a fundamental element for everyone's daily life: it is necessary to make them accessible to everyone.

GOAL 9 - Building resistant infrastructures, promoting sustainable and inclusive industrialization and promoting innovation: both infrastructure and industry are important to support the entire economic development and our well-being by becoming sustainable and reliable with technological development and research.

GOAL 11 - Make cities and communities safe, inclusive, resistant and sustainable. The environment that surrounds us affects our habits and lifestyles: the improvement from a sustainable point of view of our living spaces is important.

GOAL 12 - Ensuring sustainable consumption and production models. Our planet needs to be respected and safeguarded: waste and chemicals released above all by large multinational companies must be reduced through sustainable policies based on product recycling.

GOAL 13 - Take urgent action to combat climate change and its impact: the issue of climate change must be addressed with sustainable global policies and strategies in order to stem environmental risks and actual natural disasters.

GOAL 14 - Safeguarding the oceans, seas and marine resources for their sustainable development: the conservation and sustainable exploitation of the oceans, seas and all those resources within them are important for our life and for our health.

GOAL 15 - Protect, re-establish and promote the sustainable use of terrestrial ecosystems, the sustainable management of forests, combat desertification, stop and reverse the degradation of the territory and stop the loss of biodiversity.: preserving our planet is a task entrusted to all and for this reason it is necessary to pursue a joint action to protect, restore and promote the sustainable use of the terrestrial ecosystem.

GOAL 17 - Reinforce the meanings of implementation and revitalize global collaborations for sustainable development: strong involvement of all components of society, from businesses to the public sector, from civil society to philanthropic institutions, from universities and research centers to information and culture operators.

The GreenMetric user interface has been recognized globally as the only simple and accessible sustainability ranking and acts as a reference point, in particular to help universities in developing countries to create sustainability, promote research and development technology and accelerate development in all research areas, through various technological improvements.

The significant impact of the GreenMetric UI ranking is shown in the growing number of participants covering regions of North America, South America, Europe, Africa, Asia and Australia and Oceania [3, 4].

2   UI Greenmetric World University Ranking

2.1   Short Presentation of the Survey

UI GreenMetric World University Ranking, promoted by the University of Indonesia (UI), is a world ranking of universities launched in 2010 with the aim of creating an increasingly extensive sustainable environment.

This ranking offers the opportunity to compare the environmental sustainability of the various university campuses through scores assigned to certain indicators that measure the efforts made in developing an ecological infrastructure that involves education, research and the surrounding environment and stimulates improvement actions.

The initiative has always been appreciated by academic communities all over the world, recording over the years a continuous increase in universities and in the different participating countries (in 2019 the number of universities that joined the ranking is 8 times higher than in 2010). Over the years more and more attention has been shown not only to the environmental aspect but also to the economic and social aspect, comparing itself more fully on the concept of global sustainability (Fig. 1).

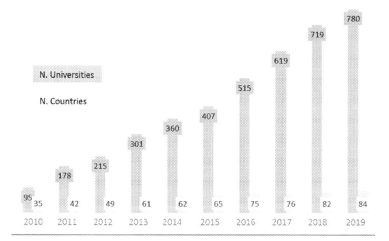

Fig. 1. Number of universities and countries belonging to the UI GreenMetric. Source: Our elaboration on Green Metric data

2.2 Methodology of Green Metric

The Green Metric working group, made up of experts in cross-cutting areas, selected 6 categories deemed most important for the assessment of universities that deal with sustainability using certain criteria. These criteria allow universities to be grouped according to their size, location (urban, suburban, rural), degree of green space available, electricity consumption and carbon emissions, transport used and related policies, water resource management, waste management. The information required to construct the sustainability indicators is therefore placed in the respective categories chosen and, in some case, they are transversal to several categories. Each category is assigned a specific weight and the indicators are assigned a certain score useful for the construction of the ranking [5].

Of course, over the ten years the methodology used has changed, adapting to different needs. In the first version used in 2010, 23 indicators were used within five categories; in 2011 the indicators were increased and brought to 34; in 2012 only one indicator was eliminated. Subsequently, the introduction of a new category related to education and research on sustainability was considered. In 2015, the methodology improves with the inclusion of two questions relating to carbon footprints in the "energy and climate change" section and some sub-indicators in the "water" and "transport" section. A big change in the methodology is made in 2016 considering the new sustainability trends. In 2017, documentation was requested to demonstrate what was declared [6].

The six categories identified and currently used are [5]:

- Setting and Infrastructure (15%): the category of indicators focuses on the type of Campus and its infrastructure by providing basic information on the impact of the university on the environment. This allows the participating universities to reflect on the possibility of providing more space for greenery and protecting the environment, as well as investing in the development of sustainable energy to be considered Green Campus.
- Energy and climate change (21%): the category of indicators focuses universities on energy issues and climate change by verifying the use of energy efficient appliances, the policy of using renewable energy, the energy saving program, green building, adaptation to climate change, the policy of reducing greenhouse gas emissions, for their greater commitment in this direction in buildings. Carbon footprint measurement was introduced in 2015 to induce universities to look into problems within their campuses.
- Waste (18%): the recycling of toxic waste, the treatment of organic waste, the treatment of inorganic waste, the disposal of water waste, the policy to reduce the use of paper and plastic are important factors since university staff and students on campus contribute to the production of a large number of waste. So, it is important to monitor and measure waste treatment and recovery activities.
- Water (10%): universities must reduce water consumption and increase the water conservation program through the use of water-efficient equipment and habitat protection.

Table 1. Indicators by category and maximum score

Categories	Indicators	Max score
Setting and Infrastructure (SI)	The ratio of open space area to total area	300
	Total area on campus covered in forest vegetation	200
	Total area on campus covered in planted vegetation	300
	Total area on campus for water absorption besides the forest and planted vegetation	200
	The total open space area divided by total campus population	300
	Percentage of university budget for sustainability efforts within a year	200
Energy and Climate Change (EC)	Energy efficient appliances usage	200
	Smart building implementation	300
	Number of renewable energy sources in campus	300
	The total electricity usage divided by total campus population (kWh per person)	300
	The ratio of renewable energy production divided by total energy usage per year	200
	Elements of green building implementation as reflected in all construction and renovation policies	300
	Greenhouse gas emission reduction program	200
	The total carbon footprint divided by total campus population (metric tons per person)	300
Waste (WS)	Recycling program for university waste	300
	Program to reduce the use of paper and plastic on campus	300
	Organic waste treatment	300
	Inorganic waste treatment	300
	Toxic waste treatment	300
	Sewage disposal	300
Water (WR)	Water conservation program implementation	300
	Water recycling program implementation	300
	Water efficient appliances usage	200
	Treated water consumed	200
Transportation (TR)	The total number of vehicles (cars and motorcycles) divided by total campus population	200
	Shuttle services	300
	Zero Emission Vehicles (ZEV) policy on campus	200
	The total number of Zero Emission Vehicles (ZEV) divided by total campus population	200
	The ratio of the parking area to total campus area	200
	Transportation program designed to limit or decrease the parking area on campus for the last 3 years (from 2016 to 2018)	200
	Number of transportation initiatives to decrease private vehicles on campus	200
	Pedestrian path policy on campus	300
Education and Research (ED)	The ratio of sustainability courses to total courses/subjects	300
	The ratio of sustainability research funding to total research funding	300
	Number of scholarly publications on sustainability	300
	Number of events related to sustainability	300
	Number of student organizations related to sustainability	300
	University-run sustainability website	200
	Sustainability report	100

Source: Our elaboration on Green Metric data

- Transportation (18%): transport policy is important for limiting the number of private cars on campus, encouraging the use of environmentally friendly public transport, cycling or walking to reduce the carbon footprint on campus and create an environment healthier.
- Education (18%): the university plays an important role in training new generations on issues related to sustainability. Therefore, the offer of training in this sense and any other available means of awareness (events, publications, web) have a fundamental educational role.

Specific indicators are required within the 6 categories. Most of them have an associated score to be able to draw up the ranking (Table 1).

3   Main Results

3.1   International Ranking

It is possible to calculate some significant statistical indicators in the period from 2010 to 2019 based on the total scores from general classification.

The number of universities that participated in the ranking over the years has increased, starting from 95 universities in 2010 up to 780 international universities in 2019 (last year of the survey).

The total average score since 2010 has decreased until 2015, except for 2014 and then increases again until 2019. The minimum and maximum scores oscillate between 550 and 9,125 in the 10 years considered (Table 2).

Table 2. Some descriptive statistics about total score (2010–2019)

Year	N. Universities	N. Countries	Total score			
			Average	Min	Max	Std dev.
2010	95	35	5,464	2,153	8,213	1,324
2011	178	42	5,423	2,185	8,034	1,298
2012	215	49	4,980	1,857	7,569	1,127
2013	301	61	4,945	1,807	7,521	1,138
2014	360	62	5,169	1,957	7,803	1,208
2015	407	65	4,351	1,449	7,267	1,278
2016	515	75	4,425	805	8,398	1,360
2017	619	76	4,426	1,413	7,552	1,185
2018	719	82	4,810	1,025	9,125	1,523
2019	780	84	4,990	550	9,075	1,656

Source: Our elaboration on Green Metric data.

These universities are located all over the world; in 2019 most of them (about 80%) are in the following countries (Fig. 2), with a prevalence in Indonesia (9.2% of the total), Russia, Iraq, Colombia and Turkey (5%).

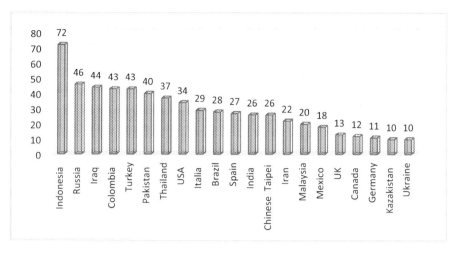

Fig. 2. Number of universities in the ranking by country (2019). Source: Our elaboration on Green Metric data

The universities classified in the Top 25 (Fig. 3) are mainly located in the United Kingdom (32%), United States of America, Netherland and Ireland (12%).

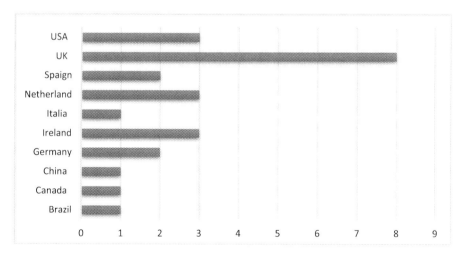

Fig. 3. The Top 25 university in 2019 by nationality (2019). Source: Our elaboration on Green Metric data

The top 10 universities of 2019 are instead located mainly in the United Kingdom (Fig. 4).

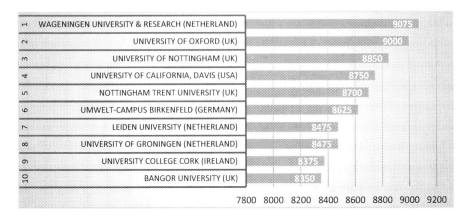

Fig. 4. The Top 10 university in 2019. Source: Our elaboration on Green Metric data

In the 2018 edition we find all the top 10 of 2019 but with different positioning, except for Leiden University which is replaced by the University of Connecticut. The University of Wageningen is confirmed first in the two editions.

In the last four years (2016–2019) at least 5 universities remain in the top 10. In the other years the number decreases but the presence of other universities intensifies. The University of Nottingham is always present among the top 10 in the whole decade.

3.2   The Rankings for Categories of Indicators

In addition to the general classification it is possible to draw up a sub-ranking: ranking by Indicators (for 6 categories), ranking by Campus Setting for the type of Campus (urban, suburban, rural, in the city center - high rise building), ranking by Country.

Analyzing the ranking of the Top 10 by categories of Indicators (Table 3) it is clear how the overall score hides the peculiarity of the results achieved in the different categories, since it is made up of the sum of several indicators (as indicated in the previous paragraph). In fact, it can be noted that universities that have not obtained the highest score have excellent ratings in certain categories. For example, Bangor University, which is positioned in tenth position, has very high values in the categories "Transportation" and "Education and research". On the contrary, Wageningen University & Research (Netherland), which is in first position in the ranking with 9,075 total points, presents lower values than the universities classified in the following positions as regards the category "Setting and Infrastructure" and "Transportation".

This further classification helps universities to understand for which categories of indicators it is necessary to intervene more in order to improve both the ranking and the living conditions of their university.

Table 3. The Top 10 university per categories of indicators in 2019

	University	Setting and infrastructure	Energy and climate change	Wast	Water	Transportation	Education and research	Total score
1	Wageningen University & Research	1,125	1,800	1,800	1,000	1,550	1,800	**9,075**
2	University of Oxford	1,200	1,800	1,800	1,000	1,625	1,575	**9,000**
3	University of California, Davis	1,300	1,650	1,725	1,000	1,525	1,650	**8,850**
4	University of Nottingham	1,250	1,525	1,800	925	1,525	1,725	**8,750**
5	Nottingham Trent University	1,200	1,675	1,800	700	1,525	1,800	**8,700**
6	Umwelt-Campus Birkenfeld	975	1,775	1,500	1,000	1,650	1,725	**8,625**
7	Leiden University	650	1,850	1,725	1,000	1,750	1,500	**8,475**
8	University of Groningen	925	1,575	1,800	1,000	1,525	1,650	**8,475**
9	University College Cork	900	1,750	1,725	775	1,650	1,575	**8,375**
10	Bangor University	1,025	1,675	1,725	425	1,700	1,800	**8,350**

Source: Our elaboration on Green Metric data.

3.3 Classification for the Type of Campus Setting

Regarding the classification for the type of Campus Setting there are 5 different types:

- rural: located in the countryside
- suburban: located in extremely peripheral area with respect to the city
- urban: located in a large urban area
- in city center: located in the city center
- high rise building: consisting of a single tall building

In 2019 they belong to the first group, i.e. the **Rural**, 64 universities mainly located in the USA (9 universities) and India (7 universities). The top 10 of these categories is represented by the following universities (Fig. 5) among which there is the sixth classified in the general top.

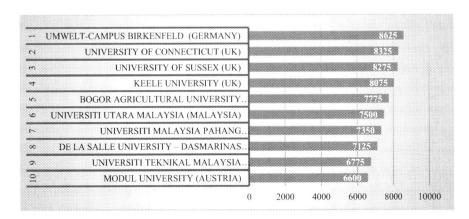

Fig. 5. The Top 10 Rural university in 2019. Source: Our elaboration on Green Metric data

Also, in the 2019 ranking belong to the group of **Suburban** 183 universities mainly located in Indonesia (19 universities) and Thailand (18 universities). The top 10 of these categories is represented by the following universities (Fig. 6) among which there are the first, third and tenth classified in the general top.

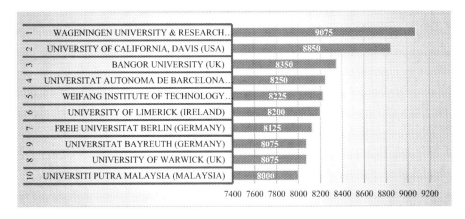

Fig. 6. The Top 10 Suburban university in 2019. Source: Our elaboration on Green Metric data

In the **Urban** ranking we find 385 universities mainly located in Indonesia (35 universities) and Colombia (30 universities). In the top 10 we find several universities in the United Kingdom and the fourth, fifth and eighth universities of the top 10 general (Fig. 7).

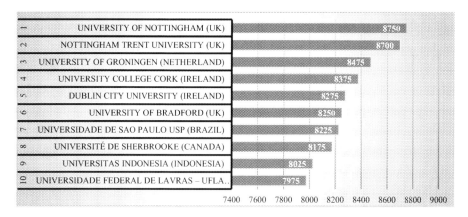

Fig. 7. The Top 10 Urban university in 2019 Source: Our elaboration on Green Metric data

In the ranking of **in city center** we find 142 universities mainly located in Iraq (16 universities) and Russia (15 universities). In the top 10 we find two large Italian universities (Bologna and Turin) and the second and seventh universities of the top 10 general (Fig. 8).

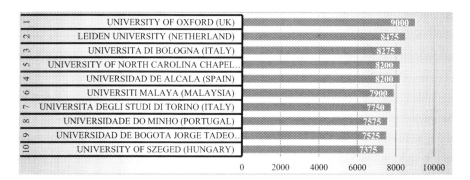

Fig. 8. The Top 10 In City Center university in 2019. Source: Our elaboration on Green Metric data

Last category is the **High-Rise Building** (Fig. 9) which includes only 5 universities occupying very low positions in the general ranking (starting from position 89).

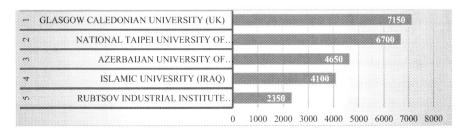

Fig. 9. The Top 10 High Rise Building in 2019. Source: Our elaboration on Green Metric data

Comparing the average scores of all types of universities (Table 4), it emerges that the "In City center" category presents values below the average in all the Categories of Indicators, unlike the universities present in the ranking of "Suburban" which have values above the average in all the Categories.

As for universities classified as "High rise building", they have the lowest average values for Setting and infrastructure and the highest ones for Energy and climate change, Wast and Transportation.

Universities classified as "Rural" have higher average values for Setting and infrastructure and lower average values for Education and research.

Values very close to the general average are found for universities classified as "Urban".

Table 4. Average of the scores by Category of indicators and by type of Campus setting

	Total score	Setting and infrastructure	Energy and climate change	Wast	Water	Transportation	Education and research
General	4,990	735	974	892	475	944	971
Rural	5,115	830	997	884	486	975	942
Suburban	5,328	832	1,002	975	534	992	994
Urban	5,002	721	976	892	478	938	997
In City center	4,456	615	915	783	381	880	882
High rise building	4,990	495	1,070	1,005	505	1,005	910

Source: Our elaboration on Green Metric data.

4   Conclusions

The construction of the ranking is based on a combination of parameters, indicators, properly weighted, chosen by the promoters of the rankings and are often the subject of some controversies due to the subjectivity in the choice of the indicators, arbitrary attribution of weights, variability in the definition of the classification of the results [7–15].

Therefore, a continuous review of indicators and weight systems is always desirable, which often are not univocally applicable to the specificity of universities. To this end, the established networks (UIGWURN network) provide an excellent tool to improve.

The indicators are fed on the basis of the data provided directly by the universities being assessed and therefore the methods of construction and attribution of the scores are subject to attention by the participants thanks also to the return of a fact file with the individual scores obtained by the team Green Metric.

In this way, universities can improve in individual categories with a greater comparison, for example, with universities located in the same geographical position or with a similar structure (campus setting).

The results reported in this study are undoubtedly encouraging and show how universities are increasingly willing to participate in order to achieve good ranking levels and improve in the field of sustainability.

References

1. Altbach, P.G.: The globalization of college and university rankings. Change Mag. High. Learn. **44**(1), 26–31 (2012)
2. Altbach, P.G.: The dilemmas of ranking. Int. High. Educ. **42**, 2–3 (2006)
3. Suwartha, N., Mohammed Ali Berawi, M.A.: The role of UI Greemetric as a global sustainable ranking for higher education institutions. Int. J. Technol. **10**(5), 862–865, 862–863 (2019). ISSN 2086-9614
4. The 17 sustainable development objectives. Blogs and news Save the Children. https://www.savethechildren.it/blog-notizie/i-17-obiettivi-di-sviluppo-sostenibile?utm_source=google&utm_medium=grants&utm_campaign=cmpntvblog&causale=14400&gclid=EAIaIQobChMI5NnD7a76AIVhc13Ch0LHQnnEAAYASAAEgJt3fD_BwE
5. Guideline UI GreenMetric World University Rankings 2019: Sustainable University in a Changing World: Lessons, Challenges and Opportunities. http://greenmetric.ui.ac.id/guidelines-page/
6. Sari, R.F., Suwartha N., Junaidi: UI green metric world university rankings: background of the ranking. In: Proceeding of the 5th International Workshop on UI GreenMetric World University Rankings (IWGM 2019)-Sustainable University in a Changing World: Lessons, Challenges and Opportunities, Cork, Ireland, 14–16 April 2019, p. viii (2019). ISBN 978-979-456-799-9
7. Dwork, C., Kumar, R., Naor, M., Sivakumar, D.: Rank aggregation methods for the Web. In: Proceedings of the 10th World Wide Web Conference, Hong-Kong, pp. 613–622, May 2001
8. Dill, D.D., Soo, M.: Academic quality, league tables and public policy: a cross national analysis of university ranking systems. High. Educ. **49**, 499–533 (2005)
9. Frey, B.S., Rost, K.: Do rankings reflect research quality? J. Appl. Econ. **13**(1), 1–38 (2010)
10. Proulx, R.: Higher education ranking and leagues tables: lessons learned from benchmarking. High. Educ. Europe **32**(1), 71–82 (2007)
11. Saisana, M., d'Hombres, B., Saltelli, A.: Rickety numbers: volatility of university rankings and policy implications. Res. Policy **40**(1), 165–177 (2011)
12. Taylor, P., Braddock, R.: International university ranking systems and the idea of university excellence. J. High. Educ. Policy Manag. **29**(3), 245–260 (2007)

13. Tofallis, C.: A different approach to university rankings. High. Educ. **63**(1), 1–18 (2012)
14. Federkeil, G.: Some aspects of ranking methodology, the CHE ranking of German universities. High. Educ. Europe **27**(4), 389–397 (2002)
15. Fidler, B., Parsons, C.: World university ranking methodologies: stability and variability. High. Educ. Rev. **40**(3), 15–34 (2008)

Real Estate Values, Tree Cover, and Per-Capita Income: An Evaluation of the Interdependencies in Buffalo City (NY)

Antonio Nesticò[1]([✉]) [ID], Theodore Endreny[2] [ID],
Maria Rosaria Guarini[3] [ID], Francesco Sica[4] [ID], and Debora Anelli[4] [ID]

[1] Department of Civil Engineering, University of Salerno,
84084 Fisciano, SA, Italy
anestico@unisa.it

[2] Department of Environmental Resources Engineering, SUNY ESF,
Syracuse, NY 13210, USA
te@esf.edu

[3] Department of Architecture and Design, Sapienza University of Rome,
00196 Rome, Italy
mariarosaria.guarini@uniroma1.it

[4] Doctoral School of Architecture and Construction, Department of Architecture
and Design, Sapienza University of Rome, 00196 Rome, Italy
{francesco.sica,debora.anelli}@uniroma1.it

Abstract. The variables that influence the price formation mechanisms of urban real estate units concern both the socio-economic both the infrastructural and environmental system of the city. In literature, the links between real estate and territory are not widely investigated. This is especially with reference to the correlation levels between Real Estate Prices and the provision of urban greening. In this perspective, aim of the research is to examine functional relations between property prices, environmental factors, and socio-economic parameters.

The applied methodology is based on statistical correlation analysis. This supports the construction of an innovative model for the estimation of the function that explains the dependence of property values on social and environmental factors that characterize the city. The elaborations concern data derived with the Tool *i-Tree Landescape*. Through Geographic Information Systems (GIS), these data are useful to obtain thematic maps illustrating the spatial distribution of Real Estate Prices, Tree Cover, and Per-Capita Income. The surveyed geographical units are the Census Blocks of Buffalo City in New York State (USA).

Keywords: Urban real estate · Tree cover · Per-Capita Income · Geographic Information Systems (GIS) · Correlation analysis · Optimization algorithm

---

The five authors contributed equally to this work.

O. Gervasi et al. (Eds.): ICCSA 2020, LNCS 12251, pp. 913–926, 2020.
https://doi.org/10.1007/978-3-030-58808-3_65

1 Introduction

The planning of city's development cannot be separated from the analysis of its socio-economic, productive and environmental system. It is essential, in fact, to define the territory, its boundaries and functional character in the perspective of «... economic development, social inclusion and environmental protection» [1].

Since 2007, the European Commission has been promoting design initiatives based on «eco-logically sound urban design practices» , i.e. interventions that provide for the creation of new green areas with which to raise the environmental level, social, and cultural quality, as well as the income capacity of the city [2–5].

The multiple effects generated by renewal urban projects nature-based are generally expressed in terms of eco-system services, in other words «... direct and indirect contributions of ecosystems to human well-being» [6]. From literature, the main urban ecosystem services are: pollutant removal, carbon storage and sequestration, microclimate adaption, landscape improvement, biodiversity, and soil protection, water resource regulation [7, 8]. It is proven that natural resources provide life-support, cultural and aesthetic benefits [9], as well as increase in the market value of urban property [10].

Millennium Ecosystem Assessment (2005) classifies ecosystem services in four categories: provisioning, regulating, cultural, and supporting services [11]. Each category pursues a specific objective, and is a function of reference territory size, which may coincide with the individual building, the block in which the building itself is located, but also with city portions [12].

According to the geographical and urban features of study area and MEA category, the ecosystem services «... can systematically influence, according to a medium-long term logic, the orientation of a specific urban environment in terms of development strategies» [13]. In this regard, Glaeser (2008) reports that the presence of natural elements and/or green areas in built-up areas can significantly affect the growth of the local economic-productive system, as well as the urban real estate market [14, 15].

It is well known that the main extrinsic variables that can affect the market value of urban buildings include the urban quality (quality of roads, buildings, squares), the green plots (public parks, and gardens), the social and demographic context, the public transport, the proximity to the Central Business District (CBD), the landscape views, the historical significance of the area, the air pollution level [16–18]. In addition to these also «... factors on the urban heat island, distribution of greenery, building density and geometry, and air quality» [19].

The literature explores the relationship between Real Estate Prices (REP) and socioeconomic parameters of the territory. In many cases the correlation between REP, per capita income of the inhabitants, and population density is highlighted [20, 21]. Some studies estimate the dependence between REP and distance of the property on both urban parks and water bodies (lakes, rivers) [22–25], both from mobility services [26–28]. There are still few scientific investigations concerning the correlation of REPs with both the distribution of greenery and air quality in cities [29].

To establish the logical-functional dependence between variables, analyses developed with different techniques can be conducted. It should mention: regression models, neural networks, time series, and hedonic pricing method. This last one evaluates quantitative values for environmental or ecosystem services that directly affect market prices for residential building, isolating with multivariate regression techniques the contribution by interest attributes on observed prices [30, 31].

2 Work Aim

Objective is to investigate the correlation levels between Real Estate Prices and the socio-economic and environmental city parameters. The considered attributes that influence property value are Tree Cover and Per-Capita Income.

The relationship between Real Estate Prices, Tree Cover, and Per-Capita Income is estimated using a linear regression model between the variables. The methodological approach on the basis of model consists of the following steps: preparation of source data and geographical maps on data distribution; analysis of spatial structure data; development of regression model describing the correlations between prices and explicative variables.

The study model is implemented with reference to Buffalo City in New York State (USA).

According to literature, linear regression models can be solved both through statistical analysis tools and mathematical programming algorithms [32]. In the present manuscript, Operational Research algorithms able to solve real linear mathematical systems set as optimization problems are used. The optimization algorithms for the estimation of correlation between multiple variables are reflected in studies concerning land-use, urban planning [33–39], and urban real estate appraisal [40–44].

The mathematical programming environment used for writing the regression system is A Mathematical Programming Language (AMPL). It is a simple and intuitive tool used for structuring mathematical programming problems. Resolutions can be developed through specific software. Some examples are CPLEX, FortMP, MINOS, and KNITRO.

The paper is divided into four sections: Sect. 3, which describes the regression model and reports the algorithms written in AMPL; Sect. 4, case study, which describes the City of Buffalo (4.1), collects the data of the study variables (4.2), and estimates the correlation coefficients (4.3); Sect. 5, which discusses the results; Sect. 6, which presents conclusions and research perspectives.

3 Model Description

In order to detect the logical-functional dependencies between study variables, a multivariate statistical analysis is used. This allows both to identify the relationships between the parameters, and to establish the terms useful for spatial correlation between them.

Methods of multivariate statistics are: (i) Principal Component Method, in the case of variables measured on qualitative scale; (ii) Correspondence Method, where both quantitative and qualitative variables are concerned; (iii) Factor-Real Analysis Method, where all variables are expressed on quantitative metric.

In the present work, a Factorial Analysis Method is used, according to which the regression function defines the dependence of Y on the x_j variables by means of the β_j coefficients (with $j = 1, \ldots, k$).

In this case, which considers the two independent variables x_1 = Tree Cover (TC) and x_2 = Per-Capita Income (PCI), the multiple regression function is of the type:

$$Y = \beta_0 + \beta_1 x_1 + \beta_2 x_2 + \varepsilon \tag{1}$$

where the variable Y = Real Estate Prices (REP) depends on x_1 and x_2 through the coefficients β_0, β_1, β_2; ε indicates the random variable, i.e. the error.

The coefficients β_0, β_1, β_2 are estimated by solving the minimisation function:

$$b := \min \sum_i \varepsilon_i^2 := \min \sum_i (y_i - \beta_0 - \beta_1 x_{1i} - \beta_2 x_{2i})^2 \tag{2}$$

where β_0 is the known term of the regression equation; β_1 and β_1 express the coefficients of x_1 and x_2 of the i-th analysis unit.

Equation (2) can be solved by using the programming algorithms of Operational Research, capable of solving systems of linear equations in order to maximize and/or minimize an objective function. In the case of (2), Quadratic Linear Programming algorithms are used.

The AMPL software is used to structure the model. In practice, the use of the AMPL programming environment allows to write the regression problem through the following steps:

1. Identification of the problem elements as SETs;
2. Specification of the parameters (PARAM) to be included in the system to be solved;
3. Definition of the variables' value (VAR);
4. Structuring of the objective function, of maximization (MAXIMIZE) or minimization (MINIMIZE), as a linear algebraic expression.

These steps define the structure of a model in parametric form (.mod file) to which the problem data are associated with a separately written .dat file.

With regard to the correlation of Real Estate Prices (REP) with Tree Cover (TC) and Per-Capita Income (PCI), the analysis units, i.e. the census areas into which the study territory is divided (set UNITS), are evaluated according to three parameters (param REP, TC, PCI). The objective is to estimate the coefficients β_0, β_1, β_2 by means of a function aimed at minimizing the summa of the quadratic difference among the values of the parameters.

The regression model written in the AMPL programming environment is in Table 1. In compliance with AMPL syntactic rules, the β_0, β_1, β_2 coefficients, i.e. the unknowns of the problem, are indicated with the letters y, x_1, x_2 respectively.

Table 1. The regression model written in A Mathematical Programming Language (AMPL) software (.mod file).

Multiple linear regression analysis model
SETS
set UNITS;
PARAMETERS
param REP {i in UNITS};
param TC {i in UNITS};
param PCI {i in UNITS };
VARIABLES
var y ≥ 0;
var x1 ≥ 0;
var x2 ≥ 0;
OBJECTIVE FUNCTION
minimize *objective*: sum {i in UNITS} (REP[i] − x1·TC[i] − x2·PCI[i] − y)^2;

The analysis units (set UNITS) are described according to three factors (set PARAMETERS): Real Estate Price, Tree Cover, and Per-Capita Income.

The unknowns (y, x_1, x_2) are continuous. The objective function is:

$$\text{minimize objective: sum } \{i \text{ in UNITS} \} \, (REP[i] - x1 \cdot TC[i] - x2 \cdot PCI[i] - y)^2$$

The MINOS optimization software was used as a solver implementing Quadratic algorithm for the problem of linear regression considered.

4 Application of the Model to Buffalo City (NY)

4.1 Study Area

Buffalo is the second largest city in the U.S. State of New York. The city is the county seat of Erie County, and a major gateway for commerce and travel across the Canada–United States border, forming part of the bi-national Buffalo Niagara Region.

It is located on the eastern shore of Lake Erie, at the head of the Niagara River, 16 miles (26 km) south of Niagara Falls. According to the United States Census Bureau, the city has an area of 52.5 square miles (136 km^2), of which 40.6 square miles (105 km^2) is land and the rest water. The total area is 22.66% water. The city consists of 286 Census Blocks with an average extension of 0.39 km^2.

Buffalo's economy has begun to see significant improvements since the early 2010s. In 2016, the U.S. Bureau of Economic Analysis valued the Buffalo area's economy at $54.9 billion. In the same year, the city's median household income is $24,536.00 and the median family income is $30,614.00.

As regards the environmental-forestry component, the Buffalo parks system has over 20 parks with several green areas accessible from any part of the city. The Olmsted Park and Parkway System is the hallmark of Buffalo's many green spaces [45].

4.2   Data Collection and Statistical Analysis

In order to investigate the correlations between socio-economic factors and ecosystem services, the data set is constructed with the values of the variables considered, Real Estate Prices, Tree Cover, Per-capita Income. The data refer to the Census Blocks Tracks of city studied considered. The values of the variables are collected using *i-Tree Landscape* Tool that gives information on tree cover, land use, and basic demographic characteristics of the census areas in United States of America. Canopy Cover data are taken directly from 2011 National Land Cover Data (NLCD), while U.S. Per capita Income and Real Estate Prices from 2011 U.S. Census Bureau data.

The statistic description of variables referring to Buffalo City is in Table 2.

Table 2.   List of variables and its statistic description.

Variables	Descriptive statistics			
	Mean	Standard deviation	Minimum	Maximum
Real Estate Prices [$]	75,140.85	61,357.07	225.0	368,900.00
Tree Cover [%]	4.82	5.77	0.03	59.20
Per-Capita Income [$]	19,190.63	10,696.74	2,434.00	76,395.00

A thematic map representative of corresponding numerical values is constructed for each variable (Fig. 1). This is done through Geographical Information Systems (GIS). The realization of the thematic maps allows a first comparison among values, useful to highlight the correlation levels.

From the comparison between maps in Fig. 1, it is evident the correlation between the variables. In particular, there is a strong correlation among those at North of Buffalo. The Census Blocks falling in this portion of territory are characterized by high values of Real Estate Prices, Tree Cover and Per-Capita Income. Here there is the presence of the Delaware Park and the Urban Zoo.

Differently for the central and southern areas of the study city. These zones are characterized by the Museum of Science, Medical Center, and Cimitery. These are areas crossed by the Buffalo River, as well as road infrastructures connecting Buffalo with nearby urban conglomerates such as Rochester. The presence of both anthropic and natural elements in Buffalo causes the formation of surfaces with Real Estate Prices, Tree Cover, and Per-Capita Income values that are not homogeneously distributed. This results in lower levels of correlation between variables.

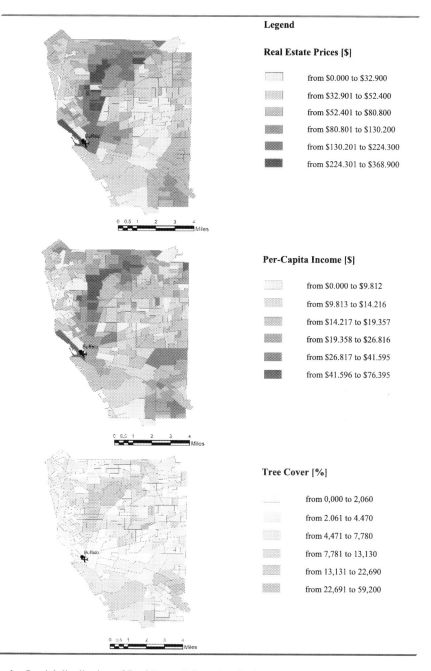

Legend

Real Estate Prices [$]

from $0.000 to $32.900
from $32.901 to $52.400
from $52.401 to $80.800
from $80.801 to $130.200
from $130.201 to $224.300
from $224.301 to $368.900

Per-Capita Income [$]

from $0.000 to $9.812
from $9.813 to $14.216
from $14.217 to $19.357
from $19.358 to $26.816
from $26.817 to $41.595
from $41.596 to $76.395

Tree Cover [%]

from 0,000 to 2,060
from 2.061 to 4.470
from 4,471 to 7,780
from 7,781 to 13,130
from 13,131 to 22,690
from 22,691 to 59,200

Fig. 1. Spatial distribution of Real Estate Prices, Pre-Capita Income, and Tree Cover in Buffalo (NY).

4.3   Estimation of Correlation Coefficients Between Variables

For the application of the proposed regression model, 287 census areas in the City of Buffalo (NY) are examined. The estimation of the correlation levels is performed by analysing census areas with the same level of infrastructural endowment and biome. For each census area of analysis is detected the:

a. Geographical extent expressed in square kilometres;
b. Median Real Estate Price in $;
c. Tree Cover in %;
d. Per-capita Income in $.

Table 3 show data-set excerpt with the values of the three parameters. On the data of Table 3 is implemented the regression model in Table 2. Thus, the regression model written in AMPL (.mod file) is in Table 4.

The .mod file in Table 4 is associated with the .dat file in Table 5.

The .mod and .dat files are called in the AMPL command line, specifying the solver that implements the algorithm:

```
ampl: reset;
ampl: model FILE.mod;
ampl: data FILE.dat;
ampl: option solver Minos;
ampl: solve.
```

The values of $y = \beta_0$, $x_1 = \beta_1$, $x_2 = \beta_2$ obtained by implementing the regression model with MINOS software are:

$$\beta_0 := 24451.548780$$
$$\beta_1 := 0.668540$$
$$\beta_2 := 0.862783$$

Table 3. Data matrix.

Census blocks	a. Area [Km²]	b. Median real estate price [$]	c. Tree cover [%]	d. Per-capita income [$]
0	0.15	32,200.00	4.01	9,205.00
1	0.48	35,417.00	3.21	12,708.00
2	0.44	69,924.00	9.22	52,698.00
3	0.22	38,060.00	4.17	15,767.00
4	0.13	44,489.00	0.00	23,225.00
5	0.14	34,949.00	2.76	12,166.00

(continued)

Table 3. (*continued*)

Census blocks	a. Area [Km^2]	b. Median real estate price [$]	c. Tree cover [%]	d. Per-capita income [$]
6	0.41	36,938.00	2.04	14,471.00
7	0.24	35,768.00	6.97	13,111.00
8	0.22	42,110.00	2.20	20,466.00
9	0.19	37,040.00	4.60	14,588.00
10	0.29	45,989.00	1.04	24,963.00
11	0.18	34,461.00	3.91	11,599.00
12	0.32	34,977.00	9.23	12,193.00
13	0.12	33,988.00	1.30	11,053.00
14	0.18	41,478.00	9.58	19,727.00
15	0.20	47,589.00	2.69	26,816.00
16	0.23	33,651.00	3.42	10,660.00
...	...	...	...	...
143	0.29	35,834.00	2.80	13,191.00
144	0.32	37,155.00	3.71	14,721.00
...	...	...	...	...
270	0.32	37,924.00	1.51	15,614.00
271	0.23	34,208.00	1.88	11,307.00
272	0.10	33,553.00	2.69	10,547.00
273	0.18	32,430.00	1.83	9,246.00
274	0.23	52,024.00	13.87	31,947.00
275	0.15	66,425.00	10.96	48,641.00
276	0.25	42,982.00	5.25	21,474.00
277	0.17	39,274.00	2.21	17,179.00
278	0.10	37,795.00	3.56	15,463.00
279	0.47	35,405.00	1.56	12,695.00
280	0.24	39,264.00	6.50	17,164.00
281	0.94	37,285.00	3.06	14,873.00
282	0.42	38,803.00	6.98	16,629.00
283	0.17	37,018.00	0.98	14,565.00
284	1.06	38,363.00	2.15	16,123.00
285	0.26	34,953.00	4.55	12,169.00
286	3.03	24,456.00	7.43	0.00

Table 4. Regression model in AMPL (file .mod).

```
## Census Blocks definition ##
set BLOCKS;

## parameters of the regression problem ##
param MREP {i in BLOCKS};
param TC   {i in BLOCKS};
param PCI  {i in BLOCKS};

## explanation of variables ##
var y  {i in BLOCKS} ≥0;
var X1 {i in BLOCKS} ≥0;
var X2 {i in BLOCKS} ≥0;

### objective function ###
minimize objective: sum{i in BLOCKS}(MREP[i] - TC*x1[i] -PCI*x2[i]- y)^2
```

Table 5. File.dat.

```
set BLOCKS := 0 1 … 144 … 286;

param MREP :=
    0      32396
    1      35419
    2      69924
    3      38060
    4      44489
    5      34949
    6      36938
    7      32396
    8      35419
   …        …
  286      24456


param TC :=
    0       4.01
    1       3.21
    2       9.22
    3       4.17
    4       0.00
    5       2.76
    6       2.04
    7       4.01
    8       3.21
   …        …
  286       7,43


param PCI :=
    0       9205
    1       1270
    2       5269
    3      15767
    4      23225
    5      12166
    6      14471
    7        920
    8      12708
   …        …
  286       0,00
```

5   Discussions

The correlation function is derived from the processing carried out:

$$REP = 24,451.55 + 0.668540 \cdot TC + 0.862783 \cdot PCI \qquad (3)$$

Equation (3) highlights the functional link of Real Estate Prices (REP) with the environmental variable Tree Cover (TC) and the socio-economic variable Per-Capita Income (PCI). The tests conducted show weak multicollinearity between the TC and PCI variables, as demonstrated by the 0.21 value of the Pearson index.

Obviously, (3) refers to a typical real estate unit, whose market value depends exclusively on TC and PCI.

From the obtained results derives a significant correlation between REP, TC, and PCI. The Census Blocks with higher Per-Capita Income and Tree Cover values are characterized by high Real Estate Prices. On the other hand, the less wealthy parts of cities have less green areas and lower Real Estate Prices. These results are in line with other U.S. cities, where the richest urban neighbourhoods, i.e., with higher PCI and REP, have a high percentage of Tree Cover [46, 47].

6   Conclusions

Socio-economic and environmental parameters of the city influence the urban real estate market. In special way, the natural elements, such as single trees and urban forests, provide multiple effects. Among these, the increase of residential building stock values in the city.

Despite in the reference literature it is recognized the relation among ecological factors and market parameters, there are few examples that examine their functional bond. In fact, the knowledge of the interdependencies including multiple variables is useful for the planning of urban sustainable actions in the respect of the reference context features.

In order to investigate the link including multiple variables, indicators representative the socio-economic and environmental urban system were identified. From the study on the urban forest benefits of Nowak D. and John F. Dwyer (2007), Tree Cover and Per-Capita Income parameters were considered as variables to be correlated to Real Estate Prices.

The novelty of the present work is characterized by the methodology implemented for the variables correlation analysis. The estimation of the relationship between Real Estate Prices, Tree Cover, and Per-Capita Income is conducted both through the construction of thematic maps with Geographic Information Systems (GIS) and by implementing statistical methodologies and mathematical programming tools. Thematic maps give the graphic representation of the values distribution. The correlation analysis then returns the function that quantitatively expresses the dependence of Real Estate Prices from Tree Cover and Per-Capita Income. For the construction of the function is defined an innovative model that implements optimization algorithms.

A case study is developed with reference to datasets taken from official sources for the City of Buffalo, in the New York State (USA). The elaborations on 287 Census Blocks demonstrate that the Real Estate Prices depend on the Tree Cover and Per-Capita Income levels. In particular, the model provides the β_1 and β_2 coefficients of the function: $\beta_1 = 0.668540$ (TC); $\beta_2 = 0.862783$ (PCI).

Research perspectives concern, on the one hand, the application of the investigation protocol to other urban realities, and, on the other one, the characterization of a function able to explain the mechanisms of formation of urban real estate values in the light not only of the socio-economic and environmental variables of the study area, but also in relation to the positional and intrinsic property characteristics.

References

1. https://ec.europa.eu/regional_policy/it/information/publications/brochures/2014/integrated-sustainable-urban-development. Accessed 07 Mar 2020
2. Haughton, G., Hunter, C.: Sustainable Cities. Routledge, London (2004)
3. European Commission: Il futuro sostenibile dell'Europa: prossime tappe. L'azione europea a favore della sostenibilità (2016)
4. Tyrväinen, L., Pauleit, S., Seeland, K., de Vries, S.: Benefits and uses of urban forests and trees. In: Konijnendijk, C., Nilsson, K., Randrup, T., Schipperijn, J. (eds.) urban forests and trees, pp. 81–114. Springer, Berlin, Heidelberg (2005). https://doi.org/10.1007/3-540-27684-X_5
5. Shackleton, S., Chinyimba, A., Hebinck, P., Shackleton, C., Kaoma, H.: Multiple benefits and values of trees in urban landscapes in two towns in northern South Africa. Landscape Urban Plan. **136**, 76–86 (2015)
6. Biodiversity Information System for Europe. https://biodiversity.europa.eu/topics/ecosystem-services. Accessed 25 Feb 2020
7. Baró, F., Chaparro, L., Gómez-Baggethun, E., Langemeyer, J., Nowak, D.J., Terradas, J.: Contribution of ecosystem services to air quality and climate change mitigation policies: the case of urban forests in Barcelona. Ambio **43**(4), 466–479 (2014)
8. Coutts, A.M., Tapper, N.J., Beringer, J., Loughnan, M., Demuzere, M.: Wateringour cities: the capacity for water sensitive urban design to support urban cooling and improve human thermal comfort in the Australian context. Prog. Phys. Geogr. **37**(1), 2–28 (2013)
9. Daily, G.C.: Nature's Services: Societal Dependence on Natural Ecosystems. Island Press, Washington, DC (1997)
10. Poudyal, N.C., Hodges, D.G., Merrett, C.D.: A hedonic analysis of the demand for and benefits of urban recreation parks. Land Use Policy **26**, 975–983 (2006)
11. Millennium Ecosystem Assessment: Ecosystems and Human Well-being: Synthesis. Island Press, Washington, DC (2005)
12. Baró, F., Palomo, I., Zulian, G., Vizcaino, P., Haase, D., Gómez-Baggethun, E.: Mapping ecosystem service capacity, flow and demand for landscape and urban planning a case study in the Barcelona metropolitan region. Land Use Policy **57**, 405–417 (2016). https://doi.org/10.1016/j.landusepol.2016.06.006
13. Bartoli, F.: Tecniche e strumenti per l'analisi economica-finanziaria, Franco Angeli, Milano (2006)
14. Glaeser, E.: Cities, Agglomeration and Spatial Equilibrium. Oxford University Press, New York (2008)

15. Glaeser, E., Kahn, M., Rappaport, J.: Why do the poor live in cities? The role of public transportation. J. Urban Econ. **63**(1), 1–24 (2008)
16. Bateman, I.J., Day, B., Lake, I., Lovett, A.A.: The effect of road traffic on residential property values: A literature review and hedonic pricing study. Edinburgh: Scottish Office and the Stationary Office (2001)
17. Benson, E.D., Hansen, J.L., Schwartz, Jr., A.L., Smersh, G.T.: Pricing residential amenities: the value of a view. J. Real Estate Financ. Econ. **16**, 55–73 (1998)
18. Bolitzer, B., Netusil, N. R.: The impact of open spaces on property values in Portland, Oregon. J. Environ. Manage. **59**, 185–193 (2000)
19. Nichol, J., Wong, M.S.: Modeling urban environmental quality in a tropical city. Landscape Urban Plan. **73**(1), 49–58 (2005)
20. Bencardino, M., Granata, M.F., Nesticò, A., Salvati, L.: Urban growth and real estate income. a comparison of analytical models. In: Gervasi, O., et al. (eds.) ICCSA 2016. LNCS, vol. 9788, pp. 151–166. Springer, Cham (2016). https://doi.org/10.1007/978-3-319-42111-7_13
21. Bencardino, M., Nesticò, A.: Urban sprawl, labor incomes and real estate values. In: Gervasi, O., et al. (eds.) ICCSA 2017. LNCS, vol. 10405, pp. 17–30. Springer, Cham (2017). https://doi.org/10.1007/978-3-319-62395-5_2
22. Tajima, K.: New estimates of the demand for urban green space: implications for valuing the environmental benefits of Boston's big dig project. J. Urban Aff. **25**(5), 641–655 (2003)
23. Troy, A., Grove, J.M.: Property values, parks, and crime: a hedonic analysis in Baltimore. Landscape Urban Plan. **87**, 233–245 (2008)
24. Poudyal, N.C., Hodges, D.G., Merrett, C.D.: A hedonic analysis of the demand for and benefits of urban recreation parks. Land Use Policy **26**, 975–983 (2009)
25. Wu, J.J., Adams, R.M., Plantinga, A.J.: Amenities in an urban equilibrium model: Residential development in Portland, Oregon. Land Econ. **80**(1), 19–32 (2004)
26. Debrezion, G., Pels, E., Rietveld, P.: The impact of railway stations on residential and commercial property value: a meta-analysis. J. Real Estate Financ. Econ. **35**, 161–180 (2007)
27. Schaerer, C., Baranzini, B., Ramirez, J.V., Thalmann, P.: Using the hedonic approach to value natural land uses in an urban area: an application to Geneva and Zurich. Économie publique/Public Econ. **20**, 147–167 (2007)
28. Jim, C.Y., Chen, W.Y.: Value of scenic views: hedonic assessment of private housing in Hong Kong. Landscape Urban Plan. **91**, 226–234 (2009)
29. Nichol, J., Wong, M.S.: Modeling urban environmental quality in a tropical city. Landscape Urban Plan. **73**(1), 49–58 (2005)
30. Garrod, G., Willis, K.G.: Valuing goods characteristics: an application of the hedonic price method to environmental attributes. J. Environ. Manage. **34**, 59–76 (1992)
31. Powe, N.A., Garrod, G.D., Willis, K.G.: Valuation of urban amenities using an hedonic price model. J. Prop. Res. **12**, 137–147 (1995)
32. Giorgetti, M., Massaro, D.: Ricerca e percorsi di analisi dati con SPSS. Pearson Italia Spa, (2007)
33. Anselin, L.: Spatial dependence and spatial structural instability in applied regression analysis. J. Reg. Sci. **30**(2), 185–207 (1990)
34. Feng, Y., Yan, L., Xiaohua, T., Miaolong, L., Susu, D.: Modeling dynamic urban growth using cellular automata and particle swarm optimization rules. Landscape Urban Plan. **102**(3), 188–196 (2011)
35. Morano, P., Tajani, F.: Saving soil and financial feasibility. a model to support public-private partnerships in the regeneration of abandoned areas. Land Use Policy **73**, 40–48 (2018)

36. Nesticò, A., Morano, P., Sica, F.: A model to support the public administration decisions for the investments selection on historic buildings. J. Cult. Herit. **33**, 201–207 (2018). https://doi.org/10.1016/j.culher.2018.03.008

37. Della Spina, L.: A multi-level integrated approach to designing complex urban scenarios in support of strategic planning and urban regeneration. In: Calabrò, F., Della Spina, L., Bevilacqua, C. (eds.) ISHT 2018. SIST, vol. 100, pp. 226–237. Springer, Cham (2019). https://doi.org/10.1007/978-3-319-92099-3_27

38. Nesticò, A., Guarini, M.R., Morano, P., Sica, F.: An economic analysis algorithm for urban forestry projects. Sustainability **11**(2), 314 (2019). https://doi.org/10.3390/su11020314

39. Ribera, F., Nesticò, A., Cucco, P., Maselli, G.: A multicriteria approach to identify the Highest and Best Use for historical buildings. J. Cult. Herit. **41**, 166–177 (2020). https://doi.org/10.1016/j.culher.2019.06.004

40. Borst, R.A., McCluskey, W.J.: Using geographically weighted regression to detect housing submarkets: modeling large-scale spatial variations in value. J. Prop. Tax Assess. Adm. **5**(1), 21–54 (2008)

41. De Mare, G., Nesticò, A., Tajani, F.: The rational quantification of social housing. In: Murgante, B., et al. (eds.) ICCSA 2012. LNCS, vol. 7334, pp. 27–43. Springer, Heidelberg (2012). https://doi.org/10.1007/978-3-642-31075-1_3

42. Bidanset, P.E., Lombard, J.R.: Evaluating spatial model accuracy in mass real estate appraisal: a comparison of geographically weighted regression and the spatial lag model. Cityscape **16**(3), 169–182 (2014)

43. Morano, P., Tajani, F., Di Liddo, F., Anelli, D.: A feasibility analysis of the refurbishment investments in the italian residential market. Sustainability **12**(6), 2503 (2020)

44. Nesticò, A., Maselli, G.: A protocol for the estimate of the social rate of time preference: the case studies of Italy and the USA. J. Econ. Stud. **47**(3), 527–545 (2020). https://doi.org/10.1108/JES-02-2019-0081

45. Buffalo economy news. https://www.buffalony.gov/. Accessed 07 Mar 2020

46. Nowak, D.J., Dwyer, J.F.: Understanding the benefits and costs of urban forest ecosystems. In: Kuser, J.E. (ed.) Urban and Community Forestry in the Northeast, pp. 25–46. Springer, Dordrecht (2007). https://doi.org/10.1007/978-1-4020-4289-8_2

47. Schwarz, K., et al.: Trees grow on money: urban tree canopy cover and environmental justice. PLoS ONE **10**(4) (2015) 42*

COVID 19: Health, Statistical and Constitutional Aspects

Francesco Perchinunno[1], Luigia Stefania Stucci[2],
and Paola Perchinunno[3(✉)]

[1] Ionic Department of Mediterranean Legal and Economic Systems Society
Environment Cultures, University of Bari "Aldo Moro", Bari, Italy
francesco.perchinunno@uniba.it
[2] Medical Oncology Unit, Policlinico Hospital, Bari,
University of Bari "Aldo Moro", Bari, Italy
stefania.stucci@uniba.it
[3] Department of Economics, Management and Business,
University of Bari "Aldo Moro", Bari, Italy
paola.perchinunno@uniba.it

Abstract. Starting from the assumption that "*health is the primary good*" and although the humanitarian character of the epidemic remains the most urgent aspect to be treated, unfortunately it is not the only one. As acknowledged by the World Health Organization on 11 March, the COVID-19 epidemic has become a worldwide pandemic. The response to the emergency connected with the spread of the virus, to limit its negative impact on the economic system, must consist in a rapid and targeted intervention. The aim of this paper is to analyse the main aspects of Covid 19 from the point of view of health, constitutional and statistical, highlighting the evolution of the/phenomenon and its territorial diffusion, with reference to the spread of the virus in Italy.

Keywords: Covid 19 · Constitutional aspects · Statistical data · Public health

1 Introduction

Starting from the assumption that "*health is the primary good*" and although the humanitarian character of the epidemic remains the most urgent aspect to be treated, unfortunately it is not the only one.

The restrictive measures aimed at containing the virus have a negative economic effect both through the closure of commercial activities and changes in household spending decisions, and through the blocking of certain industrial activities, with interruptions even in global value chains.

It's not easy to quantify the effects of these measures on the economic system as they will depend on a number of factors such as duration, the spread of contagion in the

---

The contribution is the result of joint reflections by the authors, with the following contributions attributed to F. Perchinunno (chapters 1, 3 and 5), L S. Stucci (chapter 2) and to P. Perchinunno (chapter 4).

territorial context and the rigidity in the implementation of containment measures in other countries, especially those linked to trade relations with ours.

The negative effects on the economic system are propagated through a double shock, of the demand (reduction of the consumption of the families, decrease of the tourist flows, decrease of the external demand) and of the supply (closing of the activities, lack of intermediate assets of production) which is widespread in all productive sectors (wholesale and retail trade, transport, accommodation and catering services, real estate).

As acknowledged by the World Health Organization on 11 March, the COVID-19 epidemic has become a worldwide pandemic.

The response to the emergency connected with the spread of the virus, to limit its negative impact on the economic system, must consist in a rapid and targeted intervention.

The aim of this paper is to analyze the main aspects of Covid 19: that is, to analyze the phenomenon from the point of view of health, constitutional and statistical, highlighting the evolution of the phenomenon and its territorial diffusion, with particular reference to the spread of the virus in Italy, a heavily affected nation.

2   Clinical and Health-Related Features of COVID-19 Infection

2.1   Background

Coronaviruses (CoVs) are a family of respiratory viruses that induce cause mild to moderate diseases simulating flu syndrome or severe respiratory syndromes [1–3]. Past years, two respiratory infections by coronavirus have been described such the severe acute respiratory syndrome (SARS) and Middle East respiratory syndrome (MERS).

SARS is knowns as a virus developed in southern China in November 2002 resulting in 774 deaths in in the 2002–2003 period [4]; MERS is detected for the first time in Saudi Arabia in 2012 and produced 858 deaths [5]. In December 2019, viral pneumoniae were linked to a new coronavirus in Wuhan and in following days a new coronavirus, known as 2019-nCoV, was isolated [6]. On February 11, 2020, the World Health Organization (WHO) formally designed the new coronavirus 2019 infection Coronavirus Disease 2019 (COVID-19). The increase in the number of cases and expansion of geographical areas induced WHO to declare pandemic disease on March 11, 2020 [7].

2.2   Clinical Aspects

SARS-CoV-2 infection has been associated with different several clinical manifestations ranging from mild disease to death, which occurs from sepsis and/or Acute respiratory distress syndrome (ARDS). Early diagnosis of the clinical symptoms is recommended for promptly starting the adequate restrictive measures as social quarantine and treatments to avoid spreading of the virus and complications in patients [8]. Fever, cough and dyspnea are the most common symptoms reported whereas those patients who develop pneumonia and ground-glass opacity in radiological scans have higher risk of complications [8, 9].

Moreover, elderly patients with comorbidities can develop acute respiratory distress syndrome more rapidly and die of multiple organ failure. The case fatality rate (CFR) ranged from 10.5% to and 5.6% for cardiovascular diseases, diabetes, chronic respiratory diseases, hypertension and tumors, respectively [10]. Rare manifestations induced by COVID-19 are gastrointestinal disorders such as vomiting, diarrhea, and abdominal pain, as observed in about 2–10% of patients and nausea and diarrhea could arise early compared to respiratory signs and symptoms [11]. Myalgia, pharyngodynia and anorexia were observed but rarely early recognized. Many patients developed multiorgan dysfunction such as liver deficiency and acute renal failure and furthermore mental confusion [12].

Neurological symptoms of COVID-19 infection are common and often severe such as dizziness, loss of smell (anosmia), loss of taste (ageusia), muscle pain and weakness, impaired consciousness, and cerebrovascular complications. Furthermore, we do not yet know the acute or long-term consequences of the COVID-19 lockdown and social isolation as well as increasing unemployment, financial insecurity, and poverty on mental health [13].

2.3   Impact on Public Health

At the beginning of COVID-19 disease, restriction measures have been issued by Chinese government for limiting the spread of the virus, such as suspension of public transport, closure of airports, cancellation of celebrations and closure of parks and cinemas [14, 15].

Subsequently, more drastic containment measures were applied in China up to the suspension of all non-essential activities resulting in decrease of COVID-19 spreading progression, as observed in Wuhan. First, the WHO and the Emergency Committee under International Health Regulations have stressed the need to screen travelers from affected areas. After, the WHO has developed a strategy to contain the impact of COVID-19 that included blocking the chain of transmission (working on patients as well as close contacts), to avoid the health system disruption and to limit the impact of the pandemic as much as possible. Isolation of symptomatic subjects is mandatory to reduce transmission as well as social distancing for healthy people is essential. Recommendations about common practices of daily life included rigorous hand hygiene with soap and water or alcohol-based solutions, avoiding coughing and/or sneezing without covering the mouth [16]. Relatively to use of surgical face masks, there is no evidence of the usefulness of face masks by healthy subjects [17]. Moreover, some countries such as Italy implemented relevant social measures as school closures, smart working and cancelling meetings, sports and cultural events [18].

3 Constitutional Aspects

In the current dramatic epidemiological framework marked by the undeniable priority need for containment of contagion, the new health safety parameters that have emerged seem to be aligned in a worrying framework that, whereas, on the one hand, there is a change in the way we live and live together, on the other, it seems to seriously jeopardize the modern democratic society in which we live and the system of values and freedoms on which our constitutional order is based, constant source of inspiration of exegetical work aimed at the reconciliation of the fundamental values of the person. The need for multi-level protection of the right to health, since the post-war period, is evident first of all from the International Health Conference (New York, 1946) and the World Health Organization (WHO) and in the definition of health as A state of complete physical, mental, social well-being and does not consist only in the absence of disease or infirmity. The possession of the best state of health that can be achieved constitutes one of the fundamental rights of every human being, whatever his race, his religion, his political opinions, his economic and social condition. Governments have the responsibility for the health of his peoples: they must take the appropriate health and social measures to be part of them. For an examination of the balancing work carried out by the Constitutional Court [19–21].

An undeniable fact is that it is an emergency event (more or less intercepted in advance, given the gradual propagation in individual countries) empirically identified and scientifically proven, so as to seriously endanger the life and health of people, in which, while acknowledging that the counter-measures must in any event be classified within the framework of the extraordinary measures dictated by the emergency, it was feared that (provocatively, although we cannot be sure it is really a totally unfounded fear and unnecessarily alarmist) the risk of the restoration of a state-centric conception, so much opposed by the Constituent Fathers [22–24].

The fact that this is not, unfortunately, a short transitional period makes a systematic preliminary question legitimate: to what extent can the situation of emergency justify the suppression of fundamental human rights? We can say, in terms of certainty, that the mentioned and indispensable work of balancing between freedom and constitutional values is the inspiring criterion underlying the recent containment measures and the emergency regulatory framework put in place by the institutions with respect to the epidemiological emergency ongoing? [25–27].

Faced with these and many other questions, the community of jurists has certainly not intended to spare its efforts by highlighting, Clearly, the reasons behind the alarm cry that is being widely used to stand in the way of repeated attempts to break down the barriers set up to defend the stronghold of our constitutional guarantee system. But we can really believe that the measures implemented to contain and combat the spread of the virus throughout the national territory, have been characterized by the necessary effectiveness and that the proclaimed intention to protect the fundamental right of everyone to public health has been marked by full effectiveness?

The range of issues that would deserve rigorous and timely deepening is undoubtedly vast and there is no claim to carry out a thorough and complete analysis of the many critical profiles emerged from the examination of the measures recently adopted in the form varied (decrees of the law, decrees of the President of the Council of Ministers, decrees of ministers, provisions of regional presidents or mayors).

It should be possible to give a brief example here. On this point, it has been widely highlighted that to guarantee the fundamental right to health of citizens, the required emergency regulatory framework has resulted in a clear and persistent imbalance to the detriment of other rights and freedoms. Relatively to "freedom to circulate" for "security and people health" (art. 16, paragraph 1) it provides for a prohibition «for proven reasons of safety or public safety», but only if meetings take place in a public place, while the measures adopted have been strictly prohibited even all meetings in private places, in addition to the known restrictions relating to religious faith and the exercise of the right to work.

Strongly affected and in some cases entirely compressed was personal freedom [28], both with the «application of the precautionary quarantine measure to persons who have had close contact with confirmed cases of spreading infectious disease or who fall within areas, located outside the Italian territory» (art. 1, paragraph 1 Legislative Decree 25 March 2020, n.19) both with the «absolute prohibition to move away from one's home or dwelling for persons subject to the quarantine measure because they were positive to the virus» (art. 1, paragraph 1 Legislative Decree n. 19), not to mention the further derogations. disposed by art. 14 Legislative Decree 9 March 2020, n.14 to the legislation on the processing of personal data, with the related and frequent violations perpetrated. Equally well known is the complex question of the abolition of the right of access for those subject to restrictions on personal freedom, then merged into equally known phenomena of violent reaction.

4   Statistical Analysis of Covid 19 Trend

4.1   Introduction

"Torture numbers, and they'll confess to anything". This sentence of Gregg Easterbrook, American journalist and writer, summarizes the questions that arise in these hours many people, daily bombarded by the "numbers of contagion", percentages and graphs on the course of the pandemic from Covid-19 in our country.

For example, the precise and certified number of infections of patients with Covid-19 does not exist and may never exist and in any case this value is closely related to the number of swabs performed. In other words, the number of new positives recorded in one day does not give much information unless it is accompanied by data on the number of tests carried out.

But we still try to work on the data trying to consider the overall values certified and not the single daily evolution.

4.2   Situation in the World

The World Health Organization data show that on May 7 the number of confirmed cases worldwide is 3,726,292 with deaths number equal to 257,405 and with 215 Countries, areas or territories with cases (Fig. 1).

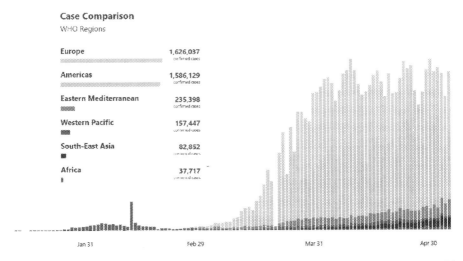

Fig. 1. Number of confirmed cases from 11/01/2020 to 07/05/2020 per Region *Source*: World Health Organization.

In particular, the countries most affected by the virus are the United States (1.2 million cases), Spain (219.329 cases) and Italy (214.457 cases). These countries are also at the top of the list for the number of deceased (Table 1). However, this situation should be compared with the percentage of infections per inhabitant and the mortality rate. Data from Johns Hopkins University show that the countries with the highest percentages of infections, as a percentage of the population, are Spain (0.469%), Belgium (0.436%), followed by the United States, Italy and Switzerland (about 0.35%). The highest mortality rate is found in Belgium (15.87%) and the United Kingdom (15.03%), followed by Italy and the Netherlands.

Table 1. Number of confirmed cases per region

Rank	Countries	Total cases	% of Covid infections per population	Number of deaths	Mortality rate
1	United States	**1,257,023**	0.38%	75,662	6.02%
2	Spaign	**221,447**	0.47%	26,070	11.77%
3	Italy	**215,858**	0.36%	29,958	13.88%
4	United Kingdom	**207,977**	0.31%	30,689	14.76%
5	Russia	**177,160**	0.12%	1,625	0.92%
6	France	**174,918**	0.27%	2,990	14.86%
7	Germany	**169,430**	0.20%	7,392	4.36%
8	Brazil	**135,773**	0.06%	9,190	6.77%
9	Turchia	**133,721**	0.16%	3,641	2.72%

(continued)

Table 1. (*continued*)

Rank	Countries	Total cases	% of Covid infections per population	Number of deaths	Mortality rate
10	Iran	**103,135**	0.12%	6,486	6.29%
11	China	**83,975**	0.01%	4,637	5.52%
12	Canada	**66,201**	0.18%	4,541	6.86%
13	Perù	**58,526**	0.18%	1,627	2.78%
14	India	**56,351**	0.00%	1,889	3.35%
15	Belgium	**51,420**	0.44%	8,415	16.37%
16	Netherlands	**41,973**	0.25%	5,306	12.64%
17	Saudi Arabia	**33,731**	0.10%	219	0.65%
18	Ecuador	**30,298**	0.17%	1,654	5.46%
19	Swiss	**30,126**	0.35%	1,810	6.01%
20	Mexico	**29,616**	0.02%	2,961	10.00%

Source: Johns Hopkins University (07/05/2020)

4.3   The Current Situation in Italy

In Italy the total cases confirmed on 7 May are 214,457, as already mentioned in the previous paragraph, with a deaths number equal to 29,684.

The trend of the pandemic since February 25 shows an exponential curve with regard to the number of total cases and a decreasing curve with regard to the number of cases currently positive (Fig. 2).

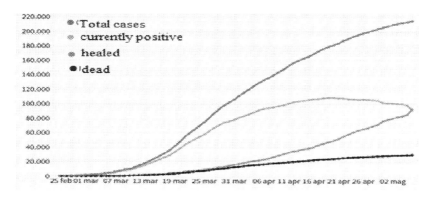

Fig. 2. Pandemic curves in Italy *Source*: https://lab24.ilsole24ore.com/coronavirus/

Going to analyze the situation in the different Italian regions we see that more than half of the positive cases are concentrated in three regions: 37% of cases in Lombardia and more than 10% in Piemonte and Emilia Romagna (Fig. 3).

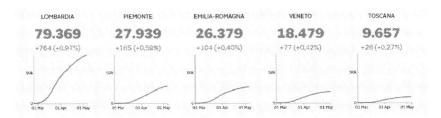

Fig. 3. Pandemic curves in some Italian regions *Source*: https://lab24.ilsole24ore.com/corona virus/

The curve trend in some Italian provinces is also interesting. At the top of the ranking we find the province of Milan, which presents on May 7 20,711 cases, followed by Turin with 14,116 cases and Brescia with 13,267 (Fig. 4).

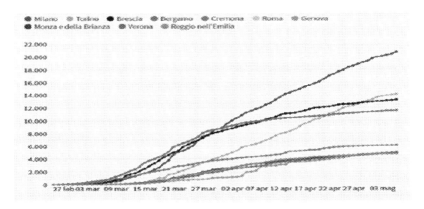

Fig. 4. Pandemic curves in some Italian regions *Source*: https://lab24.ilsole24ore.com/corona virus/

In this case, however, by comparing the number of cases with the resident population, it emerges that the provinces with the highest incidence are Cremona with 1.7% of total cases, Piacenza, 1.5% and Lodi 1.4%, the province from which the infection initially spread.

4.4   An Application of Cluster Analysis on the Italian Provinces

Based on the results of a multitude of scientific work and reports of the Intergovernmental Panel on Climate Change (IPCC) it is clear that the air quality we breathe affects our health. Climate change acts directly and indirectly in determining a wide variety of diseases, fostering new ones and acting as a multiplying force for many of the existing problems. There are still many unanswered questions, for example on the causes, certainly many, of the high lethality of Covid-19 in Italy, especially in some provinces.

In order to respond to the many studies that have been carried out on this possible link, which has been debated worldwide, The Higher Institute of Health and the Higher Institute for Environmental Protection and Research with the National System for Environmental Protection have launched an epidemiological study at national level to assess whether and to what extent the levels of air pollution are associated with the health effects of the epidemic. What is certain at the moment is that air pollution increases the risk of infections of the lower respiratory tract, especially in vulnerable individuals, elderly and people with previous diseases, conditions that also characterize the epidemic of Covid-19.

In addition, the many data collected shows that Covid-19 was particularly severe for people with other diseases, such as cardiovascular diseases, diabetes, chronic respiratory diseases, hypertension and cancer. In the people affected and without concomitant diseases the lethality was found of the average 1.4% (Case Fatality ratio, that is the deaths on the total of the reported cases), while it is superior to the average in persons with other pathologies (7.6% in cancer patients and 13.2% in heart patients).

The present work analyzes the relationships between the environmental quality and health status indicators identified by Istat (Italian National Institute of Statistics) through the Report on Fair and Sustainable Wellness (BES). Therefore, in this work we have tried to verify if the spread of the virus in some provinces can be linked to the presence in the territory of other aspects such as:

1. High mortality rates due to cancer or nervous system diseases;
2. Environmental factors (lack of green, low waste collection capacity, low use of renewable energy).

On the basis of the mix of variables considered, a cluster analysis was carried out on the data of the 207 Italian provinces in order to create homogeneous groups.

Cluster analysis is very advantageous since it provides clusters, each consisting of units with a high degree of "natural association", that are "relatively distinct" among them (i.e. external heterogeneous) and relatively similar within themselves (i.e. internal homogeneous).

The TwoStep procedure, highly efficient for large datasets, is a cluster analysis scalar algorithm and is able to simultaneously treat variables or categorical and continuous attributes. This is achieved through two steps. In the first step, defined as precluster, records are pre-classified into many small sub-clusters.

In the second step the sub-clusters produced in the pre-clusters are further classified. In this second stage, given the modest dimensions, traditional methods of clustering can also prove effective. Given a mixed model with p continuous variables x_j and q categorical variables a_j, the Bayesian Information Criterion (useful for model selection among a finite set of models, because the one with the lowest BIC has to be preferred) is defined as [29]:

$$BIC_k = r_k \ln n - 2l_k$$

The results of the cluster led to the identification of 4 clusters, of which two more numerous and two less numerous (Fig. 5). In particular, Cluster 4 includes only 8 provinces (Bergamo, Brescia, Cremona, Lodi, Milan, Piacenza, Pavia and Turin) which are by far the provinces most affected by Covid 19.

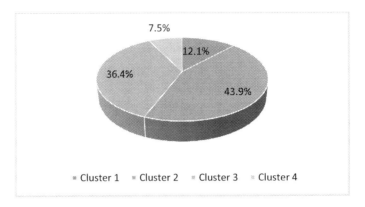

Fig. 5. Percentage composition of single cluster *Source*: Our elaboration on Two Step cluster analysis.

The average values of the individual indicators show that the first and second clusters contain those provinces poorly affected by Covid, with low number of cases (Table 2). More interesting are certainly the results of Cluster 3 and especially cluster 4 that refer to the provinces heavily affected by Covid with a very high total of cases but especially with high cancer mortality rate (8.78) or due to the nervous system (33.39). Equally high are environmental values such as the absence of green (73.59) and the absence of renewable sources (71.56).

Table 2. Average values of the variables and indicators.

	Number of provinces	Positive cases of Covid 19	Percentage cases on population	Cancer mortality rate (20–64 years)	Mortality rate due to the nervous system (over 65 years)	Absence of green	Absence of renewable sources
Cluster 1	13	824.38	0.27	8.42	27.60	12.15	34.13
Cluster 2	47	657.04	0.14	8.89	29.03	83.10	50.52
Cluster 3	39	235.,62	0.49	8.21	34.04	69.25	82.17
Cluster 4	8	9,737.88	1.09	8.78	33.39	73.59	71.56

This situation appears much more evident in the following Fig. 6, where we find the average values of the different indices. It is more evident in this figure that cluster 3 and 4 have values much higher than the average, except for the cancer mortality rate that for Cluster 4 is lower than for Cluster 2.

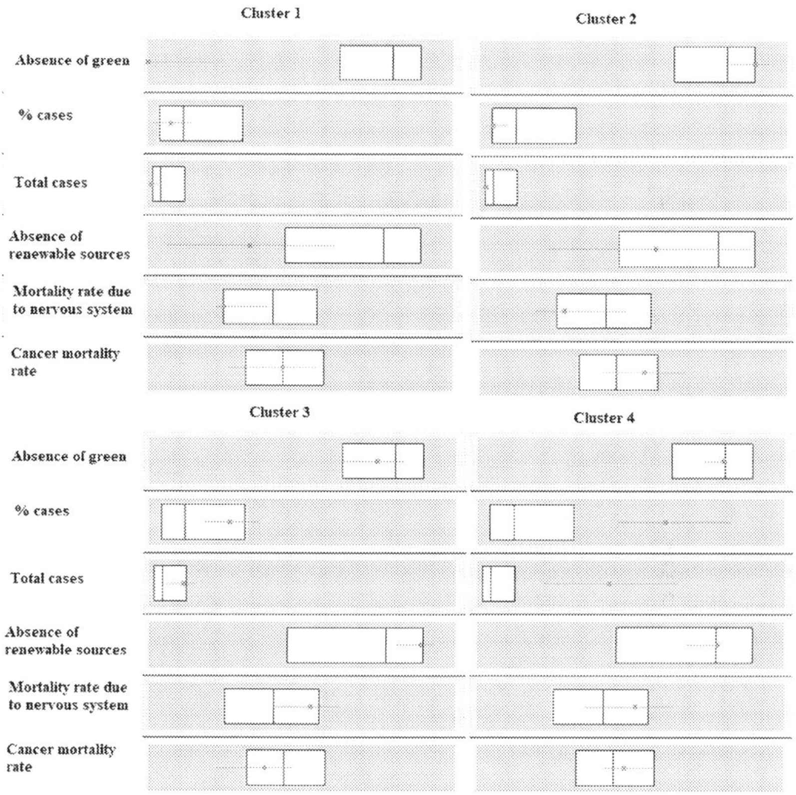

Fig. 6. Average values of the variables and indicators. *Source*: Our elaboration on Two Step cluster analysis.

5 Conclusions

As recognized by the World Health Organization on March 11, the COVID-19 epidemic has become a global pandemic, present in 215 countries.

Coronaviruses (CoVs) are a family of respiratory viruses that induce cause mild to moderate diseases simulating flu syndrome or severe respiratory syndromes. SARS-CoV-2 infection has been associated with different Several clinical manifestations ranging from mild disease to death, which occurs from Acute respiratory distress syndrome (ARDS). The case fatality rate (CFR) range from 10.5% to and 5.6% for cardiovascular diseases, diabetes, chronic respiratory diseases, hypertension and tumors, respectively.

The need for multi-level protection of the right to health, since the post-war period, is evident first of all from the International Health Conference (New York, 1946) and the World Health Organization (WHO) and in the definition of health as A state of complete physical, mental, social well-being and does not consist only in the absence of disease or infirmity.

The possession of the best state of health that can be achieved constitutes one of the fundamental rights of every human being, whatever his race, his religion, his political opinions, his economic and social condition. Strongly affected and in some cases entirely compressed was personal freedom, both with the «application of the precautionary quarantine measure to persons who have had close contact with confirmed cases of spreading infectious disease or who fall within areas.

Precise and certified number of infections of patients with Covid-19 does not exist and may never exist. The data show different trends according to the territories analyzed and the timing of spread of the virus

On the basis of the mix of variables considered, a cluster analysis was carried out on the data of the 208 Italian provinces in order to create homogeneous groups. This application was carried out in order to demonstrate as already anticipated in other studies conducted by The Higher Institute of Health and the Higher Institute for Environmental Protection and Research with the National System for Environmental Protection that the virus develops in relation to other factors. In fact, the results showed that the Italian provinces in which a greater number of cases occurred, also have high mortality rates from cancer and diseases of the nervous system, as well as problems related to environmental aspects.

Further developments will be made in the coming months based on the trend of the phenomenon, unfortunately still in the evolution phase.

We could hypothesize that a multidisciplinary approach including medical figures, epidemiologists, researchers, jurists, associations of patients and government workers is advisable to identify the best strategy to prevent other hypothetical emergencies.

References

1. Giovanni, G., d'Anchera, E., Sandri, F., Savio, M., Stefanati, A.: Coronavirus: update related to the current outbreak of COVID-19. Infect. Dis. Ther. **9**, 241–253 (2020)
2. Chen, N., Zhou, M., Dong, X., et al.: Epidemiological and clinical characteristics of 99 cases of 2019 novel coronavirus pneumonia in Wuhan, China: a descriptive study. Lancet **395**, 507–513 (2020)
3. ISS. Focolaio di infezione da un nuovo coronavirus (2019-nCoV). https://www.epicentro.iss.it/coronavirus/2019-nCoV. Accessed 10 Feb 2020
4. ECDC. Severe acute respiratory syndrome (SARS) (2016). https://www.ecdc.europa.eu/sites/default/files/documents/AER_for_2015-SARS.pdf. Accessed 12 Feb 2020
5. ECDC. Risk assessment guidelines for infectious diseases transmitted on aircraft (RAGIDA) Middle East Respiratory Syndrome Coronavirus (MERS-CoV) (2020). https://www.ecdc.europa.eu/sites/default/files/documents/infectious-diseases-transmitted-on-aircrafts-ragida-risk-assessment-guidelines.pdf. Accessed 12 Feb 2020

6. Lu, R., Zhao, X., Li, J., et al.: Genomic characterisation and epidemiology of 2019 novel coronavirus: implications for virus origins and receptor binding. Lancet **395**, 565–574 (2020)

7. WHO. Novel Coronavirus (2019-nCoV) Situation Report—22. https://www.who.int/docs/default-source/coronaviruse/situation-reports/20200211-sitrep-22-ncov.pdf?sfvrsn=fb6d49b1_2. Accessed 12 Feb 2020

8. Wang, D., et al.: Clinical characteristics of 138 hospitalized patients with 2019 novel coronavirus-infected pneumonia in Wuhan, China. JAMA **323**(11), 1061 (2020)

9. Ciotti, M., et al.: COVID-19 outbreak: an overview. Chemotherapy **64**, 215–223 (2020)

10. Wu, Z., McGoogan, J.M.: Characteristics of and important lessons from the coronavirus disease 2019 (COVID-19) outbreak in China: summary of a report of 72,314 cases from the Chinese Center for Disease Control and Prevention. JAMA **2019**, 24–27 (2020)

11. Chen, N., et al.: Epidemiological and clinical characteristics of 99 cases of 2019 novel coronavirus pneumonia in Wuhan, China: a descriptive study. Lancet **395**(10223), 507–513 (2020)

12. Holshue, M.L., DeBolt, C., Lindquist, S., et al.: First case of 2019 novel coronavirus in the United States. N. Engl. J. Med. **382**(10), 929–936 (2020)

13. Holmes, E.A., et al.: Multidisciplinary research priorities for the COVID-19 pandemic: a call for action for mental health science. Lancet Psychiatr. **7**, 547–560 (2020)

14. Wang, C., Horby, P.W., Hayden, F.G., Gao, G.F.: A novel coronavirus outbreak of global health concern. Lancet **395**, 470–473 (2020). https://www.thelancet.com/journals/lancet/article/PIIS0140-6736(20)30185-9/fulltext. Accessed 30 Jan 2020

15. WHO. Updated WHO recommendations for inter- national traffic in relation to COVID-19 outbreak. https://www.who.int/news-room/articles-detail/updated-who-recommendations-for-international-traffic-in-relation-to-covid-19-outbreak. Accessed 11 Mar 2020

16. WHO. Infection prevention and control. https://www.who.int/emergencies/diseases/novel-coronavirus-2019/technical-guidance/infection-prevention-and-control. Accessed 13 Mar 2020

17. Gazzetta Ufficiale Della Repubblica Italiana. G.U. Serie Generale n. 62 del 09 marzo 2020. https://www.trovanorme.salute.gov.it/norme/dettaglioAtto?id=73629. Accessed 10 Mar 2020

18. Gazzetta Ufficiale della Repubblica Italiana. G.U. Serie Generale, n. 59 del 08 marzo 2020. https://www.trovanorme.salute.gov.it/norme/dettaglioAtto?id=73594. Accessed 9 Mar 2020

19. Bin, R.: Diritti e argomenti. Il bilanciamento degli interessi nella giurisprudenza costituzionale, pp. 32–35. Giuffrè (1992)

20. Cartabia, M.: I principi di ragionevolezza e proporzionalità nella giurisprudenza costituzionale italiana, relazione nell'ambito della Conferenza trilaterale delle Corte costituzionali italiana, portoghese e spagnola, Roma, Palazzo della Consulta, 24-26 ottobre 2013. www.cortecostituzionale.it9ss

21. Morrone, A.: voce "Bilanciamento" (Giustizia costituzionale) in Enciclopedia del diritto", Annali, Milano, 2008, vol. II, tomo II, pp. 185- 204; Id., Il bilanciamento nello stato costituzionale: teoria e prassi delle tecniche di giudizio nei conflitti tra diritti e interessi costituzionali, Torino, Giappichelli, VIII, pp.1–149 (2014)

22. Celotto, A.: L' « abuso » del decreto-legge, vol.1, Padova (1997)

23. D'aloia, A.: Costituzione ed emergenza. L'esperienza del Coronavirus, in questa rivista, Online First - BLJ 2/2020; Ivi

24. Fabiano, L.: La catena della normativa emergenziale in risposta alle minacce di diffusione del COVID 19. Biodiritto 2/2020 (2020)

25. Morrone, A.: Il custode della ragionevolezza. Giuffrè, Milano (2001)

26. Penasa, S.: La "ragionevolezza scientifica" delle leggi nella giurisprudenza costituzionale, in Quaderni costituzionali, p. 817 ss (2009)
27. Morrone, A., Minni, F.: Il diritto alla salute nella giurisprudenza della Corte costituzionale italiana, in Rivista AIC, n.3/2013 (2013)
28. Pugh, J.: The United Kingdom's Coronavirus Act, Deprivations of Liberty, and The Right to Liberty and Security of the Person. Journal of Law and the Biosciences (2020)
29. Chiu, T., Fang, D., Chen, J., Wang, Y., Jeris, C.: A robust and scalable clustering algorithm for mixed type attributes in large database environment. In: Proceedings of the Seventh ACM SIGKDD International Conference on Knowledge Discovery and Data Mining, vol. 263, pp. 263–268 (2001)

A Model to Support the Investment Decisions Through Social Impact Bonds as Effective Financial Instruments for the Enhancement of Social Welfare Policies

Francesco Tajani[1] , Pierluigi Morano[2] , Debora Anelli[1(✉)] ,
and Carmelo Maria Torre[2]

[1] Department of Architecture and Design, "La Sapienza" University of Rome,
00196 Rome, Italy
{francesco.tajani,debora.anelli}@uniroma1.it
[2] Department of Sciences of Civil Engineering and Architecture, Polytechnic
University of Bari, 70125 Bari, Italy
{pierluigi.morano,carmelomaria.torre}@poliba.it

Abstract. Despite the widespread opinion that the traditional finance is exclusively interested in the monetary return, in the last ten years this sector has been affected by a contamination of the public principles related to the social impacts. The global development and the spread of "win-win" financial instruments such as the Social Impact Bonds (SIBs) outlines a growing interest in making an investment that aims at generating benefits for all the subjects involved, always guaranteeing a monetary return to the private investor. The complexity of identifying the social impact sectors to be preferred, in a context characterized by different social needs, represents a critical issue in the SIBs investment. This research defines a model that can constitute a decision support tool for the public and private subjects in the preliminary phases concerning the resource allocation for a social program. The proposed algorithm allows to define a temporal priority of the social impact sectors that are simultaneously able to maximize the conveniences for all the subjects involved. Through the model, the public and private subjects will be able to determine the best allocation of financial resources according to the real social needs, contributing to an effective spread of SIBs both in Italy and abroad.

Keywords: Social impact bond · Social investments · Public-private partnership · Operational research · Decision support tools

1 Introduction

In the context of Sustainable and Responsible Investment (SRI), a sustainable finance approach born after the 2007 financial crisis, named the Social Impact Investments (SII), represents a new generation of impact investing. Defined as a wide range of investments based on the assumption that private capital can intentionally contribute to positive social impacts and, at the same time, financial returns, the SII stand out for the

© Springer Nature Switzerland AG 2020
O. Gervasi et al. (Eds.): ICCSA 2020, LNCS 12251, pp. 941–951, 2020.
https://doi.org/10.1007/978-3-030-58808-3_67

proactive intention with which the investor pursues the social purpose, together with the financial return, differently from the profit-only approach of traditional finance tools. In particular, the SII consist of a transaction between a private investor and a social enterprise, aimed at beneficiaries in difficulty. Therefore, the characteristics of the SII can be summarized as follow:

- *intentionality*, that is, the willing of the investor to generate a positive impact through the investment;
- *expectations of financial returns*, differentiating themselves from donations for which investors do not expect a monetary return;
- *flexibility of the expected rate of return*, which may be below the average market level or align with the average or higher yield;
- *measurability of the impact* generated through assessment techniques that can translate impacts into monetary terms;
- *breadth of investment instruments*, through the use of different asset classes, ranging from debt to pure equity, with different yield and risk profiles.

On the demand side, the SII promote the growth process of social enterprises, the ability to propose suitable responses to the complexity of new social needs and, at the same time, an economic and financial feasibility of intervention models. On the supply side, the SII stimulate the growth of a new generation of responsible investors, helping to spread the outcome-oriented approach that forms the theoretical basis on which the different social investment instruments are based.

The social investors, called *Impact First*, are solidarity organizations whose priority is to achieve the established social or environmental purpose, to the detriment of financial returns (e.g. charitable organizations, venture capital and venture philanthropy funds). It often happens that they find themselves working together with *Financial First* investors, who are influenced mainly by the financial return they can obtain, constituting the so called *Layered Structures*, which allows the Financial First to obtain the expected returns at market rates and Impact First to be able to make the most of their investment capital, thus achieving a more significant impact than that obtained if they had independently operated. The most common motivations for private investors to carry out impact investing concern *i)* the willingness of organizations to be responsible investors (85%), *ii)* the dedication to their missions (84%), *iii)* the search for efficient modalities to achieve their impact investment objectives (71%). A further important factor for the impact investing is represented by the respect of the United Nations sustainable development goals set in the 2030 Agenda for Sustainable Development or in the climate agreement held in Paris in 2015 [7].

There is ample evidence that the growth of the social sector stimulates the growth of the economic sector; social development closely linked to the government's practices for public welfare whose austerity policies, caused by the 2007 financial crisis, have imposed a constraint on the budget of public finances [8]. The constraint on public expenditure on welfare is probably a powerful driver of the diffusion of the SII tools, therefore, the public-private partnership acquires an important role [1–3, 17]. The involvement of private financial resources for public purposes has shown its effectiveness in numerous initiatives [19, 20, 22]. In doing so, for the financial private world the welfare deficit appears as a new market, while for the policy maker's point of view the public-private partnership is a way to address the lack of public money, in order to provide the necessary resources for supporting the weaker of the community.

The Governments have been keen to support alternative service delivery as a way to shrink public sector costs and responsibilities [5] and they have begun experimenting with tools to encourage alternative service funding through innovations such as social impact bonds (SIBs). The SIBs are a financial tool used to encourage private, philanthropic or public investors to provide upfront capital to support project-oriented service delivery by private sector social investors.

2 Aim

Taking into account the framework outlined, in this research a decision support model that allows to define investment priorities among different social impact sectors has been developed, by borrowing the principles and theory of Operational Research. The model adopts the branch and bound algorithm for identifying the priority investment alternatives in the different social impact sectors considered. Each investment alternative is distinguished by several weighted criteria that define its main characteristics and represent the aspects that the social investor and the Public Administration should take into account for the adoption of the SIBs.

The model will serve the social investor and the Public Administration to carry out a screening of the types of social impact sectors in which to invest primarily for the activation of a social program using the Social Impact Bonds. The adoption of the proposed model must provide a simple and immediate support to the decisions to be taken in the preliminary phase of identifying the social issues to be addressed. The model should also support decisions to be taken for both small scale and large scale social issues.

The rest of the paper is organized as follows. In Sect. 3 the current diffusion of the SIBs has been outlined. In Sect. 4 the algorithm of the model proposed has been defined: the variables, the objective function and the constraints to be considered for the implementation of the model have been specified. In Sect. 5 the potentialities and the limitations of the proposed model have been discussed.

3 The Social Impact Bonds

Starting from 2010, the empirical literature that focuses on aspects of the SIBs implementation is growing. This consists of reports commissioned by governments and consultancies seeking to highlight the practical advantages and disadvantages to be learned from SIBs adoption [4, 18, 21]. These reports are almost exclusively qualitative and highlight many of the practical challenges involved in initiating SIBs such as high transaction costs, complexity in contractual negotiations, uncertainty relating to the risk and the return deriving from the (entrepreneurial and managerial) capacity of social enterprises to provide services related to the results of the program, difficulty in evaluating and quantifying the expected results, limitation of the budgets that can be spent by the local public administration funds. A part of this literature focuses instead on the motivations that could lead the actors to decide not to pursue a SIB, mainly identified in a lack of sharing of the interests of the subjects involved [6, 13, 16].

The SIB's strategic ambiguity has advantages for both their proponents and critics, and they are lauded for their potential:

- to foster innovation [10];
- to "scale-up" evidence-based programs [12];
- to enable flexible and personalized services tailored to service users [9];
- to improve productivity and value for money or promote more measurement and accountability both for privates and public sector [11].

Thus, there are interesting questions emerging about whether the SIBs challenge or reinforce existing governance logics for actors, institutions and networks involved in experimenting with this new commissioning tool. Developing a SIB at the local level frequently involves the development of new relationships between the actors and the subjects that provide services in the social sector in response to existing and emerging policy problems. This, in turn, may strain (or conversely, it may strengthen) existing relationships and system governance.

Since the launch of the first SIB in 2010 in England, there has been a significant spread of the phenomenon with 137 cases implemented worldwide for a total number of beneficiaries amounting to 1,711,130 (Table 1).

Table 1. Number of SIBs implemented by geographical area, capital raised and number of beneficiaries (source: own elaboration based on https://sibdatabase.socialfinance.org.uk/).

	Number of launched sibs	Capital raised	Number of beneficiaries
United Kingdom	47	£ 44,7 M	54,233
United States	26	$ 219 M	659,482
Australia	10	AUD 57,2 M	8,280
Netherlands	11	€ 11,5 M	2,850
Canada	4	C$ 4,30 M	7,201
France	5	€ 1,95 M	14,030
Israel	3	€ 22,4 M	4,050
Portugal	4	€ 1,72 M	611
Japan	3	–	–
India	3	$ 14,8 M	918,000
South Africa	1	$ 0.540 M	1,000
Finland	2	€ 10,7 M	4,300
Colombia	1	$ 0.750 M	511
Belgium	2	€ 0.230 M	450
Uganda	1	$ 2,40 M	13,830
Cameroon	1	$ 250 M	18,000
Germany	3	€ 0.300 M	148
New Zeland	2	NZ$ 7,50 M	2,700
South Korea	2	₩ 110 M	100

(*continued*)

Table 1. (*continued*)

	Number of launched sibs	Capital raised	Number of beneficiaries
Sweden	1	SEK 10,0 M	60
Switzerland	1	CHF 0.720 M	120
Argentina	1	AR$ 40,0 M	1,000
Congo	1	$ 27,0 M	0
Austria	1	€ 0.800 M	75
Peru	1	$ 0.110 M	99
Total	–	–	**1,711,130**

What is shown in Table 1 highlights the adoption of SIBs in different geographical areas of the world whose entire, characterized by a series of social needs expressed by local populations, allows to define the extent of the impact generated and consequently the system of needs expressed by populations. Generally, there are seven main macro-categories that represent the social sectors of interest of SIBs: *i)* Child and family welfare; *ii)* Criminal Justice; *iii)* Education; *iv)* Environment; *v)* Health; *vi)* Homelessness; *vii)* Workforce development. The typology of needs expressed, and the importance attributed to a certain sector through the social programs launched by local governments is highlighted by the distribution of the SIBs in each country according to the respective social impact sector (Table 2).

Table 2. Number of SIBs implemented in each social impact sector (own elaboration based on data from: https://sibdatabase.socialfinance.org.uk/)

Country	Social issue						
	Child and family welfare	Criminal justice	Education	Environment	Health	Homelessness	Workforce development
United Kingdom	6	1	1	–	8	16	15
United States	6	8	2	1	3	4	2
Australia	4	1	–	–	1	3	1
Netherlands	–	1	–	1	1	–	8
Canada	1	–	1	–	1	–	1
France	–	–	1	–	–	–	4
Israel	–	–	1	–	1	–	1
Portugal	1	–	1	–	–	–	2
Japan	–	–	–	–	3	–	–
India	–	–	2	–	1	–	–
South Africa	1	–	–	–	–	–	–

(*continued*)

Table 2. (*continued*)

Country	Social issue						
	Child and family welfare	Criminal justice	Education	Environment	Health	Homelessness	Workforce development
Finland	–	–	–	–	–	–	2
Colombia	–	–	–	–	–	–	1
Belgium	–	–	–	–	–	–	2
Uganda	–	–	–	1	–	–	–
Cameroon	–	–	–	–	1	–	–
Germany	–	–	2	–	–	–	1
New Zeland	–	1	–	–	1	–	–
South Korea	–	–	1	–	–	–	1
Sweden	–	–	1	–	–	–	–
Switzerland	–	–	–	–	–	–	1
Argentina	–	–	–	–	–	–	1
Congo	–	–	–	–	1	–	–
Austria	1	–	–	–	–	–	–
Peru	–	–	–	–	–	–	1
Subtotals	**20**	**12**	**13**	**3**	**22**	**23**	**44**

The data show a high number of interventions (44) in the Workforce development sector for almost all countries, whereas it is evident that the number of Environment initiatives is limited (3). The interventions included in the social environmental impact sector are not very frequent because they are considered complex due to the vast territorial scale in which they are carried out and the multiple aspects and subjects involved (such as the construction of a water and sewage system for the most poor cities or reforestation projects). This condition is also confirmed by the number of funds that decide to invest pursuing an environmental focus or a social focus. From the data present in the ImpactBase database (source: www.impactbase.org/) it appears in fact that, until 2017, there is a greater number of funds involved in social initiatives rather than environmental ones, also due to the target return respectively obtainable (Fig. 1).

FUNDS INVOLVED IN THE IMPACT INVESTING SECTOR

■ Social focus   ■ Environmental focus

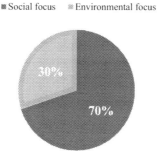

Fig. 1. Percentage of funds involved in initiatives with a social and environmental focus (source: www.impactbase.org/)

Moreover, England holds the record for the largest number of SIBs implemented (47), especially in the Homelessness (16) and Workforce development sectors (15), evidence in accordance with the reasons that led to the birth of SIBs. Except for the United States, Netherlands and Australia where the most widespread SIBs concern the sectors of Workforce Development, Criminal Justice and Child and family welfare, in the remaining geographical contexts the SIB have not had a development comparable to the previous ones. The subdivision of the projects implemented through the adoption of SIBs into macro categories of intervention is necessary to understand the relevance of the social issues felt by local governments, which as a result, carry out appropriate intervention.

4   The Model

The proposed model is based on the principles and theory of the Operational Research, a methodology of considerable flexibility and efficiency that is frequently used in the field of project evaluation and feasibility territorial investments [19, 20] for the resolution of a goal programming problem. The generic problem of goal programming can be traced back to the determination of the excellent use of scarce resources usable in alternative modalities. For the specific field of interest, the goal programming problem is characterized as follows:

- resources available in limited quantity are represented by the different social impact sectors;
- the alternative uses correspond the various weighted criteria, identifying the important aspects considered by the subjects involved (social investors, Public Administrations etc.);
- the constraints that translate into mathematical terms the temporal priority of social impact sectors taking into account the weighted criteria identified;

- the objective functions that reflect the goals pursued by the subjects involved in the social initiative, in terms of maximization (or minimization) of the respective expected outputs.

In this research, the proposed model applies the branch and bound algorithm [14, 15]. Based on a partition and branching mechanism of the sets of solutions and the calculation of a limit value of the objective function, the branch and bound algorithm proceeds for a partial exploration of the feasible solutions. In this case, clarified that, increasing the number of the n social impact sectors and the number of the m weighted criteria, the combinations to be examined become so numerous that the enumeration of the total number of solutions could be impossible, the branch and bound algorithm decomposes the domain solutions into subsets whose intersection is null and whose union coincides with the starter set. The best combination is sought in the subsets obtained through a strategy that aims to verify whether the partition should be further subdivided or may be excluded from further analysis.

The implementation of the branch and bound algorithm provides the introduction of a binary variable k_i, associated to the i-th ($i = 1, \ldots, n$) social impact sectors. Therefore, k_i represents the branching variable, which assumes a value "1" if the i-th social impact sector satisfies the objective function, and a value "0" otherwise.

With the aim of defining a temporal priority order of the investment sectors according to the different weighted criteria, it is evident that the first selection of the i-th investment sector - which is therefore the first in order of realization - excludes the other $n-1$ sectors. This will be followed by $n-2$ elaborations, aimed at identifying, in each of them, the i-th investment sector which, in terms of capacity to satisfy the intended objective function, follows that generated in the previous processing. Therefore, the first constraint of the model can be defined through Eq. (1), in order to ensure that the branching variable k_i takes the value "1" for only one of the j social impact sectors, with j = n, n − 1, …, 2 repeating this process:

$$\sum_{i=1}^{j} k_i = 1 \qquad j = n, n - 1, \ldots, 2 \tag{1}$$

The second constraint concerns the definition, for each l-th selection criterion, of the capacity (G_l) of the investment sectors to satisfy the considered constraint. By indicating with v_{il} the score of the i-th investment sector in satisfying the l-th selection criterion, the second constraint of the model is expressed by Eq. (2):

$$\sum_{i=1}^{j} k_i \cdot v_{il} = G_l \begin{cases} j = n, n - 1, \ldots, 2 \\ l = 1, \ldots, m \end{cases} \tag{2}$$

The objective function of the proposed model pursues the maximization (or the minimization) of the outputs expected by the subjects involved in the process, i.e. the sum of the capacities (G_l) of the social impact sectors to satisfy the weighted criteria considered. Taking into account that the weight attributed to each criterion of the social impact sectors could be different, thus indicating with p_l the weight attributed to the l-th selection criterion, the objective function will be expressed by Eq. (3):

$$\max/\min\left(\sum_{l=1}^{m} p_l \cdot G_l\right) \tag{3}$$

In Table 3 the algorithm of the proposed model has been reported, whereas in Table 4 the meaning of each term of the algorithm has been explained.

Table 3. Algorithm of the proposed model

Variable	k_i
Objective function	Max/min $(\sum_{l=1}^{m} p_l \cdot G_l)$
Constraints	$\sum_{i=1}^{j} k_i = 1 \quad j = n, n-1, \ldots, 2$
	$\sum_{i=1}^{j} k_i \cdot v_{il} = G_l \begin{cases} j = n, n-1, \ldots, 2 \\ l = 1, \ldots, m \end{cases}$

Table 4. Variable and exogenous parameters of the proposed model

k_i	Binary variable associated to the i-th social impact sectors
v_{il}	Score attributed to the i-th investment sector in meeting the l-th criterion
G_l	Capacity of all social impact sectors to suit the l-th criterion
p_l	Weight attributed to the l-th criterion

In the present research the branch and bound algorithm will thus be able to provide the outputs, in order of temporal priority, of social impact sectors as outputs based on the values assumed by the return function of the model. In this way, the public subject and the social investor will be able to determine the best allocation of financial resources according to the real social needs and the expected financial returns.

5   Conclusions

In recent years, traditional financial instruments have been weak in investments aimed at solving complex social problems. New innovative models have also emerged in the SRI field to respond to the growing social and environmental challenges. The SII represent one of the most interesting SRI fields due to the willing to generate value in terms of both a measurable positive social impact and a financial return. The development and global spread of "win – win" financial instruments such as SIB are proof of a growing interest in the realization of an investment that aims at generating benefits for all the subjects involved. Taking into account the frequent low effectiveness of numerous social welfare policies and the reduction of financial resources available to the public sector, the investments in SIB can constitute a useful tool for governments. Private sector investments in SIB would not only be able to help public social programs, but they could also provide such programs to a scale that would include a larger amount of beneficiaries.

However, the spread and the development of SIBs did not occur with the same force in all the countries. The complexity connected to the identification of the social impact sector in which investing in a territorial context characterized by a plurality of different social needs is one of the critical points highlighted in the literature for the development of the SIBs. The proposed operational research model has been defined to overcome this critical issue and to contribute to the spread of SIBs both in Italy and abroad.

Future developments of the present research may concern the verification of the validity and the reliability of the proposed model through the application to a real situation that can characterize, for example, a municipal area. Moreover, the definition of specific criteria that allow to represent the variety of aspects that the decision makers, according to their objectives and roles, evaluate for the activation of a social program, will allow to highlight how the process of identifying the social impact sector in which investing is closely related to a plurality of factors and requires an accurate assessment.

Aknowledgements. The authors declare that the contents of the paper are based on data and information developed within the "Innonets" Project founded by the Interreg Greece-Italy Program and are aimed at promoting the Project itself. For more information see http://interreginnonets.eu/en.

References

1. Barreca, A., Curto, R., Rolando, D.: Assessing social and territorial vulnerability on real estate submarkets. Buildings **7**(4), 94 (2017)
2. Del Giudice, V., De Paola, P., Forte, F.: Valuation of historical, cultural and environmental resources, between traditional approaches and future perspectives. In: Mondini, G., Fattinnanzi, E., Oppio, A., Bottero, M., Stanghellini, S. (eds.) SIEV 2016. GET, pp. 177–186. Springer, Cham (2018). https://doi.org/10.1007/978-3-319-78271-3_14
3. De Paola, P., Del Giudice, V., Massimo, D.E., Forte, F., Musolino, M., Malerba, A.: Isovalore maps for the spatial analysis of real estate market: a case study for a central urban area of reggio Calabria, Italy. In: Calabrò, F., Della Spina, L., Bevilacqua, C. (eds.) ISHT 2018. SIST, vol. 100, pp. 402–410. Springer, Cham (2019). https://doi.org/10.1007/978-3-319-92099-3_46
4. Disley, E., Giacomantonio, C., Kruithof, K., Sim, M.: The payment by results Social Impact Bond pilot at HMP Peterborough: final process evaluation report. Ann. Rev. Policy Des. **7**(1), 1–20 (2019)
5. Evans, B., Richmond, T., Shields, J.: Structuring neoliberal governance: the nonprofit sector, emerging new modes of control and the marketization of service delivery. Policy Soc. **24**(1), 73–97 (2005)
6. Giacomantonio, C.: Grant-maximizing but not money-making: a simple decision-tree analysis for social impact bonds. J. Soc. Entrepreneurship **8**(1), 47–66 (2017)
7. Global Impact Investing Network. https://theiin.org/asets/GIIN_2019%20Annual%20Impact%20Investor%20Survey_webfile.pdf.Accessed 21 Nov 2019

8. Morano, P., Tajani, F., Di Liddo, F., Anelli, D.: A feasibility analysis of the refurbishment investments in the Italian residential market. Sustainability **12**(6), 2503 (2020)
9. Jackson, E.T.: Evaluating social impact bonds: questions, challenges, innovations, and possibilities in measuring outcomes in impact investing. Commun. Dev. **44**(5), 608–616 (2013)
10. Leventhal, R.: Effecting Progress: Using Social Impact Bonds to Finance Social Services. N. Y. Univ. J. Law Bus. **9**, 511 (2012)
11. Liebman, J.B.: Social Impact Bonds: A Promising New Financing Model to Accelerate Social Innovation and Improve Government Performance. Center for American Progress, Boston, vol. 9 (2011)
12. Maier, F., Meyer, M.: Social impact bonds and the perils of aligned interests. Adm. Sci. **7**(3), 24 (2017)
13. McKay, K.: Evaluating social impact bonds as a new re-entry financing mechanism: a case study on re-entry programming in Maryland, Department of Legislative Services Office of Policy Analysis Annapolis (2013)
14. Nemhauser, G.L., Wolsey, L.A.: Integer and Combinatorial Optimization. Wiley, New York (1988)
15. Parker, R.G., Rardin, R.L.: Discrete Optimization. Elsevier, New York (2014)
16. Pauly, M., Swanson. A.: Social Impact Bonds in Nonprofit Health Care: New Product or New Package?. National Bureau of Economic Research (2013)
17. Pontrandolfi, P., Scorza, F.: sustainable urban regeneration policy making: inclusive participation practice. In: Gervasi, O., et al. (eds.) ICCSA 2016. LNCS, vol. 9788, pp. 552–560. Springer, Cham (2016). https://doi.org/10.1007/978-3-319-42111-7_44
18. Rudd, T., Nicoletti, E., Misner, K., Bonsu, J.: Financing Promising Evidence-Based Programs: Early Lessons from the New York City Social Impact Bond. MDRC, New York (2013)
19. Tajani, F., Morano, P.: Evaluation of vacant and redundant public properties and risk control: A model for the definition of the optimal mix of eligible functions. J. Property Invest. Financ. **35**(1), 75–100 (2017)
20. Tajani, F., Morano, P.: Risk management and goal programming for feasible territorial investments. In: Mondini, G., Fattinnanzi, E., Oppio, A., Bottero, M., Stanghellini, S. (eds.) SIEV 2016. GET, pp. 123–132. Springer, Cham (2018). https://doi.org/10.1007/978-3-319-78271-3_10
21. Tan, S., et al.: An evaluation of Social Impact Bonds in Health and Social Care: Interim Report. Policy Innovation Research Unit (2015)
22. Viglianisi, A., Rugolo, A., Calabrò, J., Della Spina, L.: Villa san giovanni transport hub: a public-private partnership opportunity. In: Calabrò, F., Della Spina, L., Bevilacqua, C. (eds.) ISHT 2018. SIST, vol. 101, pp. 211–221. Springer, Cham (2019). https://doi.org/10.1007/978-3-319-92102-0_23

Assessing the Interstitial Rent: The Effects of Touristification on the Historic Center of Naples (Italy)

Maria Cerreta$^{(\boxtimes)}$, Fernanda Della Mura, and Giuliano Poli

Department of Architecture, University of Naples Federico II, via Toledo 402, 80134 Naples, Italy
{maria.cerreta,giuliano.poli}@unina.it,
fernanda.dm91@gmail.com

Abstract. The paper explains a research methodology to understand the impacts of touristification and overtourism on the urban dimension. A block of the Naples historic downtown (Italy) has been selected as a sample to verify the transformations induced by the airification process of the city since 2015 on the social and urban environment, investigating the influence on real estate dynamics, and their impacts related to economic, social, cultural and urban criteria.

The mapping of Airbnb listings on the block has allowed to analyse the economic convenience of renting short-term housing, identifying an income surplus defined as "interstitial rent", which represents the economic advantage achieved. The analyses carried out on the average market values of the properties on Airbnb have revealed the tendency to redevelop the "waste building stock", which is unattractive for the ordinary market, in view of its turbo-valorisation by converting it into accommodation. The attractiveness characteristics of the properties were compared to the average market values to investigate the housing stock that can be removed from the ordinary residence market in view of its value enhancement. This research has highlighted the progressive gentrification of the historic center and the relocation of the permanent inhabitants outside it. Therefore, investigating the prevailing sentiment that agitates the social organisations of Europe, we have come to provide an intervention proposal for the regulation of post covid-19 housing policy, which tends to co-participated formulation of strategic choices, involving all interested parties, in order to find a shared application.

Keywords: Urban rent · Sharing economy · Short-term rental market · Web-scraping technique · Multiple regression analysis

1 Introduction

The touristification indicates the whole process of transformation of the urban space that shapes the city exclusively to make it attractive to the ordinary tourist. According to Renau [1], touristification is defined as the result of a relatively spontaneous and unplanned process of tourism development, which leads to the transformation into a

O. Gervasi et al. (Eds.): ICCSA 2020, LNCS 12251, pp. 952–967, 2020.
https://doi.org/10.1007/978-3-030-58808-3_68

tourist commodity of the space on which it insists. Some authors identify touristification as the process that transforms residential neighbourhoods into tourist areas [2, 3].

The transformation process can lead to a radical change in the appearance of the city, fuelled by that process which has been called the "Bilbao/Guggenheim effect", i.e. the possibility of changing the pre-existing aspect by investing in projects of great urban impact aimed at the search for a new image as a tourist destination [4]. After the deindustrialisation process, that took place in the last decades of the twentieth century or even after the economic crisis of 2008, to "survive" many cities have staked everything on the tourism industry, exploiting a natural vocation or creating the conditions. In all the cases observed, a common denominator, capable of being a powerful tourist magnet, seems to be the rediscovery or, in some cases, the invention of old traditional or typical atmospheres sold to tourists as authentic experiences [5].

Since touristification is a direct consequence of the emergence of mass tourism, it is the result not only of the needs but also of the trends and tastes that can make the city a "destination-city". These transformations generally do not recover anything authentic. On the contrary, they deplete the meaning of places and profoundly modify their image, homologating the destinations given daily needs related to the use (metro, pub, street food, souvenir shops) and, in the most severe cases, the creation of fakes leading to the Disneyfication [6] of cities. This is what has happened in some areas of Seoul, Korea, where some local media [7] threats such as Bukchon Hanok Village in the south of the city, famous for its traditional Korean houses, and Ihwa Mural Village in the Jongno district, in the centre of the city, known for its colourful murals, ended up becoming a sort of "theme parks", making it necessary to affix signs that invite silence during the visit of the neighbourhood, making the coexistence between tourists and residents conflicting. Other authors note that mass tourism in a historic city, in addition to causing conflicts between visitors and residents, also produces positive effects such as job creation, cosmopolitanism or the expansion of the cultural offer [8] and transforms urban spaces and neighbourhoods into recreational areas with substantial repercussions on residents [9]. Gurran et al. [2] identify in the shift of low-income residents due to the increase in the market values of the houses, the most significant social impact that tourism in the historic centre can produce. Touristification inevitably collides with the carrying capacity of tourist cities.

The United World Tourism Organisation [10, 11] defines carrying capacity as «the maximum number of people that may visit a tourist destination at the same time, without causing destruction of the physical, economic, socio-cultural environment and an unacceptable decrease in the quality if visitors' satisfaction». Carrying capacity cannot be measured by placing emphasis only on the size of tourist flows since it is influenced by the specificity of the context that is considered and in particular by two factors which are the characteristics of tourists and the features of the destination and its residents [12]. Touristification is inextricably linked to neologism "overtourism" which means "tourist overcrowding". The term was coined in 2016 by Rafat Ali [13], and according to him «Overtourism represents a potential hazard to popular destinations worldwide, as the dynamic forces that power tourism often inflicts unavoidable negative consequences if not managed well». The term was introduced in Collins in 2018 [14, 15], and became the word of the year in 2018 [16].

The issue is very close to Italy with Venice which in 2018 was elected world capital of overtourism. The repercussions of overtourism can be analysed on different levels: economic, social and environmental. It is necessary to investigate these three fields to arrive at the structuring of a methodology suitable for the study of the phenomenon and the consequent formulation of a more sustainable future scenario. According to Tomaselli [17], the tourism industry is an extractive industry, precisely like the oil industry and therefore, does not produce goods, but extracts value from "deposits" rich in added value. Furthermore, according to some authors [18], the development of a tourism industry does not guarantee to solve the local socio-economic problems faced by some destinations. Thanks to the emergence of sharing tools, such as Airbnb, the season of the so-called "platform capitalism" has been inaugurated [19–23]. According to Bernardi [24], indeed, since 2008 the traditional postulates of economic and social growth, and the entire capitalist system, have been called into question, prompting people to ask themselves what they needed and how to extract value from the things/capacities possessed. It is precisely in this context that the awareness has been acquired that every property can turn into a potential profit, every person into a potential entrepreneur. The competition in the panorama of Airbnb hosts is very high. Therefore, the reliability of the hosts is calculated through reputational algorithms, which collect opinions and feedback on the experience and disposability of the guest. In this way, each host is associated with a rating that serves to give an approximate measure of the trust that the digital tourist community places in it. What we are witnessing is the formation of "symbolic capital" [25, 26], that is a "cognitive" capital: what others think, the way they judge you, takes on value, regardless of the truthfulness of these conjectures or the existence of certain behaviours or qualities of the social actor. This symbolic capital, thanks to what we can define as a digital reputation economy, is converted into economic capital: having a good reputation will imply a higher flow of earnings [24]. Indeed, Airbnb has created new paradigms, bringing the home-sharing economy from a hidden economy to a new model of experiential tourism accessible to everyone, radically changing the hospitality market.

Research by the Ladest Laboratory [27], studying the penetration of Airbnb in thirteen Italian cities has shown how extremely unequal the distribution of rental income is. Indeed, the local administrations focus everything on tourism as a tool for the revitalisation of the economy, not only as a source to replenish the tax funds, but also to face social problems (high rate of unemployment, especially of young people). At the same time, the local communities live daily all the difficulties that uncontrolled tourism brings. It is not only private individuals or municipalities that benefit from tourism. In 2017, the diffusion of Airbnb, supported by the National Association of Italian Municipalities (ANCI) and by the Ministry for cultural goods and activities (Mibact) [28], took the opportunity of tourism to draw up a program. This program provided for the enhancement of over 40 villages, spread throughout the country, through a range of different projects with the collaboration of Airbnb for the recovery of the public spaces of three villages [29]. An effect was an out of control tourism, made up of more and more people who pass through a location for a short period, determining an environmental and social "pressure" on the context. A more significant influx of vehicles and a greater presence of people inevitably define an increase in the production of waste and an increase in the consumption of water and electricity.

The European Environment Agency [30] calculated that each tourist consumes a daily quantity of water 3 or 4 times higher than a permanent resident, with non-tourist water consumption of between 100 and 200 litres per day a person throughout Europe [31]. Furthermore, the problem of transport intervenes, according to estimates by the United Nations World Tourism Organization (UNWTO) according to which in 2017 57% of international tourism used the air mode [32]. However, according to the European Environment Agency [33], if air transport is responsible for the largest share of greenhouse gas emissions associated with tourism, while cruises remain the mode of transportation that generates the most greenhouse gas emissions per kilometre travelled. Besides, most cruises start with a flight to the port of departure, and this increases the total emissions produced by the journey by 10–30%.

In 2014 Hollenhorst et al. [34] had already revealed the unsustainability of tourism when they unmasked the image of tourism as a false embodiment of sustainability since it is one of the most carbon-emitting industries and resource depletion factor. According to some studies, indeed, the tourism sector would be responsible for the emission of 4.5 Gigaton of CO_2 in 2013, corresponding to the 8% of global carbon footprint and increasing four times more than the previous year estimation [35]. In this perspective, a general analysis allows having a clear picture of the saturation situation of a specific tourist resort at a particular time t. Nevertheless, long-term sustainability of tourist flows has to be assessed [36]. Due to the advent of mass tourism, the social mixite of consolidated urban centres has changed a lot. Overtourism, boosted by platform capitalism, has propelled existing phenomena i.e. the exploitation of labour and the gentrification of entire neighborhoods, amplifying the old social problems closely linked to the contradictions of capitalism [1].

The houses, often old and in poor condition, have been renovated and the same fate has fallen on disused industrial buildings which have been transformed into apartments and, in some cases, new buildings are built. This whole operation was possible thanks to the emergence of a rent gap [37]. Capital is invested in the neighbourhoods where the annuity values are depreciated due to the widespread degradation given an economic advantage that resides in the potential annuity due to the interventions [38]. The intervention of investors in neighborhoods brings the value of the capitalisation annuity in line with that of the potential annuity. As a direct consequence, we are witnessing the increase in the market value of the properties. Whether the cause of gentrification is tourism or the opportunity for the upper-middle class to be closer to city attractions, it is considered essential to preserve the existing social stratification to preserve what Ruth Glass called the "social personality" of a neighborhood [39].

The progressive tourist saturation of cities is reflected only on the shortage of housing, but also on the inefficiency of services that generate stressful situations in the resident population, as well as competition in the use of services in the city [40, 41]. In Spain, in Barcelona, we have witnessed several times in recent years, the reaction of citizens who have occupied the streets and squares of the city by ordering tourists to go home [42]. Because uncontrolled tourism not only saturates cities, it also brings with it environmental repercussions and transformation of places to respond to the hunger for the authenticity of the ordinary tourist, depleting the resources of the territory and expelling the most fragile and weakest part of the population "authentic". Within the above-mentioned background, the paper shows a methodology design allowing to emphasise the economic convenience of renting short-term housing, through mapping of listings and detecting of income surplus referring to as "interstitial rent" in terms of achieved economic profit.

The paper's structure proceeds as follows. The first part (Sect. 2) explains the methodological approach by describing methods and different research phases; the second (Sect. 3) identifies the case study related to the historic centre of Naples; the third (Sect. 4) analyses the transformation processes; the fourth (Sect. 5) shows the discussion of the data processed; and the last (Sect. 6) discusses the conclusions about the whole process.

2   Materials and Methods

"The first research conference on tourism and SDGs" [43] highlighted the relationships between tourism and SDGs objectives, eliciting the positive and negative impacts of tourism in terms of sustainability. In a nutshell, the SDGs consider the idea that tourism, and more generally, the activities that generate profit, must deal with the environmental contexts in which they are inserted, creating positive impacts through social responsibility programs. Therefore, the United Nations program calls for respecting the seventeen goals to capitalise while contributing to more sustainable development. Taking into account the SDGs contents, the research has focused on the study of a block in the historic centre of Naples (Italy) which takes on the features of an example to evaluate the effects of touristification in the downtown. The methodological process considers the following four main steps (Fig. 1): 1. Knowledge; 2. Data processing; 3. Assessment; 4. Outcome.

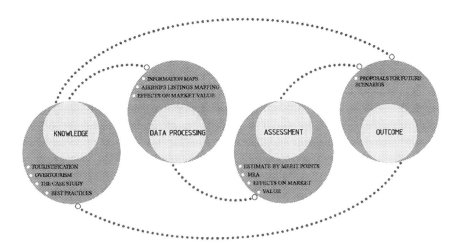

Fig. 1. The methodological framework

In the assessment phase, the different data were elaborated by multiple regression analysis. All the assessments produced to contribute to the structuring of proposals for future more sustainable intervention scenarios for the inhabitants and the city, also in light of the recent covid-19 pandemic which has generated the impossibility of moving people, changing the dynamics of uncontrolled growth of overtourism.

3   The Case Study

The case study selected a block in the heart of Naples historic downtown, located between the sections of San Lorenzo and Pendino (Fig. 2), between the fourth and second municipalities of Naples. The block is located on the ruins of the ancient agora [44], enclosed between the lower decuman south of via San Biagio dei Librai and the middle decuman north of Via dei Tribunali. Nowadays the Naples historic downtown is configured as the stratification of transformations undergone by its Greek foundation to today.

Fig. 2. The study area: on a block of Naples historic downtown (Italy)

The methodological process focuses on:

- the analysis of the transformations induced by touristification in urban dynamics;
- the study of the accommodations on Airbnb;
- the average market value of the buildings in the block on Airbnb;
- the identification of the most influential characteristics that determine the market value.

Thanks to the study of the Airbnb listings, it is possible to determine the extent of the economic advantage of renting under a short-term lease. Instead, the analysis focused on market value serves to understand what are the characteristics that properties must have more than others, to be attractive for the short-term rental market.

4   Transformation Processes: Approaches and Tools

The touristification involves several spheres of reality; therefore, to understand the transformation processes of the observed block, it has been necessary to construct a knowledge framework through the collection and processing of hard data.

The data collected refer to four criteria: economic, social, cultural and urban. The indicators matrix has been elaborated by selecting the indicators based on the survey criteria (Fig. 3), indicating for each one of them: Criterion; Indicator; Unit of measurement; Territorial coverage; Source; Reference year.

Criteria	Indicator	Unit of measurement	Territorial coverage	Source	Reference year
Economic	Marke Value - Housing	€/smq	Area OMI B8, B9	OMI	From 2014 to 2019
	Rental Value -Housing	€/smq	Area OMI B8, B9	OMI	From 2014 to 2019
	Economic activities	Number	Cadastral section	Istat	1991-2001-2011
	Restaurants and pizzeria	Number	Cadastral section	Open Street Map	2019
	Bars	Number	Cadastral section	Open Street Map	2019
	Take away	Number	Cadastral section	Open Street Map	2019
	Pastry shops	Number	Cadastral section	G.M. / on-site	2019
	Local food markets	Attendance	Neighborhood	G.M. / on-site	2019
	Local craft markets	Attendance	Neighborhood	G.M. / on-site	2019
	Supermarket	Number	Cadastral section	Open Street Map	2019
	Clothing stores	Number	Cadastral section	Open Street Map	2019
	Bookshops	Number	Cadastral section	Open Street Map	2019
	Pharmacies	Number	Cadastral section	Open Street Map	2019
	B&Bs	Number	Neighborhood	G.M. / on-site	2019
	Hotels	Number	Neighborhood	G.M. / on-site	2019
		Number	Cadastral section	Open Street Map	2019
Social	Residents total	Number	Cadastral section	Istat	1991-2001-2011
	Foreign residents	Number	Cadastral section	Istat	1991-2001-2011
	Age of the resident population	Number	Cadastral section	Istat	1991-2001-2011
	Education of residents	Number	Cadastral section	Istat	1991-2001-2011
	Occupied resident population	Number	Cadastral section	Istat	1991-2001-2011
	Unemployed resident population	Number	Cadastral section	Istat	1991-2001-2011
	Active resident population	Number	Cadastral section	Istat	1991-2001-2011
Cultural	Churches and convents	Number	Cadastral section	Open Street Map	2019
		Number	Neighborhood	G.M. / on-site	2019
	Historical buildings	Number	Neighborhood	G.M. / on-site	2019
	School buildings	Number	Neighborhood	G.M. / on-site	2019
	Castles, walls (doors) and towers	Number	Cadastral section	Open Street Map	2019
	Theaters and Cinemas	Number	Cadastral section	Open Street Map	2019
	Museums and archaeological sites	Number	Cadastral section	Open Street Map	2019
		Number	Neighborhood	On-site research	2019
	Archives	Number	Cadastral section	Open Street Map	2019
	Cloisters of religious buildings and universities	Number	Cadastral section	Open Street Map	2019
	Squares	Number	Cadastral section	Open Street Map	2019
		Number	Neighborhood	G.M. / on-site	2019
	Cultural and social associations	Number	Cadastral section	G.M.	2019
	Neighborhood events	Attendance	Cadastral section	G.M. / on-site	2019
Urban	Unoccupied buildings	Number	Cadastral section	Istat	1991-2001-2011
	Institution	Number	Cadastral section	Open Street Map	2019
	Station	Number	Cadastral section	G.M. / on-site	2019

Fig. 3. Matrix of indicators for the economic, social, cultural and urban criteria

The institutional data referring to property valuations, returned as an aggregate data that refers to a larger area than that of the investigation since the survey block is a portion of the space between different homogeneous territorial areas (OMI), therefore in the subsequent evaluations, the average values have been considered. Furthermore, the existing institutional data have a temporal limit, which cannot capture the most recent transformations. This critical point has been overcome by proceeding with the development of an integrative mapping, with field surveys (on-site) and the support of the open-source data of the Google Maps and the Open Street Map. The indicators matrix has been the starting point for the elaboration of information maps thanks to the spatial representation of the selected indicators, referring them to the census tracts indicated by Istat [45], to compare the data and provide a common basis to the different maps. The resulting maps enable us to describe the existing relationships in the selected area spatially and to quantify and compare them with the transformations ongoing induced by the growth in the number of b&bs. The number of accommodations on Airbnb has been acquired by scraping with an open-source software "Airbnb Data Collection" [46]. Data have been extracted monthly, from June 2018 to May 2019. Through their comparison with field surveys and research on home-sharing websites, it has been possible to map the Airbnb listings spatially. The mapping has made it possible to observe the relationships that the growing accommodation urban function has woven with the other features, serving as a basis for identifying new urban scenarios.

5   Results and Discussion

According to the cadastral data, 655 buildings were officially designed for housing in the block. This official number excludes most of the buildings used as dwellings called "bassi" which were once unhealthy homes or artisan workshops on the ground floor. Istat data from 2011 showed that there were 1039 residents on the block under analysis (Fig. 4).

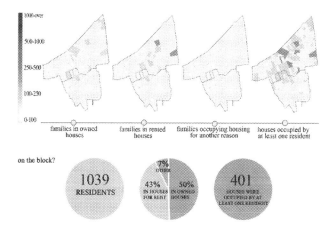

Fig. 4. Families who occupied the block in 2011

Although the city is attractive from a tourist point of view, the database of the observatory on property valuations provided by the revenue agency, underlined that, from 2014 to 2019, the market value of the property is slightly decreasing, as well as also the rental value (Fig. 5).

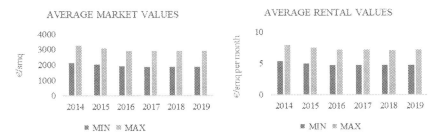

Fig. 5. Real estate dynamics from 2014 to 2019

In 2016, when Airbnb began to establish itself in the Neapolitan landscape, the average market values began to fall. This decrease also affected the average rental prices, which from 2016 onwards have undergone minimal fluctuations, with the first significant increase in the number of listings in the historical centre 197% compared to 2015 (Fig. 6).

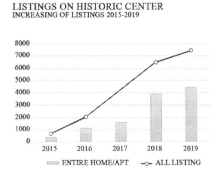

Fig. 6. Increasing of listings from 2015 to 2019 on the Naples historic centre

The average market values observed indicate that they have decreased from 2695 €/sqm in 2014 to 2380 €/sqm in 2019, while for the average rental values it went from 6.68 €/sqm per month in 2014 to 5,97 €/sqm per month in 2019.

The institutional data refer to aggregate data that consider the fluctuations of property valuations relating to three leading real estate categories: noble, civil and low-cost. In particular, the noble type housing (cadastral category A1) from 2016 onwards show constants market values in a progressive decline from 2014 going from 3900 €/smq to 3300 €/smq, the same lowering occurs for leases that decrease from 8,78 €/smq per month to 7,45 €/smq per month. The decline of property valuations is also reflected in the other cadastral categories observed.

For civil housing (cat. A2) from 2014 to 2019, the average market value decreases from 2825 €/sqm to 2537 €/sqm and the average rental value decreases from 7 €/sqm per month to 6,4 €/sqm per month.

For low-cost housing (cat. A3) from 2014 to 2019 the average market value decreases from 1962 €/sqm to 1762 €/sqm, and the average rental value decreases from 5,22 €/sqm per month to 4,72 €/sqm per month.

Through field analysis, it has been estimated that the drop in prices is often the consequence of the increasingly degraded conditions of the properties placed on the market. Degradation conditions not only concern real estate but often refer to the dilapidated context in which they are inserted. The mapping of the properties on Airbnb highlights the presence of 39 entire apartments, 37 single rooms and 0 shared rooms, for a total of 310 beds (Fig. 7).

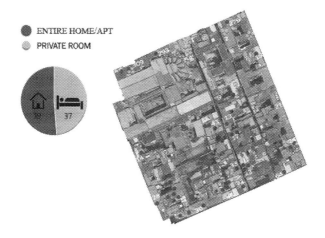

Fig. 7. Mapping of Airbnb listings

The study of the individual listings has made it possible to skim the data and to detect that 99% of the listings refer to an entire apartment, for a total of 41 real estate units present on the platform. The presence of 38% of hosts that publish multiple listings also notices the growing trend to value the second home on the short-term rental market.

The data of the 41 real estate units have been crossed with the cadastral data to obtain information relating to: the building category, the number of rooms and the income.

The cadastral research shows that the 41 real estate units 61% are popular housing (A4), 20% ultra-popular housing (A5) and 19% civil housing (A2).

From direct observation, however, it emerges that the maintenance status of the buildings is mainly average (64%), and only a small percentage (12%) has an excellent maintenance status. The average market value of these real estate units is 1,562 €/sqm. It has been calculated as a weighted average between the value inferred from the summary estimate by merit points and the calculation of the normal unit value.

Through multiple regression analysis [47], it has been possible to study all those characteristics that determine the observed market value. In this case study, the dependent variable Y is the Market Value (MV). At the same time, the predictors consist of 42 indicators at two main scales of investigation referring to: (i) the analysed property and (ii) the urban and socio-economic context.

The SPSS software provided by IBM has been used to lead data analysis and statistics. The predictors most influencing the MV - ordered by most significant influence - are the following: surface; elevator; finishes.

The market value is low compared to the position in the historic centre and the ability to extract buildings through the platform. In fact, by comparing the 2018 turnover data of the hosts with those of the taxation regime to which the properties are subject, it is shown that the properties under observation produce a surplus of income, definable as interstitial income, equal to 707%. From the observed data, it emerges that the most attractive properties considering their potential reconversion for accommodation use are those of popular category. Indeed, considering the residential building stock of the block, excluding the 0,5 and 1 room properties, those of category A2 and those already present on the platform, it has been observed that 62% of residential real estate units (408) are susceptible to turbo-valorisation. Therefore, Airbnb becomes a tool to make income from building waste, which is not attractive for the ordinary market, generating an "interstitial rent". In this context, the property rental market can choose a third way: not the lease to residents or students, but the short contract to tourists that allows to exploit the advantage of a disadvantageous starting condition and transform it into an opportunity by pursuing a higher turnover.

6   Conclusions

The research shows that degraded contexts can become an opportunity for platforms like Airbnb to relaunch the "authentic" live like a local [48].

The difference between this new accommodation and traditional tourism spaces is that their function is based on relationality rather than on visual consumption or any type of traditional tourist gaze [49, 50].

According to Dredge and Gyimóthy [51], tourism understood in this way needs to rely on new skills such as relationality, so that a new resource (private houses) can be made available to tourists. This process is leading to the creation of new meanings, where the tourists can live the experience as a local.

This process in turn transforms tourism and the space in which tourism is produced, consumed and executed. The analysis of the feedback about the hosts shows that the guests are happy to be in contact with a more authentic and raw reality of the city. For tourists, it is not a problem to walk four floors to reach the apartment, because this too is a characteristic of the real way of life, in a historic building.

The "napolitaness" is pursued, which takes shape through the imagination of tourists and materialises through the expressions of city culture and history, which from the semantic analysis are related to: art, church, tribunal, old, popular, typical.

The increase in tourist presence in the city has pointed out the gradual replacement of neighborhood businesses in favour of those intended for tourists, contributing to the change in urban balances.

In the context analysed, it is clear that it is possible to turbo-valorise a property only in particular conditions that depend on the characteristics of the property itself but also on the context in which it localised. Therefore, in this case, it is possible thanks to the natural attraction of the historic centre of Naples, combined with the marketing ability of the host who online must make his property attractive and must be able to build and maintain an excellent virtual reputation. The possibility of buying properties at a low price compared to the profit potential has generated a series of "make-up" interventions on them. These are often conducted only in terms of finishes, to give a new image to the apartments and make them increasingly attractive to tourists, thus generating the expected profits. Indeed, compared to the announcements observed, a furniture homologation emerges that can be defined as an "IKEA effect": low-cost furniture to furnish spaces according to contemporary tastes. Given the earnings prospects, a real risk, already manifested in Italy in cities like Venice [52, 53], is the bulk purchase of properties managed by holding companies, which would thus erode the already limited availability of housing in the historic centre of Naples.

Due to the progressive increase in properties destined for the short-term rental market, weaker economic groups have already been marginalised, increasing the evictions not only of residents but especially of off-site students.

The gentrification in the Naples historic downtown can be read above all in the impossibility for students to occupy properties near the numerous university sites in the historic centre. This phenomenon of expulsion from the city centre also affected other Italian cities and especially Bologna [54], where the student groups gathered in university associations and social spaces such as Link – Studenti Indipendenti and the Arci Ritmolento circle, which they are asked about the need for concrete action that would allow on the one hand to create connections and links between competing subjects, and on the other to put pressure on the political decision-maker to take radical action to put an end to the airification of the city.

The problems related to airification cities have responded with various actions from below to deal with the lack of housing, such as "solidarity couchsurfing" [55] created to give hospitality to off-site young people looking for a room to rent.

The current condition of the global pandemic of covid-19 has shown all the fragility of an economic system that foresee the tourism monoculture for the revitalisation of the cities. Indeed, the impossibility of moving freely has paralysed this economic sector, and it is, therefore, necessary to rethink the method of use of properties sold on the short-term rental market. In this circumstance, it may be essential to activate synergies between public and private resources to pursue not only economic but also social development objectives. It is desirable that the central government encourages the reintroduction of this building heritage into the medium-long term residence circuit, contributing to the resolution of the housing problem concerning those who do not have a fixed home and those who need isolation from their family.

In the long run, this partnership could be the key to the solution of urban conflicts generated by touristification and promote a greater balance in the distribution of the population residing in the city, drastically reducing the number of tourist apartments, and returning them to the stock of rental housing in agreement with what is required by the SET network [56] and the Manifesto for the Reorganisation of the City after covid19 [57, 58]. Therefore, a policy is set up, which enables a virtuous use of the waste real estate assets. The key points of the policy are identified in: tourist degrowth; integrated housing policy; economic degrowth. The tourist decreasing is necessary so that citizens can regain possession of the contended spaces [59–61] of the city and contrast its commercialisation. Only with a reduction of the tourist phenomenon can we also pursue concomitant objectives, such as mitigation of environmental impacts and precarious jobs that gravitate around the tourism industry. The ongoing global pandemic has made housing inequalities even more evident, leaving those who do not have a home entirely on the sidelines. Therefore, the vision of an integrated housing policy, which also involves private individuals in a joint effort, would allow on the one hand private individuals to obtain a profit, albeit less than the pre-covid 19 situation.

On the other hand, governments can deal with more tools for the housing emergency, promoting the social function of public and private real estate assets. An economic degrowth that supports the social and local economy, the cooperative and sharing economy, and small businesses become necessary. The current circumstances have shown that mutual support networks are indispensable especially in emergencies. Touristification and gentrification have fragmented and weakened entire neighborhoods, determining an inability to better respond to the primary needs of the resident population. A new policy has to consider the goal to calm the unlimited consumption of goods, resources and territory so that a balanced condition is reached. This process which does not exclude any segment of the market in the future, but sets guidelines for an inclusive use of the city that it is not a source of urban tensions, breaking down all forms of tourist hegemony.

References

1. Del Romero Renau, L.: Touristification, sharing economies and the new geography of urban conflicts. Urban Sci. 2(4), 104 (2018)
2. Gurran, N., Zhang, Y., Shrestha, P.: 'Pop-up'tourism or 'invasion'? Airbnb in coastal Australia. Ann. Tourism Res. 81, 102845 (2020)
3. Sequera, J., Nofre, J.: Shaken, not stirred: new debates on touristification and the limits of gentrification. City 22(5–6), 843–855 (2018)
4. Marchi, M.: Declino e rigenerazione urbana in Europa: i casi di Glasgow e Lipsia. In: Dini, F., Randelli, F. (eds.): Oltre la globalizzazione: le proposte della geografia economica. University Press, Firenze (2012)
5. Gilli, M.: Autenticità e interpretazione nell'esperienza turistica. Franco Angeli, Milano (2009)
6. D'Eramo, M.: Il selfie del mondo. Indagine sull'età del turismo. Feltrinelli, Milano (2017)
7. Korea JongAng Daily (2016). http://koreajoongangdaily.joins.com/news/article/article.aspx?aid=3023439

8. Bujalance, S.G., Barrera-Fernández, D., Scalici, M.: Touristification in historic cities. Reflections on Malaga. Revista de Turismo Contemporâneo 7(1), 93–115 (2019)
9. Lim, S.E.Y., Bouchon, F.: Blending in for a life less ordinary? Off the beaten track tourism experiences in the global city. Geoforum 86, 13–15 (2017)
10. UNWTO, Saturation of Tourist Destinations: Report of the Secretary General, World Tourism Organisation, Madrid (1981)
11. Kennell, J.: Carrying capacity, tourism. In: Jafari, J., Xiao, H. (eds.) Encyclopaedia of Tourism. Springer, Cham (2014)
12. O'Reilly, A.M.: Tourism carrying capacity: concept and issues. Tourism Manag. 7, 254–258 (1986)
13. Ali, R.: Foreword: the coming perils of overtourism (2016). https://skift.com/iceland-tourism/
14. Singh, T.: Is over-tourism the downside of mass tourism? Tourism Recreation Res. 43(4), 415–416 (2018)
15. Dickinson, G.: Dear dictionaries, this is why 'overtourism' should be your 2018 word of the year (2018). https://www.telegraph.co.uk/travel/comment/overtourism-word-of-the-year/
16. Dickinson, G.: 'Overtourism' shortlisted as Word of the Year following Telegraph Travel campaign (2018). https://www.telegraph.co.uk/travel/news/overtourism-word-of-the-year-oxford/
17. Tomaselli, E.: La metafora del petrolio. Dove porta la turistificazione, Wsimag (2018)
18. Sanchez, P.M., Adams, K.M.: The Janus-faced character of tourism in Cuba. Ann. Tourism Res. 35(1), 27–46 (2008)
19. Olma, S.: Never Mind the Sharing Economy: Here's Platform Capitalism. Institute of Network Culture (2014)
20. Martin, K., Zysman, J.: The rise of the platform economy. Issues Sci. Technol. 32(3), 61 (2016)
21. Langley, P., Leyshon, A.: Platform capitalism: the intermediation and capitalisation of digital economic circulation. Finan. Soc. 3(1), 11–31 (2017)
22. Srnicek, N.: Platform Capitalism. Polity Press, Cambridge (2017)
23. Celata, F.: Territorio 86, 48–56 (2018)
24. Bernardi, M.: Un'introduzione alla Sharing Economy. Fondazione GianGiacomo Feltrinelli, Laboratorio Expo KEYWORDS, Milano (2015)
25. Bourdieu, P.: Raison pratiques. Sur la théorie de l'action, Éditions du Seuil, Paris 1994; trad. it. Ragioni pratiche, Il Mulino, Bologna (1995)
26. Mele, V.: Capitale simbolico e stile della vita in Pierre Bourdieu, Sociologia e ricerca sociale, n. 84, Franco Angeli, Milano (2007)
27. Picascia, S., Romano, A., Teobaldi, M.: The airification of cities. Making sense of the impact of peer to peer short term letting on urban functions and economy. In: Proceedings of the Annual Congress of the Association of European Schools of Planning, pp. 2212–2223 (2017)
28. MIBACT. https://www.beniculturali.it/mibac/export/MiBAC/sito-MiBAC/Contenuti/visualizza_asset.html_664421166.html. Accessed 10 May 2020
29. AIRBNB. https://italianvillages.byairbnb.com/it/. Accessed 10 May 2020
30. Agenzia Europea dell'Ambiente, https://www.eea.europa.eu/soer-2015/europe/tourism. Accessed 10 May 2020
31. EEA: Water resources across Europe - confronting water scarcity and drought, EEA Report No 2/2009. European Environment Agency, Copenhagen, Denmark (2009)
32. UNTWO: http://people.unica.it/calamassidda/files/2017/06/UNWTO_TourismHighlights_2017.pdf. Accessed 10 May 2020

33. European Environment Agency. https://www.eea.europa.eu/it/segnali/segnali-2016/articoli/i-trasporti-aerei-e-marittimi. Accessed 10 May 2020
34. Hollenhorst, S.J., Houge-Mackenzie, S., Ostergren, D.M.: The trouble with tourism. Tourism Recreation Res. **39**(3), 305–319 (2014)
35. Lenzen, M., Sun, Y., Faturay, F., et al.: The carbon footprint of global tourism. Nat. Clim Change **8**, 522–528 (2018)
36. Tajani, F., et al.: Data-driven techniques for mass appraisals. Applications to the residential market of the city of Bari (Italy). Int. J. Bus. Intell. Data Min. **11**(2), 109–129 (2016)
37. Slater, T.: Clarifying neil smith's rent gap theory of gentrification. Tracce Urbane Rivista Italiana Transdisciplinare di Studi Urbani. **1**(1), 83–101 (2017)
38. Diappi, L., Bolchi, P.; L'effetto della rendita nelle dinamiche locali di degrado e ristrutturazione edilizia: un modello Multi Agente. In: Bruzzo, A., Ocelli, S. (eds.) Le relazioni tra conoscenza e innovazione nello sviluppo dei territori. Franco Angeli, Milano (2004)
39. Glass, R.: London: Aspects of Change. MacGibbon & Kee, London (1964)
40. Nocifora, E.: Turismo, società, patrimonio, Cedam (2019)
41. Séraphin, H., Sheeran, P., Pilato, M.: Over-tourism and the fall of Venice as a destination. J. Destination Mark. Manag. **9**, 374–376 (2018)
42. Hughes, N.: 'Tourists go home': anti-tourism industry protest in Barcelona. Soc. Mov. Stud. **17**(4), 471–477 (2018)
43. UNTWO. http://tourism4sdgs.org/tourism-for-sdgs/tourism-and-sdgs/. Accessed 10 May 2020
44. Ferraro, I.: Napoli; Atlante della città storica. Centro antico. Clean, Napoli (2002)
45. Istat - Istituto Nazionale di Statistica, Dati e indicatori (2011). https://www.istat.it/. Accessed 10 May 2020
46. GitHub. https://github.com/tomslee/airbnb-data-collection. Accessed 10 May 2020
47. Simonotti, M.: L'Analisi di regressione nelle valutazioni immobiliari. Stud. di Economia e Diritto **3**, 369–401 (1988)
48. Benner, K.: Airbnb Wants Travelers to 'Live Like a Local' With Its App. The New York Times (2016). https://www.nytimes.com/2016/04/20/technology/airbnb-wants-travelers-to-live-like-a-local-with-its-app.html
49. Urry, J.: The Tourist Gaze: Leisure and Travel in Contemporary Societies. Sage, London (1990)
50. Richards, G.: Creativity and tourism in the city. Curr. Issues Tourism **17**(2), 119–144 (2013)
51. Dredge, D., Gyimóthy, S.: Collaborative Economy and Tourism: Perspectives, Politics, Policies and Prospects. Tourism on the Verge. Springer (2017)
52. Hardy, P.: Sinking city: how Venice is managing Europe's worst tourism crisis. The Guardian (2019)
53. Cerreta, M., Cannatella, D., Poli, G., Sposito, S.: Climate change and transformability scenario evaluation for venice (Italy) port-city through ANP method. In: Gervasi, O., et al. (eds.) ICCSA 2015. LNCS, vol. 9158, pp. 50–63. Springer, Cham (2015). https://doi.org/10.1007/978-3-319-21410-8_4
54. Semi, G.: Gentrification. Tutte le città come Disneyland? Il Mulino, Bologna (2015)
55. Arci. https://www.arci.it/appuntamento/welcome-fuorisede-il-couchsurfing-solidale/. Accessed 10 May 2020
56. Emergenza Cultura. https://emergenzacultura.org/2020/04/29/la-rete-set-il-covid-19-e-le-sue-conseguenze/?fbclid=IwAR1FnQbD7ctD8aqDhCw_W00CMneS5ZOTx2ZRqUE6yui-FlvNFf4t0ShpbFM. Accessed 10 May 2020
57. Manifiesto por la reorganización de la ciudad tras el covid19. https://manifiesto.perspectivasanomalas.org/. Accessed 10 May 2020

58. Argemi, A.: Por una Barcelona menos mercantilizada y más humana Más de mil personas firman el 'Manifiesto por la reorganización de la ciudad tras la covid19'. El Pais (2020)
59. Ruggiero, L., Scrofani, L.: Turismo e competitività urbana. FrancoAngeli, Milano (2011)
60. Cusimano, G.: Spazi contesi spazi condivisi. Geografie dell'interculturalità, Pàtron, Bologna (2010)
61. Tajani, F., Torre, C.M., Di Liddo, F.: Financial feasibility assessment of public property assets valorization: a case study in rome (Italy). In: Misra, S., et al. (eds.) ICCSA 2019. LNCS, vol. 11620, pp. 82–93. Springer, Cham (2019). https://doi.org/10.1007/978-3-030-24296-1_9

Sustainable Redevelopment: The Cost-Revenue Analysis to Support the Urban Planning Decisions

Pierluigi Morano[1] , Maria Rosaria Guarini[2] ,
Francesco Tajani[2] , and Debora Anelli[2(✉)]

[1] Department of Sciences of Civil Engineering and Architecture,
Polytechnic University of Bari, 70125 Bari, Italy
pierluigi.morano@poliba.it
[2] Department of Architecture and Design, "La Sapienza" University of Rome,
00196 Rome, Italy
{mariarosaria.guarini,francesco.tajani,
debora.anelli}@uniroma1.it

Abstract. The sustainable development of cities, by focusing on abandoned or degraded areas, represent the key driver that is able to remedy to damages that are produced by uncontrolled urban growth. The aim of the work is to highlight usefulness of the Cost-Revenue Analysis (CRA) in decision-making process concerning the enhancement of a military complex located in the city of Rome (Italy). In particular, existence of conditions that allow the Public Administration (PA) to request private entrepreneur additional operations to those established by the law is intended to examine, ensuring financial convenience of both subjects involved. The results achieved demonstrate feasibility of initiative, but also existence of wide margins for financial convenience of private entrepreneur that allow the PA to request additional financial resources or the direct realization of the initiatives to be intended to public functions. Furthermore, the influence of combination of the intended uses on urban planning loads and financial budget of the initiative is tested by analyzing the variation of share of gross floor area that can be attributed to each intended use. The adoption of the CRA can be implemented by subjects involved in the early stage of redevelopment initiatives in order to provide the size of the amount of financial resources that are necessary.

Keywords: Public-private partnership · Financial analysis · Decision support tools · Abandoned properties · Saving soil · Urban redevelopment

1 The Sustainable Urban Redevelopment as a "Zero Consumption" Soil Saving Strategy

The reduction of the land take environmental impacts through the sustainable development of the cities is currently an open topic in the worldwide research debate. An efficient and shared strategy by guaranteeing the sustainable recovery of our territories is the early step in order to achieve the European goal of "no net land take by 2050".

© Springer Nature Switzerland AG 2020
O. Gervasi et al. (Eds.): ICCSA 2020, LNCS 12251, pp. 968–980, 2020.
https://doi.org/10.1007/978-3-030-58808-3_69

In Italy, in the last 50 years the reduction of the Agricultural Used Area, equal to 6 million hectares, is pointed out by the analysis of the soil sealing, the uncontrolled urban growth and the landscape transformations dynamics [13]. The data considered, in fact, show that the spread of phenomena such as the diffusion, the dispersion and the urban decentralization have caused the loss of a large amount of the natural surfaces within the cities. Some of these phenomena can affect entire communities or individual urban districts which thus giving rise to "cities within the cities" that are configured as urban voids, also due to the particular social, economic and territorial realities [7].

The problems related to the urban voids undermine the sustainable development, discourage the investment and the economic growth, promote the deterioration of the natural and the built environment [17, 20]. Instead, the point of view that must be taken appreciate the degraded areas as a resource with a great potential, for public and private subjects, since they represent the opportunity to active regeneration interventions in order to remedy to the disfigurement that are produced over the years by the uncontrolled soil sealing processes [11, 12]. Therefore, the urban redevelopment, if it is intended as the "Planning activity, aimed at recovering a valid qualitative and functional dimension in urban and/or building structures, as a whole or in single parts, compromised by obsolescence or degradation" [1], represents an efficient and shared strategy which allows to provide a "zero consumption" saving soil solution [14].

In the current context, urban land use planning plays a crucial role in the development of support systems able to regulate the dynamics of the cities [6, 9]. If the national urban planning law (Law no.1150/1942) had the goal of regulating the urban development on a territorial scale - with the institution of the General Regulatory Plan (GRP) -, since the end of the '70s the goal of the urban planning regulation tools became the recovery and the enhancement of the degraded areas and the abandoned properties, largely.

The most critical issues of soil sealing processes regard preservation of natural resources, employment of sustainable materials for buildings construction, reclaim of contaminated or abandoned industrial areas, enhancement of properties in disuse. With the lack of a national regulatory framework, that currently has not yet been approved by the Parliament, many of the Italian regions have demonstrated an increasing attention to soil saving issue by drafting several requirements specifically dedicated to the containment of land take and highlighting the potential contribution given by renewal initiatives. Some regions, in fact, have established requirements for adoption of measures aimed at enhancing the existing abandoned heritage as the most effective solution to reduce land take [5].

The limited availability of public financial resources and the need to redevelop the city are the reasons of growing participation of private investors in urban planning initiatives of disused or degraded areas by financing the solutions capable of spreading the burdens and the risks thus giving the possibility, to the initiatives, to be carried out in compliance with the established aims [18]. Recently, the governmental institutions have focused their attention to the contribution made by the urban redevelopment operations in public-private partnership (PPP). Some experiences - such as those promoted by the European Commission in 2006, by the National Council of Conservative Landscape Architects in 2012 and by the National Urban Planning Institute in 2017 - show that success mainly depends on: robustness of public direction; adoption of the most suitable forms of PPP for each initiative; creation of dedicated financial funds in order to avoid exacerbating the public financial resources.

Furthermore, in Italy the economic potential impact induced by a widespread urban redevelopment campaign is attested equal to 327 million euros and the amount of the concession fees collected by the Public Administration (PA) is about of 17 million euros [3]. The PA has thus an additional incentive, of a financial type, for compensating the complex management burden related to singular interests of several stakeholders involved and the great number of uncertain variables. In order to overcome such burdens the public and private subjects can recourse to the use of suitable urban planning tools in order to : *i)* provide a support for planning and managing the financial feasibility of the initiative; *ii)* take into account the needs and the requirements of the subject involved in determining the balance sheet of their conveniences; *iii)* define the goals and evaluate the effects of the project both *ex-ante* and *ex-post*; *iv)* adequately allocate the initiative's financial and failure risks among the subjects.

2   The Contribution of the Assessment Tools in the Urban Redevelopment Processes

The complexity that characterizes themes and tools used to decide, plan and implement the initiatives is one of the most common feature of the urban redevelopment projects, according to the contingent presence of public and private functions to be established, heterogeneity of subjects involved (PA, private entrepreneurs, land owners, etc.), financial resources and multiplicity of funding channels. It is clear, therefore, that each initiative requires evaluation and control of a large number of variables that, in a different way, concern the stakeholders interested and affect the feasibility of the initiative [19]. The most crucial variables may be:

- the location of areas and their accessibility, which affect the choice of the most appropriate intended use and present and future property values;
- the maintenance state of area and intended use of the site;
- the structure of property, fragmented or not, which can invalidate the feasibility of the initiative;
- the GRP requirements and constraints related to the area;
- the policies and the aims of the PA, which can influence timing and any facilitations (tax, etc.);
- the local demand of similar available areas;
- the social and economic effects produced by the intervention on the surrounding areas;
- the legal and fiscal aspects, which vary according to the subjects involved;
- the demands of production system;
- the dynamics of local real estate market.

Starting from these variables, the decision-making process must reach the balance of conveniences of subjects involved, while ensuring the feasibility of the initiative [8].

The adoption of the financial and economic evaluations allows to formulate value judgments and economic convenience ones, useful for integrating and systemizing the variables that affect the feasibility of the initiative [10, 21].

The relationships that are established among different subjects involved in the initiatives mainly depend on the structure of the properties and the onerousness of the investment [15, 16]. In particular, a negotiation relationship is established between PA and private subject as owner of the availability of the property or as a property developer [4]. The interests involved are, therefore, of two types: the first one, for private subject, expressed by maximizing the revenues deriving from the realization of buildings and minimizing incurred costs; the second one, for public subject, that deal with the economic sphere [24]. Depending on the point of view considered, different analysis can be carried out: for private investor, the Cost-Revenue Analysis (CRA) - among the several financial assessment methodologies - is a technique widely spread in the context of urban projects *ex-ante* evaluation. It is a financial analysis that does not take into account qualitative effects of the project but is aimed at assessing the ability of the project to generate revenues to a greater extent than costs necessary for its realization. The time factor, in the "instant" variant, can be excluded in order to avoid complications deriving from use of discount rate. The CRA application requires an accuracy level of the amount of financial quantities analyzed such that of a broad estimate, therefore, in line with the level of detail of initial stages of the project. Furthermore, compared to other economic and multidimensional approaches, does not require specific professional skills or knowledge of particular private software. This feature allows both technicians of the PA and real estate developer involved to obtain a summary appraisal of the cost and revenue items required by the initiative. Another reason linked to the CRA application's utility, is that it is the starting point for implementing a Cost-Benefit Analysis (CBA) of the project, thus improving the accuracy of the *ex-ante* evaluation carried out. In fact, if the point of view is that of the public subject, the adoption of the CBA widens evaluation of the project to the effects that can generates on the community as a whole, positive and negative ones, by comparing costs and economic benefits that can be monetized [22]. Both the aforementioned methodologies are useful reference for the decision-makers to order the different project solutions on the basis of the respective synthetic indicator's information - i.e. the Net Present Value (NPV), the Internal Rate of Return (IRR), the Cost-Benefit Ratio (C/B) and the Pay Back Period (PBP) - provided for each solution.

It should also be noted that in the context of the urban transformations planning, the use of the multi-criteria techniques to analyze the conflicting qualitative and quantitative decision-making problems is increasingly widespread [23]. Furthermore, the growing involvement of inhabitants in urban redevelopment projects requires adoption of techniques able to objectify as much as possible opinions collected in the community [2].

In the context of the urban renovation projects conducted in the PPP form, the assessment tools have the role to support the choices in order to be effective and efficient, i.e. that are suitable of guaranteeing the economic and the financial convenience of the subjects involved by respecting the urban and the natural environment.

3   Aim of the Work

The aim of the work is to highlight the usefulness of the assessment tools to support the decision-making process with reference to a real case study concerning the enhancement of a military abandoned complex located in a central area of the city of Rome (Italy). In particular, in order to verify the possibility for the PA to request additional operations from the private entrepreneur more than the minimum ones required by the law, the CRA is implemented, on the basis of the real physical, urban and market data and in line with the current GRP provisions.

The assessments relating to the different aspects of the initiative can lead the PA to simulate the expected financial quantities required (costs/ revenues) for the initiative, which is assumed to be realized by a generic private entrepreneur, and to evaluate its financial convenience. By this way, the public operator will be able to correctly calibrate - ensuring the financial feasibility of the initiative - the extent and the type of the requests that can be required to the private entrepreneur, in relation to the type of the initiative, its complexity and the level of demand that the redeveloped properties will have on the local real estate market.

The private entrepreneur, instead, can use the CRA to verify its own financial balance sheet, among which cost items will also include the amounts of the realization of the public buildings, including the related burdens. The adoption of the CRA allows thus the private entrepreneur to broaden the vision of the critical issues related to the initiative under analysis, highlighting the strengths and the uncertain points by recognizing the risks.

The paper is structured as follows: in the Sect. 4 the case study is described, the cost and the revenue items considered for the implementation of the CRA are presented; in Sect. 5 the conclusions and the future insights of the work are drawn.

4   Case Study

The case study is the PPP initiative for the enhancement of the *Triumphal Fortress,* that is a disused military complex composed by abandoned buildings and degraded spaces located in the North-West of the city of Rome. Following the decommissioning by the Ministry of the Defense, in the 2007 it was declared "of particularly important interest and therefore subjected to all the protection provisions contained in the current legislation" and, subsequently, it was acquired by the PA. The properties and the bordering areas are one of the six military complexes owned by the PA and which are part of the well-known "Entrenched Field" system: the 15 military fortress and the 4 artillery built between 1877 and 1891. The assets of the system are disused structures, in extreme degradation conditions, since they have not been in use for decades, whereas, by the virtue of their strategic location, they could represent an important redevelopment opportunity for the entire city. When the Triumphal Fortress was built, the countryside was the only thing that surrounded it and only after the XX century urban expansion was incorporated into the built environment. According to the current GRP's provisions, the area has now a strategic role for the public reuse focused on offering the urban-level service functions which are required by the district.

Overall, the *Entrenched Field* system represents an evident discrepancy between the cultural and the identity value that is recognized by the respective local communities and the impossibility of enjoying it. In order to pursue the enhancement of the *Entrenched Field's* properties the goals set out by the GRP's provisions foreseen are: *i)* the recovery and the redevelopment of the urban morphology's traces that are represented by the Fortresses; *ii)* the strengthening of the specific local identities of the territorial zones surrounding each buildings by including the cultural functions; *iii)* the improvement of the sustainable mobility in order to create an environmental network for each building. Among the general purposes, as regards the area of the Triumphal Fortress a public service hub is planned to be accessible both from the district in which is located and the users of the neighboring urban areas.

The buildings in the area and their original intended uses are described below and reported in the Fig. 1:

Fig. 1. The currently existing buildings in the area of the Triumphal Fortress. (Color figure online)

- the Fortress, consisting of all the buildings once used for the so-called "Ulivelli Barracks", built in 1886 and now in disuse (beige color);
- a prestigious property once destinated for the airship shelter, the so-called "Hangar of the Ulivelli Barracks", dating back to the 1930s and actually abandoned (blue color);
- other buildings of lesser architectural importance (gray color);
- an uncultivated green area (green color).

Given the purposes of the PA related to redevelopment and functional recovery of historic buildings to be used for public facilities (such as the head office of the Municipality XIV), and to the transformation of the adjacent areas for the residential and tertiary functions, the perimeter of the area which include the above described properties was defined as a "Recovery Area". Furthermore, three land sectors have been established (Fig. 2) - i.e. *Sub. 1*, *Sub. 2* and *Sub. 3* - that can be sold to the private entrepreneur in order to be redeveloped. The programmatic tools establish to build a total Gross Floor Surface (GFS) of no more than 8,200 m^2, with the following intended uses:

- Sub.1: 5,000 m^2 are the maximum GFS and residential (*res*) and tertiary (*off*) units are allowed. The intervention categories for aforementioned intended uses, established by the Art. 9 of the Technical Implementation Standards (TIS) of the city of Rome, provide demolition and reconstruction with an increase in the GFS and in the above-ground volume/new construction;
- Sub.2: 1,800 m^2 are the maximum GFS are allowed and the intended uses include residential (*res*) and tertiary (*off*) units. The intervention categories for aforementioned intended uses, governed by the Art. 9 of the TIS of Rome, provide demolition and reconstruction with an increase in the GFS and in the above-ground volume/new construction;
- Sub.3: 1,400 m^2 are the maximum GFS and the intended uses consist of residential (*res*) and tertiary (*off*) units. For the intended uses the same categories of intervention valid for Sub.1 and Sub.2 are applied, according to the TIS.

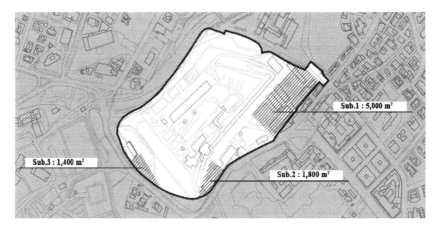

Fig. 2. - Identification of the three land sectors (Sub.1-Sub.2-Sub.3) within the area of the Triumphal Fortress.

4.1   Assumptions

The tested hypothesis is that redevelopment initiatives of the three land sectors are managed only by one general private entrepreneur. In this case, the relationship between public and private subjects can be summarized through the following exchange mechanisms:

– The PA gives building rights to private entrepreneur in order to build residential and tertiary units in the three land sectors;
– The private entrepreneur:
  • realizes the GFS quantity that is destined to the residential and tertiary units allowed by building rights and sells them on local real estate market, obtaining the monetary revenues that are produced by sales;
  • is required to pay primary, secondary and construction urbanization fees (Law no.10/1977) depending on the intended uses to be realized;
  • is obligated to reclaim areas and to transfer them to the PA freely of charge for primary and secondary urbanization operations;
  • reclaims additional areas for an amount of 4,810 m^2.

The definition of margins of agreement between subjects involved plays a decisive role since the absence of the balance of economic and financial conveniences between PA and private entrepreneur affects the feasibility of the initiative. Therefore, in order to verify the risks and the advantages of the parties involved it is necessary to adopt adequate assessment tools during this preliminary stage.

In this relationship, the real estate market has an important role linked to the absorption capacity and the phase of the local real estate market (expansion, crisis, recovery, etc.). on which the units of the building process will be sold, as established by the agreements between PA and private entrepreneur. It should be noted that in official consulted documents there are no limits on the maximum quantity of residential or tertiary units to be build. In other words, the distribution of the total GFS in shares of residential and tertiary units is defined in accordance with PA and private entrepreneur. For this reason, in the present work the hypothesis that in each of the three land sectors the 80% of the total GFS is destined to residential units and the 20% is for tertiary ones is firstly sets out. Moreover, without prejudice to legal obligations deriving from the Ministerial Decree no.1444/68, the freely transfer to the PA of additional areas destined for public neighborhood services for no less than 4,810 m^2 is envisaged, in order to satisfy provisions of urban planning standards for the area.

4.2   Cost-Revenue Analysis

In order to verify the financial feasibility of the initiative, the CRA is implemented by instantly comparing the expected financial items considered: all costs and revenues that are generated by the initiative occur at the same time, without considering the distribution that can reasonably have over time and effects - related to discount operations - that may be affect the results. This assessment methodology is well-suited with the accuracy that characterizes the preliminary phases of the initiative, for which indications required concern only the extent of amounts involved.

Specifically, the simulation of the financial balance sheet of redevelopment project managed by the generic private entrepreneur requires collection of data related to costs and selling prices to be implemented in financial balance sheets. In the present case study, these data (shown in Table 1) are identified by carrying out local real estate market analysis and by consulting construction companies operating in the area.

Table 1. Economic parameters of private investor's balance sheet.

Cost items	Unit cost/percentage incidence
Purchase of the area	200 €/m²
Registration tax and notary fees	11%
Primary urbanization fees (res)	23.82 €/m³
Primary urbanization fees (off)	39.31 €/m³
Secondary urbanization fees (res)	19.82 €/m³
Demolition of the existing volumes	35 €/m³
Construction costs (res)	1,000 €/m²
Construction costs (off)	850 €/m²
Construction charges (res)	12.50%
Construction charges (off)	12.50%
Realization of the parking spaces (res)	48 €/m²
Realization of the parking spaces (off)	48 €/m²
Realization of the green spaces (res)	32.55 €/m²
Realization of access roads	93 €/m²
Private entrepreneur's normal profit	20%
Technical and general expenses	9%
Financial fees	10%
Revenue items	*Unit price*
Residential sale	3,950 €/m²
Residential parking sale	925 €/m²
Tertiary sale	2,925 €/m²

By assigning the parametric amounts listed in Table 1 to the quantities foreseen by transformation hypotheses for each of the three land sector, the financial balance sheet of private entrepreneur is obtained, as described in Table 2.

The application of the CRA by instantly comparing costs and revenues assumed for the project has enabled to define the effective margins of financial convenience of private entrepreneur. The initiative is convenient on all the three land sectors. In particular, the evaluation tool adopted has allowed to determine the amount of extra-profit (see Net Values in Table 2) generated by the entire redevelopment project equal to $NV_1 + NV_2 + NV_3 = 5,926,864$ € (NV_t). This is the amount that the private entrepreneur earns from the whole initiative, in addition to normal profit included in cost items (shown in Table 3) to the extent of 20% of expected revenues.

Table 2. - The private investor's balance sheet for each land sector.

Sub.1	
Items	*Quantities*
Total land surface	13,074 m^2
Total costs	16,471,473 €
Total revenues	19,095,000 €
Net value (NV$_1$)	2,623,527 €

Sub.2	
Items	*Quantities*
Total land surface	2,676 m^2
Total costs	5,185,167 €
Total revenues	6,874,200 €
Net value (NV$_2$)	1,689,033 €

Sub.3	
Items	*Quantities*
Total land surface	939 m^2
Total costs	3,732,296 €
Total revenues	5,346,600 €
Net value (NV$_3$)	1,614,304 €

Table 3. - The normal profit of private investor for each land sector's interventions.

Normal profit	
Items	*Quantities*
NP$_1$	3,819,000 €
NP$_2$	1,374,840 €
NP$_3$	1,069,320 €
Total (NP$_t$)	6,263,160 €

The total normal profit (NP$_t$) quantity is equal to NP$_1$ + NP$_2$ + NP$_3$ = 6,263,160 € (see Normal Profit in Table 3). Definitely, by adding normal profit to extra-profit, private entrepreneur gains a total profit of NP$_t$ + NV$_t$ = 12,190,024 €. This output proves that there are wide margins for further requests that can be elicited by the PA.

Another information that can be obtained with the CRA adoption, concerns the variation of the NV$_t$ of the initiative linked to different combinations of shares of dwellings and tertiary units on the total extent allowed. The graph of Fig. 3 represents the relationship that occurs between the NV$_t$, i.e. the amount of extra-profit that belongs to private entrepreneur, and the percentage of total GFS that is differently destined to each intended use.

As can be seen from the graph, by varying the percentage incidence on the total of GFS that is intended for residential units (or symmetrically of the GFS that is intended

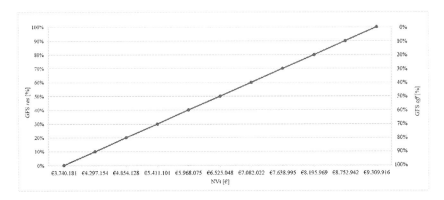

Fig. 3. NV$_t$ of the initiative related to the percentage of the total GFS attributed to the residential or tertiary units.

for tertiary ones), different margins of convenience are obtained. By increasing residential GFS and reducing that for tertiary units, the amount of extra-profit generated is considerably higher than the one of opposite situation, i.e. when tertiary GFS increases and residential one decreases. This analysis shows up that the functional mix to be realizing is a delicate issue that the PA musts take into account, not only for urban weights that this mix determines, but also for effects it generates on financial balance sheet of initiative.

5 Conclusions

The growing establishment of the awareness to mitigate the impacts produced by soil sealing due to uncontrolled urban growth through sustainable urban redevelopment policies, is felt by all the authorities. The urban redevelopment policies, in fact, impose territorial transformation actions that allow to restore identity values to urban spaces characterized by physical and social degradation. Since the complexity is a common feature of urban renewal processes, the assessment tools are necessary to provide a useful support for making effective and efficient choices with the aim of guaranteeing economic and financial convenience for the subjects involved, by preserving natural environment.

This research is linked to issues outlined. In particular, the aim has been to highlight usefulness of the CRA assessment technique to support decision-making process of enhancement of abandoned Triumphal Fortress military complex located in a central area of the city of Rome (Italy). In particular, in order to verify possibility for PA to request additional operations to private entrepreneur more than the minimum ones required by the law, the CRA has been implemented by considering real physical, urban and market data and by applying current GRP provisions. The implementation of the CRA with instant comparison of expected financial quantities has allowed to determine balance sheets of PA and private entrepreneur and to highlight the existence of wide bargaining margins which, if not modified, leave the private developer with a substantial profit. The achieved results enable the PA to request the private entrepreneur i) the further financial resources or ii) the direct realization of the operations that are intended to the public services. In other words, the use of the proposed assessment technique has allowed the subjects involved to quantify cost and revenue items relating to transaction so that, in compliance with constraints of financial convenience, the fair distribution of costs and advantages according to the functional mix has been guaranteed. This is an output that otherwise would not have been available.

The financial analysis conducted could represent a first step of a broader methodological approach that provides in a second phase the analysis of discounted cash flows, for a more accurate financial feasibility check, and the determination of GFS quantity which represents the break-even point of the initiative. Operating in this way, the entire financial soundness of the project to enhance the Triumphal Fortress would be probed.

Future improvements of the research may also concern the application of Operational Research algorithms for determining the specific urban parameters' quantity able to define the maximum reduction of land take impacts, by guarantying the financial interests of subjects involved.

Aknowledgements. The authors declare that the contents of the paper are based on data and information developed within the "Innonets" Project founded by the Interreg Greece-Italy Program and are aimed at promoting the Project itself. For more information see http://interreginnonets.eu/en.

References

1. Borri, D.: Lessico urbanistico, annotato e figurato. Edizioni Dedalo (1985)
2. Butt, M.A., Li, S.: Usability evaluation of collaborative PPGIS-GeoCWMI for supporting public participation during municipal planning and management services. Appl. Geomatics 7(3), 139–161 (2014). https://doi.org/10.1007/s12518-014-0141-0
3. Centro Studi Sogeea: Primo rapporto sulla rigenerazione urbana in Italia (2018)
4. Copiello, S.: Progetti urbani in partenariato. Studi di fattibilità e piano economico finanziario, vol. 2. Alinea Editrice (2011)
5. CUFAA, S.P.: Il Consumo di suolo, dinamiche territoriali e servizi ecosistemici. https://www.snpambiente.it/wp-content/uploads/2019/09/Rapporto_consumo_di_suolo_20190917-1.pdf. Accessed 20 Dec 2019
6. Dostatni, E., Diakun, J., Grajewski, D., Wichniarek, R., Karwasz, A.: Multi-agent system to support decision-making process in ecodesign. In: 10th International Conference on Soft Computing Models in Industrial and Environmental Applications, pp. 463–474. Springer, Cham (2015)
7. Gargiulo, C., Davino, A.: Processi di rivitalizzazione e riqualificazione urbana: dalla pianificazione del recupero all'attuazione degli interventi (2000)
8. Gazzetta Ufficiale della Repubblica Italiana: Relazione sulle attività concernenti il Partenariato pubblico-privato (PPP) 2017–2018 (2019)
9. Ghavami, S.M., Taleai, M., Arentze, T.: An intelligent spatial land use planning support system using socially rational agents. Int. J. Geogr. Inf. Sci. 31(5), 1022–1041 (2017)

10. Guarini, M.R., D'Addabbo, N., Morano, P., Tajani, F.: Multi-criteria analysis in compound decision processes: the ahp and the architectural competition for the chamber of deputies in Rome (Italy). Buildings **7**, 38 (2017)
11. Guarini, M.R., Morano, P., Sica, F.: Integrated ecosystem design: an evaluation model to support the choice of eco-compatible technological solutions for residential building. Energies **12**, 2659 (2019)
12. Indovina, F.: Governare la città con l'urbanistica. Guida agli strumenti di pianificazione urbana e del territorio. Maggioli Editore, Rimini (2005)
13. ISPRA-SNPA: Consumo di suolo, dinamiche territoriali e servizi ecosistemici - Edizione 2018. Rapporti 288/2018 (2018)
14. Las Casas, G., Scorza, F., Murgante, B.: New urban agenda and open challenges for urban and regional planning. In: Calabrò, F., Della Spina, L., Bevilacqua, C. (eds.) ISHT 2018. SIST, vol. 100, pp. 282–288. Springer, Cham (2019). https://doi.org/10.1007/978-3-319-92099-3_33
15. Liang, Q., Hu, H.: Uncertainty analysis of value for money assessment for public-private partnership projects. J. Shanghai Jiaotong Univ. (Sci.) **22**(6), 672–681 (2017). https://doi.org/10.1007/s12204-017-1889-2
16. Luque, J.P., Ikromov, N., Noseworthy, W.B.: Financial feasibility analysis: planning for the possible. Affordable Housing Development, pp. 81–87. Springer, Cham (2019). https://doi.org/10.1007/978-3-030-04064-2_6
17. Magnolfi, N.: Rigenerazione urbana come paradigma del piano e del progetto urbanistico: contenuti e strumenti per innescare il processo. Doctoral dissertation, Politecnico di Torino (2018)
18. Mattia, S.: Governo del territorio e ruolo della valutazione. In: Miccoli, S. (a cura di) Rinnovo urbano e valutazioni integrate, pp. 33–68. DEI-Tipografia del Genio Civile, Roma (2009)
19. Morano, P.: Un modello di perequazione urbanistico-estimativo. Graffiti, Napoli (1999)
20. Morano, P., Guarini, M.R., Tajani, F., Di Liddo, F., Anelli, D.: Incidence of different types of urban green spaces on property prices. A case study in the Flaminio district of Rome (Italy). In: Misra, S., et al. (eds.) ICCSA 2019. LNCS, vol. 11622, pp. 23–34. Springer, Cham (2019). https://doi.org/10.1007/978-3-030-24305-0_3
21. Morano, P., Tajani, F., Di Liddo, F., Guarnaccia, C.: The value of the energy retrofit in the Italian housing market: two case-studies compared. WSEAS Trans. Bus. Econ. **15**, 249–258 (2018)
22. Nas, T.F.: Cost-Benefit Analysis: Theory and Application. Lexington Books (2016)
23. Omidipoor, M., Jelokhani-Niaraki, M., Moeinmehr, A., Sadeghi-Niaraki, A., Choi, S.M.: A GIS-based decision support system for facilitating participatory urban renewal process. Land Use Policy **88**, 104150 (2019)
24. Stanghellini, S.: Il negoziato pubblico privato nei progetti urbani: principi, metodi e tecniche di valutazione. DEI-Tipografia del Genio civile (2012)

A Procedure to Evaluate the Extra-Charge of Urbanization

Maria Rosaria Guarini[1]([✉]) [iD], Pierluigi Morano[2] [iD],
and Alessandro Micheli[3] [iD]

[1] Department of Architecture and Design, Sapienza University of Rome,
00196 Rome, Italy
mariarosaria.guarini@uniroma1.it
[2] Department of Science of Civil Engineering and Architecture,
Polytechnic University of Bari, 70125 Bari, Italy
pierluigi.morano@poliba.it
[3] Doctoral School of Architecture and Construction, Department of Architecture
and Design, Sapienza University of Rome, 00196 Rome, Italy
alessandro.micheli@uniroma1.it

Abstract. The Public-Private Partnership is also used, at international level, to carry out interventions in urban variant on areas or buildings that, by enhancing the value of (private) real estate properties, guarantee an economic return for the local community through an extraordinary contribution of urbanization (ECU) paid to the Administration by the private entity. The Italian Regions and Municipalities that have regulated the ECU calculation adopt a procedure of transformation value that does not specify either the duration or the risk-related rate of return of the real estate operation, which are essential factors for its balanced quantification. The paper defines a procedure that integrates methodologies, diffused in practice and literature, with official datasets to determine objectively the ECU, also evaluating its sensitivity to the critical variables of the transformative process. The proposed procedure is easy to apply and adaptable to a wide range of interventions and to different phases of the public-private (re) negotiation. It is in line with the Italian case and it is consistent with the international evaluation standards (IVS), having a general validity in the estimation field.

Keywords: Extra charge of urbanization · Public–private partnership · Discount rate · Capital gain appraisal · Land value capture · Real estate development · Discounted cash flow analisys

Conceptualization, M.R.G., P.M. and A.M.; Data curation, A.M.; Formal analysis, M.R.G., P.M. and A.M.; Investigation, M.R.G. and A.M.; Methodology, M.R.G., P.M. and A.M.; Resources, M.R. G. and A.M.; Supervision, M.R.G. and P.M.; Validation, M.R.G. and P.M.; Visualization, M.R.G., P.M. and A.M.; Writing—original draft, M.R.G., P.M. and A.M.; Writing—review & editing, M.R. G., P.M. and A.M.

© Springer Nature Switzerland AG 2020
O. Gervasi et al. (Eds.): ICCSA 2020, LNCS 12251, pp. 981–999, 2020.
https://doi.org/10.1007/978-3-030-58808-3_70

1 Introduction

At least since the second half of the 20th century, the Public Private Partnership (PPP) has been a very present topic in the international scientific debate, also in reference to urban settlement transformation projects. Both contractual and institutionalized PPP [18], concerns a bundle of legal institutions that Member States have transposed from the European Policy Framework. It is recognised that the quality of urban processes derives from the capacity for interaction between the public and private sectors, according to advanced cooperative forms [11].

The current scarcity of resources increases the need for the Public Administration (PA) to finance, including from private funding, the public investments that make the city live without increasing public debt [12]. Urban land income is a logical, rather than economic, approach to the issue: it can be at least partially taxed because its value depends significantly on the investments (and decisions) that the PA and other private entities make in/around the city [5].

The purpose of taxation is not necessarily the values of the land and property stock (already strongly affected) but concerns the emerging or extra capital gain (ECG), i.e. its variation (capitalized in the value/price of the land) generated by the processes of real estate (re)development or urban rigeneration and transformation of land uses, from greenfiels to brownfields sites or from blighted areas to renewal ones [26].

In most advanced countries, the private actor is called to contribute to the regeneration and development costs of the city, sharing with the public sector the surplus value generated by urban transformation, through various modalities (generally combined and used with different intensities) [7, 14]. The French *taxe d'aménagement*, the Spanish *cargas de urbanizaciòn*, the American *impact fees* and the English *development permit* are tools for this purpose [8].

Even in less developed and developing countries, agencies and large international study centres are increasingly oriented towards differentiated practices and forms of *value recapture* - i.e. the recovery (in favour of the community) of surplus values created directly by public action or investment - and *value sharing* - i.e. the fair distribution between public and private of surplus values in urban transformation (from agricultural use to residential and productive land use) - through national and local rules or negotiated agreements between PA and real estate operators [8].

Among the most common practices in Europe, the partial recovery by the public of the capital gains of private urban transformation is typical of the Spanish and Italian case. In Italy this extraordinary charge of urbanisation (ECU) is calculated by the municipal administration as an additional planning permission burden; it must beat least equal to 50% of the ECG produced by interventions on buildings in urban variant which, in general, presupposes a negotiated agreement between the parties.

Operationally and in literature, the procedure of the Transformation Value (TV) is used to quantify the ECG; this criterion requires a duration of the initiative and a Discount Rate (DR) to be estimated in a prudential way because it is crucial to assess the sustainability of the project [26] also using the Discounted Cash Flow (DCF).

The most used techniques to determine the DR are the Weighted Average Cost of Capital (WACC), the Build-Up Method and the Capital Asset Pricing Model (CAPM) [17, 20].

It should be noted that the analysis of medium/long-term public-private relationships (at least 5–10 years) cannot be limited to the start of the implementation phase and it shoud overcome the conviction that the initial commitments and balances must be preserved even in subsequent variants [26]: in fact, the implementation of the contracted projects may require changes due to: renewed market needs, objective construction difficulties (not attributable to developers), etc.

It therefore seems appropriate to focus on the transformation capital gains that emerge in the valuation processes, ex-*ante* and ex-*post*, about the initial and final net property values, and how these are calculated [26].

The definition of a method that makes this capital gain assessable, with effective and conditional parameters, can be a valid support also to re-negotiate the economic-financial relationships, in the final stage or working progress [26].

The paper defines a relatively straightforward method that, integrating commonly used techniques and data easily available (online) from authoritative sources, allows the determination of the DR and the ECU of a real estate initiative. This is in order to: i) contribute to the discussion of the real estate development and urban regeneration initiatives, also in PPPs, focusing on the partial recovery to the community - as a *benefit sharing* [30] - of the properties transformation value (lands or buildings) because their added value can be understood as *public value* [25]; ii) investigate how the economic-financial evaluation can seek a satisfactory balance between public benefits and private convenience also in view of the extraordinary contribution (ECU) envisaged - at European level [8] - for the private players; iii) limit the subjectivity in public-private negotiations which often have very different outcomes depending on the actors involved and their contractual capacity [26].

The following are presented and discussed in the next section: 2) the regulation of the ECU, the procedure to estimate the TV and the techniques to determine the DR; 3) the proposed procedure to quantify the ECU congruently, correlating it to its sensitivity to critical variables such as time, market values and DR; 4) conclusions and future prospects of the research undertaken.

2   Materials and Methods

2.1   Preamble

Considering that the regulation of PPPs and the application of the ECU in Italy are in line with European and international practice, this section is a preamble to the proposed procedure and includes: an excursus on the regulation and calculation methods of the ECU that are based on the TV criterion but without explicating the 'time' and 'risk' factors (although these significantly affect the DR, the consequent TV and thus the ECU); a description of the main methodologies and techniques used to determine the TV and the DR.

2.2   Extraordinary Charge of Urbanization (ECU)

Law no. 164/2014, which has regulated the payment of the ECU in Italy, is inspired by the most advanced international laws on the subject and allocates the private resources of the revenue produced to carry out public works or services (urban regeneration, environmental protection, social reform) [12]. This legislative (and not regulatory) [1] provision treats the ECU as a consensual and negotiated planning permission burden - in addition to the primary and secondary urbanisation costs - in relation to the increase in real estate value due to urban variant (UV), derogations from the current General Plan (GP) or changes in use.

However, the building and urban planning matter concerns the "shared" legislative power of the State-Regions, but more than five years (2019) since the approval of the national law, as a general principle, only 4 out of 20 Regions have legislated on the subject (Emilia-Romagna, Liguria, Piemonte, Puglia) and only Piemonte and Puglia seem to have fully implemented its content [12] while Liguria (LR 16/2008, as amended by LR 41/2014) has not specified the calculation method. The deliberations of these Regions show differences in the calculation methodology of the ECU to be adopted, the way it is applied and the provision of incentives or disincentives [12].

Operationally, the analytical and synthetic methods are identified. The Regions Piemonte (DGR 222974/2016), Puglia (LR 18/2019), Emilia-Romagna (DGR 186/2018) and many municipalities apply the first one according to the Formula 1:

$$ECU = 50\% \, ECG = 50\% \, [TV1 - TV0]$$
$$= 50\% \, [(MV1 - K1) - (MV0 - K0)] \qquad (1)$$

where TV-Transformation Value of the property is given by the difference between the Market Value of the goods produced by transformation (MV), inferred from the quotations published by the Italian Tax Agency's Real Estate Market Observatory (OMI), and all transformation costs (K) (see as example Table 1).

The second one is provided by Emilia-Romagna for interventions located outside the urbanised territory and without any project complexity; it calculates the ECG (Formula 2) as the difference between the market values after (V *post*) and before (V *ante*) the transformation generated by the UV. These values are deduced from market analysis or from building areas value for municipal tax purposes and, for V *ante*, also from the "average farm values" of the Region.

$$ECU = 50\% \, ECG = 50\% \, [V\, post - V\, ante] \qquad (2)$$

Formulas 1 and 2 do not consider the discount factor and even the analytical method, although more defined than the synthetic one, presents many inconsistencies [13].

A survey conducted online to identify the municipal regulations on the ECU, showed that, as at 31 March 2020, 37 municipalities have adopted, in essence, one of these two methods but with different variants and various coefficients; few municipalities set the ECU in a percentage (on the ECG) higher than the minimum (50%), including the Municipality of Rome which was among the first administrations to regulate the calculation of the ECU, adopting a measure of 66% (Rome Assembly Resolution 128/2014).

In the Regions that have ignored or opposed national law, such as Lombardia [9] and Veneto, the municipalities have not transposed it or have adopted targets and methods for determining the ECU with their own criteria. Many regional and municipal measures have then tried to decrease the potential of the rule by reducing, during the evaluation, the surplus value generated by the interventions [12] through deductible charges, overestimations of the initial land value, overestimations of works or land disposals, miscellaneous interest and extra-profits that are abnormal and not-ordinary [8].

The heterogeneity and obvious contradictions of the calculation mechanism adopted by most Regions and municipalities hinders the effective application of the law [12].

The evaluation models of the ECU found in the literature [4, 26], referring to the same scheme of the TV, calculate the riskiness of the intervention within the percentage of the Promoter's profit (empirically established) and determine the discount factor only over its entire theoretical duration (5 years) on the basis of a predetermined interest rate [26].

2.3   Methodologies for Determining the Transformation Value (TV)

In the Italian estimative tradition, TV is applied both as an estimative procedure - valid for the generality of market operators - and as a criterion of estimation or economic aspect (derivative), referring to a particular operator and expression of economic convenience judgements.

The IVS distinguish the MV and values different from it: they are not logically different from the "derivative" and typical economic aspects of the Italian school such as TV [23]. Internationally, TV is part of the concepts of *highest and best use* and *value in use* [2, 17].

In the Italian estimative tradition, the analytical criterion for determining the TV of an asset is given by the difference between the revenues obtained from the marketing of goods produced by transformation (Market Value post-transformation - MV_P) and the transformation costs (K), direct (DC: technical construction cost) and indirect (IC: charges and other costs necessary to start and carry out the transformation), including the promoter's profit (Up) to remunerate his investment on the basis of its risk. These quantities are discounted with a discount factor $1/q^n = 1/(1 + r)^n$ which is a function of the discount rate r and the time of the normal transformation n (Formula 3):

$$TV = MV_P/q^n - (K + Up)/q^n \qquad (3)$$

As already pointed out, in practice Italian PAs use an 'atemporal' formula (without q^n) to determine TV and thus quantifying the ECU amount.

In order to estimate the MV_P (such as revenues from: sale, rent, management; etc.) it is possible to use the analysis of [15]: a) historical data (income achieved in the recent past), b) potential income (fully operational), c) market data (through goods comparable to those in question), d) average income (considering all above points).

The MV is usually determined synthetically or parametrically by reference to the target market data from official sources or direct market research.

In the absence of a detailed design, direct costs (DC) can be deducted from: a) similar interventions through the synthetic-comparative procedure, which identifies a parametric unit cost (€/sqm, €/cm); b) commercial publications (e.g. Italian publisher Maggioli, DEI) by means of statistically surveyed and, if necessary, adjusted parametric costs; c) summary cost estimate, generally attached to the feasibility study or to the general project.

Indirect costs (IC) are generally derived as a percentage of MV_P and DC: Italian technical literature and municipal regulations for calculating the ECU use different terms and percentages, but the cost items are substantially the same. By way of example, Table 1 shows those contained in the Resolution of the Municipal Administration of Rome.

The DCF [21] considers a more detailed distribution of costs /revenues and determines TV as Net Present Value (NPV), given by the algebraic sum (Σ) of costs (excluding Up) and revenues (expected cash flows (Ft)), discounted using a discount factor $(1/q^t)$ according to their time sequence (t) within a suitable duration (n) (Formula 4).

$$TV = \text{NPV} = \sum_{t=1}^{n} \frac{F_t}{q^t} = \sum_{t=1}^{n} \frac{F_t}{(1+r)^t} \tag{4}$$

Table 1. Methods of calculating transformation costs (Rome assembly resolution no. 128/2014)

Cost items	Calculation mode
DC - Direct Costs	
C_0 – technical construction cost;	Parametrically[a]
IC - Indirect Costs	
C_1 – cost of area adaptation and connection to utilities/networks;	2–5% C_0
C_2 – urbanization charges (Presidential Decree no. 380/2001, art.16)	10% C_0
C_3 – professional and supplementary charges and contingencies;	8–12% $(C_0 + C_1)$
C_4 – marketing charges;	2–3% MV_P
C_5 – borrowing costs on debt D, with $q^n = (1 + i)^n$ and interest rate i = Spread + EurIRS or Euribor (loan duration 15 years)[b]	$D\%(C_0 + C_1 + C_2 + C_3) * (q^n-1)$
C_6 – Promoter's profit (gross) Up	15–25% MV_P
K - Total cost of transformation	$C_0 + C_1 + C_2 + C_3 + C_4 + C_5 + C_6$

(a) calculated using the €/sqm values of the "Prezziario del Collegio Ingegneri e Architetti di Milano" (DEI, latest edition available at the time of the estimate) for specific uses;
(b) only a first pre-amortisation period of 5 years (n = 1, 2, ..., 5) is considered as follows: 10% (1st year); 30% (2nd year); 40% (3rd year); 20% (4th year); 0% (5th year).

In both cases, the DR (which is difficult to deduce from the market) is decisive because the TV varies considerably even at small rate deviations and can therefore influence the amount of the ECU. The construction of the discount rate (DR) in the DCF is based on the concept of risk-return, also taking into account debt capital [3, 29].

In general, Ft are gross of interest payable on debt and subjective income taxes as these costs would lead to different values for the same property as the financing choices and subjective tax situation change.

The exclusion of the tax variable does not influence or weaken the methodology.

The possible growth rate of Ft can be appropriately analyzed according to the expected trend of the market or business sector concerned.

Due to the uncertainty of price development trends, a constant price exercise is frequently used, generally adopting the last data availability. While this allows to focus on quantity forecasts, price changes and inflation can have a major impact on the project's valuation. [19]. Alternatively, to use current discount and interest rates, both price and quantity forecasts should be made by assumptions about trend increases in price/cost and price/revenue rather than applying a uniform expected inflation rate (probably elusive for periods longer than 3–5 years) [19].

Assuming the expected inflation, it is theoretically equivalent to express Ft at nominal or real value but they must be consistent with DR [22]. Assuming that a discount rate before interest is used, in fact, part of the real loss in value of capital is incorporated [19].

The minimum duration of the analysis period (n) is 5 years but in particular cases it may exceed 10 years, depending on the duration of the lease contracts or the forecast of significant transformations over a long period [15, 26].

2.4   Discount Rate (DR)

The direct research of DR requires the detection of a sufficiently large and homogeneous sample of contracts, which is often difficult. In the absence of income and market prices in the specific real estate segment (or to verify the estimated rate), its indirect estimation is used by comparing the operators expectation in this sector with those found in other markets, considering that the expected return on the investment is proportional to its risk degree. The most widely used models for this purpose are:

- WACC, derived from the corporate field and widespread in the practice of the DCF [11, 15, 16, 26, 27, 32], is based on the assumption that the invested capital derives from a mix of financial sources, debt (D) and equity (E), according to the Formula 5:

$$WACC = kd * (1 - t) * D/(E + D) + ke * E/(E + D) \qquad (5)$$

where Kd is the cost of debt, Ke is the cost of equity, and t is the tax rate.

In general, Kd is assumed to be equal to the EurIRS increased by the Spread (remuneration for the credit institution granting the loan) [15, 32], inferred from the main Central National Banks (Deutsch Bank, BNP Paribas, Credit Agricole etc.), and the costs related to the loan structuring (Literature Fees) [11].

- The Build-Up Approach [15], an "additive method" that identifies the expected return on investments by adding the different economic return differentials (dx) (Formula 6). These differentials d represent the risk within the variability of the yield (ymin = minimum yield; ymax = maximum yield) specific to each factor defining the risk of a real estate iniziative, such that: ymin < d (n) < ymax.

$$r = d(1) + d(2) + \ldots + d(n) = n\sum x = 1\, d(x) \qquad (6)$$

- CAPM [28] based on the idea of systematic risk (non-diversifiable) and market risk premium MRP as compensation for the investor given by the Rm return of a market portfolio minus the Rrf return of a risk free bond; it is extended to real estate to determine Ke [11, 15, 16, 26, 27, 32] according to the Formula 7:

$$Ke = Rrf + (\beta_{imm} * MRP) + \gamma = Rrf + [\beta_{imm} * (Rm - Rrf)] + \gamma \quad (7)$$

where: β_{imm} (or β) is the covariance index of the real estate sector and measures the sensitivity of the return on investment to movements of the entire market; γ is the premium for the additional risk component depending on the characteristics of the specific real estate project (location, type, size, etc.).

To determine Rrf, β_{imm}, MRP it is possible to refer to official sources or online datasets used in international evaluation practice [32] which make available freely numerous data in the economic and financial field (Bloomberg, Damodaran etc.) (Table 2).

In general, the γ estimate requires a market survey of similar real estate development projects already carried out, in order to gather their most relevant data considering their specific characteristics. The Italian Tax Agency (ITA) has proposed a model of real estate risk assessment γ based on the dataset of 42 real estate development projects carried out in Rome and analyzed by means of several concatenated statistical-economic estimation methods (DCSEOMI Methodology) [10]. This model, in relation to the market segment investigated and the information available, selects the investment risk factors from those indicated in TEGoVA's reports (2003, 2010) and describes each one by a level/nomenclature associated with a scale of conventional importance (the level associated with a higher rate of return corresponds to the highest score). Having defined the various differentials characterising each factor and assigned the specific value (risk level) of each differential, the expected γ of the initiative is given by the sum of the values corresponding to the individual risk components (Table 3).

The γ of real estate development projects may therefore vary within these values:

$$\gamma_{min} = 0,18\% + 0,03\% + 0,06\% + 0,06\% + 0,12\% = 0,45\% \quad (8)$$

$$\gamma_{max} = 9,35\% + 1,58\% + 2,87\% + 2,94\% + 6,20\% = 22,94\% \quad (9)$$

Table 2. Official sources and data platforms to determine the components of: a) Ke; b) Kd

Factor	Source	Benchmark
a) components of Ke		
Rrf	National Central Bank; Bloomberg	Nominal yields on government bonds with a maturity comparable to the duration of the operation (average reference period)
MRP	Bloomberg	Inferable values for the national market concerned
	Damodaran	Equity Risk Premium (ERP) for the country concerned
β	Damodaran	Beta of the sector: *Real Estate (Development)* - Western Europe
b) components of Kd		
EurIrs (o Euribor)	www.Euribor.it (or Il Sole 24Ore, for Italy)	Average of the values recorded backwards for a period representative of the market trend
Spread	Deutsch Bank, BNP Paribas, Credit Agricole, ecc	Percentage value determined by the main European Banks (estimated at around 2%)
	Damodaran	Annual published value for "Western Europe"
Success + arranging fee	Technical literature	Costs related to the structuring of funding quantifiable respectively in the order of 0.7% + 0.9% = 1.6% (Literature Fees)

Table 3. Model proposed by the Italian RA to assess the specific risk γ of property investments

Specific risk factor levels	Location	Dimensional factors	Building equipment	Property size	Presence of competitors	
	0,0948316	0,0160615	0,0291632	0,0298292	0,062866	Weights
	40,74%	6,90%	12,53%	12,82%	27,01%	% on the total
1	0,18%	0,03%	0,06%	0,06%	0,12%	
2	2,01%	0,65%	1,18%	0,82%	2,54%	
3	3,83%	1,58%	2,87%	1,78%	6,20%	
4	6,59%			2,94%		
5	9,35%					

3 Proposed Procedure for Determining the ECU

3.1 Structure and Steps of the Procedure

The proposed procedure to calculate the ECU of a real estate development intervention in urbanistic variant (UV), public or private initiative referable to building land or consolidated urban planning, considers the most representative variables of the transformative process, declined both in the scenario 'before' and 'after' the Event or Variant (*with-without-principle*), including factors that normally are not considered in the methodologies used by the Italian PA, such as duration and riskiness of the investment.

In the role of PAs, this tool can be used to evaluate a hypothesis of urban transformation in order to: i) establish the fundamental parameters and essential contents of the possible urban planning programme (as a basis of the tender); ii) verify the data proposed by the Promoter, compared to those considered "ordinary" by the PA with reference to the sources established in the procedure; iii) determine and/or verify a fair distribution of the costs and benefits of the initiative, ex-*ante* ex-*post* or in progress.

Given the elements of uncertainty and the temporal distance of the assumptions necessary for the procedure, it is important for the PA not to apply it uncritically. Here it is assumed that the Promoter formulates a proposal to the PA articulated in the two intervention scenarios (*ante/post* UV) according to the data reported in Table 4.

Table 4. Input data of the proposed model to enhance a free area in derogation of the GP

Data typology	Parameters per scenario (S)	Description /formula /note
Data urban planning	Typology intervention	New building, urban renovation, etc.
	Land Area (of intervention)	LA (mq), (for area)
	Land Use Index	LUI_S (smq/sqm or cm/sqm)
	Gross Floor Area (building)[a]	$GFA_S = LUI_S * LA$
	Building Types Areas	$(BTAi)_S = [(\%i)*GFA]_S$
Timing (analysis period)	Time span transformation	n_S (in years); t = generic year of duration n
	Time distribution MV, K	$[(\%MV)t, (\%DC)t, (\%ICK)t, (\%ICR)t]_S$
Revenue of transformation	Market Value (MV)	$(MV)_S = \Sigma(MVi)_S = \Sigma(UMVi)_S*(BTAi)_S$
	Average Unit MV	$UMV_S = [\Sigma(UMVi*BTAi)_S] / GFA_S$
Cost of transformation $(K)_S = DC_S + IC_S$	Direct Cost (DC)	$DC_S [= \Sigma(UDCi_S*BTAi_S)]$
	average Unit DC	$UDC_S = [\Sigma(UDCi_S*BTAi_S)] / GFA_S$
	Indirect Costs IC = $ICK_S + ICR_S$	$ICK_S = [\Sigma(\%_K)*DC]_S$, $ICR_S = [(\%_R)*MV]_S$
	Extra Charge ofUrbanization	ECU
Financial parameters	Cost of financing/funding	Ke (cost of Equity), Kd (cost of Debt)
	Leverage ratio	D/E

([a]) according to the rules of Presidential Decree 138/98, also used to determine OMI quotations

The procedure is structured in 6 phases; their logical-operational steps are illustrated in detail in the following paragraphs and summarized in Fig. 1.

The data proposed by the Promoter in the initiative (private) or in an interlocutory phase of the public evidence procedure, are re-determined by the PA (step 1) and compared with each other (step 2) in order to select the most advantageous values for the PA, in the *ante* and *post* scenarios, as input data of the calculation system (according to the DCF) to determine the ECU ("reference value") as the difference of the *post/ante*-UV NPVs (step 3). The variation tables of the ECU, constructed as UMV and DR vary - starting from their reference values (step 4) - and extending the duration of *m* years (step 5), highlight its sensitivity to these critical variables and therefore the randomness degree of the results. This *what-if-analisys* can lead to confirm the "reference value" or to repeat the calculation (step 6), modifying the input data (step 1 and 2).

Given its repetitive nature, the procedure maintains its validity also by diversifying the characteristics of the scenarios. This makes it suitable to a wide range of interventions and to the various phases of the iniziative, starting from the pre-contractual one.

Step 1 – Determination and Verification of the Data. The PA verifies the urban planning data proposed by the Promoter for the *ante*-UV scenario, in compliance with the current GP, and assesses at its discretion the eligibility of: those proposed for the *post*-UV scenario, as an exception to the GP; duration of the operation (n) and time distribution of the expected flows, proposed by the Promoter (which is responsible for the organization of the overall transformation process) assuming them if they are consistent with the governance strategies of the territory. Instead, PA redetermines for each scenario, the other data of the proposal, i.e.: i) estimates the DCs, using parametric unit costs (UDCi) derived from landmark sector publications or from online applications (provided by organizations of proven reliability), possibly by applying corrective coefficients in relation to location and environmental context, time of estimation, consistency and coherence of the comparative work categories (quality level); ii) quantify the MV_P revenues using the average quotations (usually expressed in €/sqm) published by the OMI for the area concerned (or neighbouring /similar areas) and for each building type envisaged by the transformation (or assimilable) in the assumed state of maintenance. The new building values, if not specified in OMI, are assumed to be equal to the maximum value of the "normal" state of maintenance increased by 30%; iii) identifies the economic-financial parameters to quantify the DR, consulting official sources and online datasets/platforms commonly used at international level, also by reworkable excel files (Table 5).

Step 2 – Data Comparison. The comparison between the data calculated by the PA (^A) and those proposed by the Promoter (^P) is aimed at selecting the values that (tend to) maximize the ECG of the initiative as input of the DCF (Table 6). Indirect costs (IC) are determined at the same time and proportionally (in percentage terms) to the input parameters (MV_P, DC) according to the specific municipal regulation. The PA then determines the DR indirectly (WACC method) by the Formula 10 which, prudentially (in relation to the calculation of NPV and ECU), does not report the tax rate *t* deductible from the *Kd*.

$$DR = (ke)/[(D/E) + 1] + (kd) * (D/E)/[(D/E) + 1]$$
$$= (Rrf + \beta * MRP + \gamma)/[(D/E) + 1] + (Eurirs + Spread + Fee) * (D/E)/[(D/E) + 1]$$

$$(10)$$

Table 5. Technical-financial parameters to quantify the DR and their sources/references.

Index	Font	Reference
Rft	Central National Bank	Average yield on government bonds - by residual life span comparable to the analysis period of the transformation - in the year closest to the time of the estimate for a period sufficiently representative of their performance
		es. for Italy: https://www.bancaditalia.it/compiti/operazioni-mef/rendistato-rendiob/
β	Damodaran	Factor "Beta" of the "Real Estate (Development)" (Industry name)
		http://people.stern.nyu.edu/adamodar/ link "Europe" of Current data set "Levered and Unlevered Betas by Industry" (Topic: "Discount Rate Estimation"), downloading Regional datasets (Excel)
MRP	Damodaran	The most recent "Equity Risk Premium" value assigned to the country concerned
		http://people.stern.nyu.edu/adamodar/New_Home_Page/datafile/ctryprem.html or http://people. stern.nyu.edu/adamodar/ click on "Current Data" – "Risk Premiums for Other Markets"
D/E	Damodaran	"D/E Ratio" of the "Real Estate (Development)" (Industry name)
		Download the Regional datasets from http://people.stern.nyu.edu/adamodar/ (see β)
Eurirs	European Banking Federation	Average of the reference interbank rate recorded backwards from the time of the estimate and with a maturity (years) comparable to the analysis period.
		https://www.ebf.eu (data are published in the main newspapers/magazines specialized in the financial sector (e.g. in Italy: ilsole24ore) or can be inferred from other sources/bank institutes etc.)
Spread	Damodaran	Last percentage of "Adj. Default Spread" assigned to the country concerned
		http://people.stern.nyu.edu/adamodar/New_Home_Page/datafile/ctryprem.html
γ	Italian Tax Agency	DCSEOMI Methodology (see par. 2.4, Table 3)
		https://www.agenziaentrate.gov.it/portale/web/guest/schede/fabbricatiterreni/omi/pubblicazioni/ quaderni-osservatorio
Factor	*Location*	*Dimensional Factors*	*Building Equipment*	*Property Size*	*Presence of Competitors*	*Total*
Level	L_1 (from 1 to 5)	L_2 (from 1 to 3)	L_3 (from 1 to 3)	L_4 (from 1 to 4)	L_5 (from 1 to 3)	$\Sigma(L)_i$

Table 6. Data comparison and DCF input parameter selection (^A-administration,^P-promoter)

Scenario	Unit Direct Cost (UD^A, UD^P)	Unit Market Value (UMV^A, UMV^P)	Cost of Equity (Ke^A, Ke^P)	Borrowing Cost (Kd^A, Kd^P)	Ratio Leverage (D/E^A, D/E^P)
Post UV	MIN	MAX	MIN	MIN	MAX
Ante UV	MAX	MIN	MAX	MAX	MIN

Step 3 - Implementation of Calculation Scheme. By entering the input data in the calculation scheme (DCF), the PA determines the Ft of each scenario as the algebraic sum of costs and revenues (MV-ΣCi)t for each year (t) of the analysis period (n) and discount them by the factor 1/(1 + DR)t based on their time allocation within that duration. The difference (NPV*post*-NPV*ante*) returns the ECG potential of the initiative and the ECU ($\geq$ 50%ECG) as "reference value" - obtained by applying the procedure - that already allows to assess the adequacy of the ECU offered by the Promoter. This "reference value" will then be confirmed or not as a result of the following two steps.

The procedure also aims to limit the number and discretion of the choices necessary for its application but it is clear that the DR and all the hypotheses, concerning the amount of Ft and the transformation timing, have a strong impact on the amount of NPV and ECU.

Indeed, the subsequent sensitivity analysis (*What-if-analisys*) concerns the impact and variation of critical inputs on the ECU and in particular the DR/UMV pair (step 4) and the duration *n* (step 5). This is intended to structure a framework of risk-price-duration variability to validate the reference value that, within this framework, can guarantee a fair distribution between public and private costs and benefits.

Step 4 - What if Analisys: ECUas DR and UMV Changes. The ECU variation is determined in relation to a range of DR and UMV variation (Table 7) by processing 4 double input matrices having the same structure (Table 8): the first two contain, for each scenario, the multiple NPVs corresponding to each DR/UMV pair which vary – starting from the reference values - within the ranges and according to the percentage unit of increase/decrease (Δ) indicated in Table 7.

Table 7. DR and UMV variation range (UMV-related ICs will also vary)

Factor	Min	Max	Range (max-min)	Range discretization	Δ%
DRi	DR (γ_{min}) **	DR (γ_{max}) **	δDR = + DR (γ = 0,45%) – DR (γ = 22,94%)	2x = (max – min)/ΔR	ΔR = ± 0,50%
UMVj	−20% UMV*	+20% UMV*	δMV = |40% UMV|	2y = (max – min)/ΔV	ΔV = ± 2,50%

Note. if DRP $\neq$ DRA *the %*ΔR *will be an appropriate (sub)multiple of* \|DRP-DRA\| *and will vary within a narrower common range:* δ'DR = DRA(γ-*min*)-DRP(γ-*max*) *if* DRA > DRP *(and vice versa) with* 2x' = δ'DR/ΔR

(*) equal to twice the commonly accepted alea estimative; (**) see Sect. 2.3.

Table 8. ECU values (or similarly ECG, NPVs) as DR-UMV changes

Scenario POST	DR$_{P-MIN}$	DR$_P$-pΔR%	DR$_P$	DR$_P$+pΔR%	DR$_{P-MAX}$	
0,80*UMV$_P$	ECU(min,min)	(...)	ECU(med,min)	(...)	ECU(max,min)	0,80*UMV$_A$
(1-uΔV)*UMV$_P$	(...)	(...)	(...)	(...)	(...)	(1-uΔV)*UMV$_A$
UMV$_P$	ECU(min,med)	(...)	**ECU(med,med)**	(...)	ECU(max,med)	UMV$_A$
(1+uΔV)*UMV$_P$	(...)	(...)	(...)	(...)	(...)	(1+uΔV)*UMV$_A$
1,20*UMV$_P$	ECU(min,max)	(...)	ECU(med,max)	(...)	ECU(max,max)	1,20*UMV$_A$
	DR$_{A-MIN}$	DR$_A$-aΔR%	**DR$_A$**	DR$_A$+aΔR%	DR$_{A-MAX}$	**Scenario ANTE**

Note. u *varies between 1 and y*; a *and* p *vary between: 1 and x, if* DR$_P$=DR$_A$, *1 and x' if* DR$_P$≠DR$_A$.

The 3rd matrix contains the multiple ECGijs obtained by subtracting from each NPV of the *post*-UV scenario the analogous NPV of the *ante*-UV scenario; the 4th matrix contains therefore the corresponding ECUijs, calculated in percentage on the ECG.

If the two scenarios present a different level of risk (DR$_P$ ≠ DR$_A$) and/or market (UMV$_P$ ≠ UMV$_A$), the difference between the multiple NPVPs contained in the two variation tables/matrices (NPV$_P$-NPV$_A$) will occur between the values corresponding to the pairs (DRi, UMVj) that have the same gap, namely |UMV$_P$-UMV$_A$| e/o |DRi$_P$-DRi$_A$|.

Step 5 - What if Analisys: ECU as the Time (*n*) Changes. The elaborations of step 4 are replicated as *n* is increased, by moving (delaying) the start of the operation up to - at least - 3 years (m). A reasonable discretization of the ranges established for DR and UMV can facilitate the computational phase and a better clarity of presentation.

Step 6 - End or Reiteration. The procedure ends if the "reference value" of the ECU (step 3) is confirmed as a balance between risks and opportunities of Promoter and PA in the overall frame emerging from step 4–5; otherwise it needs to be repeated from step 1 (by varying the data).

The greater awareness of the PA about the possible options and the main aspects of the investment (e.g. timing) may correspond to a greater confidence of the investor which, in any case, may be based on a high level of sharing choices

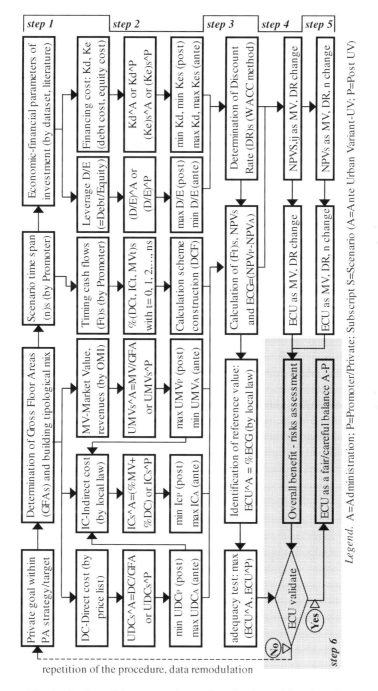

Fig. 1. Outline of the proposed procedure for determining the ECU

3.2   Utilities of the Procedure

The procedure described allows a relatively simple use also for PA (with a possible predefined excel sheet) and a more detailed-accurate determination of the ECU than the simplified methods used by Italian municipalities, including the critical factors in the quantification of TV (and the ECU itself). It can be a potential analysis and evaluation tool which, besides the verification of the data proposed by the Promoter, offers the PA various utilities in the different phases of a partnership agreement (Table 9). In times of economic and financial instability or crisis, this procedure can introduce elements of flexibility that make it possible to extend the amounts due over time [12] and to include new updated information, monitoring the variables (to be pre-established) that have an impact on the profitability of the real estate investment (e.g. prices, time) and simulating alternative intervention scenarios to reduce uncertainty about its possible outcomes.

Table 9. Utility and support of the proposed procedure regarding the transformation process

Phase	Utilities and PA support
Preliminary (pre-contractual)	Aware choice of parameters (UMV etc.) and/or valid references to: i) activate the renegotiation of the agreement, also in relation to any pre-defined value thresholds (e.g. % lower revenues, longer authorization process); ii) renegotiate the contents of the agreement
Negotiation	To pursue the most advantageous balance by repeating the procedure as the inputs vary - GFA, UMV, typological mix (e.g. presence of social housing) etc. - and public "guarantees" regarding one or more components of real estate risk (and consequent financial benefits).
Contractual (implementation)	To monitor factors of economic-financial equilibrium and facilitate the contractual revision by objective/measurable variation of pre-established parameters (price level, difficulty in accessing credit, etc.) within a framework of less legal/factual uncertainty
Closure	To find the contractual terms/conditions - parameterised in advance, from pre-established sources - also to establishing the postponement and/or the amount of any final rate of the ECU
Check (final)	To verify the compliance of the completed intervention with the agreed conditions (e.g. price/revenue of goods produced/sold, price/construction cost, also with the aid of BIM technology

4   Conclusions and Perspectives

The law that introduced the ECU in Italy is in line with the advanced international legislation on the subject, but it has been ignored and weakened (2019): in practice, a 'fair' distribution between public and private of the surplus values emerging from urban transformations is far from being achieved; the current distribution rewards the private, land income or real estate finance to the detriment of public investment in the city [6].

The paper illustrates an experimental procedure conformed on the Italian case to quantify the ECU owed by the private actor to the PA, on the basis of the respective economic and financial advantages, and can be extended, also in international contexts, where applicable: contractors' contributions for urbanization and territorial infrastructure; urban requalification of lands or buildings (with or without change of use) or valorisation of agricultural areas; programmes/projects (also in PPP) that presuppose an *ante/post* transformative events of the *status quo*.

Compared to the methods used in practice and in the literature, the procedure: (i) duly considers time and risk factors of the initiative, allowing the DR of the investment to be easily estimated; (ii) 'measures' the level of uncertainty inherent in the final result through sensitivity analysis versus *MV, DR, n* as performance indicators - or their combinations - that can be easily interpreted by the parties involved. This also in order to strengthen: the awareness of operators and PA about the most critical project drivers; fairness and transparency of public action towards investors and the community.

The socio-economic context becomes increasingly uncertain and complex, making investments more difficult and risky; therefore the use of the proposed procedure may support a different culture of "governing by contract" [31] that goes beyond the traditional antinomy of the parties and is based on the transparency of public-private relations, on effective and loyal cooperation, and on the awareness of the real mutual convenience for the success of the agreement established.

The potential of the proposed procedure can be further tested with regard to: i) more types of interventions, related to buildings or areas (including services); ii) different phases of the transformation process, ex *ante*/ex *post*; iii) various combinations of subjects (promoter/builder/owner/PA); iv) quick check of more complex analyses. It could be usefull to use: methodologies that also take into account the correlations between the critical factors (e.g. Montecarlo); parameters that make the sensitive aspects of the procedure objectifiable (e.g. random real estate market coefficient) [24] or that allow the inclusion of the inflation and/or revaluation/evaluation parameter of real estate values; further and better sources or reliable datasets to easily quantify the economic and financial parameters, including the ICs of the intervention.

References

1. ANCE, Il contributo straordinario per maggior valore nelle regole di regioni e comuni. http://www.anceaies.it/wp-content/uploads/2016/03/Dossier-Ance-sulcontributo-straordinario-per-maggior-valore-Allegato-contributostraordinario__marzo2016.pdf. Accessed 05 Apr 2020
2. Boyce, B.N.: Real Estate Appraisal Terminology. Bullinger Publishing Company, Cambridge (1975)
3. Cacciamani, C. (ed.): Real Estate, 3rd edn. Egea, Milan (2012)

4. Calabrò, F., Della, S.L.: The public-private partnerships in buildings regeneration: a model appraisal of the benefits and for land value capture. Adv. Mater. Res. **931–932**, 555–559 (2014)

5. Camagni, R.: Il finanziamento della città pubblica. In: Baioni, M. (ed.) La costruzione della città pubblica. Alinea, Firenze (2008)

6. Camagni, R.: Verso una riforma della governance territoriale: area vasta e controllo della rendita fondiaria e immobiliare (2011). https://storicamente.org/quadterr2/camagni.htm. Accessed 05 May 2020

7. Camagni, R.: Risorse per la qualità urbana. In: Cecchini, D., Castelli, G. (eds.) Scenari, risorse, metodi e realizzazioni per città sostenibili, pp. 39–52, Gangemi, Roma (2012)

8. Camagni, R.: La riforma della fiscalità urbanistica. In: Petretto, A., Lattarulo, P. (ed.) Contributi sulla riforma dell'imposizione locale in Italia, Carocci, Roma (2016)

9. Camagni, R.: Il mistero del "contributo straordinario (2019). https://www.arcipelagomilano. org/archives/51773. Accessed 05 Apr 2020

10. Caravella, G., Lisi, G., Pizzirani, F.: Costruzione di un modello di valutazione del rischio immobiliare: fondamenti teorici e prime evidenze empiriche, Quaderni dell'Osservatorio, Direzione Centrale Osservatorio del Mercato Immobiliare e Servizi Estimativi (DCSEOMI) - Agenzia delle Entrate (2019)

11. Copiello, S.: Progetti urbani in partenariato. Studi di fattibilità e piano economico finanziario. Alinea, Firenze (2011)

12. Crupi, F.: Rigenerazione urbana e città pubblica. Il Contributo Straordinario di Urbanizzazione nelle leggi regionali. In: XVI edn Urbanpromo, "Città Contemporanea Gigante dai Piedi d'Argilla, Torino (2019)

13. Crupi, F.: Fiscalità urbana e governo del territorio. Tra integrazione delle riforme e complementarietà degli strumenti. In: Talia, M. (a cura di) Il bisogno di giustizia nella città che cambia. Planum Publisher, Roma-Milano (2018)

14. Curti, F.: Le condizioni di sostenibilità del welfare urbano. In: Curti, F. (ed.) Lo scambio leale, Officina edizioni, Roma (2006)

15. Direzione Centrale Osservatorio del Mercato Immobiliare e Servizi Estimativi - Agenzia del Territorio, Manuale Operativo delle Stime Immobiliari, FrancoAngeli, Milano (2011)

16. Del Giudice, V.: Estimo e valutazione economica dei progetti. Loffredo, Napoli (2015)

17. Dotzour, M.G., Grissom, T.V., Liu, C.H., Pearson, T.: Highest and best use: the evolving paradigm. J. Real Estate Res. **5**(1), 17–32 (1990)

18. European Commission, Green Paper on public-private partnerships and Community law on public contracts and concessions, COM 327 (2004)

19. Florio, M.: La valutazione degli investimenti pubblici. I progetti di sviluppo nell'Unione Europea e nell'esperienza internazionale, vol. 1, 2nd edn., Franco Angeli, Milano (2003)

20. Hoesli, M., Morri, G.: Investimento immobiliare. Hoepli, Milano (2010)

21. International Valuation Standards Council – IVSC, International Valuation Standards, London (2013)

22. Lanzavecchia, A.: L'analisi finanziaria degli investimenti immobiliari. In: Cacciamani, C. (ed.) Real Estate. 3rd edn., pp. 331–352, Egea, Milan (2012)

23. Manganelli, B.: A proposal for a synopsis in real estate appraisal between the Italian doctrine and international valuation standards. Valori e valutazioni **18**, 9–16 (2017)

24. Manganelli, B., Morano, P., Tajani, F.: Risk assessment in estimating the capitalization rate. WSEAS Trans. Bus. Econ. **11**, 199–208 (2014)

25. Micelli, E.: Partenariato e Urbanistica. Il cofinanziamento dei servizi e infrastrutture per mezzo degli strumenti di piano. In: Codecasa, G. (a cura di) Governare il Partenariato pubblico e privato nei progetti urbani, Politecnica Maggioli, città (2010)

26. Modigliani, D., Camagni, R., Dongarrà, A., Fonti, R., Messina, G., Tamburini, M.: Definizioni e determinanti della rendita urbana: metodologia di calcolo e sue applicazioni alla Città Metropolitana di Roma. INU Edizioni, Roma (2016)
27. Rubeo, F.: Partenariato pubblico-privato per lo sviluppo urbano, 3rd edn. Politecnica Maggioli, Santarcangelo di Romagna (RN) (2012)
28. Sharpe, W.F.: Capital asset prices: a theory of market equilibrium under conditions of risk. J. Financ. **19**, 425–442 (1964)
29. Tajani, F., Morano, P., Di Liddo, F., Locurcio, M.: An innovative interpretation of the DCFA evaluation criteria in the public-private partnership for the enhancement of the public property assets. Smart Innov. Syst. Technol. **100**, 305–313 (2019)
30. Togni, R.M., Ferrante, G.: Clausole di benefit sharing: equità ed efficienza nell'affidamento dei contratti di PPP. In: Ricchi, M., Cartei, G.F. (ed.) Finanza di progetto e partenariato pubblico-privato 2015. Editoriale scientifica, Napoli (2015)
31. Valaguzza, S.: Collaborare nell'interesse pubblico. Editoriale Scientifica, Napoli (2019)
32. Vecchi, V.: Le analisi economico-finanziarie per la valutazione e gestione di progetti di PPP. In: Ricchi, M., Cartei, G.F. (a cura di) Finanza di progetto e partenariato pubblico-privato 2015. Editoriale scientifica, Napoli (2015)

The Effects of Urban Transformation Projects on the Real Estate Market: A Case Study in Bari (Italy)

Pierluigi Morano[1], Francesco Tajani[2], Felicia Di Liddo[1(✉)], Carmelo Maria Torre[1], and Marco Locurcio[1]

[1] Department of Science of Civil Engineering and Architecture, Polytechnic University of Bari, 70126 Bari, Italy
felicia.diliddo@poliba.it
[2] Department of Architecture and Design, "Sapienza" University of Rome, 00196 Rome, Italy

Abstract. In the present research the effects of urban transformation projects on the housing prices have been analyzed. With reference to a redevelopment initiative in the peripheral area of the city of Bari (Southern Italy), the study has been carried out on a sample of two hundred residential properties, sold in the period 2017–2019 and located in the peripheral district. The main influencing factors considered by sellers and buyers in the negotiation phase have been collected. The application of a data-driven technique has allowed to identify a statistically reliable model through which the functional relationships between the variables considered and the selling prices in the current state have been detected, and the new market values generated by the redevelopment project have been then determined, in order to compare them with those relating to the current state. The results obtained point out an increase in the values of the sample properties as a result of the redevelopment project, confirming the empirically expected outputs.

Keywords: Urban redevelopment · Property enhancement · Urban quality · Housing market · Residential selling prices

1 Introduction

The urban transformation processes play an important role in the context of public policies aimed at promoting the renewal of cities and satisfying the community needs. The issue of recovery and functional reconversion of degraded urban areas and/or public properties has been widely dealt with in the reference literature and in the urban planning practice [1, 4, 5, 11, 13, 14, 16, 22].

The property enhancement initiatives, intended as actions for the renovation of the existing property assets [8], on the one hand, and for the redevelopment of abandoned public spaces in terms of urban regeneration, determine environmental, economic and social externalities that are reflected on the qualitative levels of the urban system [2, 6, 20]. In this framework, the assessment of the effects deriving from the implementation of an urban transformation project in quantitative terms aims at *i)* weighing

© Springer Nature Switzerland AG 2020
O. Gervasi et al. (Eds.): ICCSA 2020, LNCS 12251, pp. 1000–1015, 2020.
https://doi.org/10.1007/978-3-030-58808-3_71

the quality level of urban spaces, *ii)* supporting the decision-making processes in the territorial redevelopment initiatives and *iii)* verifying their effectiveness. Currently, in fact, most urban systems have wide degraded areas. It's about the degraded housing, insufficient or inadequate facilities, ineffective public transport, large abandoned former industrial areas, environmental risks and problems, unattractive and disconnected urban services, widespread unemployment and relevant social problems, such as poverty, low levels of education, aging populations, etc. The decentralization process of the production system from central to more peripheral areas is the main cause of the "emptying" and consequent abandonment of wide urban areas. The recovery initiatives of these spaces allow both to reuse entire portions of cities no longer used for their original functions, and to start wider regeneration and reorganization processes of the urban system. Currently, the public policies of urban territory government, which have no longer been engaged in the strategies for expanding borders, focus on the planning of effective redevelopment initiatives in degraded and/or abandoned areas of cities, in order to improve the level of urban quality, in terms of "appropriate" equipment and infrastructure, adequate to the effective demand of the population who lives in the requalification territories [19]. So far, these areas have been considered places of social unease maximum concentration. Currently, instead, the degraded areas are considered a resource for public and private operators: for the public subject, in fact, they represent an opportunity for a growth of the city image and to demonstrate the skills and efficiency of the administration and to attract capital and private investors; for the private subject, on the other hand, a degraded urban area represents the opportunity for financial earnings deriving from the enhancement of the existing property asset and/or transformation of urban spaces.

The recovery of the underused urban areas represents an important stimulus for a new economic development of the city, not only for quantity of surface but also for physical and environmental qualities, due to the often strategically relevant locations that these areas occupy for the organization and the proper functioning of the entire urban system.

2 Aim

The present research concerns the framework outlined. The aim regards the analysis of the effects that urban transformation projects - property enhancement or public spaces redevelopment initiatives - produce on the real estate market in terms of variation in the housing prices.

In particular, the study intends to define a model for the *ex ante* assessment of the variation in the housing prices produced by an urban transformation project. The evaluation tool will be able to constitute a procedure to support the decision-making processes in the redevelopment of property assets or the re-functionalization of public spaces and it can be used on an urban scale by the several subjects involved in the decision phases of urban planning. Furthermore, on a micro-urban scale, the model can be focus on a single initiative or different projects. The model proposed will address, on an urban scale, to Public Administrations or, in the case of cooperation between the public and private sectors (Public Private Partnership - PPP), to private investors, with the aim of defining the most suitable place, among the various identified, where to concentrate the available financial resources.

The analysis is applied to the city of Bari (Southern Italy), with reference to an urban redevelopment project currently in progress.

The paper is structured as follows. In Sect. 3 the case study, related to the initiative for the regeneration of the San Girolamo waterfront located in the peripheral area of the city of Bari (Southern Italy), has been described and the variables considered in the model have been identified. In Sect. 4, the methodology implemented has been explained and the main criteria used to assess the reliability of the returned models have been specified. In Sect. 5, the implementation of the method to the case study related to the current state has been illustrated and the results obtained in terms of specific statistic performances and empirical reliability of the functional relationships returned by the technique have been presented. In Sect. 6, with reference to the situation post intervention, some considerations on the results obtained have been discussed. Finally in Sect. 7 the conclusions of the work have been reported.

3   Case Study

3.1   Current Use

The case study concerns the urban redevelopment of the San Girolamo waterfront located in the peripheral area of the city of Bari (Southern Italy). Figure 1 shows the San Girolamo waterfront localization in the urban context of the city.

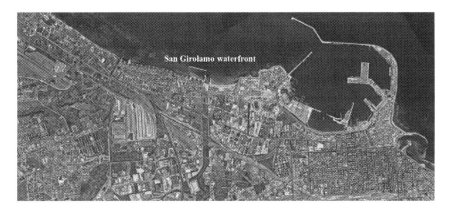

Fig. 1. San Girolamo waterfont localization (broken red line) in the urban context of the city of Bari (Color figure online)

The San Girolamo waterfront, in the homonymous district north-west of the city of Bari, extends for over a kilometer between Lama Lamasinata and Lama Balice.

With reference to the neighborhood and to the area overlooking the coast, there is a relevant demand for public services not satisfied, above all due to the lack of collective spaces such as "places for socializing". In particular, currently the commercial intended use located on the buildings ground floors along the coastline are often unused and the promenade is considered exclusively as a fast crossing on the edge rather than as a service axis of the neighborhood. Finally, especially in the west area, on the border with Lama Balice, there are ruins and old and unused buildings characterized by widespread environmental degradation. In Fig. 2 San Girolamo waterfront current state is shown.

Fig. 2. San Girolamo waterfront current state

3.2   The Urban Redevelopment Project

The urban redevelopment project considered provides for the recovery and the functional transformation of the San Girolamo waterfront (Fig. 3).

The intervention aims to determinate a new landscape and environmental configuration overlooking the sea and new ways to use the spaces, by involving the economic and social sectors of the neighborhood in consistency with the sustainability principles and taking into account the environmental conditions.

The project, currently in progress, intends to give a new image of the San Girolamo district, through the redevelopment of the relationship between the urban asset and the sea, by creating new urban spaces and using the natural element of water to enhance the urban landscape.

Furthermore, the redevelopment initiative will encourage economic and social activities for the purpose of a wider socio-economic regeneration of the neighborhood, also through the enhancement of the current public areas characterized by a high degradation state.

The redevelopment of the waterfront provides for the pedestrianization of the road axis with the creation of an "square on the sea" of 8,000 square meters on two levels with about 600 seats facing the sea, and the introduction of new urban functions currently not sufficiently enhanced and equipped such as: service activities, places for leisure, sport, swimming, socialization.

The project proposal aims at introducing implications on a territorial scale, locating on the sea a structure able to attract, due to its exceptional nature, also scientific and tourist interests.

In addition, the project intends to act at the landscape level, in order to protect the seafront from the physical aggression of the water, which progressively causes the coast erosion, through protective elements physically and biologically compatible with the environment and using local natural materials (limestone boulders).

With reference to the coastline, some operations are planned to create new spaces for bathing - in particular four new beaches.

Among the objectives of the project, the rationalization of the mobility system in the neighborhood should be highlighted, through the proposal for a more efficient roadway with a prevailing pedestrianization and cycling on the seafront and with the provision of a new urban water transport which connect the urban areas of Torre a Mare, Palese and Santo Spirito and the cruise terminal of the city of Bari.

Finally, the urban redevelopment project involves the introduction of a structure - the Aquarium - which will constitute a strong architectural sign towards the sea, integrated with the nautical activities and over the water.

The new San Girolamo waterfront will transform the appearance of this part of the coast, becoming a relevant example of urban redevelopment between the city and the sea.

Fig. 3. San Girolamo waterfront redevelopment project

3.3   Variables of the Model

With reference to the peripheral urban area of the city of Bari, in which the project considered in the present research is located, the study sample consists of two hundred residential properties sold in the two-year period 2017–2019 (in particular from the second half of 2017 to the first half of 2019).

For each property the total selling price, expressed in euro (P - dependent variable), and the factors most influential on the residential prices (independent variables) have been collected. As confirmed by the real estate agents operating in the specific housing market, the factors identified represent the characteristics considered by buyers and sellers of residential properties in the negotiation phase. Thus, several real estate operators located in the peripheral urban area of the city of Bari have been consulted.

A list of different intrinsic (relating to the property) and extrinsic (relating to the urban context) factors has been shown to them and it has been asked to indicate the most requested ones by potential buyers and/or to add any missing variables.

In particular, the independent variables considered in the model have been as follows:

1. the total surface of the property (S), expressed in square meters of gross floor area;
2. the number of bathrooms in the building (B);
3. the presence of the lift in the building where the property is located (A);
4. the quality of the maintenance conditions of the property (Sc), taken as a qualitative variable and differentiated, through a synthetic evaluation, by the scores 1, 3 and 5, respectively corresponding to the categories "to be restructured", "fit for habitation" and "restructured";
5. the maintenance conditions of the public spaces adjacent to the property (Sa), assessed through a scale of scores (1, 3, 5) attributed by panels of experts (sociologists, landscape architects, etc.), where the score "1" indicates a bad maintenance conditions of the public spaces, the score "3" a good state and the score "5" an excellent state;
6. the road private and public vehicular traffic (buses) level of the building area (T), assessed by a team of experts (sociologists, landscapers, architects, engineers, etc.) through a scale of scores defined as follows: score 1 indicates a road characterized by high traffic intensity, score 3 indicates a medium traffic intensity, score 5 indicates a road characterized by low traffic congestion;
7. the distance of the property from the nearest railway station (Ds), calculated in kilometers it takes to walk to it;
8. the perceived quality of public space level in the area in which the property is located (Qp), assessed on the basis of affirmative or negative items, to which a sufficiently representative sample of users assigns a numerical score that ranges from "1" (disagreement with the verbal expression) to "5" (agreement with the verbal expression). This perceived quality assessment technique has been borrowed from the studies carried out by Fornara, Bonaiuto and Bonnes [3, 9] for the evaluation of the perceived residential urban quality. In particular, the items considered are the following:

– you can meet bad people,
– people often behave uncivilly,
– late in the evening there is the risk of dangerous encounters,
– the streets are safe enough;

9. the perceived environmental quality level of the area in which the property is located (Qn), assessed through numerical judgments on the basis of the scores scale from "1" (disagreement with the verbal expression) to "5" (agreement with the verbal expression), assigned by a sample of users sufficiently representative of the urban area. The items considered are:

– this neighborhood is generally not polluted,
– this is a quiet neighborhood,
– residents' health is threatened by pollution,
– he heavy traffic in this neighborhood is very annoying,
– there are green areas for relaxing,
– going to a park means travelling to other parts of the city,
– the green areas are well-equipped;

10. the perceived urbanistic quality level of the area in which the property is located (Qu). This variable is determined through six items, to which a sample of individuals assigns a numerical judgment on the basis of their subjective perception of the quality level. The selected items are:

– this neighborhood is too cut-off from the rest of the city,
– his neighborhood is well-connected with important parts of the city,
– the city-center can be easily reached from this neighborhood,
– in the neighborhood there are enough green areas,
– it is easy to cycle around,
– going into this neighborhood means going round in circles;

11. the perceived social quality level of the area in which the property is located (Qs), assessed by means of an articulated system of items. In particular, a subjective judgment is assigned by a sample of individuals, expressed in numerical terms from "1" (total dissent from the verbal expression) to "5" (total consent). The items considered are:

– this neighborhood is well-served with stores,
– there are all kinds of stores,
– stores do not provide a wide range of products,
– stores are not well-distributed,
– stores selling the most needed products can be easily reached,
– this neighborhood is well-equipped with sports grounds,
– various sports can be played,
– sports grounds are insufficient,
– this neighborhood is not well-equipped to host cultural events,
– there are often cultural events,
– libraries are adequate for residents' needs,
– this neighborhood has good school facilities,
– schools can be easily reached on foot,
– schools are located in bad-quality buildings,
– children and teenagers are forced to move from this neighborhood to go to school,
– social services are inadequate,
– the local health service is satisfactory,
– the local health service is inadequate.

It should be highlighted that the variables Qp, Qn, Qu, Qs concern the perceived urban quality, assessed through subjective judgments of an interview sample. In particular, these factors have been included among those most influential on the residential prices, as they consider the opinion of the population and/or frequent and occasional users of the urban area and they transform the expressed opinions in quantitative terms. Thus, these variables take into account the community perceptions of the urban quality into the selling price formation.

4   The Method

The methodological approach applied in the present research is the Evolutionary Polynomial Regression (EPR), which integrates the best features of numerical regression with genetic programming [10]. EPR is a hybrid data-driven technique that uses a multi-objective Genetic Algorithm in order *i)* to combine numerical and symbolic regression methods using polynomial structures, *ii)* to search those model expressions that simultaneously maximize the accuracy of the data and the parsimony of the final mathematical functions.

Set the dependent variable *(Y)* and the independent variables *(Xi)*, defined the parameters useful to return the function form able to define the functional relationship $Y = f(Xi)$, the generic expression of the non-linear model implemented in EPR is summarized by Eq. (1):

$$Y = a_0 + \sum_{i=1}^{n} \left[a_i \cdot (X_1)^{(i,1)} \cdot \ldots \cdot (X_j)^{(i,j)} \cdot f((X_1)^{(i,j+1)} \cdot \ldots \cdot (X_j)^{(i,2j)}) \right] \qquad (1)$$

where a_0 is an optional bias, n is the number of additive terms, the length of the polynomial expression (bias excluded), a_i represents numeric parameters to be identified, X_i are the explanatory variables candidate to be selected by the model, (i, l) - with $l = (1,\ldots, 2j)$ - is the exponent of the *l-th* input variable within the *i-th* term, f is a function chosen by the user among a set of possible mathematical expressions. The exponents (i, l) are also selected by the user in a range of possible real numbers. The parameters a_i are evaluated by the Least Squares Method.

The EPR implementation involves the selection and generation of a series of different models whose functional form is the best combination of the input variables *Xi*, identifying for each one the exponents *(i, l)* and the numerical coefficients a_i.

The EPR main advantage is that the genetic algorithm underlying the procedure does not require the exogenous definition of the mathematical expression and of the number of parameters that fit better the data collected, but the iterative process itself returns the best solution [21]. Thus, the EPR overcomes the classical multiple regression method, as it selects only the "good" solutions and rejects the "bad" ones in order to obtain the best performance of the final results. Moreover, EPR applies an evolutionary multi-objective genetic algorithm as an optimization strategy based on the Pareto dominance criterion aimed at *i)* maximizing the model accuracy through the satisfaction of appropriate statistical criteria for the verification of the equation; *ii)* maximizing the model's parsimony through the minimization of the number of terms (a_i); *iii)* reducing the complexity of the model through the minimization of the number of the explanatory variables (Xi) of the final equation.

The key idea of the EPR method concerns the search of the best functional form of the price function in which each term is a combination of the independent variables with a numerical multiplier coefficient and a proper exponent.

The accuracy of each algebraic expression generated by the EPR technique implementation is checked by the Coefficient of Determination (CoD), defined in Eq. (2):

$$COD = 1 - \frac{N-1}{N} \cdot \frac{\sum_N (y_{detected} - y_{estimated})^2}{\sum_N (y_{detected} - mean(y_{detected}))^2} \quad (2)$$

where $y_{estimated}$ are the values of the dependent variable estimated by the methodology, $y_{detected}$ are the collected values of the dependent variable, N is the sample size in analysis. The CoD value varies between 0% and 100%. The closer the CoD value is to 100%, the higher the statistical performance of the model returned by EPR. The technique EPR returns a set of mathematical expressions characterized by a different level of statistical accuracy and a different complexity level of the algebraic structure. The analysis of the compromise solutions between the statistical performance and the complexity of the expression allows to select the most suitable models according to the specific application. With reference to the real estate sector, so far EPR has been generally used to determine the price function $Y = f(x_1; x_2; ... x_n)$ in order to identify the explanatory factors most influential in the mechanisms for the formation of the housing prices in different territorial contexts and to analyze the marginal contribution of each of them on the prices [15, 17, 18].

5   Application of the Method of the Case Study (*Ante-project* Situation)

With reference to the current state (*ante-project* situation), the EPR method has been implemented considering the structure of the generic model identified in Eq. (1) without function *f* selected and with the dependent variable Y as the natural logarithm of the total selling price ($Y = ln(P)$) [7, 12]. Each additive monomial term of the mathematical expression is assumed as a combination of the inputs - the explanatory variables X_i - raised to the proper numerical exponents. In particular, in order to have a wide range of models, the candidate exponents belong to the set (0; 0.5; 1; 2) and the maximum number n of additive terms in final expressions is assumed to be eight.

The implementation of the econometric method has generated several solutions, each one characterized by a different number of additive terms, combinations of the variables and a different level of CoD. With reference to the peripheral urban area of the city of Bari where the study sample properties considered in the research are located, the model defined by Eq. (3) has been selected as able to explain the mechanism of formation of selling prices in the current situation in the specific urban area.

This model is characterized by a high CoD value (+79.48%) and it considers all the explanatory variables considered in the analysis.

$$
\begin{aligned}
Y = &+ 1.69 \cdot Ds^{0.5} \cdot Qu^{0.5} \cdot Qn^{0.5} + 0.35 \cdot Sc + 0.65 \cdot A^{0.5} \cdot Ds^{0.5} \cdot Qp^{0.5} \\
&+ 5.66 \cdot B^{0.5} \cdot Sa^{0.5} \cdot T^{0.5} Qn^2 \cdot Qs^2 + 6.34 \cdot S^{0.5} - 2.76 \cdot S + 8.32
\end{aligned}
$$
$$(3)$$

The mathematical expression of the model of Eq. (3) does not allow to immediately verify the empirical coherence of the coefficients signs of the explanatory variables selected, being the same variable present in more terms and/or combined within the same term with other factors. In the present research an empirical approach has been used, in order to verify the empirical consistency of the functional relationships returned by the EPR implementation on the study sample and to define the marginal contribution of each factor selected by the model. In particular, the procedure adopted constitutes a simplified exogenous approach which, instead of determining the partial derivative of the dependent variable with respect to the i-th variable, provides for the variation of the i-th variable analyzed in the variation interval in the observed sample, keeping the mathematical terms of the other variables are constant - i.e. equal to the average value for the quantitative variables and 1 for the other dummy variables. Figure 4 shows the functional relationships obtained for the study sample considered.

6 Determination of the Housing Prices After the Redevelopment Initiative (*Post-project* Situation)

The proposed model for the assessment of the effects of the urban redevelopment initiative considered in the city of Bari on the housing prices provides for this mandatory assumption: the function that links the selling prices and the intrinsic and extrinsic factors that contribute to their formation after the initiative (*post-project* situation) remains the model obtained by the implementation of the algorithm EPR in

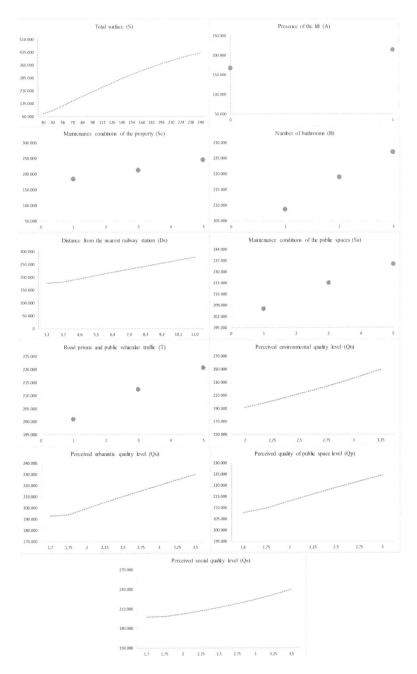

Fig. 4. Functional relationship between the explanatory variables selected by the model and the housing prices in the current state (*ante-project* situation)

order to identify the functional relationships between prices and explanatory variables in the current state (*ante-project* situation), i.e. the model of Eq. (3).

It is assumed that in the *post-project* situation the structural dynamics of the real estate market and, in particular, of the residential sector are the same of the current state. Therefore, the most influencing factors for the sellers and the buyers in the processes of selling prices do not change compared to those considered in the *ante-project* situation. The values of the extrinsic variables for which a variation is expected have been replaced in the model, using the same operating methods implemented for the definition of the same variables in the current state. With reference to the intrinsic and extrinsic variables that are not modified in the *post-project* situation, the values of these characteristics previously determined and/or calculated do not vary. Therefore, it is possible to determine the new selling prices of the sample properties, assessed as a result of the project realization. The results obtained in terms of comparison between the estimated *ante-project* prices and the prices assessed after the implementation of the redevelopment initiative, confirm the outputs empirically expected in terms of growth in the market values of the properties located in the area of the redevelopment. In particular, the average price increase assessed is approximately equal to +32%.

The propagation effect of the redevelopment initiative on the housing prices highlights a direct functional link between the proximity of the property to the requalified waterfront and the positive variation of the expected values. In fact, it should be noted that the residential properties of the study sample that overlook the new waterfront are those most involved.

With reference to the nearest properties to the urban redevelopment project, significant variations of selling prices have been assessed, also equal to +140%. Therefore, this confirms the strong relevance that the San Girolamo waterfront initiative could be on the urban context in which it is located, in terms of urban quality level improvement. In fact, the project solution is aimed to rehabilitate the waterfront portion and the neighborhood by overcoming the public services lack, the current collective spaces degradation state and by contrasting the social exclusion.

The new public urban space realization will allow the promotion of economic and social activities able to trigger off wider urban development processes, among which the increase in value of existing property asset represents an important factor of the more general economic recovery.

With reference to the case study considered in the present research, in Fig. 5 a flow chart shows the main steps implemented in order to analyze the effects of San Girolamo waterfront redevelopment project on the housing prices in the peripheral area of the city of Bari.

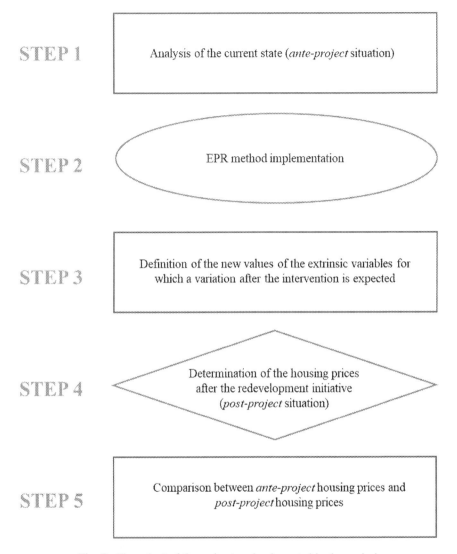

Fig. 5. Flow chart of the main steps implemented in the analysis

7   Conclusions

The processes of urban transformation - property enhancement or public spaces redevelopment initiatives - constitute operations of public spaces recovery on different scale, modifying the physical and functional structure of the urban system and producing several effects on the surrounding context and, in a gradually lower form, on the city.

With reference to the current need to support the public administrations and the private investors through valid evaluation tools in order to guide urban planning processes towards effective choices in the medium and long term, in the present research a model for the *ex ante* assessment of the variation in the housing prices that an urban transformation project generates, has been developed. Referring to an urban redevelopment project currently in progress in the peripheral area of the city of Bari (Southern Italy), a study sample of two-hundred residential properties has been collected. The methodology EPR has been implemented in order to identify the main factors that are influent in the phenomena of housing price formation in the current state (*ante-project* situation). In order to analyze the effects of San Girolamo waterfront redevelopment project on the housing prices, the new values of the extrinsic variables selected for which a variation is expected as result of the redevelopment initiative, have been assessed. Using the model generated by the implementation of the EPR technique related to the current state, for each property the new prices have been determined and they have been compared with those assessed in the *ante-project* situation.

The analysis has been limited to the geographic context of the peripheral area of the city of Bari in which the project is located.

Further developments may be carried out to study the "complementary" and additional effects of San Girolamo waterfront redevelopment project on the housing prices in the other city areas and, more generally, on the entire urban system of the city of Bari. The new waterfront could constitute a landmark not only for the peripheral San Girolamo district, but also for the entire city of Bari.

The present work is part of a research of high and current interest. In particular, the proposed model aimed at assessing the effects of an urban transformation initiative on the real estate market will be able to support the decision-making processes and to integrate the traditional financial and economic tools currently implemented for the feasibility analysis of the projects. Finally, further insights may address *i)* the application of the proposed procedure in other national and international territorial contexts and the experimentation of the same technique for the study of the complementary effects of different contemporary transformation projects, *ii)* the comparison of the models obtained by the EPR implementation with the output generated by other techniques (e.g. Artificial Neural Networks, Cellular Automata, spatial analysis, ecc.).

Aknowledgements. The authors declare that the contents of the paper are based on data and information developed within the "Innonets" Project founded by the Interreg Greece-Italy Program and are aimed at promoting the Project itself. For more information see http://interreginnonets.eu/en.

References

1. Adair, A., Berry, J., McGreal, S., Deddis, B., Hirst, S.: The financing of urban regeneration. Land Use Policy **17**(2), 147–156 (2000)
2. Agnoletti, C., Bocci, C.: Gli effetti economici e distributivi degli interventi di riqualificazione urbana. XVII Congresso nazionale Associazione Italiana di Valutazione – AIV -, Napoli, 10–11 aprile 2014 (2014)

3. Bonaiuto, M., Fornara, F., Bonnes, M.: Indexes of perceived residential environment quality and neighbourhood attachment in urban environments: a confirmation study on the city of Rome. Landscape Urban Plann. **65**(1–2), 41–52 (2003)
4. Bottero, M., D'Alpaos, C., Oppio, A.: Decision-making for urban planning and regional development. In: Advances in Operations Research 2019 (2019)
5. Bottero, M., D'Alpaos, C., Oppio, A.: Ranking of adaptive reuse strategies for abandoned industrial heritage in vulnerable contexts: a multiple criteria decision aiding approach. Sustainability **11**(3), 785 (2019)
6. Calabrò, F., Della Spina, L.: The cultural and environmental resources for sustainable development of rural areas in economically disadvantaged contexts - economic-appraisals issues of a model of management for the valorization of public assets. Adv. Mater. Res. **869**, 43–48 (2013)
7. Cassel, E., Mendelsohn, R.: The choice of functional forms for hedonic price equations: comment. J. Urban Econ. **18**(2), 135–142 (1985)
8. Della Spina, L., Calabrò, F.: Decision support model for conservation, reuse and valorization of the historic cultural heritage. In: Gervasi, O., et al. (eds.) ICCSA 2018. LNCS, vol. 10962, pp. 3–17. Springer, Cham (2018). https://doi.org/10.1007/978-3-319-95168-3_1
9. Fornara, F., Bonaiuto, M., Bonnes, M.: Indicatori di qualità urbana residenziale percepita (IQURP). Manuale d'uso di scale psicometriche per scopi di ricerca e applicativi. Milano: Franco Angeli (2010)
10. Giustolisi, O., Savic, D.: Advances in data-driven analyses and modelling using EPR-MOGA. J. Hydroinformatics **11**(3–4), 225–236 (2009)
11. Itard, L., Meijer, F., Vrins, E., Hoiting, H.: Building renovation and modernisation in Europe: state of the art review. Final Report Erabuild, Delft. 31 (2008)
12. Lynch, A.K., Rasmussen, D.W.: Proximity, neighbourhood and the efficacy of exclusion. Urban Stud. **41**(2), 285–298 (2004)
13. Mangialardo, A., Micelli, E.: Reconstruction or reuse? How real estate values and planning choices impact urban redevelopment. Sustainability **12**(10), 4060 (2020)
14. Micelli, E., Mangialardo, A.: Recycling the city new perspective on the real-estate market and construction industry. In: Bisello, A., Vettorato, D., Stephens, R., Elisei, P. (eds.) SSPCR 2015. GET, pp. 115–125. Springer, Cham (2017). https://doi.org/10.1007/978-3-319-44899-2_8
15. Morano, P., Guarini, M.R., Tajani, F., Di Liddo, F., Anelli, D.: Incidence of different types of urban green spaces on property prices. a case study in the flaminio district of Rome (Italy). In: Misra, S., et al. (eds.) ICCSA 2019. LNCS, vol. 11622, pp. 23–34. Springer, Cham (2019). https://doi.org/10.1007/978-3-030-24305-0_3
16. Morano, P., Locurcio, M., Tajani, F., Guarini, M.R.: Fuzzy logic and coherence control in multi-criteria evaluation of urban redevelopment projects. Int. J. Bus. Intell. Data Min. 14, **10**(1), 73–93 (2015)
17. Morano, P., Rosato, P., Tajani, F., Di Liddo, F.: An analysis of the energy efficiency impacts on the residential property prices in the city of Bari (Italy). In: Mondini, G., Oppio, A., Stanghellini, S., Bottero, M., Abastante, F. (eds.) Values and Functions for Future Cities. GET, pp. 73–88. Springer, Cham (2020). https://doi.org/10.1007/978-3-030-23786-8_5
18. Morano, P., Tajani, F., Guarini, M.R., Di Liddo, F., Anelli, D.: A multivariate econometric analysis for the forecasting of the interdependences between the housing prices and the socio-economic factors in the city of Barcelona (Spain). In: Misra, S., et al. (eds.) ICCSA 2019. LNCS, vol. 11622, pp. 13–22. Springer, Cham (2019). https://doi.org/10.1007/978-3-030-24305-0_2
19. Ombuen, S., Ricci, M., Segnalini, O.: I programmi complessi. Il Sole 24 ore, Milano (2000)

20. Rosato, P., Breil, M., Giupponi, C., Berto, R.: Assessing the impact of urban improvement on housing values: a hedonic pricing and multi-attribute analysis model for the historic centre of Venice. Buildings **7**(4), 112 (2017)
21. Tajani, F., Morano, P., Torre, C.M., Di Liddo, F.: An analysis of the influence of property tax on housing prices in the Apulia region (Italy). Buildings **7**(3), 67 (2017)
22. Wanner, P., Tricarico, L.: La carta AUDIS della Rigenerazione Urbana e il dibattito europeo sulla qualità urbana. Documento d'inquadramento e di verifica delle basi culturali della Carta e della Matrice della Qualità Urbana AUDIS, Venezia (2012). http://www.audis.it

Circular Enhancement of the Cultural Heritage: An Adaptive Reuse Strategy for Ercolano Heritagescape

Maria Cerreta[(✉)] and Valentina Savino

Department of Architecture, University of Naples Federico II, via Toledo 402,
Naples 80134, Italy
maria.cerreta@unina.it, valentina.savino93@gmail.com

Abstract. The circular economy paradigm identifies the need for rational use and adequate reuse of all resources, including cultural heritage. This study explores the opportunity to apply a circular economy model to culture-led regeneration processes oriented to cultural heritage valorisation. The methodological process identified tries to structure an integrated approach, able to combine the tools of building renovation and that of multidimensional evaluation to define a circular enhancement strategy for cultural heritage. Starting from the local cultural values and the changing uses of urban spaces related to the case study of the municipality of Ercolano (Italy), the structured decision-making process analyse how to optimise tangible/intangible cultural resources for local and sustainable development; to generate values and activate the engagement of communities through new sustainable uses; to build widespread and capillary complex networks among people, values, and spaces. In this interdisciplinary approach, the role of cultural and creative industries (CCI) is ever more significant, with the ability to create communication between management models, sustainability assessments, deliberative approaches and conservation strategies, regeneration and enhancement of cultural heritage.

Keywords: Circular economy · Culture-led regeneration · Multi-criteria analysis · PROMETHEE-Gaia method

1 Introduction

With the rapid growth of the developing world, the demand for natural resources continues to develop exponentially in the coming decades. As a result, there is an increase in environmental and climate impacts. In this context, the diffusion of a new circular model of production and consumption constitutes an element of strategic importance for achieving the Sustainable Development Goals (SDGs) purposes [1]. It represents, at the same time, a factor in relaunching the country competitiveness. The meaning of "waste" is no longer the traditional one to which we are used concerning production processes, but *waste* can arise from the human transformations implemented in some areas. At the same time, the cultural abandoned heritage can be conceived like a *waste* and can be triggered in a different reconversion process according to a particular value chain [2, 3]. Cultural heritage is, in its different expressions, an integral

© Springer Nature Switzerland AG 2020
O. Gervasi et al. (Eds.): ICCSA 2020, LNCS 12251, pp. 1016–1033, 2020.
https://doi.org/10.1007/978-3-030-58808-3_72

part of the cultural landscape, increasingly expressed through the identification of new values, not only linked to the aesthetic-naturalistic aspect, but also to the conception of the landscape itself as a cultural and historical symbol [4]. In the scientific literature, Mary-Catherine Garden had employed the term "heritagescape" to reference "landscapes of heritage" [5], heritage sites that are understood to exist within wider physical and social expanses. This neologism thus is both a descriptor and a methodology for parsing the living processes that circulate at these historic places: «In thinking of heritage sites as heritagescapes - i.e. as landscapes - it draws attention to their qualities as dynamic, changing spaces. It also offers the opportunity to locate sites in the context of their larger environment and draws attention to the importance of the setting. Accepting the heritage site as a landscape locates these places in their rightful place as a fluid, changing space with which people regularly interact» [5]. Heritagescapes are about the visible, physical place and also about the experiences that people have with that place [6].

Cultural heritage can be seen as a driver for vital and active transformations, stimulating new perspectives for the local community development, using the resources of the environment and nature, regulating human and social relationships [7, 8]. Culture and creativity are essential to promote the innovation capacity of local stakeholders (citizens and public employees, public and private actors, profit and nonprofit organisations, etc.) [9]. They are strictly connected to the concept of resilience [10], because the innovative potential is re-defined as "recovery capacity", and considered as the ability to absorb, adapt, transform and prepare for past and future shocks and stresses to ensure sustainable development, well-being and inclusive growth [11]. This interpretation of the resilience concept is related to a selection of criteria provided by the OECD. The choice could be applied to perceive the potential for innovation within cities: robustness, redundancy, reflexivity, flexibility, resourcefulness, inclusiveness, integration [12]. The relationship between resilience and circularity has been highlighted by numerous authors [13], that identified circularity as an archetype of sustainable business models [14]. Indeed, circularity can be identified as one of several options for promoting the sustainability of a system. These options are seen as beneficial in principle and can also be combined to add earnings or achieve synergies [15], describe circular strategies as an option to increase efficiency or dematerialisation [16], and to generate new values. Value-focused thinking, according to Keeney's theories [17], helps to distinguish between alternatives and values, stating that alternatives are the means to achieve the most fundamental values, while values are the principles used for evaluation. Thinking about values allows for discovering hidden goals [18]. In the framework proposed by Keeney, the strategic objectives, the fundamental objectives and the decision-making context are defined. According to this suggestion, a hierarchy of primary objectives for the circular economy model can be developed by "expanding" the purposes to the three dimensions of sustainability, social, economic and environmental. The integration of the approaches, circularity and sustainability, can be useful in defining a cognitive framework for determining a strategy of urban regeneration of cultural heritage (material and immaterial), able to ensure an integrated decision-making process (Fig. 1). The first level is *Increase the sustainable creation of value*, consisting of the triple-bottom-line subdivision on the second hierarchy level and afterwards unfolding down to level 5. The sorting in this hierarchy is unambiguous, meaning that each objective has only one relation going out to the next hierarchy level, and there are no relations within a hierarchy level. The increase of the economic performance of the system means to *Increase the decoupling of economic growth and the use of scarce resources as third-level objectives.* The second-level goal to *Increase the ecologic performance of the system* means to *Increase regeneration/valorisation of products, parts and material.* The second-level objective to *Increase human welfare* consists of *an increase in the number of jobs* and the *increase* of *the social standard.* In this perspective, the circular economy [19] can represent a tool for achieving and implementing sustainable development, minimising negative impacts, and producing benefits, related to environmental, economic, and socio-cultural systems [20].

The overall outcome of the paper highlighted that through procedural rationality and control systems based on the multi-level strategies could be tested and calibrated in terms of long-term sustainability and resilience.

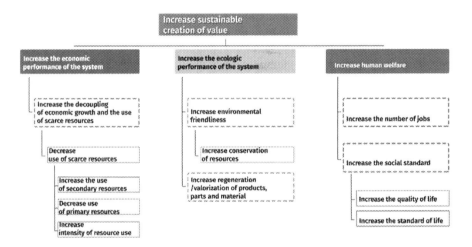

Fig. 1. Fundamental objectives for the circular economy aimed at cultural heritage

To analyse the interaction between cultural heritage enhancement and circular economy, the first part of the paper (Sect. 2) shows the purpose of research and the methodological approach; the second one (Sect. 3) describes the case study for which the methodology has been elaborated; the third one (Sect. 4) analyses the results, and the last part (Sect. 5) concerns with discussion and conclusions about the issues afforded.

2   Culture, Economies, and Creative Processes for Urban Regeneration

Cultural and creative industries [21, 22], with their multidisciplinary nature, can be seen as enabling contexts for the enhancement of cultural heritage. The Creation of Shared Value (CVC) is directly functional to the competitive position and ability to obtain company profits. Optimising and using specific resources and skills, the CVC builds economic value through the creation of social value [23], generates employment opportunities as well as continuous innovation and considers an advanced form of shared responsibility. The latest, from Corporate Social Responsibility (CSR) and Community Social Responsibility (RSC), is configured as Social Responsibility of the Territory (RTD) [24], capable of generating complex social value [7]. Although the role of cultural heritage in sustainable development has been recognised in the international debate (SDGs), it is explicitly mentioned only once in the Goal 11 (Target 11.4) *Strengthen efforts to protect and safeguard the world's cultural and natural heritage* and, to date, only one indicator has been identified about it, as the result of a series of public consultations with agencies and organisations. The selected indicator of Target 11.4.1 aiming at illustrating how financial efforts/actions made by public authorities, at different levels (local, national, and international levels), alone or in partnership (for example, with civil society organisations and private sector), to protect and safeguard cultural heritage, have a direct impact in making cities and human settlements more sustainable. The circular economy approach becomes relevant for cultural heritage as a driver for urban regeneration, increasing the heritage life cycle, creating new value and promoting local development. The Italian Atlas of Circular Economy indicates dimensions, criteria and indicators deemed to be fundamental in the evaluation of the production processes analysed and in which the practices of valorisation of the abandoned or "discarded" cultural heritage play a significant role, implementing the enabling factors recognised by the European Environmental Agency [25]: eco-design, recovery, renewal and regeneration; recycle; economic incentives and financing, business models, eco-innovation, governance, skills and knowledge. Cultural capital is the driver of a regeneration process where the transversal interconnections between the production cycles of the regeneration are related to the management phase and configure a circular process of multidimensional production of value. Indeed, the circular economy model is based on the processes of reuse, recycling, restoration, regeneration of resources, to avoid any form of waste and underutilisation [26]. This is the general perspective in which to place the proposal for reuse and restoration, generating new use and not use values [27]. Reuse is defined as the set of construction and/or reuse interventions of a building and a building system designed to meet new conditions and new requirements [28]. Over time, it has gradually been enriched with arguments relating to cultural, socio-economic and ecological objectives which underline its potential as a driver of more comprehensive strategies for regenerating the context, reducing land consumption and contrasting urban expansion processes. The decision-making process plays a key role and characterises the reuse initiatives [29], also, in the light of an arena that is enriched with new actors and new systems of relationships, new reflections urging on the need for multi-level governance, capable of supporting these initiatives with a view to territorial regeneration and cultural and social innovation. For the success of the culture-led processes [30], the integration of top-down and bottom-up approaches is necessary, which makes dialogue, mutual exchange and cooperation possible. Thus, a "hybrid" approach emerges [29], which seeks to consider the complexity of a culture-led development process by facing it and managing it through interdisciplinary tools.

3   The Methodological Approach

The methodological process is outlined within a place-based approach [31, 32], to include a plurality of local actors in the decision-making process and building a sustainable and circular development strategy, that recognises the central role of social capital and tangible and intangible cultural capital of the territory. This approach underlies a systemic interpretation of the territory, aimed at a non-punctual but widespread enhancement, which identifies as an essential prerequisite the strengthening of the relationships between the physical components and the cultural, social and economic components to trigger new development trajectories. The first part, therefore, proposes a data collection approach (Fig. 2), which connects culture, economies and reactive processes for urban regeneration. The second part explains an ex-post evaluation of some creative experiences, applying the multi-criteria method PROMETHEE-GAIA to analyse the results from different perspectives and find a balanced decision system that considers the role of each criterion. The third section concerns the elaboration of a meta-strategy through the re-elaboration of potentials and critical aspects of the different strategies. The meta-strategy has been elaborated for the Ercolano case study. Subsequently, through the ex-ante evaluation, it has been possible to move from the methodology identification to its application to the case study.

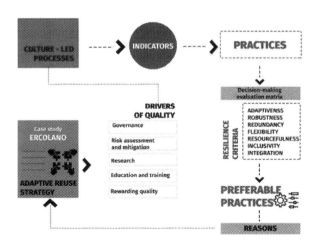

Fig. 2. The methodological workflow

The ex-post evaluation is interpreted as an objective and systematic assessment of an ongoing or completed project, practice, programme or policy, its design, implementation, and results [33]. This approach is comprehensive, and it relates to many types of assessment, from socio-economic evaluations to business-value evaluations, and from holistic approaches to performance measurement evaluations [34]; some examples include: ex-post recalculations of ex-ante cost-benefit analyses; assessments based on the principles of corporate finance; multi-criteria evaluations as a combination of quantitative/qualitative approaches. On the other hand, usually, the ex-ante evaluation is applied before the ex-post one, as the ex-ante evaluation is a broad initial assessment aimed at identifying which alternative will yield the most significant benefit from an intended investment. Indeed, in an ex-ante evaluation, much must be based on assumptions because fewer facts are available [35]. For this reason, concerning the Ercolano case study, the ex-post evaluation has been used before, as its goal is first and most importantly, to assess the lessons learned in an undertaking. The motivation for using ex-post assessment is principal that it contributes to double-loop learning. Finally, in this case, to strengthen the methodological process, the two approaches have been merged, as the information deriving from the ex-post evaluation reinforces the structure of the ex-ante evaluation defined for Ercolano.

4 From an Ex-Post Evaluation to an Ex-Ante Evaluation for the Heritagescape of Ercolano

4.1 Learning from the Best Practices

Taking into account the considerations made so far, an ex-post evaluation of five creative practices of regeneration and adaptive reuse process throughout Italy has been implemented in order to identify some relevant criteria, able to put some theoretical concepts into practice and making them explicit. An inventory of the different types of approaches that could be used in similar contexts concerning all the issues addressed was structured, analysing the critical aspects and the potentials related to the decision-making process, actors involved, economic and human resources and results. The five practices were chosen considering three examples from Southern Italy (two in Naples and one in Puglia), an example from Central Italy, and an example from Northern Italy: the FOQUS "Spanish Quarters Foundation" in Naples; the Catacombs of San Gennaro in Naples; the Ex Fadda, Urban Laboratory in San Vito dei Normanni (Br); the Oz "Officine ZERO" in Rome; the Basic Project - Ex Ansaldo in Milan.

The selection of the above practices was based on the following criteria deemed significant:

− *Characteristics of the context*: the initiatives took place in degraded contexts, with great potential linked to cultural heritage and relations between the inhabitants, thanks to the presence of associations, widespread talents and deep-rooted traditions.
− *Component of social and cultural innovation:* in the context of the selected initiatives, essential elements of cultural innovation must be considered: artistic

sensitivity, managerial know-how, social responsibility, while for social innovation: promotion of actions based on principles of cooperation, practical sustainability, civic entrepreneurship, assumption of responsibility, knowledge and training.
– *Circularity and sustainability component:* practices are focused on the reuse and use of all "waste resources" in the area, resulting in a "drag and drop" effect that favours the evolution and development of the social and cultural capital of the community.

From the analysis and evaluation of good practices and by using criteria and indicators selected from the scientific literature, an evaluation matrix has been elaborated (Table 1). It considers the seven resilience criteria reported by the OECD [10] as criteria to assess the sustainability of the processes (applying the previous considerations on the circular economy and sustainable development). Based on their definitions, the indicators, divided by typologies, have been inserted for each criterion, considering the cultural dimension of sustainable development, also taking into account the circularity objectives (Fig. 1) [18], linking cultural activities with social, economic and cultural impacts. These indicators concern cultural inclusion, resources, skills and cultural policies.

Table 1. Evaluation matrix

Criteria	Characteristics	Categories of indicators	Indicator	UM
Robustness	Robustness depends on a system which is well-designed, built and managed to absorb the impact of a shock and continue to operate	Type of partnership	Private	Yes/No
			Social	Yes/No
			Public	Yes/No
			Public-social	Yes/No
			Private-social	Yes/No
			Public-private	Yes/No
		Type of management industry	Social industry	Yes/No
			Cultural and creative industry	Yes/No
			Hybrid industry	Yes/No
Redundancy	An urban system able to meet the need for spare capacity when faced with unexpected demand, a disruptive event or extreme pressure	Finance research methods	Private financing	Yes/No
			Public financing	Yes/No
			Self-financing	Yes/No
		Cost of the investment	Construction costs	Euro
		Financial intermediaries	Agreements made with attentive parties	Number

(*continued*)

Table 1. (*continued*)

Criteria	Characteristics	Categories of indicators	Indicator	UM
Adaptiveness	An adaptive urban system manages uncertainty by evolving, modifying standards, norms or past behaviour, using evidence to identify solutions and applying the knowledge gained from experience when making decisions about the future	Tables for monitoring changes	Tables for monitoring changes	Number
		Development oriented policies	New enterprises	Number
			Private enterprises involved	Number
			Public enterprises involved	Number
			Social enterprises involved	Number
			New jobs	Number
		Capacity building	Training courses for economic and social development	Number/Year
			Participants in training courses	Number/Year
			New highly specialised jobs	Number
Flexibility	A flexible urban system allows individuals, households, businesses, communities and government to adjust behavior or actions in order to rapidly respond to change	Recovery of abandoned building heritage	Building units recovered	Number
			Percentage of spaces used	%
			Surface of buildings recovered	Square meters
		Recovery of the environment	Green surface recovered	Square meters
			Meeting places integrated into the green system	Number
		Tourist offer	Receptive exercises	Number

(*continued*)

Table 1. (*continued*)

Criteria	Characteristics	Categories of indicators	Indicator	UM
Resourcefulness	A resourceful urban system can effectively and quickly restore the functionality of essential services and systems in a crisis or under highly constrained conditions	Initiatives enhancing the cultural heritage	Occasional events	Number/Year
			Training courses	Number/Year
			People involved	Number/Day
		Communication strategies	Social media management	Number
			Social accounts	Number
Inclusivity	An inclusive urban system ensures that diverse actors and communities are fully consulted, engaged and empowered in the policy process	Type of decision-making process	Top-down	Yes/No
			Bottom-up	Yes/No
			Hybrid process	Yes/No
		Local wealth through social inclusion	Volunteers involved	Number/Year
			Disadvantaged people involved	Number/Year
Integration	An integrated urban system promotes a co-operative and a collaborative approach to policymaking and programming that transcends sectoral and administrative boundaries	Differentiation of the actors	Private actors	Number
			Social actor	Number
			Public actor	Number

Data collection was obtained through the interview of referents of each practice and from the sources of the web (websites, newspaper articles, etc.). Once the matrix has been completed for each practice, the evaluation has been carried out through a multi-criteria analysis [36], implementing the Preference Ranking Organisation METHOD for Enriched Evaluation (PROMETHEE-Gaia) method [37]. The PROMETHEE-GAIA method is based on the computation of pair-wise preference degrees (scored between 0 and 1), which rank all the alternatives from best to worst for the decision-maker. The pair-wise comparisons of the alternatives are based on three preference flows for consolidating the results: Phi+ (f+): the positive flow; Phi− (f−): the negative flow; Phi (f): the net flow. The PROMETHEE-GAIA points out how the decision-maker perceives the difference between the objective evaluations (often measured) on every criterion. The preference degree is computed for each criterion, rescaling or enriching the assessment of the actions through preference information.

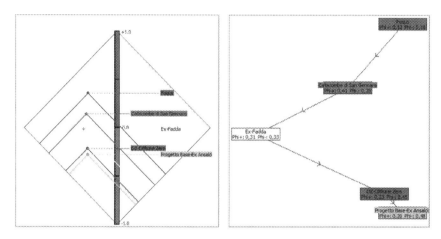

Fig. 3. Evaluation of alternatives (a) PROMETHEE Diamond (b) PROMETHEE Network (source: authors' elaboration of PROMETHEE results)

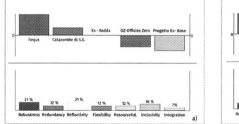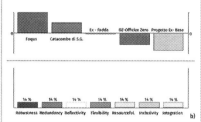

Fig. 4. Sensitivity analysis – Walking weights for: (a) Practices' criteria weights

By the implementation of the PROMETHEE-GAIA, the selected practices were first analysed overall to have a general ranking and, then, the individual profile of each practice. The PROMETHEE Diamond (Fig. 3), a two-dimensional representation of two classifications (partial and complete), where every alternative is represented as a point in the plane (Phi+, Phi−) angled at 45°. In the PROMETHEE Network, actions are instead symbolised by nodes and arrows drawn from emerging preferences.

The results of the two analysis show that FOQUS and Catacombs of San Gennaro are preferable, while Ex-Fadda occupies a position in the middle of the ranking. The final ranking was verified through the sensitivity analysis too, by Walking weights tool (Fig. 4), and underlines that FOQUS is characterised by a preferable strategy for a culture-led regeneration process. The study of the general classification data suggests that FOQUS has a preferred strategy for a culture-led regeneration, in particular from *reflexivity*, *inclusivity* and *integration*. The example of the Catacombs of San Gennaro is better for *flexibility*, *inclusivity* and *integration*. The preferable characteristics of Ex Fadda are *resourcefulness*, *redundancy* and *inclusivity*. Officine Zero compared to other practices, have not found good results; their strategy has focused on enhancing *flexibility* more. The preferable characteristics of the profile of the Ex-Ansaldo BASE Project are *resourcefulness* and *robustness*. In light of the results obtained, we can determine the meta-strategy.

4.2   A Culture-Led Regeneration Process for Ercolano

The area of interest, Ercolano, lies at the western foot of Mount Vesuvius, on the Bay of Naples, just southeast of the city of Naples. Figure 5 frames the focus area, which extends over the coast of the Tyrrhenian Sea.

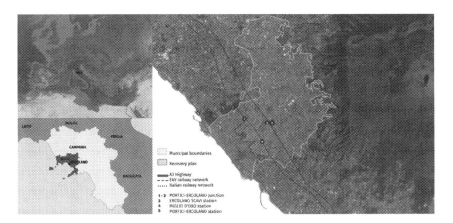

Fig. 5. The area of Ercolano

Ercolano is well known for one of the most important archaeological sites in Italy (Archeological Site of Herculaneum), for the National Park of Mt. Vesuvius created in 1995, for the Miglio d'Oro, the Corso Resina (the old Strada Regia per le Calabrie) and the peculiar street market of Pugliano. However, most resources are not fully enhanced, and others are abandoned. The inhabitants are 53,843, and population density is approximately 2,700/kmqs. The unemployment rate is 27.3%, while the young unemployment rate is 60.6%. During this phase, the methodological process consists of soft and hard data collection to know the settlement system. The *meta-strategy* scheme (Fig. 6) is divided into strategic actions, and punctual actions declined in *who, how, and what* and *duration* of the action. This scheme applied to the case study has resulted in an adaptive reuse strategy for Ercolano, taking into account what emerged from the knowledge of the settlement system. In analysing the settlement system, we classified cultural heritage based on its potential for reuse. There is a type of *heritage* [9, 10] that is enhanced, and this category includes the Archaeological Park with the Excavations; however, underutilised compared to their potential, and other particularly significant resources used on rare occasions such as the Villas of Miglio d'Oro, mostly on events such as festivals. However, they are resources that, although not or inadequately valued, have their own precise identity, due to their cultural and symbolic value. Then a large number of historic buildings, public places, green areas are concretely abandoned, without intended use. The territory affected by our analysis is made up of many *waste-assets*, of a *heritage*, which for the most part, is abandoned or not used to its full potential.

WHAT: STRATEGIC ACTIONS	HOW: TO REACH THE ACTION	TYPE OF ACTION	WHO: ACTORS	DURATION
RECOVERY AND REUSE OF BUILDINGS DISPOSED OF IN THE AREA AS CATALYSTS OF THE PROCESS	Establishment and placement of a management company / entity operating in the cultural and creative sector on the hybrid model	Process actions / Governance		Long term
	Public-private-social partnership between the actors operating in the area in dialogue with the institutions	Process actions / Governance	ENTITIES AND ASSOCIATIONS THAT OPERATE IN THE AREA WITH THE INSTITUTIONS	Long term
	Establishment of permanent tables for the evaluation and monitoring of changes	Process actions / Governance	INDUSTRY EXPERTS	Long term
	Establishment of financial agreements	Process actions / Governance	FINANCIAL BETWEEN PARTIES LOOK TO PROMOTE EMPLOYMENT IN AREAS OF DEVELOPMENT OF SUSTAINABILITY / SOCIAL RESPONSIBILITY	Long term
	Adjustment of the project site	Material actions	INDUSTRY EXPERTS	Middle term
PROMOTION OF CAPACITY BUILDING AND CREATION OF SPECIALIZATION OPPORTUNITIES	Location of training courses	Material actions	INSTITUTIONS AND ASSOCIATIONS OPERATING IN THE AREA UNDER THE CULTURAL AND CREATIVE	Middle term
	Organization of training workshops for the creation of opportunities for economic and social development	Intangible actions	ENTITIES AND ASSOCIATIONS THAT OPERATE IN THE TERRITORY IN THE FIELD OF TRAINING	Middle term
ENHANCEMENT OF THE COMMUNITY AND CULTURAL HERITAGE	Organization of cultural events and manifestations	Intangible actions		Short term
	Activation and training courses (periodicals) for the promotion and enhancement of the community and cultural heritage	Intangible actions	INSTITUTIONS AND ASSOCIATIONS OPERATING ON THE TERRITORY IN THE FIELD OF EDUCATION	Middle term
INCLUSION OF DISADVANTAGED PERSONS	Organization of activities carried out periodically for social inclusion	Intangible actions		Short term
	Use of the share capital of Ercolano in the new established activities (agricultural, productive-textile, tourist)	Intangible actions	INSTITUTIONS AND ASSOCIATIONS OPERATING IN THE AREA IN THE SOCIAL FIELD	Long term
OPTIMIZATION OF RESOURCES IN THE AREA	Courses in the field of tourism training	Intangible actions	ORGANIZATION MANAGEMENT FORMAT BY ASSOCIATIONS IN THE AREA	Middle term
	Activation of communication campaigns	Intangible actions		Short term
RECOVERY OF GREEN AREAS	Organization of activities and events related to the recovery of the environment	Intangible actions		Short term
	Creation of meeting places	Material actions	INSTITUTIONS AND ASSOCIATIONS OPERATING IN THE FIELD OF SOCIAL AND CULTURAL	Long term
	Recovery of urban greenery through planting	Material actions		Long term

Fig. 6. Meta-strategy scheme

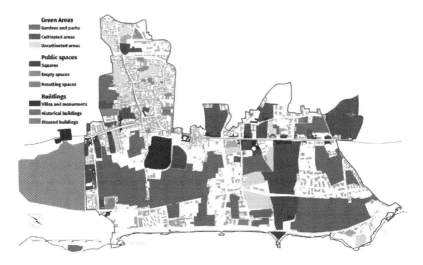

Fig. 7. Map of the settlement system resources

These resources have great potential to activate regeneration processes which, by putting into play the components analysed so far, can generate economic and non-economic values, improve commitment and involve communities through new sustainable uses by building complex capillary networks between people, values and spaces. The culture-led regeneration and adaptive reuse are linked to the fact that there is a significant amount of *abandoned landscape*, which can become drivers of vital and attractive transformations. This is classified in the map (Fig. 7) that shows the state of affairs in categories, such as green areas, buildings and public spaces.

The following map (Fig. 8) shows the classification of the different resources according to their potential as use and non-use allow us to identify opportunities for an adaptive reuse strategy and therefore able to develop new uses.

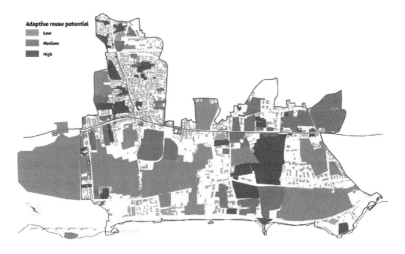

Fig. 8. Map of the potential for adaptive reuse

After the phase of the settlement system knowledge, through the collection of data, the detection of preferences, a fairly clear picture of the state of affairs is obtained. The adaptive reuse proposal for Ercolano derive from the information relating to the settlement system and the results obtained through the evaluation of the good practices analysed previously.

An incremental strategy has been developed in order to create an interpersonal network spread over the territory, reinforcing the urban fabric.

The main functions of this strategy start from a cultural hub (Palazzo Tarascone), where a managing entity is located, continuing to incorporate other resources, promoting their circularity and sustainability. Palazzo Tarascone is the ideal starting point of a cultural hub because it is situated in a strategic position concerning some points of great interest such as the Archaeological Park, the MAV, the Vesuvian Villas, the Reggia di Portici, as well as the Pezze di Pugliano market and other waste resources with great potential for reactivation. The strategic actions previously identified have been applied to the entire settlement system (Fig. 9), creating a network of actors and functions (Fig. 10).

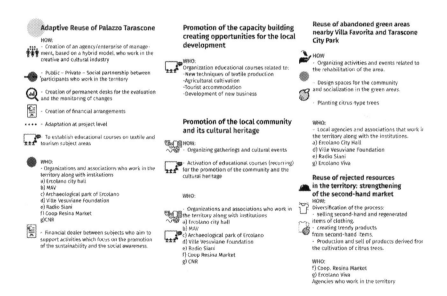

Fig. 9. Actions of the strategy: how and who

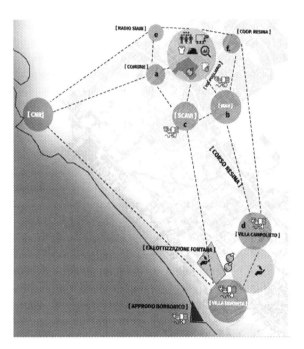

Fig. 10. The localisation of the strategy actions

5 Conclusions

The identification of a system of values, models and indicators has made it possible to select practices of culture-led urban regeneration where culture is conceived as a driver for circular and sustainable transformations. The elaboration of an adaptive reuse strategy for the city of Ercolano has been possible through the analysis and the ex post evaluation of selected practices. The knowledge of the settlement system and the detection of preferences have been proved to be essential elements to authentically describe the context and specifically decline the adaptive reuse strategy for Ercolano.

This methodological process enabled the use of an adaptive territorial strategy that connects the various resources and cultural heritage, developing in a comprehensive manner throughout the territory. A strategy that exploits the analysed components has been reconstructed backwards once the theoretical issues had been identified through the study of the literature and the analysis of the good practices of adaptive reuse. Through the practices, some typologies have been identified by making theoretical concepts explicit. Based on this, an inventory of the different types of approaches that could be used in similar contexts has been found. In fact, within this kind of "inventory", just the factors already tested have been kept to be eventually used in future strategies. This kind of adaptive reuse strategy keeps all the components analysed, and it is very useful for a process of regeneration of cultural heritage as sustainable as possible. So, it has been possible to apply this strategy to the case study Ercolano. In this kind of context, those *waste heritagescapes* have been identified to systematise all the issues of the settlement system. Therefore, through this strategy, it would be possible to enhance not only a degraded context as Ercolano, but also other sites in similar conditions to promote the economic, social and cultural development to the benefit of the entire community. The promotion and strengthening of cultural and creative clusters, through the spatial concentration of cultural and creative industries (CCI), could constitute an effective approach in support of policies for urban regeneration and the attraction of talent and intellectual capital, as well as forms of capacity building in historic cities and metropolitan areas. At the same time, there is a need to activate creative governance processes, intended as a tool capable of supporting not only metropolitan cities but also small towns. The interaction between research, experimentation and learning in the different levels of governance, allows combining strategic, tactical and operational approaches through a systematic and integrated process in the fields of institutional innovation, social cohesion, promotion of new local economies, by supporting the ability to self-organise the community and reducing the distances between people and institutions.

Acknowledgement. The work on the case study was undertaken during the internship carried out within the PRIN project (Projects of Relevance of National Interest) Metropolitan cities: economic-territorial strategies, financial constraints and circular regeneration at the Department of Architecture, University of Naples Federico II, February–July 2018, in collaboration with the Municipality of Ercolano. The methodological process it is part of the Master degree thesis in Architecture of Valentina Savino, Department of Architecture (DiARC), University of Naples Federico II, tutor prof. Maria Cerreta, co-tutor prof. Maria Rita Pinto and prof. Serena Viola.

References

1. United Nations (UN): Transforming our World: The 2030 Agenda for Sustainable Development. United Nations, New York (2015)
2. Fusco Girard, L., Baycan, T., Nijkamp, P. (eds.): Sustainable City and Creativity: Promoting Creative Urban Initiatives. Routledge, London (2012)
3. Fusco Girard, L., Nijkamp, P.: Le valutazioni per lo sviluppo sostenibile della città e del territorio. Franco Angeli, Milano (2012)
4. ICOMOS: Guidance on Heritage Impact Assessments for Cultural World Heritage Properties (2011). https://www.icomos.org/world_heritage/HIA_20110201.pdf. Accessed 03 Feb 2020
5. Garden, M.-C.: The heritagescape: looking at landscapes of the past. Int. J. Heritage Stud. **12**(5), 394–411 (2006)
6. Di Giovine, M.: The Heritage-Scape: UNESCO, World Heritage, and Tourism. Lexington Books, Lanham (2009)
7. Cerreta, M., Daldanise, G., Sposito, S.: Public spaces culture-led regeneration: monitoring complex values networks in action. Urbani izziv **29**, 9–28 (2018)
8. CHCfE: Cultural Heritage Counts for Europe (2015). http://blogs.encatc.org/culturalheritagecountsforeurope/outcomes/. Accessed 03 Feb 2020
9. KEA European Affairs: The economy of culture in Europe, European Commission Directorate-General for Education and Culture, Brussels (2016)
10. OECD: Guidelines for Resilience Systems Analysis. OECD Publishing, Paris (2014)
11. Sugahara, M., Bermont, L.: Energy and Resilient Cities, OECD Regional Development Working Papers, No. 2016/05, OECD Publishing, Paris (2016)

12. Concilio, G., Tosoni, I. (eds.): Innovation Capacity and the City. SAST. Springer, Cham (2019). https://doi.org/10.1007/978-3-030-00123-0
13. Fabbricatti, K., Biancamano, P.: Circular economy and resilience thinking for historic urban landscape regeneration: the case of Torre Annunziata, Naples. Sustainability 11(12), 3391 (2019). https://doi.org/10.3390/su11123391
14. Bocken, N.M.P., Short, S.W., Rana, P., Evans, S.: A literature and practice review to develop sustainable business model archetypes. J. Clean. Prod. 65, 42–56 (2014)
15. Weissbrod, I., Bocken, N.M.P.: Developing sustainable business experimentation capability. A case study. J. Cleaner Prod. 142, 2663–2676 (2017)
16. Geissdoerfer, M., Savaget, P., Evans, S.: The Cambridge business model innovation process. In: 14th Global Conference on Sustainable Manufacturing, GCSM, Stellenbosch, 3–5 October 2016 (2016)
17. Keeney, R.L.: Value-focused thinking: identifying decision opportunities and creating alternatives. Eur. J. Oper. Res. 92(3), 537–549 (1996)
18. Velte, C.J., Scheller, K., Steinhilpera, R.: Circular economy through objectives – development of a proceeding to understand and shape a circular economy using value-focused thinking. In: 25th CIRP Life Cycle Engineering (LCE) Conference, Copenhagen, Denmark, 30 April–2 May 2018, Procedia CIRP, vol. 69, pp. 775–780 (2018)
19. Preston, F.: A Global Redesign? Shaping the Circular Economy; Chatham House Briefing Paper, EERG BP 2012/02. Chatham House, London (2012)
20. Cerreta, M., Panaro, S., Poli, G.: A knowledge-based approach for the implementation of a SDSS in the Partenio Regional Park (Italy). In: Gervasi, O., et al. (eds.) ICCSA 2016. LNCS, vol. 9789, pp. 111–124. Springer, Cham (2016). https://doi.org/10.1007/978-3-319-42089-9_8
21. Micelli, E.: Modelli ibridi di partnership pubblico-privato nei progetti urbani. Scienze Regionali 8(2), 97–112 (2009)
22. Solima, L.: L'impresa culturale: processi e strumenti di gestione. Carocci, Roma (2005)
23. Porter, M.E., Kramer, M.R.: Creare valore condiviso. Harvard Bus. Rev. Italia 1(2), 68–84 (2011)
24. Del Baldo, M., Demartini, P.: Responsabilità sociale di territorio, network sinergici e governo locale. Piccola impresa/Small Bus. 3, 1–28 (2016)
25. EEA – European Environment Agency: Circular Economy in Europe. Developing the knowledge base. EU Low and Publications (2016)
26. Boulding, K.E.: The economics of the Coming Speeship Earth. In: Henry, J. (ed.) Environmental Quality Issues in a Growing Economy, pp. 3–14. John Hopkins Press, Baltimore (1966)
27. Cerreta, M., Daldanise, G.: Community branding (Co-Bra): a collaborative decision making process for urban regeneration. In: Gervasi, O., et al. (eds.) ICCSA 2017. LNCS, vol. 10406, pp. 730–746. Springer, Cham (2017). https://doi.org/10.1007/978-3-319-62398-6_52
28. Hasnain, H., Mohseni, F.: Creative ideation and adaptive reuse: a solution to sustainable urban heritage conservation. IOP Conference Series Earth and Environmental Science, vol. 126, 012075, pp 1–8 (2018). https://doi.org/10.1088/1755-1315/126/1/012075
29. Misirlisoy, D., Günçe, K.: Assessment of the adaptive reuse of castles as museums: case of Cyprus. Int. J. Sustain. Dev. Plann. 11(2), 147–159 (2016). https://doi.org/10.2495/SDP-V11-N2-147-159
30. Sacco, P.L., Pedrini, S.: Il distretto culturale: mito o opportunità. Il Risparmio 51(3), 101–155 (2003)
31. Huggins, R., Clifton, N.: Competitiveness, creativity and place-based development. Environ. Plann. A 43, 1341–1362 (2011)

32. Pugalis, L., Bentley, G.: (Re)appraising place-based economic development strategies. Local Econ. **29**, 273–282 (2014)
33. Samset, K.: Project Evaluation: Making Investments Succeed. Tapir Academic Press, Trondheim (2003)
34. Olsson, N.O.E., Krane, H.P., Rolstadås, A., Veiseth, M.: Influence of reference points in ex-post evaluations of rail infrastructure projects. Transp. Policy **17**(4), 251–258 (2010)
35. Samset, K., Christensen, T.: Ex ante project evaluation and the complexity of early decision-making. Public Organ. Rev. **17**(1), 1–17 (2015). https://doi.org/10.1007/s11115-015-0326-y
36. Torre, C.M., Attardi, R., Sannicandro, V.: Integrating financial analysis and decision theory for the evaluation of alternative reuse scenarios of historical buildings. In: Gervasi, O., et al. (eds.) ICCSA 2016. LNCS, vol. 9789, pp. 177–190. Springer, Cham (2016). https://doi.org/10.1007/978-3-319-42089-9_13
37. Brans, J.P., Mareschal, B.: The PROMETHEE methods for MCDM. The PROMCALC, GAIA and BANKADVISER software. In: Bana e Costa, C.A. (ed.) Readings in Multiple Criteria Decision Aid, pp. 216–252. Springer, Heidelberg (1990). https://doi.org/10.1007/978-3-642-75935-2_10

Author Index

Aleixo, Sandra M.   157
Altafini, Diego   291
Andrade, Marina A. P.   429, 446, 481
Anelli, Debora   913, 941, 968
Antunes, Ana Rita   247
Aragona, Francesco   776
Azevedo, Olga   415
Azevedo, Ramoni Z. S.   455

Balucani, Nadia   707, 717, 730, 744
Battisti, Fabrizio   869
Bennouar, Djamel   89
Bicho, Estela   415
Bleil de Souza, Clarice   324
Bogdanov, Alexander   58, 72
Bollini, Letizia   805
Bougherara, Maamar   89
Braga, Ana Cristina   185, 199, 231, 247
Braga, José   415

Calhamonas, Gabriel   446
Camarda, Domenico   340
Campo, Orazio   869
Caracciolo, Adriana   744
Carrasco, Alexandre   481
Carrasqueira, Pedro   552
Carriço, Nelson   429
Carvalho, Óscar   231
Casavecchia, Piergiorgio   744
Cazzolle, Monica   899
Ceccarelli, Cecilia   730
Cerreta, Maria   855, 952, 1016
Chioni, Chiara   324
Clua, Esteban W. G.   121
Collazos-Morales, Carlos   134
Costa, Lino   274
Costa, M. Fernanda P.   538, 606, 664
Cunha, Ângela   231
Cutini, Valerio   291, 324

D'Apuzzo, Mauro   367
da Silva Junior, Ayres Nishio   121

de Aragão, Emília Valença Ferreira   717
de Azevedo, Victor Ribeiro   103
De Lotto, Roberto   759
de Macedo Mourelle, Luiza   89, 103
De Mare, Gianluigi   836
De Paola, Pierfrancesco   884
De Ruggiero, Manuela   776
de Souza, Julio C. Stacchini   121
Degtyarev, Alexander   58, 72
Di Liddo, Felicia   1000
Di Pinto, Valerio   307
Dias, Joana   552, 564, 576
Díaz-Charris, Luis David   134
Dimas, Isabel Dórdio   214
Do Coutto Filho, Milton B.   121
Dolores, Luigi   836
Dostov, Victor   3

Endreny, Theodore   913
Erlhagen, Wolfram   415
Esposito, Dario   340
Evangelisti, Azzurra   367

Faginas-Lago, Noelia   707
Falcinelli, Stefano   693, 744
Faria, Susana   470
Fatkina, Anna   48
Fazia, Irene Della   805
Fernandes, Carlos   415
Fernandes, Edite M. G. P.   538, 606, 664
Fernandes, Paula O.   521
Fernandes, Sérgio   429
Ferreira, Ângela P.   261, 521
Ferreira, Brígida   552
Ferreira, Fátima   147
Ferreira, Flora   415
Ferreira, Rita   231
Ferrero, Stefano   683
Filipe, Vítor   620
Filomena Teodoro, M.   481
Flores, Paulo   231
Freitas, Adelaide   214

Gago, Miguel F.   415
Gaspar-Cunha, A.   664
Godinho, Pedro   576
González, Ramón E. R.   134
Grela, Karolina   787
Gremmelspacher, Jonas Manuel   352
Gruetzmacher, Sarah B.   261
Guarini, Maria Rosaria   913, 968, 981

Hanzl, Małgorzata   787
Hendrix, Eligius M. T.   590
Hiremath, Sandesh Athni   648
Hortencio, Hanna Pamplona   632

Iakushkin, Oleg   48

Jiménez-Cabas, Javier   134
Jiménez-Vilcherrez, Judith Keren   383

Kalay, Yehuda E.   340
Kemcha, Rebiha   89
Khvatov, Valery   58, 72
Kirillov, Denis   25
Korkhov, Vladimir   25
Kulabukhova, Nataliia   14
Kyazhin, Sergey   37

Lampreia, Suzana   492
Locurcio, Marco   1000
Lopes, Maria do Carmo   552
Lourenço, Paulo Renato   214

Macchiaroli, Maria   836
Machado e Costa, Francisco   185, 199
Makarov, Mikhail   25
Malevanniy, Daniil   48
Mancini, Luca   707, 717, 730
Manganelli, Benedetto   884
Marques, Ana   231
Marques, Mário J. Simões   446
Martínez-Bachs, Berta   683
Maselli, Gabriella   836
Matos, Marina A.   664
Meléndez-Pertuz, Farid   134
Merino Gómez, Elena   836
Micheli, Alessandro   981
Minucci, Amanda S.   521
Misiak, Jakub   787
Monteiro, Francisca   231

Montrone, Silvestro   899
Morano, Pierluigi   941, 968, 981, 1000
Muccio, Eugenio   855
Mura, Fernanda Della   952
Muts, Pavlo   590

Naboni, Emanuele   352
Nedjah, Nadia   89, 103
Nelas, José   564
Nesticò, Antonio   836, 913
Nicolosi, Vittorio   367
Nik, Vahid M.   352
Nowak, Ivo   590
Nunes, Isabel   446

Pacheco, António   147
Pacifici, Leonardo   744
Pavlova, Ekaterina   48
Perchinunno, Francesco   927
Perchinunno, Paola   899, 927
Pereira, Ana I.   507
Petrunin, Vadim   25
Pezzica, Camilla   324
Pietra, Caterina   759
Pirani, Fernando   693
Poli, Giuliano   855, 952
Popov, Vladimir   37

Quintas, José Pedro   199

Rebelo, José B.   492
Rebelo, Teresa   214
Regalbuto, Stefania   855
Ribeiro, Helena   147
Rimola, Albert   683
Rinaldi, Antonio M.   307
Rocha, Ana Maria A. C.   538, 606, 664
Rocha, Humberto   552
Romano, Francesca   855
Ronconi, Débora Pretti   632
Rosa, Álvaro   481
Rosi, Marzio   693, 707, 717, 730, 744

Salgado, Carla   470
Salvo, Francesca   776
Santilli, Daniela   367
Santos, Francisco   274
Santos, Isaac P.   455
Savino, Valentina   1016

Schaumann, Davide   340
Semenova, Elizaveta A.   395
Shchegoleva, Nadezhda   58, 72
Shust, Pavel   3
Sica, Francesco   913
Silva, Alberto   214
Silva, Carina   173
Silva, Filipe   231
Silva, José   231
Silva, Mariana   231
Sivolova, Julija   352
Skouteris, Dimitrios   707, 717, 730, 744
Sousa, Lisete   173
Sousa, Nuno   415
Stucci, Luigia Stefania   927

Tajani, Francesco   941, 968, 1000
Tataranna, Sabina   884
Tavano, Daniela   776
Teixeira, Ana   507, 620
Teixeira, Pedro   620

Teles, Júlia   157
Teodoro, M. Filomena   429, 446, 492
Torre, Carmelo Maria   941, 1000
Torres, Pedro   576
Trindade, Bruno   231
Turkman, Maria Antónia Amaral   173

Uteshev, Alexei Yu.   395
Uteshev, Alexey   72

Valença Ferreira de Aragão, Emília   707
Vaz, Clara B.   261
Vazart, Fanny   730
Vecchiocattivi, Franco   693
Venco, Elisabetta Maria   759
Ventura, Tiago   552

Živković, Ljiljana   820
Zvyagintsev, Mikhail   72

Printed in the United States
By Bookmasters